Springer Series in Astrophysics and Cosmology

The series covers all areas of astrophysics and cosmology, including theory, observations, and instrumentation. It publishes monographs and edited volumes. All books are authored or edited by leading experts in the field and are primarily intended for researchers and graduate students.

Christian Byrnes · Gabriele Franciolini ·
Tomohiro Harada · Paolo Pani · Misao Sasaki
Editors

Primordial Black Holes

Springer

Editors
Christian Byrnes
Department of Physics and Astronomy
University of Sussex
Brighton, UK

Gabriele Franciolini
Department of Theoretical Physics
CERN
Geneva, Switzerland

Tomohiro Harada
Department of Physics
Rikkyo University
Tokyo, Japan

Paolo Pani
Department of Physics
Sapienza University of Rome
Rome, Italy

Misao Sasaki
Kavli Institute for the Physics
and Mathematics of the Universe
University of Tokyo
Kashiwa, Chiba, Japan

ISSN 2731-734X ISSN 2731-7358 (electronic)
Springer Series in Astrophysics and Cosmology
ISBN 978-981-97-8886-6 ISBN 978-981-97-8887-3 (eBook)
https://doi.org/10.1007/978-981-97-8887-3

This Springer imprint is published by the registered company Springer Nature Singapore Pte Ltd.
The registered company address is: 152 Beach Road, #21-01/04 Gateway East, Singapore 189721, Singapore

To the next-generation cosmologists, particle physicists, and astrophysicists.

Foreword

In early February 2016, some colleagues and I were having lunch in the cafeteria of the Space Telescope Science Institute, across the street from the Johns Hopkins physics building, speculating about rumors that LIGO was going to announce the discovery of the merger of two 30-solar-mass black holes. The discovery of any gravitational-wave signal at all would have been revolutionary, but the high masses were particularly intriguing. A prevailing pre-LIGO mindset was that neutron-star mergers would, given the higher abundance of neutron stars, be more numerous than black-hole mergers. We also thought that if/when black-hole mergers were seen, they would involve—given the steeply falling stellar initial mass function—black holes in the 5–10 $M_\odot$ range, not $\sim 30\, M_\odot$ makes sense given the greater accessible volume for more massive black holes, but our understanding of this point came just a bit later.

I had worked on the periphery of gravitational-wave science, but none of us at the table were compact-object nor gravitational-wave specialists. We were primarily focused on cosmology and so, as unlikely as any connection may have seemed, we brainstormed for implications of the impending LIGO announcement for cosmology. After tossing around several *really* crazy ideas, we came across the notion that 30 $M_\odot$ black holes may have something to do with dark matter. In further discussions that afternoon we realized with order-of-magnitude estimates that if the dark matter were composed of 30 $M_\odot$ black holes, they would merge with a rate consistent with that implied by that one LIGO event. In the subsequent days, we checked these estimates more carefully and concluded, after scouring the literature, that there were no robust empirical arguments that would rule out the hypothesis that dark matter was made of 30 $M_\odot$ black holes. As nutty as this idea seemed, the coincidence between the merger rate of such black holes and the LIGO merger rate was too intriguing to ignore, and so we wrote a paper. We then found shortly thereafter that others had also been bouncing similar ideas around.

We got very deeply involved in this work as well as that we reported in several subsequent papers. The LIGO discovery was a remarkably profound discovery and a triumph for the half century of work and hundreds (thousands?) of careers devoted to the project, and we jumped on the bandwagon, as did many others. What's more, though, is that theoretical astrophysics allows one to be a generalist, employ a variety of tools in theoretical physics, and study a number of interesting physical systems. Our work on PBH dark matter involved galactic and stellar dynamics, gravitational dynamics, accretion physics, CMB spectral distortions and fluctuations, large-scale clustering, black-hole populations, gravitational microlensing, fast radio bursts, gravitational waves, and gamma-ray bursts—i.e., many (though not all—I don't think we ever thought about early-Universe mechanisms to produce PBH dark matter) of the subjects discussed in this book.

Of course, the idea of primordial black holes, and the possibility that they might be dark matter, was not new. But I would have been horrified had you told me just a few years earlier I would write a paper on PBH dark matter. I came of age in physics at a time when most theorists supposed that dark matter had to be WIMPs, or maybe axions. For a long time, it felt, with these two seemingly natural possibilities,that all other dark-matter candidates, like PBHs, were wacky. But with increasingly stringent null results from direct searches, and then the absence of any hints of new physics from the LHC, the back flips and leaps of faith required to make the WIMP work made us all far more willing to entertain a more diverse range of dark-matter contenders. This can be seen with citation counts for the classic 1974 Carr-Hawking paper.

The renewed interest in PBHs, since LIGO, has spawned a broad range of novel and important work. Although I may be a bit more skeptical than some contributors to this volume that PBHs will be found to be dark matter, I don't know that I am any more skeptical of this particular dark-matter candidate than any other. Moreover,the study of PBHs is valuable (as is the study of plenty of other skeptical ideas in cosmology) because it leads to novel questions in astrophysics, motivates new types of experiments, measurements, observations, and analyses, and "stress tests" our understanding of fundamental physics and astrophysical systems. For example, I have always been intrigued by the notion that the primordial curvature perturbation predicted by inflation is scale-invariant, adiabatic, and gaussian over ~ 30 orders of magnitude in distance scale; work on PBHs has advanced ideas for augmenting and testing this prediction. As seen in this volume, there have also been angles provided on other mysteries in astrophysics, like the formation of the first stars and the seeds for supermassive black holes. Interest in PBHs has helped motivate new types of gravitational-wave analyses and pushed others into parameter regimes that would have otherwise been ignored. And PBHs may also still be lurking out there, even if they are not the primary component of dark matter. In short, although this is a book about PBHs, it's also a book about a broader range of interesting issues in physics, astrophysics.

Baltimore, US
September 2024

Marc Kamionkowski

Preface

Primordial Black Holes (PBHs) were proposed more than 50 years ago as black holes potentially formed across a vast mass range in the early universe, long before the first stars. They represent a unique probe for accessing information about the primordial era of the universe and, potentially, cosmological inflation. Furthermore, their existence has been connected to the dark matter puzzle, and a vast body of research has been devoted to testing this hypothesis. While at least one mass window is still not ruled out, it is nonetheless exciting that even a subdominant population of PBHs could contribute to solving long-standing problems in modern physics and astrophysics. These problems include the formation of supermassive black hole seeds at high redshift, the generation of some of the gravitational-wave events detected so far, and the potential to be associated with novel gravitational-wave sources detectable with current and future instruments.

The quest to detect PBHs has proved to be extremely challenging. Extensive research has primarily set a variety of constraints on the fraction of dark matter composed of these objects, although some potential hints of their existence may be lurking behind recent observations.

Despite the impressive amount of research work and excellent reviews on this subject, there was no textbook aimed at introducing the multifaceted aspects of PBH physics in a pedagogical manner. This is particularly urgent given the multidisciplinarity of the subject, merging different communities and expertise. The scope of this book is to fill this gap by providing an up-to-date and comprehensive overview of PBH physics, with a pedagogical outlook, covering details on PBH formation scenarios, their cosmological evolution, astrophysical implications, connections with gravitational-wave astronomy, and a critical discussion of the latest and future constraints.

Unlike the many outstanding reviews on this subject, this book is also addressed to graduate students and researchers who may not be familiar with the incredibly vast and diverse range of topics connected by this interdisciplinary class of objects. It provides detailed, accessible explanations of important key results rather than presenting a purely technical collection of the latest literature. Our goal was to compile a textbook that could remain as a useful reference in the years to come, despite the fast developments that this field will certainly have.

The topic is inherently cross-cutting, connecting areas as diverse as cosmology, particle physics, high-energy astrophysics, classical gravitation, gravitational-wave astronomy, and numerical simulations. Since the inception of this field of research, the investigation of PBHs has proven to be an almost unmatched playground for young and ambitious physicists to explore diverse aspects of modern theoretical physics. To reflect this diversity, the book includes 27 chapters written by key researchers working in these different areas, divided into 6 parts. Thus, in addition to being the first textbook on the subject, this is intended to be the first community-driven comprehensive review on PBH physics, also reflecting different points of view, open issues, and ongoing debates. In particular, the reader might recognize different attitudes in presenting the most recent observational data and current constraints in Part V. Some authors consider certain recent observations as hints of PBH detection, while others are more cautious also because of the astrophysical uncertainties, the reach of which is discussed in detail in this book. We tried to provide a faithful account of this ongoing debate, which should certainly be clarified in the near future.

The first and last parts of the book (Part I and Part VI, respectively) highlight the historical perspective, including a description of how the field has evolved since its early days, and provide an outlook on how future research is expected to address the most pressing questions.

In Part II, we discuss the formation of PBHs. Most chapters focus on PBHs formed from large amplitude density perturbations soon after horizon entry during radiation domination, which is widely considered the least exotic formation scenario. This is introduced carefully in Chap. 2 but the final chapter of this part (Chap. 9) highlights alternative formation mechanisms. Chapter 3 explains how inflation may generate the required large amplitude peak of the primordial power spectrum as well as contains a section on recent and less settled research about loop corrections which may challenge the viability of the tree-level predictions of some inflationary models which generate PBHs. Exactly how large the amplitude of a perturbation has to be in order to collapse to form PBHs can only be determined using numerical simulations, and this topic is covered in Chap. 4. How this is related to the required amplitude of the power spectrum and the PBH mass function is outlined in Chap. 6 assuming Gaussian perturbations and radiation domination. The extension to phase transitions when the equation of state drops below the value during radiation domination is considered in Chap. 5. There are three relevant phase transitions within the standard model of particle physics. The extension to non-Gaussian fluctuations is studied in Chap. 7 with one motivation for non-Gaussian perturbations being provided by Chap. 8, which studies quantum diffusion during inflation.

In Part III, we discuss the cosmological evolution and astrophysical implications of the PBH scenario. Chapter 10 deals with an overview of the evolution of the early universe in the presence of PBHs. Chapter 11 deals with the impact of PBHs on the formation of the first stars and galaxies, whereas Chap. 12 provides a complementary view of the same problem in terms of known physics to explain black-hole seeds and their growth. Chapter 13 describes the role of accretion on the PBH mass and spin evolution. Chapter 14 describes the astrophysical implications of scenarios in which PBHs co-exist with other dark-matter candidates (which must be the case if not all the DM is in the form of PBHs), and a particularly interesting way to study this mixed dark matter model is via dense haloes around PBHs.

Part IV is devoted to a recent development in the field: testing the PBH scenario with gravitational-wave observations. Chapter 15 gives an overview on the subject, presenting current gravitational-wave searches for PBHs and opportunities with future detectors. Chapter 16 discusses the formation of PBH binaries and their merger rates, which is a crucial ingredient for gravitational-wave searches. Chapter 17 is devoted to the various gravitational-wave backgrounds associated with PBHs, while Chap. 18 is focused on the gravitational-wave signatures of subsolar mass objects, which are a potential smoking gun signature for PBHs.

Part V addresses the many observational constraints on PBHs, especially limiting the fraction of dark matter that can be composed of these objects. Chapter 19 presents an overview on the subject, including an up-to-date summary plot of the most relevant constraints, whereas Chap. 20 proposes a positivist view, which interprets some existing observations as indications for the existence of PBHs and suggests opportunities for their discovery. Although the latter is a topic of heated debate in recent literature, skeptics might see it as a collection of conjectured areas with significant potential for groundbreaking discoveries. Chapters 21 and 22 are devoted to constraints in particularly interesting and challenging mass windows, respectively, the ultralight range where Hawking radiation of PBHs is relevant and the asteroid-mass range where PBHs could comprise the entirety of the dark matter. Chapter 23 is devoted to discuss constraints coming from lensing, while Chap. 24 deals with constraints coming from measurements of the cosmic microwave background, which can be affected by PBH accretion. Chapters 25 and 26 are devoted to gravitational-wave constraints, respectively, from merger detections by current and future ground-based interferometers and from searches for the stochastic background associated with PBH formation and binaries.

Finally, the last part (Part VI) discusses the future of PBH physics, listing urgent open questions and proposing a roadmap for the exciting years ahead. In short, we hope this book will provide a unique reference for both newcomers to the topic and those seeking to learn about specific specialized sub-areas.

Brighton, UK	Christian Byrnes
Geneva, Switzerland	Gabriele Franciolini
Tokyo, Japan	Tomohiro Harada
Rome, Italy	Paolo Pani
Kashiwa, Japan	Misao Sasaki
July 2024	

Acknowledgements

We are indebted to Prof. Cosimo Bambi, who suggested the need for this book. We extend our gratitude to the outstanding authors who accepted the invitation to contribute to this collective effort. Their expertise and dedication have made this comprehensive resource possible.

Finally, an equally important thanks goes to the scientists who kindly accepted our invitation to act as reviewers for portions of the book. We list their names here, in alphabetical order: Jamie Bamber, Albert Escrivà, Daniele Gaggero, Simona Gallerani, Andrew Gow, Güntherr Hasinger, Christopher Hirata, Joaquim Iguaz Juan, Kaloian Lozanov, David Mulryne, Kai Murai, Ken K. Y. Ng, Alexander Nitz, Mauro Pieroni, Yuuiti Sendouda, Yuichiro Tada, Marco Taoso, Eemeli Tomberg, Caner Unal, and Shuichiro Yokoyama. We also thank the other reviewers who preferred to remain anonymous, and other colleagues and students who provided important suggestions and improvements: Amirah Aljazaeri, Hayami Iizuka, Xavier Pritchard, and Koichiro Uehara.

Contents

Editors and Contributors

About the Editors

Christian Byrnes is professor of theoretical cosmology at the University of Sussex where he held the Royal Society University Research Fellowship from 2012 for seven years. He received his Ph.D. in 2007 from the Institute of Cosmology and Gravitation in Portsmouth and has also been a fellow in CERN and held postdoctoral positions in the German universities at Heidelberg and Bielefeld. He has researched many areas of primordial black hole and early universe physics over the last decade.

Gabriele Franciolini is a senior postdoctoral fellow at CERN, where he specializes in theoretical physics with a focus on gravitational waves, black holes, dark matter, and the early universe. He earned his Ph.D. from the University of Geneva, concentrating on the study of primordial black holes, and has since become a key contributor to major international collaborations, including the Einstein Telescope, the European Pulsar Timing Array, and LISA. Franciolini received the prestigious Buchalter Cosmology Prize in 2024 and has co-authored over 70 publications in leading scientific journals.

Tomohiro Harada is a professor of physics at Rikkyo University. He belongs to the Institute of Theoretical Physics, Department of Physics. He received his Ph.D. from Kyoto University in 1999. Right after his Ph.D., he joined the same institute as a postdoctoral fellow. Prior to moving to Rikkyo University, he held postdoctoral positions at Waseda University, Queen Mary University of London, and Kyoto University. He took up a lectureship at Rikkyo University in 2006. He was promoted to associate professor in 2008 and to full professor in 2015. He has authored or co-authored over 140 publications in leading scientific journals in the research field of general relativity, relativistic astrophysics, and cosmology. His research interests include black hole physics, gravitational waves, primordial black holes, gravitational collapse, spacetimes singularities, and dark energy.

Paolo Pani is a professor of Theoretical Physics and coordinator of the gravity theory group at Sapienza University of Rome, where he has been the principal investigator of several European projects. He received the INFN Fubini Prize and the SIGRAV Prize. He is a co-author of the book "Superradiance," of the textbook "General Relativity and its Applications," and of over 180 scientific publications on black-hole physics and gravitational-wave phenomenology, and their connections to fundamental physics.

Misao Sasaki is a professor emeritus at Kyoto University and currently a project professor at Kavli Institute for the Physics and Mathematics of the Universe, University of Tokyo. His areas of expertise are general relativity and theoretical cosmology. He is an author of the textbook "General Relativity" (in Japanese), co-author of the book "The δN Formalism in Cosmological Perturbation Theory," and has published over 300 scientific articles in peer-reviewed journals.

The Editorial Board is formed by well-known experts in the field, with diverse and complementary expertise covering all topics of the book, and who will serve as a bridge across the different communities addressed by this project.

Contributors

Nicola Bellomo Texas Center for Cosmology and Astroparticle Physics, Weinberg Institute, Department of Physics, The University of Texas at Austin, Austin, TX, USA;
Dipartimento di Fisica e Astronomia G. Galilei, Università degli Studi di Padova, Padova, Italy

Volker Bromm Department of Astronomy, University of Texas, Austin, TX, USA

Bernard J. Carr School of Physics and Astronomy, Queen Mary University of London, London, UK

Zu-Cheng Chen Department of Physics, Hunan Normal University, Yuelu District, China

Philippa S. Cole Dipartimento di Fisica "G. Occhialini", Universitá degli Studi di Milano-Bicocca, Milano, Italy;
INFN, Sezione di Milano-Bicocca, Milano, Italy

Monica Colpi Department of Physics G. Occhialini, University of Milano-Bicocca, Milano, Italy

Valerio De Luca Center for Particle Cosmology, Department of Physics and Astronomy, University of Pennsylvania, Philadelphia, PA, USA

Guillem Domènech Institute for Theoretical Physics, Leibniz University Hannover, Hannover, Germany

Marcos M. Flores Laboratoire de Physique de l'École Normale Supérieure, ENS, Université PSL, CNRS, Sorbonne Université, Université Paris Cité, Paris, France

Anne M. Green School of Physics and Astronomy, University of Nottingham, Nottingham, UK

Alex Hall Institute for Astronomy, University of Edinburgh, Edinburgh, UK

Francesco Iacovelli Département de Physique Théorique, Université de Genève, Genève 4, Switzerland;
Gravitational Wave Science Center (GWSC), Université de Genève, Geneva, Switzerland

Karsten Jedamzik Laboratoire de Univers et Particules de Montpellier, UMR5299-CNRS, Universite de Montpellier, Montpellier, France

Jorge Jiménez-Vicente Departamento de Física Teórica y del Cosmos, Universidad de Granada, Granada, Spain;
Instituto Carlos I de Física Teórica y Computacional, Universidad de Granada, Granada, Spain

Kazunori Kohri Division of Science, National Astronomical Observatory of Japan (NAOJ), and SOKENDAI, Mitaka, Tokyo, Japan;
Theory Center, IPNS, KEK, Tsukuba, Ibaraki, Japan;
QUP (WPI), KEK, Tsukuba, Ibaraki, Japan;
Kavli IPMU (WPI), University of Tokyo, Kashiwa, Chiba, Japan

Ely D. Kovetz Department of Physics, Ben-Gurion University of the Negev, Be'er Sheva, Israel

Jason Kristiano Department of Physics and Research Center for the Early Universe (RESCEU), Graduate School of Science, The University of Tokyo, Tokyo, Japan

F. Kühnel Max-Planck-Institut für Physik, Garching, Germany;
Arnold Sommerfeld Center, Ludwig-Maximilians-Universität, München, Germany

Sachiko Kuroyanagi Instituto de Física Teórica UAM-CSIC, Universidad Autónoma de Madrid, Madrid, Spain;
Department of Physics, Nagoya University, Furo-cho Chikusa-ku, Nagoya, Japan

Alexander Kusenko Department of Physics and Astronomy, University of California, Los Angeles, CA, USA;
Kavli Institute for the Physics and Mathematics of the Universe (WPI), UTIAS, The University of Tokyo, Kashiwa, Chiba, Japan

Boyuan Liu Institute of Astronomy, University of Cambridge, Cambridge, CB3 0HA, UK

Alessandro Lupi Department of Science and Advanced Technology, University of Insubria, Como, Italy

Michele Maggiore Département de Physique Théorique, Université de Genève, Genève 4, Switzerland;
Gravitational Wave Science Center (GWSC), Université de Genève, Geneva, Switzerland

Evencio Mediavilla Instituto de Astrofísica de Canarias, Vía Láctea S/N, La Laguna, Tenerife, Spain;
Departamento de Astrofísica, Universidad de la Laguna, La Laguna, Tenerife, Spain

Andrew L. Miller Nikhef—National Institute for Subatomic Physics, XG, Amsterdam, The Netherlands;
Institute for Gravitational and Subatomic Physics (GRASP), Utrecht University, CC, Utrecht, The Netherlands

Ilia Musco Dipartimento di Fisica, Sapienza Università di Roma, Roma, Italy;
INFN, Sezione di Roma, Roma, Italy

Shi Pi Institute of Theoretical Physics, Chinese Academy of Sciences, Beijing, China;
Center for High Energy Physics, Peking University, Beijing, China;
Kavli Institute for the Physics and Mathematics of the Universe (WPI), The University of Tokyo, Kashiwa, Chiba, Japan

Stefano Profumo University of California, Santa Cruz, and Santa Cruz Institute for Particle Physics, Santa Cruz, CA, USA

Martti Raidal Keemilise ja Bioloogilise Füüsika Instituut, Tallinn, Estonia

Antonio Riotto Département de Physique Théorique, Université de Genève, Geneva, Switzerland;
Gravitational Wave Science Center (GWSC), Université de Genève, Geneva, Switzerland

Alba Romero-Rodríguez Chargèe de recherche du Centre National de La Recherche Scientifique (CNRS), France, LAPP, IN2P3 (Institut de Physique Nuclèaire et de Physique des Particules), Univ. Savoie Mont Blanc, CNRS, Laboratoire d' Annecy de Physique des Particules - IN2P3, Annecy, France

Pasquale Dario Serpico LAPTh, CNRS and University of Savoie Mont Blanc, Annecy, France

Joseph Silk Institut d'Astrophysique de Paris, Paris, France;
Department of Physics and Astronomy, The Johns Hopkins University, Baltimore, MD, USA;
Beecroft Institute for Particle Astrophysics and Cosmology, Department of Physics, University of Oxford, Oxford, UK

Teruaki Suyama Department of Physics, Tokyo Institute of Technology, Meguro-ku, Tokyo, Japan

Peter Tinyakov Université Libre de Bruxelles (ULB), Brussels, Belgium

Ville Vaskonen Keemilise ja Bioloogilise Füüsika Instituut, Tallinn, Estonia;
Dipartimento di Fisica e Astronomia, Università degli Studi di Padova, Padova, Italy;
Istituto Nazionale di Fisica Nucleare, Sezione di Padova, Padova, Italy

Hardi Veermäe Keemilise ja Bioloogilise Füüsika Instituut, Tallinn, Estonia

Vincent Vennin Laboratoire de Physique de l'École Normale Supérieure, ENS, CNRS, Université PSL, Sorbonne Université, Université Paris Cité, Paris, France

David Wands Institute of Cosmology & Gravitation, University of Portsmouth, Portsmouth, UK

Jun'ichi Yokoyama Department of Physics and Research Center for the Early Universe (RESCEU), Graduate School of Science, The University of Tokyo, Tokyo, Japan;
Kavli Institute for the Physics and Mathematics of the Universe (Kavli IPMU), WPI, UTIAS, The University of Tokyo, Kashiwa, Chiba, Japan;
Trans-Scale Quantum Science Institute, The University of Tokyo, Tokyo, Japan

Chul-Moon Yoo Division of Particle and Astrophysical Science, Graduate School of Science, Nagoya University, Nagoya, Japan

Sam Young Lorentz Institute for Theoretical Physics, University of Leiden, Leiden, CA, The Netherlands

Conventions and Acronyms

Unfortunately, primordial black hole physics connects various communities that historically use different mathematical conventions. Unless otherwise stated, the book uses a mostly plus signature convention, with the Minkowski metric being $\eta_{\mu\nu} = \text{diag}(-1, 1, 1, 1)$. Most chapters use natural units ($\hbar = c = 1$) but others use geometrized units ($G = c = 1$). This choice, and the corresponding definition of the Planck mass, $M_P = \sqrt{\hbar c/G}$, and of the reduced Planck mass, $m_P = \sqrt{\hbar c/(8\pi G)}$, will be clarified in each chapter when needed.

While each chapter will introduce its own list of acronyms the first time they are encountered, for the reader's convenience we list here the main ones which are typically used throughout the book:

3G	Third Generation
ABH	Astrophysical Black Hole
ALP	Axion-Like Particle
BBH	Binary Black Hole
BBN	Big Bang Nucleosynthesis
BH	Black Hole
BNS	Binary Neutron Star
BPL	Broken Power-Law
BSM	Beyond Standard Model
CBC	Compact Binary Coalescence
CE	Cosmic Explorer
CI	Credible Interval
CMB	Cosmic Microwave Background
DM	Dark Matter
dS	de Sitter
EFT	Effective Field Theory
EM	Electromagnetic
EMRI	Extreme Mass Ratio Inspiral
ET	Einstein Telescope
FAR	False Alarm Rate

FLRW	Friedmann–Lemaître–Robertson–Walker
GR	General Relativity
GW	Gravitational Wave
GWTC	Gravitational-Wave Transient Catalog
IMBH	Intermediate Mass Black Hole
ISCO	Innermost Stable Circular Orbit
JWST	James Webb Space Telescope
KAGRA	Kamioka Gravitational-Wave Detector
LIGO	Laser Interferometer Gravitational-Wave Observatory
LISA	Laser Interferometer Space Antenna
LMC	Large Magellanic Cloud
LVK	Ligo/Virgo/Kagra
MACHO	Massive Compact Halo Object
MW	Milky Way
NFW	Navarro–Frenk–White
NR	Numerical Relativity
NS	Neutron Star
PBH	Primordial Black Hole
PNG	Primordial non-Gaussianity
RD	Radiation Domination
SGWB	Stochastic Gravitational-Wave Background
SIGWB	Scalar-Induced Gravitational-Wave Background
SM	Standard Model
SMBH	Super Massive Black Hole
SMPBH	Super Massive Primordial Black Hole
SNR	Signal-to-Noise Ratio
SR	Slow Roll
UFD	Ultra-Faint Dwarf Galaxy
USR	Ultra Slow Roll
WIMP	Weakly Interacting Massive Particle

Part I
Overview

Chapter 1
The History of Primordial Black Holes

Bernard J. Carr and Anne M. Green

Abstract We overview the history of primordial black hole (PBH) research from the first papers around 50 years ago to the present epoch. The history may be divided into four periods, the dividing lines being marked by three key developments: inflation on the theoretical front and the detection of microlensing events by the MACHO project and gravitational waves by the LIGO/Virgo/KAGRA project on the observation front. However, they are also characterised by somewhat different focuses of research. The period 1967–1980 covered the groundbreaking work on PBH formation and evaporation. The period 1980–1996 mainly focussed on their formation, while the period 1996–2016 consolidated the work on formation but also collated the constraints on the PBH abundance. In the period 2016–2024 there was a shift of emphasis to the search for evidence for PBHs and—while opinions about the strength of the purported evidence vary—this has motivated more careful studies of some aspects of the subject. Certainly the soaring number of papers on PBHs in this last period indicates a growing interest in the topic.

1.1 Introduction

General relativity predicts that a region of mass M forms a black hole (i.e. a region where the gravitational field is so strong that not even light can escape) if it falls within its Schwarzschild radius $R_S \equiv 2GM/c^2$. Black holes could exist over a wide range of mass scales, although astrophysical processes could only produce them above a solar mass. Those larger than several solar masses would form at the endpoint of evolution of ordinary stars and there should be billions of these even in the disc of our own Galaxy. "Intermediate Mass Black Holes" (IMBHs) would derive from stars bigger than $100 M_\odot$, which are radiation-dominated and collapse due to an

B. J. Carr
School of Physics and Astronomy, Queen Mary University of London, London, UK
e-mail: b.j.carr@qmul.ac.uk

A. M. Green (✉)
School of Physics and Astronomy, University of Nottingham, Nottingham, UK
e-mail: anne.green@nottingham.ac.uk

C. Byrnes et al. (eds.), *Primordial Black Holes*, Springer Series in Astrophysics and Cosmology, https://doi.org/10.1007/978-981-97-8887-3_1

instability during oxygen-burning, and the first stars may have been in this range. "Supermassive Black Holes" (SMBHs), with masses from $10^6 M_\odot$ to $10^{10} M_\odot$, are thought to reside in galactic nuclei, with our own Galaxy harbouring one of mass $4 \times 10^6 M_\odot$ and quasars being powered by ones with mass of around $10^8 M_\odot$. There is now overwhelming evidence for these astrophysical types of black holes, but they can only provide a small fraction of the dark matter density.

Black holes could also have formed in the early Universe and these are termed "primordial". Comparing the cosmological density at a time t after the Big Bang and the density required for a region of mass M to fall within its Schwarzschild radius implies that primordial black holes (PBHs) would initially have around the cosmological horizon mass $M \sim c^3 t/G$ at formation. So they would have the Planck mass ($M_{\rm Pl} \sim 10^{-5}$ g) if they formed at the Planck time (10^{-43} s), $1 M_\odot$ if they formed at the quantum chromodynamics (QCD) epoch (10^{-5} s) and $10^5 M_\odot$ if they formed at $t \sim 1$ s. Therefore PBHs could span an enormous mass range and are the only black holes which could be smaller than a solar mass. In particular, only PBHs could be light enough for Hawking radiation to be important, those lighter than the Earth being hotter than the cosmic microwave background (CMB) and those lighter than 10^{15} g evaporating within the current age of the Universe.

The wide range of masses of black holes and their crucial rôle in linking macrophysics and microphysics is summarised in Fig. 1.1. The edge of the orange circle can be regarded as a sort of "clock" in which the scale changes by a factor of 10

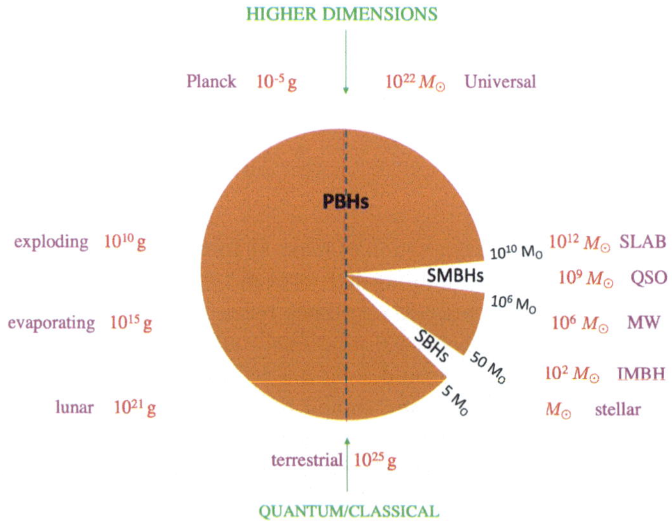

Fig. 1.1 This diagram indicates the mass and size of various types of black holes, with the division between the micro and macro domains being indicated by the vertical line. QSO stands for "Quasi-Stellar Object", MW for "Milky Way", IMBH for "Intermediate Mass Black Hole", SLAB for "Stupendously Large Black Hole". Stellar black holes (SBHs) and supermassive black holes (SMBHs) occupy only small slivers, whereas PBHs occupy a much wider range (shown in orange)

for each minute, from the Planck scale at the top left to the scale of the observable Universe at the top right. The top itself corresponds to the Big Bang because at the horizon distance one is peering back to an epoch when the Universe was very small, so the very large meets the very small there. The various types of black holes are labelled by their mass and positioned according to their Schwarzschild radius. On the right are the astrophysical black holes, with the well-established stellar and supermassive ones corresponding to the segments between 5 and 50 $M_{\odot}$ and between 10^6 and $10^{10} M_{\odot}$, respectively. On the left—and possibly extending somewhat to the right—are the more speculative PBHs.

The vertical line between the bottom (planetary-mass black holes) and the top (Planck-mass black holes) provides a convenient division between the microphysical and macrophysical domains. Quantum emission is suppressed by accretion of the CMB to the right of the bottom point, so this might be regarded as the transition between quantum and classical black holes. The effects of extra dimensions could be important at the top, especially if they are compactified on a scale much larger than the Planck length. In this context, there is a sense in which the whole Universe might be regarded as a PBH; this is because in brane cosmology (in which one extra dimension is extended) the Universe can be regarded as emerging from a five-dimensional black hole.

The study of PBHs provides a unique probe of four areas of physics: (1) the early Universe ($M < 10^{15}$ g); (2) gravitational collapse ($M > 10^{15}$ g); (3) high energy physics ($M \sim 10^{15}$ g); and (4) quantum gravity ($M \sim 10^{-5}$ g). Although we still cannot be certain that PBHs formed in *any* of these mass ranges, these many interesting applications suggest that nature would be remiss if their existence were precluded.

From a historical perspective, it should be stressed that PBHs have attracted increasing attention in recent years. Following the founding papers in the 1970s, there were only a dozen or so publications per year for the next two decades, although Hawking radiation obviously attracted attention. The rate rose to around a hundred per year after the MACHO microlensing results in 1996 but the most dramatic rise occurred after the first LIGO-Virgo gravitational wave events in 2016. This is illustrated in Fig. 1.2.

The structure of this chapter is primarily chronological, with successive sections covering the periods 1967–1980, 1980–1996, 1996–2016 and 2016–2024. The first period covered the initial groundbreaking work on PBH formation and evaporation. The second period focussed mainly on their formation, in particular from the collapse of large inflationary density perturbations, so we take this period to start with the inflationary proposal in 1980. The third period continued the studies of formation but also saw much work on PBH constraints, this being intensified after the detection of microlensing events by the MACHO project in 1996. The fourth period was initiated by the LIGO/Virgo detection in 2016 of gravitational waves from the mergers of multi-solar mass black holes, some of which could be PBHs, but it also saw claims of positive evidence for PBHs from several other observations. The authors of this chapter have somewhat different opinions about the strength of this purported evidence but this merely reflects the different opinions of the PBH community as a whole. So this chapter may be more balanced than it might have been otherwise.

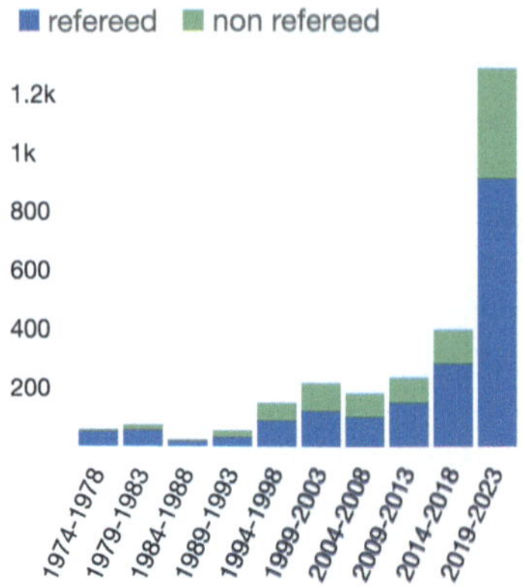

Fig. 1.2 Number of articles in SAO/NASA Astrophysics Data System with "Primordial Black Hole" in title in four-year bins

1.2 Early History (1967–1980)

In this section, we describe the early history of PBHs. The first paper on the topic was by Zeldovich and Novikov [202] in 1967 but they concluded (incorrectly) that PBHs would have accreted catastrophically and therefore could not have formed. This did not deter Hawking [104] from proposing a specific scenario in 1971. This was also partly flawed but it ultimately led to his discovery of Hawking radiation in 1974 [100]. This prompted more general interest in PBHs but it also implied observational constraints which suggested their existence was less likely.

1.2.1 Formation and Accretion

Since the cosmological density at a time t after the Big Bang is $\rho \sim 1/(Gt^2)$ and the density required for a region of mass M to fall within its Schwarzschild radius is $\rho \sim c^6/(G^3M^2)$, PBHs would initially have a mass

$$M \sim \frac{c^3 t}{G} \sim 10^{15} \left(\frac{t}{10^{-23}\,\mathrm{s}}\right) \mathrm{g}\,. \tag{1.1}$$

This is just the cosmological horizon mass at their formation epoch. The first proposal for the existence of PBHs was in a paper by Hawking just over 50 years ago [104]. He argued that PBHs of the Planck mass would be electrically charged and thereby capture electrons or protons to form "atoms". These could then leave tracks in bubble chambers and collections of them might accumulate in the centres of stars. This might explain the low flux of neutrinos coming from the Sun (which was then unexplained). Later it was realised that such small black holes would lose their charge through quantum effects.

In fact, the first discussion of PBHs, including Eq. (1.1) for the mass, was in a paper by Zeldovich and Novikov [202] several years before Hawking's paper. However, they concluded that the existence of PBHs was unlikely on the basis of a Bondi accretion analysis. This suggested that the PBH mass would increase according to

$$M = \frac{\eta c^3 t/G}{1 + (t/t_1)(\eta c^3 t_f/GM_f - 1)} \approx \begin{cases} M_f & (M_f \ll \eta c^3 t_f/G) \\ \eta t_f & (M_f \sim \eta c^3 t_f/G)\,, \end{cases} \tag{1.2}$$

where M_f is the formation mass and η is a constant of order unity. Thus PBHs with initial size comparable to the horizon (as expected) should grow as fast as the horizon and reach a mass of $10^{17} M_\odot$ by the end of the radiation-dominated era. Since the existence of such huge black holes is precluded, this might suggest that PBHs never formed. However, this argument neglects the cosmic expansion, which is important for PBHs with the horizon size and would inhibit accretion, and in 1974 Carr and Hawking showed that there is no self-similar solution in general relativity in which a back hole formed from local collapse can grow as fast as the horizon [52]. Furthermore, the black hole would soon become much smaller than the horizon, at which point Eq. (1.2) should apply, so one would not expect much growth at all. This removed the concerns raised by Zeldovich–Novikov and reinvigorated PBH research.

At this time more detailed studies of PBH formation were initiated. Only overdensities larger than the Jeans length at maximum expansion can collapse against the pressure and this is around $\sqrt{w}$ times the horizon size for $p = w\rho$. If PBHs formed from Gaussian density perturbations of root-mean-square amplitude $\epsilon(M)$ at the horizon epoch, then the fraction of the Universe collapsing into PBHs of mass M should be [40]

$$\beta(M) \sim \epsilon(M) \exp\left[-\frac{w^2}{2\epsilon(M)^2}\right]. \tag{1.3}$$

At that time it was expected that $\epsilon(M)$ should be scale-invariant, corresponding to a Harrison–Zeldovich spectrum [98, 203]. In this case, β is also scale-invariant and the PBH mass function should have the form [40]

$$dN/dM \propto M^{-\left(\frac{1+3w}{1+w}\right)-1}\,, \tag{1.4}$$

which just falls as $M^{-5/2}$ for a radiation-dominated universe ($w = 1/3$), as expected. However, this does not apply for the fluctuations expected in the inflationary scenario since (as discussed later) these are not exactly scale-invariant. The first numerical studies of PBH formation were carried out by Nadezhin et al. [156], modelling overdense regions as $k = +1$ FRW models matched to a $k = 0$ background by a vacuum region. These roughly confirmed the analytic prediction but included the effect of pressure gradients, resulting in PBHs somewhat smaller than the horizon.

On cosmological scales the amplitude of the fluctuations at horizon crossing is only around 10^{-5}, so the exponential dependence in Eq. (1.3) implies that the collapse fraction should be tiny if ϵ is scale-independent. Observations also require this since, if the current density parameter of PBHs which form at redshift z is $\Omega_{\rm PBH}$, in units of the critical density, then the initial collapse fraction is [40]

$$\beta = \frac{\Omega_{\rm PBH}}{\Omega_{\rm R}}(1+z)^{-1} \approx 10^{-6}\Omega_{\rm PBH}\left(\frac{t}{\rm s}\right)^{1/2} \approx 10^{-18}\Omega_{\rm PBH}\left(\frac{M}{10^{15}\,{\rm g}}\right)^{1/2}, \quad (1.5)$$

where $\Omega_{\rm R} \approx 10^{-4}$ is the current density parameter of radiation and we have used Eq. (1.1) at the last step. The $(1+z)$ factor arises because the radiation density scales as $(1+z)^4$, whereas the PBH density scales as $(1+z)^3$. So β must be tiny even if PBHs provide all of the dark matter. This is a potential criticism of the PBH dark matter proposal, since it requires fine-tuning of the collapse fraction and even greater fine-tuning of the density fluctuations. There is also the puzzling feature that the PBH and baryon densities are so close if PBHs provide the dark matter. However, as discussed later, there is one scenario in which this arises naturally.

1.2.2 Evaporation and Constraints

The realisation that PBHs might be small prompted Hawking to study their quantum properties. This led to his famous discovery [100] that black holes radiate thermally with a temperature

$$T_{\rm BH} = \frac{\hbar c^3}{8\pi G M k_{\rm B}} \approx 10^{-7}\left(\frac{M}{M_\odot}\right)^{-1} {\rm K}\,, \quad (1.6)$$

so they evaporate on a timescale

$$\tau(M) \approx 8000\frac{G^2M^3}{\hbar c^4} \approx 10^{67}\left(\frac{M}{Msun}\right)^3 {\rm yr}\,. \quad (1.7)$$

The numerical factor in the second expression would be 5120π if the radiation were exactly black-body but this does not apply because the spectrum also depends upon the mass and spin of the particles being emitted.

PBHs with initial mass $M_* \sim 10^{15}$ g, which formed at 10^{-23} s and had the size of a proton, would be evaporating now and lighter ones would have evaporated at an earlier epoch. However, evaporation would be suppressed for PBHs heavier than the Earth, 10^{27} g, since they would be cooler than the cosmic microwave background (CMB) and so would accrete rather than evaporate.

Hawking's discovery has not yet been confirmed experimentally and there remain major conceptual puzzles associated with the process. Nevertheless, it is generally recognised as one of the key developments in 20th-century physics because it beautifully unifies general relativity, quantum mechanics and thermodynamics. The fact that Hawking was only led to this discovery through contemplating the properties of PBHs illustrates that it has been useful to study them even if they do not exist. However, at first sight it was bad news for PBH enthusiasts. Since PBHs with a mass of 10^{15} g would be producing photons with energy of order 100 MeV at the present epoch, the observational limit on the γ-ray background intensity at 100 MeV immediately implied that their density could not exceed 10^{-8} times the critical density [168]. This implied that there was little chance of detecting black hole explosions at the present epoch, which would have confirmed the existence of both PBHs and Hawking radiation.

Nevertheless, the evaporation of PBHs smaller than 10^{15} g could still have many interesting cosmological consequences [37], each associated with the different types of particle emitted, and it also implied interesting constraints on the collapse fraction $\beta(M)$. These constraints were first brought together by Novikov et al. [163] in 1979. The strongest one is the γ-ray limit associated with the 10^{15} g PBHs evaporating at the present epoch. Others are associated with the generation of entropy and modifications to the cosmological production of light elements, PBHs with $M \sim 10^{10}$ g having a lifetime $\tau \sim 10^3$ s and therefore evaporating at the big bang nucleosynthesis (BBN) epoch. Injection of high-energy neutrinos and antineutrinos [192] changes the weak interaction freeze-out time and hence the neutron-to-proton ratio at the onset of BBN, which changes ^{4}He production. PBHs with $M = 10^{10}$–10^{13} g evaporate after BBN, increasing the baryon-to-entropy ratio at nucleosynthesis and resulting in overproduction of ^{4}He and underproduction of D [148]. Emission of high-energy nucleons and antinucleons [201] increases the primordial deuterium abundance due to the capture of free neutrons by protons and spallation of ^{4}He. The emission of photons by PBHs with $M > 10^{10}$ g dissociates the deuterons produced in nucleosynthesis [136]. The limits associated with these effects are shown in the top panel of Fig. 1.3. The equivalent figure 20 years later is shown on the bottom and is not very different.

PBHs were also invoked to explain certain observations. For example, evaporating PBHs of around 10^{15} g might explain the 511 keV annihilation line radiation from the Galactic centre [165] or antiprotons in cosmic rays [37]. PBHs more massive than 10^{15} g might provide the dark matter since these are unaffected by Hawking radiation, with Chapline [55] suggesting this in 1975. In this case, as pointed out by Mészáros [146] in the same year, sufficiently large ones could generate cosmic structures through the $\sqrt{N}$ Poisson effect.

1.3 Studies of PBH Formation (1980–1996)

This section overviews developments in the 1980s and early 1990s, in particular studies of PBH formation. This period includes the development of inflationary theory and, since inflation generates density fluctuations, there was interest in whether these could produce PBHs. Indeed, constraints can be imposed on inflationary models by

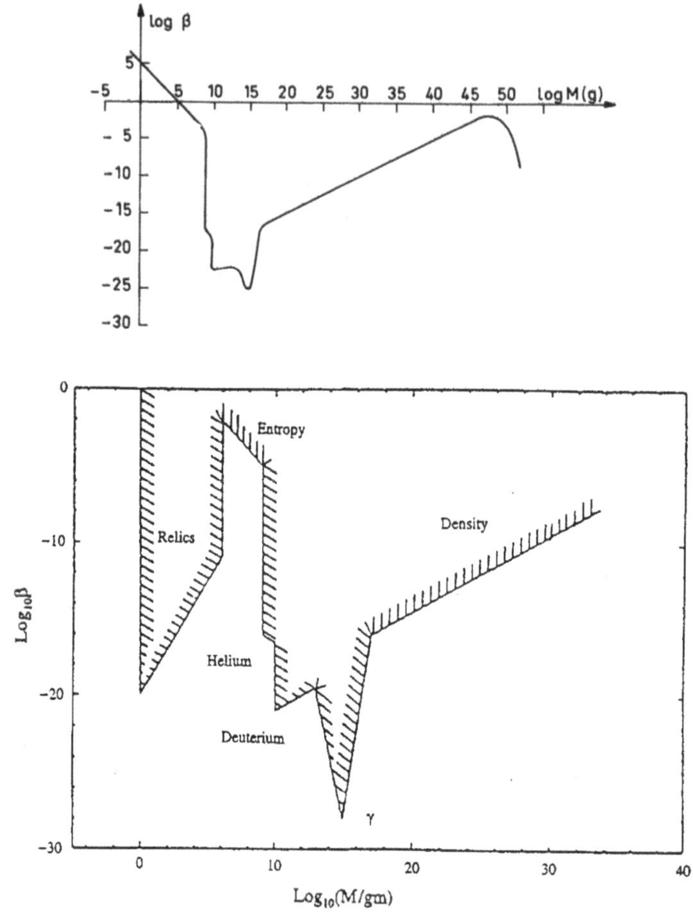

Fig. 1.3 Evaporation constraints on $\beta(M)$, the fraction of Universe collapsing into PBHs of mass M, from Ref. [163] in 1979 (top) and Ref. [41] in 1994 (bottom)

the requirement that they do not overproduce them. Other PBH formation mechanisms were also proposed, and improved calculations of the evaporation of PBHs led to refinements in abundance constraints.

1.3.1 Formation from Inflationary Density Perturbations

Inflation, a period of accelerated expansion in the early Universe, was proposed in the early 1980s [6, 93, 181, 185]. It was quickly realised that quantum fluctuations in the scalar field driving inflation would lead to density perturbations from which structures can form [94, 102, 135, 186], so there was naturally interest in whether these fluctuations could be large enough to generate an interesting abundance of PBHs. Also, since inflation exponentially reduces the number density of any PBHs formed earlier, the horizon mass at the end of inflation sets a lower limit on the mass of PBHs which are subsequently observable. The tensor contribution to the CMB temperature anisotropies on large angular scales limits the reheat temperature to be less than 10^{16} GeV, which corresponds to a horizon mass of ~ 1 g [53].

The amplitude of the density perturbations on CMB scales is of order $\sim 10^{-5}$ [183, 194]. From Eq. (1.3) this amplitude is too small to generate even one PBH per current horizon volume. To produce a non-negligible abundance, the amplitude must be far larger on some small scale. In single-field slow-roll inflation models the amplitude of the density perturbation is $\epsilon \propto V^{3/2}/V'$ where V is the potential and V' is its slope. One requires either a blue spectrum, where the primordial perturbations increase monotonically on smaller scales, or a spike on some scale. Both of these are possible with single-field inflation but one needs to go beyond the slow-roll approximation. In 1993 Carr and Lidsey [53] studied PBH formation in models with blue spectra. In this case, PBHs could form abundantly and avoiding their overproduction placed a tighter constraint on the spectral index of the primordial perturbations than CMB observations did at that time [41]. In 1994 Ivanov et al. pointed out that a plateau in the potential, with $V' \to 0$, could generate large PBH-forming perturbations [113]. However, more complicated models were also proposed, with García-Bellido et al. [82] invoking hybrid inflation; this has two fields, with one having large quantum fluctuations as it undergoes a phase transition which ends inflation. It was pointed out by Dolgov and Silk [69] that PBHs with a lognormal mass function could be formed from baryon isocurvature fluctuations generated by various inflation models.

1.3.2 PBH Formation During Matter-Domination

It is usually assumed that the Universe is radiation-dominated from matter-radiation equality ($t_{eq} = 1.7 \times 10^{12}$ s) back to very early times. There could, however, be an early period of matter domination before BBN due, for instance, to long-lived non-relativistic particles dominating the Universe and then decaying. In the 1980s Khlopov and Polnarev [124, 173] studied PBH formation during such a period of early matter domination. A matter-like phase might also arise at the end of inflation if reheating is slow [41]. In these cases, the criteria for PBH formation from the collapse of density perturbations are very different from those during radiation-domination (as covered in Sect. 1.2.1). A perturbation must be close to spherically symmetric to form a PBH, rather than collapsing to form a pancake-like or cigar-like configuration [124]. It must also collapse within its Schwarzschild radius before a caustic can form at its centre [173].

1.3.3 PBH Formation from Phase Transitions

In the early 1980s, various authors showed that PBHs, with mass of order the horizon mass, could form from the collisions of bubbles formed during a first-order phase transition [63, 103, 126]. However, to form a non-negligible PBH abundance the bubble nucleation rate had to be fine-tuned, so that bubble collisions occur but without the phase transition completing too quickly.

Phase transitions in the early Universe can also lead to the formation of a network of one-dimensional topological defects, known as cosmic strings [125]. These include oscillating loops that are smaller than the Hubble radius and in the late 1980s it was shown a PBH can form from the collapse of a loop if it becomes smaller than its Schwarzschild radius [101, 174]. The PBH mass is proportional to the loop mass, which is proportional to the horizon mass, and the rate at which loops collapse to form PBHs is independent of time. Therefore, as for PBHs formed from scale-invariant density perturbations, the number density of PBHs formed is $\mathrm{d}n/\mathrm{d}M \propto M^{-5/2}$ [141]. As indicated by numerical simulations in the early 1990s, the fraction of loops collapsing to PBHs is very sensitive to the string tension $(G\mu)$ [7, 11, 33] and there were already strong constraints on this parameter from other observations.

1.3.4 Observational Probes

In the early 1990s MacGibbon and Webber [139, 140] studied the evaporation of PBHs in more detail. Since black holes emit all elementary particles with rest mass less than the black hole temperature, T_{BH}, they argued that PBHs with mass $M \lesssim 2 \times 10^{14}$ g, which are hotter than than the QCD confinement scale, should emit quark-gluon jets which then fragment. The total emission is thus the combination of the primary and secondary emission. MacGibbon and Carr [142] studied the observational consequences of this, in particular their possible contribution to cosmic rays and the extragalactic gamma-ray background. They placed constraints on the contribution from PBHs, and hence their abundance, but they did not claim positive evidence for them.

In 1985 Lacey and Ostriker argued that the observed puffing of the Galactic disc could be due to halo black holes with mass around $10^6\,M_\odot$ [132], older stars being heated more than younger ones. Although dynamical constraints now exclude such large PBHs providing the dark matter, clusters of smaller PBHs could also explain disc heating [44]. However, heating by a combination of spiral density waves and giant molecular clouds may also explain the data [131]. The suggestion that sufficiently massive PBHs could generate cosmic structures through the 'seed' or 'Poisson' effect, as first pointed out by Meszaros [146], was explored in detail in several papers during the 1980s [45, 46, 77]. Bond and Carr [26] also considered the possibility that LIGO could detect the bursts and gravitational wave background generated by binary black holes, although they did not require them to be primordial.

In the early 1990s Hawkins [105] argued that observed variations in the brightness of quasars indicate a large abundance of compact objects.

1.4 Studies of PBH Formation and Constraints (1996–2016)

Until the mid-1990s there was no evidence for PBHs and no compelling reason to associate them with the dark matter problem. However, there was a flurry of excitement in 1996, when microlensing searches for massive compact halo objects (MACHOs) suggested that the dark matter could be black holes of mass $0.5M_\odot$ [8]. Alternative microlensing candidates could be excluded and it was realised that PBHs of this mass might naturally form at the quark-hadron phase transition [114]. With further data, the MACHO collaboration found that solar-mass compact objects could comprise only 20% of the dark matter [10] and other microlensing observations during this period subsequently excluded compact objects in the mass range $10^{-7}M_\odot$ to $10M_\odot$ from providing all of it [191]. Putative evidence for PBHs thus became a constraint on their abundance. Indeed, much of the focus of research in this period was on the collation of constraints over a huge range of mass scales from 10^{-5} g to $10^{12}M_\odot$. These constraints were associated with numerous astrophysical and cosmological effects—related to evaporation, microlensing, dynamics, accretion and gravitational waves—and expressed as upper limits on $f_{\rm PBH}(M)$, the fraction of the dark matter in PBHs with mass M. They are summarised briefly below and discussed in more detail in Part V. See Refs. [39, 43] for the status of constraints in 2010 and 2016, respectively. However, much of the literature in this period continued the study of PBH formation, so we will start by reviewing this.

1.4.1 Further Studies of Formation from Inflation

It was pointed out in the late 1990s that since PBHs form from rare large perturbations, their abundance depends sensitively on the shape of the large-amplitude tail of the probability distribution of the perturbations [29, 112]. For the flat inflaton potential required to generate large perturbations in single-field models (see Sect. 1.3.1), it was also argued that quantum fluctuations play a significant role in the dynamics of the field and can generate a non-Gaussian probability distribution for the fluctuations [198].

Various inflation models that can produce large, PBH-forming perturbations were explored, such as hilltop inflation (where the field evolves away from a local maximum) [127] and double inflation (with two periods of inflation) [122]. It was also realised that the reheating era, at the end of inflation, could produce large perturbations [21, 91]. Several authors pointed out that the amplitude of large density perturbations could be indirectly constrained via limits on the gravitational waves they induce [13, 22, 178] (see Chap. 17). In the mid-2000s, Green et al. [92] calculated the abundance of PBHs using the classic Bardeen, Bond, Kaiser and Szalay (BBKS) [20] peaks theory rather than the traditional Press-Schechter approach which leads to Eq. (1.3). This is discussed in more detail in Chap. 6 and is not only relevant to inflation.

1.4.2 Formation from Other Processes

It was realised in the 1990s that due to critical phenomena [57], the mass of a black hole depends on the amplitude of the perturbation from which it forms: $M \propto M_{\mathrm{H}}(\delta - \delta_{\mathrm{c}})^{\gamma}$, where the scaling exponent γ is constant for a given equation of state. Niemeyer and Jedamzik [161] pointed out that this applies to PBH formation and calculated the resulting PBH mass function (assuming all PBHs form at the same time). They also carried out simulations of PBH formation for different shaped perturbations [160]. Subsequent work by Musco and his collaborators [153–155] in this period explored critical collapse further, in particular verifying the mass scaling for small $(\delta - \delta_{\mathrm{c}})$ and studying a range of equation of state parameters $(0 < w < 0.6)$. See Chap. 4 for further details.

Jedamzik pointed out that the reduction in pressure at the QCD phase transition leads to enhanced PBH formation at this time, with the horizon mass (and hence PBH mass) being around $1\,M_{\odot}$ [115]. There were also improvements in specifying the criteria for PBH formation, with Ref. [157] showing that the threshold depends on the shape of the perturbation. Numerical studies of PBH formation during matter-domination found that the initial mass fraction of PBHs in this case is given by $\beta \approx 0.21\epsilon^{13/2}$ for $\epsilon \ll 1$ [96]. Since angular momentum is significant for PBHs formed during matter-domination, they have larger spins than those formed during radiation-domination [97].

1.4.3 Evaporation Constraints

In a series of papers starting in 1997 Cline and colleagues argued that exploding PBHs could explain some short-duration gamma-ray bursts [60]. Unlike cosmological gamma-ray bursts, these would be located within the Galactic halo and therefore anisotropically distributed. If true, this would require some new effect when the black hole temperature reaches the QCD scale, possibly associated with the formation of a photosphere due to electron-positron or QCD interactions [106]. However, in 2008 MacGibbon, Carr and Page [143] showed that evaporating black holes do not form photospheres. They also found that a more precise value for the initial mass of a PBH evaporating today is $M_{*} \approx 5 \times 10^{14}$ g. The constraints on $\beta(M)$ for PBHs with initial mass $M \lesssim 10^{17}$ g were extended and improved by numerous studies of the

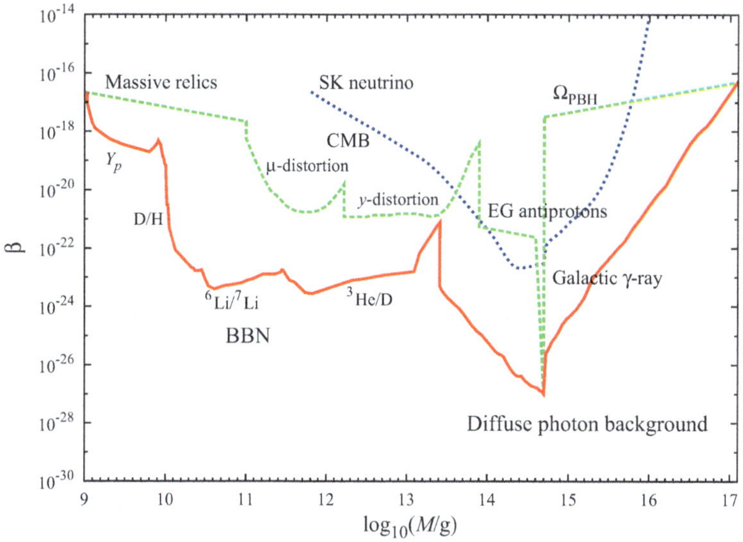

Fig. 1.4 Evaporation constraints on $\beta(M)$, the fraction of Universe collapsing into PBHs of mass M, from Ref. [39] in 2010

effects of PBH evaporations on the early Universe. These constraints were summarised in Ref. [39], which also updated the cosmological nucleosynthesis bound, and are shown in Fig. 1.4.

If PBHs of mass M_* are clustered inside the Galactic halo, as expected, then their quantum evaporation should generate a Galactic γ-ray background. Since this would be anisotropic, it should be separable from the extragalactic γ-ray background. While the dominant contribution to the latter comes from the time-integrated emission of PBHs with initial mass M_*, the Galactic background is dominated by the instantaneous emission of those with initial mass slightly larger than M_* and current mass below M_*. In 1996 Wright [195] claimed that a Galactic background had been detected in EGRET observations between 30 and 120 MeV and attributed this to PBHs. A later analysis of EGRET data in 2009, assuming a variety of PBH distributions, was given by Lehoucq et al. [134]. In 2016 Ref. [38] reassessed this limit by including a realistic model for the PBH mass spectrum and a more precise relationship between the initial and current PBH mass. Specifically, the PBHs which generate the Galactic background have an initial mass $M_*(1+\mu)$ and a current mass $(3\mu)^{1/3}M_*$ with $\mu \ll 1$.

1.4.4 Lensing and Gravitational-Wave Constraints

Stellar microlensing, the temporary achromatic amplification of a star which occurs when a compact object crosses the line of sight to a star [167], is a long-standing probe of the abundance of planetary and solar-mass compact objects. In the 1990s the EROS, MACHO and OGLE projects started microlensing surveys of the Milky Way. In particular, they monitored millions of stars in the Magellanic Clouds to probe compact objects in our halo. In their 2-year results [8] the MACHO collaboration found significantly more events than expected from known stellar populations. Their number and durations were consistent with roughly half of our halo being in compact objects with mass of around $0.5 M_{\odot}$.

Baryon budget arguments excluded such a large population of astrophysical objects (e.g. white dwarfs) [74] and PBHs were a plausible explanation for these events. However, in the MACHO project's subsequent (5.7 year) results, the best-fit halo fraction dropped by a factor of roughly two, with the absence of long-duration events excluding compact objects with mass $(1-30) M_{\odot}$ from comprising all of the dark matter [9]. Subsequent results from EROS and OGLE, under the assumption that the halo is an isothermal sphere with a flat rotation curve, excluded planetary and stellar mass compact objects making up more than about 10% of the halo [190, 196].

Nakamura et al. [111, 158] pointed out that if PBHs make up a significant fraction of the dark matter, then PBH binaries could form at early times. Pairs of PBHs that happen to be close together would decouple from the expansion of the Universe before matter-radiation equality, with the torque from nearby PBHs leading to the formation of binaries with high eccentricity. The coalescence times of these binaries could be comparable to the age of the Universe, so gravitational waves from their mergers would be detectable by interferometers such as LIGO providing the binaries survive within halos to late times.

1.4.5 Dynamical and Accretion Constraints

Carr and Sakellariadou [54] provided a comprehensive collection of dynamical constraints on dark compact objects. Encounters of compact objects with mass $M \gtrsim 10 M_{\odot}$ with wide binary stars will increase the energy of those binaries, increasing their separation and disrupting the widest ones [17]. Limits on the fraction of the MW halo in PBHs can therefore be obtained by comparing observed and simulated wide binary separation distributions [149, 176, 197]. Reliable constraints require a large sample of confirmed halo binaries and also accurate modelling of their initial properties [176]. Afshordi et al. [4] showed that the Poisson fluctuations in the PBH distribution would lead to the formation of PBH clusters shortly after matter-radiation equality and the enhanced clustering on subgalactic scales would have important implications for observations of Lyman-alpha clouds. It was also suggested that asteroid-mass PBHs (10^{17} g $\lesssim M \lesssim 10^{22}$ g) could be probed via the consequences of their interactions with stars [34, 88, 170].

Gas accretion onto PBHs with $M \gtrsim 10 M_{\odot}$ has potentially observable consequences. The resulting emission can modify the recombination history of the Universe [36] and Ricotti and collaborators calculated the constraints from the anisotropies and spectral features of the CMB [177]. For more details see Ref. [150] and Chap. 11. Mack et al. [144] showed that if dark matter is composed of a mixture of PBHs and particles, then PBHs will accrete halos of particle dark matter which have a steep density profile as discussed in Chap. 14. Lacki and Beacom [133] pointed out that PBHs and WIMPs cannot both make up a significant fraction of the dark matter as WIMP annihilation in these halos would produce a higher than observed flux of γ−rays.

1.5 Improved Calculations and Search for Evidence (2016–2024)

The detection of gravitational waves from the mergers of black holes with masses of $10 - 100\, M_{\odot}$ by LIGO/Virgo has generated a huge wave of interest in PBHs as the origin of some of these events and as dark matter. Indeed, perhaps the most significant change of emphasis in this fourth period has been the search for *evidence* for PBHs. Partly prompted by this development, there have also been significant improvements in calculations of the abundance and mass function of PBHs formed from large density perturbations, and other formation mechanisms have been advocated. New probes of the PBH abundance have also been proposed and improvements made to existing constraints and signatures (see Chaps. 19 and 20). The status of the constraints in 2021 was reviewed in Refs. [48, 90] and a recent form of the $f_{\rm PBH}(M)$ diagram is shown in Fig. 1.5. The shift towards the search for positive evidence for PBHs was signalled by the 2018 paper of Clesse and García-Bellido [58]. This was followed by their 2021 paper with Carr and Kuhnel, in which much of the claimed evidence was shown to be compatible with a 'natural' thermal history model in which the sound speed dips at various times and most strikingly at the QCD transition at 10^{-5} MeVs [49]. More recently the claimed evidence has been reviewed in Ref. [50], which promotes what has been termed a 'positivist' approach. Some of the arguments are mentioned below but they are discussed in more detail in Chap. 20.

1.5.1 Further Work on PBH Formation

Byrnes et al. [31] calculated the mass function for PBHs formed at the QCD epoch, using an accurate form for the softened equation of state then. In general the threshold for PBH formation, δ_c, will decrease at any phase transition where there is a reduction

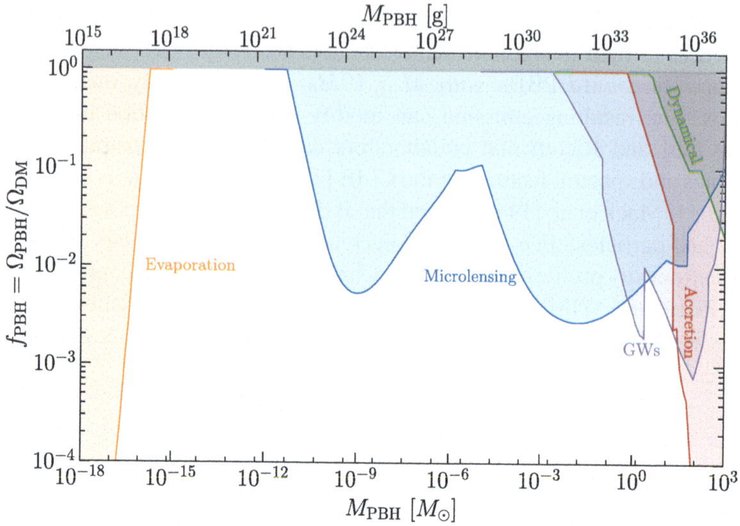

Fig. 1.5 Constraints on the fraction of DM in the form of PBHs, f_{PBHs}, as a function of mass, M_{PBH}, assuming all PBHs have the same mass. The bounds shown are (left to right) from evaporation (orange), microlensing (blue), gravitational waves (purple), accretion (red) and dynamical effects (green). For each type of bound the tightest constraint at each mass is shown and the shaded regions are excluded under standard assumptions. Figure created using Kavanagh's PBHbounds code [121], which is regularly updated to include the latest constraints

in pressure, so the PBH mass function will have peaks at the corresponding horizon mass (see Chap. 5). Carr et al. [49] extended this idea by pointing out that the thermal history of the Universe produces four dips in the sound speed in the period $(10^{-8} - 10)$ s, leading to corresponding bumps in the PBH mass function if the amplitude of the primordial perturbations is large across the corresponding range of scales.

There has been further extensive work exploring which inflation models can generate large PBH-forming density perturbations, while also satisfying the constraints on the amplitude and scale dependence of perturbations on cosmological scales [5]. This is not possible in single-field slow-roll models (e.g. Refs. [19, 107]) but can be achieved with a feature in the potential, such as an inflection point (e.g. Refs. [19, 83]) or a small dip/peak (e.g. Refs. [107, 147]), which leads to a period of ultra-slow-roll inflation. Other methods for generating large inflationary density perturbations have been proposed, such as multi-field models with rapid turns in field space [79, 169]); for an overview see Ref. [166]. New mechanisms for the production of PBH-forming fluctuations have also been proposed, for instance, Q-balls produced by the fragmentation of scalar field condensate [62, 70], long-range forces mediated by scalar fields [12, 75] and the collapse of cosmic string cusps [116]. For further details see Chap. 9.

It was pointed out in Ref. [86] that the shape of density perturbations, and hence the threshold for collapse, depends on the form of the primordial power spectrum.

Subsequently, Escrivá et al. [72] showed that the threshold is universal if specified in terms of the average of the compaction function, which is similar to the Schwarzschild gravitational potential. See Chaps. 2 and 6 for further details. It was realised that even if the curvature perturbations are Gaussian, the non-linear relationship between density and curvature perturbations means that the distribution of large density perturbations is inevitably non-Gaussian [66, 123, 200]. Chapter 7 provides a detailed discussion of non-Gaussianity and an approach to calculating the probability distribution of the primordial fluctuations for stochastic inflation is described in Chap. 8.

The issue of the fine-tuning required to generate an interesting PBH abundance has also been addressed. Fine-tuning is required in three different senses: the collapse fraction β must be tiny even if they constitute all of the dark matter; if PBHs form from the collapse of density perturbations, their amplitude must be even more fine-tuned; if the perturbations are generated by inflation, the parameters of the inflation model must also be fine-tuned. The first problem has been addressed by García-Bellido et al. [85], who point out that PBHs which form at the QCD epoch would necessarily have $\beta \sim \eta$, where η is the cosmological baryon-to-photon ratio, if they provide the dark matter. They propose a baryogenesis scenario in which the PBH formation at the QCD epoch naturally generates this relation and this also explains why the PBH and baryon densities are comparable. The second problem is then viewed from an anthropic perspective. The third problem has been discussed by Cole et al. [61].

1.5.2 Extended Mass Functions and Clustering

Constraints on the PBH abundance are usually calculated assuming that PBHs have a smooth density distribution and a single mass (i.e. a delta-function or monochromatic mass function). However, most formation mechanisms produce PBHs that have enhanced clustering on small scales and an extended mass function. The latter problem was first addressed in Refs. [43, 89] and a general method for applying constraints for a monochromatic mass function to an extended mass function was presented in Ref. [51]. The fraction of dark matter in PBHs is constrained to be less than 1 over a wider range of masses than for a monochromatic mass function [89] but the tightest value of each constraint is weakened [51].

The clustering of PBHs as a result of their Poisson fluctuations has featured in all the historical periods but several groups studied this effect more extensively in this fourth one. In particular, the fact that the first baryonic clouds could form earlier than in the standard scenario, would have interesting observational consequences, as stressed by Kashlinsky [118]. Later Inman and Ali-Haïmoud carried out numerical simulations of the clusters that form shortly after matter-radiation equality if PBHs make up a fraction of the dark matter [109]. This clustering modifies some of the PBH constraints but in different ways. For the diffuse clusters formed from Gaussian perturbations the change in the microlensing constraints is small [87, 172]. However for compact clusters (which could be formed from non-Gaussian perturbations) the microlensing constraints would be weakened [32], while other constraints would be tightened [67, 199].

Adamek et al. [3] studied the ultracompact minihalos that form around PBHs when dark matter is a mixture of PBHs and WIMPs. The limits for such mixed dark matter models were updated in Ref. [42], using the most recent and detailed simulations of the particle dark matter halos around PBHs [27]. As also emphasised in Ref. [23], definitive evidence for the existence of PBHs would rule out WIMPs making up a significant fraction of the dark matter. See Chaps. 14 and 20 for more details.

1.5.3 Evidence and Constraints from Gravitational Waves and Microlensing

In 2016 the LIGO-Virgo collaboration announced the discovery of gravitational waves from mergers of black holes [1]. Soon afterwards several papers suggested that these could be PBHs and also make up some of the dark matter [24, 59, 179]. Whereas Ref. [24] focussed on PBH binaries formed within dark matter halos at late times, Ref. [179] focussed on binaries forming at early times, but the latter process is probably dominant. If all of the dark matter is in such PBHs, then (provided the orbits of the binaries are not changed significantly during structure formation) the expected present-day merger rate is several orders of magnitude larger than that measured by LIGO-Virgo [179]. However, a subdominant PBH component of the dark matter could still be responsible for some of the observed events [65, 76, 84, 95, 108, 180] and this is discussed further in Chap. 25. Although the mainstream view is that all the LIGO-Virgo-KAGRA detections derive from astrophysical black holes, the masses are larger than initially expected [129] and some lie in mass gaps where stellar remnants should not be found [50]. Also, the observations suggest smaller spins than would be expected for astrophysical black holes [84]. The PBH origin proposal is still controversial but future observations of the mass, spin and redshift distribution of the objects should clarify the issue soon. Further details and more recent developments are covered in Part IV.

In 2017 Niikura et al. [162] carried out a microlensing survey of the Andromeda galaxy (M31) which was sensitive to short-duration events and hence lighter compact objects. They found only one candidate event and extended the range of masses constrained by microlensing down to $\sim 10^{22}$ g [64, 162, 184]. Lighter PBHs cannot be probed by optical observations of main-sequence stars; both the finite size of the source stars and wave-optics effects (from the PBH Schwarzschild radius being comparable with the wavelength of light) reduce the microlensing event rate (see Refs. [150, 188] and references therein).

The Milky Way microlensing constraints have now been extended to larger masses, $M \sim 10^3 M_\odot$, by combining data from different surveys to obtain sensitivity to longer duration events [25, 151]. However, Ref. [81] argues that these limits can be weakened when one allows for the extended PBH mass function expected in the thermal history model and the falling rotation curve indicated by recent observations of the Milky Way.

Various other microlensing constraints on stellar mass PBHs have been studied, including the magnification distribution of type 1a supernovae (SNe) [205], super-magnified stars such as Icarus [152, 164] (where a background star at a relatively high redshift passes close to a caustic in a galaxy cluster and is magnified by a huge factor [193]) and quasar microlensing [73]. See Chap. 23 for details.

1.5.4 Evidence and Constraints from Evaporation, Dynamical, Accretion and Nucleosynthetic Effects

In 2019 Arbey and Auffinger publicly released the `BlackHawk` code which calculates the evaporation spectra produced by any population of PBHs [14, 15]. Various observations (e.g. of MeV gamma-rays) were used to tightly constrain the abundance of PBHs lighter than 10^{17} g and limits on evaporating PBHs were overviewed in Refs. [48] in 2021, [16] in 2023. These results are discussed further in Chap. 19.

Also in 2019 Tian et al. [189] found that the separation distribution of halo-like wide binaries from GAIA-DR2 steepens at large radii, but is not sharply truncated, as would be expected if they are perturbed by compact objects. They argued that this, combined with the uncertainty in the initial separation distribution, makes it hard to probe PBHs using wide binaries. However, constraints were placed on heavier PBHs from their dynamical effects on stars in dwarf galaxies. Two-body interactions equalize the kinetic energies of populations, so if a significant fraction of the dark matter in dwarf galaxies is composed of PBHs with $M \gtrsim 10 M_\odot$, the stars there will gain energy and their distribution will expand [28, 187, 204]. In 2017 the effects of PBH accretion on the recombination history of the Universe were refined [175] and other present-day accretion probes were proposed [80]. In particular, accretion of interstellar gas onto PBHs in the Milky Way would lead to observable X-ray and radio emission [110, 145] and heat the gas in dwarf galaxies [138]. See Chaps. 13 and 24 for more details.

Fuller et al. [78] showed that some r-process elements (i.e. those generated by fast nuclear reactions) can be produced by the interaction of PBHs with neutron stars if they have $f > 0.01$ in the mass range $10^{-14} - 10^{-8}\, M_\odot$. Abramowicz and Bejger [2] argued that collisions of neutron stars with PBHs of mass 10^{23} g may explain the millisecond durations and large luminosities of fast radio bursts.

1.5.5 Evidence and Constraints from Cosmic Structure, Dwarf Galaxies and Supermassive Black Holes

An updated discussion of the effects of PBHs on the generation of cosmic structure, through the seed and Poisson effects was provided in Ref. [47]. The mass binding at redshift z_B is $4000\,M\,z_B^{-1}\,M_\odot$ for the seed effect and $10^7\,f\,M\,z_B^{-2}\,M_\odot$ for the Poisson effect. Having $f = 1$ requires $M < 10^3\,M_\odot$ and so the Poisson effect could only bind a scale below $10^{10}\,z_B^{-2}\,M_\odot$, which is necessarily subgalactic. However, providing the mass and dark matter fraction of the PBHs are large enough, this effect could generate the first bound objects at a much earlier epoch than in the standard cosmological scenario, with interesting observational consequences. This may explain the cosmic-infrared-background fluctuation excess first detected in Spitzer data [117], with a consistency check being provided by correlations of the source-subtracted infrared and X-ray cosmic backgrounds. This line of evidence has been studied extensively by Kashlinsky and colleagues [119, 120, 159] and more recently by Hasinger et al. [99] and Cappelutti et al. [35].

There are problems with the standard CDM scenario which Silk claims can be resolved with PBHs in the intermediate mass range [182]. In particular, he has argued that such PBHs could be ubiquitous in early dwarf galaxies, being mostly passive today but active in their gas-rich past. This is suggested by observations of AGN [18, 128, 171] and early feedback from these objects could provide a unified explanation for many dwarf galaxy anomalies. Besides providing a phase of early galaxy formation and seeds for SMBHs at high z, they could: (1) suppress the number of luminous dwarfs; (2) generate cores in dwarfs by dynamical heating; (3) resolve the "too big to fail" problem; (4) create bulgeless disks; (5) form ultra-faint dwarfs and ultra-diffuse galaxies; (6) reduce the baryon fraction in Milky-Way-type galaxies; (7) explain ultra-luminous X-ray sources in the outskirts of galaxies; (8) trigger star formation in dwarfs via AGN.

The mainstream view is that the SMBHs in galactic nuclei form from dynamical processes *after* galaxies, but this proposal is becoming increasingly challenged by the high mass and redshift of some SMBHs. For example, there is a population of red galaxies at $7.4 < z < 9.1$ with central SMBHs [130] and ALMA has observed an AGN at $z \sim 7$ whose large luminosity suggests a $10^9 M_\odot$ black hole [71]. It is unclear that such objects can form quickly enough in the standard model, so this suggests that the SMBHs—or at least their seeds—could form *before* galaxies [68]. Thus Liu and Bromm [137] have argued that unusually massive galaxies at $z > 10$ could be evidence that structure formation is accelerated by $10^9 M_\odot$ PBHs making up $10^{-6} - 10^{-3}$ of the dark matter (see Chap. 11). Reference [49] claims that the $10^6 M_\odot$ PBHs formed at the electron-positron annihilation epoch could provide the required seeds but Byrnes et al. [30] have argued that is precluded by limits on the μ-distortion in the CMB [56].

1.6 Conclusion

We are at a tantalizing point in the history of PBHs since their existence is likely to be either confirmed or excluded within the next decade. The community has a range of views as to which outcome is most likely, and even the authors of this chapter have different expectations. However, there is no doubt that there has been a huge surge of interest in PBHs in recent years and this is important because it has prompted more sophisticated studies of their expected characteristics and consequences. For researchers who advocate PBHs as a solution of the dark matter problem, it should be stressed that there is a division between those who favour PBHs in the asteroidal and solar mass ranges. The former view has the advantage that it is not excluded by current observations. The latter view is driven by the claimed evidence but is more controversial due to the constraints in this mass region.

References

1. B.P. Abbott et al., Observation of gravitational waves from a binary black hole merger. Phys. Rev. Lett. **116**(6), 061102 (2016). https://doi.org/10.1103/PhysRevLett.116.061102. arXiv: 1602.03837 [gr-qc]
2. M. Abramowicz, M. Bejger, M. Wielgus, Collisions of neutron stars with primordial black holes as fast radio bursts engines. Astrophys. J. **868**(1), 17 (2018). https://doi.org/10.3847/1538-4357/aae64a. arXiv: 1704.05931 [astro-ph.HE]
3. J. Adamek et al., WIMPs and stellar-mass primordial black holes are incompatible. Phys. Rev. D **100**(2), 023506 (2019). https://doi.org/10.1103/PhysRevD.100.023506. arXiv:1901.08528 [astro-ph.CO]
4. N. Afshordi, P. McDonald, D.N. Spergel, Primordial black holes as dark matter: the power spectrum and evaporation of early structures. Astrophys. J. Lett. **594**, L71–L74 (2003). https://doi.org/10.1086/378763. arXiv: astro-ph/0302035
5. Y. Akrami et al., Planck 2018 results. X. Constraints on inflation. Astron. Astrophys. **641**, A10 (2020). https://doi.org/10.1051/0004-6361/201833887. arXiv: 1807.06211 [astro-ph.CO]
6. A. Albrecht, P. Steinhardt, Cosmology for grand unified theories with radiatively induced symmetry breaking. Phys. Rev. Lett. **48**, 1220–1223 (1982). https://doi.org/10.1103/PhysRevLett.48.1220
7. A. Albrecht, N. Turok, Evolution of cosmic string networks. Phys. Rev. D **40**, 973–1001 (1989). https://doi.org/10.1103/PhysRevD.40.973. https://link.aps.org/doi/10.1103/PhysRevD.40.973
8. C. Alcock et al., The MACHO project LMC microlensing results from the first two years and the nature of the galactic dark halo. Astrophys. J. **486**, 697–726 (1997). https://doi.org/10.1086/304535. arXiv: astro-ph/9606165
9. C. Alcock et al., The MACHO project: microlensing results from 5.7 years of LMC observations. Astrophys. J. **542**, 281–307 (2000). https://doi.org/10.1086/309512. arXiv: astro-ph/0001272
10. C. Alcock et al., The MACHO project: Microlensing results from 5.7 years of LMC observations. Astrophys. J. **542**, 281–307 (2000). https://doi.org/10.1086/309512. arXiv: astro-ph/0001272
11. B. Allen, E.P.S. Shellard, Cosmic-string evolution: a numerical simulation. Phys. Rev. Lett. **64**, 119–122 (1990). https://doi.org/10.1103/PhysRevLett.64.119. https://link.aps.org/doi/10.1103/PhysRevLett.64.119

12. L. Amendola, J. Rubio, C. Wetterich, Primordial black holes from fifth forces. Phys. Rev. D **97**(8), 081302 (2018). https://doi.org/10.1103/PhysRevD.97.081302. arXiv: 1711.09915 [astro-ph.CO]
13. K.N. Ananda, C. Clarkson, D. Wands, The Cosmological gravitational wave background from primordial density perturbations. Phys. Rev. D **75**, 123518 (2007). https://doi.org/10.1103/PhysRevD.75.123518. arXiv: gr-qc/0612013
14. A. Arbey, J. Auffinger, BlackHawk: a public code for calculating the Hawking evaporation spectra of any black hole distribution. Eur. Phys. J. C **79**(8), 693 (2019). https://doi.org/10.1140/epjc/s10052-019-7161-1. arXiv: 1905.04268 [gr-qc]
15. A. Arbey, J. Auffinger, Physics beyond the standard model with BlackHawk v2.0. Eur. Phys. J. C **81**, 910 (2021). https://doi.org/10.1140/epjc/s10052-021-09702-8. arXiv: 2108.02737 [gr-qc]
16. J. Auffinger, Primordial black hole constraints with Hawking radiation–a review. Prog. Part. Nucl. Phys. **131**, 104040 (2023). https://doi.org/10.1016/j.ppnp.2023.104040. arXiv: 2206.02672 [astro-ph.CO]
17. J.N. Bahcall, P. Hut, S. Tremaine, Maximum mass of objects that constitute unseen disk material. Astrophys. J **290**, 15–20 (1985). https://doi.org/10.1086/162953
18. V. Baldassare et al., X-ray and ultraviolet properties of AGN in nearby Dwarf Galaxies. Astrophys. J. **836**(1), 20 (2017). https://doi.org/10.3847/1538-4357/836/1/20. arXiv: 1609.07148 [astro-ph.HE]
19. G. Ballesteros, M. Taoso, Primordial black hole dark matter from single field inflation. Phys. Rev. D **97**(2), 023501 (2018). https://doi.org/10.1103/PhysRevD.97.023501. arXiv: 1709.05565 [hep-ph]
20. J.M. Bardeen et al., The statistics of peaks of Gaussian random fields. Astrophys. J. **304**, 15–61 (1986). https://doi.org/10.1086/164143
21. B.A. Bassett, S. Tsujikawa, Inflationary preheating and primordial black holes. Phys. Rev. D **63**, 123503 (2001). https://doi.org/10.1103/PhysRevD.63.123503. arXiv: hep-ph/0008328
22. D. Baumann et al., Gravitational wave spectrum induced by primordial scalar perturbations. Phys. Rev. D **76**, 084019 (2007). https://doi.org/10.1103/PhysRevD.76.084019. arXiv: hep-th/0703290
23. G. Bertone et al., Primordial black holes as silver bullets for new physics at the weak scale. Phys. Rev. D **100**(12), 123013 (2019). https://doi.org/10.1103/PhysRevD.100.123013. arXiv: 1905.01238 [hep-ph]
24. S. Bird et al., Did LIGO detect dark matter? Phys. Rev. Lett. **116**(20), 201301 (2016). https://doi.org/10.1103/PhysRevLett.116.201301. arXiv: 1603.00464 [astro-ph.CO]
25. T. Blaineau et al., New limits from microlensing on Galactic black holes in the mass range 10 $M_\odot < M < 1000\ M_\odot$. Astron. Astrophys. **664**, A106 (2022). https://doi.org/10.1051/0004-6361/202243430. arXiv: 2202.13819 [astro-ph.GA]
26. R. Bond, B. Carr, Gravitational waves from a population of binary black holes. Mon. Not. Roy. astron. Soc. **207**, 585–609 (1984). https://doi.org/10.1093/mnras/207.3.585
27. M. Boudaud et al., In-depth analysis of the clustering of dark matter particles around primordial black holes I: density profiles (2021). arXiv: 2106.07480 [astro-ph.CO]
28. T.D. Brandt, Constraints on MACHO dark matter from compact stellar systems in ultra-faint Dwarf Galaxies. Astrophys. J. Lett. **824**(2), L31 (2016). https://doi.org/10.3847/2041-8205/824/2/L31. arXiv: 1605.03665 [astro-ph.GA]
29. J.S. Bullock, J.R. Primack, NonGaussian fluctuations and primordial black holes from inflation. Phys. Rev. D **55**, 7423–7439 (1997). https://doi.org/10.1103/PhysRevD.55.7423. arXiv: astro-ph/9611106
30. C. Byrnes, P. Cole, S. Patil, Steepest growth of the power spectrum and primordial black holes. JCAP **2019**, 028 (1906). https://doi.org/10.1088/1475-7516/2019/06/028. arXiv: 1811.11158 [astro-ph.CO]
31. C.T. Byrnes et al., Primordial black holes with an accurate QCD equation of state. JCAP **08**, 041 (2018). https://doi.org/10.1088/1475-7516/2018/08/041. arXiv: 1801.06138 [astro-ph.CO]

32. J. Calcino, J. Garcia-Bellido, T.M. Davis, Updating the MACHO fraction of the Milky Way dark halowith improved mass models. Mon. Not. Roy. Astron. Soc. **479**(3), 2889–2905 (2018). https://doi.org/10.1093/mnras/sty1368. arXiv: 1803.09205 [astro-ph.CO]
33. R.R. Caldwell, B. Allen, Cosmological constraints on cosmic string gravitational radiation. Phys. Rev. D **45**, 3447–3468 (1992). https://doi.org/10.1103/PhysRevD.45.3447
34. F. Capela, M. Pshirkov, P. Tinyakov, Constraints on primordial black holes as dark matter candidates from capture by neutron stars. Phys. Rev. D **87**(12), 123524 (2013). https://doi.org/10.1103/PhysRevD.87.123524. arXiv: 1301.4984 [astro-ph.CO]
35. N. Cappelluti, G. Hasinger, P. Natarajan, Exploring the high-redshift PBH-ΛCDM Universe: early black hole seeding, the first stars and cosmic radiation backgrounds. Astrophys. J. **926**(2), 205 (2022). https://doi.org/10.3847/1538-4357/ac332d. arXiv: 2109.08701 [astro-ph.CO]
36. B.J. Carr, Pregalactic black hole accretion and the thermal history of the universe. Mon. Not. Roy. Astron. Soc. **194**, 639–668 (1981). https://doi.org/10.1093/mnras/194.3.639
37. B.J. Carr, Some cosmological consequences of primordial black-hole evaporations. Astrophys. J. **206**, 8–25 (1976). https://doi.org/10.1086/154351
38. B.J. Carr et al., Constraints on primordial black holes from the Galactic gamma-ray background. Phys. Rev. D **94**(4), 044029 (2016). https://doi.org/10.1103/PhysRevD.94.044029. arXiv: 1604.05349 [astro-ph.CO]
39. B.J. Carr et al., New cosmological constraints on primordial black holes. Phys. Rev. D **81**, 104019 (2010). https://doi.org/10.1103/PhysRevD.81.104019. arXiv: 0912.5297 [astro-ph.CO]
40. B. Carr, The Primordial black hole mass spectrum. Astrophys. J. **201**, 1–19 (1975). https://doi.org/10.1086/153853
41. B. Carr, J. Gilbert, J. Lidsey, Black hole relics and inflation: limits on blue perturbation spectra. Phys. Rev. D **50**, 4853–4867 (1994). https://doi.org/10.1103/PhysRevD.50.4853. arXiv: astro-ph/9405027
42. B. Carr, F. Kuhnel, L. Visinelli, Black holes and WIMPs: all or nothing or something else. Mon. Not. Roy. Astron. Soc. **506**(3), 3648–3661 (2021). https://doi.org/10.1093/mnras/stab1930. arXiv: 2011.01930 [astro-ph.CO]
43. B. Carr, F. Kühnel, M. Sandstad, Primordial black holes as dark matter. Phys. Rev. D **94**(8), 083504 (2016). https://doi.org/10.1103/PhysRevD.94.083504
44. B. Carr, C. Lacey, Dark clusters in galactic halos? Astrophys. J. **316**, 23–35 (1987). https://doi.org/10.1086/165176
45. B. Carr, M.J. Rees, Can pregalactic objects generate galaxies? Mon. Not. Roy. Astron. Soc. **206**, 801–818 (1984). https://doi.org/10.1093/mnras/206.4.801
46. B. Carr, J. Silk, Can graininess in the early universe make galaxies? Astrophys. J. **268**, 1–16 (1983). https://doi.org/10.1086/160924
47. B. Carr, J. Silk, Primordial black holes as generators of cosmic structures. Mon. Not. Roy. Astron. Soc. **478**(3), 3756–3775 (2018). https://doi.org/10.1093/mnras/sty1204
48. B. Carr et al., Constraints on primordial black holes. Rept. Prog. Phys. **84**(11), 116902 (2021). https://doi.org/10.1088/1361-6633/ac1e31. arXiv: 2002.12778 [astro-ph.CO]
49. B. Carr et al., Cosmic conundra explained by thermal history and primordial black holes. Phys. Dark Univ. **31**, 100755 (2021). https://doi.org/10.1016/j.dark.2020.100755. arXiv: 1906.08217 [astro-ph.CO]
50. B. Carr et al., Observational evidence for primordial black holes: a positivist perspective. Phys. Rept. **1054**, 1–68 (2024). https://doi.org/10.1016/j.physrep.2023.11.005. arXiv: 2306.03903 [astro-ph.CO]
51. B. Carr et al., Primordial black hole constraints for extended mass functions. Phys. Rev. D **96**(2), 023514 (2017). https://doi.org/10.1103/PhysRevD.96.023514. arXiv: 1705.05567 [astro-ph.CO]
52. B.J. Carr, S.W. Hawking, Black holes in the early Universe. Mon. Not. Roy. Astron. Soc. **168**, 399–415 (1974). https://doi.org/10.1093/mnras/168.2.399

53. B.J. Carr, J.E. Lidsey, Primordial black holes and generalized constraints on chaotic inflation. Phys. Rev. D **48**, 543–553 (1993). https://doi.org/10.1103/PhysRevD.48.543
54. B.J. Carr, M. Sakellariadou, Dynamical constraints on dark compact objects. Astrophys. J. **516**, 195–220 (1999). https://doi.org/10.1086/307071
55. G. Chapline, Cosmological effects of primordial black holes. Nature (London) **253**, 251–+ (1975). https://doi.org/10.1038/253251a0
56. J. Chluba, A.L. Erickcek, I. Ben-Dayan, Probing the inflaton: small-scale power spectrum constraints from measurements of the CMB energy spectrum. Astrophys. J. **758**, 76 (2012). https://doi.org/10.1088/0004-637X/758/2/76. arXiv: 1203.2681 [astro-ph.CO]
57. M.W. Choptuik, Universality and scaling in gravitational collapse of a massless scalar field. Phys. Rev. Lett. **70**, 9–12 (1993). https://doi.org/10.1103/PhysRevLett.70.9
58. S. Clesse, J. García-Bellido, Seven hints for primordial black hole dark matter. Phys. Dark Univ. **22**, 137–146 (2018). https://doi.org/10.1016/j.dark.2018.08.004. arXiv: 1711.10458 [astro-ph.CO]
59. S. Clesse, J. García-Bellido, The clustering of massive primordial black holes as dark matter: measuring their mass distribution with advanced LIGO. Phys. Dark Univ. **15**, 142–147 (2017). https://doi.org/10.1016/j.dark.2016.10.002. arXiv: 1603.05234 [astro-ph.CO]
60. D.B. Cline, D.A. Sanders, W. Hong, Further evidence for gamma ray bursts consistent with primordial black hole evaporation. Astrophys. J. **486**, 169–178 (1997). https://doi.org/10.1086/304480
61. P.S. Cole et al., Primordial black holes from single-field inflation: a fine-tuning audit. JCAP **08**, 031 (2023). https://doi.org/10.1088/1475-7516/2023/08/031 . arXiv: 2304.01997 [astro-ph.CO]
62. E. Cotner, A. Kusenko, Primordial black holes from supersymmetry in the early universe. Phys. Rev. Lett. **119**(3), 031103 (2017). https://doi.org/10.1103/PhysRevLett.119.031103. arXiv: 1612.02529 [astro-ph.CO]
63. M. Crawford, D.N. Schramm, Spontaneous generation of density perturbations in the early universe. Nature **298**, 538–540 (1982). https://doi.org/10.1038/298538a0
64. D. Croon et al., Subaru-HSC through a different lens: microlensing by extended dark matter structures. Phys. Rev. D **102**(8), 083021 (2020). https://doi.org/10.1103/PhysRevD.102.083021. arXiv: 2007.12697 [astro-ph.CO]
65. V. De Luca et al., Primordial black holes confront LIGO/Virgo data: current situation. JCAP **06**, 044 (2020). https://doi.org/10.1088/1475-7516/2020/06/044. arXiv: 2005.05641 [astro-ph.CO]
66. V. De Luca et al., The ineludible non-Gaussianity of the primordial black hole abundance. JCAP **07**, 048 (2019). https://doi.org/10.1088/1475-7516/2019/07/048. arXiv: 1904.00970 [astro-ph.CO]
67. V. De Luca et al., Ruling out initially clustered primordial black holes as dark matter. Phys. Rev. Lett. **129**(19), 191302 (2022). https://doi.org/10.1103/PhysRevLett.129.191302. arXiv: 2208.01683 [astro-ph.CO]
68. A. Dolgov, Tension between HST/JWST and ΛCDM cosmology, PBH, and antimatter in the galaxy, in *14th Frascati workshop on Multifrequency Behaviour of High Energy Cosmic Sources* (2023). arXiv: 2310.00671 [astro-ph.CO]
69. A. Dolgov, J. Silk, Baryon isocurvature fluctuations at small scales and baryonic dark matter. Phys. Rev. D **47**, 4244–4255 (1993). https://doi.org/10.1103/PhysRevD.47.4244. https://link.aps.org/doi/10.1103/PhysRevD.47.4244
70. G. Dvali, F. Kühnel, M. Zantedeschi, Vortices in black holes. Phys. Rev. Lett. **129**(6), 061302 (2022). https://doi.org/10.1103/PhysRevLett.129.061302. arXiv: 2112.08354 [hep-th]
71. R. Endsley et al., ALMA confirmation of an obscured hyperluminous radio-loud AGN. Monthly Notices R. Astron. Soc. **520**(3), 4609–4620 (2023). https://doi.org/10.1093/mnras/stad266. https://doi.org/10.1093%20%2Fmnras%2Fstad266
72. A. Escrivà, C. Germani, R.K. Sheth, Universal threshold for primordial black hole formation. Phys. Rev. D **101**(4), 044022 (2020). https://doi.org/10.1103/PhysRevD.101.044022. arXiv: 1907.13311 [gr-qc]

73. A. Esteban-Gutiérrez et al., Constraints on the abundance of primordial black holes from x-ray quasar microlensing observations: substellar to planetary mass range. Astrophys. J. **954**(2), 172 (2023). https://doi.org/10.3847/1538-4357/ace62f. arXiv: 2307.07473 [astro-ph.CO]
74. B.D. Fields, K. Freese, D.S. Graff, Chemical abundance constraints on white dwarfs as halo dark matter. Astrophys. J. **534**, 265–276 (2000). https://doi.org/10.1086/308727. arXiv: astro-ph/9904291
75. M.M. Flores, A. Kusenko, Primordial black holes from long-range scalar forces and scalar radiative cooling. Phys. Rev. Lett. **126**(4), 041101 (2021). https://doi.org/10.1103/PhysRevLett.126.041101. arXiv: 2008.12456 [astro-ph.CO]
76. G. Franciolini et al., Searching for a subpopulation of primordial black holes in LIGO-Virgo gravitational-wave data. Phys. Rev. D **105**(8), 083526 (2022). https://doi.org/10.1103/PhysRevD.105.083526. arXiv: 2105.03349 [gr-qc]
77. K. Freese, R. Price, D. Schramm, Formation of population III stars and galaxies with primordial planetary-mass black holes. Astrophys. J. **275**, 405–412 (1983). https://doi.org/10.1086/161542
78. G. Fuller, A. Kusenko, V. Takhistov, Primordial black holes and r-process nucleosynthesis. Phys. Rev. Lett. **119**(6), 061101 (2017). https://doi.org/10.1103/PhysRevLett.119.061101. arXiv: 1704.01129 [astro-ph.HE]
79. J. Fumagalli et al., Turning in the landscape: a new mechanism for generating primordial black holes. Phys. Lett. B **841**, 137921 (2023). https://doi.org/10.1016/j.physletb.2023.137921. arXiv: 2004.08369 [hep-th]
80. D. Gaggero et al., Searching for primordial black holes in the radio and x-ray sky. Phys. Rev. Lett. **118**(24), 241101 (2017). https://doi.org/10.1103/PhysRevLett.118.241101. arXiv: 1612.00457 [astro-ph.HE]
81. J. Garcia-Bellido, M. Hawkins, Reanalysis of the MACHO constraints on PBH in the light of Gaia DR3 data (2024). arXiv: 2402.00212 [astro-ph.GA]
82. J. Garcia-Bellido, A.D. Linde, D. Wands, Density perturbations and black hole formation in hybrid inflation. Phys. Rev. D **54**, 6040–6058 (1996). https://doi.org/10.1103/PhysRevD.54.6040. arXiv: astro-ph/9605094
83. J. Garcia-Bellido, E.R. Morales, Primordial black holes from single field models of inflation. Phys. Dark Univ. **18**, 47–54 (2017). https://doi.org/10.1016/j.dark.2017.09.007. arXiv: 1702.03901 [astro-ph.CO]
84. J. García-Bellido, Massive primordial black holes as dark matter and their detection with gravitational waves. J. Phys. Conf. Ser. **840**(1), 012032 (2017). https://doi.org/10.1088/1742-6596/840/1/012032. arXiv: 1702.08275 [astro-ph.CO]
85. J. García-Bellido, B. Carr, S. Clesse, Primordial black holes and a common origin of baryons and dark matter. Universe **8**(1), 12 (2021). https://doi.org/10.3390/universe8010012. arXiv: 1904.11482 [astro-ph.CO]
86. C. Germani, I. Musco, Abundance of primordial black holes depends on the shape of the inflationary power spectrum. Phys. Rev. Lett. **122**(14), 141302 (2019). https://doi.org/10.1103/PhysRevLett.122.141302. arXiv: 1805.04087 [astro-ph.CO]
87. M. Gorton, A.M. Green, Effect of clustering on primordial black hole microlensing constraints. JCAP **08**(08), 035 (2022). https://doi.org/10.1088/1475-7516/2022/08/035. arXiv: 2203.04209 [astro-ph.CO]
88. P.W. Graham, S. Rajendran, J. Varela, Dark matter triggers of supernovae. Phys. Rev. D **92**(6), 063007 (2015). https://doi.org/10.1103/PhysRevD.92.063007. arXiv: 1505.04444 [hep-ph]
89. A.M. Green, Microlensing and dynamical constraints on primordial black hole dark matter with an extended mass function. Phys. Rev. D **94**(6), 063530 (2016). https://doi.org/10.1103/PhysRevD.94.063530. arXiv: 1609.01143 [astro-ph.CO]
90. A.M. Green, B.J. Kavanagh, Primordial Black Holes as a dark matter candidate. J. Phys. G **48**(4), 043001 (2021). https://doi.org/10.1088/1361-6471/abc534. arXiv: 2007.10722 [astro-ph.CO]

91. A.M. Green, K.A. Malik, Primordial black hole production due to preheating. Phys. Rev. D **64**, 021301 (2001). https://doi.org/10.1103/PhysRevD.64.021301. arXiv: hep-ph/0008113
92. A.M. Green et al., A New calculation of the mass fraction of primordial black holes. Phys. Rev. D **70**, 041502 (2004). https://doi.org/10.1103/PhysRevD.70.041502. arXiv: astro-ph/0403181
93. A.H. Guth, The inflationary universe: a possible solution to the horizon and flatness problems. Phys. Rev. D **23**, 347–356 (1981). Ed. by Li-Zhi Fang, R. Ruffini. https://doi.org/10.1103/PhysRevD.23.347
94. A.H. Guth, S.Y. Pi, Fluctuations in the new inflationary universe. Phys. Rev. Lett. **49**, 1110–1113 (1982). https://doi.org/10.1103/PhysRevLett.49.1110
95. A. Hall, A.D. Gow, C.T. Byrnes, Bayesian analysis of LIGO-Virgo mergers: primordial vs. astrophysical black hole populations. Phys. Rev. D **102**, 123524 (2020)
96. T. Harada et al., Primordial black hole formation in the matter-dominated phase of the Universe. Astrophys. J. **833**(1), 61 (2016). https://doi.org/10.3847/1538-4357/833/1/61. arXiv: 1609.01588 [astro-ph.CO]
97. T. Harada et al., Spins of primordial black holes formed in the matter-dominated phase of the Universe. Phys. Rev. D **96**(8), 083517 (2017). [Erratum: Phys. Rev. D **99**, 069904 (2019)]. https://doi.org/10.1103/PhysRevD.96.083517. arXiv: 1707.03595 [gr-qc]
98. E.R. Harrison, Fluctuations at the threshold of classical cosmology. Phys. Rev. D **1**(10), 2726–2730 (1970). https://doi.org/10.1103/PhysRevD.1.2726
99. G. Hasinger, Illuminating the dark ages: cosmic backgrounds from accretion onto primordial black hole dark matter. JCAP **07**, 022 (2020). https://doi.org/10.1088/1475-7516/2020/07/022. arXiv: 2003.05150 [astro-ph.CO]
100. S.W. Hawking, Black hole explosions. Nature **248**, 30–31 (1974). https://doi.org/10.1038/248030a0
101. S.W. Hawking, Black holes from cosmic strings. Phys. Lett. B **231**, 237–239 (1989). https://doi.org/10.1016/0370-2693(89)90206-2
102. S.W. Hawking, The development of irregularities in a single bubble inflationary universe. Phys. Lett. B **115**, 295 (1982). https://doi.org/10.1016/0370-2693(82)90373-2
103. S.W. Hawking, I.G. Moss, J.M. Stewart, Bubble collisions in the very early universe. Phys. Rev. D **26**, 2681 (1982). https://doi.org/10.1103/PhysRevD.26.2681
104. S. Hawking, Gravitationally collapsed objects of very low mass. Mon. Not. Roy. Astron. Soc. **152**, 75 (1971)
105. M.R.S. Hawkins, Gravitational microlensing, quasar variability and missing matter. Nature **366**(6452), 242–245 (1993). https://doi.org/10.1038/366242a0
106. A.F. Heckler, On the formation of a Hawking radiation photosphere around microscopic black holes. Phys. Rev. D **55**, 480–488 (1997). https://doi.org/10.1103/PhysRevD.55.480. arXiv: astro-ph/9601029
107. M.P. Hertzberg, M. Yamada, Primordial black holes from polynomial potentials in single field inflation. Phys. Rev. D **97**(8), 083509 (2018). https://doi.org/10.1103/PhysRevD.97.083509. arXiv: 1712.09750 [astro-ph.CO]
108. G. Hütsi et al., Two populations of LIGO-Virgo black holes. JCAP **03**, 068 (2021). https://doi.org/10.1088/1475-7516/2021/03/068. arXiv: 2012.02786 [astro-ph.CO]
109. D. Inman, Y. Ali-Haïmoud, Early structure formation in primordial black hole cosmologies. Phys. Rev. D **100**(8), 083528 (2019). https://doi.org/10.1103/PhysRevD.100.083528. arXiv: 1907.08129 [astro-ph.CO]
110. Y. Inoue, A. Kusenko, New X-ray bound on density of primordial black holes. JCAP **10**, 034 (2017). https://doi.org/10.1088/1475-7516/2017/10/034. arXiv: 1705.00791 [astro-ph.CO]
111. K. Ioka et al., Black hole binary formation in the expanding universe: three body problem approximation. Phys. Rev. D **58**, 063003 (1998). https://doi.org/10.1103/PhysRevD.58.063003. arXiv: astro-ph/9807018
112. P. Ivanov, Nonlinear metric perturbations and production of primordial black holes. Phys. Rev. D **57**, 7145–7154 (1998). https://doi.org/10.1103/PhysRevD.57.7145. arXiv: astro-ph/9708224

113. P. Ivanov, P. Naselsky, I. Novikov, Inflation and primordial black holes as dark matter. Phys. Rev. D **50**, 7173–7178 (1994). https://doi.org/10.1103/PhysRevD.50.7173
114. K. Jedamzik, Could MACHOs be primordial black holes formed during the QCD epoch? Phys. Rept. **307**, 155–162 (1998). https://doi.org/10.1016/S0370-1573(98)00067-2. arXiv: astro-ph/9805147 [astro-ph]
115. K. Jedamzik, Primordial black hole formation during the QCD epoch. Phys. Rev. D **55**, 5871–5875 (1997). https://doi.org/10.1103/PhysRevD.55.R5871. arXiv: astro-ph/9605152
116. A.C. Jenkins, M. Sakellariadou, Primordial black holes from cusp collapse on cosmic strings (2020). arXiv: 2006.16249 [astro-ph.CO]
117. A. Kashlinsky et al., Tracing the first stars with fluctuations of the cosmic infrared background. Nature **438**(7064), 45–50 (2005). https://doi.org/10.1038/nature04143. arXiv: astro-ph/0511105 [astro-ph]
118. A. Kashlinsky, LIGO gravitational wave detection, primordial black holes, and the near-IR cosmic infrared background anisotropies. Astrophys. J. Lett. **823**(2), L25 (2016). https://doi.org/10.3847/2041-8205/823/2/L25. arXiv: 1605.04023 [astro-ph.CO]
119. A. Kashlinsky et al., Looking at cosmic near-infrared background radiation anisotropies. Reviews of Modern Physics **90**(2), 025006 (2018). https://doi.org/10.1103/RevModPhys.90.025006. arXiv: 1802.07774 [astro-ph.CO]
120. A. Kashlinsky et al., Probing the cross-power of unresolved cosmic infrared/X-ray backgrounds with upcoming space missions. Astrophys. J. Lett. **871**(1), L6 (2019). https://doi.org/10.3847/2041-8213/aafaf6. arXiv: 1812.01535 [astro-ph.CO]
121. B.J. Kavanagh, *PBHbounds, 10.5281/zenodo.3538998*. https://doi.org/10.5281/zenodo.3538998
122. M. Kawasaki, N. Sugiyama, T. Yanagida, Primordial black hole formation in a double inflation model in supergravity. Phys. Rev. D **57**, 6050–6056 (1998). https://doi.org/10.1103/PhysRevD.57.6050. arXiv: hep-ph/9710259
123. M. Kawasaki, H. Nakatsuka, Effect of nonlinearity between density and curvature perturbations on the primordial black hole formation. Phys. Rev. D **99**(12), 123501 (2019). https://doi.org/10.1103/PhysRevD.99.123501. arXiv: 1903.02994 [astro-ph.CO]
124. M.Y. Khlopov, A.G. Polnarev, Primordial black holes as a cosmological test of grand unification. Phys. Lett. B **97**, 383–387 (1980). https://doi.org/10.1016/0370-2693(80)90624-3
125. T.W.B. Kibble, Topology of cosmic domains and strings. J. Phys. A **9**, 1387–1398 (1976). https://doi.org/10.1088/0305-4470/9/8/029
126. H. Kodama, M. Sasaki, K. Sato, Abundance of primordial holes produced by cosmological first order phase transition. Prog. Theor. Phys. **68**, 1979 (1982). https://doi.org/10.1143/PTP.68.1979
127. K. Kohri, D.H. Lyth, A. Melchiorri, Black hole formation and slow-roll inflation. JCAP **04**, 038 (2008). https://doi.org/10.1088/1475-7516/2008/04/038. arXiv: 0711.50060 [hep-ph]
128. J. Kormendy, L.C. Ho, Coevolution (or not) of supermassive black holes and host galaxies. Ann. Rev. Astron. Astrophys. **51**(1), 511–653 (2013). https://doi.org/10.1146/annurev-astro-082708-101811. arXiv: 1304.7762 [astro-ph.CO]
129. E. Kovetz, Probing primordial-black-hole dark matter with gravitational waves. Phys. Rev. Lett. **119**(13), 131301 (2017). https://doi.org/10.1103/PhysRevLett.119.131301. arXiv: 1705.09182 [astro-ph.CO]
130. I. Labbé et al., A population of red candidate massive galaxies $\sim$600 Myr after the Big Bang. Nature **616**(7956), 266–269 (2023). https://doi.org/10.1038/s41586-023-05786-2.https://doi.org/10.1038%20%2Fs41586-023-05786-2
131. C.G. Lacey, The disk heating problem. Dynamics of disc galaxies, in ed. by B. Sundelius (1991), p. 257
132. C. Lacey, J. Ostriker, Massive black holes in galactic halos? Astrophys. J. **299**, 633–652 (1985). https://doi.org/10.1086/163729
133. B.C. Lacki, J.F. Beacom, Primordial black holes as dark matter: almost all or almost nothing. Astrophys. J. Lett. **720**, L67–L71 (2010). https://doi.org/10.1088/2041-8205/720/1/L67. arXiv: 1003.3466 [astro-ph.CO]

134. R. Lehoucq et al., New constraints on the primordial black hole number density from Galactic γ-ray astronomy. Astron. Astrophys. **502**, 37–43 (2009). https://doi.org/10.1051/0004-6361/200911961. arXiv: 0906.1648 [astro-ph.HE]
135. A.D. Linde, Scalar field fluctuations in expanding universe and the new inflationary universe scenario. Phys. Lett. B **116**, 335–339 (1982). https://doi.org/10.1016/0370-2693(82)90293-3
136. D. Lindley, Primordial black holes and the deuterium abundance. Mon. Not. Roy. Astron. Soc. **193**, 593–601 (1980). https://doi.org/10.1093/mnras/193.3.593
137. B. Liu, V. Bromm, Accelerating early massive galaxy formation with primordial black holes. Astrophys. J. Lett. **937**(2), L30 (2022). https://doi.org/10.3847/2041-8213/ac927f. arXiv: 2208.13178 [astro-ph.CO]
138. P. Lu et al., Constraining primordial black holes with Dwarf Galaxy heating. Astrophys. J. Lett. **908**(2), L23 (2021). https://doi.org/10.3847/2041-8213/abdcb6. arXiv: 2007.02213 [astro-ph.CO]
139. J.H. MacGibbon, B.R. Webber, Quark and gluon jet emission from primordial black holes: the instantaneous spectra. Phys. Rev. D **41**, 3052–3079 (1990). https://doi.org/10.1103/PhysRevD.41.3052
140. J.H. MacGibbon, Quark and gluon jet emission from primordial black holes. 2. The Lifetime emission. Phys. Rev. D **44**, 376–392 (1991). https://doi.org/10.1103/PhysRevD.44.376
141. J.H. MacGibbon, R.H. Brandenberger, U.F. Wichoski, Limits on black hole formation from cosmic string loops. Phys. Rev. D **57**, 2158–2165 (1998). https://doi.org/10.1103/PhysRevD.57.2158. arXiv: astro-ph/9707146
142. J.H. MacGibbon, B.J. Carr, Cosmic rays from primordial black holes. Astrophys. J. **371**, 447–469 (1991). https://doi.org/10.1086/169909
143. J.H. MacGibbon, B.J. Carr, D.N. Page, Do evaporating black holes form photospheres? Phys. Rev. D **78**, 064043 (2008). https://doi.org/10.1103/PhysRevD.78.064043. arXiv: 0709.2380 [astro-ph]
144. K.J. Mack, J.P. Ostriker, M. Ricotti, Growth of structure seeded by primordial black holes. Astrophys. J. **665**, 1277–1287 (2007). https://doi.org/10.1086/518998. arXiv: astro-ph/0608642
145. J. Manshanden et al., Multi-wavelength astronomical searches for primordial black holes. JCAP **06**, 026 (2019). https://doi.org/10.1088/1475-7516/2019/06/026. arXiv: 1812.07967 [astro-ph.HE]
146. P. Mészáros, Primeval black holes and galaxy formation. Astron. Astrophys. **38**, 5–13 (1975)
147. S.S. Mishra, V. Sahni, Primordial Black Holes from a tiny bump/dip in the Inflaton potential. JCAP **04**, 007 (2020). https://doi.org/10.1088/1475-7516/2020/04/007. arXiv: 1911.00057 [gr-qc]
148. S. Miyama, K. Sato, The upper bound of the number density of primordial black holes from the big bang nucleosynthesis. Prog. Theor. Phys. **59**, 1012 (1978). https://doi.org/10.1143/PTP.59.1012
149. M.A. Monroy-Rodríguez, C. Allen, The end of the MACHO Era, revisited: new limits on MACHO masses from Halo wide binaries. Astrophys. J. **790**(2), 159 (2014). https://doi.org/10.1088/0004-637X/790/2/159. arXiv: 1406.5169 [astro-ph.GA]
150. P. Montero-Camacho et al., Revisiting constraints on asteroid-mass primordial black holes as dark matter candidates. JCAP **08**, 031 (2019). https://doi.org/10.1088/1475-7516/2019/08/031. arXiv: 1906.05950 [astro-ph.CO]
151. P. Mroz et al., No massive black holes in the Milky Way halo (2024). arXiv: 2403.02386 [astro-ph.GA]
152. C.V. Müller, J. Miralda-Escudé, Limits on dark matter compact objects implied by supermagnified stars in lensing clusters (2024). arXiv: 2403.16989 [astro-ph.CO]
153. I. Musco, J.C. Miller, Primordial black hole formation in the early universe: critical behaviour and self-similarity. Class. Quant. Grav. **30**, 145009 (2013). https://doi.org/10.1088/0264-9381/30/14/145009. arXiv: 1201.2379 [gr-qc]

154. I. Musco, J.C. Miller, A.G. Polnarev, Primordial black hole formation in the radiative era: investigation of the critical nature of the collapse. Class. Quant. Grav. **26**, 235001 (2009). https://doi.org/10.1088/0264-9381/26/23/235001. arXiv: 0811.14520 [gr-qc]
155. I. Musco, J.C. Miller, L. Rezzolla, Computations of primordial black hole formation. Class. Quant. Grav. **22**, 1405–1424 (2005). https://doi.org/10.1088/0264-9381/22/7/013. arXiv: gr-qc/0412063
156. D.K. Nadezhin, I.D. Novikov, A.G. Polnarev, The hydrodynamics of primordial black hole formation. Sov. Astron. **22**, 129–138 (1978)
157. T. Nakama et al., Identifying the most crucial parameters of the initial curvature profile for primordial black hole formation. JCAP **01**, 037 (2014). https://doi.org/10.1088/1475-7516/2014/01/037. arXiv: 1310.3007 [gr-qc]
158. T. Nakamura et al., Gravitational waves from coalescing black hole MACHO binaries. Astrophys. J. Lett. **487**, L139–L142 (1997). https://doi.org/10.1086/310886. arXiv: astro-ph/9708060
159. N. Cappelluti et al., Cross-correlating cosmic infrared and x-ray background fluctuations: evidence of significant black hole populations among the CIB sources. Astrophys. J. **769**(1), 68 (2013). https://doi.org/10.1088/0004-637X/769/1/68. arXiv: 1210.5302 [astro-ph.CO]
160. J.C. Niemeyer, K. Jedamzik, Dynamics of primordial black hole formation. Phys. Rev. D **59**, 124013 (1999). https://doi.org/10.1103/PhysRevD.59.124013. arXiv: astro-ph/9901292
161. J.C. Niemeyer, K. Jedamzik, Near-critical gravitational collapse and the initial mass function of primordial black holes. Phys. Rev. Lett. **80**, 5481–5484 (1998). https://doi.org/10.1103/PhysRevLett.80.5481. arXiv: astro-ph/9709072
162. H. Niikura et al., Microlensing constraints on primordial black holes with Subaru/HSC Andromeda observations. Nature Astron. **3**(6), 524–534 (2019). https://doi.org/10.1038/s41550-019-0723-1. arXiv: 1701.02151 [astro-ph.CO]
163. I.D. Novikov et al., Primordial black holes. Astron. Astrophys. **80**, 104–109 (1979)
164. M. Oguri et al., Understanding caustic crossings in giant arcs: characteristic scales, event rates, and constraints on compact dark matter. Phys. Rev. D **97**(2), 023518 (2018). https://doi.org/10.1103/PhysRevD.97.023518. arXiv: 1710.00148 [astro-ph.CO]
165. P.N. Okele, M.J. Rees, Observational consequences of positron production by evaporating black holes. Astron. Astrophys. **81**, 263–+ (1980)
166. O. Özsoy, G. Tasinato, Inflation and primordial black holes. Universe **9**(5), 203 (2023). https://doi.org/10.3390/universe9050203. arXiv: 2301.03600 [astro-ph.CO]
167. B. Paczynski, Gravitational microlensing by the galactic halo. Astrophys. J. **304**, 1–5 (1986). https://doi.org/10.1086/164140
168. D. Page, S. Hawking, Gamma rays from primordial black holes. Astrophys. J. **206**, 1–7 (1976). https://doi.org/10.1086/154350
169. G.A. Palma, S. Sypsas, C. Zenteno, Seeding primordial black holes in multifield inflation. Phys. Rev. Lett. **125**(12), 121301 (2020). https://doi.org/10.1103/PhysRevLett.125.121301. arXiv: 2004.06106 [astro-ph.CO]
170. P. Pani, A. Loeb, Tidal capture of a primordial black hole by a neutron star: implications for constraints on dark matter. JCAP **06**, 026 (2014). https://doi.org/10.1088/1475-7516/2014/06/026. arXiv: 1401.3025 [astro-ph.CO]
171. K. Pardo et al., X-ray detected active galactic nuclei in Dwarf Galaxies at 0. Astrophys. J. **831**, 203 (2016). https://doi.org/10.3847/0004-637X/831/2/203. arXiv: 1603.01622
172. M. Petač, J. Lavalle, K. Jedamzik, Microlensing constraints on clustered primordial black holes. Phys. Rev. D **105**(8), 083520 (2022). https://doi.org/10.1103/PhysRevD.105.083520. arXiv: 2201.02521 [astro-ph.CO]
173. A.G. Polnarev, M.Y. Khlopov, The stage of superheavy particle dominance in the universe and primordial black holes. Astron. Zh. **58**, 706–716 (1981)
174. A. Polnarev, R. Zembowicz, Formation of primordial black holes by cosmic strings. Phys. Rev. D **43**, 1106–1109 (1991). https://doi.org/10.1103/PhysRevD.43.1106
175. V. Poulin et al., CMB bounds on disk-accreting massive primordial black holes. Phys. Rev. D **96**(8), 083524 (2017). https://doi.org/10.1103/PhysRevD.96.083524. arXiv: 1707.04206 [astro-ph.CO]

176. D.P. Quinn et al., On the reported death of the MACHO era. Mon. Not. Roy. Astron. Soc. **396**(1), L11–L15 (2009). https://doi.org/10.1111/j.1745-3933.2009.00652.x. arXiv: 0903.1644 [astro-ph.GA]
177. M. Ricotti, J.P. Ostriker, K.J. Mack, Effect of primordial black holes on the cosmic microwave background and cosmological parameter estimates. Astrophys. J. **680**, 829 (2008). https://doi.org/10.1086/587831. arXiv: 0709.0524 [astro-ph]
178. R. Saito, J. Yokoyama, Gravitational wave background as a probe of the primordial black hole abundance. Phys. Rev. Lett. **102**, 161101 (2009). [Erratum: Phys.Rev.Lett. 107, 069901 (2011)]. https://doi.org/10.1103/PhysRevLett.102.161101. arXiv: 0812.4339 [astro-ph]
179. M. Sasaki et al., Primordial black hole scenario for the gravitational-wave event GW150914. Phys. Rev. Lett. **117**(6), 061101 (2016). [Erratum: Phys.Rev.Lett. 121, 059901 (2018)]. https://doi.org/10.1103/PhysRevLett.117.061101. arXiv: 1603.08338 [astro-ph.CO]
180. M. Sasaki et al., Primordial black holes–perspectives in gravitational wave astronomy. Class. Q. Grav. **35**(6), 063001 (2018). https://doi.org/10.1088/1361-6382/aaa7b4. arXiv: 1801.05235 [astro-ph.CO]
181. K. Sato, First order phase transition of a vacuum and expansion of the universe. Mon. Not. Roy. astron. Soc. **195**, 467–479 (1981)
182. J. Silk, Feedback by massive black holes in gas-rich Dwarf Galaxies. Astrophys. J. Lett. **839**(1), L13 (2017). https://doi.org/10.3847/2041-8213/aa67da. arXiv: 1703.08553 [astro-ph.GA]
183. G.F. Smoot et al., Structure in the COBE differential microwave radiometer first year maps. Astrophys. J. Lett. **396**, L1–L5 (1992). https://doi.org/10.1086/186504
184. N. Smyth et al., Updated constraints on asteroid-mass primordial black holes as dark matter. Phys. Rev. D **101**(6), 063005 (2020). https://doi.org/10.1103/PhysRevD.101.063005. arXiv: 1910.01285 [astro-ph.CO]
185. A.A. Starobinsky, A new type of isotropic cosmological models without singularity. Phys. Lett. B **91**, 99–102 (1980). Ed. by I.M. Khalatnikov, V.P. Mineev. https://doi.org/10.1016/0370-2693(80)90670-X
186. A.A. Starobinsky, Dynamics of phase transition in the new inflationary universe scenario and generation of perturbations. Phys. Lett. B **117**, 175–178 (1982). https://doi.org/10.1016/0370-2693(82)90541-X
187. J. Stegmann et al., Improved constraints from ultra-faint dwarf galaxies on primordial black holes as dark matter. Mon. Not. Roy. Astron. Soc. **492**(4), 5247–5260 (2020). https://doi.org/10.1093/mnras/staa170. arXiv: 1910.04793 [astro-ph.GA]
188. S. Sugiyama, T. Kurita, M. Takada, On the wave optics effect on primordial black hole constraints from optical microlensing search. Mon. Not. Roy. Astron. Soc. **493**(3), 3632–3641 (2020). https://doi.org/10.1093/mnras/staa407. arXiv: 1905.06066 [astro-ph.CO]
189. H.-J. Tian et al., The separation distribution of ultrawide binaries across galactic populations. Astrophys. J. Suppl. Ser. **246**(1), 4 (2019). ISSN: 1538-4365. https://doi.org/10.3847/1538-4365/ab54c4. http://dx.doi.org/10.3847/1538-4365/ab54c4
190. P. Tisserand et al., Limits on the Macho content of the galactic halo from the EROS-2 survey of the magellanic clouds. Astron. Astrophys. **469**, 387–404 (2007). https://doi.org/10.1051/0004-6361:20066017. arXiv: astro-ph/0607207
191. P. Tisserand et al., Limits on the Macho content of the galactic halo from the EROS-2 survey of the magellanic clouds. Astron. Astrophys. **469**, 387–404 (2007). https://doi.org/10.1051/0004-6361:20066017. arXiv: astro-ph/0607207
192. B.V. Vainer, P.D. Naselskii, Possible cosmological consequences of primordial black hole evaporation. Astron. Zh. **55**, 231–235 (1978). [Sov. Astron. **22**, 138 (1978).]
193. T. Venumadhav, L. Dai, J. Miralda-Escudé, Microlensing of extremely magnified stars near caustics of galaxy clusters. Astrophys. J. **850**(1), 49 (2017). https://doi.org/10.3847/1538-4357/aa9575. arXiv: 1707.00003 [astro-ph.CO]
194. E.L. Wright et al., Interpretation of the cosmic microwave background radiation anisotropy detected by the COBE differential microwave radiometer. Astrophys. J. Lett. **396**, L13–L18 (1992). https://doi.org/10.1086/186506

195. E.L. Wright, On the density of primordial black holes in the Galactic halo. Astrophys. J. **459**, 487–490 (1996). https://doi.org/10.1086/176910. arXiv: astro-ph/9509074
196. L. Wyrzykowski et al., The OGLE view of microlensing towards the magellanic clouds. IV. OGLE-III SMC data and final conclusions on MACHOs. Mon. Not. Roy. Astron. Soc. **416**, 2949 (2011). https://doi.org/10.1111/j.1365-2966.2011.19243.x. arXiv: 1106.2925 [astro-ph.GA]
197. J. Yoo, J. Chaname, A. Gould, The end of the MACHO era: limits on halo dark matter from stellar halo wide binaries. Astrophys. J. **601**, 311–318 (2004). https://doi.org/10.1086/380562. arXiv: astro-ph/0307437
198. S. Young, C. Byrnes, Signatures of non-gaussianity in the isocurvature modes of primordial black hole dark matter. JCAP **1504**(04), 034 (2015). https://doi.org/10.1088/1475-7516/2015/04/034. arXiv: 1503.01505 [astro-ph.CO]
199. S. Young, C.T. Byrnes, Initial clustering and the primordial black hole merger rate. JCAP **03**, 004 (2020). https://doi.org/10.1088/1475-7516/2020/03/004. arXiv: 1910.06077 [astro-ph.CO]
200. S. Young, I. Musco, C.T. Byrnes, Primordial black hole formation and abundance: contribution from the non-linear relation between the density and curvature perturbation. JCAP **11**, 012 (2019). https://doi.org/10.1088/1475-7516/2019/11/012. arXiv: 1904.00984 [astro-ph.CO]
201. Y.B. Zel'dovich et al., Primordial black holes and the deuterium problem. Pisma Astron. Zh. **3**, 208–211 (1977). [Sov. Astron. Lett. **3**, 110 (1977).]
202. Y. Zel'dovich, I. Novikov, The hypothesis of cores retarded during expansion and the hot cosmological model. Sov. Astron. **10**, 602 (1967)
203. Y. Zeldovich, A hypothesis, unifying the structure and the entropy of the universe. Mon. Not. R. Astron. Soc. **160**, 1P (1972). https://doi.org/10.1093/mnras/160.1.1P
204. Q. Zhu et al., Primordial black holes as dark matter: constraints from compact ultra-faint Dwarfs. Mon. Not. Roy. Astron. Soc. **476**(1), 2–11 (2018). https://doi.org/10.1093/mnras/sty079. arXiv: 1710.05032 [astro-ph.CO]
205. M. Zumalacarregui, U. Seljak, Limits on stellar-mass compact objects as dark matter from gravitational lensing of type Ia supernovae. Phys. Rev. Lett. **121**(14), 141101 (2018). https://doi.org/10.1103/PhysRevLett.121.141101. arXiv: 1712.02240 [astro-ph.CO]

Part II
Primordial Black Hole Formation

Chapter 2
Overall Picture: A Beginner's Guide to Primordial Black Hole Formation

Teruaki Suyama and Chul-Moon Yoo

Abstract The aim of this chapter is to offer an accessible overview for beginners seeking a foundational understanding of Primordial Black Holes (PBHs), encompassing their formation and abundance, without delving into extensive and complex equations. It is important to note that this chapter is designed to be introductory, striving to be self-contained so that all explanations and equations provided herein can be comprehended or derived by readers with a basic grasp of general relativity and cosmology, without reliance on external literature. Our goal is not solely to present quantitatively accurate facts that demand heavy computation or simulation for verification and to guide readers towards many reference papers, but rather to foster a clear physical understanding at the qualitative level. The price to pay is that this chapter is neither exhaustive nor deeply quantitative; any numerical values and associated equations should be interpreted as rough estimates.

2.1 Basic Picture

Among the various mechanisms for PBH formation, we narrow our focus to the most standard scenario: the direct gravitational collapse of a Hubble-sized overdense region in the radiation dominated (RD) universe. This choice is made to ensure the chapter remains self-contained within the confines of space limitations (see Chap. 9 for alternatives). This scenario naturally aligns with the framework of inflationary cosmology, where classical primordial perturbations arise from the quantum fluctuations of scalar fields during inflation. The spatial size of such perturbations is super-Hubble at the end of inflation. Subsequently, during the RD era, these perturbations

T. Suyama (✉)
Department of Physics, Tokyo Institute of Technology, 2-12-1 Ookayama, Meguro-ku, Tokyo 152-8551, Japan
e-mail: suyama@phys.titech.ac.jp

C.-M. Yoo
Division of Particle and Astrophysical Science, Graduate School of Science, Nagoya University, Nagoya 464-8602, Japan
e-mail: yoo.chulmoon.k6@f.mail.nagoya-u.ac.jp

C. Byrnes et al. (eds.), *Primordial Black Holes*, Springer Series in Astrophysics and Cosmology, https://doi.org/10.1007/978-981-97-8887-3_2

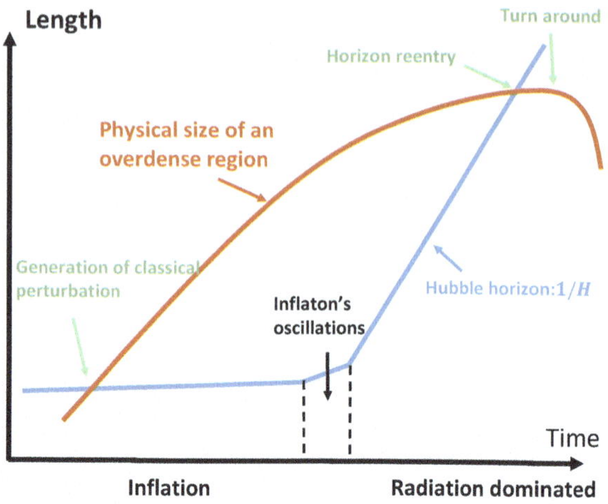

Fig. 2.1 Schematic diagram showing a relation between the physical size of an overdense region and the Hubble horizon

re-enter the Hubble horizon.[1] If a particular region exhibits sufficient overdensity at the time of the horizon re-entry, it quickly stops its expansion and undergoes gravitational collapse, resulting in the formation of a black hole (see Fig. 2.1).

To understand this picture in more detail, let us examine an overdense region whose spatial extent remains within the super-Hubble regime during the RD era. In what follows, we assume that all elementary particles present during the RD era are in a state of thermal equilibrium, devoid of any chemical potential. As a result, the system can be fully characterized by a single variable—the radiation temperature—where the pressure P is uniquely fixed by the energy density ρ. Consequently, the perturbation has only one degree of freedom, referred to as an adiabatic perturbation. The spacetime metric covering the super-Hubble overdense region can be expressed as a perturbation on top of the Friedmann–Lemaître–Robertson–Walker (FLRW) metric:

$$ds^2 = -\mathcal{N}^2(t, \boldsymbol{x})dt^2 + a^2(t)e^{2\zeta(t,\boldsymbol{x})}d\boldsymbol{x}^2. \tag{2.1}$$

Here $\zeta(t, \boldsymbol{x})$ is the curvature perturbation. Using the above metric, the time component of the conservation law $\nabla_\mu T^\mu_\nu = 0$, where T^μ_ν is the energy-momentum tensor of radiation, becomes

$$\dot{\zeta} = -\frac{\dot{a}}{a} - \frac{\dot{\rho}}{3(\rho + P)}, \tag{2.2}$$

where $\dot{} \equiv \frac{\partial}{\partial t}$. Now let us choose a time slice (=a three dimensional hypersurface where t is constant) on which the energy density ρ is spatially uniform. Because of the adiabaticity condition, the pressure P is also uniform on this slice. Thus, the

[1] Some short-wavelength perturbations may re-enter the Hubble horizon during the inflaton's oscillating phase in which the universe approximately looks like the matter-dominated era. Because of smaller pressure, the conditions for PBH formation in this phase is less restrictive than during the RD era. Although very interesting, this case is not covered in this chapter, but see Chap. 5.

right-hand side of Eq. (2.2) is a function of t only. However, a part of ζ which depends only on t is physically meaningless (i.e., it can be absorbed into the redefinition of the scale factor $a(t)$), we deduce that $\dot{\zeta} = 0$ on the uniform density slice. Furthermore, on this slice, each Hubble patch evolves in the same manner, given that the energy density serves as the sole characteristic of the system, with no distinguishing factors among various Hubble patches. This implies that $\mathcal{N}$ is also independent of $\boldsymbol{x}$ and can be taken to be 1 by a change of time variable. More detailed arguments regarding this paragraph are given in [4].

Having established the conservation of ζ on the uniform density slice, let us consider the perturbed region on this slice for which the perturbation is described by the curvature of the three-dimensional hypersurface. For simplicity, we assume this region is spherically symmetric with a comoving radius r_s. Accordingly, the metric may be written as

$$ds^2 = -dt^2 + a^2(t)e^{2\zeta(r)}\left(dr^2 + r^2 d\Omega^2\right), \tag{2.3}$$

where we impose $\zeta = 0$ for $r > r_s$ so that the metric reduces to the FLRW metric outside the perturbed region. For a reason that will become clear shortly, we assume the three-dimensional region is positively curved. The 3-curvature $R^{(3)}$ (which equates to $\frac{1}{6}$ times the three-dimensional Ricci scalar) is expressed in terms of ζ as

$$R^{(3)} = -\frac{1}{3a^2 e^{2\zeta(r)} r}\left(4\zeta' + r\zeta'^2 + 2r\zeta''\right), \tag{2.4}$$

where $' \equiv \frac{d}{dr}$. Therefore, for $R^{(3)}$ to be positive, ζ must be a decreasing function with respect to the comoving radius, as depicted in Fig. 2.2.

While the curvature perturbation ζ is widely used, particularly in computing observables of various inflation models, it is more convenient here to adopt a new radial coordinate R defined as $R = re^{\zeta(r)}$, where $a(t)R$ represents the proper area radius. In terms of R, the metric (2.3) transforms into

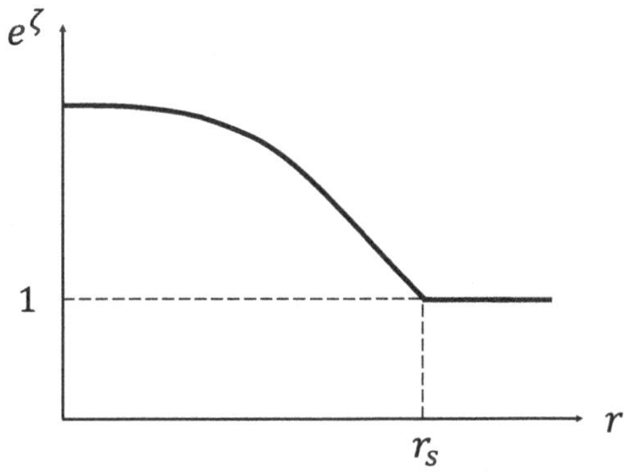

Fig. 2.2 Schematic profile of $e^{\zeta(r)}$ as a function of r

$$ds^2 = -dt^2 + a^2(t)\left(\frac{dR^2}{1 - K(R)R^2} + R^2 d\Omega^2\right), \tag{2.5}$$

where $K(R)$ is defined as

$$K(R) = -\frac{2 + r\zeta'}{re^{2\zeta}}\zeta'. \tag{2.6}$$

This quantity takes a positive value at $R = 0$ and gradually varies as R changes. As a zeroth-order approximation, let us assume this variation remains small up to r_s and treat K as constant. In other words, in the expression of the 3-curvature:

$$R^{(3)} = \frac{1}{a^2}\left(K + \frac{R}{3}\frac{dK}{dR}\right), \tag{2.7}$$

we ignore the second term. Consequently, the Hamiltonian constraint becomes

$$H^2 = \frac{8\pi G}{3}\rho - R^{(3)} \simeq \frac{8\pi G}{3}\rho - \frac{K}{a^2}, \tag{2.8}$$

where H is the Hubble parameter of the perturbed region. Then, the perturbed region evolves akin to a part of the closed FLRW universe characterized by a comoving curvature radius $1/\sqrt{K}$. For illustrative purposes, Fig. 2.3 presents a schematic depiction of the embedding of the closed FLRW universe, where the three-dimensional hypersurface is substituted with a two-dimensional surface.

In the scenario where the perturbed region is modeled akin to a closed FLRW universe, the perturbed region reaches maximal expansion ($a = a_{\rm max}$) at which the right-hand side of Eq. (2.8) vanishes, leading to a subsequent contraction phase where the closed universe ends with a big crunch. For an observer outside the perturbed region, this contraction phase manifests as gravitational collapse followed by the formation of a black hole. However, in reality, the propagation of perturbations as sound waves, not included in the preceding discussion, will impede gravitational collapse if these waves traverse the entire perturbed region by the time of maximal expansion. In other words, in order for the BH formation to occur, the size of the

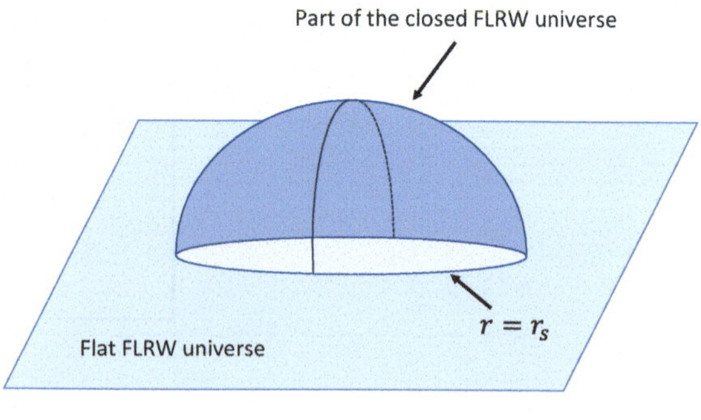

Fig. 2.3 Schematic diagram showing that the perturbed region can be regarded as a part of the closed FLRW universe

perturbed region at the time of maximal expansion must be larger than the Jeans length (which corresponds to the maximum length scale on which pressure forces can resist gravitational collapse):

$$a_{\max} r_s > \frac{c_s}{\sqrt{G\rho_{\max}}}, \tag{2.9}$$

where $c_s = 1/\sqrt{3}$ is the sound speed of radiation. By making use of the Hamiltonian constraint, this condition can be reformulated as

$$r_s > \frac{c_s}{\sqrt{K}}. \tag{2.10}$$

Given that $1\sqrt{K}$ is the curvature radius of the closed Universe and $c_s \lesssim 1$, this criterion tells that BH formation occurs only when the comoving size of the perturbed region is comparable to the total comoving size (i.e., curvature radius) of the corresponding closed universe.

This conceptual framework for PBH formation leads us to examine an intriguing scenario. Consider a situation where perturbed regions, each with a curvature radius comparable to their size of the regions (depicted as regions labeled A and B in the left side of Fig. 2.4) are superposed on another much bigger perturbed region, also possessing a curvature radius comparable to its spatial extent. In the context of Fig. 2.4, PBHs A and B on the right side form immediately after regions A and B re-enter the Hubble horizon. Around this time, the significantly larger perturbed region remains super-Hubble and does not play any role in the formation of PBHs A and B: From the perspective of regions A and B, it is indistinguishable whether they sit on top of a much larger perturbed region or not- a concept akin to our inability to discern the Earth's curvature in daily life. For a while afterwards, PBHs A and B remain isolated in the FLRW background until the larger perturbed region re-enters the Hubble horizon. Right after this time, the larger region transforms into a PBH. Since PBHs A and B reside within the larger region, they are engulfed by it during its

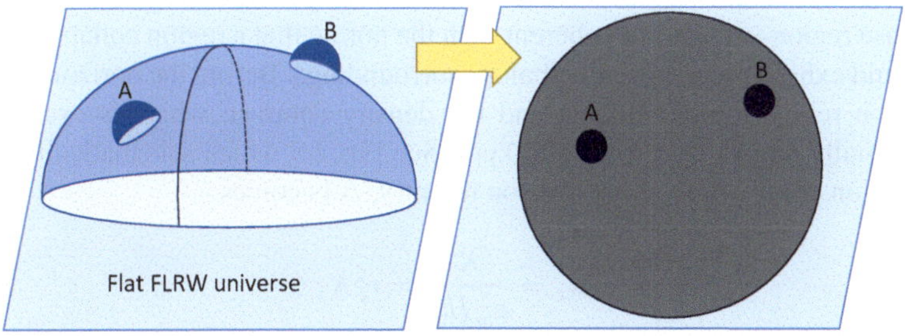

Fig. 2.4 Left figure: two perturbed regions as sections of the closed FLRW universe (labeled A and B) on top of another section of larger closed FLRW universe. Right figure: schematic picture showing that two PBHs formed from the regions A and B are absorbed by the bigger PBH later

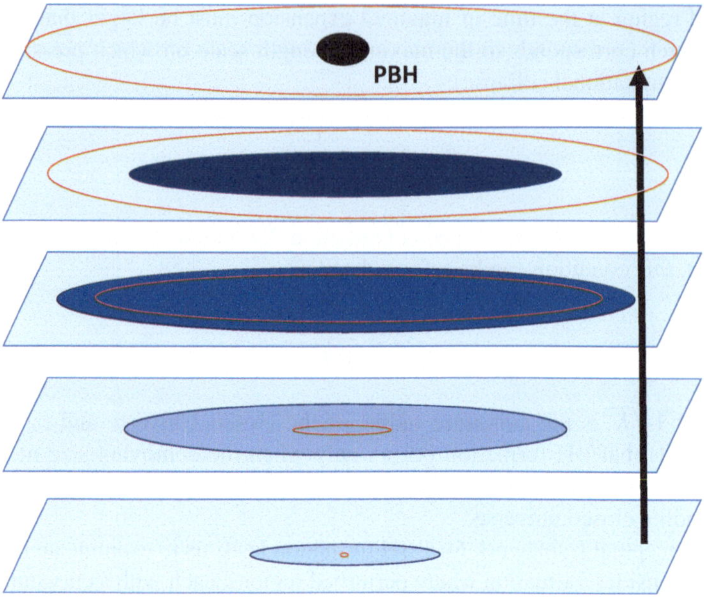

Fig. 2.5 Time evolution of the density contrast Δ represented as variation of the color intensity. The arrow represents the direction of time. The red circle at each time shows the Hubble horizon

formation. Consequently, at the conclusion of this process, what remains is a single, larger PBH formed from the initial, larger perturbed region.

At this stage, it is pertinent to consider the aforementioned perturbation on the uniform expansion hypersurface, where the expansion rate is not perturbed and the perturbation is characterized by the perturbation of radiation energy density Δ:

$$\Delta = \frac{\rho - \bar{\rho}}{\bar{\rho}} = \frac{K}{a^2 H^2}. \tag{2.11}$$

Since $K > 0$, this quantity is positive and hence the perturbed region represents an overdense region—a concept coherent with the notion that a region collapsing into a BH should exhibit greater density than its surroundings. Before the horizon re-entry, the region remains super-Hubble and the density contrast, which is exceedingly small initially, grows as a^2 in the RD era. See Fig. 2.5 which schematically shows the situation. At the time of the horizon re-entry, Δ becomes

$$\Delta_{\rm hr} = \frac{K}{a_{\rm hr}^2 H_{\rm hr}^2} = r_s^2 K, \tag{2.12}$$

where in the last equality we have used the relation $a_{\rm hr} r_s = 1/H_{\rm hr}$. Combining this result with Eq. (2.10) yields $\Delta_{\rm hr} > c_s^2$. In light of our current rough estimation, the right-hand side of this condition should not be taken literally. Rather, the essence of this crude analysis lies in recognizing that the overdense region evolves into a BH whenever the density contrast at the time of horizon re-entry approaches unity.

Given that a BH forms via direct gravitational collapse of the overdense region extending across the Hubble horizon, the mass of the BH at the formation time is approximately equal to the horizon mass:

$$M_{\rm PBH} \simeq \frac{4\pi}{3} H^{-3} \times \rho = \frac{1}{2GH}, \tag{2.13}$$

where in the last equality we have used the Friedmann equation. Precisely speaking, the right-hand side overestimates the PBH mass by a factor of $\mathcal{O}(1)$ since BH formation occurs slightly after horizon re-entry.

Considering that the BH is surrounded by radiation, its mass increases through accretion. Assuming radial infall of radiation onto the BH at the event horizon, we estimate the growth rate as

$$\frac{dM_{\rm PBH}}{dt} = 4\pi r_g^2 \rho_{\rm rad}, \tag{2.14}$$

where $r_g = 2GM_{\rm PBH}$ is the Schwarzschild radius of the BH and $\rho_{\rm rad}$ is the radiation energy density around the BH. Approximating $\rho_{\rm rad}$ to be the background cosmological radiation density, integration of the above differential equation yields the BH mass at time t as

$$M_{\rm PBH}(t) = \frac{M_{\rm PBH,ini}}{1 - 3GM_{\rm PBH,ini} H_{\rm ini} \left(1 - \frac{t_{\rm ini}}{t}\right)}, \tag{2.15}$$

where the subscript "ini" means the corresponding quantity is evaluated at the initial time $t_{\rm ini}$. Thus, if the Schwarzschild radius of the initial BH is smaller than the Hubble horizon, which has been confirmed by numerical simulations, the denominator of Eq. (2.15) quickly converges to a value not significantly different from unity. In essence, a PBH does not significantly augment its mass after its formation as long as it is immersed in the cosmological environment. Thus, we assume Eq. (2.13) gives a reasonable estimate of the PBH mass not only at the formation time but also at the present epoch.

Given that the mass of a PBH is essentially determined by the horizon mass at the time of horizon re-entry, there is a one-to-one correspondence between the comoving radius of the overdense region and the resultant PBH mass. To elucidate this relationship, let us first observe that $H_{\rm hr} \propto 1/a_{\rm hr}^2$ during the RD era. Combining this with the relation $a_{\rm hr} r_s = 1/H_{\rm hr}$ gives a scaling $H_{\rm hr} \propto 1/r_s^2$. Thus, the PBH mass directly is proportional to r_s^2. Taking the proportionality coefficient into account, Eq. (2.13) becomes

$$M_{\rm PBH} \sim 10\, M_\odot \left(\frac{r_s}{1\ {\rm pc}}\right)^2, \tag{2.16}$$

where we have assumed $g_* = 100$ for the relativistic degrees of freedom, defined by $\rho_{\rm rad} = \frac{\pi^2}{30} g_* T^4$.[2]

Since the minimum length scale of the perturbation in the context of inflation is about $H_0^{-1} e^{-60} \sim 1$ m (assuming inflation lasted for $N = 60$ number of e-folds after our observable universe exited the Hubble horizon during inflation), the smallest PBH mass could be as small as 10 g. This characteristic is pivotal and sets PBHs apart from astrophysical BHs: while sub-solar mass BHs cannot originate from the gravitational collapse of massive stars, there exists no such lower limit on the PBH mass. As it is clear from the preceding description of how PBHs can be formed in the RD universe, the only requisite condition for PBH formation is a sufficiently large overdensity. Once this condition is met, the formation of PBH becomes an inevitable consequence, irrespective of the PBH mass.

It is also noteworthy to consider the potential maximum mass of PBHs formed during the RD era in the standard Big Bang model, specifically just before the epoch of matter-radiation equality. With a Hubble horizon of 100 Mpc at the time of matter-radiation equality, we calculate the corresponding PBH mass to be approximately $10^{17}\ M_\odot$. This figure significantly surpasses the observed maximum mass of supermassive black holes ($\sim 10^{10}\ M_\odot$), suggesting that PBHs could serve as the progenitors of supermassive black holes, the origins of which remain enigmatic (see Chaps. 11 and 12).

Before closing this section, it is imperative to acknowledge that we have disregarded the effect of Hawking radiation in evaluating the time variation of PBH mass. By modeling the BH as a black body with the radius r_g and temperature $T_{\rm BH} = 1/(8\pi G M)$ (i.e., Hawking temperature), emitting only photons, the presence of Hawking radiation introduces a term $-4\pi r_g^2 \sigma T_{\rm BH}^4$ on the right-hand side of Eq. (2.14), where $\sigma = \frac{\pi^2}{60}$ is the Stefan-Boltzmann constant in the natural units. Integrating the modified Eq. (2.14) without including the effect of accretion (i.e., $\rho_{\rm rad} = 0$), it can be verified that the BH completely depletes its mass after a duration τ given by

$$\tau = 5120\pi G^2 M_{\rm PBH}^3 \simeq 2 \times 10^{10}\ {\rm yr} \left(\frac{M_{\rm PBH}}{2 \times 10^{14}\ {\rm g}} \right)^3 . \tag{2.17}$$

This estimated lifetime excludes other types of (nearly) massless particles such as neutrinos and gravitons and overestimates the BH's lifetime of the BH by a factor of $\mathcal{O}(10)$ (more detailed calculations are given in [6]). Nonetheless, this simple calculation demonstrates that PBHs lighter than $\sim 10^{15}$ g evaporate before the present time, rendering their abundance zero in the present universe, even if produced in the early universe. Conversely, PBHs significantly heavier than $\sim 10^{15}$ g have lifetimes far surpassing the age of the universe, with Hawking radiation exerting negligible impact on the time variation of their mass. In what follows, we exclusively consider massive PBHs for which the influence of Hawking radiation can be safely neglected.

---

[2] Using the thermodynamic relation $\rho_{\rm rad} = \frac{\pi^2}{30} g_* T^4$, the conservation of entropy, and $a_{\rm hr} r_s = 1/H_{\rm hr}$, it can be verified that the dependence of $M_{\rm PBH}$ on g_* is weak: $M_{\rm PBH} \propto g_*^{-1/6}$.

2.2 Criterion for PBH Formation

Let us discuss the criterion for PBH formation a bit more carefully, but based on an intuitive idea. Roughly speaking, in an asymptotically flat case, a black hole forms when the mass M gets compacted into a sufficiently small region which can be enclosed by the circumference $2\pi r_g$, where $r_g = 2GM$ is the Schwarzschild radius (hoop conjecture [8]). Therefore the compactness $GM/R_{\rm area}$ at a certain characteristic time would be useful to judge the possibility of PBH formation, where $R_{\rm area}$ is the area radius of the region concerned. It would be quite natural to evaluate the compactness at the horizon entry time for PBH formation. However, for the background universe, the horizon entry condition may give

$$R_{\rm area} = H^{-1} \Rightarrow r_g = 2GM = \frac{8\pi G}{3} R_{\rm area}^3 \rho = H^2 R_{\rm area}^3 = R_{\rm area}. \tag{2.18}$$

That is, for the FLRW background, in which no PBH formation occurs, the area radius is equivalent to the Schwarzschild radius of the mass inside it at the horizon entry time. Apparently, the criterion for the asymptotically flat situation cannot be applied and we need some modification.

Since overdense regions are relevant to PBH formation, one simple and natural quantity is not the total mass but the mass excess δM divided by the area radius, which is called compaction function [7]. Here the mass excess is the excess above the background value with the same area radius. The exact definition and related issues can be found in Refs. [2, 3]. Here we derive the useful expression of the compaction function $\mathcal{C}$ based on the qualitative discussion in the last section as follows[3]:

$$\begin{aligned}\mathcal{C}(r) := \frac{G\delta M}{aR} = (\rho - \bar{\rho})\frac{4\pi G}{3} a^2 R^2 = \left(\rho - \frac{3H^2}{8\pi G}\right)\frac{4\pi G}{3} a^2 R^2 \\ \simeq \frac{1}{2} K R^2 = \frac{1}{2}\left[1 - (1 + r\zeta')^2\right],\end{aligned} \tag{2.19}$$

where we have implicitly assumed the uniform Hubble gauge in the second equality. It should be noted that the expression has no time dependence and the compaction function is a conserved quantity in the super-Hubble limit. In Ref. [3], it has been shown that, at the leading order of the long-wavelength limit, the compaction function is equivalent to the density perturbation averaged inside the radius at the horizon entry. Therefore using the compaction function for PBH formation criterion is consistent with the discussion of the Jeans criterion given in the last section.

In Ref. [7], the following criterion is introduced

$$\mathcal{C}_{\rm max} := \mathcal{C}(r_{\rm m}) > \mathcal{C}_{\rm th}, \tag{2.20}$$

[3] Technically, this expression corresponds to the Shibata-Sasaki compaction function defined in Eq. (3.3) in Ref. [2].

where $r_{\rm m}$ is the radius at which the compaction function takes the maximum value $\mathcal{C}_{\rm max}$, and $\mathcal{C}_{\rm th}$ is a certain threshold value. This criterion has been compared with spherically symmetric simulations in, e.g., Refs. [3, 7]. Then the value of $\mathcal{C}_{\rm th}$ has been estimated as $\sim$0.4 with about 10% error depending on the profile of the initial density perturbation for moderate profiles (see also Ref. [5] for the profile dependence). We note that a more accurate criterion, whose accuracy is about 2% for moderate profiles, has been proposed in Ref. [1] also based on the compaction function. See Chap. 4 for details about numerical simulations of PBH formation.

2.3 Abundance of PBHs

In the preceding section, we explored an ideal scenario where there is only one overdense region with a fixed amplitude of perturbation. However, in reality, perturbations are randomly distributed throughout the universe. In such a scenario, one might intuitively expect that a particular region would form a PBH if the overdensity $\Delta_{\rm hr}$, obtained by averaging the density contrast over that region at the time of horizon re-entry, exceeds the threshold. In what follows, we adopt this simplified picture.[4] For notational simplicity, we omit the subscript "hr" and instead put "M" to emphasize that it represents the overdensity corresponding to the horizon mass M. Since the perturbations are stochastic in nature, Δ_M for any given region is also a stochastic variable governed by a specific probability distribution function $P(\Delta_M)$. Then, the probability β_M that a given region forms a PBH is given by

$$\beta_M = \int_{\Delta_{\rm th}}^{\infty} P(\Delta_M) d\Delta_M. \tag{2.21}$$

If the perturbations large enough to form PBHs predominantly consist of modes with nearly the same comoving wavelength, then this quantity represents the fraction of PBHs, all with masses close to M, in the total energy density:

$$\beta_M = \frac{\rho_{\rm PBH}}{\rho}. \tag{2.22}$$

It is important to note that this fraction is evaluated right after PBH formation. Afterwards, $\rho_{\rm PBH}$ decays like $1/a^3$ while ρ, dominated by radiation, decays like $1/a^4$ if we ignore the time variation of the relativistic degrees of freedom. Thus, $\Omega_{\rm PBH}$ at the present time can be estimated as

$$\Omega_{\rm PBH} \simeq \Omega_m \frac{a_{\rm eq}}{a_M} \beta_M, \tag{2.23}$$

[4] To preserve the conceptual clarity of this chapter, we continue to employ $\Delta_{\rm hr}$ in our discussions, recognizing that the compaction function primarily serves as a technical measure.

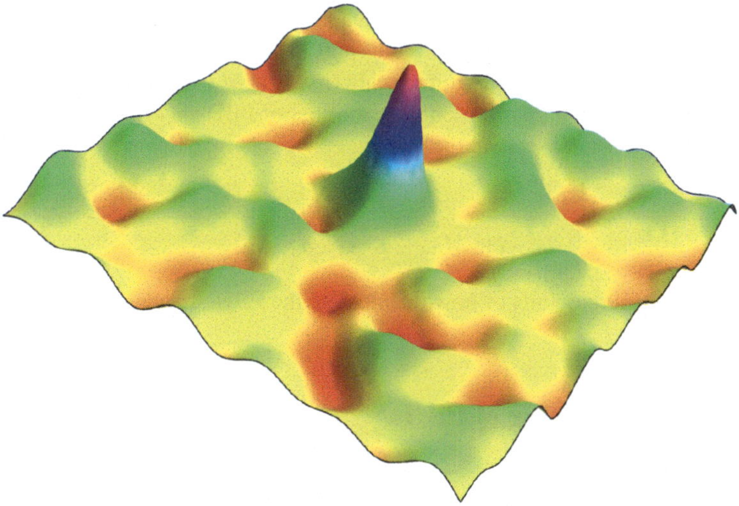

Fig. 2.6 Random distribution of Δ in two-dimensional space. Only the regions in which Δ is exceptionally larger than the standard deviation are the site of PBH formation

where $\Omega_m \simeq 0.2$ is the matter fraction, and $a_{\rm eq}$ and a_M denote the scale factors at matter-radiation equality and at PBH formation, respectively. By definition, $\Omega_{\rm PBH} \le \Omega_m$. Furthermore, since PBHs form in the deep RD era, $a_M \ll a_{\rm eq}$. Putting these together, we conclude that $\beta_M \ll 1$. Thus, PBH formation is an exceeding rare process: only a region where Δ accidentally takes a very large value evolves into a BH (see Fig. 2.6 which schematically shows the situation). To gain a more quantitative perspective, we can reformulate the above equation to derive:

$$\beta_M \simeq \left(\frac{M}{10^{18} M_\odot}\right)^{\frac{1}{2}} f_{\rm PBH}, \tag{2.24}$$

where $f_{\rm PBH} \equiv \Omega_{\rm PBH}/\Omega_m$ represents the fraction of PBH in dark matter.[5] Thus, for instance, for solar-mass PBHs, at most only one-billionth of the Hubble patches in the observable universe are allowed to transition into PBHs. The allowed probability of PBH formation decreases for smaller PBH masses. This arises because the smaller the PBH mass, the earlier the formation time, affording more time for the PBH energy density to become relatively significant compared to the radiation energy density.

The fraction (2.21) gives a reasonable estimate of the fraction of PBHs if all PBHs have the same mass M. However, if the perturbations relevant for PBH formation consist of Fourier modes covering a broad range of (comoving) wavelengths, different modes re-enter the Hubble horizon at varying times, resulting in the formation of

[5] It should be stated here that PBHs satisfy the conditions for them to be dark matter: PBHs are non-relativistic objects and have only gravitational interactions.

PBHs with different masses. In such a scenario, as β_M defined by Eq. (2.21) is a dimensionless quantity, it is natural to interpret it as the fraction of PBHs within an infinitesimal mass range in logarithmic scale $(\ln M, \ln M + d\ln M)$. Then, the PBH mass function $\frac{dn}{d\ln M}$, which represents the number of PBHs per unit volume and per unit mass in logarithmic scale at the present epoch, is given by

$$\frac{dn}{d\ln M} = \frac{3H_0^2\Omega_m}{8\pi G}\frac{a_{\rm eq}}{a_M}\frac{\beta_M}{M}. \tag{2.25}$$

See Chap. 6 for more details of how to calculate the PBH abundance.

2.4 Initial Clustering of PBHs

As a final topic of our chapter, let us investigate the circumstances under which PBHs can exhibit clustering from the moment of their formation (i.e., on comoving length scales larger than the comoving size of regions that give rise to PBHs). In what follows, for simplicity, we assume that all PBHs form simultaneously (i.e., they have the same mass). To start with, let us go back to the uniform density slice, where perturbations are represented by the curvature of three-dimensional hypersurfaces. To understand the essential part, we suppose that the curvature perturbation $\zeta(\boldsymbol{x})$ comprises only modes with a comoving length scale corresponding to PBHs (λ_S) and with a comoving length scale λ_L significantly larger than λ_S. In this scenario, ζ at any given point is a superposition of ζ_L, which varies only on the length scale λ_L, and ζ_S, which varies on the length scale λ_S.

Now, let us first consider a case where ζ_L and ζ_S are not correlated. One simple example belonging to this case is when ζ obeys Gaussian statistics, for which there are no correlations between modes of different comoving lengths. However, it is important to note that uncorrelation between ζ_L and ζ_S can also occur in cases of strong non-Gaussianity limited to modes with similar length scales. This situation is depicted in the bottom-left figure of Fig. 2.7 which schematically represents the uniform density hypersurfaces. The small semi-spheres represent perturbed regions with a length scale λ_S, while the color modulation denotes the variation of the value of ζ_L. According to our assumption of no correlation between ζ_S and ζ_L, the small semi-spheres are uniformly distributed irrespective of the background color. In other words, there is no clustering of the semi-spheres over the length scale λ_L. Those semi-spheres become PBHs after the horizon re-entry, and the spatial distribution of PBHs traces the original distribution of the semi-spheres. Hence, PBHs are not clustered on the scale λ_L.[6]

[6] After the mean distance of neighboring PBHs becomes smaller than the Hubble horizon, gravitational force between PBHs sets in and leads to clustering of PBHs at late epoch. Such clustering effect is not covered in this chapter.

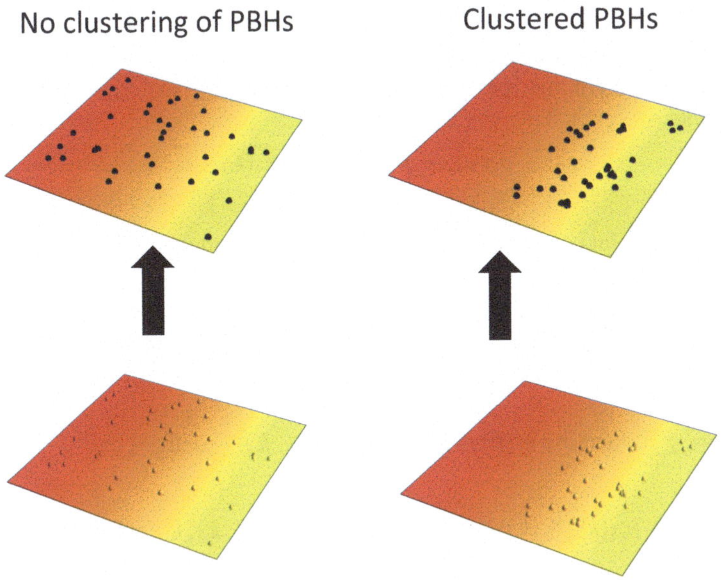

Fig. 2.7 Left(right) figure shows a case where ζ_L is uncorrelated(correlated) with ζ_S for which PBHs are uniformly distributed (clustered) on large scales. Small semi-spheres on the bottom figures are perturbed regions with length scale λ_S and the color modulation represents the variation of the value of ζ_L

Let us next explore a scenario where ζ_S is correlated with ζ_L, as illustrated in the bottom-right figure in Fig. 2.7. Specifically, we consider $\langle \zeta_S^2 \zeta_L \rangle$ to be non-zero, where $\langle \cdots \rangle$ is the ensemble average over perturbations. This correlation implies that the local variance of ζ_S is correlated with ζ_L. In the current example, semi-spheres exist only in regions where ζ_L takes values corresponding to the yellow color. In other words, the probability distribution of ζ_S depends on the value of ζ_L. As in the first case, spatial ditribution of PBHs is the same as that of the semi-spheres. Thus, PBHs exhibit clustering on the scale λ_L right after their formation, as depicted in the top-right figure.

In the above discussion, it is the correlation between ζ_L and the variance of ζ_S that induces clustering of PBHs on the large scale λ_L. However, in a broader context, the large-scale component need not be ζ_L but could be other types of perturbations that do not contribute to the curvature perturbation, known as isocurvature perturbation. Denoting such large-scale perturbation as S_L, clustering of PBHs requires the non-vanishing of $\langle S_L \zeta_S^2 \rangle$. If such a correlation exists, the local variance of ζ_S varies over the scale λ_L in conjunction with S_L. In other words, the two-point function of the local variance of ζ_S is non-vanishing at the scale λ_L:

$$\xi(r) \equiv \langle \zeta_S^2(\boldsymbol{x}) \zeta_S^2(\boldsymbol{y}) \rangle \neq 0, \tag{2.26}$$

for $r = |\boldsymbol{x} - \boldsymbol{y}| = \lambda_L$. Therefore, it is the four-point function of the curvature perturbation that produces the clustering of PBHs.

To gain an alternative perspective, let us express ζ_S in terms of its Fourier components

$$\zeta_S(\boldsymbol{x}) = \int \frac{d\boldsymbol{k}}{(2\pi)^3} e^{i\boldsymbol{k}\cdot\boldsymbol{x}} \tilde{\zeta}_S(\boldsymbol{k}). \tag{2.27}$$

Then, the two-point correlation function $\xi(r)$ can be written as

$$\xi(r) = \int \prod_{i=1}^{4} \frac{d\boldsymbol{k}_i}{(2\pi)^3} e^{i(\boldsymbol{k}_1+\boldsymbol{k}_2)\cdot(\boldsymbol{x}-\boldsymbol{y})} (2\pi)^3 T(\boldsymbol{k}_1, \boldsymbol{k}_2, \boldsymbol{k}_3, \boldsymbol{k}_4) \delta(\boldsymbol{k}_t). \tag{2.28}$$

Here $\boldsymbol{k}_t = \sum_{i=1}^{4} \boldsymbol{k}_i$ and the trispectrum $T(\boldsymbol{k}_1, \boldsymbol{k}_2, \boldsymbol{k}_3, \boldsymbol{k}_4)$ is defined by

$$\langle \tilde{\zeta}_S(\boldsymbol{k}_1)\tilde{\zeta}_S(\boldsymbol{k}_2)\tilde{\zeta}_S(\boldsymbol{k}_3)\tilde{\zeta}_S(\boldsymbol{k}_4)\rangle = (2\pi)^3 T(\boldsymbol{k}_1, \boldsymbol{k}_2, \boldsymbol{k}_3, \boldsymbol{k}_4) \delta(\boldsymbol{k}_t), \tag{2.29}$$

where the Dirac's delta function appears because of the translation invariance of the correlation functions. In the limit $\lambda_S/\lambda_L \to 0$, the integral of Eq. (2.28) will be dominated by modes satisfying $\boldsymbol{k}_1 + \boldsymbol{k}_2 \approx \boldsymbol{0}$, where length of all the wavevectors is concentrated about λ_S^{-1}. Thus, the large-scale clustering of PBHs is sourced by the so-called collapsed limit of the trispectrum (see Fig. 2.8 for the diagram showing the collapsed limit). See Chap. 7 for more about the impact of non-Gaussianities.

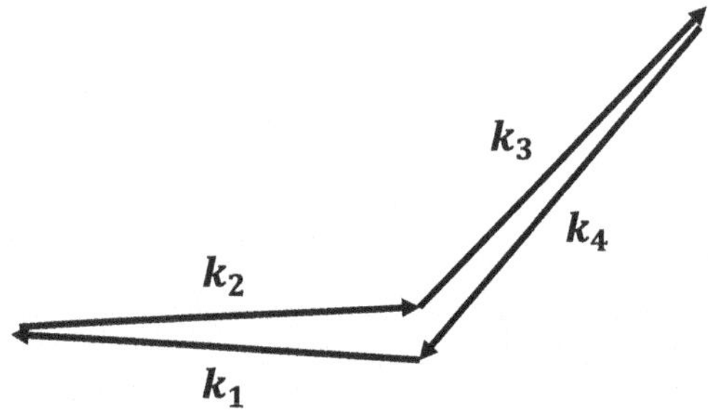

Fig. 2.8 A quadrilateral corresponding to the collapsed limit of the trispectrum of the curvature perturbation. Each side represents the wavevector $\boldsymbol{k}_i$ $(i = 1, 2, 3, 4)$. $|\boldsymbol{k}_1 + \boldsymbol{k}_2|(= |\boldsymbol{k}_3 + \boldsymbol{k}_4|$ is much smaller than $|\boldsymbol{k}_i|$ in the collapsed limit

2.5 Summary

This chapter offered an accessible overview for beginners seeking a foundational understanding of PBHs regarding their formation and abundance. A rudimentary grasp of general relativity and cosmology is adequate for understanding all explanations and equations provided, facilitating the acquisition of the underlying simple physical pictures. Readers are encouraged to explore additional resources such as the review paper [9] and other chapters within this book for further investigation.

Acknowledgements We extend our gratitude to Tomotaka Kuroda for his valuable insights that have enhanced the clarity and readability of this chapter.

References

1. A. Escrivà, C. Germani, R.K. Sheth, Universal threshold for primordial black hole formation. Phys. Rev. D **101**(4), 044022 (2020). https://doi.org/10.1103/PhysRevD.101.044022. arXiv:1907.13311 [gr-qc]
2. T. Harada, C.-M. Yoo, Y. Koga, Revisiting compaction functions for primordial black hole formation. Phys. Rev. D **108**(4), 043515 (2023). https://doi.org/10.1103/PhysRevD.108.043515. arXiv:2304.13284 [gr-qc]
3. T. Harada et al., Cosmological long-wavelength solutions and primordial black hole formation. Phys. Rev. D **91**(8), 084057 (2015). https://doi.org/10.1103/PhysRevD.91.084057. arXiv: 1503.03934 [gr-qc]
4. D.H. Lyth, K.A. Malik, M. Sasaki, A General proof of the conservation of the curvature perturbation. JCAP **05**, 004 (2005). https://doi.org/10.1088/1475-7516/2005/05/004. arXiv:astro-ph/0411220
5. I. Musco, Threshold for primordial black holes: dependence on the shape of the cosmological perturbations. Phys. Rev. D **100**(12), 123524 (2019). https://doi.org/10.1103/PhysRevD.100.123524. arXiv:1809.02127 [gr-qc]
6. D.N. Page, Particle emission rates from a black hole: massless particles from an uncharged, nonrotating hole. Phys. Rev. D **13**, 198–206 (1976). https://doi.org/10.1103/PhysRevD.13.198
7. M. Shibata, M. Sasaki, Black hole formation in the Friedmann universe: formulation and computation in numerical relativity. Phys. Rev. D **60**, 084002 (1999). https://doi.org/10.1103/PhysRevD.60.084002. arXiv: gr-qc/9905064
8. K.S. Thorne, *Magic Without Magic*. Ed. by Klauder, J. San Francisco: Freeman (1972)
9. C.-M. Yoo, The basics of primordial black hole formation and abundance estimation. Galaxies **10**(6), 112 (2022). https://doi.org/10.3390/galaxies10060112. arXiv: 2211.13512 [astro-ph.CO]

Chapter 3
Generating Large Primordial Fluctuations in Single-Field Inflation for Primordial Black Hole Formation

Jason Kristiano and Jun'ichi Yokoyama

Abstract In order to produce appreciable amount of primordial black holes (PBHs), the squared amplitude of the curvature perturbation must take a large value of $\mathcal{O}(0.01)$, namely, seven digits larger than the value observed by cosmic microwave background radiation (CMB) on large scales. Such a large fluctuation can be achieved by violating the slow-roll (SR) condition within a short duration. The best known of such possibilities is the ultraslow-roll (USR) inflation. We calculate the power spectrum of curvature perturbation in a simple single-field inflation model which evolves through the SR-USR-SR regimes so that both large-amplitude small-scale fluctuation for PBH formation and small-amplitude large-scale fluctuation as observed by CMB are realized. We further calculate the bispectrum and one-loop correction to the power spectrum induced by the third-order action of curvature perturbation as the beginning of precision cosmology on small scales. As a result, we argue that single-field inflationary models realizing PBH formation can be constrained by the quantum correction.

3.1 Introduction

PBHs are formed when an overdense region with a magnitude of density fluctuation close to unity enters the Hubble horizon during radiation domination [26, 64, 65, 154]. The most popular formation mechanism of such a large

J. Kristiano (✉) · J. Yokoyama
Department of Physics and Research Center for the Early Universe (RESCEU), Graduate School of Science, The University of Tokyo, Tokyo 113-0033, Japan
e-mail: jkristiano@resceu.s.u-tokyo.ac.jp

J. Yokoyama
e-mail: junichi.yokoyama@ipmu.jp

J. Yokoyama
Kavli Institute for the Physics and Mathematics of the Universe (Kavli IPMU), WPI, UTIAS, The University of Tokyo, Kashiwa, Chiba 277-8568, Japan

Trans-Scale Quantum Science Institute, The University of Tokyo, Tokyo 113-0033, Japan

C. Byrnes et al. (eds.), *Primordial Black Holes*, Springer Series in Astrophysics and Cosmology, https://doi.org/10.1007/978-981-97-8887-3_3

perturbation on small scales relevant to PBHs is quantum effects during cosmic inflation in the very early universe [63, 133, 139] (see [134] for a review). In this chapter, we focus on single-field inflation models that can produce large fluctuations on small scales relevant to PBH formation with particular attention to our recent finding [99, 100] that higher-order quantum effects can restrict the feasibility of such models, even though they are observationally possible because there are no stringent observational constraints on primordial fluctuations on small scales [33, 34, 74, 81, 91, 94, 114, 115, 149] unlike on large scales where observations of CMB tightly constrain the fluctuation spectrum [1, 3, 4].

The simplest inflation model consistent with CMB observation is canonical SR inflation, where the scalar field which drives the inflation, called the inflaton, is described by a canonical kinetic term and a potential. At a certain range, the potential has to be slightly tilted to realize the SR condition, so it can explain CMB observations. Beyond such a range, one can construct a theory that violates the SR condition, which usually results in the amplification of fluctuation on small scales [132, 152] due to rapid slowdown of the inflaton's velocity. The simplest example of a violation of the SR condition is the USR condition [95, 96, 106, 109, 153], which can be generalized to the constant-roll condition [110–112]. It occurs when the inflaton enters an extremely flat region in the potential [78], where the first derivative of the potential vanishes. Peaks of the enhanced fluctuation can collapse into PBHs after such a fluctuation reenters the horizon. An example potential that realizes the formation of PBHs while satisfying the observational constraint on large scales is shown in Fig. 3.1.

In this chapter, we explain how violation of the SR approximation can lead to amplification of primordial fluctuation on certain scales. Although we choose to focus on USR inflation, the method presented in this chapter can be easily generalized to other mechanism that yields rapid slowdown of the inflaton's velocity. In Sect. 3.2, we introduce two-point functions or the power spectrum of the fluctuations in SR inflation, USR inflation, and USR to SR transition. In Sect. 3.3, we introduce three-point functions of fluctuations in the same setup as in Sect. 3.2. In Sect. 3.4, we work out an inflation model with SR-USR-SR transitions. In this model, fluctuations are amplified on scales corresponding to the time of the USR phase. In Sect. 3.5, we discuss other PBH formation mechanisms within a single-field inflation framework and summarize our review.

3.2 Two-Point Functions

The action of canonical inflation is given by

$$S = \frac{1}{2} \int \mathrm{d}^4 x \sqrt{-g} \left[m_\mathrm{p}^2 R - (\partial_\mu \phi)^2 - 2V(\phi) \right], \tag{3.1}$$

where m_p is reduced Planck scale, $g = \det g_{\mu\nu}$, $g_{\mu\nu}$ and R are metric tensor and its Ricci scalar. Consider a spatially flat, homogeneous, and isotropic background,

$$\mathrm{d}s^2 = -\mathrm{d}t^2 + a^2(t)\mathrm{d}\boldsymbol{x}^2 = a^2(\tau)(-\mathrm{d}\tau^2 + \mathrm{d}\boldsymbol{x}^2), \tag{3.2}$$

where τ is conformal time. Equations of motion for the scale factor $a(t)$ and the homogeneous part of the inflaton $\phi(t)$ are the Friedmann equations

$$H^2 = \frac{1}{3m_p^2}\left(\frac{1}{2}\dot{\phi}^2 + V(\phi)\right), \quad \dot{H} = -\frac{\dot{\phi}^2}{2m_p^2}, \tag{3.3}$$

with $H = \dot{a}/a$ being the Hubble parameter, and the Klein-Gordon equation

$$\ddot{\phi} + 3H\dot{\phi} + \frac{\mathrm{d}V}{\mathrm{d}\phi} = 0. \tag{3.4}$$

Here, a dot denotes time derivative.

During inflation, the evolution of the Hubble parameter is parameterized by the slow-roll (SR) parameters. The SR parameters are defined as

$$\epsilon_{n+1} = \frac{\dot{\epsilon}_n}{\epsilon_n H}, \quad \epsilon_1 \equiv -\frac{\dot{H}}{H^2}, \tag{3.5}$$

where n is a positive integer. In case the Hubble parameter H is almost a constant, we have $\epsilon_1 \ll 1$. This condition is called the quasi-de Sitter (quasi-dS) approximation. Thus, the scale factor can be approximated as

$$a \approx -\frac{1}{H\tau} \propto e^{Ht}. \tag{3.6}$$

Another stronger condition is called the SR approximation, which is defined as $|\epsilon_n| \ll 1$ for every n. Therefore, the SR approximation implies a quasi-dS approximation, while the converse statement is not true. Violation of the SR approximation does not imply a breakdown of the quasi-dS approximation.

Small perturbation from the homogeneous part, $\phi(t)$, of the inflaton $\phi(\boldsymbol{x}, t)$ and metric can be expressed as

$$\begin{aligned} \phi(\boldsymbol{x}, t) &= \phi(t) + \delta\phi(\boldsymbol{x}, t), \\ \mathrm{d}s^2 &= -N^2\mathrm{d}t^2 + \gamma_{ij}(\mathrm{d}x^i + N^i\mathrm{d}t)(\mathrm{d}x^j + N^j\mathrm{d}t), \end{aligned} \tag{3.7}$$

where γ_{ij} is the three-dimensional metric on slices of constant t, N is the lapse function, and N^i is the shift vector. We choose comoving gauge condition

$$\delta\phi(\boldsymbol{x}, t) = 0, \ \gamma_{ij}(\boldsymbol{x}, t) = a^2(t)e^{2\zeta(\boldsymbol{x},t)}\delta_{ij}, \tag{3.8}$$

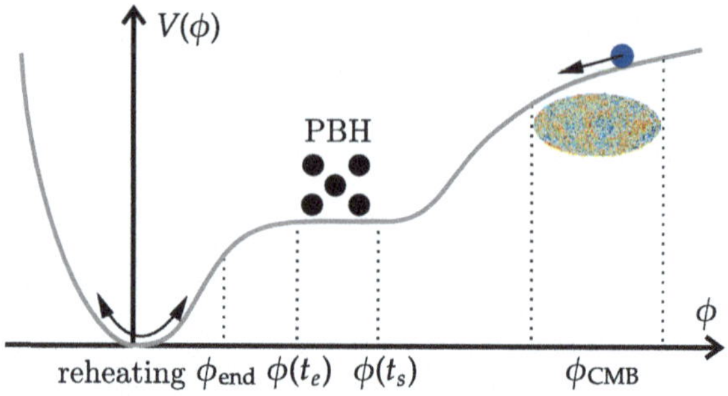

Fig. 3.1 Schematic picture of the inflaton potential realizing PBH formation from USR condition. When the inflaton is around $\phi_{\rm CMB}$, scales probed by CMB observations leave the horizon and it is in the SR regime. It enters an extremely flat region at $t = t_s$ undergoing an USR period. It enters the SR period again at $t = t_e$ until $\phi_{\rm end}$, the end of inflation

where $\zeta(\boldsymbol{x}, t)$ is the comoving curvature perturbation. N and N^i are obtained by solving the constraint equations. In this review, we do not consider tensor perturbations.

Expanding the action (3.1) up to the second order of the curvature perturbation yields

$$S^{(2)}[\zeta] = m_{\rm p}^2 \int {\rm d}t\, {\rm d}^3x\, a^3\epsilon_1 \left[\dot{\zeta}^2 - \frac{1}{a^2}(\partial_i\zeta)^2\right]. \tag{3.9}$$

Introducing the Mukhanov–Sasaki variable by $v = zm_{\rm p}\zeta$, where $z = a\sqrt{2\epsilon_1}$, the action becomes canonically normalized

$$S^{(2)}_{\rm can}[v] = \frac{1}{2}\int {\rm d}\tau\, {\rm d}^3x \left[(v')^2 - (\partial_i v)^2 + \frac{z''}{z}v^2\right], \tag{3.10}$$

where a prime denotes derivative with respect to τ. We can see that the quantity z''/z behaves as an effective time-dependent mass squared. In momentum space, quantization is performed by promoting the Mukhanov–Sasaki variable to an operator

$$\hat{v}_{\boldsymbol{k}}(\tau) = m_{\rm p}z(\tau)\hat{\zeta}_{\boldsymbol{k}}(\tau) = v_k(\tau)\hat{a}_{\boldsymbol{k}} + v_k^*(\tau)\hat{a}^\dagger_{-\boldsymbol{k}}, \tag{3.11}$$

where the mode function $v_k(\tau)$ satisfies the Mukhanov–Sasaki equation

$$v_k'' + \left(k^2 - \frac{z''}{z}\right)v_k = 0. \tag{3.12}$$

In terms of the SR parameters, the effective mass term can be written as

$$\frac{z''}{z} = (aH)^2\left(2 - \epsilon_1 + \frac{3}{2}\epsilon_2 - \frac{1}{2}\epsilon_1\epsilon_2 + \frac{1}{4}\epsilon_2^2 + \frac{1}{2}\epsilon_2\epsilon_3\right). \tag{3.13}$$

From (3.10) we find that $\pi^v = v'$ is momentum conjugate to v, so canonical quantization imposes a commutator

$$[\hat{v}_{\boldsymbol{k}}(\tau), \hat{\pi}^v_{\boldsymbol{k}'}(\tau)] = i\hbar\delta(\boldsymbol{k} - \boldsymbol{k}'), \tag{3.14}$$

and the operators satisfy $[\hat{a}_{\boldsymbol{k}}, \hat{a}^\dagger_{\boldsymbol{k}'}] = (2\pi)^3\delta(\boldsymbol{k} - \boldsymbol{k}')$ under the normalization condition

$$v_k'^* v_k - v_k' v_k^* = i\hbar. \tag{3.15}$$

The two-point functions of curvature perturbation and power spectrum can be written as

$$\langle\zeta_{\boldsymbol{k}}(\tau)\zeta_{\boldsymbol{k}'}(\tau)\rangle \equiv (2\pi)^3\delta(\boldsymbol{k} + \boldsymbol{k}')\,\langle\!\langle\zeta_{\boldsymbol{k}}(\tau)\zeta_{-\boldsymbol{k}}(\tau)\rangle\!\rangle\,, \tag{3.16}$$

$$\Delta_s^2(k,\tau) \equiv \frac{k^3}{2\pi^2}\langle\!\langle\zeta_{\boldsymbol{k}}(\tau)\zeta_{-\boldsymbol{k}}(\tau)\rangle\!\rangle, \tag{3.17}$$

where the bracket $\langle\cdots\rangle = \langle 0|\cdots|0\rangle$ denotes the vacuum expectation value (often called the VEV), the double bracket $\langle\!\langle\cdots\rangle\!\rangle$ means the VEV after taking out the Dirac-delta function of total momentum, and $\Delta_s^2(k,\tau)$ is the power spectrum multiplied by the phase space density.

3.2.1 *Slow-Roll Approximation*

Using the SR approximation, the effective mass becomes

$$\frac{z''}{z} = (aH)^2\left[2 - \epsilon_1 + \frac{3}{2}\epsilon_2 + O(\epsilon^2)\right]. \tag{3.18}$$

Up to first-order corrections in the SR approximation, aH as a function of τ reads

$$aH = -\frac{1}{\tau(1-\epsilon_1)}. \tag{3.19}$$

Then, the effective mass can be expressed in terms of the conformal time as

$$\frac{z''}{z} = \frac{1}{\tau^2}\left(2 + 3\epsilon_1 + \frac{3}{2}\epsilon_2\right) = \frac{1}{\tau^2}\left(\nu^2 - \frac{1}{4}\right), \tag{3.20}$$

where ν is defined as

$$\nu = \frac{3}{2} + \epsilon_1 + \frac{1}{2}\epsilon_2. \tag{3.21}$$

At early time, $\tau \to -\infty$, the Mukhanov–Sasaki equation approaches

$$v_k'' + k^2 v_k = 0, \tag{3.22}$$

which is the equation of a simple harmonic oscillator. Requiring minimum energy state and normalization (3.15), the solution of the Mukhanov–Sasaki equation at early time is given by

$$v_k(\tau \to -\infty) = \frac{1}{\sqrt{2k}} e^{-ik\tau}. \tag{3.23}$$

This minimum energy state corresponds to the initial vacuum state called Bunch–Davies vacuum, defined as $\hat{a}_{\boldsymbol{k}}|0\rangle = 0$.

At general times, the solution of the Mukhanov–Sasaki equation is given by

$$v_k(\tau) = \sqrt{-k\tau}\left[c_1 H_\nu^{(1)}(-k\tau) + c_2 H_\nu^{(2)}(-k\tau)\right], \tag{3.24}$$

where c_1 and c_2 are coefficients to be fixed to a boundary condition. Given the asymptotic limit of Hankel functions

$$\lim_{x\to\infty} H_\nu^{(1,2)}(x) = \sqrt{\frac{2}{\pi x}} \exp\left[\pm i\left(x - \frac{\nu\pi}{2} - \frac{\pi}{4}\right)\right], \tag{3.25}$$

we can match the general solution to the early time limit (3.23) that leads to

$$c_1 = \sqrt{\frac{\pi}{4k}} e^{i\frac{\pi}{4}(2\nu+1)} \quad \text{and} \quad c_2 = 0. \tag{3.26}$$

Then, the mode function of the Mukhanov–Sasaki variable becomes

$$v_k(\tau) = \sqrt{\frac{\pi}{4}} e^{i\frac{\pi}{4}(2\nu+1)} \sqrt{-\tau} H_\nu^{(1)}(-k\tau). \tag{3.27}$$

To obtain the mode function of the curvature perturbation, we need to know the explicit function of $z(\tau)$. Starting from

$$\frac{z'}{z} = aH\left(1 + \frac{1}{2}\epsilon_2\right), \tag{3.28}$$

we can substitute (3.19) to obtain

$$\begin{aligned} \int \frac{\mathrm{d}z}{z} &= -\int \frac{\mathrm{d}\tau}{\tau}\left(1 + \epsilon_1 + \frac{1}{2}\epsilon_2\right) \\ \frac{z(\tau)}{z_\star} &= \left(\frac{\tau}{\tau_\star}\right)^{-(1+\epsilon_1+\frac{1}{2}\epsilon_2)}, \end{aligned} \tag{3.29}$$

where $\tau_\star$ is an arbitrary time reference and $z_\star = z(\tau_\star)$. Choosing $\tau_\star$ as horizon crossing time $k = a(\tau_\star)H(\tau_\star)$ so

$$\tau_\star = -\left[\frac{1}{aH(1-\epsilon_1)}\right]_\star = -\frac{1}{k(1-\epsilon_1)_\star}, \tag{3.30}$$

$$z_\star = \left(a\sqrt{2\epsilon_1}\right)_\star = \left(\frac{k}{H}\sqrt{2\epsilon_1}\right)_\star, \tag{3.31}$$

then the explicit function of $z(\tau)$ is

$$z(\tau) = \left(\frac{k}{H}\sqrt{2\epsilon_1}\right)_\star [-k\tau(1-\epsilon_1)_\star]^{-(1+\epsilon_1+\frac{1}{2}\epsilon_2)}. \tag{3.32}$$

Then, the mode function of curvature perturbation reads

$$\begin{aligned}\zeta_k(\tau) &= \frac{v_k(\tau)}{m_\mathrm{p} z(\tau)} \\ &= \left(\frac{H}{m_\mathrm{p} k\sqrt{2\epsilon_1}}\right)_\star \sqrt{\frac{\pi}{4}}(-k\tau)^{(1+\epsilon_1+\frac{1}{2}\epsilon_2)} e^{i\frac{\pi}{4}(2\nu+1)}\sqrt{-\tau}H_\nu^{(1)}(-k\tau).\end{aligned} \tag{3.33}$$

With another asymptotic limit of Hankel function

$$\lim_{x\to 0} H_\nu^{(1)}(x) = -\frac{i}{\pi}\Gamma(\nu)\left(\frac{x}{2}\right)^{-\nu}, \tag{3.34}$$

the curvature perturbation at the end of inflation $\tau = \tau_0(\to 0)$, approaches

$$\zeta_k(\tau_0) = \left(\frac{iH}{m_\mathrm{p}\sqrt{2k^3\epsilon_1}}\right)_\star \frac{1}{\sqrt{4\pi}} 2^{3/2}\Gamma(3/2). \tag{3.35}$$

Also, the power spectrum reads

$$\Delta_s^2(k,\tau_0) = \left(\frac{H^2}{8\pi^2 m_\mathrm{p}^2\epsilon_1}\right)_\star, \tag{3.36}$$

with a weak k-dependence due to the horizon crossing condition manifested in the spectral tilt

$$n_s(k,\tau_0) - 1 = \frac{\mathrm{d}\log\Delta_s^2(k,\tau_0)}{\mathrm{d}\log k} = (-2\epsilon_1 - \epsilon_2)_\star. \tag{3.37}$$

Another way to derive power spectrum in the SR approximation is by looking at the exact de Sitter case. In exact de Sitter, the effective mass is

$$\frac{z''}{z} = \frac{2}{\tau^2}. \tag{3.38}$$

The solution of the Mukhanov–Sasaki equation with Bunch–Davies initial vacuum is

$$v_k(\tau) = \frac{1}{\sqrt{2k}}\left(1 - \frac{i}{k\tau}\right)e^{-ik\tau}. \tag{3.39}$$

When we proceed to derive the curvature perturbation, we run into a trouble because in exact de Sitter we find $\epsilon_1 = 0$. However, as a first-order approximation of the SR parameters, we can simply consider the finite value of ϵ_1. Also, the Hubble and SR parameters have a weak time dependence.

Numerical calculation of $\zeta_k(\tau)$ shows that evaluating the proportionality coefficient $1/z$ to $v_k(\tau)$ at the horizon crossing time $\tau = \tau_\star$ yields an excellent fit. So the curvature perturbation (3.33) can be approximated as

$$\zeta_k(\tau) = \left(\frac{iH}{2m_\mathrm{p}\sqrt{k^3\epsilon_1}}\right)_\star (1 + ik\tau)e^{-ik\tau}. \tag{3.40}$$

Substituting $\tau \to 0$ leads to the power spectrum (3.36).

The power spectrum (3.36) can be expressed as a function of the potential. We can rewrite the Friedmann equation as

$$V = H^2 m_\mathrm{p}^2 (3 - \epsilon_1). \tag{3.41}$$

Taking a derivative with respect to ϕ yields

$$V_{,\phi} \equiv \frac{\mathrm{d}V}{\mathrm{d}\phi} = \frac{1}{\sqrt{2}} H^2 m_\mathrm{p} \sqrt{\epsilon_1}(6 - 2\epsilon_1 + \epsilon_2). \tag{3.42}$$

At leading order in the SR approximation, the first SR parameter can be written as

$$\epsilon_1 \simeq \frac{m_\mathrm{p}^2}{2}\left(\frac{V_{,\phi}}{V}\right)^2. \tag{3.43}$$

Therefore, ϵ_1 has an implicit time dependence because ϕ depends on time.

From (3.36), we can see that the power spectrum at the wavenumber p corresponds to ϵ_1 evaluated at $\tau = -1/p$. This implies that observation of the power spectrum at the wavenumber p constrains $V_{,\phi}(\phi(\tau))$ at $\tau = -1/p$. To solve the horizon problem, the fluctuations in the CMB must be around 60 e-folds before the end of inflation. Moreover, fluctuations are observed within the finite range of wavenumber $0.005\,\mathrm{Mpc}^{-1} < p < 0.2\,\mathrm{Mpc}^{-1}$ by CMB. This means that CMB observations constrain only a finite region of the potential around 60 e-folds before the end of inflation. Beyond such a region, there are much looser observational constraints on the potential. Therefore, it is possible to have an inflation model in which the

SR approximation is violated after the horizon-crossing time of the CMB scales and before the end of inflation.

Comparison with observations can be made if and only if we are convinced that quantum fluctuations generated during inflation behave as classical statistical fluctuations to leave observable traces in our universe. The following intuitive argument can be developed for the classicalization of curvature fluctuations. In the superhorizon regime, $k \ll aH$, the mode function behaves as

$$\zeta_k(\tau) = \left(\frac{iH}{2m_\mathrm{p}\sqrt{k^3\epsilon_1}}\right)_\star \left[1 + O\left(\frac{k^2}{a^2H^2}\right)\right], \tag{3.44}$$

which means $\zeta_k^*(\tau) = -\zeta_k(\tau)$ to the leading order. So we find $\hat{\zeta}_{\boldsymbol{k}}(\tau) = \zeta_k(\tau)(\hat{a}_{\boldsymbol{k}} - \hat{a}^\dagger_{-\boldsymbol{k}})$ and its conjugate momentum $\hat{\pi}^\zeta_{\boldsymbol{k}}(\tau) = m_\mathrm{p}^2 z^2(\tau)\zeta_k'(\tau)(\hat{a}_{\boldsymbol{k}} - \hat{a}^\dagger_{-\boldsymbol{k}})$ have the same operator dependence, and apparently commute with each other, behaving classically [125]. This is why long-wave quantum fluctuations behave as if they are classical statistical fluctuations to provide the origin of large-scale structures and CMB anisotropies. The more precise statement, however, is that the quantum commutator

$$\left[\hat{\zeta}_{\boldsymbol{k}}(\tau), \hat{\pi}^\zeta_{\boldsymbol{k}'}(\tau)\right] = i\hbar\delta(\boldsymbol{k} - \boldsymbol{k}') \tag{3.45}$$

always holds, and it is the commutator

$$\left[\hat{\zeta}_{\boldsymbol{p}}(\tau), \hat{\zeta}'_{\boldsymbol{q}}(\tau)\right] = \frac{i\hbar}{2m_\mathrm{p}^2\epsilon_1(\tau)a^2(\tau)}\delta(\boldsymbol{p} - \boldsymbol{q}) \tag{3.46}$$

that decreases exponentially during the standard slow-roll inflation.

Observation of the spectral tilt $n_s = 0.9649 \pm 0.0042$ on the CMB scale implies that the SR parameters ϵ_1, $|\epsilon_2|$ at $\tau = -1/p$ have a value $O(0.01)$, based on (3.37). If the first SR parameter decreases rapidly at a later time $\tau \gg -1/p$, the power spectrum is amplified at wavenumber $k \gg p$. This can be achieved by violating the SR approximation at a later time. In the next subsections, we will show some inflation models that violate the SR approximation.

3.2.2 Ultraslow-Roll Inflation

In this subsection, we consider a regime where the potential is constant, called ultra-slow-roll inflation (USR) [95, 96, 106, 109, 153]. Because $\mathrm{d}V/\mathrm{d}\phi = 0$, the Klein-Gordon equation (3.4) becomes $\ddot{\phi} = -3H\dot{\phi}$, so $\dot{\phi} \propto a^{-3}$. This makes ϵ_1 strongly time-dependent and extremely small as

$$\epsilon_1 = \frac{\dot{\phi}^2}{2m_{\rm p}^2 H^2} \propto a^{-6}, \tag{3.47}$$

and the second SR parameter becomes

$$\epsilon_2 \equiv \frac{\dot{\epsilon}_1}{\epsilon_1 H} = 2\epsilon_1 + 2\frac{\ddot{\phi}}{\dot{\phi} H} \simeq -6, \tag{3.48}$$

which breaks the SR approximation. The effective mass can be written as

$$\frac{z''}{z} = (aH)^2 \left[2 + \frac{3}{2}\epsilon_2 + \frac{1}{4}\epsilon_2^2 + O(\epsilon_1) \right] = \frac{2}{\tau^2}, \tag{3.49}$$

which is the same as in the exact de Sitter case. Thus, the solution of the Mukhanov–Sasaki equation with Bunch–Davies initial vacuum is the same as (3.39). Then, the curvature perturbation is given by

$$\zeta_k(\tau) = \frac{iH}{2m_{\rm p}\sqrt{k^3 \epsilon_1(\tau)}} (1 + ik\tau) e^{-ik\tau}. \tag{3.50}$$

Here, ϵ_1 has a strong dependence on time, which can be expressed as

$$\epsilon_1(\tau) = \epsilon_1(\tau_\star) \left(\frac{a(\tau)}{a(\tau_\star)} \right)^{-6}, \tag{3.51}$$

where $\tau_\star$ is the horizon-crossing time of a characteristic wavenumber $k_\star$. Therefore, the curvature perturbation becomes

$$\zeta_k(\tau) = \frac{iH}{2m_{\rm p}\sqrt{k^3 \epsilon_1(\tau_\star)}} \left(\frac{a(\tau)}{a(\tau_\star)} \right)^{3} (1 + ik\tau) e^{-ik\tau}. \tag{3.52}$$

At the end of inflation, the power spectrum is

$$\Delta_s^2(k, \tau_0) = \frac{H^2}{8\pi^2 m_{\rm p}^2 \epsilon_1(\tau_\star)} \left(\frac{a(\tau_0)}{a(\tau_\star)} \right)^{6}. \tag{3.53}$$

To solve the horizon problem, CMB-scale fluctuations must cross the horizon at $N_{\rm CMB} \approx 60$ e-folds before the end of inflation. It means that power spectrum of CMB fluctuations can be written as

$$\Delta_s^2(p, \tau_0) = \frac{H^2}{8\pi^2 m_{\rm p}^2 \epsilon_1(\tau_\star)} e^{6N_{\rm CMB}}. \tag{3.54}$$

To explain the observed $\Delta_s^2(p, \tau_0) = 2.1 \times 10^{-9}$, for a reasonable value of $\epsilon_1(\tau_\star)$, an extremely small value of H/m_p is required. Because H/m_p characterizes the energy scale of inflation, an extremely small value of H/m_p can reach the energy scale of the Big Bang nucleosynthesis. Hence, inflation with only USR period cannot reach the required e-folds number to solve the horizon problem.

3.2.3 Constant-Roll Inflation

We can generalize the USR condition to $\epsilon_2 = \text{constant}$ [110–112]. By taking reference at time $\tau_\star$, the first SR parameter can be written as

$$\epsilon_1(\tau) = \epsilon_1(\tau_\star)\left(\frac{a(\tau)}{a(\tau_\star)}\right)^{\epsilon_2}, \tag{3.55}$$

Rewriting the second SR parameter as $\epsilon_2 = -2(3+\alpha)$, the effective mass can be expressed as

$$\frac{z''}{z} = \frac{1}{\tau^2}(2+\alpha)(1+\alpha) = \frac{1}{\tau^2}\left(\nu^2 - \frac{1}{4}\right), \tag{3.56}$$

where ν is defined as

$$\nu = \left|\alpha + \frac{3}{2}\right|. \tag{3.57}$$

The solution of the Mukhanov–Sasaki equation with Bunch–Davies initial condition is the same as (3.27). The power spectrum is given by

$$\Delta_s^2(k, \tau_0) = \frac{H^2}{8\pi^2 m_\mathrm{p}^2 \epsilon_1(\tau_0)} \frac{1}{2\pi} 2^{2\nu}[\Gamma(\nu)]^2 \left(\frac{k}{aH}\right)^{3-2\nu}. \tag{3.58}$$

We can read that the spectral tilt is $n_s - 1 = 3 - 2\nu$. Therefore, there are two possible solutions for α, given by

$$\alpha_\pm = -\frac{3}{2} \pm \frac{1}{2}(4 - n_s). \tag{3.59}$$

However, we have to choose a solution that does not grow outside the horizon. Recall the second-order action of ζ that reads

$$S^{(2)}[\zeta] = m_\mathrm{p}^2 \int \mathrm{d}\tau\, \mathrm{d}^3x\, a^2 \epsilon_1 \left[(\zeta')^2 - (\partial_i \zeta)^2\right]. \tag{3.60}$$

The equation of motion is given by

$$\zeta_k'' + \frac{(a^2\epsilon_1)'}{a^2\epsilon_1}\zeta_k' + k^2\zeta_k = 0 \tag{3.61}$$

At superhorizon scale, we can neglect the last term in the equation of motion. Hence, evolution of the curvature perturbation outside the horizon is

$$\zeta_k(\tau_0) \to A_k + B_k \int_{\tau_\star}^{\tau_0} \frac{d\tau}{a^2(\tau)\epsilon_1(\tau)}, \tag{3.62}$$

where A_k and B_k are time-independent coefficients. The second term determines whether $\zeta_k(\tau_0)$ grows or decays. Substituting (3.55), the integral becomes

$$\int_{\tau_\star}^{\tau_0} \frac{d\tau}{a^2(\tau)\epsilon_1(\tau)} \sim \tau_0^{\epsilon_2+3}. \tag{3.63}$$

This shows that curvature perturbation grows outside the horizon if $\epsilon_2 < -3$ or equivalently $\alpha > -3/2$. It was pointed out in [132] that the curvature perturbation grows outside the horizon if $3 - \epsilon_1 + \epsilon_2 < 0$ is satisfied. Clearly, USR condition satisfies this condition since $\epsilon_2 = -6$.

From observation of spectral tilt $n_s = 0.9649 \pm 0.0042$ on CMB scale, we obtain $\alpha_+ \approx 0$ and $\alpha_- \approx -3$. Solution $\alpha_+ \approx 0$ corresponds to $\epsilon_2 \approx -6$, which approximately equal to USR inflation. As explained in the previous subsection, although it yields an almost scale-invariant power spectrum, this solution cannot explain the required number of e-folds of inflation because the curvature perturbation grows outside the horizon. Therefore, we have to choose α_- as the solution because it corresponds to the decaying curvature perturbation. With this solution, constant-roll inflation can explain the observed CMB fluctuations.

3.2.4 *Ultraslow-Roll to Slow-Roll Transition*

In this subsection, we consider an inflation model in which the second parameter SR evolves from the USR to the SR period at conformal time τ_s. For $\tau < \tau_s$, the inflaton is in the USR period. At time $\tau \geq \tau_s$, the inflaton undergoes a transition from the USR period to the SR period. In the next part, we consider two different evolutions of the second SR parameter: smooth and sharp transitions. The former means that the second SR parameter evolves as a continuous function of time, while in the latter it evolves as a step function of time. We follow the parameterization of $\epsilon_2(\tau)$ in [24], which is also used in many literatures. The sketches of $\epsilon_2(\tau)$ for both cases are shown by the gray and blue curves in Fig. 3.2.

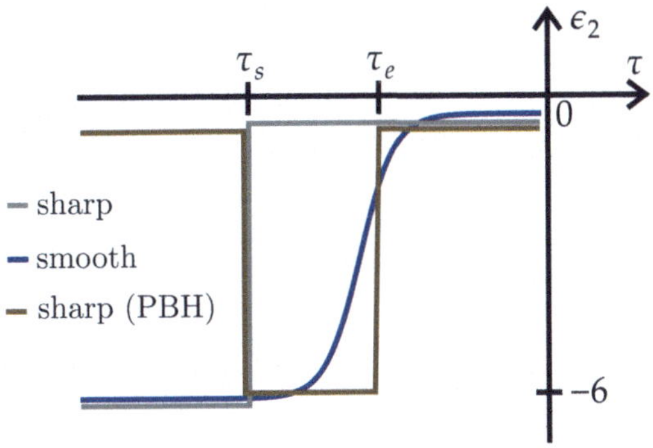

Fig. 3.2 Sketch of $\epsilon_2(\tau)$ for sharp and smooth transition from USR to SR period are shown by the gray and blue curve, respectively. Brown curve shows sharp transition from SR-USR-SR period, which will be discussed in Sect. 3.4

3.2.4.1 Smooth Transition

At time $\tau > \tau_s$, consider evolution of the first and second SR parameter as

$$\epsilon_1(\tau) = \epsilon_1(\tau_s)\frac{3\left[\left(\frac{\tau_s}{\tau}\right)^s (s-3)+s+3\right]^2}{2s^2(s+3)}\left(\frac{\tau}{\tau_s}\right)^{s+3}, \tag{3.64}$$

$$\epsilon_2(\tau) = s-3+\frac{2s(s+3)}{\left(\frac{\tau_s}{\tau}\right)^s (s-3)+s+3}, \tag{3.65}$$

where $s = \sqrt{9-12\eta_V} \simeq 3-2\eta_V$ and $\eta_V \to 0$. The parameter η_V corresponds to the second derivative of the potential at the transition point. In this model, the second SR parameter evolves from USR period $\epsilon_2(\tau_s) = -6$ to SR period $\epsilon_2(\tau_0) = -2\eta_V$. With this parametrization, the effective mass becomes

$$\frac{z''}{z} = \frac{1}{\tau^2}(2-3\eta_V) = \frac{1}{\tau^2}\left(\nu^2-\frac{1}{4}\right), \tag{3.66}$$

where

$$\nu^2 = \frac{9}{4}-3\eta_V. \tag{3.67}$$

Because $\eta_V \to 0$, the solution of the Mukhanov–Sasaki equation with Bunch–Davies initial vacuum is the same as (3.39). Then, the curvature perturbation is given by

$$\zeta_k(\tau) = \frac{iH}{2m_p\sqrt{k^3\epsilon_1(\tau)}}(1+ik\tau)e^{-ik\tau}, \tag{3.68}$$

and power spectrum at the end of inflation becomes

$$\Delta_s^2(k,\tau_0) = \frac{H^2}{8\pi^2 m_{\rm p}^2 \epsilon_1(\tau_0)}. \tag{3.69}$$

3.2.4.2 Sharp Transition

Consider the evolution of the second SR parameter as a step function of time

$$\epsilon_2(\tau) = -\Delta\epsilon_2 \left[1 - \theta(\tau - \tau_s)\right], \tag{3.70}$$

where $\Delta\epsilon_2 = 6$. At time $\tau < \tau_s$, the inflaton is in the USR period with the Bunch–Davies initial vacuum. In this period, the curvature perturbation and its time derivative are given by

$$\begin{aligned}\zeta_k(\tau) &= \frac{iH}{2m_{\rm p}\sqrt{k^3\epsilon_1(\tau)}} e^{-ik\tau}(1+ik\tau) \\ &= \frac{iH}{2m_{\rm p}\sqrt{k^3\epsilon_1(\tau_s)}} \left(\frac{\tau_s}{\tau}\right)^3 e^{-ik\tau}(1+ik\tau),\end{aligned} \tag{3.71}$$

$$\zeta_k'(\tau) = \frac{iH}{2m_{\rm p}\sqrt{k^3\epsilon_1(\tau_s)}} \left(\frac{\tau_s}{\tau}\right)^3 e^{-ik\tau} \left(k^2\tau - \frac{3}{\tau} - 3ik\right). \tag{3.72}$$

At time $\tau \geq \tau_s$, the inflaton is in the SR period. In this period, the curvature perturbation and its time derivative are given by

$$\zeta_k(\tau) = \frac{iH}{2m_{\rm p}\sqrt{k^3\epsilon_1(\tau_s)}} \left[\mathcal{A}_k e^{-ik\tau}(1+ik\tau) - \mathcal{B}_k e^{ik\tau}(1-ik\tau)\right], \tag{3.73}$$

$$\zeta_k'(\tau) = \frac{iH}{2m_{\rm p}\sqrt{k^3\epsilon_1(\tau_s)}} k^2\tau \left(\mathcal{A}_k e^{-ik\tau} - \mathcal{B}_k e^{ik\tau}\right), \tag{3.74}$$

where $\mathcal{A}_k$ and $\mathcal{B}_k$ are general coefficients to be determined from boundary conditions [21, 46, 83, 101, 102, 123, 140, 143]. This time, we have to consider both positive and negative frequency solutions because the boundary condition at $\tau = \tau_s$ does not correspond to Bunch–Davies vacuum. Solutions of the coefficients by requiring continuity of $\zeta_k(\tau)$ and $\zeta_k'(\tau)$ at transition $\tau = \tau_s$ are

$$\mathcal{A}_k = 1 + \frac{3(1+k^2\tau_s^2)}{2ik^3\tau_s^3}, \tag{3.75}$$

$$\mathcal{B}_k = \frac{3(1+ik\tau_s)^2}{2ik^3\tau_s^3} e^{-2ik\tau_s}. \tag{3.76}$$

At the end of inflation, the power spectrum becomes

$$\Delta_s^2(k,\tau_0) = \frac{k^3}{2\pi^2}|\zeta_k(\tau_0)|^2 = \frac{H^2}{8\pi^2 m_\mathrm{p}^2 \epsilon_1(\tau_s)}|\mathcal{A}_k - \mathcal{B}_k|^2 . \tag{3.77}$$

For a very long-wavelength modes, $|k\tau_s| \ll 1$, the power spectrum approaches

$$\Delta_s^2\left(k \ll \frac{1}{|\tau_s|}, \tau_0\right) = 4 \times \frac{H^2}{8\pi^2 m_\mathrm{p}^2 \epsilon_1(\tau_s)}. \tag{3.78}$$

3.3 Three-Point Functions

So far, we have considered the second-order action of the curvature perturbation that leads to two-point functions or the power spectrum at the tree level in the language of quantum field theory. Within the cosmological perturbation theory framework, higher-order correlation functions resulting from higher-order expansion of the action are expected to be small. However, because we are interested in PBH formation, where the curvature perturbation is large on certain small scales, higher-order correlation functions can be quite large (see also Chap. 7).

In this section, we discuss the third-order action that yields three-point functions of the curvature perturbation. Expanding (3.1) to third-order of ζ yields the interaction action [105]

$$S^{(3)}[\zeta] = S_\mathrm{bulk}[\zeta] + S_\mathrm{B}[\zeta] + m_\mathrm{p}^2 \int \mathrm{d}t\, \mathrm{d}^3x\; 2f(\zeta)\left(\frac{\delta L}{\delta \zeta}\right)_1 . \tag{3.79}$$

The bulk interaction $S_\mathrm{bulk}[\zeta]$ reads

$$\begin{aligned} S_\mathrm{bulk}[\zeta] = m_\mathrm{p}^2 \int \mathrm{d}t\, \mathrm{d}^3x\; a^3 \Bigg[& \epsilon_1^2 \dot{\zeta}^2 \zeta + \frac{1}{a^2}\epsilon_1^2 (\partial_i \zeta)^2 \zeta - 2\epsilon_1 \dot{\zeta}\, \partial_i \zeta\, \partial_i \chi \\ & -\frac{1}{2}\epsilon_1^3 \dot{\zeta}^2 \zeta + \frac{1}{2}\epsilon_1 \zeta (\partial_i \partial_j \chi)^2 + \frac{1}{2}\epsilon_1 \dot{\epsilon}_2 \dot{\zeta} \zeta^2 \Bigg], \end{aligned} \tag{3.80}$$

where $\chi = \epsilon_1 \partial^{-2} \dot{\zeta}$. $S_\mathrm{B}[\zeta]$ is the surface action given by the total derivatives [7, 20]. The last term is an interaction term proportional to the equation of motion in the lowest order,

$$\left(\frac{\delta L}{\delta \zeta}\right)_1 = \frac{\mathrm{d}}{\mathrm{d}t}(\epsilon_1 a^3 \dot{\zeta}) - \epsilon_1 a \partial^2 \zeta . \tag{3.81}$$

The function $f(\zeta)$ is explicitly given by

$$f(\zeta) = \frac{\epsilon_2}{4}\zeta^2 + \frac{\dot{\zeta}}{H}\zeta + \frac{1}{4a^2H^2}[-(\partial_i\zeta)^2 + \partial^{-2}\partial_i\partial_j(\partial_i\zeta\partial_j\zeta)] + \frac{1}{2H}[\partial_i\zeta\partial_i\chi - \partial^{-2}\partial_i\partial_j(\partial_i\zeta\partial_j\chi)]. \tag{3.82}$$

Performing field redefinition $\zeta = \boldsymbol{\zeta} + f(\boldsymbol{\zeta})$ generates third-order terms from the second-order action (3.9) that cancel relevant surface terms containing $\dot{\zeta}$. As a result, we find a useful relation [7, 20, 105]

$$S^{(2)}[\zeta] + S^{(3)}[\zeta] = S^{(2)}[\boldsymbol{\zeta}] + S_{\rm bulk}[\boldsymbol{\zeta}]. \tag{3.83}$$

In terms of $\boldsymbol{\zeta}$, the total action is simply given by the second-order action (3.9) and bulk interaction (3.80).

Three-point functions of ζ can be written schematically as

$$\langle\zeta_{k_1}(\tau)\zeta_{k_2}(\tau)\zeta_{k_3}(\tau)\rangle = \langle\boldsymbol{\zeta}_{k_1}(\tau)\boldsymbol{\zeta}_{k_2}(\tau)\boldsymbol{\zeta}_{k_3}(\tau)\rangle + \text{redefinition terms}, \tag{3.84}$$

where the first term is generated by bulk interaction (3.80) and the second term is boundary contribution at τ from field redefinition. Higher-order correction to the expectation value of an operator $O(\tau)$ is calculated by the in-in perturbation theory

$$\langle O(\tau)\rangle = \left\langle \left[\bar{\rm T}\exp\left(i\int_{-\infty}^{\tau} d\tau' H_{\rm int}(\tau')\right)\right]\hat{O}(\tau)\left[{\rm T}\exp\left(-i\int_{-\infty}^{\tau} d\tau' H_{\rm int}(\tau')\right)\right]\right\rangle, \tag{3.85}$$

where T and $\bar{\rm T}$ denote time and antitime ordering. In this case, the interaction Hamiltonian is simply $H_{\rm int} = -\int d^3x\mathcal{L}_{\rm bulk}$, where $\mathcal{L}_{\rm bulk}$ is the integrand of $S_{\rm bulk}$. For three-point functions, the operator is $\boldsymbol{\zeta}_{k_1}(\tau)\boldsymbol{\zeta}_{k_2}(\tau)\boldsymbol{\zeta}_{k_3}(\tau)$. First-order expansion of (3.85) reads

$$\langle O(\tau)\rangle = 2\int_{-\infty}^{\tau} d\tau_1 \,{\rm Im}\langle\hat{O}(\tau)H_{\rm int}(\tau_1)\rangle. \tag{3.86}$$

Bispectrum $\langle\langle\zeta_{k_1}(\tau)\zeta_{k_2}(\tau)\zeta_{k_3}(\tau)\rangle\rangle$ is defined as

$$\langle\zeta_{k_1}(\tau)\zeta_{k_2}(\tau)\zeta_{k_3}(\tau)\rangle = (2\pi)^3\delta(\boldsymbol{k}_1+\boldsymbol{k}_2+\boldsymbol{k}_3)\langle\langle\zeta_{k_1}(\tau)\zeta_{k_2}(\tau)\zeta_{k_3}(\tau)\rangle\rangle. \tag{3.87}$$

A bispectrum has local type if it can be expressed as

$$\begin{aligned}\langle\langle\zeta_{k_1}(\tau_0)\zeta_{k_2}(\tau_0)\zeta_{k_3}(\tau_0)\rangle\rangle = \frac{6}{5}f_{\rm NL}\Big[&\left|\zeta_{k_1}(\tau_0)\right|^2\left|\zeta_{k_2}(\tau_0)\right|^2 + \left|\zeta_{k_2}(\tau_0)\right|^2\left|\zeta_{k_3}(\tau_0)\right|^2 \\ &+ \left|\zeta_{k_1}(\tau_0)\right|^2\left|\zeta_{k_3}(\tau_0)\right|^2\Big],\end{aligned} \tag{3.88}$$

where the coefficient $f_{\rm NL}$ is a non-linearity parameter.

3.3.1 Slow-Roll Approximation

In the SR approximation, the first line in cubic self-interaction (3.80) has $O(\epsilon^2)$ coefficients, while the second line has $O(\epsilon^3)$ coefficients. At the end of inflation, when the relevant modes are in the superhorizon limit, the leading term in the field redefinition (3.82) is the first term, because the other terms decay outside the horizon. Therefore, the third-order action becomes

$$S_{\text{bulk}}[\boldsymbol{\zeta}] = m_{\text{p}}^2 \int \mathrm{d}t\, \mathrm{d}^3x\, a^3 \left[\epsilon_1^2 \dot{\boldsymbol{\zeta}}^2 \boldsymbol{\zeta} + \frac{1}{a^2} \epsilon_1^2 (\partial_i \boldsymbol{\zeta})^2 \boldsymbol{\zeta} - 2\epsilon_1^2 \dot{\boldsymbol{\zeta}} \partial_i \boldsymbol{\zeta} \partial_i \partial^{-2} \dot{\boldsymbol{\zeta}} + O(\epsilon^3) \right], \tag{3.89}$$

with $\zeta = \boldsymbol{\zeta} + \epsilon_2 \boldsymbol{\zeta}^2/4$. Using in-in perturbation theory (3.86), we obtain the bispectrum of $\boldsymbol{\zeta}$ as

$$\langle\langle \boldsymbol{\zeta}_{k_1}(\tau_0) \boldsymbol{\zeta}_{k_2}(\tau_0) \boldsymbol{\zeta}_{k_3}(\tau_0) \rangle\rangle_{O(\epsilon^2)} = \frac{H^4}{32 m_{\text{p}}^4 \epsilon_1^2 (k_1 k_2 k_3)^3} \tag{3.90}$$

$$\times \left[-\epsilon_{1\star} \sum_i k_i^3 + \epsilon_{1\star} \sum_{i \neq j} k_i k_j^2 + 8\epsilon_{1\star} \frac{\sum_{i>j} k_i^2 k_j^2}{\sum_i k_i} \right].$$

To arrive at the bispectrum of ζ, we have to include contributions from the field redefinition

$$\langle\langle \zeta_{k_1}(\tau_0) \zeta_{k_2}(\tau_0) \zeta_{k_3}(\tau_0) \rangle\rangle = \langle\langle \boldsymbol{\zeta}_{k_1}(\tau_0) \boldsymbol{\zeta}_{k_2}(\tau_0) \boldsymbol{\zeta}_{k_3}(\tau_0) \rangle\rangle \tag{3.91}$$

$$+ \frac{1}{4} \epsilon_2(\tau_0) \times 2 \left[\left|\zeta_{k_1}(\tau_0)\right|^2 \left|\zeta_{k_2}(\tau_0)\right|^2 + \left|\zeta_{k_2}(\tau_0)\right|^2 \left|\zeta_{k_3}(\tau_0)\right|^2 + \left|\zeta_{k_1}(\tau_0)\right|^2 \left|\zeta_{k_3}(\tau_0)\right|^2 \right].$$

However, assuming ϵ_3 as a constant, $\epsilon_2(\tau_0)$ diverges as $\tau_0 \to 0$. The explicit time dependence is given by

$$\epsilon_2(\tau_0) = \epsilon_{2\star} \left(\frac{\tau_0}{\tau_\star} \right)^{-\epsilon_{3\star}} \simeq \epsilon_{2\star} \left(1 - \epsilon_{3\star} \log \frac{\tau_0}{\tau_\star} \right). \tag{3.92}$$

At leading order of $\epsilon_2(\tau_0)$, the bispectrum of ζ is

$$\langle\langle \zeta_{k_1}(\tau_0) \zeta_{k_2}(\tau_0) \zeta_{k_3}(\tau_0) \rangle\rangle_{O(\epsilon^2)} = \frac{H^4}{32 m_{\text{p}}^4 \epsilon_1^2 (k_1 k_2 k_3)^3} \tag{3.93}$$

$$\times \left[(\epsilon_2 - \epsilon_1)_\star \sum_i k_i^3 + \epsilon_{1\star} \sum_{i \neq j} k_i k_j^2 + 8\epsilon_{1\star} \frac{\sum_{i>j} k_i^2 k_j^2}{\sum_i k_i} \right].$$

In the squeezed limit, where one of the wavenumbers, k_1, is much smaller than the others, $k_1 \to 0$, the bispectrum becomes

$$\langle\langle \zeta_{k_1}(\tau_0)\zeta_{k_2}(\tau_0)\zeta_{-k_2}(\tau_0)\rangle\rangle = (2\epsilon_1 + 4\epsilon_2)_\star \frac{H^4}{32m_{\rm p}^4\epsilon_1^2(k_1k_2)^3}. \tag{3.94}$$

In terms of the mode function and spectral tilt, the squeezed bispectrum can be written as

$$\langle\langle \zeta_{k_1}(\tau_0)\zeta_{k_2}(\tau_0)\zeta_{-k_2}(\tau_0)\rangle\rangle = -(n_s(k_2,\tau_0) - 1)\left|\zeta_{k_1}(\tau_0)\right|^2 \left|\zeta_{k_2}(\tau_0)\right|^2, \tag{3.95}$$

which is called Maldacena's theorem. Here, we have shown Maldacena's theorem by explicit derivation. It can also be proven by a model-independent method by considering the evolution of the small-scale perturbation in a rescaled dS background due to the long-wavelength perturbation [40].

Another important point discussed in [105] is the bispectrum induced by the last term in (3.80)

$$S_{\rm bulk}[\boldsymbol{\zeta}] \supset \frac{1}{2}m_{\rm p}^2 \int {\rm d}t\, {\rm d}^3x\, a^3\epsilon_1\dot{\epsilon}_2\dot{\boldsymbol{\zeta}}\boldsymbol{\zeta}^2. \tag{3.96}$$

Although it is SR-suppressed compared to (3.89), such term induces bispectrum that evolves outside the horizon.

$$\langle\langle \boldsymbol{\zeta}_{k_1}(\tau_0)\boldsymbol{\zeta}_{k_2}(\tau_0)\boldsymbol{\zeta}_{k_3}(\tau_0)\rangle\rangle_{O(\epsilon^3)} = \left(\epsilon_{2\star}\epsilon_{3\star}\log\frac{\tau_0}{\tau_\star}\right)\frac{H^4}{32m_{\rm p}^4\epsilon_1^2(k_1k_2k_3)^3}(k_1^3+k_2^3+k_3^3), \tag{3.97}$$

where $\star$ denotes horizon crossing of the total wavenumber $k_1 + k_2 + k_3$. It shows that the bispectrum of $\boldsymbol{\zeta}$ evolves outside the horizon due to $\log \tau_0$ dependence. However, for the bispectrum of ζ, there is an additional contribution from the field redefinition (3.91). After summing with the second term in (3.92), the $\log \tau_0$ dependence cancels out. Therefore, the bispectrum of ζ does not evolve outside the horizon.

3.3.2 *Ultraslow-Roll Inflation*

During USR period, ϵ_1 becomes extremely small because it scales as a^{-6}. Also, $\dot{\epsilon}_2 = 0$ because ϵ_2 is a constant. Therefore, the contribution of the bulk interaction to the bispectrum is negligible and the leading contribution comes from the field redefinition. The relevant terms in the field redefinition are [117]

$$f(\boldsymbol{\zeta}) = \frac{\epsilon_2}{4}\boldsymbol{\zeta}^2 + \frac{\dot{\boldsymbol{\zeta}}}{H}\boldsymbol{\zeta}, \tag{3.98}$$

which contribute

$$\langle\langle \zeta_{k_1}(\tau_0)\zeta_{k_2}(\tau_0)\zeta_{k_3}(\tau_0)\rangle\rangle = 2\left(\frac{\epsilon_2}{4}+3\right)\left[\left|\zeta_{k_1}(\tau_0)\right|^2\left|\zeta_{k_2}(\tau_0)\right|^2+\left|\zeta_{k_2}(\tau_0)\right|^2\left|\zeta_{k_3}(\tau_0)\right|^2\right.$$
$$\left.+\left|\zeta_{k_1}(\tau_0)\right|^2\left|\zeta_{k_3}(\tau_0)\right|^2\right]. \tag{3.99}$$

The factor of 3 comes from $\dot{\zeta}_k(\tau_0) = 3H\zeta_k(\tau_0)$, which can be derived from the USR mode function (3.52). Hence, we can read the local non-Gaussianity as

$$\frac{6}{5}f_{\rm NL} = 3 \Longrightarrow f_{\rm NL} = \frac{5}{2}. \tag{3.100}$$

In the squeezed limit, the bispectrum becomes

$$\langle\langle \zeta_{k_1}(\tau_0)\zeta_{k_2}(\tau_0)\zeta_{-k_2}(\tau_0)\rangle\rangle = 6\left|\zeta_{k_1}(\tau_0)\right|^2\left|\zeta_{k_2}(\tau_0)\right|^2, \tag{3.101}$$

which clearly violates Maldacena's theorem (3.95). Indeed, for a shift-symmetric action of the inflaton, Maldacena's theorem is generalized in [18, 49]. The squeezed bispectrum (3.101) can be reproduced from a generalized Maldacena's theorem after substituting the USR condition.

3.3.3 *Ultraslow-Roll to Slow-Roll Transition*

In case there is a transition of the second SR parameter, $\dot{\epsilon_2}$ becomes large during the transition. Therefore, the leading term in the third-order action is [22, 24, 27, 43, 117]

$$S_{\rm bulk}[\boldsymbol{\zeta}] = m_{\rm p}^2\int {\rm d}\tau\,{\rm d}^3x\,a^2\left[\frac{1}{2}\epsilon_1\epsilon_2'\boldsymbol{\zeta}'\boldsymbol{\zeta}^2+O(\epsilon^2)\right]. \tag{3.102}$$

Because we are interested in transition from USR to SR period, relation between ζ and $\boldsymbol{\zeta}$ at the end of inflation is $\zeta = \boldsymbol{\zeta}+\epsilon_2\boldsymbol{\zeta}^2/4$. However, because $|\epsilon_2(\tau_0)| \ll 1$, contribution from field redefinition to the bispectrum is much smaller than the bulk interaction (3.102). Thus, the bispectrum is given by

$$\langle\langle \zeta_{k_1}(\tau_0)\zeta_{k_2}(\tau_0)\zeta_{k_3}(\tau_0)\rangle\rangle = -2m_{\rm p}^2\int_{-\infty}^{\tau_0}{\rm d}\tau\,\epsilon_1(\tau)\epsilon_2'(\tau)a^2(\tau) \tag{3.103}$$
$$\times{\rm Im}\left[\zeta_{k_1}(\tau_0)\zeta_{k_2}(\tau_0)\zeta_{k_3}(\tau_0)\zeta_{k_1}^*(\tau)\zeta_{k_2}^*(\tau)\zeta_{k_3}'^*(\tau)+{\rm perm}\right].$$

In the next part, we will evaluate the time integral for sharp and smooth transition of $\epsilon_2(\tau)$. Similarly as in Sect. 3.2.4, we follow the parametrization of $\epsilon_2(\tau)$ in [24].

3.3.3.1 Sharp Transition

Consider evolution of the second SR parameter as step function, given by (3.70), its time derivative is a Dirac-delta function

$$\epsilon_2'(\tau) = \Delta\epsilon_2 \delta(\tau - \tau_s). \tag{3.104}$$

Substituting it to the time integral in (3.103) leads to

$$\begin{aligned} \langle\langle \zeta_{\mathbf{k}_1}(\tau_0)\zeta_{\mathbf{k}_2}(\tau_0)\zeta_{\mathbf{k}_3}(\tau_0)\rangle\rangle &= -2m_{\mathrm{p}}^2\epsilon_1(\tau_s)a^2(\tau_s)\Delta\epsilon_2 \\ &\times \mathrm{Im}\left[\zeta_{k_1}(\tau_0)\zeta_{k_2}(\tau_0)\zeta_{k_3}(\tau_0)\zeta_{k_1}^*(\tau_s)\zeta_{k_2}^*(\tau_s)\zeta_{k_3}'^*(\tau_s) + \mathrm{perm}\right]. \end{aligned} \tag{3.105}$$

Then, we can substitute $\zeta_k(\tau)$ and $\zeta_k'(\tau)$ given in Sect. 3.2.4.2 to obtain the bispectrum. Although the general form of bispectrum is complicated, in the limit $k_1, k_2, k_3 \ll k_s$ the bispectrum becomes

$$\begin{aligned} \langle\langle \zeta_{\mathbf{k}_1}(\tau_0)\zeta_{\mathbf{k}_2}(\tau_0)\zeta_{\mathbf{k}_3}(\tau_0)\rangle\rangle &= 2m_{\mathrm{p}}^2\epsilon_1(\tau_s)a^2(\tau_s)\Delta\epsilon_2\left(\frac{H^2}{4m_{\mathrm{p}}^2\epsilon_1(\tau_s)}\right)^3 \\ &\times \frac{4}{k_s^2}\left(\frac{1}{k_1^3k_2^3} + \frac{1}{k_2^3k_3^3} + \frac{1}{k_1^3k_3^3}\right), \end{aligned} \tag{3.106}$$

which is a local type bispectrum. Thus, we can read the nonlinear parameter $f_{\mathrm{NL}} = 5/8$.

3.3.3.2 Smooth Transition

Consider evolution of the second SR parameter given by (3.65), its time derivative is

$$\epsilon_2'(\tau) = \frac{-2s^2(s-3)(s+3)}{\tau\left[\left(\frac{\tau_s}{\tau}\right)^s(s-3)+s+3\right]^2}\left(\frac{\tau_s}{\tau}\right)^s. \tag{3.107}$$

Because the second SR parameter is constant for $\tau < \tau_s$, the domain of time integral in (3.103) is simplified to

$$\begin{aligned} \langle\langle \zeta_{\mathbf{k}_1}(\tau_0)\zeta_{\mathbf{k}_2}(\tau_0)\zeta_{\mathbf{k}_3}(\tau_0)\rangle\rangle &= -2m_{\mathrm{p}}^2\int_{\tau_s}^{\tau_0} \mathrm{d}\tau\ \epsilon_1(\tau)\epsilon_2'(\tau)a^2(\tau) \\ &\times \mathrm{Im}\left[\zeta_{k_1}(\tau_0)\zeta_{k_2}(\tau_0)\zeta_{k_3}(\tau_0)\zeta_{k_1}^*(\tau)\zeta_{k_2}^*(\tau)\zeta_{k_3}'^*(\tau) + \mathrm{perm}\right]. \end{aligned} \tag{3.108}$$

Unfortunately, the general form of bispectrum is complicated. However, it becomes simple in the squeezed limit $k_1 \ll k_2 = k_3$ for perturbation modes that cross the horizon much before the transition $k_1, k_2, k_3 \ll k_s$. Substituting (3.64), (3.107), and (3.68) to the time integral leads to

$$
\begin{aligned}
\langle\langle \zeta_{k_1}(\tau_0)\zeta_{k_2}(\tau_0)\zeta_{-k_2}(\tau_0)\rangle\rangle = & \left|\zeta_{k_1}(\tau_0)\right|^2 \left|\zeta_{k_2}(\tau_0)\right|^2 \\
& \times \int_{\tau_s}^{\tau_0} \mathrm{d}\tau \sqrt{\frac{\epsilon_1(\tau_0)}{\epsilon_1(\tau)}} \epsilon_2'(\tau) \left(1 + \frac{1}{2}\epsilon_2(\tau)\right) \\
= & - 2\eta_V \sqrt{\frac{\epsilon_1(\tau_0)}{\epsilon_1(\tau_s)}} \left|\zeta_{k_1}(\tau_0)\right|^2 \left|\zeta_{k_2}(\tau_0)\right|^2 . \qquad (3.109)
\end{aligned}
$$

Thus, we can read off the nonlinear parameter

$$
\frac{12}{5} f_{\mathrm{NL}} = -2\eta_V \sqrt{\frac{\epsilon_1(\tau_0)}{\epsilon_1(\tau_s)}}, \qquad (3.110)
$$

which is suppressed by SR parameter η_V. This means that the bispectrum generated by the transition (3.65) is negligible. In [98], it is argued that suppression in the bispectrum is a result of (3.65) satisfying Wands duality condition [147], which makes the integrand of (3.102) a total time derivative. The squeezed bispectrum in both smooth and sharp transition cases can be reproduced from the generalized Maldacena's theorem by considering only background evolution [116] following [40].

3.4 Generation of Large Fluctuation

In Sect. 3.2.2, we have shown that violation of the SR approximation, especially the USR condition, leads to the growth of curvature perturbation outside the horizon [132, 152]. Although the USR period cannot realize the required e-folds number of inflation, it can be useful for other purposes. If the inflaton undergoes USR period only for a short duration, the curvature perturbation will be enhanced only within that duration. At the end of inflation, it corresponds to amplification of the curvature perturbation at a range of wavenumbers that cross the horizon during the USR period.

Assuming sharp transitions, evolution of the second SR parameter corresponding to the potential in Fig. 3.1 is shown by the brown curve in Fig. 3.2. The inflaton undergoes USR period in the flat region of potential from time t_s to t_e. In order to explain the observed CMB fluctuations, at early time, $t \lesssim t_s$, the inflation was in SR period with slightly tilted potential. At late time, $t \gtrsim t_e$, there is little observational constraint on the potential. A potential is viable as long as the curvature perturbation does not grow during the final period.

In this section, we study the power spectrum and bispectrum of curvature perturbation in such a setup.

3.4.1 Power Spectrum

At early time, $t \lesssim t_s$, the inflaton was in SR period with Bunch–Davies initial vacuum, so the curvature perturbation is given by

$$\zeta_k(\tau) = \left(\frac{iH}{2m_{\mathrm{p}}\sqrt{k^3\epsilon_{1s}}} \right)_\star e^{-ik\tau}(1+ik\tau), \tag{3.111}$$

where ϵ_{1s} is the first SR parameter during this SR period.

At $t_s \lesssim t \lesssim t_e$, the inflaton is in the intermediate USR period. We define τ_s and τ_e as the conformal time corresponding to t_s and t_e, respectively. The first SR parameter can be written as $\epsilon_1(\tau) = \epsilon_{1s}(\tau/\tau_s)^6$ based on proportionality in (3.47). Therefore, the curvature perturbation becomes

$$\zeta_k(\tau) = \left(\frac{iH}{2m_{\mathrm{p}}\sqrt{k^3\epsilon_{1s}}} \right)_\star \left(\frac{\tau_s}{\tau}\right)^3 \left[\mathcal{A}_k e^{-ik\tau}(1+ik\tau) - \mathcal{B}_k e^{ik\tau}(1-ik\tau)\right], \tag{3.112}$$

where coefficients $\mathcal{A}_k$ and $\mathcal{B}_k$ are determined by matching to the SR solution (3.111) at the boundary. Solutions of the coefficients by requiring continuity of $\zeta_k(\tau)$ and $\zeta_k'(\tau)$ at transition $\tau = \tau_s$ are

$$\mathcal{A}_k = 1 - \frac{3(1+k^2\tau_s^2)}{2ik^3\tau_s^3}, \tag{3.113}$$

$$\mathcal{B}_k = -\frac{3(1+ik\tau_s)^2}{2ik^3\tau_s^3} e^{-2ik\tau_s}. \tag{3.114}$$

At late time, $t \gtrsim t_e$, the inflaton goes back to SR dynamics. The curvature perturbation can be written as

$$\zeta_k(\tau) = \left(\frac{iH}{2m_{\mathrm{p}}\sqrt{k^3\epsilon_{1s}}} \right)_\star \left(\frac{\tau_s}{\tau_e}\right)^3 \left[\mathcal{C}_k e^{-ik\tau}(1+ik\tau) - \mathcal{D}_k e^{ik\tau}(1-ik\tau)\right], \tag{3.115}$$

where coefficients $\mathcal{C}_k$ and $\mathcal{D}_k$ are determined by matching to the USR solution (3.112) at the boundary. Solutions of the coefficients by requiring continuity of $\zeta_k(\tau)$ and $\zeta_k'(\tau)$ at transition $\tau = \tau_e$ are

$$\begin{aligned} \mathcal{C}_k = \frac{-1}{4k^6\tau_s^3\tau_e^3} & \left\{9(k\tau_s - i)^2(k\tau_e + i)^2 e^{2ik(\tau_e-\tau_s)} \right. \\ & \left. - \left[k^2\tau_s^2(2k\tau_s+3i)+3i\right]\left[k^2\tau_e^2(2k\tau_e-3i)-3i\right]\right\}, \end{aligned} \tag{3.116}$$

$$\mathcal{D}_k = \frac{3}{4k^6\tau_s^3\tau_e^3}\{e^{-2ik\tau_s}[3+k^2\tau_e^2(3-2ik\tau_e)](k\tau_s-i)^2 \\ +ie^{-2ik\tau_e}\left[3i+k^2\tau_s^2(2k\tau_s+3i)\right](k\tau_e-i)^2\}. \tag{3.117}$$

As before, k_s and k_e are wavenumbers which cross the horizon at τ_s and τ_e, respectively. At the end of inflation, τ_0 ($\to 0$), the tree-level power spectrum is

$$\Delta^2_{s(0)}(k,\tau_0) = \frac{k^3}{2\pi^2}|\zeta_k(\tau_0)|^2 = \left(\frac{H^2}{8\pi^2 m_\mathrm{p}^2\epsilon_{1s}}\right)_\star \left(\frac{k_e}{k_s}\right)^6 |\mathcal{C}_k - \mathcal{D}_k|^2. \tag{3.118}$$

On large scale, the power spectrum approaches an almost scale-invariant limit

$$\Delta^2_{s(\mathrm{SR})}(k) \equiv \Delta^2_{s(0)}(k \ll k_s,\tau_0) = \left(\frac{H^2}{8\pi^2 m_\mathrm{p}^2\epsilon_{1s}}\right)_\star \left[1+\mathcal{O}\left(\frac{k^2}{k_s^2}\right)\right], \tag{3.119}$$

which is consistent with CMB observation. Assuming $k_e \gg k_s$, the next-order approximation is

$$\Delta^2_{s(0)}(k,\tau_0) \simeq \left(\frac{H^2}{8\pi^2 m_\mathrm{p}^2\epsilon_{1s}}\right)_\star \left[1-\frac{8}{5}\left(\frac{k_e}{k_s}\right)^3\left(\frac{k}{k_s}\right)^2 \\ +\frac{24}{35}\left(\frac{k_e}{k_s}\right)^3\left(\frac{k}{k_s}\right)^4+\mathcal{O}\left(\frac{k^6}{k_s^6}\right)\right]. \tag{3.120}$$

The minus sign in the coefficient of $(k/k_s)^2$ is important because it can explain a dip in the power spectrum before amplification start. As k approaches k_s, the k^2-term grows until it reaches unity. At this point, the k^4-term is not large enough yet, so the power spectrum almost vanishes. This yields a dip in the power spectrum, which happens at the wavenumber

$$k_\mathrm{dip} \simeq k_s\left[\frac{5}{8}\left(\frac{k_s}{k_e}\right)^3\right]^{1/2}. \tag{3.121}$$

It has been argued that this dip may lead to an interesting observational consequence [10].

At $k_\mathrm{dip} < k < k_s$, the k^4-term grows while the k^6-term is still small, which explains the apparent k^4-growth in the power spectrum [21]. On a small scale with a larger wavenumber, $k_s \lesssim k \lesssim k_e$, the power spectrum is oscillating around

$$4\times\Delta^2_{s(\mathrm{PBH})} \equiv 4\times\Delta^2_{s(\mathrm{SR})}(k_s)\left(\frac{k_e}{k_s}\right)^6, \tag{3.122}$$

whose high-density peak may collapse into PBHs. The numerical coefficient of 4 appears for the same reason as (3.78). It is amplified by a factor of $(k_e/k_s)^6$ compared to the CMB-scale power spectrum. To generate a significant abundance of PBHs, typically peak with $\mathcal{O}(0.01)$ on small-scale power spectrum is required, see Chaps. 2 and 6. This means that the power spectrum is amplified $\mathcal{O}(10^7)$ compared to CMB-scale. In terms of the number of e-folds during the USR period $N_{\rm USR} = \log[a(\tau_e)/a(\tau_s)]$, we need to have

$$N_{\rm USR} = \log \frac{\tau_s}{\tau_e} = \log \frac{k_e}{k_s} \simeq \frac{1}{6} \log 10^7 = 2.68, \tag{3.123}$$

to achieve a significant abundance of PBHs (Fig. 3.3).

In this model, we can see that the power spectrum remains amplified even for $k > k_e$, which oscillates around $\Delta^2_{s({\rm PBH})}$. This happens because the inflaton goes directly to the SR period after the USR period. In case the inflaton undergoes a constant-roll period with $\epsilon_2 > 0$ after the USR period, the power spectrum at the end of inflation will decrease for $k > k_e$.

In this review, we choose an USR period for the intermediate phase as an example. In principle, the intermediate phase can be generalized to a constant-roll period as long as the curvature perturbation grows outside the horizon. The requirement was discussed below Eq. (3.63). However, for a general constant-roll period, the mode function becomes the Hankel function, which is given by (3.24) with (3.57). Technically, it becomes more difficult to match the coefficient of constant-roll period with the SR period because it involves derivative of Hankel function. In [143], a method to approximate the mode function is proposed by utilizing the fact that the duration of the SR violating regime is extremely short compared to the total duration

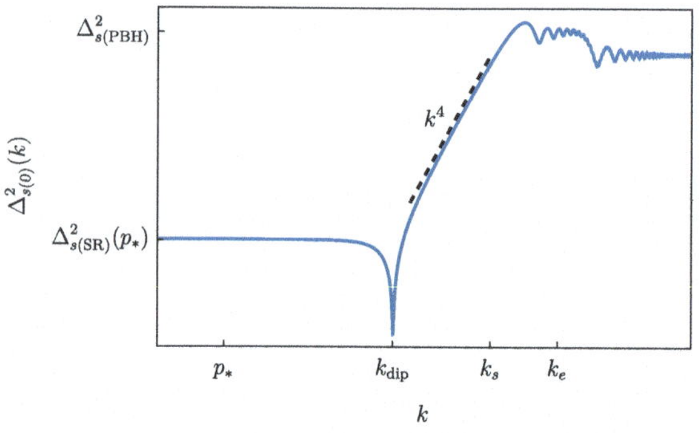

Fig. 3.3 Power spectrum of the curvature perturbation. At CMB pivot scale, $p_* = 0.05$ Mpc^{-1}, the power spectrum is almost scale invariant with amplitude $\Delta^2_{s({\rm SR})}(p_*) = 2.1 \times 10^{-9}$, based on observational result [4]. At $k_{\rm dip} < k < k_s$, the power spectrum grows with function k^4 until it reaches the peak. At $k_s < k < k_e$, the power spectrum is oscillating around $4 \times \Delta^2_{s({\rm PBH})}$. Similar behavior at $k > k_e$ because SR period is imposed after the USR period

of inflation. An inflation model with two periods of SR violation is also introduced in [143] that yields a brief period of k^8 growth in the power spectrum, which is much steeper than k^4 growth.

3.4.2 Bispectrum

Consider the evolution of the second SR parameter shown by the brown curve in Fig. 3.2. Its time derivative can be modelled as Dirac-delta function

$$\epsilon_2'(\tau) = \Delta\epsilon_2[-\delta(\tau-\tau_s)+\delta(\tau-\tau_e)]. \tag{3.124}$$

Substituting it to the time integral in (3.103) leads to

$$\begin{aligned}&\langle\langle\zeta_{\boldsymbol{k}_1}(\tau_0)\zeta_{\boldsymbol{k}_2}(\tau_0)\zeta_{\boldsymbol{k}_3}(\tau_0)\rangle\rangle = \\ &2m_{\rm p}^2\epsilon_1(\tau_s)a^2(\tau_s)\Delta\epsilon_2\mathrm{Im}\left[\zeta_{k_1}(\tau_0)\zeta_{k_2}(\tau_0)\zeta_{k_3}(\tau_0)\zeta_{k_1}^*(\tau_s)\zeta_{k_2}^*(\tau_s)\zeta_{k_3}'^*(\tau_s)\right] \\ &-2m_{\rm p}^2\epsilon_1(\tau_e)a^2(\tau_e)\Delta\epsilon_2\mathrm{Im}\left[\zeta_{k_1}(\tau_0)\zeta_{k_2}(\tau_0)\zeta_{k_3}(\tau_0)\zeta_{k_1}^*(\tau_e)\zeta_{k_2}^*(\tau_e)\zeta_{k_3}'^*(\tau_e)\right] + \text{perm}.\end{aligned} \tag{3.125}$$

In the squeezed limit, the bispectrum becomes

$$\begin{aligned}\langle\langle\zeta_{\boldsymbol{k}_1}(\tau_0)\zeta_{\boldsymbol{k}_2}(\tau_0)\zeta_{-\boldsymbol{k}_2}(\tau_0)\rangle\rangle &= -\left|\zeta_{k_1}(\tau_0)\right|^2\left|\zeta_{k_2}(\tau_0)\right|^2 \\ \times\Bigg\{4\Delta\epsilon_2 m_{\rm p}^2\epsilon_1(\tau_e)a^2(\tau_e)\mathrm{Im}&\left[\frac{\zeta_{k_2}^2(\tau_0)}{\left|\zeta_{k_2}(\tau_0)\right|^2}\zeta_{k_2}^*(\tau_e)\zeta_{k_2}'^*(\tau_e)\right] \\ -4\Delta\epsilon_2 m_{\rm p}^2\epsilon_1(\tau_s)a^2(\tau_s)\mathrm{Im}&\left[\frac{\zeta_{k_2}^2(\tau_0)}{\left|\zeta_{k_2}(\tau_0)\right|^2}\zeta_{k_2}^*(\tau_s)\zeta_{k_2}'^*(\tau_s)\right]\Bigg\}.\end{aligned} \tag{3.126}$$

We define the term inside parenthesis as $C_0(k)$, so the bispectrum can be written as

$$\langle\langle\zeta_{\boldsymbol{k}_1}(\tau_0)\zeta_{\boldsymbol{k}_2}(\tau_0)\zeta_{-\boldsymbol{k}_2}(\tau_0)\rangle\rangle = -C_0(k_2)\left|\zeta_{k_1}(\tau_0)\right|^2\left|\zeta_{k_2}(\tau_0)\right|^2. \tag{3.127}$$

We can show that $C_0(k) = n_s(k,\tau_0) - 1$ analytically, which confirms Maldacena's theorem explicitly [99] (Fig. 3.4).

At the end of inflation, we have shown that only the bulk interaction (3.102) contributes to the bispectrum. During the USR period, the bispectrum also satisfies Maldacena's theorem, where both the bulk interaction (3.102) and field redefinition (3.98) are important. Therefore, in general time, the squeezed bispectrum is

$$\langle\langle\zeta_{\boldsymbol{k}_1}(\tau)\zeta_{\boldsymbol{k}_2}(\tau)\zeta_{-\boldsymbol{k}_2}(\tau)\rangle\rangle = -(n_s(k_2,\tau)-1)\left|\zeta_{k_1}(\tau)\right|^2\left|\zeta_{k_2}(\tau)\right|^2. \tag{3.128}$$

We can read that spectral tilt of the short-scale perturbation, depends on time even at superhorizon scale, specifically around the transition time. A plot of the spectral tilt

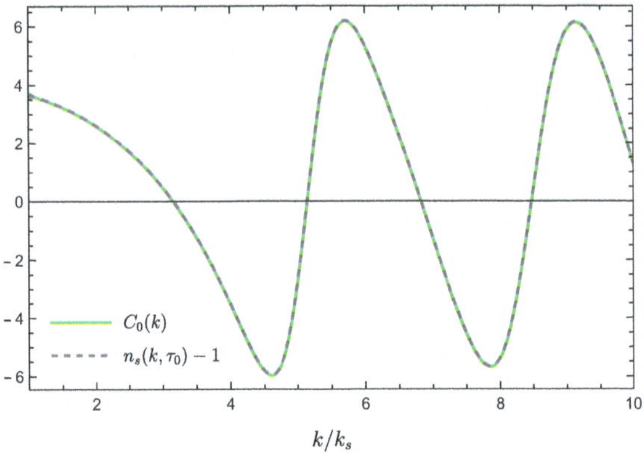

Fig. 3.4 Plot of $C_0(k)$ and $n_s(k, \tau_0) - 1$. We choose $k_e/k_s = 10$ only for illustrative purposes

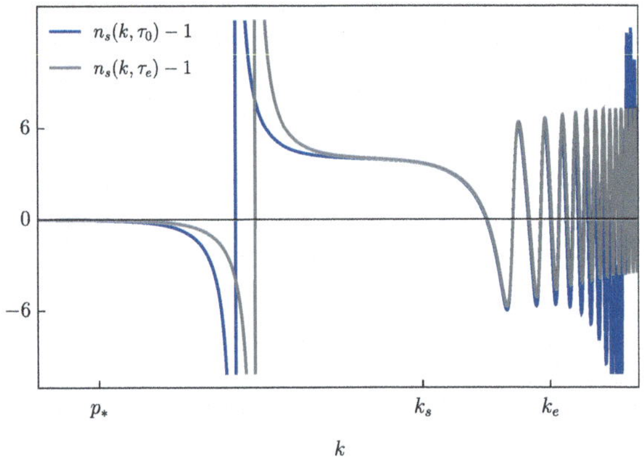

Fig. 3.5 Plot of $n_s - 1$ at the end of USR period and at the end of inflation. For a range of wavenumber between p_* and k_s, the spectral tilt evolves, although such perturbations already far outside the horizon at the end of USR period

at the end of USR period and at the end of inflation are shown in Fig. 3.5. Thus, in contrast to the case of the SR inflation in Sect. 3.3.1 where time dependence $\log \tau_0$ is cancelled, the squeezed bispectrum evolves outside the horizon in this model. In case all wavenumbers have magnitude around the peak scale, the bispectrum is expected to deviate significantly from Maldacena's limit.

In the presence of peaks and troughs in the power spectrum, Maldacena's theorem is generalized in [118]. In addition, the consistency relation (3.128) fixes the sign of $f_{\rm NL}$ at the peak, which has an implication for the abundances of PBH [54].

3.4.3 Beyond the Tree-Level Contribution

So far, we have discussed two-point and three-point functions of the curvature perturbation at tree-level approximation. We have shown that the sharp transition of the second SR parameter generates a $\mathcal{O}(1)$ contribution to the nonlinear parameter $f_{\rm NL}$. It is a large contribution to the bispectrum, which implies that a higher-order correction than first-order perturbation theory can also be large.

The second-order expansion of in-in perturbation theory reads

$$\langle O(\tau)\rangle_{(1)} = \langle O(\tau)\rangle^{\dagger}_{(0,2)} + \langle O(\tau)\rangle_{(1,1)} + \langle O(\tau)\rangle_{(0,2)}, \tag{3.129}$$

$$\langle O(\tau)\rangle_{(1,1)} = \int_{-\infty}^{\tau} \mathrm{d}\tau_1 \int_{-\infty}^{\tau} \mathrm{d}\tau_2 \left\langle H_{\text{int}}(\tau_1)\hat{O}(\tau)H_{\text{int}}(\tau_2)\right\rangle,$$

$$\langle O(\tau)\rangle_{(0,2)} = -\int_{-\infty}^{\tau} \mathrm{d}\tau_1 \int_{-\infty}^{\tau_1} \mathrm{d}\tau_2 \left\langle \hat{O}(\tau)H_{\text{int}}(\tau_1)H_{\text{int}}(\tau_2)\right\rangle.$$

In the presence of a cubic self-interaction in the Hamiltonian, the second-order perturbation contributes to a one-loop correction to the power spectrum and tree-level contribution to the trispectrum. For an inflation model with sharp transition of the second SR parameter, a one-loop correction to the large-scale power spectrum was first derived in [100]. In this case, the operator is $\zeta_{\boldsymbol{p}}(\tau)\zeta_{-\boldsymbol{p}}(\tau)$, where $\mathbf{p}$ is the CMB scale wavevector, evaluated at $\tau = \tau_0$ ($\rightarrow 0$).

At the end of inflation, there is no contribution from the field redefinition because it is evaluated at the SR period. Only the bulk interaction (3.102) contributes to the one-loop correction. Substituting the corresponding Hamiltonian of (3.102) to (3.129) yields

$$\begin{aligned}\langle\zeta_{\boldsymbol{p}}(\tau_0)\zeta_{-\boldsymbol{p}}(\tau_0)\rangle_{(1,1)} &= \frac{1}{4}m_{\text{p}}^4 \int_{-\infty}^{\tau_0} \mathrm{d}\tau_1\, a^2(\tau_1)\epsilon_1(\tau_1)\epsilon_2'(\tau_1)\\ \times \int_{-\infty}^{\tau_0} \mathrm{d}\tau_2\, a^2(\tau_2)\epsilon_1(\tau_2)\epsilon_2'(\tau_2) &\int \prod_{a=1}^{6}\left[\frac{\mathrm{d}^3k_a}{(2\pi)^3}\right]\delta(\boldsymbol{k}_1+\boldsymbol{k}_2+\boldsymbol{k}_3)\delta(\boldsymbol{k}_4+\boldsymbol{k}_5+\boldsymbol{k}_6)\\ \times \left\langle\zeta'_{\boldsymbol{k}_1}(\tau_1)\zeta_{\boldsymbol{k}_2}(\tau_1)\zeta_{\boldsymbol{k}_3}(\tau_1)\right.&\left.\zeta_{\boldsymbol{p}}(\tau_0)\zeta_{-\boldsymbol{p}}(\tau_0)\zeta'_{\boldsymbol{k}_4}(\tau_2)\zeta_{\boldsymbol{k}_5}(\tau_2)\zeta_{\boldsymbol{k}_6}(\tau_2)\right\rangle,\end{aligned} \tag{3.130}$$

$$\begin{aligned}\langle\zeta_{\boldsymbol{p}}(\tau_0)\zeta_{-\boldsymbol{p}}(\tau_0)\rangle_{(0,2)} &= -\frac{1}{4}m_{\text{p}}^4 \int_{-\infty}^{\tau_0} \mathrm{d}\tau_1\, a^2(\tau_1)\epsilon_1(\tau_1)\epsilon_2'(\tau_1)\\ \times \int_{-\infty}^{\tau_1} \mathrm{d}\tau_2\, a^2(\tau_2)\epsilon_1(\tau_2)\epsilon_2'(\tau_2) &\int \prod_{a=1}^{6}\left[\frac{\mathrm{d}^3k_a}{(2\pi)^3}\right]\delta(\boldsymbol{k}_1+\boldsymbol{k}_2+\boldsymbol{k}_3)\delta(\boldsymbol{k}_4+\boldsymbol{k}_5+\boldsymbol{k}_6)\\ \times \left\langle\zeta_{\boldsymbol{p}}(\tau_0)\zeta_{-\boldsymbol{p}}(\tau_0)\zeta'_{\boldsymbol{k}_1}(\tau_1)\right.&\left.\zeta_{\boldsymbol{k}_2}(\tau_1)\zeta_{\boldsymbol{k}_3}(\tau_1)\zeta'_{\boldsymbol{k}_4}(\tau_2)\zeta_{\boldsymbol{k}_5}(\tau_2)\zeta_{\boldsymbol{k}_6}(\tau_2)\right\rangle.\end{aligned} \tag{3.131}$$

Performing Wick contraction and summing both terms lead to

$$\begin{aligned}\langle\!\langle\zeta_{\boldsymbol{p}}(\tau_0)\zeta_{-\boldsymbol{p}}(\tau_0)\rangle\!\rangle_{(1)} &= \frac{1}{4}m_{\text{p}}^4\epsilon_1^2(\tau_e)a^4(\tau_e)(\Delta\epsilon_2)^2\\ \times \int \frac{\mathrm{d}^3k}{(2\pi)^3}&\left\{\zeta_p(\tau_0)\zeta_p^*(\tau_0)\left(4\zeta_p'\zeta_p'^*\zeta_q\zeta_q^*\zeta_k\zeta_k^* + 8\zeta_p'^*\zeta_p\zeta_q'\zeta_q^*\zeta_k\zeta_k^* + 8\zeta_p'\zeta_p^*\zeta_q'^*\zeta_q\zeta_k\zeta_k^*\right.\right.\\ &\left.+16\zeta_p\zeta_p^*\zeta_q'\zeta_q'^*\zeta_k\zeta_k^*\right)_{\tau=\tau_e} - \mathrm{Re}\left[\zeta_p(\tau_0)\zeta_p(\tau_0)\left(4\zeta_p'^*\zeta_p'^*\zeta_q\zeta_q^*\zeta_k\zeta_k^* + 8\zeta_p'^*\zeta_p^*\zeta_q'\zeta_q^*\zeta_k\zeta_k^*\right.\right.\\ &\left.\left.\left.+8\zeta_p'^*\zeta_p^*\zeta_q'^*\zeta_q\zeta_k\zeta_k^* + 16\zeta_p^*\zeta_p^*\zeta_q'\zeta_q'^*\zeta_k\zeta_k^*\right)_{\tau=\tau_e}\right]\right\}.\end{aligned} \tag{3.132}$$

Here, we only show the leading contribution to the one-loop correction that comes from integral of Dirac-delta function at $\tau = \tau_e$. The contributions of $\tau = \tau_s$ are subleading, which are explicitly evaluated in [99].

For $p \ll k$, the one-loop correction can be simplified to

$$\begin{aligned}\langle\langle \zeta_p(\tau_0)\zeta_{-p}(\tau_0)\rangle\rangle_{(1)} =& \frac{1}{4}m_{\rm p}^4(\Delta\epsilon_2)^2 \left|\zeta_p(\tau_0)\right|^2 \\ &\times \int \frac{{\rm d}^3k}{(2\pi)^3} 16\left[\epsilon_1^2 a^4 \left|\zeta_k\right|^2 {\rm Im}(\zeta_p\zeta_p'^*){\rm Im}(\zeta_k\zeta_k'^*)\right]_{\tau=\tau_e}.\end{aligned} \tag{3.133}$$

We find that the one-loop correction is exactly proportional to the quantum commutator (3.46), which implies the importance of decaying mode even for superhorizon mode function. If we only consider the constant mode of curvature perturbation at superhorizon scale, we will obtain an incorrect result $\left[\hat{\zeta}_p(\tau), \hat{\zeta}_q'(\tau)\right] = 0$. From commutator (3.46), we find

$$ {\rm Im}(\zeta_k(\tau)\zeta_k'^*(\tau)) = \frac{1}{4m_{\rm p}^2\epsilon(\tau)a^2(\tau)}, \tag{3.134}$$

which does not depend on wavenumber k. As we have already seen, in the standard SR inflation this quantity decreases exponentially as a function of conformal time. However, during USR regime, since ϵ decreases much more rapidly than a^{-2}, the right-hand-side of (3.46) and (3.134) *increases* exponentially, which makes the one-loop correction to the power spectrum large even on the large scales probed by CMB.

Then, substituting (3.134) into (3.133), we obtain a one-loop correction to the power spectrum

$$ \Delta^2_{s(1)}(p,\tau_0) = \frac{1}{4}(\Delta\epsilon_2)^2\Delta^2_{s(0)}(p,\tau_0)\int_{k_{\rm IR}}^{k_{\rm UV}} \frac{{\rm d}k}{k}\Delta^2_{s(0)}(k,\tau_e). \tag{3.135}$$

Because we are interested in the effect of large fluctuation on small scales related to PBH formation to the fluctuations on large scales observed by CMB, a natural choice of cutoffs are $k_{\rm IR} = k_s$ and $k_{\rm UV} = k_e$. For $k_e/k_s \ll 1$, evaluating the wavenumber integral with small-scale mode function (3.112) yields

$$ \Delta^2_{s(1)}(p,\tau_0) = \frac{1}{4}(\Delta\epsilon_2)^2\Delta^2_{s(0)}(p,\tau_0)\Delta^2_{s({\rm PBH})}\left(1.1 + \log\frac{k_e}{k_s}\right). \tag{3.136}$$

Requiring the one-loop correction to be much smaller than tree-level contribution yields an upper-bound on small-scale power spectrum

$$ \Delta^2_{s({\rm PBH})} \ll \frac{1}{(\Delta\epsilon_2)^2} \simeq 0.03, \tag{3.137}$$

where the numerical value is obtained by substituting $\Delta\epsilon_2 = 6$ for transition from USR to SR period. Such a bound suggests that sharp transition of the second SR parameter cannot generate large fluctuation on small scales, otherwise perturbation theory breaks down. This result should be understood under the caveat that there are other contributions from the quartic Hamiltonian, both induced by the third-order action (3.102) and fourth-order action.

3.5 Discussion and Summary

3.5.1 Discussion

Following [83], PBH formation models in single-field inflation can be classified into two categories. The first category is models with features in the inflationary potential or nonminimally coupled inflaton with potential defined in the Einstein frame. Models with an extremely flat feature or an inflection point in the potential fall in this category [14, 15, 31, 32, 38, 39, 41, 42, 47, 48, 57, 61, 62, 78, 89, 90, 92, 103, 119, 122, 124, 126, 127, 138, 150]. These models can be derived from high-energy theories such as supergravity [61, 89, 90, 92, 119, 138, 150], axion monodromy [14, 31], scalaron in R^2-gravity [124], α-attractor [41, 42, 57, 103], and string theory [38, 39, 122]. Also, there are models from Higgs inflation which do not require theories beyond the standard model [15, 32, 47, 48, 127]. Other examples of features are a tiny bump or dip [9, 73, 108, 128, 148, 155], an upward or downward step [23, 75, 76, 85, 93, 110, 120], polynomial shape [16, 45, 66, 82], and Coleman-Weinberg potential [19, 84, 132, 151]. In these examples, modification of potential makes a sharp transition of the second SR parameter on the inflaton dynamics.

The second category is models with modified gravity or beyond nonminimally coupled inflaton. Examples for this category are models based on k [5] or G [97] inflation [67, 69, 70, 136, 137, 146], the effective field theory of inflation [8, 11, 12], $f(R)$ gravity [56], a nonminimal derivative coupling [58, 68, 71], Gauss-Bonnet inflation [86, 88], and bumpy axion inflation [121]. In these examples, amplification of small-scale perturbation can be realized by a sharp transition of the second SR parameter and/or other parameters. In [11, 12], amplification of small-scale perturbation is caused by a sharp transition of the sound speed, c_s, a quantity that parametrizes deviation from a canonical kinetic term. In this case, coupling ϵ_2' in cubic self-interaction (3.102) is modified to $(\epsilon_2/c_s^2)'$ [28, 135]. For this category, rapid slowdown of the inflaton's velocity might not be the only reason of amplification of the curvature perturbation.

There have been some works following [99, 100], most of them derive and qualitatively confirm the one-loop correction by other methods, although quantitatively the loop amplitude has not been settled yet in the community. Despite that, our claim [100] was criticized by the author of [130] who attempted to evaluate the same correction with a different method. In case of sharp transition of the second SR parameter, we have shown in [99] that after more careful and proper calculation, one can reproduce our result even if the method proposed in [130] is used. In case the second SR parameter evolves continuously with function (3.107), the bispectrum is suppressed by a SR parameter η_V [129]. In [98], we have argued that suppression on the bispectrum and one-loop correction occur only for a specific smooth transition that satisfies Wands duality condition [147].

Indeed, there are issues related to regularization and renormalization of the divergence. In [100], we show that constraint (3.137) corresponds to the requirement to perform the renormalization order by order in perturbation theory. Another method for performing regularization, renormalization, and resummation of loop corrections is proposed in [35–37], where the sound speed of the curvature perturbation is introduced within the effective field theory of inflation framework. Numerical integration of the one-loop correction has been performed in [44, 55], confirming our analytical result in the limit of the sharp transition. Also, Ref. [55] pointed out that a dip in the power spectrum (3.121) is a tree-level artifact, which disappears after adding a one-loop correction. One-loop correction to the power spectrum around the peak scale has been investigated in [60]. Specifically, for an inflationary potential with oscillatory features, it has been calculated in [25, 77].

Moreover, the δN formalism and stochastic inflation in the presence of a sharp transition of the second SR parameter have been investigated in [79, 107]. The one-loop correction has also been derived from the δN formalism in [53, 72]. The extension from USR to constraint-roll period has been derived in [113], which yields a constraint involving a constant second SR parameter and small-scale power spectrum. In the limit $|\epsilon_2| \to \infty$, the bispectrum and one-loop correction has been investigated in [144, 145], based on a method that approximates the curvature perturbation in the presence of a very short period where the SR approximation is violated [143]. In addition to cubic self-interaction, quartic self-interaction [6, 80] in the Hamiltonian can also contribute to the one-loop correction, in first-order perturbation theory. Its contribution has been investigated, either by direct expansion of the quartic derivative of the potential [104] or by an effective field theory approach [2, 51].

However, there are two works that completely disagree with our conclusions [59, 142]. In their approach, instead of performing field redefinition, they calculate a one-loop correction directly from third-order action with total time derivative terms and terms proportional to the equation of motion, as shown in (3.79). Currently, they have been criticized by [52]. Other perspectives on the effect of the total time derivative on the correlation function have been presented in [17, 87]. We hope to solve this discrepancy by pointing out what is missing in [59, 142]. Beyond the power spectrum, one-loop correction to the bispectrum has been explored in [50].

While we have been performing computations in the comoving gauge, some authors do so adopting the flat-slicing gauge. In this gauge, perturbations are represented by the inflaton's fluctuation $\delta\phi$, with vanishing curvature perturbation. Prior to our work, the quantum correction to $\delta\phi$ was investigated in [29, 141] and followed up in [30], which is consistent with our claim. The effect of a smooth transition of the second SR parameter in the flat-slicing gauge has been explored in [13, 131].

3.5.2 Summary

In this chapter, we have explained the mechanism to generate large primordial fluctuation in single-field inflation. In order to amplify the curvature perturbation, violation of the SR approximation is needed. We have discussed the simplest example of an inflation model that violates the SR approximation, called USR inflation. We have shown that USR inflation leads to growth of the curvature perturbation outside the horizon. The consequence is that USR inflation cannot explain the required e-folds number to solve the horizon problem.

Nevertheless, one can utilize the properties of the USR period for another purpose. If the USR period is sandwiched by two SR periods, the curvature perturbation grows temporarily during the USR period. At the end of inflation, the power spectrum of the curvature perturbation is amplified on small scales corresponding to the time scale of the USR period, while keeping the small amplitude on large scale consistent with CMB observations.

Moreover, we have explained the nonlinear evolution of the curvature perturbation induced by cubic self-interaction, which yields three-point functions or bispectrum. The cubic self-interaction consists of bulk interaction and field redefinition. In USR inflation, only the field redefinition contributes to the bispectrum. When a transition from USR to SR period exists, the bulk interaction that is proportional to the time derivative of the second SR parameter becomes important. If the transition is sharp, the nonlinear parameter of the bispectrum has an $\mathcal{O}(1)$ value. This means that the bispectrum is large enough, which indicates that the nonlinear effect cannot be neglected. At the same time, such a nonlinear effect gives backreaction to the power spectrum as a one-loop correction that leads to a constraint on the small-scale power spectrum associated with transition of the second SR parameter.

Acknowledgements J. K. acknowledges the support from JSPS KAKENHI Grants No. 22KJ1006 and No. 22J20289. J. K. is also supported by the Global Science Graduate Course (GSGC) program of The University of Tokyo. J. Y. is supported by JSPS KAKENHI Grant No. 20H05639.

References

1. N. Aghanim et al., Planck 2018 results. I. Overview and the cosmological legacy of Planck. Astron. Astrophys. **641**, A1 (2020). https://doi.org/10.1051/0004-6361/201833880. arXiv: 1807.06205 [astro-ph.CO]
2. M. Akhshik, H. Firouzjahi, S. Jazayeri, Effective field theory of non-attractor inflation. JCAP **07**, 048 (2015). https://doi.org/10.1088/1475-7516/2015/07/048. arXiv: 1501.01099 [hep-th]
3. Y. Akrami et al., Planck 2018 results. IX. Constraints on primordial non-Gaussianity. Astron. Astrophys. **641**, A9 (2020). https://doi.org/10.1051/0004-6361/201935891. arXiv: 1905.05697 [astro-ph.CO]
4. Y. Akrami et al., Planck 2018 results. X. Constraints on inflation. Astron. Astrophys. **641**, A10 (2020). https://doi.org/10.1051/0004-6361/201833887. arXiv: 1807.06211 [astro-ph.CO]

5. C. Armendariz-Picon, T. Damour, V.F. Mukhanov, k - inflation. Phys. Lett. B **458**, 209–218 (1999). https://doi.org/10.1016/S0370-2693(99)00603-6. arXiv: hep-th/9904075
6. F. Arroja, K. Koyama, Non-Gaussianity from the trispectrum in general single field inflation. Phys. Rev. D **77**, 083517 (2008). https://doi.org/10.1103/PhysRevD.77.083517. arXiv: 0802.1167 [hep-th]
7. F. Arroja, T. Tanaka, A note on the role of the boundary terms for the non-Gaussianity in general k-inflation. JCAP **05**, 005 (2011). https://doi.org/10.1088/1475-7516/2011/05/005. arXiv: 1103.1102 [astro-ph.CO]
8. A. Ashoorioon, A. Rostami, J.T. Firouzjaee, EFT compatible PBHs: effective spawning of the seeds for primordial black holes during inflation. JHEP **07**, 087 (2021). https://doi.org/10.1007/JHEP07(2021)087. arXiv: 1912.13326 [astro-ph.CO]
9. V. Atal, J. Garriga, A. Marcos-Caballero, Primordial black hole formation with non-Gaussian curvature perturbations. JCAP **09**, 073 (2019). https://doi.org/10.1088/1475-7516/2019/09/073. arXiv: 1905.13202 [astro-ph.CO]
10. S. Balaji et al., Observing nulling of primordial correlations via the 21-cm signal. Phys. Rev. Lett. **129**(26), 261301 (2022). https://doi.org/10.1103/PhysRevLett.129.261301. arXiv: 2206.06386 [astro-ph.CO]
11. G. Ballesteros, J.B. Jimenez, M. Pieroni, Black hole formation from a general quadratic action for inflationary primordial fluctuations. JCAP **06**, 016 (2019). https://doi.org/10.1088/1475-7516/2019/06/016. arXiv: 1811.03065 [astro-ph.CO]
12. G. Ballesteros, S. Céspedes, L. Santoni, Large power spectrum and primordial black holes in the effective theory of inflation. JHEP **01**, 074 (2022). https://doi.org/10.1007/JHEP01(2022)074. arXiv: 2109.00567 [hep-th]
13. G. Ballesteros, J.G. Egea, One-loop power spectrum in ultra slow-roll inflation and implications for primordial black hole dark matter. JCAP **07**, 052 (2024). https://doi.org/10.1088/1475-7516/2024/07/052. arXiv: 2404.07196 [astro-ph.CO]
14. G. Ballesteros, J. Rey, F. Rompineve, Detuning primordial black hole dark matter with early matter domination and axion monodromy. JCAP **06**, 014 (2020). https://doi.org/10.1088/1475-7516/2020/06/014. arXiv: 1912.01638 [astro-ph.CO]
15. G. Ballesteros, M. Taoso, Primordial black hole dark matter from single field inflation. Phys. Rev. D **97**(2), 023501 (2018). https://doi.org/10.1103/PhysRevD.97.023501. arXiv: 1709.05565 [hep-ph]
16. G. Ballesteros et al., Primordial black holes as dark matter and gravitational waves from single-field polynomial inflation. JCAP **07**, 025 (2020). https://doi.org/10.1088/1475-7516/2020/07/025. arXiv: 2001.08220 [astro-ph.CO]
17. M. Braglia, L. Pinol, No time to derive: unraveling total time derivatives in in-in perturbation theory. J. High Energy Phys. Vol 2024, Article Number 68 (2024). arXiv: 2403.14558 [astro-ph.CO]
18. R. Bravo et al., A generalized non-Gaussian consistency relation for single field inflation. JCAP **05**, 024 (2018). https://doi.org/10.1088/1475-7516/2018/05/024. arXiv: 1711.02680 [astro-ph.CO]
19. E. Bugaev, P. Klimai, Large curvature perturbations near horizon crossing in single-field inflation models. Phys. Rev. D **78**, 063515 (2008). https://doi.org/10.1103/PhysRevD.78.063515. arXiv: 0806.4541 [astro-ph]
20. C. Burrage, R.H. Ribeiro, D. Seery, Large slow-roll corrections to the bispectrum of noncanonical inflation. JCAP **07**, 032 (2011). https://doi.org/10.1088/1475-7516/2011/07/032. arXiv: 1103.4126 [astro-ph.CO]
21. C. Byrnes, P. Cole, S. Patil, Steepest growth of the power spectrum and primordial black holes. JCAP **1906**, 028 (2019). https://doi.org/10.1088/1475-7516/2019/06/028. arXiv: 1811.11158 [astro-ph.CO]
22. Y.-F. Cai et al., Features from the non-attractor beginning of inflation. JCAP **10**, 017 (2016). https://doi.org/10.1088/1475-7516/2016/10/017. arXiv: 1607.07872 [astro-ph.CO]
23. Y.-F. Cai et al., One small step for an inflaton, one giant leap for inflation: a novel non-Gaussian tail and primordial black holes. Phys. Lett. B **834**, 137461 (2022). https://doi.org/10.1016/j.physletb.2022.137461. arXiv: 2112.13836 [astro-ph.CO]

24. Y.-F. Cai et al., Revisiting non-Gaussianity from non-attractor inflation models. JCAP **05**, 012 (2018). https://doi.org/10.1088/1475-7516/2018/05/012. arXiv: 1712.09998 [astro-ph.CO]
25. A. Caravano, K. Inomata, S. Renaux-Petel, The inflationary butterfly effect: non-perturbative dynamics from small-scale features. Phys. Rev. Lett. **133**, 151001 (2024). arXiv: 2403.12811 [astro-ph.CO]
26. B.J. Carr, S.W. Hawking, Black holes in the early Universe. Mon. Not. Roy. Astron. Soc. **168**, 399–415 (1974). https://doi.org/10.1093/mnras/168.2.399
27. X. Chen et al., In-in and δN calculations of the bispectrum from non-attractor single-field inflation. JCAP **12**, 039 (2013). https://doi.org/10.1088/1475-7516/2013/12/039. arXiv: 1308.5341 [astro-ph.CO]
28. X. Chen et al., Observational signatures and non-Gaussianities of general single field inflation. JCAP **01**, 002 (2007). https://doi.org/10.1088/1475-7516/2007/01/002. arXiv: hep-th/0605045
29. S.-L. Cheng, D.-S. Lee, K.-W. Ng, Power spectrum of primordial perturbations during ultra-slow-roll inflation with back reaction effects. Phys. Lett. B **827**, 136956 (2022). https://doi.org/10.1016/j.physletb.2022.136956. arXiv: 2106.09275 [astro-ph.CO]
30. S.-L. Cheng, D.-S. Lee, K.-W. Ng, Primordial perturbations from ultra-slow-roll single-field inflation with quantum loop effects (2023). arXiv: 2305.16810 [astro-ph.CO]
31. S.-L. Cheng, W. Lee, K.-W. Ng, Primordial black holes and associated gravitational waves in axion monodromy inflation. JCAP **07**, 001 (2018). https://doi.org/10.1088/1475-7516/2018/07/001. arXiv: 1801.09050 [astro-ph.CO]
32. D.Y. Cheong, S.M. Lee, S.C. Park, Primordial black holes in Higgs-R^2 inflation as the whole of dark matter. JCAP **01**, 032 (2021). https://doi.org/10.1088/1475-7516/2021/01/032. arXiv: 1912.12032 [hep-ph]
33. J. Chluba et al., New horizons in cosmology with spectral distortions of the cosmic microwave background. Exper. Astron. **51**(3), 1515–1554 (2021). https://doi.org/10.1007/s10686-021-09729-5. arXiv: 1909.01593 [astro-ph.CO]
34. J. Chluba, A.L. Erickcek, I. Ben-Dayan, Probing the inflaton: small-scale power spectrum constraints from measurements of the CMB energy spectrum. Astrophys. J. **758**, 76 (2012). https://doi.org/10.1088/0004-637X/758/2/76. arXiv: 1203.2681 [astro-ph.CO]
35. S. Choudhury, M.R. Gangopadhyay, M. Sami, No-go for the formation of heavy mass primordial black holes in single field inflation. Eur. Phys. J. C **84**, 884 (2024). https://doi.org/10.1140/epjc/s10052-024-13218-2. arXiv: 2301.10000 [astro-ph.CO]
36. S. Choudhury, S. Panda, M. Sami, PBH formation in EFT of single field inflation with sharp transition. Phys. Lett. B **845**, 138123 (2023). https://doi.org/10.1016/j.physletb.2023.138123. arXiv: 2302.05655 [astro-ph.CO]
37. S. Choudhury, S. Panda, M. Sami, Quantum loop effects on the power spectrum and constraints on primordial black holes. JCAP **11**, 066 (2023). https://doi.org/10.1088/1475-7516/2023/11/066. arXiv: 2303.06066 [astro-ph.CO]
38. M. Cicoli, V.A. Diaz, F.G. Pedro, Primordial black holes from string inflation. JCAP **06**, 034 (2018). https://doi.org/10.1088/1475-7516/2018/06/034. arXiv: 1803.02837 [hep-th]
39. M. Cicoli, F.G. Pedro, N. Pedron, Secondary GWs and PBHs in string inflation: formation and detectability. JCAP **08**(08), 030 (2022). https://doi.org/10.1088/1475-7516/2022/08/030. arXiv: 2203.00021 [hep-th]
40. P. Creminelli, M. Zaldarriaga, Single field consistency relation for the 3-point function. JCAP **10**, 006 (2004). https://doi.org/10.1088/1475-7516/2004/10/006. arXiv: astro-ph/0407059
41. I. Dalianis, A. Kehagias, G. Tringas, Primordial black holes from α-attractors. JCAP **01**, 037 (2019). https://doi.org/10.1088/1475-7516/2019/01/037. arXiv: 1805.09483 [astro-ph.CO]
42. I. Dalianis, G. Tringas, Primordial black hole remnants as dark matter produced in thermal, matter, and runaway-quintessence postinflationary scenarios. Phys. Rev. D **100**(8), 083512 (2019). https://doi.org/10.1103/PhysRevD.100.083512. arXiv: 1905.01741 [astro-ph.CO]

43. M.W. Davies, P. Carrilho, D.J. Mulryne, Non-Gaussianity in inflationary scenarios for primordial black holes. JCAP **06**(06), 019 (2022). https://doi.org/10.1088/1475-7516/2022/06/019. arXiv: 2110.08189 [astro-ph.CO]
44. M.W. Davies, L. Iacconi, D.J. Mulryne, Numerical 1-loop correction from a potential yielding ultra-slow-roll dynamics. JCAP **04**, 050 (2024). https://doi.org/10.1088/1475-7516/2024/04/050. arXiv: 2312.05694 [astro-ph.CO]
45. H. Di, Y. Gong, Primordial black holes and second order gravitational waves from ultra-slow-roll inflation. JCAP **07**, 007 (2018). https://doi.org/10.1088/1475-7516/2018/07/007. arXiv: 1707.09578 [astro-ph.CO]
46. G. Domènech, G. Vargas, T. Vargas, An exact model for enhancing/suppressing primordial fluctuations. JCAP **03**, 002 (2024). https://doi.org/10.1088/1475-7516/2024/03/002. arXiv: 2309.05750 [astro-ph.CO]
47. M. Drees, Y. Xu, Overshooting, critical Higgs inflation and second order gravitational wave signatures. Eur. Phys. J. C **81**(2), 182 (2021). https://doi.org/10.1140/epjc/s10052-021-08976-2. arXiv: 1905.13581 [hep-ph]
48. J.M. Ezquiaga, J. García-Bellido, E.R. Morales, Primordial black hole production in critical Higgs inflation. Phys. Lett. B **776**, 345–349 (2018). https://doi.org/10.1016/j.physletb.2017.11.039. arXiv: 1705.04861 [astro-ph.CO]
49. B. Finelli et al., Soft theorems for shift-symmetric cosmologies. Phys. Rev. D **97**(6), 063531 (2018). https://doi.org/10.1103/PhysRevD.97.063531. arXiv: 1711.03737 [hep-th]
50. H. Firouzjahi, Loop corrections in the bispectrum in ultraslow-roll inflation with PBHs formation. Phys. Rev. D **110**, 043519 (2024). https://doi.org/10.1103/PhysRevD.110.043519. arXiv: 2403.03841 [astro-ph.CO]
51. H. Firouzjahi, One-loop corrections in power spectrum in single field inflation. JCAP **10**, 006 (2023). https://doi.org/10.1088/1475-7516/2023/10/006. arXiv: 2303.12025 [astro-ph.CO]
52. H. Firouzjahi, Revisiting loop corrections in single field ultraslow-roll inflation. Phys. Rev. D **109**(4), 043514 (2024). https://doi.org/10.1103/PhysRevD.109.043514. arXiv: 2311.04080 [astro-ph.CO]
53. H. Firouzjahi, A. Riotto, Primordial Black Holes and loops in single-field inflation. JCAP **02**, 021 (2024). https://doi.org/10.1088/1475-7516/2024/02/021. arXiv: 2304.07801 [astro-ph.CO]
54. H. Firouzjahi, A. Riotto, The sign of non-Gaussianity and the primordial black holes abundance (2023). arXiv: 2309.10536 [astro-ph.CO]
55. G. Franciolini et al., Perturbativity in the presence of ultraslow-roll dynamics. Phys. Rev. D **109**, 123550 (2024). https://doi.org/10.1103/PhysRevD.109.123550. arXiv: 2305.03491 [astro-ph.CO]
56. D. Frolovsky, S.V. Ketov, S. Saburov, Formation of primordial black holes after Starobinsky inflation. Mod. Phys. Lett. A **37**(21), 2250135 (2022). https://doi.org/10.1142/S0217732322501358. arXiv: 2205.00603 [astro-ph.CO]
57. C. Fu, S.-J. Wang, Primordial black holes and induced gravitational waves from double-pole inflation. JCAP **06**, 012 (2023). https://doi.org/10.1088/1475-7516/2023/06/012. arXiv: 2211.03523 [astro-ph.CO]
58. C. Fu, P. Wu, H. Yu, Primordial black holes from inflation with nonminimal derivative coupling. Phys. Rev. D **100**(6), 063532 (2019). https://doi.org/10.1103/PhysRevD.100.063532. arXiv: 1907.05042 [astro-ph.CO]
59. J. Fumagalli, Absence of one-loop effects on large scales from small scales in non-slow-roll dynamics (2023). arXiv: 2305.19263 [astro-ph.CO]
60. J. Fumagalli et al., One-loop infrared rescattering by enhanced scalar fluctuations during inflation (2023). arXiv: 2307.08358 [astro-ph.CO]
61. T.-J. Gao, Z.-K. Guo, Primordial black hole production in inflationary models of supergravity with a single chiral superfield. Phys. Rev. D **98**(6), 063526 (2018). https://doi.org/10.1103/PhysRevD.98.063526. arXiv: 1806.09320 [hep-ph]

62. J. Garcia-Bellido, E.R. Morales, Primordial black holes from single field models of inflation. Phys. Dark Univ. **18**, 47–54 (2017). https://doi.org/10.1016/j.dark.2017.09.007. arXiv: 1702.03901 [astro-ph.CO]
63. A.H. Guth, The inflationary universe: a possible solution to the horizon and flatness problems. Phys. Rev. D **23**, 347–356 (1981). Ed. by L.-Z. Fang, R. Ruffini. https://doi.org/10.1103/PhysRevD.23.347
64. S.W. Hawking, Black hole explosions. Nature **248**, 30–31 (1974). https://doi.org/10.1038/248030a0
65. S. Hawking, Gravitationally collapsed objects of very low mass. Mon. Not. Roy. Astron. Soc. **152**, 75 (1971)
66. M.P. Hertzberg, M. Yamada, Primordial black holes from polynomial potentials in single field inflation. Phys. Rev. D **97**(8), 083509 (2018). https://doi.org/10.1103/PhysRevD.97.083509. arXiv: 1712.09750 [astro-ph.CO]
67. S. Heydari, K. Karami, Primordial black holes and secondary gravitational waves from generalized power-law non-canonical inflation with quartic potential. Eur. Phys. J. C **84**(2), 127 (2024). https://doi.org/10.1140/epjc/s10052-024-12489-z. arXiv: 2310.11030 [gr-qc]
68. S. Heydari, K. Karami, Primordial black holes ensued from exponential potential and coupling parameter in nonminimal derivative inflation model. JCAP **03**(03), 033 (2022). https://doi.org/10.1088/1475-7516/2022/03/033. arXiv: 2111.00494 [gr-qc]
69. S. Heydari, K. Karami, Primordial black holes generated by fast-roll mechanism in non-canonical natural inflation (2024). arXiv: 2405.08563 [gr-qc]
70. S. Heydari, K. Karami, Primordial black holes in non-canonical scalar field inflation driven by quartic potential in the presence of bump. JCAP **02**, 047 (2024). https://doi.org/10.1088/1475-7516/2024/02/047. arXiv: 2309.01239 [astro-ph.CO]
71. S. Heydari, K. Karami, Primordial black holes in nonminimal derivative coupling inflation with quartic potential and reheating consideration. Eur. Phys. J. C **82**(1), 83 (2022). https://doi.org/10.1140/epjc/s10052-022-10036-2. arXiv: 2107.10550 [gr-qc]
72. L. Iacconi, D. Mulryne, D. Seery, Loop corrections in the separate universe picture. JCAP **06**, 062 (2024). https://doi.org/10.1088/1475-7516/2024/06/062. arXiv: 2312.12424 [astro-ph.CO]
73. L. Iacconi et al., Revisiting small-scale fluctuations in α-attractor models of inflation. JCAP **06**(06), 007 (2022). https://doi.org/10.1088/1475-7516/2022/06/007. arXiv: 2112.05092 [astro-ph.CO]
74. K. Inomata, M. Kawasaki, Y. Tada, Revisiting constraints on small scale perturbations from big-bang nucleosynthesis. Phys. Rev. D **94**(4), 043527 (2016). https://doi.org/10.1103/PhysRevD.94.043527. arXiv: 1605.04646 [astro-ph.CO]
75. K. Inomata, E. McDonough, W. Hu, Amplification of primordial perturbations from the rise or fall of the inflaton. JCAP **02**(02), 031 (2022). https://doi.org/10.1088/1475-7516/2022/02/031. arXiv: 2110.14641 [astro-ph.CO]
76. K. Inomata, E. McDonough, W. Hu, Primordial black holes arise when the inflaton falls. Phys. Rev. D **104**(12), 123553 (2021). https://doi.org/10.1103/PhysRevD.104.123553. arXiv: 2104.03972 [astro-ph.CO]
77. K. Inomata et al., Questions on calculation of primordial power spectrum with large spikes: the resonance model case. JCAP **04**, 011 (2023). https://doi.org/10.1088/1475-7516/2023/04/011. arXiv: 2211.02586 [astro-ph.CO]
78. P. Ivanov, P. Naselsky, I. Novikov, Inflation and primordial black holes as dark matter. Phys. Rev. D **50**, 7173–7178 (1994). https://doi.org/10.1103/PhysRevD.50.7173
79. J.H.P. Jackson et al., The separate-universe approach and sudden transitions during inflation. JCAP **05**, 053 (2024). https://doi.org/10.1088/1475-7516/2024/05/053. arXiv: 2311.03281 [astro-ph.CO]
80. P.R. Jarnhus, M.S. Sloth, de Sitter limit of inflation and nonlinear perturbation theory. JCAP **02**, 013 (2008). https://doi.org/10.1088/1475-7516/2008/02/013. arXiv: 0709.2708 [hep-th]
81. D. Jeong et al., Silk damping at a redshift of a billion: a new limit on small-scale adiabatic perturbations. Phys. Rev. Lett. **113**, 061301 (2014). https://doi.org/10.1103/PhysRevLett.113.061301. arXiv: 1403.3697 [astro-ph.CO]

82. K. Kannike et al., Single field double inflation and primordial black holes. JCAP **09**, 020 (2017). https://doi.org/10.1088/1475-7516/2017/09/020. arXiv: 1705.06225 [astro-ph.CO]
83. A. Karam et al., Anatomy of single-field inflationary models for primordial black holes. JCAP **03**, 013 (2023). https://doi.org/10.1088/1475-7516/2023/03/013. arXiv: 2205.13540 [astro-ph.CO]
84. A. Karam et al., Primordial black holes and inflation from double-well potentials. JCAP **09**, 002 (2023). https://doi.org/10.1088/1475-7516/2023/09/002. arXiv: 2305.09630 [astro-ph.CO]
85. R. Kawaguchi, T. Fujita, M. Sasaki, Highly asymmetric probability distribution from a finite-width upward step during inflation. JCAP **11**, 021 (2023). https://doi.org/10.1088/1475-7516/2023/11/021. arXiv: 2305.18140 [astro-ph.CO]
86. R. Kawaguchi, S. Tsujikawa, Primordial black holes from Higgs inflation with a Gauss-Bonnet coupling. Phys. Rev. D **107**(6), 063508 (2023). https://doi.org/10.1103/PhysRevD.107.063508. arXiv: 2211.13364 [astro-ph.CO]
87. R. Kawaguchi, S. Tsujikawa, Y. Yamada, Roles of boundary and equation-of-motion terms in cosmological correlation functions. Phys. Lett. B **856** (2024). arXiv: 2403.16022 [hep-th]
88. S. Kawai, J. Kim, Primordial black holes from Gauss-Bonnet-corrected single field inflation. Phys. Rev. D **104**(8), 083545 (2021). https://doi.org/10.1103/PhysRevD.104.083545. arXiv: 2108.01340 [astro-ph.CO]
89. M. Kawasaki, N. Sugiyama, T. Yanagida, Primordial black hole formation in a double inflation model in supergravity. Phys. Rev. D **57**, 6050–6056 (1998). https://doi.org/10.1103/PhysRevD.57.6050. arXiv: hep-ph/9710259
90. M. Kawasaki, T. Yanagida, Primordial black hole formation in supergravity. Phys. Rev. D **59**, 043512 (1999). https://doi.org/10.1103/PhysRevD.59.043512. arXiv: hep-ph/9807544
91. M. Kawasaki, H. Nakatsuka, K. Nakayama, Constraints on small-scale primordial density fluctuation from cosmic microwave background through dark matter annihilation. JCAP **03**(03), 061 (2022). https://doi.org/10.1088/1475-7516/2022/03/061. arXiv: 2110.12620 [astro-ph.CO]
92. M. Kawasaki et al., Primordial black holes as dark matter in supergravity inflation models. Phys. Rev. D **94**(8), 083523 (2016). https://doi.org/10.1103/PhysRevD.94.083523. arXiv: 1606.07631 [astro-ph.CO]
93. K. Kefala et al., Features of the inflaton potential and the power spectrum of cosmological perturbations. Phys. Rev. D **104**(2), 023506 (2021). https://doi.org/10.1103/PhysRevD.104.023506. arXiv: 2010.12483 [astro-ph.CO]
94. R. Kimura et al., Reconstruction of primordial power spectrum of curvature perturbation from the merger rate of primordial black hole binaries. JCAP **04**, 031 (2021). https://doi.org/10.1088/1475-7516/2021/04/031. arXiv: 2102.05280 [astro-ph.CO]
95. W.H. Kinney, A Hamilton-Jacobi approach to nonslow roll inflation. Phys. Rev. D **56**, 2002–2009 (1997). https://doi.org/10.1103/PhysRevD.56.2002. arXiv: hep-ph/9702427
96. W.H. Kinney, Horizon crossing and inflation with large eta. Phys. Rev. D **72**, 023515 (2005). https://doi.org/10.1103/PhysRevD.72.023515. arXiv: gr-qc/0503017
97. T. Kobayashi, M. Yamaguchi, J. Yokoyama, Generalized G-inflation: inflation with the most general second-order field equations. Prog. Theor. Phys. **126**, 511–529 (2011). https://doi.org/10.1143/PTP.126.511. arXiv: 1105.5723 [hep-th]
98. J. Kristiano, J. Yokoyama, Comparing sharp and smooth transitions of the second slow-roll parameter in single-field inflation. JCAP **10**, 036 (2024). https://doi.org/10.1088/1475-7516/2024/10/036. arXiv: 2405.12145 [astro-ph.CO]
99. J. Kristiano, J. Yokoyama, Note on the bispectrum and one-loop corrections in single-field inflation with primordial black hole formation. Phys. Rev. D **109**, 103541 (2024). https://doi.org/10.1103/PhysRevD.109.103541. arXiv: 2303.00341 [hep-th]
100. J. Kristiano, J. Yokoyama, Constraining primordial black hole formation from single-field inflation. Phys. Rev. Lett. **132**, 221003 (2024). https://doi.org/10.1103/PhysRevLett.132.221003. arXiv: 2211.03395 [hep-th]

101. S.M. Leach et al., Enhancement of superhorizon scale inflationary curvature perturbations. Phys. Rev. D **64**, 023512 (2001). https://doi.org/10.1103/PhysRevD.64.023512. arXiv: astro-ph/0101406
102. J. Liu, Z.-K. Guo, R.-G. Cai, Analytical approximation of the scalar spectrum in the ultraslow-roll inflationary models. Phys. Rev. D **101**(8), 083535 (2020). https://doi.org/10.1103/PhysRevD.101.083535. arXiv: 2003.02075 [astro-ph.CO]
103. R. Mahbub, Primordial black hole formation in inflationary α-attractor models. Phys. Rev. D **101**(2), 023533 (2020). https://doi.org/10.1103/PhysRevD.101.023533. arXiv: 1910.10602 [astro-ph.CO]
104. S. Maity et al., Loop contributions to the scalar power spectrum due to quartic order action in ultra slow roll inflation. J. Cosmology Astroparticle Phys. (2023). arXiv: 2307.13636 [astro-ph.CO]
105. J.M. Maldacena, Non-Gaussian features of primordial fluctuations in single field inflationary models. JHEP **05**, 013 (2003). https://doi.org/10.1088/1126-6708/2003/05/013. arXiv: astro-ph/0210603
106. J. Martin, H. Motohashi, T. Suyama, Ultra slow-roll inflation and the non-Gaussianity consistency relation. Phys. Rev. D **87**(2), 023514 (2013). https://doi.org/10.1103/PhysRevD.87.023514. arXiv: 1211.0083 [astro-ph.CO]
107. S.S. Mishra, E.J. Copeland, A.M. Green, Primordial black holes and stochastic inflation beyond slow roll. Part I. Noise matrix elements. JCAP **09**, 005 (2023). https://doi.org/10.1088/1475-7516/2023/09/005. arXiv: 2303.17375 [astro-ph.CO]
108. S.S. Mishra, V. Sahni, Primordial black holes from a tiny bump/dip in the inflaton potential. JCAP **04**, 007 (2020). https://doi.org/10.1088/1475-7516/2020/04/007. arXiv: 1911.00057 [gr-qc]
109. H. Motohashi, W. Hu, Primordial black holes and slow-roll violation. Phys. Rev. D **96**(6), 063503 (2017). https://doi.org/10.1103/PhysRevD.96.063503. arXiv: 1706.06784 [astro-ph.CO]
110. H. Motohashi, S. Mukohyama, M. Oliosi, Constant roll and primordial black holes. JCAP **03**, 002 (2020). https://doi.org/10.1088/1475-7516/2020/03/002. arXiv: 1910.13235 [gr-qc]
111. H. Motohashi, A.A. Starobinsky, Constant-roll inflation: confrontation with recent observational data. EPL **117**(3), 39001 (2017). https://doi.org/10.1209/0295-5075/117/39001. arXiv: 1702.05847 [astro-ph.CO]
112. H. Motohashi, A.A. Starobinsky, J. Yokoyama, Inflation with a constant rate of roll. JCAP **09**, 018 (2015). https://doi.org/10.1088/1475-7516/2015/09/018. arXiv: 1411.5021 [astro-ph.CO]
113. H. Motohashi, Y. Tada, Squeezed bispectrum and one-loop corrections in transient constant-roll inflation. JCAP **08**, 069 (2023). https://doi.org/10.1088/1475-7516/2023/08/069. arXiv: 2303.16035 [astro-ph.CO]
114. T. Nakama, T. Suyama, J. Yokoyama, Reheating the universe once more: the dissipation of acoustic waves as a novel probe of primordial inhomogeneities on even smaller scales. Phys. Rev. Lett. **113**, 061302 (2014). https://doi.org/10.1103/PhysRevLett.113.061302. arXiv: 1403.5407 [astro-ph.CO]
115. T. Nakama et al., Constraints on small-scale primordial power by annihilation signals from extragalactic dark matter minihalos. Phys. Rev. D **97**(2), 023539 (2018). https://doi.org/10.1103/PhysRevD.97.023539. arXiv: 1712.08820 [astro-ph.CO]
116. M.H. Namjoo, One consistency relation for all single-field inflationary models. JCAP **05**, 041 (2024). https://doi.org/10.1088/1475-7516/2024/05/041. arXiv: 2311.12777 [astro-ph.CO]
117. M.H. Namjoo, H. Firouzjahi, M. Sasaki, Violation of non-Gaussianity consistency relation in a single field inflationary model. EPL **101**(3), 39001 (2013). https://doi.org/10.1209/0295-5075/101/39001. arXiv: 1210.3692 [astro-ph.CO]
118. M.H. Namjoo, B. Nikbakht, Non-Gaussianity consistency relations and their consequences for the peaks. arXiv: 2401.12958 [astro-ph.CO]

119. D.V. Nanopoulos, V.C. Spanos, I.D. Stamou, Primordial black holes from no-scale supergravity. Phys. Rev. D **102**(8), 083536 (2020). https://doi.org/10.1103/PhysRevD.102.083536. arXiv: 2008.01457 [astro-ph.CO]
120. K.-W. Ng, Y.-P. Wu, Constant-rate inflation: primordial black holes from conformal weight transitions. JHEP **11**, 076 (2021). https://doi.org/10.1007/JHEP11(2021)076. arXiv: 2102.05620 [astro-ph.CO]
121. O. Özsoy, Z. Lalak, Primordial black holes as dark matter and gravitational waves from bumpy axion inflation. JCAP **01**, 040 (2021). https://doi.org/10.1088/1475-7516/2021/01/040. arXiv: 2008.07549 [astro-ph.CO]
122. O. Özsoy et al., Mechanisms for primordial black hole production in string theory. JCAP **07**, 005 (2018). https://doi.org/10.1088/1475-7516/2018/07/005. arXiv: 1803.07626 [hep-th]
123. S. Pi, J. Wang, Primordial black hole formation in Starobinsky's linear potential model. JCAP **06**, 018 (2023). https://doi.org/10.1088/1475-7516/2023/06/018. arXiv: 2209.14183 [astro-ph.CO]
124. S. Pi et al., Scalaron from R^2-gravity as a heavy field. JCAP **05**, 042 (2018). https://doi.org/10.1088/1475-7516/2018/05/042. arXiv: 1712.09896 [astro-ph.CO]
125. D. Polarski, A.A. Starobinsky, Semiclassicality and decoherence of cosmological perturbations. Class. Quant. Grav. **13**, 377–392 (1996). https://doi.org/10.1088/0264-9381/13/3/006. arXiv: gr-qc/9504030
126. H.V. Ragavendra et al., Primordial black holes and secondary gravitational waves from ultra-slow roll and punctuated inflation. Phys. Rev. D **103**(8), 083510 (2021). https://doi.org/10.1103/PhysRevD.103.083510. arXiv: 2008.12202 [astro-ph.CO]
127. S. Rasanen, E. Tomberg, Planck scale black hole dark matter from Higgs inflation. JCAP **01**, 038 (2019). https://doi.org/10.1088/1475-7516/2019/01/038. arXiv: 1810.12608 [astro-ph.CO]
128. K. Rezazadeh et al., Non-Gaussianity and secondary gravitational waves from primordial black holes production in α-attractor inflation. Eur. Phys. J. C **82**(8), 758 (2022). https://doi.org/10.1140/epjc/s10052-022-10735-w. arXiv: 2110.01482 [gr-qc]
129. A. Riotto, The primordial black hole formation from single-field inflation is still not ruled out (2023). arXiv: 2303.01727 [astro-ph.CO]
130. A. Riotto, The primordial black hole formation from single-field inflation is not ruled out (2023). arXiv: 2301.00599 [astro-ph.CO]
131. S. Saburov, S.V. Ketov, Quantum loop corrections in the modified gravity model of Starobinsky inflation with primordial black hole production (2024). arXiv: 2402.02934 [gr-qc]
132. R. Saito, J. Yokoyama, R. Nagata, Single-field inflation, anomalous enhancement of superhorizon fluctuations, and non-Gaussianity in primordial black hole formation. JCAP **06**, 024 (2008). https://doi.org/10.1088/1475-7516/2008/06/024. arXiv: 0804.3470 [astro-ph]
133. K. Sato, First order phase transition of a vacuum and expansion of the universe. Mon. Not. Roy. astron. Soc. **195**, 467–479 (1981)
134. K. Sato, J. Yokoyama, Inflationary cosmology: first 30+ years. Int. J. Mod. Phys. D **24**(11), 1530025 (2015). https://doi.org/10.1142/S0218271815300256
135. D. Seery, J.E. Lidsey, Primordial non-Gaussianities in single field inflation. JCAP **06**, 003 (2005). https://doi.org/10.1088/1475-7516/2005/06/003. arXiv: astro-ph/0503692
136. M. Solbi, K. Karami, Primordial black holes and induced gravitational waves in k-inflation. JCAP **08**, 056 (2021). https://doi.org/10.1088/1475-7516/2021/08/056. arXiv: 2102.05651 [astro-ph.CO]
137. M. Solbi, K. Karami, Primordial black holes formation in the inflationary model with field-dependent kinetic term for quartic and natural potentials. Eur. Phys. J. C **81**(10), 884 (2021). https://doi.org/10.1140/epjc/s10052-021-09690-9. arXiv: 2106.02863 [astro-ph.CO]
138. I.D. Stamou, Mechanisms of producing primordial black holes by breaking the $SU(2,1)/SU(2) \times U(1)$ symmetry. Phys. Rev. D **103**(8), 083512 (2021). https://doi.org/10.1103/PhysRevD.103.083512. arXiv: 2104.08654 [hep-ph]

139. A.A. Starobinsky, A new type of isotropic cosmological models without singularity. Phys. Lett. B **91**, 99–102 (1980). Ed. by I.M. Khalatnikov, V.P. Mineev. https://doi.org/10.1016/0370-2693(80)90670-X
140. A.A. Starobinsky, Spectrum of adiabatic perturbations in the universe when there are singularities in the inflation potential. JETP Lett. **55**, 489–494 (1992)
141. W.-C. Syu, D.-S. Lee, K.-W. Ng, Quantum loop effects to the power spectrum of primordial perturbations during ultra slow-roll inflation. Phys. Rev. D **101**(2), 025013 (2020). https://doi.org/10.1103/PhysRevD.101.025013. arXiv: 1907.13089 [gr-qc]
142. Y. Tada, T. Terada, J. Tokuda, Cancellation of quantum corrections on the soft curvature perturbations. JHEP **01**, 105 (2024). https://doi.org/10.1007/JHEP01(2024)105. arXiv: 2308.04732 [hep-th]
143. G. Tasinato, An analytic approach to non-slow-roll inflation. Phys. Rev. D **103**(2), 023535 (2021). https://doi.org/10.1103/PhysRevD.103.023535. arXiv: 2012.02518 [hep-th]
144. G. Tasinato, Large —η— approach to single field inflation. Phys. Rev. D **108**(4), 043526 (2023). https://doi.org/10.1103/PhysRevD.108.043526. arXiv: 2305.11568 [hep-th]
145. G. Tasinato, Non-Gaussianities and the large —η— approach to inflation. Phys. Rev. D **109**(6), 063510 (2024). https://doi.org/10.1103/PhysRevD.109.063510. arXiv: 2312.03498 [hep-th]
146. Z. Teimoori et al., Mechanism of primordial black holes production and secondary gravitational waves in α-attractor Galileon inflationary scenario. JCAP **2021**(10), 018 (2021). https://doi.org/10.1088/1475-7516/2021/10/018. arXiv: 2107.07620 [astro-ph.CO]
147. D. Wands, Duality invariance of cosmological perturbation spectra. Phys. Rev. D **60**, 023507 (1999). https://doi.org/10.1103/PhysRevD.60.023507. arXiv: gr-qc/9809062
148. Q. Wang et al., Primordial black holes from the perturbations in the inflaton potential in peak theory. Phys. Rev. D **104**(8), 083546 (2021). https://doi.org/10.1103/PhysRevD.104.083546. arXiv: 2111.10028 [astro-ph.CO]
149. X. Wang et al., Reconstruction of power spectrum of primordial curvature perturbations on small scales from primordial black hole binaries scenario of LIGO/VIRGO detection. Sci. China Phys. Mech. Astron. **66**(6), 260462 (2023). https://doi.org/10.1007/s11433-023-2091-x. arXiv: 2209.12911 [astro-ph.CO]
150. L. Wu, Y. Gong, T. Li, Primordial black holes and secondary gravitational waves from string inspired general no-scale supergravity. Phys. Rev. D **104**(12), 123544 (2021). https://doi.org/10.1103/PhysRevD.104.123544. arXiv: 2105.07694 [gr-qc]
151. J. Yokoyama, Chaotic new inflation and formation of primordial black holes. Phys. Rev. D **58**, 083510 (1998). https://doi.org/10.1103/PhysRevD.58.083510. arXiv: astro-ph/9802357
152. J. Yokoyama, Chaotic new inflation and primordial spectrum of adiabatic fluctuations. Phys. Rev. D **59**, 107303 (1999). https://doi.org/10.1103/PhysRevD.59.107303
153. J. Yokoyama, S. Inoue, Curvature perturbation at the local extremum of the inflaton's potential. Phys. Lett. B **524**, 15–20 (2002). https://doi.org/10.1016/S0370-2693(01)01369-7. arXiv: hep-ph/0104083
154. Y.B. Zel'dovich, I.D. Novikov, The hypothesis of cores retarded during expansion and the Hot cosmological model. Sov. Astron. **10**, 602–603 (1967)
155. R. Zheng, J. Shi, T. Qiu, On primordial black holes and secondary gravitational waves generated from inflation with solo/multi-bumpy potential. Chin. Phys. C **46**(4), 045103 (2022). https://doi.org/10.1088/1674-1137/ac42bd. arXiv: 2106.04303 [astro-ph.CO]

Chapter 4
Numerical Simulations of Primordial Black Holes

Ilia Musco

Abstract The aim of this chapter is to present a consistent description of the mathematical and numerical investigation on primordial black hole formation from the collapse of adiabatic cosmological perturbations performed in the last 25 years, describing the mechanism of gravitational collapse. Despite the presence of some technicalities, at the end of this chapter the reader will should have a consistent scenario, knowing how the results of numerical simulations depend from the initial conditions described by the power spectrum of cosmological perturbations. This will provide the basic knowledge required to take the first steps in the research of primordial black hole formation.

4.1 Introduction

Numerical simulations have played a crucial role in the study of primordial black holes (PBHs), in particular to understand the formation mechanism by gravitational collapse of cosmological perturbations within an expanding early Universe. This is a fully non-linear problem needing the solution of the Einstein equations with numerical simulations, computing the value of the (collapse) threshold δ_c and the corresponding mass distribution characterised by critical collapse, two fundamental aspects to evaluate the cosmological impact of PBHs.

In this chapter we will first review the mathematical formalism necessary to set up the initial condition of numerical simulations, following the long wave-length approximation when cosmological perturbations are still growing with the expansion of the Universe, well before the onset of the collapse. Starting from this mathematical framework, using the basic assumption of spherical symmetry, we will understand how the numerical results obtained at the end of numerical simulations, when PBHs are formed from the collapse of cosmological perturbations, relates to the initial conditions of the early Universe. Although non spherical studies has recently been

I. Musco (✉)
Dipartimento di Fisica, Sapienza Università di Roma, Piazzale Aldo Moro 5, 00185 Roma, Italy

INFN, Sezione di Roma, Piazzale Aldo Moro 2, 00185 Roma, Italy
e-mail: ilia.rmusco@roma1.infn.it

C. Byrnes et al. (eds.), *Primordial Black Holes*, Springer Series in Astrophysics and Cosmology, https://doi.org/10.1007/978-981-97-8887-3_4

performed, these have not yet provided enough results to understand to which extent the results in spherical symmetry are a good approximation. In the next future we should expect the research on PBH formation to develop in this direction.

The chapter is intended as a self consistent review to allow the reader to understand the basic principle of this topic, without going into all the technical details that the interested reader can investigate further by looking at the list of references provided. Throughout geometrical units with $c = G = 1$ are used.

4.2 Mathematical Formulation

In this chapter we will consider the content of the early Universe described by a perfect fluid, with the stress energy tensor defined as

$$T^{\mu\nu} = (\rho + p)u^{\mu}u^{\nu} + pg^{\mu\nu}, \tag{4.1}$$

where ρ and p are the fluid energy density and the pressure measured in the comoving frame of the fluid, $g^{\mu\nu}$ is the metric tensor, while u^{μ} is the four-velocity of the fluid normalized such that $u^{\mu}u_{\mu} = -1$.

Historically simulations of PBH formation have been performed by different groups [2, 28], and more recently by Shibata and Sasaki [34] using the *constant mean curvature gauge*, and by Niemeyer and Jedamzik [15] using the *comoving gauge* which is going to be used also in the following discussion. The covoimg gauge has the advantage to be more intuitive to set up the initial conditions, and for this reason is a convenient and pedagogical choice to start getting familiar with numerical simulations of PBHs. The relation between different gauges has been analysed extensively in [12, 18].

In the assumption of spherical symmetry the explicit form of the Einstein equations in the comoving gauge is based on the following diagonal form of the metric [20]

$$ds^2 = -A^2(r,t)\mathrm{d}t^2 + B^2(r,t)\mathrm{d}r^2 + R^2(r,t)\mathrm{d}\Omega^2 \tag{4.2}$$

where the radial coordinate r is taken to be comoving with the fluid, i.e. the four-velocity of the fluid is equal to the unit normal vector orthogonal to the hypersurface of constant time t, usually referred to as *cosmic time*. This metric corresponds to an orthogonal comoving foliation of the spacetime where the shift (i.e. the $drdt$-term) has been imposed to be equal to zero. This leaves three non zero coefficients of the metric, A, B and R, which are positive functions of r and t; where R in particular is the *areal radius*, and $d\Omega = d\theta^2 + \sin^2\theta\, d\phi^2$ is the element of a 2-sphere of symmetry. When the Universe is homogeneous and isotropic the metric (4.2) corresponds to the FLRW metric

$$ds^2 = -\mathrm{d}t^2 + a^2(t)\left[\frac{\mathrm{d}r^2}{1 - Kr^2} + r^2\mathrm{d}\Omega^2\right] \tag{4.3}$$

where $a(t)$ is the scale factor and $K = 0, \pm 1$ being the spatial curvature for a flat, closed and open Universe.

4.2.1 The Misner–Sharp–Hernandez Equations

Starting from the metric (4.2) it is convenient to introduce two differential operators:

$$D_t \equiv \frac{1}{A}\frac{\partial}{\partial t} \quad \text{and} \quad D_r \equiv \frac{1}{B}\frac{\partial}{\partial r}, \tag{4.4}$$

which represents, respectively, derivatives with respect to *proper time* and radial *proper distance* in the comoving frame of the fluid. Applying them to the areal radius R one can define

$$U \equiv D_t R = \frac{1}{A}\frac{\partial R}{\partial t} \quad \text{and} \quad \Gamma \equiv D_r R = \frac{1}{B}\frac{\partial R}{\partial r}, \tag{4.5}$$

where U is measuring the velocity of the fluid with respect to the centre of coordinates (i.e. the Eulerian non comoving frame where R is used as the radial coordinate), that in the homogeneous and isotropic FLRW Universe is simply given by the Hubble law $U = HR$ with $R(r,t) = a(t)r$ and $H(t) \equiv \dot{a}(t)/a(t)$. The quantity Γ, instead, gives a measure of the spatial curvature, that in the FLRW becomes $\Gamma^2 = 1 - Kr^2$, being just a constant ($\Gamma = 1$) when the Universe is homogeneous, isotropic and spatially flat.

These two quantities are related to the so called *Misner–Sharp–Hernandez mass* M by the *constraint equation*

$$\Gamma^2 = 1 + U^2 - \frac{2M}{R}, \tag{4.6}$$

which in spherical symmetry is a unique and well defined measure of the mass within a sphere of radius R, and assuming a perfect fluid defined in Eq. (4.1) the mass M can simply be written as

$$M = \int_0^R 4\pi R^2 \rho \, dR. \tag{4.7}$$

The Misner–Sharp–Hernandez hydrodynamic equations obtained from the Einstein equations and the conservation of the stress energy tensor (see [19, 20] for the details of the derivation) are:

$$D_t U = -\frac{\Gamma}{\rho + p} D_r p - \frac{M}{R^2} - 4\pi R p, \tag{4.8}$$

$$D_t \rho_0 = -\frac{\rho_0}{\Gamma R^2} D_r (R^2 U), \tag{4.9}$$

$$D_t \rho = \frac{\rho + p}{\rho_0} D_t \rho_0 \,, \tag{4.10}$$

$$D_r A = -\frac{A}{\rho + p} D_r p \,, \tag{4.11}$$

$$D_r M = 4\pi R^2 \Gamma \rho \,, \tag{4.12}$$

where ρ_0 in Eqs. (4.9) and (4.10) is the rest mass density (or understood as the copmpression factor, being proportional to the particle number density, even if the fluid consists of massless particles). Applying the time operator D_t to the constraint equation (4.6) we can obtain two more, although not linearly independent, useful equations:

$$D_t \Gamma = -\frac{U}{e + p} D_r p \,, \tag{4.13}$$

$$D_t M = -4\pi R^2 U p \,. \tag{4.14}$$

To solve this set of equations we need to close the system specifying also an equation of state, which gives the relation between the pressure and the components of the energy density. For the cosmological fluid the standard choice is to take a constant ratio between pressure and energy density, such as

$$p = w\rho \,. \tag{4.15}$$

In particular $w = 0$ corresponds to a pressureless fluid (often refereed to as "dust") while $w = 1/3$ corresponds to a radiation dominated fluid (normally called as "radiation"). The equation of state can be more complicated during particular epochs of the Universe such as the the QCD phase transition, where $w = w(T)$ is a function of the temperature T and the equation of state of the Universe goes through a temporary softening (i.e. $0.23 \lesssim w < 1/3$) (see Chap. 5).

4.2.2 Gradient Expansion and the Quasi-Homogeneous Solution

Here I am going to review the explicit solution of the Misner–Sharp–Hernandez set of equations in the long wavelength approximation, when the perturbation length scale is much larger than the cosmological horizon (from now on identified as the super-horizon regime). In this regime the curvature profile is a key time independent quantity describing, from a geometrical point of view, the origin of adiabatic cosmological perturbations. Using the comoving gauge it can be conveniently described by a function $K(r)$, replacing the constant curvature parameter of the FLRW metric (4.3), or as a perturbation of the scale factor, writing

$$ds^2 = \begin{cases} -\,\mathrm{d}t^2 + a^2(t)\left[\dfrac{\mathrm{d}r^2}{1-K(r)r^2} + r^2\mathrm{d}\Omega^2\right] \\ -\,\mathrm{d}t^2 + a^2(t)e^{2\zeta(\hat{r})}\left[\mathrm{d}\hat{r}^2 + \hat{r}^2\mathrm{d}\Omega^2\right]. \end{cases} \tag{4.16}$$

These two expressions are asymptotic solutions of the Einstein equations in the limit of $t \to 0$, while the full solution is given by the *quasi-homogeneous solution* obtained with a gradient expansion, when the pressure gradients are small.

The coordinate transformation between $K(r)$ and $\zeta(\hat{r})$ is obtained equating separately the radial and angular components of the two asymptotic metrics [14]

$$\begin{cases} r = \hat{r}e^{\zeta(\hat{r})} \\ \dfrac{\mathrm{d}r}{\sqrt{1-K(r)r^2}} = e^{\zeta(\hat{r})}\mathrm{d}\hat{r}\,. \end{cases} \tag{4.17}$$

From the first expression one gets the relation between $\hat{r}$ and r

$$\frac{\mathrm{d}r}{\mathrm{d}\hat{r}} = e^{\zeta(\hat{r})}\left(1 + \hat{r}\zeta'(\hat{r})\right), \tag{4.18}$$

and inserting this into the second expression one finally has

$$K(r)r^2 = -\hat{r}\zeta'(\hat{r})\left[2 + \hat{r}\zeta'(\hat{r})\right]. \tag{4.19}$$

For more details, please see [12, 21].

To perform the gradient expansion it is necessary to define a small parameter $\epsilon \equiv 1/H(t)L \ll 1$ where $L \propto a(t)$ is the perturbation length scale and $R_{\mathrm{H}} \equiv 1/H$ is the cosmological horizon. The time evolution of the scale factor and the Hubble parameter in the Einstein De-Sitter model (i.e. homogeneous and isotropic with $K = 0$)

$$a(t) \propto t^{\frac{2}{3(1+w)}} \text{ and } H(t) \propto \frac{1}{t} \;\Rightarrow\; \epsilon \propto t^{\frac{1+3w}{3(1+w)}} \tag{4.20}$$

shows that $\epsilon(t)$ is parameterisation of the time evolution of the Universe: for matter with $w > -1/3$ (like dust and radiation) the limit of $\epsilon \to 0$ corresponds to $t \to 0$.

For pure growing modes the asymptotic values of the metric tensor is $O(\epsilon^2)$, plus higher order terms that can be neglected when $\epsilon \ll 1$ (see [18] for the general proof), and one can write the components of the cosmic time metric defined in (4.2) as

$$A = 1 + \epsilon^2\tilde{A} \tag{4.21}$$

$$B = \frac{a(t)}{\sqrt{1-K(r)r^2}}(1+\epsilon^2\tilde{B}) = a(t)e^{\zeta(\hat{r})}(1+\epsilon^2\tilde{B}) \tag{4.22}$$

$$R = a(t)r(1+\epsilon^2\tilde{R}) = a(t)e^{\zeta(\hat{r})}\hat{r}(1+\epsilon^2\tilde{R}) \tag{4.23}$$

while, using the same approach, the hydrodynamical variables can be expanded as

$$\rho = \rho_b(t)(1 + \epsilon^2\tilde{\rho}) \tag{4.24}$$
$$U = H(t)R(1 + \epsilon^2\tilde{U}) \tag{4.25}$$
$$M = \frac{4\pi}{3}\rho_b(t)R^3(1 + \epsilon^2\tilde{M}) \tag{4.26}$$

where the pressure is calculated with the equation of state given by Eq. (4.15). Writing the constraint Eq. (4.6) using the expansion in ϵ^2, one obtains

$$K(r)r^2 = (aHr)^2\epsilon^2\left(\tilde{M} - 2\tilde{U}\right) \tag{4.27}$$

and looking at this expression one can appreciate why in (4.25) and (4.26) we have put R instead of $R_b = a(t)r$ outside the parenthesis.This simplifies the calculation allowing decomposition of the perturbation of M and U into the fundamental components.

In the super-horizon regime the perturbation profiles are sourced by the curvature profile $K(r)$ or $\zeta(\hat{r})$, while the time evolution is tracked by ϵ^2, with a clear separation of variables between time and space dependence. This is because in the gradient expansion approximation the pressure gradients are negligible, and the time evolution of ϵ^2 is equivalent to a pure growing mode in the linear theory.

According to the quasi-homogeneous solution, the energy density contrast $\delta\rho/\rho_b$ for adiabatic perturbations takes the following expression

$$\frac{\delta\rho}{\rho_b}(r,t) = \Phi\frac{\left[K(r)r^3\right]'}{3r^2} = -\frac{4}{3}\Phi\left(\frac{1}{aH}\right)^2 e^{-5\zeta(\hat{r})/2}\nabla^2 e^{\zeta(\hat{r})/2}, \tag{4.28}$$

where the function $\Phi(t)$ is a coefficient depending on the equation of state of the Universe, obtained by solving the following first order differential equation

$$R_H\frac{d\Phi}{dR_H} + \frac{5+3w}{3(1+w)}\Phi - 1 = 0 \tag{4.29}$$

integrated from past infinity (i.e. t, $R_H \to 0$) to the time (i.e. the value of ϵ) when the perturbation is computed. When the equation of state is characterised by a constant value $\bar{w}$ of w, one has $d\Phi/dt = 0$, and the equation (4.29) is analytically solved by

$$\bar{\Phi} = \frac{3(1+\bar{w})}{(5+3\bar{w})}, \tag{4.30}$$

yielding $\bar{\Phi} = 2/3$ for a radiation fluid with $\bar{w} = 1/3$. The value of Φ is measuring to which extent the curvature profile appearing in the left hand side of Einstein equations, is affecting the stress energy tensor on the right hand side. Note that $\Phi = 0$ if $w = -1$, consistent with the cosmological constant which cannot be perturbed.[1]

The explicit expression of the quasi-homogeneous solution can be written as a linear combination of energy density and the velocity perturbations

$$\tilde{\rho} = \begin{cases} \dfrac{3(1+w)}{5+3w}\left[K(r) + \dfrac{r}{3}K'(r)\right] r_k^2 \\ -\dfrac{2(1+w)}{5+3w}\dfrac{e^{2\zeta(\hat{r}_k)}}{e^{2\zeta(\hat{r})}}\left[\zeta''(\hat{r}) + \zeta'(\hat{r})\left(\dfrac{2}{r} + \dfrac{1}{2}\zeta'(\hat{r})\right)\right]\hat{r}_k^2 \end{cases} \tag{4.31}$$

$$\tilde{U} = \begin{cases} -\dfrac{1}{5+3w}K(r)r_k^2 \\ \dfrac{1}{5+3w}\dfrac{e^{2\zeta(\hat{r}_k)}}{e^{2\zeta(\hat{r})}}\zeta'(\hat{r})\left[\dfrac{2}{\hat{r}} + \zeta'(\hat{r})\right]\hat{r}_k^2 \end{cases} \tag{4.32}$$

$$\tilde{A} = -\frac{w}{1+w}\tilde{\rho} \tag{4.33}$$

$$\tilde{M} = -3(1+w)\tilde{U} \tag{4.34}$$

$$\tilde{R} = -\frac{w}{(1+3w)(1+w)}\tilde{\rho} + \frac{1}{1+3w}\tilde{U} \tag{4.35}$$

$$\tilde{B} = \frac{w}{(1+3w)(1+w)}r\frac{d\tilde{\rho}}{dr} \tag{4.36}$$

where

$$r\frac{\mathrm{d}}{\mathrm{d}r} = \frac{\hat{r}}{1+\hat{r}\zeta'(\hat{r})}\frac{\mathrm{d}}{\mathrm{d}\hat{r}},$$

and consistently with a pure growing solution (i.e. characterised by one degree of freedom), $\tilde{\rho}$ and $\tilde{U}$ are related to each other as

$$\tilde{\rho} = -(1+w)\frac{1}{r^2}\frac{\mathrm{d}}{\mathrm{d}r}\left(r^3\tilde{U}\right) \tag{4.37}$$

$$\tilde{U} = -\frac{1}{(1+w)}\frac{1}{r^3}\int \tilde{\rho}\, r^2 dr\,. \tag{4.38}$$

The term $\tilde{B}$ can be interpreted as the seeds of pressure gradients, breaking the self similar behaviour of the quasi-homogeneous solution when ϵ is $O(1)$.

At this point we need to specify the values of the background quantities: the energy density $\rho_b(t)$, the Hubble parameter $H(t)$ and the scale factor $a(t)$, describing the background solution of a FLRW Universe. This allows one to write $\epsilon(t)$ as

---

[1] The function Φ is analogous to the coefficient appearing in the Bardeen potential in cosmological linear perturbation theory, where the Bardeen Potential Ψ on super-horizon scales is given by $\Psi(r) = -\frac{3(1+w)}{5+3w}\zeta(r)$.

$$\epsilon(t) = \frac{1}{a(t)H(t)r_{\mathrm{m}}} \tag{4.39}$$

where $r_{\mathrm{m}} \equiv L/a(t)$ is the comoving perturbation length scale.[2]

The boundary condition at infinity of the curvature profile distinguishes between an energy density profile that is compensated, i.e. if the integral of $\delta\rho/\rho_{\mathrm{b}}$ over an infinite volume is equal to zero, and a non compensated profiles which does not satisfy this condition. This implies that for a compensated profile we have

$$\int_0^\infty 4\pi r^2 \tilde{\rho} \mathrm{d}r = 0 \quad \Rightarrow \quad \begin{cases} \lim\limits_{r\to\infty} K(r)r^3 = 0 \\ \lim\limits_{\hat{r}\to\infty} \zeta(\hat{r})\hat{r} = 0 \end{cases} \tag{4.40}$$

Using (4.27) in the definition of Γ (see (4.5)) one obtains that the lowest (zeroth order) term is time independent, consistent with the interpretation of Γ as a measure of the spatial curvature.

$$\Gamma^2 = 1 - K(r)r^2 = \left(1 + \hat{r}\zeta'(\hat{r})\right)^2 . \tag{4.41}$$

It is interesting to notice that when $K(r)r^2 = 1$ we meet a coordinate singularity in the first expression of the asymptotic metric seen before in (4.16). As was also pointed out in [17] this can be removed by a coordinate transformation, as in (4.17), distinguishing between PBHs of type I ($K(r)r^2 < 1$) and PBHs of type II, where $K(r)r^2 = 1$ is reached (see [3] for more details). To avoid 0/0 terms in the Misner–Sharp–Hernandez equations we need to have $\Gamma > 0$, which ensure a monotonic behaviour of the areal radius R as function of the comoving coordinate r, while to study PBHs of Type II, characterised by a region where $\Gamma < 0$, a different gauge like the spatially flat coordinates has to be used. Type II initial configurations, which gives always to the formation of PBHs, have been carefully analysed in [17, 35].

4.2.3 Compaction Function and the Perturbation Amplitude

The averaged overdensity $\delta(r,t)$ is defined as the mass excess of the energy density averaged over the spherical volume of radius $R(r,t)$, that is

$$\delta(r,t) \equiv \frac{4\pi}{V}\int_0^R \frac{\delta\rho}{\rho_{\mathrm{b}}} R^2 \mathrm{d}R \simeq \frac{3}{\tilde{r}^3}\int_0^{\tilde{r}} \frac{\delta\rho}{\rho_{\mathrm{b}}} r^2 \mathrm{d}r , \tag{4.42}$$

where the second equality is obtained by neglecting the higher–order terms in ϵ, approximating $R \simeq a(t)\tilde{r}$, which allows making a simple integration over the

[2] The quasi-homogeneous solution including higher–order terms in ϵ has been studied in [32].

comoving volume of radius r. The perturbation amplitude δ is then measured at the maximum of the compaction function [34] defined here as

$$\mathcal{C} \equiv 2\frac{M(r,t) - M_{\mathrm{b}}(r,t)}{R(r,t)}\,, \tag{4.43}$$

where the numerator is given by the difference between the Misner–Sharp mass within a sphere of radius $R(r,t)$, and the background mass $M_{\mathrm{b}}(r,t) = 4\pi\rho_{\mathrm{b}}(r,t)R^3(r,t)/3$ computed within the same areal radius, but calculated with respect to a spatially flat FLRW metric. The compaction function was originally introduced by [34] in a different gauge, although there was some confusion in their definition, which does not affect its significance. At super-horizon scales, the compaction function and the averaged overdensity defined in (4.42) are equivalent. The Shibata-Sasaki compaction function and the definition given here in (4.43) using the comoving gauge are related up to a constant factor. They both can be written in terms of the first spatial derivative of the curvature perturbation ζ in the long-wavelength limit, see [10, 12, 21] for more details.

According to the quasi-homogeneous solutions, on super-horizon scales (i.e. $\epsilon \ll 1$) the compaction function is time independent, and is given by

$$\mathcal{C}(r) = \Phi K(r) r^2 = -\Phi \hat{r}\zeta'(\hat{r})\left[2 + \hat{r}\zeta'(r)\right]\,, \tag{4.44}$$

with the comoving length scale of the perturbation being defined by $r = r_{\mathrm{m}}$, where the compaction function reaches its maximum (i.e. $\mathcal{C}'(r_{\mathrm{m}}) = 0$), which implies

$$K(r_{\mathrm{m}}) + \frac{r_{\mathrm{m}}}{2}K'(r_{\mathrm{m}}) = 0\,, \qquad \zeta'(\hat{r}_{\mathrm{m}}) + r_{\mathrm{m}}\zeta''(\hat{r}_{\mathrm{m}}) = 0\,. \tag{4.45}$$

The compaction function is then related to the energy density profile by

$$\mathcal{C}(r) = \frac{\tilde{r}^2}{\tilde{r}_{\mathrm{m}}^2}\delta(r, t_{\mathrm{H}})\,, \tag{4.46}$$

where t_{H} is defined as the cosmological horizon crossing time, when $aHr_{\mathrm{m}} = 1$ ($\epsilon = 1$). Although in this regime the gradient expansion approximation is not very accurate, and the horizon crossing defined in this way is only a linear extrapolation, this provides a well defined criterion for consistently measuring the perturbation amplitude, understanding how the threshold is varying in terms of the shape of the initial curvature profile (see [21] for more details).

The perturbation amplitude δ of the perturbation is defined as the excess of mass averaged over a spherical volume of radius R_{m} using a top-hat window function, computed at the time of the cosmological horizon crossing t_{H}. This quantity is equivalent to the peak amplitude of the compaction function measured on super- horizon scales.

Inserting the expression for $\delta\rho/\rho_{\mathrm{b}}$ given by (4.28) into (4.42), a simple calculation seen in [21] gives the fundamental relation

$$\delta = 3\frac{\delta\rho}{\rho_b}(r_\text{m}, t_H) \tag{4.47}$$

independent from the specific shape of the curvature profile, consistent with identifying the maximum of the compaction function as the length scale of the perturbation.

Looking at (4.44) the perturbation amplitude δ can be written in terms of a variable with Gaussian statistics linearly related to the curvature perturbation $\delta_l \equiv -2\Phi\, r\zeta'(r_m)$

$$\delta = \delta_l\left(1 - \frac{1}{4\Phi}\delta_l\right). \tag{4.48}$$

This expression is very useful to compute the mass distribution of PBHs, as explained in Chap. 6. From expression (4.48) there are two different values of δ_l corresponding to the same value of δ, with a maximum value of $\delta = \Phi$ for $\delta_l = 2\Phi$. This degeneracy in the perturbation amplitude measured with δ is related to the difference between cosmological perturbations of Type I and Type II mentioned earlier.

4.3 Initial Conditions

As we will see in the next section the threshold for PBHs depends on the shape of the cosmological perturbation used to set up the initial conditions, characterised by the width of the peak of the compaction function. This shape dependence can be measured by a dimensionless parameter defined as

$$\alpha = -\frac{\mathcal{C}''(r_\text{m})r_\text{m}^2}{4\mathcal{C}(r_\text{m})}, \tag{4.49}$$

introduced by [5, 21], where the family of curvature profiles $K(r)$ can be parameterised in terms of an exponential as

$$K(r) = \mathcal{A}\exp\left[-\frac{1}{\alpha}\left(\frac{r}{r_\text{m}}\right)^{2\alpha}\right]. \tag{4.50}$$

This identifies a basis of centrally peaked profiles, describing the main features of all of the possible shapes. Although this profile has a spiky behaviour in the centre for $\alpha < 0.5$, where the first derivative diverges, the consistency of the numerical evolution is guaranteed by the pressure gradients smoothing the profile.[3]

---

[3] Alternative families of profiles, like the polynomial one, has been used in [5]. This is not showing a divergence in the central value of the first derivative, but is not always centrally peaked. Both choices in the end gives consistent results.

In Fig. 4.1 one can see the energy density profile $\delta\rho/\rho_b(r, t_H)$ normalised at horizon crossing time t_H ($aHr_m = 1$), and the compaction function $\mathcal{C}(r)$, both plotted against r/r_m for different values of α, obtained by inserting (4.50) into (4.28) and (4.44) respectively. The profile of the energy density becomes more peaked for $\alpha < 1$ (red lines) while the compaction function profile becomes broader, whereas the dashed line describes the typical Mexican-hat profile ($\alpha = 1$). On the contrary the shape of the compaction function is more peaked for values of $\alpha > 1$ (blue lines), corresponding to broad profiles of the energy density. Because we are considering a pure growing mode with one degree of freedom, the shape of the compaction function and the energy density are specified by the choice of the time independent curvature profile $K(r)$ where the shape parameter α is appearing explicitly within the argument of the exponential.

It is important to appreciate that, when replacing (4.50) or any other K-profile into (4.49), the value of α is independent of the perturbation amplitude $\delta = \mathcal{C}(r_m)$ which is varying with the peak of the curvature profile $\mathcal{A}$, which is an overall factor simplifying in the ratio between the second derivatives and the value of the compaction function computed at its maximum. *The parameter α is distinguishing between different shapes of the perturbation measured at the peak of the compactness, when the curvature profile is expressed in terms of $K(r)$.*

A correlation arises instead, when the shape parameter is measured in terms of $\zeta(\hat{r})$: using the coordinate transformation of (4.17) one obtains that

$$\alpha = -\frac{\mathcal{C}''(\hat{r}_m)\hat{r}_m^2}{4\mathcal{C}(\hat{r}_m)\left[1 - \frac{3}{2}\mathcal{C}(\hat{r}_m)\right]}, \tag{4.51}$$

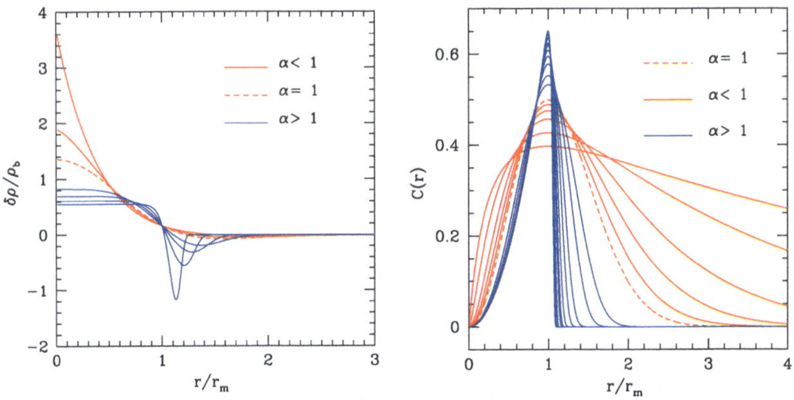

Fig. 4.1 This figure shows the behavior of $\delta\rho/\rho_b(r, t_H)$ and $\mathcal{C}(r)$ plotted against r/r_m, when $aHr_m = 1$, for $\alpha = 0.5, 0.75, 1, 2, 3, 5, 10$. The profiles with $\alpha \leq 1$ are plotted with a red line (a dashed line for $\alpha = 1$) while blue lines are used for profiles with $\alpha > 1$. The left and right plots are taken from [26, 27], respectively

where the term proportional to $C'(\hat{r})$ appearing during the calculation is equal to zero when calculated at $\hat{r}_{\mathrm{m}}$ because of (4.45). This expression shows that the peak of the compaction function does not cancel out with the peak of the second derivative, when computed with respect to $\hat{r}$ instead of r. This is related to the exponential term of the coordinate transformation $r = \hat{r}e^{\zeta(\hat{r})}$. In general, when varying the amplitude of the perturbation, the values of $C''(\hat{r}_{\mathrm{m}})$ and $C(\hat{r}_{\mathrm{m}})$ are not independent, but correlated, changing according to the given value of α.

4.4 Numerical Simulations

The results for PBH formation presented in this section have been obtained using an explicit Lagrangian hydrodynamic code developed in [24, 25], designed for calculations in an expanding cosmological background. The reader is referred to those papers, and reference therein, for more details of the methods used. This is similar in most respect to the method used by Niemeyer and Jedamzik in [16], following from the Baumgarte et al code [1]. In order to achieve a good resolution an adaptive mesh refinement has been implemented, with the grid spacing increasing exponentially going outwards. The initial condition are specified in the linear regime, using the gradient expansion approach described above, with the perturbation length scales at least ten times larger than the cosmological horizon radius R_{H}, and the grid reaching out to $10R_{\mathrm{H}}$. This requires between 1000 and 2000 grid points.

For collapses leading to black-hole formation, the calculations proceed in two stages: first, initial data is specified on a spacelike slice at constant cosmic time, specifying the time independent curvature profile from which one can compute the energy density profile ρ, the four velocity component U, and the areal radius R at initial time t_{i}. The rest of the initial data are computed as functions of these quantities using the Misner-Sharp-Hernandez equations, used also to evolve and generate a second set of initial data on a null slice (at constant observer time). To do this, an outgoing radial light ray is traced out from the centre and parameter values are stored as it passes the boundary of each grid zone. This second set of initial data is then evolved using the set of Hernandez–Misner equations [25].

The unperturbed background model is taken as a spatially flat FLRW model, for which $\Gamma = 1$, with $\rho(r, t)$ being a constant at any particular cosmic time t. The perturbations of ρ and U are then superimposed on this background in the way described in the previous section. During the following evolution, the lapse functions, both for cosmic time and observer time, are set equal to unity at the outer boundary of the grid, synchronising the cosmic and observer times with local clock time there as measured by the local co-moving FLRW fundamental observer.

4.4.1 The Threshold for PBHs

To distinguish whether a cosmological perturbation is able or not to form a PBH depends on the perturbation amplitude δ: if it is larger than a threshold δ_c, an apparent horizon will form, satisfying the condition for the formation of a marginally trapped surface $R(r, t) = 2M(r, t)$. From Eq. (4.6) this corresponds to the condition $\Gamma^2 = U^2$, which has two possible solutions:

- $\Gamma = U$: this is the condition for the cosmological horizon $R_H = 1/H$, which is also an apparent horizon, within an expanding region of the FLRW Universe ($U > 0$).
- $\Gamma = -U$: this is the condition for the formation of the apparent horizon for a black hole, within a collapsing region ($U < 0$).

The mathematical properties of a marginally trapped surface have been discussed in detail in [13].

In Fig. 4.2 one can see the evolution of the ratio $2M/R$ for two particular but representative cases in a radiation dominated Universe ($w = 1/3$): in the left plot ($\delta > \delta_c$) the formation of an apparent horizon for a black hole is obtained after a close equilibrium between gravitational force and pressure gradients, while in the right plot ($\delta < \delta_c$) the collapse is bouncing, and the perturbation is dispersed into the surrounding medium. In the observer time reference frame the formation of the apparent horizon is obtained asymptotically for the null time $u \to \infty$ synchronised with the condition $R(r, t) \to 2M(r, t)$, avoiding the formation of the singularity inside the black hole. In practice the simulations are stopped when $(1 - 2M/R) < 10^{-3}$, and further evolution is negligible. This allows one to compute the mass of the black hole with very good accuracy.

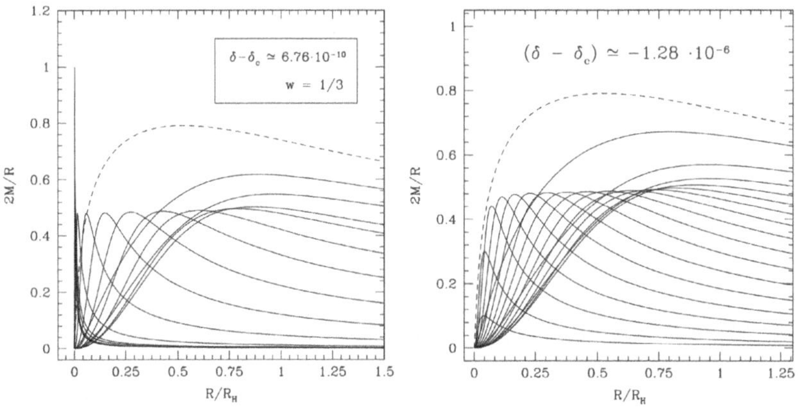

Fig. 4.2 These plots show the behavior of the threshold $2M/R$ against R/R_H when $w = 1/3$. The left panel corresponds to a case when $\delta > \delta_c$ and a PBH is formed with the formation of an apparent horizon $R = 2M$, while the right panel represent a case when $\delta < \delta_c$ and the perturbation is bounced and dispersed into the surrounding medium. The right plot is taken from [24]

The threshold for PBH formation has been originally estimated in [4] using a nonrelativistic Jeans scale argument, showing that the threshold is of the order of the sound speed squared, i.e. $\delta_c \sim w$ for $p = w\rho$. This has been generalised in [11] including the relativistic correction, which for $w = 1/3$ gives in the comoving gauge $\delta_c \simeq 0.4$. For a more accurate evaluation, necessary to compute the abundance of PBHs, numerical simulations of general relativistic nonlinear dynamics of PBH formation are necessary. In the work by Shibata and Sasaki [34], where the threshold is computed with the peak of the compaction function using the constant mean curvature gauge, it was found a value of the peak approximately equal to 0.4 that in the comoving gauge, using the definition of (4.43), gives 0.53. (There is a factor of 2 difference between the two definitions and the coefficient 3(1+w)/(5+3w) depending on the equation of state to multiply when going from the constant mean curvature gauge to the comoving one, i.e. for radiation 2 * 2/3 * 0.4 = 0.53). This differs from the valued obtained in [16], where the threshold was computed with the averaged density contrast $\delta_c \simeq 0.7$. It was then shown in [25, 33] that a consistent value of $\delta_c \simeq 0.5$ would be obtained if the initial conditions of the numerical simulation were imposed on super-horizon scales.

More recently, the shape dependence of the threshold has been analysed in [5, 21, 27, 29]. Given the equation of state, the value of the threshold for PBHs depends on the shape of cosmological perturbations. For a radiation dominated Universe $2/5 \leq \delta_c \leq 2/3$, with the corresponding threshold for the Gaussian component δ_l within the range $0.49 \lesssim \delta_{lc} \leq 4/3$. The shape dependence is parameterised by the dimensionless parameter α defined above, measuring the width of the compaction function at the peak, where the apparent horizon is going to form: for larger values of α the peak of the compaction function becomes narrower, with a sharp transition from the density of the central region within r_m and the outer region, whereas for smaller values of α the transition is smoother. This affects the efficiency of the pressure gradients trying to prevent the black hole forming, and explains why δ_c increases for larger values of α, as can be seen in Fig. 4.3. In practice, the collapse is mainly

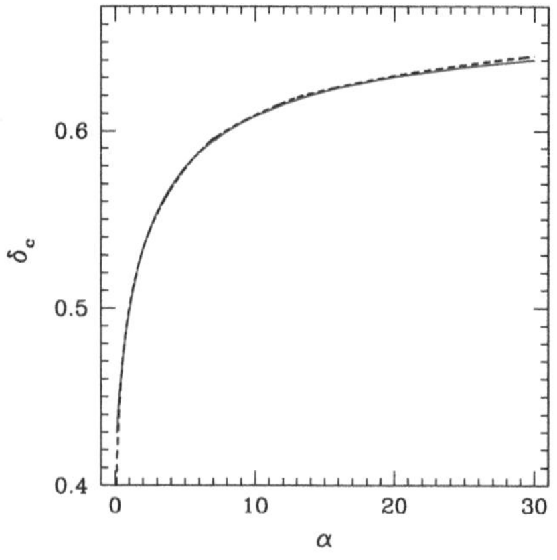

Fig. 4.3 The threshold for PBHs during the radiation dominated Universe. These plots show the behavior of the threshold δ_c when $w = 1/3$, as functions of the shape parameter α. The blue line represents the numerical results while the analytic fit given by (4.53) is plotted with a dashed line. Taken from [22]

affected by the matter distribution inside the region forming the black hole, while the corrections induced by the particular configuration of the tail outside this region affect the threshold values up to 1–2%.

According to the numerical simulations, in a radiation dominated Universe the threshold for PBH formation can be written as a function of the shape parameter α using the analytic expression derived in [5]

$$\delta_c \simeq \frac{4}{15} e^{-\frac{1}{\alpha}} \frac{\alpha^{1-\frac{5}{2\alpha}}}{\Gamma\left(\frac{5}{2\alpha}\right) - \Gamma\left(\frac{5}{2\alpha}, \frac{1}{\alpha}\right)}, \tag{4.52}$$

obtained from the interesting property that the average of the compaction function for the threshold δ_c is shape independent, i.e. $\overline{\mathcal{C}}(r_m) \simeq 0.4$. An alternative analytical expression for (4.52) is given by the numerical fit obtained in [27]:

$$\delta_c = \begin{cases} \alpha^{0.047} - 0.50 & 0.1 \lesssim \alpha \lesssim 7 \\ \alpha^{0.035} - 0.475 & 7 \lesssim \alpha \lesssim 13 \\ \alpha^{0.026} - 0.45 & 13 \lesssim \alpha \lesssim 30 \end{cases} \tag{4.53}$$

and in Fig. 4.3 we show the numerical behaviour of δ_c obtained by [21], plotted with a blue line, which is very well fitted by (4.53), plotted with a dashed line.

Following the decomposition of (4.48), it is possible to write the shape parameter α as

$$\alpha = \frac{\alpha_G}{\left(1 - \frac{1}{4}\delta_l\right)\left(1 - \frac{1}{2}\delta_l\right)} \tag{4.54}$$

where

$$\alpha_G = -\frac{\delta_l'' r_m^2}{4\delta_l} \tag{4.55}$$

is the Gaussian shape parameter obtained in the linear approximation ($\delta_l \ll 1$), independent of the amplitude of δ_l. Using this decomposition the typical shape parameter of the average profile shape can be computed from the shape of the power spectrum of the cosmological perturbation if ζ is a Gaussian variable, apart from some possible variation depending on the effects of the sub-horizon modes that could affect the collapse. Using this in the context of peak theory (see Chap. 6), it is possible to compute the threshold δ_c from the shape of the Gaussian power spectrum following a simple prescription. The algorithm accounts for both the nonlinearities associated with the relation between the Gaussian curvature perturbation and the perturbation energy density profile. The full derivation of the prescription can be found in [27], here we provide the reader with a schematic overview summarised in few simple steps:

1. **The power spectrum of the curvature perturbation**: take the primordial power spectrum $\mathcal{P}_\zeta$ of the Gaussian curvature perturbation and compute, on super-horizon scales, its convolution with the transfer function $T(k,\eta)$

$$P_\zeta(k,\eta) = \frac{2\pi^2}{k^3}\mathcal{P}_\zeta(k)T^2(k,\eta).$$

2. **The comoving length scale** $\hat{r}_\mathrm{m}$ of the perturbation is related to the characteristic scale k_* of the power spectrum P_ζ. Compute the value of $k_*\hat{r}_\mathrm{m}$ by solving the following integral equation

$$\int \mathrm{d}k k^2 \left[(k^2\hat{r}_m^2 - 1)\frac{\sin(k\hat{r}_m)}{k\hat{r}_m} + \cos(k\hat{r}_m) \right] P_\zeta(k,\eta) = 0\,.$$

3. **The shape parameter:** compute the corresponding shape parameter α of the collapsing perturbation, including the correction from the non linear effects, by solving the following equation

$$F(\alpha)\left[1+F(\alpha)\right]\alpha = -\frac{1}{2}\left[1 + \hat{r}_m \frac{\int \mathrm{d}k k^4 \cos(k\hat{r}_m) P_\zeta(k,\eta)}{\int \mathrm{d}k k^3 \sin(k\hat{r}_m) P_\zeta(k,\eta)}\right]$$

$$F(\alpha) = \sqrt{1 - \frac{2}{5} e^{-\frac{1}{\alpha}} \frac{\alpha^{1-\frac{5}{2\alpha}}}{\Gamma\left(\frac{5}{2\alpha}\right) - \Gamma\left(\frac{5}{2\alpha}, \frac{1}{\alpha}\right)}}\,.$$

4. **The threshold** δ_c is computed as a function of α using (4.53).

The threshold δ_c for PBH formation defined here is computed at the cosmological horizon crossing by making a linear extrapolation from the super-horizon regime, where the curvature is time independent, imposing $aHr_m = 1$, where the cosmological horizon $R_\mathrm{H} = 1/H$ is defined with respect to the background. In this way one is extending the validity of the gradient expansion approximation up to $\epsilon = 1$. Although this is not very accurate, it represents a well defined criterion for measuring the perturbation amplitude, and has been widely used in the literature. This definition of δ_c does not take into account the non linear effects of the higher orders in the gradient expansion approach that at horizon crossing would not be negligible: the analysis done in [27] has found that the value of the threshold computed at the non linear horizon crossing is about two times larger than the one computed with an extrapolation from the super-horizon regime. Understanding to which extent this has relevant implications in the computation of the abundance of PBHs it is still matter of investigation, and is beyond the scope of the discussion of this chapter.

4.4.2 *Critical Collapse*

The final aspect of PBH formation that it is important to discuss before concluding this chapter is the mechanism of critical collapse ([8, 9] and references therein), necessary for a proper computation of the mass distribution of PBHs. It was shown in [16, 31] that the mass spectrum of PBHs is characterised by the scaling law relation given by

$$M_{\rm PBH} = K\, M_{\rm H}\, (\delta - \delta_{\rm c})^{\gamma} \ . \tag{4.56}$$

where the cosmological horizon mass $M_{\rm H}$ identifies the epoch of the Universe when the cosmological perturbations collapsing into PBHs are crossing the cosmological horizon, while γ is a parameter depending only on the equation of state [30], with $\gamma \simeq 0.36$ when $w = 1/3$. The other two parameters K and $\delta_{\rm c}$ also depend on the effects of pressure gradients, i.e. the equation of state, described by the value of w and $c_{\rm s}^2$, and the shape of the initial configuration, identified by the shape parameter α.

The nature of critical collapse in the context of PBHs has been then intensively investigated by the author of this chapter [23, 24] and the left panel of Fig. 4.4 shows the scaling law behaviour for different values of α: for $M_{\rm PBH} < M_{\rm H}$ the exponent $\gamma \simeq 0.36$ is constant, while for larger values deviation from scaling are visible. This is because the critical collapse is characterised by a self similar behaviour [23], which is scale free. The right panel of Fig. 4.4 shows how K is varying with α (a similar behaviour has been observed also in [6]).

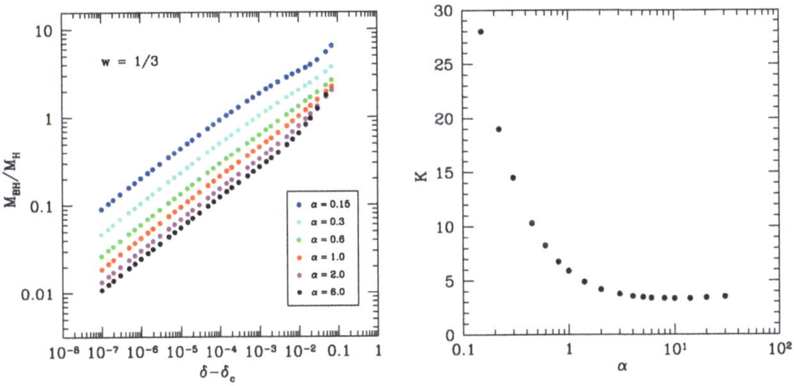

Fig. 4.4 The mass of PBHs for a radiation dominated Universe. This figure shows the critical collapse behaviour of equation (4.56) for $w = 1/3$: in the left panel we plot the scaling laws for different values of α obtained from the results of the numerical simulations, while in the right panel we show the corresponding dependence of the scaling law parameter K as function of α. Taken from [22]

Critical collapse arises naturally in the context of PBH formation, also when the equation of state is different from the simple $p = w\rho$, as in the context of PBH formation during the QCD phase transition (see Chap. 5). This case characterised by an intrinsic scale, and the scaling law behaviour is modified with the γ, δ_c and K being a function of the equation of state varying with the horizon crossing [22]. The formation of PBHs during the QCD phase transition could have important implications for solar mass black holes forming binary systems that during the merging emit gravitation waves. The possibility that some of the sources detected by LIGO/VIRGO has a primordial origin is matter of an interesting debate [7]—see Chap. 25.

4.5 Summary

In this chapter we have explained the mechanism of PBH formation from the gravitational collapse of cosmological perturbation in the very early Universe, presenting in particular numerical results when the Universe is radiation dominated. Starting from the mathematical framework of the long wave-length approximation, when cosmological perturbations are super-horizon, we have seen how to specify initial condition of numerical simulations using the gradient expansion approach, where all geometrical and hydrodynamical variables are functions of a time independent curvature perturbation, describing the geometrical origin of adiabatic perturbations.

Assuming spherical symmetry, which is well justified in the context of PBH formation, because we are dealing with cosmological perturbations with high peaks, general relativistic numerical simulations allow one to compute the threshold δ_c as a function of a simple shape parameter, corresponding to the shape of a Gaussian power spectrum for cosmological perturbations.

Finally we have seen that the masses of PBHs follow the scaling law relation of critical collapse, a mechanism arising naturally in the context of PBH formation. These are the fundamental ingredients that can be used to compute the abundance of PBHs in peak theory, understanding to which extent PBHs could play an important role in the current Universe, explaining for example a non negligible component of dark matter or the origin of binary systems merging with the emission of gravitational waves. All these are challenging questions in modern Cosmology.

References

1. T.W. Baumgarte, S.L. Shapiro, S.A. Teukolsky, Computing supernova collapse to neutron stars and black holes. **443**, 717 (1995). https://doi.org/10.1086/175563
2. G. V. Bicknell, R. N. Henriksen, Formation of primordial black holes. Astrophys. J. Part 1 **232**, 670 (1979)
3. B.J. Carr, T. Harada, Separate universe problem: 40 years on. Phys. Rev. D **91**(8), 084048 (2015). https://doi.org/10.1103/PhysRevD.91.084048. arXiv: 1405.3624 [astro-ph.CO]

4. B. Carr, The Primordial black hole mass spectrum. Astrophys. J. **201**, 1–19 (1975). https://doi.org/10.1086/153853
5. A. Escrivà, C. Germani, R.K. Sheth, Universal threshold for primordial black hole formation. Phys. Rev. D **101**(4), 044022 (2020). https://doi.org/10.1103/PhysRevD.101.044022. arXiv: 1907.13311 [gr-qc]
6. A. Escrivà, A.E. Romano, Effects of the shape of curvature peaks on the size of primordial black holes. JCAP **05**, 066 (2021). https://doi.org/10.1088/1475-7516/2021/05/066. arXiv: 2103.03867 [gr-qc]
7. G. Franciolini et al., From inflation to black hole mergers and back again: gravitational-wave data-driven constraints on inflationary scenarios with a first-principle model of primordial black holes across the QCD epoch. Phys. Rev. D **106**(12), 123526 (2022). https://doi.org/10.1103/PhysRevD.106.123526. arXiv: 2209.05959 [astro-ph.CO]
8. C. Gundlach, J.M. Martin-Garcia, Critical phenomena in gravitational collapse. Living Rev. Rel. **10**, 5 (2007). https://doi.org/10.12942/lrr-2007-5. arXiv: 0711.4620 [gr-qc]
9. T. Harada, H. Maeda, Convergence to a selfsimilar solution in general relativistic gravitational collapse. Phys. Rev. D **63**, 084022 (2001). https://doi.org/10.1103/PhysRevD.63.084022. arXiv: gr-qc/0101064
10. T. Harada, C.-M. Yoo, Y. Koga, Revisiting compaction functions for primordial black hole formation. Phys. Rev. D **108**(4), 043515 (2023). https://doi.org/10.1103/PhysRevD.108.043515. arXiv: 2304.13284 [gr-qc]
11. T. Harada, C.-M. Yoo, K. Kohri, Threshold of primordial black hole formation. Phys. Rev. D **88**(8), 084051 (2013). [Erratum: Phys. Rev. D **89**, 029903 (2014)]. https://doi.org/10.1103/PhysRevD.88.084051. arXiv: 1309.4201 [astro-ph.CO]
12. T. Harada et al., Cosmological long-wavelength solutions and primordial black hole formation. Phys. Rev. D **91**(8), 084057 (2015). https://doi.org/10.1103/PhysRevD.91.084057. arXiv: 1503.03934 [gr-qc]
13. A. Helou, I. Musco, J.C. Miller, Causal nature and dynamics of trapping horizons in black hole collapse. Class. Quant. Grav. **34**(13), 135012 (2017). https://doi.org/10.1088/1361-6382/aa6d8f. arXiv: 1601.05109 [gr-qc]
14. J.C. Hidalgo, A.G. Polnarev, Probability of primordial black hole formation and its dependence on the radial profile of initial configurations. Phys. Rev. D **79**, 044006 (2009). https://doi.org/10.1103/PhysRevD.79.044006. arXiv: 0806.2752 [astro-ph]
15. K. Jedamzik, J.C. Niemeyer, Primordial black hole formation during first order phase transitions. Phys. Rev. D **59**, 124014 (1999). https://doi.org/10.1103/PhysRevD.59.124014. arXiv: astro-ph/9901293
16. C. Jens, K. Niemeyer, Jedamzik, Dynamics of primordial black hole formation. Phys. Rev. D **59**, 124013 (1999). https://doi.org/10.1103/PhysRevD.59.124013. arXiv: astro-ph/9901292
17. M. Kopp, S. Hofmann, J. Weller, Separate universes do not constrain primordial black hole formation. Phys. Rev. D **83**, 124025 (2011). https://doi.org/10.1103/PhysRevD.83.124025. arXiv: 1012.4369 [astro-ph.CO]
18. D.H. Lyth, K.A. Malik, M. Sasaki, A general proof of the conservation of the curvature perturbation. JCAP **05**, 004 (2005). https://doi.org/10.1088/1475-7516/2005/05/004. arXiv: astro-ph/0411220
19. M.M. May, R.H. White, Hydrodynamic calculations of general-relativistic collapse. Phys. Rev. **141**, 1232–1241 (1966). https://doi.org/10.1103/PhysRev.141.1232
20. C.W. Misner, D.H. Sharp, Relativistic equations for adiabatic, spherically symmetric gravitational collapse. Phys. Rev. **136**, B571–B576 (1964). https://doi.org/10.1103/PhysRev.136.B571
21. I. Musco, Threshold for primordial black holes: dependence on the shape of the cosmological perturbations. Phys. Rev. D **100**(12), 123524 (2019). https://doi.org/10.1103/PhysRevD.100.123524. arXiv: 1809.02127 [gr-qc]
22. I. Musco, K. Jedamzik, S. Young, Primordial black hole formation during the QCD phase transition: threshold, mass distribution and abundance (2023). arXiv: 2303.07980 [astro-ph.CO]

23. I. Musco, J.C. Miller, Primordial black hole formation in the early universe: critical behaviour and self-similarity. Class. Quant. Grav. **30**, 145009 (2013). https://doi.org/10.1088/0264-9381/30/14/145009. arXiv: 1201.2379 [gr-qc]
24. I. Musco, J.C. Miller, A.G. Polnarev, Primordial black hole formation in the radiative era: investigation of the critical nature of the collapse. Class. Quant. Grav. **26**, 235001 (2009). https://doi.org/10.1088/0264-9381/26/23/235001. arXiv: 0811.1452 [gr-qc]
25. I. Musco, J.C. Miller, L. Rezzolla, Computations of primordial black hole formation. Class. Quant. Grav. **22**, 1405–1424 (2005). https://doi.org/10.1088/0264-9381/22/7/013. arXiv: gr-qc/0412063
26. I. Musco, T. Papanikolaou, Primordial black hole formation for an anisotropic perfect fluid: Initial conditions and estimation of the threshold. Phys. Rev. D **106**(8), 083017 (2022). https://doi.org/10.1103/PhysRevD.106.083017. arXiv: 2110.05982 [gr-qc]
27. I. Musco et al., Threshold for primordial black holes. II. A simple analytic prescription. Phys. Rev. D **103**(6), 063538 (2021). https://doi.org/10.1103/PhysRevD.103.063538. arXiv: 2011.03014 [astro-ph.CO]
28. D. K. Nadezhin, I. D. Novikov, A. G. Polnarev, The hydrodynamics of primordial black hole formation. Sov. Astron. **22**, 129 (1978)
29. T. Nakama et al., Identifying the most crucial parameters of the initial curvature profile for primordial black hole formation. JCAP **01**, 037 (2014). https://doi.org/10.1088/1475-7516/2014/01/037. arXiv: 1310.3007 [gr-qc]
30. D.W. Neilsen, M.W. Choptuik, Critical phenomena in perfect fluids. Class. Quant. Grav. **17**, 761–782 (2000). https://doi.org/10.1088/0264-9381/17/4/303. arXiv: gr-qc/9812053
31. J.C. Niemeyer, K. Jedamzik, Near-critical gravitational collapse and the initial mass function of primordial black holes. Phys. Rev. Lett. **80**, 5481–5484 (1998). https://doi.org/10.1103/PhysRevLett.80.5481. arXiv: astro-ph/9709072
32. A.G. Polnarev, T. Nakama, J. Yokoyama, Self-consistent initial conditions for primordial black hole formation. JCAP **09**, 027 (2012). https://doi.org/10.1088/1475-7516/2012/09/027. arXiv: 1204.6601 [gr-qc]
33. A.G. Polnarev, I. Musco, Curvature profiles as initial conditions for primordial black hole formation. Class. Quant. Grav. **24**, 1405–1432 (2007). https://doi.org/10.1088/0264-9381/24/6/003. arXiv: gr-qc/0605122
34. M. Shibata, M. Sasaki, Black hole formation in the Friedmann universe: formulation and computation in numerical relativity. Phys. Rev. D **60**, 084002 (1999). https://doi.org/10.1103/PhysRevD.60.084002. arXiv: gr-qc/9905064
35. K. Uehara et al., Numerical simulation of type II primordial black hole formation (2024). arXiv: 2401.06329 [gr-qc]

Chapter 5
Primordial Black Hole Formation During Cosmic Phase Transitions

Karsten Jedamzik

Abstract Primordial black holes (PBHs) may be part of the dark matter. It is shown here that PBHs form more easily during cosmic phase transitions. For approximately scale-invariant primordial curvature fluctuations the PBH mass function may therefore leave a record of the thermal history of the early Universe. In particular a peak is expected on the $1.9 M_\odot$ scale due to the cosmic QCD transition.

5.1 General Considerations

Primordial black hole (PBH) formation during cosmic phase transitions may be particularly efficient. In what follows we only discuss the case of PBH formation due to pre-existing curvature fluctuations produced during inflation, as described in Chap. 2. We omit a discussion of other scenarios [7, 10] where substantial sub-horizon size density fluctuations are generated due to the dynamics of the phase transition itself. Consider a pre-existing slightly non-linear density fluctuation when it crosses into the horizon, i.e. of size $r \approx r_H$, where $r_H = 2ct = c/H$, is the horizon length during a radiation dominated Universe with c the speed of light, t cosmic time, and H the Hubble expansion rate. Approximate the overdense region to be in vacuum, i.e. not having cosmic matter around it. One can then compute the Schwarzschild radius of the overdense region

$$r_S = \frac{2GM}{c^2} = \frac{2G}{c^2}\frac{4\pi}{3}\rho r_H^3 = \frac{2G}{c^2}\frac{4\pi}{3}\rho\left(\frac{c}{H}\right)^3 = \frac{2G}{c^2}\frac{c^3}{2HG} = \frac{c}{H} = r_H\,, \quad (5.1)$$

where in the second equality we have used the horizon mass and in the fourth equality the Friedman equation $H^2 = 8\pi G\rho/3$. Here G and ρ are the gravitational constant and energy density, respectively. It is seen that the overdense matter contained on the horizon scale is close to its own Schwarzschild radius and therefore close to

K. Jedamzik (✉)
Laboratoire de Univers et Particules de Montpellier, UMR5299-CNRS, Universite de Montpellier, 34095 Montpellier, France
e-mail: karsten.jedamzik@umontpellier.fr

C. Byrnes et al. (eds.), *Primordial Black Holes*, Springer Series in Astrophysics and Cosmology, https://doi.org/10.1007/978-981-97-8887-3_5

forming a PBH. It is known that the overdensity of a sub-horizon region does not further grow in a radiation dominated Universe , such that a PBH will either form at horizon crossing, or never, since $r_S < r_H$ during the subsequent evolution. What is the role of pressure forces during PBH formation? From astrophysics we are used to the concept that an overdense region may only gravitationally collapse if its size is large enough, larger than the Jeans length. This is the case if the pressure response time, i.e. the time for a pressure (sound) wave to travel over the entire overdense region, $t_p = r/c_s$, with c_s the speed of sound, is larger than the gravitational collapse time scale, $t_g \approx 1/\sqrt{G\rho}$. Using the speed of sound $c_s = c/\sqrt{3}$ of a relativistic gas, as appropriate in a radiation dominated Universe, one may easily derive the Jeans length

$$r_J \approx c_s/\sqrt{G\rho} = \frac{\sqrt{8\pi}}{3}\frac{c}{H} \approx 1.67 r_H \tag{5.2}$$

It is seen that forces due to pressure gradients are important. The PBH formation process is thus a competition between gravity and pressure forces. This implies that during any epochs in the early Universe where pressure forces are reduced due to a varying equation of state, PBHs may form more easily than during a purely radiation dominated Universe. As PBH formation is a rare process, occurring on an exponential tail (cf. Chaps. 2, 6 and 7), the slightest reduction of the overdensity required to form a PBH, may lead to a very significant enhancement of PBH formation. For an approximately scale-invariant spectrum of pre-existing inflationary curvature perturbations, the PBH mass function should therefore be peaked at the approximate horizon masses during epochs where a softening of the equations of state occurs.

5.2 Generic First-Order Phase Transitions

Consider a generic first order phase transition. During such a transition there is a low-energy phase and a high-energy phase of the matter. These phases may coexist at coexistence temperature T_c. In thermodynamic equilibrium the quantities $p = -(\partial E/\partial V)_S$ and $T = (\partial E/\partial S)_V$, which define pressure p and temperature T, are continuous at temperature T_c. Here E, V, and S are energy, volume, and entropy, respectively. During the phase transition, when the system "cools", a high-energy phase with energy density $\rho_h = \rho_l + L$ is converted to a low-energy phase, while the system stays at the same temperature T_c. Here L is latent heat. Only when all high-energy phase is converted to low-energy phase, will the temperature drop below T_c. Borrowing the parametrisation of the MIT model (developed to describe a first order QCD transition) one may write

$$p_l(T) = \frac{1}{3} g_l \frac{\pi^2}{30} T^4 \quad ; \quad p_h(T) = \frac{1}{3} g_h \frac{\pi^2}{30} T^4 - B \tag{5.3}$$

$$\rho_l(T) = g_l \frac{\pi^2}{30} T^4 \quad ; \quad \rho_h(T) = g_h \frac{\pi^2}{30} T^4 + B \tag{5.4}$$

$$s_l(T) = \frac{4}{3} g_l \frac{\pi^2}{30} T^3 \quad ; \quad s_h(T) = \frac{4}{3} g_h \frac{\pi^2}{30} T^3 \quad . \tag{5.5}$$

In the above s is entropy density and g are statistical weights (i.e. essentially the number of relativistic species in a phase). Furthermore the "bag constant" B acts essentially as a vacuum energy density for the high energy phase. In the QCD case it was introduced to account for the interaction energy associated within the strongly interacting quark-gluon plasma. During the middle of a phase transition the plasma consists of bubbles of one phase surrounded by the other phase. Over sufficiently large regions, i.e. $l >> r_b$, where r_b is the typical size of a bubble, one may define an average density and entropy

$$\langle p \rangle = p_l(T_c) = p_h(T_c) \tag{5.6}$$

$$\langle \rho \rangle = f_l \rho_l(T_c) + (1 - f_l)\rho_h(T_c) \tag{5.7}$$

$$\langle s \rangle = f_l s_l(T_c) + (1 - f_l) s_h(T_c), \tag{5.8}$$

where f_l is the fraction of volume existing in the low-energy phase. During the phase transition the pressure and temperature stay constant and the evolution of average energy- and entropy- density is governed by the evolution of the volume fraction f_l. It is therefore clear that over sufficiently large regions the effective speed of sound is zero

$$c_S^{eff} = \sqrt{\left(\frac{\partial \langle p \rangle}{\partial \langle \rho \rangle}\right)_S} \approx 0 \tag{5.9}$$

These arguments of course only apply in thermodynamic equilibrium, and not for super-cooled or -heated phase transitions. Using the constancy for entropy during adiabatic evolution (i.e. sa^3 = constant) as well as Eqs. 5.4, 5.5, 5.7, and 5.8 one may derive the evolution of average energy density with scale factor a during a first-order phase transition

$$\langle \rho \rangle (a) = \left(\frac{a_0}{a}\right)^3 [\rho_h + \frac{1}{3}\rho_l] - \frac{1}{3}\rho_l \, . \tag{5.10}$$

for $a_b < a < a_e$ with a_b and a_e the scale factor at the beginning and the end of the phase transition, respectively. It is interesting to note that it is more akin of a dust like phase (i.e. $\rho \sim 1/a^3$) than a radiation dominated phase (i.e. $\rho \sim 1/a^4$). During dust like phases density perturbations can grow unhindered by pressure forces. We therefore expect PBH formation during first-order phase transitions to be facilitated due to the absence of pressure forces.

The PBH formation process during a short first-order phase transition has been numerically investigated in [12]. This general relativistic hydrodynamic simulation of the evolution of a spherical symmetric overdensity shows the formation of a PBH. In Fig. 5.1 the evolution of the density profile as a function of radial coordinate is shown. A first-order phase transition with $c_s^{eff} \approx 0$ has been assumed in the energy density range $\rho = 0.5 - 1$, as indicated by the dashed lines, and $c_s = 1/\sqrt{3}$ for $\rho > 1$

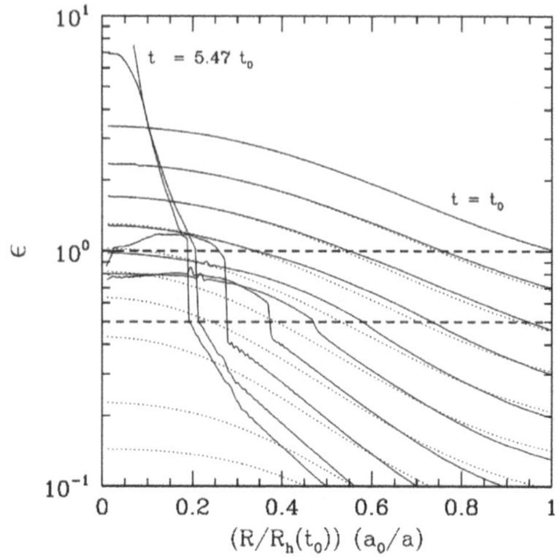

Fig. 5.1 The evolution of energy density as a function of radial coordinate for a density fluctuation generated during inflation and re-entering the horizon after inflation. Results for two scenarios are shown. **a** a first-order phase transition in the density range $\rho = 0.5 - 1$ (solid) and **b** a radiation dominated Universe without phase transition (dotted). Figure taken from [12]

and $\rho < 0.5$ has been adopted. Note that only the inner region of the simulation is shown. It is seen that at late times when the fluctuation is collapsing any matter which reaches $\rho_l(T_c)$ from below during the collapse, will go through the entire phase transition, until $\rho \geq \rho_h(T_c)$ as there are no pressure forces to prevent this collapse. This bias gives the fluctuation the extra kick to form a PBH. In comparison, the dotted lines show the evolution of the density profile in the absence of a first-order phase transition, where for the same initial fluctuation no PBH is formed.

5.3 QCD Phase Transition

In the seventies when nuclear physicists discovered a large number of nuclear resonances, it was speculated that the associated softening of the equation of state could lead to an enhancement of PBH formation [6]. With the discovery of quantum chromo-dynamics it became clear that nuclear resonances were not the fundamental particles, but were made up of quarks and gluons. For a long time it was thought that the necessary transition from hadronic matter (mesons and baryons) to their constituents (quarks and gluons) at $T \sim 100\,\text{MeV}$ could be a first-order phase transition. It was speculated that the softening of the equation of state could lead to an enhancement of solar mass PBHs and potentially explain the MACHO signal of some compact dark matter in the Milky Way halo [13, 14].[1] Detailed lattice gauge simulations in this millenium [2, 3] seem to have established that the QCD transition

[1] The question if there are some massive compact Milky Way halo objects is still under debate, as the EROS [18] and MACHO [1] collaborations find different results. It can, however, be excluded that the Milky way halo is entirely made up of solar mass PBHs, as they would violate microlensing constraints. This conclusion can not be circumvented when the PBHs are members of clusters [17].

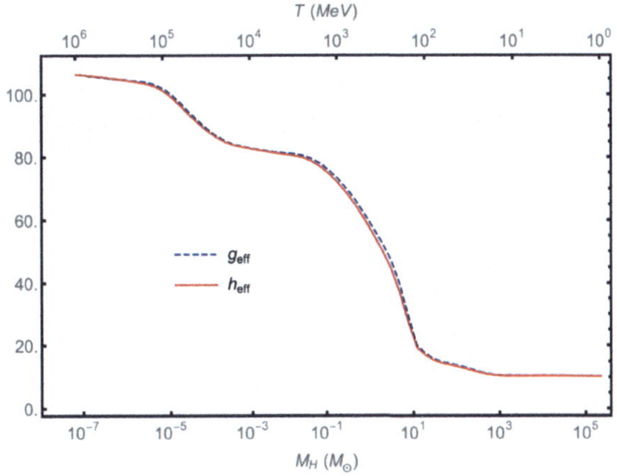

Fig. 5.2 The statistical weights for energy density $g_{\text{eff}} = 30\rho/(\pi^2 T^4)$ and entropy density $h_{\text{eff}} = 45s/(2\pi^2 T^3)$ as a function of cosmic temperature T and horizon mass $M_h(T)$. Taken from Ref. [11]. Kindly provided by the authors of [11, 15]

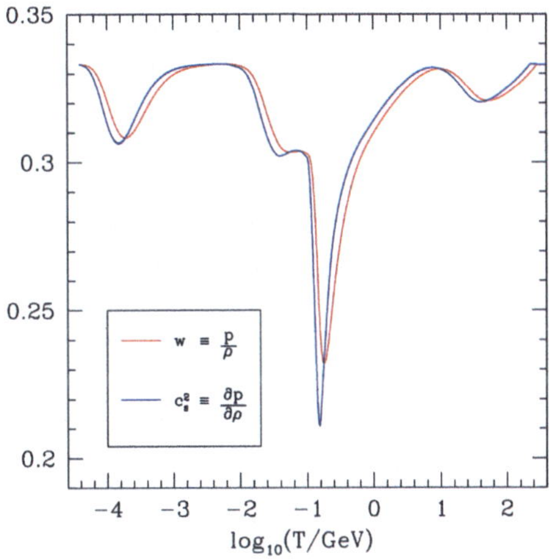

Fig. 5.3 The speed of sound squared c_s^2 and $w = p/\rho$ in the early Universe for a wide range of temperatures. Taken from [16]

is no phase transition but a cross-over. There is nevertheless a reduction of speed of sound during the crossover. Figure 5.2 shows the relativistic degrees of freedom g for cosmic temperatures which include the electroweak- and QCD- phase transitions. It also associates the horizon, i.e. the approximated PBH mass, with cosmic temperature. It is seen that during the QCD transition g drops from $g \approx 80$ to $g \approx 11$. In Fig. 5.3 the square of the speed of sound, c_s^2, and $w = p/\rho$ are shown for a wide range of temperatures between $T \approx 25$ keV and $T \approx 400$ GeV. It is seen that, though the reduction of c_s^2 from a value of $1/3$ is not as drastic as for a first-order phase transition where $c_s^2 \approx 0$, that there are several epochs where c_s^2 is reduced. Most notable is the reduction during the QCD transition at $T \approx 100$ MeV. It is relatively large due to the large number of degrees of freedom involved. Smaller reductions occur during the electroweak transition at $T \approx 100$ GeV, the pion annihilation epoch at $T \approx 50$ GeV, and the e^+e^- annihilation epoch at $T \approx 200$ keV. During epochs where the plasma contains particles with mass $M \sim T$, a generic reduction in c_s^2 and w occurs, as rest mass energy does not contribute to pressure. Only when T has dropped significantly below M, thereby provoking the annihilation of massive particles with their anti-particles, will c_s^2 bounce back to its radiation dominated value. This is the case for the massive gauge bosons W and Z during the electroweak transition, massive pions shortly after the QCD transition and massive electrons and positrons during Big Bang nucleosynthesis.

The detailed equation of state has been used to compute the density (or curvature) threshold δ_c at each epoch required for PBH formation. Pre-existing fluctuations with $\delta \geq \delta_c$ form PBHs upon horizon crossing, whereas fluctuations with $\delta < \delta_c$ are dispersed by pressure forces (cf. Chap. 4). These studies have been done by either approximate analytical collapse models [4, 5] or by full numerical simulation [8, 16]. In Fig. 5.4 the PBH formation threshold δ_c relative to the threshold in a purely radiation dominated Universe is shown. The underlying study [16] used fully general-relativistic hydrodyamics simulations of spherical fluctuations with different fluctuation profiles (given by the value α). Only the QCD epoch is shown, leading, however, to a comparatively large range in horizon masses (~PBH masses). It is seen that the ~30% reduction in c_s^2 leads to a ~10% reduction in threshold δ_c. This study finds that for a wide range of curvature fluctuations, the apparent horizon always occurs first in conditions when matter is at the minimum c_s^2. We have observed similar results for first order phase transitions in the previous section. It is also observed that critical scaling phenomena (see also Chap. 4) hold even in the QCD case with a varying equation of state, albeit with a different exponent than that in a plasma with a purely radiation dominated Universe.

The results shown in Fig. 5.4 may be used to compute the QCD PBH mass function as a function of spectral index and fluctuation shape α. The calculations take full account of critical phenomena and apply peaks theory (Chap. 6) for the computation

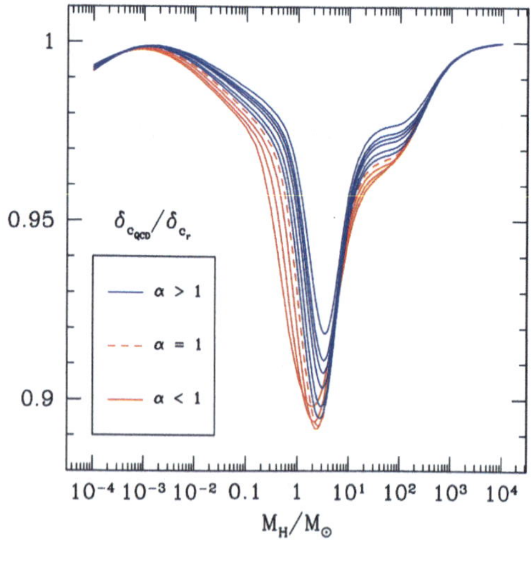

Fig. 5.4 The critical threshold for PBH formation during the QCD epoch normalized by the value of a radiation dominated Universe. The parameter α quantifies the shape of the fluctuation. Taken from [16]

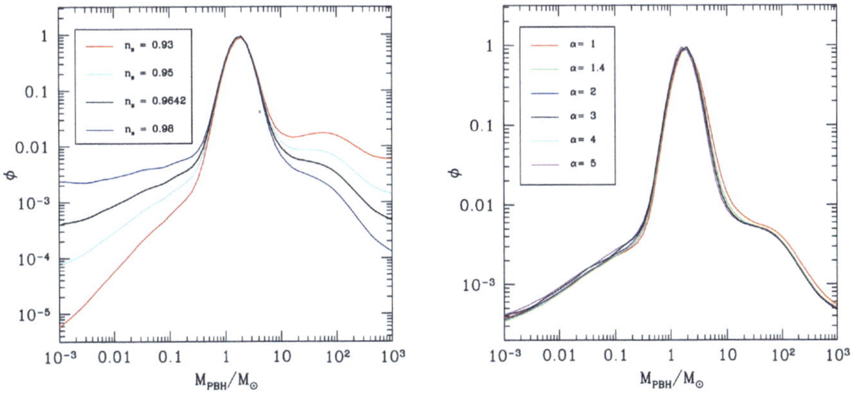

Fig. 5.5 The QCD PBH mass function $\Phi(M_{\rm PBH})$. Left panel: The mass function for varying spectral index n_s and $\alpha = 3$. Right panel: The mass function for varying α and $n_s = 0.9642$. Taken from [16]

of the PBH mass function. Results are shown in Fig. 5.5 with the assumption of pre-existing Gaussian fluctuations. It is seen that the resulting peak in the mass function, which is at $M \approx 1.9 M_\odot$, is surprisingly independent of the fluctuation shape parameter α and spectral index n_s of the underlying fluctuations. Of course the latter holds only when the fluctuations are approximately scale-invariant. For very blue spectra $n_s > 1$ PBH formation is dominated at the smallest scales, and vice versa for very red spectra $n_s < 1$. In such cases the QCD peak would not be very important. But for approximately scale-invariant spectra, PBH formation on the QCD scale $M_{\rm pbh} \approx 1.9 M_\odot$ is a factor thousand more probable than on other scales. It is also noted that the pion annihilation epoch leads to a small enhancement on mass scales $M_{\rm pbh} > 10 M_\odot$. Approximately one per cent of all QCD PBHs are thus massive. If some fraction of the observed LIGO/Virgo/Kagra $M > M_\odot$ events are PBHs [9, 15], then gravitational wave detectors have only observed the tip of the iceberg, as there would be a factor ~100 more low mass PBHs. Of course, the obvious smoking gun for the existence of PBH dark matter in our Universe would be the discovery of PBHs of sub-Chandrasekar mass. Such PBHs are formed in QCD enhanced PBH formation scenarios in quite large abundance.

5.4 The e^+e^- Annihilation Epoch

It has been noted [5, 14], that the decrease of w and c_s^2 during the e^+e^- annihilation epoch could lead to an enhancement of the abundance of PBHs on the $M_{\rm pbh} \sim 10^5 M_\odot$ scale. However this may not necessarily be the case, as the cosmic e^+e^- annihilation epoch is quite different from the QCD epoch and the electroweak epoch. At temperatures $T \approx 1 - 2\,{\rm MeV}$, shortly before the e^+e^- annihilation, with a maximum reduction of w at $T \approx 200\,\mathrm{keV}$, neutrinos decouple from the Universe. In particular, since their interactions with the rest of the plasma freeze out, neutrinos can free-stream out of overdense regions, effectively reducing the overdensity. One may schematically write for the overdensity

$$\frac{\delta\rho}{\rho} = \frac{\sum\limits_i \delta\rho_i}{\sum\limits_i \langle\rho_i\rangle}, \tag{5.11}$$

where brackets denote cosmic average and the index i runs over particle species. For adiabatic perturbations one has $\delta\rho_i = K\langle\rho_i\rangle$ with K a quantity independent of species, such that $\delta\rho/\rho = K$. However, the free-streaming of the neutrinos destroys the adiabaticity of the perturbation since $\delta\rho_\nu \approx 0$. Having $\langle\rho_\nu\rangle/\langle\rho_{\mathrm{tot}}\rangle = g_\nu/g_{\mathrm{tot}} = 5.25/10.75 \approx 0.5$ one may estimate that the original perturbation has only approximately half of the overdensity after neutrino free-streaming. On the other hand, the critical threshold for PBH formation will not reduce by as much as a factor of two due to the e^+e^- equation of state, such that the argument would imply that formation of PBHs during any epoch after neutrino coupling is highly suppressed for scale invariant perturbation spectra. Seen from the opposite point of view, even if neutrinos are initially homogeneous, the density perturbation in photons and e^+e^- would gravitationally attract neutrinos. Since those neutrinos do not even exert pressure, a further reduction of w to $w \sim 0.33/2$ would occur, favoring PBH formation. Reference [16] argues that the first argument may be dominant, leading to a suppression of PBH formation. However, only a dedicated simulation of PBH formation with a fluid and free-streaming component could definitely answer this question. The formation of PBHs during the e^+e^- annihilation is therefore an open question.

5.5 Conclusions

In this short chapter we have considered the process of PBH formation from pre-existing inflationary curvature fluctuations during cosmic phase transitions. We have presented analytical arguments which show that due the reduction of pressure forces during cosmic phase transitions and particle annihilation periods PBH formation is facilitated. These analytical arguments have been confirmed by fully non-relativistic numerical simulations. We have found that for approximately scale-invariant spectra of the pre-existing curvature fluctuations a drastic enhancement of the PBH mass function occurs on the approximate horizon mass scale at the phase transition. For the QCD transition this corresponds to a well pronounced peak at $M \approx 1.9M_\odot$ in the

PBH mass function. We have argued that PBH formation during the e^+e^- annihilation may be either reduced or enhanced due to the particularity of neutrino decoupling. If PBHs of mass $M \sim 10^5 M_\odot$ form more easily during the e^+e^- annihilation epoch is currently an open question.

References

1. C. Alcock et al., The MACHO project: microlensing results from 5.7 years of LMC observations. Astrophys. J. **542**, 281–307 (2000). https://doi.org/10.1086/309512. arXiv:astro-ph/0001272
2. T. Bhattacharya et al., QCD phase transition with chiral quarks and physical quark masses. Phys. Rev. Lett. **113**(8), 082001 (2014). https://doi.org/10.1103/PhysRevLett.113.082001. arXiv:1402.5175 [hep-lat]
3. Sz. Borsanyi et al., Calculation of the axion mass based on high-temperature lattice quantum chromodynamics. Nature **539**(7627), 69–71 (2016). https://doi.org/10.1038/nature20115. arXiv:1606.07494 [hep-lat]
4. C.T. Byrnes et al., Primordial black holes with an accurate QCD equation of state. JCAP **08**, 041 (2018). https://doi.org/10.1088/1475-7516/2018/08/041. arXiv:1801.06138 [astro-ph.CO]
5. B. Carr et al., Cosmic conundra explained by thermal history and primordial black holes. Phys. Dark Univ. **31**, 100755 (2021). https://doi.org/10.1016/j.dark.2020.100755. arXiv:1906.08217 [astro-ph.CO]
6. G.F. Chapline, Hadron physics and primordial black holes. Phys. Rev. D **12**, 2949–2954 (1975). https://doi.org/10.1103/PhysRevD.12.2949
7. M. Crawford, D.N. Schramm, Spontaneous generation of density perturbations in the early universe. Nature **298**, 538–540 (1982). https://doi.org/10.1038/298538a0
8. A. Escrivà, E. Bagui, S. Clesse, Simulations of PBH formation at the QCD epoch and comparison with the GWTC-3 catalog. JCAP **05**, 004 (2023). https://doi.org/10.1088/1475-7516/2023/05/004. arXiv: 2209.06196 [astro-ph.CO]
9. G. Franciolini et al., From inflation to black hole mergers and back again: gravitational-wave data-driven constraints on inflationary scenarios with a first-principle model of primordial black holes across the QCD epoch. Phys. Rev. D **106**(12), 123526 (2022). https://doi.org/10.1103/PhysRevD.106.123526. arXiv: 2209.05959 [astro-ph.CO]
10. L.J. Hall, S. Hsu, Cosmological production of black holes. Phys. Rev. Lett. **64**, 2848–2851 (1990). https://doi.org/10.1103/PhysRevLett.64.2848
11. J.I. Juan, Ruling out QCD phase transition as a PBH origin of LIGO/Virgo events. SciPost Phys. Proc. **12**, 052 (2023). https://doi.org/10.21468/SciPostPhysProc.12.052
12. K. Jedamzik, J.C. Niemeyer, Primordial black hole formation during first order phase transitions. Phys. Rev. D **59**, 124014 (1999). https://doi.org/10.1103/PhysRevD.59.124014. arXiv: astro-ph/9901293
13. K. Jedamzik, Could MACHOs be primordial black holes formed during the QCD epoch? Phys. Rept. **307**, 155–162 (1998). https://doi.org/10.1016/S0370-1573(98)00067-2. arXiv: astro-ph/9805147 [astro-ph]
14. K. Jedamzik, Primordial black hole formation during the QCD epoch. Phys. Rev. D **55**, 5871–5875 (1997). https://doi.org/10.1103/PhysRevD.55.R5871. arXiv: astro-ph/9605152
15. J.I. Juan, P.D. Serpico, G.F. Abellán, The QCD phase transition behind a PBH origin of LIGO/Virgo events? JCAP **07**(07), 009 (2022). https://doi.org/10.1088/1475-7516/2022/07/009. arXiv: 2204.07027 [astro-ph.CO]
16. I. Musco, K. Jedamzik, S. Young, Primordial black hole formation during the QCD phase transition: threshold, mass distribution and abundance (2023). arXiv: 2303.07980 [astro-ph.CO]

17. M. Petač, J. Lavalle, K. Jedamzik, Microlensing constraints on clustered primordial black holes. Phys. Rev. D **105**(8), 083520 (2022). https://doi.org/10.1103/PhysRevD.105.083520. arXiv: 2201.02521 [astro-ph.CO]
18. P. Tisserand et al., Limits on the Macho content of the Galactic Halo from the EROS-2 Survey of the magellanic clouds. Astron. Astrophys. **469**, 387–404 (2007). https://doi.org/10.1051/0004-6361:20066017. arXiv: astro-ph/0607207

Chapter 6
Computation of the Abundance of Primordial Black Holes

Sam Young

Abstract An accurate calculation of their abundance is crucial for numerous aspects of cosmology related to primordial black holes (PBHs). For example, placing constraints on the primordial power spectrum from constraints on the abundance of PBHs (or vice-versa), calculating the mass function observable today, or predicting the merger rate of (primordial) black holes observable by gravitational wave observatories such as LIGO, Virgo and KAGRA. In this chapter, we will discuss the different methods used for the calculation of the abundance of PBHs forming from large-amplitude cosmological perturbations, assuming only a minimal understanding of modern cosmology. Different parameters to describe cosmological perturbations will be considered (including different choices for the window function), and it will be argued that the compaction is typically the most appropriate choice. Different methodologies for calculating the abundance and mass function are explained, including *Press-Schechter*-type and peaks theory approaches.

6.1 Introduction

Primordial black holes (PBHs) have a wealth of potential implications for cosmology—from placing unique constraints on the primordial power spectrum, solving numerous unexplained phenomena (such as the early formation of supermassive black holes (SMBHs)), and the observed presence of dark matter in the Universe. Drawing robust conclusions in such scenarios typically relies on a calculation of quantities such as the abundance of PBHs from given initial conditions, the masses at which they form, and their initial clustering. Making this calculation as accurate and reliable as possible is therefore crucial.

In this chapter, we will discuss the calculation of the abundance of PBHs from the collapse of large density fluctuations during radiation domination (as in Chap. 2,

S. Young (✉)
Lorentz Institute for Theoretical Physics, University of Leiden, Niels Bohrweg 2, Leiden 2333, CA, The Netherlands
e-mail: young@lorentz.leidenuniv.nl

C. Byrnes et al. (eds.), *Primordial Black Holes*, Springer Series in Astrophysics and Cosmology, https://doi.org/10.1007/978-981-97-8887-3_6

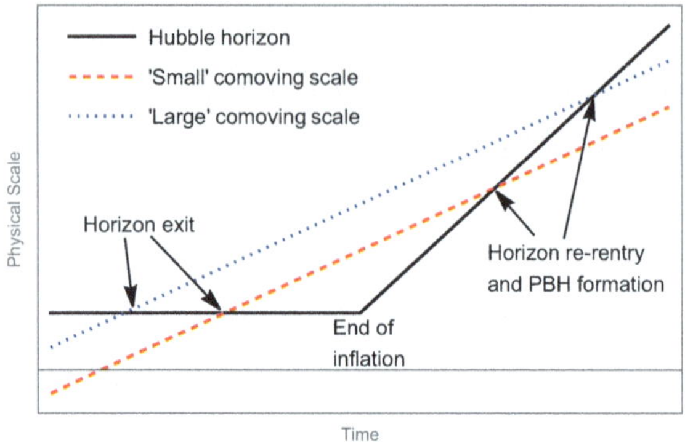

Fig. 6.1 *Horizon exit and re-entry.* The black line shows the Hubble horizon as a function of time, remaining constant during (quasi-)exponential inflation, before growing during the radiation dominated epoch following reheating. The dashed red and dotted blue lines show how perturbations of a small and large scale, respectively, first exit the horizon during inflation before re-entering at a later time. PBHs can form upon horizon re-entry if the amplitude of a perturbation is large enough

see Chap. 9 for alternatives), but let us first turn our attention to the cosmological setting in which such PBHs form. During (exponential) inflation, the physical Hubble horizon is constant, and quantum fluctuations become classicial density perturbations as they exit the horizon as the universe expands. Once inflation ends and the universe becomes radiation dominated, the Hubble horizon starts to grow, and modes which had previously exited the horizon re-enter. Figure 6.1 shows a schematic plot of how perturbations exit and re-enter the horizon. Larger-scale perturbations exit the horizon earlier, and then re-enter the horizon later. Once a perturbation re-enters the horizon, if the amplitude of the perturbation is large enough, gravity can overcome pressure forces, causing it to quickly collapse and form a PBH [5, 18, 40], with a mass of approximately the horizon mass, or else be quickly damped out if the perturbation is not dense enough. This means that larger-scale perturbations correspond to earlier times during inflation, and more massive PBHs which form at later times.

Once an abundance of PBHs has formed, they behave on large scales like a pressure-less fluid, and the PBH density grows inversely proportional to the cube of the scale factor,

$$\rho_{\mathrm{PBH}} \propto a^{-3}, \tag{6.1}$$

i.e. the PBH density on large scales evolves like matter, whilst the density of the radiation fluid grows inversely to the fourth power of the scale factor,

$$\rho_{\mathrm{r}} \propto a^{-4}. \tag{6.2}$$

Although not relevant for our discussion here, the final quantity which makes up a significant fraction of the universe today is dark energy, which we will assume is due to a cosmological constant Λ. The density of the cosmological constant remains constant over time,

$$\rho_\Lambda \propto \text{const}. \tag{6.3}$$

The density parameter Ω_i describes the density of a particular fluid ρ_i relative to the total density ρ_T,

$$\Omega_i = \frac{\rho_i}{\rho_T}. \tag{6.4}$$

Fig. 6.2 shows how the density parameter changes over time for matter, radiation and the cosmological constant. PBHs form very early during radiation domination, at a time when the (relative) density of matter is very low. Because the abundance of PBHs cannot be larger than the abundance of matter, with the exception of very light PBHs which evaporate very quickly, the abundance of PBHs in the universe at the time of formation is typically assumed to be very small: PBHs are extremely rare at the time of formation.

For PBHs to form, significantly larger perturbations are needed than are observed on large scales in e.g. the cosmic microwave background radiation (CMB). Perturbations are required to be approximately 50% more dense than the background [25, 26, 29, 30]. A consequence of this is that the primordial power spectrum must be orders of magnitude larger on small scales than the value of $\sim 2 \times 10^{-9}$ observed in the CMB [1], with the typical values quoted in the literature around $\mathcal{O}(10^{-2})$ (see e.g. [6]). However, there can be significant deviation in this value depending on the method used (e.g. [16]).

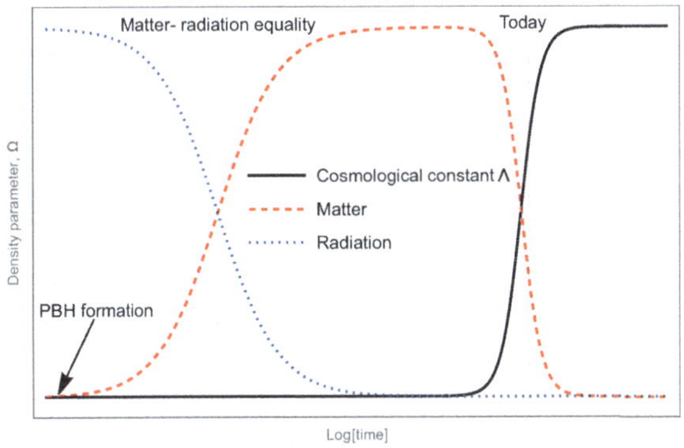

Fig. 6.2 *The evolution of cosmological densities.* The figure show the relative energy density of different species within the universe. Initially, following reheating, the energy density of the universe was dominated by radiation. As the universe expands, the radiation density decreases faster than the matter density, and the universe enter a matter dominated epoch. Finally, the universe becomes dominated by the cosmological constant, whose energy density remains constant. Because PBHs act like matter, and form very early during radiation domination, the fraction of the energy of the universe contained in PBHs is typically expected to be very small (except in models where such PBHs evaporate extremely rapidly)

In this chapter, we will first discuss how cosmological perturbations can be described and how they evolve over time. We will then discuss the typical profile shapes of perturbations, and how this can affect the calculation of the PBH abundance. We will then explain the different methods which can be used to calculate the PBH abundance, beginning with a simple *Press-Schechter*-type approach, before moving on to the theory of peaks and its extensions. Finally, we will consider the effect of non-linearities in the density, which become important as large perturbations are required for PBH formation.

6.2 Describing Cosmological Perturbations

Many models of inflation can predict a large number of PBHs to form in the early universe, including the running mass model [9], axion inflation [3, 31], or a waterfall transition during hybrid inflation [4, 13, 21], and many others. The curvature perturbation ζ is typically used to study cosmological perturbations generated with the different models and to predict their observable consequences (NB. Often $\mathcal{R}$ is used instead to denote the curvature perturbation, whilst ζ is used to describe the curvature perturbation only on a uniform density slicing), and is also what is typically constrained by large-scale observations, such as the *Planck* satellite observations of the CMB [1]. The curvature perturbation ζ appears as a perturbative quantity in the Friedmann-Lemaitre-Robertson-Walker (FLRW) metric in the comoving uniform-density gauge as

$$\mathrm{d}s^2 = -\mathrm{d}t^2 + a^2(t)\exp(2\zeta)\mathrm{d}\mathbf{X}^2, \tag{6.5}$$

where t is time, $a(t)$ is the scale factor and $\mathbf{X}$ represents the three comoving spatial coordinates. To use constraints on PBHs to constrain models of inflation (or to make predictions of PBH abundances from a given model), we therefore need to relate the primordial curvature perturbation power spectrum $\mathcal{P}_\zeta(k)$ to the abundance of PBHs of different masses.

However, the curvature perturbation itself is not a suitable parameter to determine PBH formation. Whilst it is possible to state a threshold value for ζ above which a fluctuation will collapse to form a PBH [33], this assumes that the curvature perturbation goes to zero outside the scale of the perturbation. This is not generally the case: small-scale perturbations will typically be superposed on top of large-scale perturbations. Naively using the curvature perturbation as the formation threshold would therefore suggest that larger-scale perturbations have a significant impact on PBH formation. However, precisely because such perturbations are well outside the horizon at the time of PBH formation, the separate universe approach suggests that it should not affect the local evolution of the universe [34]. This can be seen by separating the curvature perturbation into a small horizon-scale perturbation ζ_s and a large super-horizon perturbation ζ_l, and then treating the large-scale super-horizon mode as a constant:

$$\zeta = \zeta_s + \zeta_l. \tag{6.6}$$

We then consider the term in the FLRW metric where the curvature perturbation appears,

$$a(t)\exp(\zeta) \to a(t)\exp(\zeta_s + \zeta_l) = \left[a(t)\exp(\zeta_l)\right]\exp(\zeta_s). \tag{6.7}$$

If we then consider that ζ_l is (very close to) constant on the horizon scale of ζ_s, we can treat that term as simply a rescaling of the scale factor, corresponding to a time shift t' in the background evolution of the universe,

$$a(t) \to a(t)\exp(\zeta_l) = a(t + t'). \tag{6.8}$$

We conclude that, super-horizon modes in ζ should not affect whether a PBH forms or not, but can have an effect on the time at which a PBH would form.

Seemingly the most obvious parameter to use is then the density contrast δ, defined as

$$\delta(t, \mathbf{x}) \equiv \frac{\rho - \bar{\rho}}{\bar{\rho}} \tag{6.9}$$

which, at first order,[1] is related to ζ as

$$\delta(t, \mathbf{x}) = -\frac{2(1+\omega)}{5+3\omega}\left(\frac{1}{aH}\right)^2 \nabla^2\zeta(\mathbf{x}), \tag{6.10}$$

where $\mathbf{x}$ is the spatial coordinate, ρ is the density, $\bar{\rho}$ is its background value, ω is the equation of state (we will consider that $\omega = 1/3$ during radiation domination for the rest of this chapter), $H(t) = \dot{a}(t)/a(t)$ is the Hubble parameter, and $\nabla^2 = \frac{\partial^2}{\partial x^2} + \frac{\partial^2}{\partial y^2} + \frac{\partial^2}{\partial z^2}$ is the Laplacian. The comoving Hubble horizon is given by $R_{\mathrm{H}} = (aH)^{-1}$. In Fourier space, this relation is given by

$$\delta(t, k) = \frac{2(1+\omega)}{5+3\omega}\left(\frac{k}{aH}\right)^2 \zeta(t, k). \tag{6.11}$$

On super-horizon scales, ζ is constant over time, $\zeta(t, k) \to \zeta(k)$, and so the term $\left(\frac{k}{aH}\right)^2$ contains the time-dependance of the density contrast: perturbations becoming larger over time as the horizon grows, and larger-scale modes are suppressed by a factor k^2.

On sub-horizon scales, it is necessary to account for the evolution of perturbations, which, at linear order, is easily accounted for in Fourier space by including a transfer function $T(R_{\mathrm{H}}, k)$ which includes the time-dependence,

[1] We will consider the effect of non-linearities arising from higher order terms in Sect. 6.5.

$$\delta(R_{\rm H}, k) = \frac{2(1+\omega)}{5+3\omega}\left(\frac{k}{aH}\right)^2 \zeta(k) T(R_{\rm H}, k), \tag{6.12}$$

with the transfer function given by [19],

$$T(R_{\rm H}, k) = 3\frac{\sin\left(\frac{kR_{\rm H}}{\sqrt{3}}\right) - \frac{kR_{\rm H}}{\sqrt{3}}\cos\left(\frac{kR_{\rm H}}{\sqrt{3}}\right)}{\left(\frac{kR_{\rm H}}{\sqrt{3}}\right)^3}, \tag{6.13}$$

where the Hubble horizon scale $R_{\rm H}(t)$ is a function of time. However, it will be more convenient for our purposes to simply use $R_{\rm H}$ itself as the time parameter. On super-horizon scales, $k \ll R_{\rm H}$, and the transfer function approach unity, recovering Eq. (6.11), and it is highly oscillatory on sub-horizon scales.

It is common to use the density contrast as the formation criterion: for any perturbations where the density peaks above some threshold δ_c, it is considered that a PBH will form. However, simulations (see Chap. 4) reveal that the formation (or not) of a PBH is best determined by the average density within the horizon, rather than the peak value (likely at the center of the perturbation) [23], which brings us to the discussion of smoothing functions (also referred to as window functions, or simply windows).

6.2.1 Smoothing Functions

A smoothing function can be considered as an averaging function—it returns the average value of a quantity within a specified volume, centred on a certain coordinate. In order to calculate the smoothed density contrast δ_R, a smoothing function $W(R_s, \mathbf{x})$ is convolved with the density contrast:

$$\delta_R(t, \mathbf{x}, R_s) = \int \mathrm{d}^3\mathbf{y}\,\delta(t, \mathbf{x}-\mathbf{y}) W(R_s, \mathbf{y}), \tag{6.14}$$

where R_s is the smoothing scale, and the integration is performed over all space (or at least, the full extent of the smoothing function). In Fourier space, the density is expressed simply as a multiplication with the smoothing function:

$$\delta_R(t, k) = \delta(t, k)\tilde{W}(k, R_s), \tag{6.15}$$

where the tilde denotes a Fourier transform.

A smoothing function will typically define a volume of space to be integrated over (potentially giving different weight to different regions), divided by the volume. Conceptually, the simplest smoothing function is the (real space) top-hat window function $T_{\rm TH}$,

$$W_{\rm TH}(\mathbf{x}, R_s) = \frac{3}{4\pi R_s^3}\Theta_H(R - x), \tag{6.16}$$

where $\Theta_H(x)$ is the Heaviside theta function, and $x = |\mathbf{x}|$. We can insert this into Eq. (6.14) to obtain

$$\delta_R(t, \mathbf{x}, R_s) = \frac{3}{4\pi R_s^3}\int \mathrm{d}^3\mathbf{y}\delta(t, \mathbf{x} - \mathbf{y})\Theta_H(R - y). \tag{6.17}$$

We further simplify by choosing to center our coordinates on $\mathbf{x}$ and assuming spherical symmetry around this point,

$$\delta_R(t, 0, R_s) = \frac{3}{4\pi R_s^3}\int_0^{R_s} \mathrm{d}\mathbf{r}4\pi r^2\delta(t, r), \tag{6.18}$$

where we have used $\mathrm{d}^3y = 4\pi r^2\mathrm{d}r$, which is valid under spherical symmetry. It is clear, here, that the smoothed value δ_R is simply δ integrated over a sphere of radius R_s, divided by the volume of the sphere i.e. the volume-averaged value.

In addition to the top-hat smoothing function, another common choice is the Gaussian smoothing function,

$$W_G(\mathbf{x}, R_s) = \frac{1}{\left(\pi R_s^2\right)^{3/2}} \exp\left(\frac{-x^2}{R_s^2}\right). \tag{6.19}$$

We note that, often, there is an additional factor of $1/2$ included in the exponent of this function which has been neglected here. This is a free choice, but is made such that the characteristic scales of the Gaussian and top-hat window functions coincide (i.e. the integrand $4\pi r^2 W(r, R_s)$ peaks at R_s in both cases).

The Fourier transforms of these smoothing functions are given by

$$\tilde{W}_{\rm TH}(k, R_s) = 3\frac{\sin(kR_s) - \frac{kR_s}{\sqrt{3}}\cos\left(\frac{kR_s}{\sqrt{3}}\right)}{\left(\frac{kR_s}{\sqrt{3}}\right)^3}, \tag{6.20}$$

$$\tilde{W}_{\rm G}(k, R_s) = \exp\left(\frac{-k^2R_s^2}{4}\right). \tag{6.21}$$

Figure 6.3 shows the real- and Fourier-space transforms of the top-hat and Gaussian smoothing functions. Which smoothing function to use is the subject of ongoing debate, but *a priori* there is currently no reason to choose one smoothing over another—although both presented here have their own advantages and disadvantages which will be discussed later.

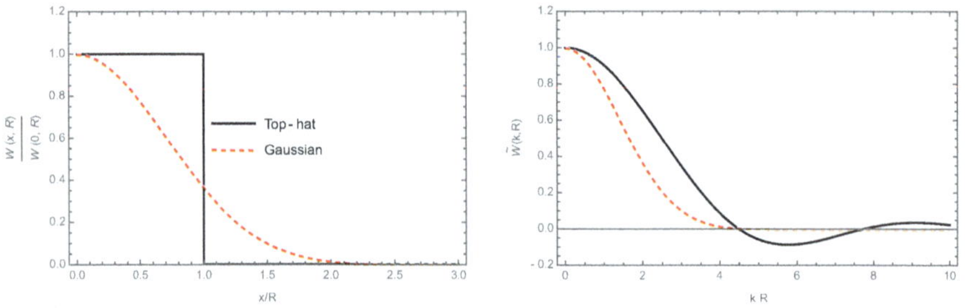

Fig. 6.3 *Smoothing functions.* The shape of top-hat and Gaussian smoothing functions are shown in real-space (left) and Fourier-space (right)

The use of a smoothing function has several benefits compared to using the unsmoothed density contrast. As previously mentioned, the threshold for collapse is more accurately stated in terms of the smoothed density contrast. Whilst the exact value depends on the shape of a the specific perturbation being considered, the critical value varies in the range $0.41 \lesssim \delta_{R,c} \lesssim 0.67$ for extreme shapes of perturbations. However, the critical value for the peak value of the (unsmoothed) density contrast can, in principle vary between positive and negative infinity (e.g. for perturbations with a very sharp peak or trough at the centre) [23]. In addition, smoothing the distribution on different scales means that the abundance of PBHs forming on different scales (and, therefore, of different masses) can be calculated [36].

6.2.2 The Compaction Function

Very closely related to the smoothed density contrast, the compaction function C has, in recent years, become the standard parameter used to describe perturbations, and is defined as

$$C(r, t, \mathbf{x}) = \frac{2\delta M(r, t, \mathbf{x})}{r}, \tag{6.22}$$

where $\delta M(r, \mathbf{x}) = M(r, t, \mathbf{x}) - \bar{M}(r, t)$ is the mass excess within a sphere of areal radius r centred on spatial coordinate $\mathbf{x}$; $M(r, t, \mathrm{x})$ is the Misner-Sharp mass and the bar denotes the background value in a region of unperturbed space. As we will see soon, the compaction is constant on superhorizon scales, and the time dependance is typically dropped. The mass excess δM can be expressed as an integral over the density

$$\delta M(r, t, \mathbf{x}) = \int \mathrm{d}^3\mathbf{y} \left[\rho(t, \mathbf{x} - \mathbf{y}) - \bar{\rho}(t)\right] \Theta_H(r - y). \tag{6.23}$$

Using natural units, $c = G = 1$, and assuming a flat universe, the Friedmann equation gives an expression for the background density,

$$\bar{\rho}(t) = \frac{3\,H^2(t)}{8\pi}. \tag{6.24}$$

Using the definition for the density contrast, Eq. (6.9), allows us to rewrite Eq. (6.23) as

$$\delta M(r, t, \mathbf{x}) = \frac{3H^2}{4\pi} \int \mathrm{d}^3\mathbf{y}\delta(t, \mathbf{x} - \mathbf{y})\Theta_H(r - y). \tag{6.25}$$

The compaction function can then be expressed as,

$$C(r, t, \mathbf{x}) = \frac{r^2}{r_\mathrm{H}^2} \int \mathrm{d}^3\mathbf{y}\delta(t, \mathbf{x} - \mathbf{y})\frac{3}{4\pi r^3}\Theta_H(r - y), \tag{6.26}$$

where $r_\mathrm{H} = H^{-1}$ is the physical Hubble radius. The term inside the integral is the top-hat smoothing function, Eq. (6.16), so finally we can write

$$C_\mathrm{TH}(r, t, \mathbf{x}) = \frac{r^2}{R_\mathrm{H}^2} \int \mathrm{d}^3\mathbf{y}\delta(t, \mathbf{x} - \mathbf{y}) W_\mathrm{TH}(r, \mathbf{y}), \tag{6.27}$$

which is exactly the same as the smoothed density contrast, Eq. (6.14), where r has taken the role of the smoothing scale, and multiplied by r^2/r_H^2. The compaction function can also be defined using a Gaussian smoothing:

$$C_\mathrm{G}(r, t, \mathbf{x}) = \frac{r^2}{R_\mathrm{H}^2} \int \mathrm{d}^3\mathbf{y}\delta(t, \mathbf{x} - \mathbf{y}) W_\mathrm{G}(r, \mathbf{y}). \tag{6.28}$$

The compaction function is therefore extremely closely related to the smoothed density contrast, and indeed, the two are often used interchangeably in the literature (see e.g. reference [36]). This is especially true as the smoothed density contrast is often evaluated at horizon entry, such that $r/r_\mathrm{H} = 1$. However, the compaction is typically preferred as it is conserved on super-horizon scales. This can be seen by inserting Eq. (6.9), and replacing the physical radius $r = aR$ with the comoving radius R:

$$C(R, \mathbf{x}) = -R^2\frac{2(1+\omega)}{5+3\omega} \int \mathrm{d}^3\mathbf{y}\nabla^2\zeta(\mathbf{x} - \mathbf{y}) W(R, \mathbf{y}), \tag{6.29}$$

where the time-dependance has been dropped, because ζ is conserved on super-horizon scales.

It is typically simpler to leave this integral as it appears, as the expressions in Fourier space are quite simple (e.g. Eq. (6.15)). However, it will be useful later if we perform the integration analytically for the case of the top-hat smoothing, which can be completed under the assumption of spherical symmetry. This assumption is typically justified because, as discussed in Sect. 6.1, PBHs are extremely rare at the time of formation—which therefore means that the perturbations from which they form are very rare, due to the large amplitude required. The theory of peaks tells us that, the larger a perturbation is, the more spherical symmetry it will have [2].[2] Under the assumption of spherical symmetry, the Laplacian $\nabla^2\zeta$ becomes

$$\nabla^2\zeta = \zeta''(r) + \frac{2}{r}\zeta'(r), \tag{6.30}$$

where r is here the radial coordinate, and the prime denotes a derivative with respect to r. The compaction can then be expressed as

$$C(R) = -\frac{2(1+\omega)}{5+3\omega}\frac{3}{R}\int_0^R drr^2\left[\zeta''(r) + \frac{2}{r}\zeta'(r)\right] = -\frac{6(1+\omega)}{5+3\omega}R\zeta'(R). \tag{6.31}$$

To avoid confusion, we stress here that, the ζ' term has arisen as the surface term of an integral over a spherical volume—therefore corresponding to ζ' smoothed over a spherical shell.

6.3 Perturbation Profiles

The single most important parameter when considering the abundance of PBHs which will form from the collapse of cosmological perturbations is the threshold amplitude of the perturbations required for collapse, C_c. Carr's initial order-of-magnitude estimate, calculated using a Jeans' Length argument, was that the threshold for collapse is equal to the equation of state, $C_c \sim \omega$ [5, 7], which is equal to $1/3$ during radiation domination. Since then, many more complex methods have been developed, utilising either analytic methods (e.g. [11, 17]), or numerical simulations (e.g. [10, 12, 25, 26, 28, 32]). The precise value depends on the initial profile shape of the perturbation which forms a PBH, which is a subject of immense scrutiny. In this section, we will briefly turn our attention towards the profile shape one might therefore expect for cosmological perturbations.

We will first consider how a single perturbation may be described using the various parameters discussed in Sect. 6.2, and we will consider only the initial perturbation on super-horizon scales, before any complicated evolution close to horizon entry. We therefore set the transfer function equal to unity, $T(t,k) = 1$. We will take a spherically symmetric perturbation in ζ in an otherwise empty universe, given by a Gaussian profile:

$$\zeta(r) = A\exp\left(-\frac{r^2}{r_p^2}\right), \tag{6.32}$$

[2] However, we note that it has been pointed out that, whilst the perturbations to the density (or compaction) are large and rare, this does not necessarily correspond to large, rare perturbations in ζ [35].

where A describes the amplitude of the perturbation, r_p determines the scale of the perturbation, and r is the comoving radial coordinate. The same perturbation can be described in terms of the density contrast δ and results in the ubiquitous Mexican-hat profile,

$$\delta(r) = -\frac{4(1+\omega)}{5+3\omega}\left(\frac{1}{aH}\right)^2 A\frac{\exp\left(-\frac{r^2}{r_p^2}\right)\left(2r^2-3r_p^2\right)}{r_p^4}, \tag{6.33}$$

as well as the compaction C (for a top-hat smoothing),

$$C(r) = -\frac{12(1+\omega)}{5+3\omega}A\frac{r^2}{r_p^2}\exp\left(-\frac{r^2}{r_p^2}\right). \tag{6.34}$$

Figure 6.4 shows the profile shapes in each of these parameters, where we have made the following choices, $\left[A, r_p, (aH)^{-1}\right] = [0.5, 1, 1]$. A is chosen to correspond approximately to the amplitude for PBH formation, whilst the choice of r_p is arbitrary, and $(aH)^{-1}$ is chosen so that the amplitude of the density is similar to the others (technically, in the super-horizon limit, we should have chosen $(aH)^{-1} \ll r_p$). For ζ and δ, the profile simply represents the value of that parameter at the coordinate r, and this is also often considered the case for C. However, it is more accurate to consider the shown profile which corresponds to the value of the compaction at the centre, when a smoothing scale $R_s = r$ has been used.

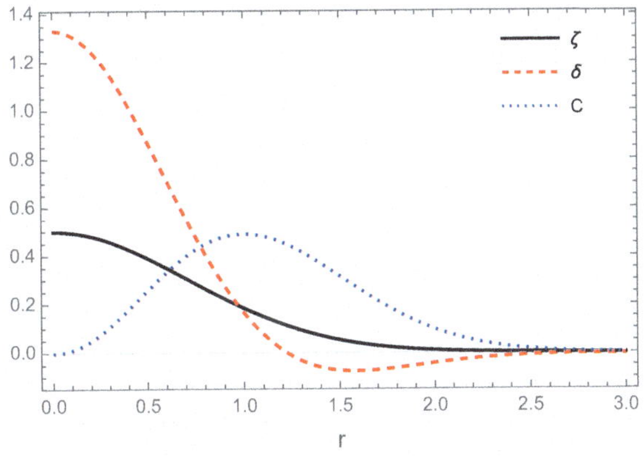

Fig. 6.4 *Parameterising a perturbation.* The same perturbation is shown, described with several different parameters: the curvature perturbation ζ, the density contrast δ and the compaction C. For ζ and δ the plot shows the value at the radial coordinate r. For the compaction, the plot shows the amplitude of the compaction at the centre of the perturbation, when smoothed with a smoothing scale r

For any profile shape, the compaction is generically zero at $r \to 0$ and $r \to \infty$, and peaks at some value of r. The value at which this occurs is used to define the characteristic scale r_m of the perturbation: $C'(r_m) \equiv 0$, and can be found by solving

$$\zeta'(r_m) + r_m \zeta''(r_m) = 0. \tag{6.35}$$

The characteristic scale r_m will be important later as it corresponds to the scale (and mass) at which a PBH forms. In this case, with the profile given in Eq. (6.32), we find that r_m coincides with our choice of parameter for the scale, $r_m = r_p$.

We will now turn our attention to the consideration of what profile shapes we might expect to see in the density in different scenarios—as this will affect the amplitude a perturbation needs to have in order to form a PBH. The expected profile shape can be calculated by considering the amplitude of different Fourier modes in the universe [2], which is described by the power spectrum. The power spectrum $P_\zeta(k)$ is the Fourier transform of the 2-point correlation function,

$$\langle \zeta(\mathbf{k}_1)\zeta(\mathbf{k}_2)\rangle = (2\pi)^3 \delta_D^{(3)}(\mathbf{k}_1 + \mathbf{k}_2) P_\zeta(k), \tag{6.36}$$

where $\delta_D^{(3)}$ is the (3D) Dirac-delta function. We will consider the dimensionless power spectrum $\mathcal{P}_\zeta(k)$, given by

$$\mathcal{P}_\zeta(k) = \frac{k^3}{2\pi^2} P_\zeta(k). \tag{6.37}$$

The effect of the power spectrum on perturbation profiles is considered in detail in the context of PBH formation by references [14, 36]. In this case, we will consider density profiles, and so consider the density power spectrum, $\mathcal{P}_\delta(k)$, which is related to the curvature perturbation power spectrum, $\mathcal{P}_\zeta(k)$, as

$$\mathcal{P}_\delta(R_{\mathrm{H}}, k) = \frac{4(1+\omega)^2}{(5+3\omega)^2} (k R_{\mathrm{H}})^4 \mathcal{P}_\zeta(k) T^2(R_{\mathrm{H}}, k), \tag{6.38}$$

where, again, we use the comoving Hubble horizon R_{H} as the parameter for time, and will here consider the initial perturbations long before horizon entry, such that $k \ll R_{\mathrm{H}}$, and $T(R_{\mathrm{H}}, k) \approx 1$.

Assuming spherical symmetry, the profile shape close to the centre of a perturbation can be stated in terms of the radial coordinate r,

$$\delta(R_{\mathrm{H}}, r) = \delta(R_{\mathrm{H}}, 0) \frac{\xi(R_{\mathrm{H}}, r)}{\xi(R_{\mathrm{H}}, 0)}, \tag{6.39}$$

where $\delta(R_{\mathrm{H}}, 0)$ is the amplitude at the centre of the perturbation. The function $\xi(R_{\mathrm{H}}, r)$, which describes the average profile shape, is given by

$$\xi(R_{\mathrm{H}}, t) = \frac{1}{(2\pi)^3} \int \frac{\mathrm{d}k}{k} \frac{\sin(kr)}{kr} \mathcal{P}_\delta(R_{\mathrm{H}}, k), \tag{6.40}$$

where Eq. (6.38) relates the density and curvature power spectra. We will consider several forms for the curvature perturbation power spectrum, which cover a wide range of models in the literature:

- the (Dirac-)delta power spectrum

$$\mathcal{P}_\zeta = \mathcal{A}\delta_{\rm D}(k - k_*), \tag{6.41}$$

  where $\delta_{\rm D}$ is the Dirac-delta function. It is notable that this doesn't correspond to a realistic or physical power spectrum, but it is often used as a reasonable approximation to sufficiently narrow spectra.
- the narrow-peak power spectrum

$$\mathcal{P}_\zeta = \mathcal{A}\exp\left(-\frac{(k - k_*)^2}{2\Delta^2}\right), \tag{6.42}$$

  which contains a Gaussian-shaped peak. It is also possible to utilise a lognormal peak instead.
- The power-law power spectrum

$$\mathcal{P}_\zeta = \mathcal{A}\left(\frac{k}{k_*}\right)^{n_s-1}\Theta_H(k - k_{min})\Theta_H(k_* - k), \tag{6.43}$$

  where the Heaviside functions provide both a UV and IR cut-off. We will here consider the scale-invariant power spectrum, $n_s = 1$.

Whilst these describe very different power spectra, one thing that they all share in common is a cut-off at some scale in both the UV and IR. Without the UV cut-offs, the integral in Eq. (6.40) diverges—and, as we will see later, the variance of the distribution also diverges. The IR cut-off is not so significant, as the k^4 term when converting to the density power spectrum, as in Eq. (6.38), naturally provides a sharp IR cut-off.

Figure 6.5 shows the resulting density power spectra, using Eq. (6.38), and average density profiles, calculated using Eq. (6.39). We can see that, whilst the curvature perturbation power spectra are very different, the density power spectrum are well characterised by a sharp peak. As a result, the average density profile in each case is also very similar (although note that the characteristic scale r_m of the average profiles does vary slightly, and it has been normalised to 1 in the figure), and matches closely to the Mexican-hat profile already considered, Eq. (6.33). Because of this, the threshold amplitude for collapse also shows little variation, and we can safely use the threshold for the Mexican-hat profile shape, $C_c \approx 0.5$. There is, however, some dependance on the shape of the power spectrum, which is discussed in reference [14].

We note that this critical value corresponds to a top-hat smoothing, and, as we will see later, the choice of smoothing function can have a significant impact on the calculated PBH abundance. If the choice is made to use a Gaussian smoothing function for the calculation, then the critical value for the Gaussian-smoothed compaction should

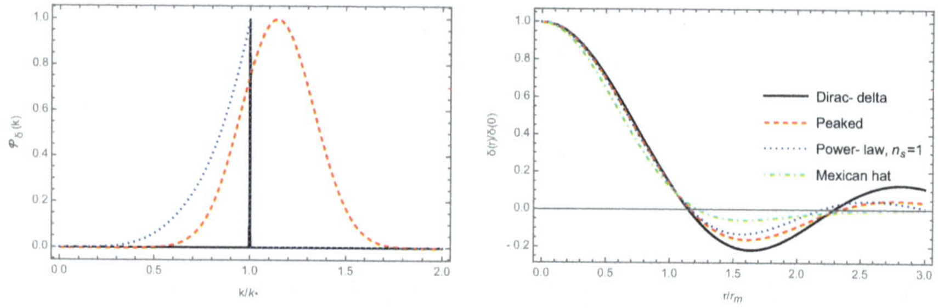

Fig. 6.5 *Power spectra and average profile shape.* The left plot shows several different possible choices for the density power spectrum, normalised to have an amplitude of unity at the peak value. The right plot shows that the average profile associated with such power spectrum, with the Mexican hat profile shape also included with the green dotted line

be used: $C_{G,c} \approx 0.18$. When properly accounted for, this introduces an uncertainty in the amplitude of the power spectrum required to form a given abundance of PBHs of order 10% [36].

6.4 Methods for the Calculation

We will first discuss some generalities with the calculation, before discussing the simplest method, a *Press-Schechter*-type approach, and finally the theory of peaks and its extensions. Throughout this chapter, we will assume a Gaussian distribution for the curvature perturbation ζ. At first-order, the density δ, compaction C and curvature perturbation ζ are linearly related, implying both δ and C also follow a Gaussian distribution. Since we will use C for the calculation, we will consider the probability density function (PDF) of C given by,

$$P(C) = \frac{1}{\sqrt{2\pi\sigma_0^2}} \exp\left(-\frac{C^2}{2\sigma_0^2}\right), \tag{6.44}$$

where σ_0^2 is its variance. The variance can be calculated by taking the zeroth-order (n = 0) moment of the power spectrum,

$$\begin{aligned}\sigma_n^2(R_s, R_\mathrm{H}) &= \int_0^\infty \frac{\mathrm{d}k}{k} k^{2n} \mathcal{P}_C(R_s, R_\mathrm{H}, k) \\ &= \int_0^\infty \frac{\mathrm{d}k}{k} k^{2n} \frac{4(1+\omega)^2}{(5+3\omega)^2} (kR_s)^4 \mathcal{P}_\zeta(k) \tilde{W}^2(R_s, k) T^2(R_\mathrm{H}, k),\end{aligned} \tag{6.45}$$

where R_S is the smoothing scale, and R_H is the Hubble horizon. For the *Press-Schechter*-type approach, we will only need σ_0^2, whilst higher order terms will be needed for the theory of peaks.

Several things are note-worthy at this point. The first is that, whilst PBH formation is an extremely non-linear process, $T(R_H, k)$ is the linear transfer function. It is, therefore, preferable to complete calculations in the super-horizon limit $R_s \gg R_H$, such that $\delta \ll 1$, and $T(R_H, k) \approx 1$. The second is that the integrand contains the factor k^4, which can cause the integral to diverge if there is no UV cut-off sharper than k^{-4}. This is the case for a Gaussian smoothing, Eq. (6.21), which has an exponential cut-off, but not necessarily for the top-hat smoothing, Eq. (6.20), which (though highly oscillatory) drops as k^{-4}. Therefore, when the top-hat function is used, the calculation is often performed at horizon crossing, $R_S = R_H$ on a somewhat *ad hoc* basis, and the effect of the transfer function then guarantees that the integral converges. For further reading on this topic, see more details in [36], and the effect of non-linearities at horizon crossing has been studied in [8].

The initial abundance of PBHs is typically stated in terms of β, the energy fraction of the universe collapsing into black holes at a given time t,

$$\beta(t) = \Omega_{\rm PBH}(t)\big|_{\rm formation} = \left.\frac{\rho_{\rm PBH}}{\rho_r}\right|_{\rm formation}, \tag{6.46}$$

where $\rho_{\rm PBH}$ is the mean energy density of PBHs, ρ_r is the mean radiation density, and we are assuming complete radiation domination. It will be helpful to consider β as the fraction of the universe collapsing to form PBHs at a time t when the Hubble horizon is R_s and the horizon mass is M_H, where these 3 parameters can be used interchangeably. Note that the smoothing scale R_s is used here, rather than the (actual) Hubble horizon R_H at the time of calculation. This is because the calculation typically works by comparing the *initial* amplitude of perturbations to the critical amplitude, long before such perturbations enter the horizon. The scale of the perturbations forming PBHs, corresponding to the smoothing scale, is therefore much larger than the Hubble scale at this time. However, as discussed in the previous paragraph, it is often necessary to set $R_s = R_H$.

Throughout this chapter, we will assume that PBHs form spontaneously upon horizon re-entry. Accounting for the fact that PBHs take some time to form after horizon re-entry would mean a larger value for β (because ρ_r drops faster than $\rho_{\rm PBH}$ as the universe expands), which would then be cancelled out by the relative changes in the PBH and radiation densities as the universe expands when the final PBH abundance is calculated at a later time. For simplicity, this is therefore neglected here, but may be important if one is considering the formation of very light PBHs which evaporate extremely rapidly.

6.4.1 *The* Press-Schechter-*type approach*

The idea behind the *Press-Schechter*-type approach is that we can calculate β, by calculating the fraction of the volume of the universe where the compaction is above the threshold value. This gives us the collapse constraint

$$n(\bar{\nu}) = \delta_{\rm D}\left(\nu - \bar{\nu}\right)\Theta_H(\nu - \nu_c), \tag{6.47}$$

where $\delta_{\rm D}$ is the Dirac-delta function, $\nu_c = C_c/\sigma_0$, and

$$\nu(\mathbf{x}) = \frac{C(\mathbf{x})}{\sigma_0}, \tag{6.48}$$

has unit variance. This returns the number of regions (either 0 or 1) where ν takes a value between $\bar{\nu}$ and $\bar{\nu} + \mathrm{d}\nu$, above the critical value at point $\mathbf{x}$. To calculate β, we will want the expectation value of n:

$$\mathcal{N}(\bar{\nu}) = \langle n\rangle = \int_{-\infty}^{\infty} \mathrm{d}\nu n(\bar{\nu})P(\nu) = \begin{cases} P(\bar{\nu}) & \bar{\nu} \geq \nu_c \\ 0 & \bar{\nu} < \nu_c \end{cases}. \tag{6.49}$$

We can then calculate β as

$$\beta(t) = 2\int_{\nu_c}^{\infty} \mathrm{d}\nu \frac{M_{\rm PBH}}{M_{\rm H}} \frac{1}{\sqrt{2\pi}} \exp\left(-\frac{\nu^2}{2}\right), \tag{6.50}$$

where the fraction $M_{\rm PBH}/M_{\rm H}$ gives the fraction of each Hubble patch which collapses to form a PBH. The time dependance is somewhat hidden here,

The simplest scenario is to assume that PBHs form with a fixed fraction of the horizon mass, $M_{\rm PBH}/M_{\rm H} = \alpha$, whilst simulations of PBH formation show that the PBH mass depends on both the scale and amplitude of the perturbation from which it formed, which will be discussed in Sect. 6.4.1.1. The factor 2 arises as a "fudge-factor" from requiring that, for $\nu_c \ll 1$, all of the energy universe should collapse to form PBHs, $\beta = 1$. Without the addition of the factor of 2, this integral would give $\beta = 0.5$ (for $\alpha = 1$).

For $M_{\rm PBH}/M_{\rm H} = \alpha$, the expression for β is simply the complementary error function erfc,

$$\beta(t) = \alpha\, \mathrm{erfc}\left(\frac{\nu_c}{\sqrt{2}}\right). \tag{6.51}$$

Using the large x limit of erfc(x) gives,

$$\beta(t) \approx \frac{\alpha}{\sqrt{2\pi}\,\nu_c} \exp\left(-\frac{\nu_c^2}{2}\right). \tag{6.52}$$

Whilst not very accurate, this analytic expression does give an order-of-magnitude estimate for the amplitude of the power spectrum required to form a given abundance of PBHs, and illustrate the exponential dependence of β on σ_0 (and thus, on the power spectrum). Because PBHs form from rare perturbations in the extreme tail of the distribution, $\nu_c \gg 1$, small changes in the power spectrum result in large changes to β.

6.4.1.1 The Critical Scaling Relationship

When a perturbation collapses to form a PBH, the final mass depends on how much of the surrounding matter has been pulled in. For this to happen, gravitational attraction towards the centre needs to overcome pressure forces and the initial Hubble expansion of the surrounding material. The larger the density at the centre, the greater the force of gravity, and the more material is pulled into the final PBH. The result is that PBH mass depends on the amplitude of the perturbation from which it formed, as well as the scale. Simulations of PBH formation reveal that the PBH follows a power law known as the critical scaling relation [24, 25, 38],

$$M_{\rm PBH} = \mathcal{K} M_{\rm H} \left(C - C_c\right)^{\gamma} , \tag{6.53}$$

where the exact values depend on the specific profile considered, but we will take $\mathcal{K} = 4$, and $\gamma = 0.36$ as representative values during radiation domination. The horizon mass when the comoving Hubble horizon is R_H, assuming radiation domination until the time of matter-radiation equality, is given by

$$M_{\rm H}(R_{\rm H}) = M_{\rm eq}\left(\frac{g_*}{g_{\star,\rm eq}}\right)^{-1/6}\left(\frac{R_{\rm H}}{R_{\rm eq}}\right)^2 M_{\rm eq} \approx 0.5\left(\frac{g_*}{10.75}\right)^{-1/6}\left(\frac{R_{\rm H}}{10^{-6}{\rm Mpc}}\right)^2 M_\odot, \tag{6.54}$$

where g_* is the number of relativistic degrees of freedom (although the dependance on this is extremely weak and is typically neglected), the subscript eq denotes the value at matter-radiation equality, and $M_\odot$ is a solar mass. For the parameter choices in [27], $M_{\rm eq} = 2.8 \times 10^{17} M_\odot$. Here, we have given the horizon mass $M_{\rm H}$ as a function of the Hubble horizon R_H at a given time, but for our purposes it will be more convenient to consider what the horizon mass will be at the time when the smoothing scale enters the horizon, which is achieved simply by replacing $R_{\rm H}$ with R_s. Therefore, the expression we will use for the horizon mass is therefore

$$M_{\rm H}(M_{\rm H}) \approx 0.5\left(\frac{g_*}{10.75}\right)^{-1/6}\left(\frac{R_s}{10^{-6}{\rm Mpc}}\right)^2 M_\odot. \tag{6.55}$$

Utilising Eq. (6.53), the expression for the initial PBH abundance becomes

$$\beta(t) = 2\int_{\nu_c}^{\infty} \mathrm{d}\nu \mathcal{K}\left(\nu\sigma_0 - \nu_c\sigma_0\right)^{\gamma} \frac{1}{\sqrt{2\pi}} \exp\left(-\frac{\nu^2}{2}\right), \tag{6.56}$$

where the integral is now performed numerically.

Whilst not needed at this point, it will be helpful later to express β as an integral over the PBH mass. Equations (6.53) and (6.48) can be used to express ν as a function of the PBH mass

$$\nu(M_{\mathrm{PBH}}) = \frac{C(M_{\mathrm{PBH}})}{\sigma_0} = \frac{1}{\sigma_0}\left(\frac{M_{\mathrm{PBH}}}{\mathcal{K}M_{\mathrm{H}}}\right)^{1/\gamma} + \nu_c, \tag{6.57}$$

and differentiated to give

$$\frac{\mathrm{d}\nu}{\mathrm{d}M_{\mathrm{PBH}}} = \frac{1}{\gamma\mathcal{K}M_{\mathrm{H}}\sigma_0}\left(\frac{M_{\mathrm{PBH}}}{\mathcal{K}M_{\mathrm{H}}}\right)^{\frac{1}{\gamma}-1}. \tag{6.58}$$

Performing a change of variable, we can then express β as

$$\beta(t) = 2\int_{0}^{\infty} \mathrm{d}M_{\mathrm{PBH}}\left[\frac{1}{\gamma\mathcal{K}M_{\mathrm{H}}\sigma_0}\left(\frac{M_{\mathrm{PBH}}}{\mathcal{K}M_{\mathrm{H}}}\right)^{\frac{1}{\gamma}-1}\right]\frac{M_{\mathrm{PBH}}}{M_{\mathrm{H}}}\frac{1}{\sqrt{2\pi}} \exp\left(-\frac{\nu^2(M_{\mathrm{PBH}})}{2}\right). \tag{6.59}$$

Including the effect of the critical scaling relationship means that, even when only considering perturbations of a single scale (i.e. PBHs forming at a single time), PBHs form with a range of masses. It also means that, in addition to affecting the total abundance of PBHs which form, the amplitude of the power spectrum also affects the masses at which they form. Figure 6.6 shows the abundance of PBH forming at different scales, normalised by their overall abundance, with values for the variance $\sigma_0^2 = \{0.003, 0.0065, 0.01\}$. These were chosen to correspond approximately with typical values for the abundance, and give $\beta \approx \{4\times10^{-20}, 4\times10^{-10}, 5\times10^{-7}\}$. For larger σ_0^2, the distribution peaks at larger masses—although still very similar to the horizon mass. NB. This plot is independent of the shape of the power spectrum. In the *Press-Schechter* formalism, the power spectrum only enters into the calculation via the σ_0^2 term—which we have here simply chosen values for.

6.4.1.2 The Total Primordial Black Hole Abundance

We have so far considered the abundance of PBHs at a single time of formation, β. In order to calculate the total abundance of PBHs at a later time (e.g. today), we need to sum the amount of PBHs forming at all times, and account for the evolution of their density. Assuming that the universe behaves purely as radiation domination until the

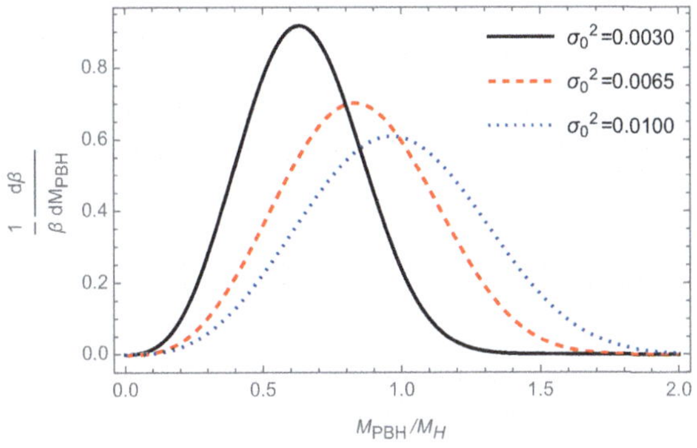

Fig. 6.6 *Critical scaling and extended mass functions.* The figure shows the masses of PBHs forming at a specific time (or, equivalently, at a specific scale), normalised by the total abundance. The critical scaling relationship between the PBH mass and the amplitude of perturbations means that PBHs form at a range of masses, even when they only form at one specific time. The larger the amplitude of perturbations, the larger the mass of the PBHs which form

time of matter-radiation equality, then we can express the abundance of PBHs which formed when the horizon mass was M_H as a function of β,

$$\beta_\mathrm{eq}(M_\mathrm{H}) = \frac{a_\mathrm{eq}}{a(M_\mathrm{H})}\beta(M_\mathrm{H}) = \left(\frac{M_\mathrm{eq}}{M_\mathrm{H}}\right)^{1/2}\beta(M_\mathrm{H}). \tag{6.60}$$

The total PBH abundance at matter-radiation equality is therefore found by integrating over all times (scales) at which PBHs form,

$$\Omega_\mathrm{PBH} = \int_{M_\mathrm{min}}^{M_\mathrm{max}} \frac{\mathrm{d}M_\mathrm{H}}{M_\mathrm{H}}\left(\frac{M_\mathrm{eq}}{M_\mathrm{H}}\right)^{1/2}\beta(M_\mathrm{H}). \tag{6.61}$$

The integral should technically be performed over all scales, but in practice it is only necessary between the minimum and maximum scales being considered. Typically, this range is set by the power spectrum under consideration, where PBH formation occurs over a limited range of scales near the peak of the power spectrum (or, between UV and IR cut-offs).

The total PBH abundance is often also expressed in terms of the fraction of (cold) dark matter composed of PBHs

$$f_\mathrm{PBH} = \frac{\Omega_\mathrm{PBH}}{\Omega_\mathrm{CDM}}, \tag{6.62}$$

where Ω_CDM is the dark matter density parameter. This is also useful because it is a time-independent quantity (after PBHs have stopped forming).

6.4.1.3 The Mass Function

In addition to the total abundance, a key potential observable of PBHs is the mass function $\psi(M_{\rm PBH})$, which describes how many PBHs there are at different masses. There are several different ways in which this can be parameterised, but we will use the definition,

$$\psi(M_{\rm PBH}) = \frac{1}{\Omega_{\rm CDM}} \frac{{\rm d}\Omega_{\rm PBH}}{{\rm d}(\ln M_{\rm PBH})} = M_{\rm PBH} \frac{{\rm d}f_{\rm PBH}}{{\rm d}M_{\rm PBH}}, \tag{6.63}$$

which satisfies the normalisation condition,

$$\int {\rm d}(\ln M_{\rm PBH}) \psi(M_{\rm PBH}) = f_{\rm PBH}. \tag{6.64}$$

Using Eqs. (6.59), (6.61) and (6.62) gives us the final expression for the mass function

$$\psi(M_{\rm PBH}) = \frac{1}{\Omega_{\rm CDM}} \int_{M_{\rm min}}^{M_{\rm max}} \frac{{\rm d}M_{\rm H}}{M_{\rm H}} \left(\frac{M_{\rm eq}}{M_{\rm H}}\right)^{1/2} \frac{2}{\gamma \mathcal{K} M_{\rm H} \sigma_0} \left(\frac{M_{\rm PBH}}{\mathcal{K} M_{\rm H}}\right)^{\frac{1}{\gamma}-1} \times \frac{M_{\rm PBH}}{M_{\rm H}} \frac{1}{\sqrt{2\pi}} \exp\left(-\frac{\nu^2(M_{\rm PBH})}{2}\right), \tag{6.65}$$

where the expression for $\nu(M_{\rm PBH})$ is given in Eq. (6.57). Recall that the variance σ_0^2 is a function of the smoothing scale, which gives the integrand a dependence on the horizon mass $M_{\rm H}$.

6.4.2 *The Theory of Peaks*

The theory of peaks, developed by Bardeen, Bond, Kaiser, and Szalay (BBKS) [2], extends upon the *Press-Schechter* formalism by stating that compact objects (PBHs in this case) will form at peaks in the density above some threshold value. Our collapse constraint becomes

$$n(\mathbf{x}) = \delta_D^{(3)}(\eta) \Theta_H(\lambda_3) \delta_{\rm D}^{(3)}(\nu - \bar{\nu})\, \Theta_H(\nu - \nu_c), \tag{6.66}$$

where the index (3) indicates a 3-dimensional function, σ_n are moments of the power spectrum given by Eq. (6.45), η is given by

$$\eta(\mathbf{x}) = \frac{1}{\sigma_1} \nabla C(\mathbf{x}), \tag{6.67}$$

ζ_{ij} is given by

$$\zeta_{ij}(\mathbf{x}) = \frac{1}{\sigma_2}\frac{\partial^2 C((x))}{\partial x_i \partial x_j}, \tag{6.68}$$

and finally, λ_3 is the smallest eigenvalue of $-\zeta_{ij}$. The collapse constraint $n(\mathbf{x})$ returns the number of peaks (either 0 or 1) of height between $\bar{\nu}$ and $\bar{\nu} + \mathrm{d}\nu$, above the critical value ν_c in the infinitesimal volume d^3x, divided by d^3x. This is analogous to the collapse constraint used in the *Press-Schechter*-type approach, where the additional $\delta_D^{(3)}(\eta)$ term constrains the spatial derivative of C to be zero, whilst the $\Theta_H(\lambda_3)$ term specifies peaks rather than troughs.

The collapse constraint depends on the spatial coordinate $\mathbf{x}$ only through a set of continuous Gaussian fields evaluated at the same position, which we will denote $\mathbf{y}$,

$$\mathbf{y} = \{\nu, \eta_i, \zeta_{ij}\}. \tag{6.69}$$

In order to calculate the abundance of PBHs, we need to calculate the average value for n, calculated by integrating over the PDF $P(\mathbf{y})$

$$\mathcal{N} = \int \mathrm{d}\mathbf{y}\frac{\sigma_2^3}{\sigma_1^3}\left|\det\zeta_{ij}\right| nP(\mathbf{y}). \tag{6.70}$$

The term $\frac{\sigma_2^3}{\sigma_1^3}\left|\det\zeta_{ij}\right|$ is included to account for the transformation from the Gaussian variables of the constraint to the physical coordinates $\mathbf{x}$. Whilst not straightforward to calculate, the integral gives the result

$$\begin{aligned}\mathcal{N}(\nu) = {} & \frac{\sigma_2^3}{3^{3/2}(2\pi)^2\sigma_1^3}\exp\left(-\frac{\nu^2}{2}\right)\Theta_H(\nu-\nu_c)\\ & \times\int_0^\infty \mathrm{d}u f(u)\frac{1}{\sqrt{2\pi(1-\gamma_{02}^2)}}\exp\left(-\frac{(u-\gamma_{02}\nu)^2}{2(1-\gamma_{02}^2)}\right),\end{aligned} \tag{6.71}$$

where $\gamma_{02} = \sigma_1^2/\sigma_0\sigma_2$ is the correlation function, and the function $f(u)$ is given by

$$\begin{aligned}f(u) = {} & \sqrt{\frac{2}{5\pi}}\left[\left(\frac{u^2}{2}-\frac{8}{5}\right)\exp\left(-\frac{5u}{2}\right)+\left(\frac{31u^2}{4}+\frac{8}{5}\right)\exp\left(-\frac{5u}{8}\right)\right]\\ & +\frac{1}{2}\left(u^3-3u\right)\left[\mathrm{erf}\left(\sqrt{\frac{5}{2}}u\right)+\mathrm{erf}\left(\sqrt{\frac{5}{2}}\frac{u}{2}\right)\right],\end{aligned} \tag{6.72}$$

where $\mathrm{erf}(x)$ is the error function

$$\text{erf}(x) = \frac{2}{\pi} \int_0^x \text{d}t \exp(-t^2). \tag{6.73}$$

For the full derivation, the interested reader is directed towards the original BBKS paper [2], as well as section II and III of [20].

This expression is somewhat complicated, but fortunately can be greatly simplified when one considers that PBHs form from high peaks, and one can then make the assumption $\upsilon \gg 1$, which is known as the high-peak limit. The integrand appearing in Eq. (6.71) is the function $f(u)$ multiplied by a Gaussian PDF with mean $\gamma_{02}\nu$ and variance $(1-\gamma_{02}^2)$, where we generically have $\gamma_{02}\nu \gg (1-\gamma_{02}^2)$. This means that the PDF can be well approximated by a Dirac-delta function, $\delta_D(u-\gamma_{02}\nu)$. Under this approximation, the integral is then equal to $f(\gamma_{02}\nu)$, with $\gamma_{02}\nu \gg 1$. The exponential terms in Eq. (6.72) are then very small, whilst the erf are close to unity. We then have

$$f(u) \approx u^3 - 3u \approx u^3. \tag{6.74}$$

Inserting this into Eq. (6.71) gives the us the well-known (and much simpler) number density of peaks in the high-peak limit:

$$\mathcal{N}_{hi-pk}(\nu) = \frac{1}{3^{3/2}(2\pi)^2} \frac{\sigma_1^3}{\sigma_0^3} \nu^3 \exp\left(-\frac{\nu^2}{2}\right). \tag{6.75}$$

From this expression, we can easily see that the factor σ_1^3/σ_0^3 gives the expression in units of $k^3 \propto 1/\text{volume}$, since $\mathcal{N}$ is a number density (rather than simply a number, as in Sect. 6.4.1).

We can now proceed with the calculation of the PBH abundance, which proceeds in much the same way as before, except that, since $\mathcal{N}$ now gives a number density, we now need to multiply by a volume to find the fraction of the universe collapsing into PBHs at a given time (or equivelently, scale or horizon mass):

$$\beta(R_s) = V(R_s) \int_{\nu_c}^{\infty} \text{d}\nu \frac{M_{\text{PBH}}}{M_{\text{H}}} \frac{1}{3^{3/2}(2\pi)^2} \frac{\sigma_1^3}{\sigma_0^3} \nu^3 \exp\left(-\frac{\nu^2}{2}\right), \tag{6.76}$$

where, again, the critical scaling relationship, Eq. (6.53), can be used to relate the PBH mass to the amplitude of the perturbation which formed it. The volume term $V(R_s)$ is the volume of the window function being used,

$$V(R_s) = \begin{cases} (2\pi)^{3/2} R_s^3 & \text{Gaussian smoothing,} \\ \frac{4\pi}{3} R_s^3 & \text{Top}-\text{hat smoothing.} \end{cases} \tag{6.77}$$

The total abundance and mass function can then be calculated using the same method as in Sect. 6.4.1.

It is noteworthy that the dominant term in the expression for β remains the unchanged exponential term. However, using the theory of peaks, rather than *Press-Schechter* generally predicts a higher abundance of PBHs, due to the additional ν^3 term appearing in the expression. However, when one calculates the amplitude of the power spectrum required to produce a specified PBH abundance, the difference is typically only of order 10%, although significant changes can be seen in the mass function [16, 37].

6.4.3 Extensions to the Theory of Peaks

When studying the large-scale structure of the universe, a similar approach is taken to that considered here. Initial fluctuations in the density of the universe grow over time and provide the seeds for the formation of compact objects (such as galaxies and galaxy clusters, in this case), and one can calculate the later abundance of e.g. galaxies by considering the initial distribution of perturbations and the amplitude required for collapse. A common problem encountered in this case is the cloud-in-cloud phenomenon, which arises when a small-scale compact object is calculated to form inside a large-scale compact object, resulting in double counting. A common method used to solve this problem for large-scale structure is the excursion set approach, which works by adding the stipulation that haloes are only considered to collapse to form a compact object if there are no haloes on larger scales which will form compact objects at that scale.

The excursion set has been considered in the case of PBH formation, and it is found that the cloud-in-cloud phenomenon, whereby a smaller mass PBH would be absorbed by bigger mass PBHs forming later, is basically absent for a Gaussian distribution [22]. This is due to the rarity of PBHs at formation, meaning that only a very small fraction of small mass PBHs form at locations where larger mass PBHs later form.[3]

However, a related problem to the cloud-in-cloud problem can be relevant for PBHs, which is most easily seen if we consider a single perturbation, such as that shown in Fig. 6.4. For a smoothing scale $R_s = r = 0$, the compaction C is also zero, and as the smoothing scale increases, C peaks at the characteristic scale r_m before dropping back to zero. If the peak value of C is above the threshold for collapse, then the perturbation will collapse to form a PBH, with a mass determined by the amplitude of C at the peak and the scale r_m (as in Eq. (6.53)). However, this does mean that the compaction is above the threshold for a range of smoothing scales (unless the peak occurs at exactly $C = C_c$, in which case a vanishingly small PBH forms). The calculations that we have considered thus far would, therefore, consider

[3] However, this may be different for non-Gaussian distributions, which could mean that PBHs form at small masses at a location where there is a large-scale perturbation which will collapse to a large mass PBH (although this does not occur for non-Gaussianities arising as a result of non-linearities of the type considered in Sect. 6.5).

that PBHs form at each of these scales, and sum over the relevant range of scales. However, in reality, only a single PBH would form, corresponding to the scale at which C peaks.

This can be accounted for by extending the theory of peaks to stipulate that PBHs form not only at locations where the compaction peaks, but also at the (smoothing) scale at which the compaction peaks:

$$n(\mathbf{x}) = \delta_D\left(\frac{\partial C}{\partial R}\right)\Theta_H\left(\frac{\partial^2 C}{\partial C^2}\right)\delta_D^{(3)}(\eta)\Theta_H^{(3)}(\lambda_3)\delta_{\rm D}^{(3)}(\nu-\bar{\nu})\,\Theta_H(\nu-\nu_c). \quad (6.78)$$

The calculations necessary were developed simultaneously by Germani & Sheth [15] and Young & Musso [39]. The effect of accounting for the improved the theory of peaks calculated was considered by Gow et al. [16], and it was found that the change to power spectrum amplitude to produce a given PBH abundance only varies by a few percent, and that the calculated mass function is also largely unchanged. For this reason, the extensions to the theory of peaks are typically not included in calculations.

6.5 The Effect of Non-linearities

Until this point, we have been considering a linear relation between the curvature perturbation ζ and the density contrast δ, resulting also in a linear relation to the compaction C. However, for PBHs to form, large amplitude perturbations are needed, and it is therefore necessary to go beyond linear order [38].

The first effect we need to account for is a coordinate transform between the comoving uniform density gauge in which we have defined ζ and the comoving synchronous gauge in which we calculate C and determine PBH formation. We will still consider that PBHs form at high peaks which are spherically symmetric, and therefore only consider the radial coordinate. The radial coordinate $\hat{r}$ in the comoving synchronous gauge is then related to the radial coordinate r appearing in the comoving uniform density gauge by

$$\hat{r} = a(t)\exp\left(\zeta(r)\right)r. \quad (6.79)$$

For small amplitude perturbations, $\hat{r}$ increases monotonically with r, and these are known as type I perturbations. However, as larger ζ is considered, there comes a point where this is no longer true, meaning that increasing coordinate radius r does not coincide with increasing physical distance. Perturbations where this occurs are known as type II perturbations, and their evolution (and their subsequent collapse to PBHs) is not currently well understood. We can find the point at which the switch between type I and II perturbations occurs by searching for a stationary point $\mathrm{d}\hat{r}/\mathrm{d}r = 0$. Stationary points occur when we have

$$-r\zeta'(r) = 1. \quad (6.80)$$

For any given perturbation, this function peaks at the scale r_m (see the discussion in Sect. 6.3), and we therefore have the condition for type II perturbations,

$$-r_m\zeta'(r_m) > 1. \tag{6.81}$$

The second effect we need to consider is the non-linear relationship between ζ and δ, and we will again assume that ζ follows a Gaussian distribution. In the super-horizon regime, δ can be expressed as a function of ζ as

$$\delta(\hat{r}) = -\frac{4(1+\omega)}{5+3\omega}\left(\frac{1}{aH}\right)^2 \exp\left(-\frac{5\zeta(r)}{2}\right)\nabla^2 \exp\left(\frac{\zeta(r)}{2}\right), \tag{6.82}$$

where $\hat{r}$ is the radial coordinate in the comoving synchronous gauge, and in spherical symmetry

$$\nabla^2 = \frac{\partial^2}{\partial r^2} + \frac{2}{r}\frac{\partial}{\partial r}. \tag{6.83}$$

We note that Eq. (6.82) is exact to all orders in ζ, but is correct to second order in $\epsilon = 1/HL$, where L is the length scale characterizing the perturbation, and this expression is therefore valid in the super-horizon regime.

Combining these equations and inserting them into the equation for the compaction with a top-hat window function, Eq. (6.27) gives

$$C(r) = -\frac{2}{ae^{\zeta(r)}r}\frac{3H^2}{8\pi}\int_0^r \mathrm{d}\left(ae^{\zeta(r)}r\right)\left[4\pi\left(ae^{\zeta(r)}r\right)^2\right]$$
$$\times \frac{2(1+\omega)}{5+3\omega}\left(\frac{1}{aH}\right)^2 e^{-2\zeta(r)}\left[\zeta''(r) + \frac{2}{r}\zeta'(r) + \frac{1}{2}\left(\zeta'(r)\right)^2\right], \tag{6.84}$$

where the prime denotes a derivative with respect to r. Fortunately, this complicated looking integral has a simple analytic solution:

$$C(r) = -\frac{3(1+\omega)}{5+3\omega}\left[2r\zeta'(r) - \left(r\zeta'(r)\right)^2\right]. \tag{6.85}$$

The linear component of this expression,

$$C_1(r) = -\frac{6(1+\omega)}{5+3\omega}r\zeta'(r), \tag{6.86}$$

is exactly the same as the expression for the compaction seen in Eq. (6.31), and the variance $\langle C_1^2\rangle = \sigma_0^2$ is calculated exactly as before, as in Eq. (6.45). Setting $\omega = 1/3$ during radiation domination, we have a simple quadratic relation between the compaction C and its linear component C_1:

$$C = C_1 - \frac{3}{8}C_1^2, \tag{6.87}$$

which takes a maximum value $C = 2/3$ when $C_1 = -(4/3)r\zeta'(r) = 4/3$. If we consider the characteristic scale r_m, where the compaction peaks, this corresponds to the switch between type I and II perturbations, seen in Eq. (6.81).

For the large, rare perturbations which form PBHs, peaks in the non-linear compaction field will correspond to peaks in the linear compaction field [38], and we can therefore simply apply the same calculation methods we have used before, with some small modifications. We will consider that PBHs form for type I perturbations with $C > C_c$. Solving Eq. (6.87) for C_1 gives

$$C_{1\pm} = \frac{4}{3}\left(1 \pm \sqrt{\frac{2-3C}{2}}\right), \tag{6.88}$$

which gives two solutions for any value $C < 2/3$, where C_{1-} and C_{1+} correspond to type I and II perturbations respectively. Going forwards, we will consider only type I solutions, and will drop the minus sign in the subscript. Taking $C_c \approx 0.5$ (for Mexican hat-type perturbations) means that we will consider PBHs to form in the range,

$$C_{1,c} \approx 0.67 < C < \frac{4}{3}. \tag{6.89}$$

Due to the exponential suppression in the abundance of large peaks, the upper limit will have hardly any effect on the total abundance of PBHs (although we will need to account for it in the calculation nonetheless, to avoid the appearance of imaginary solutions).

Neglecting the upper bound for the moment, therefore, we can obtain a simple estimate for the effect of non-linearities using Eq. (6.52), using $C_{1,c} \approx 0.67$ instead of $C_c = 0.5$. The result is a severe reduction in the number of PBHs forming, typically by numerous orders of magnitude. In order to achieve the same PBH abundance as before, then σ_0^2 and the amplitude of the power spectrum needs to be larger by a factor,

$$\left(\frac{C_{1,c}}{C_c}\right)^2 = \left(\frac{4}{3C_c}\left(1 \pm \sqrt{\frac{2-3C_c}{2}}\right)\right)^2 \approx 1.78, \tag{6.90}$$

where we have used $C_c = 0.5$ in the final approximation on the right-hand side. This means that, when constraints on the PBH abundance are translated into constraints on the primordial power spectrum, they are almost a factor of 2 weaker than if non-linearities were ignored.

The main differences we will need to make in the calculation of the PBH abundance concerns the critical scaling relationship, and the limits of several integrals. Inserting Eq. (6.87) into the equation for the critical scaling relation, Eq. (6.53), gives the PBH mass as a function of C_1

$$M_{\rm PBH} = \mathcal{K} M_{\rm H} \left(C_1 - \frac{3}{8} C_1^2 - C_c \right)^{\gamma}, \tag{6.91}$$

where the values are unchanged (we previously chose $\mathcal{K} = 4$ and $\gamma = 0.36$ as representative values during radiation domination). Because the compaction takes a maximum value C_{max} at $C_{1,max} = 4/3$, we now have a maximum mass PBH which can form for a given horizon mass,

$$M_{\rm PBH,max} = \mathcal{K} M_{\rm H} \left(C_{max} - C_c \right)^{\gamma} \approx 2.1 M_{\rm H}, \tag{6.92}$$

or, equivalently, the minimum horizon mass which can form a PBH of a certain mass,

$$M_{\rm H,min} = \frac{M_{\rm PBH}}{\mathcal{K} \left(C_{max} - C_c \right)^{\gamma}} \approx 0.48 M_{\rm PBH}. \tag{6.93}$$

Inverting Eq. (6.91) gives us ν as a function of the PBH mass,

$$\nu(M_{\rm PBH}) = \frac{C_1(M_{\rm PBH})}{\sigma_0} = \frac{2}{3\sigma_0} \left(2 - \sqrt{4 - 6C_c - 6 \left(\frac{M_{\rm PBH}}{\mathcal{K} M_{\rm H}} \right)^{1/\gamma}} \right), \tag{6.94}$$

where we have kept only the solution for type I perturbations, and σ_n is given by Eq. (6.45). We will also require the derivative,

$$\frac{{\rm d}\nu}{{\rm d}M_{\rm PBH}} = \frac{\sqrt{2} \left(\frac{M_{\rm PBH}}{\mathcal{K} M_{\rm H}} \right)^{\frac{1}{\gamma} - 1}}{\gamma \mathcal{K} M_{\rm H} \sqrt{2 - 3C_c - 3 \left(\frac{M_{\rm PBH}}{\mathcal{K} M_{M_{\rm H}}} \right)^{\frac{1}{\gamma}}}}. \tag{6.95}$$

Using the theory of peaks, the initial abundance of PBHs collapsing to form PBHs is given by,

$$\beta(r) = \frac{4\pi r^3}{3} \int_{\nu_c}^{\frac{4}{3\sigma_0}} {\rm d}\nu \mathcal{K} \left(\nu\sigma_0 - \frac{3}{8} (\nu\sigma_0)^2 - C_c \right)^{\gamma} \frac{1}{3^{3/2}(2\pi)^2} \frac{\sigma_1^3}{\sigma_0^3} \exp\left(-\frac{\nu^2}{2} \right), \tag{6.96}$$

We have here included the upper limit for PBH formation, corresponding to $C_{\rm max} = 4/3$, although this typically has a minimal effect on the PBH abundance.

Again, the total PBH abundance $f_{\rm PBH}$ and the mass function $\psi_{\rm PBH}$ can then be obtained following the same method as described in Sect. 6.4.1;

$$f_{\rm PBH} = \frac{\Omega_{\rm PBH}}{\Omega_{\rm CDM}} = \frac{1}{\Omega_{\rm CDM}} \int_{M_{\rm min}}^{M_{\rm max}} \frac{{\rm d}M_{\rm H}}{M_{\rm H}} \left(\frac{M_{\rm eq}}{M_{\rm H}} \right)^{1/2} \beta(M_{\rm H}), \tag{6.97}$$

where Eq. (6.54) relates the horizon mass $M_{\rm H}$ to the (smoothing) scale r. The integral runs over the entire range of horizon masses for which PBH formation occurs.

The PBH mass function is, likewise, calculated in the same fashion as previously,

$$\psi(M_{\rm PBH}) = M_{\rm PBH}\frac{{\rm d}f_{\rm PBH}}{{\rm d}M_{\rm PBH}} = \frac{M_{\rm PBH}}{\Omega_{\rm CDM}}\int\limits_{M_{\rm H,min}}^{M_{\rm max}}\frac{{\rm d}M_{\rm H}}{M_{\rm H}}\left(\frac{M_{\rm eq}}{M_{\rm H}}\right)^{1/2}$$
$$\times\frac{4\pi r^3(M_{\rm H})}{3}\frac{{\rm d}\nu}{{\rm d}M_{\rm PBH}}\frac{M_{\rm PBH}}{M_{\rm H}}\frac{1}{3^{3/2}(2\pi)^2}\frac{\sigma_1^3}{\sigma_0^3}\exp\left(-\frac{\nu^2(M_{\rm PBH})}{2}\right). \quad (6.98)$$

It is here important to note that we now integrate from the minimum horizon mass which forms a PBH of mass $M_{\rm PBH}$, given by Eq. (6.93). It is now important not to neglect this limit, as the peak height ν becomes imaginary for smaller values.

6.6 Summary

We have discussed how cosmological perturbations collapse to form PBHs at horizon entry, with larger mass PBHs corresponding to larger scale perturbations formed at earlier times during cosmological inflation. Because PBHs form deep in the radiation dominated epoch, and are constrained to be less abundant than dark matter, they are very rare upon formation, making up only a small fraction of the energy content of the universe. The abundance of PBHs depends upon the amplitude of perturbations in the early universe, with larger amplitude perturbations resulting in a higher PBH abundance. This means that the PBH abundance can be calculated from the primordial power spectrum, providing a unique method to constrain the early universe via constraints on the PBH abundance. An accurate calculation of the PBH abundance also means that predictions can be made for other interesting factors, such as the mass function and initial clustering.

There are numerous ways in which perturbations can be parameterised, including the curvature perturbation and the density contrast, but in the context of PBH formation, it is typically most useful to consider perturbations described by the compaction. A perturbation is considered to collapse to form a PBH if the compaction peaks above a certain threshold value, C_c. The compaction function is conserved on super-horizon scales, and the statistics are relatively simple, even when non-linearities are accounted for. However, as perturbations re-enter the horizon, their evolution becomes significantly more complicated—although a linear transfer function is often used to describe the evolution on horizon and sub-horizon scales, the validity of which can be questioned [8]. It is therefore preferable to calculate the PBH abundance in the super-horizon regime by comparing the initial amplitude of perturbations (long before horizon entry) to the critical value. However, depending on the power spectrum being considered, this is not always possible.

The most important parameter for determining the PBH abundance is the threshold value for collapse. The critical value is sufficiently large that no PBHs are expected to form in the universe, if the power spectrum is not significantly larger on small scales than is visible in the CMB. The exact value of the threshold depends upon the profile shape of a given perturbation, but since perturbations are generally expected to be close to the Mexican-hat profile, it is safe to use the critical value corresponding to that profile, $C_c = 0.5$.

Several methods for calculating the PBH abundance and mass function have been described, with the theory of peaks being the most common method. Extensions to the theory of peaks do exist, but these typically do not offer a significant improvement in the accuracy of such calculations, and can be far more complex.

Throughout this chapter, we have assumed an initial Gaussian distribution for the curvature perturbation ζ (see Chap. 7 for an extension to non-Gaussianity), but in Sect. 6.5 the effect of non-linearity between ζ and the density contrast is considered. This non-linearity results in the compaction following a non-Gaussian distribution function, which significantly increases the amplitude of the power spectrum required to form a given abundance of PBHs.

References

1. Y. Akrami et al., Planck 2018 results. X. Constraints on inflation. Astron. Astrophys. **641**, A10 (2020). https://doi.org/10.1051/0004-6361/201833887. arXiv: 1807.06211 [astro-ph.CO]
2. J.M. Bardeen et al., The statistics of peaks of Gaussian random fields. Astrophys. J. **304**, 15–61 (1986). https://doi.org/10.1086/164143
3. E. Bugaev, P. Klimai, Axion inflation with gauge field production and primordial black holes. Phys. Rev. D **90**(10), 103501 (2014). https://doi.org/10.1103/PhysRevD.90.103501. arXiv: 1312.7435 [astro-ph.CO]
4. E. Bugaev, P. Klimai, Formation of primordial black holes from non-Gaussian perturbations produced in a waterfall transition. Phys. Rev. **D85**, 103504 (2012). https://doi.org/10.1103/PhysRevD.85.103504. arXiv: 1112.5601 [astro-ph.CO]
5. B. Carr, The primordial black hole mass spectrum. Astrophys. J. **201**, 1–19 (1975). https://doi.org/10.1086/153853
6. B. Carr et al., Constraints on primordial black holes. Rept. Prog. Phys. **84**(11), 116902 (2021). https://doi.org/10.1088/1361-6633/ac1e31. arXiv: 2002.12778 [astro-ph.CO]
7. B.J. Carr, S.W. Hawking, Black holes in the early universe. Mon. Not. Roy. Astron. Soc. **168**, 399–415 (1974). https://doi.org/10.1093/mnras/168.2.399
8. V. De Luca, A. Kehagias, A. Riotto, How well do we know the primordial black hole abundance: The crucial role of nonlinearities when approaching the horizon. Phys. Rev. D **108**(6), 063531 (2023). https://doi.org/10.1103/PhysRevD.108.063531. arXiv: 2307.13633 [astro-ph.CO]
9. M. Drees, E. Erfani, Running-mass inflation model and primordial black holes. JCAP **1104**, 005 (2011). https://doi.org/10.1088/1475-7516/2011/04/005. arXiv: 1102.2340 [hep-ph]
10. A. Escrivà, Simulation of primordial black hole formation using pseudo-spectral methods. Phys. Dark Univ. **27**, 100466 (2020). https://doi.org/10.1016/j.dark.2020.100466. arXiv: 1907.13065 [gr-qc]

11. A. Escrivà, C. Germani, R.K. Sheth, Analytical thresholds for black hole formation in general cosmological backgrounds. JCAP **01**, 030 (2021). https://doi.org/10.1088/1475-7516/2021/01/030. arXiv: 2007.05564 [gr-qc]
12. A. Escrivà, C. Germani, R.K. Sheth, Universal threshold for primordial black hole formation. Phys. Rev. D **101**(4), 044022 (2020). https://doi.org/10.1103/PhysRevD.101.044022. arXiv: 1907.13311 [gr-qc]
13. J. García-Bellido, A.D. Linde, D. Wands, Density perturbations and black hole formation in hybrid inflation. Phys. Rev. D **54**, 6040–6058 (1996). https://doi.org/10.1103/PhysRevD.54.6040. arXiv: astro-ph/9605094
14. C. Germani, I. Musco, Abundance of primordial black holes depends on the shape of the inflationary power spectrum. Phys. Rev. Lett. **122**(14), 141302 (2019). https://doi.org/10.1103/PhysRevLett.122.141302. arXiv: 1805.04087 [astro-ph.CO]
15. C. Germani, R.K. Sheth, Nonlinear statistics of primordial black holes from Gaussian curvature perturbations. Phys. Rev. D **101**(6), 063520 (2020). https://doi.org/10.1103/PhysRevD.101.063520. arXiv: 1912.07072 [astro-ph.CO]
16. A.D. Gow et al., The power spectrum on small scales: robust constraints and comparing PBH methodologies. JCAP **02**, 002 (2021). https://doi.org/10.1088/1475-7516/2021/02/002. arXiv: 2008.03289 [astro-ph.CO]
17. T. Harada, S. Jhingan, Spherical and nonspherical models of primordial black hole formation: exact solutions. PTEP **2016**(9), 093E04 (2016). https://doi.org/10.1093/ptep/ptw123. arXiv: 1512.08639 [gr-qc]
18. S. Hawking, Gravitationally collapsed objects of very low mass. Mon. Not. Roy. Astron. Soc. **152**, 75 (1971)
19. A.S. Josan, A.M. Green, K.A. Malik, Generalised constraints on the curvature perturbation from primordial black holes. Phys. Rev. D **79**, 103520 (2009). https://doi.org/10.1103/PhysRevD.79.103520. arXiv: 0903.3184 [astro-ph.CO]
20. T. Lazeyras, M. Musso, V. Desjacques, Lagrangian bias of generic large-scale structure tracers. Phys. Rev. D **93**(6), 063007 (2016). https://doi.org/10.1103/PhysRevD.93.063007. arXiv: 1512.05283 [astro-ph.CO]
21. D.H. Lyth, The hybrid inflation waterfall and the primordial curvature perturbation. JCAP **05**, 022 (2012). https://doi.org/10.1088/1475-7516/2012/05/022. arXiv: 1201.4312 [astro-ph.CO]
22. A. Moradinezhad Dizgah, G. Franciolini, A. Riotto, Primordial black holes from broad spectra: abundance and clustering. JCAP **11**, 001 (2019). https://doi.org/10.1088/1475-7516/2019/11/001. arXiv: 1906.08978 [astro-ph.CO]
23. I. Musco, Threshold for primordial black holes: dependence on the shape of the cosmological perturbations. Phys. Rev. D **100**(12), 123524 (2019). https://doi.org/10.1103/PhysRevD.100.123524. arXiv: 1809.02127 [gr-qc]
24. I. Musco, J.C. Miller, Primordial black hole formation in the early universe: critical behaviour and self-similarity. Class. Quant. Grav. **30**, 145009 (2013). https://doi.org/10.1088/0264-9381/30/14/145009. arXiv: 1201.2379 [gr-qc]
25. I. Musco, J.C. Miller, A.G. Polnarev, Primordial black hole formation in the radiative era: investigation of the critical nature of the collapse. Class. Quant. Grav. **26**, 235001 (2009). https://doi.org/10.1088/0264-9381/26/23/235001. arXiv: 0811.1452 [gr-qc]
26. I. Musco, J.C. Miller, L. Rezzolla, Computations of primordial black hole formation. Class. Quant. Grav. **22**, 1405–1424 (2005). https://doi.org/10.1088/0264-9381/22/7/013. arXiv: gr-qc/0412063
27. T. Nakama, J. Silk, M. Kamionkowski, Stochastic gravitational waves associated with the formation of primordial black holes. Phys. Rev. D **95**(4), 043511 (2017). https://doi.org/10.1103/PhysRevD.95.043511. arXiv: 1612.06264 [astro-ph.CO]
28. T. Nakama et al., Identifying the most crucial parameters of the initial curvature profile for primordial black hole formation. JCAP **01**, 037 (2014). https://doi.org/10.1088/1475-7516/2014/01/037. arXiv: 1310.3007 [gr-qc]

29. T. Nakama et al., Investigating formation condition of primordial black holes for generalized initial perturbation profiles, in 10th international symposium on cosmology and particle astrophysics (2014). arXiv: 1401.7740 [gr-qc]
30. T. Nakama et al., Numerical simulation of primordial black hole formation. JPS Conf. Proc. **1**, 013115 (2014). https://doi.org/10.7566/JPSCP.1.013115
31. O. Özsoy et al., Mechanisms for primordial black hole production in string theory. JCAP **07**, 005 (2018). https://doi.org/10.1088/1475-7516/2018/07/005. arXiv: 1803.07626 [hep-th]
32. A.G. Polnarev, I. Musco, Curvature profiles as initial conditions for primordial black hole formation. Class. Quant. Grav. **24**, 1405–1432 (2007). https://doi.org/10.1088/0264-9381/24/6/003. arXiv: gr-qc/0605122
33. M. Shibata, M. Sasaki, Black hole formation in the Friedmann universe: formulation and computation in numerical relativity. Phys. Rev. D **60**, 084002 (1999). https://doi.org/10.1103/PhysRevD.60.084002. arXiv: gr-qc/9905064
34. D. Wands et al., A new approach to the evolution of cosmological perturbations on large scales. Phys. Rev. D **62**, 043527 (2000). https://doi.org/10.1103/PhysRevD.62.043527. arXiv: astro-ph/0003278
35. S. Young, Peaks and primordial black holes: the effect of non-Gaussianity. JCAP **05**(05), 037 (2022). https://doi.org/10.1088/1475-7516/2022/05/037. arXiv: 2201.13345 [astro-ph.CO]
36. S. Young, The primordial black hole formation criterion re-examined: parametrisation, timing and the choice of window function. Int. J. Mod. Phys. D **29**(02), 2030002 (2019). https://doi.org/10.1142/S0218271820300025. arXiv: 1905.01230 [astro-ph.CO]
37. S. Young, C.T. Byrnes, M. Sasaki, Calculating the mass fraction of primordial black holes. JCAP **07**, 045 (2014). https://doi.org/10.1088/1475-7516/2014/07/045. arXiv: 1405.7023 [gr-qc]
38. S. Young, I. Musco, C.T. Byrnes, Primordial black hole formation and abundance: contribution from the non-linear relation between the density and curvature perturbation. JCAP **11**, 012 (2019). https://doi.org/10.1088/1475-7516/2019/11/012. arXiv: 1904.00984 [astro-ph.CO]
39. S. Young, M. Musso. Application of peaks theory to the abundance of primordial black holes. JCAP **11**, 022 (2020). https://doi.org/10.1088/1475-7516/2020/11/022. arXiv: 2001.06469 [astro-ph.CO]
40. Y.B. Zel'dovich, I.D. Novikov, Soviet Astron. **10**, 602 (1967)

Chapter 7
Non-Gaussianities and Primordial Black Holes

Shi Pi

Abstract The most promising mechanism of generating PBHs is by the enhancement of power spectrum of the primordial curvature perturbation, which is usually accompanied by the the enhancement of non-Gaussianity that crucially changes the abundance of PBHs. In this chapter I will discuss how non-Gaussianity is generated in single field inflation as well as in the curvaton scenario, and then discuss how to calculate PBH mass function and induced gravitational waves (GWs) with such non-Gaussianities. When the PBH abundance is fixed, non-Gaussianity only has mild impact on the spectral shape of the induced GWs, which gives relatively robust predictions in mHz and nHz GW experiments.

7.1 Overview

PBHs are formed when the density contrast $\delta\rho/\rho$ exceeds some critical value $\delta_{\rm th} \sim 0.4$ when the perturbation reenter the Hubble horizon [59, 65, 125, 143, 174, 270]. The peak mass of PBHs depends on the horizon mass at the reentry moment of the over-densed region, while the abundance of the PBHs depends on the number of such Hubble patches which can satisfy the condition of PBH formation (like $\delta\rho/\rho > \delta_{\rm th}$). Motivated by the Press-Schechter formalism in calculating the halo mass function in galaxy formation [209], the PBH abundance was calculated by integrating the probability density function (PDF) of the density contrast $\mathbb{P}(\delta\rho/\rho)$ from the threshold. Later, it was realized that the criterion for PBH formation should be set on the compaction function $\mathscr{C}(t,r) = 2G\delta M(t,r)/r$ [229], which is a direct application of the hoop conjecture [175]. The calculation based on

S. Pi (✉)
Institute of Theoretical Physics, Chinese Academy of Sciences, Beijing 100190, China
e-mail: shi.pi@itp.ac.cn

Center for High Energy Physics, Peking University, Beijing 100871, China

Kavli Institute for the Physics and Mathematics of the Universe (WPI), The University of Tokyo, Kashiwa, Chiba 277-8583, Japan

C. Byrnes et al. (eds.), *Primordial Black Holes*, Springer Series in Astrophysics and Cosmology, https://doi.org/10.1007/978-981-97-8887-3_7

the PDF of compaction function $\mathscr{C}$ is called the Press-Schechter-type method, which is reviewed in Chap. 6 under the assumption that the curvature perturbation is Gaussian. By the observation of the cosmic microwave background (CMB) anisotropies and the large scale structure, we already confirm the primordial curvature perturbation on comoving slices, $\mathcal{R}$, is nearly scale invariant and Gaussian, with a power spectrum $\mathcal{P}_\mathcal{R} \sim 10^{-9}$ on scales larger than about 1 Mpc [9]. However, this is far too small to generate PBHs (to play the role say, as microlensing events, black hole merger events, with detectable stochastic GW from the mergers, scalar induced GW, etc.), and a rough estimate show that the power spectrum of the curvature perturbation should be enhanced to $\mathcal{P}_\mathcal{R} \sim 10^{-2}$. If it is still Gaussian, the number of Hubble patches of PBH formation is solely determined by the power spectrum and the threshold of the compaction function,[1] as is shown in Sect. 6.4.1.

A related phenomenon is the stochastic gravitational wave background induced by the enhanced curvature perturbation. In General Relativity, the curvature perturbation $\mathcal{R}$ and tensor perturbation h_{ij} are only coupled at nonlinear order, and the tensor-scalar-scalar type coupling $\sim h_{ij}\partial_i\mathcal{R}\partial_j\mathcal{R}$ can induce second-order GWs [16, 39, 58, 168–170, 190, 196, 197].[2] Therefore, the curvature perturbation can source the GWs after it reenters the Hubble horizon, which is called scalar induced secondary GWs. The peak frequency, as well as the PBH mass generated by such a perturbation, are determined by the horizon scale at the reentry moment, which are connected by [220]

$$f_{\rm IGW} \approx 3{\rm Hz}\left(\frac{M_{\rm PBH}}{10^{16}{\rm g}}\right)^{-\frac{1}{2}}. \tag{7.1}$$

At the peak scale, the spectral energy density of the induced GWs we observe today (normalized by the critical density), is roughly $\Omega_{\rm GW,0} \equiv \rho_{\rm cr}^{-1}\left({\rm d}\rho_{\rm GW}/{\rm d}\ln k\right)_{t_0} \sim 10^{-6}\mathcal{P}_\mathcal{R}^2$, where $\mathcal{P}_\mathcal{R}$ is the primordial power spectrum of $\mathcal{R}$. The nearly scale-invariant power spectrum extrapolated from CMB scales predicts a scale-invariant induced GW with an amplitude of $\Omega_{\rm GW,0} \sim 10^{-24}$, far less than the observational bound [39]. However, for the enhanced curvature perturbation with abundant PBH formation ($\mathcal{P}_\mathcal{R} \sim \mathcal{O}(10^{-2})$), the induced GW is also enhanced to $\Omega_{\rm GW,0} \sim 10^{-10}$, which reaches the detectable range of many experiments like the next-generation ground- and space-borne interferometers and pulsar timing arrays. As both $\Omega_{\rm GW,0}$ and PBH abundance $f_{\rm PBH}$ depend on $\mathcal{P}_\mathcal{R}$, they can be cross-checked to probe the primordial power spectrum on small scales [25, 46, 47, 220, 221]. For more discussion on the induced GWs, see Chap. 17.

To enhance the power spectrum of the curvature perturbation, usually a deviation from the slow-roll attractor is required. Depending on the model, this not only changes the shape and amplitude of the enhanced power spectrum, but also the statistics of

[1] Besides the variance and the threshold, the PBH abundance also depends on the choice of window functions. This is especially important when the power spectrum is broad. We ignore this dependence for simplicity. For more discussions, see Sect. 6.2.1.

[2] For GWs induced by other couplings like scalar-tensor-tensor or tensor-tensor-tensor, see for instance Refs. [35, 67, 207, 268].

the curvature perturbation. For discussion on such inflation models, see Chap. 3. Even a small non-Gaussianity can change the relation between the PBH abundance and induced GW spectrum significantly, as the induced GW depends mainly on the variance of the PDF of the curvature perturbation, while the PBH abundance depends on the high-σ tail of the PDF. A small change in the PDF, i.e. deviation from the Gaussianity, can give rise to a huge enhancement/suppression in the PBH abundance [28, 43, 48, 50, 93, 119, 126, 132, 142, 144, 164, 171, 208, 217, 226, 239, 256, 257, 260, 264, 267], but only changes the amplitude of the induced GW spectrum mildly [2, 4, 56, 112, 113, 188, 246, 247].

On large scales, non-Gaussianity is often described by the perturbative series [148]

$$\mathcal{R} = \mathcal{R}_g + \frac{3}{5} f_{\rm NL} (\mathcal{R}_g^2 - \langle \mathcal{R}_g \rangle^2) + \cdots , \tag{7.2}$$

which was used to describe the curvature perturbation on small scales for PBH formation [17, 28, 50, 105, 199, 238, 262, 264, 267] as well as the induced GWs [56, 112, 188, 246, 247]. However, in many models, the non-linear parameter $f_{\rm NL}$ reaches $O(1)$ [69, 80, 167, 178, 191–193], so the higher orders in (7.2) might be important, which are highly model-dependent. These higher order terms must be taken into account in the PBH formation. Fortunately, by using the δN formalism, a fully nonlinear curvature perturbation can be derived for single field inflation and for the curvaton scenario, of which the PDF of $\mathcal{R}$ can be calculated analytically. Various methods are developed to calculate the PBH abundance for such a PDF, including the Press-Schechter-type formalism [43, 98, 119] and the theory of peaks [27, 30, 114, 120, 144, 217, 239, 258–260, 266]. The Press-Schechter-type formalism for Gaussian curvature perturbation is thoroughly reviewed in Chap. 6, which is straightforward to be further extended to the non-Gaussian case in this chapter.

Considering non-Gaussianity in the PBH formation and induced GW generation is important in many theoretical and observational issues in cosmology.

- PBHs might be the supermassive or stupendously large BHs which seed the galaxy or structure formation [40, 61, 62, 140]. To satisfy the stringent constraints from CMB μ- and y-distortion [145], large non-Gaussianities must be included [29, 41, 51, 62, 117, 127, 159, 186, 187, 189].
- According to the current observational constraints, asteroid-mass PBH can be all the dark matter, of which the accompanying induced GWs are detectable by space-borne interferometers LISA [24, 31, 32, 34], Taiji [216], and TianQin [156]. As positive non-Gaussianity can enhance the PBH abundance greatly, the required power spectrum of the curvature perturbation for $\Omega_{\rm PBH} = \Omega_{\rm DM}$ is reduced with positive non-Gaussianity, and the induced GW spectrum is also reduced, yet it is still above the LISA/Taiji/TianQin sensitivity curve [37, 56, 246]. The large non-Gaussian limit with $f_{\rm NL} \gg 1$ (i.e. the χ^2-distribution) can be realized in the curvaton model when the curvaton is negligible at its decay [36, 88, 98, 205, 224].
- In 2023, the Pulsar Timing Array collaborations NANOGrav [7, 8], EPTA combined with InPTA [20–22], PPTA [214, 215, 272], CPTA [254], as well as IPTA

[6], announced the strong evidence of nHz stochastic GW background, while the induced GW interpretation is favored [5]. However, as the amplitude of the nHz GW spectrum is relatively high ($\sim 10^{-9}$), sub-solar-mass PBHs might be overproduced, i.e. $\Omega_{\rm PBH} > \Omega_{\rm DM}$. This tension can be alleviated by introducing a negative non-Gaussianity with $f_{\rm NL} \lesssim -1$ to suppress the PBH formation [73, 107, 160]. Such a large and negative $f_{\rm NL}$ should be seen as the leading order of a fully non-linear curvature perturbation, which can be realized in the curvaton scenario when the curvaton dominates at its decay moment [68, 98, 205].

- It was shown that PBHs obey a Poisson distribution when the density perturbation is Gaussian [12, 85], while non-Gaussianity can make PBHs cluster initially on small scales [82, 105, 236, 238, 262, 263], which enhances the merger rate of PBHs significantly [75, 213]. Such clustered PBHs can arguably escape from the microlensing detections, which assume Poisson distribution [111].
- When the power spectrum of the curvature perturbation is enhanced on small scales, the loop corrections on CMB scales might be large, which may jeopardize the mechanism of enhancing the power spectrum on small scales to generate PBHs [151, 152]. It was soon realized that the loop correction can be suppressed when the transition to slow-roll inflation is smooth [102–104, 218]. For recent discussions, see [33, 70, 72, 81, 101, 106, 108, 109, 130, 131, 150, 165, 179, 237, 241] and see Chap. 3.

For more discussion on the observational implications, see Part V and the references therein, for instance the review papers [60, 63, 64].

This chapter is organized as follows. In Sect. 7.2 we review the calculation of PBH mass function, emphasizing on when and how the non-Gaussianity comes into play. In Sect. 7.3, we study some typical inflation models which can generate large non-Gaussianity: the ultra-slow-roll inflation, constant-roll inflation with a bumpy potential, single-field inflation with piecewise quadratic potential, and the curvaton scenario. After getting the non-Gaussian curvature perturbations in these models, we calculate the PBH mass function and abundance by the Press-Schechter-type formalism in Sect. 7.4, as well as the scalar-induced GWs in Sect. 7.5. We conclude by discussing some interesting topics to be studied in the future in Sect. 7.6.

7.2 A Brief Review of PBH Formation

7.2.1 *PBH Abundance in the Gaussian Case*

In this section we will briefly review how to calculate the PBH mass function and abundance by the Press-Schechter formalism, emphasizing on where and how the non-linearities and primordial non-Gaussianities enter (see Chaps. 2 and 6). Our starting point is the power spectrum of the primordial curvature perturbation on comoving slices, $\mathcal{P}_{\mathcal{R}}(k)$, calculated at the end of inflation. In the simplest Press-

Schechter formalism, the density contrast on comoving slices is used, which is related to $\mathcal{R}$ by the Poisson equation

$$\delta \equiv \left(\frac{\delta\rho}{\rho_0}\right)_c \approx -\frac{2(1+w)}{5+3w}\frac{1}{H^2a^2}\nabla^2\mathcal{R}. \tag{7.3}$$

In the radiation dominated case ($w = 1/3$), the prefactor is 4/9.

The density contrast threshold $\delta_{\rm th}$ is to determine whether a PBH could form in a specific Hubble patch or not, which was argued by the Jeans instability to be approximately $\delta_{\rm th} \approx w$ [59]. A recent analytical result is the Harada-Yoo-Kohri (HYK) threshold [122]:

$$\delta_{\rm th} = \frac{3(1+w)}{5+3w}\sin^2\left(\frac{\pi\sqrt{w}}{1+3w}\right) \xrightarrow{w\to 1/3} 0.41. \tag{7.4}$$

Given the PDF of the curvature perturbation $\mathbb{P}(\mathcal{R})$, the PDF of the density contrast is given by

$$\mathbb{P}(\delta) = \mathbb{P}(\mathcal{R})\left|\frac{{\rm d}\mathcal{R}}{{\rm d}\delta}\right|, \tag{7.5}$$

where the Jacobian $|{\rm d}\mathcal{R}/{\rm d}\delta|$ can be easily calculated from the linear Poisson equation (7.3). With these quantities, we can calculate the PBH abundance in the following way:

$$\left.\begin{array}{r}\mathcal{R} \xrightarrow{\text{Eq. (7.3)}} \delta \\ \mathbb{P}(\mathcal{R}) \xrightarrow{\text{Eq. (7.5)}} \mathbb{P}(\delta)\end{array}\right\} \xrightarrow[\text{Window function}]{\delta_{\rm th}=0.41(\text{HYK limit})} \beta = \int_{\delta_{\rm th}} \mathbb{P}(\delta)\frac{M(\delta)}{M_H}\,{\rm d}\delta. \tag{7.6}$$

We will not discuss issues about the window function in this chapter. The key point of the above calculation is that, if the curvature perturbation $\mathcal{R}$ is Gaussian, the linear approximation (7.3) tells us the density contrast is also Gaussian. Therefore the integration in (7.6) is simply a complementary error function, which gives a Gaussian suppression $\sim\exp(-\delta_{\rm th}^2/2\sigma_\delta^2)$ in the far tail. See (6.52) in Chap. 6. This simple estimation is frequently quoted in the literature. However, it depends crucially on the Gaussianity of $\mathcal{R}$ and the linear approximation (7.3), both of which are doubtful when $\delta > \delta_{\rm th}$.

7.2.2 *Press-Schechter-Type Formalism*

The estimation (7.6) neglects some important facts in the calculation of PBH abundance, which might change the result significantly.

7.2.2.1 Compaction Function

The criterion of PBH formation is not simply a threshold of the density contrast $\delta\rho/\rho$. As discussed in Chap. 4, numerical studies on the black hole collapse show that the criterion should be put on the averaged mass excess inside a sphere of a fixed areal radius, dubbed the compaction function [229]. In a reduced-circumference polar coordinates, we have[3]

$$\mathscr{C}(t,\bar{r},\mathbf{x}) \equiv \frac{2\delta M(t,\bar{r},\mathbf{x})}{\bar{r}} = \frac{3}{\bar{r}}(aH)^2\int_0^{\bar{r}} \mathrm{d}\varrho\, \varrho^2\delta(t,\varrho,\mathbf{x}), \tag{7.9}$$

where we set $G = 1$. We can see that the compaction function is the averaged density contrast inside a sphere of radius $\bar{r}$, which is conserved on superhorizon scales. For details of how to introduce the compaction function, see Sect. 6.2.2. For the discussion of gauge issue, see [123]. It seems that, if we substitute the linear relation (7.3) into (7.9), we have (See also (6.31))

$$\mathscr{C}_\ell = -\frac{6(1+w)}{5+3w} r\frac{\partial\mathcal{R}}{\partial r} \xrightarrow{w=1/3} -\frac{4}{3}r\frac{\partial\mathcal{R}}{\partial r}. \tag{7.10}$$

However, linear relation (7.3) is only the leading-order approximation of [83, 123, 141, 265]

$$\delta = -\frac{4}{3}\cdot\frac{3(1+w)}{5+3w}\left(\frac{1}{aH}\right)^2 e^{-5\mathcal{R}/2}\nabla^2 e^{\mathcal{R}/2}, \tag{7.11}$$

Therefore we should use

$$\mathscr{C} = \mathscr{C}_\ell\left(1-\frac{5+3w}{12(1+w)}\mathscr{C}_\ell\right) \xrightarrow{w=1/3} \mathscr{C} = \mathscr{C}_\ell - \frac{3}{8}\mathscr{C}_\ell^2. \tag{7.12}$$

Such a nonlinear relation will naturally bring non-Gaussianity to the calculation of PBH formation even if $\mathcal{R}$ is Gaussian. For more details, see Sect. 6.5. From now on, we will focus on radiation-dominated universe with $w = 1/3$.

---

[3] In this chapter, following Refs. [92, 115, 116], we write the spatial metric as

$$\mathrm{d}\Sigma^2 = e^{2\mathcal{R}(r)}\left(\mathrm{d}r^2 + r^2\mathrm{d}\Omega^2\right) = \frac{\mathrm{d}\bar{r}^2}{1-K(\bar{r})\bar{r}^2} + \bar{r}^2\mathrm{d}\Omega^2 \tag{7.7}$$

where $\mathrm{d}\Omega^2 = \mathrm{d}\theta^2 + \sin^2\theta\mathrm{d}\phi^2$ is the two-dimensional line element on a sphere. $\bar{r}$ is called the reduced-circumference radius, which is related to r by the local expansion:

$$\bar{r} = e^{\mathcal{R}(r)}r. \tag{7.8}$$

7.2.2.2 Threshold

Around a density peak, the compaction function $\mathscr{C}(\bar{r})$ is a function of the radius $\bar{r}$. Numerical calculations tell us that whether an overdense region can collapse into a black hole depends on whether the maximum of the compaction function, $\mathscr{C}(\bar{r}_m)$, exceeds the critical value $\mathscr{C}_{\rm th}$, which varies from 0.4 to 0.67, depending on the density profile of the overdensed region [184]. It was further found that only the width of the compaction function peak, parameterized by $q \equiv -\mathscr{C}''(\bar{r}_m)\bar{r}_m^2/(4\mathscr{C}(\bar{r}_m))$, is important [91]. Large q corresponds to a narrow $\mathscr{C}(\bar{r}_m)$, given by a top-hat density profile which drops rapidly around $\bar{r}_m$. Strong pressure gradients at $\bar{r}_m$ can resist the gravitational collapse, thus $\mathscr{C}_{\rm th}$ reaches its maximum 2/3 in this limit, as is considered in Chap. 6. On the contrary, small q corresponds to a broad $\mathscr{C}(\bar{r}_m)$, given by a density profile highly peaked at the center. As the pressure gradients can be neglected, it reduces to the HYK limit $\mathscr{C}_{\rm th} \approx \delta_{\rm th} \approx 0.4$. In principle the density profiles and their statistical properties should be input if we want to calculate PBH abundance appropriately. Some hypothetical profiles are given by fitting the numerical results [91, 184, 261]. In the theory of peaks, the profiles can be derived by the multipole moments of the power spectrum, provided the curvature perturbation is Gaussian [27, 30, 144, 258, 259].

By using (7.12), $\mathscr{C}'(r) = 0$ gives $\mathscr{C}_\ell'(r) = 0$ or $\mathscr{C}_\ell(r) = 4/3$. The former condition determines the extreme of $\mathscr{C}(r)$ at r_m, where its second derivative is

$$\mathscr{C}''(r_m) = \left(1 - \frac{3}{4}\mathscr{C}_\ell(r_m)\right)\mathscr{C}_\ell''(r_m). \tag{7.13}$$

We see that depending on whether $\mathscr{C}_\ell < 4/3$ or $\mathscr{C}_\ell > 4/3$, the maximum $\mathscr{C}(r_m)$ corresponds to the maximum or minimum of $\mathscr{C}_\ell$. Only the former case, dubbed type I fluctuation [149], is consistent with the linear approximation. Type II fluctuations with $\mathscr{C}_\ell > 4/3$ are rarer, which we do not consider in this chapter. Given the profile-dependent threshold on $\mathscr{C}$, the corresponding threshold on the linear compaction function $\mathscr{C}_\ell$ can be solved by (7.12)

$$\mathscr{C}_{\ell,\rm th} = \frac{4}{3}\left(1 - \sqrt{1 - \frac{3}{2}\mathscr{C}_{\rm th}}\right). \tag{7.14}$$

As two examples, the threshold of a Maxican-hat profile considered in Chap. 6 gives $\mathscr{C}_{\rm th} \approx 1/2$ and $\mathscr{C}_{\ell,\rm th} \approx 2/3$. In the theory of peaks, a monochromatic power spectrum gives $\mathscr{C}_{\rm th} \approx 0.587$, thus $\mathscr{C}_{\ell,\rm th} \approx 0.872$ [144], which will be used in this chapter.

When the power spectrum of $\mathcal{R}$ is broad, for every scale we study, we should add a window function to smooth out the unwanted short-wavelength fluctuations. This is highly non-trivial and there are uncertainties from the choice of the window function, which will not be discussed in this chapter.

7.2.2.3 Primordial Non-Gaussianity

The primordial non-Gaussianity of $\mathcal{R}$ (if any) should be taken into account when deriving the PDF of $\mathscr{C}$ from the PDF of $\mathcal{R}$, which is the main topic of this chapter.

We already know from Sect. 7.2.2.1 that the compaction function $\mathscr{C}$ is a quadratic function of $\mathscr{C}_\ell$. In the presence of primordial non-Gaussianity in $\mathcal{R}$, even $\mathscr{C}_\ell$ is non-Gaussian. As we will see, for local non-Gaussianities, $\mathcal{R}$ is a function of a Gaussian-distributed variable, $\mathcal{R}_{\rm g}$, which obeys the Gaussian PDF

$$\mathbb{P}(\mathcal{R}_{\rm g}) = \frac{1}{\sqrt{2\pi\sigma_{\mathcal{R}}^2}} \exp\left(-\frac{\mathcal{R}_g^2}{2\sigma_{\mathcal{R}}^2}\right), \tag{7.15}$$

with

$$\sigma_{\mathcal{R}}^2 = \int \frac{{\rm d}k}{k} \tilde{W}^2(k, r) \mathcal{P}_{\mathcal{R}_{\rm g}}(k). \tag{7.16}$$

Local-type non-Gaussianity gives

$$\mathcal{R} = \mathscr{F}[\mathcal{R}_{\rm g}(r)], \qquad \mathcal{R}_{\rm g} = \mathscr{G}[\mathcal{R}(r)], \tag{7.17}$$

where $\mathscr{G} \equiv \mathscr{F}^{-1}$ with $\mathscr{F}(0) = 0$ and $\mathscr{F}'(0) = 1$, such that (7.17) can be expanded as $\mathscr{F}(\mathcal{R}_{\rm g}) \approx \mathcal{R}_{\rm g}$ when $\mathcal{R}_{\rm g}$ is small. Concrete examples of such nonlinear relations (7.17) will be discussed in the next section. From (7.15), the corresponding PDF of $\mathcal{R}$ is

$$\mathbb{P}(\mathcal{R}) = \mathbb{P}_{\rm G}(\mathcal{R}_{\rm g}) \cdot \left|\frac{{\rm d}\mathscr{F}}{{\rm d}\mathcal{R}_{\rm g}}\right|^{-1} = \mathbb{P}_{\rm G}(\mathcal{R}_{\rm g}) \cdot \left|\frac{{\rm d}\mathscr{G}}{{\rm d}\mathcal{R}}\right|. \tag{7.18}$$

This brings non-Gaussianity to the linear compaction function by

$$\mathscr{C}_\ell = -\frac{4}{3} r \frac{{\rm d}\mathscr{F}}{{\rm d}\mathcal{R}_{\rm g}} \frac{\partial \mathcal{R}_{\rm g}(r)}{\partial r}. \tag{7.19}$$

7.2.3 PBH Abundance with Primordial Non-Gaussianity

According to the discussion above, the PBH abundance should be calculated as following [43, 98, 119]

$$\left.\begin{array}{l} \mathcal{R} \xrightarrow{\text{Eq. (7.10)}} \mathscr{C}_\ell \xrightarrow{\text{Eq. (7.12)}} \mathscr{C}(\mathscr{C}_\ell) \\ \begin{pmatrix}\text{PNG from}\\ \text{Inflation}\end{pmatrix} \to \mathbb{P}(\mathcal{R}) \xrightarrow{\text{Eq. (7.18)}} \mathbb{P}(\mathscr{C}_\ell) \end{array}\right\} \xrightarrow[\text{Window function}]{\mathscr{C}_{\rm th}(\text{profile})} \beta = \int_{\mathscr{C}_{\ell,{\rm th}}}^{4/3} \mathbb{P}(\mathscr{C}_\ell) \frac{M(\mathscr{C}_\ell)}{M_H} {\rm d}\mathscr{C}_\ell, \tag{7.20}$$

where PNG stands for primordial non-Gaussianity of the curvature perturbation generated during inflation, which will be discussed in detail in the next section. Calculations following the routines in (7.20) are called extended Press-Schechter or Press-Schechter-type formalism, which is in principle different from another frequently used method, *the theory of peaks* [27, 30, 114, 120, 144, 217, 239, 258–260, 266], by two main points. In Press-Schechter-type formalism, the PBH abundance is calculated by integrating the PDF of $\mathscr{C}_\ell$ from the threshold given by $\mathscr{C}_\ell \geq \mathscr{C}_{\ell,\mathrm{th}}$, while in the theory of peaks the PBH abundance is given by counting the number of peaks n_{pk} where the averaged compaction function in a sphere of radius r_m, $\langle\mathscr{C}\rangle_{r\leq r_m}$, exceeds a universal threshold 2/5 [91].

7.3 Primordial Non-Gaussianities of the Curvature Perturbation

Usually, the primordial curvature perturbation on comoving slices $\mathcal{R}$ originates from the quantum fluctuations of the inflaton φ during inflation [11, 45, 121, 158, 183, 230]. In linear perturbation theory, $\mathcal{R}_g = -(H/\dot{\varphi})\delta\varphi$ [182, 222], where $\delta\varphi$ is the field perturbation on spatially-flat slices. The non-Gaussian curvature perturbation is usually parameterized as the perturbative series of (7.17) [148],

$$\mathcal{R} = \mathcal{R}_g + \frac{3}{5} f_{\mathrm{NL}} \left(\mathcal{R}_g^2 - \langle\mathcal{R}_g^2\rangle\right) + \cdots . \tag{7.21}$$

In slow-roll inflation, the nonlinear parameter f_{NL} is related to the spectral tilt by the consistency relation $f_{\mathrm{NL}} = (5/12)(1 - n_s)$ [78, 166], which is suppressed by slow-roll parameters as $n_s - 1 = -2\epsilon_H - \eta_H$ with $\epsilon_H \equiv -\dot{H}/H^2$ and $\eta_H \equiv \dot{\epsilon}_H/(H\epsilon_H)$. On scales larger than 1 Mpc, the observational constraint is $f_{\mathrm{NL}} = -0.9 \pm 5.1$ [1, 10].

As is commented above, the PBH abundance depends sensitively on the non-Gaussianities, thus even a small non-Gaussianity can change the PBH abundance significantly. Furthermore, as the power spectrum is enhanced rapidly (usually as power-law, $k^{\mathcal{O}(1)}$) before the peak scale, an intuitive guess based on the consistency relation $f_{\mathrm{NL}} \sim 1 - n_s$ is that non-Gaussianity will also be enhanced to $\mathcal{O}(1)$. For instance, in ultra-slow-roll inflation, $\epsilon_H \lll 1$ and $\eta_H = -6$. If we only focus on the slow-roll attractor mode, we have

$$f_{\mathrm{NL}} = \frac{5}{12}(1 - n_s) = \frac{5}{12}(2\epsilon_H + \eta_H) \approx \frac{5}{12} \times (-6) = -\frac{5}{2}. \tag{7.22}$$

Accidentally, this rough estimate gives the correct absolute value with an opposite sign [192]. Anyway this indicates that the non-Gaussianity may reach $\mathcal{O}(1)$ when the power spectrum increases rapidly.

In principle, when $|f_{\rm NL}| \gtrsim 1$, the higher-order terms include an infinite number of parameters, i.e. $g_{\rm NL}$, $h_{\rm NL}$, $i_{\rm NL}$, etc., which makes it impossible to have a model-independent analysis. Fortunately, in some typical models, the non-Gaussian curvature perturbation can be calculated neatly by the δN formalism, based on the *separate universe* approach [162, 223, 250]. During inflation, the gradient terms decay rapidly on superhorizon scales, which guarantees that in a region slightly larger than the Hubble patch, the inflaton field takes a homogeneous value, which varies from region to region. The initial condition for φ in the numerous separate universes obeys a Gaussian distribution, which evolves independently on superhorizon scales. The δN formalism tells us that the final curvature perturbation in such a patch is the difference of total local expansion, i.e. the e-folding number N, between this patch to the fiducial $\langle N\rangle$, evaluated from an initial spatially-flat slice to the final comoving slice:

$$\mathcal{R} = \delta N \equiv N(\varphi, \dot{\varphi}) - \langle N(\langle\varphi\rangle, \langle\dot{\varphi}\rangle)\rangle. \tag{7.23}$$

This formula does not require a decomposition of a background and a perturbation, thus can be used for large perturbations. We will use it to study the fully nonlinear curvature perturbation in the following sections. In Chap. 8, (7.23) will be used to study quantum diffusion.

7.3.1 Ultra-Slow-Roll Inflation on a Flat Potential

When $|f_{\rm NL}| \gtrsim 1$, we should take into account all the higher-order terms, which is only possible for some specific models. The simplest example is the so-called ultra-slow-roll inflation. Consider a plateau of potential $V(\varphi) = V_0$ for $\varphi_i < \varphi < \varphi_t$. On the plateau, the inflaton with an initial velocity moves toward the positive-φ direction. The equation of motion of φ is $\ddot{\varphi} + 3H\dot{\varphi} = 0$, where $H \approx \sqrt{V_0}/(\sqrt{3}M_{\rm Pl})$ is nearly a constant on the plateau. Conventionally, we define the e-folding number N, counted backward in time from $N(\varphi_t) = 0$, such that $\mathrm{d}N = -H\mathrm{d}t$. The equation of motion becomes

$$\frac{\mathrm{d}^2\varphi}{\mathrm{d}N^2} - 3\frac{\mathrm{d}\varphi}{\mathrm{d}N} = 0. \tag{7.24}$$

Two general solutions to (7.24) are $\varphi =$ constant and $\varphi \propto e^{3N}$. The former static solution is the attractor, but it should not dominate at the classical level, otherwise the inflaton gets stuck on the plateau, which can only end inflation with the help of quantum diffusion. The non-attractor decelerating solution $\varphi \propto e^{3N}$ is necessary as it helps the inflaton to classically shoot out of the potential plateau with large enough initial velocity. We will see that the excitation of such non-attractor solutions is the origin of the large non-Gaussianity. Together with the boundary condition at the endpoint φ_t of the plateau, the special solution is

$$\varphi(N) = \varphi_t + \frac{\pi_t}{3}\left(1 - e^{3N}\right), \tag{7.25}$$

where π_t is the velocity $\pi \equiv -\mathrm{d}\varphi/\mathrm{d}N$ at the endpoint where $N = 0$, such that $\varphi(0) = \varphi_t$, $\pi(0) = \pi_t$. On the plateau, we have $\pi(N) = \pi_t e^{3N}$ as expected, and (7.25) can be rewritten as a "conservation law"

$$\pi(N) + 3\varphi(N) = \pi_t + 3\varphi_t, \tag{7.26}$$

which is valid for all the trajectories at any e-folding number. For a perturbed universe, (7.26) gives $\tilde{\pi}(\tilde{N}) + 3\tilde{\varphi}(\tilde{N}) = \tilde{\pi}_t + 3\varphi_t$, which in turn gives

$$\delta\pi + 3\delta\varphi = \delta\pi_t. \tag{7.27}$$

We defined $\delta\varphi \equiv \tilde{\varphi} - \varphi$, $\delta\pi \equiv \tilde{\pi} - \pi$ and $\delta\pi_t \equiv \tilde{\pi}_t - \pi_t$, which are Gaussian variables as φ is a massless field. The physical meaning of (7.26) and (7.27) can be well understood in the schematic phase portrait Fig. 7.1.

From $\pi(N) = \pi_t e^{3N}$, we have

$$N = \frac{1}{3}\ln\frac{\pi}{\pi_t} = -\frac{1}{3}\ln\left(1 + \frac{3(\varphi - \varphi_t)}{\pi}\right), \tag{7.28}$$

where the second equality comes from (7.26), which is useful to determine the equal-N lines in Fig. 7.1: They are radial lines from $(\pi = 0,\ \varphi = \varphi_t)$.

In the separate universe approach, another remote and independent "perturbed universe" happens to have another e-folding number of $\tilde{N} = (1/3)\ln\tilde{\pi}/\tilde{\pi}_t$, with

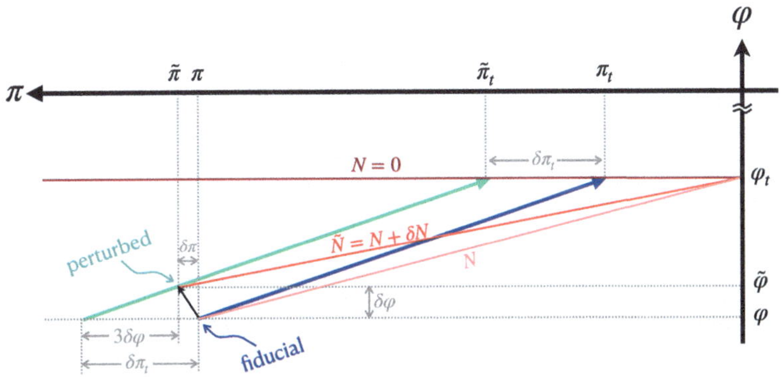

Fig. 7.1 The schematic phase portrait of the ultra-slow-roll inflation on a flat plateau. The blue and green parallel lines with arrows are the inflaton trajectories in the background universe and in a perturbed Hubble patch, respectively. The red lines are equal-N lines, determined by the conservation law (7.26), while $(\tilde{\pi}, \tilde{\varphi})$ at the perturbed trajectory corresponds to an e-folding number $\tilde{N} = N + \delta N$. Such a perturbed point $(\tilde{\pi}, \tilde{\varphi})$ satisfies (7.27), as is clearly shown geometrically. For more discussion on such a phase portrait, see [199]

$\tilde{\pi} = \pi + \delta\pi$, and $\tilde{\pi}_t = \pi_t + \delta\pi_t$. In the δN formalism, the late-time curvature perturbation in this perturbed universe is the difference between these e-folding numbers

$$\delta N = \tilde{N} - N = \frac{1}{3}\ln\left(\frac{\tilde{\pi}}{\tilde{\pi}_t}\frac{\pi_t}{\pi}\right) = \frac{1}{3}\ln\left(1 - \frac{\delta\pi}{\pi}\right) - \frac{1}{3}\ln\left(1 - \frac{\delta\pi_t}{\pi_t}\right). \tag{7.29}$$

Note that from from (7.27) we have $\delta\pi \sim \delta\pi_t$, so unless $N \approx 0$, the first term in (7.29) is always negligible compared to the second term as $\pi = \pi_t e^{3N} \gg \pi_t$. If the later evolution does not contribute significantly to δN,[4] the final curvature perturbation is

$$\mathcal{R} \approx -\frac{1}{3}\ln\left(1 - \frac{\delta\pi_t}{\pi_t}\right), \tag{7.30}$$

which is highly non-Gaussian with a nominal nonlinear parameter

$$f_{\rm NL} = \frac{5}{2}. \tag{7.31}$$

Apparently, the non-Gaussianity expressed in (7.30) can not be simply described by quadratic expansion with $f_{\rm NL} = 5/2$. We should emphasize that (7.30) is only valid when the contribution from the later evolution to δN is negligible, the meaning of which will be discussed in the subsection below.

7.3.2 Constant-Roll on a Quadratic Potential

In the previous section we see how the non-attractor solution is important in enhancing the power spectrum and generating large non-Gaussianities. Such a conclusion is general, valid also for other types of potentials. In this subsection we will consider another analytically solvable model – constant-roll inflation on a quadratic potential, which can be inserted as an intermediate stage between two slow-roll stages to enhance the curvature perturbation. We consider the following potential

$$V(\varphi) = V_0 + \frac{1}{2}m^2\varphi^2, \tag{7.32}$$

shown in Fig. 7.2. We assume $V_0 \gg |m^2\varphi^2|$, such that the Hubble parameter can still be approximated by a constant $H \approx \sqrt{V_0}/\sqrt{3}M_{\rm Pl}$, while the second slow-roll parameter $\eta \equiv m^2/3H^2$ is also a constant. The equation of motion is

[4] This requirement is not trivial. It is equivalent to require that the inflaton accelerates to the later slow-roll attractor after it classically shoots out of the flat pleateau. On the other hand, if φ decelerates, the δN contributed from the later stage is dominant, which is usually Gaussian and thus the entire curvature perturbation is nearly Gaussian. See Sect. 7.3.3 for a detailed discussion.

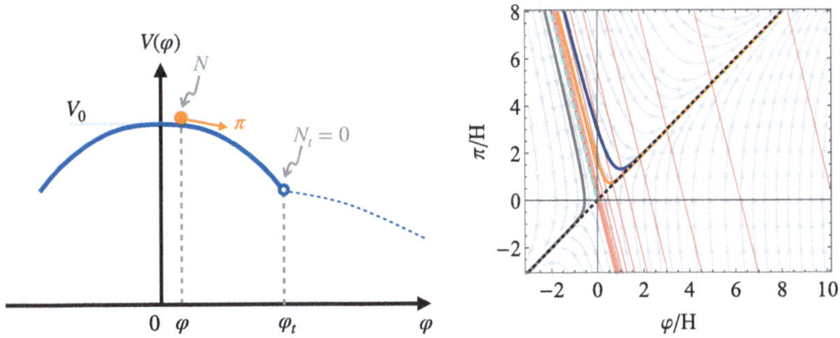

Fig. 7.2 Inflation potential (7.32), and the phase portrait of (7.33) with $\eta = -4/3$. The initial conditions for the blue, orange, cyan, and gray curves are $\pi_i = 14, 13, 12, 11$ at $\varphi_i = -3$ with $\varphi_t = 100$ (in the unit of H). These solutions are not on the attractor $\pi + \lambda_- \varphi = 0$ initially, which is displayed by the diagonal dotted black line. The equal-N lines, based on (7.37), are shown in thin red lines with equal difference of $\Delta N = 0.5$. It is clearly shown that for the same $\delta\varphi$, the largest δN comes from around the critical trajectory (cyan dashed) in the constant-roll stage

$$\frac{d^2\varphi}{dN^2} - 3\frac{d\varphi}{dN} + 3\eta\varphi = 0, \tag{7.33}$$

which is written in the e-folding number N, counted backwards in time from a junction point φ_t far enough from the bump, such that the solution is already an attractor there.[5] Setting $\varphi \propto e^{\lambda N}$, the characteristic root λ of (7.33) is

$$\lambda^2 - 3\lambda + 3\eta = 0, \quad \Longrightarrow \quad \lambda_\pm = \frac{3 \pm \sqrt{9 - 12\eta}}{2}. \tag{7.34}$$

For $\eta < 3/4$, we have two real solutions with $\lambda_- < \lambda_+$, which means φ has two independent solutions: the attractor solution $\varphi \propto e^{\lambda_- N}$ and the non-attractor solution $\varphi \propto e^{\lambda_+ N}$. Again, defining the velocity $\pi(N) = -d\varphi/dN$, the attractor and non-attractor solutions satisfy

$$\pi + \lambda_\mp \varphi = 0. \tag{7.35}$$

Considering the boundary conditions $\pi(0) = \pi_t$, $\varphi(0) = \varphi_t$, the special solution is

$$\varphi(N) = -\frac{\pi_t + \lambda_- \varphi_t}{\lambda_+ - \lambda_-} e^{\lambda_+ N} + \frac{\pi_t + \lambda_+ \varphi_t}{\lambda_+ - \lambda_-} e^{\lambda_- N}. \tag{7.36}$$

[5] For the curvature perturbation which exits the horizon during the constant-roll stage, there will be still some e-folding number from φ_t to the end of inflation, say φ_e. However, if the trajectories we study here already merged into the attractor solution before φ_t, these Hubble patches evolves uniquely after φ_t, thus the contribution to δN from φ_t to φ_e is negligible.

Therefore, we can solve these equations for N:

$$N = \frac{1}{\lambda_\pm} \ln \frac{\pi + \lambda_\mp \varphi}{\pi_t + \lambda_\mp \varphi_t}, \tag{7.37}$$

The two different expressions for N in (7.37) are equivalent under the "conservation law"

$$\left(\frac{\pi + \lambda_+\varphi}{\pi_t + \lambda_+\varphi_t}\right)^{\lambda_+} = \left(\frac{\pi + \lambda_-\varphi}{\pi_t + \lambda_-\varphi_t}\right)^{\lambda_-}, \tag{7.38}$$

where both sides equal to $e^{\lambda_+\lambda_- N} = e^{3\eta N}$, a direct consequence of the solution (7.36). This is the fiducial e-folding number. For a perturbed universe where $\tilde{\varphi} = \varphi + \delta\varphi$, $\tilde{\pi} = \pi + \delta\pi$, $\tilde{\pi}_t = \pi_t + \delta\pi_t$, $\tilde{N} = N + \delta N$, we have

$$N + \delta N = \frac{1}{\lambda_\pm} \ln \frac{\pi + \delta\pi + \lambda_\mp \varphi}{\pi_t + \delta\pi_t + \lambda_\mp \varphi_t}. \tag{7.39}$$

Subtracting (7.37) from (7.39), we have

$$\mathcal{R} = \delta N = \frac{1}{\lambda_\pm} \ln\left(1 + \frac{\delta\pi + \lambda_\mp\delta\varphi}{\pi + \lambda_\mp\varphi}\right) - \frac{1}{\lambda_\pm} \ln\left(1 + \frac{\delta\pi_t}{\pi_t + \lambda_\mp\varphi_t}\right). \tag{7.40}$$

We see that the curvature perturbation has two equivalent expressions, which is called *logarithmic duality* in Ref. [204]. Apparently it comes from the quadratic nature of the potential, which is a good approximation when the field excursion we study is not too long. The equivalence of the upper and lower formulas in (7.40) can be easily shown by the perturbed version of conservation law (7.38):

$$\left(1 + \frac{\delta\pi_t}{\pi_t + \lambda_+\varphi_t}\right)^{-\lambda_+} \left(1 + \frac{\delta\pi_t}{\pi_t + \lambda_-\varphi_t}\right)^{\lambda_-} = \left(1 + \frac{\delta\pi + \lambda_-\delta\varphi}{\pi + \lambda_-\varphi}\right)^{\lambda_-} \left(1 + \frac{\delta\pi + \lambda_+\delta\varphi}{\pi + \lambda_+\varphi}\right)^{-\lambda_+}. \tag{7.41}$$

In principle, the perturbation at the boundary $\delta\pi_t$ induced by the initial perturbation $(\delta\varphi, \delta\pi)$ should be solved by (7.41). However, in this case we do not need it after assuming that the inflaton is already the attractor solution at φ_t, i.e. $\pi_t + \lambda_-\varphi_t \approx 0$ and $\delta\pi_t \approx 0$. Therefore, it is more convenient to use the lower line of (7.40), as the second logarithm is negligible:

$$\mathcal{R} \approx \frac{1}{\lambda_-} \ln\left(1 + \frac{\delta\pi + \lambda_+\delta\varphi}{\pi + \lambda_+\varphi}\right) = \frac{1}{\lambda_-} \ln\left(1 - \lambda_- \left.\frac{(\delta\varphi)_k}{\pi}\right|_{a \gg k/H}\right). \tag{7.42}$$

In the second step we use the fact that $\delta\varphi$ and φ obeys the same equation of motion (7.33) on superhorizon scales (i.e. $k^2 \ll H^2a^2$), when the potential is quadratic.[6]

[6] This is a special case of the well know proportionality $\delta\varphi \propto \dot{\varphi}/H$ on superhorizon scales for *any* potential in single-field inflation.

Therefore, $(\delta\pi + \lambda_+\delta\varphi)/(\pi + \lambda_+\varphi)$ is a constant along each trajectory, which can be evaluated at a much later time after horizon exit (marked by $a \gg k/H$) when φ is already in the attractor (but still before the boundary) [27]. Note that $(\delta\varphi)_k/\pi$ in (7.42) was stretched out of the horizon at $a_k = k/H$, which is now on a much larger scale of $a/k \gg 1/H$. It can be calculated by evolving the solution of the Mukhanov-Sasaki equation to late time $a \gg k/H$, which should not be confused with the "new" perturbations exiting the horizon at that moment. The phase diagram is shown in Fig. 7.2.

Slow-roll inflation and ultra-slow-roll inflation discussed in Sect. 7.3.1 are special cases of (7.40). Slow roll is the attractor solution when $\lambda_- \ll 1, \lambda_+ \approx 3$. Expanding (7.42), we get (7.21) with $\mathcal{R}_g = -\delta\varphi/\pi = H\delta\varphi/\dot{\varphi}$ and $f_{\rm NL} = -(5/6)\lambda_-$. In ultra-slow-roll inflation, the non-attractor solution dominates. Substituting $\lambda_- = 0, \lambda_+ = 3$ into the upper line of Eq. (7.40), we can get (7.29).

7.3.3 A General Analysis on Single-Field Inflation

We have shown how a quadratic piece of the potential (7.32) can give rise to the logarithmic relation (7.40) of $\mathcal{R}(\delta\varphi, \delta\pi)$. The Gaussian variables $(\delta\varphi, \delta\pi)$ originate from quantum fluctuations on subhorizon scales, which should be evaluated on the initial spatially-flat surface after horizon exit. In general, such a potential (7.32) is a segment of a global potential. In a typical three-segment model, the initial condition of (φ, π) are usually provided by a preceding (slow-roll) stage, while after the bump/plateau, the potential is usually followed by another slow-roll potential to end inflation. Before and after the junction point, the attractor solutions are usually different. This can excite the non-attractor solution transiently, which contribute to the total δN as we discussed above. All of such contributions should be taken into account to calculate the final δN.

We consider a piecewise potential consisting of two parabolas:

$$V(\varphi) = \begin{cases} V_0 + \dfrac{m^2}{2}\left(\varphi^2 - \varphi_t^2\right), & \text{for } \varphi \le \varphi_t; \\ V_0 - \dfrac{m'^2}{2}\left(\varphi_t - \varphi_m\right)^2 + \dfrac{m'^2}{2}\left(\varphi - \varphi_m\right)^2, & \text{for } \varphi > \varphi_t, \end{cases} \tag{7.43}$$

where φ_t is the junction point of the two quadratic potentials with $V(\varphi_t) = V_0$, and φ_m is the minimum of $V(\varphi > \varphi_t)$. For simplicity, we set the origin to make $\varphi = 0$ the maximum (minimum) of $V(\varphi < \varphi_t)$, when $m^2 < 0\,(m^2 > 0)$. This is an extension of the Starobinsky model with a hinged linear potential [232] to quadratic level, which can be further extended to a multi-exponential potential [87]. Inflation ends at φ_e in the second stage. A schematic figure of this piecewise potential is shown in Fig. 7.3. $V(\varphi)$ is continuous, but $V'(\varphi)$ has a jump at φ_t. This means φ' is continuous, but a change in the slope means that the attractor solution for $\varphi < \varphi_t$ is not the attractor solution for $\varphi > \varphi_t$, thus φ will decelerate/accelerate to reach the later attractor

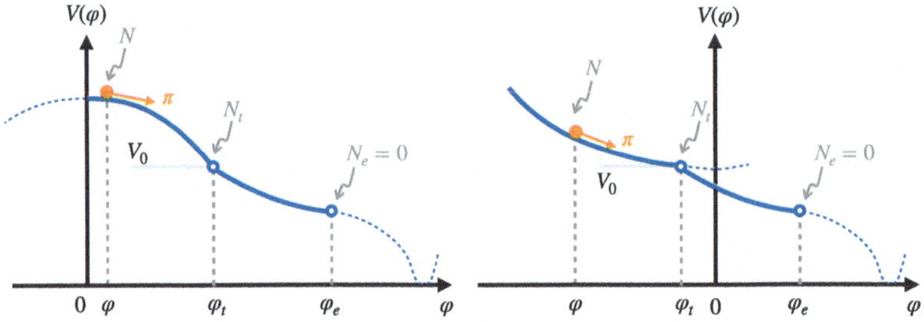

Fig. 7.3 Schematic pictures of the piecewise quadratic potential $V(\varphi)$ in (7.43), with $m^2 < 0$ (left) and $m^2 > 0$ (right). The potential is continuous at the transition point φ_t, where $V(\varphi_t) = V_0$, but $V'(\varphi_t)$ has a jump. The origin is chosen at the local maximum (left panel) or minimum (right panel) of $V(\varphi < \varphi_t)$, for $m^2 < 0$ or $m^2 > 0$, respectively. The e-folding numbers are labeled at φ, φ_t, and φ_e

solution, during which the curvature perturbation will be enhanced/suppressed, and large non-Gaussianity could be generated.

By the method used in Sect. 7.3.2, we can solve δN contributed in all the stages, and add together to get the final curvature perturbation [204]:

$$\mathcal{R} = \delta N_1 \left(\{\varphi, \pi\} \to \{\varphi_t, \pi_t\}\right) + \delta N_2 \left(\{\varphi_t, \pi_t\} \to \{\varphi_e, \pi_e\}\right), \tag{7.44}$$

$$\delta N_1 = \frac{1}{\lambda_\pm} \ln\left(1 + \frac{\delta\pi + \lambda_\mp \delta\varphi}{\pi + \lambda_\mp \varphi}\right) - \frac{1}{\lambda_\pm} \ln\left(1 + \frac{\delta\pi_t}{\pi_t + \lambda_\mp \varphi_t}\right), \tag{7.45}$$

$$\delta N_2 = \frac{1}{\lambda'_\pm} \ln\left(1 + \frac{\delta\pi_t}{\pi_t + \lambda'_\mp(\varphi_t - \varphi_m)}\right) - \frac{1}{\lambda'_\pm} \ln\left(1 + \frac{\delta\pi_e}{\pi_e + \lambda'_\mp(\varphi_e - \varphi_m)}\right), \tag{7.46}$$

where $\lambda_\pm$ and $\lambda'_\pm$ are the characteristic roots of each piece, defined by (7.34). The equivalence of upper and lower expressions of each δN_i are guaranteed by the perturbed conservation law (7.41) for each piece. Here, $\delta\varphi$ and $\delta\pi$ are the inflaton perturbation and its momentum (i.e. time derivative with respect to $-N$) on an initial spatially-flat slice around (φ, π), which should be solved from the equation of motion for $\delta\varphi$, and evaluated at a moment slightly later than the horizon exit. In slow-roll stage, typically $\delta\varphi \approx H/(2\pi)$ and $\delta\pi \approx 0$ as $\delta\pi$ decays exponentially after horizon exit, yet it is not so simple for non-attractor stages. For a concrete example of calculating $\delta\varphi$ and $\delta\pi$ in ultra-slow-roll inflation, see [87, 133]. After we know $\delta\varphi$ and $\delta\pi$, the corresponding perturbations on all the later boundaries, i.e. $\delta\pi_t$, $\delta\pi_e$, etc., can be solved by the perturbed conservation law (7.41) for each piece, although this equation can only be solved numerically except for a few special values of λ_- listed in Ref. [204]. As we have shown before, finally all the different trajectories induced by the initial perturbation $(\delta\pi, \delta\varphi)$ merge to the unique attractor trajectory. Since then, the e-folding number only depends on φ, and δN in this stage only depends on the time difference when φ falls on to the attractor, which is usually Gaussian. It is straightforward to extend (7.44) to include more segments of quadratic potentials, and for every segment of the quadratic piece, two logarithms from the early and late boundaries are present, similar to (7.46).

As a simple example, let's discuss ultra-slow-roll inflation on a flat potential connected to a slow-roll potential. Then (7.45) reduces to (7.30), while (7.46) reduces to (7.42):

$$\mathcal{R} \approx -\frac{1}{3}\ln\left(1+\frac{\delta\pi_t}{\pi_t}\right)+\frac{1}{\lambda'_-}\ln\left(1+\frac{\delta\pi_t}{\pi_t+\lambda'_+(\varphi_t-\varphi_m)}\right). \tag{7.47}$$

In general, the final curvature perturbation is a sum of these two terms, and the PDF of $\mathcal{R}$ can only be solved numerically when two logarithmic functions are equally important. However, it can be solved analytically when one of the logarithms dominates. When $\pi_t \gg \lambda'_-(\varphi_t-\varphi_m)$, the velocity at the end of the flat plateau surpasses the attractor velocity of the slow-roll potential, thus a transient deceleration stage is followed. The second term of (7.47) dominates, so the final curvature perturbation is nearly Gaussian with $f_{\rm NL}=-(6/5)\lambda'_- \approx -(6/5)\eta'$. On the contrary, when $\pi_t \ll \lambda'_-(\varphi_t-\varphi_m)$, the velocity at the end of flat plateau is not yet the attractor of the slow-roll potential, thus a transient acceleration stage is followed. The first term of (7.47) dominates, so the final curvature perturbation is non-Gaussian with $f_{\rm NL}=5/2$. These two limits are called "smooth transition" and "sharp transition" respectively in [54], where a parameter $h \equiv 6|\lambda'_-(\varphi_t-\varphi_m)/\pi_t|$ was defined to compare the slow-roll attractor velocity $-\lambda'_-(\varphi_t-\varphi_m)$ and the realistic end-of-ultra-slow-roll velocity π_t at the transition point φ_t. To avoid confusion, we should emphasize that the comparison is between the velocities, but not between the slopes of the potentials.

This result is easy to understand by looking at the phase portrait. For the "smooth transition" case, although the potential behaves differently for $\varphi<\varphi_t$ and $\varphi>\varphi_t$, but the non-attractor solution ($\pi=-3\varphi$) in the flat plateau is almost the same as the non-attractor solution in a slow-roll potential ($\pi=-\lambda'_+\varphi \approx (3-\eta')\varphi$). Therefore, this case is similar to the bumpy potential we studied in Sect. 7.3.2, and (7.42) tells us $\mathcal{R}$ is nearly Gaussian with $f_{\rm NL}=-(6/5)\lambda'_-$. The phase portrait Fig. 7.4 looks similar to that of a slow-roll inflation, as the ultra-slow-roll trajectories are similar to the fast-roll trajectories in an ordinary slow-roll inflation, which are parallel to the equal-N lines. The difference in the e-folding number for two nearby trajectories are proportional to the difference in φ, therefore $\delta N \propto \delta\varphi$ which preserves Gaussianity. The well studied Starobinsky model [42, 87, 154, 206, 232, 240] belongs to this limit.

For the sharp transition case, as is shown in Fig. 7.5, the inflaton experiences an acceleration after the ultra-slow-roll stage, therefore the nearby trajectories, slightly shifted by the velocity difference $\delta\pi_t$ generated during the ultra-slow-roll stage, can also induce difference in e-folding number δN in the later slow-roll stage. However, as $\pi_t \ll 1$, trajectories with different π_t's near the origin are almost on the same fast-roll trajectory in the slow-roll era for $\varphi>\varphi_t$, as the trajectories near the φ-axis are perpendicular to it. On this trajectory, the difference in N caused by $\delta\pi_t$ is

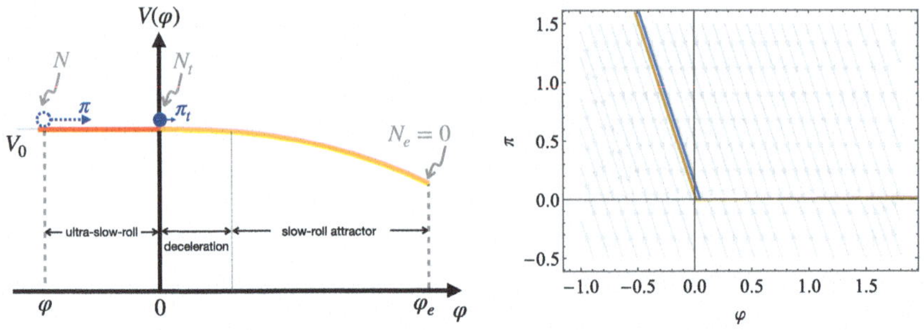

Fig. 7.4 The schematic potential (left panel) and the phase portrait (right panel) of the "smooth transition" from ultra-slow-roll to slow-roll, with $\eta' = -0.01$, $\varphi_m = 0$. For the fiducial (blue) and perturbed (orange) trajectories we set $\pi = 9.15,\ 9.05$ at $\varphi = -3$ (in the unit of H), respectively. To show the difference of the trajectories clearly, $\delta\pi$ is exaggerated. Equal-N lines are parallel to the fast-roll streams (i.e. $\pi \propto -3\varphi$) with almost equal intervals, which are not drawn for visual clearance. This clearly shows that the contribution to δN from the ultra-slow-roll stage is negligible, and the contribution from the later stage preserves slow-roll-type Gaussianity

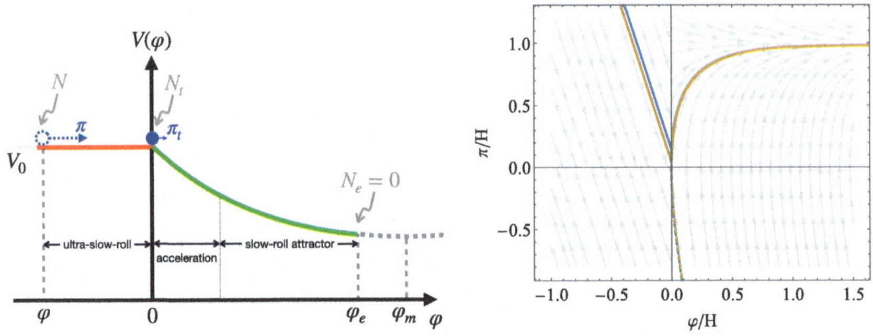

Fig. 7.5 The schematic potential (left panel) and the phase portrait (right panel) of the "sharp transition" from ultra-slow-roll to slow-roll, with $\eta' = 0.01$, $\varphi_m = 100$ (in the unit of H). For the fiducial (blue) and perturbed (orange) trajectories we set $\pi = 9.15,\ 9.05$ at $\varphi = -3$, respectively. To show the difference of the trajectories clearly, $\delta\pi$ is exaggerated. Near the origin, the trajectories merge to a unique fast-roll trajectory with different π, which contribute negligibly to δN in the subsequent evolution. The total δN comes mainly from the ultra-slow-roll stage, given by (7.30)

negligible, and $\varphi = \varphi_t$ is approximately an equal-N surface. Following the discussion in Sect. 7.3.1, we will arrive at the non-Gaussian relation (7.30) straightforwardly.

7.3.4 The Curvaton Scenario

The curvaton scenario is a special two-field inflation model, of which the curvature perturbation is generated by the density perturbation of the curvaton field χ at its decay after inflation. This was first proposed in [89, 163, 177] as an alternative method of generating the curvature perturbation, which was usually referred to as the outcome of the quantum fluctuation of the inflaton field φ during inflation. In the simplest curvaton models the power spectrum is nearly scale-invariant. To generate PBHs, the power spectrum should be enhanced on a specific scale, which can be realized in the axion-curvaton model [18, 134, 138, 139], non-minimal curvaton scenario [68, 173, 205], or the multi-curvaton model [235]. In the cuvaton scenario, it was shown that non-Gaussianity is inevitably enhanced compared to the slow-roll inflation, and the non-linear parameter is [36, 88, 224][7]

$$f_{\rm NL} = \frac{5}{4r}, \qquad r \equiv \frac{3\Omega_{\chi,\rm dec}}{4-\Omega_{\chi,\rm dec}}, \tag{7.48}$$

where r is roughly the cosmological density parameter of the curvaton χ at its decay. Apparently such a nominal $f_{\rm NL}$ is not enough to calculate the PBH abundance, as $f_{\rm NL}$ is always larger than 1.

We will skip the model building and the dynamics of the curvaton field, but focus on its fully non-Gaussian curvature perturbation on the uniform-density slice, which is derived in Ref. [224] for a Gaussian curvaton field perturbation in the sudden decay approximation:

$$\zeta = \ln\left(K^{1/2}\frac{1+\sqrt{ArK^{-3/2}-1}}{(3+r)^{1/3}}\right), \tag{7.49}$$

where r is defined in (7.48), and

$$A = \left(1+\frac{\delta\chi}{\chi}\right)^2, \quad \text{(on spatially-flat slicing)} \tag{7.50}$$

$$P = (Ar)^2 + \sqrt{(Ar)^4 + (3+r)(1-r)^3}, \tag{7.51}$$

$$K = \frac{1}{2}\left[P^{1/3} + (r-1)(3+r)^{1/3}P^{-1/3}\right]. \tag{7.52}$$

This fully non-linear curvature perturbation ζ can be used to calculate the PBH formation, following the procedure we show in Eq. (7.20) [97, 98, 118].

The fully non-Gaussian curvature perturbation (7.49) seems complicated, but it has two interesting limiting cases for $r \ll 1$ and $r \to 1$, both of which have fruitful physical implications. In the former limit, the curvaton energy density is negligible at its decay, and (7.49) can be expanded for small r:

$$\zeta = \frac{r}{3}(A-1) - \frac{r^2}{9}(A-1)^2 + \cdots = \frac{2}{3}r\frac{\delta\chi}{\chi} + \frac{r}{3}\left(1-\frac{4r}{3}\right)\left(\frac{\delta\chi}{\chi}\right)^2 + \cdots, \tag{7.53}$$

[7] For simplicity, we only consider a quadratic potential for the curvaton χ, and the decay is supposed to be instantaneous. For more details of these assumptions, see Ref. [224].

where dots are higher-order terms of $O(r^3)$. We see that ζ has the familiar form of the quadratic local non-Gaussianity (7.29), with $\zeta_g = (2/3)r\delta\chi/\chi$ and $f_{\rm NL} = 5/(4r)$. Cubic and higher-order terms are suppressed when $r \ll 1$, which makes it possible to truncate (7.53) at quadratic level even when $f_{\rm NL} \gg 1$. Therefore, counterintuitively, such a quadratic expression (7.53) holds for large $f_{\rm NL}$, towards the limit of a χ^2-distribution.

The opposite limit $r \to 1$ is more complicated and attracts much attention recently. In this limit, we simply have $P = 2(Ar)^2$, $K = (Ar/2)^{2/3}$, and

$$\zeta = \frac{1}{3}\ln A = \frac{2}{3}\ln\left|1+\frac{\delta\chi}{\chi}\right|. \tag{7.54}$$

A simple derivation of (7.54) is as follows. As curvaton dominates the universe at its decay, the curvature perturbation on the uniform-curvaton slice ζ_χ equals to the curvature perturbation on uniform density slice ζ, which gives [205]

$$\zeta = \zeta_\chi(t,\mathbf{x}) = \int_{\rho_\chi(t)}^{\rho_\chi(t,\mathbf{x})}\frac{\mathrm{d}\widetilde{\rho}_\chi}{3\widetilde{\rho}_\chi} = \frac{1}{3}\ln\frac{\rho_\chi(t,\mathbf{x})}{\rho_\chi(t)} = \frac{2}{3}\ln\left|1+\frac{\delta\chi}{\chi}\right|, \tag{7.55}$$

where in the last step we used $\rho_\chi = \frac{1}{2}m_\chi^2\chi^2$. Therefore, in the $r \to 1$ limit, the curvature perturbation ζ has a logarithmic dependence on the curvaton perturbation $\delta\chi$, which is similar to (7.42), but with a negative nonlinear parameter

$$f_{\rm NL} = -\frac{5}{4}. \tag{7.56}$$

This means that the curvaton scenario can suppress the formation of PBHs in the $r \to 1$ limit, which is a natural model to avoid the overproduction of PBHs when interpreting the nHz stochastic GWs as induced GWs [107].

7.4 PBH Abundance with Non-Gaussian Curvature Perturbation

The non-Gaussian curvature perturbation discussed in the previous section may have a significantly different tail from the Gaussian tail in the PDF in the large $\mathcal{R}$ regime. As we commented, the quantum fluctuations of the inflaton field on subhorizon scales naturally provide a random variable $\delta\varphi$ which has a Gaussian distribution among all the Hubble patches, as the quantum fluctuation is Gaussian when the potential is approximately quadratic. In most of the cases, the Gaussian component of the curvature perturbation $\mathcal{R}_{\rm g}$ can be defined as proportional to the inflaton field perturbation $\delta\varphi$ which recovers the leading order Gaussian relation $\mathcal{R}_{\rm g} \equiv -(H/\dot\varphi)\delta\varphi$. As we have discussed in Sect. 7.2.2, the nonlinear relation $\mathcal{R} = \mathscr{F}(\mathcal{R}_{\rm g})$ can bring non-Gaussianities to the PDF of $\mathcal{R}$, as is shown in (7.17). Then the PBH abundance should be calculated following (7.20), by transferring the PDF $\mathbb{P}(\mathcal{R})$ to that of the linear compaction function $\mathbb{P}(\mathscr{C}_\ell)$.

For instance, consider the following logarithmic relation of $\mathcal{R}$, (7.30):

$$\mathcal{R} = \mathscr{F}(\mathcal{R}_\mathrm{g}) = \frac{1}{\lambda}\ln\left(1+\lambda\mathcal{R}_\mathrm{g}\right), \tag{7.57}$$

where $\mathcal{R}_\mathrm{g}$ is Gaussian. This formula represents ultra-slow-roll inflation ($\lambda = -3$), constant-roll inflation ($\lambda = \lambda_-$ defined in (7.34)), and the $r \to 1$ limit of the curvaton scenario ($\lambda = 3/2$). Knowing the PDF of $\mathcal{R}_\mathrm{g}$ and the Jacobian (7.18), the PDF of $\mathcal{R}$ is given by

$$\mathbb{P}(\mathcal{R}) = \frac{e^{\lambda\mathcal{R}}}{\sqrt{2\pi}\,\sigma_{\mathcal{R}\mathrm{g}}}\exp\left[-\frac{\left(e^{\lambda\mathcal{R}}-1\right)^2}{2\lambda^2\sigma^2_{\mathcal{R}\mathrm{g}}}\right], \tag{7.58}$$

where $\sigma_{\mathcal{R}\mathrm{g}}$ is the root mean square of the Gaussian perturbation $\mathcal{R}_\mathrm{g}$, defined in (7.16). For a negative λ (including $\lambda = -3$), it goes as $\exp(\lambda\mathcal{R})$ when $\mathcal{R} \gg \mathcal{O}(0.1)$, so is called *exponential tail* PDF. Such a PDF is less suppressed compared to the Gaussian tail, thus the PBH abundance will be greatly enhanced. Although $\mathbb{P}(\mathcal{R})$ is not directly used in the calculation of the PBH mass function in the Press-Schechter-type formalism, it is straightforward to see the origin of such enhancement by the tail of $\mathbb{P}(\mathcal{R})$ when $\mathcal{R} \gg \sigma_{\mathcal{R}_\mathrm{g}}$. On the contrary, for a positive λ, the double exponential suppression is dominant, $\mathbb{P} \propto \exp(-c^2 e^{\lambda\mathcal{R}})$, called *Gumbel-distribution-like* PDF.

We will show how to calculate the PBH abundance for non-Gaussian curvature perturbation (7.57) in the Press-Schechter-type formalism, following [119]. By using (7.19), the linear compaction function $\mathscr{C}_\ell$ can be written as

$$\mathscr{C}_\ell \equiv \frac{X}{Y}, \qquad \text{with} \quad X \equiv -\frac{4}{3}r\mathcal{R}_\mathrm{g}'(r), \quad Y \equiv 1+\lambda\mathcal{R}_\mathrm{g}. \tag{7.59}$$

The covariance matrix is $\mathbf{\Sigma} = \begin{pmatrix} \sigma_X^2 & \varrho\sigma_X\sigma_Y \\ \varrho\sigma_X\sigma_Y & \sigma_Y^2 \end{pmatrix}$, with

$$\sigma_X^2 = \langle XX\rangle = \left(-\frac{4}{3}\right)^2 \int (kr)^2 \mathcal{P}_{\mathcal{R}\mathrm{g}}(k)\left(\frac{\mathrm{d}j_0}{\mathrm{d}z}(kr)\right)^2 \mathrm{d}\ln k, \tag{7.60}$$

$$\varrho\sigma_X\sigma_Y = \langle XY\rangle = \left(-\frac{4}{3}\right)\lambda\int (kr)\mathcal{P}_{\mathcal{R}\mathrm{g}}(k)\, j_0(kr)\frac{\mathrm{d}j_0}{\mathrm{d}z}(kr)\mathrm{d}\ln k, \tag{7.61}$$

$$\sigma_Y^2 = \langle YY\rangle = \lambda^2\int \mathcal{P}_{\mathcal{R}\mathrm{g}}(k)\, j_0^2(kr)\mathrm{d}\ln k, \tag{7.62}$$

where ϱ is the correlation between X and Y, $j_0(z) = \sin z/z$ is the zeroth order spherical Bessel function, and we assume a spherical ζ profile of

$$\mathcal{R}_g(r) = \frac{1}{\sqrt{2\pi}} \int \mathrm{d}k\, k j_0(kr)\mathcal{R}_k, \qquad \langle \mathcal{R}_k \mathcal{R}_p \rangle = \delta(k-p)\frac{2\pi^2}{k^3}\mathcal{P}_{\mathcal{R}g}(k). \tag{7.63}$$

The joint PDF of X and Y becomes

$$\begin{aligned}\mathbb{P}(X, Y) &= \frac{1}{2\pi\sqrt{\det\mathbf{\Sigma}}} \exp\left[-\frac{1}{2}(X, Y-1)\mathbf{\Sigma}^{-1}\begin{pmatrix} X \\ Y-1 \end{pmatrix}\right] \\ &= \frac{1}{2\pi\sigma_X\sigma_Y\sqrt{1-\varrho^2}} \exp\left(-\frac{X^2}{2\sigma_X^2}\right) \exp\left[-\frac{1}{2\left(1-\varrho^2\right)}\left(\varrho\frac{X}{\sigma_X} - \frac{Y-1}{\sigma_Y}\right)^2\right],\end{aligned} \tag{7.64}$$

which gives the PDF of $\mathscr{C}_\ell$,

$$\mathbb{P}(\mathscr{C}_\ell) = \int \mathrm{d}X\mathrm{d}Y\, \mathbb{P}(X, Y)\delta\left(\mathscr{C}_\ell - \frac{X}{Y}\right) = \int \mathrm{d}X\, |Y| \cdot \mathbb{P}\left(X \to \mathscr{C}_\ell Y, Y\right). \tag{7.65}$$

The mass of the PBH obeys a power-law scaling from the critical collapse [71, 94, 124, 147, 185, 194]

$$\frac{M(\mathscr{C}_\ell)}{M_H} \sim K\left(\mathscr{C} - \mathscr{C}_{\rm th}\right)^\gamma = K\left[\left(\mathscr{C}_\ell - \frac{3}{8}\mathscr{C}_\ell^2\right) - \mathscr{C}_{\rm th}\right]^\gamma, \tag{7.66}$$

where $\gamma \approx 0.36$ and $K \sim 1$. Solving this quadratic algebraic equation of $\mathscr{C}_\ell$, we have

$$\mathscr{C}_\ell = \frac{4}{3}\left(1 - \sqrt{1 - \frac{3}{2}\mathscr{C}_{\rm th} - \frac{3}{2}\left(\frac{M}{KM_H}\right)^{1/\gamma}}\right). \tag{7.67}$$

Another solution with $\mathscr{C}_\ell > 4/3$ is discarded as it is for the Type II fluctuation. With (7.66), to change the variable from $\mathscr{C}_\ell$ to M, we have

$$\frac{\mathrm{d}\mathscr{C}_\ell}{\mathrm{d}\ln M} = \frac{1}{\gamma}\frac{\mathscr{C}_\ell - \frac{3}{8}\mathscr{C}_\ell^2 - \mathscr{C}_{\rm th}}{1 - \frac{3}{4}\mathscr{C}_\ell}. \tag{7.68}$$

Therefore, substituting (7.66), (7.68) together with (7.65) back into (7.20), we get the PBH mass function at the formation time

$$\beta(M)\mathrm{d}\ln M \approx K\frac{\left(\mathscr{C}_\ell - \frac{3}{8}\mathscr{C}_\ell^2 - \mathscr{C}_{\rm th}\right)^{\gamma+1}}{\gamma\left(1 - \frac{3}{4}\mathscr{C}_\ell\right)}\mathbb{P}(\mathscr{C}_\ell)\mathrm{d}\ln M, \tag{7.69}$$

where $\mathscr{C}_\ell$ should be expressed as a function of M, given by (7.67). The PBH mass function today is redshifted from the horizon reentry to equality, as the PBH energy density decays as a^{-3} in the radiation dominated background $\sim a^{-4}$:

$$f_{\rm PBH}(M) \approx 3.81 \times 10^8 \left(\frac{g_{*i}}{106.75}\right)^{-1/4} \left(\frac{h}{0.67}\right)^{-2} \left(\frac{M_\odot}{M_H}\right)^{1/2} \left(\frac{M}{M_H}\right)^{-1/2} \beta(M). \tag{7.70}$$

The PBH abundance today $f_{\rm PBH} = \int f_{\rm PBH}(M) {\rm d}\ln M$, is the integration of the mass function over the entire mass range. For PBHs to be all the dark matter requires $f_{\rm PBH} = 1$.

For simplicity, we only consider the monochromatic power spectrum

$$\mathcal{P}_{\mathcal{R}g} = \mathcal{A}_{\mathcal{R}} \delta(\ln k - \ln k_*). \tag{7.71}$$

From (7.60) to (7.62), we have

$$\sigma_X = \frac{4}{3}(k_* r)\mathcal{A}_{\mathcal{R}}^{1/2} \left|\frac{{\rm d}j_0}{{\rm d}z}(k_* r)\right|, \qquad \sigma_Y = |\lambda| \mathcal{A}_{\mathcal{R}}^{1/2} \left|j_0(k_* r)\right|, \tag{7.72}$$

and $\varrho = {\rm sgn}(-\lambda j_0(k_* r) j_0'(k_* r))$. For simplicity, we choose r_m to be the same as the Gaussian case, $k_* r_m \approx 2.74$, which gives $\varrho = {\rm sgn}(\lambda)$. With $\varrho^2 = 1$, the PDF (7.65) reduces to be one-dimensional, as X and Y are fully correlated. Together with $X = \mathscr{C}_\ell Y$, we have

$$X = \frac{\sigma_X \mathscr{C}_\ell}{\sigma_X - {\rm sgn}(\lambda)\sigma_Y \mathscr{C}_\ell}, \qquad Y = \frac{\sigma_X}{\sigma_X - {\rm sgn}(\lambda)\sigma_Y \mathscr{C}_\ell}. \tag{7.73}$$

And the degenerated PDF of $\mathscr{C}_\ell$ becomes

$$\mathbb{P}(\mathscr{C}_\ell) = \frac{1}{\sqrt{2\pi\sigma_X^2}} \left(\frac{\sigma_X}{\sigma_X - {\rm sgn}(\lambda)\sigma_Y \mathscr{C}_\ell}\right)^2 \exp\left[-\frac{1}{2}\left(\frac{\mathscr{C}_\ell}{\sigma_X - {\rm sgn}(\lambda)\sigma_Y \mathscr{C}_\ell}\right)^2\right]. \tag{7.74}$$

Apparently for $\lambda \to 0$, it degenerates to the Gaussian PDF. Roughly speaking, the variance of PDF (7.74) increases for a negative λ (i.e. positive nominal $f_{\rm NL}$), which enhances PBH formation. As $\mathscr{C}_\ell$ increases, the PDF (7.74) finally approaches the Cauchy distribution $\mathbb{P}(\mathscr{C}_\ell) \propto \mathscr{C}_\ell^{-2}$, which, however, is far from the Type-I-PBH formation region of $0.872 < \mathscr{C}_\ell < 4/3$ [43]. We calculated the PBH mass function $f_{\rm PBH}(M)$ for such a monochromatic power spectrum (7.71) in the ultra-slow-roll inflation ($\lambda = -3$), shown in the left panel of Fig. 7.6. We choose a typical PBH peak mass of $M_{k^*} = 10^{-12}\, M_\odot$, and normalize the amplitude of the power spectrum $\mathcal{A}_{\mathcal{R}}$ to be 4.151077×10^{-3}, such that $f_{\rm PBH} = 1$, i.e. PBHs are all the dark matter. Interestingly, as is clearly seen in Fig. 7.6, for the narrow peak case, the critical infrared scaling $f_{\rm PBH}(M) \propto M^{1+1/\gamma} \sim M^{3.78}$ which originates from the universal behavior is almost parallel to the constraints from the 21-cm signal which is proportional to M^4 [176]. So the entire asteroid-mass window is open for narrow-peak power spectrum even with non-Gaussianity.

Another example is the perturbative series of (7.21). Similarly we can define

$$\mathscr{C}_\ell = XZ, \quad X \equiv -\frac{4}{3} r\mathcal{R}_{\rm g}'(r), \quad Z = 1 + \frac{6}{5} f_{\rm NL}\mathcal{R}_{\rm g}. \tag{7.75}$$

By the same method, for a narrow spectrum, we can get the PBH abundance $f_{\rm PBH}$ as a function of $\mathcal{A}_\mathcal{R}$ for different $f_{\rm NL}$'s, given by (7.20) with the PDF of $\mathscr{C}_\ell$

$$\mathbb{P}(\mathscr{C}_\ell) = \sum_{i=+,-} \frac{|Z_i|}{\sqrt{2\pi}|\sigma_X Z_i^2 + \mathrm{sgn}(f_{\rm NL})\mathscr{C}_\ell \sigma_Z|} \exp\left(-\frac{\mathscr{C}_\ell^2}{2\sigma_X^2 Z_i^2}\right), \tag{7.76}$$

with $Z_\pm = 1/2 \pm \sqrt{1/4 + \mathrm{sgn}(f_{\rm NL})(\sigma_Z/\sigma_X)\mathscr{C}_\ell}$, and

$$\sigma_X = \frac{4}{3}(k_* r)\mathcal{A}_\mathcal{R}^{1/2}\left|\frac{\mathrm{d} j_0}{\mathrm{d}z}(k_* r)\right|, \qquad \sigma_Z = \frac{6}{5}\mathcal{A}_\mathcal{R}^{1/2}\left|f_{\rm NL} j_0(k_* r)\right|. \tag{7.77}$$

When $f_{\rm NL} \to 0$, the "+" branch goes back to Gaussian PDF, while the "−" branch disappears. Roughly speaking, for a positive $f_{\rm NL}$, the variance of $\mathbb{P}(\mathscr{C}_\ell)$ increases, which enhances PBH production. In the large-$f_{\rm NL}$ limit, the PDF reduces to χ^2-distribution (a special form of gamma distribution):

$$\mathbb{P}(\mathscr{C}_\ell) \approx \frac{1}{2\sqrt{2\pi\sigma_X\sigma_Z\mathscr{C}_\ell}} \exp\left(-\frac{\mathscr{C}_\ell}{2\sigma_X\sigma_Z}\right). \tag{7.78}$$

Besides its form, the key point is that in this limit only the product $\sigma_X\sigma_Z \propto \mathcal{A}_\mathcal{R} f_{\rm NL}$ is relevant, and numerical integration shows that $f_{\rm PBH} = 1$ requires $\mathcal{A}_\mathcal{R} f_{\rm NL} \sim 0.07$.

We can calculate the total abundance $f_{\rm PBH}$ as a function of $\mathcal{A}_\mathcal{R}$ following (7.20), which is shown in the right panel of Fig. 7.6, together with the result from ultra-slow-roll inflation and curvaton scenario with $r \to 1$. As a positive $f_{\rm NL}$ enhances the PBH abundance, the $f_{\rm PBH}(\mathcal{A}_\mathcal{R})$ curve moves upward for positive $f_{\rm NL}$'s, which makes the required $\mathcal{A}_\mathcal{R}$ for $f_{\rm PBH} = 1$ smaller, as we described in the introduction. Also, because of the higher-order terms, the logarithmic relation (7.57) always generates more PBHs than the quadratic expansion with the same $f_{\rm NL}$, which requires a slightly smaller $\mathcal{A}_\mathcal{R}$ for the same abundance of PBH. This is clearly shown for two cases: $f_{\rm NL} = 5/2$ with ultra-slow-roll inflation, and $f_{\rm NL} = -5/4$ with curvaton scenario with $r \to 1$. The values of $\mathcal{A}_\mathcal{R}$ required to realize PBH dark matter for $M_{k_*} = 10^{-12}\,M_\odot$ are listed in Table 7.1, which can be well fitted by

$$\mathcal{A}_\mathcal{R} \approx \frac{6.6\times 10^{-3}}{1 + 0.1 f_{\rm NL}}, \tag{7.79}$$

which recovers $\mathcal{A}_\mathcal{R} f_{\rm NL} \to 0.07$ when $f_{\rm NL} \gg 1$.

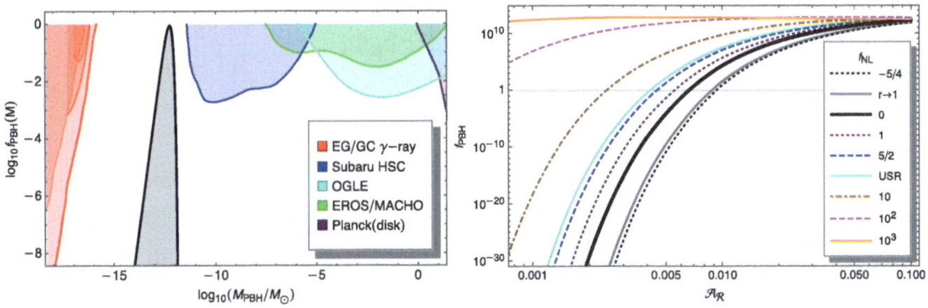

Fig. 7.6 Left: The PBH mass function $f_{\rm PBH}(M)$ given by (7.70) with a monochromatic power spectrum (7.71) in a ultra-slow-roll inflation with a sharp end, together with the some observational constraints. From left to right: extra-galactic γ-rays [63], the 511 keV line from galactic center [84, 153], 21-cm signal from EDGES [176], Subaru HSC [195], EROS [244], OGLE [180, 181], and CMB anisotropy with disk accretion [227]. The total power and central mass are $\mathcal{A}_{\mathcal{R}} = 4.151 \times 10^{-3}$ and $M_{k_*} = 10^{-12} M_\odot$, such that $\int f_{\rm PBH}(M) \mathrm{d} \ln M = 1$. Right: The total PBH abundance $f_{\rm PBH}$ as a function of the total power $\mathcal{A}_{\mathcal{R}}$, for a monochromatic power spectrum (7.71). From right to left, the curves are for $f_{\rm NL} = -5/4$ (black dotted), curvaton with $r \to 1$ (gray), Gaussian case ($f_{\rm NL} = 0$) (black thick), $f_{\rm NL} = 1$ (purple dotted), $f_{\rm NL} = 5/2$ (blue dashed), ultra-slow-roll inflation (cyan), $f_{\rm NL} = 10$ (brown dot-dashed), $f_{\rm NL} = 10^2$ (magenta dashed), and $f_{\rm NL} = 10^3$ (orange), respectively

Table 7.1 The total power $\mathcal{A}_{\mathcal{R}}$ for a monochromatic power spectrum (7.71) needed to make PBH to be all the dark matter at $M_{k_*} = 2 \times 10^{21}$ g, with different $f_{\rm NL}$'s. "$r \to 1$" and "USR" stand for curvaton model with $r \to 1$ and ultra-slow-roll inflation, described by a logarithmic $\mathcal{R}(\mathcal{R}_{\rm g})$ relation (7.57) with $\lambda = 3/2$ ($f_{\rm NL} = -5/4$) and $\lambda = -3$ ($f_{\rm NL} = 5/2$), respectively

$f_{\rm NL}$	$\mathcal{A}_{\mathcal{R}}$	$f_{\rm NL}$	$\mathcal{A}_{\mathcal{R}}$
$-5/4$	9.328243×10^{-3}	5/2	4.479141×10^{-3}
$r \to 1$	8.808810×10^{-3}	USR	4.151077×10^{-3}
0	6.635506×10^{-3}	10	2.456822×10^{-3}
0.1	6.498845×10^{-3}	10^2	4.618021×10^{-4}
1	5.519888×10^{-3}	10^3	5.690648×10^{-5}

7.5 Induced Gravitational Waves

The curvature perturbation can source gravitational waves via nonlinear coupling terms like scale-scalar-tensor. In this paper, we only briefly discuss the impact of the non-Gaussian curvature perturbation on the amplitude (via the normalization by the PBH abundance) as well as the shape of the induced GW spectrum. For a general discussion of scalar-induced GWs, see [86, 253] as well as Chap. 17.

For each polarization, the tensor perturbation obeys the following equation of motion [16, 39, 197]

$$h''_{\mathbf{k}} + 2\mathcal{H}h'_{\mathbf{k}} + k^2 h_{\mathbf{k}} = \int \frac{d^3p}{(2\pi)^{3/2}} e^{ij}_{\mathbf{k}} p_i p_j \left[2\Phi_{\mathbf{p}}\Phi_{\mathbf{k-p}} + \frac{4\left(a\Phi_{\mathbf{p}}\right)'\left(a\Phi_{\mathbf{k-p}}\right)'}{3(1+w)a^2\mathcal{H}^2} \right], \tag{7.80}$$

where $\Phi = \frac{2}{3}\mathcal{R}$ is the curvature perturbation in longitudinal gauge calculated in the radiation dominated era, and $e^{ij}_{\mathbf{k}}$ is one of the polarization tensor. After solving the equation of motion (7.80), we can calculate the energy spectrum of the GWs generated at the horizon reentry, and evaluate it until today. For a simple spectral shape of $\mathcal{R}$ like a δ-function peak [146], lognormal peak [203], flat plateau [221], etc., the GW spectrum can be calculated semi-analytically. Suppose the non-Gaussian curvature perturbation given by the local form (7.17), $\mathcal{R} = \mathscr{F}(\mathcal{R}_g)$, can be expanded as

$$\mathcal{R} = \mathcal{R}_g + F_{\rm NL}\left(\mathcal{R}_g^2 - \langle \mathcal{R}_g^2\rangle\right) + G_{\rm NL}\mathcal{R}_g^3 + H_{\rm NL}\left(\mathcal{R}_g^4 - \langle \mathcal{R}_g^4\rangle\right) + I_{\rm NL}\mathcal{R}_g^5 + \cdots \tag{7.81}$$

where $F_{\rm NL} = (3/5)f_{\rm NL}$, $G_{\rm NL} = (9/25)g_{\rm NL}$, etc. The GW spectrum today is given by a redshifted GW spectrum from the matter-radiation equality [56, 246]

$$\Omega_{\rm GW}(f, \eta_0)h^2 = 1.6\times 10^{-5}\left(\frac{g_{*s}(\eta_k)}{106.75}\right)^{-1/3}\left(\frac{\Omega_{r,0}h^2}{4.1\times 10^{-5}}\right)\Omega_{\rm GW,eq}(f), \tag{7.82}$$

where

$$\begin{aligned}
\Omega_{\rm GW,eq}(k) = 3 \int_0^\infty \mathrm{d}v \int_{|1-v|}^{1+v} \mathrm{d}u \frac{1}{4u^2v^2}\left[\frac{4v^2-(1+v^2-u^2)^2}{4uv}\right]^2 \left(\frac{u^2+v^2-3}{2uv}\right)^4 \\
\cdot \left[\left(\ln\left|\frac{3-(u+v)^2}{3-(u-v)^2}\right| - \frac{4uv}{u^2+v^2-3}\right)^2 + \pi^2\Theta\left(u+v-\sqrt{3}\right)\right] \\
\cdot \left(\mathcal{P}_{\mathcal{R}g}(uk) + F_{\rm NL}^2 \int_0^\infty \mathrm{d}\nu \int_{|1-\nu|}^{1+\nu} \mathrm{d}\mu \, \frac{\mathcal{P}_{\mathcal{R}g}(\mu uk)\mathcal{P}_{\mathcal{R}g}(\nu uk)}{\mu^2\nu^2} + \cdots\right) \\
\cdot \left(\mathcal{P}_{\mathcal{R}g}(vk) + F_{\rm NL}^2 \int_0^\infty \mathrm{d}\lambda \int_{|1-\lambda|}^{1+\lambda} \mathrm{d}\rho \, \frac{\mathcal{P}_{\mathcal{R}g}(\lambda vk)\mathcal{P}_{\mathcal{R}g}(\rho vk)}{\lambda^2\rho^2} + \cdots\right) + \cdots .
\end{aligned} \tag{7.83}$$

Here for simplicity we omit the higher-order terms as well as the contribution from the connected part $\langle \mathcal{R}^4\rangle_c$, which are negligibly small when $F_{\rm NL} \lesssim 10$. For the full expression, see Ref. [4] for quadratic expansion (up to $F_{\rm NL}$), Refs. [155, 269] for cubic expansion (up to $G_{\rm NL}$), Ref. [202] for quintic expansion (7.81) (up to $I_{\rm NL}$), and Ref. [2] for ultra-slow-roll inflation up to $\mathcal{A}_{\mathcal{R}}^4$.

The energy spectrum of the induced GW (7.83) can only be calculated for perturbative series like (7.81). Fortunately, because the induced GWs mainly depends on the power spectrum of the curvature perturbation $\mathcal{P}_{\mathcal{R}}$ which is at most $\sim 10^{-2}$ for an enhanced peak, the higher order terms like $G_{\rm NL}\mathcal{R}^3$, $H_{\rm NL}\mathcal{R}^4$ contribute negligibly to the GW spectrum, as long as these coefficients $G_{\rm NL}$'s are not too large. Loose bounds for them are $G_{\rm NL} \ll O(10^2)F_{\rm NL}$, $H_{\rm NL} \ll O(10^4)F_{\rm NL}$, etc., which are easy to be satisfied for most of the models.

There are two important examples. In the curvaton scenario with negligible curvaton energy density at its decay, the perturbative series of (7.21) is valid even if $f_{\rm NL} \gg 1$, as the higher order terms ($G_{\rm NL}\mathcal{R}^2$ etc.) are always strongly suppressed. In the ultra-slow-roll inflation with a sharp end, $\mathcal{R}(\mathcal{R}_{\rm g})$ is given by (7.57), of which the nonlinear parameters are

$$F_{\rm NL} = 3/2, \quad G_{\rm NL} = 3, \quad H_{\rm NL} = 27/4, \quad \cdots . \tag{7.84}$$

This surely guarantees that higher-order contributions are negligible, which was shown explicitly up to $\mathcal{A}_{\mathcal{R}}^4$ in Ref. [2].

According to the current observational constraints, asteroid mass PBHs can serve as all the dark matter, of which the scalar induced GWs are in the milliHz to deciHz band, to be explored by the space-borne interferometers. As an concrete example, we calculate the induced GW spectra for monochromatic power spectra (7.71) with different $f_{\rm NL}$'s, normalized such that PBHs of $M_{k_*} \approx 10^{-12}M_\odot$ are all the dark matter. The resultant induced GWs peak at $k_* = 6.73 \times 10^{-3}$ Hz, which are shown in Fig. 7.7 together with the power-law integrated sensitivity curves [243] of the space-borne interferometers LISA [13–15, 34], Taiji [219], TianQin [161], and DECIGO [136, 137]. The induced GW from Gaussian curvature perturbation with a

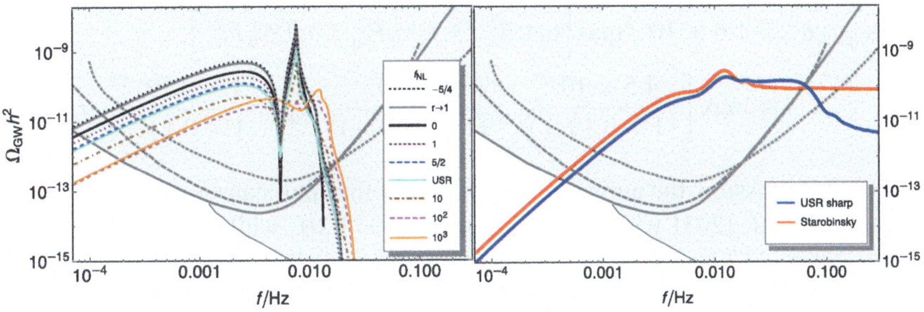

Fig. 7.7 Left: The spectrum of GWs induced by the enhanced narrow-peak power spectrum (7.71). The amplitude $\mathcal{A}_{\mathcal{R}}$ for different parameters follows Table 7.1, such that PBHs $\sim 10^{-12}M_\odot$ are all the dark matter, shown in Fig. 7.6. From right to left, the orange, magenta dashed, brown dot-dashed, cyan thick, blue dashed, purple dotted, black thick, gray, and black dotted curves are for $f_{\rm NL} = 10^3$, $f_{\rm NL} = 10^2$, $f_{\rm NL} = 10$, ultra-slow-roll, $f_{\rm NL} = 5/2$, $f_{\rm NL} = 1$, Gaussian case ($f_{\rm NL} = 0$), curvaton scenario with $r \to 1$, and $f_{\rm NL} = -5/4$, respectively. The curves are drawn by [3], following [4], together with the designed power-law integrated sensitivity curve of space-borne interferometers LISA (gray thick) [38], Taiji (gray dashed) [251], TianQin (gray dotted) [156], and DECIGO (gray thin) [136, 137]. Right: Two more realistic signals, the spectrum of GWs induced by ultra-slow-roll inflation with a sharp transition to slow-roll (blue), and a smooth transition to slow-roll, i.e. Starobinsky's linear potential model (red). The right panel was drawn by (7.83) with the help of the SIGWfast [252] package

monochromatic power spectrum displays an infrared scaling of $\propto k^2$, a dip of zero at $\sqrt{2/3}k_*$, a resonant peak of logarithmic divergence at $(2/\sqrt{3})k_*$, and a sharp cutoff beyond $2k_*$ [146]. This infrared scaling is a consequence of the infinitesimal scale of the source [55], which is preserved even in the non-Gaussian case [4]. As we commented before, positive $f_{\rm NL}$ can enhance the PBH formation, which requires less $\mathcal{A}_\mathcal{R}$ for a fixed PBH abundance (e.g. $f_{\rm PBH}=1$), thus smaller GW spectrum is displayed. However, when $f_{\rm NL}\lesssim 10$, the suppression is within one order of magnitude, which makes the amplitude and shape of the GW spectrum similar to the Gaussian one and difficult to be distinguished. In this case, the GWs beyond the cutoff frequency $2k_*$ is purely from the non-Gaussian part, which is a smoking gun of primordial non-Gaussianity [2, 56]. Although these contributions are tiny compared with the main peak, they could be probed by the deci-hertz GW detector like BBO (0.1 to 1 Hz) [77, 79] or DECIGO (10^{-2} to 1 Hz) [136, 137].

As $f_{\rm NL}$ increases, the main peak contributed by the Gaussian part decreases due to the lower requirement of $\mathcal{A}_\mathcal{R}$ because of the enhanced productivity of PBHs. However, the non-Gaussian contribution increases slightly, and finally dominates the GW spectrum and changes its shape significantly. The non-Gaussian source is roughly a convolution of two spectra with different wavenumbers, which fills the dip at $\sqrt{2/3}k_*$. Also, $\Omega_{\rm GW}^{\rm (NG)}$ has a mild bump instead of a sharp peak at $\sim\sqrt{3}k_*$, of which the amplitude is $\sim F_{\rm NL}^4\mathcal{A}_\mathcal{R}^4$. It becomes visible by LISA for $f_{\rm NL}\gtrsim 20$.

A very important issue is that when PBHs constitute all the dark matter, $\Omega_{\rm GW}$ has a lower bound which is detectable by the space-borne interferometers [56]. To see it clearly, note that the peak value of $\Omega_{\rm GW}$ is given by the maximum of the three peaks contributed by $O(\mathcal{A}_\mathcal{R}^2)$-, $O(\mathcal{A}_\mathcal{R}^3 f_{\rm NL}^2)$-, and $O(\mathcal{A}_\mathcal{R}^4 f_{\rm NL}^4)$-terms:

$$\begin{aligned}\Omega_{\rm GW,peak}h^2 &\approx 1.6\times10^{-5}\max\left[6.4\mathcal{A}_\mathcal{R}^2,\ 3.7\mathcal{A}_\mathcal{R}^3F_{\rm NL}^2,\ 3.9\mathcal{A}_\mathcal{R}^4F_{\rm NL}^4\right],\\ &\approx \max\left[\frac{4.5\times10^{-9}}{(1+0.1f_{\rm NL})^2},\ \frac{6.1\times10^{-12}}{(1+0.1f_{\rm NL})^3}f_{\rm NL}^2,\ \frac{1.5\times10^{-14}}{(1+0.1f_{\rm NL})^4}f_{\rm NL}^4\right].\end{aligned}$$

In the first expression, the numerical coefficient of the Gaussian contribution ("6.4") comes from Ref. [203] with smoothing scale $\Delta=0.01$, while the coefficients of the $F_{\rm NL}^2$-term ("3.7") and $F_{\rm NL}^4$-term ("3.9") come from Ref. [4] (also with a width of $\Delta=0.01$). In the second equality, we use (7.79) for PBH as all the dark matter. From the above fitting formula, it is easy to see that $\Omega_{\rm GW,peak}h^2$ has a minimum of $\Omega_{\rm GW,peak}^{\rm (min)}h^2\sim 8.4\times10^{-11}$ when $f_{\rm NL}\sim 63$. After that, $\Omega_{\rm GW,peak}h^2$ only increase slightly, and finally approaches a constant of 1.1×10^{-10}. This minimum is still well above the sensitivity curves of space-borne interferometers. Therefore, the detectability of scalar-induced GWs is robust against the nonlinear parameter $f_{\rm NL}$, λ_- (see (7.34) and (7.42)), or r (see (7.48) and (7.49)), which made it an important scientific goal of LISA [24, 31, 32, 34], Taiji [216], and TianQin [156] to probe the PBH dark matter.

We should caution that the δ-function peak we considered above is not physical. For more realistic models, the enhancement and decay of the power spectrum around the peak are milder, usually obeying power-law. Especially, in the typical model of ultra-slow-roll inflation, $\mathcal{P}_{\mathcal{R}}(k)$ increases as $\propto k^4$ before it reaches the maximum [49]. There might be modulated oscillations of period $\sim \pi$ if the transition from slow-roll to ultra-slow-roll is abrupt [42, 76, 206, 232]. The UV behavior of the power spectrum depends on the shape of the second slow-roll potential, usually displaying another power-law $\mathcal{P}_{\mathcal{R}} \propto k^{n_{s,2}-1}$. If the ultra-slow-roll stage ends sharply, the power spectrum drops off a little bit with new modulated oscillations with a larger period of $\sim \pi e^{N_{\rm USR}}$ [206]. According to our discussion in Sect. 7.3.3, these oscillations mark the sharp end of ultra-slow-roll stage, which are accompanied with the logarithmic relation (7.30). Based on the analytical expression of $\mathcal{P}_{\mathcal{R}}(k)$ (cf. Fig. 3.4 in Chap. 3) as well as the statistical properties, we can draw the GW spectrum associated with a fixed PBH abundance. Currently, there is no satisfactory method of calculating PBH abundance for such a broad spectrum with non-Gaussianity, so we choose the same parameters as in the narrow peak case (marked "USR" in Table 7.1), aiming to demonstrate the feature but not the exact amplitude of the induced GW spectrum in the ultra-slow-roll inflation. This is shown in the right panel of Fig. 7.7. For comparison, we also draw the induced GW from Starobinsky's linear potential model (with the narrow-peak amplitude), which represents the smooth end of ultra-slow-roll stage with negligible non-Gaussianity. We can see that similar to the δ-function peak case, non-Gaussianity enhances the UV part of the induced GW spectrum, which makes the primary peak of the GW spectrum obscure. A drop off with modulated oscillations can be recognized as a smoking gun for the non-Gaussianity in the ultra-slow-roll inflation shown in (7.30), which are detectable by DECIGO and BBO when PBHs constitute all the dark matter.

When the transition from ultra-slow-roll to the second slow-roll stage is smooth, the curvature perturbation is dominated by the δN contributed by the second slow-roll stage, which has the normal consistency relation $f_{\rm NL} = (5/12)(1 - n_{s,2})$ with $n_{s,2}$ the UV tilt of $\mathcal{P}_{\mathcal{R}}(k)$. In such induced GW spectra, there will be no modulated oscillations with period of $\sim \pi e^{N_{\rm USR}}$, but $f_{\rm NL}$ is determined by the UV slope of the GW spectrum, as $\Omega_{\rm GW}(k) \sim k^{2(n_{s,2}-1)}$. This can also be used to probe non-Gaussianity [26].

In summary, it is clearly shown in Figs. 7.6 and 7.7 that the dependence of the induced GW amplitude on the non-Gaussianity is mild once the PBH abundance is fixed. However, the non-Gaussian impact on the spectral shape could be detected in the future.

7.6 Conclusion and Discussion

In this chapter we review the impact of the primordial non-Gaussinity in the curvature perturbation on PBH formation as well as on the generation of scalar-induced GWs. We discussed some concrete inflation models to realize these interesting non-Gaussian curvature perturbations, and demonstrate how to calculate the PBH abundance as well as the induced GWs when non-Gaussianity is taken into account. Positive/negative non-Gaussianity can greatly enhance/suppress the PBH abundance, as they change the tail of the PDF significantly. However, the induced GW is mainly determined by the power spectrum, thus non-Gaussianity can not significantly change the induced GW spectrum. When fixing the PBH abundance, positive/negative non-Gaussianity only mildly suppresses/enhances the induced GWs. Therefore, the amplitude of the induded GW spectrum in the presence of abundant PBHs is a relatively robust prediction, which is an important scientific goal for many current and future experiments.

We only focus on the simplest local-type non-Gaussianity, which gives $\mathcal{R}(\mathcal{R}_g)$ as a function of Gaussian variable $\mathcal{R}_g$. Scale dependence of such a relation is not considered here [198, 199, 210, 211, 242]. Especially, various types of non-Gaussianities like equilateral, orthogonal, or folded shapes are also important, not only on the abundance but also on the clustering [82, 172, 212, 228, 236, 238, 263, 267]. Recently, Ref. [171] studied the bispectrum and trispectrum in the PBH formation systematically, and shows that in the narrow power spectrum case, the nonlinear parameter $f_{\rm NL}$ entering the calculation of PBH formation should be a linear combination of $f_{\rm NL}$ of local, equilateral, folded, and orthogonal shapes:

$$f_{\rm NL}^{\rm eff} \equiv f_{\rm NL}^{\rm loc} - 3 f_{\rm NL}^{\rm eq} + 3 f_{\rm NL}^{\rm fo} - 9 f_{\rm NL}^{\rm ort}, \tag{7.85}$$

which implies that other shapes of non-Gaussianity should also be taken into account when calculating the PBH abundance. Such non-Gaussianities are important for non-canonical kinetic terms, like for instance k-essence or G-inflation [74, 110, 157, 255, 271].

Our starting point is the δN formalism, which assumes that the Hubble patches we study are far enough from each other, such that each patch can be seen as a local Friedmann universe where the field value (and other perturbed quantities) takes a random initial value and evolves independently in the later stage. This is called the *separate universe approach*, which works quite well in the slow-roll regime. However, in the beginning of ultra-slow-roll stage (marked by $\varphi_{\rm i}$), the gradient terms of both growing mode and decaying mode are rather important, which give rise to the well-known k^4 growth [49, 76, 154]. Such a k-dependence cannot be recovered by separate universe approach and the classical δN formalism based on it [133]. Fortunately, such a failure is only serious for $\varphi \lesssim \sqrt{\pi}\varphi_{\rm i}$ [87], which makes the separate universe approach and the δN formalism still valid at the peak of the enhanced power spectrum, which is located at $\varphi_{\rm p} \approx \pi\varphi_{\rm i}$ [206]. Apparently, if we extend the δN formalism to include the spatial curvature on the initial slice taken slightly after horizon exit, the leading-order correction of the gradient expansion can be taken into account [23].

We did not discuss the interesting case of negative non-Gaussianity, which is connected to Type II PBHs. Simple estimate shows that negative non-Gaussianity will significantly suppress the PBH abundance, which was calculated in a perturbative way [264]. For a fully nonlinear curvature perturbation with a logarithmic dependence on the Gaussian variable, the PDF of $\mathcal{R}$ displays a Gumbel-like distribution [204] (also called double exponential suppression [129]):

$$\mathbb{P}(\mathcal{R}) \sim \exp\left(-c^2 e^{2\lambda_- \mathcal{R}}\right). \tag{7.86}$$

This can be realized, for instance, in the $r \to 1$ limit of the curvaton scenario which corresponds to $\lambda_- = 3/2$. Direct calculation shows that it can significantly suppress PBH formation. This suppression was utilized to avoid the overproduction of sub-solar-mass PBHs when interpreting the nHz stochastoc GW background as the scalar-induced GWs [73, 107, 160]. Recent study by theory of peaks indicates that the PBH formation for negative f_{NL} is quite difficult [144], and numerical simulations show that there is a lower bound $f_{\mathrm{NL}} \gtrsim -1.2$ below which the Type I PBHs can not form, based on a monochromatic power spectrum [93]. Around this regime the profiles of the density peaks become complicated, and the Type II PBH are crucial [245], which we did not touch here.

In many models, the non-attractor behavior is excited by the sudden change of the slop of the potential, which itself is continuous. This actually means that the change of slope is realized in an energy scale much larger than H, i.e. $V'' \gg H^2$ at φ_t. Another class of models display a discontinuity or a bump in the potential, which causes a loss of kinetic energy. Large non-Gaussianity can also be generated in such a case, which gives a highly asymmetric PDF of $\mathcal{R}$ [44, 52, 53, 135]. If the initial velocity is not enough, the inflaton may not overshoot the step/bump in some Hubble patches. Recent study shows that such a Hubble patch finally collapses to a PBH, as the energy density in the surrounding area become smaller than this patch and starts to compress it. When $f_{\mathrm{NL}} \gtrsim 2.6$, the PBHs from such bubbles are dominant [57, 90].

In principle, for a non-quadratic potential, the equation of motion for the inflaton cannot be solved analytically, and the analysis based on quadratic potentials in Sect. 7.3.3 can not apply. However, inspired by the logarithmic relation of $\mathcal{R}(\mathcal{R}_{\mathrm{g}})$ and its exponential-tail PDF (7.58), it is reasonable to conjecture a more general PDF of $\mathcal{R}$ directly [128, 129, 189]. For instance, a simple guess is

$$\mathbb{P}(\mathcal{R}) \propto \exp\left(-c\,|\mathcal{R}|^p\right), \tag{7.87}$$

where c is a positive constant. The PDF with $0 < p < 1$ is called heavy tailed, which is more effective in generating PBHs than the exponential-tail PDF. This extra efficiency is needed to generate enough supermassive PBHs to seed early galaxies, as the power spectrum of $\mathcal{R}$ on that scale is strongly constrained by μ- and y-distortions, $\mathcal{P}_{\mathcal{R}} \lesssim 10^{-5}$ [29, 51, 62, 117, 145, 159, 186]. Such a heavy-tail PDF can arise from self-interacting curvaton scenario [127], while a lighter tail of $p = 3/2$ can be generated in the non-perturbative regime of inflation model with a $\dot{\mathcal{R}}^4$ interaction [66].

Up to now, all the discussions above are based on the classical δN formalism, which allows us to divide the field value φ in a Hubble patch into "background" φ_0 and "perturbation" $\delta\varphi$. Such a split becomes difficult (if not completely impossible) when the perturbations are large and overwhelm the background value. As we commented, this happens when the inflaton is nearly static on a flat plateau due to the Hubble friction, and quantum diffusion must be taken into account [231, 233, 234]. Although the exponential-tail PDF like (7.58) appears as the leading order of the PDF in the semi-classical limit, more complicated and bizarre PDF arises in the quantum regime [19, 48, 95, 96, 99, 100, 200, 201, 248, 249]. This is the topic of the next chapter.

Acknowledgements I would like to thank Diego Cruces, Guillem Domènech, Jaume Garriga, Cristiano Germani, Cristian Joana, Misao Sasaki, Teruaki Suyama, Takahiro Tanaka, Dong-Gang Wang, and Jianing Wang for their comments and discussions, and to thank Ryoto Inui and Ao Wang for helping me draw Fig. 7.7. I also thank the hospitality of YITP, Kavli IPMU, IBS-CTPU, and KIAS during my visit when this paper was finalized, and thank all the participants of the 2nd International Workshop on Gravitational Waves and the Early Universe for the intriguing discussions. This work is supported in part by the National Key Research and Development Program of China Grant No. 2021YFC2203004, by Project No. 12475066 and No. 12047503 of the National Natural Science Foundation of China, by the JSPS Grant-in-Aid for Early-Career Scientists No. JP20K14461, by JSPS KAKENHI No. JP24K00624, and by the World Premier International Research Center Initiative (WPI Initiative), MEXT, Japan.

References

1. T.M.C. Abbott et al., Dark energy survey year 3 results: cosmological constraints from galaxy clustering and weak lensing. Phys. Rev. D **105**(2), 023520 (2022). https://doi.org/10.1103/PhysRevD.105.023520. arXiv: 2105.13549 [astro-ph.CO]
2. K.T. Abe et al., Primordial black holes and gravitational waves induced by exponential-tailed perturbations (2022). arXiv:2209.13891 [astro-ph.CO]
3. P. Adshead, K. Lozanov, Z. Weiner, *ngsigw-results*. Version v2023.1 (2023). https://doi.org/10.5281/zenodo.8176108
4. P. Adshead, K.D. Lozanov, Z.J. Weiner, Non-Gaussianity and the induced gravitational wave background. JCAP **10**, 080 (2021). https://doi.org/10.1088/1475-7516/2021/10/080. arXiv: 2105.01659 [astro-ph.CO]
5. A. Afzal et al., The NANOGrav 15 yr data set: search for signals from new physics. Astrophys. J. Lett. **951**(1), L11 (2023). https://doi.org/10.3847/2041-8213/acdc91. arXiv: 2306.16219 [astro-ph.HE]
6. G. Agazie et al., Comparing recent PTA results on the nanohertz stochastic gravitational wave background (2023). arXiv: 2309.00693 [astro-ph.HE]
7. G. Agazie et al., The NANOGrav 15 yr data set: evidence for a gravitational-wave background. Astrophys. J. Lett. **951**(1), L8 (2023). https://doi.org/10.3847/2041-8213/acdac6. arXiv: 2306.16213 [astro-ph.HE]
8. G. Agazie et al., The NANOGrav 15 yr data set: observations and timing of 68 millisecond pulsars. Astrophys. J. Lett. **951**(1), L9 (2023). https://doi.org/10.3847/2041-8213/acda9a. arXiv: 2306.16217 [astro-ph.HE]
9. Y. Akrami et al., Planck 2018 results. X. Constraints on inflation (2018). arXiv: 1807.06211 [astro-ph.CO]
10. Y. Akrami et al., Planck 2018 results. X. Constraints on inflation. Astron. Astrophys. **641**, A10 (2020). https://doi.org/10.1051/0004-6361/201833887. arXiv: 1807.06211 [astro-ph.CO]
11. A. Albrecht, P. Steinhardt, Cosmology for grand unified theories with radiatively induced symmetry breaking. Phys. Rev. Lett. **48**, 1220–1223 (1982). https://doi.org/10.1103/PhysRevLett.48.1220
12. Y. Ali-Haïmoud, Correlation function of high-threshold regions and application to the initial small-scale clustering of primordial black holes. Phys. Rev. Lett. **121**(8), 081304 (2018). https://doi.org/10.1103/PhysRevLett.121.081304. arXiv: 1805.05912 [astro-ph.CO]
13. P. Amaro-Seoane et al., eLISA/NGO: astrophysics and cosmology in the gravitational-wave millihertz regime. GW Notes **6**, 4–110 (2013). arXiv: 1201.3621 [astro-ph.CO]

14. P. Amaro-Seoane et al., Laser interferometer space antenna (2017). arXiv: 1702.00786 [astro-ph.IM]
15. P. Amaro-Seoane et al., Low-frequency gravitational-wave science with eLISA/NGO. Class. Quant. Grav. **29**, 124016 (2012). Ed. by Mark Hannam et al. https://doi.org/10.1088/0264-9381/29/12/124016. arXiv: 1202.0839 [gr-qc]
16. K.N. Ananda, C. Clarkson, D. Wands, The cosmological gravitational wave background from primordial density perturbations. Phys. Rev. D **75**, 123518 (2007). https://doi.org/10.1103/PhysRevD.75.123518. arXiv: gr-qc/0612013
17. K. Ando, M. Kawasaki, H. Nakatsuka, Formation of primordial black holes in an axionlike curvaton model. Phys. Rev. D **98**(8), 083508 (2018). https://doi.org/10.1103/PhysRevD.98.083508. arXiv: 1805.07757 [astro-ph.CO]
18. K. Ando et al., Primordial black holes for the LIGO events in the axionlike curvaton model. Phys. Rev. D **97**(12), 123512 (2018). https://doi.org/10.1103/PhysRevD.97.123512. arXiv: 1711.08956 [astro-ph.CO]
19. C. Animali, V. Vennin, Primordial black holes from stochastic tunnelling (2022). arXiv: 2210.03812 [astro-ph.CO]
20. J. Antoniadis et al., The second data release from the European pulsar timing array I. The dataset and timing analysis (2023). https://doi.org/10.1051/0004-6361/202346841. arXiv: 2306.16224 [astro-ph.HE]
21. J. Antoniadis et al., The second data release from the European pulsar timing array III. Search for gravitational wave signals. Astron. Astrophys. **678**, A50 (2023). https://doi.org/10.1051/0004-6361/202346844. arXiv: 2306.16214 [astro-ph.HE]
22. J. Antoniadis et al., The second data release from the European Pulsar Timing Array: V. Implications for massive black holes, dark matter and the early Universe (2023). arXiv: 2306.16227 [astro-ph.CO]
23. D. Artigas, S. Pi, T. Tanaka, Extended δN formalism (2024). arXiv: 2408.09964 [astro-ph.CO]
24. K.G. Arun et al., New horizons for fundamental physics with LISA. Living Rev. Rel. **25**(1), 4 (2022). https://doi.org/10.1007/s41114-022-00036-9. arXiv: 2205.01597 [gr-qc]
25. H. Assadullahi, D. Wands, Constraints on primordial density perturbations from induced gravitational waves. Phys. Rev. D **81**, 023527 (2010). https://doi.org/10.1103/PhysRevD.81.023527. arXiv: 0907.4073 [astro-ph.CO]
26. V. Atal, G. Domènech, Probing non-Gaussianities with the high frequency tail of induced gravitational waves. JCAP **06**, 001 (2021). https://doi.org/10.1088/1475-7516/2021/06/001. arXiv: 2103.01056 [astro-ph.CO]
27. V. Atal, J. Garriga, A. Marcos-Caballero, Primordial black hole formation with non-Gaussian curvature perturbations. JCAP **09**, 073 (2019). https://doi.org/10.1088/1475-7516/2019/09/073. arXiv: 1905.13202 [astro-ph.CO]
28. V. Atal, C. Germani, The role of non-Gaussianities in primordial black hole formation. Phys. Dark Univ. **24**, 100275 (2019). https://doi.org/10.1016/j.dark.2019.100275. arXiv: 1811.07857 [astro-ph.CO]
29. V. Atal, A. Sanglas, N. Triantafyllou, NANOGrav signal as mergers of stupendously large primordial black holes (2020). arXiv: 2012.14721 [astro-ph.CO]
30. V. Atal et al., PBH in single field inflation: the effect of shape dispersion and non-Gaussianities. JCAP **05**, 022 (2020). https://doi.org/10.1088/1475-7516/2020/05/022. arXiv: 1908.11357 [astro-ph.CO]
31. P. Auclair et al., Cosmology with the laser interferometer space antenna (2022). arXiv: 2204.05434 [astro-ph.CO]
32. E. Bagui et al., Primordial black holes and their gravitational-wave signatures (2023). arXiv: 2310.19857 [astro-ph.CO]
33. G. Ballesteros, J. Gambín Egea, One-loop power spectrum in ultra slow-roll inflation and implications for primordial black hole dark matter (2024). arXiv: 2404.07196 [astro-ph.CO]

34. E. Barausse et al., Prospects for fundamental physics with LISA. Gen. Rel. Grav. **52**(8), 81 (2020). https://doi.org/10.1007/s10714-020-02691-1. arXiv: 2001.09793 [gr-qc]
35. P. Bari et al., Gravitational waves induced by scalar-tensor mixing. Phys. Rev. D **109**(2), 023509 (2024). https://doi.org/10.1103/PhysRevD.109.023509. arXiv: 2307.05404 [astro-ph.CO]
36. N. Bartolo, S. Matarrese, A. Riotto, On nonGaussianity in the curvaton scenario. Phys. Rev. D **69**, 043503 (2004). https://doi.org/10.1103/PhysRevD.69.043503. arXiv: hep-ph/0309033
37. N. Bartolo et al., Primordial black hole dark matter: LISA serendipity. Phys. Rev. Lett. **122**(21), 211301 (2019). https://doi.org/10.1103/PhysRevLett.122.211301. arXiv: 1810.12218 [astro-ph.CO]
38. N. Bartolo et al., Science with the space-based interferometer LISA. IV: probing inflation with gravitational waves. JCAP **12**, 026 (2016). https://doi.org/10.1088/1475-7516/2016/12/026. arXiv: 1610.06481 [astro-ph.CO]
39. D. Baumann et al., Gravitational wave spectrum induced by primordial scalar perturbations. Phys. Rev. D **76**, 084019 (2007). https://doi.org/10.1103/PhysRevD.76.084019. arXiv: hep-th/0703290
40. R. Bean, J. Magueijo, Could supermassive black holes be quintessential primordial black holes? Phys. Rev. D **66**, 063505 (2002). https://doi.org/10.1103/PhysRevD.66.063505. arXiv: astro-ph/0204486
41. M. Biagetti, G. Franciolini, A. Riotto, High-redshift JWST observations and primordial Non-Gaussianity. Astrophys. J. **944**(2), 113 (2023). https://doi.org/10.3847/1538-4357/acb5ea. arXiv: 2210.04812 [astro-ph.CO]
42. M. Biagetti et al., Primordial black holes from inflation and quantum diffusion. JCAP **07**, 032 (2018). https://doi.org/10.1088/1475-7516/2018/07/032. arXiv: 1804.07124 [astro-ph.CO]
43. M. Biagetti et al., The formation probability of primordial black holes. Phys. Lett. B **820**, 136602 (2021). https://doi.org/10.1016/j.physletb.2021.136602. arXiv: 2105.07810 [astro-ph.CO]
44. V. Briaud, V. Vennin, Uphill inflation. JCAP **06**, 029 (2023). https://doi.org/10.1088/1475-7516/2023/06/029. arXiv: 2301.09336 [astro-ph.CO]
45. R. Brout, F. Englert, E. Gunzig, The creation of the universe as a quantum phenomenon. Annals Phys. **115**, 78 (1978). https://doi.org/10.1016/0003-4916(78)90176-8
46. E. Bugaev, P. Klimai, Constraints on the induced gravitational wave background from primordial black holes. Phys. Rev. D **83**, 083521 (2011). https://doi.org/10.1103/PhysRevD.83.083521. arXiv: 1012.4697 [astro-ph.CO]
47. E. Bugaev, P. Klimai, Induced gravitational wave background and primordial black holes. Phys. Rev. D **81**, 023517 (2010). https://doi.org/10.1103/PhysRevD.81.023517. arXiv: 0908.0664 [astro-ph.CO]
48. J.S. Bullock, J.R. Primack, NonGaussian fluctuations and primordial black holes from inflation. Phys. Rev. D **55**, 7423–7439 (1997). https://doi.org/10.1103/PhysRevD.55.7423. arXiv: astro-ph/9611106
49. C. Byrnes, P. Cole, S. Patil, Steepest growth of the power spectrum and primordial black holes. JCAP **1906**, 028 (2019). https://doi.org/10.1088/1475-7516/2019/06/028. arXiv: 1811.11158 [astro-ph.CO]
50. C.T. Byrnes, E.J. Copeland, A.M. Green, Primordial black holes as a tool for constraining non-Gaussianity. Phys. Rev. D **86**, 043512 (2012). https://doi.org/10.1103/PhysRevD.86.043512. arXiv: 1206.4188 [astro-ph.CO]
51. C.T. Byrnes, J. Lesgourgues, D. Sharma, Robust μ-distortion constraints on primordial supermassive black holes from non-Gaussian perturbations (2024). arXiv: 2404.18475 [astro-ph.CO]
52. Y.-F. Cai et al., Highly non-Gaussian tails and primordial black holes from single-field inflation (2022). arXiv: 2207.11910 [astro-ph.CO]
53. Y.-F. Cai et al., One small step for an inflaton, one giant leap for inflation: a novel non-Gaussian tail and primordial black holes. Phys. Lett. B **834**, 137461 (2022). https://doi.org/10.1016/j.physletb.2022.137461. arXiv: 2112.13836 [astro-ph.CO]

54. Y.-F. Cai et al., Revisiting non-Gaussianity from non-attractor inflation models. JCAP **05**, 012 (2018). https://doi.org/10.1088/1475-7516/2018/05/012. arXiv: 1712.09998 [astro-ph.CO]
55. R.-G. Cai, S. Pi, M. Sasaki, Universal infrared scaling of gravitational wave background spectra. Phys. Rev. D **102**(8), 083528 (2020). https://doi.org/10.1103/PhysRevD.102.083528. arXiv: 1909.13728 [astro-ph.CO]
56. R.-G. Cai, S. Pi, M. Sasaki, Gravitational waves induced by non-Gaussian scalar perturbations. Phys. Rev. Lett. **122**(20), 201101 (2019). https://doi.org/10.1103/PhysRevLett.122.201101. arXiv: 1810.11000 [astro-ph.CO]
57. A. Caravano, K. Inomata, S. Renaux-Petel, The inflationary butterfly effect: non-perturbative dynamics from small-scale features (2024). arXiv: 2403.12811 [astro-ph.CO]
58. C. Carbone, S. Matarrese, A unified treatment of cosmological perturbations from super-horizon to small scales. Phys. Rev. D **71**, 043508 (2005). https://doi.org/10.1103/PhysRevD.71.043508. arXiv: astro-ph/0407611
59. B. Carr, The primordial black hole mass spectrum. Astrophys. J. **201**, 1–19 (1975). https://doi.org/10.1086/153853
60. B. Carr, F. Kuhnel, Primordial black holes as dark matter: recent developments. Ann. Rev. Nucl. Part. Sci. **70**, 355–394 (2020). https://doi.org/10.1146/annurev-nucl-050520-125911. arXiv: 2006.02838 [astro-ph.CO]
61. B. Carr, F. Kuhnel, L. Visinelli, Constraints on stupendously large black holes. Mon. Not. Roy. Astron. Soc. **501**, 2 (2021). https://doi.org/10.1093/mnras/staa3651. arXiv: 2008.08077 [astro-ph.CO]
62. B. Carr, J. Silk, Primordial black holes as generators of cosmic structures. Mon. Not. Roy. Astron. Soc. **478**(3), 3756–3775 (2018). https://doi.org/10.1093/mnras/sty1204
63. B. Carr et al., Constraints on primordial black holes. Rept. Prog. Phys. **84**(11), 116902 (2021). https://doi.org/10.1088/1361-6633/ac1e31. arXiv: 2002.12778 [astro-ph.CO]
64. B. Carr et al., Observational evidence for primordial black holes: a positivist perspective. Phys. Rept. **1054**, 1–68 (2024). https://doi.org/10.1016/j.physrep.2023.11.005. arXiv: 2306.03903 [astro-ph.CO]
65. B.J. Carr, S.W. Hawking, Black holes in the early Universe. Mon. Not. Roy. Astron. Soc. **168**, 399–415 (1974). https://doi.org/10.1093/mnras/168.2.399
66. M. Celoria et al., Beyond perturbation theory in inflation. JCAP **06**, 051 (2021). https://doi.org/10.1088/1475-7516/2021/06/051. arXiv: 2103.09244 [hep-th]
67. Z. Chang, X. Zhang, J.-Z. Zhou, Gravitational waves from primordial scalar and tensor perturbations. Phys. Rev. D **107**(6), 063510 (2023). https://doi.org/10.1103/PhysRevD.107.063510. arXiv: 2209.07693 [astro-ph.CO]
68. C. Chen et al., Growth of curvature perturbations for PBH formation & detectable GWs in non-minimal curvaton scenario revisited. JCAP **08**, 041 (2023). https://doi.org/10.1088/1475-7516/2023/08/041. arXiv: 2305.12325 [astro-ph.CO]
69. X. Chen et al., A single field inflation model with large local non-Gaussianity. EPL **102**(5), 59001 (2013). https://doi.org/10.1209/0295-5075/102/59001. arXiv: 1301.5699 [hep-th]
70. S.-L. Cheng, D.-S. Lee, K.-W. Ng, Primordial perturbations from ultra-slow-roll single-field inflation with quantum loop effects (2023). arXiv: 2305.16810 [astro-ph.CO]
71. M.W. Choptuik, Universality and scaling in gravitational collapse of a massless scalar field. Phys. Rev. Lett. **70**, 9–12 (1993). https://doi.org/10.1103/PhysRevLett.70.9
72. S. Choudhury, S. Panda, M. Sami, Quantum loop effects on the power spectrum and constraints on primordial black holes. JCAP **11**, 066 (2023). https://doi.org/10.1088/1475-7516/2023/11/066. arXiv: 2303.06066 [astro-ph.CO]
73. S. Choudhury et al., Primordial non-Gaussianity as a saviour for PBH overproduction in SIGWs generated by pulsar timing arrays for Galileon inflation (2023). arXiv: 2310.11034 [astro-ph.CO]
74. S. Choudhury et al., Primordial non-Gaussianity from ultra slow-roll Galileon inflation. JCAP **01**, 012 (2024). https://doi.org/10.1088/1475-7516/2024/01/012. arXiv: 2306.12334 [astro-ph.CO]

75. S. Clesse, J. García-Bellido, The clustering of massive primordial black holes as dark matter: measuring their mass distribution with advanced LIGO. Phys. Dark Univ. **15**, 142–147 (2017). https://doi.org/10.1016/j.dark.2016.10.002. arXiv: 1603.05234 [astro-ph.CO]
76. P.S. Cole et al., Steepest growth re-examined: repercussions for primordial black hole formation (2022). arXiv: 2204.07573 [astro-ph.CO]
77. V. Corbin, N.J. Cornish, Detecting the cosmic gravitational wave background with the big bang observer. Class. Quant. Grav. **23**, 2435–2446 (2006). https://doi.org/10.1088/0264-9381/23/7/014. arXiv: gr-qc/0512039
78. P. Creminelli, M. Zaldarriaga, Single field consistency relation for the 3-point function. JCAP **10**, 006 (2004). https://doi.org/10.1088/1475-7516/2004/10/006. arXiv: astro-ph/0407059
79. J. Crowder, N.J. Cornish, Beyond LISA: exploring future gravitational wave missions. Phys. Rev. D **72**, 083005 (2005). https://doi.org/10.1103/PhysRevD.72.083005. arXiv: gr-qc/0506015
80. M.W. Davies, P. Carrilho, D.J. Mulryne, Non-Gaussianity in inflationary scenarios for primordial black holes. JCAP **06**(06), 019 (2022). https://doi.org/10.1088/1475-7516/2022/06/019. arXiv: 2110.08189 [astro-ph.CO]
81. M.W. Davies, L. Iacconi, D.J. Mulryne, Numerical 1-loop correction from a potential yielding ultra-slow-roll dynamics. JCAP **04**, 050 (2024). https://doi.org/10.1088/1475-7516/2024/04/050. arXiv: 2312.05694 [astro-ph.CO]
82. V. De Luca, G. Franciolini, A. Riotto, Constraining the initial primordial black hole clustering with CMB distortion. Phys. Rev. D **104**(6), 063526 (2021). https://doi.org/10.1103/PhysRevD.104.063526. arXiv: 2103.16369 [astro-ph.CO]
83. V. De Luca et al., The ineludible non-Gaussianity of the primordial black hole abundance. JCAP **07**, 048 (2019). https://doi.org/10.1088/1475-7516/2019/07/048. arXiv: 1904.00970 [astro-ph.CO]
84. W. DeRocco, P.W. Graham, Constraining primordial black hole abundance with the galactic 511 keV line. Phys. Rev. Lett. **123**(25), 251102 (2019). https://doi.org/10.1103/PhysRevLett.123.251102. arXiv: 1906.07740 [astro-ph.CO]
85. V. Desjacques, A. Riotto, Spatial clustering of primordial black holes. Phys. Rev. D **98**(12), 123533 (2018). https://doi.org/10.1103/PhysRevD.98.123533. arXiv: 1806.10414 [astro-ph.CO]
86. G. Domènech, Scalar induced gravitational waves review. Universe **7**(11), 398 (2021). https://doi.org/10.3390/universe7110398. arXiv: 2109.01398 [gr-qc]
87. G. Domènech, G. Vargas, T. Vargas, An exact model for enhancing/suppressing primordial fluctuations. JCAP **03**, 002 (2024). https://doi.org/10.1088/1475-7516/2024/03/002. arXiv: 2309.05750 [astro-ph.CO]
88. K. Enqvist, S. Nurmi, Non-Gaussianity in curvaton models with nearly quadratic potential. JCAP **10**, 013 (2005). https://doi.org/10.1088/1475-7516/2005/10/013. arXiv: astro-ph/0508573
89. K. Enqvist, M.S. Sloth, Adiabatic CMB perturbations in pre - big bang string cosmology. Nucl. Phys. B **626**, 395–409 (2002). https://doi.org/10.1016/S0550-3213(02)00043-3. arXiv: hep-ph/0109214
90. A. Escrivà, V. Atal, J. Garriga, Formation of trapped vacuum bubbles during inflation, and consequences for PBH scenarios. JCAP **10**, 035 (2023). https://doi.org/10.1088/1475-7516/2023/10/035. arXiv: 2306.09990 [astro-ph.CO]
91. A. Escrivà, C. Germani, R.K. Sheth, Universal threshold for primordial black hole formation. Phys. Rev. D **101**(4), 044022 (2020). https://doi.org/10.1103/PhysRevD.101.044022. arXiv: 1907.13311 [gr-qc]
92. A. Escrivà, F. Kuhnel, Y. Tada, Primordial black holes (2022). arXiv: 2211.05767 [astro-ph.CO]
93. A. Escrivà et al., Simulation of primordial black holes with large negative non-Gaussianity. JCAP **05**(05), 012 (2022). https://doi.org/10.1088/1475-7516/2022/05/012. arXiv: 2202.01028 [astro-ph.CO]

94. C.R. Evans, J.S. Coleman, Observation of critical phenomena and selfsimilarity in the gravitational collapse of radiation fluid. Phys. Rev. Lett. **72**, 1782–1785 (1994). https://doi.org/10.1103/PhysRevLett.72.1782. arXiv: gr-qc/9402041
95. J.M. Ezquiaga, J. García-Bellido, V. Vennin, The exponential tail of inflationary fluctuations: consequences for primordial black holes. JCAP **03**, 029 (2020). https://doi.org/10.1088/1475-7516/2020/03/029. arXiv: 1912.05399 [astro-ph.CO]
96. J.M. Ezquiaga, J. García-Bellido, V. Vennin, Massive galaxy clusters like El Gordo hint at primordial quantum diffusion. Phys. Rev. Lett. (2022). arXiv:2207.06317. arXiv: 2207.06317 [astro-ph.CO]
97. G. Ferrante et al., Primordial black holes in the curvaton model: possible connections to pulsar timing arrays and dark matter. JCAP **06**, 057 (2023). https://doi.org/10.1088/1475-7516/2023/06/057. arXiv: 2305.13382 [astro-ph.CO]
98. G. Ferrante et al., Primordial non-Gaussianity up to all orders: theoretical aspects and implications for primordial black hole models. Phys. Rev. D **107**(4), 043520 (2023). https://doi.org/10.1103/PhysRevD.107.043520. arXiv: 2211.01728 [astro-ph.CO]
99. D.G. Figueroa et al., Implications of stochastic effects for primordial black hole production in ultra-slow-roll inflation. JCAP **05**(05), 027 (2022). https://doi.org/10.1088/1475-7516/2022/05/027. arXiv: 2111.07437 [astro-ph.CO]
100. D.G. Figueroa et al., Non-Gaussian tail of the curvature perturbation in stochastic ultraslow-roll inflation: implications for primordial black hole production. Phys. Rev. Lett. **127**(10), 101302 (2021). https://doi.org/10.1103/PhysRevLett.127.101302. arXiv: 2012.06551 [astro-ph.CO]
101. H. Firouzjahi, Loop corrections in bispectrum in USR inflation with PBHs formation (2024). arXiv: 2403.03841 [astro-ph.CO]
102. H. Firouzjahi, One-loop corrections in power spectrum in single field inflation. JCAP **10**, 006 (2023). https://doi.org/10.1088/1475-7516/2023/10/006. arXiv: 2303.12025 [astro-ph.CO]
103. H. Firouzjahi, Revisiting loop corrections in single field ultraslow-roll inflation. Phys. Rev. D **109**(4), 043514 (2024). https://doi.org/10.1103/PhysRevD.109.043514. arXiv: 2311.04080 [astro-ph.CO]
104. H. Firouzjahi, A. Riotto, Primordial black holes and loops in single-field inflation (2023). arXiv: 2304.07801 [astro-ph.CO]
105. G. Franciolini et al., Primordial black holes from inflation and non-Gaussianity. JCAP **03**, 016 (2018). https://doi.org/10.1088/1475-7516/2018/03/016. arXiv: 1801.09415 [astro-ph.CO]
106. G. Franciolini et al., One loop to rule them all: Perturbativity in the presence of ultra slow-roll dynamics (2023). arXiv: 2305.03491 [astro-ph.CO]
107. G. Franciolini et al., The recent gravitational wave observation by pulsar timing arrays and primordial black holes: the importance of non-Gaussianities (2023). arXiv: 2306.17149 [astro-ph.CO]
108. J. Fumagalli, Absence of one-loop effects on large scales from small scales in non-slow-roll dynamics (2023). arXiv: 2305.19263 [astro-ph.CO]
109. J. Fumagalli et al., One-loop infrared rescattering by enhanced scalar fluctuations during inflation (2023). arXiv: 2307.08358 [astro-ph.CO]
110. Q. Gao, Primordial black holes and secondary gravitational waves from chaotic inflation. Sci. China Phys. Mech. Astron. **64**(8), 280411 (2021). https://doi.org/10.1007/s11433-021-1708-9. arXiv: 2102.07369 [gr-qc]
111. J. García-Bellido, S. Clesse, Constraints from microlensing experiments on clustered primordial black holes. Phys. Dark Univ. **19**, 144–148 (2018). https://doi.org/10.1016/j.dark.2018.01.001. arXiv: 1710.04694 [astro-ph.CO]
112. J. García-Bellido, M. Peloso, C. Unal, Gravitational Wave signatures of inflationary models from primordial black hole dark matter. JCAP **09**, 013 (2017). https://doi.org/10.1088/1475-7516/2017/09/013. arXiv: 1707.02441 [astro-ph.CO]

113. S. Garcia-Saenz et al., No-go theorem for scalar-trispectrum-induced gravitational waves (2022). arXiv: 2207.14267 [astro-ph.CO]
114. C. Germani, I. Musco, Abundance of primordial black holes depends on the shape of the inflationary power spectrum. Phys. Rev. Lett. **122**(14), 141302 (2019). https://doi.org/10.1103/PhysRevLett.122.141302. arXiv: 1805.04087 [astro-ph.CO]
115. C. Germani, R.K. Sheth, Nonlinear statistics of primordial black holes from Gaussian curvature perturbations. Phys. Rev. D **101**(6), 063520 (2020). https://doi.org/10.1103/PhysRevD.101.063520. arXiv: 1912.07072 [astro-ph.CO]
116. C. Germani, R.K. Sheth, The statistics of primordial black holes in a radiation-dominated universe: recent and new results. Universe **9**(9), 421 (2023). https://doi.org/10.3390/universe9090421. arXiv: 2308.02971 [astro-ph.CO]
117. Y. Gouttenoire et al., Scrutinizing the primordial black holes interpretation of PTA gravitational waves and JWST early galaxies (2023). arXiv: 2307.01457 [astro-ph.CO]
118. A.D. Gow, T. Miranda, S. Nurmi, Primordial black holes from a curvaton scenario with strongly non-Gaussian perturbations. JCAP **11**, 006 (2023). https://doi.org/10.1088/1475-7516/2023/11/006. arXiv: 2307.03078 [astro-ph.CO]
119. A.D. Gow et al., Non-perturbative non-Gaussianity and primordial black holes. EPL **142**(4), 49001 (2023). https://doi.org/10.1209/0295-5075/acd417. arXiv: 2211.08348 [astro-ph.CO]
120. A.M. Green et al., A New calculation of the mass fraction of primordial black holes. Phys. Rev. D **70**, 041502 (2004). https://doi.org/10.1103/PhysRevD.70.041502. arXiv: astro-ph/0403181
121. A.H. Guth, The inflationary universe: a possible solution to the horizon and flatness problems. Phys. Rev. D **23**, 347–356 (1981). Ed. by L.-Z. Fang, R. Ruffini. https://doi.org/10.1103/PhysRevD.23.347
122. T. Harada, C.-M. Yoo, K. Kohri, Threshold of primordial black hole formation. Phys. Rev. D **88**(8), 084051 (2013). [Erratum: Phys. Rev. D **89**, 029903 (2014)]. https://doi.org/10.1103/PhysRevD.88.084051. arXiv: 1309.4201 [astro-ph.CO]
123. T. Harada et al., Cosmological long-wavelength solutions and primordial black hole formation. Phys. Rev. D **91**(8), 084057 (2015). https://doi.org/10.1103/PhysRevD.91.084057. arXiv: 1503.03934 [gr-qc]
124. I. Hawke, J.M. Stewart, The dynamics of primordial black hole formation. Class. Quant. Grav. **19**, 3687–3707 (2002). https://doi.org/10.1088/0264-9381/19/14/310
125. S. Hawking, Gravitationally collapsed objects of very low mass. Mon. Not. Roy. Astron. Soc. **152**, 75 (1971)
126. J.C. Hidalgo, The effect of non-Gaussian curvature perturbations on the formation of primordial black holes (2007). arXiv: 0708.3875 [astro-ph]
127. D. Hooper et al., Supermassive primordial black holes from inflation (2023). arXiv: 2308.00756 [astro-ph.CO]
128. S. Hooshangi, M.H. Namjoo, M. Noorbala, Rare events are nonperturbative: primordial black holes from heavy-tailed distributions. Phys. Lett. B **834**, 137400 (2022). https://doi.org/10.1016/j.physletb.2022.137400. arXiv: 2112.04520 [astro-ph.CO]
129. S. Hooshangi, M.H. Namjoo, M. Noorbala, Tail diversity from inflation. JCAP **09**, 023 (2023). https://doi.org/10.1088/1475-7516/2023/09/023. arXiv: 2305.19257 [astro-ph.CO]
130. L. Iacconi, D. Mulryne, D. Seery, Loop corrections in the separate universe picture. arXiv: 2312.12424 [astro-ph.CO]
131. K. Inomata, Curvature Perturbations Protected Against One Loop (2024). arXiv: 2403.04682 [astro-ph.CO]
132. P. Ivanov, Nonlinear metric perturbations and production of primordial black holes. Phys. Rev. D **57**, 7145–7154 (1998). https://doi.org/10.1103/PhysRevD.57.7145. arXiv: astro-ph/9708224
133. J.H.P. Jackson et al., The separate-universe approach and sudden transitions during inflation (2023). arXiv: 2311.03281 [astro-ph.CO]

134. S. Kasuya, M. Kawasaki, Axion isocurvature fluctuations with extremely blue spectrum. Phys. Rev. D **80**, 023516 (2009). https://doi.org/10.1103/PhysRevD.80.023516. arXiv: 0904.3800 [astro-ph.CO]
135. R. Kawaguchi, T. Fujita, M. Sasaki, Highly asymmetric probability distribution from a finite-width upward step during inflation. JCAP **11**, 021 (2023). https://doi.org/10.1088/1475-7516/2023/11/021. arXiv: 2305.18140 [astro-ph.CO]
136. S. Kawamura et al., The Japanese space gravitational wave antenna DECIGO. Class. Quant. Grav. **23**, S125–S132 (2006). Ed. by N. Mio. https://doi.org/10.1088/0264-9381/23/8/S17
137. S. Kawamura et al., The Japanese space gravitational wave antenna: DECIGO. Class. Quant. Grav. **28**, 094011 (2011). Ed. by Sasha Buchman and Ke-Xun Sun. https://doi.org/10.1088/0264-9381/28/9/094011
138. M. Kawasaki, N. Kitajima, T.T. Yanagida, Primordial black hole formation from an axionlike curvaton model. Phys. Rev. D **87**(6), 063519 (2013). https://doi.org/10.1103/PhysRevD.87.063519. arXiv: 1207.2550 [hep-ph]
139. M. Kawasaki, N. Kitajima, S. Yokoyama, Gravitational waves from a curvaton model with blue spectrum. JCAP **08**, 042 (2013). https://doi.org/10.1088/1475-7516/2013/08/042. arXiv: 1305.4464 [astro-ph.CO]
140. M. Kawasaki, A. Kusenko, T.T. Yanagida, Primordial seeds of supermassive black holes. Phys. Lett. B 711 (2012), 1–5. https://doi.org/10.1016/j.physletb.2012.03.056. arXiv: 1202.3848 [astro-ph.CO]
141. M. Kawasaki, H. Nakatsuka, Effect of nonlinearity between density and curvature perturbations on the primordial black hole formation. Phys. Rev. D **99**(12), 123501 (2019). https://doi.org/10.1103/PhysRevD.99.123501. arXiv: 1903.02994 [astro-ph.CO]
142. A. Kehagias, I. Musco, A. Riotto, Non-Gaussian formation of primordial black holes: effects on the threshold. JCAP **12**, 029 (2019). https://doi.org/10.1088/1475-7516/2019/12/029. arXiv: 1906.07135 [astro-ph.CO]
143. M. Khlopov, B.A. Malomed, I.B. Zeldovich, Gravitational instability of scalar fields and formation of primordial black holes. Mon. Not. Roy. Astron. Soc. **215**, 575–589 (1985)
144. N. Kitajima et al., Primordial black holes in peak theory with a non-Gaussian tail. JCAP **10**, 053 (2021). https://doi.org/10.1088/1475-7516/2021/10/053. arXiv: 2109.00791 [astro-ph.CO]
145. K. Kohri, T. Nakama, T. Suyama, Testing scenarios of primordial black holes being the seeds of supermassive black holes by ultracompact minihalos and CMB μ-distortions. Phys. Rev. D **90**(8), 083514 (2014). https://doi.org/10.1103/PhysRevD.90.083514. arXiv: 1405.5999 [astro-ph.CO]
146. K. Kohri, T. Terada, Semianalytic calculation of gravitational wave spectrum nonlinearly induced from primordial curvature perturbations. Phys. Rev. D **97**(12), 123532 (2018). https://doi.org/10.1103/PhysRevD.97.123532. arXiv: 1804.08577 [gr-qc]
147. T. Koike, T. Hara, S. Adachi, Critical behavior in gravitational collapse of radiation fluid: a renormalization group (linear perturbation) analysis. Phys. Rev. Lett. **74**, 5170–5173 (1995). https://doi.org/10.1103/PhysRevLett.74.5170. arXiv: gr-qc/9503007
148. E. Komatsu, D.N. Spergel, Acoustic signatures in the primary microwave background bispectrum. Phys. Rev. D **63**, 063002 (2001). https://doi.org/10.1103/PhysRevD.63.063002. arXiv: astro-ph/0005036
149. M. Kopp, S. Hofmann, J. Weller, Separate universes do not constrain primordial black hole Formation. Phys. Rev. D **83**, 124025 (2011). https://doi.org/10.1103/PhysRevD.83.124025. arXiv: 1012.4369 [astro-ph.CO]
150. J. Kristiano, J. Yokoyama, Comparing sharp and smooth transitions of the second slow-roll parameter in single-field inflation (2024). arXiv: 2405.12145 [astro-ph.CO]
151. J. Kristiano, J. Yokoyama, Response to criticism on "Ruling out primordial black hole formation from single-field inflation": a note on bispectrum and one-loop correction in single-field inflation with primordial black hole formation (2023). arXiv: 2303.00341 [hep-th]
152. J. Kristiano, J. Yokoyama, Ruling out primordial black hole formation from single-field inflation (2022). arXiv: 2211.03395 [hep-th]

153. R. Laha, Primordial black holes as a dark matter candidate are severely constrained by the galactic center 511 keV γ -Ray Line. Phys. Rev. Lett. **123**(25), 251101 (2019). https://doi.org/10.1103/PhysRevLett.123.251101. arXiv: 1906.09994 [astro-ph.HE]
154. S.M. Leach et al., Enhancement of superhorizon scale inflationary curvature perturbations. Phys. Rev. D **64**, 023512 (2001). https://doi.org/10.1103/PhysRevD.64.023512. arXiv: astro-ph/0101406
155. J.-P. Li et al., Complete analysis of scalar-induced gravitational waves and primordial non-Gaussianities f_{NL} and g_{NL} (2023). arXiv: 2309.07792 [astro-ph.CO]
156. Z.-C. Liang et al., Science with the TianQin observatory: Preliminary results on stochastic gravitational-wave background (2021). arXiv: 2107.08643 [astro-ph.CO]
157. J. Lin et al., Primordial black holes and secondary gravitational waves from k and G inflation. Phys. Rev. D **101**(10), 103515 (2020). https://doi.org/10.1103/PhysRevD.101.103515. arXiv: 2001.05909 [gr-qc]
158. A.D. Linde, A new inflationary universe scenario: a possible solution of the horizon, flatness, homogeneity, isotropy and primordial monopole problems. Phys. Lett. B **108**, 389–393 (1982). Ed. by L.-Z. Fang, R. Ruffini. https://doi.org/10.1016/0370-2693(82)91219-9
159. B. Liu, V. Bromm, Accelerating early massive galaxy formation with primordial black holes. Astrophys. J. Lett. **937**(2), L30 (2022). https://doi.org/10.3847/2041-8213/ac927f. arXiv: 2208.13178 [astro-ph.CO]
160. L. Liu, Z.-C. Chen, Q.-G. Huang, Implications for the non-Gaussianity of curvature perturbation from pulsar timing arrays (2023). arXiv: 2307.01102 [astro-ph.CO]
161. J. Luo et al., TianQin: a space-borne gravitational wave detector. Class. Quant. Grav. **33**(3), 035010 (2016). https://doi.org/10.1088/0264-9381/33/3/035010. arXiv: 1512.02076 [astro-ph.IM]
162. D.H. Lyth, K.A. Malik, M. Sasaki, A General proof of the conservation of the curvature perturbation. JCAP **05**, 004 (2005). https://doi.org/10.1088/1475-7516/2005/05/004. arXiv: astro-ph/0411220
163. D.H. Lyth, C. Ungarelli, D. Wands, The primordial density perturbation in the curvaton scenario. Phys. Rev. D **67**, 023503 (2003). https://doi.org/10.1103/PhysRevD.67.023503. arXiv: astro-ph/0208055
164. R. Mahbub, Impact of nonlinear overdensity statistics on primordial black hole abundance. Phys. Rev. D **102**(2), 023538 (2020). https://doi.org/10.1103/PhysRevD.102.023538. arXiv: 2005.03618 [astro-ph.CO]
165. S. Maity et al., Loop contributions to the scalar power spectrum due to quartic order action in ultra slow roll inflation (2023). arXiv: 2307.13636 [astro-ph.CO]
166. J.M. Maldacena, Non-Gaussian features of primordial fluctuations in single field inflationary models. JHEP **05**, 013 (2003). https://doi.org/10.1088/1126-6708/2003/05/013. arXiv: astro-ph/0210603
167. J. Martin, H. Motohashi, T. Suyama, Ultra slow-roll inflation and the non-Gaussianity consistency relation. Phys. Rev. D **87**(2), 023514 (2013). https://doi.org/10.1103/PhysRevD.87.023514. arXiv: 1211.0083 [astro-ph.CO]
168. S. Matarrese, S. Mollerach, M. Bruni, Second order perturbations of the Einstein-de Sitter universe. Phys. Rev. D **58**, 043504 (1998). https://doi.org/10.1103/PhysRevD.58.043504. arXiv: astro-ph/9707278
169. S. Matarrese, O. Pantano, D. Saez, A general relativistic approach to the nonlinear evolution of collisionless matter. Phys. Rev. D **47**, 1311–1323 (1993). https://doi.org/10.1103/PhysRevD.47.1311
170. S. Matarrese, O. Pantano, D. Saez, General relativistic dynamics of irrotational dust: cosmological implications. Phys. Rev. Lett. **72**, 320–323 (1994). https://doi.org/10.1103/PhysRevLett.72.320. arXiv: astro-ph/9310036
171. T. Matsubara, M. Sasaki, Non-Gaussianity effects on the primordial black hole abundance for sharply-peaked primordial spectrum. JCAP **10**, 094 (2022). https://doi.org/10.1088/1475-7516/2022/10/094. arXiv: 2208.02941 [astro-ph.CO]

172. T. Matsubara et al., Clustering of primordial black holes formed in a matter-dominated epoch. Phys. Rev. D **100**(12), 123544 (2019). https://doi.org/10.1103/PhysRevD.100.123544. arXiv: 1909.04053 [astro-ph.CO]
173. D.-S. Meng, C. Yuan, Q.-G. Huang, Primordial black holes generated by the non-minimal spectator field. Sci. China Phys. Mech. Astron. **66**(8), 280411 (2023). https://doi.org/10.1007/s11433-022-2095-5. arXiv: 2212.03577 [astro-ph.CO]
174. P. Meszaros, The behaviour of point masses in an expanding cosmological substratum. Astron. Astrophys. **37**, 225–228 (1974)
175. C.W. Misner, K.S. Thorne, J.A. Wheeler, *Gravitation*. (W. H. Freeman, San Francisco, 1973). ISBN: 978-0-7167-0344-0, 978-0-691-17779-3
176. S. Mittal et al., Constraining primordial black holes as dark matter using the global 21-cm signal with X-ray heating and excess radio background (2021). arXiv: 2107.02190 [astro-ph.CO]
177. T. Moroi, T. Takahashi, Effects of cosmological moduli fields on cosmic microwave background. Phys. Lett. B **522**, 215–221 (2001). [Erratum: Phys. Lett. B **539**, 303 (2002)]. https://doi.org/10.1016/S0370-2693(01)01295-3. arXiv: hep-ph/0110096
178. H. Motohashi, A.A. Starobinsky, J. Yokoyama, Inflation with a constant rate of roll. JCAP **09**, 018 (2015). https://doi.org/10.1088/1475-7516/2015/09/018. arXiv: 1411.5021 [astro-ph.CO]
179. H. Motohashi, Y. Tada, Squeezed bispectrum and one-loop corrections in transient constant-roll inflation. JCAP **08**, 069 (2023). https://doi.org/10.1088/1475-7516/2023/08/069. arXiv: 2303.16035 [astro-ph.CO]
180. P. Mroz et al., Microlensing optical depth and event rate toward the Large Magellanic Cloud based on 20 years of OGLE observations (2024). arXiv: 2403.02398 [astro-ph.GA]
181. P. Mroz et al., No massive black holes in the Milky Way halo (2024). arXiv: 2403.02386 [astro-ph.GA]
182. V.F. Mukhanov, Gravitational instability of the universe filled with a scalar field. JETP Lett. **41**, 493–496 (1985)
183. V.F. Mukhanov, G.V. Chibisov, Quantum fluctuations and a nonsingular universe. JETP Lett. **33**, 532–535 (1981). [Pisma Zh. Eksp. Teor. Fiz. **33**, 549 (1981)]
184. I. Musco, Threshold for primordial black holes: dependence on the shape of the cosmological perturbations. Phys. Rev. D **100**(12), 123524 (2019). https://doi.org/10.1103/PhysRevD.100.123524. arXiv: 1809.02127 [gr-qc]
185. I. Musco, J.C. Miller, A.G. Polnarev, Primordial black hole formation in the radiative era: Investigation of the critical nature of the collapse. Class. Quant. Grav. **26**, 235001 (2009). https://doi.org/10.1088/0264-9381/26/23/235001. arXiv: 0811.1452 [gr-qc]
186. T. Nakama, B. Carr, J. Silk, Limits on primordial black holes from μ distortions in cosmic microwave background. Phys. Rev. D **97**(4), 043525 (2018). https://doi.org/10.1103/PhysRevD.97.043525. arXiv: 1710.06945 [astro-ph.CO]
187. T. Nakama, K. Kohri, J. Silk, Ultracompact minihalos associated with stellar-mass primordial black holes. Phys. Rev. D **99**(12), 123530 (2019). https://doi.org/10.1103/PhysRevD.99.123530. arXiv: 1905.04477 [astro-ph.CO]
188. T. Nakama, J. Silk, M. Kamionkowski, Stochastic gravitational waves associated with the formation of primordial black holes. Phys. Rev. D **95**(4), 043511 (2017). https://doi.org/10.1103/PhysRevD.95.043511. arXiv: 1612.06264 [astro-ph.CO]
189. T. Nakama, T. Suyama, J. Yokoyama, Supermassive black holes formed by direct collapse of inflationary perturbations. Phys. Rev. D **94**(10), 103522 (2016). https://doi.org/10.1103/PhysRevD.94.103522. arXiv: 1609.02245 [gr-qc]
190. K. Nakamura, Second-order gauge invariant cosmological perturbation theory: Einstein equations in terms of gauge invariant variables. Prog. Theor. Phys. **117**, 17–74 (2007). https://doi.org/10.1143/PTP.117.17. arXiv: gr-qc/0605108
191. M.H. Namjoo, One consistency relation for all single-field inflationary models. JCAP **05**, 041 (2024). https://doi.org/10.1088/1475-7516/2024/05/041. arXiv: 2311.12777 [astro-ph.CO]

192. M.H. Namjoo, H. Firouzjahi, M. Sasaki, Violation of non-Gaussianity consistency relation in a single field inflationary model. EPL **101**(3), 39001 (2013). https://doi.org/10.1209/0295-5075/101/39001. arXiv: 1210.3692 [astro-ph.CO]
193. M.H. Namjoo, B. Nikbakht, Non-Gaussianity consistency relations and their consequences for the peaks. arXiv: 2401.12958 [astro-ph.CO]
194. J.C. Niemeyer, K. Jedamzik, Near-critical gravitational collapse and the initial mass function of primordial black holes. Phys. Rev. Lett. **80**, 5481–5484 (1998). https://doi.org/10.1103/PhysRevLett.80.5481. arXiv: astro-ph/9709072
195. H. Niikura et al., Microlensing constraints on primordial black holes with Subaru/HSC Andromeda observations. Nature Astron. **3**(6), 524–534 (2019). https://doi.org/10.1038/s41550-019-0723-1. arXiv: 1701.02151 [astro-ph.CO]
196. H. Noh, J.-C. Hwang, Second-order perturbations of the Friedmann world model. Phys. Rev. D **69**, 104011 (2004). https://doi.org/10.1103/PhysRevD.69.104011
197. B. Osano et al., Gravitational waves generated by second order effects during inflation. JCAP **04**, 003 (2007). https://doi.org/10.1088/1475-7516/2007/04/003. arXiv: gr-qc/0612108
198. O. Özsoy, G. Tasinato, Consistency conditions and primordial black holes in single field inflation. Phys. Rev. D **2022**(2), 023524 (2022). https://doi.org/10.1103/PhysRevD.105.023524. arXiv: 2111.02432 [astro-ph.CO]
199. S. Passaglia, W. Hu, H. Motohashi, Primordial black holes and local non-Gaussianity in canonical inflation. Phys. Rev. D **99**(4), 043536 (2019). https://doi.org/10.1103/PhysRevD.99.043536. arXiv: 1812.08243 [astro-ph.CO]
200. C. Pattison et al., Quantum diffusion during inflation and primordial black holes. JCAP **10**, 046 (2017). https://doi.org/10.1088/1475-7516/2017/10/046. arXiv: 1707.00537 [hep-th]
201. C. Pattison et al., Ultra-slow-roll inflation with quantum diffusion. JCAP **04**, 080 (2021). https://doi.org/10.1088/1475-7516/2021/04/080. arXiv: 2101.05741 [astro-ph.CO]
202. G. Perna et al., Fully non-Gaussian scalar-induced gravitational waves (2024). arXiv: 2403.06962 [astro-ph.CO]
203. S. Pi, M. Sasaki, Gravitational waves induced by scalar perturbations with a lognormal peak. JCAP **09**, 037 (2020). https://doi.org/10.1088/1475-7516/2020/09/037. arXiv: 2005.12306 [gr-qc]
204. S. Pi, M. Sasaki, Logarithmic duality of the curvature perturbation. Phys. Rev. Lett. **131**(1), 011002 (2023). https://doi.org/10.1103/PhysRevLett.131.011002. arXiv: 2211.13932 [astro-ph.CO]
205. S. Pi, M. Sasaki, Primordial black hole formation in non-minimal curvaton scenario (2021). arXiv: 2112.12680 [astro-ph.CO]
206. S. Pi, J. Wang, Primordial black hole formation in Starobinsky's linear potential model. JCAP **06**, 018 (2023). https://doi.org/10.1088/1475-7516/2023/06/018. arXiv: 2209.14183 [astro-ph.CO]
207. R. Picard, K.A. Malik, Induced gravitational waves: the effect of first order tensor perturbations (2023). arXiv: 2311.14513 [astro-ph.CO]
208. P.P. Avelino, Primordial black hole constraints on non-Gaussian inflation models. Phys. Rev. D **72**, 124004 (2005). https://doi.org/10.1103/PhysRevD.72.124004. arXiv: astro-ph/0510052
209. W.H. Press, P. Schechter, Formation of galaxies and clusters of galaxies by selfsimilar gravitational condensation. Astrophys. J. **187**, 425–438 (1974). https://doi.org/10.1086/152650
210. H. V. Ragavendra, Accounting for scalar non-Gaussianity in secondary gravitational waves. Phys. Rev. D **105**(6), 063533 (2022). https://doi.org/10.1103/PhysRevD.105.063533. arXiv: 2108.04193 [astro-ph.CO]
211. H.V. Ragavendra, L. Sriramkumar, Observational imprints of enhanced scalar power on small scales in ultra slow roll inflation and associated non-Gaussianities. Galaxies **11**(1), 34 (2023). https://doi.org/10.3390/galaxies11010034. arXiv: 2301.08887 [astro-ph.CO]
212. H.V. Ragavendra et al., Primordial black holes and secondary gravitational waves from ultra-slow roll and punctuated inflation. Phys. Rev. D **103**(8), 083510 (2021). https://doi.org/10.1103/PhysRevD.103.083510. arXiv: 2008.12202 [astro-ph.CO]

213. M. Raidal, V. Vaskonen, H. Veermäe, Gravitational waves from primordial black hole mergers. JCAP **09**, 037 (2017). https://doi.org/10.1088/1475-7516/2017/09/037. arXiv: 1707.01480 [astro-ph.CO]
214. D.J. Reardon et al., Search for an isotropic gravitational-wave background with the parkes pulsar timing array. Astrophys. J. Lett. **951**(1), L6 (2023). https://doi.org/10.3847/2041-8213/acdd02. arXiv: 2306.16215 [astro-ph.HE]
215. D.J. Reardon et al., The gravitational-wave background null hypothesis: characterizing noise in millisecond pulsar arrival times with the parkes pulsar timing array. Astrophys. J. Lett. **951**(1), L7 (2023). https://doi.org/10.3847/2041-8213/acdd03. arXiv: 2306.16229 [astro-ph.HE]
216. Z. Ren et al., Taiji data challenge for exploring gravitational wave universe. Front. Phys. (Beijing) **18**(6), 64302 (2023). https://doi.org/10.1007/s11467-023-1318-y. arXiv: 2301.02967 [gr-qc]
217. F. Riccardi, M. Taoso, A. Urbano, Solving peak theory in the presence of local non-Gaussianities. JCAP **08**, 060 (2021). https://doi.org/10.1088/1475-7516/2021/08/060. arXiv: 2102.04084 [astro-ph.CO]
218. A. Riotto, The primordial black hole formation from single-field inflation is still not ruled out (2023). arXiv: 2303.01727 [astro-ph.CO]
219. W.-H. Ruan et al., Taiji program: gravitational-wave sources. Int. J. Mod. Phys. A **35**(17), 2050075 (2020). https://doi.org/10.1142/S0217751X2050075X. arXiv: 1807.09495 [gr-qc]
220. R. Saito, J. Yokoyama, Gravitational wave background as a probe of the primordial black hole abundance. Phys. Rev. Lett. **102**, 161101 (2009). [Erratum: Phys. Rev. Lett. **107**, 069901 (2011)]. https://doi.org/10.1103/PhysRevLett.102.161101. arXiv: 0812.4339 [astro-ph]
221. R. Saito, J. Yokoyama, Gravitational-wave constraints on the abundance of primordial black holes. Prog. Theor. Phys. **123**, 867–886 (2010). [Erratum: Prog. Theor. Phys. **126**, 351–352 (2011)]. https://doi.org/10.1143/PTP.126.351. arXiv: 0912.5317 [astro-ph.CO]
222. M. Sasaki, Large scale quantum fluctuations in the inflationary universe. Prog. Theor. Phys. **76**, 1036 (1986). https://doi.org/10.1143/PTP.76.1036
223. M. Sasaki, E.D. Stewart, A general analytic formula for the spectral index of the density perturbations produced during inflation. Prog. Theor. Phys. **95**, 71–78 (1996). https://doi.org/10.1143/PTP.95.71. arXiv: astro-ph/9507001
224. M. Sasaki, J. Valiviita, D. Wands, Non-Gaussianity of the primordial perturbation in the curvaton model. Phys. Rev. D **74**, 103003 (2006). https://doi.org/10.1103/PhysRevD.74.103003. arXiv: astro-ph/0607627
225. K. Schmitz, New sensitivity curves for gravitational-wave signals from cosmological phase transitions. JHEP **01**, 097 (2021). https://doi.org/10.1007/JHEP01(2021)097. arXiv: 2002.04615 [hep-ph]
226. D. Seery, J.C. Hidalgo, Non-Gaussian corrections to the probability distribution of the curvature perturbation from inflation. JCAP **07**, 008 (2006). https://doi.org/10.1088/1475-7516/2006/07/008. arXiv: astro-ph/0604579
227. P.D. Serpico et al., Cosmic microwave background bounds on primordial black holes including dark matter halo accretion. Phys. Rev. Res. **2**(2), 023204 (2020). https://doi.org/10.1103/PhysRevResearch.2.023204. arXiv: 2002.10771 [astro-ph.CO]
228. S. Shandera et al., Number counts and non-Gaussianity. Phys. Rev. D **88**(10), 103506 (2013). https://doi.org/10.1103/PhysRevD.88.103506. arXiv: 1211.7361 [astro-ph.CO]
229. M. Shibata, M. Sasaki, Black hole formation in the Friedmann universe: formulation and computation in numerical relativity. Phys. Rev. D **60**, 084002 (1999). https://doi.org/10.1103/PhysRevD.60.084002. arXiv: gr-qc/9905064
230. A.A. Starobinsky, A new type of isotropic cosmological models without singularity. Phys. Lett. B **91**, 99–102 (1980). Ed. by I.M. Khalatnikov, V.P. Mineev. https://doi.org/10.1016/0370-2693(80)90670-X
231. A.A. Starobinsky, Dynamics of phase transition in the new inflationary universe scenario and generation of perturbations. Phys. Lett. B **117**, 175–178 (1982). https://doi.org/10.1016/0370-2693(82)90541-X

232. A.A. Starobinsky, Spectrum of adiabatic perturbations in the universe when there are singularities in the inflation potential. JETP Lett. **55**, 489–494 (1992)
233. A.A. Starobinsky, Stochastic de sitter (inflationary) stage in the early universe. Lect Notes Phys **246**, 107–126 (1986). https://doi.org/10.1007/3-540-16452-9_6
234. A.A. Starobinsky, J. Yokoyama, Equilibrium state of a selfinteracting scalar field in the De Sitter background. Phys. Rev. D **50**, 6357–6368 (1994). https://doi.org/10.1103/PhysRevD.50.6357. arXiv: astro-ph/9407016
235. T. Suyama, J. Yokoyama, Temporal enhancement of super-horizon curvature perturbations from decays of two curvatons and its cosmological consequences. Phys. Rev. D **84**, 083511 (2011). https://doi.org/10.1103/PhysRevD.84.083511. arXiv: 1106.5983 [astro-ph.CO]
236. T. Suyama, S. Yokoyama, Clustering of primordial black holes with non-Gaussian initial fluctuations. PTEP **2019**(10), 103E02 (2019). https://doi.org/10.1093/ptep/ptz105. arXiv: 1906.04958 [astro-ph.CO]
237. Y. Tada, T. Terada, J. Tokuda, Cancellation of quantum corrections on the soft curvature perturbations. JHEP **01**, 105 (2024). https://doi.org/10.1007/JHEP01(2024)105. arXiv: 2308.04732 [hep-th]
238. Y. Tada, S. Yokoyama, Primordial black holes as biased tracers. Phys. Rev. D **91**(12), 123534 (2015). https://doi.org/10.1103/PhysRevD.91.123534. arXiv: 1502.01124 [astro-ph.CO]
239. M. Taoso, A. Urbano, Non-Gaussianities for primordial black hole formation. JCAP **08**, 016 (2021). https://doi.org/10.1088/1475-7516/2021/08/016. arXiv: 2102.03610 [astro-ph.CO]
240. G. Tasinato, An analytic approach to non-slow-roll inflation. Phys. Rev. D **103**(2), 023535 (2021). https://doi.org/10.1103/PhysRevD.103.023535. arXiv: 2012.02518 [hep-th]
241. G. Tasinato, Large —η— approach to single field inflation. Phys. Rev. D **108**(4), 043526 (2023). https://doi.org/10.1103/PhysRevD.108.043526. arXiv: 2305.11568 [hep-th]
242. G. Tasinato, Non-Gaussianities and the large —η— approach to inflation. Phys. Rev. D **109**(6), 063510 (2024). https://doi.org/10.1103/PhysRevD.109.063510. arXiv: 2312.03498 [hep-th]
243. E. Thrane, J.D. Romano, Sensitivity curves for searches for gravitational-wave backgrounds. Phys. Rev. D **88**(12), 124032 (2013). https://doi.org/10.1103/PhysRevD.88.124032. arXiv: 1310.5300 [astro-ph.IM]
244. P. Tisserand et al., Limits on the Macho content of the galactic halo from the EROS-2 survey of the magellanic clouds. Astron. Astrophys. **469**, 387–404 (2007). https://doi.org/10.1051/0004-6361:20066017. arXiv: astro-ph/0607207
245. K. Uehara et al., Numerical simulation of type II primordial black hole formation (2024). arXiv: 2401.06329 [gr-qc]
246. C. Unal, Imprints of primordial non-Gaussianity on gravitational wave spectrum. Phys. Rev. D **99**(4), 041301 (2019). https://doi.org/10.1103/PhysRevD.99.041301. arXiv: 1811.09151 [astro-ph.CO]
247. C. Ünal, E.D. Kovetz, S.P. Patil, Multimessenger probes of inflationary fluctuations and primordial black holes. Phys. Rev. D **2021**(6), 063519 (2021). https://doi.org/10.1103/PhysRevD.103.063519. arXiv: 2008.11184 [astro-ph.CO]
248. V. Vennin, Stochastic inflation and primordial black holes. Other thesis (2020). arXiv: 2009.08715 [astro-ph.CO]
249. V. Vennin, A.A. Starobinsky, Correlation functions in stochastic inflation. Eur. Phys. J. C **75**, 413 (2015). https://doi.org/10.1140/epjc/s10052-015-3643-y. arXiv: 1506.04732 [hep-th]
250. D. Wands et al., A new approach to the evolution of cosmological perturbations on large scales. Phys. Rev. D **62**, 043527 (2000). https://doi.org/10.1103/PhysRevD.62.043527. arXiv: astro-ph/0003278
251. G. Wang, W.-B. Han, Alternative LISA-TAIJI networks: detectability to isotropic stochastic gravitational wave background (2021). arXiv: 2108.11151 [gr-qc]

252. L.T. Witkowski, SIGWfast: a python package for the computation of scalar-induced gravitational wave spectra (2022). arXiv: 2209.05296 [astro-ph.CO]
253. L.T. Witkowski et al., Expansion history-dependent oscillations in the scalar-induced gravitational wave background. JCAP **05**(05), 028 (2022). https://doi.org/10.1088/1475-7516/2022/05/028. arXiv: 2110.09480 [astro-ph.CO]
254. H. Xu et al., Searching for the Nano-Hertz stochastic gravitational wave background with the Chinese pulsar timing array data release I. Res. Astron. Astrophys. **23**(7), 075024 (2023). https://doi.org/10.1088/1674-4527/acdfa5. arXiv: 2306.16216 [astro-ph.HE]
255. Z. Yi et al., Primordial black holes and scalar-induced secondary gravitational waves from inflationary models with a noncanonical kinetic term. Phys. Rev. D **103**(6), 063534 (2021). https://doi.org/10.1103/PhysRevD.103.063534. arXiv: 2011.10606 [astro-ph.CO]
256. J. Yokoyama, Cosmological constraints on primordial black holes produced in the near critical gravitational collapse. Phys. Rev. D **58**, 107502 (1998). https://doi.org/10.1103/PhysRevD.58.107502. arXiv: gr-qc/9804041
257. C.-M. Yoo, J.-O. Gong, S. Yokoyama, Abundance of primordial black holes with local non-Gaussianity in peak theory. JCAP **09**, 033 (2019). https://doi.org/10.1088/1475-7516/2019/09/033. arXiv: 1906.06790 [astro-ph.CO]
258. C.-M. Yoo et al., Abundance of primordial black holes in peak theory for an arbitrary power spectrum. PTEP **2021**(1), 013E02 (2021). https://doi.org/10.1093/ptep/ptaa155. arXiv: 2008.02425 [astro-ph.CO]
259. C.-M. Yoo et al., Primordial black hole abundance from random Gaussian curvature perturbations and a local density threshold. PTEP **2018**(12), 123E01 (2018). https://doi.org/10.1093/ptep/pty120. arXiv: 1805.03946 [astro-ph.CO]
260. S. Young, Peaks and primordial black holes: the effect of non-Gaussianity. JCAP **05**(05), 037 (2022). https://doi.org/10.1088/1475-7516/2022/05/037. arXiv: 2201.13345 [astro-ph.CO]
261. S. Young, The primordial black hole formation criterion re-examined: parametrisation, timing and the choice of window function. Int. J. Mod. Phys. D **29**(02), 2030002 (2019). https://doi.org/10.1142/S0218271820300025. arXiv: 1905.01230 [astro-ph.CO]
262. S. Young, C. Byrnes, Signatures of non-Gaussianity in the isocurvature modes of primordial black hole dark matter. JCAP **1504**(04), 034 (2015). https://doi.org/10.1088/1475-7516/2015/04/034. arXiv: 1503.01505 [astro-ph.CO]
263. S. Young, C.T. Byrnes, Initial clustering and the primordial black hole merger rate. JCAP **03**, 004 (2020). https://doi.org/10.1088/1475-7516/2020/03/004. arXiv: 1910.06077 [astro-ph.CO]
264. S. Young, C.T. Byrnes, Primordial black holes in non-Gaussian regimes. JCAP **08**, 052 (2013). https://doi.org/10.1088/1475-7516/2013/08/052. arXiv: 1307.4995 [astro-ph.CO]
265. S. Young, I. Musco, C.T. Byrnes, Primordial black hole formation and abundance: contribution from the non-linear relation between the density and curvature perturbation. JCAP **11**, 012 (2019). https://doi.org/10.1088/1475-7516/2019/11/012. arXiv: 1904.00984 [astro-ph.CO]
266. S. Young, M. Musso, Application of peaks theory to the abundance of primordial black holes. JCAP **11**, 022 (2020). https://doi.org/10.1088/1475-7516/2020/11/022. arXiv: 2001.06469 [astro-ph.CO]
267. S. Young, D. Regan, C.T. Byrnes, Influence of large local and non-local bispectra on primordial black hole abundance. JCAP **02**, 029 (2016). https://doi.org/10.1088/1475-7516/2016/02/029. arXiv: 1512.07224 [astro-ph.CO]
268. Y.-H. Yu, S. Wang, Primordial gravitational waves assisted by cosmological scalar perturbations (2023). arXiv: 2303.03897 [astro-ph.CO]
269. C. Yuan, D.-S. Meng, Q.-G. Huang, Full analysis of the scalar-induced gravitational waves for the curvature perturbation with local-type non-Gaussianities. JCAP **12**, 036 (2023). https://doi.org/10.1088/1475-7516/2023/12/036. arXiv: 2308.07155 [astro-ph.CO]
270. B. Ya, I.D. Novikov Zel'dovich, The hypothesis of cores retarded during expansion and the hot cosmological model. Soviet Astron. AJ (Engl. Transl.) **10**, 602 (1967)

271. F. Zhang et al., Primordial non-Gaussianity from G-inflation. JCAP **04**, 045 (2021). https://doi.org/10.1088/1475-7516/2021/04/045. arXiv: 2012.06960 [astro-ph.CO]
272. A. Zic et al., The parkes pulsar timing array third data release (2023). arXiv: 2306.16230 [astro-ph.HE]

Chapter 8
Quantum Diffusion and Large Primordial Perturbations from Inflation

Vincent Vennin and David Wands

Abstract Quantum diffusion describes the inflow of vacuum quantum fluctuations as they get amplified by gravitational instability, and stretched to large distances during inflation. In this picture, the dynamics of the universe's expansion becomes stochastic, and the statistics of the curvature perturbation is encoded in the distribution of the duration of inflation. This provides a non-perturbative framework to study cosmological fluctuations during inflation, which is well-suited to the case of primordial black holes since they originate from large fluctuations. We show that standard, perturbative expectations for the primordial black hole abundance can be significantly modified by quantum-diffusion effects, and we identify a few open challenges.

8.1 Introduction

PBHs may form in the early universe when large density fluctuations are produced and gravitationally collapse into black holes. Although different mechanisms have been proposed to amplify cosmological fluctuations above the required threshold, one possibility is that the amplification takes place during inflation, see Chap. 3. If large curvature perturbations are produced during inflation, they may no longer be described by perturbative frameworks, and in this chapter we show how the stochastic-inflation formalism, in combination with the δN formalism, can be used as a non-perturbative scheme to keep track of large fluctuations. In this approach, quantum fluctuations

In memory of Alexei Starobinsky (1948–2023), the founding father of stochastic inflation, and of so much more.

V. Vennin (✉)
Laboratoire de Physique de l'École Normale Supérieure, ENS, CNRS, Université PSL, Sorbonne Université, Université Paris Cité, 75005 Paris, France
e-mail: vincent.vennin@ens.fr

D. Wands
Institute of Cosmology & Gravitation, University of Portsmouth, Dennis Sciama Building, Burnaby Road, Portsmouth PO1 3FX, UK
e-mail: david.wands@port.ac.uk

C. Byrnes et al. (eds.), *Primordial Black Holes*, Springer Series in Astrophysics and Cosmology, https://doi.org/10.1007/978-981-97-8887-3_8

provide random kicks to the large-scale dynamics of the universe, which can thus be described by means of stochastic systems. Different regions of space inflate for different amounts of time, hence the number of inflationary e-folds becomes itself a stochastic quantity, which can be identified with the curvature perturbation. We show how its statistics can be studied using first-passage-time techniques, and discuss the expected consequences for the abundance of PBHs from inflation. This abundance is strongly enhanced compared to perturbative predictions due to heavy tails in the probability distribution, which are non-perturbative in nature and inevitable. We also discuss the validity of the separate-universe approach, on which this framework rests, in particular in the context of slow-roll violations, which commonly arise in PBH-producing models. Finally, we identify a few open challenges that remain to be addressed.

8.2 Stochastic Inflation in a Nutshell

The stochastic formalism of inflation [77, 79] is an effective theory for the long-wavelength part of quantum fields living on an inflating background. It can be obtained formally by integrating out the small-wavelength part of these fields, using the language of effective-field theories and the Schwinger-Keldysh formalism [31, 52, 54, 63, 83, 84]. Here, we present a heuristic derivation that makes physical interpretation easier.

We consider a single scalar field ϕ, minimally coupled to gravity in a 4-dimensional curved space-time with metric $g_{\mu\nu}$, described by the action

$$S = \int \mathrm{d}^4x\sqrt{-g}\left[\frac{M_P^2}{2}R - \frac{1}{2}g^{\mu\nu}\partial_\mu\phi\partial_\nu\phi - V(\phi)\right]. \tag{8.1}$$

In this expression, R is the Ricci scalar curvature and $V(\phi)$ is the potential energy of the scalar field.

8.2.1 Background

At the background level, space-time is homogeneous and isotropic, so the line-element is of the Friedmann-Lemaître-Robertson-Walker (FLRW) form, $\mathrm{d}s^2 = -\mathrm{d}t^2 + a^2(t)\mathrm{d}x^2$. Here, t is cosmic time, a is the scale factor and x^i are spatial coordinates comoving with the expansion. The Einstein equations yield a constraint equation, called the Friedmann equation,

$$H^2 = \frac{V(\phi) + \frac{\dot{\phi}^2}{2}}{3M_P^2} \tag{8.2}$$

where $H = \dot{a}/a$ is the Hubble expansion rate, a dot denotes a derivative with respect to cosmic time and $M_P = \sqrt{\hbar c/(8\pi G)}$ is the reduced Planck mass. The dynamical equations for the inflaton field, ϕ, can be written as two coupled first-order equations

$$\begin{aligned} \frac{\mathrm{d}\phi}{\mathrm{d}N} &= \frac{\Pi}{a^3 H}, \\ \frac{\mathrm{d}\Pi}{\mathrm{d}N} &= -\frac{a^3}{H} V'(\phi), \end{aligned} \tag{8.3}$$

where Π is the momentum conjugate to ϕ, V' is the first derivative of the function $V(\phi)$, and time is labeled by the number of e-folds $N \equiv \ln(a)$ for future convenience. When the potential function is flat enough, $M_P|V'/V| \ll 1$ and $M_P^2|V''/V| \ll 1$, a phase of slow-roll inflation takes place where $\ddot{a} > 0$ and H varies slowly.

8.2.2 Linear Perturbations

Let us now introduce small fluctuations over the homogeneous and isotropic background, by allowing ϕ to depend on space, $\phi(\boldsymbol{x}, t) = \bar{\phi}(t) + \delta\phi(\boldsymbol{x}, t)$, where $\bar{\phi}$ follows the background dynamics given above and $|\delta\phi| \ll |\bar{\phi}|$.[1] Inserting this decomposition into Eq. (8.1), at leading order in perturbation theory one has

$$\begin{aligned} \frac{\mathrm{d}\phi_k}{\mathrm{d}N} &= \frac{\pi_k}{a^3 H}, \\ \frac{\mathrm{d}\pi_k}{\mathrm{d}N} &= -\frac{a^3}{H} V''(\bar{\phi})\phi_k - \frac{a}{H} k^2 \phi_k . \end{aligned} \tag{8.4}$$

These equations are written in Fourier space, i.e.

$$\delta\phi(\boldsymbol{x}, t) = (2\pi)^{-3/2} \int \mathrm{d}\boldsymbol{k} \left[a_{\boldsymbol{k}} \phi_k(t) e^{-i\boldsymbol{k}\cdot\boldsymbol{x}} + a_{\boldsymbol{k}}^{\dagger} \phi_k^*(t) e^{i\boldsymbol{k}\cdot\boldsymbol{x}} \right] \tag{8.5}$$

with a similar expression for $\delta\pi(\boldsymbol{x}, t)$. Here, the field $\delta\phi(\boldsymbol{x}, t)$ is quantised, so it is expanded onto the creation and annihilation operators $a_{\boldsymbol{k}}^{\dagger}$ and $a_{\boldsymbol{k}}$, which satisfy the usual commutation relations $[a_{\boldsymbol{k}}, a_{\boldsymbol{k}'}^{\dagger}] = \delta^3(\boldsymbol{k} - \boldsymbol{k}')$ and $[a_{\boldsymbol{k}}, a_{\boldsymbol{k}'}] = [a_{\boldsymbol{k}}^{\dagger}, a_{\boldsymbol{k}'}^{\dagger}] = 0$. They are independent of time, contrary to the mode functions $\phi_k(t)$ which satisfy Eq. (8.4). On isotropic backgrounds, they depend only on the modulus $k = |\boldsymbol{k}|$ of the wave-vector, and they must satisfy $\phi_{-\boldsymbol{k}} = \phi_{\boldsymbol{k}}^{\star}$ for the quantum field $\delta\phi(\boldsymbol{x}, t)$ to be Hermitian (i.e. for the classical field to be real). In the asymptotic past, $k \gg aH$ and the above reduces to the dynamics of a massless field on a Minkowski background. In this sub-Hubble regime, each Fourier mode is described by a harmonic oscillator,

[1] Metric perturbations are not included at this stage. The conditions under which this is a valid approximation when time is labeled with N are clarified in Sect. 8.3.

the ground state of which can serve as an initial condition. This is the so-called Bunch-Davies vacuum, which corresponds to setting $\phi_k = e^{-ik\eta}/(a\sqrt{2k})$ and $\pi_k = -ika^2\phi_k$ as $k\eta \to -\infty$, where η is the conformal time, related to the cosmic time t via $\mathrm{d}t = a\mathrm{d}\eta$. Together with Eq. (8.4), this specifies the state of the perturbations at any time. In particular, this is a Gaussian state, given that the ground state of the harmonic oscillator is Gaussian and that the evolution is linear. Non-Gaussianities arise at non-linear orders, and one of our goals is to describe them non perturbatively.

8.2.3 Coarse-Grained Fields

In practice, cosmological perturbation theory provides an accurate description of cosmological fluctuations so long as their amplitude remains small. In the context of PBHs, one is precisely interested in situations where the rare, high peaks of cosmological perturbations grow on super-Hubble scales and become sizeable, see Chap. 3, hence non-perturbative frameworks are required at large scales. One such framework is the stochastic-inflation formalism, which describes the dynamics of fields coarse-grained at the σ-Hubble radius, $(\sigma H)^{-1}$. Here, σ is a fixed parameter that sets the scale at which quantum fluctuations backreact onto the local FLRW geometry. In practice, one must set $\sigma \ll 1$ in order for gradient terms not to affect the dynamics of the coarse-grained field, see Sect. 8.3. One thus decomposes

$$\begin{aligned} \phi &= \tilde{\phi} + \phi_Q \\ \Pi &= \tilde{\pi} + \pi_Q \end{aligned} \tag{8.6}$$

where $\tilde{\phi}$ and $\tilde{\pi}$ are coarse-grained fields, and

$$\begin{aligned} \phi_Q &= \int_{\mathbb{R}^3} \frac{\mathrm{d}^3\boldsymbol{k}}{(2\pi)^{3/2}} W\left(\frac{k}{k_\sigma}\right)\left[a_{\boldsymbol{k}}\,\phi_k(\tau)e^{-i\boldsymbol{k}\cdot\boldsymbol{x}} + a_{\boldsymbol{k}}^\dagger\,\phi_k^\star(\tau)e^{i\boldsymbol{k}\cdot\boldsymbol{x}}\right] \\ \pi_Q &= \int_{\mathbb{R}^3} \frac{\mathrm{d}^3\boldsymbol{k}}{(2\pi)^{3/2}} W\left(\frac{k}{k_\sigma}\right)\left[a_{\boldsymbol{k}}\,\pi_k(\tau)e^{-i\boldsymbol{k}\cdot\boldsymbol{x}} + a_{\boldsymbol{k}}^\dagger\,\pi_k^\star(\tau)e^{i\boldsymbol{k}\cdot\boldsymbol{x}}\right] \end{aligned} \tag{8.7}$$

are their small-wavelength complements. Here, $k_\sigma = \sigma a H$, and the window function W is such that $W \simeq 0$ for $k \ll k_\sigma$ and $W \simeq 1$ for $k \gg k_\sigma$. The equations of motion for the coarse-grained fields are given by

$$\begin{aligned} \frac{\mathrm{d}\tilde{\phi}}{\mathrm{d}N} &= \frac{\tilde{\pi}}{a^3 H} - \frac{\mathrm{d}\phi_Q}{\mathrm{d}N} + \frac{\pi_Q}{Ha^3}, \\ \frac{\mathrm{d}\tilde{\pi}}{\mathrm{d}N} &= -\frac{a^3}{H}V'(\tilde{\phi}) - \frac{\mathrm{d}\pi_Q}{\mathrm{d}N} - \frac{a^3}{H}V''(\tilde{\phi})\phi_Q + \frac{a}{H}\Delta\phi_Q. \end{aligned} \tag{8.8}$$

In these expressions, the Laplacian of $\tilde{\phi}$ has been dropped since it is suppressed by σ, and a linear expansion in the small-wavelength perturbations ϕ_Q and π_Q has been performed. Replacing ϕ_Q and π_Q by Eq. (8.7), and making use of the fact that the mode functions ϕ_k and π_k satisfy Eq. (8.4), the above can be rewritten as [27, 28, 57, 68, 85, 90]

$$\begin{aligned}\frac{\mathrm{d}\tilde{\phi}}{\mathrm{d}N} &= \frac{\tilde{\pi}}{a^3 H} + \xi_\phi(N)\,,\\ \frac{\mathrm{d}\tilde{\pi}}{\mathrm{d}N} &= -\frac{a^3}{H}V'(\tilde{\phi}) + \xi_\pi(N)\,,\end{aligned} \tag{8.9}$$

where the quantum sources ξ_ϕ and ξ_π arise from the time dependence of k_σ in the window functions in Eq. (8.7) and are given by

$$\begin{aligned}\xi_\phi &= -\int_{\mathbb{R}^3} \frac{\mathrm{d}^3\boldsymbol{k}}{(2\pi)^{3/2}} \frac{\mathrm{d}W}{\mathrm{d}N}\left(\frac{k}{k_\sigma}\right)\left[a_{\boldsymbol{k}}\phi_k(\tau)e^{-i\boldsymbol{k}\cdot\boldsymbol{x}} + a^\dagger_{\boldsymbol{k}}\phi^\star_k(\tau)e^{i\boldsymbol{k}\cdot\boldsymbol{x}}\right],\\ \xi_\pi &= -\int_{\mathbb{R}^3} \frac{\mathrm{d}^3\boldsymbol{k}}{(2\pi)^{3/2}} \frac{\mathrm{d}W}{\mathrm{d}N}\left(\frac{k}{k_\sigma}\right)\left[a_{\boldsymbol{k}}\pi_k(\tau)e^{-i\boldsymbol{k}\cdot\boldsymbol{x}} + a^\dagger_{\boldsymbol{k}}\pi^\star_k(\tau)e^{i\boldsymbol{k}\cdot\boldsymbol{x}}\right].\end{aligned} \tag{8.10}$$

They act as source terms in the equations of the motion for the coarse-grained fields. Since they only involve fluctuations at the σ-Hubble scale, they can be evaluated in cosmological perturbation theory where they obey Gaussian statistics. Their properties are thus entirely captured by their two-point functions,

$$\left\langle \xi_f(\boldsymbol{x}_1, N)\xi_g(\boldsymbol{x}_2, N')\right\rangle = \frac{\mathrm{d}\ln k_\sigma}{\mathrm{d}N}\mathcal{P}_{fg}\left[k_\sigma(N), N\right]\mathrm{sinc}\left[k_\sigma(N)|\boldsymbol{x}_2 - \boldsymbol{x}_1|\right]\delta\left(N - N'\right) \tag{8.11}$$

where f and g denote either ϕ or π. Here, we have assumed that the window function is sharp in Fourier space, $W(k/k_\sigma) = \Theta(k/k_\sigma - 1)$. This explains why the quantum noises are white, i.e. uncorrelated over time. In the above expression, we have also introduced the reduced power spectrum,

$$\mathcal{P}_{fg}(k, N) = \frac{k^3}{2\pi^2} f_k(N) g^\star_k(N)\,. \tag{8.12}$$

8.2.4 *Langevin Equations*

The next step involves describing spatial hypersurfaces of the universe as ensembles of σ-Hubble patches. Between two such patches, the correlations between the quantum sources are suppressed by the cardinal sine function appearing in Eq. (8.11). Since gradient interactions have been already dropped when discarding the Laplacian of $\tilde{\phi}$ in Eq. (8.8), such correlations must be neglected at the order at which the calculation is performed. This implies that σ-Hubble patches evolve independently, which is usually referred to as the "separate-universe" picture.

Before elaborating more on the validity of this assumption in Sect. 8.3, let us note that, in this approach, each patch is endowed with a local FLRW geometry, in which the (local) expansion rate is still given by the Friedmann equation (8.2), where the right-hand side must now be evaluated with the (local) values of $\tilde{\phi}$ and $\tilde{\pi}$. Such an equation may however seem problematic since, although H is a geometrical classical quantity, at this stage $\tilde{\phi}$ and $\tilde{\pi}$ are still quantum fields. The present approach thus requires $\tilde{\phi}$ and $\tilde{\pi}$ to be seen as classical, random fields, rather than quantum fields. The quantum sources ξ_ϕ and ξ_π become stochastic noises, with statistical correlation functions that are identified with the quantum expectation values (8.11).

Since ξ_ϕ and ξ_π are Hermitian quantum fields placed in a Gaussian state, and given that they enter linearly in the equations of motion (8.9), they can be equivalently described by classical random variables sharing the same two-point functions, and the two descriptions are perfectly equivalent [2, 27, 50]. In contrast, the coarse-grained fields $\tilde{\phi}$ and $\tilde{\pi}$ may be placed in non-Gaussian states as an effect of non-linear evolution, and in this case the stochastic and quantum descriptions are not equivalent. It is indeed well-known that quantum systems can display correlations that cannot be accounted for using classical setups, a famous example being the celebrated Bell inequalities [9]. In fact, even if $\tilde{\phi}$ and $\tilde{\pi}$ remain in Gaussian states, non-classical correlations appear when considering "improper" combinations,[2] for which Bell inequalities can be explicitly violated [3, 51]. In practice however, these quantum signatures are often hidden in the amplitude of the so-called decaying mode [36, 42, 49, 64], which is again gradient suppressed, and may be further concealed by quantum decoherence [13, 18, 47, 48].

When such a "quantum-to-classical" transition takes place, the stochastic approach is valid, and Eq. (8.9) may be seen as Langevin equations [71]. The white Gaussian noises ξ_ϕ and ξ_π model the inflow of sub-Hubble quantum fluctuations into the super-Hubble, coarse-grained sector of the theory, which behaves as classical random fields.

8.3 Inhomogeneous Spacetime

Thus far we have considered the evolution of inhomogeneous scalar field fluctuations in an unperturbed FLRW spacetime. However, since it is the energy density of the scalar field that determines the Hubble expansion during inflation, see Eq. (8.2), then we should also consider inhomogeneous perturbations of the spacetime, consistently perturbing the metric as well as the field content. Traditionally when studying the evolution of inhomogeneous fluctuations in cosmology, one allows for arbitrary, but small perturbations about the homogeneous and isotropic background and then systematically expands order-by-order in a small parameter controlling the deviation from FLRW symmetries [46]. However this perturbative approach is liable to break

[2] In this context, "improper" denotes operators for which the Wigner-Weyl transform takes values outside their spectrum [67].

down where fluctuations become large, either due to the accumulation of many small perturbations over a long time period, e.g., due to quantum diffusion, or a sudden large enhancement in the variance of those perturbations at a particular time or scale, for example, due to a localised feature in the scalar field potential.

In practice the coarse-graining scale, k_σ introduced in Sect. 8.2.3, allows us to adopt a different non-perturbative approach to model the evolution of potentially large inhomogeneities on large scales, $k < k_\sigma$, while retaining a perturbative approach to quantum fluctuations on small scales. This is known as the separate universe approach [75, 89] and it is an implicit, but essential part of the stochastic framework outlined above. It then allows us to reconstruct the inhomogeneous spacetime metric at the end of a period of stochastic inflation in terms of the perturbation of the local integrated expansion, δN [43, 45, 75].

8.3.1 Separate-Universe Approach

An important simplification occurs when studying inhomogeneous perturbations on sufficiently large scales where anisotropy and spatial gradients can be neglected. In that case the evolution of the fields locally can be described using ordinary differential equations for the time-dependence of effectively homogeneous and isotropic fields in an FLRW cosmology, considerably simplifying the calculation, and enabling us to track the nonlinear as well as linear evolution. This is the separate-universe approach [75, 89]. Formally we can interpret this as the leading-order behaviour in a gradient expansion [40, 69, 72, 81].

For example, if we consider linear perturbations about an FLRW cosmology of the scalar field (as already introduced in Sect. 8.2.2) and in addition perturbing the lapse function (allowing for local dilation of the proper time, τ relative to the background cosmic time t):

$$\phi \to \bar{\phi} + \delta\phi\,, \quad \partial/\partial t \to \partial/\partial\tau = (1 - A)\partial/\partial t\,, \tag{8.13}$$

then, at first order in the perturbations, the Klein-Gordon equation for the scalar field at each point can be written as [60]

$$\frac{\partial^2}{\partial\tau^2}(\bar{\phi} + \delta\phi) + 3(\bar{H} + \delta H)\frac{\partial}{\partial\tau}(\bar{\phi} + \delta\phi) + V_{,\phi}(\bar{\phi} + \delta\phi) = \nabla^2(\delta\phi)\,, \tag{8.14}$$

where $\bar{H} + \delta H$ is the local expansion rate. Up to the spatial gradient term, $\nabla^2\delta\phi$, Eq. (8.14) has exactly the same form as the Klein-Gordon equation for a homogeneous scalar field $\tilde{\phi} = \bar{\phi} + \delta\phi$ in an FLRW spacetime with Hubble rate $H = \bar{H} + \delta H$. The separate-universe approach is expected be valid on scales larger than the Hubble radius, H^{-1}, such that spatial gradients of the fields are small relative to the expansion rate. Thus we identify $\tilde{\phi}$ and H appearing in the Langevin equations (8.9) with the field and expansion rate coarse-grained on scales larger than $(\sigma H)^{-1}$ where $\sigma \ll 1$, and H is given by the local Friedmann equation

$$H^2 = \frac{1}{3M_P^2}\left[V(\tilde{\phi}) + \frac{1}{2}\left(\frac{\mathrm{d}\tilde{\phi}}{\mathrm{d}\tau}\right)^2\right]. \tag{8.15}$$

The separate universe approach turns out to be particularly powerful during inflation, since the accelerated expansion of the universe ensures that the spatial gradients of comoving wavemodes decrease more quickly than the Hubble rate ($H > |\dot{H}/H|$). Thus spatial gradients rapidly decrease as modes are stretched beyond the Hubble scale.

Intuitively we consider the coarse-grained field in each super-Hubble sized patch of the universe to be evolving like a separate FLRW universe, where the fields may take different values in different patches, but are treated as homogeneous within each patch. After sewing back together the different patches, we can reconstruct the inhomogeneous evolution of our entire observable universe above the coarse-graining scale. To relate local quantities to a global coordinate system amounts to a choice of gauge. This cannot be determined by the local FLRW quantities but requires the use of additional constraint equations from cosmological perturbation theory. For example in the spatially-flat gauge the momentum constraint imposes a simple relation between the lapse function and the scalar-field perturbation [38, 46]

$$A = \frac{\dot{\phi}}{2M_P^2 H}\delta\phi .$$

In the slow-roll limit $\dot{\phi} \to 0$ and the local proper time coincides with the background cosmic time, but beyond slow roll one must consistently account for local variations in the proper time interval in different patches to be able to relate the separate-universe evolution locally to the evolution of inhomogeneous perturbations about the global background [60].

Since the Langevin equations (8.9) use the logarithmic expansion, N, to describe the evolution of the scalar field and its canonical momentum in local patches, when we describe the ensemble of many different patches at a given N, we are implictly using a gauge in which the integrated local expansion

$$N \equiv \int (\bar{H} + \delta H)(1 + A)\,\mathrm{d}t , \tag{8.16}$$

remains unperturbed. Quantum field fluctuations crossing outside the Hubble scale during inflation are commonly calculated using the Sasaki-Mukhanov [55, 73] mode equation for scalar field fluctuations in the spatially-flat gauge [38, 46]. In slow-roll inflation the integrated local expansion is unperturbed in the spatially-flat gauge, but in general there is a gauge transformation required to consistently describe scalar field fluctuations in a uniform-N gauge [60].

While the separate universe approach works well on super-Hubble scales in many situations, it has recently been shown [34] that it breaks down on a finite range of super-Hubble scales in inflationary models with features, such as an inflection point, leading to a sudden, non-adiabatic transition from slow roll to a transient, ultra-slow-roll phase. This boosts the amplitude of density perturbations on small scales, which exit the Hubble-scale after the transition, which could lead to the formation of primordial black holes after inflation, or a stochastic background of gravitational waves induced at second-order in perturbation theory. But sudden transitions can also lead to spatial gradients generating non-adiabatic perturbations on super-Hubble scales at the transition [41], breaking the usual assumption that the separate universe approach can safely be used soon after Hubble exit. In such a situation one should be cautious about applying the usual stochastic-inflation formalism, for example, by waiting until after a sudden transition [34], or until the transient, non-adiabatic perturbations have decayed, before using the Langevin equations [86].

8.3.2 δN Formalism

Primordial density perturbations at the end of inflation are commonly characterised in terms of the gauge-invariant variable, ζ, which corresponds to the scalar metric perturbation, $-\psi$, on uniform-density hypersurfaces ($\delta\rho = 0$). At first-order this is written in terms of the curvature and density perturbations in an arbitrary gauge as [46]

$$\zeta \equiv -\psi - \frac{H\delta\rho}{\dot{\rho}} . \tag{8.17}$$

In particular it can be identified with the dimensionless density perturbation on spatially-flat hypersurfaces

$$\zeta = -\left.\frac{H\delta\rho}{\dot{\rho}}\right|_{\psi=0} . \tag{8.18}$$

It can be shown that ζ is conserved for adiabatic perturbations, irrespective of the field or matter content of the universe, in the large-scale limit where spatial gradients can be neglected [89, 91]. Hence it is a particularly useful tool to link scalar-field fluctuations during inflation to perturbations in the primordial plasma some time after inflation [7].

More generally ζ can be identified with the nonlinearly perturbed expansion between a spatially-flat hypersurface ($\psi = 0$) and the uniform-density hypersurface [43, 45, 75]

$$\zeta = \delta N \equiv N\left[\boldsymbol{\Phi}(\boldsymbol{x})\right] - \langle N(\boldsymbol{\Phi})\rangle , \tag{8.19}$$

where the local expansion N was defined in equation (8.16). In our definition of ζ in (8.19) we require that the lower limit of the integral (8.16) is a spatially-flat hypersurface, where the initial field-phase space vector at some point is given by

$\boldsymbol{\Phi}(\boldsymbol{x})$, and the upper limit of the integral is a uniform-density hypersurface at the end of inflation corresponding to the field-phase space vector $\boldsymbol{\Phi}_{\text{end}}$ at the end of inflation.

Thus we are able to calculate the distribution of primordial density perturbations after inflation in terms of the distribution of values of the perturbed local expansion, N, from an initial spatially-flat hypersurface during inflation to a uniform-density hypersurface at the end of inflation. If we employ the separate-universe approach to determine the nonlinear evolution on large scales, then we can use the background FLRW equations of motion to determine local expansion, $N(\boldsymbol{\Phi})$, as a function of the local coarse-grained field-phase space values $\boldsymbol{\Phi}(\boldsymbol{x})$ [44, 74, 77, 78].

In what we will call the *classical* δN formalism the fields are coarse-grained on one initial hypersurface such that all comoving scales of interest are above the Hubble scale at that time. If we further restrict ourselves to slow-roll evolution where the local field momenta are a function of the local field values, we can then formally expand Eq. (8.19) to give [44]

$$\zeta = \sum_I \frac{\partial N}{\partial \phi^I}\delta\phi_I + \frac{1}{2}\sum_{I,J}\frac{\partial^2 N}{\partial \phi^I \partial \phi^I}\delta\phi_I\delta\phi_J + \dots . \tag{8.20}$$

Furthermore for single-field slow-roll inflation we have $\partial N/\partial\phi = -H/\dot{\phi}$ and hence Eq. (8.20) reduces to Eq. (8.18) at first order, where $\rho = \rho(\phi)$.

The classical δN formalism assumes that the subsequent evolution above one fixed comoving coarse-graining scale is independent of further fluctuations on smaller scales and can be described by the unperturbed background equations. In the *stochastic* δN formalism [88] we choose a time-dependent coarse-graining scale, determined by the physical Hubble scale at any given time, thus incorporating the effect of quantum diffusion due to fluctuations on all scales that cross the coarse-graining scale before the end of inflation. The integrated expansion up to the end of inflation becomes a stochastic variable, $\mathcal{N}[\boldsymbol{\Phi}(\boldsymbol{x})]$, whose statistical properties are a function of the initial field-phase space vector.

8.4 First-Passage-Time Analysis

The Langevin equations (8.9) can be written more generally as

$$\frac{\mathrm{d}\Phi^i}{\mathrm{d}N} = F^i(\boldsymbol{\Phi}) + G^i_j(\boldsymbol{\Phi})\xi^j(N)\,, \tag{8.21}$$

where $\boldsymbol{\Phi}$ is the field-phase space vector that comprises all fields and conjugate momenta. The classical, homogeneous equations of motion are contained in $\boldsymbol{F}$, while $\boldsymbol{G}$ is obtained by evolving the field fluctuations from the Bunch-Davies vacuum to the σ-Hubble crossing time. The white Gaussian noises ξ_i are normalised such that $\langle\xi^i(N)\xi^j(N')\rangle = \delta^{i,j}\delta(N-N')$. In the case of single-field inflation with a canonical kinetic term, $\boldsymbol{F}$ and $\boldsymbol{G}$ can be read off Eq. (8.9). Along the slow-roll attractor, quantum diffusion only occurs along the classical flow lines [27], hence field phase space reduces to $\boldsymbol{\Phi} = (\phi)$ and one has $F = -V'/(3H^2)$ and $G = H/(2\pi)$, with $H^2 = V/(3M_P^2)$. For now, let us keep the generic form (8.21), since it allows us to describe any setup, including multi-fields models in non-trivial field-space geometries [63].

8.4.1 Fokker-Planck Equation

From the Langevin equation (8.21), one can derive a Fokker-Planck equation that drives the probability $P(\boldsymbol{\Phi}, N|\boldsymbol{\Phi}_{\text{in}}, N_{\text{in}})$ to find the configuration $\boldsymbol{\Phi}$ at time N knowing that the system was initiated as $\boldsymbol{\Phi}_{\text{in}}$ at time N_{in}. It is given by [71]

$$\begin{aligned}\frac{\partial}{\partial N}P(\boldsymbol{\Phi}, N|\boldsymbol{\Phi}_{\text{in}}, N_{\text{in}}) =& \left(-\frac{\partial}{\partial \Phi^i}F^i + \frac{1}{2}\frac{\partial^2}{\partial \Phi^i \partial \Phi^j}G^j_k G^{ki}\right) P(\boldsymbol{\Phi}, N|\boldsymbol{\Phi}_{\text{in}}, N_{\text{in}})\\ \equiv& \mathcal{L}(\boldsymbol{\Phi}) \cdot P(\boldsymbol{\Phi}, N|\boldsymbol{\Phi}_{\text{in}}, N_{\text{in}}),\end{aligned} \tag{8.22}$$

where we use Itô's discretisation convention. This defines the Fokker-Planck operator $\mathcal{L}(\boldsymbol{\Phi})$, which is a differential operator acting in the $\boldsymbol{\Phi}$ direction. The Fokker-Planck equation needs to be solved from the initial condition $P(\boldsymbol{\Phi}, N_{\text{in}}|\boldsymbol{\Phi}_{\text{in}}, N_{\text{in}}) = \delta(\boldsymbol{\Phi} - \boldsymbol{\Phi}_{\text{in}})$, and with an absorbing boundary condition on the end-of-inflation hypersurface $\mathcal{C}_{\text{end}}$, such that $P(\boldsymbol{\Phi}, N|\boldsymbol{\Phi}_{\text{in}}, N_{\text{in}}) = 0$ when $\boldsymbol{\Phi} \in \mathcal{C}_{\text{end}}$, where $\mathcal{C}_{\text{end}}$ is of uniform energy density.

The Langevin equation (8.21) describes a Markovian process, hence for any fixed intermediate time $\bar{N}$, using the rules of conditional probabilities one has

$$P\left(\boldsymbol{\Phi}, N|\boldsymbol{\Phi}_{\text{in}}, N_{\text{in}}\right) = \int_\Omega \mathrm{d}\bar{\boldsymbol{\Phi}}\, P\left(\boldsymbol{\Phi}, N|\bar{\boldsymbol{\Phi}}, \bar{N}\right) P\left(\bar{\boldsymbol{\Phi}}, \bar{N}|\boldsymbol{\Phi}_{\text{in}}, N_{\text{in}}\right), \tag{8.23}$$

where Ω is the inflating domain of field-phase space, bounded by $\mathcal{C}_{\text{end}}$. By differentiating both sides of this relation with respect to $\bar{N}$, one obtains

$$\frac{\partial}{\partial \bar{N}}P\left(\boldsymbol{\Phi}, N|\bar{\boldsymbol{\Phi}}, \bar{N}\right) = \mathcal{L}^\dagger(\bar{\boldsymbol{\Phi}}) \cdot P\left(\boldsymbol{\Phi}, N|\bar{\boldsymbol{\Phi}}, \bar{N}\right), \tag{8.24}$$

where $\mathcal{L}^\dagger$ is adjoint to the Fokker-Planck operator $\mathcal{L}$ in the sense that, for any two field-phase space functions $f(\boldsymbol{\Phi})$ and $g(\boldsymbol{\Phi})$, $\int \mathrm{d}\boldsymbol{\Phi} f(\boldsymbol{\Phi})[\mathcal{L}(\boldsymbol{\Phi}) \cdot g(\boldsymbol{\Phi})] = \int \mathrm{d}\boldsymbol{\Phi}[\mathcal{L}^\dagger(\boldsymbol{\Phi}) \cdot f(\boldsymbol{\Phi})]g(\boldsymbol{\Phi})$. This is why Eq. (8.24) is called the adjoint Fokker-Planck equation, or Kolmogorov backward equation, where, using integration by parts

$$\mathcal{L}^\dagger(\boldsymbol{\Phi}) = F^i \frac{\partial}{\partial \Phi^i} + \frac{1}{2}G^i_k G^{ki}\frac{\partial^2}{\partial \Phi^i \partial \Phi^j}. \tag{8.25}$$

Our goal is to reconstruct the distribution of first-passage times $\mathcal{N}$ through the end-of-inflation hypersurface $\mathcal{C}_{\text{end}}$, starting from a given field configuration $\boldsymbol{\Phi}_{\text{in}}$, and identify it with the coarse-grained curvature perturbation using Eq. (8.19). This is the so-called stochastic-δN program [17, 25, 59, 88]. The first-passage time distribution is denoted $P_{\text{FPT},\boldsymbol{\Phi}_{\text{in}}}(\mathcal{N})$, and is a central topic in stochastic analysis, see for instance Ref. [66]. Here, we only give a few key results, following Refs. [20, 59, 88] in the context of inflationary cosmology.

A direct way to reconstruct the distribution of first passage times is through numerical sampling of the Langevin equation (8.9): for each realisation, record the duration of inflation $\mathcal{N}$, and draw a histogram (or use more advanced kernel reconstruction methods). This approach is numerically expensive, since a large number of realisations needs to be simulated to reach satisfactory levels of statistical noise, especially in the tail of the first-passage-time distribution where the statistics is scarce. Importance sampling can be used to circumvent this issue, see Ref. [33], or methods based on the formalism of stochastic excursions, see Ref. [82]. However, for analytical insight it is useful to derive a driving equation for $P_{\text{FPT},\boldsymbol{\Phi}_{\text{in}}}(\mathcal{N})$, analogous to the Fokker-Planck equation for $P(\boldsymbol{\Phi}, N|\boldsymbol{\Phi}_{\text{in}}, N_{\text{in}})$. This is the goal of this section.

8.4.2 A Warm Up: Free Diffusion

Let us start with the simple toy model consisting of a scalar field ϕ diffusing along a constant potential, along the slow-roll attractor. Upon introducing the rescaled variable $x = 2\pi(\phi - \phi_{\text{end}})/H$, the Langevin equation reads $\mathrm{d}x/\mathrm{d}N = \xi(N)$, and the Fokker-Planck equation is given by

$$\frac{\partial}{\partial N} P(x, N|x_{\text{in}}, N_{\text{in}}) = \frac{1}{2}\frac{\partial^2}{\partial x^2} P(x, N|x_{\text{in}}, N_{\text{in}}) , \tag{8.26}$$

which admits Gaussian solutions of the form

$$f_{x_0,\sigma_0}(x, N) = \frac{e^{-\frac{(x-x_0)^2}{2\sigma^2(N)}}}{\sqrt{2\pi\sigma^2(N)}} \quad \text{with} \quad \sigma^2(N) = \sigma_0^2 + N - N_{\text{in}} . \tag{8.27}$$

Since it is linear, any linear combination of such functions is still a solution. The linear combination that allows one to satisfy both the initial condition $P(x, N_{\text{in}}|x_{\text{in}}, N_{\text{in}}) = \delta(x - x_{\text{in}})$ and the absorbing condition $P(0, N|x_{\text{in}}, N_{\text{in}}) = 0$ is given by

$$P(x, N|x_{\text{in}}, N_{\text{in}}) = \left[f_{x_{\text{in}},0}(x, N) - f_{-x_{\text{in}},0}(x, N)\right]\Theta(x) , \tag{8.28}$$

where the Heaviside function restricts the solution to the inflating domain $x > 0$. This solution is obtained by adding a (negative) image of the unconstrained solution mirrored through the absorbing boundary, which is often referred to as the "method of images".

From the solution to the Fokker-Planck equation, one can compute the survival probability $S_{x_\mathrm{in}}(N)$, i.e. the probability for the field to be still inflating at time N

$$S_{x_\mathrm{in}}(N) = \int_0^\infty \mathrm{d}x\, P(x, N|x_\mathrm{in}, N_\mathrm{in}) = \mathrm{erf}\left[\frac{x_\mathrm{in}}{\sqrt{2(N-N_\mathrm{in})}}\right]. \tag{8.29}$$

Alternatively, this can be written as the probability that, starting from x_in, the first-passage time through the end of inflation x_end exceeds $N - N_\mathrm{in}$, hence

$$S_{x_\mathrm{in}}(N) = \int_{N-N_\mathrm{in}}^\infty P_{\mathrm{FPT},x_\mathrm{in}}(\mathcal{N})\mathrm{d}\mathcal{N}. \tag{8.30}$$

By equating the two expressions above, and differentiating with respect to N, one obtains $P_{\mathrm{FPT},x_\mathrm{in}}(\mathcal{N}) = -\left.\partial S_{x_\mathrm{in}}/\partial N\right|_{N=N_\mathrm{in}+\mathcal{N}}$, hence

$$P_{\mathrm{FPT},x_\mathrm{in}}(\mathcal{N}) = \frac{x_\mathrm{in}}{\sqrt{2\pi}\mathcal{N}^{3/2}}\exp\left(-\frac{x_\mathrm{in}^2}{2\mathcal{N}}\right). \tag{8.31}$$

This is called a Lévy distribution. One can see that it features a very heavy tail, since $P_{\mathrm{FPT},x_\mathrm{in}}(\mathcal{N}) \propto \mathcal{N}^{-3/2}$ at large $\mathcal{N}$. This implies that the probability to realise large excursion times $\mathcal{N}$, hence large curvature perturbations according to Sect. 8.3, is much larger than what would be obtained with Gaussian statistics. Although this conclusion is obtained with a toy, arguably unrealistic model, it is in fact entirely generic, as we shall now see.

8.4.3 Adjoint Fokker-Planck Equation

In general, the survival probability starting from a given configuration $\boldsymbol{\Phi}_\mathrm{in}$ can be defined in a way analogous to Eq. (8.29), i.e.

$$S_{\boldsymbol{\Phi}_\mathrm{in}}(N) = \int_\Omega \mathrm{d}\boldsymbol{\Phi}\, P\left(\boldsymbol{\Phi}, N|\boldsymbol{\Phi}_\mathrm{in}, N_\mathrm{in}\right). \tag{8.32}$$

Since $P(\boldsymbol{\Phi}, N|\boldsymbol{\Phi}_\mathrm{in}, N_\mathrm{in})$ satisfies the adjoint Fokker-Planck equation (8.24), so does the survival probability, $(\partial/\partial N_\mathrm{in})S_{\boldsymbol{\Phi}_\mathrm{in}}(N) = \mathcal{L}^\dagger(\boldsymbol{\Phi}_\mathrm{in}) \cdot S_{\boldsymbol{\Phi}_\mathrm{in}}(N)$. Since $\boldsymbol{F}(\boldsymbol{\Phi})$ and $\boldsymbol{G}(\boldsymbol{\Phi})$ do not depend on time explicitly, the survival probability only depends on the time difference $N - N_\mathrm{in}$, hence $(\partial/\partial N_\mathrm{in})S_{\boldsymbol{\Phi}_\mathrm{in}}(N) = -(\partial/\partial N)S_{\boldsymbol{\Phi}_\mathrm{in}}(N)$. Moreover, similarly to Eq. (8.30) one can write $S_{\boldsymbol{\Phi}_\mathrm{in}}(N) = \int_{N-N_\mathrm{in}}^\infty P_{\mathrm{FPT},\boldsymbol{\Phi}_\mathrm{in}}(\mathcal{N})\mathrm{d}\mathcal{N}$. Combining these results, one obtains

$$\frac{\partial}{\partial \mathcal{N}} P_{\mathrm{FPT},\boldsymbol{\Phi}_{\mathrm{in}}}(\mathcal{N}) = \mathcal{L}^\dagger(\boldsymbol{\Phi}_{\mathrm{in}}) \cdot P_{\mathrm{FPT},\boldsymbol{\Phi}_{\mathrm{in}}}(\mathcal{N}) , \tag{8.33}$$

i.e. the first-passage time distribution obeys the adjoint Fokker-Planck equation. The only difference with the Kolmogorov backward equation (8.24) is the boundary conditions: here, Eq. (8.33) needs to be solved with $P_{\mathrm{FPT},\boldsymbol{\Phi}_{\mathrm{in}}}(\mathcal{N}) = \delta(\mathcal{N})$ when $\boldsymbol{\Phi}_{\mathrm{in}} \in \mathcal{C}_{\mathrm{end}}$. In the free-diffusion problem, one can check that, indeed, the Lévy distribution (8.31) satisfies the adjoint Fokker-Planck equation, which in that case coincides with the Fokker-Planck equation (8.26) since $F = 0$ and G is uniform.

8.5 Heavy Tails and Non-perturbative Non-Gaussianities

The first-passage-time distribution depends on the details of the inflationary model under consideration, and a variety of profiles can be obtained, see for instance Refs. [1, 6, 12, 14, 20, 21, 23, 30, 33, 35, 37, 58, 59, 61, 70, 87]. In all cases however, the tail of the distribution, which is relevant for the formation of PBHs, is heavier than the Gaussian behaviour predicted by linear perturbation theory. In this section, we briefly review why.

8.5.1 Characteristic Function

The adjoint Fokker-Planck equation (8.33) is a linear partial differential equation, hence it can be turned into a set of ordinary differential equations by introducing the characteristic function

$$\chi\left(t, \boldsymbol{\phi}\right) = \left\langle e^{it\mathcal{N}} \right\rangle = \int_{-\infty}^{\infty} \mathrm{d}\mathcal{N} e^{it\mathcal{N}} P_{\mathrm{FPT},\boldsymbol{\phi}}(\mathcal{N}) . \tag{8.34}$$

This is nothing but the Fourier transform of the first-passage-time distribution, which is a function of the dummy parameter t. Using Eq. (8.33), it satisfies the differential equation

$$\mathcal{L}^\dagger\left(\boldsymbol{\Phi}\right) \cdot \chi(t, \boldsymbol{\Phi}) = -it\chi\left(t, \boldsymbol{\Phi}\right), \tag{8.35}$$

together with the boundary condition $\chi(t, \boldsymbol{\Phi}) = 1$ when $\boldsymbol{\Phi} \in \mathcal{C}_{\mathrm{end}}$.

8.5.2 Pole Expansion

In general, the presence of non-trivial boundary conditions results in the existence of poles, and the characteristic function can be expanded as

$$\chi\,(t,\boldsymbol{\Phi}) = f\,(t,\boldsymbol{\Phi}) + \sum_n \frac{a_n(\boldsymbol{\Phi})}{\Lambda_n - it}, \tag{8.36}$$

where f is a regular function, and we have assumed that the poles form a discrete set. The form (8.36) follows from the structure of the Fokker-Planck operator and can be justified formally, see Ref. [20], where the parameters Λ_n are shown to correspond to the eigenvalues of the adjoint Fokker-Planck operator, hence they do not depend on $\boldsymbol{\Phi}$.

Let us illustrate this decomposition with the example of free diffusion studied in Sect. 8.4.2, where an additional reflective boundary is added at ϕ_r. Such a model is referred to as a "flat well". This boundary can be removed by evaluating the formulas below in the limit $\phi_\mathrm{r} \to \infty$ but at this stage ϕ_r needs to be kept finite, since only when adding the boundary condition $(\partial/\partial\phi)\chi(t,\phi)|_{\phi=\phi_\mathrm{r}} = 0$ is the solution to Eq. (8.35) fully specified, and reads

$$\chi(t,\phi) = \frac{\cos\left[\sqrt{it}\mu\left(\frac{\phi}{\phi_\mathrm{r}} - 1\right)\right]}{\cos\left(\sqrt{it}\mu\right)}, \quad \text{where} \quad \mu^2 = \frac{24\pi^2 M_P^2 \phi_\mathrm{r}^2}{V} \tag{8.37}$$

and where we have set $\phi_\mathrm{end} = 0$ without loss of generality. The partial fraction decomposition of $\chi(t,\phi)$ is of the form (8.36), with $\Lambda_n = (\pi/\mu)^2(n+1/2)^2$ and $a_n(\phi) = (-1)^n\pi(2n+1)\cos[\frac{\pi}{2}(2n+1)(\frac{\phi}{\phi_\mathrm{r}} - 1)]/\mu^2$.

The inverse Fourier transform that is required to derive the first-passage-time distribution can be performed by means of the residue theorem, and one obtains

$$P_{\mathrm{FPT},\boldsymbol{\Phi}}\,(\mathcal{N}) = \sum_n a_n(\boldsymbol{\Phi}) e^{-\Lambda_n \mathcal{N}}. \tag{8.38}$$

On the tail, i.e. at large values of $\mathcal{N}$, the lowest eigenvalues Λ_n dominate, hence if the poles are ordered such that $\Lambda_0 < \Lambda_1 < \cdots$, the pole expansion is essentially a tail expansion.

8.5.3 Tail Behaviour

In Fig. 8.1, we display the first-passage-time distribution obtained in a tilted-inflection point model, for a few initial conditions. The solid lines correspond to the numerical solutions of Eq. (8.33) (in practice, we solved Eq. (8.35) on a grid of t values and inverse-Fourier transformed the result), while the dotted lines correspond to the prediction of linear cosmological perturbation theory. In perturbation theory, one obtains Gaussian distributions, which provide good fits to the full distributions close to the maximum. This is because the statistics of the most likely fluctuations are almost Gaussian. However, in the tail of distribution, the Gaussian

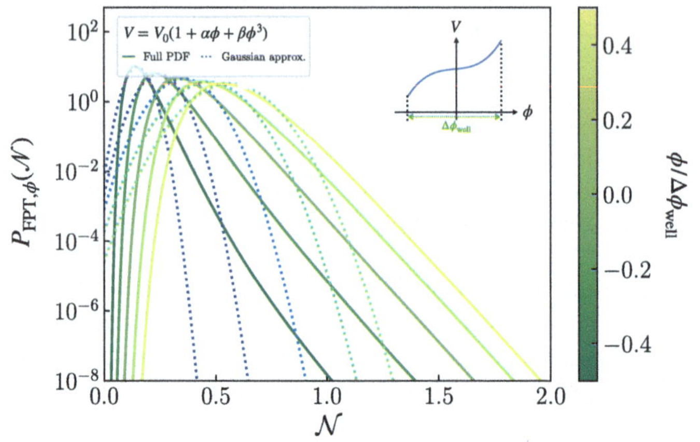

Fig. 8.1 Probability density functions of the number of e-folds generated by an inflection-point inflationary potential (see inset, with $V_0/(24\pi^2 M_P^4) = 10^{-3}$, $\alpha = 0.24/M_P$, $\beta = 9M_P^3$ and $\Delta\phi_{\rm well} = 5.83M_P$–the value for V_0 is unrealistically large but it is used for convenience of the illustration). Solid lines correspond to the full distribution functions computed by mean of the stochastic-δN formalism, where different colours correspond to different initial conditions. The dotted lines correspond to the Gaussian approximation provided by linear perturbation theory, which provides a good fit for the maximum of the distribution but however fails to describe the heavy tails, where PBHs are nonetheless produced. Adapted from Ref. [20]

approximation breaks down. Indeed, when $\mathcal{N}$ becomes large, Eq. (8.38) reduces to $P_{\rm FPT,\boldsymbol{\Phi}}(\mathcal{N}) \simeq a_0(\phi)e^{-\Lambda_0\mathcal{N}}$, i.e. the tail has an exponential profile. Compared to Gaussian suppression, $P_{\rm FPT,\boldsymbol{\Phi}}(\mathcal{N}) \propto e^{-\alpha\mathcal{N}^2}$, the tail is much heavier, and since large values of $\mathcal{N}$ correspond to large values of ζ, this can have a strong impact on the abundance of PBHs [26].

If ζ has Gaussian statistics, the probability to realise $\zeta > \zeta_{\rm c} \sim 1$ is of order $p_{\rm c} = \mathrm{erfc}[\zeta_{\rm c}/(\sqrt{2\mathcal{P}_\zeta})]/2 \propto e^{-\zeta_{\rm c}^2/(2\mathcal{P}_\zeta)}$, where erfc is the complementary error function, which we have expanded in the regime $\zeta_{\rm c} \gg \sqrt{\mathcal{P}_\zeta}$. PBH dark-matter typically requires the PBH mass fraction to be of order $p_{\rm c} \sim 10^{-20}$ at formation, see Part V, which thus corresponds to $\mathcal{P}_\zeta \simeq 10^{-2}$. For this reason it is generally considered that, for dark matter to be comprised of PBHs, such large values of the power spectrum are required, which can be challenging to achieve from a model-building perspective. However, in the presence of exponential tails, the situation is quite different. For instance, consider the case where the PDF of ζ is Gaussian across $n_{\rm SD}$ standard deviations around its maximum, and then exponential. In that case $p_{\rm c} \simeq e^{-n_{\rm SD}^2/2-\Lambda_0}/(\Lambda_0\sqrt{2\pi\mathcal{P}_\zeta})$, which clearly signals that cosmologically-relevant values of $p_{\rm c}$ do not require large values of the power-spectrum. This is why it is crucial to take quantum diffusion into account when it comes to computing the abundance and statistics of PBHs.

Finally, let us note that a compact inflating domain is required for the eigenvalues Λ_n to form a discrete set. Indeed, separated poles in the characteristic function follow from imposing boundary conditions at finite field distances, in the same way that standing waves can assume only certain frequencies in a finite-sized cavity. If the cavity becomes infinite, any frequency is allowed, and likewise, if the inflating domain is not bounded, n becomes a continuous rather than discrete index in Eq. (8.38). An explicit example of this phenomenon is given by the pure-diffusion example: in that case, $\Lambda_{n+1} - \Lambda_n = 2(n+1)\pi^2/\mu^2$, which goes to zero, since $\mu^2 \to \infty$, when $\phi_r \to \infty$, hence the set of poles becomes continuous. When this occurs the tail of $P_{\mathrm{FPT},\boldsymbol{\Phi}}(\mathcal{N})$ can no longer be approximated by $a_0(\boldsymbol{\Phi})e^{-\Lambda_0\mathcal{N}}$, hence it becomes even heavier than exponential. In the free-diffusion example, in the limit $\phi_{\mathrm{r}} \to \infty$ we found a Lévy distribution, see Eq. (8.31), for which the tail is indeed monomial rather than exponential. This makes the conclusions derived above even more relevant in that case.

8.5.4 Non-perturbative Non-Gaussianities

The importance of non-Gaussianities in shaping the detailed PBH statistics is well-known, but so far it has been mostly investigated using perturbative techniques, such as f_{NL}-type parametrisations. These expansions are perturbative in nature, i.e. they expand around Gaussian statistics, introducing corrections arising from the bispectrum (the three-point correlation function), the trispectrum (the four-point correlation function), etc. For instance, if $\mathcal{N}_{\mathrm{G}}$ is a Gaussian random variable, the local-f_{NL} parametrisation corresponds to the approximation

$$\mathcal{N} \simeq \mathcal{N}_{\mathrm{G}} + f_{\mathrm{NL}}\left(\mathcal{N}_{\mathrm{G}}^2 - \langle\mathcal{N}_{\mathrm{G}}^2\rangle\right). \tag{8.39}$$

The parameter f_{NL} can be extracted from the third moment of $\mathcal{N}$, which consists of Taylor expanding the PDF of $\mathcal{N}$ around its maximum. This approximation is compared to the full distribution in Fig. 8.2 in the free-diffusion toy model. This illustrates that, although the f_{NL} approximation improves the quality of the fit close to the maximum, it fails to capture the tail behaviour, which is crucial for the PBH abundance.

The reason for this failure is that the non-Gaussianities produced by quantum diffusion are non-perturbative in nature. Thus, they require non-perturbative parametrisations. A straightforward non-perturbative generalisation of Eq. (8.39) consists of treating the local-f_{NL} parametrisation as a Taylor expansion of an underlying local ansatz $\mathcal{N}(\boldsymbol{x}) = f[\mathcal{N}_{\mathrm{G}}(\boldsymbol{x}), \boldsymbol{x}]$, where f is any (in general non-linear) function. This approach was recently followed in Refs. [26, 37]. It has the advantage of allowing one to derive the statistics of the compaction function, hence the mass fraction of PBHs, in a consistent non-perturbative way, which is otherwise difficult (see Sect. 8.6.2 below). However, it is not yet clear whether or not non-Gaussianities inherited from quantum diffusion are well-approximated by the above local ansatz, and further investigations are required.

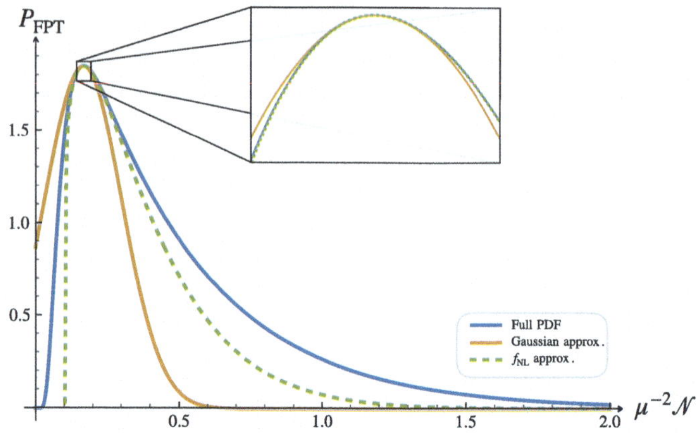

Fig. 8.2 Probability density function of the number of e-folds generated by a flat well starting from $\phi = \phi_{\rm r}$, see Eqs. (8.37)–(8.38). The orange line corresponds to the Gaussian approximation, while the dashed green line stand for the local $f_{\rm NL}$ expansion, see Eq. (8.39). The inset zooms in on the maximum of the distribution, where the Gaussian and quasi-Gaussian approximations provide good fits. However, they both fail to properly describe the heavy tails

8.6 Prospects

Let us end this chapter by reviewing a few of the open challenges that remain to be addressed to better understand the role of quantum diffusion in shaping the statistics of PBHs. Our goal is not to list all of them, but simply to highlight a few conceptual and technical issues that we think deserve further attention in the years to come.

8.6.1 *The Stochastic-δN Formalism Beyond Slow Roll*

Most applications of the stochastic-δN formalism to-date have been restricted to slow-roll inflation in which the coarse-grained fields follow an attractor trajectory where the fields' momenta are uniquely determined by the field values, reducing the phase-space to 1D for a single scalar field, for example. However models of inflation where there is a significant enhancement of primordial density perturbations on small scales necessarily break the single-field and/or slow-roll assumptions and include transient, non-attractor behaviour, such as ultra-slow roll.

Extending the phase-space beyond the slow-roll trajectory has several important consequences. Firstly, and most obviously, it raises the dimensionality of the probability distribution function $P(\boldsymbol{\Phi}, N|\boldsymbol{\Phi}_{\rm in}, N_{\rm in})$ and the Fokker-Planck equation (8.22), and hence makes more challenging the first-passage time problem to find $P_{\mathrm{FPT},\boldsymbol{\Phi}_{\rm in}}(\mathcal{N})$. In the case of a single field, for example, extending the phase space beyond slow roll [27] leads to the the characteristic function (8.34) obeying a partial differential equation [61] rather than an ordinary differential equation as in the case of slow roll for a single field [20, 59]. The first-passage-time problem in 2D has been studied in the ultra-slow roll limit, assuming the usual Bunch-Davies vacuum state for a massless field in de Sitter [24, 61]. In the small-velocity (diffusion-dominated) expansion one recovers at leading order the previous slow-roll results [59], but at next-to-leading order we find an additional series of poles in the characteristic function. This does not affect the lowest pole, and hence the leading exponential term in the tail of the probability distribution for the first passage time, but it can suppress the amplitude of the tail.

There are several other challenges. As we move away from the massless limit the power spectrum for the stochastic noise due to quantum fluctuations crossing outside the coarse-graining scale deviates from the simple result for a massless field in de Sitter, $\langle \xi_\varphi^2 \rangle = (H/2\pi)^2$, and one must also include the noise affecting the field momentum, ξ_π [16, 23, 27, 32, 53]. In slow roll, the fluctuations are overdamped on super-horizon scales leading to a squeezed state that can be treated as classical noise, but in the absence of an overdamped attractor then the question arises as to whether a classical treatment remains sufficient. Moreover, away from slow roll the covariance of the noise does not only depend on the instantaneous value of H, but it is also affected by the details of the prior background evolution. This implies that the field mode equation should be solved along each realisation of the Langevin equation, giving rise to non-Markovian processes that are challenging to investigate [22, 23].

Finally, we note that single-field models leading to a rapid rise in the primordial power spectrum on small scales require a sudden, non-adiabatic transition to force the background field out of slow-roll into an ultra-slow-roll phase. The sudden transition leads to deviations from the Bunch-Davies vacuum state (i.e. particle production) on sub-Hubble scales, and non-adiabatic perturbations, sourced by residual spatial gradients, on a finite range of super-Hubble scales at the transition [41]. These gradient terms are not included in the usual separate-universe approximation and require a re-evaluation of the classical and stochastic δN formalisms at a sudden transition [34].

8.6.2 From the Curvature Perturbation to the Compaction Function

So far in this chapter, the production of PBHs has been attributed to large local values of ζ_R, the curvature perturbation coarse-grained at the scale R, where the mass of the resulting black hole scales with the Hubble mass at the time the scale R re-enters the Hubble radius. This is however only an approximation, and modern PBH formation criteria rather depend on the smoothed density contrast or the compaction function [29, 56, 76] to determine whether a PBH forms, and use critical scaling to infer its mass [15, 19, 39], see Chaps 2, 4 and 6.

The curvature perturbation is a direct outcome of the stochastic δN formalism but the statistics of other relevant cosmological fields, at the non-perturbative level, remain unexplored. One reason why the curvature perturbation is not ideal to predict the formation of PBHs at the scale R is that it is sensitive to power at larger scales. Indeed, if ζ_R denotes the curvature perturbation coarse-grained at the scale R, from the coincident two-point function, $\langle \zeta_R^2(\boldsymbol{x}) \rangle = \int \mathrm{d} \ln k \mathcal{P}_\zeta(k) \mathrm{sinc}(kR/a)$, one can see that power at scales $k \ll a/R$ may contribute significantly to ζ_R, depending on the power spectrum. However, at large scales, ζ is conserved and may be interpreted as a renormalisation of the scale factor seen by a local observer, hence it should not affect the collapse dynamics of a local over-density. In contrast, at the linear level the comoving density contrast is related to the curvature perturbation via

$$\delta_{\mathrm{lin}} = \frac{2(1+w)}{5+3w} \frac{k^2}{a^2 H^2} \zeta , \tag{8.40}$$

where w is the equation-of-state parameter. The density contrast thus features an additional k^2 suppression at large scales, which ensures that only the scales $k \sim a/R$ contribute to the formation of black holes with size R. If one accounts for the non-linear relationship between the curvature perturbation and the density contrast, one is specifically led to consider the compaction function, but the argument remains the same.

One possible approach to compute the statistics of the density contrast or the compaction function is to consider the coarse-shelled curvature perturbation, where instead of averaging ζ over a sphere of radius R, one considers its average between two spheres of radii R_1 and R_2, $\Delta\zeta_{R_1,R_2}(\boldsymbol{x}) = \zeta_{R_2}(\boldsymbol{x}) - \zeta_{R_1}(\boldsymbol{x})$. This selects scales in the interval $a/R_2 < k < a/R_1$, so by adjusting the parameters R_1 and R_2 around R, one can try to mimic $\delta_R(\boldsymbol{x})$. Since the coarse-shelled curvature perturbation can be written as the difference between two coarse-grained curvature perturbations, its statistics can be computed in the stochastic-δN formalism, and in this way one may approximate the statistics of the comoving density contrast. This approach is developed in Ref. [80], where it is also applied to the compaction function. This requires us to compute the two-point distribution of the curvature perturbation, $P[\zeta_{R_1}(\boldsymbol{x}), \zeta_{R_2}(\boldsymbol{x})]$, but no higher statistics.

In those rare cases where all n-point functions are known, exact calculations can also be performed. For instance, under the generalised local ansatz mentioned in Sect. 8.5.4, $\zeta(\boldsymbol{x}) = f[\zeta_{\mathrm{G}}(\boldsymbol{x}), \boldsymbol{x}]$ where ζ_{G} is a Gaussian field, the statistics of the compaction function $\mathcal{C}_R$ can be derived using

$$\mathcal{C}_R = \frac{2}{3}\left[1 - \left(1 + R\zeta_R'\right)^2\right] , \tag{8.41}$$

where a prime denotes differentiation with respect to R. This approach was followed in Refs. [26, 37], where the importance of heavy tails was confirmed. Similarly, in the setup recently considered in Ref. [65], an explicit local expression is given for $\zeta_R(\boldsymbol{x})$

in terms of a set of Gaussian fields, which makes the computation of the compaction function tractable.

Beyond those simple cases however, the derivation of the compaction function in the stochastic-δN formalism remains an open issue. The use of more advanced structure-formation tools, such as the excursion set [10, 11, 62] or peak theory [8], together with critical scaling, also remains to be explored.

8.6.3 Volume Weighting and Backward Observables

In the standard picture of cosmological perturbation theory, along a given field-phase space classical trajectory, there is a one-to-one relationship between comoving scales k and the system's configuration $\boldsymbol{\Phi}_*(k)$ when that scale crosses out the Hubble radius during inflation. Therefore, by performing measurements of the statistics at the scale k, one is able to probe the inflationary setup (the potential function, the field geometry, etc.) only around that location $\boldsymbol{\Phi}_*$. In the stochastic picture however, such a one-to-one relationship does not exist. Assuming that H is quasi-constant during inflation, a given scale k exits the Hubble radius at the time $N_*(k) = N_{\mathrm{end}} - \ln(k_{\mathrm{end}}/k)$, where k_{end} denotes the comoving Hubble scale at the end of inflation. Therefore, if $\boldsymbol{\Phi}_*$ denotes the system's configuration at the time N_*, then it is a random quantity, and its distribution function is given by [4]

$$P_{\mathrm{bw}}\left(\boldsymbol{\Phi}_*\right) = P_{\mathrm{FPT},\boldsymbol{\Phi}_*}\left(N_{\mathrm{bw}}\right) \frac{\int_0^\infty P\left(\boldsymbol{\Phi}_*, N|\boldsymbol{\Phi}_{\mathrm{in}}, N_{\mathrm{in}}\right) \mathrm{d}N}{\int_{N_{\mathrm{bw}}}^\infty P_{\mathrm{FPT},\boldsymbol{\Phi}_{\mathrm{in}}}\left(N_{\mathrm{tot}}\right) \mathrm{d}N_{\mathrm{tot}}} . \tag{8.42}$$

Here, $N_{\mathrm{bw}} = \ln(k_{\mathrm{end}}/k)$ is the number of e-folds corresponding to the Hubble exit of k, counted backwards from the end of inflation. When the model under consideration features regions with large quantum diffusion, this distribution function may have a wide support, which implies that a broad region of field-phase space can be probed even if a limited range of scales is observationally accessible. Note that this is the case even if the scales being measured emerge prior to the diffusion-dominated region, which allows one to place constraints on PBH-compliant models from CMB measurements only [4].

In practice, the relationship between field values and physical distances is even more involved, and is encoded in the recursive geometry of a stochastically inflating universe. Consider indeed the situation sketched in Fig. 8.3. Here, each disk stands for a Hubble patch, and as inflation proceeds each patch gives rise to new Hubble patches, until the end-of-inflation hypersurface $\mathcal{C}_{\mathrm{end}}$ is being crossed. In this picture, the inflating space-time has the structure of a tree in the language of graph theory. From a given initial patch, denoted $\mathcal{P}_*$ in the figure and with field value $\boldsymbol{\Phi}_*$, a certain volume V is produced on the final hypersurface, which corresponds to the number of leaves originating from the root $\mathcal{P}_*$. The distribution function $P(V|\boldsymbol{\Phi}_*)$ is not straightforward to compute, since all branches originating from the root $\mathcal{P}_*$

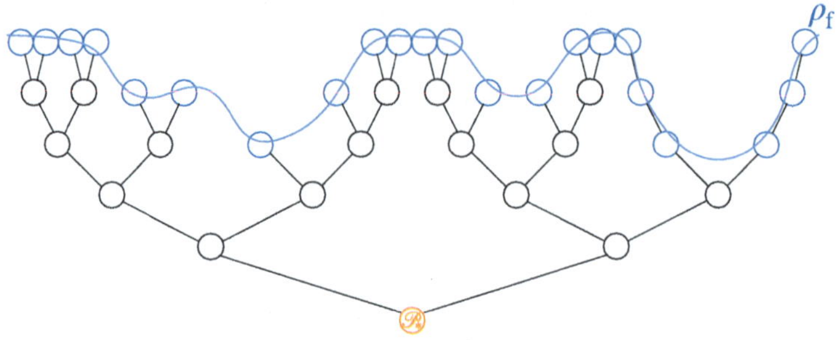

Fig. 8.3 Space-time structure of a stochastically-inflating universe (adapted from Ref. [5])

coincide until they split, hence the leaves cannot be seen as the end-point of independent stochastic realisations of the Langevin equation. Methods can nonetheless be developed to reconstruct the distribution of the final volume [5], which contains the relevant information to match distances as measured by local observers on the final hypersurface to field configurations when the corresponding spatial region emerges during inflation.

Subtleties may however arise when considering volume weighting. Consider indeed a local observer on the final hypersurface of the tree in Fig. 8.3, and task them with measuring the mean density on that hypersurface. What they would do is to measure the density in each leaf, and compute the ensemble average. As mentioned above, ensemble averages over subsets of leaves correspond to volume averages from the viewpoint of the root patch $\mathcal{P}_*$. The reason is that regions of the universe that inflate more give rise to a larger number of leaves on the final hypersurface, hence they contribute more to ensemble averages computed by local observers. As a consequence, when it comes to observable quantities, distributions should be volume-weighted by $e^{3\mathcal{N}}$. For instance, for the curvature perturbation coarse-grained at R_{end}, the Hubble radius at the end of inflation, one has

$$P\left(\zeta_{R_{\mathrm{end}}}\right) \propto P_{\mathrm{FPT},\boldsymbol{\Phi}_{\mathrm{in}}}\left(\left\langle\mathcal{N}_{\boldsymbol{\Phi}_{\mathrm{in}}}\right\rangle_V + \zeta_{R_{\mathrm{end}}}\right) e^{3\zeta_{R_{\mathrm{end}}}} , \tag{8.43}$$

where the overall factor is set such that the distribution is normalised, and $\langle\cdot\rangle_V$ is the volume-weighted stochastic average, i.e.

$$\left\langle\mathcal{N}_{\boldsymbol{\Phi}_{\mathrm{in}}}\right\rangle_V = \frac{\int P_{\mathrm{FPT},\boldsymbol{\Phi}_{\mathrm{in}}}(\mathcal{N}) e^{3\mathcal{N}} \mathcal{N} \mathrm{d}\mathcal{N}}{\int P_{\mathrm{FPT},\boldsymbol{\Phi}_{\mathrm{in}}}(\mathcal{N}) e^{3\mathcal{N}} \mathrm{d}\mathcal{N}} . \tag{8.44}$$

This is what a local observer would measure by recording $\mathcal{N}_{\boldsymbol{\Phi}_{\mathrm{in}}}$ in each leaf and computing the ensemble average. Recalling that the first-passage-time distribution has an exponential tail, $P_{\mathrm{FPT},\boldsymbol{\Phi}_{\mathrm{in}}}(\mathcal{N}) \propto e^{-\Lambda_0\mathcal{N}}$, the above formula fails to converge when $\Lambda_0 \leq 3$. This is the case where "eternal inflation" takes place, and the mean volume originating from a given patch $\mathcal{P}_*$ diverges. Note that this issue cannot be avoided by starting inflation at sufficiently low energy: as mentioned in Sect. 8.5.3,

Λ_0 is independent of $\boldsymbol{\Phi}$, hence if volume divergences arise somewhere then they arise everywhere. Note also that, as mentioned in Sect. 8.5.3, when the field-phase space is not compact, the tail of the first-passage time distribution is even heavier than exponential, hence volume divergences necessarily occur. This is the case for most inflationary models considered in the literature, hence deriving properly volume-weighted observables in this context remains an important challenge to be addressed.

Acknowledgements We would like to thank all the colleagues with whom we have investigated quantum diffusion and primordial black holes over the past few years, in particular Kenta Ando, Chiara Animali, Hooshyar Assadullahi, Vadim Briaud, Jose-María Ezquiaga, Juan García-Bellido, Andrew Gow, Joseph Jackson, Kazuya Koyama, Chris Pattison, Yuichiro Tada, Koki Tokeshi and Eemeli Tomberg.

References

1. A. Achucarro et al., The hand-made tail: non-perturbative tails from multifield inflation. JHEP **05**, 052 (2022). https://doi.org/10.1007/JHEP05(2022)052. arXiv:2112.14712 [hep-th]
2. G. Adesso, S. Ragy, A.R. Lee, Continuous variable quantum information: gaussian states and beyond. Open. Syst. Inf. Dyn. **21**, 1440001 (2014). arXiv e-prints. https://doi.org/10.48550/arXiv.1401.4679. arXiv:1401.4679 [quant-ph]
3. K. Ando, V. Vennin, Bipartite temporal Bell inequalities for two-mode squeezed states. Phys. Rev. A **102**(5), 052213 (2020). https://doi.org/10.1103/PhysRevA.102.052213. arXiv:2007.00458 [quant-ph]
4. K. Ando, V. Vennin, Power spectrum in stochastic inflation. JCAP **04**, 057 (2021). https://doi.org/10.1088/1475-7516/2021/04/057. arXiv:2012.02031 [astro-ph.CO]
5. C. Animali, V. Vennin, Clustering of primordial black holes from quantum diffusion during inflation. J. Cosmol. Astropart. Phys. (2024). https://iopscience.iop.org/article/10.1088/1475-7516/2024/08/026
6. C. Animali, V. Vennin, Primordial black holes from stochastic tunnelling (2022). https://iopscience.iop.org/article/10.1088/1475-7516/2023/02/043
7. J.M. Bardeen, P.J. Steinhardt, M.S. Turner, Spontaneous creation of almost scale—free density perturbations in an inflationary universe. Phys. Rev. D **28**, 679 (1983). https://doi.org/10.1103/PhysRevD.28.679
8. J.M. Bardeen et al., The statistics of peaks of gaussian random fields. Astrophys. J. **304**, 15–61 (1986). https://doi.org/10.1086/164143
9. J.S. Bell, On the Einstein-Podolsky-Rosen paradox. Physics Physique Fizika **1**, 195–200 (1964). https://doi.org/10.1103/PhysicsPhysiqueFizika.1.195
10. J.R. Bond et al., Excursion set mass functions for hierarchical Gaussian fluctuations. Astrophys. J. **379**, 440 (1991). https://doi.org/10.1086/170520
11. R.G. Bower, The Evolution of groups of galaxies in the Press-Schechter formalism. Mon. Not. Roy. Astron. Soc. **248**, 332 (1991)
12. V. Briaud, V. Vennin, Uphill inflation. JCAP **06**, 029 (2023). https://doi.org/10.1088/1475-7516/2023/06/029. arXiv:2301.09336 [astro-ph.CO]
13. C.P. Burgess et al., Minimal decoherence from inflation. JCAP **07**, 022 (2023). https://doi.org/10.1088/1475-7516/2023/07/022. arXiv:2211.11046 [hep-th]
14. Y.-F. Cai et al., Highly non-Gaussian tails and primordial black holes from single-field inflation. JCAP **12**, 034 (2022). arXiv:2207.11910 [astro-ph.CO]
15. M.W. Choptuik, Universality and scaling in gravitational collapse of a massless scalar field. Phys. Rev. Lett. **70**, 9–12 (1993). https://doi.org/10.1103/PhysRevLett.70.9

16. D. Cruces, C. Germani, Stochastic inflation at all order in slow-roll parameters: foundations. Phys. Rev. D **105**(2), 023533 (2022). https://doi.org/10.1103/PhysRevD.105.023533. arXiv:2107.12735 [gr-qc]
17. K. Enqvist et al., On the divergences of inflationary superhorizon perturbations. JCAP **04**, 025 (2008). https://doi.org/10.1088/1475-7516/2008/04/025. arXiv:0802.0395 [astro-ph]
18. L. Espinosa-Portalés, V. Vennin, Real-space Bell inequalities in de Sitter. JCAP **07**(07), 037 (2022). https://doi.org/10.1088/1475-7516/2022/07/037. arXiv:2203.03505 [quant-ph]
19. C.R. Evans, J.S. Coleman, Observation of critical phenomena and selfsimilarity in the gravitational collapse of radiation fluid. Phys. Rev. Lett. **72**, 1782–1785 (1994). https://doi.org/10.1103/PhysRevLett.72.1782. arXiv:gr-qc/9402041
20. J.M. Ezquiaga, J. García-Bellido, V. Vennin, The exponential tail of inflationary fluctuations: consequences for primordial black holes. JCAP **03**, 029 (2020). https://doi.org/10.1088/1475-7516/2020/03/029. arXiv:1912.05399 [astro-ph.CO]
21. J.M. Ezquiaga, J. García-Bellido, V. Vennin, Massive galaxy clusters like El Gordo hint at primordial quantum diffusion. Phys. Rev. Lett. (2022). arXiv:2207.06317 [astro-ph.CO]
22. D.G. Figueroa et al., Implications of stochastic effects for primordial black hole production in ultra-slow-roll inflation. JCAP **05**(05), 027 (2022). https://doi.org/10.1088/1475-7516/2022/05/027. arXiv:2111.07437 [astro-ph.CO]
23. D.G. Figueroa et al., Non-Gaussian tail of the curvature perturbation in stochastic ultraslow-roll inflation: implications for primordial black hole production. Phys. Rev. Lett. **127**(10), 101302 (2021). https://doi.org/10.1103/PhysRevLett.127.101302. arXiv:2012.06551 [astro-ph.CO]
24. H. Firouzjahi, A. Nassiri-Rad, M. Noorbala, Stochastic Ultra Slow Roll Inflation. JCAP **01**, 040 (2019). https://doi.org/10.1088/1475-7516/2019/01/040. arXiv:1811.02175 [hep-th]
25. T. Fujita et al., A new algorithm for calculating the curvature perturbations in stochastic inflation. JCAP **12**, 036 (2013). https://doi.org/10.1088/1475-7516/2013/12/036. arXiv:1308.4754 [astro-ph.CO]
26. A.D. Gow et al., Non-perturbative non-Gaussianity and primordial black holes. EPL **142**(4), 49001 (2023). https://doi.org/10.1209/0295-5075/acd417. arXiv:2211.08348 [astro-ph.CO]
27. J. Grain, V. Vennin, Stochastic inflation in phase space: is slow roll a stochastic attractor? JCAP **05**, 045 (2017). https://doi.org/10.1088/1475-7516/2017/05/045. arXiv:1703.00447 [gr-qc]
28. S. Habib, Stochastic inflation: the quantum phase space approach. Phys. Rev. D **46**, 2408–2427 (1992). https://doi.org/10.1103/PhysRevD.46.2408. arXiv:gr-qc/9208006
29. T. Harada et al., Cosmological long-wavelength solutions and primordial black hole formation. Phys. Rev. D **91**(8), 084057 (2015). https://doi.org/10.1103/PhysRevD.91.084057. arXiv:1503.03934 [gr-qc]
30. S. Hooshangi, M.H. Namjoo, M. Noorbala, Tail diversity from inflation. JCAP **09**, 023 (2023). https://doi.org/10.1088/1475-7516/2023/09/023. arXiv:2305.19257 [astro-ph.CO]
31. B.L. Hu, S. Sinha, A fluctuation—dissipation relation for semiclassical cosmology. Phys. Rev. D **51**, 1587–1606 (1995). https://doi.org/10.1103/PhysRevD.51.1587. arXiv:gr-qc/9403054
32. L. Iacconi, D.J. Mulryne, Multi-field inflation with large scalar fluctuations: non-Gaussianity and perturbativity. JCAP **09**, 033 (2023). https://doi.org/10.1088/1475-7516/2023/09/033. arXiv:2304.14260 [astro-ph.CO]
33. J.H.P. Jackson et al., Numerical simulations of stochastic inflation using importance sampling. JCAP **10**, 067 (2022). https://doi.org/10.1088/1475-7516/2022/10/067. arXiv:2206.11234 [astro-ph.CO]
34. J.H.P. Jackson et al., The separate-universe approach and sudden transitions during inflation. J. Cosmol. Astropart. Phys. (2023). https://iopscience.iop.org/article/10.1088/1475-7516/2024/05/053
35. R. Kawaguchi, T. Fujita, M. Sasaki, Highly asymmetric probability distribution from a finite-width upward step during inflation. JCAP **11**, 021 (2023). https://doi.org/10.1088/1475-7516/2023/11/021. arXiv:2305.18140 [astro-ph.CO]
36. C. Kiefer, D. Polarski, Why do cosmological perturbations look classical to us? Adv. Sci. Lett. **2**, 164–173 (2009). https://doi.org/10.1166/asl.2009.1023. arXiv:0810.0087 [astro-ph]

37. N. Kitajima et al., Primordial black holes in peak theory with a non-Gaussian tail. JCAP **10**, 053 (2021). https://doi.org/10.1088/1475-7516/2021/10/053. arXiv:2109.00791 [astro-ph.CO]
38. H. Kodama, M. Sasaki, Cosmological perturbation theory. Prog. Theor. Phys. Suppl. **78**, 1–166 (1984). https://doi.org/10.1143/PTPS.78.1
39. T. Koike, T. Hara, S. Adachi, Critical behavior in gravitational collapse of radiation fluid: a Renormalization group (linear perturbation) analysis. Phys. Rev. Lett. **74**, 5170–5173 (1995). https://doi.org/10.1103/PhysRevLett.74.5170. arXiv:gr-qc/9503007
40. Y.L. Launay, G.I. Rigopoulos, E.P.S. Shellard, Stochastic inflation in general relativity. Phys. Rev. D **109**, 123523 (2024). arXiv:2401.08530 [gr-qc]
41. S.M. Leach et al., Enhancement of superhorizon scale inflationary curvature perturbations. Phys. Rev. D **64**, 023512 (2001). https://doi.org/10.1103/PhysRevD.64.023512. arXiv:astro-ph/0101406
42. J. Lesgourgues, D. Polarski, A.A. Starobinsky, Quantum to classical transition of cosmological perturbations for nonvacuum initial states. Nucl. Phys. B **497**, 479–510 (1997). https://doi.org/10.1016/S0550-3213(97)00224-1. arXiv:gr-qc/9611019
43. D.H. Lyth, K.A. Malik, M. Sasaki, A General proof of the conservation of the curvature perturbation. JCAP **05**, 004 (2005). https://doi.org/10.1088/1475-7516/2005/05/004. arXiv:astro-ph/0411220
44. D.H. Lyth, Y. Rodriguez, The Inflationary prediction for primordial non-Gaussianity. Phys. Rev. Lett. **95**, 121302 (2005). https://doi.org/10.1103/PhysRevLett.95.121302. arXiv:astro-ph/0504045
45. D.H. Lyth, D. Wands, Conserved cosmological perturbations. Phys. Rev. D **68**, 103515 (2003). https://doi.org/10.1103/PhysRevD.68.103515. arXiv:astro-ph/0306498
46. K.A. Malik, D. Wands, Cosmological perturbations. Phys. Rept. **475**, 1–51 (2009). https://doi.org/10.1016/j.physrep.2009.03.001. arXiv:0809.4944 [astro-ph]
47. J. Martin, V. Vennin Real-space entanglement in the cosmic microwave background. JCAP **10**, 036 (2021). https://doi.org/10.1088/1475-7516/2021/10/036. arXiv:2106.15100 [gr-qc]
48. J. Martin, A. Micheli, V. Vennin, Discord and decoherence. JCAP **04**(04), 051 (2022). https://doi.org/10.1088/1475-7516/2022/04/051. arXiv:2112.05037 [quant-ph]
49. J. Martin, V. Vennin, Obstructions to bell CMB experiments. Phys. Rev. D **96**(6), 063501 (2017). https://doi.org/10.1103/PhysRevD.96.063501. arXiv:1706.05001 [astro-ph.CO]
50. J. Martin, V. Vennin, Quantum discord of cosmic inflation: can we show that cmb anisotropies are of quantum-mechanical origin? Phys. Rev. D **93**(2), 023505 (2016). https://doi.org/10.1103/PhysRevD.93.023505. arXiv:1510.04038 [astro-ph.CO]
51. J. Martin, V. Vennin, Bell inequalities for continuous-variable systems in generic squeezed states. Phys. Rev. A **93**(6), 062117 (2016). https://doi.org/10.1103/PhysRevA.93.062117. arXiv:1605.02944 [quant-ph]
52. S. Matarrese, M.A. Musso, A. Riotto, Influence of superhorizon scales on cosmological observables generated during inflation. JCAP **05**, 008 (2004). https://doi.org/10.1088/1475-7516/2004/05/008. arXiv:hep-th/0311059
53. S.S. Mishra, E.J. Copeland, A.M. Green, Primordial black holes and stochastic inflation beyond slow roll. Part I. Noise matrix elements. JCAP **09**, 005 (2023). https://doi.org/10.1088/1475-7516/2023/09/005. arXiv:2303.17375 [astro-ph.CO]
54. M. Morikawa, Dissipation and fluctuation of quantum fields in expanding universes. Phys. Rev. D **42**, 1027–1034 (1990). https://doi.org/10.1103/PhysRevD.42.1027
55. V.F. Mukhanov, Quantum theory of gauge invariant cosmological perturbations. Sov. Phys. JETP **67**, 1297–1302 (1988)
56. I. Musco, Threshold for primordial black holes: dependence on the shape of the cosmological perturbations. Phys. Rev. D **100**(12), 123524 (2019). https://doi.org/10.1103/PhysRevD.100.123524. arXiv:1809.02127 [gr-qc]
57. K. Nakao, Y. Nambu, M. Sasaki, Stochastic dynamics of new inflation. Prog. Theor. Phys. **80**, 1041 (1988). https://doi.org/10.1143/PTP.80.1041
58. G. Panagopoulos, E. Silverstein, Primordial black holes from non-gaussian tails (2019). arXiv:1906.02827 [hep-th]

59. C. Pattison et al., Quantum diffusion during inflation and primordial black holes. JCAP **10**, 046 (2017). https://doi.org/10.1088/1475-7516/2017/10/046. arXiv:1707.00537 [hep-th]
60. C. Pattison et al., Stochastic inflation beyond slow roll. JCAP **07**, 031 (2019). https://doi.org/10.1088/1475-7516/2019/07/031. arXiv:1905.06300 [astro-ph.CO]
61. C. Pattison et al., Ultra-slow-roll inflation with quantum diffusion. JCAP **04**, 080 (2021). https://doi.org/10.1088/1475-7516/2021/04/080. arXiv:2101.05741 [astro-ph.CO]
62. J.A. Peacock, A.F. Heavens, Alternatives to the Press-Schechter cosmological mass function. Mon. Not. Roy. Astron. Soc. **243**, 133–143 (1990)
63. L. Pinol, S. Renaux-Petel, Y. Tada, A manifestly covariant theory of multifield stochastic inflation in phase space: solving the discretisation ambiguity in stochastic inflation. JCAP **04**, 048 (2021). https://doi.org/10.1088/1475-7516/2021/04/048. arXiv:2008.07497 [astro-ph.CO]
64. D. Polarski, A.A. Starobinsky, Semiclassicality and decoherence of cosmological perturbations. Class. Quant. Grav. **13**, 377–392 (1996). https://doi.org/10.1088/0264-9381/13/3/006. arXiv:gr-qc/9504030
65. S. Raatikainen, S. Rasanen, E. Tomberg, Primordial black hole compaction function from stochastic fluctuations in ultra-slow-roll inflation. Phys. Rev. Lett. **133**, 121403 (2023). arXiv:2312.12911 [astro-ph.CO]
66. S. Redner, *A Guide to First-Passage Processes* (Cambridge University Press, 2001)
67. M. Revzen, The Wigner function as distribution function. Found. Phys. **36**(4), 546–562 (2006). https://doi.org/10.1007/s10701-005-9037-5
68. G.I. Rigopoulos, E.P.S. Shellard, B.J.W. van Tent, Non-linear perturbations in multiple-field inflation. Phys. Rev. D **73**, 083521 (2006). https://doi.org/10.1103/PhysRevD.73.083521. arXiv:astro-ph/0504508
69. G.I. Rigopoulos, E.P.S. Shellard, The separate universe approach and the evolution of nonlinear superhorizon cosmological perturbations. Phys. Rev. D **68**, 123518 (2003). https://doi.org/10.1103/PhysRevD.68.123518. arXiv:astro-ph/0306620
70. G. Rigopoulos, A. Wilkins, Computing first-passage times with the functional renormalisation group. JCAP **04**, 046 (2023). https://doi.org/10.1088/1475-7516/2023/04/046. arXiv:2211.09649 [astro-ph.CO]
71. H. Risken, T. Frank, *The Fokker-Planck Equation: Methods of Solution and Applications*. Springer Series in Synergetics (Springer, Berlin Heidelberg, 1996)
72. D.S. Salopek, J.R. Bond, Nonlinear evolution of long wavelength metric fluctuations in inflationary models. Phys. Rev. D **42**, 3936–3962 (1990). https://doi.org/10.1103/PhysRevD.42.3936
73. M. Sasaki, Large scale quantum fluctuations in the inflationary universe. Prog. Theor. Phys. **76**, 1036 (1986). https://doi.org/10.1143/PTP.76.1036
74. M. Sasaki, E.D. Stewart, A General analytic formula for the spectral index of the density perturbations produced during inflation. Prog. Theor. Phys. **95**, 71–78 (1996). https://doi.org/10.1143/PTP.95.71. arXiv:astro-ph/9507001
75. M. Sasaki, T. Tanaka, Super horizon scale dynamics of multi scalar inflation. Prog. Theor. Phys. **99**, 763–782 (1998). https://doi.org/10.1143/PTP.99.763. arXiv:gr-qc/9801017
76. M. Shibata, M. Sasaki, Black hole formation in the Friedmann universe: formulation and computation in numerical relativity. Phys. Rev. D **60**, 084002 (1999). https://doi.org/10.1103/PhysRevD.60.084002. arXiv:gr-qc/9905064
77. A.A. Starobinsky, Dynamics of phase transition in the new inflationary universe scenario and generation of perturbations. Phys. Lett. B **117**, 175–178 (1982). https://doi.org/10.1016/0370-2693(82)90541-X
78. A.A. Starobinsky, Multicomponent de Sitter (inflationary) stages and the generation of perturbations. JETP Lett. **42**, 152–155 (1985)
79. A.A. Starobinsky, Stochastic de Sitter (inflationary) stage in the early universe. Lecture Notes in Physics, vol. 246 (1986), pp. 107–126. https://doi.org/10.1007/3-540-16452-9_6
80. Y. Tada, V. Vennin, Statistics of coarse-grained cosmological fields in stochastic inflation. JCAP **02**(02), 021 (2022). https://doi.org/10.1088/1475-7516/2022/02/021. arXiv:2111.15280 [astro-ph.CO]

81. Y. Tanaka, M. Sasaki, Gradient expansion approach to nonlinear superhorizon perturbations. II. A single scalar field. Prog. Theor. Phys. **118**, 455–473 (2007). https://doi.org/10.1143/PTP.118.455. arXiv:0706.0678 [gr-qc]
82. K. Tokeshi, V. Vennin, Why does inflation look single field to us? Phys. Rev. Lett. **132**, 251001 (2023). arXiv:2310.16649 [astro-ph.CO]
83. J. Tokuda, T. Tanaka, Can all the infrared secular growth really be understood as increase of classical statistical variance? JCAP **11**, 022 (2018). https://doi.org/10.1088/1475-7516/2018/11/022. arXiv:1806.03262 [hep-th]
84. J. Tokuda, T. Tanaka, Statistical nature of infrared dynamics on de Sitter background. JCAP **02**, 014 (2018). https://doi.org/10.1088/1475-7516/2018/02/014. arXiv:1708.01734 [gr-qc]
85. A.J. Tolley, M. Wyman, Stochastic inflation revisited: non-slow roll statistics and DBI inflation. JCAP **04**, 028 (2008). https://doi.org/10.1088/1475-7516/2008/04/028. arXiv:0801.1854 [hep-th]
86. E. Tomberg, Stochastic constant-roll inflation and primordial black holes. Phys. Rev. D **108**(4), 043502 (2023). https://doi.org/10.1103/PhysRevD.108.043502. arXiv:2304.10903 [astro-ph.CO]
87. V. Vennin, Stochastic inflation and primordial black holes. Other thesis, 2020. arXiv:2009.08715 [astro-ph.CO]
88. V. Vennin, A.A. Starobinsky, Correlation functions in stochastic inflation. Eur. Phys. J. C **75**, 413 (2015). https://doi.org/10.1140/epjc/s10052-015-3643-y. (arXiv:1506.04732 [hep-th])
89. D. Wands et al., A new approach to the evolution of cosmological perturbations on large scales. Phys. Rev. D **62**, 043527 (2000). https://doi.org/10.1103/PhysRevD.62.043527. arXiv:astro-ph/0003278
90. J. Weenink, T. Prokopec, On decoherence of cosmological perturbations and stochastic inflation (2011). arXiv:1108.3994 [gr-qc]
91. S. Weinberg, Adiabatic modes in cosmology. Phys. Rev. D **67**, 123504 (2003). https://doi.org/10.1103/PhysRevD.67.123504. arXiv:astro-ph/0302326

Chapter 9
New Ideas on the Formation and Astrophysical Detection of Primordial Black Holes

Marcos M. Flores and Alexander Kusenko

Abstract Recently, a number of novel scenarios for primordial black hole (PBH) formation have been discovered. Some of them require very minimal new physics, some others require no new ingredients besides those already present in commonly considered models, such as supersymmetry. At the same time, new strategies have emerged for detection of PBHs. For example, an observation of an orphan kilonova unaccompanied by the gravitational waves signal of merging neutron stars, but associated with a fast radio burst, could be a smoking gun of PBH dark matter. We review some new ideas for PBH formation and detection.

9.1 Introduction

Black holes are known to exist in nature, and can be produced in stellar explosions. However, the dense early Universe could have provided an additional channel for producing black holes, including supermassive black holes, as well as those with masses much smaller than a solar mass. The latter is a particularly exciting possibility because in the range of masses $10^{17} - 10^{23}$ g, primordial black holes (PBHs) could explain the entirety of dark matter. The fact that black holes are known to exist makes PBHs a very appealing dark matter candidate [1–39].

A very popular scenario for PBH formation involves production of relatively large density perturbations on specific small scales [7, 16, 22, 27, 38–41]. When these perturbations re-enter the horizon, black holes form with masses of the order of the horizon size.

M. M. Flores
Laboratoire de Physique de l'École Normale Supérieure, ENS, Université PSL, CNRS, Sorbonne Université, Université Paris Cité, 75005 Paris, France
e-mail: marcos.flores@phys.ens.fr

A. Kusenko (✉)
Department of Physics and Astronomy, University of California, Los Angeles, CA 90095-1547, USA
e-mail: kusenko@ucla.edu

Kavli Institute for the Physics and Mathematics of the Universe (WPI), UTIAS, The University of Tokyo, Kashiwa, Chiba 277-8583, Japan

C. Byrnes et al. (eds.), *Primordial Black Holes*, Springer Series in Astrophysics and Cosmology, https://doi.org/10.1007/978-981-97-8887-3_9

An alternative class of scenarios involves overdensities on subhorizon scales, which can become black holes with masses much smaller than the horizon at the time of formation. It was realized relatively recently that many models of particle physics beyond the Standard Model already have all the ingredients for PBH formation. For example, the widely studied field of supersymmetry (including its minimal version, the Minimal Supersymmetric Standard Model) has PBHs as one of its generic dark matter candidates, thanks to the flat directions predicted within its framework [42].

Another important and recent development is a new understanding of strategies to search for dark matter in the form of PBHs, including through microlensing observations and using neutron stars as detectors.

We will review some of the recently proposed PBH formation scenarios, as well as new and promising detection techniques.

9.2 PBHs from Yukawa Interactions

We begin with a remarkably simple scenario which involves only one fermion and one boson in a hypothesized dark sector. If these two fields interact via a Yukawa coupling, this can potentially lead to PBH formation.

9.2.1 Primordial Structure Formation

Black hole formation, in the astrophysical context, arises from the evolution and collapse of heavy stars. Under the standard assumptions of ΛCDM, the formation of structure only occurs after matter-radiation equality. To see this, note that the matter density perturbations, $\delta_m \equiv (\rho_m - \bar{\rho}_m)/\bar{\rho}_m$, as described by cosmological perturbation theory, evolve via the differential equation

$$\ddot{\delta}_m + 2H\dot{\delta}_m = 4\pi G a^2 \bar{\rho}_m \delta_m \tag{9.1}$$

where $\dot{} \equiv d/dt$, a is the scale factor, $H \equiv \dot{a}/a$ is the Hubble parameter and $\bar{\rho}_m$ background value of the matter density. In the early, radiation dominated Universe $a \propto t^{1/2}$ and $\bar{\rho}_m \sim 0$ enabling us to conclude that growing mode

$$\delta_m(a) \propto \ln a, \qquad \text{(gravity only)}. \tag{9.2}$$

The standard conclusion of this result is that matter perturbations cannot grow during radiation domination. Naturally, this result implies any non-trivial structure like stars, galaxies, etc., cannot form at these early stages in the evolution of the Universe. Without these objects, the formation black holes in a fashion similar to astrophysical black hole formation appears out of reach.

The inability for structure form during radiation domination is related, in part, to the feeble strength of gravity. Therefore, enabling a phase of *primordial structure formation* requires the introduction of an additional, stronger, force [43–50]. The natural candidate for such a force is a scalar-mediated Yukawa interaction:

$$\mathcal{L} \supset \frac{1}{2}m_\varphi^2\varphi^2 + (m_\psi + y\varphi)\bar{\psi}\psi. \tag{9.3}$$

Scalar-mediated Yukawa interactions lead to forces which are always attractive. Generically, these forces are also always stronger than gravity. This becomes apparent by comparing the couplings of the two forces,

$$\beta \equiv y\left(\frac{M_{\text{Pl}}}{m_\psi}\right) \tag{9.4}$$

where M_{Pl} is the reduced Planck mass $M_{\text{Pl}} = 1/\sqrt{8\pi G} \sim 2.4 \times 10^{18}$ GeV. The appearance of M_{Pl} in the above ratio implies that generically, $\beta \gg 1$. The well-known Yukawa interaction potential,

$$V(r) = -\frac{y^2}{4\pi r}e^{-m_\varphi r} \tag{9.5}$$

illustrates that for length scales $r < m_\varphi^{-1}$, the Yukawa potential is Coulomb-like. In the cosmological context, this fact is particularly applicable, especially when the horizon size is smaller than the Compton wavelength of the mediator mass, or equivalently, $m_\varphi < H$.

Before discussing the details of primordial structure formation, we will first specify a set of conditions required before any appreciable structure can form. First, we will require that $\bar{\psi}\psi \leftrightarrow \varphi\varphi$ interactions freeze-out. This is to establish a fixed population of fermions out of which to generate nontrivial structure. Second, we will require that the ψ fermions are non-relativistic. Lastly, we will demand that φ-radiation pressure is negligible or, in other words, that the scalar mean free path is comparable to the Hubble radius. These conditions are satisfied when the temperature of the fermionic dark sector is $T \sim m_\psi$.

With these conditions in mind, we will proceed by using cosmological perturbation theory to describe the growth of matter perturbations in the presence of a long-range scalar-mediated force. The first thing to note is that scalar forces couple to number density rather than energy density, as is the case with gravity. Therefore it is more useful to examine the *number* density contrast, $\delta_\psi = \delta n_\psi / n_\psi$, as opposed to the energy density contrast. The key equation which describes the growth of overdensities in the ψ fluid is [47],

$$\delta_\psi'' + \frac{2+3x}{2x(1+x)}\delta_\psi' = \frac{3}{2x(1+x)}\delta_\psi f_\psi\left[1 + \frac{2\beta^2}{1+(k/m_\varphi)^{-2}}\right] \tag{9.6}$$

where $' \equiv d/dx$ with $x = a/a_{\rm eq}$, $a_{\rm eq}$ being the scale factor when $\rho_r = \rho_m + \rho_\psi$ and $f_\psi = \rho_\psi/(\rho_m + \rho_\psi) \leq 1$. For readers familiar with cosmological perturbation theory, this equation is similar to the Meszaros equation, with an explicit dependence on the scale k. During the radiation dominated era, i.e., $x \ll 1$ then

$$\delta_\psi = c_1 I_0(\sqrt{6\alpha x}) + c_2 K_0(\sqrt{6\alpha x}), \qquad \alpha = f_\psi \left[1 + \frac{2\beta^2}{1 + (k/m_\varphi)^{-2}}\right], \tag{9.7}$$

where I_0 and K_0 are the zeroth-order modified Bessel functions. For $6\alpha x \gg 1$, the growing mode becomes exponential

$$\delta_\psi \sim I_0(\sqrt{6\alpha x}) \sim \frac{e^{\sqrt{6\alpha x}}}{\sqrt{2\pi\sqrt{6\alpha x}}}. \tag{9.8}$$

When the characteristic timescale associated with the growth of perturbations, $\delta_\psi/\dot{\delta}_\psi$, is shorter than the Hubble time, then overdensities will rapidly form. Furthermore, one can see that in the limit that $\beta \to 0$ and $x \ll 1$ then $\delta_\psi \sim$ const., as is expected.

Once the perturbations reach $\delta_\psi \sim 1$, the dynamics of structure formation become non-linear and Eq. (9.6) no longer holds. Physically, this corresponds to the formation of virialized ψ halos whose eventual collapse will lead to the formation of black holes.

The breakdown of linear perturbation theory sets the stage for numerical studies. Before discussing recent numerical work related to primordial structure formation, we will point out that the preceding analysis provides a simplistic picture, which in actuality, could be more complicated. For example, the fermion ψ acquires an effective mass $m_{\rm eff}(t) = m_\psi + y\varphi(t)$ when φ is non-zero. This is particularly relevant when φ self-interactions are included in the theory, Eq.(9.3), which can induce non-trivial effects. For a more comprehensive discussion, we point the reader to Refs. [46, 47].

Figure 9.1 illustrates the growth of structure throughout a radiation dominated era. As is typical in structure formation, halos form as nodes within a network of filaments. However, as time evolves the halos become more concentrated with their maximum radius being determined by the inverse mass of the mediator φ.

9.2.2 *Energy Dissipation and Collapse*

Yet another advantage of the long-range Yukawa interaction is its ability to facilitate dissipation through scalar radiation. Just as accelerating electric charges cause the emission of electromagnetic waves, accelerating fermions, coupled to a scalar mediator, emit scalar waves. Without dissipation, the ψ halos formed via long-range scalar interactions would simply remain as stable virialized halos, assuming the constitute particles are stable [50]. While this alone is an interesting possibility, dissipation is an unavoidable consequence of Yukawa interactions.

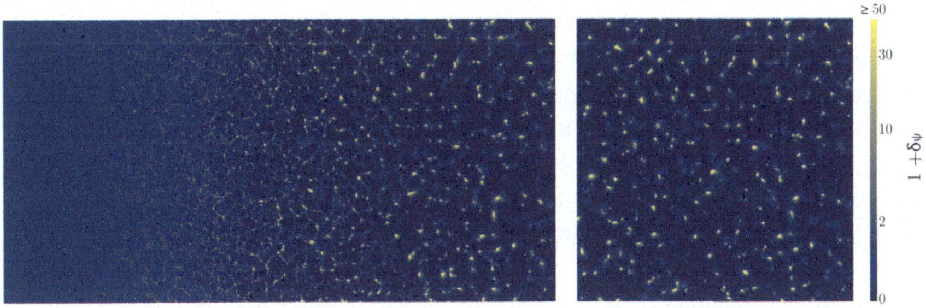

Fig. 9.1 Numerical simulations of primordial structure formation. Starting with small perturbations, the above figure demonstrates the growth of nonlinear structure during a radiation dominated era. Adapted from Ref. [47]

The dynamics of scalar radiation are complicated and heavily dependent on the characteristics of individual halos. For simplicity, we will outline the general history of a given ψ halo. The details of this evolution are an area of active research, and future analytic and numerical work will hopefully shed light on the complex dynamics which can occur as scalar radiation becomes important.

There are various different radiation channels available for scalar emission, which can broadly be classified into oscillatory and non-oscillatory channels. Oscillatory channels are related to the rotation of a given halo. Coherent oscillations correspond to rotation of the entire halo with one single frequency ω. The power radiated through coherent motion is simply related to the coupling y and the number of particles within the halo N_h through $P_{\text{coh}} \propto y^2 N_h^2$. Incoherent oscillations correspond to rotation of different particles, or subhalos, with different oscillation frequencies ω. The energy radiated in this case instead scales as $P_{\text{incoh}} \propto y^2 N_h$, as is required by the low number density limit.

The non-oscillatory emission channels involve microscopic phenomena which produce scalar quanta. The first non-oscillatory channel we will discuss is a consequence of pair-wise interactions of fermions, which lead to scalar emission, i.e., bremsstrahlung. This process is similar to free-free emission of photons from plasma [51, 52]. However, unlike the plasmas considered in traditional astrophysical contexts, which contain oppositely charged particles, our system contains only identical particles. This being the case, bremsstrahlung radiation due to two-particle collisions is quadrupole rather than dipole and analogous to the *e*-*e* component of free-free emission from plasma [51, 52]. Similar in spirit to bremsstrahlung is bound state formation, which also leads to the emission of scalar radiation. Bound state formation will be important in the early stages of halo formation, when the interacting particles are non-relativistic and the relative kinetic energy between particles is sufficiently low compared to the binding energy $|E_{\text{bind}}| \sim m_\psi y^4$. However, as the halo collapses, high-energy scalars could re-ionize any ψ bound states which may have formed at the early stages of cooling.

The last channel we will discuss deserves special attention because it is the inevitable fate of any collapsing halo. At a given point, scalar radiation will become trapped as the optical depth of φ particles increases within the increasingly dense halo. Once this stage of the evolution is reached, radiation can only escape from a thin surface layer of the collapsing halo. Cooling from the surface will drive less dense, higher energy ψ plasma upward. Simultaneously the cooler, denser, outer layer will fall inward. This drives convection currents which rapidly cool the fireball, leading the halo toward total collapse.

The emission channels discussed above involve complicated, non-linear dynamics which require further investigation to fully understand. Despite the complicated physics involved it is possible to demonstrate that the timescales associated with the processes discussed above are sufficiently short to facilitate efficient collapse. Associated with a given halo is its radiative cooling timescale τ_{cool} which is defined as

$$\tau_{\text{cool}}(R, M_h) \equiv \frac{E}{dE/dt} = \frac{E}{P_{\text{brem}} + P_{\text{surf}} + \cdots}. \tag{9.9}$$

As indicated above, the cooling rate depends on the halo mass and its radius. In general, the radius is a monochromatically decreasing function of time. For the dissipation channels we discussed above, the cooling rate decreases as $R(t) \to 0$, implying that collapse is a runaway process [48].

Once a given halo has reached a point where its cooling timescale is smaller than the Hubble time H^{-1}, the collapse of the halo will commence rapidly on cosmological timescales. The implications of this collapse are wide-ranging [53–55], but we will focus only on the possibility of PBH formation.

9.2.3 PBH Formation

The collapse process discussed above will increase the inter-halo particle interaction rate. If the fermions involved are symmetric, this increased number density could lead to particle annihilations that would destroy any structure which may have formed. While this process itself has interesting cosmological implications, it acts as a roadblock for the formation of PBHs.

To avoid this issue, we will require a particle asymmetry in the fermion sector, i.e.,

$$\eta_\psi \equiv \frac{n_\psi - n_{\bar{\psi}}}{s} \tag{9.10}$$

where s is the comoving entropy number density. Requiring a particle asymmetry is motivated by existing asymmetric dark matter models, e.g., [56, 57], as well as the existence of the asymmetry in the Standard Model sector. The inclusion of an asymmetry in the ψ fermion sector ensures that annihilations will not disrupt the collapse of previously formed ψ halos, and provides a path toward PBH formation.

The last remaining potential roadblock is Fermi degeneracy pressure. Earlier examinations of scalar-mediated fermionic dark matter have demonstrated that larger bound states can form and act as macroscopic dark matter [49, 58–61]. For the formation of large PBHs Fermi degeneracy pressure can be ignored. To see this, we compare the average distance between ψ fermions within a given halo to the Compton wave length of the fermions. This leads to an expression analogous to the Chandrasekhar limit,

$$M_{\rm PBH} \geq \frac{(3/4\pi)^2}{(2G)^{3/2} m_\psi^2} \simeq 1.3 \times 10^{-3}\, M_\odot \left(\frac{5\ \text{GeV}}{m_\psi}\right)^2 . \tag{9.11}$$

The above expression tells us that for PBH masses exceeding $\sim 10^{-3}\ M_\odot$, Fermi degeneracy pressure never comes into play. For smaller masses, one must consider the fermionic pressure which may inhibit the formation of a PBH. In order to ensure that Fermi degeneracy pressure is not an issue, we require that the Fermi energy be small compared to the potential energy $y^2 N/R$ as $R \to R_S$. This condition is similar to the Chandrasekhar limit but is modified by a factor of $\beta^{-3/2}$. For the parameters relevant for PBH formation, this condition will not hamper the formation of PBHs relevant for dark matter [48].

The abundance of PBHs will be determined by the initial number density of the ψ fluid. Given that we are considering asymmetric fermions, the number density is simply given by $\eta_\psi T^3$. Under the assumption that all of the fermions end up in halos and therefore fall into PBHs, the PBH abundance $f_{\rm PBH}$ will be given by

$$f_{\rm PBH} \equiv \frac{\Omega_{\rm PBH}}{\Omega_{\rm DM}} = 0.2 \frac{m_\psi}{m_p} \frac{\eta_\psi}{\eta_{\rm B}} = \left(\frac{m_\psi}{5\ \text{GeV}}\right) \left(\frac{\eta_\psi}{10^{-10}}\right) . \tag{9.12}$$

Another advantage of using an asymmetric fermion to generate PBHs comes directly from the above formula. In particular, this scenario inherits the motivating feature behind asymmetric dark matter, namely an explanation for the factor five difference between the dark matter and visible matter energy densities.

The mass distribution of PBHs generated via primordial structure formation will be an extended distribution. An absolute lower limit can be placed on the minimum size of a PBH formed via this mechanism. In particular, the Compton wavelength of the ψ particles must fit within the Schwarzschild radius of their prospective PBH. This gives

$$M_{\rm PBH}^{\rm min} \geq \frac{4\pi m_P^2}{m_\psi} . \tag{9.13}$$

The maximum PBH size is determined by the total ψ mass within the Hubble volume at the time of formation. A precise understanding of the mass distribution of ψ halos, and therefore PBHs, has yet to be established. Following Ref. [48], we will adopt a distribution motivated by the Press-Schechter formalism. In particular, we will assume that the halo mass distribution is described by

$$M^2 \frac{dN_h}{dM} \propto \frac{1}{\sqrt{\pi}} \left(\frac{M}{M_*}\right)^{1/2} e^{-M/M_*} \tag{9.14}$$

where the overall normalization is set by requiring that the integrated mass function gives f_{PBH} as determined by Eq. (9.12). For details about the precise nature of the mass function, we point the reader to Ref. [48].

Figure 9.2 illustrates the mass function for two set of parameters. Taking $m_\psi = 5$ GeV and $\eta_\psi = 10^{-10}$ (which is comparable to the observed value of the baryon asymmetry) gives $f_{\text{PBH}} = 1$ and produces a distribution which sits within the unconstrained dark matter window. Alternatively, one can adjust the parameters to $\eta_\psi = 10^{-9}$ and $m_\psi = 5$ MeV which leads to a population of PBHs which would act as a subcomponent of dark matter, but might be relevant for present-day gravitational wave experiments.

9.2.4 *Observational Signatures*

Primordial structure formation in general provides a rich phenomenology with many possible observable signals. For example, the introduction of a light degree of freedom though the scalar φ could potentially modify ΔN_{eff}. Not only is this allowed, but it may be favorable in addressing the so-called Hubble tension [65–73]. Apart from this, two particular observables are of interest in the context of PBHs, namely the expected spin distribution and the emission of gravitational waves.

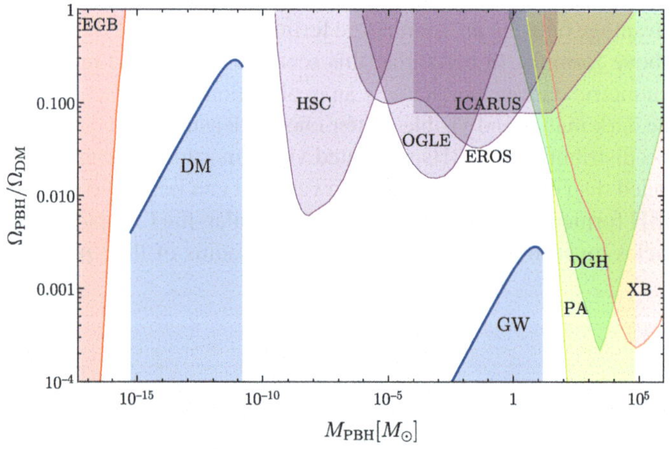

Fig. 9.2 The PBH abundance within the primordial structure formation framework. The dark matter (DM) curve corresponds to $\eta_\psi \sim \eta_B \sim 10^{-10}$ with $m_\psi = 5$ GeV. Alternatively, the gravitational wave curve (GW) relevant to LIGO, Virgo, and KAGRA, corresponds to $\eta_\psi = 10^{-9}$ and $m_\psi = 5$ MeV. See Ref. [48] for further details. The constraints plotted are reproduced from Refs. [62, 63] and the references within, e.g., [64]

9.2.4.1 Spin Distribution

An advantage of using primordial structure formation and scalar radiation for the generation of PBHs is the efficient removal of both energy and angular momentum. In particular, the efficient removal of angular momentum avoids the need to generate spherical configurations of matter which are required for PBH formation in the gravity-only context.

The removal of angular momentum of a spinning fermion halo is due to two possible effects. First, rotation of the halo as a whole will lead to emission of energy, which is directly related to the rate of angular momentum loss. Specifically, if decomposed in a spherical basis one can demonstrate that, [74]

$$\frac{dJ_{\ell m}}{dt} = \frac{m}{\omega}\left(\frac{dE_{\ell m}}{dt}\right) \tag{9.15}$$

where ℓ and $m = -\ell, \ldots, \ell$ are the spherical expansion indices and ω is the oscillation frequency. Alternatively, one can examine the emission of scalar quanta due to pair-wise scattering. In that case, the loss of angular momentum can be seen to be driven from blueshifting or redshifting of radiation as it is being emitted in a direction parallel or antiparallel relative to the tangential velocity of the rotating halo. This leads to an angular momentum loss rate which is also proportional to the energy loss, but also depends on the tangential velocity in a nontrivial fashion, [74]

$$\frac{dJ}{dt} = -R\frac{dE}{dt}f(v), \qquad f(v) = \frac{3\pi^2}{4}\left[\frac{(1+v^2)\tanh^{-1}v - v}{v^2}\right] \tag{9.16}$$

where the function $f(v)$ accounts for the red(blue)shifting of the scalar quanta. For small velocities $f(v \ll 1) \propto v$.

As demonstrated in Ref. [74], the characteristic timescale associated with angular momentum loss $\tau_J \equiv J/(dJ/dt)$ is smaller than the cooling timescale, $\tau_{\rm cool}$, for the parameters relevant for PBH formation. This implies that angular momentum losses are rapid, and any halo which may have formed initially spinning will rapidly shed its angular momentum. Given that this is the case, any PBH which may have formed through primordial structure formation and collapse should be spinless at the time of formation. Naturally, accretion and mergers can alter the PBH spin distribution after formation. We will leave discussion of these effects to Ref. [75].

9.2.4.2 Gravitational Waves

The key ingredient for the generation of gravitational waves is a time-changing mass quadrupole moment. Given that the ψ perturbations will, in general, collapse in a nonspherical fashion it is natural to expect that gravitational waves will be emitted. Since conventional wisdom requires a matter dominated era to facilitate collapse, early works such as Refs. [76–79] have previously examined gravitational waves in

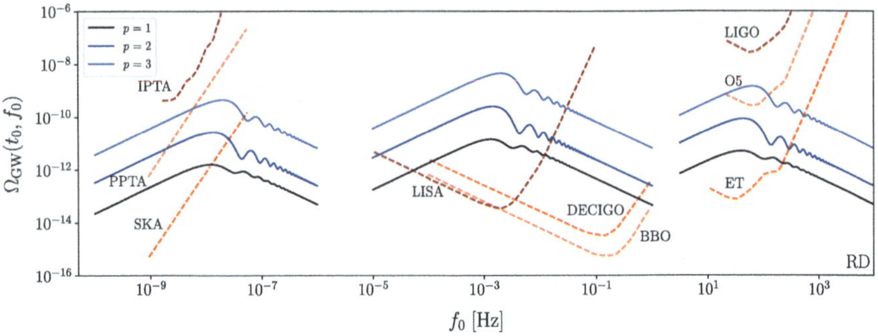

Fig. 9.3 Predicted gravitational wave signals due to the formation of structure due to a long-range interaction. The distribution of masses of the overdensities are assumed to be monochromatic. Here, the masses of the overdensities are $10^{-2}\ M_\odot$ (left), $10^{-12}\ M_\odot$ (center), and $10^{-21}\ M_\odot$ (right). For further details, see [84] and the references within

this context. The generation of gravitational waves from isocurvature perturbations in a radiation dominated era have also been examined in the literature [80–83]. The majority of the literature referenced uses cosmological perturbation theory to determine the gravitational wave signal. While this should also be possible within the framework of primordial structure formation, the details are nontrivial. The one difficulty arises from the fact that the scalar mediator couples to the fermionic number density as opposed to the energy density. Analysis of the growth of perturbations, including the discussion above, is always performed in the Newtonian limit. More work needs to be done to describe the growth of overdensities in a fully relativistic framework.

To avoid these issues, Ref. [84] generalized the framework of Ref. [77] which utilizes the Zeldovich approximation [85]. The Zeldovich approximation allows us to examine the time-changing mass quadrupole moment of individual matter perturbations. By doing so, we can avoid using cosmological perturbation theory and also examine dynamics beyond the traditional regime permissible in a linear framework.

The details of the calculations involved are quite complicated and are presented in Refs. [55, 77, 86]. To parameterize the growth of additional structure, Ref. [84] assumed that fermion perturbations grow as

$$\delta_\psi \propto a^p \tag{9.17}$$

where p parametrizes the strength of long-range scalar interaction. In Fig. 9.3, we present the signal of gravitational waves for a broad range of frequencies. The peak of the distribution is determined by the time in which the δ_ψ enters into the horizon. This time is associated with the mass of the perturbation, with $f_{\rm max} \propto M^{-1/2}$. It should also be noted that the signals presented in Fig. 9.3 assume a monochromatic distribution of overdensities.

Outside of the context of PBHs, gravitational waves produced through primordial structure formation offer an opportunity to examine presence of long-range interactions in the early Universe. This effect, combined with traditional particle physics detection methods, offers a unique, multi-approached strategy for probing physics beyond the Standard Model.

9.2.5 *Future Directions*

Primordial structure formation remains an active field of research which requires multidisciplinary collaboration to tackle many of the remaining outstanding questions. Numerical simulations will continue to probe the formation of structure during radiation domination. At the present moment, these simulations do not include dissipation due to scalar radiation. Dissipation could dramatically impact the mass distribution of halos formed, and is an important factor for PBH formation. Numerical simulations will also be important in understanding the later stages of collapse, particularly during the surface cooling phase. These open questions, and others, continue to propel our exploration of the early Universe toward new and exciting directions.

9.3 PBHs as a Generic Dark Matter Candidate in MSSM and Other Models with Supersymmetry

It is important to realize that PBH production is a fairly generic consequence of supersymmetry. For example, the Minimal Supersymmetric Standard Model (MSSM) has all the ingredients necessary for PBH production. Whether or not PBHs are produced and the abundance of PBHs depend on the details of supersymmetry breaking.

9.3.1 *Flat Directions, Q-Balls, and PBHs*

What makes the production of PBHs possible is the existence of flat directions in the MSSM and other supersymmetric generalizations of the Standard Model. These flat directions ϕ correspond to zeros of the scalar potential: $U(\phi) = 0$ in the limit of unbroken supersymmetry [42]. Supersymmetry breaking lifts the flat direction in a model-dependent way, but the lifted flat directions remain D-flat, so that they grow logarithmically or at most quadratically. Such directions in the scalar potential develop large vacuum expectation values (VEV) in the early Universe. This is due to two reasons.

First, the effective minimum of the potential in de Sitter space can be displaced because of the terms $\sim cH^2\phi^2$ that come from the Kaehler potential [87, 88]. If $c < 0$, the field acquires a nonzero VEV.

Second, regardless of where the minimum is, the field does not remain in that minimum everywhere in space during de Sitter expansion. Instead, the field undergoes fluctuations on superhorizon scales [89–94]. The results of these papers can be summarized as follows: in de Sitter space, each scalar field is displaced from the minimum, and each degree of freedom, on average, carries the amount of energy density of the order of H_{infl}^4, where H_{infl} is the Hubble parameter during inflation. (Since every degree of freedom carries the average energy $\sim H_{\text{infl}}$, one can think of H_{infl} as the effective temperature of de Sitter space for the purposes of scalar fluctuations.) For some simple potentials, the shift in the minimum due to first effect and the average deviation from that minimum due to second effect can be of the same order of magnitude.

An initially homogeneous condensate can fragment into stable or relatively stable Q-balls [95], which become the building blocks of PBHs. The population of Q-balls behaves as matter, but it is matter with relatively few particles per horizon. Therefore, the Poisson fluctuations are unsuppressed, and some fluctuations can be big enough to collapse to black holes [8, 9, 11].

Supersymmetry provides a natural framework for PBH formation from scalar field fragmentation. It is particularly interesting that electroweak-scale supersymmetry naturally results in the PBH masses for which there are no observational constraints and which can account for all dark matter. While Q-ball formation is possible in different models of supersymmetry breaking, let us consider gauge-mediated supersymmetry breaking at the scale $M_{\text{SUSY}} \sim 100$ TeV. At the time of fragmentation, when the energy density of the condensate is a fraction $f \sim r_f^{-1/2} \lesssim 1$ of the total energy density ($\rho_\phi \sim M_{\text{SUSY}}^4 \equiv f \times \rho_{\text{tot}}$), the mass inside the horizon is

$$M_{\text{hor}} \sim r_f^{-1/4}\left(\frac{M_{\text{Pl}}^3}{M_{\text{SUSY}}^2}\right) \sim 10^{22}\text{g}\left(\frac{100\ \text{TeV}}{M_{\text{SUSY}}}\right)^2 . \tag{9.18}$$

Let the number of Q-balls per the horizon be $N_H \sim 100$. A typical PBH results from coalescence of 10 to 100 Q-balls, and the mass of a typical PBH is

$$M_{\text{PBH}} \sim r_f^{-1/4} \times 10^{22}\text{g}\left(\frac{100\ \text{TeV}}{M_{\text{SUSY}}}\right)^2 , \tag{9.19}$$

which is consistent with the open window for dark matter in the form of PBH at masses

$$10^{17}\text{g} \lesssim M_{\text{PBH}} \lesssim 10^{22}\text{g} . \tag{9.20}$$

It is intriguing that supersymmetry just above the electroweak scale predicts the masses of PBHs consistent with the current bounds for dark matter.

9.3.2 Long-Range Interactions Mediated by Flat Directions

The existence of flat directions allows for another PBH formation scenario, which also starts with fragmentation of the scalar condensate into Q-balls but then proceeds in a different dynamical fashion. The previously mentioned scenario requires the introduction of an intermediate matter dominated era to facilitate the collapse of overdensities via gravity. This requirement is again a consequence of the fact that traditionally, matter perturbations only grow during a matter dominated phase. Given that gravity is relatively inefficient in removing energy and angular momentum, only rare, spherical configurations can ultimately collapse into black holes. This limitation led the authors of Ref. [96] to pursue a less limited framework.

With the framework of primordial structure formation in mind, Ref. [96] included an additional scalar force which mediates interactions between Q-balls. Just as in the case of primordial structure formation, this scalar interaction can lead to the coalescence and collapse of overdensities, even during radiation domination. As discussed before, the removal of energy and angular momentum is very efficient for scalar waves, thus alleviating the need for special spherical configurations of particles.

As with primordial structure formation, the dynamics in this scenario are complicated, and numerical simulations will play a key role in future investigations. However, given the size and strengths of the forces involved, the expectation is that after scalar fragmentation large systems of charges will be drawn together. Ultimately, these charges will start to collide and merge, resulting in a population of larger and larger charges. For Q-balls formed from the fragmentation of flat directions, their mass and radius are dictated by, [97]

$$M_Q = M_{\rm SUSY} Q^{3/4}, \qquad R_Q = \frac{Q^{1/4}}{M_{\rm SUSY}} \tag{9.21}$$

where Q is the $U(1)$ charge associated with the flat direction which permits the formation of Q balls. Setting $R_Q = R_S = 2GM_Q$ establishes threshold where a charge will eventually fit within its own Schwarzschild radius. This leads to a predicted PBH mass which depends only on the SUSY breaking scale,

$$M_{\rm PBH} \sim 10^{22}\ {\rm g} \left(\frac{100\ {\rm TeV}}{M_{\rm SUSY}}\right)^2. \tag{9.22}$$

Compared to the previous, gravity-only mechanism, the mass dependence is nearly exactly the same. The formation of PBHs via supersymmetry and long-range scalar forces has the ability to produce populations of PBHs which can be the totality of dark matter, or which may explain microlensing observations by HSC and OGLE [98–100].

9.3.2.1 Tests for High-Scale Supersymmetry

The fact that the PBH mass is inversely proportional to the scale of supersymmetry breaking leads to the opportunity to examine high-scale supersymmetry by studying the implications of light PBHs. Although PBHs with masses below $\lesssim 10^{17}$ g would have evaporated by the present-day, the implications of a light population of PBHs could still be observable today. In particular, if these light PBHs had the opportunity to dominate the energy density of the Universe, then a detectable gravitational wave signal can be produced via the *poltergeist mechanism* [101–112].

The poltergeist phenomenon is a result of the fact that curvature perturbations source gravitational waves at second order in perturbation theory [76, 113, 114]. Two features of an intermediate matter dominated era help in generating a larger gravitational wave signal. First, unlike during radiation domination, the gravitational potential during a matter dominated era does not decay, even on subhorizon scales. Second, if the transition from intermediate matter domination to radiation domination is rapid, then there is a resonant enhancement. This phenomenon arises from the quick conversion of matter overdensities into relativistic sound waves, and the enhancement of the comoving gravitational waves that travel with these sound waves.

The frequency of the gravitational waves produced by the poltergeist mechanism is determined by the wavelength of the fluctuations reentering the horizon at the time of the PBH evaporation [108], not by the size of an individual PBH. The corresponding temperature is $T_R \propto M_{PBH}^{-3/2}$. For the relevant range of masses $M_{PBH} \sim 1 - 100$ kg, the gravitational waves can be probed by the future GW observations.

The strength of the gravitational wave signal produced from an intermediate matter dominated era and PBH evaporation depends on the length of the matter dominated era, and the speed of the transition from matter domination to radiation domination. Within the framework for supersymmetric PBH formation we have discussed, there exists a large parameter space which could generate PBHs with the masses required to produce the necessary conditions for generating a detectable gravitational wave signal. As highlighted in Ref. [105], a wide range of supersymmetric breaking scales can be explored with many of proposed future gravitational wave detectors.

9.4 Neutron Stars as a PBH Detector

If PBHs comprise dark matter, their number density is very small for the allowed mass window. Indeed, no more than one dark-matter black hole could be found in the vicinity of the solar system. A dark matter-black hole, with an asteroid-scale mass, would have a Schwarzschild radius which is smaller than the wavelength of visible light. If such a black hole passes through a star or a planet, the drag forces are not sufficient to stop the black hole. However, PBHs can get captured by neutron stars, thanks to their higher density. This will occur in regions with high dark matter densities, like the galactic center. Once a PBH settles inside the neutron star, it consumes the neutron star from the inside, turning it into a 1–2 $M_\odot$ black hole [36,

115–120]. This offers an exciting possibility to use neutron stars as a PBH detectors. One may wonder if the disruptions of neutron stars by PBHs could explain the missing pulsar problem [121], but the rates in the central parsec may not be high enough [122]. The evidence of a neutron star disruption by a PBH consists of several observable effects: a kilonova explosion [117], a fast radio burst [117, 123], and a population of 1–2 $M_\odot$ black holes which can be detected by LIGO/VIRGO, KAGRA, etc. [36, 117, 119, 120]

The average life-time of a neutron star in a PBH rich environment is determined by three physical processes. The first, and most important is the PBH capture rate

$$F \equiv f_{\rm PBH} \times F_0^{\rm MW}, \tag{9.23}$$

where $F_0^{\rm MW}$ is the Milky Way capture rate given by, [124]

$$F_0^{\rm MW} = \sqrt{6\pi}\,\frac{\rho_{\rm DM}}{M_{\rm PBH}\bar{v}}\left(\frac{2GM_{\rm NS}R_{\rm NS}}{1-2GM_{\rm NS}/R_{\rm NS}}\right)\left(1-e^{-3E_{\rm loss}/(M_{\rm PBH}\bar{v}^2)}\right), \tag{9.24}$$

where $\bar{v}$ is the dark matter velocity dispersion, $\rho_{\rm DM}$ is the dark matter energy density and

$$E_{\rm loss} \approx 58.8\frac{G^2M_{\rm PBH}^2M_{\rm NS}}{R_{\rm NS}^2} \tag{9.25}$$

is the average interaction loss energy during a neutron star-PBH interaction. The capture rate may also be enhanced by considering the neutron star velocity dispersion [117]. However, for the parameter space we consider this effect will be small. For our analysis, we consider neutron stars with masses $M_{\rm NS} = 1.5\ M_\odot$ and radii $R_{\rm NS} = 10$ km.

Once a PBH is gravitationally bound to a given neutron star, multiple collisions with its host neutron star are required before the PBH finally settles into the neutron star core. This settling time scale has been estimated to be [124]

$$t_{\rm set} \simeq 1.3\times 10^9\ {\rm yr}\left(\frac{M_{\rm PBH}}{10^{19}\ {\rm g}}\right)^{-3/2}. \tag{9.26}$$

Once settled at the neutron star core, the sublunar PBH will begin to consume the surrounding neutron rich material. Therefore, the final contribution to the average neutron star lifetime is an estimation of the accretion rate. For simplicity, we will assume spherical Bondi accretion,

$$\dot{M}_{\rm PBH} = 4\pi\lambda_s G^2 M_{\rm PBH}^2 \rho_{\rm NS}/v_s^3, \tag{9.27}$$

where $M_{\rm PBH}$ is the time evolving mass of the growing black hole, v_s is the sound speed, $\rho_{\rm NS}$ is the neutron star density, and λ_s is an order one parameter depending on the neutron star equation of state. For a neutron star described by an $n = 3$ polytrope, $v_s = 0.17$, $\rho_{\rm NS} = 10^{15}$ g cm^{-3}, and $\lambda_s = 0.707$ [125], we find

$$t_{\mathrm{acc}} \equiv \frac{M_{\mathrm{PBH}}}{dM_{\mathrm{PBH}}/dt} \simeq 10\ \mathrm{yr}\left(\frac{10^{19}\ \mathrm{g}}{M_{\mathrm{PBH}}}\right). \tag{9.28}$$

Though we are examining incredibly small black holes, the traditional Bondi accretion rate is still reliable for the PBHs in the dark matter mass range [126].

The average neutron star lifetime will be given by

$$\langle t_{\mathrm{NS}}\rangle = F^{-1} + t_{\mathrm{set}} + t_{\mathrm{acc}}. \tag{9.29}$$

For the majority of parameter space, the interaction timescale F^{-1} is the main contribution to $\langle t_{\mathrm{NS}}\rangle$. Assuming the capture and conversion of PBHs follows a Poisson process, the number of converted neutron stars as a function of time is

$$\Upsilon(t) = 1 - \exp(-t/\langle t_{\mathrm{NS}}\rangle). \tag{9.30}$$

For the neutron star parameters mentioned above and $f_{\mathrm{PBH}} = 1$, PBHs with masses between 10^{17} to 10^{25} g lead to an average lifetime of $\langle t_{\mathrm{NS}}\rangle \lesssim 10^{12}$ yr [117]. Such a lifetime suggests that $\mathcal{O}(1-10)\%$ of pulsars in a PBH-rich environment will be consumed within the lifetime of the galaxy. This is consistent with observations which suggest an underabundance of millisecond pulsars within the regions closest to the Milky Way's central supermassive black hole, Sagittarius A* (Sgr A*) [121].

In general, the destruction of a neutron star is a dramatic event [117]. If the neutron star hosting a black hole is a rapidly rotating millisecond pulsar, the equatorial regions move close to the speed of light. During the last milliseconds of the neutron star's demise, its radius decreases dramatically, and, by angular momentum conservation, the star spins up, leading to release of neutron rich material. This produces a kilonova explosion with similar optical properties as the merger of neutron stars.

Kilonova explosions are a likely site of the r-process nucleosynthesis. The only kilonova observed so far was a neutron star merger detected by LIGO. This detection triggered quick follow-up optical observations, which were generally consistent with a r-process nucleosynthesis site. However, the rates of neutron star mergers are not sufficient to explain all r-process nucleosynthesis by neutron star collisions [127]. A disruption of a neutron star by PBH would provide an additional contribution to synthesis of heavy elements.

When a PBH settles in the center of a neutron star and consumes it from the inside, the system does not generate the gravitational waves signal similar to the neutron star merger. In fact, due to an approximate axial symmetry, any gravitational waves radiation is relatively weak. In the past, observation of a kilonova event without a signal from LIGO was incredibly improbable. However, the advent of Rubin LSST, ZTF and future surveys open the opportunity to detect tens of kilonovae per year without the need for a gravitational waves trigger. An additional signature of PBH induced kilonova is a fast radio burst [117, 123]. During the last milliseconds of the neutron stars collapse, its magnetic field changes rapidly, generating a strong and observable radio signal [117, 123].

Therefore, observation of an orphan kilonova unaccompanied by the gravitational waves signature of a neutron star merger, but accompanied by a fast radio burst would be a smoking-gun indicator of PBH-induced destruction of a neutron star [117].

9.4.1 G Objects

Neutron star destruction via PBHs also offers the opportunity to explain another peculiarity in the galactic center, namely the presence of mysterious class of objects called *G* objects [128]. Located within the central inner most region of the galactic center, six *G* objects have been identified. Dynamically, *G* objects act as stellar mass objects. Thermally however, these same objects appear to radiate in a similar fashion to dust and ionized gas.

The two most studied *G* objects, named G1 and G2, have also demonstrated a surprising resilience to tidal disruption. When passing close to Sgr A* both G1 and G2 remained intact, despite the fact that both objects experienced large tidal forces during their passage. This observation seems to indicate that *G* objects are not simply clouds of gas, but instead contain a stellar-mass core which provides gravitational stability. Many attempts have been made to explain the nature of *G* objects. Most models have postulated that the central stellar-mass core is a star. For example, the central star may be a young, low-mass star that has retained a protoplanetary disk [129] or that generated a mass-loss envelope [130]. Alternative models of *G* object formation have viewed the *G* objects as a merger product of a binary system [131–133].

In Ref. [134], the authors offered an origin for *G* objects which relies on the interactions between PBHs and neutron stars. As indicated before, the destruction of neutron stars will occur primarily in the inner most regions of the galaxy. Furthermore, as stated before, the final product of PBH-induced neutron star destruction will be stellar-mass black hole surrounded by an atmosphere of dust and gas. These objects can be interpreted as the observed population of *G* objects. As discussed in Ref. [134], the number of expected converted neutron stars are consistent with the number of observed *G* objects. Additionally, the luminosity of a solar mass black hole cloaked by an atmosphere are consistent with the luminosities inferred from *G* object observations. These two coincidences point to the possibility that neutron star destruction via sub-lunar PBHs may account for the appearance of these mysterious objects within our galaxy. It is through this lens that the existence of G object-like systems provide another pathway for testing PBHs as a dark matter candidate.

9.5 Other Scenarios

The formation of PBHs has continued to remain a topic of much interest within the literature. There is a vast and ever-growing set of models which can accommodate the generation of black holes at the earliest stages of the history of the Universe. Given the enormity of available models, we will be unable to mention all of the possibilities. However, in order to provide the reader with an overview of the field we will include a broad summary of the landscape of non-inflationary PBH formation mechanisms.

PBHs from phase transitions: Phase transitions have long-since been considered a possible formation channel for PBHs. A variety of physical phenomena associated with phase transitions can lead to the formation of black holes. Early work considered the collision of bubble walls [135] or the in-fall of the domain wall which separates true- from false-vacuum regions [136–138]. The latter scenario has the unique feature of forming worm-hole solutions which appear as PBHs to an outside observer, while internal observers would perceive themselves to be in a separate and expanding Universe. This possibility has also been examined within the context of inflation, particularly with vacuum bubbles which may have nucleated during the inflationary phase of the early Universe [32, 139–142]. Recently, there has also been considerable interest in PBHs from supercooled phase transitions [136–138, 143–152].

Particles which are massive in the true vacuum state could become trapped in the false vacuum region. These heavy particles would not have the energy to penetrate the bubble walls separating the two regions, and be compressed into a black hole [143, 153–155].

Phase transitions, like the QCD phase transition, are associated with large changes in the effective number of degrees of freedom. As such, phase transitions have the ability to depress the sound speed. Since the critical collapse criteria in the canonical inflationary scenario is related to the sound speed, PBH production may be enhanced at the scales associated with a given phase transition [156–158].

PBHs from topological defects: Alongside phase transitions, the formation of topological defects, which act as the building blocks of PBHs, may also lead to the formation of PBHs. This includes the collapse of cosmic string loops [159–164], or the collapse of domain walls [165–168].

Other scenarios and future directions: Besides those based on phase transitions and topological defects, there are also many other possible PBH formation mechanisms, for example, those based on quark confinement [169], formation from isocurvature perturbations [170] or PBH formation from stochastic tunnelling [171].

Short-lived black holes and supermassive black holes: The James Webb Space Telescope allowed for the discovery of supermassive black holes at very high red-shifts, when the Universe was less than a billion years old [172]. Small black holes with lifetimes too short to be dark matter could form and decay during the dark ages, after recombination. Evaporation of such black holes could provide the source of heat necessary for direct collapse of large gas clouds to supermassve black holes [173].

Recently, there has been interest in *nonprimordial* light black holes forming in the present epoch [174–177]. In particular, small black holes forming at present time in the galactic center could simultaneously explain the GeV excess, as well as the flux of cosmic-ray antiprotons and in a few tentative antihelium events reported by the anti-matter spectrometer AMS-02 [174, 177].

Acknowledgements The work of M.M.F was supported by the European Union's Horizon 2020 research and innovation programme under grant agreement No 101002846, ERC CoG CosmoChart.

A.K. was supported by the U.S. Department of Energy (DOE) Grant No. DE-SC0009937; by World Premier International Research Center Initiative (WPI), MEXT, Japan; and by Japan Society for the Promotion of Science (JSPS) KAKENHI Grant No. JP20H05853.

References

1. K. Ando et al., Primordial black holes for the LIGO events in the axionlike curvaton model. Phys. Rev. D **97**(12), 123512 (2018). https://doi.org/10.1103/PhysRevD.97.123512. arXiv:1711.08956 [astro-ph.CO]
2. U. Banik et al., Constraining the mass density of free-floating black holes using razor-thin lensing arcs. Mon. Not. R. Astron. Soc. **483**(2), 1558–1573 (2019). https://doi.org/10.1093/mnras/sty3267. (arXiv:1811.00637 [astro-ph.CO])
3. M. Biagetti et al., The formation probability of primordial black holes. Phys. Lett. B **820**, 136602 (2021). https://doi.org/10.1016/j.physletb.2021.136602. arXiv:2105.07810 [astro-ph.CO]
4. B.J. Carr et al., New cosmological constraints on primordial black holes. Phys. Rev. D **81**, 104019 (2010). https://doi.org/10.1103/PhysRevD.81.104019. arXiv:0912.5297 [astro-ph.CO]
5. B. Carr, F. Kühnel, M. Sandstad, Primordial black holes as dark matter. Phys. Rev. D **94**(8), 083504 (2016). https://doi.org/10.1103/PhysRevD.94.083504
6. B. Carr, J. Silk, Primordial black holes as generators of cosmic structures. Mon. Not. R. Astron. Soc. **478**(3), 3756–3775 (2018). https://doi.org/10.1093/mnras/sty1204
7. B.J. Carr, S.W. Hawking, Black holes in the early Universe. Mon. Not. R. Astron. Soc. **168**, 399–415 (1974). https://doi.org/10.1093/mnras/168.2.399
8. E. Cotner, A. Kusenko, Primordial black holes from scalar field evolution in the early universe. Phys. Rev. D **96**(10), 103002 (2017). https://doi.org/10.1103/PhysRevD.96.103002. arXiv:1706.09003 [astro-ph.CO]
9. E. Cotner, A. Kusenko, Primordial black holes from supersymmetry in the early universe. Phys. Rev. Lett. **119**(3), 031103 (2017). https://doi.org/10.1103/PhysRevLett.119.031103. arXiv:1612.02529 [astro-ph.CO]
10. E. Cotner, A. Kusenko, V. Takhistov, Primordial black holes from inflaton fragmentation into oscillons. Phys. Rev. D **98**(8), 083513 (2018). https://doi.org/10.1103/PhysRevD.98.083513. arXiv:1801.03321 [astro-ph.CO]
11. E. Cotner et al., Analytic description of primordial black hole formation from scalar field fragmentation. JCAP **10**, 077 (2019). https://doi.org/10.1088/1475-7516/2019/10/077. arXiv:1907.10613 [astro-ph.CO]
12. A. Dolgov, J. Silk, Baryon isocurvature fluctuations at small scales and baryonic dark matter. Phys. Rev. D **47**, 4244–4255 (1993). https://doi.org/10.1103/PhysRevD.47.4244
13. A. Escrivà, C. Germani, R.K. Sheth, Universal threshold for primordial black hole formation. Phys. Rev. D **101**(4), 044022 (2020). https://doi.org/10.1103/PhysRevD.101.044022. arXiv:1907.13311 [gr-qc]
14. P.H. Frampton et al., Primordial black holes as all dark matter. JCAP **04**, 023 (2010). https://doi.org/10.1088/1475-7516/2010/04/023. arXiv:1001.2308 [hep-ph]
15. J.A. de Freitas Pacheco, J. Silk, Primordial rotating black holes. Phys. Rev. D **101**(8), 083022 (2020). https://doi.org/10.1103/PhysRevD.101.083022. arXiv:2003.12072 [astro-ph.CO]
16. J. García-Bellido, A.D. Linde, D. Wands, Density perturbations and black hole formation in hybrid inflation. Phys. Rev. D **54**, 6040–6058 (1996). https://doi.org/10.1103/PhysRevD.54.6040. arXiv:astro-ph/9605094
17. J. García-Bellido, M. Peloso, C. Unal, Gravitational wave signatures of inflationary models from primordial black hole dark matter. JCAP **09**, 013 (2017). https://doi.org/10.1088/1475 7516/2017/09/013. arXiv:1707.02441 [astro-ph.CO]

18. J. Georg, S. Watson, A preferred mass range for primordial black hole formation and black holes as dark matter revisited. JHEP **09**, 138 (2017). https://doi.org/10.1007/JHEP09(2017)138. arXiv:1703.04825 [astro-ph.CO]
19. C. Germani, I. Musco, Abundance of primordial black holes depends on the shape of the inflationary power spectrum. Phys. Rev. Lett. **122**(14), 141302 (2019). https://doi.org/10.1103/PhysRevLett.122.141302. arXiv:1805.04087 [astro-ph.CO]
20. C. Germani, R.K. Sheth, Nonlinear statistics of primordial black holes from Gaussian curvature perturbations. Phys. Rev. D **101**(6), 063520 (2020). https://doi.org/10.1103/PhysRevD.101.063520. arXiv:1912.07072 [astro-ph.CO]
21. A.M. Green et al., A new calculation of the mass fraction of primordial black holes. Phys. Rev. D **70**, 041502 (2004). https://doi.org/10.1103/PhysRevD.70.041502. arXiv:astro-ph/0403181
22. S. Hawking, Gravitationally collapsed objects of very low mass. Mon. Not. R. Astron. Soc. **152**, 75 (1971)
23. F. Hoyle, R.A. Lyttleton, The effect of interstellar matter on climatic variation. Proc. Camb. Philos. Soc. **35**(3), 405 (1939). https://doi.org/10.1017/S0305004100021150
24. K. Inomata et al., Double inflation as a single origin of primordial black holes for all dark matter and LIGO observations. Phys. Rev. D **97**(4), 043514 (2018). https://doi.org/10.1103/PhysRevD.97.043514. arXiv:1711.06129 [astro-ph.CO]
25. K. Inomata et al., Inflationary primordial black holes as all dark matter. Phys. Rev. D **96**(4), 043504 (2017). https://doi.org/10.1103/PhysRevD.96.043504. arXiv:1701.02544 [astro-ph.CO]
26. K. Inomata et al., Inflationary primordial black holes for the LIGO gravitational wave events and pulsar timing array experiments. Phys. Rev. D **95**(12), 123510 (2017). https://doi.org/10.1103/PhysRevD.95.123510. arXiv:1611.06130 [astro-ph.CO]
27. M. Kawasaki, N. Sugiyama, T. Yanagida, Primordial black hole formation in a double inflation model in supergravity. Phys. Rev. D **57**, 6050–6056 (1998). https://doi.org/10.1103/PhysRevD.57.6050. arXiv:hep-ph/9710259
28. M. Kawasaki et al., Primordial black holes as dark matter in supergravity inflation models. Phys. Rev. D **94**(8), 083523 (2016). https://doi.org/10.1103/PhysRevD.94.083523. arXiv:1606.07631 [astro-ph.CO]
29. M. Khlopov, B.A. Malomed, I.B. Zeldovich, Gravitational instability of scalar fields and formation of primordial black holes. Mon. Not. R. Astron. Soc. **215**, 575–589 (1985)
30. M.Y. Khlopov, Primordial black holes. Res. Astron. Astrophys. **10**, 495–528 (2010). https://doi.org/10.1088/1674-4527/10/6/001. arXiv:0801.0116 [astro-ph]
31. B. Kocsis et al., Hidden universality in the merger rate distribution in the primordial black hole scenario. Astrophys. J. **854**(1), 41 (2018). https://doi.org/10.3847/1538-4357/aaa7f4. arXiv:1709.09007 [astro-ph.CO]
32. A. Kusenko et al., Exploring primordial black holes from the multiverse with optical telescopes. Phys. Rev. Lett. **125**(18), 18 (2020). https://doi.org/10.1103/PhysRevLett.125.181304. arXiv:2001.09160 [astro-ph.CO]
33. S. Pi et al., Scalaron from R^2-gravity as a heavy field. JCAP **05**, 042 (2018). https://doi.org/10.1088/1475-7516/2018/05/042. arXiv:1712.09896 [astro-ph.CO]
34. A. Riotto, J. Silk, The future of primordial black holes: open questions and roadmap (2024). arXiv:2403.02907 [astro-ph.CO]
35. M. Sasaki et al., Primordial black holes–perspectives in gravitational wave astronomy. Class. Quant. Grav. **35**(6), 063001 (2018). https://doi.org/10.1088/1361-6382/aaa7b4. arXiv:1801.05235 [astro-ph.CO]
36. V. Takhistov, G.M. Fuller, A. Kusenko, Test for the origin of solar mass black holes. Phys. Rev. Lett. **126**(7), 071101 (2021). https://doi.org/10.1103/PhysRevLett.126.071101. arXiv:2008.12780 [astro-ph.HE]
37. E.L. Wright, On the density of pbh's in the galactic halo. Astrophys. J. **459**, 487 (1996). https://doi.org/10.1086/176910. arXiv:astro-ph/9509074
38. J. Yokoyama, Formation of MACHO primordial black holes in inflationary cosmology. Astron. Astrophys. **318**, 673 (1997). arXiv:astro-ph/9509027

39. Y.B. Zel'dovich, I.D. Novikov, Sov. Astron. **10**, 602 (1967)
40. B. Carr, The Primordial black hole mass spectrum. Astrophys. J. **201**, 1–19 (1975). https://doi.org/10.1086/153853
41. M. Kawasaki, A. Kusenko, T.T. Yanagida, Primordial seeds of supermassive black holes. Phys. Lett. B **711**, 1–5 (2012). https://doi.org/10.1016/j.physletb.2012.03.056. arXiv:1202.3848 [astro-ph.CO]
42. T. Gherghetta, C.F. Kolda, S.P. Martin, Flat directions in the scalar potential of the supersymmetric standard model. Nucl. Phys. B **468**, 37–58 (1996). https://doi.org/10.1016/0550-3213(96)00095-8. arXiv:hep-ph/9510370 [hep-ph]
43. L. Amendola, J. Rubio, C. Wetterich, Primordial black holes from fifth forces. Phys. Rev. D **97**(8), 081302 (2018). https://doi.org/10.1103/PhysRevD.97.081302. arXiv:1711.09915 [astro-ph.CO]
44. Y. Ayaita, M. Weber, C. Wetterich, Neutrino lump fluid in growing neutrino quintessence. Phys. Rev. D **87**(4), 043519 (2013). https://doi.org/10.1103/PhysRevD.87.043519. arXiv:1211.6589 [astro-ph.CO]
45. S. Casas, V. Pettorino, C. Wetterich, Dynamics of neutrino lumps in growing neutrino quintessence. Phys. Rev. D **94**(10), 103518 (2016). https://doi.org/10.1103/PhysRevD.94.103518. arXiv:1608.02358 [astro-ph.CO]
46. G. Domènech, M. Sasaki, Cosmology of strongly interacting fermions in the early universe. JCAP **06**, 030 (2021). https://doi.org/10.1088/1475-7516/2021/06/030. arXiv:2104.05271 [hep-th]
47. G. Domènech et al., Halo formation from Yukawa forces in the very early Universe. Phys. Rev. D **108**(10), 103543 (2023). https://doi.org/10.1103/PhysRevD.108.103543. arXiv:2304.13053 [astro-ph.CO]
48. M.M. Flores, A. Kusenko, Primordial black holes from long-range scalar forces and scalar radiative cooling. Phys. Rev. Lett. **126**(4), 041101 (2021). https://doi.org/10.1103/PhysRevLett.126.041101. arXiv:2008.12456 [astro-ph.CO]
49. M.M. Flores, L. Yifan, A. Kusenko, Structure formation after reheating: supermassive primordial black holes and Fermi ball dark matter. Phys. Rev. D **108**(12), 123511 (2023). https://doi.org/10.1103/PhysRevD.108.123511. arXiv:2308.09094 [astro-ph.CO]
50. S. Savastano et al., Primordial dark matter halos from fifth forces. Phys. Rev. D **100**(8), 083518 (2019). https://doi.org/10.1103/PhysRevD.100.083518. arXiv:1906.05300 [astro-ph.CO]
51. M.S. Maxon, E.G. Corman, Electron-Electron Bremsstrahlung from a Quantum Plasma ($Z = 1$). Phys. Rev. **163**, 156–162 (1967). https://doi.org/10.1103/PhysRev.163.156. https://link.aps.org/doi/10.1103/PhysRev.163.156
52. S. Maxon, Bremsstrahlung rate and spectra from a hot gas ($Z = 1$). Phys. Rev. A **5**, 1630–1633 (1972). https://doi.org/10.1103/PhysRevA.5.1630. https://link.aps.org/doi/10.1103/PhysRevA.5.1630
53. R. Durrer, A. Kusenko, Magnetogenesis from early structure formation due to Yukawa forces. JCAP **11**, 002 (2023). https://doi.org/10.1088/1475-7516/2023/11/002. arXiv:2209.13313 [astro-ph.CO]
54. M.M. Flores, C. Kouvaris, A. Kusenko, Defrosting and Blast Freezing Dark Matter. Phys. Rev. D **108**(10), 103545 (2023). https://doi.org/10.1103/PhysRevD.108.103545. arXiv:2306.04056 [hep-ph]
55. M.M. Flores et al., Fireball baryogenesis from early structure formation due to Yukawa forces. Phys. Rev. D **108**(9), L091705 (2023). https://doi.org/10.1103/PhysRevD.108.L091705. arXiv:2208.09789 [hep-ph]
56. K. Petraki, R.R. Volkas, Review of asymmetric dark matter. Int. J. Mod. Phys. A **28**, 1330028 (2013). https://doi.org/10.1142/S0217751X13300287. arXiv:1305.4939 [hep-ph]
57. K.M. Zurek, Asymmetric dark matter: theories, signatures, and constraints. Phys. Rept. **537**, 91–121 (2014). https://doi.org/10.1016/j.physrep.2013.12.001. arXiv:1308.0338 [hep-ph]
58. M.I. Gresham, H.K. Lou, K.M. Zurek, Early Universe synthesis of asymmetric dark matter nuggets. Phys. Rev. D **97**(3), 036003 (2018). https://doi.org/10.1103/PhysRevD.97.036003. arXiv:1707.02316 [hep-ph]

59. M.I. Gresham, H.K. Lou, K.M. Zurek, Nuclear structure of bound states of asymmetric dark matter. Phys. Rev. D **96**(9), 096012 (2017). https://doi.org/10.1103/PhysRevD.96.096012. arXiv:1707.02313 [hep-ph]
60. M.I. Gresham, K.M. Zurek, Asymmetric dark stars and neutron star stability. Phys. Rev. D **99**(8), 083008 (2019). https://doi.org/10.1103/PhysRevD.99.083008. arXiv:1809.08254 [astro-ph.CO]
61. M.B. Wise, Y. Zhang, Stable bound states of asymmetric dark matter. Phys. Rev. D **90**(5), 055030 (2014). [Erratum: Phys. Rev. D **91**, 039907 (2015)]. https://doi.org/10.1103/PhysRevD.90.055030. arXiv:1407.4121 [hep-ph]
62. B. Carr, F. Kuhnel, Primordial black holes as dark matter: recent developments. Ann. Rev. Nucl. Part. Sci. **70**, 355–394 (2020). https://doi.org/10.1146/annurev-nucl-050520-125911. arXiv:2006.02838 [astro-ph.CO]
63. B. Carr et al., Constraints on primordial black holes. Rept. Prog. Phys. **84**(11), 116902 (2021). https://doi.org/10.1088/1361-6633/ac1e31. arXiv:2002.12778 [astro-ph.CO]
64. P. Mroz et al., No large population of unbound or wide-orbit Jupiter-mass planets. Nature **548**(7666), 183–186 (2017). https://doi.org/10.1038/nature23276. arXiv:1707.07634 [astro-ph.EP]
65. L.A. Anchordoqui, S.E.P. Bergliaffa, Hot thermal universe endowed with massive dark vector fields and the Hubble tension. Phys. Rev. D **100**(12), 123525 (2019). https://doi.org/10.1103/PhysRevD.100.123525. arXiv:1910.05860 [astro-ph.CO]
66. J.L. Bernal, L. Verde, A.G. Riess, The trouble with H_0. JCAP **10**, 019 (2016). https://doi.org/10.1088/1475-7516/2016/10/019. arXiv:1607.05617 [astro-ph.CO]
67. M. Escudero, S.J. Witte, A CMB search for the neutrino mass mechanism and its relation to the Hubble tension. Eur. Phys. J. C **80**(4), 294 (2020). https://doi.org/10.1140/epjc/s10052-020-7854-5. arXiv:1909.04044 [astro-ph.CO]
68. G.B. Gelmini, A. Kusenko, V. Takhistov, Possible hints of sterile neutrinos in recent measurements of the Hubble parameter. JCAP **06**, 002 (2021). https://doi.org/10.1088/1475-7516/2021/06/002. arXiv:1906.10136 [astro-ph.CO]
69. G.B. Gelmini et al., Big Bang Nucleosynthesis constraints on sterile neutrino and lepton asymmetry of the Universe. JCAP **09**, 051 (2020). https://doi.org/10.1088/1475-7516/2020/09/051. arXiv:2005.06721 [hep-ph]
70. L. Perivolaropoulos, F. Skara, Challenges for ΛCDM: an update. New Astron. Rev. **95**, (2022). https://doi.org/10.1016/j.newar.2022.101659. arXiv:2105.05208 [astro-ph.CO]
71. S. Vagnozzi, New physics in light of the H_0 tension: an alternative view. Phys. Rev. D **102**(2), 023518 (2020). https://doi.org/10.1103/PhysRevD.102.023518. arXiv:1907.07569 [astro-ph.CO]
72. K. Vattis, S.M. Koushiappas, A. Loeb, Dark matter decaying in the late Universe can relieve the H0 tension. Phys. Rev. D **99**(12), 121302 (2019). https://doi.org/10.1103/PhysRevD.99.121302. arXiv:1903.06220 [astro-ph.CO]
73. K.C. Wong et al., H0LiCOW–XIII. A 2.4 per cent measurement of H0 from lensed quasars: 5.3 σ tension between early- and late-Universe probes. Mon. Not. R. Astron. Soc. **498**(1), 1420–1439 (2020). https://doi.org/10.1093/mnras/stz3094. arXiv:1907.04869 [astro-ph.CO]
74. M.M. Flores, A. Kusenko, Spins of primordial black holes formed in different cosmological scenarios. Phys. Rev. D **104**(6), 063008 (2021). https://doi.org/10.1103/PhysRevD.104.063008. arXiv:2106.03237 [astro-ph.CO]
75. V. De Luca, N. Bellomo, Accretion, emission, mass and spin evolution (2023). arXiv:2312.14097 [astro-ph.CO]
76. H. Assadullahi, D. Wands, Gravitational waves from an early matter era. Phys. Rev. D **79**, 083511 (2009). https://doi.org/10.1103/PhysRevD.79.083511. arXiv:0901.0989 [astro-ph.CO]
77. I. Dalianis, C. Kouvaris, Gravitational waves from density perturbations in an early matter domination era (2020). arXiv:2012.09255 [astro-ph.CO]
78. K. Jedamzik, M. Lemoine, J. Martin, Generation of gravitational waves during early structure formation between cosmic inflation and reheating. JCAP **04**, 021 (2010). https://doi.org/10.1088/1475-7516/2010/04/021. arXiv:1002.3278 [astro-ph.CO]

79. T. Nakama, Stochastic gravitational waves associated with primordial black holes formed during an early matter era. Phys. Rev. D **101**(6), 063519 (2020). https://doi.org/10.1103/PhysRevD.101.063519
80. K.N. Ananda, C. Clarkson, D. Wands, The Cosmological gravitational wave background from primordial density perturbations. Phys. Rev. D **75**, 123518 (2007). https://doi.org/10.1103/PhysRevD.75.123518. arXiv:gr-qc/0612013
81. D. Baumann et al., Gravitational wave spectrum induced by primordial scalar perturbations. Phys. Rev. D **76**, 084019 (2007). https://doi.org/10.1103/PhysRevD.76.084019. arXiv:hep-th/0703290
82. V. De Luca et al., Gravitational waves from peaks. JCAP **09**, 059 (2019). https://doi.org/10.1088/1475-7516/2019/09/059. arXiv:1905.13459 [astro-ph.CO]
83. G. Domènech, S. Passaglia, S. Renaux-Petel, Gravitational waves from dark matter isocurvature. JCAP **03**(03), 023 (2022). https://doi.org/10.1088/1475-7516/2022/03/023. arXiv:2112.10163 [astro-ph.CO]
84. M.M. Flores, A. Kusenko, M. Sasaki, Gravitational waves from rapid structure formation on microscopic scales before matter-radiation equality. Phys. Rev. Lett. **131**(1), 011003 (2023). https://doi.org/10.1103/PhysRevLett.131.011003. arXiv:2209.04970 [astro-ph.CO]
85. Y.B. Zel'dovich, Gravitational instability: an approximate theory for large density perturbations. Astron. Astrophys. **5**, 84–89 (1970)
86. T. Harada et al., Primordial black hole formation in the matter-dominated phase of the Universe. Astrophys. J. **833**(1), 61 (2016). https://doi.org/10.3847/1538-4357/833/1/61. arXiv:1609.01588 [astro-ph.CO]
87. I. Affleck, M. Dine, A new mechanism for baryogenesis. Nucl. Phys. B **249**, 361–380 (1985). https://doi.org/10.1016/0550-3213(85)90021-5
88. M. Dine, L. Randall, S.D. Thomas, Supersymmetry breaking in the early universe. Phys. Rev. Lett. **75**, 398–401 (1995). https://doi.org/10.1103/PhysRevLett.75.398. arXiv:hep-ph/9503303
89. T.S. Bunch, P.C.W. Davies, Quantum field theory in de sitter space: renormalization by point splitting. Proc. R. Soc. Lond. A **360**, 117–134 (1978). https://doi.org/10.1098/rspa.1978.0060
90. N.A. Chernikov, E.A. Tagirov, Quantum theory of scalar fields in de Sitter space-time. Ann. Inst. H. Poincare Phys. Theor. A **9**, 109 (1968)
91. K.-M. Lee, E.J. Weinberg, Decay of the true vacuum in curved space-time. Phys. Rev. D **36**, 1088 (1987). https://doi.org/10.1103/PhysRevD.36.1088
92. A.D. Linde, Scalar field fluctuations in expanding universe and the new inflationary universe scenario. Phys. Lett. B **116**, 335–339 (1982). https://doi.org/10.1016/0370-2693(82)90293-3
93. A.A. Starobinsky, J. Yokoyama, Equilibrium state of a selfinteracting scalar field in the De Sitter background. Phys. Rev. D **50**, 6357–6368 (1994). https://doi.org/10.1103/PhysRevD.50.6357. arXiv:astro-ph/9407016
94. E.A. Tagirov, Consequences of field quantization in de Sitter type cosmological models. Ann. Phys. **76**, 561–579 (1973). https://doi.org/10.1016/0003-4916(73)90047-X
95. A. Kusenko, M.E. Shaposhnikov, Supersymmetric Q balls as dark matter. Phys. Lett. B **418**, 46–54 (1998). https://doi.org/10.1016/S0370-2693(97)01375-0. arXiv:hep-ph/9709492 [hep-ph]
96. M.M. Flores, A. Kusenko, Primordial black holes as a dark matter candidate in theories with supersymmetry and inflation. JCAP **05**, 013 (2023). https://doi.org/10.1088/1475-7516/2023/05/013. arXiv:2108.08416 [hep-ph]
97. M. Dine, A. Kusenko, The Origin of the matter—antimatter asymmetry. Rev. Mod. Phys. **76**, 1 (2003). https://doi.org/10.1103/RevModPhys.76.1. arXiv:hep-ph/0303065 [hep-ph]
98. H. Niikura et al., Microlensing constraints on primordial black holes with Subaru/HSC Andromeda observations. Nat. Astron. **3**(6), 524–534 (2019). https://doi.org/10.1038/s41550-019-0723-1. arXiv:1701.02151 [astro-ph.CO]
99. H. Niikura et al., Constraints on Earth-mass primordial black holes from OGLE 5-year microlensing events. Phys. Rev. D **99**(8), 083503 (2019). https://doi.org/10.1103/PhysRevD.99.083503. arXiv:1901.07120 [astro-ph.CO]

100. S. Sugiyama, M. Takada, A. Kusenko, Possible evidence of axion stars in HSC and OGLE microlensing events. Phys. Lett. B **840**, 137891 (2023). https://doi.org/10.1016/j.physletb.2023.137891. arXiv:2108.03063 [hep-ph]
101. N. Bhaumik, R.K. Jain, Small-scale induced gravitational waves from primordial black holes, a stringent lower mass bound, and the imprints of an early matter to radiation transition. Phys. Rev. D **104**(2), 023531 (2021). https://doi.org/10.1103/PhysRevD.104.023531. arXiv:2009.10424 [astro-ph.CO]
102. G. Domènech, C. Lin, M. Sasaki, Gravitational wave constraints on the primordial black hole dominated early universe. JCAP **04**, 062 (2021). https://doi.org/10.1088/1475-7516/2021/04/062. arXiv:2012.08151 [gr-qc]
103. G. Domènech, M. Sasaki, Gravitational wave hints black hole remnants as dark matter. Class. Quant. Grav. **40**(17), 177001 (2023). https://doi.org/10.1088/1361-6382/ace493. arXiv:2303.07661 [gr-qc]
104. G. Domènech, V. Takhistov, M. Sasaki, Exploring evaporating primordial black holes with gravitational waves. Phys. Lett. B **823**, 136722 (2021). https://doi.org/10.1016/j.physletb.2021.136722. arXiv:2105.06816 [astro-ph.CO]
105. M.M. Flores et al., Testing high scale supersymmetry via second order gravitational waves. Phys. Rev. D **108**(12), 123002 (2023). https://doi.org/10.1103/PhysRevD.108.123002. arXiv:2308.15522 [hep-ph]
106. K. Harigaya, K. Inomata, T. Terada, Gravitational wave production from axion rotations right after a transition to kination. Phys. Rev. D **108**(8), L081303 (2023). https://doi.org/10.1103/PhysRevD.108.L081303. arXiv:2305.14242 [hep-ph]
107. K. Inomata et al., Enhancement of gravitational waves induced by scalar perturbations due to a sudden transition from an early matter era to the radiation era. Phys. Rev. D **100**(4), 043532 (2019). https://doi.org/10.1103/PhysRevD.100.043532. arXiv:1904.12879 [astro-ph.CO]
108. K. Inomata et al., Gravitational wave production right after a primordial black hole evaporation. Phys. Rev. D **101**(12), 123533 (2020). https://doi.org/10.1103/PhysRevD.101.123533. arXiv:2003.10455 [astro-ph.CO]
109. K. Inomata et al., Gravitational waves induced by scalar perturbations during a gradual transition from an early matter era to the radiation era. JCAP **10**, 071 (2019). https://doi.org/10.1088/1475-7516/2019/10/071. arXiv:1904.12878 [astro-ph.CO]
110. T. Papanikolaou, V. Vennin, D. Langlois, Gravitational waves from a universe filled with primordial black holes. JCAP **03**, 053 (2021). https://doi.org/10.1088/1475-7516/2021/03/053. arXiv:2010.11573 [astro-ph.CO]
111. M. Pearce et al., Gravitational wave signals from early matter domination: interpolating between fast and slow transitions (2023). arXiv:2311.12340 [astro-ph.CO]
112. G. White et al., Detectable gravitational wave signals from Affleck-Dine baryogenesis. Phys. Rev. Lett. **127**(18), 181601 (2021). https://doi.org/10.1103/PhysRevLett.127.181601. arXiv:2105.11655 [hep-ph]
113. L. Alabidi et al., Observable induced gravitational waves from an early matter phase. JCAP **05**, 033 (2013). https://doi.org/10.1088/1475-7516/2013/05/033. arXiv:1303.4519 [astro-ph.CO]
114. K. Kohri, T. Terada, Semianalytic calculation of gravitational wave spectrum nonlinearly induced from primordial curvature perturbations. Phys. Rev. D **97**(12), 123532 (2018). https://doi.org/10.1103/PhysRevD.97.123532. arXiv:1804.08577 [gr-qc]
115. T.W. Baumgarte, S.L. Shapiro, Primordial black holes captured by neutron stars: relativistic point-mass treatment (2024). arXiv:2404.08735 [gr-qc]
116. T.W. Baumgarte, S.L. Shapiro, Primordial black holes captured by neutron stars: simulations in general relativity (2024). arXiv:2405.10365 [gr-qc]
117. G. Fuller, A. Kusenko, V. Takhistov, Primordial black holes and r-process nucleosynthesis. Phys. Rev. Lett. **119**(6), 061101 (2017). https://doi.org/10.1103/PhysRevLett.119.061101. arXiv:1704.01129 [astro-ph.HE]
118. C. Kouvaris, P. Tinyakov, Growth of black holes in the interior of rotating neutron stars. Phys. Rev. D **90**(4), 043512 (2014). https://doi.org/10.1103/PhysRevD.90.043512. arXiv:1312.3764 [astro-ph.SR]

119. V. Takhistov, Positrons from primordial black hole microquasars and gamma-ray bursts. Phys. Lett. B **789**, 538–544 (2019). https://doi.org/10.1016/j.physletb.2018.12.043. arXiv:1710.09458 [astro-ph.HE]
120. V. Takhistov, Transmuted gravity wave signals from primordial black holes. Phys. Lett. B **782**, 77–82 (2018). https://doi.org/10.1016/j.physletb.2018.05.026. arXiv:1707.05849 [astro-ph.CO]
121. J. Dexter, R.M. O'Leary, The peculiar pulsar population of the central parsec. Astrophys. J. Lett. **783**(1), L7 (2014). https://doi.org/10.1088/2041-8205/783/1/L7
122. R. Caiozzo, G. Bertone, F. Kühnel, Revisiting primordial black hole capture by neutron stars (2024). arXiv:2404.08057 [astro-ph.HE]
123. M. Abramowicz, M. Bejger, M. Wielgus, Collisions of neutron stars with primordial black holes as fast radio bursts engines. Astrophys. J. **868**(1), 17 (2018). https://doi.org/10.3847/1538-4357/aae64a. arXiv:1704.05931 [astro-ph.HE]
124. F. Capela, M. Pshirkov, P. Tinyakov, Constraints on primordial black holes as dark matter candidates from capture by neutron stars. Phys. Rev. D **87**(12), 123524 (2013). https://doi.org/10.1103/PhysRevD.87.123524. arXiv:1301.4984 [astro-ph.CO]
125. S.L. Shapiro, S.A. Teukolsky, *Black Holes, White Dwarfs, and Neutron Stars: the Physics of Compact Objects* (John Wiley & Sons, 2008)
126. P. Giffin et al., Neutron star quantum death by small black holes. Phys. Rev. D **105**(12), 123030 (2022). https://doi.org/10.1103/PhysRevD.105.123030. arXiv:2105.06504 [hep-ph]
127. C. Kobayashi, A.I. Karakas, M. Lugaro, The origin of elements from carbon to uranium. Astrophys. J. **900**(2), 179 (2020). https://doi.org/10.3847/1538-4357/abae65. arXiv:2008.04660 [astro-ph.GA]
128. A. Ciurlo et al., A population of dust-enshrouded objects orbiting the Galactic black hole. Nature **577**(7790), 337–340 (2020). https://doi.org/10.1038/s41586-019-1883-y
129. R.A. Murray-Clay, A. Loeb, Disruption of a proto-planetary disc by the black hole at the milky way centre. Nat. Commun. **3**(1), 1–5 (2012)
130. N. Scoville, A. Burkert, The galactic center cloud G2–a young low-mass star with a stellar wind. Astrophys. J. **768**(2), 108 (2013)
131. S. Prodan, F. Antonini, H.B. Perets, Secular evolution of binaries near massive black holes: formation of compact binaries, merger/collision products and G2-like objects. Astrophys. J. **799**(2), 118 (2015)
132. A.P. Stephan et al., Merging binaries in the galactic center: the eccentric Kozai–Lidov mechanism with stellar evolution. Mon. Not. R. Astron. Soc. **460**(4), 3494–3504 (2016)
133. A.P. Stephan et al., The fate of binaries in the galactic center: the mundane and the exotic. Astrophys. J. **878**(1), 58 (2019). https://doi.org/10.3847/1538-4357/ab1e4d. arXiv:1903.00010 [astro-ph.SR]
134. M.M. Flores et al., G objects and primordial black holes. Phys. Rev. D **108**(6), L061301 (2023). https://doi.org/10.1103/PhysRevD.108.L061301. arXiv:2308.08623 [astro-ph.CO]
135. S.W. Hawking, I.G. Moss, J.M. Stewart, Bubble collisions in the very early universe. Phys. Rev. D **26**, 2681 (1982). https://doi.org/10.1103/PhysRevD.26.2681
136. S.K. Blau, E.I. Guendelman, A.H. Guth, The dynamics of false vacuum bubbles. Phys. Rev. D **35**, 1747 (1987). https://doi.org/10.1103/PhysRevD.35.1747
137. H. Kodama, M. Sasaki, K. Sato, Abundance of primordial holes produced by cosmological first order phase transition. Prog. Theor. Phys. **68**, 1979 (1982). https://doi.org/10.1143/PTP.68.1979
138. K. Sato et al., Multiproduction of universes by first order phase transition of a vacuum. Phys. Lett. B **108**, 103–107 (1982). https://doi.org/10.1016/0370-2693(82)91152-2
139. H. Deng, A. Vilenkin, Primordial black hole formation by vacuum bubbles. JCAP **12**, 044 (2017). https://doi.org/10.1088/1475-7516/2017/12/044. arXiv:1710.02865 [gr-qc]
140. J. Garriga, A. Vilenkin, J. Zhang, Black holes and the multiverse. JCAP **02**, 064 (2016). https://doi.org/10.1088/1475-7516/2016/02/064. arXiv:1512.01819 [hep-th]
141. H.-L. Huang, Y.-S. Piao, Towards supermassive primordial black holes from inflationary bubbles (2023). arXiv:2312.11982 [astro-ph.CO]

142. H.-L. Huang et al., Supermassive primordial black holes in multiverse: for nano-Hertz gravitational wave and high-redshift JWST galaxies (2023). arXiv:2306.17577 [gr-qc]
143. M.J. Baker et al., Detailed calculation of primordial black hole formation during first-order cosmological phase transitions (2021). arXiv:2110.00005 [astro-ph.CO]
144. I. Baldes, M.O. Olea-Romacho, Primordial black holes as dark matter: interferometric tests of phase transition origin (2023). arXiv:2307.11639 [hep-ph]
145. A. Conaci et al., Slaying axion-like particles via gravitational waves and primordial black holes from supercooled phase transition (2024). arXiv:2401.09411 [astro-ph.CO]
146. M.M. Flores, A. Kusenko, M. Sasaki, Revisiting formation of primordial black holes in a supercooled first-order phase transition (2024). arXiv:2402.13341 [hep-ph]
147. Y. Gouttenoire, T. Volansky, Primordial black holes from supercooled phase transitions (2023). arXiv:2305.04942 [hep-ph]
148. R. Jinno, J. Kume, M. Yamada, Super-slow phase transition catalyzed by BHs and the birth of baby BHs (2023). arXiv:2310.06901 [hep-ph]
149. K. Kawana, T. Kim, P. Lu, PBH formation from overdensities in delayed vacuum transitions (2022). arXiv:2212.14037 [astro-ph.CO]
150. M. Lewicki, P. Toczek, V. Vaskonen, Black holes and gravitational waves from slow phase transitions (2024). arXiv:2402.04158 [astro-ph.CO]
151. M. Lewicki, P. Toczek, V. Vaskonen, Primordial black holes from strong first-order phase transitions. JHEP **09**, 092 (2023). https://doi.org/10.1007/JHEP09(2023)092. arXiv:2305.04924 [astro-ph.CO]
152. J. Liu et al., Primordial black hole production during first-order phase transitions. Phys. Rev. D **105**(2), L021303 (2022). https://doi.org/10.1103/PhysRevD.105.L021303. arXiv:2106.05637 [astro-ph.CO]
153. M.J. Baker et al., Primordial black holes from first-order cosmological phase transitions (2021). arXiv:2105.07481 [astro-ph.CO]
154. K. Kawana, L. Philip, K.-P. Xie, First-order phase transition and fate of false vacuum remnants. JCAP **10**, 030 (2022). https://doi.org/10.1088/1475-7516/2022/10/030. arXiv:2206.09923 [astro-ph.CO]
155. K. Kawana, K.-P. Xie, Primordial black holes from a cosmic phase transition: the collapse of Fermi-balls. Phys. Lett. B **824**, 136791 (2022). https://doi.org/10.1016/j.physletb.2021.136791. arXiv:2106.00111 [astro-ph.CO]
156. B. Carr et al., Cosmic conundra explained by thermal history and primordial black holes. Phys. Dark Univ. **31**, 100755 (2021). https://doi.org/10.1016/j.dark.2020.100755. arXiv:1906.08217 [astro-ph.CO]
157. M. Crawford, D.N. Schramm, Spontaneous generation of density perturbations in the early universe. Nature **298**, 538–540 (1982). https://doi.org/10.1038/298538a0
158. K. Jedamzik, Primordial black hole formation during the QCD epoch. Phys. Rev. D **55**, 5871–5875 (1997). https://doi.org/10.1103/PhysRevD.55.R5871. arXiv:astro-ph/9605152
159. R.R. Caldwell, P. Casper, Formation of black holes from collapsed cosmic string loops. Phys. Rev. D **53**, 3002–3010 (1996). https://doi.org/10.1103/PhysRevD.53.3002. arXiv:gr-qc/9509012
160. J. Garriga, M. Sakellariadou, Effects of friction on cosmic strings. Phys. Rev. D **48**, 2502–2515 (1993). https://doi.org/10.1103/PhysRevD.48.2502. arXiv:hep-th/9303024
161. S.W. Hawking, Black holes from cosmic strings. Phys. Lett. B **231**, 237–239 (1989). https://doi.org/10.1016/0370-2693(89)90206-2
162. A.C. Jenkins, M. Sakellariadou, Primordial black holes from cusp collapse on cosmic strings (2020). arXiv:2006.16249 [astro-ph.CO]
163. J.H. MacGibbon, R.H. Brandenberger, U.F. Wichoski, Limits on black hole formation from cosmic string loops. Phys. Rev. D **57**, 2158–2165 (1998). https://doi.org/10.1103/PhysRevD.57.2158. arXiv:astro-ph/9707146
164. A. Polnarev, R. Zembowicz, Formation of primordial black holes by cosmic strings. Phys. Rev. D **43**, 1106–1109 (1991). https://doi.org/10.1103/PhysRevD.43.1106

165. G.B. Gelmini, A. Simpson, E. Vitagliano, Catastrogenesis: DM, GWs, and PBHs from ALP string-wall networks. JCAP **02**, 031 (2023). https://doi.org/10.1088/1475-7516/2023/02/031. arXiv:2207.07126 [hep-ph]
166. G.B. Gelmini et al., Primordial black hole dark matter from catastrogenesis with unstable pseudo-Goldstone bosons. JCAP **06**, 055 (2023). https://doi.org/10.1088/1475-7516/2023/06/055. arXiv:2303.14107 [hep-ph]
167. Y. Gouttenoire, E. Vitagliano, Primordial black holes and wormholes from domain wall networks. Phys. Rev. D **109**(12), 123507 (2024). https://doi.org/10.1103/PhysRevD.109.123507. arXiv:2311.07670 [hep-ph]
168. S.G. Rubin, A.S. Sakharov, M.Y. Khlopov, The Formation of primary galactic nuclei during phase transitions in the early universe. J. Exp. Theor. Phys. **91**, 921–929 (2001). https://doi.org/10.1134/1.1385631. arXiv:hep-ph/0106187
169. G. Dvali, F. Kühnel, M. Zantedeschi, Primordial black holes from confinement. Phys. Rev. D **104**(12), 123507 (2021). https://doi.org/10.1103/PhysRevD.104.123507. arXiv:2108.09471 [hep-ph]
170. S. Passaglia, M. Sasaki, Primordial black holes from CDM isocurvature perturbations. Phys. Rev. D **105**(10), 103530 (2022). https://doi.org/10.1103/PhysRevD.105.103530. arXiv:2109.12824 [astro-ph.CO]
171. C. Animali, V. Vennin, Primordial black holes from stochastic tunnelling (2022). arXiv:2210.03812 [astro-ph.CO]
172. A.J. Bunker et al., JADES NIRSpec Spectroscopy of GN-z11: Lyman- α emission and possible enhanced nitrogen abundance in a z = 10.60 luminous galaxy. Astron. Astrophys. **677**, A88 (2023). https://doi.org/10.1051/0004-6361/202346159. arXiv:2302.07256 [astro-ph.GA]
173. Y. Lu, Z.S.C. Picker, A. Kusenko, Feeding plankton to whales: high-redshift supermassive black holes from tiny black hole explosions (2023). arXiv:2312.15062 [astro-ph.GA]
174. M. Korwar, S. Profumo, Late-forming black holes and the antiproton, gamma-ray, and anti-helium excesses (2024). arXiv:2403.18656 [astro-ph.CO]
175. L. Philip, K. Kawana, A. Kusenko, Late-forming primordial black holes: beyond the CMB era. Phys. Rev. D **107**(10), 103037 (2023). https://doi.org/10.1103/PhysRevD.107.103037. arXiv:2210.16462 [astro-ph.CO]
176. Z.S.C. Picker, A. Kusenko, Constraints on late-forming exploding black holes. Phys. Rev. D **108**(2), 023012 (2023). https://doi.org/10.1103/PhysRevD.108.023012. arXiv:2305.13429 [astro-ph.CO]
177. Z.S.C. Picker, A. Kusenko, Explaining the GeV excess with exploding black holes. Phys. Lett. B **845**, 138175 (2023). https://doi.org/10.1016/j.physletb.2023.138175. arXiv:2305.13434 [astro-ph.HE]

Part III
Cosmological Evolution and Astrophysics Implications

Chapter 10
Cosmological Evolution in a Universe with Primordial Black Holes: Overall Picture

Ely D. Kovetz

Abstract We present a lightning overview of the implications of primordial black holes on cosmological and astrophysical evolution, from the early to the late observable Universe, with emphasis on the cosmic dawn and reionization epochs and on hierarchical large-scale structure formation. We discuss how primordial black holes may affect dark-matter halo evolution, star formation, supermassive-black-hole formation, and how they may interact with particle dark matter. Meanwhile, the primordial black holes themselves evolve via accretion, emission and mergers.

10.1 Introduction

To put things into context, this book is split into six parts. Following the introduction (Part I), Parts II, III, IV, and V cover different aspects of the primordial black hole (PBH) scenario. As illustrated in Fig. 10.1, we can use an analogy with the different search methods for particle dark matter to understand the logic behind this partition. The illustration shows a Feynman-like diagram with two legs connected to PBHs and the other two connected to "standard-model" astrophysical objects. The arrows, indicating different ways to follow the diagram, correspond to the production (or formation) of PBHs (discussed in Part II of this book), their annihilation via mergers (which sources gravitational waves, see Part IV) and their interaction with standard astrophysics and cosmology, which imposes constraints on the PBH abundance (see Part V) and also has implications for cosmic evolution - the topic of Part III below. The main focus in this Part is on stellar-mass (or more massive) PBHs and their effects on structure evolution, supermassive black hole formation, their accretion and merger histories, etc.

E. D. Kovetz (✉)
Department of Physics, Ben-Gurion University of the Negev, 84105 Be'er Sheva, Israel
e-mail: kovetz@bgu.ac.il

C. Byrnes et al. (eds.), *Primordial Black Holes*, Springer Series in Astrophysics and Cosmology, https://doi.org/10.1007/978-981-97-8887-3_10

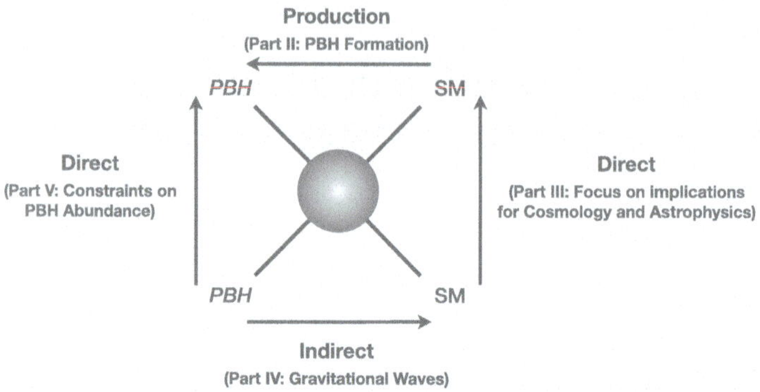

Fig. 10.1 Illustration, adapted from [29]. In analogy to search methods for (particle) dark matter: *production* in colliders, *direct detection* via interactions with standard model particles and *indirect detection* via products of dark matter annihilation; we show a Feynman diagram connecting two PBHs with two "standard-model" astrophysical objects. This book covers PBH *formation* in Part II, abundance *constraints* in Part V, *gravitational wave* signatures in Part IV, and in Part III: *implications* on cosmology and astrophysics

10.2 Early-Universe Implications

10.2.1 Initial Conditions for Structure Formation

As explained in Parts I and II, PBHs are formed in the very early Universe, deep in the radiation dominated era under the standard scenario, or in an early matter dominated era in more exotic models. They source additional density perturbations through gravity, which affect the process of large-scale structure formation and modify the halo mass distribution [11]. The affected mass range depends on the scales of the enhanced perturbations associated with the PBHs (see Fig. 10.2).

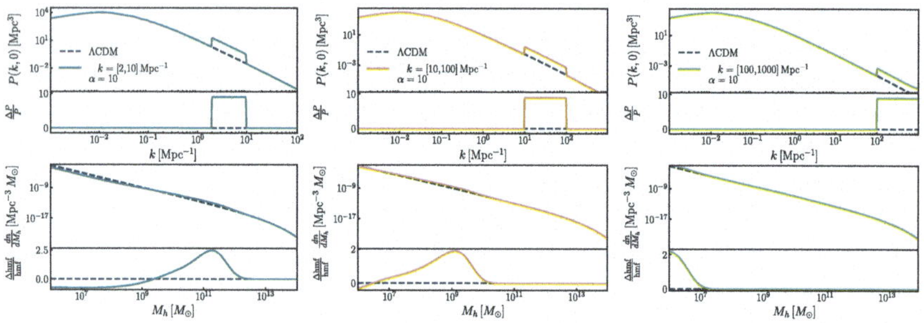

Fig. 10.2 Enhanced matter fluctuations on different scales translate into excess halos formation in corresponding mass ranges. Adapted from [31]. Credit: Sarah Libanore

The effect of PBHs on structure formation can be divided to their local effect on their (individual) small-scale surrounding (where they act as seeds for the gravitational evolution), and their overall statistical effect on the clustering properties of the fluctuations, as they source random Poisson fluctuations in addition to the standard adiabatic ones [24]. As will be discussed next in Chap. 11, the seed effect in general is more important when the PBH abundance is small, while the random Poisson clustering effect is more important when PBHs account for a large fraction of the dark matter. Dedicated N-body simulations can be used to explore these different regimes and account for the non-linear evolution of structure formation in the presence of PBHs [24, 32].

10.2.2 First Stars and Galaxies

Naively, enhanced fluctuations from PBHs will lead to excess structure growth and increased galaxy and star formation at cosmic dawn. This will have direct effects on the early-Universe 21cm signal, on signals related to the epoch of reionization such as the CMB optical depth, the Lyman-alpha forest of high-redshift quasars (which is sensitive to the HI neutral fraction), and on the high-redshift ultraviolet luminosity functions. Predicting the effects on these signals may seem straightforward. However, there are non-trivial feedback effects resulting from PBH accretion from the primordial plasma [1, 3, 9, 17, 23, 27, 36, 37, 39, 41, 42, 46]. Energy injection from this process changes the thermal and ionization states of the intergalactic medium (IGM) and the interstellar medium (ISM) [43, 48].

To determine the effect on star formation, for example, we should distinguish between PopIII stars, formed in molecular-cooling galaxies that reside in dark matter halos with masses in the range $10^5 - 10^8\, M_\odot$ [28, 44], and PopII stars that are formed in atomic-cooling galaxies in $\gtrsim 10^8\, M_\odot$ massive halos [10]. The latter are metal-enriched and more rare at very high redshift ($z \lesssim 30$).

While the presence of PBHs results in higher abundance of halos that can host PopIII stars, the heating that results from their accretion increases the molecular cooling threshold and acts to inhibit the formation of stars and galaxies in these halos. These competing effects, which are also strongly dependent on the PBH mass spectrum, have been estimated for example using the Press-Schechter formalism [5]. More careful studies, that relax the assumption of Gaussian density fluctuations, which in particular is inaccurate in the presence of PBHs [15], require dedicated simulations [12, 32]. As will be shown in the next chapter, stellar-mass PBHs allowed by existing observational constraints can alter the shape of the halo density profiles, the thermal and chemical properties of the baryonic gas, and have (limited) impact on PopIII star formation (Figs. 10.3 and 10.4).

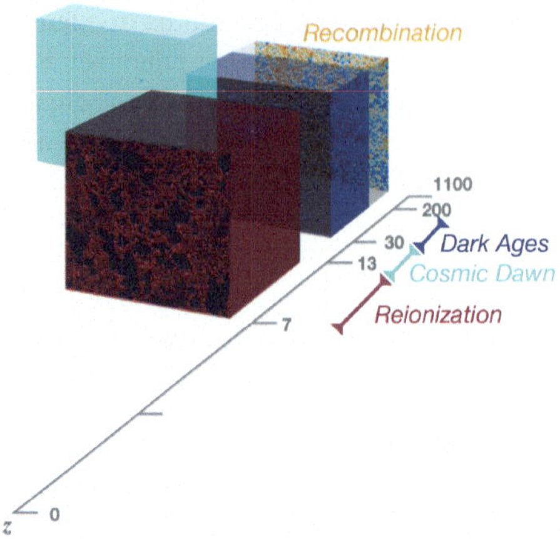

Fig. 10.3 Epochs in the high-redshift Universe which are affected by the presence of PBHs, through their effects on density fluctuations and their collapse into halos, feedback on star formation at cosmic dawn, and injection of energy and ionizing photons as byproducts of accretion from their surroundings and their signatures in recombination and reionization

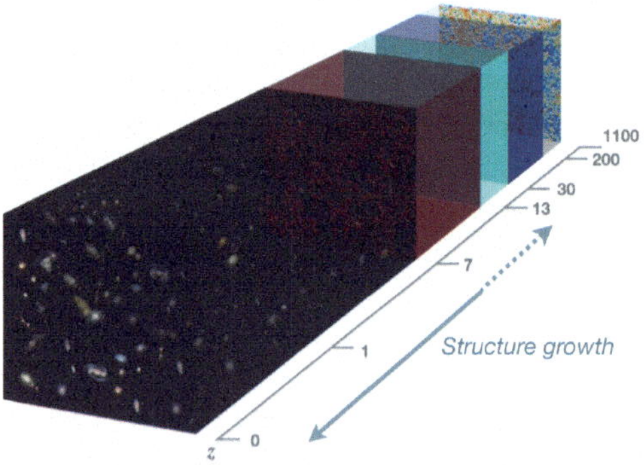

Fig. 10.4 The hierarchical growth of structure in the Universe can be affected by PBHs. For example, PBHs can provide seed for supermassive black holes. They can also impact the dynamics in dense environments like globular clusters and galactic centers. The properties of the PBHs themselves also evolve with time, through accretion, emission and mergers

10.3 Late-Time Implications

10.3.1 PBHs as Massive Black Hole Seeds

An important effect of PBHs is their potential role as seeds of massive black holes. In particular, stellar-mass PBHs can lead to direct collapse of gas clumps as large as $10^4 - 10^6\,M_\odot$, if Lyman-Werner radiation from their accretion disks is strong enough to dissociate H2 in metal-poor halos and suppress molecular cooling and fragmentation of the gas, allowing it to directly collapse into black holes [45]. These can provide the seeds for supermassive black holes (SMBHs) observed at high redshift, requiring more modest growth via accretion and mergers, without the need for super-Eddington accretion. Meanwhile, a small abundance of supermassive PBHs, allowed by existing constraints on PBH dark matter (constraints can be more easily evaded in models with non-Gaussian primordial fluctuations [26, 35]), can directly account for the seed SMBH population. An enhanced population of $\sim 10^7 - 10^{11}\,M_\odot$ SMBHs (e.g. in a mixed scenario with astrophysical and primordial SMBHs) can also potentially be related to the stochastic gravitational-wave background signal recently observed by pulsar-timing arrays [2, 20]. Chapter 12 presents a comprehensive review of the growth channels of SMBHs with emphasis on alternatives to a primordial seed origin.

10.3.2 Evolution of PBH Properties

In addition to the evolution of dark matter halos and the galaxies and stars they host, the properties of the PBHs themselves evolve with time. In particular, accretion onto PBHs can result in significant increase of their mass and spin [14, 38], while binary inspiral and merger can affect their spin and eccentricity. The effects of mergers on the gravitational wave signal are covered extensively in Part IV of this book. Meanwhile, Chap. 13 discusses the accretion process in detail, and also accounts for scenarios with dark matter "dressing" (these assume PBHs comprise only a small fraction of the total dark matter, and are typically engulfed in a halo made up of the dominant dark matter component), which exhibit very different accretion rates [41].

In order to accurately calculate the accretion process, the resulting energy injection into the surrounding environment has to be taken into account, as it can generate important feedback effects, such as collisional or photo ionization and outflows (winds, jets) that produce radiative feedback on the accreting gas [3]. In addition, while its effect is not necessarily large, the geometry (specifically the deviation from spherical symmetry) of the accretion has to be modeled as well.

As will be shown in Chap. 13, gas accretion can result in mass increase of several orders of magnitude for stellar-mass PBHs, with larger growth in binary systems (that also tend to drive the binary mass ratio to unity) [13]. While natal PBH spins are generally expected to be very small [22], accretion can potentially enhance them to non-negligible levels [14].

Overall, the effects of mass, spin as well as orbital and merger characteristics of annihilating PBH binaries can be studied with gravitational-wave measurements [18] (see Part IV).

10.4 Implications on Particle Dark Matter

In light of the numerous constraints on the PBH abundance (typically quantified as their fraction of the total dark matter energy density), if they exist, they only account for a small fraction of the total dark matter [21] outside the asteroid-mass window [34]. Therefore, it is interesting to investigate the possible interactions of PBHs with popular particle dark matter models, such as weakly-interacting massive particles (WIMPs) [4] or axion-like particles (ALP) [33]. The former interact with standard model particles via gravity and the weak force (or new mediators in typical extensions of the original WIMP scenario), and are the target of active searches in colliders as well as direct and indirect detection experiments. The latter are mostly searched for indirectly, exploiting their coupling to photons, e.g. by looking for axion-photon conversion in the presence of magnetic fields or considering their possible decay channels into photons.

Chapter 14 discusses in detail the prospects of detecting mixed dark matter scenarios, some of which predict very clear signatures. For example, the presence of WIMPs around PBHs will lead to enhanced annihilation signals, which place strong constraints on the mixed PBH-WIMP scenario [8]. Meanwhile, ALP will form clouds around PBHs with increased density compared to the standard dark matter spikes expected to form around black holes, which could lead to various detectable signatures [6, 7, 16, 19, 25, 30, 40, 47].

References

1. M.A. Abramowicz, P.C. Fragile, Foundations of black hole accretion disk theory. Living Rev. Rel. **16**, 1 (2013). https://doi.org/10.12942/lrr-2013-1. arXiv:1104.5499 [astro-ph.HE]
2. G. Agazie et al., The NANOGrav 15 yr data set: constraints on supermassive black hole binaries from the gravitational-wave background. Astrophys. J. Lett. **952**(2), L37 (2023). https://doi.org/10.3847/2041-8213/ace18b. arXiv:2306.16220 [astro-ph.HE]
3. Y. Ali-Haimoud, M. Kamionkowski, Cosmic microwave background limits on accreting primordial black holes. Phys. Rev. D **95**(4), 043534 (2017). https://doi.org/10.1103/PhysRevD.95.043534. arXiv:1612.05644 [astro-ph.CO]
4. G. Arcadi et al., The waning of the WIMP? A review of models, searches, and constraints. Eur. Phys. J. C **78**(3), (2018). ISSN: 1434-6052. https://doi.org/10.1140/epjc/s10052-018-5662-y. http://dx.doi.org/10.1140/epjc/s10052-018-5662-y
5. F. Atrio-Barandela, The effect of primordial black holes and streaming motions on structure formation. Astrophys. J. **939**(2), 69 (2022). https://doi.org/10.3847/1538-4357/ac983e. arXiv:2209.04737 [astro-ph.CO]
6. G. Bertone, D. Merritt, Time-dependent models for dark matter at the Galactic Center. Phys. Rev. D **72**, 103502 (2005). https://doi.org/10.1103/PhysRevD.72.103502. arXiv:astro-ph/0501555
7. G. Bertone, A.R. Zentner, J. Silk, A new signature of dark matter annihilations: gamma-rays from intermediate-mass black holes. Phys. Rev. D **72**, 103517 (2005). https://doi.org/10.1103/PhysRevD.72.103517. arXiv:astro-ph/0509565

8. G. Bertone et al., Primordial black holes as silver bullets for new physics at the weak scale. Phys. Rev. D **100**(12), 123013 (2019). https://doi.org/10.1103/PhysRevD.100.123013. arXiv:1905.01238 [hep-ph]
9. V. Bosch-Ramon, N. Bellomo, Mechanical feedback effects on primordial black hole accretion. Astron. Astrophys. **638**, A132 (2020). https://doi.org/10.1051/0004-6361/202037537. arXiv:2004.11224 [astro-ph.CO]
10. V. Bromm, N. Yoshida, The first galaxies. Annual Rev. Astron. Astrophys. **49**, 373–407 (2011). https://doi.org/10.1146/annurev-astro-081710-102608. arXiv:1102.4638 [astro-ph.CO]
11. B. Carr, J. Silk, Primordial black holes as generators of cosmic structures. Mon. Not. Roy. Astron. Soc. **478**(3), 3756–3775 (2018). https://doi.org/10.1093/mnras/sty1204
12. P.E. Colazo, F. Stasyszyn, N. Padilla, Structure formation with primordial black holes to alleviate early star formation tension revealed by JWST. Astron. Astrophys. **685**, L8 (2024). https://doi.org/10.1051/0004-6361/202449565. arXiv:2404.13110 [astro-ph.CO]
13. V. De Luca et al., Primordial black holes confront LIGO/Virgo data: current situation. JCAP **06**, 044 (2020). https://doi.org/10.1088/1475-7516/2020/06/044. arXiv:2005.05641 [astro-ph.CO]
14. V. De Luca et al., The evolution of primordial black holes and their final observable spins. JCAP **04**, 052 (2020). https://doi.org/10.1088/1475-7516/2020/04/052. arXiv:2003.02778 [astro-ph.CO]
15. V. Desjacques, A. Riotto, Spatial clustering of primordial black holes. Phys. Rev. D **98**(12), 123533 (2018). https://doi.org/10.1103/PhysRevD.98.123533. arXiv:1806.10414 [astro-ph.CO]
16. K. Eda et al., New probe of dark-matter properties: gravitational waves from an intermediate-mass black hole embedded in a dark-matter minispike. Phys. Rev. Lett. **110**(22), 221101 (2013). https://doi.org/10.1103/PhysRevLett.110.221101. (arXiv:1301.5971 [gr-qc])
17. X. Fan, E. Bañados, R.A. Simcoe, Quasars and the intergalactic medium at cosmic dawn. Annu. Rev. Astron. Astrophys. **61**, 373–426 (2023)
18. G. Franciolini et al., How to assess the primordial origin of single gravitational-wave events with mass, spin, eccentricity, and deformability measurements. Phys. Rev. D **105**(6), 063510 (2022). https://doi.org/10.1103/PhysRevD.105.063510. arXiv:2112.10660 [astro-ph.CO]
19. P. Gondolo, J. Silk, Dark matter annihilation at the galactic center. Phys. Rev. Lett. **83**, 1719–1722 (1999). https://doi.org/10.1103/PhysRevLett.83.1719. arXiv:astro-ph/9906391
20. Y. Gouttenoire et al., Scrutinizing the primordial black holes interpretation of PTA gravitational waves and JWST early galaxies (2023). arXiv:2307.01457 [astro-ph.CO]
21. A.M. Green, B.J. Kavanagh, Primordial Black Holes as a dark matter candidate. J. Phys. G **48**(4), 043001 (2021). https://doi.org/10.1088/1361-6471/abc534. arXiv:2007.10722 [astro-ph.CO]
22. T. Harada et al., Spins of primordial black holes formed in the radiation-dominated phase of the universe: first-order effect. Astrophys. J. **908**(2), 140 (2021). https://doi.org/10.3847/1538-4357/abd9b9. arXiv:2011.00710 [astro-ph.CO]
23. G. Hasinger, Illuminating the dark ages: cosmic backgrounds from accretion onto primordial black hole dark matter. JCAP **07**, 022 (2020). https://doi.org/10.1088/1475-7516/2020/07/022. arXiv:2003.05150 [astro-ph.CO]
24. D. Inman, Y. Ali-Haimoud, Early structure formation in primordial black hole cosmologies. Phys. Rev. D **100**(8), 083528 (2019). https://doi.org/10.1103/PhysRevD.100.083528. arXiv:1907.08129 [astro-ph.CO]
25. B.J. Kavanagh et al., Detecting dark matter around black holes with gravitational waves: effects of dark-matter dynamics on the gravitational waveform. Phys. Rev. D **102**(8), 083006 (2020). https://doi.org/10.1103/PhysRevD.102.083006. arXiv:2002.12811 [gr-qc]
26. M. Kawasaki, K. Murai, Formation of supermassive primordial black holes by Affleck-Dine mechanism. Phys. Rev. D **100**(10), 103521 (2019). https://doi.org/10.1103/PhysRevD.100.103521. arXiv:1907.02273 [astro-ph.CO]
27. A. King, K. Pounds, Powerful outflows and feedback from active galactic nuclei. Ann. Rev. Astron. Astrophys. **53**, 115–154 (2015). https://doi.org/10.1146/annurev-astro-082214-122316. arXiv:1503.05206 [astro-ph.GA]

28. R.S. Klessen, S.C.O. Glover, The first stars: formation, properties, and impact. Annu. Rev. Astron. Astrophys. **61**, (2023)
29. E. Kovetz, Probing primordial-black-hole dark matter with gravitational waves. Phys. Rev. Lett. **119**(13), 131301 (2017). https://doi.org/10.1103/PhysRevLett.119.131301. arXiv:1705.09182 [astro-ph.CO]
30. R. Lasenby, Black hole superradiance as a probe of ultra-light new particles. Proc. Int. Astron. Union **12**(S324), 273–278 (2016). https://doi.org/10.1017/S1743921317002319
31. S. Libanore et al., Unveiling cosmological information on small scales with line intensity mapping. Phys. Rev. D **106**(12), 123512 (2022). https://doi.org/10.1103/PhysRevD.106.123512. arXiv:2208.01658 [astro-ph.CO]
32. B. Liu, S. Zhang, V. Bromm, Effects of stellar-mass primordial black holes on first star formation. Mon. Not. Roy. Astron. Soc. **514**(2), 2376–2396 (2022). https://doi.org/10.1093/mnras/stac1472. arXiv:2204.06330 [astro-ph.GA]
33. D.J.E. Marsh, Axions and ALPs: a very short introduction, in *13th Patras Workshop on Axions, WIMPs and WISPs* (2018), pp. 59–74. https://doi.org/10.3204/DESY-PROC-2017-02/marsh_david. arXiv:1712.03018 [hep-ph]
34. P. Montero-Camacho et al., Revisiting constraints on asteroid-mass primordial black holes as dark matter candidates. J. Cosmol. Astropart. Phys. **2019**(08), 031 (2019). ISSN: 1475-7516. https://doi.org/10.1088/1475-7516/2019/08/031. http://dx.doi.org/10.1088/1475-7516/2019/08/031
35. T. Nakama, B. Carr, J. Silk, Limits on primordial black holes from μ distortions in cosmic microwave background. Phys. Rev. D **97**(4), 043525 (2018). https://doi.org/10.1103/PhysRevD.97.043525. arXiv:1710.06945 [astro-ph.CO]
36. L. Piga et al., The effect of outflows on CMB bounds from Primordial Black Hole accretion. JCAP **12**, 016 (2022). https://doi.org/10.1088/1475-7516/2022/12/016. arXiv:2210.14934 [astro-ph.CO]
37. V. Poulin et al., CMB bounds on disk-accreting massive primordial black holes. Phys. Rev. D **96**(8), 083524 (2017). https://doi.org/10.1103/PhysRevD.96.083524. arXiv:1707.04206 [astro-ph.CO]
38. M. Ricotti, Bondi accretion in the early universe. Astrophys. J. **662**, 53–61 (2007). https://doi.org/10.1086/516562. arXiv:0706.0864 [astro-ph]
39. M. Ricotti, J.P. Ostriker, K.J. Mack, Effect of primordial black holes on the cosmic microwave background and cosmological parameter estimates. Astrophys. J. **680**, 829 (2008). https://doi.org/10.1086/587831. arXiv:0709.0524 [astro-ph]
40. L. Sadeghian, F. Ferrer, C.M. Will, Dark matter distributions around massive black holes: a general relativistic analysis. Phys. Rev. D **88**(6), 063522 (2013). https://doi.org/10.1103/PhysRevD.88.063522. arXiv:1305.2619 [astro-ph.GA]
41. P.D. Serpico et al., Cosmic microwave background bounds on primordial black holes including dark matter halo accretion. Phys. Rev. Res. **2**(2), 023204 (2020). https://doi.org/10.1103/PhysRevResearch.2.023204. arXiv:2002.10771 [astro-ph.CO]
42. J. Silk, Unleashing positive feedback: linking the rates of star formation, supermassive black hole accretion and outflows in distant galaxies. Astrophys. J. **772**, 112 (2013). https://doi.org/10.1088/0004-637X/772/2/112. arXiv:1305.5840 [astro-ph.CO]
43. V. Takhistov et al., Interstellar gas heating by primordial black holes. JCAP **03**(03), 017 (2022). https://doi.org/10.1088/1475-7516/2022/03/017. arXiv:2105.06099 [astro-ph.GA]
44. M. Trenti, M. Stiavelli, The formation rates of population III stars and chemical enrichment of halos during the reionization era. Astrophys. J. **694**, 879–892 (2009). https://doi.org/10.1088/0004-637X/694/2/879. arXiv:0901.0711 [astro-ph.CO]
45. J.H. Wise et al., Formation of massive black holes in rapidly growing pre-galactic gas clouds. Nature **566**, 85–88 (2019). https://doi.org/10.1038/s41586-019-0873-4. arXiv:1901.07563 [astro-ph.GA]

46. F. Yuan, R. Narayan, Hot accretion flows around black holes. Ann. Rev. Astron. Astrophys. **52**, 529–588 (2014). https://doi.org/10.1146/annurev-astro-082812-141003. arXiv:1401.0586 [astro-ph.HE]
47. H.-S. Zhao, J. Silk, Mini-dark halos with intermediate mass black holes. Phys. Rev. Lett. **95**, 011301 (2005). https://doi.org/10.1103/PhysRevLett.95.011301. arXiv:astro-ph/0501625
48. F. Ziparo et al., Cosmic radiation backgrounds from primordial black holes. Mon. Not. Roy. Astron. Soc. **517**(1), 1086–1097 (2022). https://doi.org/10.1093/mnras/stac2705. arXiv:2209.09907 [astro-ph.CO]

Chapter 11
Impact of Primordial Black Holes on the Formation of the First Stars and Galaxies

Boyuan Liu and Volker Bromm

Abstract Recent gravitational-wave observations of binary black-hole mergers and the stochastic gravitational-wave background have triggered renewed interest in primordial black holes in the stellar-mass ($\sim 10 - 100\ M_{\odot}$) and supermassive regimes ($\sim 10^7 - 10^{11}\ M_{\odot}$). Although only a small fraction ($\lesssim 1\%$) of dark matter in the form of primordial black holes is required to explain such observations, these primordial black holes may play important roles in early structure/star/galaxy formation. In this chapter, we combine semi-analytical analysis and cosmological simulations to explore the possible impact of primordial black holes on the formation of the first stars and galaxies, taking into account two (competing) effects of primordial black holes: acceleration of structure formation and gas heating by black hole accretion feedback. We find that the impact of stellar-mass primordial black holes (allowed by existing observational constraints) on primordial star formation is likely minor, although they do alter the properties of the first star-forming halos/clouds and can potentially trigger the formation of massive black holes, while supermassive primordial black holes serve as seeds of massive structures that can explain the apparent overabundance of massive galaxies in recent observations. Our tentative models and results call for future studies with improved modeling of the interactions between primordial black holes, particle dark matter, and baryons to better understand the impact of primordial black holes on early star/galaxy/structure formation and their imprints in high-redshift observations.

B. Liu (✉)
Institute of Astronomy, University of Cambridge, Madingley Road, Cambridge, CB3 0HA, UK
e-mail: bl527@cam.ac.uk

V. Bromm
Department of Astronomy, University of Texas, Austin, TX 78712, USA
e-mail: vbromm@astro.as.utexas.edu

C. Byrnes et al. (eds.), *Primordial Black Holes*, Springer Series in Astrophysics and Cosmology, https://doi.org/10.1007/978-981-97-8887-3_11

11.1 Introduction

Recent gravitational-wave (GW) observations provide novel hints on the presence of primordial black holes (PBHs) [26, 55] as an important component of dark matter (DM). In particular, mergers of stellar-mass ($\sim$10–100 $M_{\odot}$) PBHs can potentially explain the entire population of the binary black hole (BBH) mergers detected by the LIGO-Virgo-KAGRA Scientific Collaboration [1], if such PBHs make up a small mass fraction of DM $f_{\rm PBH}$$\sim$$0.1-1\%$ [8, 9, 62, 99, 121]. Although it is unlikely that all BBH mergers originate from PBHs [51], a non-negligible contribution from PBH mergers (with $f_{\rm PBH}\sim 0.01-0.1\%$) is favored by the observed rate of mass-gap events[1] like GW190521 [34, 41] (see also Part IV and Chap. 25). On the other hand, the SGWB recently detected by pulsar timing arrays [5] can also be explained by mergers of supermassive ($\sim 10^7-10^{11}\,M_{\odot}$) PBHs given a similar PBH mass fraction in DM of $f_{\rm PBH}\sim 0.1-1\%$ [46, 59], although such a high abundance of supermassive PBHs (SMPBHs) is (marginally) in tension with local observational constraints that require $f_{\rm PBH}\lesssim 10^{-3}$ (independent of PBH formation mechanisms) [25], even if we ignore the high-z CMB μ-distortion constraint that rules out PBHs with $m_{\rm PBH}\gtrsim 10^4\,M_{\odot}$ formed by the spherical collapse of overdensities in a Gaussian density field (see Sect. 11.5 and Part V of this book). Moreover, even if PBHs in these two mass regimes only account for a small fraction of DM, their existence may have interesting implications for various astrophysical processes, in particular structure/star/galaxy formation in the first billion years after the Big Bang (the so-called Cosmic Dawn). Here, we review from a theoretical standpoint how PBHs may impact the formation of the first stars and galaxies via their gravitational and non-gravitational effects.

Formed in the very early universe, PBHs are sources of gravity that produce additional perturbations in the matter density field, resulting in accelerated structure formation with respect to the standard ΛCDM case. In general, PBHs accelerate structure formation individually through the 'seed' effect and collectively via the 'Poisson' effect [24]. The 'seed' effect dominates when the (mass) fraction of PBHs in DM is very small ($f_{\rm PBH}\to 0$), such that non-linear structures (i.e., DM halos) form around individual PBHs and do not interact with each other. In the idealized case where a single PBH with an initial mass $m_{\rm PBH}$ resides within a uniform background of particle DM (PDM), the mass of the halo seeded by the PBH follows a simple scaling law $M_{\rm B}\sim m_{\rm PBH}a/a_{\rm eq}$ in the matter-dominated epoch [84], where $a=1/(1+z)$ is the current scale-factor and $a_{\rm eq}\sim 1/3400$ the scale factor at matter-radiation equality.[2] The 'Poisson' effect dominates in the opposite limit $f_{\rm PBH}\to 1$, as the random

[1] Here, 'mass gap' refers to the mass range ($\sim 50-130\,M_{\odot}$) unachievable for a remnant black hole within standard stellar evolution theory [87].

[2] This idealized picture of isolated growth of halos around PBHs will break down at late epochs when large-scale adiabatic perturbations have grown enough to drive clustering/interactions of PBH-seeded structures and/or when competition of accretion between individual PBH-seeded halos occurs at $a\gtrsim a_{\rm crit}\equiv a_{\rm eq}/f_{\rm PBH}$ with the critical scale factor $a_{\rm crit}$ defined by $M_{\rm B}(a_{\rm crit})=m_{\rm PBH}/f_{\rm PBH}$.

distribution of PBHs at small scales produce isocurvature Poisson perturbations in the primordial density field on top of the standard adiabatic/isentropic perturbations. The interplay between these two effects involves complex nonlinear dynamics in the intermediate case when PBHs do not make up all DM but are sufficiently abundant to allow interactions/mergers between DM halos induced by individual PBHs, which is still an active area of research (see, e.g., [63, 80, 124] for the most recent progress with cosmological simulations). To zeroth order, the 'seed' effect dominates at mass scales smaller than $\sim M_{\rm B}$, whereas the 'Poisson' effect is important at mass scales around $\sim m_{\rm PBH}/f_{\rm PBH}$.

In addition to the gravitational effects on DM structures that provide the gravitational potential wells for gas (baryonic matter) to condense and form stars/galaxies, PBHs also affect star/galaxy formation via the accretion of gas and the corresponding feedback. When gas falls onto a BH, part of its gravitational energy is converted into radiative, magnetic, and kinetic energy that can impact the surrounding medium through, e.g., ionization, heating, and shocks. This process of BH (accretion) feedback in general involves complex physics at multiple scales (from the Schwarzschild radius to a few Mpc), and it remains under vigorous theoretical and observational investigations (see, e.g., [2, 39, 40, 53, 73, 109, 122]). Such 'non-gravitational' effects should also be taken into account in a complete theoretical model of early star/galaxy formation under the influence of PBHs.

In this chapter, we introduce the key theoretical ingredients required to study the impact of PBHs on early star/galaxy formation and demonstrate their applications with examples of idealized PBH models and the relevant astrophysical context. The aim is to pave the way for future work that can further refine and extend these theoretical tools to consider more general PBH models and a broader range of astrophysical phenomena. For simplicity, we assume that all PBHs have the *same* (initial) mass $m_{\rm PBH}$, and we focus on the stellar-mass ($\sim 10-100\,{\rm M}_\odot$) and supermassive ($\sim 10^7-10^{11}\,{\rm M}_\odot$) regimes motivated by recent GW observations and JWST results. We refer the reader to Part II of the book for discussions on the PBH mass spectra from different formation mechanisms, some of which can indeed produce highly enhanced peaks like the ones assumed here (e.g., [23, 57]). We also ignore any additional clustering of PBHs beyond the small-scale random distribution at birth. Such clustering can occur in certain PBH formation mechanisms [13] (see Chap. 20), with non-trivial implications for the effects and observational signatures of PBHs [16, 35, 36] (see also Chaps. 16 and 23).

This chapter is organized as follows. In Sect. 11.2, we outline a heuristic theoretical framework for structure formation with PBHs, introducing the additional terms from PBHs in the linear power spectrum (Sect. 11.2.1) and a numerical scheme to generate initial conditions for cosmological simulations including PBHs (Sect. 11.2.2). In Sect. 11.3, we describe a simple model for PBH accretion/feedback in terms of gas heating. In Sect. 11.4, based on the tools developed in previous sections, we combine semi-analytical analysis and cosmological simulations to investigate the effects of stellar-mass PBHs on primordial star formation, focusing on the cosmic stellar mass density (Sect. 11.4.1), the internal structure of the first star-forming clouds/halos (Sect. 11.4.2), and the formation of massive BH seeds (Sect. 11.4.3). In Sect. 11.5, we briefly discuss how the 'seed' effect of SMPBHs could explain the unusually massive galaxy candidates observed by JWST with a back-of-the-envelope derivation. Finally, in Sect. 11.6, we summarize our key findings and discuss the directions for future work.

Throughout this chapter, unless otherwise noted, we adopt the cosmological parameters $\Omega_m = 0.3089$, $\Omega_b = 0.04864$, $\Omega_r = 9.12 \times 10^{-5}$, $n_s = 0.9667$, $\sigma_8 = 0.8159$, and $h \equiv H_0/(100\ \mathrm{km\ s^{-1}\ Mpc^{-1}}) = 0.6774$ measured by the *Planck* satellite for the standard ΛCDM cosmology [4], which gives $a_{\mathrm{eq}} = \Omega_r/\Omega_m = 2.952 \times 10^{-4}$.

11.2 Structure Formation Within PBH Cosmologies

In this section, we assemble a heuristic theoretical framework for cosmic structure formation in the presence of PBHs. For simplicity and in the absence of a better theory, we follow the formalism of linear perturbation theory developed for the standard ΛCDM case and introduce terms/treatments for the effects of PBHs. We first introduce the linear power spectrum of DM density fluctuations (extrapolated to the present time, $a = 1$) that includes the 'Poisson' effect from PBHs as additional isocurvature and correlation terms (Sect. 11.2.1). Then we describe a tentative numerical scheme to generate initial conditions for cosmological simulations with PBHs, designed to capture both 'Poisson' and 'seed' effects (Sect. 11.2.2).

! Attention
Due to the discrete and nonlinear nature of PBHs as point sources, linear perturbation theory is only strictly valid at large scales where we can treat PBHs as a quasi-homogeneous ideal pressure-less fluid like PDM.

11.2.1 Power Spectrum of DM Density Fluctuations with PBHs

Following the analysis in [63], we assume that both PDM and PBHs are subject to the same primordial adiabatic perturbations at all scales that cause the primordial overdensity δ^0_{ad}, while the primordial overdensity of PBHs has an additional term δ^0_{iso} for isocurvature perturbations due to their discreteness that are important at small scales and assumed to be (initially) *uncorrelated* with the adiabatic piece δ^0_{ad}, so that the primordial overdensities of PDM and PBHs can be written as $\delta_{\mathrm{PDM}}(a \to 0) \equiv \delta^0_{\mathrm{PDM}} = \delta^0_{\mathrm{ad}}$ and $\delta_{\mathrm{PBH}}(a \to 0) \equiv \delta^0_{\mathrm{PBH}} = \delta^0_{\mathrm{ad}} + \delta^0_{\mathrm{iso}}$. The total DM overdensity is defined as $\delta_{\rm DM} \equiv (1 - f_{\rm PBH})\delta_{\rm PDM} + f_{\rm PBH}\delta_{\rm PBH}$, whose evolution in the linear regime, starting from the initial condition $\delta^0_{\rm DM} = \delta^0_{\rm ad} + f_{\rm PBH}\delta^0_{\rm iso}$, follows

$$\delta_{\rm DM}(a) = T_{\rm ad}(a)\delta^0_{\rm ad} + f_{\rm PBH}T_{\rm iso}(a)\delta^0_{\rm iso} \,. \tag{11.1}$$

Here, $T_{\rm ad}(a)$ and $T_{\rm iso}(a)$ are the linear transfer functions for the adiabatic and isocurvature modes, respectively, which are normalized to unity at $a \to 0$. To derive the evolution of the sub-components $\delta_{\rm PDM}$ and $\delta_{\rm PBH}$, we further define the difference field $\delta_- \equiv \delta_{\rm PBH} - \delta_{\rm PDM}$. For simplicity, we assume that PBHs form with negligible velocities relative to PDM, such that $\delta_- = \delta^0_- = \delta^0_{\rm iso}$ remains constant according to the linearized continuity equations [63]. In this way, we have

$$\delta_{\rm PDM}(a) = \delta_{\rm DM}(a) - f_{\rm PBH}\delta^0_{\rm iso} = T_{\rm ad}(a)\delta^0_{\rm ad} + [T_{\rm iso}(a) - 1]f_{\rm PBH}\delta^0_{\rm iso} \,, \tag{11.2}$$

$$\delta_{\rm PBH}(a) = \delta_{\rm PDM}(a) + \delta^0_{\rm iso} = T_{\rm ad}(a)\delta^0_{\rm ad} + T^{\rm PBH}_{\rm iso}(a)\delta^0_{\rm iso} \,, \tag{11.3}$$

where $T^{\rm PBH}_{\rm iso}(a) \equiv 1 + [T_{\rm iso}(a) - 1]\, f_{\rm PBH}$.

To appreciate the effects of PBHs on structure formation given the above formalism, let us focus on the total DM density fluctuations $\delta_{\rm DM}(a)$, whose statistical properties are well captured by the linear power spectrum $P_{\rm DM}(k, a)$, which in turn is the Fourier transform of the autocorrelation function. This can be written as

$$P_{\rm DM}(k, a) = T^2_{\rm ad}(a)P^0_{\rm ad}(k) + f^2_{\rm PBH}T^2_{\rm iso}(a)P^0_{\rm iso}(k) \;, \tag{11.4}$$

ignoring any correlation between the adiabatic and isocurvature modes. Here, the first term is identical to the linear power spectrum in the standard ΛCDM case, which is well constrained at large ($k \lesssim 3h$ Mpc^{-1}) scales by observations (see Fig. 19 in [6]), while the second term captures the effects of PBHs. Since the PBH 'seed' effect is by definition nonlinear, in this linear theory, we only consider the 'Poisson' effect with

$$P^0_{\rm iso}(k) \equiv 2\pi^2\Delta^2_{\rm Poisson}(k)/k^3 = \bar{n}^{-1}_{\rm PBH} \,, \quad k > k_{\rm eq} \,, \tag{11.5}$$

where $\Delta^2_{\rm Poisson}(k) = (k/k_\star)^3$ is the variance of primordial Poisson perturbations given $k_\star = (2\pi^2\bar{n}_{\rm PBH})^{1/3}$ as the critical scale around the mean separation between PBHs and $\bar{n}_{\rm PBH} = f_{\rm PBH}[3H_0^2(\Omega_m - \Omega_b)]/(8\pi G m_{\rm PBH})$ as the cosmic average (comoving) number density of PBHs. For simplicity, perturbations from PBHs are only included at scales smaller than the horizon scale at matter-radiation equality, defined by $k_{\rm eq}{\sim}0.01$ Mpc^{-1}.

The linear power spectra can be used to calculate halo abundances semi-analytically using the Press-Schechter (PS) formalism [92, 98], which will be discussed in Sect. 11.4 below. We note that in Eq. 11.4, we have ignored the correlation between isocurvature and adiabatic perturbations, so that the halo abundances derived in this way should be regarded as *conservative* estimates. Such correlation can be significant when PBHs fall into large structures due to adiabatic perturbations and meanwhile induce/disrupt PDM structures around themselves on small scales. Exploring this, cosmological simulations (see Sect. 4 in [80]) have shown that the PS formalism based on the linear power spectrum without correlation (Eq. 11.4) indeed predicts less abundant halos than those in simulations in the halo mass range $0.5\mathrm{m_{PBH}/f_{PBH}} \leq \mathrm{M_h} \leq 20\mathrm{m_{PBH}/f_{PBH}}$. A heuristic term that captures the correlation can be included to achieve better agreements between the results of the PS formalism and simulations for $m_{\rm PBH} \sim 30\ \mathrm{M_\odot}$, $f_{\rm PBH} \sim 10^{-4} - 0.1$:

$$P_{\rm DM}(k,a) = T_{\rm ad}^2(a)P_{\rm ad}^0(k) + f_{\rm PBH}^2 T_{\rm iso}^2(a)P_{\rm iso}^0(k) + P_{\rm corr}(k,a)\ , \tag{11.6}$$

$$P_{\rm corr}(k,a) = f_{\rm PBH}T_{\rm iso}(a)\Delta_{\rm Poisson}^2(k)T_{\rm ad}^2(a)P_{\rm ad}^0(k),\quad k_{\rm eq} < k < 3k_\star\ . \tag{11.7}$$

Here the correlation term (Eq. 11.7) is restricted to large scales ($k < 3k_\star$), because the small-scale behavior should be dominated by the Poisson noise in the PBH distribution. In fact, the effect of correlation is expected to be only important for $f_{\rm PBH} \lesssim 10^{-2}$ at intermediate scales where PBHs are clustered by adiabatic perturbations but still have significant Poisson noise in their distribution, and the gravity from PBHs do affect clustering of PDM but is not strong enough to seed or completely disrupt structures. More work needs to be done to validate this approach given the numerical difficulties in cosmological simulations (see the subsection below). Here we simply use the linear power spectrum with the heuristic correlation term (Eq. 11.6) to derive *optimistic* estimates of the halo abundances under the effects of PBHs.

Considering the self-similar nature of perturbation growth in the linear regime, it is common to evaluate the (extrapolated) linear power spectrum at $a = 1$. To do so, we use the `python` package COLOSSUS [37] to derive the adiabatic term $T_{\rm ad}^2(a=1)P_{\rm ad}^0(k)$ based on *Planck* observations [4], and calculate $T_{\rm iso}^0(a=1)$ with $T_{\rm iso}(a) = D(a)$ considering only the growing modes of isocurvature perturbations, where $D(a)$ is the linear growth rate given by the approximation formula from [63]:

$$D(a) \approx \left(1 + \frac{3\gamma a}{2a_- a_{\rm eq}}\right)^{a_-}\ ,\quad \gamma = \frac{\Omega_m - \Omega_b}{\Omega_m}\ ,\quad a_- = \frac{1}{4}\left(\sqrt{1+24\gamma} - 1\right)\ . \tag{11.8}$$

As examples, in Fig. 11.1 we show the linear power spectra at $a = 1$ for the reference standard ΛCDM case ($f_{\rm PBH} = 0$, long-dashed) and 4 PBH models with $m_{\rm PBH} = 33\ \mathrm{M_\odot}$, $f_{\rm PBH} = 10^{-4}$ (solid), 10^{-3} (dashed), 0.01 (dashed-dotted) and 0.1 (dotted). For each PBH model, two versions of the linear power spectrum with and without the heuristic correlation term are plotted with the thin and thick curves, respectively. It is evident that the Poisson noise term arising from PBHs overwhelms the intrinsic adiabatic term at small scales ($k \to \infty$), implying that structure formation will be accelerated by PBHs, especially at small scales and at early epochs.

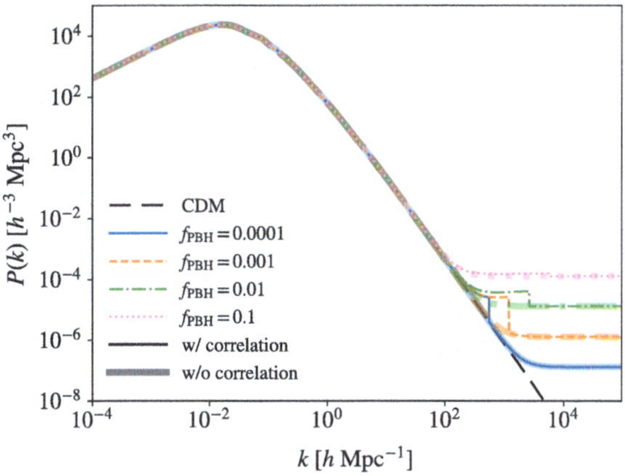

Fig. 11.1 Linear power spectra of DM density fluctuations (extrapolated to $a = 1$) for the PBH models with $m_{\rm PBH} = 33\ \rm M_\odot$, $f_{\rm PBH} = 10^{-4}$ (solid), 10^{-3} (dashed), 0.01 (dashed-dotted) and 0.1 (dotted), in comparison with the reference power spectrum for the standard ΛCDM case (long-dashed) provided by the `python` package COLOSSUS [37] based on *Planck* observations [4], where two versions of the linear power spectrum with and without the heuristic correlation term (Eq. 11.7) are shown for each PBH model with the thin and thick curves, respectively

11.2.2 Initial Conditions for Cosmological Simulations with PBHs

The linear power spectrum derived in the last subsection (Eqs. 11.4–11.7) enables us to study the 'Poisson' effect of PBHs on structure formation in a simple semi-analytical manner, as will be discussed in Sect. 11.4. However, this simple approach cannot capture the non-linear dynamics that shape the internal structure of and interactions between halos containing PBHs. One can model such nonlinear effects with cosmological (hydrodynamic) N-body simulations, in which gas can further be included in addition to DM to follow star and galaxy formation, as well as BH accretion and feedback (see, e.g., [10, 106, 119] for reviews).

In order to utilize such advanced numerical machinery for our purposes, we must generate initial conditions for cosmological N-body simulations that include the effects of PBHs properly. In particular, we need to know the initial positions and velocities of PBHs and simulation particles that embody (ensembles of) PDM. However, this is a challenging task, since state-of-the-art cosmological (hydrodynamic) simulation codes for star/galaxy formation (e.g., GIZMO [58], AREPO [110] and RAMSES [115]) are designed to start in the matter-dominated epoch with initial redshifts of a few hundred when small-scale isocurvature perturbations of PBHs are already highly nonlinear while existing initial condition generators (e.g., MUSIC [50]) only consider adiabatic perturbations that remain in the (quasi-)linear regime in this epoch. In this section, we outline a tentative numerical scheme for generating cosmological initial conditions with PBHs based on the exploratory work by [80]. We highlight the difficulties of modeling the nonlinear effects of PBHs in the initial conditions and discuss possible solutions that are yet to be improved in future studies.

11.2.2.1 Initial Distribution of PBHs

We start with an initial simulation box at $a_{\rm ini}$ produced for the standard ΛCDM cosmology that only contains PDM (simulation) particles whose initial density and velocity fields reflect the standard adiabatic perturbations embodied by $T_{\rm ad}(a)\delta^0_{\rm ad}$ in Eqs. 11.1–11.3. We then place PBHs into the box according to Eq. 11.3 in two steps:

1. We superimpose a Cartesian grid on the simulation box, whose cell size is chosen to have around one PBH in each cell on average, and calculate the local overdensity of DM in each cell i as δ_i. The cell size is meant to reflect the scale below which the discreteness of PBHs is important (see the next step).
2. For each cell i, we draw the number of PBHs contained in it, N_i, from a Poisson distribution $p_N = \lambda_i^N e^{-\lambda_i}/(N!)$ with $\lambda_i = (\delta_i + 1) f_{\rm PBH} M_{\rm cell}/m_{\rm PBH}$, where $M_{\rm cell}$ is the average mass of DM per cell. Then we place N_i PBHs randomly in the cell.
3. Once the positions of PBHs are known, we assign a velocity to each PBH as the velocity of its nearest-neighbor PDM particle, assuming that PBHs form with negligible velocities relative to PDM. Finally, the mass of each PDM particle is reduced by a fraction of $f_{\rm PBH}$ for mass conservation.

In this way, the fact that λ_i is proportional to δ_i+1 and PBHs have the same velocity field as PDM captures the large-scale adiabatic perturbations on PBHs, i.e., the term $T_{\rm ad}(a)\delta^0_{\rm ad}$ in Eq. 11.3, while the random draw of N_i and random spatial distribution of PBHs in each cell capture the isocurvature term $T_{\rm iso}^{\rm PBH}(a)\delta^0_{\rm iso}$.

11.2.2.2 Perturbations of PBHs on PDM and Gas

After addressing the PBH component, let us focus on the PDM component, for which adiabatic perturbations are already in place, and we only need to further implement the effects of PBHs on PDM denoted by the second term $[T_{\rm iso}(a) - 1] f_{\rm PBH}\delta^0_{\rm iso}$ in Eq. 11.2. To do so, we first follow the Zel'dovich approximation [92, 123] (which is commonly-used to produce initial conditions for the standard adiabatic perturbations) to associate the (co-moving) displacement $\boldsymbol{\psi}$ and (physical) velocity $\Delta\mathbf{v}$ fields produced by PBHs with their (co-moving) acceleration field $\nabla\phi_{\rm iso}(\mathbf{x})$:

$$\boldsymbol{\psi}(\mathbf{x}) = -\frac{2D_{\rm PBH}(a_{\rm ini})}{3\Omega_m H_0^2}\nabla\phi_{\rm iso}(\mathbf{x})\ ,\quad \Delta\mathbf{v}(\mathbf{x}) = \frac{a_{\rm ini}\dot{D}_{\rm PBH}(a_{\rm ini})}{D_{\rm PBH}(a_{\rm ini})}\boldsymbol{\psi}(\mathbf{x})\ , \tag{11.9}$$

$$\nabla\phi_{\rm iso}(\mathbf{x}) = 4\pi G m_{\rm PBH}\sum_i \frac{\mathbf{x}-\mathbf{x}_i}{|\mathbf{x}-\mathbf{x}_i|^3}\ , \tag{11.10}$$

given the (co-moving) coordinates of PBH particles[3] $\mathbf{x}_i$ at $a_{\rm ini}$. Here $D_{\rm PBH}(a_{\rm ini})$ is a growth factor, and we have $D_{\rm PBH}(a_{\rm ini}) = T_{\rm iso}(a_{\rm ini}) - 1 = D(a_{\rm ini}) - 1$ (considering only the growing modes) according to the linear theory (Eq. 11.2). Naively, with these equations, one can apply position and velocity changes to each PDM particle j with original coordinate $\mathbf{x}_j$ and velocity $\mathbf{v}_j$ as $\mathbf{x}_j = \mathbf{x}_j + \boldsymbol{\psi}(\mathbf{x}_j)$ and $\mathbf{v}_j = \mathbf{v}_j + \Delta\mathbf{v}(\mathbf{x}_j)$. This simple approach is only strictly valid in the linear regime when $|\boldsymbol{\psi}|$ is small, but perturbations to particles very close to PBHs are nonlinear (i.e., the 'seed' effect). Considering a single PBH for simplicity, a natural definition of the nonlinear scale is $d_{\rm nl} = [D_P BH(a_i ni) f_P BH/\bar{n}_{\rm PBH}]^{1/3}$, since when the distance between a PDM particle to the PBH r is below $d_{\rm nl}$, we will have $|\boldsymbol{\psi}| \gtrsim r$ according to Eqs. 11.9 and 11.10. If we start the simulation deep in the radiation-dominated regime ($a_{\rm ini} \ll a_{\rm eq}$) when $D_{\rm PBH} \approx 1.5\gamma a_{\rm ini}/a_{\rm eq}$ (see Eq. 11.8) is very small, this is not a serious problem as $d_{\rm nl}$ will usually be much smaller than the (initial) resolution of the simulation.

However, to study star/galaxy formation with existing simulation codes, we want to start the simulation in the matter-dominated epoch with $a_{\rm ini} \sim 0.001 - 0.01$, because it is computationally intensive to simulate baryonic physics at very high redshifts. In this case, the exact form of $D_{\rm PBH}(a_{\rm ini})$ can be different from what is expected/extrapolated from the linear theory, and $d_{\rm nl}$ can be larger than the simulation resolution or even larger than the structures that we are interested in. Here, we introduce two numerical treatments to address these issues targeting the 'Poisson' and 'seed' effects, respectively, which will be combined in the end to construct a tentative scheme that captures both effects.

First, if we are only concerned with formation of structures much larger than $d_{\rm nl}$ at late epochs under the collective influence of multiple PBHs (i.e., the 'Poisson' effect), we can insure linearity everywhere in the initial setup, by applying a truncation on the displacement with [63, 80]

$$\boldsymbol{\psi} = \min\left(1, d_{\rm PDM}/|\boldsymbol{\psi}|\right)\boldsymbol{\psi}\ , \tag{11.11}$$

given $d_{\rm PDM}$ as the average distance between PDM particles, which is a measure of the (initial) resolution of the simulation. Here, the truncation will only delay structure formation around PBHs at small scales (i.e., the 'seed' effect), while the perturbations from PBHs that affect large structures are still small in the initial condition and are therefore unaffected by the truncation.

Next, to capture the nonlinear structures around individual PBHs (i.e., the 'seed' effect), we often need stronger perturbations than allowed by the truncation (Eq. 11.11) in the region very close to a PBH (i.e., within $d_{\rm nl}$). Considering a PDM particle perturbed by a single PBH at a distance r away from it, the displacement $\boldsymbol{\psi}$ points from the PDM particle to the BH, and we have $|\boldsymbol{\psi}| \gtrsim r$ for $r \lesssim d_{\rm nl}$, which indicates that nonlinear dynamics (e.g., shell crossing) is already important in this region. The final outcome of such nonlinear dynamics is a virialized halo of mass

[3] Since we are using the positions of PBHs at $a_{\rm ini}$ (rather than $a \to 0$) that are already affected by large-scale adiabatic perturbations, the maximal correlation between adiabatic and isocurvature perturbations is automatically taken into account here.

$M_B \sim m_{PBH}a/a_{eq}$ [84] with an overdensity of $\Delta_{vir} = 18\pi^2 \sim 200$ according to the spherical collapse theory (see Chap. 5 in [92]), which is also what we want to produce in an idealized simulation of one single PBH. In light of this, we can apply the following truncation

$$\boldsymbol{\psi} = \min\left(1, f_{shrink}r/|\boldsymbol{\psi}|\right)\boldsymbol{\psi}\ , \quad f_{shrink} = 1 - \Delta_{vir}^{-1/3} \approx 0.822\ , \tag{11.12}$$

to create a uniform sphere of PDM around the PBH with an overdensity of Δ_{vir} and radius of $\sim d_{nl}$, mimicking the non-linear structure seeded by the PBH (see [67] for another approach to creating nonlinear structures around point-sources of gravity in the initial condition based on virialization). We find by test simulations of a single PBH starting from $a_{ini} \in [a_{eq}, 0.01]$ with the codes GIZMO [58] and AREPO [110] (whose gravity solvers are both based on GADGET- 2 [111]) that the initial uniform sphere will quickly collapse and virialize within a local Hubble time to become a halo[4] whose mass grows linearly with a as $M_h \sim (a/a_{ini})(4\pi/3)D_{PBH}(a_{ini})m_{PBH}$. This linear relation with a is consistent with $M_B \sim m_{PBH}a/a_{eq}$ predicted by the spherical collapse theory [84], but the normalization is off by a factor of a few if we derive the growth factor from the linear growth rate $D_{PBH}(a_{ini}) = D(a_{ini}) - 1$ (see Eq. 11.8). This implies that the strength of perturbations is overestimated by the (extrapolated) linear theory or there are numerical issues in our simulations. In fact, even if we do not consider the nonlinear structure initially and use the stronger truncation in Eq. 11.11, the halo mass will converge to $M_h \sim (a/a_{ini})(4\pi/3)D_{PBH}(a_{ini})m_{PBH}$ at late epochs. In light of this, we adopt $D_{PBH}(a_{ini}) \sim 3a_{ini}/(4\pi a_{eq})$ to ensure $M_h \sim M_B$. As $D_{PBH}(a) \propto a$, we have $\dot{D}_{PBH}(a)/D_{PBH}(a) = \dot{a}/a = H(a)$, which can be used to derive $\Delta\mathbf{v}$ (Eq. 11.9).

Finally, we combine the two prescriptions by dividing the perturbation of PBHs on a PDM particle into two pieces $\boldsymbol{\psi} = \boldsymbol{\psi}_{seed} + \boldsymbol{\psi}_{Poisson}$. Here $\boldsymbol{\psi}_{seed}$ arises from the 'seed' effect of the nearest PBH, while $\boldsymbol{\psi}_{Poisson}$ captures the collective 'Poisson' effect of all the other PBHs. In practice, we only consider the nearest PBH in Eq. 11.10 to calculate $\boldsymbol{\psi}_{seed}$, which is subject to the truncation based on halo overdensity (Eq. 11.12). For $\boldsymbol{\psi}_{Poisson}$, we add up the accelerations from all the other PBHs in Eq. 11.10 and use the truncation based on resolution (Eq. 11.11). When gas particles are included in the initial condition, we apply PBH perturbations to them following the same procedure developed for PDM (Eq. 11.9–11.12) but with a smaller growth factor $D_{PBH}(a_{ini}) \sim 3a_{ini}/(4\pi a_{rec})$, assuming that isocurvature perturbations in gas only starts to grow after photon-gas decoupling (i.e., recombination) at $a_{rec} = 1/(1+z_{rec}) \simeq 1/1100$.

We find by dark-matter-only test simulations with GIZMO [58] for $f_{PBH} \leq 0.1$ that the relation $M_h \sim M_B \sim m_{PBH}a/a_{eq}$ holds at early stages dominated by the 'seed' effect, while the mass distributions of halos from simulations are consistent with the

[4] We use the code ROCKSTAR [12] to identify halos in our simulations and calculate halo (virial) masses and radii.

predictions from the PS formalism[5] based on the linear power spectrum (Eq. 11.6) including the correlation term (Eq. 11.7) at late epochs dominated by the 'Poisson' effect for haloes with $M_{\rm h} \sim (0.5 - 20) m_{\rm PBH}/f_{\rm PBH}$.

11.3 PBH Accretion and Feedback

Beyond structure formation, the evolution of gas and radiation at Cosmic Dawn is also regulated by PBHs via BH accretion and feedback. Current studies focus on the radiation and heating produced by accreting PBHs. The former contributes to a variety of cosmic radiation backgrounds ([3, 20, 54, 86, 91, 104, 125, 126]) and affects the thermal and ionization histories of the intergalactic medium (IGM), which provides important evidence/constraints for PBHs (see Part V of this book). The latter alters the thermal and chemical properties of the interstellar medium (ISM, [28, 76, 83, 114]) that are crucial for star formation. In this section, we describe a simple model to calculate the heating rate from an accreting PBH as a function of environmental parameters, which will be used in Sect. 11.4 to evaluate the effects of PBHs on the first star-forming halos. The reader is referred to Chap. 13 (see also Chap. 24) for a broader and more detailed discussion on PBH accretion and feedback, in particular their impact on the evolution of PBHs themselves.

The mass accretion rate $\dot{m}_{\rm acc}$ of a BH with a mass of $m_{\rm BH}$ surrounded by a medium with gas density $\rho_{\rm gas}$ is given by the Bondi-Hoyle-Lyttleton formula

$$\dot{m}_{\rm acc} = 4\pi r_{\rm B}^2 \tilde{v} \rho_{\rm gas} = \frac{4\pi (G m_{\rm BH})^2 \rho_{\rm gas}}{\tilde{v}^3} , \tag{11.13}$$

where $r_{\rm B} = G m_{\rm BH}/\tilde{v}^2$ is the Bondi radius, and $\tilde{v} = \sqrt{v_{\rm rel}^2 + c_s^2}$ is the characteristic velocity. Here $v_{\rm rel}$ is the relative velocity between the BH and the medium, and c_s is the gas sound speed.

The bolometric luminosity $L_{\rm BH}$ of the accretion flow is then related to the accretion rate by $L_{\rm BH} = \epsilon_{\rm EM} \dot{m}_{\rm acc} c^2$, where $\epsilon_{\rm EM}$ is the radiation efficiency which depends on the detailed properties of the accretion flow and to the first order is a function of accretion rate and BH mass. For simplicity, we use the formula of $\epsilon_{\rm EM}$ in [96] to

[5] However, the abundance of smaller ($M_{\rm h} \lesssim 0.5 m_{\rm PBH}/f_{\rm PBH}$) halos in PBH simulations is significantly lower than that predicted by the PS formalism (by up to two orders of magnitude), even lower than the abundance in the standard ΛCDM case. The reason is that the formation and growth of small halos without PBHs are suppressed by the disruption and engulfment of nearby PBH-induced halos [124]. The PS formalism also overproduces large ($M_{\rm h} \gtrsim 20 m_{\rm PBH}/f_{\rm PBH}$) halos by up to a factor of 10 for $f_{\rm PBH} \gtrsim 0.01$. The assembly of such massive halos containing many PBHs are likely delayed by the tightly bound (sub)structures seeded by PBHs. This is consistent with the finding in [63] that the non-linear growth of large-scale fluctuations caused by PBH is slower than expected from the linear theory extrapolation. These discrepancies (see Fig. 14 in [80]) indicate that the PS formalism cannot fully capture the non-linear dynamics and non-Gaussianity caused by PBHs in structure formation.

capture the transition from optically thick and geometrically thin, radiatively efficient accretion disks, to optically thin, geometrically thick, radiatively inefficient advection dominated accretion flows:

$$\epsilon_{\rm EM} = \frac{\epsilon_0 A\eta}{1+A\eta}\ , \quad \eta \equiv \dot{m}_{\rm acc}/\dot{m}_{\rm Edd}\ , \tag{11.14}$$

where $A = 100$, $\epsilon_0 = 0.057$ is the radiative efficiency for non-rotating Schwarzschild BHs (assuming negligible spins of PBHs for simplicity), and η is the Eddington ratio defined by the Eddington accretion rate $\dot{m}_{\rm Edd}$, which for primordial gas made of $\sim 75\%$ hydrogen and $\sim 25\%$ helium (mass fractions) is

$$\dot{m}_{\rm Edd} \simeq 2.7\times 10^{-6}\ {\rm M_\odot\ yr^{-1}} \left(\frac{m_{\rm BH}}{100\ {\rm M_\odot}}\right)\left(\frac{\epsilon_0}{0.1}\right)^{-1}\ . \tag{11.15}$$

It is shown in [80] that Eq. 11.14 is consistent with the results of detailed calculations of the spectra of BH accretion flows based on [114].

Now, given Eqs. 11.13–11.15, we can derive the heating rate of an accreting BH as

$$P(m_{\rm BH}, \rho_{\rm gas}, \tilde{v}) = f_{\rm abs} f_h \epsilon_{\rm EM}(m_{\rm BH}, \dot{m}_{\rm acc})\dot{m}_{\rm acc}(m_{\rm BH}, \rho_{\rm gas}, \tilde{v})c^2\ , \tag{11.16}$$

where $f_{\rm abs}$ is the fraction of radiation energy (carried by ionizing photons) absorbed by the system that we are concerned with (e.g., primordial star-forming halos), and f_h is the fraction of energy deposited into the ISM as heat. We assume a constant absorption faction $f_{\rm abs} = 0.66$ for simplicity according to radiative transfer calculations for typical primordial star-forming clouds in [80], and adopt $f_h = 1/3$ following [83, 114].

11.4 Primordial Star Formation Under the Influence of Stellar-Mass PBHs

In this section, we explore the effects of stellar-mass PBHs on primordial star formation. We focus on the particular case with an initial PBH mass of $m_{\rm PBH} = 33\ {\rm M_\odot}$. We choose this mass because it is the location of the Gaussian peak in the best-fit Power-Law + Peak model of the mass distribution of BHs detected in GW observations of binary BH mergers by the LIGO-Virgo-KAGRA Scientific Collaboration [1]. Varying $m_{\rm PBH}$ by a factor of a few does not change our conclusions. We focus on PBH (initial) mass fractions in the range $f_{\rm PBH} \in [10^{-4} - 0.1]$ with $f_{\rm PBH} = 10^{-3}$ being the fiducial case, motivated by existing constraints (and their uncertainties) on PBH abundance [25] (see also Part V of the book). Here $f_{\rm PBH} \gtrsim 0.01$ is strongly disfavored for $m_{\rm PBH} \sim 30\ {\rm M}_\odot$ by observations from GWs, the 21-cm signal and CMB. We still consider the extreme regime of $f_{\rm PBH} \sim 0.01 - 0.1$ for the sake of theoretical exploration.

In the standard ΛCDM cosmology, the first generation of stars, the so-called Population III (Pop III), form typically in minihalos of $M_{\rm h} \sim 10^5 - 10^8\ {\rm M}_\odot$ at $z \sim 10 - 30$ where *primordial* (pure H/He) gas clouds collapse via cooling by molecular hydrogen H_2 in the absence of metals (see, e.g., [17, 74] for reviews). Here the minimum halo mass required to form Pop III stars is $M_{\rm mol} \sim 10^5 - 10^7\ {\rm M}_\odot$ [107, 116], which is determined by the criterion of efficient molecular cooling and is a function of redshift and environmental parameters (see below). The upper bound reflects the fact that massive halos ($M_{\rm h} \gtrsim 10^8\ {\rm M}_\odot$) tend to be metal-enriched and form second-generation stars [19]. The abundance of such massive halos is also much lower than that of low-mass Pop III star-forming halos at high z. Therefore, we focus on $M_{\rm h} < 10^8\ {\rm M}_\odot$ when considering Pop III star formation.

In our case with $m_{\rm PBH} = 33\ {\rm M}_\odot$, the abundance of halos that host Pop III stars in the standard ΛCDM cosmology is enhanced by PBHs due to their 'Poisson' effect (Sect. 11.2.1), while the 'seed' effect is expected to be unimportant because the mass of an isolated halo seeded by a PBH is $M_{\rm B} \lesssim 10^4\ {\rm M}_\odot$ at $z \gtrsim 10$, much smaller than $M_{\rm mol}$. On the other hand, PBHs accrete and heat the surrounding gas, such that stronger (molecular) cooling is required for the clouds to collapse and form stars, leading to higher $M_{\rm mol}$, which tends to reduce the abundance of Pop III star-forming halos. Here we first consider these two competing effects self-consistently in a semi-analytical analysis to evaluate the global impact of PBHs on the cosmic mass density of Pop III stars ever formed in the universe (Sect. 11.4.1). Then we look into the local effects of PBHs on the internal structure of Pop III star-forming halos and clouds using the results from the cosmological zoom-in simulations in [80] (Sect. 11.4.2). Finally, we briefly discuss the potential impact of PBHs on the formation of massive BH seeds in more massive halos ($M_{\rm h} \gtrsim 10^8\ {\rm M}_\odot$) in Sect. 11.4.3.

11.4.1 Cosmic Stellar Mass Density of Pop III Stars

To appreciate the acceleration of structure formation by the 'Poisson' effect of PBHs, we calculate the halo mass function $dn_{\rm h}/dM_{\rm h}$ (co-moving number density of halos per unit halo mass) with the PS formalism (see Chap. 7 of [92] for details) using the linear power spectrum with enhanced small-scale power by PBHs (Eqs. 11.4–11.7). We adopt the commonly-used top-hat window function to derive the density variance at a given mass scale (see Sec. 6.1.3 in [92]) and consider the correction for ellipsoidal dynamics (Eq. 7.67 in [92]) that is required for good agreements between the predictions of the PS formalism and N-body simulations. The results are shown in Fig. 11.2 for $f_{\rm PBH} = 10^{-4}$ (solid), 10^{-3} (dashed), 0.01 (dashed-dotted) and 0.1 (dotted), compared with the standard ΛCDM results (long-dashed) at $z = 30$ (left) and $z = 15$ (right). Similar to Fig. 11.1, the halo mass functions for two versions of the linear power spectrum with and without the heuristic correlation term (Eq. 11.7) are plotted with the thin and thick curves (as optimistic and conservative estimates) for each PBH model, respectively. Clearly, compared with the standard ΛCDM case (long-dashed), the abundances of halos are boosted by PBHs by up to ~ 3 orders of magnitude for $M_{\rm h} \sim 10^5\ \rm M_\odot$ at $z = 30$ in the extreme model with $f_{\rm PBH} = 0.1$. The effect is stronger at higher redshift. Taking into account the correlation term (Eq. 11.7) in the linear power spectrum (thick curves) significantly increases the halo abundance at $M_{\rm h} \gtrsim M_{\rm B} \sim$ a few thousand $\rm M_\odot$ for $f_{\rm PBH} \lesssim 0.01$.

! Attention

The Press-Schechter formalism assumes that density fluctuations are Gaussian. However, in the presence of PBHs, the density field is non-Gaussian, especially at small scales dominated by isocurvature perturbations from the Poisson distributed PBHs [30, 36, 63]. Our calculation of the halo mass function with the Press-Schechter formalism here is an approximation commonly used in previous studies [11, 20, 70, 71]. Cosmological simulations with proper initial conditions are required to accurately derive the halo mass functions in PBH cosmologies (see Sect. 11.2.2).

To interpret these results in the context of Pop III star formation, in Fig. 11.2 we also plot the halo mass thresholds of Pop III star formation $M_{\rm mol}$ in the standard ΛCDM cosmology for 2 situations: First (dot-dash-dotted), in the idealized case without any environmental effects (e.g., halo assembly history, metal enrichment, radiative feedback, cosmic rays and streaming motion between DM and gas), we have

$$M_{\rm mol} \sim 1.54 \times 10^5\ \rm M_\odot[(1+z)/31]^{-2.074} \tag{11.17}$$

according to the simple analysis of primordial chemistry and cooling in [116]. Second (densely-dashed), we consider the environmental effects of Lyman-Werner (LW) radiation and streaming motion[6] (SM) between DM and baryons with the fit formula of $M_{\rm mol}(J_{21}, v_{\rm bc})$ derived from the cosmological simulations in [107] (see their Eqs. 11–13):

$$\begin{aligned}
&\log(M_{\rm mol}/\rm M_\odot) = \log(M_0/\rm M_\odot) + b \times v_{\rm bc}/\sigma_{\rm rms}\ ,\\
&\log(M_0/\rm M_\odot) = 5.562 \times (1 + 0.279\sqrt{J_{21}})\ , \quad b = 0.614 \times (1 - 0.56\sqrt{J_{21}})\ ,
\end{aligned} \tag{11.18}$$

[6] With PBHs, the effects of gas-DM streaming are weaker than those in the standard CDM case due to structures induced by PBHs, which can facilitate gas collapse in star-forming halos for $f_{\rm PBH} m_{\rm PBH} \gtrsim 3\ \rm M_\odot$ [70]. Here, we ignore such effects for simplicity, as we are mostly concerned with PBH models with $f_{\rm PBH} m_{\rm PBH} \lesssim 3\ \rm M_\odot$, although similar trends are also observed in cosmological zoom-in simulations for $f_{\rm PBH} = 0.001$ and $m_{\rm PBH} = 33\ \rm M_\odot$ [80] (see also [11]).

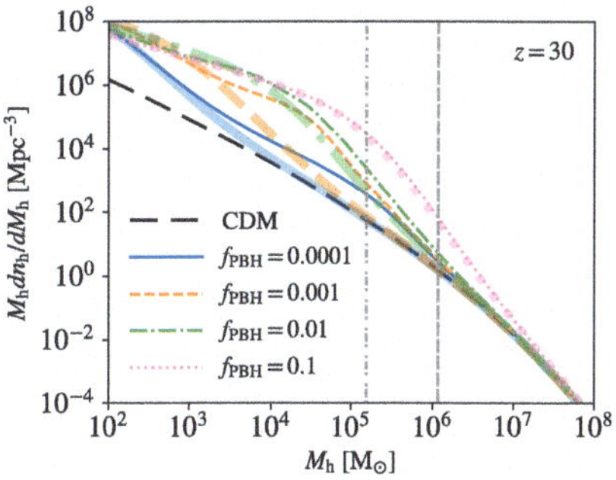

Fig. 11.2 Halo mass functions for the PBH models with $m_{\rm PBH} = 33\ {\rm M}_\odot$, $f_{\rm PBH} = 10^{-4}$ (solid), 10^{-3} (dashed), 0.01 (dashed-dotted) and 0.1 (dotted), compared with the ΛCDM reference mass function (long-dashed), at $z = 30$ (*left*) and $z = 15$ (*right*), where two versions of the halo mass function with and without the heuristic correlation term in the linear power spectrum (Eq. 11.7) are shown for each PBH model with the thin and thick curves, and the dot-dash-dotted and densely dashed vertical lines label the mass thresholds of Pop III star formation via molecular cooling in the standard ΛCDM case with and without the environmental effects of LWB and SM [107, 116], respectively

where J_{21} is the intensity of LW background (LWB) as a function of redshift in units of 10^{-21} erg s^{-1} cm^{-2} Hz^{-1} sr^{-2}, and $v_{\rm bc}/\sigma_{\rm rms}$ is the relative velocity between DM and gas in units of the SM root-mean-square velocity $\sigma_{\rm rms}$. We adopt $J_{21} = 10^{2-z/5}$ from [48] and $v_{\rm bc}/\sigma_{\rm rms} = 0.8$ as the most common SM velocity [107] to capture the typical levels of LWB and SM. Such environmental effects increase $M_{\rm mol}$ by a factor of ~ 8 (5) at $z = 30$ (15) compared with the idealized case (Eq. 11.17) because the LWB dissociates H_2 and the SM engenders additional turbulence energy in the primordial cloud, thus delaying collapse. We find that for $f_{\rm PBH} \sim 10^{-4} - 0.01$ (allowed by observational constraints) the abundance of halos above $M_{\rm mol}$ is significantly enhanced only in the idealized case (Eq. 11.17), and only when the correlation term is considered for $f_{\rm PBH} \sim 10^{-4} - 10^{-3}$. The effects of PBHs are much weaker for more massive halos above the mass threshold subject to the environmental effects of LWB and SM (Eq. 11.18), such that only the extreme model with $f_{\rm PBH} = 0.1$ exhibits significant differences.

To quantify the effects of PBHs, let us consider the impact of PBH accretion heating on the mass threshold of Pop III star formation. For simplicity, we focus on the idealized case without environmental effects. To the first order, the mass threshold $M_{\rm mol}$ corresponds to the minimum halo mass that satisfies the Rees-Ostriker-Silk cooling criterion [100, 108] $t_{\rm cool} \leq t_{\rm ff}$, where $t_{\rm cool}$ is the cooling timescale, and $t_{\rm ff}$ is the free-fall timescale:

$$t_{\rm cool} = \frac{(3/2)k_B T_{\rm vir}}{\Lambda(T_{\rm vir}, n_{\rm H})x_{\rm H_2} - \Gamma(m_{\rm PBH}, f_{\rm PBH}, \rho_{\rm gas}, \tilde{v})}\,, \quad (11.19)$$

$$t_{\rm ff} = \sqrt{\frac{3\pi}{32G\rho_{\rm gas}}}\,. \quad (11.20)$$

Here $T_{\rm vir}$ is the virial temperature of the halo (see Eq. 2 in [116]), $x_{\rm H_2}$ is the maximum abundance of H_2 (i.e., ratio of number densities between H_2 and hydrogen nuclei), $\rho_{\rm gas} = X^{-1} m_{\rm H} n_{\rm H} = \Delta_{\rm gas}\bar{\rho}_{\rm gas}$ is the characteristic (physical) density of gas defined by the overdensity parameter Δ and the cosmic average gas density $\bar{\rho}_{\rm gas}$, $n_{\rm H}$ is the number density of hydrogen nuclei related to $\rho_{\rm gas}$ by the primordial hydrogen mass fraction $X \simeq 0.76$ and proton mass $m_{\rm H}$, Λ is the cooling rate per H_2, and Γ is the heating rate per baryon from PBH accretion.

To evaluate the cooling rate, we adopt the approximation formulae of $x_{\rm H_2}$ and Λ in [116] for the standard ΛCDM case, and further consider the enhancement of H_2 formation due to ionization of gas by accreting PBHs found in cosmological simulations [80] (captured by the multiplication term with $f_{\rm PBH}$ in Eq. 11.21):

$$x_{\rm H2} \simeq 3.5 \times 10^{-4} \left(\frac{T_{\rm vir}}{1000\ {\rm K}}\right)^{1.52} \times \max[1, 3(f_{\rm PBH}/0.1)^{0.15}] \,, \tag{11.21}$$

$$\Lambda(T_{\rm vir}, n_{\rm H}) \simeq 10^{-31.6}\ {\rm erg\ s^{-1}} \left(\frac{T}{100\ {\rm K}}\right)^{3.4} \left(\frac{n_{\rm H}}{10^{-4}\ {\rm cm^{-3}}}\right) . \tag{11.22}$$

The heating rate per baryon is related to the heating rate per BH (Eq. 11.16) by

$$\Gamma = \frac{f_{\rm PBH}\mu m_{\rm H}(\Omega_m - \Omega_b)}{m_{\rm PBH}\Omega_b} P(m_{\rm PBH}, \rho_{\rm gas}, \tilde{v}) \,, \tag{11.23}$$

where $\mu \simeq 1.22$ is the mean molecular weight of primordial gas, and we estimate the characteristic velocity between PBHs and gas as

$$\tilde{v} \sim \sqrt{\frac{GM_{\rm h}}{R_{\rm vir}}} \sim 5.4\ {\rm km\ s^{-1}} \left(\frac{M_{\rm h}}{10^6\ {\rm M_\odot}}\right)^{\frac{1}{3}} \left(\frac{21}{1+z}\right)^{\frac{1}{2}} . \tag{11.24}$$

Here for simplicity, we use $m_{\rm BH} = m_{\rm PBH}$ to evaluate the heating rate per BH (Eq. 11.16) considering that mass growth by accretion is minor ($\lesssim 10\%$) at $z \gtrsim 10$ for our stellar-mass PBHs with an initial mass of $m_{\rm PBH} = 33\ {\rm M_\odot}$, as shown in [80] (see also Chap. 13 and [44]). Finally, we set the overdensity parameter as $\Delta_{\rm gas} = 1300$ to reproduce the results from [116] for the ΛCDM case (Eq. 11.17) with $f_{\rm PBH} = 0$.

The results for $f_{\rm PBH} = 10^{-4}$ (solid), 10^{-3} (dashed), 0.01 (dashed-dotted), 0.1 (dotted), 10^{-5} (long-dashed) and 10^{-6} (dot-dash-dotted) are shown in Fig. 11.3, compared with the standard ΛCDM case (thick solid). As expected, $M_{\rm mol}$ increases with $f_{\rm PBH}$, by up to a factor of $\sim 3-5$ at $z \sim 20-40$ for $f_{\rm PBH} = 0.1$, compared with the ΛCDM case (thick solid). The effect of PBHs is stronger at higher z and the mass threshold converges to the ΛCDM value at low z. These results are generally consistent (within a factor of 2) with the cosmological simulations in [80]. Interestingly, we find that even for the extreme model with $f_{\rm PBH} = 0.1$, the PBH-enhanced halo mass threshold without environmental effects remains below that with environmental effects of LWB and SM (Eq. 11.18). This indicates that the heating effect of PBHs on $M_{\rm mol}$ is likely unimportant in most regions of the universe as it is overwhelmed by

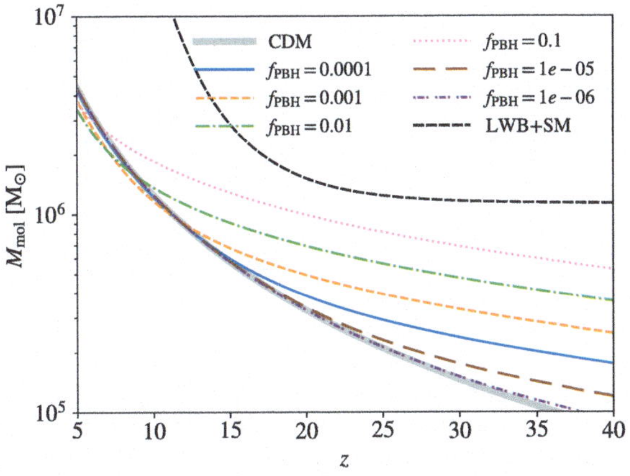

Fig. 11.3 Halo mass thresholds of Pop III star formation via molecular cooling as functions of redshift without environmental effects, for the PBH models with $m_{\rm PBH} = 33\ {\rm M}_\odot$, $f_{\rm PBH} = 10^{-4}$ (solid), 10^{-3} (dashed), 0.01 (dashed-dotted), 0.1 (dotted), 10^{-5} (long-dashed) and 10^{-6} (dot-dash-dotted), compared with the standard ΛCDM case (thick solid, Eq. 11.17), and the halo mass threshold under the environmental effects of LWB and SM (densely dashed, Eq. 11.18)

the effects of LWB and SM. In the above analysis, we have ignored PDM substructures around PBHs which can reduce the central concentration of PDM and delay cloud collapse (see Sect. 11.4.2), leading to higher $M_{\rm mol}$. This (dynamical) effect is only important for $f_{\rm PBH} \gtrsim 0.01$ according to [80], and remains minor compared with the environmental effects of LWB and SM (Eq. 11.18). Here, we have also ignored the local LW radiation from accreting PBHs, which can potentially reduce $x_{\rm H_2}$ and increase $M_{\rm mol}$. This effect is sensitive to the spectra of accreting PBHs and gas densities around PBHs and may even be stronger than the PBH heating and environmental effects (see Sect. 11.4.3). Besides, the X-rays from accreting PBHs also regulate the thermochemistry of primordial gas via ionization and heating, which can either facilitate or delay Pop III star formation (e.g., [61, 97, 103]). Future studies should employ (radiative) hydrodynamic simulations to thoroughly assess the impact of local LW and X-ray feedback from PBHs on first star formation.

Finally, we combine the halo mass function $dn_{\rm h}/dM_{\rm h}$ and mass threshold $M_{\rm mol}$ of Pop III star formation (as functions of $f_{\rm PBH}$ and z) to estimate the cosmic mass density of Pop III stars ever formed. For simplicity, we assume that gas is converted into stars at a constant star formation efficiency $\epsilon_{\star,\rm III}$ in Pop III star-forming halos with masses $M_{\rm h} \in [M_{\rm mol}, 10^8\ {\rm M}_\odot]$, independent of PBH parameters (which is justified in Sect. 11.4.2), so that the cosmic (co-moving) mass density of Pop III stars (ever formed) can be written as

$$\rho_{\star,\rm III} = \epsilon_{\star,\rm III}\frac{\Omega_b}{\Omega_m}\int_{M_{\rm mol}}^{10^8\ {\rm M}_\odot}\frac{dn_{\rm h}}{dM_{\rm h}}M_{\rm h}dM_{\rm h}\ . \tag{11.25}$$

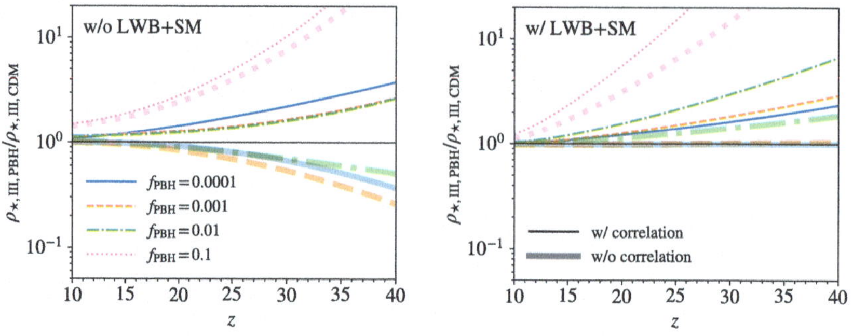

Fig. 11.4 Ratios of the mass densities of Pop III stars ever formed in halos of $M_{\mathrm{h}} \in [M_{\mathrm{mol}}, 10^8\ \mathrm{M}_{\odot}]$ between the PBH and standard ΛCDM cases, for the PBH models with $m_{\mathrm{PBH}} = 33\ \mathrm{M}_{\odot}$, $f_{\mathrm{PBH}} = 10^{-4}$ (solid), 10^{-3} (dashed), 0.01 (dashed-dotted) and 0.1 (dotted), where the *left panel* shows the results for the halo mass threshold M_{mol} without the environmental effects (see Fig. 11.3), while the *right panel* shows the results with the typical environmental effects of LWB and SM (Eq. 11.18), and the results with and without the heuristic correlation term (Eq. 11.7) in the calculation of the halo mass function are shown for each PBH model with the thin and thick curves, respectively

Since we are concerned with the difference made by PBHs, we focus on the ratio of $\rho_{\star,\mathrm{III}}$ between the PBH and standard ΛCDM models, i.e., $\rho_{\star,\mathrm{III,PBH}}/\rho_{\star,\mathrm{III,CDM}}$.

Figure 11.4 displays the results for $f_{\mathrm{PBH}} = 10^{-4}$ (solid), 10^{-3} (dashed), 0.01 (dashed-dotted) and 0.1 (dotted), where we consider 2 models of $dn_{\mathrm{h}}/dM_{\mathrm{h}}$ with and without the correlation term in the linear power spectrum (Eq. 11.7), denoted by the thin and thick curves, respectively. On the left panel, we show the results for the idealized case of M_{mol} without environmental effects (i.e., considering only the PBH heating effect on M_{mol}). It turns out that for $f_{\mathrm{PBH}} \sim 10^{-4} - 0.01$, $\rho_{\star,\mathrm{III}}$ is reduced by PBHs for the conservative estimate of $dn_{\mathrm{h}}/dM_{\mathrm{h}}$ without the correlation term in the linear power spectrum (Eq. 11.7) by up to a factor ~ 2 at $z \lesssim 30$ (when most Pop III stars form). On the contrary, given the optimistic estimate of $dn_{\mathrm{h}}/dM_{\mathrm{h}}$ with the correlation term, $\rho_{\star,\mathrm{III}}$ is increased by up to a factor ~ 2. In both cases, the effects of PBHs are stronger at higher z and show weak, non-monotonic dependence on f_{PBH}. On the right panel, we show the results for the mass threshold M_{mol} under the influence of typical environmental effects of LWB and SM which overwhelm the heating effect of PBHs. Here the effects of PBH are negligible for $f_{\mathrm{PBH}} \lesssim 10^{-3}$ without the correlation term, and $\rho_{\star,\mathrm{III}}$ increases monotonically with f_{PBH}, reaching up to ~ 3 times the ΛCDM value at $z \lesssim 30$ for $f_{\mathrm{PBH}} \sim 10^{-4} - 0.01$ when the optimistic estimate of $dn_{\mathrm{h}}/dM_{\mathrm{h}}$ is adopted. For all the 4 combinations of M_{mol} and $dn_{\mathrm{h}}/dM_{\mathrm{h}}$ models, $\rho_{\star,\mathrm{III}}$ is significantly enhanced by more than one order of magnitude at $z \gtrsim 30$ in the extreme case of $f_{\mathrm{PBH}} = 0.1$. On the other hand, for $f_{\mathrm{PBH}} \sim 10^{-4} - 0.01$ allowed by existing constraints, the variation of $\rho_{\star,\mathrm{III}}$ caused by PBHs is within a factor of 3, much smaller than the large (~ 1.5 dex) uncertainties in current theoretical predictions on the cosmic Pop III star formation history in the standard ΛCDM cosmology (see, e.g., Fig. 5 in [79]). Therefore, we conclude that the global impact of stellar-mass PBHs allowed by existing observational constraints on Pop III star formation is minor.

11.4.2 Internal Structure of Pop III Star-Forming Halos and Clouds

Although it is shown by the semi-analytical analysis above that the cosmic mass density of Pop III stars is not changed greatly by stellar-mass PBHs with $f_{\rm PBH} \lesssim 0.01$, such PBHs may still have interesting local effects on the internal structure of star-forming halos and clouds. In this subsection, we use the results from the high-resolution cosmological zoom-in simulations in [80] as examples to demonstrate the local effects of PBHs at the halo and cloud scales. In particular, we consider the simulations for $f_{\rm PBH} = 0$ (CDM), 0.001, 0.01 and 0.1 (with $m_{\rm PBH} = 33\ {\rm M}_{\odot}$) in their Case A zoom-in region with a mass resolution for PDM (gas) of ~ 2 (0.4) $\rm M_{\odot}$. The runs start at $a_{\rm ini} = 1/301$ from cosmological initial conditions generated by a method similar to that[7] described in Sect. 11.2.2 with the resolution-based truncation (Eq. 11.11), suitable for the 'Poisson' effect that is relevant for Pop III star formation. The initial adiabatic density perturbations in the initial conditions are enhanced by adopting $\sigma_8 = 2$ rather than the cosmic mean $\sigma_8 = 0.8159$ [4] to mimic the typical overdense regions where the first stars form at $z \sim 30$. These simulations are also equipped with a non-equilibrium primordial chemistry and cooling network and sub-grid models for BH accretion, dynamical friction and feedback (which is consistent with the model in Sect. 11.3) to self-consistently capture the gravitational and heating effects of PBHs on the nonlinear dynamics of halo assembly and cloud collapse as well as the thermal and chemical evolution of primordial gas. The reader is referred to Sec. 2 of [80] for details on the numerical methods and setup of the simulations.[8]

[7] Since the gravity of PBHs outside the zoom-in region is not considered, the enhancement of PBH perturbations (at large scales) by PBH clustering and mode correlation can be overestimated if all PBHs in the zoom-in region are used to evaluate the acceleration field (Eq. 11.10). Therefore, [80] restrict PBH perturbations to small scales by using at most the 64 nearest PBH particles within $2d_{\rm PBH}$ around the PDM/gas particle when evaluating Eq. 11.10, where $d_{\rm PBH}$ is the average separation between PBHs. Besides, $D_{\rm PBH}(a_{\rm ini}) = T_{\rm iso}(a_{\rm ini}) - 1 = D(a_{\rm ini}) - 1$ is used in [80], without the correction motivated by the 'seed' effect (see Sect. 11.2.2), so that the PBH perturbations in their simulations may be overestimated. Nevertheless, it is verified in [80] (see their Appendix A) that the conclusions made in this section on the internal structure are still valid despite such complications.

[8] In fact, when environmental effects are ignored, first star formation is delayed for $f_{\rm PBH} \gtrsim 0.01$ in the zoom-in simulations from [80], which seems in conflict with the prediction by Eq. 11.25 using the optimistic estimate of the halo mass function from the PS formalism with the correction term (Eq. 11.7). On the other hand, star formation is accelerated in the zoom-in simulations for $f_{\rm PBH} \lesssim 0.01$, inconsistent with the semi-analytical prediction without the correction term. It is difficult to interpret these discrepancies as the trends obtained from zoom-in simulations are sensitive to the specific halo assembly histories in the small, biased zoom-in regions (i.e., cosmic variance) and cannot be naively generalized. Cosmological simulations in larger volumes are required to validate/calibrate the PS formalism.

We focus on the snapshot when the maximum hydrogen number density reaches $10^5\ \mathrm{cm}^{-3}$, which marks the onset of Pop III star formation. At this moment, a dense ($n_{\mathrm{H}} \gtrsim 10^4\ \mathrm{cm}^{-3}$) cold ($T \lesssim 10^3$ K) gas clump of $\sim 10^3\ \mathrm{M}_{\odot}$ has formed by run-away collapse with efficient molecular cooling in the central parsec of the most massive halo in each simulation, which is the typical environment for Pop III star formation. Such a dense gas clump is expected to further collapse and form Pop III stars in a very short (free-fall) timescale of $t_{\mathrm{ff}} \sim 0.5$ Myr. The simulations stop at this moment because higher resolution is required to follow the subsequent collapse. The fact that such typical Pop III star-forming gas clumps occur in all simulations with different PBH parameters implies that the standard picture of Pop III star formation at sub-parsec scales is not changed by the stellar-mass PBHs considered here. Nevertheless, PBHs can have noticeable effects in the outer regions of the halo via their heating effect and the PDM substructures around them, as shown in Fig. 11.5.

To better appreciate this point, we plot the density profiles of the halos hosting the Pop III star-forming clumps in Fig. 11.6 for $f_{\mathrm{PBH}} = 0$ (CDM, black), 0.001 (orange) and 0.01 (green) and 0.1 (pink). Here the halo/cloud center is defined as the gas density peak. The left panel shows the hydrogen number density profiles $n_{\mathrm{H}}(r)$, and the right panel shows the enclosed mass profiles $M_{\mathrm{enc}}(r)$ of all components: gas (solid), PDM (dashed) and PBHs (dotted). We find that indeed the gas density profiles are almost identical in the central region ($\lesssim 1$ pc) occupied by the star-forming gas clump for all simulations. As f_{PBH} increases, the gas distribution in the outer part of the halo (1 pc $\lesssim r \lesssim R_{\mathrm{vir}}$) becomes more clumpy, and the density is higher for $r \sim 1 - 20$ pc, showing that gas is pilled up in this region. This is likely caused by the PBH heating and PDM substructures around PBHs that can slow down cloud collapse. Nevertheless, as $r \to R_{\mathrm{vir}}$, the gas density profile converges to the power-law form, $n_{\mathrm{H}}(r) \propto r^{-2}$, of an isothermal distribution, which is a typical feature of Pop III star-forming halos (see, e.g., [56]). The density profile of PDM also becomes shallower with higher f_{PBH}, especially in the central region ($r \lesssim 10$ pc) where the

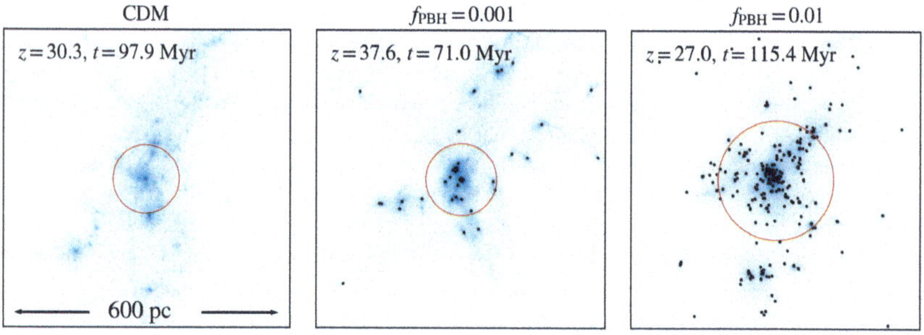

Fig. 11.5 Projected distributions of PDM and PBHs at the onset of star formation in a data slice with a (physical) extent of 600 pc and a thickness of 300 pc, for the reference ΛCDM simulation (*left*) and PBH simulations with $m_{\mathrm{PBH}} = 33\ \mathrm{M}_{\odot}$, $f_{\mathrm{PBH}} = 0.001$ (*middle*) and 0.01 (*right*), in the Case A zoom-in region from [80], where PBHs are plotted with black dots, and the circles indicate the halo virial radii

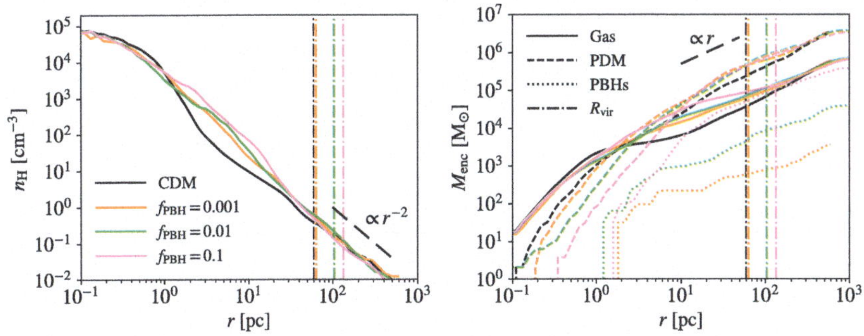

Fig. 11.6 Internal structure of the star-forming halo at the onset of star formation in terms of the hydrogen number density profile (*left*) and enclosed mass profiles (*right*) for gas (solid), PDM (dashed) and PBHs (dotted), for the reference ΛCDM (CDM) simulation (black) and PBH simulations with $m_{PBH} = 33\ M_\odot$, $f_{PBH} = 0.001$ (orange) and 0.01 (green) and 0.1 (pink), in the Case A zoom-in region from [80], where the halo virial radii are shown with the vertical dashed-dotted lines

PDM density decreases with f_{PBH}. This indicates that it is more difficult for PDM to concentrate at the center when there are more PBHs, consistent with the results in [63]. The reason is that substructures of PDM around PBHs are more tightly bound and therefore more difficult to destroy/strip during virialization compared with their BH-less counterparts in the ΛCDM case (see Fig. 11.5). For PBHs, we have $M_{enc}(r) \propto r$ at $r \gtrsim 10$ pc, corresponding to an isothermal distribution independent of f_{PBH}, while the density profile is shallower in the central region with increasing f_{PBH}. In fact, only up to a few PBHs can reach $r \lesssim 2$ pc regardless of the overall PBH abundance, and there are no PBHs at $r < 1$ pc in all the 19 simulations that include PBH heating run by [80]. This interesting outcome can be interpreted with the survivor bias for gas condensation. Since we define the halo center as the gas density peak, PBHs cannot be too close to the center, because otherwise, gas will not be able to cool and condense due to the strong heating of nearby PBHs.

Next, we evaluate the impact of PBHs on the thermal and chemical properties of gas in Fig. 11.7, where we show the temperature-density behavior (left) and abundances of H_2 (dashed) and free electrons e^- (dotted) as functions of density (right) for the same ΛCDM and PBH models considered in Fig. 11.6. The gas temperature generally increases with f_{PBH} for $f_{PBH} \gtrsim 0.001$ at halo outskirts with $n_H \lesssim 1\ \mathrm{cm^{-3}}$ where cooling is relatively inefficient. The reason is that the halo mass at the onset of star formation increases with f_{PBH} due to the heating of PBHs, which results in stronger compression heating by halo assembly with a higher virial temperature. The H_2 abundance also increases with f_{PBH} by a factor almost independent of density at $n_H \gtrsim 10\ \mathrm{cm^{-3}}$, as the heating and ionization from accreting PBHs facilitate H_2 formation. This trend can be described by a simple power-law fit in Eq. 11.21. Beyond these clear trends with f_{PBH}, we find large temporal variations of temperature and electron abundance as functions of density in the central region with $n_H \gtrsim 10^2\ \mathrm{cm^{-3}}$ (i.e., $r \lesssim 10$ pc, see the left panel of Fig. 11.6), although they are always above the

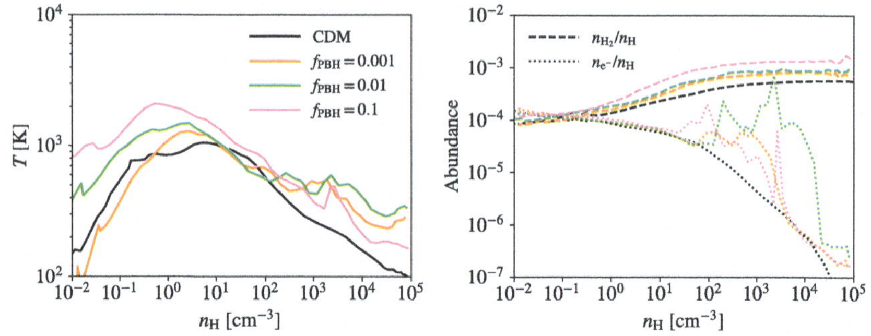

Fig. 11.7 Thermal and chemical properties of the star-forming halo at the onset of star formation in terms of the temperature-density diagram (*left*) and abundances of H_2 (dashed) and e^- (dotted) as functions of density (*right*), for the reference ΛCDM (CDM) simulation (black) and PBH simulations with $m_{PBH} = 33\ M_\odot$, $f_{PBH} = 0.001$ (orange) and 0.01 (green) and 0.1 (pink), in the Case A zoom-in region from [80]

standard ΛCDM evolution track. This shows that gas properties are sensitive to the detailed stochastic gaseous environments around a small number of PBHs in the central region.

Finally, let us consider the fate of the dense clumps formed at the end of the simulations under the influence of PBHs. These clumps are expected to collapse and form Pop III stars within a timescale of $t_{ff} \sim 0.5$ Myr, while nearby PBHs will sink to the clump center over the dynamical friction timescale t_{DF}. If a PBH can reach the densest central region ($r \lesssim 10^{-3}$ pc) with $n_H \sim 10^9\ cm^{-3}$ before completion of star formation (and evaporation of the cloud by stellar feedback), it can undergo super-Eddington accretion to grow significantly (i.e., by a factor of ~ 10 within ~ 0.01 Myr) and even suppress star formation by evaporating the cloud with accretion feedback (see Sec. 4.3 in [80]). We can compare the dynamical friction timescale t_{DF} of PBHs with the cloud collapse timescale t_{ff} to discern whether PBHs can reach the center to significantly affect the star formation process. For a BH of mass m_{BH} with initial apocentric distance r and velocity v_{BH} with respect to the center, the dynamical friction timescale can be estimated by [14]

$$\frac{t_{DF}}{\text{Myr}} \simeq \frac{340}{\ln \Lambda} \left(\frac{r}{3\ \text{pc}}\right)^2 \left(\frac{v_{BH}}{10\ \text{km s}}^{-1}\right) \left(\frac{m_{BH}}{100\ M_\odot}\right)^{-1}, \quad (11.26)$$

where $\ln \Lambda \sim 10$ is the Coulomb logarithm. In our case of $m_{BH} = 33\ M_\odot$, the simulations predict that $v_{BH} \sim 5 - 10\ \text{km s}^{-1}$ and $r \gtrsim 1$ pc (due to the survivor bias of gas condensation), so that we have $t_{DF} \gtrsim 6$ Myr $\gg t_{SF}$. This indicates that stellar-mass PBHs are unlikely to sink into Pop III star-forming disks and, therefore, cannot interact with Pop III protostars directly. Nevertheless, t_{DF} is comparable to the typical lifetimes of Pop III stars and relaxation timescales of Pop III star clusters, so that PBHs may still affect Pop III stellar evolution and dynamics of Pop III star clusters, which can be an interesting topic for future research.

11.4.3 Impact of PBHs on the Formation of Massive BH Seeds

We have shown that the impact of stellar-mass PBHs (allowed by existing observational constraints) on Pop III star formation in molecular-cooling minihalos ($M_{\rm h} \sim 10^5 - 10^8\ {\rm M}_\odot$) is likely minor both globally and locally. However, the PBH effects may still be important in larger, atomic-cooling halos ($M_{\rm h} \gtrsim 10^8\ {\rm M}_\odot$). In particular, if PBH accretion produces strong LW radiation that can efficiently dissociate H_2 in a metal-pool halo, molecular cooling and fragmentation will be suppressed, such that gas can collapse directly via atomic cooling to form a massive ($\sim 10^4 - 10^6\ {\rm M}_\odot$) BH [18]. This direct-collapse BH (DCBH) scenario (reviewed by, e.g., [49, 77]) is an important channel of forming the seeds of the supermassive BHs observed in high-z quasars, but it is only expected to occur in very rare sites with strong external LW radiation fields or high inflow rates in the standard ΛCDM cosmology [101, 118, 120]. To explore whether DCBH formation can be enhanced by PBHs, we calculate the LW intensity produced by PBHs at the halo center, $J_{21,{\rm PBH}}(r = 0)$. Here we treat the LW radiation from PBHs as an 'external' background by ignoring self-shielding, in order to compare our results with the typical critical intensity of DCBH formation $J_{21,{\rm crit}} = 1000$, defined for an external radiation field with a temperature $T_{\rm rad} \gtrsim 2 \times 10^4$ K [113]. We assume that both gas and PBHs follow isothermal distributions ($M_{\rm enc} \propto r$), based on simulation results [80, 105, 120]. The PBH distribution is further truncated at $r = 1$ pc considering the survivor bias of gas condensation. We use Eq. 11.13 to calculate the accretion rate of each PBH given the characteristic velocity $\tilde{v} \sim \sqrt{GM_{\rm h}/R_{\rm vir}}$, and derive the specific luminosity in the LW band from the accretion rate using the spectrum model in [114]. We integrate over the volume defined by the radius range $r \in [1\ {\rm pc}, R_{\rm vir}]$ to sum up the contributions from individual PBHs.

The resulting $J_{21,{\rm PBH}}(r = 0)$ as a function of halo mass at $z = 20$ for $f_{\rm PBH} = 0.001$ is shown in Fig. 11.8 as an example. In fact, we find that our results can be described by a simple power-law fit, within a factor 2 uncertainty, for $m_{\rm PBH} \sim 10 - 100\ {\rm M}_\odot$ and $M_{\rm h} \sim 10^6 - 10^{10}\ {\rm M}_\odot$ at $z \sim 10 - 30$:

$$J_{21,{\rm PBH}}(r = 0) \sim 5 \times 10^4 f_{\rm PBH} (M_{\rm h}/10^7\ {\rm M}_\odot)^{0.56} (m_{\rm PBH}/33\ {\rm M}_\odot)^{0.9}\ . \tag{11.27}$$

This implies that halos with $M_{\rm h} \gtrsim 3.3 \times 10^7\ {\rm M}_\odot\ (f_{\rm PBH}/0.01)^{-1.8}\ (m_{\rm PBH}/33\ {\rm M}_\odot)^{-1.6}$ will meet the criterion $J_{21} \gtrsim 10^3$ for DCBH formation [113]. Therefore, for $f_{\rm PBH} \sim 0.001 - 0.01$ still (marginally) allowed by observational constraints, stellar-mass PBHs can potentially trigger the formation of massive BH seeds via direct collapse in typical atomic-cooling halos ($M_{\rm h} \sim 10^8 - 10^{10}\ {\rm M}_\odot$). Radiative hydrodynamic simulations with detailed modeling of the radiation fields from accreting PBHs are required to verify this possibility.

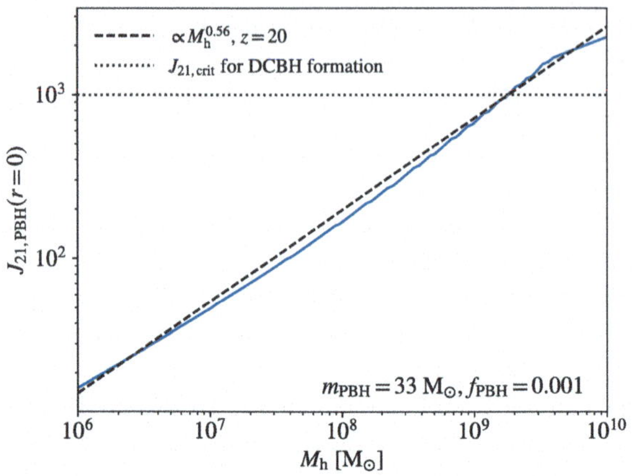

Fig. 11.8 Intensity of the LW radiation at the halo center from accreting PBHs with $m_{PBH} = 33\ M_\odot$ and $f_{PBH} = 0.001$ as a function of halo mass for $z = 20$, assuming isothermal distributions of gas and PBHs, ignoring self-shielding (i.e., treating the LW radiation as an 'external' background), where the dashed line shows the power-law fit (Eq. 11.27) that captures the results of detailed calculations (solid) within a factor of 2 uncertainty, for $m_{PBH} \sim 10 - 100\ M_\odot$ and $M_h \sim 10^6 - 10^{10}\ M_\odot$ at $z \sim 10 - 30$, and the dotted horizontal line denotes the critical intensity, $J_{21,crit} \sim 1000$, for DCBH formation [113]

11.5 Formation of Massive Galaxies Seeded by SMPBHs

JWST has recently discovered a population of unusually massive (stellar masses $\gtrsim 10^{10}\ M_\odot$) galaxy candidates formed at $z \gtrsim 8$ [21, 22, 33, 43, 75, 82, 94, 117] and a surprising ubiquity of (overmassive) SMBHs in galaxies at $z \gtrsim 5$ [7, 38, 45, 47, 52, 85, 88, 95], which is seemingly in tension with typical theoretical predictions on high-z galaxy and SMBH formation in the standard ΛCDM cosmology [15, 29, 89]. A promising explanation is SMPBHs (or other primordial massive objects like cosmic string loops [66]) as a component of DM, which can serve as seeds of both SMBHs and massive galaxies [31, 78, 112]. Such enhanced early galaxy formation by SMPBHs may also be able to explain the excess of galaxy-galaxy strong lensing probability and star formation observed in (proto-)clusters of galaxies [90, 102], and mergers of SMPBHs can also contribute to the SGWB recently discovered by pulsar timing arrays [46] (see Part IV of the book). In this section, we use the extraordinarily massive old galaxy candidate ZF-UDS-7329 detected by JWST (with a spectroscopic redshift $z = 3.21$) [43] as an example to demonstrate how the SMPBH parameters required to explain certain observations can be identified with a back-of-the-envelope calculation focusing on the 'seed' effect (see also [33, 94, 117] for JWST observations of similar massive old galaxy candidates unexpected in the standard ΛCDM cosmology).

It is found by [43] that the observed spectrum of ZF-UDS-7329 can be explained by an old stellar population formed at $z \gtrsim 11$ with a total stellar mass of $M_\star \sim$

2.5×10^{11} $M_\odot$. The co-moving number density of galaxies like ZF-UDS-7329 is inferred as $n_g \sim 6 \times 10^{-7}$ Mpc^{-3} from the survey volume. In the standard ΛCDM cosmology, such massive galaxies (and the corresponding massive halos) are not expected to occur with the inferred abundance until $z \sim 6$. Now, we assume that this galaxy resides in a halo seeded by an SMPBH with an initial mass of $m_{\rm PBH}$, so that the halo mass follows $M_{\rm h} \sim M_{\rm B}(m_{\rm PBH}, z) = m_{\rm PBH}/[(1+z)a_{\rm eq}]$ in the idealized case of isolated growth. To reproduce the observed stellar mass and abundance of ZF-UDS-7329, we simply require $\bar{n}_{\rm PBH} > n_g$ and $M_\star = \epsilon(\Omega_b/\Omega_m)M_{\rm B}$ at $z = 11$, given the star formation efficiency $\epsilon < 1$ and the average number density of SMPBHs $\bar{n}_{\rm PBH} = f_{\rm PBH}[3H_0^2(\Omega_m - \Omega_b)]/(8\pi G m_{\rm PBH})$. Considering the typical (extreme) limit $\epsilon < 0.1$ (1) for such massive galaxies, we have $m_{\rm PBH} \gtrsim 5.6 \times 10^{10\,(9)}$ $M_\odot$ and $f_{\rm PBH} \gtrsim 10^{-6\,(7)}$. We also require $f_{\rm PBH} < (1+z)a_{\rm eq} \sim 0.0035$, which is a necessary (but not sufficient) condition for isolated growth of PBH seeded halos, and $m_{\rm PBH} \lesssim 10^{11}$ $M_\odot$ considering the non-detection of BHs above 10^{11} $M_\odot$. We summarize these requirements and the resulting region in the $m_{\rm PBH}$-$f_{\rm PBH}$ parameter space in Fig. 11.9.

In fact, if PBHs form in the standard spherical collapse scenario, and primordial density fluctuations are Gaussian, such SMPBH models with $m_{\rm PBH} \sim 10^{10}$ $M_\odot$ and $f_{\rm PBH} \sim 10^{-7} - 0.0035$ are ruled out by the observed limit of CMB μ-distortion [25] (see also Part V of the book). However, the CMB μ-distortion constraint can be weakened/lifted if primordial density fluctuations are highly non-Gaussian [57, 93] or PBHs form in non-standard scenarios such as inhomogeneous baryogenesis with

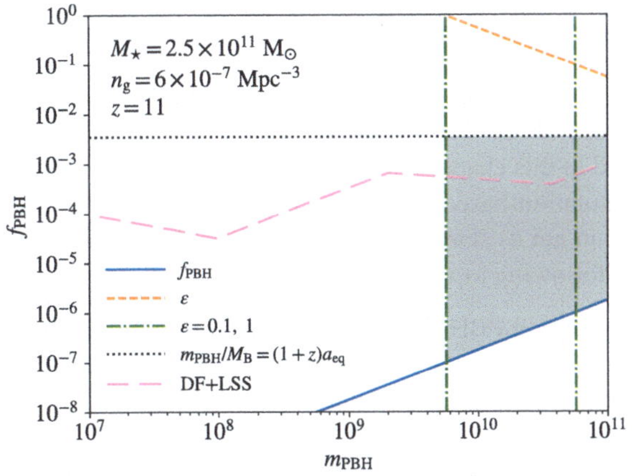

Fig. 11.9 The PBH parameters (shaded region) required to explain the extraordinarily massive galaxy candidate ZF-UDS-7329 formed at $z = 11$ [43] with the 'seed' effect of SMPBHs, where the solid line shows the lower limit of $f_{\rm PBH}$, given by the galaxy abundance requirement $\bar{n}_{\rm PBH} > n_g$, the dashed line shows the star formation efficiency ϵ required to explain the stellar mass with $M_\star = \epsilon M_{\rm B}$, the vertical dash-dotted lines label typical values of $\epsilon = 0.1$ and 1, and the horizontal dotted lines denotes the necessary (but not sufficient) condition $f_{\rm PBH} < (1+z)a_{\rm eq}$ for isolated growth, and the long-dashed curve shows the combined constraint from dynamical friction (DF) [27] and large-scale structure (LSS) [24] compiled by [25]

the modified Affleck-Dine mechanism [68, 69, 72], first-order phase transitions [32, 65, 81], and the collapse of supercritical vacuum bubbles nucleated during inflation [42, 60] (see Part II of the book). If we bypass the CMB μ-distortion constraint with these 'non-standard' scenarios, the strongest remaining constraints around this mass scale come from the infall of PBHs into the Galactic center by dynamical friction [27] and large-scale structure statistics [24], which jointly require $f_{\rm PBH} \lesssim 10^{-3}$ (see the long-dashed curve in Fig. 11.9).[9] In this case, a substantial region in the PBH parameter space ($m_{\rm PBH} \sim 10^{7} - 10^{11}$ M$_{\odot}$, $f_{\rm PBH} \sim 10^{-7} - 10^{-3}$) is still allowed to explain recent JWST observations [43, 75, 78, 112]. However, a key uncertainty in our idealized analysis is the star formation efficiency ϵ that captures the complex interplay between star formation and BH accretion feedback. It is possible that star formation will be significantly suppressed by the SMPBH, leading to extremely small ϵ that requires $m_{\rm PBH} \gtrsim 10^{11}$ M$_{\odot}$ in conflicts with observations. SMPBHs and the massive galaxies seeded by them may also violate other observational constraints from the early Universe, such as the 21 cm signal, ionization history, cosmic infrared and X-ray backgrounds [71]. Moreover, in the presence of PBH-seeded halos, first star formation can happen at very high redshifts ($z \gtrsim 100$) under peculiar conditions, leading to distinct stellar properties compared with the standard ΛCDM case [64]. Future studies with detailed modeling of the interactions between PBHs, baryons and PDM (in cosmological simulations) are required to fully understand star/galaxy formation in PBH cosmologies and evaluate their viability against observations.

11.6 Summary and Outlook

A possible PBH contribution to the cosmic DM density adds qualitatively new aspects to the physics of cosmological structure formation and early star/galaxy formation. As we have discussed in this chapter, these novel aspects present challenges to existing analytical and simulation-based treatments. We have suggested idealized, or heuristic approaches that can get us started, but need to be further developed in future work. In particular, the following topics deserve further investigation:

- Halo mass functions in PBH cosmologies from robust analytical models and cosmological simulations that capture the interplay between the 'Poisson' and 'seed' effects of PBHs on structure formation

[9] The UV luminosity functions of galaxies observed by HST at $z \sim 8$ also place constraints on PBH parameters which, compared with the constraints from large-scale structure [24], are slightly weaker in the conservative scenario of the 'Poisson' effect modeled by the PS formalism with suppressed small-scale power at $k > k_{\star}$ from nonlinear dynamics around PBHs (see Fig. 1 in [46]), and is thus not shown here. The constraints from HST observations can be stronger with more optimistic estimates of the small-scale power of PBH perturbations (see Fig. 3 in [46]), but still unable to completely rule out SMPBHs as seeds of the observed massive galaxy candidates.

- Effects of LW and X-ray feedback from PBHs on early star formation and DCBH formation triggered by PBHs
- Interplay between star formation, stellar feedback, BH accretion and feedback in structures induced by SMPBHs

The overall motivation for such dedicated effort is the importance of broadly exploring the behavior and viability of PBH dark matter, in terms of not violating existing and future empirical constraints.

The first stars and galaxies are probing cosmic structure on the smallest scales, and provide therefore an ideal laboratory for the nature of DM, both the microphysics of PDM, reflecting extensions to the standard model of particle physics, as well as the mass spectrum and abundance of PBHs, similarly depending on exotic physical processes in the ultra-early Universe. Developing this astrophysical laboratory may indeed be the only way to conquer the dark matter frontier if the production or detection of dark matter particles continues to elude experimental physics. More specifically, for PBHs, the challenge is to disentangle their properties and impact on cosmic history from the suite of astrophysical black holes. Here, again, the very onset of star and galaxy formation promises to provide unique fingerprints for black holes of such vastly different origins. We thus face an exciting theoretical and computational challenge in elucidating PBH-ΛCDM cosmology, at the very moment when frontier telescopes, such as the JWST, Euclid, and soon the Roman Space Telescope, and other observational facilities are coming online.

References

1. R. Abbott et al., Population properties of compact objects from the second LIGO-Virgo gravitational-wave transient catalog. Astrophys. J. Lett. **913**(1), L7 (2021). https://doi.org/10.3847/2041-8213/abe949. arXiv:2010.14533 [astro-ph.HE]
2. M.A. Abramowicz, P.C. Fragile, Foundations of black hole accretion disk theory. Living Rev. Rel. **16**, 1 (2013). https://doi.org/10.12942/lrr-2013-1. arXiv:2010.14533 [astro-ph.HE]
3. S.K. Acharya, J. Dhandha, J. Chluba, Can accreting primordial black holes explain the excess radio background? Mon. Not. R. Astron. Soc. **517**(2), 2454–2461 (2022). https://doi.org/10.1093/mnras/stac2739. arXiv:2010.14533 [astro-ph.CO]
4. P.A.R. Ade et al., Planck 2015 results. XIII. Cosmological parameters. Astron. Astrophys. **594**, A13 (2016). https://doi.org/10.1051/0004-6361/201525830. arXiv:2010.14533 [astro-ph.CO]
5. Gabriella Agazie et al., The NANOGrav 15 yr data set: constraints on supermassive black hole binaries from the gravitational-wave background. Astrophys. J. Lett. **952**(2), L37 (2023). https://doi.org/10.3847/2041-8213/ace18b. arXiv:2010.14533 [astro-ph.HE]
6. N. Aghanim et al., Planck 2018 results. I. Overview and the cosmological legacy of Planck. Astron. Astrophys. **641**, A1 (2020). https://doi.org/10.1051/0004-6361/201833880. arXiv:2010.14533 [astro-ph.CO]
7. H.B. Akins et al., COSMOS-Web: the over-abundance and physical nature of "little red dots"–Implications for early galaxy and SMBH assembly (2024). arXiv e-prints, arXiv:2406.10341. https://doi.org/10.48550/arXiv.2406.10341. arXiv:2010.14533 [astro-ph.GA]
8. Y. Ali-Haïmoud, E.D. Kovetz, M. Kamionkowski, Merger rate of primordial black-hole binaries. Phys. Rev. D **96**(12), 123523 (2017). https://doi.org/10.1103/PhysRevD.96.123523. arXiv:2010.14533 [astro-ph.CO]

9. M. Andrés-Carcasona et al., Constraints on primordial black holes from LIGO-Virgo-KAGRA O3 events (2024). arXiv:2010.14533 [astro-ph.CO]
10. R.E. Angulo, O. Hahn, Large-scale dark matter simulations. Living Rev. Comput. Astrophys. **8**(1), 1 (2022)
11. F. Atrio-Barandela, The effect of Primordial Black Holes and streaming motions on structure formation. Astrophys. J. **939**(2), 69 (2022). https://doi.org/10.3847/1538-4357/ac983e. arXiv:2010.14533 [astro-ph.CO]
12. Peter S. Behroozi, Risa H. Wechsler, Wu. Hao-Yi, The rockstar phase-space temporal halo finder and the velocity offsets of cluster cores. Astrophys. J. **762**, 109 (2013). https://doi.org/10.1088/0004-637X/762/2/109. arXiv:2010.14533 [astro-ph.CO]
13. K.M. Belotsky et al., Clusters of primordial black holes. Eur. Phys. J. C **79**(3), 246 (2019). https://doi.org/10.1140/epjc/s10052-019-6741-4. arXiv:2010.14533 [astro-ph.CO]
14. J. Binney, S. Tremaine, *Galactic Dynamics*, vol. 20 (Princeton University Press, 2011)
15. M. Boylan-Kolchin, Stress testing Λ CDM with high-redshift galaxy candidates. Nat. Astron. **7**(6), 731–735 (2023). https://doi.org/10.1038/s41550-023-01937-7. arXiv:2010.14533 [astro-ph.CO]
16. T. Bringmann et al., Towards closing the window of primordial black holes as dark matter: the case of large clustering. Phys. Rev. D **99**(6), 063532 (2019). https://doi.org/10.1103/PhysRevD.99.063532. arXiv:2010.14533 [astro-ph.CO]
17. V. Bromm, Formation of the first stars. Rep. Prog. Phys. **76**(112901), 112901 (2013). https://doi.org/10.1088/0034-4885/76/11/112901. arXiv:2010.14533 [astro-ph.CO]
18. V. Bromm, A. Loeb, Formation of the first supermassive black holes. Astrophys. J. **596**, 34–46 (2003). https://doi.org/10.1086/377529. arXiv:2010.14533 [astro-ph]
19. V. Bromm, N. Yoshida, The first galaxies. Annu. Rev. Astron. Astrophys. **49**, 373–407 (2011). https://doi.org/10.1146/annurev-astro-081710-102608. arXiv:2010.14533 [astro-ph.CO]
20. N. Cappelluti, G. Hasinger, P. Natarajan, Exploring the high-redshift PBH-Λ CDM universe: early black hole seeding, the first stars and cosmic radiation backgrounds. Astrophys. J. **926**(2), 205 (2022). https://doi.org/10.3847/1538-4357/ac332d. (arXiv:2010.14533 [astro-ph.CO])
21. A.C. Carnall et al., A surprising abundance of massive quiescent galaxies at 3<z<5 in the first data from JWST CEERS. Mon. Not. R. Astron. Soc. **520**(3), 3974–3985 (2023). https://doi.org/10.1093/mnras/stad369. arXiv:2010.14533 [astro-ph.GA]
22. S. Carniani et al., A shining cosmic dawn: spectroscopic confirmation of two luminous galaxies at $z \sim 14$ (2024). arXiv e-prints, arXiv:2405.18485. https://doi.org/10.48550/arXiv.2405.18485. arXiv:2010.14533 [astro-ph.GA]
23. B. Carr, F. Kuhnel, Primordial black holes with multimodal mass spectra. Phys. Rev. D **99**(10), 103535 (2019). https://doi.org/10.1103/PhysRevD.99.103535. arXiv:2010.14533 [astro-ph.CO]
24. B. Carr, J. Silk, Primordial black holes as generators of cosmic structures. Mon. Not. R. Astron. Soc. **478**(3), 3756–3775 (2018). https://doi.org/10.1093/mnras/sty1204
25. B. Carr et al., Constraints on primordial black holes. Rept. Prog. Phys. **84**(11), 116902 (2021). https://doi.org/10.1088/1361-6633/ac1e31. arXiv:2010.14533 [astro-ph.CO]
26. B.J. Carr, S.W. Hawking, Black holes in the early Universe. Mon. Not. R. Astron. Soc. **168**, 399–415 (1974). https://doi.org/10.1093/mnras/168.2.399
27. B.J. Carr, M. Sakellariadou, Dynamical constraints on dark compact objects. Astrophys. J. **516**, 195–220 (1999). https://doi.org/10.1086/307071
28. C. Casanueva-Villarreal et al., Impact of primordial black hole dark matter on gas properties at very high redshift: a semi-analytical model (2024). arXiv:2010.14533 [astro-ph.GA]
29. Y. Chen, H.J. Mo, K. Wang, Massive dark matter haloes at high redshift: implications for observations in the JWST era. Mon. Not. R. Astron. Soc. **526**(2), 2542–2559 (2023). https://doi.org/10.1093/mnras/stad2866. arXiv:2010.14533 [astro-ph.GA]
30. J.R. Chisholm, Clustering of primordial black holes: basic results. Phys. Rev. D **73**, 083504 (2006). https://doi.org/10.1103/PhysRevD.73.083504. arXiv:2010.14533
31. P.E. Colazo, F. Stasyszyn, N. Padilla, Structure formation with primordial black holes to alleviate early star formation tension revealed by JWST. Astron. Astrophys. **685**, L8 (2024). https://doi.org/10.1051/0004-6361/202449565. arXiv:2010.14533 [astro-ph.CO]

32. H. Davoudiasl, P.B. Denton, J. Gehrlein, Supermassive black holes, ultralight dark matter, and gravitational waves from a first order phase transition. Phys. Rev. Lett. **128**(8), 081101 (2022). https://doi.org/10.1103/PhysRevLett.128.081101. arXiv:2010.14533 [astro-ph.CO]
33. A. de Graaff et al., Efficient formation of a massive quiescent galaxy at redshift 4.9 (2024). arXiv e-prints, arXiv:2404.05683. https://doi.org/10.48550/arXiv.2404.05683. arXiv:2010.14533 [astro-ph.GA]
34. V. De Luca et al., Bayesian evidence for both astrophysical and primordial black holes: mapping the GWTC-2 catalog to third-generation detectors. JCAP **05**, 003 (2021). https://doi.org/10.1088/1475-7516/2021/05/003. arXiv:2010.14533 [astro-ph.CO]
35. V. De Luca et al., The clustering evolution of primordial black holes. JCAP **11**, 028 (2020). https://doi.org/10.1088/1475-7516/2020/11/028. arXiv:2010.14533 [astro-ph.CO]
36. V. Desjacques, A. Riotto, Spatial clustering of primordial black holes. Phys. Rev. D **98**(12), 123533 (2018). https://doi.org/10.1103/PhysRevD.98.123533. arXiv:2010.14533 [astro-ph.CO]
37. B. Diemer, COLOSSUS: a python toolkit for cosmology, large-scale structure, and dark matter halos. Astrophys. J. Suppl. **239**(2), 35 (2018). https://doi.org/10.3847/1538-4365/aaee8c. arXiv:2010.14533 [astro-ph.CO]
38. E. Durodola, F. Pacucci, R.C. Hickox, Exploring the AGN fraction of a sample of JWST's little red dots at $5<z<8$: overmassive black holes are strongly favored (2024). arXiv:2010.14533 [astro-ph.GA]
39. G. Fabbiano, M. Elvis, The interaction of the active nucleus with the host galaxy interstellar medium, in *Handbook of X-ray and Gamma-ray Astrophysics* (Springer, 2022), p. 92. https://doi.org/10.1007/978-981-16-4544-0_111-1
40. X. Fan, E. Bañados, R.A. Simcoe, Quasars and the intergalactic medium at cosmic dawn. Annu. Rev. Astron. Astrophys. **61**, 373–426 (2023)
41. G. Franciolini et al., Searching for a subpopulation of primordial black holes in LIGO-Virgo gravitational-wave data. Phys. Rev. D **105**(8), 083526 (2022). https://doi.org/10.1103/PhysRevD.105.083526. arXiv:2010.14533 [gr-qc]
42. J. Garriga, A. Vilenkin, J. Zhang, Black holes and the multiverse. JCAP **02**, 064 (2016). https://doi.org/10.1088/1475-7516/2016/02/064. arXiv:2010.14533 [hep-th]
43. K. Glazebrook et al., A massive galaxy that formed its stars at $z \approx 11$. Nature **628**(8007), 277–281 (2024). https://doi.org/10.1038/s41586-024-07191-9. arXiv:2010.14533 [astro-ph.GA]
44. S.T. Gordon et al., Hungry or not: how stellar-mass black holes grow (or don't) in dark matter mini-haloes at high resolution. Mon. Not. R. Astron. Soc. **529**(1), 604–627 (2024). https://doi.org/10.1093/mnras/stae566. arXiv:2010.14533 [astro-ph.GA]
45. A.D. Goulding et al., UNCOVER: the growth of the first massive black holes from JWST/NIRSpec—spectroscopic redshift confirmation of an X-Ray luminous AGN at $z = 10.1$. Astrophys. J. Lett. **955**(1), L24 (2023). https://doi.org/10.3847/2041-8213/acf7c5. arXiv:2010.14533 [astro-ph.GA]
46. Y. Gouttenoire et al., Scrutinizing the primordial black holes interpretation of PTA gravitational waves and JWST early galaxies (2023). arXiv:2010.14533 [astro-ph.CO]
47. J.E. Greene et al., UNCOVER spectroscopy confirms the surprising ubiquity of active galactic nuclei in red sources at $z > 5$. Astrophys. J. **964**(1), 39 (2024). https://doi.org/10.3847/1538-4357/ad1e5f
48. T.H. Greif, V. Bromm, Two populations of metal-free stars in the early Universe. Mon. Not. R. Astron. Soc. **373**, 128–138 (2006). https://doi.org/10.1111/j.1365-2966.2006.11017.x. arXiv:2010.14533
49. L. Haemmerlé et al., Formation of the first stars and black holes. Space Sci. Rev. **216**(4), 48 (2020). https://doi.org/10.1007/s11214-020-00673-y. arXiv:2010.14533 [astro-ph.GA]
50. O. Hahn, T. Abel, Multi-scale initial conditions for cosmological simulations. Mon. Not. R. Astron. Soc. **415**, 2101–2121 (2011). https://doi.org/10.1111/j.1365-2966.2011.18820.x. arXiv:2010.14533 [astro-ph.CO]
51. A. Hall, A.D. Gow, C.T. Byrnes, Bayesian analysis of LIGO-Virgo mergers: primordial vs. astrophysical black hole populations. Phys. Rev. D **102**, 123524 (2020). https://doi.org/10.1103/PhysRevD.102.123524. arXiv:2010.14533 [astro-ph.CO]

52. Y. Harikane et al., A JWST/NIRSpec first census of broad-line AGNs at z = 4-7: detection of 10 faint AGNs with M_{BH} 10^6-10^8 $M_\odot$ and their host galaxy properties. Astrophys. J. **959**(1), 39 (2023). https://doi.org/10.3847/1538-4357/ad029e. arXiv:2010.14533 [astro-ph.GA]
53. C.M. Harrison, Impact of supermassive black hole growth on star formation. Nat. Astron. **1**(0165), 0165 (2017). https://doi.org/10.1038/s41550-017-0165. arXiv:2010.14533 [astro-ph.GA]
54. G. Hasinger, Illuminating the dark ages: Cosmic backgrounds from accretion onto primordial black hole dark matter. JCAP **07**, 022 (2020). https://doi.org/10.1088/1475-7516/2020/07/022. arXiv:2010.14533 [astro-ph.CO]
55. S.W. Hawking, Black hole explosions. Nature **248**, 30–31 (1974). https://doi.org/10.1038/248030a0
56. S. Hirano et al., Primordial star formation under the influence of far ultraviolet radiation: 1540 cosmological haloes and the stellar mass distribution. Mon. Not. R. Astron. Soc. **448**(1), 568–587 (2015). https://doi.org/10.1093/mnras/stv044. arXiv:2010.14533 [astro-ph.GA]
57. D. Hooper et al., Supermassive primordial black holes from inflation (2023). arXiv:2010.14533 [astro-ph.CO]
58. P.F. Hopkins, A new class of accurate, mesh-free hydrodynamic simulation methods. Mon. Not. R. Astron. Soc. **450**(1), 53–110 (2015). https://doi.org/10.1093/mnras/stv195. arXiv:2010.14533 [astro-ph.CO]
59. H.-L. Huang, Y.-S. Piao, Towards supermassive primordial black holes from inflationary bubbles (2023). arXiv e-prints, arXiv:2312.11982. https://doi.org/10.48550/arXiv.2312.11982. arXiv:2010.14533 [astro-ph.CO]
60. H.-L. Huang et al., Supermassive primordial black holes in multiverse: for nano-Hertz gravitational wave and high-redshift JWST galaxies (2023). arXiv e-prints, arXiv:2306.17577. https://doi.org/10.48550/arXiv.2306.17577. arXiv:2010.14533 [gr-qc]
61. J.A. Hummel et al., The first stars: formation under X-ray feedback. Mon. Not. R. Astron. Soc. **453**(4), 4136–4147 (2015). https://doi.org/10.1093/mnras/stv1902. arXiv:2010.14533 [astro-ph.GA]
62. G. Hütsi et al., Two populations of LIGO-Virgo black holes. JCAP **03**, 068 (2021). https://doi.org/10.1088/1475-7516/2021/03/068. arXiv:2010.14533 [astro-ph.CO]
63. D. Inman, Y. Ali-Haïmoud, Early structure formation in primordial black hole cosmologies. Phys. Rev. D **100**(8), 083528 (2019). https://doi.org/10.1103/PhysRevD.100.083528. arXiv:2010.14533 [astro-ph.CO]
64. M. Ito, K. Omukai, First star formation in extremely early epochs (2024). arXiv:2010.14533 [astro-ph.GA]
65. K. Jedamzik, J.C. Niemeyer, Primordial black hole formation during first order phase transitions. Phys. Rev. D **59**, 124014 (1999). https://doi.org/10.1103/PhysRevD.59.124014. arXiv:2010.14533
66. H. Jiao, R. Brandenberger, A. Refregier, Early structure formation from cosmic string loops in light of early JWST observations. Phys. Rev. D **108**(4), 043510 (2023). https://doi.org/10.1103/PhysRevD.108.043510. arXiv:2010.14533 [astro-ph.CO])
67. H. Jiao, R. Brandenberger, A. Refregier, N-body simulation of early structure formation from cosmic string loops. Phys. Rev. D **109**(12), 123524 (2024). https://doi.org/10.1103/PhysRevD.109.123524. arXiv:2010.14533 [astro-ph.CO]
68. K. Kasai, M. Kawasaki, K. Murai, Revisiting the Affleck-Dine mechanism for primordial black hole formation. JCAP **10**, 048 (2022). https://doi.org/10.1088/1475-7516/2022/10/048. arXiv:2010.14533 [astro-ph.CO]
69. Kentaro Kasai et al., Supermassive black hole formation from Affleck-Dine mechanism with suppressed clustering on large scales (2024). arXiv:2010.14533 [astro-ph.CO]
70. A. Kashlinsky. Cosmological advection flows in the presence of primordial black holes as dark matter and formation of first sources. Phys. Rev. Lett. **126**(1), 011101 (2021). https://doi.org/10.1103/PhysRevLett.126.011101. arXiv:2010.14533 [astro-ph.CO]
71. A. Kashlinsky, LIGO gravitational wave detection, primordial black holes, and the near-IR cosmic infrared background anisotropies. Astrophys. J. Lett. **823**(L25), L25 (2016). https://doi.org/10.3847/2041-8205/823/2/L25. arXiv:2010.14533 [astro-ph.CO]

72. M. Kawasaki, K. Murai, Formation of supermassive primordial black holes by Affleck-Dine mechanism. Phys. Rev. D **100**(10), 103521 (2019). https://doi.org/10.1103/PhysRevD.100.103521. arXiv:2010.14533 [astro-ph.CO]
73. A. King, K. Pounds, Powerful outflows and feedback from active galactic nuclei. Ann. Rev. Astron. Astrophys. **53**, 115–154 (2015). https://doi.org/10.1146/annurev-astro-082214-122316. arXiv:2010.14533 [astro-ph.GA]
74. R.S. Klessen, S.C.O. Glover, The first stars: formation, properties, and impact. Annu. Rev. Astron. Astrophys. **61**, (2023)
75. I. Labbé et al., A population of red candidate massive galaxies 600 Myr after the Big Bang. Nature **616**, 266–269 (2023). https://doi.org/10.1038/s41586-023-05786-2. arXiv:2010.14533 [astro-ph.GA]
76. R. Laha, L. Philip, V. Takhistov, Gas heating from spinning and non-spinning evaporating primordial black holes. Phys. Lett. B **820**, 136459 (2021). https://doi.org/10.1016/j.physletb.2021.136459. arXiv:2010.14533 [astro-ph.CO]
77. M.A. Latif, Black hole formation via gas-dynamical processes, in *Formation of the First Black Holes* (World Scientific, 2019), pp. 99–113
78. B. Liu, V. Bromm, Accelerating early massive galaxy formation with primordial black holes. Astrophys. J. Lett. **937**(2), L30 (2022). https://doi.org/10.3847/2041-8213/ac927f. arXiv:2010.14533 [astro-ph.CO]
79. B. Liu, V. Bromm, When did Population III star formation end? Mon. Not. R. Astron. Soc. **497**(3), 2839–2854 (2020). https://doi.org/10.1093/mnras/staa2143. arXiv:2010.14533 [astro-ph.GA]
80. B. Liu, S. Zhang, V. Bromm, Effects of stellar-mass primordial black holes on first star formation. Mon. Not. R. Astron. Soc. **514**(2), 2376–2396 (2022). https://doi.org/10.1093/mnras/stac1472. (arXiv:2010.14533 [astro-ph.GA])
81. J. Liu et al., Primordial black hole production during first-order phase transitions. Phys. Rev. D **105**(2), L021303 (2022). https://doi.org/10.1103/PhysRevD.105.L021303. arXiv:2010.14533 [astro-ph.CO]
82. M. Lopez-Corredoira et al., Age of massive galaxies at redshift 8 (2024). arXiv:2010.14533 [astro-ph.CO]
83. L. Philip et al., Constraining primordial black holes with dwarf galaxy heating. Astrophys. J. Lett. **908**(2), L23 (2021). https://doi.org/10.3847/2041-8213/abdcb6. arXiv:2010.14533 [astro-ph.CO]
84. K.J. Mack, J.P. Ostriker, M. Ricotti, Growth of structure seeded by primordial black holes. Astrophys. J. **665**, 1277–1287 (2007). https://doi.org/10.1086/518998. arXiv:2010.14533
85. R. Maiolino et al., JADES. The diverse population of infant Black Holes at $4<z<11$: merging, tiny, poor, but mighty (2023). arXiv:2308.01230
86. D. Manzoni et al., Primordial black holes as near-infrared background sources. Mon. Not. R. Astron. Soc. **527**(2), 4153–4161 (2024)
87. M. Mapelli, *Formation Channels of Single and Binary Stellar-Mass Black Holes* (Springer, 2020), pp. 1–65
88. J. Matthee et al., Little red dots: an abundant population of faint active galactic nuclei at z sim 5 revealed by the EIGER and FRESCO JWST surveys. Astrophys. J. **963**(2), 129 (2024). https://doi.org/10.3847/1538-4357/ad2345. arXiv:2010.14533 [astro-ph.GA]
89. N. Menci et al., High-redshift galaxies from early JWST observations: constraints on dark energy models. Astrophys. J. Lett. **938**(1), L5 (2022). https://doi.org/10.3847/2041-8213/ac96e9. arXiv:2010.14533 [astro-ph.CO]
90. M. Meneghetti et al., A persistent excess of galaxy-galaxy strong lensing observed in galaxy clusters. Astron. Astrophys. **678**, L2 (2023). https://doi.org/10.1051/0004-6361/202346975. arXiv:2010.14533 [astro-ph.CO]
91. S. Mittal et al., Constraining primordial black holes as dark matter using the global 21-cm signal with X-ray heating and excess radio background (2021). arXiv:2010.14533 [astro-ph.CO]

92. H. Mo, F. Van den Bosch, S. White, *Galaxy Formation and Evolution* (Cambridge University Press, 2010)
93. T. Nakama, B. Carr, J. Silk, Limits on primordial black holes from μ distortions in cosmic microwave background. Phys. Rev. D **97**(4), 043525 (2018). https://doi.org/10.1103/PhysRevD.97.043525. arXiv:2010.14533 [astro-ph.CO]
94. T. Nanayakkara et al., A population of faint, old, and massive quiescent galaxies at 3âĽ¯¡zâĽ¯¡4 revealed by JWST NIRSpec Spectroscopy. Sci. Rep. **14**(3724), 3724 (2024). https://doi.org/10.1038/s41598-024-52585-4. arXiv:2010.14533 [astro-ph.GA]
95. P. Natarajan et al., First detection of an over-massive black hole galaxy: UHZ1–evidence for heavy black hole seeds from direct collapse? (2023). arXiv:2308.02654
96. A. Negri, M. Volonteri, Black hole feeding and feedback: the physics inside the 'sub-grid'. Mon. Not. R. Astron. Soc. **467**(3), 3475–3492 (2017). https://doi.org/10.1093/mnras/stx362. arXiv:2010.14533 [astro-ph.GA]
97. J. Park, M. Ricotti, K. Sugimura, Population III star formation in an X-ray background – I. Critical halo mass of formation and total mass in stars. Mon. Not. R. Astron. Soc. **508**(4), 6176–6192 (2021). https://doi.org/10.1093/mnras/stab2999. arXiv:2010.14533 [astro-ph.GA]
98. H.W. Press, P. Schechter, Formation of galaxies and clusters of galaxies by selfsimilar gravitational condensation. Astrophys. J. **187**, 425–438 (1974). https://doi.org/10.1086/152650
99. M. Raidal, V. Vaskonen, H. Veermäe, Gravitational waves from primordial black hole mergers. JCAP **09**, 037 (2017). https://doi.org/10.1088/1475-7516/2017/09/037. arXiv:2010.14533 [astro-ph.CO]
100. M.J. Rees, J.P. Ostriker, Cooling, dynamics and fragmentation of massive gas clouds: clues to the masses and radii of galaxies and clusters. Mon. Not. R. Astron. Soc. **179**, 541–559 (1977). https://doi.org/10.1093/mnras/179.4.541
101. J.A. Regan et al., The emergence of the first star-free atomic cooling haloes in the universe. Mon. Not. R. Astron. Soc. **492**(2), 3021–3031 (2020). https://doi.org/10.1093/mnras/staa035. arXiv:2010.14533 [astro-ph.GA]
102. R.-S. Remus, K. Dolag, H. Dannerbauer, The young and the wild: what happens to protoclusters forming at redshift z *approx* 4? Astrophys. J. **950**(2), 191 (2023). https://doi.org/10.3847/1538-4357/accb91. arXiv:2010.14533 [astro-ph.CO]
103. M. Ricotti, X-ray twinkles and population III stars. Mon. Not. R. Astron. Soc. **462**(1), 601–609 (2016). https://doi.org/10.1093/mnras/stw1672. arXiv:2010.14533 [astro-ph.GA]
104. M. Ricotti, J.P. Ostriker, K.J. Mack, Effect of primordial black holes on the cosmic microwave background and cosmological parameter estimates. Astrophys. J. **680**, 829 (2008). https://doi.org/10.1086/587831. arXiv:2010.14533 [astro-ph]
105. M. Safarzadeh, Z. Haiman, Formation of GW190521 via gas accretion onto population III stellar black hole remnants born in high-redshift minihalos. Astrophys. J. Lett. **903**(1), L21 (2020). https://doi.org/10.3847/2041-8213/abc253. arXiv:2010.14533 [astro-ph.HE]
106. L.V. Sales, A. Wetzel, A. Fattahi, Baryonic solutions and challenges for cosmological models of dwarf galaxies. Nat. Astron. **6**(8), 897–910 (2022). https://doi.org/10.1038/s41550-022-01689-w. arXiv:2010.14533 [astro-ph.GA]
107. Anna T. P. Schauer et al., The influence of streaming velocities and Lyman–Werner radiation on the formation of the first stars. Mon. Not. R. Astron. Soc. **507**(2), 1775–1787 (2021). https://doi.org/10.1093/mnras/stab1953. arXiv:2010.14533 [astro-ph.GA]
108. J. Silk, On the Fragmentation of cosmic gas clouds. 1. The formation of galaxies and the first generation of stars. Astrophys. J. **211**, 638–648 (1977). https://doi.org/10.1086/154972
109. J. Silk, Unleashing positive feedback: linking the rates of star formation, supermassive black hole accretion and outflows in distant galaxies. Astrophys. J. **772**, 112 (2013). https://doi.org/10.1088/0004-637X/772/2/112. arXiv:2010.14533 [astro-ph.CO]
110. V. Springel, E pur si muove: galiliean-invariant cosmological hydrodynamical simulations on a moving mesh. Mon. Not. R. Astron. Soc. **401**, 791 (2010). https://doi.org/10.1111/j.1365-2966.2009.15715.x. arXiv:2010.14533 [astro-ph.CO]

111. V. Springel, The cosmological simulation code GADGET-2. Mon. Not. R. Astron. Soc. **364**, 1105–1134 (2005). https://doi.org/10.1111/j.1365-2966.2005.09655.x. arXiv:2010.14533
112. B.-Y. Su, N. Li, L. Feng, An inflation model for massive primordial black holes to interpret the JWST observations (2023). arXiv:2306.05364
113. K. Sugimura, K. Omukai, A.K. Inoue, The critical radiation intensity for direct collapse black hole formation: dependence on the radiation spectral shape. Mon. Not. R. Astron. Soc. **445**(1), 544–553 (2014). https://doi.org/10.1093/mnras/stu1778. arXiv:2010.14533 [astro-ph.GA]
114. V. Takhistov et al., Interstellar gas heating by primordial black holes. JCAP **03**(03), 017 (2022). https://doi.org/10.1088/1475-7516/2022/03/017. arXiv:2010.14533 [astro-ph.GA]
115. R. Teyssier, Cosmological hydrodynamics with adaptive mesh refinement: a new high resolution code called RAMSES. Astron. Astrophys. **385**, 337–364 (2002). https://doi.org/10.1051/0004-6361:20011817. arXiv:2010.14533
116. M. Trenti, M. Stiavelli, The formation rates of population III stars and chemical enrichment of halos during the reionization era. Astrophys. J. **694**, 879–892 (2009). https://doi.org/10.1088/0004-637X/694/2/879. arXiv:2010.14533 [astro-ph.CO]
117. S.M. Urbano Stawinski et al., Spectroscopic confirmation of an ultra-massive galaxy in a protocluster at $z \sim 4.9$ (2024). arXiv e-prints, arXiv:2404.16036. arXiv:2010.14533 [astro-ph.GA]
118. E. Visbal, Z. Haiman, G.L. Bryan, Direct collapse black hole formation from synchronized pairs of atomic cooling haloes. Mon. Not. R. Astron. Soc. **445**(1), 1056–1063 (2014). https://doi.org/10.1093/mnras/stu1794. arXiv:2010.14533 [astro-ph.GA]
119. M. Vogelsberger et al., Cosmological simulations of galaxy formation. Nat. Rev. Phys. **2**(1), 42–66 (2020). https://doi.org/10.1038/s42254-019-0127-2. arXiv:2010.14533 [astro-ph.GA]
120. J.H. Wise et al., Formation of massive black holes in rapidly growing pre-galactic gas clouds. Nature **566**, 85–88 (2019). https://doi.org/10.1038/s41586-019-0873-4. arXiv:2010.14533 [astro-ph.GA]
121. K.W.K. Wong et al., Constraining the primordial black hole scenario with Bayesian inference and machine learning: the GWTC-2 gravitational wave catalog. Phys. Rev. D **103**(2), 023026 (2021). https://doi.org/10.1103/PhysRevD.103.023026. arXiv:2010.14533 [gr-qc]
122. F. Yuan, R. Narayan, Hot accretion flows around black holes. Ann. Rev. Astron. Astrophys. **52**, 529–588 (2014). https://doi.org/10.1146/annurev-astro-082812-141003. arXiv:2010.14533 [astro-ph.HE]
123. Y.B. Zeldovich, Gravitational instability: an approximate theory for large density perturbations. Astron. Astrophys. **5**, 84–89 (1970)
124. S. Zhang, V. Bromm, B. Liu, How do primordial black holes change the halo mass function and structure? (2024). arXiv:2010.14533 [astro-ph.CO]
125. S. Zhang, B. Liu, V. Bromm, Distinguishing the impact and signature of black holes from different origins in early cosmic history. Mon. Not. R. Astron. Soc. **528**(1), 180–197 (2024). https://doi.org/10.1093/mnras/stad3986. arXiv:2010.14533 [astro-ph.CO]
126. F. Ziparo et al., Cosmic radiation backgrounds from primordial black holes. Mon. Not. R. Astron. Soc. **517**(1), 1086–1097 (2022). https://doi.org/10.1093/mnras/stac2705. arXiv:2010.14533 [astro-ph.CO]

Chapter 12
Black Hole Seeds at Cosmic Dawn and Their Growth

Monica Colpi and Alessandro Lupi

Abstract The black holes of this chapter form, grow and evolve following the formation and assembly of dark matter halos and the collapse of their baryonic components during cosmic structure formation: we call them *astrophysical black holes*. These black holes weighing millions to ten billions of solar masses are sites of extreme and violent phenomena, and shape the environment with the energy they produce. Their existence is necessary to explain the emission from the population of Active Galactic Nuclei, the luminous flares of tidally disrupted stars in galactic nuclei, the fast motion of the S-stars at the Galactic Center and the long-lived jets emerging from M87-like galaxies. Supermassive black holes do not come to birth "supermassive" but must have grown from *black hole seeds* of intermediate mass $\mathcal{O}(100 - 10^5\,\mathrm{M}_\odot)$. Here we describe avenues of *seed* formation at redshift $z \lesssim 30$, and investigate on their growth using "known" physics but extrapolated to untested astrophysical regimes and environments. The nature of *seeds* and their exceptionally fast evolution toward being massive is still unconstrained, and here we explore paths alternative to a *primordial origin*.

12.1 Introduction

Astrophysical black holes (BHs, hereafter) are defined as black holes resulting from the gravitational collapse of baryonic matter during cosmic structure formation. They carry masses ranging from the minimum mass of a rotating neutron star, of about $\mathcal{O}(3\,\mathrm{M}_\odot)$, up to values as large as $\mathcal{O}(10^{10}\,\mathrm{M}_\odot)$. They live in the galaxies of the Universe, either as relics of massive stars—called *stellar black holes* —or as engines

M. Colpi (✉)
Department of Physics G. Occhialini, University of Milano-Bicocca, Milano, Italy
e-mail: monica.colpi@unimib.it

A. Lupi
Department of Science and Advanced Technology, University of Insubria, Como, Italy
e-mail: alessandro.lupi@uninsubria.it

C. Byrnes et al. (eds.), *Primordial Black Holes*, Springer Series in Astrophysics and Cosmology, https://doi.org/10.1007/978-981-97-8887-3_12

at the centers of galaxies—the *supermassive black holes*—shining as Active Galactic Nuclei (AGN) due to accretion of matter near the black hole event horizon.

Primordial black holes (PBHs hereafter) can carry any mass from $\mathcal{O}(\sim 10^{-20}\,\mathrm{M}_\odot)$ (limited by Hawking evaporation) up to $\mathcal{O}(10^{10}\,\mathrm{M}_\odot)$ or more. Therefore, there exists a mass range in which BHs do overlap with PBHs. Once they have formed, PBHs embedded in galactic structures can grow by accreting baryons or/and merge with other black holes as *astrophysical* black holes do. Thus, no distinction can be easily made between PBHs and BHs in the *overlap region* above $\mathcal{O}(3\,\mathrm{M}_\odot)$, based on the mere measurement of their mass. It is only through population model selection analysis that we can potentially infer, within a given mass range or class of sources, the contribution of PBHs relative to that of BHs (see Chap. 25).

The observation of a $\mathcal{O}(0.1-1\,\mathrm{M}_\odot)$ black hole would instead provide compelling evidence of a PBH origin, and interestingly this mass range is accessible to current ground-based gravitational-wave interferometers (see Chap. 18). Soon after the discovery of the first gravitational wave (GW) signal GW150914 [84] by the LIGO Scientific Collaboration and Virgo Collaboration, and of the signals that followed [83, 87], various groups proposed that some of the merging black holes producing these signals could have primordial origin [18, 23, 136]. As these signals originate from the coalescence of black holes with masses $\mathcal{O}(10-40\,\mathrm{M}_\odot)$ in the overlap region, they could descend either from the first metal-free stars [60] or from binaries forming across cosmic ages [98]. As of today, no sub-solar merger event has been detected, that would provide a smoking gun signature of the presence of a PBH, and an upper limit on the fraction of dark matter in PBHs in this mass range has been set around 6% [86].

Understanding the onset of star, black hole and galaxy formation at cosmic dawn, a few hundreds of Myrs after the Big Bang, is one of the key unanswered questions in contemporary astrophysics and cosmology. With the successful launch of the James Webb Space Telescope (JWST), new upcoming electromagnetic observatories as the Vera Rubin Observatory [70], the Nancy Grace Roman Telescope [58], and in the next decade with the launch of the Laser Interferometer Space Antenna (LISA) [31] and the construction of more sensitive terrestrial gravitational wave detectors as Einstein Telescope (ET) [27, 124] and Cosmic Explorer (CE) [130], this early epoch of cosmic history will become accessible to direct observations, thus finally shedding light on the origin(s) of BHs. It is pushing our boundaries to extremely high redshift, up to $z \sim 30$–40, and in a suitable range of masses, that we may learn directly about the existence of PBHs, as stars are expected to form later inside the first collapsing dark matter halos [105].

While the rest of this book focuses on the physics and potential consequences of PBHs (in particular see Chaps. 10 and 11 for the cosmological evolution and impact on the first stars and galaxies), here we provide a complementary view in terms of known physics to explain black hole seeds and their growth. We consider *astrophysical supermassive black holes*, and focus on their formation avenues below redshift $z \sim 30$. Excellent reviews have been recently published on black hole seeds, which add great value and content to this short chapter, which is not comprehensive of the vast literature. These reviews cover most if not all topics related to this rapidly evolving field [57, 69, 156]. In this book, Liu & Bromm explored the possible impact of PBHs on the formation of the first stars and galaxies, and on the role of PBHs as seeds. In this chapter, we describe seed formation mechanisms and growth in the earliest galactic halos, through stellar/gas processes in different environments, and explore their evolution.

This chapter is organised as follows: in Sect. 12.2 we briefly outline the observations of the high-redshift AGN, called quasars due to their large luminosities (in excess of $10^{13}\ L_{\odot}$), of AGN at lower redshifts, and massive dark objects in quiescent galaxies, which are all conducive to the concept of *black holes seeds*. In Sect. 12.3 we describe seeds forming from the gravitational collapse of single stars or massive gas clouds, and seeds that form via dynamical processes in dense environments, such as nuclear star clusters or through collisions and mergers. In Sect. 12.4 we describe the complexity of the evolution model for the growth of seeds over cosmic history and extract examples from the literature. Not a single process, but a variety of them with different statistical weight are likely to concur to the formation and evolution of the seeds. In Sect. 12.5 we discuss the future role of gravitational wave observations in identifying seed black holes and SMBH in binaries, in order to disentangle their origin and evolution channels.

12.2 A Universe of Supermassive Black Holes

Most of what we know about *astrophysical supermassive black holes* (SMBHs) thus far has been informed form the observation of about a million AGN covering broad ranges in redshift ($0.1 \lesssim z \lesssim 7{-}8$) and luminosities ($44 \lesssim \log(L_{\rm bol}/{\rm erg\,s^{-1}}) \lesssim 49$) [159], from the rare quasars shining when the Universe was about 700 Myrs old, during the epoch of cosmic reionization [161] to the population of SMBHs observed as *massive dark objects* in almost all bright galaxies of the today Universe, representing the endpoint of the AGN evolution [78].

In Fig. 12.1 we sketch the mass function of both supermassive and stellar BHs, and highlight the presence of a "desert" (white area) and of an unconstrained region between the two families (light grey area) where *intermediate mass* BHs (IMBHs) with masses $O(100\,{\rm M}_{\odot} - 10^5\,{\rm M}_{\odot})$ are expected to exist. These IMBHs are difficult to observe due to their weaker accretion luminosities and weaker imprint on the kinematics of orbiting stars. Nonetheless, observations are now uncovering these BHs in dwarf galaxies [104].

Whether the black hole mass function is continuous over the two families or whether the SMBH mass function displays a cut-off at a critical mass, in the IMBH window, is yet unknown. As illustrated in Fig. 12.1, stellar BHs are now being detected in galaxies as gravitational wave sources with masses $O(3{-}60\,{\rm M}_{\odot})$ with the heaviest observed so far in GW190521 [82, 85] with a mass of $\sim 150\,{\rm M}_{\odot}$. The lightest SMBH candidate, in the galaxy NGC 205 [112], weighs only 6800 ${\rm M}_{\odot}$, SgA*—the black hole in our Milky Way Galaxy—has a mass of $4.4 \times 10^6\,{\rm M}_{\odot}$ [52],

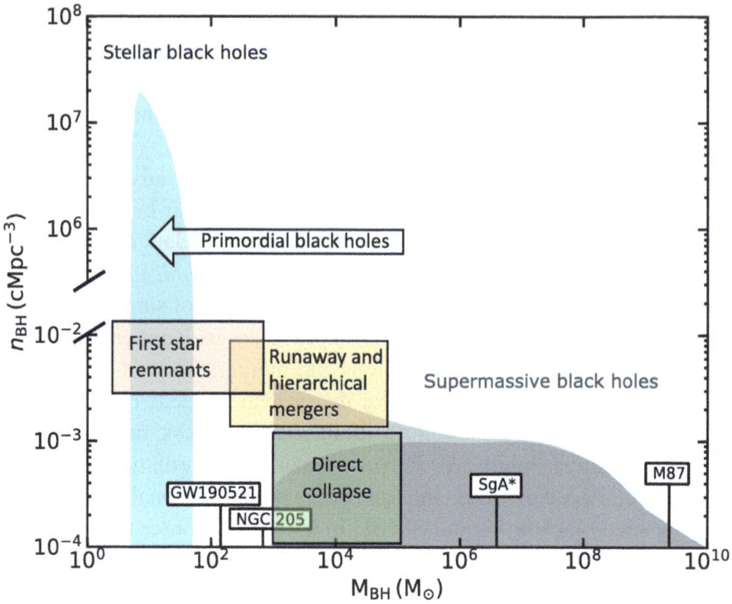

Fig. 12.1 Cartoon sketching the number density in stellar and supermassive BHs, with the light shaded grey area indicating the uncertainties in the shape of the mass function around $O(10^3 - 10^5\,M_\odot)$. The white area between the stellar and supermassive BHs highlights the zone comprising IMBHs, where observations are challenging. Boxes indicate the different channels of *black hole seed* formation. The figure is a modification of Fig. 1 in [156]

while among prototypical SMBHs we highlight the $6.5 \times 10^9\,M_\odot$ BH in the elliptical galaxy M87 [45]. Rare and distant SMBHs as heavy as a few $10^{10}\,M_\odot$ are now being discovered in large surveys [116].

A suite of observations indicates that SMBHs must have grown in mass over cosmic time mainly by gas accretion [99, 142]. This is inferred comparing the local SMBH mass-density, ρ_{BH}, derived from the observation of quiescent SMBHs in bright nearby galaxies, with the mass-density increment experienced by the accreting SMBHs when shining as AGN, $\Delta\rho_{BH}$. This results in a $\rho_{BH} \sim 2 - 5 \times 10^5\,M_\odot\,Mpc^{-3}$ in close match with $\Delta\rho_{BH} \sim 3.5 \times 10^5(\eta/0.1)^{-1}\,M_\odot\,Mpc^{-3}$ (with η denoting the radiative efficiency). The SMBH growth by accretion, as inferred by these two independent estimates, has led to the concept of the existence of *seeds*, i.e. IMBHs of yet unknown origin.

The discovery of the reionization quasars [48, 161] with masses of $O(10^9\,M_\odot)$ at $z \sim 6 - 7$ poses additional constraints on the way SMBHs have grown. It requires that *seeds* came into being at very high redshift ($z \sim 10 - 30$), when the Universe was only a few hundred Myrs old and the first dark matter halos were collapsing to spark star formations following baryon infall [106]. These *seeds* may have been born as relics of the first stars or with masses straddling the two BH families, in what is currently an observational desert. The IMBH population, filling this mass-gap, could hold the key to understanding the formation of the *seeds* and in Fig. 12.1 we indicate paths of formation, described in Sect. 12.3.

SMBHs are central actors in the ecology of galaxy formation and evolution, due to the energy they produce and deposit in the host galaxy when they shine as quasars [46]. This has led to the discovery of a relation between the SMBH mass $M_{\rm BH}$ and the stellar mass M_* and stellar velocity dispersion of the galaxy they inhabit, which indicates that SMBHs *coevolve* with their host [78]. The relation $M_{\rm BH} - M_*$, with its slope, normalization, and scatter has emerged as one of the most important scaling relations in unveiling the nature of the SMBH-galaxy coevolution. *Seed* formation and growth appear indeed to be ruled by the interplay among various feedback processes, i.e. between the BH (its emission affecting the state of the interstellar medium) and the stars (forming in cool regions and exploding in supernovae events) and the multi-phase gas (feeding the BH and being heated/cooled by radiation from the BH and stars). This can explain in a single, albeit complex, picture the relation between star formation, *seed* and AGN and galaxy evolution [145].

JWST is on the verge of revolutionizing the study of high redshift ($4 \lesssim z \lesssim 9$) AGN and the early coevolution of SMBHs with their host galaxies. The unprecedented sensitivity and resolution of NIRCam imaging have enabled the first detection of stellar light associated to AGN during the reionization epoch, leading to estimates of the stellar masses and therefore of the $M_{\rm BH} - M_*$ relation close to the seeding epochs, which may shed light on the formation mechanisms of SMBHs. See, e.g. [41, 59, 95, 96, 149] and the references therein. The observations claim SMBHs in these systems to be significantly more massive than expected from the local $M_{\rm BH} - M_*$ and a debate is ongoing on whether this is an intrinsically overmassive population or a manifestation of a selection effect [81, 117].

In addition, JWST is revealing a new population of dust-reddened broad-line AGN at redshifts $z > 5$ with BH masses of $10^7 - 10^9\,{\rm M}_\odot$ and high number densities [56]. Lastly, the discovery of a candidate SMBH at $z \sim 10$ in the lensed galaxy GHZ9 with a lower limit for its mass of $8 \times 10^7\,{\rm M}_\odot$ [77] starts constraining the *seed* origin of this SMBH. An over-abundant $z \sim 10$ SMBH population would imply a higher-than-expected seed formation efficiency, and rapid growth. This observation is contemporaneous of the finding of a smaller SMBH of $10^6\,{\rm M}_\odot$ in the galaxy GN-z11, at reshift $z \sim 11$ accreting at five times the Eddington limit [95]. In addition, SMBHs at $4 < z < 11$ are now being discovered [96], some in close pairs and with lower masses, in the range $O(10^6 - 10^8\,{\rm M}_\odot)$ [96].

12.3 Channels of Black Hole Seed Formation

Seeds are IMBHs with masses in the range $O(10^2 - 10^5\,{\rm M}_\odot)$ and are the endpoint of the unhalted *gravitational collapse* from an initial configuration that can either be (as described below) a massive star, a supermassive star, or a very massive star outcome of runaway mergers among stars, or the end-result of hierarchical sequence of black hole merges [156]. Each channel, highlighted in Fig. 12.1, identifies a preferred mass range, a specific astrophysical environment, and a formation redshift. Finally, *seeds* can have primordial origin.

We distinguish here (i) *channels linked to stars and gas* from (ii) *channels linked to stellar/black hole dynamics*. *Seeds* are transient, high-redshift objects, undergoing fast evolution to account for the observation of the high-z quasars. *Seeds* can also evolve on longer timescales to account for the lower-redshift rich population of AGN. Furthermore, *starved seeds* can be present as IMBHs in dense star clusters or/and in dwarf galaxies [104].

12.3.1 Channels Linked to Stars and Gas

Light seeds fall in the first category (i) and have nominal masses from $\sim 100\,M_\odot$ to a few $100\,M_\odot$. They are the relics of the most massive (population III) stars forming in pristine metal-free gas clouds inside dark matter halos of $\sim 10^6\,M_\odot$ collapsing at redshifts $z \sim 10$–30 [61, 69, 156]. In metal-free conditions stars originate from the fragmentation of gas clouds which cool due to the presence of molecular hydrogen [147]. Because of the fewer—collisionally excited—spectral transitions than in matter enriched by metals (i.e. the elements heavier than helium), high-z pristine star forming regions have high equilibrium temperatures T (of about 100 K) and therefore a high Jeans mass, $M_J \propto T^{3/2}$, which is the mass setting the scale of fragmentation. This is expected to lead to a rather *top-heavy* stellar initial mass function, from solar masses but up to several hundreds of solar masses. Thus, many stars end their lives as black holes, with masses comparable to that of the progenitor stars, as winds during stellar evolution are heavily suppressed due to the lack of atomic lines offering a strong interaction with the stellar radiation [143]. Metal poor/free stars with masses between about 130 and $250\,M_\odot$ are subjected to the so called *pair-instability*, triggered when copious electron-positron pairs are created in the very hot helium core of these massive stars [158]. Due to uncontrolled carbon-oxygen nuclear burning, these stars leave no remnant, and an *upper-mass gap* in the BH mass function is expected to be present between about $65\,M_\odot$ and $135\,M_\odot$. *Light* seeds are generally assumed to lie beyond this pair instability gap to guarantee their growth via accretion of gas and stars because of their deeper gravitational potential well than lower-mass BHs. Even though the physics of pristine gas is rather simple, the level of detail required to accurately determine the mass spectrum of population III stars resulted in the need of high-resolution numerical simulations able to directly track the impact of stellar radiation on the mass accretion of the protostars [62, 65, 76, 114, 148]. Despite this effort, a consensus is still missing.

Heavy seeds belong again to category (i) and are generally thought to carry masses of $10^4 - 10^5\,M_\odot$. They form from the *direct collapse* of metal-free gas clouds in dark matter "atomic-cooling" halos of $\sim 10^8\,M_\odot$ at lower redhifts, $z \sim 15$ [14, 125, 156]. In this scenario, a gas cloud contracts coherently possibly forming a star of comparable mass that collapses to a massive seed, either via an intermediate phase known as *quasistar* powered by continuing accretion onto a growing small black hole embryo, or by a *supermassive star* (SMS) driven by general relativistic instabilities and fuel exhaustion [14, 68, 153]. *Heavy* seeds are considered to be rare, due to their contrived birth environmental conditions [2, 126, 154]. They need to form in halos surrounded by companion haloes undergoing intense (and metal polluting) star formation. An intense UV radiation, from adjacent star forming regions, is requested in order to photo-dissociate molecular hydrogen and allow cooling via atomic hydrogen transitions, at temperatures of $\sim 10^4$ K, together with a large infall rate of metal-free/poor gas to suppress fragmentation of the birth cloud and feed the central object. Using a radiation hydro-dynamical simulation to describe early galaxy formation, in [157] the authors show the natural emergence of metal-free halos massive enough and with sufficiently high gas infall rates to form a SMS. They pointed further out that repeated mergers among halos, rather than a critical UV flux, may be the main driver of massive BH formation since turbulence plays a role in providing additional stabilization of the gas cloud, against fragmentation. This process is called *dynamical heating*, and it represents the PdV work and shocks associated to rapid gas infall, which is a major driver of central gas collapse, possibly as important as the dissociating radiation. Recently, in [80] it is found that heating by compression in cosmic web filaments is an important mechanism to consider. It suggests that for the formation of *heavy* seeds the environment of atomic cooling halos is more relevant than the dissociating radiation emitted by adjacent star forming regions [63, 91]. Starting from the results by [80, 128, 157] further explores the evolution of metal-free atomic-cooling, irradiated halos with extremely high-resolution simulations. The authors find that gas clouds fragment producing *seeds* with masses of a few 10^3 $M_\odot$, which are lower than those typically assumed for *heavy seeds*.

12.3.2 Channels Linked to Stellar/Black Hole Dynamics

The *runaway stellar mergers* is a scenario which belongs to category (ii), i.e. the dynamical channel [156]. Here, seeds of $O(100 - 10^3\,M_\odot)$ can form from the collapse of a *very massive star* (VMS) in the central region of a compact, metal-poor star cluster. Clusters with core's densities as large as $10^6\,M_\odot\,pc^{-3}$ undergo gravitational collapse driven by two-body relaxation processes on a timescale shorter than the life of the most massive stars [8, 36, 122, 129]. At the time of cluster core contraction, multiple stellar collisions can form a VMS. These collisions are not the end-result of head-on, hyperbolic encounters among stars. They occur gently via single-binary encounters of main sequence stars, which repeat in a runaway fashion on timescales of less than a few Myrs. The formed VMS is then collapsing into an IMBH. The IMBH can sink deep into the cluster and further grow by accreting stars [9]. We remark that stellar collisions and mergers can actually overcome the *pair-instability* by "mixing" stars in different evolutionary stages, a mechanism that allows increasing the stellar mass while keeping the helium core below the mass-threshold for explosive carbon-oxygen nuclear burning [37]. Stellar mergers of this type have proven to be a viable way to generate BHs within the *pair-instability gap* in star clusters and, in some cases, to generate IMBHs beyond that gap. IMBH formation has been investigated recently through dedicated studies in low-metallicity young and old star clusters, in light of their importance as gravitational wave sources both for ground-based and space-based detectors [9, 34, 38].

The *hierarchical black hole mergers* is an alternative scenario [7, 50, 53, 156] which belong to (ii). In star clusters of $10^4 - 10^7\,M_\odot$, stellar BHs sink rapidly (on timescales of a few Myrs) by dynamical friction forming a nested core where close, repeated encounters between binary and single BHs take place. Dynamical recoils resulting from these encounters can either heat the core which expands lowering the interaction rate, or eject the BHs, before they merge, inhibiting the sequence of hierarchical growth. Hierarchical black hole mergers have been proven to occur in a variety of direct N-body simulations but the impact of the recoil induced by the emission of gravitational waves at the time of any merger is found to inhibit hierarchical growth. This causes the sequence of mergers to terminate at the second or third generation, unless the cluster has a very high central velocity dispersion (in excess of $100\,\text{km s}^{-1}$) to retain the product of the merger [7]. Alternatively, [90] explored a model whereby a major central gas inflow occurs onto a nuclear star cluster at the center of a forming galactic disk. The inflow, triggered by disk instabilities, steepens the gravitational potential well of the star cluster to the extent that a massive *seed* of $\sim 10^3\,M_\odot$ can form through hierarchical BH mergers [35]. This process, which is independent of the metal content of the parent cluster, is competitive relative to others, and is predicted to peak at $z \sim 10$.

Merging galaxies at redshifts $z \sim 8 - 10$ can provide an alternative pathway to explain the rapid emergence of bright, rare and massive high-z quasars [102]. In this model, major mergers between the most massive galaxies, already enriched to solar metallicity values, lead to the formation of a nuclear supermassive disk of a few parsecs in size and mass $\sim 10^9\,M_\odot$ in less than a Myr, due to prodigious gas inflows. Here, dynamical heating as gas falls into the deepest part of the potential well, and stirring by supernova blast waves, generate a turbulent multi-phase interstellar medium, resulting only in a moderate fragmentation, despite the temperature falling below 10^3 K. The disk becomes Jeans-unstable as well as bar-unstable and collapses further until it develops a general-relativistic instability, directly forming a SMS, which collapses into a BH of $\mathcal{O}(10^6 - 10^8\,M_\odot)$ [101].

12.4 Seed Growth

Even though a proper assessment of the fundamental mechanisms behind seed formation is crucial to our understanding of the origins of SMBHs, an even more important aspect is represented by their subsequent growth, which is required to explain current observations of AGN and quasars. In this section, we describe shortly how BHs grow in general, from their initial state as *seeds* to their later state as SMBHs.

12.4.1 Growing a Black Hole

In the simplest picture of accretion, the rate of mass increase of a *seed* or SMBH can be estimated as follows:

$$\dot{M}_{\rm BH} = (1-\eta)\dot{M} = [(1-\eta)/\eta] f_{\rm E} L_{\rm E}/c^2, \tag{12.1}$$

where the average accretion rate $\dot{M}$ is related to the bolometric luminosity $L = \eta \dot{M} c^2$ with η representing the radiative efficiency, given by the ratio of the energy that flows out to infinity in the form of radiation to the rest mass-energy that flows in, crossing the event horizon. The luminosity L is expressed in terms of the Eddington luminosity $L_{\rm E} = 4\pi G M_{\rm BH} m_{\rm p} c/\sigma_{\rm T} = 1.3 \times 10^{38} (M_{\rm BH}/\,{\rm M}_{\odot})\,{\rm erg\,s}^{-1}$, with $f_{\rm E} \equiv L/L_{\rm E}$, the Eddington factor which accounts for emission at a sub- ($f_{\rm E} < 1$) or at a super- ($f_{\rm E} > 1$) Eddington rate. In Eq. (12.1), $(1-\eta)$ represents the fraction of rest-mass energy which is accreted by the BH and thus measures the mass-energy increment experienced by the BH in any of the accretion episodes. In case of continuous accretion, Eq. (12.1) implies an $e-$ fold growth of the seed/SMBH on a timescale (often referred to as the Salpeter timescale), $\tau_{\rm S} = 4.7 \times 10^8 \eta\, [(1-\eta) f_{\rm E}]^{-1}$yrs. For $\eta \sim 0.1$—typical of radiatively efficient accretion [140]—and $f_{\rm E} \sim 0.1$, this timescale is short (about 3 %) compared to the age of the Universe today, but can be a large fraction of the "then" age of the Universe when considering high-z quasars present a few hundreds of Myrs after the Big Bang. In Fig. 12.2, we show growth tracks as described by Eq. (12.1) for *light* and *heavy* seeds. By shifting the formation redshift to progressively higher z, *light* and *heavy* seeds can both explain the observations of the high-z quasars. But, as soon as we constrain the redshift window, (*heavy seeds* form later than *light seeds*), one has to tune the values of $f_{\rm E}$, i.e. assume super-Eddington episodes of accretion, to match observations. Thus uncertainties remain large, given the arbitrariness of the choices made and the degeneracy of the process.

Equation (12.1) is an oversimplification of the seed/SMBH growth. A first correction could be introduced to account for the difference between the radiative efficiency, now indicated as $\eta_{\rm rad}$, and the accretion efficiency that we denote here as $(1-\eta_{\rm out})$ expressing the net mass-energy crossing the BH horizon [71]. With this distinction, $\eta_{\rm rad}$ and $(1-\eta_{\rm out})$ depend on the geometry and radiative properties of the accretion disk, on the presence of outflows/winds and jets, and ultimately on the spin of the accreting BH (see Sect. 12.4.2). At any cosmic time t, the mass increase of the BH can be modelled as

$$\dot{M}_{\rm BH} = M_{\rm BH}(t) \left\{ \frac{(1-\eta_{\rm out}(t))}{\eta_{\rm rad}(t)} \frac{f_{\rm E}(t)}{t_{\rm S}} \right\}, \tag{12.2}$$

with $ts = 4.7 \times 10^8$ yrs.

The geometry and radiative properties of an accretion disk are mainly determined by the dimensionless parameter $f_{\rm E}$, by the degree of magnetization of the accreting

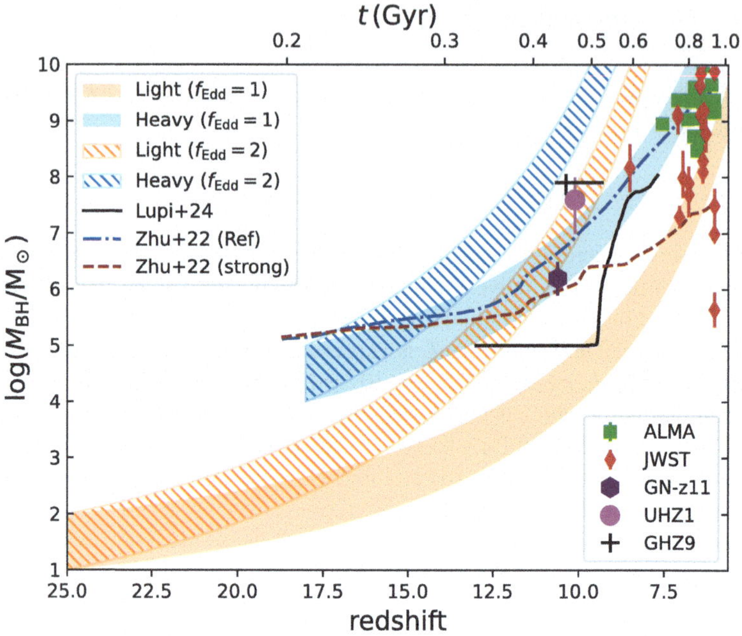

Fig. 12.2 The BH mass versus redshift z (cosmic time). The figure shows tracks of BH growth, for different initial *seed* masses assuming uninterrupted accretion with a radiative efficiency of 10%, computed from Eq. (12.1). All symbols with color and shape as in the legend refer to high-redshift ($z > 6$) observations of quasars by ALMA [110] and JWST [41, 59, 97, 149, 162]. *Light seeds* can explain these quasars and even GN-z11, UHZ1, and GHZ9 if (in these last three cases) they are accreting at a super-Eddington rate. *Heavy seeds*, forming at lower redshifts, do not necessarily need sustained super-Eddington accretion, more naturally accounting for the highest redshift quasars. There is a debate whether sustained accretion above the Eddington limit is likely in high-redshift galaxies [56, 92, 100, 127, 152]. The solid, black track is the result of a state-of-the-art high-resolution cosmological zoom-in simulations of a $z \sim 7.5$ quasar, where a *heavy seed* of $10^5\,M_\odot$ formed at $z = 10$ grows by more than two orders of magnitude in a phase of super-Eddington accretion lasting tens of Myrs. Here, disk geometry and feedback are implemented as described in Sect. 12.4.1, Eq. (12.2), including the MAD phase. The dot-dashed blue line and the brown dashed one correspond to cosmological simulations by [163] assuming Eddington-limited accretion and two different AGN feedback efficiencies, 0.1 and 0.2 respectively

plasma, and by the BH spin $J_{BH} = \chi_{BH} G M_{BH}^2/c$, where $0 \leq \chi_{BH} \leq 1$ is the dimensionless amplitude. A highly spinning BH has an innermost stable circular orbit closer to the event horizon and $0.0573(\chi_{BH=0}) < \eta_{rad} < 0.42(\chi_{BH=1})$ [94]. We can identify, for seek of simplicity, three main regimes: (1) for $f_E \lesssim 10^{-3} - 10^{-2}$ (sub-Eddington case), an advection dominated accretion flow (ADAF) [107] forms in which matter falls without radiating away much of its energy, due to inefficient cooling. The disk is geometrically thick, optically thin and ions and electrons are thermally decoupled, resulting in a two-temperature accretion flow where the radiative efficiency is low, scaling as $\eta_{rad} \propto f_E^{2/3}$ [140, 160]; (2) for $10^{-2} \lesssim f_E \lesssim 1$ the disk is optically thick and geometrically thin, the fluid velocity is nearly Keplerian and is described by the α-disk viscosity model of Shakura and Sunyaev [140] with $\eta_{\rm rad}$ depending on the BH spin and $\eta_{\rm out} \sim \eta_{\rm rad}$. Most of the AGN are found to radiate in this regime; (3) for $f_{\rm E} \gtrsim 1$, corresponding to super-Eddington accretion, the disk is described by the *slim-disk solution* in which radiation is trapped and advected inwards, leading to $(1 - \eta_{\rm out}) \sim 1$ and to values of the radiative efficiency smaller than in case (2) by a factor ~ 10. In this case Eq. (12.2) can be modelled as a function of $f_{\rm E}$ as in [1, 94]. This favors rapid BH growth.

Inside the accreting plasma, electromagnetic currents can easily develop that result in magnetic fields advected with the flow. At super-Eddington rates the disk can reach the so-called Magnetically Arrested Disk (MAD) state [19, 67, 108]. Here, the plasma drags-in a strong poloidal magnetic field which disrupts the axisymmetric accretion flow near the BH. Inside the disruption radius, the gas accretes in discrete blobs and almost the entire energy of the gas is released as magnetic energy and most importantly as mechanical energy in the form of a collimated relativistic jets [21, 134]. In the states (2) and (3) radiation and mechanical energy can affect, respectively, the surrounding gas feeding the BH. This self-regulates the accretion rate (and in turn the BH growth) setting the conditions for the BH *feedback* described in Sect. 12.4.3.

Figure 12.2 shows the growth of *light* (*heavy*) seeds from Eq. (12.1), forming at the earliest redshift $z = 25$ ($z = 17$), and accreting with $f_{\rm E} = 1$ and $f_{\rm E} = 2$ without interruption, with $\eta = 0.1$. In the figure, we report the most recent observations by ALMA [110] and JWST [41, 56, 59, 97, 149, 162], and highlight three of the most extreme AGN, GN-z11 [95], UHZ1 [24, 109] and GHZ9 [109], mentioned in Sect. 12.2. On the same plot we overlay the growth track of a *heavy* seed of $10^5\,{\rm M}_\odot$ that was let to evolve inside a high-resolution cosmological zoom-in simulation of a $z \sim 7.5$ quasar, where the accretion regimes described in Sect. 12.4.1 were self-consistently included, together with the associated radiative and kinetic feedback [92]. The seed has experienced a relatively long phase of super-Eddington that enhanced the mass by more than two orders of magnitude, after an early inactive phase. The simulation is halted at $z \sim 7.5$ but we expect that further accretion episodes at the Eddington level can further enhance the SMBH mass to reach the "clouds" of observed quasars, growing along the blue strip.

12.4.2 Spin Evolution

The spin of a BH is a key astrophysical parameter. At birth it provides information on the dynamics of gravitational collapse [115]. Instead, the spin of SMBHs charts the way accretion and mergers proceeded during their cosmic growth [15].

Measuring the spin is observationally challenging and this measure is limited to nearby and bright accreting BHs. For AGN, the spin is inferred through X-ray reflection spectroscopy and thermal continuum fitting [133]. It is found to be $0.5 \lesssim \chi_{\rm BH} \lesssim 1$. The spin of stellar BHs in X-ray binary sources is inferred using the same techniques as for AGN and is to be the interval $0.3 \lesssim \chi_{\rm BH} \lesssim 0.8$ [132]. The *effective* spin of merging stellar BH binaries—a mass-weighted value of the individual spins projected on the binary angular momentum vector—has been measured from the GW signals [87]. It is found to have very low values, hinting for evolutionary pathways different from those experienced in X-ray binaries.

For stellar BHs, spin magnitudes at birth are affected by rotation, metallicity, extent of mass losses/winds of the progenitor star and by the presence of a companion star when in a binary [55]. Hierarchically assembled BHs, which merge with spins isotropically distributed, are found to carry dimensionless spins clustered around $\chi_{BH} \sim 0.7$ [54, 66]. The spin of quasi-stars depends instead on the amount of rotation stored in the stellar envelope [14].

The spin of a *seed* BH, after birth, is subjected to changes due to accretion and mergers. In the case of accretion, matter carries across the event horizon the angular momentum it has at the innermost stable circular orbit. The BH angular momentum increases (decreases) if the disk is co-rotating (counter-rotating) relative to the BH spin direction. Changing the spin amplitude χ_{BH} from 0 to 1 (in case of equatorial, prograde accretion) or from 1 to 0 (in case of retrograde accretion) requires a mass load comparable to the BH mass [13]. As seed BHs increase their mass by many orders of magnitude, the spin undergoes large changes and information residing in the initial spin distribution is erased. Mergers also change the spin as the newly formed BH inherits non only the spin of the two BHs but acquires the residual orbital angular momentum of the binary [11, 64]. Thus accretion and mergers shape the SMBH evolution and are closely interconnected.

During the cosmic evolution of a seed/SMBH, accretion episodes are expected to repeat recursively. They can occur in a *coherent* way (i.e., keeping the BH spin orientation almost parallel to the angular momentum of the disk) or/and at random phases in a *chaotic* way (i.e., having the spin a random orientation relative to the angular momentum of the disk) [74]. In case the spin of a seed/SMBH is misaligned relative to the direction of the angular momentum of the disk, gas elements inside the disk undergo Lense-Thirring precession. Vertical shear viscosity in the differentially rotating annuli ensures that the inner disk region aligns (or anti-aligns) its orbital angular momentum with J_{BH} out to a transition radius making the disk warped [119]. At the same time, the torque provided by the warped disk makes the BH spin to precess. This configuration evolves into a state of lowest energy, owing to angular momentum conservation, in which the SMBH aligns its spin to the total angular momentum of the composite system [75, 150]. Thus, the spin orientation changes as well. Modeling phases of either chaotic or coherent accretion is a way to exploit how a seed grows, through cosmic ages. It is also a way to capture the physics of BH fueling, which is connected to the properties of the BH environment, affected by BH feedback and by the way the host galaxy evolves. The measurement of the spins of growing seeds at different redshifts will guide our understanding of the mechanism of BH evolution [138, 139], and this will be possible for SMBHs in binaries, detected as GW sources as commented in Sect. 12.5. The joint mass and spin evolution has been implemented in a number of different codes [25, 28, 29, 43, 49, 118], and included in large-scale cosmological simulations [42] and in semi-analytical models [72].

12.4.3 Radiative and Mechanical Feedback

The growth of a seed or of a SMBH is halted or/and delayed when the gas reservoir from which the BH accretes diminishes. The cause of this depletion is manyfold. In the AGN/quasar phase, large amounts of energy and momentum are released in the form or radiation, winds, fast outflows and jets from sub-pc to kpc scales. When a large scale-jet is launched it can affect the intergalactic medium in specific, anisotropic directions. The energy/momentum deposited by the seed/SMBH in its vicinity depletes the gas reservoir preventing accretion. Accretion resumes only when this effect—called *AGN feedback*—has subsided. Similarly, supernova explosions near the BH suppress accretion and the seed/SMBH growth can be delayed until the starburst has ended and gas returns to cool down for a new cycle of accretion and star formation to start. Therefore, AGN and stellar feedback can be viewed as recursive episodes of accretion-feedback-starvation along the course of evolution of a seed/SMBH that needs to be considered in any model of SMBH formation and evolution.

Note, however, that AGN feedback has much more profound impact that the mere MBH growth regulation/suppression. In facts, the feedback energy released by the accretion process also represents a unique tool to study physical processes occurring around AGN. In particular, the accreting material around MBHs is thought to typically settle in a disk-like structure, whose properties strongly depend on the inflows from large scales. The most common regime is called 'radiatively efficient' and is characterised by the geometrically thin optically thick disk described as case (2) (Sect. 12.4.1) in which thermal radiation escapes from the disk surfaces, producing a well defined electromagnetic spectrum [140]. On its way outwards, radiation is partially re-processed by intervening gas, producing specific features that are commonly employed to identify AGN, as the emission from the "broad line region" (a region of gas clouds whose orbits are fully determined by the SMBH gravitational pull) or the non-thermal X-ray emission from the corona, a hot ionised region in which electrons up-scatter the continuum photons from the accretion disk. Depending on the opacity of the gas around the accretion disk, the absorbed radiation can transfer sufficient momentum or heating to expel the gas from the galactic nucleus in a fast wind, with velocities up to 0.2c [123, 141].

An alternative mechanism to produce feedback is through magnetically-driven winds and jets. In this case, the energy source is the magnetic field of the accreting material. Magnetic fields are always present in an accretion disk and are thought to become the main energy source for radiatively-inefficient accretion flows, i.e. at very low accretion rates ($f_E \lesssim 0.01$) or close to and above the Eddington limit [135]. In these cases (as described in Sect. 12.4.1), magnetic fields can either produce a magnetocentrifugal force [20] able to expel material from the disk (when the poloidal component of the magnetic field is large) or tangle in the vicinity of the horizon up to the breaking point, i.e. when the field lines open and charged particles are released in a collimated jet [22], as the one observed in M87. In the latter case, in particular, the energy to launch the jet comes from the rotational energy of the MBH, which causes the MBH to spin down, and feedback can be anisotropic.

12.4.4 Merging Seeds

Seeds and SMBHs inhabit halos, which grow via repeated major and minor mergers with other halos. Since halo mergers may lead to the formation of binary BH (BBH) systems, BHs could be observed as GW sources at the time of their coalescence (we defer the reader to Sect. 12.5 for further details). Stellar and gaseous torques are effective in driving the BHs from the kpc scale typical of a galaxy merger down to sub-milli-parsec separations, the scale below which GW emission drives the inspiral down to coalescence [4].

The process of pairing and hardening of seeds and SMBHs shows bottlenecks due to the difficulty in accounting realistically for the different torques in extracting angular momentum [4]. The "effectiveness" of pairing is debated for the seeds, despite the fact that the halo-halo merger rate is higher at higher z, and dynamical timescales are shorter than today. Cosmological simulations show that seeds wander in the (proto-)galaxies in which they form, unless they carry masses as large as $\sim 10^5\,M_\odot$ [93, 120]. The displacement induced by their wandering in general reduces not only the amount of mass that the seed can initially accrete, but the ability to pair and form close binaries with other *seeds*. Forming a binary of two *seeds*, able to coalesce on a time shorter than the Hubble time, appears challenging. *Light seeds* may have formed in situ (as binary stars do) or they may have formed via triple interactions in the earliest merging halos [155]. After coalescence, depending on the spin direction and magnitude, seeds (as any other binary SMBH) receive a gravitational kick due to the anisotropy of the emission, as a consequence of linear momentum conservation [89]. Since kick velocities range from a few 100 km s^{-1} to 2000 km s^{-1}, merging seeds can be kicked off their host, which implies the loss of a "growing seed".

12.4.5 Growing Seeds in Numerical Simulations

Current numerical simulations of seed growth can account for both the BH and stellar feedback, included at "sub-grid" level, and modeled semi-analytically, due to huge dynamic range needed to properly resolve accretion disk scales. This limitation does not allow numerical simulations to properly account for physical processes governing the seed formation mechanism. To date, most large scale studies only include heavy seeds [40, 42, 121, 137, 151], with only a few attempts at including lower masses [17, 163]. In addition, the seeding itself is typically based on very crude assumptions, and no detailed exploration of the different formation mechanisms has been performed yet.

Regarding BH growth, the impact of BH feedback is commonly calibrated to reproduce the AGN luminosity function at low-redshift and galaxy quenching, and is only broadly connected to the physics of accretion [6, 30, 32, 33, 39, 44]. The inclusion of accretion and feedback (by stars and BHs) has showed that accretion is episodic and recursive, rather than continuous, hence far from the idealised tracks shown in Fig. 12.2.

Recently, JWST results have strengthened the need for a fast growth, even in the case of heavy seeds, challenging current simulations that, because of the intermittency of the accretion and the impact of supernova feedback in low-mass systems, struggle to reproduce the most massive and distant SMBHs observed. A possible solution to this problem comes from accretion bursts above the Eddington limit, whose plausibility has been demonstrated observationally in the case of neutron stars and stellar black holes [10, 88], while only hints of its occurrence have been found in the case of AGN, due to the uncertainties in the BH mass estimate. Theoretically, super-Eddington phases have been investigated in detail using dedicated simulations on accretion disk scales [73, 134, 135], completely disconnected from galactic scales and the cosmological structure formation. In the last few years, however, the interest in the super-Eddington regime has increased (in particular after the first JWST results), with some studies simply allowing for accretion above the Eddington limit, without any change in the "sub-grid" treatment of BHs [5, 16, 113, 163], and a few trying to even include a more physically-motivated description of BH accretion and feedback [92, 100, 127, 131]. An example of this attempt is shown in Fig. 12.2 as a black dashed line, which highlights how, despite a strong kinetic feedback, super-Eddington accretion phases can be sustained for relatively long timescales (tens of Myr) to let a *heavy seed* of $10^5\ M_{\odot}$ to grow by two or more orders of magnitude. The BH spin was shown to remain relatively low during super-Eddington phase, resulting in a lower (mechanical) feedback efficiency, and a faster growth that can explain the over-massiveness of observed high-redshift quasars. Even though it remains easier for heavy seeds to grow to the masses observed to date, the plausibility of super-Eddington accretion phases does not allow to exclude the light seed scenario, especially in the light of (1) the large abundance of SMBHs at high-redshift compared to the standard predictions for the heavy seed scenario [125], and (2) the extremely unlikely continuous growth of MBH seeds, even the most massive ones.

12.4.6 Growing Seeds in Semi-Analytical Models

Seed growth in the cosmological context of structure formation and evolution is studied in semi-analytical models, where dark matter halo formation, growth and assembly is extracted either from the Press-Schechter formalism [12] or from large-scale cosmological simulations such as the *Millennium* suite [26, 146].

The physics of *seed* formation is implemented using analytical schemes based on fundamental physical arguments or/and taking as input the outcomes of high-resolution simulations devoted to study specific processes related to *seed* formation.

As a general rule, these models are tested and constrained by astronomical observations, such as the redshift dependent luminosity function, when considering AGN. A large number of free parameters enters these models, tuned to reproduce the most updated observations. As of today, semi-analytical models can reproduce the bulk of the AGN population with *light seeds*, and the population of the heaviest high-z quasars with *heavy seeds* [144].

As an illustration of the outcome of a semi-analytical model in the upper panel of Fig. 12.3 we show the evolution of the SMBH mass function from $z = 4$ to $z = 0$ from the *L-Galaxies* semi-analytical model by [72]. By redshift $z \sim 1$ the bulk of the SMBHs is already assembled, whereas the main evolution of the population is seen at $z > 2$. Notice the progressive increase of the mass function at the higher masses with

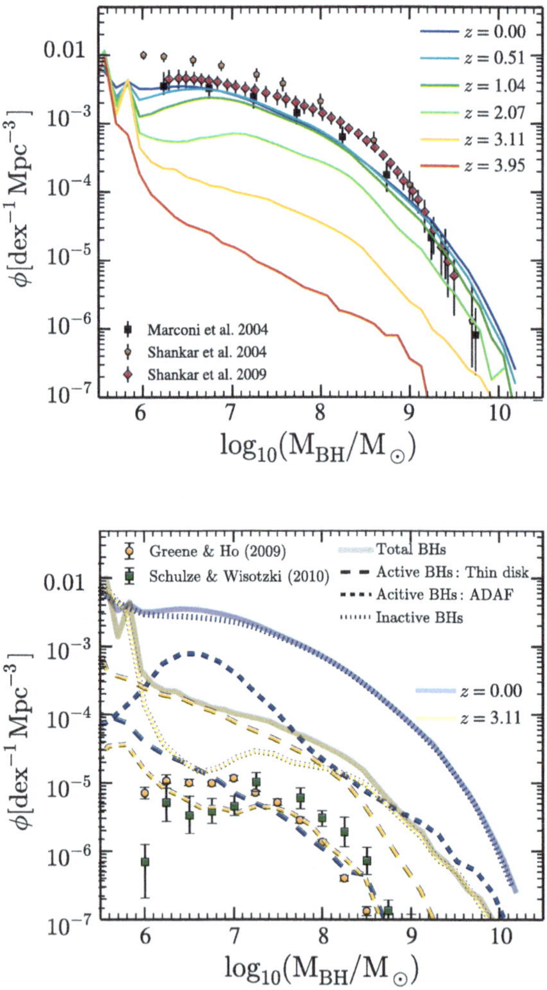

Fig. 12.3 Upper panel: Redshift evolution of the SMBH mass function from [72] compared to observational data at $z \sim 0$ (denoted with symbols). Lower panel: SMBH mass function at $z \sim 0$ (blue) and $z = 3$ (yellow), divided between active SMBHs accreting in the ADAF regime (short-dashed line) and in the geometrically thin α-disk regime (long-dashed line). At $z = 0$, the inactive SMBHs are denoted with dotted line and whole population as solid line. The figure is a modification of Fig. 4 in [72]. Courtesy of Izquierdo Villalba

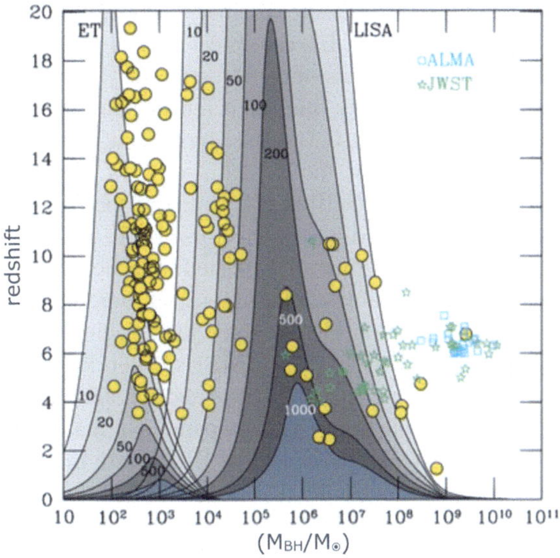

Fig. 12.4 ET and LISA cosmic horizons. The figure shows contour lines of constant SNR in the redshift z versus total mass of the BBH plane. Yellow dots denote loci of BBH mergers from [155], which models a population of *light* and *heavy binary seeds* forming via triple interaction in merging halos. Squares and stars denote the quasars observed at $4 < z < 11$, from JWST [95, 96] and ALMA [79]. Courtesy of Rosa Valiante

decreasing redshift. The lower panel of Fig. 12.3 shows the mass function of active and inactive SMBHs at $z \sim 0$ and $z \sim 3$, considering as threshold between active and non-active the Eddington ratio $f_E \sim 10^{-4}$, as below this value the emission is almost undetectable. At $z \sim 3$ the population of active SMBHs is more than one dex higher than today, indicating that accretion is declining with cosmic time. The inactive population increases from $z \sim 3$ to $z \sim 0$ with the most massive SMBHs becoming non-active at an earlier epoch. Regardless of the SMBH mass, $z \sim 3$ active SMBHs are characterized by a thin disc geometry fuelled by Eddington limited accretion flows. This picture changes at $z \sim 0$, where ADAF becomes the main accretion geometry for their feeding. The example shows the ability of semi-analytical models to explore different physical states of accretion that can be confronted with observations.

12.5 Seeds in the Gravitational Wave Universe

Third-generation ground-based interferometers and LISA in space will explore the *Gravitational Wave* Universe, in the bandwidth of $[5, 10^3]$ Hz and [0.1, 100] mHz, respectively. These GW observatories will detect the GW signals of binary BHs (BBHs) mergers over a wide mass range $O(10 - 10^7\ M_\odot)$ and redshifts $0 < z < 20$, extending the realm of observations to the time when the first stars and galaxies began to shine, and the first *seeds* began to form and grow.

The reach of ET and LISA is shown in Fig. 12.4 where contours of constant Signal-to-Noise-Ratio (SNR) are drawn in the space of redshift and binary BH total-mass. Thanks to the sensitivity of ET at frequencies down to ~ 5 Hz, the Telescope can capture GW signals from binary—population III—stars, and inform us on their initial mass function. ET can detect signals from BBHs beyond the *pair instability* mass-gap, with masses of $\mathcal{O}(100 - 400\,\mathrm{M}_\odot)$, out to $z \sim 10 - 15$ with the accuracy needed to confidently infer their redshift and masses [47]. This range of redshifts covers only the first 0.3 to 0.5 Gyrs of the age of the Universe. Therefore, these binaries could probe the existence of *light* seeds. Beyond redshift $z \sim 30$ to 40, merging PBHs of $\mathcal{O}(10\,\mathrm{M}_\odot)$, formed in the early Universe, can be detected and studied from ground-based detectors [111].

ET and CE together are expected to detect about a million of BBH mergers per year across the whole stellar mass spectrum, providing a unique view of the high-mass star formation history since their beginnings. Below redshift $z \sim 1$ ET can detect the signals from binary IMBHs with primaries in the mass range $\mathcal{O}(1000\,\mathrm{M}_\odot)$ that may form and reside in dense nuclear star clusters [51].

LISA's cosmic horizon is comparable to that of ET, as shown in Fig. 12.4. LISA can detect GW merger signals of BBHs with masses $\mathcal{O}(10^4 - 10^6\,\mathrm{M}_\odot)$ at redshifts $z \sim 10 - 15$, with confident parameter estimation [31]. These signals could represent grown *light seeds* or be *heavy* seeds. Deciphering the root can only be possible on a statistical basis [139]. With LISA and ET, the comparison between the detection of high redshift signals in the two GW bands will be instrumental in determining, at statistical level, the relative contribution of different seeding processes in building up the population of SMBHs, and the role of mergers versus accretion in determining their growth.

In Fig. 12.4 we show merger events (yellow dots) from a semi-analytical model aimed at reproducing the formation of three selected quasars at redshifts 0.2, 4, and 6 [155]. Mergers above $10^5\,\mathrm{M}_\odot$ come from at least one *heavy* growing seed in the binary, lighter mergers arise from the *light* seed population, which appears to give the dominant contribution in this model. Several assumptions underlie this model (i.e. the role of triple mergers for the formation of binary *seeds*). It is reported here to illustrates the transit of growing *seeds* in the two GW bandwidths, due to their evolution via accretion and mergers.

Finally, ET and LISA overlap in redshift and mass space below $z \sim 0.5$, just in the IMBH window. These IMBHs are therefore prime multiband sources, as they will be detected first by LISA, in their early inspiral-only phase, and later at merger in the high frequency window of terrestrial interferometers [51]. These multiband GW observations will quantify the continuity between the stellar-mass, intermediate-mass BHs and the SMBH population.

In Fig. 12.4, we plot the recent observations by JWST and ALMA. For the first time, electromagnetic observations of AGN discover *accreting* SMBH with masses $\mathcal{O}(10^6 - 10^7\,\mathrm{M}_\odot)$ up to $z \sim 9$ with a clear overlap with the LISA prospected sources. This will help further constraining the modeling of LISA sources.

LISA will discover not only merging SMBH with masses $\mathcal{O}(10^4 - 10^7\,\mathrm{M}_\odot)$ over a broad redshift range, but also *quiescent* SMBHs at the centers of galaxies, by detecting the GW signal from the so-called Extreme Mass Ratio Inspirals (EMRIs) [3] detectable out to $z \sim 3$ [31]. An EMRI is identified as a stellar BH orbiting around a SMBH on a highly relativistic, mildly eccentric orbit, displaying both in-plane and Lense-Thirring (off-plane) precession [103]. The stellar BH covers $\sim 10^4 - 10^5$ cycles before crossing the SMBH event horizon, mapping the entire spacetime structure of the Kerr SMBH. Parameter estimation of the GW signal can provide the mass and spin of the SMBH with extraordinary accuracy [31]. Therefore, the predictions made on the spins of SMBHs, by both numerical and semi-analytical models, could be tested uniquely with LISA, shedding light on the way SMBH have grown below $z < 4$. This measurement is again unique of LISA.

12.6 Conclusions

In this chapter we outlined different avenues of black hole *seed formation* within the baryonic channel, as an alternative to a primordial origin. These astrophysical avenues are not mutually exclusive given the variety of the mechanisms underlying their birth, assembly and evolution. Characterizing the different channels requires the modeling of many different astrophysical processes: from radiative, atomic and molecular physics to the physics of gravitational collapse, from the close astrophysical environment of the *seed* and of the growing SMBH to the large scale of the galaxy host, and from the galaxy host to the cosmic web scale. The intrinsic complexity and multi-dimensionality of the problem translate into a high level of uncertainty in establishing the nature of *seeds*. No clear mass scale exists for BHs in contrast to the case of neutron stars, whose basic physics has been tested on solid grounds. This implies the existence of wide range of masses for the *seeds*, which future electromagnetic and gravitational-wave observations of high-z AGN, and binary BHs are expected to constrain. Inferring the true nature of *seeds* requires observing the first stars, galaxies and BHs in the few hundreds of millions of years from the Big Bang. The repercussion of the early stages of BH formation is vast and propagates through the whole cosmic history, making this among the most fascinating problems we will be facing in the next decade.

In Chaps. 10 and 11 of this book, the reader finds an updated, critical discussion on the effects of PBHs on the generation of cosmic structures, on the BH seeding mechanism, and on the impact of PBHs on the formation of the first stars and galaxies. In this chapter we took the conservative view that *seed black holes* are the outcome of stellar evolution processes or outcome of the collapse/collision of baryonic systems. Consequently, the SMBHs that we see powering the most distant quasars are expected to form *after* stars have formed. A smoking gun observation of PBHs as *seeds* of at least a sub-population of AGN would be the discovery of an accreting black hole surrounded by little gas (to become visible) but hosted in a star-poor forming galaxy halo. Future astronomical observatories will continue to explore and pierce deep into the epoch of cosmic dawn and the increasing knowledge we are building today on AGN and GW sources is already posing new questions which call for a quantitative response.

References

1. M.A. Abramowicz et al., Slim accretion disks. **332**, 646 (1988). https://doi.org/10.1086/166683
2. B. Agarwal et al., Ubiquitous seeding of supermassive black holes by direct collapse. **425**(4), 2854–2871 (2012). https://doi.org/10.1111/j.1365-2966.2012.21651.x. arXiv: 1205.6464 [astro-ph.CO]
3. P. Amaro-Seoane, Relativistic dynamics and extreme mass ratio inspirals. Living Rev. Relativity **21**(1), 4 (2018). https://doi.org/10.1007/s41114-018-0013-8. arXiv: 1205.5240 [astro-ph.CO]
4. P. Amaro-Seoane et al., Astrophysics with the Laser Interferometer Space Antenna. Living Rev. Relativity **26**(1), 2 (2023). https://doi.org/10.1007/s41114-022-00041-y. arXiv: 2203.06016 [gr-qc]
5. D. Anglés-Alcázar et al., Cosmological Simulations of Quasar Fueling to Subparsec Scales Using Lagrangian Hyper-refinement. **917**(2), 53 (2021). https://doi.org/10.3847/1538-4357/ac09e8. arXiv: 2008.12303 [astro-ph.GA]
6. D. Anglés-Alcázar et al., Gravitational torque-driven black hole growth and feedback in cosmological simulations. **464**(3), 2840–2853 (2017). https://doi.org/10.1093/mnras/stw2565. arXiv: 1603.08007 [astro-ph.GA]
7. F. Antonini, F.A. Rasio, erging Black Hole Binaries in Galactic Nuclei: Implications for Advanced-LIGO Detections. **831**(2), 187 (2016). https://doi.org/10.3847/0004-637X/831/2/187. arXiv: 1606.04889 [astro-ph.HE]
8. M.A. Sedda et al., Breaching the limit: formation of GW190521-like and IMBH mergers in young massive clusters (2021). arXiv:2105.07003 [astro-ph.GA]
9. M.A. sedda et al., The DRAGON-II simulations - III. Compact binary mergers in clusters with up to 1 million stars: mass, spin, eccentricity, merger rate, and pair instability supernovae rate. **528**(3), 5140–5159 (2024). https://doi.org/10.1093/mnras/stad3951. arXiv: 2307.04807 [astro-ph.HE]
10. M. Bachetti et al., An ultraluminous X-ray source powered by an accreting neutron star. **514**(7521), 202–204 (2014). https://doi.org/10.1038/nature13791. arXiv: 1410.3590 [astro-ph.HE]
11. E. Barausse, L. Rezzolla, Redicting the direction of the final spin from the coalescence of two black holes. **704**(1), L40–L44 (2009). https://doi.org/10.1088/0004-637X/704/1/L40. arXiv: 0904.2577 [gr-qc]
12. E. Barausse et al., Massive black hole merger rates: the effect of kiloparsec separation wandering and supernova feedback **904**(1), 16 (2020). https://doi.org/10.3847/1538-4357/abba7f. arXiv: 2006.03065 [astro-ph.GA]
13. J.M. Bardeen, J.A. Petterson, The Lense-Thirring effect and accretion disks around Kerr black holes. **195**, L65 (1975). https://doi.org/10.1086/181711
14. M.C. Begelman, M. Volonteri, M.J. Rees. Formation of supermassive black holes by direct collapse in pre-galactic haloes. **370**(1), 289–298 (2006). https://doi.org/10.1111/j.1365-2966.2006.10467.x. arXiv: astro-ph/0602363 [astro-ph]
15. E. Berti, M. Volonteri, Cosmological Black Hole Spin Evolution by Mergers and Accretion. **684**(2), 822–828 (2008). https://doi.org/10.1086/590379. arXiv: 0802.0025 [astro-ph]
16. A.K. Bhowmick et al., Probing the $z \gtrsim 6$ quasars in a universe with IllustrisTNG physics: impact of gas-based black hole seeding models. **516**(1), 138–157 (2022). https://doi.org/10.1093/mnras/stac2238. arXiv: 2205.05717 [astro-ph.GA]
17. A.K. Bhowmick et al., Representing low-mass black hole seeds in cosmological simulations: A new sub-grid stochastic seed model. **529**(4), 3768–3792 (2024). https://doi.org/10.1093/mnras/stae780. arXiv: 2309.15341 [astro-ph.GA]
18. S. Bird et al,. Did LIGO Detect Dark Matter? **116**(20), 201301 (2016). https://doi.org/10.1103/PhysRevLett.116.201301. arXiv: 1603.00464 [astro-ph.CO]
19. G.S. Bisnovatyi-Kogan, A.A. Ruzmaikin, The accretion of matter by a collapsing star in the presence of a magnetic field. **28**(1), 45–59 (1974). https://doi.org/10.1007/BF00642237

20. R.D. Blandford, D.G. Payne, Hydromagnetic flows from accretion disks and the production of radio jets. **199**, 883–903 (1982). https://doi.org/10.1093/mnras/199.4.883
21. R.D. Blandford, R.L. Znajek, Electromagnetic extraction of energy from Kerr black holes. **179**, 433–456 (1977). https://doi.org/10.1093/mnras/179.3.433
22. R.D. Blandford, R.L. Znajek, Electromagnetic extraction of energy from Kerr black holes. **179**, 433–456 (1977). https://doi.org/10.1093/mnras/179.3.433
23. S. Blinnikov et al., Solving puzzles of GW150914 by primordial black holes. **2016**(11), 036 (2016). https://doi.org/10.1088/1475-7516/2016/11/036. arXiv: 1611.00541 [astro-ph.HE]
24. Á. Bogdán et al., Evidence for heavy-seed origin of early supermassive black holes from a $z \approx 10$ X-ray quasar. Nat. Astron. **8**, 126–133 (2024). https://doi.org/10.1038/s41550-023-02111-9. arXiv: 2305.15458 [astro-ph.GA]
25. F. Bollati et al., On the connection between AGN radiative feedback and massive black hole spin (2023). https://doi.org/10.48550/arXiv.2311.07576. arXiv: 2311.07576 [astro-ph.GA]
26. M. Boylan-Kolchin et al. Resolving cosmic structure formation with the Millennium-II Simulation. 398.3 (2009), pp. 1150–1164. https://doi.org/10.1111/j.1365-2966.2009.15191.x. arXiv: 0903.3041 [astro-ph.CO]
27. M. Branchesi et al. Science with the Einstein Telescope: a comparison of different designs. 2023.7, 068 (2023), p. 068. https://doi.org/10.1088/1475-7516/2023/07/068. arxiv:2303.15923 [gr-qc]
28. S. Bustamante, V. Springel, Spin evolution and feedback of supermassive black holes in cosmological simulations. **490**(3), 4133–4153 (2019). https://doi.org/10.1093/mnras/stz2836. arXiv: 1902.04651 [astro-ph.GA]
29. E. Cenci et al. Black hole spin evolution in warped accretion discs. **500**(3), 3719–3727 (2021). https://doi.org/10.1093/mnras/staa3449. arXiv: 2011.06596 [astro-ph.GA]
30. E. Choi et al., Radiative and momentum-based mechanical active galactic nucleus feedback in a three-dimensional galaxy evolution code. **754**(2), 125 (2012). https://doi.org/10.1088/0004-637X/754/2/125. arXiv: 1205.2082 [astro-ph.GA]
31. M. Colpi et al. LISA Definition Study Report (2024). https://doi.org/10.48550/arXiv.2402.07571. arXiv: 2402.07571 [astro-ph.CO]
32. T. Costa, D. Sijacki, M.G. Haehnelt. Feedback from active galactic nuclei: energy- versus momentum-driving. **444**(3), 2355–2376 (2014). https://doi.org/10.1093/mnras/stu1632. arXiv: 1406.2691 [astro-ph.GA]
33. T. Costa et al. AGN-driven outflows and the formation of Lyα nebulae around high-z quasars. **517**(2), 1767–1790 (2022). https://doi.org/10.1093/mnras/stac2432. arXiv: 2203.11232 [astro-ph.GA]
34. M. Dall'Amico et al., Eccentric black hole mergers via three-body interactions in young, globular, and nuclear star clusters. **683**, A186 (2024). https://doi.org/10.1051/0004-6361/202348745. arXiv: 2303.07421 [astro-ph.HE]
35. M.B. Davies, M. Coleman Miller, J.M. Bellovary. Supermassive black hole formation via gas accretion in nuclear stellar clusters. **740**(2), L42 (2011). https://doi.org/10.1088/2041-8205/740/2/L42. arXiv: 1106.5943 [astro-ph.CO]
36. B. Devecchi et al., High-redshift formation and evolution of central massive objects - II. The census of BH seeds. **421**(2), 1465–1475 (2012). https://doi.org/10.1111/j.1365-2966.2012.20406.x. arXiv: 1201.3761 [astro-ph.CO]
37. U.N. Di Carlo et al., Binary black holes in the pair instability mass gap. **497**(1), 1043–1049 (2020). https://doi.org/10.1093/mnras/staa1997. arXiv: 1911.01434 [astro-ph.HE]
38. U.N. Di Carlo et al., Intermediate-mass black holes from stellar mergers in young star clusters. **507**(4), 5132–5143 (2021). https://doi.org/10.1093/mnras/stab2390. arXiv: 2105.01085 [astro-ph.GA]
39. T. Di Matteo, V. Springel, L. Hernquist, Energy input from quasars regulates the growth and activity of black holes and their host galaxies. **433**(7026), 604–607 (2005). https://doi.org/10.1038/nature03335. arXiv: astro-ph/0502199 [astro-ph]

40. T. Di Matteo et al., The origin of the most massive black holes at high-z: BlueTides and the next quasar frontier. **467**(4), 4243–4251 (2017). https://doi.org/10.1093/mnras/stx319. arXiv: 1606.08871 [astro-ph.GA]
41. X. Ding et al., Detection of stellar light from quasar host galaxies at redshifts above 6. **621**(7977), 51–55 (2023). https://doi.org/10.1038/s41586-023-06345-5. arXiv: 2211.14329 [astro-ph.GA]
42. Y. Dubois, M. Volonteri, J. Silk, Black hole evolution - III. Statistical properties of mass growth and spin evolution using large-scale hydrodynamical cosmological simulations. **440**(2), 1590–1606 (2014). https://doi.org/10.1093/mnras/stu373. arXiv: 1304.4583 [astro-ph.CO]
43. Y. Dubois et al., Black hole evolution - II. Spinning black holes in a supernova-driven turbulent interstellar medium. **440**(3), 2333–2346 (2014). https://doi.org/10.1093/mnras/stu425. arXiv: 1401.1220 [astro-ph.GA]
44. Y. Dubois et al., Blowing cold flows away: the impact of early AGN activity on the formation of a brightest cluster galaxy progenitor. **428**(4), 2885–2900 (2013). https://doi.org/10.1093/mnras/sts224. arXiv: 1206.5838 [astro-ph.CO]
45. Event Horizon Telescope Collaboration et al., First M87 event horizon telescope results. i. the shadow of the supermassive black hole. **875**(1), L1 (2019). https://doi.org/10.3847/2041-8213/ab0ec7. arXiv: 1906.11238 [astro-ph.GA]
46. A.C. Fabian, Observational evidence of active galactic nuclei feedback. **50**, 455–489 (2012). https://doi.org/10.1146/annurev-astro-081811-125521. arXiv: 1204.4114 [astro-ph.CO]
47. S. Fairhurst et al. Identifying heavy stellar black holes at cosmological distances with next-generation gravitational-wave observatories. **529**(3), 2116–2130 (2024). https://doi.org/10.1093/mnras/stae443. arXiv: 2310.18158 [astro-ph.HE]
48. X. Fan, Evolution of high-redshift quasars **50**(9–10), 665–671 (2006). https://doi.org/10.1016/j.newar.2006.06.077
49. D. Fiacconi, D. Sijacki, J.E. Pringle. Galactic nuclei evolution with spinning black holes: method and implementation. **477**(3), 3807–3835 (2018). https://doi.org/10.1093/mnras/sty893. arXiv: 1712.00023 [astro-ph.GA]
50. G. Fragione, B. Kocsis, Black hole mergers from an evolving population of globular clusters. **121**(16), 161103 (2018). https://doi.org/10.1103/PhysRevLett.121.161103. arXiv: 1806.02351 [astro-ph.GA]
51. G. Fragione et al. Merger rates of intermediate-mass black hole binaries in nuclear star clusters. **933**(2), 170 (2022). https://doi.org/10.3847/1538-4357/ac75d0. arXiv: 2204.03745 [astro-ph.HE]
52. R. Genzel, F. Eisenhauer, S. Gillessen. The Galactic Center massive black hole and nuclear star cluster. *Reviews of Modern Physics* **82**(4), 3121–3195 (2010). https://doi.org/10.1103/RevModPhys.82.3121. arXiv: 1006.0064 [astro-ph.GA]
53. D. Gerosa, M. Fishbach, Hierarchical mergers of stellar-mass black holes and their gravitational-wave signatures. *Nature Astronomy* 5 (2021), pp. 749–760. 10.1038/s41550-021-01398-w. arXiv: 2105.03439 [astro-ph.HE]
54. D. Gerosa, M. Fishbach, Hierarchical mergers of stellar-mass black holes and their gravitational-wave signatures. Nat. Astron. **5** , 749–760 (2021). https://doi.org/10.1038/s41550-021-01398-w. arXiv: 2105.03439 [astro-ph.HE]
55. D. Gerosa et al. Spin orientations of merging black holes formed from the evolution of stellar binaries. **98**(8), 084036 (2018). https://doi.org/10.1103/PhysRevD.98.084036. arXiv: 1808.02491 [astro-ph.HE]
56. J.E. Greene et al., UNCOVER spectroscopy confirms the surprising ubiquity of active galactic nuclei in red sources at z *spsdoigtsps* 5. **964**(1), 39 (2024). https://doi.org/10.3847/1538-4357/ad1e5f. arXiv: 2309.05714 [astro-ph.GA]
57. L. Haemmerlé et al. Formation of the first stars and black holes. **216**(4), 48 (2020). https://doi.org/10.1007/s11214-020-00673-y. arXiv: 2003.10533 [astro-ph.GA]
58. Z. Haiman et al., Massive black hole binaries as LISA precursors in the roman high latitude time domain survey (2023). https://doi.org/10.48550/arXiv.2306.14990. arXiv: 2306.14990 [astro-ph.HE]

59. Y. Harikane et al. A JWST/NIRSpec First Census of Broad-line AGNs at z = 4-7: Detection of 10 Faint AGNs with M $_{BH}$ 10^6-10^8 M $_{\odot}$ and Their Host Galaxy Properties. **959**(1), 39 (2023). 10.3847/1538-4357/ad029e. arXiv: 2303.11946 [astro-ph.GA]
60. T. Hartwig et al., Gravitational waves from the remnants of the first stars. **460**(1), L74–L78 (2016). https://doi.org/10.1093/mnrasl/slw074. arXiv: 1603.05655 [astro-ph.GA]
61. A. Heger et al. How Massive Single Stars End Their Life. **591**(1), 288–300 (2003). https://doi.org/10.1086/375341. arXiv: astro-ph/0212469 [astro-ph]
62. S. Hirano et al. Primordial star formation under the influence of far ultraviolet radiation: 1540 cosmological haloes and the stellar mass distribution. **448**(1), 568–587 (2015). https://doi.org/10.1093/mnras/stv044. arXiv: 1501.01630 [astro-ph.GA]
63. S. Hirano et al. Supersonic gas streams enhance the formation of massive black holes in the early universe. *Science* **357**(6358), 1375–1378 (2017). https://doi.org/10.1126/science.aai9119. arXiv: 1709.09863 [astro-ph.CO]
64. F. Hofmann, E. Barausse, L. Rezzolla, The Final Spin from Binary Black Holes in Quasi-circular Orbits. **825**(2) L19 (2016). https://doi.org/10.3847/2041-8205/825/2/L19 arXiv: 1605.01938 [gr-qc]
65. T. Hosokawa et al., Formation of massive primordial stars: intermittent UV feedback with episodic mass accretion. **824**(2), 119 (2016). https://doi.org/10.3847/0004-637X/824/2/119. arXiv: 1510.01407 [astro-ph.GA]
66. Scott A. Hughes and Roger D. Blandford. Black Hole Mass and Spin Coevolution by Mergers. 585.2 (2003), pp. L101–L104. 10.1086/375495. arXiv: astro-ph/0208484 [astro-ph]
67. I. V. Igumenshchev, R. Narayan, M.A. Abramowicz, Three-dimensional magnetohydrodynamic simulations of radiatively inefficient accretion flows. 592(2), 1042–1059 (2003). https://doi.org/10.1086/375769. arXiv: astro-ph/0301402 [astro-ph]
68. K. Inayoshi, K. Omukai, E. Tasker, Formation of an embryonic supermassive star in the first galaxy. **445**, L109–L113 (2014). https://doi.org/10.1093/mnrasl/slu151. arXiv: 1404.4630 [astro-ph.GA]
69. K. Inayoshi, E. Visbal, Z. Haiman, The assembly of the first massive black holes. **58**, 27–97 (2020). https://doi.org/10.1146/annurev-astro-120419-014455. arXiv: 1911.05791 [astro-ph.GA]
70. Ž. Ivezić et al. LSST: from science drivers to reference design and anticipated data products. **873**(2), 111 (2019). https://doi.org/10.3847/1538-4357/ab042c. arXiv: 0805.2366 [astro-ph]
71. D. Izquierdo-Villalba et al. From galactic nuclei to the halo outskirts: tracing supermassive black holes across cosmic history and environments. **495**(4), 4681–4706 (2020). https://doi.org/10.1093/mnras/staa1399. arXiv: 2001.10548 [astro-ph.GA]
72. D. Izquierdo-Villalba et al., From galactic nuclei to the halo outskirts: tracing supermassive black holes across cosmic history and environments. **495**(4), 4681–4706 (2020). https://doi.org/10.1093/mnras/staa1399. arXiv: 2001.10548 [astro-ph.GA]
73. Y-F. Jiang, J.M. Stone, S.W. Davis, Super-eddington accretion disks around supermassive black holes. **880**(2), 67 (2019). https://doi.org/10.3847/1538-4357/ab29ff. arXiv: 1709.02845 [astro-ph.HE]
74. A.R. King,J. E. Pringle, Growing supermassive black holes by chaotic accretion. **373**(1), L90–L92 (2006). https://doi.org/10.1111/j.1745-3933.2006.00249.x. arXiv: astro-ph/0609598 [astro-ph]
75. A.R. King et al., Aligning spinning black holes and accretion discs. **363**(1), 49–56 (2005). https://doi.org/10.1111/j.1365-2966.2005.09378.x. arXiv: astro-ph/0507098 [astro-ph]
76. R.S. Klessen, S.C.O. Glover, The first stars: formation, properties, and impact. **61**, 65–130 (2023). https://doi.org/10.1146/annurev-astro-071221-053453. arXiv: 2303.12500 [astro-ph.CO]
77. D.D. Kocevski et al., Hidden little monsters: spectroscopic identification of low-mass, broad-line AGNs at z5 with CEERS. **954**(1), L4 (2023). https://doi.org/10.3847/2041-8213/ace5a0. arXiv: 2302.00012 [astro-ph.GA]

78. J. Kormendy, L.C. Ho, Coevolution (Or Not) of supermassive black holes and host galaxies **51**(1), 511–653 (2013). https://doi.org/10.1146/annurev-astro-082708-101811. arXiv: 1304.7762 [astro-ph.CO]
79. I. Labbe et al., UNCOVER: candidate red active galactic nuclei at 3¡z¡7 with JWST and ALMA. https://doi.org/10.48550/arXiv.2306.07320. arXiv: 2306.07320 [astro-ph.GA]
80. M.A. Latif et al., Turbulent cold flows gave birth to the first quasars. **607**(7917), 48–51 (2022). https://doi.org/10.1038/s41586-022-04813-y. arXiv: 2207.05093 [astro-ph.GA]
81. J. Li et al., Tip of the iceberg: overmassive black holes at 4¡z¡7 found by JWST are not inconsistent with the local $\mathcal{M}_{\rm BH}$-$\mathcal{M}_\star$ relation (2024). https://doi.org/10.48550/arXiv.2403.00074. arXiv: 2403.00074 [astro-ph.GA]
82. LIGO Scientific Collaboration and Virgo Collaboration, GW190521: a binary black hole merger with a total mass of 150 $M_\odot$. **125**(10), 101102 (2020). https://doi.org/10.1103/PhysRevLett.125.101102
83. LIGO Scientific Collaboration and Virgo Collaboration, GWTC-2: Compact Binary Coalescences Observed by LIGO and Virgo During the First Half of the Third Observing Run (2020), arXiv:2010.14527. arXiv: 2010.14527 [gr-qc]
84. LIGO Scientific Collaboration and Virgo Collaboration, Observation of Gravitational Waves from a Binary Black Hole Merger. **116**(6), 061102 (2016). https://doi.org/10.1103/PhysRevLett.116.061102. arXiv: 1602.03837 [gr-qc]
85. LIGO Scientific Collaboration and Virgo Collaboration, Properties and astrophysical implications of the 150 $M_\odot$ binary black hole merger GW190521. **900**(1), L13 (2020). https://doi.org/10.3847/2041-8213/aba493. arXiv: 2009.01190 [astro-ph.HE]
86. LIGO Scientific Collaboration and Virgo Collaboration, Search for subsolar-mass binaries in the first half of advanced LIGO's and advanced Virgo's third observing run. **129**(6), 061104 (2022). https://doi.org/10.1103/PhysRevLett.129.061104 arXiv: 2109.12197 [astro-ph.CO]
87. VIRGO Collaboration Ligo Scientific Collaboration and Kagra Collaboration, GWTC-3: Compact Binary Coalescences Observed by LIGO and Virgo during the Second Part of the Third Observing Run. Phys. Rev. X **13**(4), 041039 (2023). https://doi.org/10.1103/PhysRevX.13.041039. arXiv: 2111.03606 [gr-qc]
88. D. Lin et al. A likely decade-long sustained tidal disruption event. Nat Astron **1**, 0033 (2017). https://doi.org/10.1038/s41550-016-0033. arXiv: 1702.00792 [astro-ph.HE]
89. C.O. Lousto, Y. Zlochower, Modeling gravitational recoil from precessing highly spinning unequal-mass black-hole binaries. **79**(6), 064018 (2009). https://doi.org/10.1103/PhysRevD.79.064018. arXiv: 0805.0159 [gr-qc]
90. A. Lupi et al. Constraining the high-redshift formation of black hole seeds in nuclear star clusters with gas inflows. **442**(4), 3616–3626 (2014). https://doi.org/10.1093/mnras/stu1120. arXiv: 1406.2325 [astro-ph.GA]
91. A. Lupi, Z. Haiman, M. Volonteri, Forming massive seed black holes in high-redshift quasar host progenitors. **503**(4), 5046–5060 (2021). https://doi.org/10.1093/mnras/stab692. arXiv: 2102.05051 [astro-ph.GA]
92. A. Lupi et al., Sustained super-Eddington accretion in high-redshift quasars (2023). arXiv: 2312.08422 [astro-ph.GA]
93. L. Ma et al., Seeds don't sink: even massive black hole 'seeds' cannot migrate to galaxy centres efficiently. **508**(2), 1973–1985 (2021). https://doi.org/10.1093/mnras/stab2713. arXiv: 2101.02727 [astro-ph.GA]
94. P. Madau, F. Haardt, M. Dotti, Super-critical growth of massive black holes from stellar-mass seeds. **784**(2), L38 (2014). https://doi.org/10.1088/2041-8205/784/2/L38. arXiv: 1402.6995 [astro-ph.CO]
95. R. Maiolino et al., A small and vigorous black hole in the early Universe. **627**(8002), 59–63 (2024). https://doi.org/10.1038/s41586-024-07052-5. arXiv: 2305.12492 [astro-ph.GA]
96. R. Maiolino et al., JADES. The diverse population of infant Black Holes at 4¡z¡11: merging, tiny, poor, but mighty (2023). https://doi.org/10.48550/arXiv.2308.01230. arXiv: 2308.01230 [astro-ph.GA]

97. R. Maiolino et al., JWST-JADES. Possible population III signatures at z=10.6 in the halo of GN-z11 (2023). https://doi.org/10.48550/arXiv.2306.00953. arXiv: 2306.00953 [astro-ph.GA]
98. I. Mandel, A. Farmer, Merging stellar-mass binary black holes. **955**, 1–24 (2022). https://doi.org/10.1016/j.physrep.2022.01.003. arXiv: 1806.05820 [astro-ph.HE]
99. A. Marconi et al., Local supermassive black holes, relics of active galactic nuclei and the X-ray background. **351**(1), 169–185 (2004). https://doi.org/10.1111/j.1365-2966.2004.07765.x. arXiv: astro-ph/0311619 [astro-ph]
100. W. Massonneau et al., How the super-Eddington regime regulates black hole growth in high-redshift galaxies. **670**, A180 (2023). https://doi.org/10.1051/0004-6361/202243170. arXiv: 2201.08766
101. L. Mayer et al., Direct formation of massive black holes via dynamical collapse in metal-enriched merging galaxies at z 10: fully cosmological simulations. **961**(1), 76 (2024). https://doi.org/10.3847/1538-4357/ad11cf. arXiv: 2304.02066 [astro-ph.GA]
102. L. Mayer et al., Direct formation of supermassive black holes in metal-enriched gas at the heart of high-redshift galaxy mergers. **810**(1), 51 (2015). https://doi.org/10.1088/0004-637X/810/1/51. arXiv: 1411.5683 [astro-ph.GA]
103. D. Merritt et al., Stellar dynamics of extreme-mass-ratio inspirals. **84**(4), 044024 (2011). https://doi.org/10.1103/PhysRevD.84.044024. arXiv: 1102.3180 [astro-ph.CO]
104. M. Mezcua, Observational evidence for intermediate-mass black holes. Int. J. Mod. Phys. D **26**(11), 1730021 (2017). https://doi.org/10.1142/S021827181730021X. arXiv: 1705.09667 [astro-ph.GA]
105. H. Mo, F.C. van den Bosch, S. White, *Galaxy Formation and Evolution* (2010)
106. H. Mo, F.C. van den Bosch, S. White, *Galaxy Formation and Evolution* (2010)
107. R. Narayan, R. Mahadevan, and E. Quataert. Advection-dominated accretion around black holes. *Theory of Black Hole Accretion Disks*. Ed. by M.A. Abramowicz, G. Björnsson, and J.E. Pringle (1998), pp. 148–182. https://doi.org/10.48550. arXiv: astro-ph/9803141 [astro-ph]
108. R. Narayan, I.V. Igumenshchev, M.A. Abramowicz, Magnetically arrested disk: an energetically efficient accretion flow. **55**, L69–L72 (2003). https://doi.org/10.1093/pasj/55.6.L69. arXiv: astro-ph/0305029 [astro-ph]
109. P. Natarajan et al., First detection of an overmassive black hole galaxy uhz1: evidence for heavy black hole seed formation from direct collapse. **960**(1), L1 (2024). https://doi.org/10.3847/2041-8213/ad0e76. arXiv: 2308.02654 [astro-ph.HE]
110. M. Neeleman et al., The kinematics of z $gtrsim$ 6 quasar host galaxies. **911**(2), 141 (2021). https://doi.org/10.3847/1538-4357/abe70f. arXiv: 2102.05679 [astro-ph.GA]
111. K.K.Y. Ng et al., Constraining high-redshift stellar-mass primordial black holes with next-generation ground-based gravitational-wave detectors. **933**(2), L41 (2022). https://doi.org/10.3847/2041-8213/ac7aae. arXiv: 2204.11864 [astro-ph.CO]
112. D.D. Nguyen et al., Improved dynamical constraints on the masses of the central black holes in nearby low-mass early-type galactic nuclei and the first black hole determination for NGC 205. **872**(1), 104 (2019). https://doi.org/10.3847/1538-4357/aafe7a. arXiv: 1901.05496 [astro-ph.GA]
113. Y. Ni et al., The ASTRID simulation: the evolution of supermassive black holes. **513**(1), 670–692 (2022). https://doi.org/10.1093/mnras/stac351. arXiv: 2110.14154 [astro-ph.GA]
114. S. Nishijima, S. Hirano, H. Umeda, Low-mass Pop III star formation due to the HD-cooling induced by weak Lyman-Werner radiation (2023). https://doi.org/10.48550/arXiv.2311.10386. arXiv: 2311.10386 [astro-ph.GA]
115. E. O'Connor, C.D. Ott, Black hole formation in failing core-collapse supernovae. **730**(2), 70 (2011). https://doi.org/10.1088/0004-637X/730/2/70. arXiv: 1010.5550 [astro-ph.HE]
116. C.A. Onken et al., A thirty-four billion solar mass black hole in SMSS J2157-3602, the most luminous known quasar. **496**(2), 2309–2314 (2020). https://doi.org/10.1093/mnras/staa1635. arXiv: 2005.06868 [astro-ph.GA]

117. F. Pacucci et al., JWST CEERS and JADES active galaxies at z = 4-7 violate the local M $-M_{\star}$ Relation at ¿3σ: implications for low-mass black holes and seeding models. **957**(1), L3 (2023). https://doi.org/10.3847/2041-8213/ad0158. arXiv: 2308.12331 [astro-ph.GA]
118. S. Peirani et al., Cosmic evolution of black hole spin and galaxy orientations: Clues from the NewHorizon and Galactica simulations. **686**, A233 (2024). https://doi.org/10.1051/0004-6361/202349101. (arXiv: 2401.03712)
119. A. Perego et al., Mass and spin co-evolution during the alignment of a black hole in a warped accretion disc. **399**(4), 2249–2263 (2009). https://doi.org/10.1111/j.1365-2966.2009.15427.x. arXiv: 0907.3742 [astro-ph.CO]
120. H. Pfister et al., The erratic dynamical life of black hole seeds in high-redshift galaxies. **486**(1), 101–111 (2019). https://doi.org/10.1093/mnras/stz822. arXiv: 1902.01297 [astro-ph.GA]
121. A. Pillepich et al., Simulating galaxy formation with the IllustrisTNG model. **473**(3), 4077–4106 (2018). https://doi.org/10.1093/mnras/stx2656. arXiv: 1703.02970 [astro-ph.GA]
122. S.F. Portegies Zwart et al., Formation of massive black holes through runaway collisions in dense young star clusters. **428**(6984), 724–726 (2004). https://doi.org/10.1038/nature02448. arXiv: astro-ph/0402622 [astro-ph]
123. D. Proga, How much x-ray and uv radiation processes are coupled in accretion disks? The Active Galactic Nucl. Case **630**(1), L9–L12 (2005). https://doi.org/10.1086/462417. arXiv: astro-ph/0507152 [astro-ph]
124. M. Punturo et al., The Einstein Telescope: a third-generation gravitational wave observatory. Class. Quantum Gravity **27**(19), 194002 (2010). https://doi.org/10.1088/0264-9381/27/19/194002
125. J. Regan, M. Volonteri, Massive black hole seeds (2024). https://doi.org/10.48550/arXiv.2405.17975. arXiv: 2405.17975 [astro-ph.GA]
126. J.A. Regan et al., Rapid formation of massive black holes in close proximity to embryonic protogalaxies. Nat. Astron. **1**, 0075 (2017). https://doi.org/10.1038/s41550-017-0075. arXiv: 1703.03805 [astro-ph.GA]
127. J.A. Regan et al., Super-Eddington accretion and feedback from the first massive seed black holes. **486**(3), 3892–3906 (2019). https://doi.org/10.1093/mnras/stz1045. arXiv: 1811.04953 [astro-ph.GA]
128. J.A. Regan et al., The formation of very massive stars in early galaxies and implications for intermediate mass black holes. Open J. Astrophys. **3**(1), 15 (2020). https://doi.org/10.21105/astro.2008.08090. arXiv: 2008.08090 [astro-ph.GA]
129. B. Reinoso et al., Formation of supermassive stars in the first star clusters. **521**(3), 3553–3569 (2023). https://doi.org/10.1093/mnras/stad790. arXiv: 2303.07827 [astro-ph.GA]
130. D. Reitze et al., Cosmic explorer: The U.S. contribution to gravitational-wave astronomy beyond LIGO. Bull. Am. Astron. Soc. **51**, 35 (2019). arXiv: 1907.04833 [astro-ph.IM]
131. D. Rennehan et al., Three regimes of black hole feedback (2023). https://doi.org/10.48550/arXiv.2309.15898. arXiv: 2309.15898 [astro-ph.GA]
132. C.S. Reynolds, Measuring black hole spin using X-Ray reflection spectroscopy. **183**(1-4), 277– 294 (2014). https://doi.org/10.1007/s11214-013-0006-6. arXiv: 1302.3260 [astro-ph.HE]
133. C.S. Reynolds, The spin of supermassive black holes. Class. Quantum Gravity **30**(24), 244004 (2013). https://doi.org/10.1088/0264-9381/30/24/244004. arXiv: 1307.3246 [astro-ph.HE]
134. A. Ricarte, R. Narayan, B. Curd, Recipes for jet feedback and spin evolution of black holes with strongly magnetized super-eddington accretion disks. **954**(1), L22 (2023). https://doi.org/10.3847/2041-8213/aceda5. arXiv: 2307.04621 [astro-ph.HE]
135. A. Sadowski, R. Narayan, Three-dimensional simulations of supercritical black hole accretion discs - luminosities, photon trapping and variability. **456**(4), 3929–3947 (2016). https://doi.org/10.1093/mnras/stv2941. arXiv: 1509.03168 [astro-ph.HE]
136. M. Sasaki et al., Primordial black hole scenario for the gravitational-wave event GW150914. **117**(6), 061101 (2016). https://doi.org/10.1103/PhysRevLett.117.061101. arXiv: 1603.08338 [astro-ph.CO]

137. J. Schaye et al., The EAGLE project: simulating the evolution and assembly of galaxies and their environments. **446**(1), 521–554 (2015). https://doi.org/10.1093/mnras/stu2058. arXiv: 1407.7040 [astro-ph.GA]
138. A. Sesana et al., Linking the spin evolution of massive black holes to galaxy kinematics. **794**(2), 104 (2014). https://doi.org/10.1088/0004-637X/794/2/104. arXiv: 1402.7088 [astro-ph.CO]
139. A. Sesana et al., Reconstructing the massive black hole cosmic history through gravitational waves. **83**(4), 044036 (2011). https://doi.org/10.1103/PhysRevD.83.044036. arXiv: 1011.5893 [astro-ph.CO]
140. N.I. Shakura, R.A. Sunyaev, Black holes in binary systems. Observational appearance. **24**, 337–355 (1973)
141. I. Shlosman, P.A. Vitello, G. Shaviv, Active galactic nuclei - Internal dynamics and formation of emission clouds. **294**, 96–105 (1985). https://doi.org/10.1086/163278
142. A. Soltan, Masses of quasars. **200**, 115–122 (1982). https://doi.org/10.1093/mnras/200.1.115
143. M. Spera, M. Mapelli, A. Bressan. The mass spectrum of compact remnants from the PARSEC stellar evolution tracks. **451**(4), 4086–4103 (2015). https://doi.org/10.1093/mnras/stv1161. arXiv: 1505.05201 [astro-ph.SR]
144. D. Spinoso et al., Multiflavour SMBH seeding and evolution in cosmological environments. **518**(3), 4672–4692 (2023). https://doi.org/10.1093/mnras/stac3169. arXiv: 2203.13846 [astro-ph.GA]
145. V. Springel, T. Di Matteo, L. Hernquist, Modelling feedback from stars and black holes in galaxy mergers. **361**(3), 776–794 (2005). https://doi.org/10.1111/j.1365-2966.2005.09238.x. arXiv: astro-ph/0411108 [astro-ph]
146. V. Springel et al., Simulations of the formation, evolution and clustering of galaxies and quasars. **435**(7042), 629–636 (2005). https://doi.org/10.1038/nature03597. arXiv: astro-ph/0504097 [astro-ph]
147. A. Stacy, V. Bromm, A.T. Lee, Building up the Population III initial mass function from cosmological initial conditions. **462**(2), 1307–1328 (2016). https://doi.org/10.1093/mnras/stw1728. arXiv: 1603.09475 [astro-ph.GA]
148. A. Stacy, V. Bromm, A.T. Lee, Building up the Population III initial mass function from cosmological initial conditions. **462**(2), 1307–1328 (2016). https://doi.org/10.1093/mnras/stw1728. arXiv: 1603.09475 [astro-ph.GA]
149. M.A. Stone et al., Undermassive host galaxies of Five z $\sim$ 6 luminous quasars detected with JWST. **964**(1), 90 (2024). https://doi.org/10.3847/1538-4357/ad2a57. arXiv: 2310.18395 [astro-ph.GA]
150. K.S. Thorne et al., Rapidly rotating holes. *Black Holes: The Membrane Paradigm*. Ed. by Kip S. Thorne, Richard H. Price, and Douglas A. MacDonald (1986), pp. 67–120
151. M. Tremmel et al., The Romulus cosmological simulations: a physical approach to the formation, dynamics and accretion models of SMBHs. **470**(1), 1121–1139 (2017). https://doi.org/10.1093/mnras/stx1160. arXiv: 1607.02151 [astro-ph.GA]
152. A. Trinca et al., The low-end of the black hole mass function at cosmic dawn. **511**(1), 616–640 (2022). https://doi.org/10.1093/mnras/stac062. arXiv: 2201.02630 [astro-ph.GA]
153. H. Umeda et al., The final fates of accreting supermassive stars. **830**(2), L34 (2016). https://doi.org/10.3847/2041-8205/830/2/L34. arXiv: 1609.04457 [astro-ph.SR]
154. R. Valiante et al., From the first stars to the first black holes. **457**(3), 3356–3371 (2016). https://doi.org/10.1093/mnras/stw225. arXiv: 1601.07915 [astro-ph.GA]
155. R. Valiante et al., Unveiling early black hole growth with multifrequency gravitational wave observations. **500**(3), 4095–4109 (2021). https://doi.org/10.1093/mnras/staa3395. arXiv: 2010.15096 [astro-ph.GA]
156. M. Volonteri, M. Habouzit, M. Colpi, The origins of massive black holes. Nat. Rev. Phys. **3**(11), 732–743 (2021). https://doi.org/10.1038/s42254-021-00364-9. arXiv: 2110.10175 [astro-ph.GA]
157. J.H. Wise et al., Formation of massive black holes in rapidly growing pre-galactic gas clouds. **566**(7742), 85–88 (2019). https://doi.org/10.1038/s41586-019-0873-4. arXiv: 1901.07563 [astro-ph.GA]

158. S.E. Woosley, A. Heger, The pair-instability mass gap for black holes **912**(2), L31 (2021). https://doi.org/10.3847/2041-8213/abf2c4. arXiv: 2103.07933 [astro-ph.SR]
159. Q. Wu, Y. Shen, A catalog of quasar properties from sloan digital sky survey data release 16. **263**(2), 42 (2022). https://doi.org/10.3847/1538-4365/ac9ead. arXiv: 2209.03987 [astro-ph.GA]
160. F-G. Xie, F. Yuan, Radiative efficiency of hot accretion flows. **427**(2), 1580–1586 (2012). https://doi.org/10.1111/j.1365-2966.2012.22030.x. arXiv: 1207.3113 [astro-ph.HE]
161. J. Yang et al., Probing early supermassive black hole growth and quasar evolution with near-infrared Spectroscopy of 37 Reionization-era Quasars at 6.3 ¡ z $\leq$ 7.64 **923**(2), 262 (2021). https://doi.org/10.3847/1538-4357/ac2b32. arXiv: 2109.13942 [astro-ph.GA]
162. M. Yue et al., EIGER V. Characterizing the host galaxies of luminous quasars at $zrsim$6(2023). https://doi.org/10.48550/arXiv.2309.04614. arXiv: 2309.04614 [astro-ph.GA]
163. Q. Zhu et al., The formation of the first quasars: the black hole seeds, accretion, and feedback models. **514**(4), 5583–5606 (2022). https://doi.org/10.1093/mnras/stac1556. arXiv: 2012.01458 [astro-ph.GA]

Chapter 13
The Accretion, Emission, Mass and Spin Evolution of Primordial Black Holes

Valerio De Luca and Nicola Bellomo

Abstract Throughout the cosmic history, primordial black holes may experience an efficient phase of baryonic mass accretion from the surrounding medium. While the realm of accretion physics is marked by numerous uncertainties, and a comprehensive understanding remains elusive, recent investigations have delved into this area, exploring its implications for the cosmological evolution of these compact objects. Notably, primordial black holes could experience characteristic growths of their masses and spins, accompanied by the emission of radiation, ultimately responsible for feedback effects that could weaken the efficiency of the process. In this chapter we review the basic formalism to describe the accretion rate, luminosity function and feedback effects, in order to provide distinctive predictions for the evolution of the primordial black hole mass and spin parameters.

13.1 Accretion Model

During their cosmological evolution, primordial black holes (PBHs) can efficiently accrete particles from the medium through which they move. This process can happen either for isolated PBHs, which individually flow in the surrounding gas, or for PBHs which are assembled in binary systems from the epoch of matter-radiation equality, and it can occur both in the early stages of the universe or within the cosmological structures formed in the late universe. An efficient phase of accretion strongly impacts

V. . De Luca (✉)
Center for Particle Cosmology, Department of Physics and Astronomy, University of Pennsylvania, 209 S. 33rd St., Philadelphia, PA 19104, USA
e-mail: vdeluca@sas.upenn.edu

N. Bellomo
Texas Center for Cosmology and Astroparticle Physics, Weinberg Institute, Department of Physics, The University of Texas at Austin, Austin, TX 78712, USA
e-mail: nicola.bellomo@unipd.it

Dipartimento di Fisica e Astronomia G. Galilei, Università degli Studi di Padova, Via Marzolo 8, 35131 Padova, Italy

C. Byrnes et al. (eds.), *Primordial Black Holes*, Springer Series in Astrophysics and Cosmology, https://doi.org/10.1007/978-981-97-8887-3_13

on the evolution of these cosmic relics by increasing their mass and spin parameters, as well as converting a fraction of the accreted mass to radiation. The injection of the latter in the primordial plasma could change the thermal and ionization histories, leading to distortions in the frequency spectrum and temperature/polarization power spectra of the CMB. The corresponding constraints on a PBH population undergoing an efficient accretion phase will be thoroughly discussed in Chap. 24.

The purpose of this Section is instead to summarise the main formalism to model accretion for individual or binary PBHs, in order to set the stage for computing the evolution of their mass and spin parameters. The interested reader can find more comprehensive discussions and complementary details in the following list of Refs. [3, 16, 17, 52, 53]. In this chapter we will adopt natural units $\hbar = c = 1$.

13.1.1 Accretion Onto Individual PBHs

During the early stages of the universe, individual PBHs orbiting in the primordial gas could accrete dark matter and baryons within the so-called Bondi radius, describing the distance in which the PBH gravitational potential dominates the dynamics of the infalling particles. The main effects of this process are the consequential increase of the PBH masses and spins, as well as the amount of radiation injected in the surrounding fluid. The latter may ionize and/or heat the accreting gas, affecting the radiative output and giving rise to feedback effects. A dedicated discussion of these effects is postponed to the next Section.

In the following we provide a general description of accretion, focusing on PBHs which are either in isolation (*naked*) or have an envelope of dark matter particles surrounding them (*dressed*). The latter situation may be realised when the bulk of the dark matter is not made of PBHs, but of a dominant component that seeds, after the redshift of matter-radiation equivalence, the growth of a dark halo. Its additional contribution to the PBH gravitational potential may thus increase the gas accretion rate onto the central PBH by several orders of magnitude, affecting their cosmological evolution. Finally, we introduce the concept of luminosity, in order to quantitatively estimate the amount of radiation emitted in the medium during the accretion process.

13.1.1.1 Basic Formalism for Spherical Accretion

We consider accretion of a hydrogen gas onto an isolated point mass, bathed in the quasi-uniform CMB radiation field. Aside from radiative feedback effects, which we neglect in the first place to simplify the picture, the second aspect to consider is the geometry of the accretion flow. In particular, if the characteristic angular momentum of the accreted gas is smaller than the angular momentum at the innermost stable circular orbit, then accretion has mostly a spherical flow. Otherwise, an accretion disk may form. The latter could easily catalyse the emission of radiation by bremsstrahlung from the hot ionized plasma near the event horizon, while in the spherical case the large viscous heating required to dissipate angular momentum leads to radiating a significant fraction of the rest-mass energy [58]. Since the estimate of the gas angular momentum requires dedicated discussions on the PBH-baryon relative velocity, we first derive the most conservative accretion rate and luminosity by adopting a spherical accretion model, and postpone the discussion of the disk formation to the next Section.

In general, one should solve for the coupled, time-dependent, fluid, heat and ionization equations. However, as long as the characteristic accretion timescale is much shorter than the Hubble timescale, one can make the steady-state approximation and adopt a simplified Newtonian treatment in the region far from the BH horizon [3]. The typical accretion time scale is given in terms of the Salpeter time by

$$\tau_{\rm acc} \equiv \frac{\tau_{\rm Salp}}{\dot{m}} = \frac{\sigma_{\rm T}}{4\pi m_p}\frac{1}{\dot{m}} = \frac{4.5\times 10^8\,{\rm yr}}{\dot{m}}, \tag{13.1}$$

where $\sigma_{\rm T}$ is the Thompson cross section, m_p is the proton mass and $\dot{m}$ indicates the dimensionless accretion rate normalised to the Eddington one

$$\dot{m} = \frac{\dot{M}}{\dot{M}_{\rm Edd}} \quad \text{with} \quad \dot{M}_{\rm Edd} = 1.44\times 10^{17}\left(\frac{M}{M_\odot}\right){\rm g\,s^{-1}}. \tag{13.2}$$

The behaviour of the accretion rate $\dot{M}$ as a function of redshift z and PBH mass M will be the main focus of this Section. As showed in Refs. [52, 53], for PBH masses $M \lesssim 10^4 M_\odot$, a steady-state approximation is satisfied, so we shall limit ourselves to this mass range to simplify the time-dependent problem.

The steady-state mass and momentum equations for the fluid are

$$4\pi r^2\rho|v| = \dot{M} = {\rm const}, \tag{13.3}$$

$$v\frac{dv}{dr} = -\frac{GM}{r^2} - \frac{1}{\rho}\frac{dP}{dr} - Hv - \frac{4}{3}\frac{\overline{x}_e\sigma_{\rm T}\rho_{\rm CMB}}{m_p}v, \tag{13.4}$$

in terms of the peculiar radial velocity v of the accreted gas (i.e. the velocity with respect to the Hubble flow), its density ρ, temperature T, and pressure P, given by

$$P = \frac{\rho}{m_p}(1+\overline{x}_e)T, \tag{13.5}$$

where the ionization fraction x_e has been equated to its background value $\overline{x}_e$ in the absence of feedback effects [3]. Following the steady-state approximation, we have neglected partial time derivative $\partial v/\partial t$, as well as the self-gravity of the accreted gas [52]. Furthermore, the Hubble drag term Hv has the same form of the last term in the momentum equation, that describes the drag force due to Compton scattering of the ambient nearly homogeneous CMB photons with energy density $\rho_{\rm CMB}$, which therefore contribute as an effective viscosity term in the equation [52].

The last equation of the system is the heat equation, that reads

$$v\rho^{2/3}\frac{d}{dr}\left(\frac{T}{\rho^{2/3}}\right) = \frac{8\overline{x}_e\sigma_T\rho_{\rm CMB}}{3m_e(1+\overline{x}_e)}(T_{\rm CMB} - T)\,, \tag{13.6}$$

considering only Compton cooling by CMB photons with temperature $T_{\rm CMB}$ as a heat sink in this region [46]. In the steady-state approximation this equation enforces the condition $T(r \to \infty) = T_{\rm CMB}$. Similarly, at large distances from the PBH, the density should approach the baryonic one, $\rho(r \to \infty) = \rho_{\rm gas}$, with

$$\rho_{\rm gas} = 250\,\mu m_p\left(\frac{1+z}{1000}\right)^3 \,{\rm g\,cm^{-3}}\,, \tag{13.7}$$

for an hydrogen gas with mean molecular weight $\mu = 1.22$.

The gas pressure and density far from the point mass allow to define an effective velocity $v_{\rm eff} = \sqrt{P(r \to \infty)/\rho(r \to \infty)}$ (to be discussed below), such that an isolated PBH, moving with a relative velocity $v_{\rm rel}$ with respect to the surrounding gas with sound speed c_s, accretes particles with a rate given by the Bondi-Hoyle expression [58]

$$\dot{M}_{\rm B} = 4\pi\lambda\rho_{\rm gas}v_{\rm eff}r_{\rm B}^2\,, \tag{13.8}$$

where we have defined the Bondi radius as the region where accretion effectively occurs [8]

$$r_{\rm B} \equiv \frac{GM}{v_{\rm eff}^2} \simeq 1.3\times 10^{-4}\left(\frac{M}{M_\odot}\right)\left(\frac{v_{\rm eff}}{5.7\,{\rm km\,s^{-1}}}\right)^{-2}\,{\rm pc}\,, \tag{13.9}$$

and $v_{\rm eff} = \sqrt{v_{\rm rel}^2 + c_s^2}$. For a gas in equilibrium at the temperature of the intergalactic medium, the corresponding sound speed can be approximated as [17, 52, 53]

$$c_s = \sqrt{\frac{k_B T_{\rm gas}}{\mu m_p}} \simeq 5.7\left(\frac{1+z}{1000}\right)^{1/2}\left[\left(\frac{1+z_{\rm dec}}{1+z}\right)^{\beta} + 1\right]^{-1/2\beta}\,{\rm km\,s^{-1}}\,, \tag{13.10}$$

in terms of the Boltzmann constant k_B and the gas temperature $T_{\rm gas}$, that depends on the physics at large scales. In the second equality, by neglecting the feedback effect induced by accreting PBHs on the gas (to be discussed in the next Section), one can express it as a function of redshift $z_{\rm dec} \simeq 130(\Omega_b h^2/0.022)^{2/5}$, at which the gas temperature deviates significantly from the CMB temperature, and of the characteristic exponent $\beta = 1.72$.

The constant term in the fluid equation, expressed by the accretion parameter λ, is introduced to capture, in a phenomenological manner, the influence of various non-gravitational forces that could oppose the gravitational pull of the compact object and thereby reduce the accretion rate. For spherical accretion onto a point mass, it is of order unity and keeps into account the effects of the Hubble expansion, the coupling of the CMB radiation to the gas through Compton scattering, and the gas viscosity [52]. Its analytical approximation is given by [3]

$$\lambda = \frac{1}{\lambda_{\rm iso}} \left[\lambda_{\rm ad} + (\lambda_{\rm iso} - \lambda_{\rm ad}) \left(\frac{\gamma_{\rm c}^2}{88 + \gamma_{\rm c}^2} \right)^{0.22} \right] \times \left[\frac{1}{(\sqrt{1+\beta_{\rm c}} + 1)^2} \exp\left(\frac{9/2}{3 + \beta_{\rm c}^{3/4}} \right) \right], \tag{13.11}$$

where $\lambda_{\rm ad} = (3/5)^{3/2}/4 \approx 0.12$ and $\lambda_{\rm iso} = e^{3/2}/4 \approx 1.12$. The coefficients $\beta_{\rm c}$ and $\gamma_{\rm c}$ are the dimensionless Compton drag and cooling rates from the surrounding CMB photons respectively, and they read [3]

$$\beta_{\rm c} = \frac{4}{3} \frac{x_e \sigma_{\rm T} \rho_{\rm gas}}{m_p} \frac{r_{\rm B}}{v_{\rm eff}}, \qquad \gamma_{\rm c} = \frac{8}{3} \frac{x_e \sigma_{\rm T} \rho_{\rm gas}}{m_e (1 + x_e)} \frac{r_{\rm B}}{v_{\rm eff}}. \tag{13.12}$$

The knowledge of the exact value of λ requires computing the solution to the system of coupled equations. Some limits can however be deduced analytically. First, when $\gamma_c \gg 1$, Compton cooling efficiently maintains thermal equilibrium down to distances smaller than the Bondi radius. At that point pressure forces are negligible relative to gravity, and the temperature is no longer relevant to the other fluid variables. For this isothermal setup, the expression simplifies to

$$\lambda = \frac{1}{(\sqrt{1+\beta_{\rm c}} + 1)^2} \exp\left(\frac{9/2}{3 + \beta_{\rm c}^{3/4}} \right) \quad \text{for} \quad \gamma_c \gg 1\,. \tag{13.13}$$

On the other hand, for negligible Compton drag $\beta_c \ll 1$, the momentum equation can no longer be rewritten as a conservation equation and one must solve explicitly the coupled fluid and heat equations, and determine the parameter λ numerically. A good fit to the numerical results has been found in [3] and reads

$$\lambda \approx \lambda_{\rm ad} + (\lambda_{\rm iso} - \lambda_{\rm ad}) \left(\frac{\gamma_{\rm c}^2}{88 + \gamma_{\rm c}^2} \right)^{0.22} \quad \text{for} \quad \beta_c \ll 1\,. \tag{13.14}$$

The combination of these two results allows for the approximated expression shown in Eq. (13.11), which is strictly valid for $\beta_c \ll \gamma_c$ and for $\beta_c \ll 1, \gamma_c \gg 1$.

The final ingredient in the expression of the accretion rate is the BH peculiar velocity $v_{\rm rel}$, that has to be self-consistently accounted for to properly estimate the gas temperature and the dynamics of the accretion process. While a dedicated time-dependent hydrodynamical simulation is probably needed to provide an exact result, given the dependence of the PBH proper motion on the amplitude of the inhomogeneities of the DM and baryon fluids, some analytical insights can be derived. Following Ref. [53], and assuming PBHs to behave like DM particles, one can identify two main regimes of interest for the PBH relative velocity $v_{\rm rel}$: a Gaussian linear contribution on large scales $v_{\rm L}$, whose power spectrum and variance can be extracted from linear Boltzmann codes, and a small-scale contribution due to non-linear clustering of PBHs, $v_{\rm NL}$.

In the linear regime, gas and dark matter perturbations are coupled to each other until the decoupling redshift $z_{\rm dec} \simeq 130$ is reached. At earlier times, the Silk damping acts suppressing the growth of inhomogeneities on small scales, such that the PBH peculiar velocity is of order of the gas sound speed. On the other hand, at redshifts $z < z_{\rm dec}$ (before entering the nonlinear regime), the gas flow lags behind the DM with a relative velocity $v_{\rm L} = v_{\rm DM} - v_{\rm b}$. This velocity has a complex time and scale-dependence, since baryons undergo acoustic oscillations and the dark-matter overdensities grow (see for example Fig. 2 of [53]). Reference [19] explicitly compute $\langle v_{\rm L}^2 \rangle$ as a function of time and find that it is mostly constant for $z \gtrsim 10^3$ and decays like $1 + z$ at smaller redshifts. It therefore follows the simple dependence [19]

$$\sqrt{\langle v_{\rm L}^2 \rangle} \approx \min\left[1, z/10^3\right] \times 30\,{\rm km/s}\,. \tag{13.15}$$

As discussed above, the PBH accretion rate depends on the combination $v_{\rm eff} = \sqrt{v_{\rm rel}^2 + c_s^2}$, such that the total energy injected in the plasma is obtained by averaging over the Gaussian distribution of relative velocities. The final average reads [53]

$$\begin{aligned}
\langle v_{\rm eff} \rangle_{\rm A} &\sim c_s \left(\frac{16}{\sqrt{2\pi}} \mathcal{M}^3 \right)^{\frac{1}{6}} \theta(\mathcal{M} - 1) + c_s \left(1 + \mathcal{M}^2\right)^{\frac{1}{2}} \theta(1 - \mathcal{M}) \,, \\
\langle v_{\rm eff} \rangle_{\rm B} &\sim c_s \mathcal{M} \left[\sqrt{\frac{2}{\pi}} \ln\left(\frac{2}{e} \mathcal{M} \right) \right]^{-\frac{1}{3}} \theta(\mathcal{M} - 1) + c_s \left(1 + \mathcal{M}^2\right)^{\frac{1}{2}} \theta(1 - \mathcal{M}) \,,
\end{aligned} \tag{13.16}$$

in terms of the Heaviside step function θ and Mach number $\mathcal{M} = \langle v_{\rm rel} \rangle / c_s$, that identifies a supersonic ($\mathcal{M} > 1$) or subsonic ($\mathcal{M} < 1$) motion. Here the scenario A refers to a low efficient accretion rate, $\dot{m} < 1$, characterised by a spherical geometry, while scenario B refers to an efficient accretion rate, $\dot{m} > 1$, which, as shown below, could support the presence of an accretion disk.

In the nonlinear regime, at redshifts $z \lesssim \mathcal{O}(10)$, the growth of nonlinear perturbations dominates the proper motion of PBHs. The velocity of PBHs which are massive enough to fall into the potential wells of the first cosmic structures could be sufficiently large to stop accretion, until they came to rest at the center of the host halo. At that point, it is unclear whether they meet favorable conditions for accretion to start again. One can assume the characteristic velocity induced by nonlinear structures to be comparable with the circular velocity of virialized halos as a function of redshift. By determining their masses following the Press-Schechter approach, the circular and thus typical velocity reads [53]

$$v_{\rm eff} \simeq (8 \cdot 10^2\,{\rm km/s}) \exp[-0.6(1+z)] \sqrt{\frac{1+z}{10}}\,. \tag{13.17}$$

This estimate, based on a fluid-like approximation for the formation of dark matter halos, may suffer from few drawbacks: first, the discrete nature of PBHs, as well as their spatial distribution at formation, may change the power spectrum at small scales and invalidate the Press-Schechter assumption; second, this approach does not take into account kinetic effects, such as the thermal velocity distribution around the bulk motion velocity, and the putative presence of shocks and instabilities, which may hamper the applicability of the approach to small scales [49]. As further discussed in the next section about feedback effects, the beginning of structure formation and the associated nonlinearities provide therefore a large source of uncertainty in the efficiency and duration of the accretion process.

The setup outline above describes naked PBHs, i.e. black holes in isolation in the surrounding medium. However, if PBHs do not comprise the totality of the dark matter ($f_{\rm PBH} \lesssim 1$), an additional and dominant dark matter component would be present around them and modify the accretion picture. Current observational constraints show that PBHs with masses larger than $\mathcal{O}(M_\odot)$ can comprise only a fraction of the dark matter in the universe [12]. In this case, PBHs would attract the surrounding dark matter particles, generating an extended dark matter halo during the matter-dominated epoch with mass [1, 5, 11, 36]

$$M_h(z) = 3M\left(\frac{1+z}{1000}\right)^{-1}. \tag{13.18}$$

It grows with time as long as PBHs are in isolation, and stops when all the available dark matter has been accreted, i.e. approximately when $3f_{\rm PBH}(1+z/1000)^{-1} = 1$, or when tidal effects of nearby PBHs and other halos are efficient in destroying it.

Such a halo is characterised by a typical spherical density profile $\rho \propto r^{-\alpha}$ (with $\alpha \simeq 9/4$ [1, 36] as confirmed by N-body simulations), truncated at a radius

$$r_h \simeq 0.019\,\text{pc}\left(\frac{M}{M_\odot}\right)^{1/3}\left(\frac{1+z}{1000}\right)^{-1}. \tag{13.19}$$

While direct accretion of dark matter onto the PBH is negligible [51, 52], the halo acts as a catalyst enhancing the baryonic accretion rate.

The presence of a dark halo clothing is usually taken into account in the accretion parameter λ [52]. To estimate the hierarchy of scales in the problem, one can define the parameter

$$\kappa \equiv \frac{r_{\rm B}}{r_h} = 0.22\left(\frac{1+z}{1000}\right)\left(\frac{M_h}{M_\odot}\right)^{2/3}\left(\frac{v_{\rm eff}}{\text{km s}^{-1}}\right)^{-2}. \tag{13.20}$$

When $\kappa \geq 2$, i.e. when the typical size of the halo is smaller than the Bondi radius, the accretion rate is the same as the one for a PBH of point mass M_h. On the other hand, if $\kappa < 2$, only a fraction of the dark halo would be relevant for accretion, and one has to correct the quantities entering in the parameter λ with respect to the naked case as [52]

$$\beta_c^h \equiv \kappa^{\frac{p}{1-p}}\beta_c, \quad \lambda^h \equiv \bar{\Upsilon}^{\frac{p}{1-p}}\lambda(\beta_c^h), \quad \bar{\Upsilon} = \left(1 + 10\beta_c^h\right)^{\frac{1}{10}} \exp(2-\kappa)\left(\frac{\kappa}{2}\right)^2, \tag{13.21}$$

where $p = 2 - \alpha \simeq -1/4$. As shown in Refs. [17, 54], a dark halo dress could enhance the accretion rate of few orders of magnitude, providing an important effect to be considered in the modelling of accretion.

13.1.1.2 Beyond Spherical Symmetry

The previous results were based on the assumption of a spherically symmetric accretion flow onto the accreting PBHs. Let us now discuss the accretion model when one goes beyond this assumption. While no complete theory of disk accretion exists from first principles, the standard scenario for disk formation requires that the accreted baryonic particles carry an angular momentum larger than the one at the innermost stable circular orbit. To understand the validity of this requirement, one can start from the expression of the baryon velocity variance $\sigma_{\text{b}} = \langle v_{\text{b}} \rangle$ provided in Refs. [43, 53] as

$$\sigma_{\text{b}} \simeq (3.8 \times 10^{-7}\text{km/s})\, \xi^{-1.7}(z) \left(\frac{1+z}{1000}\right)^{-1} \left(\frac{M_h}{M_\odot}\right)^{0.85}, \tag{13.22}$$

where $\xi(z) = \text{Max}[1, \langle v_{\text{eff}} \rangle / c_s]$ describes the effect of the PBH proper motion in reducing the Bondi radius. The above criterion can be recast in a condition for the gas velocity to be larger than the Keplerian velocity close to the PBH, $\sigma_{\text{b}} \gtrsim 2D\xi^2(z)c_s^2$, in terms of a constant $D \sim O(1 \div 10)$ describing relativistic corrections. Using the expression for the dark halo mass shown in Eq. (13.18), one can estimate the minimum PBH mass for which the accreting gas acquires a disk geometry [17, 53],

$$M \gtrsim 6 \times 10^2 M_\odot\, D^{1.17} \xi^{4.33}(z) \frac{(1 + z/1000)^{3.35}}{\left[1 + 0.031\,(1 + z/1000)^{-1.72}\right]^{0.68}}. \tag{13.23}$$

It is interesting to also point out that, even without any velocity dispersion, the disk formation criterion can be satisfied if the nonlinear PBH motions at small scales are taken into account [49]. In that case, the presence of near PBH neighbors may induce the formation of binaries, with size larger than the Bondi radius, which allow for a bulk motion of the baryonic gas with respect to the PBH pair center of mass. The corresponding condition for disk formation then reads [49]

$$\sqrt{f_{\text{PBH}}} \left(\frac{M}{M_\odot}\right) \gg \left(\frac{1+z}{730}\right)^3, \tag{13.24}$$

which shows that, whenever heavy PBHs constitute a large fraction of the dark matter, disk formation may happen at large redshifts because of their nonlinear motions.

Finally, notice that, contrarily to baryonic particles, the angular momentum of accreting dark matter is typically small and thus does not lead to the formation of a disk, while it impacts on the density profile of the dark halo which envelops the PBH.

Besides the necessary condition derived in Eq. (13.23), the disk geometry also depends on the strength of the accretion rate, $\dot{m}$. Different regimes of $\dot{m}$ can be considered. First, if accretion is not efficient enough, $\dot{m} < 1$, but the geometry is non-spherical (when Eq. (13.23) is satisfied), an advection-dominated accretion flow (ADAF) may form [40]. On the other hand, when $\dot{m} \gtrsim 1$, non-spherical accretion can give rise to a geometrically thin accretion disk [55]. Finally, for an extremely large accretion rate, $\dot{m} \gg 1$, the accretion luminosity might be strong enough that the disk "puffs up" and becomes thicker. These regimes will be discussed in more details in the next Section.

For the PBH masses we are interested in, $\dot{m}$ never exceeds unit significantly (see discussion in Sect. 13.3), such that the thin-disk approximation should be reliable in the slightly super-Eddington regime of PBHs. Notice also that the condition $\dot{m} \gtrsim 1$ is always more stringent than the requirement (13.23), and it can therefore be considered as a sufficient condition for the formation of a thin disk around an isolated PBH [17].

13.1.1.3 Bolometric Luminosity

Accreting PBHs are responsible of injecting radiation into the surrounding medium. One can estimate the amount of emitted energy by computing the PBH bolometric luminosity, which is a necessary ingredient to study the role of the emitted radiation in the heating and ionization of the intergalactic medium.

The bolometric luminosity of an accreting PBH depends on the matter accreted onto the PBH per unit time $\dot{M}$, as well as on an overall efficiency factor ϵ related to the conversion of accreted matter into radiation. It reads

$$L = \epsilon \dot{M} = \epsilon \dot{m} L_{\text{Edd}}\,, \tag{13.25}$$

in terms of the Eddington luminosity L_{Edd}, that provides an upper bound at which spherical accretion is balanced by radiation pressure, and defined as

$$L_{\text{Edd}} = \frac{4\pi \mu G M m_p}{\sigma_{\text{T}}} \simeq 1.3 \cdot 10^{38} \left(\frac{M}{M_\odot}\right) \text{erg/s}\,. \tag{13.26}$$

The efficiency factor ϵ depends on the accretion rate $\dot{m}$ and on the geometry of the accretion flow, so that $\epsilon = \epsilon(\dot{m})$. Its typical value is of the order of $O(0.1)$ for spherical accretion and, in order to satisfy the Eddington bound, decreases as $\dot{M}$ increases.

Solving for the behaviour of $\epsilon(\dot{m})$ is in general a complicated radiative transfer problem. However, following the description discussed above, an approximate understanding can be outlined for different geometries of the accretion flow [53]. For spherical accretion, the dominant emission mechanism is bremsstrahlung, with most of the radiation originating from the region just outside the event horizon. In this case, the efficiency to convert rest-mass energy into radiation is proportional to $\dot{m}$ and it is given by $\epsilon = 0.011\dot{m}$ [56, 57] (even smaller values can be achieved when Compton cooling is negligible [3]).

When the gas angular momentum is non negligible (i.e. when the condition (13.23) is satisfied), different regimes arise depending on the gas cooling efficiency and accretion rate $\dot{m}$. For example, when $\dot{m} < 1$, the accretion flow is typically not dense enough for electrons to cool efficiently, such that most of the rest-mass energy is not radiated away and is advected into the central BH, giving rise to an advection-dominated accretion flow (ADAF) [30, 66]. In this case the accretion disk is hot, geometrically thick and optically thin, and the radiative efficiency is linearly suppressed and may reach values $\epsilon \simeq 0.1\dot{m}$ [40]. On the other hand, if the radiative cooling of the gas is efficient and $\dot{m} \gtrsim 1$, theoretical models suggest the formation of thin-disk geometries [55]. In this case, the maximal energy per unit mass available is uniquely determined by the binding energy at the innermost stable orbit, corresponding to a radiative efficiency approaching a constant value $\epsilon \simeq 0.1$ [53]. In this case therefore the bolometric luminosity can never exceed the Eddington limit. Finally, for larger accretion rates $\dot{m} > 1$, several phenomena may come into play. For instance, radiative and kinetic feedback can break stationary conditions, generating outflows or periods of high luminosity alternating with periods of low luminosity. Otherwise, phases with quasi-steady state, super-Eddington, mass accretion can take place, with a corresponding drop in efficiency to satisfy $L \lesssim L_{\rm Edd}$. The impact of feedback effects on the accretion rate is therefore crucial to understand the effective geometry of disks around PBHs; in particular, if feedback phenomena are efficient enough in suppressing the rate below the Eddington limit, then one would expect accreting PBHs during the dark ages to form an ADAF [2, 49]. While a complete understanding of their effects is still under investigation, we provide a more detailed discussion in the next Section and in Chap. 24.

The efficient emission of radiation in the surrounding medium could modify its properties, resulting eventually into its heating and ionization, and breaking the assumption of constant ionization fraction over the accretion flow. In particular, close enough to the PBH, the gas may eventually become fully ionized through photoionizations by the outgoing radiation field [3]. By solving the time evolution of the ionization fraction coupled to the one for the gas temperature, one could get a more precise estimate of the gas accretion rate onto PBHs and the corresponding bolometric luminosity. The interested reader can find dedicated discussions on these effects in Refs. [3, 68, 69] and Chap. 24.

13.1.2 Accretion Onto PBH Binaries

After their formation in the early universe, PBHs are dynamically coupled to the cosmic expansion, with negligible peculiar velocities. When their local density becomes comparable to the one of the surrounding radiation fluid, usually after the matter-radiation equality, they can become gravitationally bound to each other. However, because of the presence of large Poisson fluctuations at small scales, some PBHs can decouple even earlier. When they decouple, the closest PBH pairs would start to fall towards each other in a head-on collision, unless the torque caused by the gravitational field of the surrounding PBHs and other matter inhomogeneities prevents it. In that case the pair would give rise to a binary system [31, 39, 50].

From that epoch on, PBH binaries could experience a phase of baryonic accretion, as it happens for individual PBHs. In this case, however, one has to take into account both local processes, onto the individual PBHs in the binary, and global processes, i.e. on the binary as a whole. In this Section we summarise how accretion proceeds for PBH binaries following Refs. [16, 17].

Let us consider a PBH binary with total mass $M_{\rm tot} = M_1 + M_2$, mass ratio $q = M_2/M_1 \leq 1$, reduced mass $\mu = M_1 M_2/(M_1 + M_2)$, and orbital parameters a and e, describing the semi-major axis and eccentricity, respectively. The setup is schematically depicted in Fig. 13.1.

Similarly to the case of individual PBHs, one can identify a Bondi radius of the binary

$$r_{\rm B}^{\rm bin} = \frac{G M_{\rm tot}}{v_{\rm eff}^2}\,, \tag{13.27}$$

in terms of the effective velocity $v_{\rm eff} = \sqrt{c_s^2 + v_{\rm rel}^2}$, with which the binary center of mass moves relative to the surrounding gas. The presence of another scale, the binary semi-major axis a, identifies two different regimes of interest for the accretion process: for large semi-major axis, $a \gg r_{\rm B}^{\rm bin}$, accretion occurs onto the single PBHs

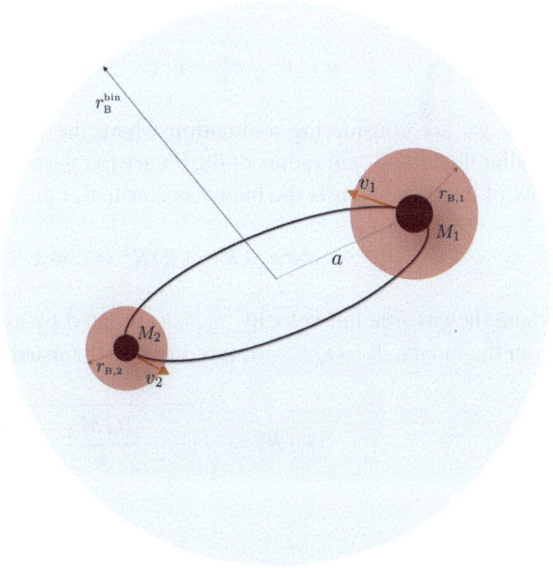

Fig. 13.1 Pictorial representation of a PBH binary undergoing a phase of baryonic accretion. The different scales of the problem are shown, including the Bondi radius of the binary which is taken to be much bigger than the orbital separation. Figure adapted from Ref. [16]

independently (each moving at a characteristic velocity that depends on the orbital one). In this case the description outlined in the previous Section would apply. On the other hand, as the binary hardens and the orbital separation decreases, the binary Bondi radius would become the dominant scale ($a \ll r_B^{bin}$) and change the properties of accretion. We will discuss this case below. Notice that this could occur much before gravitational waves emission becomes the dominant mechanism for driving the inspiral evolution.

The corresponding accretion rate for a PBH binary reads

$$\dot{M}_{bin} = 4\pi \lambda \rho_{gas} v_{eff} r_B^{bin\,2} . \tag{13.28}$$

Using the geometrical properties of the binary system, we can now derive how accretion proceeds for each binary component. The PBH positions and velocities with respect to the center of mass are given by [48]

$$r_1 = \frac{q}{1+q} r, \qquad v_1 = \frac{q}{1+q} v; \qquad r_2 = \frac{1}{1+q} r, \qquad v_2 = \frac{1}{1+q} v \tag{13.29}$$

as a function of their relative distance and velocity

$$r = a(1 - e\cos u), \qquad v = \sqrt{GM_{tot}\left(\frac{2}{r} - \frac{1}{a}\right)}, \tag{13.30}$$

and orbital angle u. The latter evolves in time following the law $\sqrt{a^3/GM_{tot}}(u(t) - e\sin u(t)) = t - C$, in terms of an integration constant C [48]. The corresponding effective velocities for each PBH component are given by

$$v_{eff,1} = \sqrt{v_{eff}^2 + v_1^2}, \qquad v_{eff,2} = \sqrt{v_{eff}^2 + v_2^2}. \tag{13.31}$$

Since we are considering a situation where the typical binary semi-axis is much smaller than the Bondi radius of the binary (as shown in Fig. 13.1), the total infalling flow of baryons towards the binary is constant, i.e.

$$4\pi \rho_{gas}(R) v_{ff}(R) R^2 = \text{const} = \dot{M}_{bin}, \tag{13.32}$$

where the gas free fall velocity, v_{ff}, is computed by assuming that at large distances from the binary, $R \sim r_B^{bin} \gg a$, it reduces to the usual effective velocity v_{eff}, i.e. [16]

$$v_{ff}(R) = \sqrt{v_{eff}^2 + \frac{2GM_{tot}}{R} - \frac{2GM_{tot}}{r_B^{bin}}}. \tag{13.33}$$

The gas density profile $\rho_{\text{gas}}(R)$ at a distance R from the binary center of mass then reads

$$\rho_{\text{gas}}(R) = \frac{\dot{M}_{\text{bin}}}{4\pi v_{\text{ff}}(R) R^2}, \tag{13.34}$$

showing that a constant flow of infalling baryons towards the binary results into an increase in the density compared to its mean cosmic value at the binary Bondi radius, ρ_{gas}.

From the definition of the gas density profile $\rho_{\text{gas}}(R)$, one can then write down the accretion rates for the individual binary components in terms of their effective Bondi radii $r_{\text{B,i}} = GM_i/v_{\text{eff,i}}^2$ as

$$\dot{M}_1 = 4\pi G^2 \rho_{\text{gas}}(r_{\text{B,1}}) v_{\text{eff,1}}^{-3} M_1^2, \qquad \dot{M}_2 = 4\pi G^2 \rho_{\text{gas}}(r_{\text{B,2}}) v_{\text{eff,2}}^{-3} M_2^2. \tag{13.35}$$

Few considerations are in order. First, one can notice that, since the local PBH velocities shown in Eq. (13.31) are of the order of the orbital velocity v, which is much larger than v_{rel} and c_s when $a \ll r_{\text{B}}^{\text{bin}}$, the individual Bondi radii $r_{\text{B,i}}$ are much smaller than the binary Bondi radius $r_{\text{B}}^{\text{bin}}$. Second, in the above expressions we have assumed the individual PBHs to be naked, fixing their accretion parameter $\lambda \approx 1$, while we have described the whole binary as clothed by a dark halo, with parameter λ that takes into account the ratio of the binary Bondi radius compared to the dark halo radius [53], $r_{\text{B}}^{\text{bin}}/r_h^{\text{bin}}$, as discussed in the previous Section. The assumption of neglecting the dark halos for single PBHs is based on the consideration that, during the orbital evolution, the halos surrounding each PBHs would get destroyed by tidal effects, leaving one single halo surrounding the binary [13, 33].

After some simplifications, the individual PBH accretion rates read

$$\dot{M}_1 = \dot{M}_{\text{bin}} \sqrt{\frac{1+\zeta+(1-\zeta)\gamma^2}{2(1+\zeta)(1+q)+(1-\zeta)(1+2q)\gamma^2}}, \tag{13.36}$$

$$\dot{M}_2 = \dot{M}_{\text{bin}} \sqrt{\frac{(1+\zeta)q+(1-\zeta)q^3\gamma^2}{2(1+\zeta)(1+q)+(1-\zeta)q^2(2+q)\gamma^2}}, \tag{13.37}$$

in terms of $\zeta = e\cos u$ and $\gamma^2 = a v_{\text{eff}}^2/\mu q$. Since the orbital period is usually much smaller than the accretion timescale τ_{acc}, one can average over the angle u, eliminating the time dependence, and also notice that the dependence on the eccentricity is negligible. Finally, since $GM_1 q^2/a \gg v_{\text{eff}}^2$ for PBHs heavier than a solar mass and for large enough semi-major axis, one can further simplify the expressions to [16]

$$\dot{M}_1 = \dot{M}_{\text{bin}} \frac{1}{\sqrt{2(1+q)}}, \qquad \dot{M}_2 = \dot{M}_{\text{bin}} \frac{\sqrt{q}}{\sqrt{2(1+q)}}, \tag{13.38}$$

which recover the expected behaviour $\dot{M}_1 = \dot{M}_2 = \dot{M}_{\text{bin}}/2$ in the limit of equal PBH masses $q \to 1$. These equations therefore provide the accretion rate for the PBH binary components. Reintroducing the Eddington normalised rates, $\dot{m}_i = \tau_{\rm Salp}\dot{M}_i/M_i$, one gets

$$\dot{m}_1 = \dot{m}_{\rm bin}\sqrt{\frac{1+q}{2}}, \qquad \dot{m}_2 = \dot{m}_{\rm bin}\sqrt{\frac{1+q}{2q}}. \tag{13.39}$$

From these equations, one can also derive the time evolution for the binary mass ratio, which reads [16]

$$\dot{q} = q\left(\frac{\dot{M}_2}{M_2} - \frac{\dot{M}_1}{M_1}\right) = \frac{q}{\tau_{\rm Salp}}\left(\dot{m}_2 - \dot{m}_1\right). \tag{13.40}$$

Since $\dot{m}_2 > \dot{m}_1$ as shown in Eq. (13.39), this equation proves that the mass ratio grows until it reaches a stationary point when $q = 1$ and $\dot{m}_1 = \dot{m}_2$. In other words, accretion onto a PBH binary pushes the binary masses to balance each other on secular time scales.

Let us close this Section by discussing the formation of an accretion disk for PBHs in binary systems. As shown above, each binary component has a velocity of the order of the orbital velocity $v_{1,2} \sim v$, which is much larger than the binary effective velocity $v_{\rm eff}$. Although the Bondi radius of the individual PBHs is much smaller than the binary Bondi radius, it is still parametrically larger than the ISCO radius, $r_{\rm B,i} > r_{\rm ISCO,i}$, such that conservation of angular momentum along the accretion flow implies that the condition for the formation of a disk is always satisfied (see Eq. (13.23)) [16]. Furthermore, the condition $\dot{m} \gtrsim 1$, to ensure the formation of a thin accretion disk, is more easily fulfilled by PBHs in binaries rather than by individual ones, since accretion for binaries is enhanced by the larger total mass of the system. In Sect. 13.3 we will show the effects of baryonic accretion on the mass and spin parameters of PBHs which are either isolated or in binaries.

13.2 Feedback Effects

So far we have considered the dynamics of an accreting PBH without considering the effect of emission on the surrounding environment. However, accurate modelling of accretion requires considering the reciprocal interaction between the PBH and the surrounding medium. Feedback effects describe this intertwined relation between the PBH accretion and emission mechanism, and the gas medium. The importance of feedback effects is well understood in other branches of Astrophysics, for instance in galaxy dynamics, where AGN feedback turns out to be a key ingredient in explaining galaxy evolution, see, e.g., Ref. [38] for a review on the subject. In this Section we will discuss the different type of feedback effects which may arise during a PBH accretion phase.

13.2.1 Local Thermal and Ionization Feedback

Local feedback effects refer to a set of physical effects the PBH has on its local environment, where "local" in this context refers to the PBH sphere of influence. The solution of Eqs. (13.3), (13.4) and (13.6) describes the temperature profile of the gas around the PBH, and in particular it describes the temperature at the Schwarzschild radius depending on the gas ionization scheme assumed [3]. Two common ionization schemes are typically assumed, "collisional ionization" and "photoionization".

In the former scenario, the temperature of the gas grows due to the adiabatic infall towards the PBH until it reaches the ionization temperature $T_{\rm ion} \approx 10^4$ K. At that stage, all the energy gained by the infall dynamics is converted into ionizing the gas, and only when the gas becomes fully ionized the temperature starts growing again. On the other hand, in the latter scenario, ionization is driven by high-energy photons emitted by ultra-relativistic electrons close to the Schwarzschild radius, thus the growth of the temperature does not need to stop while the gas is getting ionized, resulting in a higher temperature T_S of the gas close to the compact object. In both simplified pictures the temperature grows as $T(r) \propto r^{-n}$, where the adiabatic index is $n \approx 1$ ($n \approx 2/3$) when $T \lesssim m_e$ ($T \gtrsim m_e$). The photoionization scenario is characterized by a higher temperature, around $T_S \approx 10^{11}$ K, than the one in the collisional ionization scenario ($T_S \approx 10^9$ K), thus in a higher luminosity of the PBH.

In the simplified picture described above it was assumed that heating of the infalling gas is only due to compression. On the other hand, the radiation produced by the PBH can heat the surrounding material also through Compton scattering, inducing a local thermal feedback effect. It turns out that such Compton heating is comparable to the standard adiabatic heating, or to the Compton cooling to CMB photons at scales comparable to the Bondi radius [3], and it is therefore negligible. Similarly, whether the plasma around the PBH is optically thin or thick also has an impact in the transmission of heating across the surrounding environment. However, it can be showed that as long as the accretion rate is smaller than the Eddington limit, i.e., $\dot{m} \ll 1$, the plasma is optically thin and the previous modelling is not affected by this assumption [3].

Finally, let us stress that also the choice of either of the two simplifying ionization schemes cannot fully describe the realistic case. This can be easily checked by analysing the size of the Strömgren sphere around an accreting PBH [3]. In particular, the collisional ionization picture predicts a radius of such sphere that is smaller than the prediction obtained from detailed balance argument [42], while in the case of photoionization the radius is found to be larger. Thus, the realistic scenario should fall between these two possibilities, and it is heavinly impacted the feedback provided by the accreting PBH.

13.2.2 Global Feedback

Due to the supersonic motion of DM across the cosmic medium at high redshifts [65], the geometry of accretion cannot be purely spherical, even at scales of the order of the Bondi radius. Instead, an axisymmetric accretion column develops behind the PBH, through which the medium is channeled directly to the accretor. Despite this fundamental difference in the geometry, accretion still proceeds at a rate comparable to the Bondi-Hoyle-Littleton one [8, 29] described in Sect. 13.1.

Conversely, in front of the compact object, we observe the formation of a bow-shaped ionization front on scales larger than the Bondi radius for PBH masses $M_{\rm PBH} \gtrsim 10^2 M_\odot$ [45, 53, 60], which distinguish the neutral region ahead of the front from the region behind it, which embeds the PBH, where the medium has been ionized. Due to the presence of these fronts, the gas density close to the PBH is reduced, inducing a lower accretion rate than in the pure spherical case [21]. As shown in Ref. [44], this conclusion seems to be valid also for relatively slow PBHs, and even in the presence of dark matter halos or accretion disks [2], strongly impacting the efficiency and phenomenology of the accretion process. Moreover, gas inside these ionization bubbles will get heated, and as soon as its temperature reaches values higher than the background one, accretion is further reduced, especially at redshifts lower than recombination.

On the other hand, the stability of this kind of structure is non-trivial, since they can be disrupted due to the high speed motion of the compact object [60]. In that case the PBH can experience bursts in accretion due to the sudden inflow of cold gas, that can be easily accreted, even though the dynamics of this process is highly chaotic.

13.2.3 Mechanical and Radiative Feedback

Because of their interaction with the surrounding environment, PBHs can generate complex structures called outflows, winds and/or jets, depending on their level of collimation, that extend from the region close to the event horizon up to scales many times larger than the Bondi radius, see, e.g., the seminal Refs. [6, 7]. These outflows may modify the accretion efficiency both through mechanical or radiative processes, which effectively sweep the medium away from the compact object, or simply just heat it. While the role of radiative feedback is harder to model because it strongly depends on the details of the outflow and the medium, mechanical feedback processes are more robust to small changes in the details of the modelling, and they have already been studied in multiple contexts, ranging from stars, stellar mass BHs, supermassive BHs and galaxies, see, e.g., Refs. [25, 32, 34, 35, 59, 67].

On the other hand, the role of mechanical feedback on PBH accretion has been explored only recently [9, 10, 47]. In this scenario, the presence of a magnetic field and/or some excess of thermal energy in the accreted gas in the proximity of the compact object can launch supersonic outflows with different degrees of collimation. Even under the conservative assumption that these outflows carry only a small fraction of the accreted energy, such structures can escape the vicinity of the PBH, overcome the ram pressure of the inflowing gas, reach scales larger than the Bondi radius and sweep away and/or heat the gas. The level of effectiveness of this process depends significantly on the angle between the PBH direction of motion and the outflow direction, in particular it is maximised if the two are aligned. This pure mechanical effect reduces accretion because the hot medium needs time to cool down and approach again the PBH.

The dynamics of this process is highly chaotic, and numerical simulations are required to derive the exact amount of accreted gas [9, 10, 25, 35]. Nevertheless, under conservative assumptions, and for choices of gas density and PBH velocity that resemble the conditions present during the epochs after the cosmological recombination era and previous to structure formation, it was showed that mechanical feedback alone can reduce the accretion rate by an order of magnitude. Finally, the outflow itself could irradiate energy that would eventually be deposited into the surrounding medium; however, the implications of this feedback effect and its role on the accretion dynamics have been studied only in the context of PBH accreting in virialized structures, see, e.g., Ref. [63].

13.2.4 Feedback Mechanisms in the Late Universe

In summary, modelling in a self-consistent manner the accretion and subsequent emission of radiation is one of the open questions in astrophysics. This Chapter covered the basics of feedback mechanisms that have mostly been explored in the high-redshift Universe, previously to the formation of virialized structures.

At the time of their formation, a part of the PBH population would start falling in their gravitational potential wells around redshifts $z \simeq \mathcal{O}(10)$, increasing the relative velocity up to one order of magnitude [28]. This will result in a consequent suppression of the accretion rate.

Other examples of feedback have been considered in the literature, for instance concerning the gravitational role played by the gas surrounding the PBH [52, 54, 63], the possibility of accreting at super-Eddington rates [61, 62] (where the formation of thick disks may impact on the accretion luminosity and feedback), or the possibility of having PBH not evolving “in isolation” [3]. However, all these feedback mechanisms are effective only for PBHs with masses $M_{\rm PBH} \gtrsim 10^4\, M_\odot$, where the picture presented in Sect. 13.1 cannot be used to accurately describe them.

We conclude by noting that once virialized structures form, the study of PBHs in non-linear environments becomes even more challenging, and can be reasonably done only with the help of complex and model-dependent numerical simulations, without relying on simple analytic prescriptions. In this sense we parameterise the uncertainties in certain aspects of the accretion model by introducing an effective cut-off redshift $z_{\rm cut\text{-}off}$, after which PBH accretion is not efficient anymore, and limiting its effectiveness to the pre-Cosmic Dawn era [16]. The exact value of this cut-off redshift is treated as a phenomenological parameter in Sect. 13.3.

13.3 Effects of Accretion on a PBH Population

Gas accretion provides the main effect driving the mass and spin evolution of PBHs during their cosmic evolution. In particular, at redshifts $z \in (10 \div 100)$, PBHs of certain mass may experience an efficient accretion phase, provided that they are surrounded by a dark matter halo which catalyse the process, and that feedback effects are not strong enough in suppressing the rate. From the discussion in the previous Section, we will implement $z_{\text{cut-off}}$ as a parameter describing the duration and efficiency of the accretion phase. For low enough cut-off redshifts, one can show that the late time PBH mass and spin distributions might be significantly different from those at high redshifts. In the following we summarise the main effects of accretion onto a PBH population. The interested reader can find further details in Refs. [16, 17, 24].

13.3.1 PBH Mass Evolution

For PBHs in the solar mass range, baryonic accretion is mostly ineffective at redshifts $z \gtrsim 100$, because of the longer accretion timescale compared to the age of the universe at that epoch. On the other hand, at smaller redshifts, PBHs may start evolving rapidly if they are heavy enough, $M \gtrsim O(10) M_\odot$. A PBH, formed at redshift z_i with initial mass $M^\text{i} \equiv M(z_\text{i})$, may reach a final mass $M(M^\text{i}, z_{\text{cut-off}})$, obtained by solving the equation

$$\frac{dM}{dt} = \dot{M}_\text{B}(M, t) = \dot{M}_\text{Edd}\dot{m}, \quad z(t) \in [z_\text{i}; z_{\text{cut-off}}], \; M(t) \in [M^\text{i}; M(M^\text{i}, z)], \quad (13.41)$$

depending on the actual redshift $z_{\text{cut-off}}$ at which structure formation and reionization take place, as well as on feedback effects.

Following the Bondi formalism discussed in Sect. 13.1, an illustrative plot of the accretion rate $\dot{m}$ in terms of the PBH mass M and redshift z, and assuming a sharp efficiency decrease after $z_{\text{cut-off}} \sim 10$, is shown in Fig. 13.2. Based on the seminal work [53], and including the effects of a dark matter halo as thoroughly discussed in Sect. 13.1, it shows that PBH accretion may be efficient in some regions of the parameter space, $\dot{m} \gtrsim 1$, for $M \gtrsim O(30) M_\odot$, in the absence of a significant radiative feedback. For masses larger than about $10^4 M_\odot$, feedback effects are crucial in the modelling of the accretion rate, and more involved computations are needed to provide a definite behaviour. A different choice in the cut-off redshift amounts to a drastic decrease of $\dot{m}$ at that corresponding value.

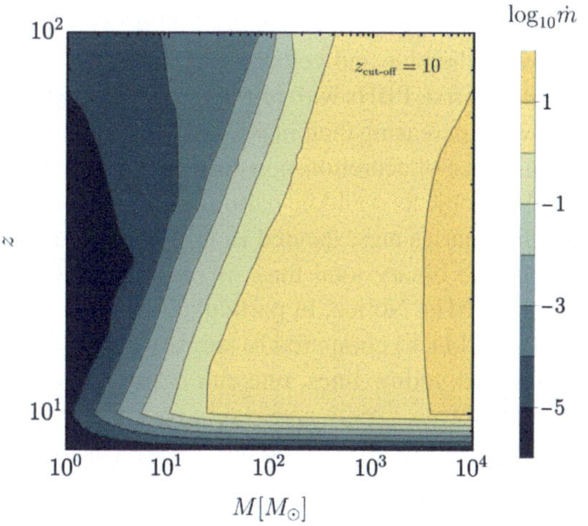

Fig. 13.2 Accretion rate $\dot{m}$ as a function of the PBH mass M and redshift z, assuming a cut-off redshift $z_{\text{cut-off}} = 10$ after which the accretion efficiency drops. Figure adapted from Ref. [17]

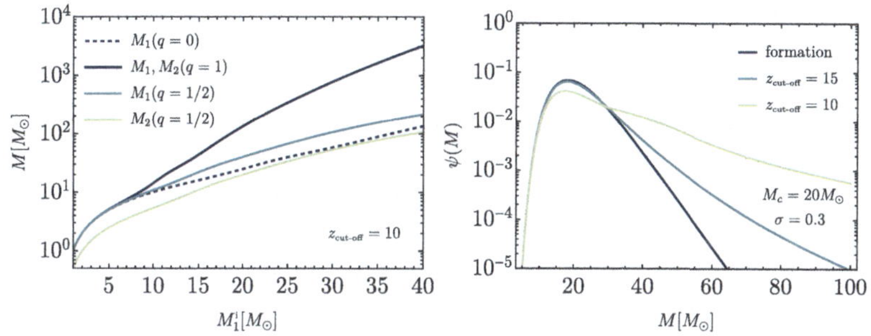

Fig. 13.3 Left panel: Behaviour of the PBH final masses M in terms of the initial ones M^{i}, assuming different values for the final mass ratio q. We have fixed the cut-off redshift to the representative value $z_{\text{cut-off}} = 10$. The dashed black line shows the evolution for a single PBH, while the solid lines hold for PBH binaries. Right panel: Evolution of the PBH mass function $\psi(M)$, assuming an initial lognormal shape with central mass $M_c = 20M_\odot$ and variance $\sigma = 0.3$, for different values of the cut-off redshift $z_{\text{cut-off}} = 15, 10$. Figure adapted from Ref. [16]

Once the accretion rate $\dot{m}$ for individual PBHs is known, and using the extended description for accretion when PBHs are assembled in binaries (see Sect. 13.1.2), one can solve Eq. (13.41) to get the evolution of the PBH mass, which is shown in the left panel of Fig. 13.3. In particular, since the initial binary parameters are unmeasurable, it is more relevant to analyse the dependence of the final masses M at redshifts after the cut-off, for each value of the initial PBH mass M^{i}. The black dashed line shows the case of an isolated PBH, while the solid lines show the final masses M_1, M_2 of a PBH binary, varying the initial primary component M_1^{i} and assuming different values for the final mass ratio, $q = M_2/M_1$. The black solid line shows the case of an equal mass binary $q = 1$, while the blue and yellow lines show the case when $q = 1/2$.

Few comments follow. First, PBHs with initial masses larger than $\mathcal{O}(10)M_\odot$ experience an efficient growth, increasing their masses of few orders of magnitude at small redshifts, under the presence of accretion-boosting dark matter halos and neglecting radiative feedback. These objects could be natural candidates for intermediate-mass BHs. Second, PBHs in binaries are expected to grow more efficiently compared to individual PBHs, since the binary total mass is responsible for a larger accretion rate compared to single PBHs. Notice, in particular, the growth of initially nearly-equal mass binaries (solid black) compared to single PBHs (dashed black). Finally, by comparing the blue and yellow lines, one can easily realise that the secondary component in the binary always experiences a relative stronger growth compared to the primary. This result gives the expectation that strongly-accreting binaries tend to have final mass ratios close to unity, see Eq. (13.40).

The evolution of PBHs under accretion affects as well their mass distribution in the universe. The mass function $\psi(M, z)$ is usually defined as the fraction of PBHs with mass in the interval $(M, M + \mathrm{d}M)$ at redshift z. For an initial $\psi(M^{\mathrm{i}}, z_{\mathrm{i}})$, its evolution is governed by [15, 17]

$$\psi(M(M^{\mathrm{i}}, z), z)\mathrm{d}M = \psi(M^{\mathrm{i}}, z_{\mathrm{i}})\mathrm{d}M^{\mathrm{i}}\,. \tag{13.42}$$

The latter is mostly dictated by isolated PBHs, which dominate the total PBH abundance in the universe. The main effect of accretion on the mass distribution is to broaden it at high masses, giving rise to a high-mass tail that can be orders of magnitude above its formation value [15]. A representative example is shown in the right panel of Fig. 13.3, where we have assumed an initial lognormal distribution with central scale M_c and variance σ (black line). Its evolution after the accretion phase, assuming different cut-off redshifts, is shown in the blue and yellow lines, highlighting the appearance of a tail at high masses.

Finally, let us mention that also the PBH abundance in the universe is affected by baryonic accretion. By assuming a non-relativistic dominant component (whose energy density scales as the inverse of the volume), one can show that [15]

$$f_{\mathrm{PBH}}(z) = \frac{\langle M(z)\rangle}{\langle M(z_{\mathrm{i}})\rangle(f^{-1}_{\mathrm{PBH}}(z_{\mathrm{i}}) - 1) + \langle M(z)\rangle}, \quad \langle M(z)\rangle = \int \mathrm{d}M M\psi(M, z)\,. \tag{13.43}$$

Because of accretion, $f_{\mathrm{PBH}}(z)$ can be significantly larger than its value at formation, $f_{\mathrm{PBH}}(z_{\mathrm{i}})$, see Ref. [16] for details.

13.3.2 PBH Spin Evolution

If baryonic accretion is not efficient enough during the cosmic history, the spin of PBHs is natal. In the standard formation scenario during the radiation era, the dimensionless spin parameter $\chi = |\mathbf{J}|/GM^2$ at formation is of percent level [18, 27, 37], although larger values can be achieved in other formation scenarios, see e.g. [14, 20, 22, 26] and Chap. 9 for further details.

On the other hand, in the case of efficient accretion, the formation of a thin disk leads to an efficient angular momentum transfer from the baryonic particles to the PBH, which could spin-up in the process. In other words, when the condition to form a thin accretion disk is satisfied, $\dot{m} \gtrsim 1$ (see discussion in Sect. 13.1.1.2), mass accretion is accompanied by an increase of the PBH spin. For PBH binaries, the necessary condition $\dot{m} \gtrsim 1$ is even more easily fulfilled, such that angular momentum transfer is more efficient [16, 17].

Since the disk is expected to be located along the equatorial plane [41, 55], the PBH spin would align perpendicularly to the disk plane. In such a configuration, one can use a geodesic model to describe the angular-momentum growth [4]. According to this model, the rate of variation of the Kerr parameter is related to the mass accretion rate $\dot{M}$ by [4, 17, 64]

$$\dot{\chi} = (\mathcal{F}(\chi) - 2\chi)\frac{\dot{M}}{M}, \tag{13.44}$$

in terms of the quantity $\mathcal{F}(\chi) \equiv L(M, J)/ME(M, J)$, which is only a function of χ and depends on

$$E(M, J) = \sqrt{1 - 2\frac{GM}{3r_{\rm ISCO}}} \quad \text{and} \quad L(M, J) = \frac{2GM}{3\sqrt{3}}\left(1 + 2\sqrt{3\frac{r_{\rm ISCO}}{GM} - 2}\right). \tag{13.45}$$

In these equations, the innermost stable circular orbit (ISCO) radius reads

$$r_{\rm ISCO}(M, J) = GM\left[3 + Z_2 - \sqrt{(3 - Z_1)(3 + Z_1 + 2Z_2)}\right], \tag{13.46}$$

with $Z_1 = 1 + \left(1 - \chi^2\right)^{1/3}\left[(1 + \chi)^{1/3} + (1 - \chi)^{1/3}\right]$ and $Z_2 = \sqrt{3\chi^2 + Z_1^2}$. One finds that, once the thin-disk hypothesis is satisfied, the PBH spin is expected to grow over a typical accretion time scale $\tau_{\rm acc}$, until it reaches the maximal value $\chi_{\rm max} = 0.998$, where radiation effects come into play and prevent reaching extremality [64].

In the left panel of Fig. 13.4, we show the final PBH spins as a function of their final masses for different combinations of binary systems and assuming a cut-off reshift $z_{\text{cut-off}} = 10$. Similarly to Fig. 13.3, the dashed black line shows the spin evolution of a single PBH, while the solid lines show the evolution of PBH spins in binaries. The plot shows a clear correlation between the final PBH mass and spin: low-mass PBHs are slowly spinning or non-spinning, whereas high-mass PBHs are rapidly

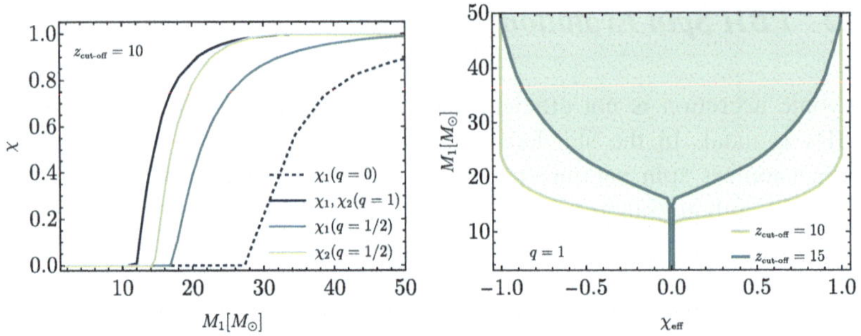

Fig. 13.4 Left panel: Behaviour of the PBH final spins χ in terms of the final masses M, assuming different values for the final mass ratio q. We have fixed the cut-off redshift to the representative value $z_{\text{cut-off}} = 10$. The dashed black line shows the evolution for a single PBH, while the solid lines hold for PBH binaries. Right panel: 99% (3σ) confidence intervals for the effective spin parameter χ_{eff} for given final PBH masses $M_1 = M_2$, for different cut-off redshifts $z_{\text{cut-off}} = 15, 10$. Figure adapted from Refs. [16, 17]

spinning. The mass scale at which this continuous transition occurs depends on the cut-off redshift and on the system properties (mass ratio etc.). For example, equal mass binaries, characterised by a stronger accretion phase, transit to a maximally spinning configuration for lower masses compared to single PBHs or to PBHs in binaries with smaller mass ratios. Finally, because of its relative larger accretion rate, the secondary binary component always spins more than the primary (see green and yellow lines in the plot).

The characteristic PBH mass-spin relations described in this Figure can be parametrically written using the expressions [24]

$$\chi_i(M_1, q, z_{\text{cut-off}}) \approx 10^{-2} + \text{Min}\left[0.988,\, 10^{f_i^b}\left(\frac{M_1}{M_\odot} - f_i^a\right)^{f_i^c}\theta\left(\frac{M_1}{M_\odot} - f_i^a\right)\right], \tag{13.47}$$

where $i = \{1, 2\}$ indicates the binary components, and the coefficients $f_i^{a,b,c}$ depend on the cut-off redshift $z_{\text{cut-off}}$ and final mass ratio q. Those functions are expanded as a polynomial series of the form [24]

$$f_i^\alpha(z_{\text{cut-off}}, q) = \alpha_i^0 + \sum_{j=1}^{3}\alpha_i^{z,j}\left(\frac{z_{\text{cut-off}}}{10}\right)^j + \sum_{j=1}^{3}\alpha_i^{q,j}q^j + \sum_{j,k=1}^{2}\alpha_i^{j,k}\left(\frac{z_{\text{cut-off}}}{10}\right)^j q^k, \tag{13.48}$$

with $\alpha = [a, b, c]$. The coefficients in the analytical expression are reported in Table 13.1.

Table 13.1 Numerical coefficients in the analytical fit described in Eq. (13.48) for the mass-spin relations of PBH binaries [24]

a_1^0	$a_1^{z,1}$	$a_1^{z,2}$	$a_1^{z,3}$	$a_1^{q,1}$	$a_1^{q,2}$	$a_1^{q,3}$	$a_1^{1,1}$	$a_1^{2,1}$	$a_1^{1,2}$	$a_1^{2,2}$
57.8531	−66.8879	43.9529	−5.46522	−56.4905	39.4605	−10.5127	37.4532	−17.5600	−17.5899	7.32670
b_1^0	$b_1^{z,1}$	$b_1^{z,2}$	$b_1^{z,3}$	$b_1^{q,1}$	$b_1^{q,2}$	$b_1^{q,3}$	$b_1^{1,1}$	$b_1^{2,1}$	$b_1^{1,2}$	$b_1^{2,2}$
2.14680	−3.65483	1.23732	−0.185276	−1.59262	−1.33445	0.940219	2.48367	0.0136971	−0.313974	−0.218091
c_1^0	$c_1^{z,1}$	$c_1^{z,2}$	$c_1^{z,3}$	$c_1^{q,1}$	$c_1^{q,2}$	$c_1^{q,3}$	$c_1^{1,1}$	$c_1^{2,1}$	$c_1^{1,2}$	$c_1^{2,2}$
0.441418	−0.738179	0.834177	−0.175491	−0.231674	2.12451	−0.787300	−0.0461876	−1.20687	−0.234563	.477210
a_2^0	$a_2^{z,1}$	$a_2^{z,2}$	$a_2^{z,3}$	$a_2^{q,1}$	$a_2^{q,2}$	$a_2^{q,3}$	$a_2^{1,1}$	$a_2^{2,1}$	$a_2^{1,2}$	$a_2^{2,2}$
44.3220	−72.7617	50.9837	−8.27027	19.8378	−33.8142	18.3605	−6.80676	1.95003	0.0581762	−0.957243
b_2^0	$b_2^{z,1}$	$b_2^{z,2}$	$b_2^{z,3}$	$b_2^{q,1}$	$b_2^{q,2}$	$b_2^{q,3}$	$b_2^{1,1}$	$b_2^{2,1}$	$b_2^{1,2}$	$b_2^{2,2}$
3.65282	−6.94442	3.55860	−0.630911	−0.474109	−0.199862	0.0523957	0.737077	0.0855668	−0.178022	−0.0212303
c_2^0	$c_2^{z,1}$	$c_2^{z,2}$	$c_2^{z,3}$	$c_2^{q,1}$	$c_2^{q,2}$	$c_2^{q,3}$	$c_2^{1,1}$	$c_2^{2,1}$	$c_2^{1,2}$	$c_2^{2,2}$
−0.189439	1.28502	−0.587638	0.0864602	−0.905386	1.25085	−0.346207	0.158765	−0.447706	−0.0109165	0.0945026

In the context of gravitational wave observations, a useful parameter entering the waveform is the effective spin parameter

$$\chi_{\rm eff} = \frac{\mathbf{J}_1/M_1 + \mathbf{J}_2/M_2}{M_1 + M_2} \cdot \hat{L}\,, \tag{13.49}$$

in terms of the direction of the orbital angular momentum $\hat{L}$. Its distribution is shown in the right panel of Fig. 13.4 where, following Refs. [16, 17], we have averaged over the angles between the total angular momentum and the individual PBH spins. The plot shows the parameter in terms of the final PBH mass M_1, for an equal mass binary $q = 1$, and for different values of the cut-off redshifts. For small PBH masses, when accretion is not efficient, the distributions peak around vanishing values, since the PBH spins kept their natal values. On the other hand, for PBHs heavier than about $\mathcal{O}(20)M_\odot$, accretion has been efficient enough in increasing the PBH spins near extremality and the distribution reaches large values. The correlation between low (high) values of the masses and low (high) spins is manifest, with a sharp transition around $M \sim \mathcal{O}(20)M_\odot$ depending on the value of the cut-off redshift. In particular, larger cut-offs imply shorter accretion phases and therefore narrower distributions. This correlation provides a distinctive feature to distinguish PBHs from other compact objects, and could be used to identify PBHs among other BHs in gravitational wave data [23, 24].

Acknowledgements V.DL. is supported by funds provided by the Center for Particle Cosmology at the University of Pennsylvania. He would like to thank his co-authors on this subject, G. Franciolini, P. Pani and A. Riotto. N.B. is supported by PRD/ARPE 2022 "Cosmology with Gravitational waves and Large Scale Structure - CosmoGraLSS" and acknowledges partial support from the National Science Foundation (NSF) under Grant No. PHY-2112884. He would also like to thank his co-authors on this subject, V. Bosch-Ramon, L. Piga, M. Lucca, S. Matarrese, A. Raccanelli, and L. Verde.

References

1. J. Adamek et al., WIMPs and stellar-mass primordial black holes are incompatible. Phys. Rev. D **100**(2), 023506 (2019). https://doi.org/10.1103/PhysRevD.100.023506. arXiv: 1901.08528 [astro-ph.CO]
2. D. Agius et al., Feedback in the dark: a critical examination of CMB bounds on primordial black holes (2024). arXiv: 2403.18895 [hep-ph]
3. Y. Ali-Haïmoud, M. Kamionkowski, Cosmic microwave background limits on accreting primordial black holes. Phys. Rev. D **95**(4), 043534 (2017). https://doi.org/10.1103/PhysRevD.95.043534. arXiv: 1612.05644 [astro-ph.CO]
4. J.M. Bardeen, W.H. Press, S.A. Teukolsky, Rotating black holes: Locally nonrotating frames, energy extraction, and scalar synchrotron radiation. Astrophys. J. **178**, 347 (1972). https://doi.org/10.1086/151796
5. V.S. Berezinsky, V.I. Dokuchaev, Y.N. Eroshenko, Formation and internal structure of superdense dark matter clumps and ultracompact minihaloes. JCAP **11**, 059 (2013). https://doi.org/10.1088/1475-7516/2013/11/059. arXiv: 1308.6742 [astro-ph.CO]

6. R.D. Blandford, D.G. Payne, Hydromagnetic flows from accretion discs and the production of radio jets. Mon. Not. Roy. Astron. Soc. **199**, 883–903 (1982). https://doi.org/10.1093/mnras/199.4.883
7. R.D. Blandford, R.L. Znajek, Electromagnetic extraction of energy from Kerr black holes. Mon. Not. Roy. Astron. Soc. **179**, 433–456 (1977). https://doi.org/10.1093/mnras/179.3.433
8. H. Bondi, F. Hoyle, On the mechanism of accretion by stars. Mon. Not. Roy. Astron. Soc. **104**, 273 (1944). https://doi.org/10.1093/mnras/104.5.273
9. V. Bosch-Ramon, 3D hydrodynamical simulations of the impact of mechanical feedback on accretion in supersonic stellar-mass black holes. Astron. Astrophys. **660**, A5 (2022). https://doi.org/10.1051/0004-6361/202142821. arXiv: 2201.09601 [astro-ph.HE]
10. V. Bosch-Ramon, N. Bellomo, Mechanical feedback effects on primordial black hole accretion. Astron. Astrophys. **638**, A132 (2020). https://doi.org/10.1051/0004-6361/202037537. arXiv: 2004.11224 [astro-ph.CO]
11. M. Boudaud et al., In-depth analysis of the clustering of dark matter particles around primordial black holes I: density profiles (2021). arXiv:2106.07480, arXiv:2106.07480. arXiv: [astro-ph.CO]
12. B. Carr et al., Constraints on primordial black holes. Rept. Prog. Phys. **84**(11), 116902 (2021). https://doi.org/10.1088/1361-6633/ac1e31. arXiv: 2002.12778 [astro-ph.CO]
13. A. Coogan et al., Measuring the dark matter environments of black hole binaries with gravitational waves. Phys. Rev. D **105**(4), 043009 (2022). https://doi.org/10.1103/PhysRevD.105.043009. arXiv: 2108.04154 [gr-qc]
14. E. Cotner, A. Kusenko, Primordial black holes from scalar field evolution in the early universe. Phys. Rev. D **96**(10), 103002 (2017). https://doi.org/10.1103/PhysRevD.96.103002. arXiv: 1706.09003 [astro-ph.CO]
15. V. De Luca et al., Constraints on primordial black holes: the importance of accretion. Phys. Rev. D **102**(4), 043505 (2020). https://doi.org/10.1103/PhysRevD.102.043505. arXiv: 2003.12589 [astro-ph.CO]
16. V. De Luca et al., Primordial black holes confront LIGO/Virgo data: current situation. JCAP **06**, 044 (2020). https://doi.org/10.1088/1475-7516/2020/06/044. arXiv: 2005.05641 [astro-ph.CO]
17. V. De Luca et al., The evolution of primordial black holes and their final observable spins. JCAP **04**, 052 (2020). https://doi.org/10.1088/1475-7516/2020/04/052. arXiv: 2003.02778 [astro-ph.CO]
18. V. De Luca et al., The initial spin probability distribution of primordial black holes. JCAP **05**, 018 (2019). https://doi.org/10.1088/1475-7516/2019/05/018. arXiv: 1903.01179 [astro-ph.CO]
19. C. Dvorkin, K. Blum, M. Kamionkowski, Constraining dark Matter-Baryon Scattering with linear cosmology. Phys. Rev. D **89**(2), 023519 (2014). https://doi.org/10.1103/PhysRevD.89.023519. arXiv: 1311.2937 [astro-ph.CO]
20. Y.N. Eroshenko, Spin of primordial black holes in the model with collapsing domain walls. JCAP **12**(12), 041 (2021). https://doi.org/10.1088/1475-7516/2021/12/041. arXiv: 2111.03403 [astro-ph.CO]
21. G. Facchinetti, M. Lucca, S. Clesse, Relaxing CMB bounds on primordial black holes: the role of ionization fronts. Phys. Rev. D **107**(4), 043537 (2023). https://doi.org/10.1103/PhysRevD.107.043537. arXiv: 2212.07969 [astro-ph.CO]
22. M.M. Flores, A. Kusenko, Spins of primordial black holes formed in different cosmological scenarios. Phys. Rev. D **104**(6), 063008 (2021). https://doi.org/10.1103/PhysRevD.104.063008. arXiv: 2106.03237 [astro-ph.CO]
23. G. Franciolini, P. Pani, Searching for mass-spin correlations in the population of gravitational-wave events: The GWTC-3 case study. Phys. Rev. D **105**(12), 23024 (2022). https://doi.org/10.1103/PhysRevD.105.123024. arXiv: 2201.13098 [astro-ph.HE]
24. G. Franciolini et al., How to assess the primordial origin of single gravitational-wave events with mass, spin, eccentricity, and deformability measurements. Phys. Rev. D **105**(6), 063510 (2022). https://doi.org/10.1103/PhysRevD.105.063510. arXiv: 2112.10660 [astro-ph.CO]

25. A. Gruzinov, Y. Levin, C.D. Matzner, Negative dynamical friction on compact objects moving through dense gas. Mon. Not. Roy. Astron. Soc. **492**(2), 2755–2761 (2020). https://doi.org/10.1093/mnras/staa013. arXiv: 1906.01186 [astro-ph.HE]
26. T. Harada et al., Spins of primordial black holes formed in the matter-dominated phase of the Universe. Phys. Rev. D **96**(8), 083517 (2017). [Erratum: Phys.Rev.D 99, 069904 (2019)]. https://doi.org/10.1103/PhysRevD.96.083517. arXiv: 1707.03595 [gr-qc]
27. T. Harada et al., Spins of primordial black holes formed in the radiation-dominated phase of the universe: first-order effect. Astrophys. J. **908**(2), 140 (2021). https://doi.org/10.3847/1538-4357/abd9b9. arXiv: 2011.00710 [astro-ph.CO]
28. G. Hasinger, Illuminating the dark ages: Cosmic backgrounds from accretion onto primordial black hole dark matter. JCAP **07**, 022 (2020). https://doi.org/10.1088/1475-7516/2020/07/022. arXiv: 2003.05150 [astro-ph.CO]
29. F. Hoyle, R.A. Lyttleton, The effect of interstellar matter on climatic variation. Math. Proc. Camb. Philos. Soc. **35**(3), 405–415 (1939). https://doi.org/10.1017/S0305004100021150
30. S. Ichimaru, Bimodal behavior of accretion disks - Theory and application to Cygnus X-1 transitions. Astrophys. J. **214**, 840–855 (1977). https://doi.org/10.1086/155314
31. K. Ioka et al., Black hole binary formation in the expanding universe: three body problem approximation. Phys. Rev. D **58**, 063003 (1998). https://doi.org/10.1103/PhysRevD.58.063003. arXiv: astro-ph/9807018
32. K. Ioka et al., GW 150914-like black holes as Galactic high-energy sources. Mon. Not. Roy. Astron. Soc. **470**(3), 3332–3345 (2017). https://doi.org/10.1093/mnras/stx1337. arXiv: 1612.03913 [astro-ph.HE]
33. B.J. Kavanagh, D. Gaggero, G. Bertone, Merger rate of a subdominant population of primordial black holes. Phys. Rev. D **98**(2), 023536 (2018). https://doi.org/10.1103/PhysRevD.98.023536. arXiv: 1805.09034 [astro-ph.CO]
34. A. Levinson, E. Nakar, Limits on the growth rate of supermassive black holes at early cosmic epochs. Mon. Not. Roy. Astron. Soc. **473**(2), 2673–2678 (2018). https://doi.org/10.1093/mnras/stx2542. arXiv: 1709.04175 [astro-ph.GA]
35. X. Li et al., Simulation of a compact object with outflows moving through a gaseous background. Mon. Not. Roy. Astron. Soc. **494**(2), 2327–2336 (2020). https://doi.org/10.1093/mnras/staa900. arXiv: 1912.06864 [astro-ph.HE]
36. Katherine J. Mack, Jeremiah P. Ostriker, Massimo Ricotti, Growth of structure seeded by primordial black holes. Astrophys. J. **665**, 1277–1287 (2007). https://doi.org/10.1086/518998. (arXiv: astro-ph/0608642)
37. M. Mirbabayi, A. Gruzinov, J. Noreña, Spin of primordial black holes. JCAP **03**, 017 (2020). https://doi.org/10.1088/1475-7516/2020/03/017. arXiv: 1901.05963 [astro-ph.CO]
38. R. Morganti, The many routes to AGN feedback. Front. Astron. Space Sci. **4**, 42 (2017). https://doi.org/10.3389/fspas.2017.00042. arXiv: 1712.05301 [astro-ph.GA]
39. Takashi Nakamura et al., Gravitational waves from coalescing black hole MACHO binaries. Astrophys. J. Lett. **487**, L139–L142 (1997). https://doi.org/10.1086/310886. (arXiv: astro-ph/9708060)
40. R. Narayan, I. Yi, Advection dominated accretion: Underfed black holes and neutron stars. Astrophys. J. **452**, 710 (1995). https://doi.org/10.1086/176343. arXiv: astro-ph/9411059
41. I.D. Novikov, K.S. Thorne, Astrophysics of black holes. Black Holes (Les Astres Occlus) (1973), pp. 343–450
42. E. Donald, *Osterbrock and Gary J* (Ferland, Astrophysics of gaseous nebulae and active galactic nuclei, 2006)
43. J.P. Ostriker, Y. Suto, The mach number of the cosmic flow: a critical test for current theories. Astrophys. J. **348**, 378 (1990). https://doi.org/10.1086/168247
44. K. Park, M. Ricotti, Accretion onto black holes from large scales regulated by radiative feedback. III. Enhanced luminosity of intermediate mass black holes moving at supersonic speeds. Astrophys. J. **767**, 163 (2013). https://doi.org/10.1088/0004-637X/767/2/163. arXiv: 1211.0542 [astro-ph.CO]

45. K. Park, M. Ricotti, Accretion onto black holes from large scales regulated by radiative feedback. iii. Enhanced luminosity of intermediate-mass black holes moving at supersonic speeds. Astrophys. J. **767**(2), 163 (2013) https://doi.org/10.1088/0004-637X/767/2/163
46. P.J.E. Peebles, Recombination of the primeval plasma. Astrophys. J. **153**, 1 (1968). https://doi.org/10.1086/149628
47. L. Piga et al., The effect of outflows on CMB bounds from Primordial Black Hole accretion. JCAP **12**, 016 (2022). https://doi.org/10.1088/1475-7516/2022/12/016. arXiv: 2210.14934 [astro-ph.CO]
48. E. Poisson, C.M. Will, *Gravity: Newtonian, Post-Newtonian. Relativistic*. (Cambridge University Press, 2014). https://doi.org/10.1017/CBO9781139507486
49. V. Poulin et al., CMB bounds on disk-accreting massive primordial black holes. Phys. Rev. D **96**(8), 083524 (2017). https://doi.org/10.1103/PhysRevD.96.083524. arXiv: 1707.04206 [astro-ph.CO]
50. M. Raidal et al., Formation and evolution of primordial black hole binaries in the early universe. JCAP **02**, 018 (2019). https://doi.org/10.1088/1475-7516/2019/02/018. arXiv: 1812.01930 [astro-ph.CO]
51. J.R. Rice, B. Zhang, Cosmological evolution of primordial black holes. JHEAp **13–14**, 22–31 (2017). https://doi.org/10.1016/j.jheap.2017.02.002. arXiv: 1702.08069 [astro-ph.HE]
52. M. Ricotti, Bondi accretion in the early universe. Astrophys. J. **662**, 53–61 (2007). https://doi.org/10.1086/516562. arXiv: 0706.0864 [astro-ph]
53. M. Ricotti, J.P. Ostriker, K.J. Mack, Effect of primordial black holes on the cosmic microwave background and cosmological parameter estimates. Astrophys. J. **680**, 829 (2008). https://doi.org/10.1086/587831. arXiv: 0709.0524 [astro-ph]
54. P.D. Serpico et al., Cosmic microwave background bounds on primordial black holes including dark matter halo accretion. Phys. Rev. Res. **2**(2), 023204 (2020). https://doi.org/10.1103/PhysRevResearch.2.023204. arXiv: 2002.10771 [astro-ph.CO]
55. N.I. Shakura, R.A. Sunyaev, Black holes in binary systems. Observational appearance. Astron. Astrophys. **24**, 337–355 (1973)
56. Stuart L. Shapiro, Accretion onto black holes: the emergent radiation spectrum. Astrophys. J. **180**, 531–546 (1973). https://doi.org/10.1086/151982
57. S.L. Shapiro, Accretion onto black holes: the emergent radiation spectrum. II. Magnetic effects. Astrophys. J. **185**, 69–82 (1973). https://doi.org/10.1086/152396
58. Stuart L. Shapiro, Saul A. Teukolsky, Black holes, white dwarfs and neutron stars. Phys. Compact Objects. (1983). https://doi.org/10.1002/9783527617661
59. N. Soker, The jet feedback mechanism (JFM) in stars, galaxies and clusters. New Astron. Rev. **75**, 1–23 (2016). https://doi.org/10.1016/j.newar.2016.08.002. arXiv: 1605.02672 [astro-ph.HE]
60. K. Sugimura, M. Ricotti, Structure and instability of the ionization fronts around moving black holes. Mon. Not. Roy. Astron. Soc. **495**(3), 2966–2978 (2020). ISSN: 0035-8711. https://doi.org/10.1093/mnras/staa1394. https://academic.oup.com/mnras/article-pdf/495/3/2966/33337508/staa1394.pdf
61. E. Takeo et al., Rapid growth of black holes accompanied with hot or warm outflows exposed to anisotropic super-Eddington radiation. Mon. Not. Roy. Astron. Soc. **476**(1), 673–682 (2018). https://doi.org/10.1093/mnras/sty264. arXiv: 1705.05382 [astro-ph.HE]
62. E. Takeo et al., Super-Eddington growth of black holes in the early universe: effects of disc radiation spectra. Mon. Not. Roy. Astron. Soc. **488**(2), 2689–2700 (2019). https://doi.org/10.1093/mnras/stz1899. arXiv: 1901.04514 [astro-ph.HE]
63. V. Takhistov et al., Impacts of Jets and winds from primordial black holes. Mon. Not. Roy. Astron. Soc. **517**(1), L1–L4 (2022). https://doi.org/10.1093/mnrasl/slac097. arXiv: 2111.08699 [astro-ph.HE]
64. K.S. Thorne, Disk accretion onto a black hole. 2. Evolution of the hole. Astrophys. J. **191**, 507–520 (1974). https://doi.org/10.1086/152991
65. D. Tseliakhovich, C. Hirata, Relative velocity of dark matter and baryonic fluids and the formation of the first structures. Phys. Rev. D **82**(8), 083520 (2010). https://doi.org/10.1103/PhysRevD.82.083520. arXiv: 1005.2416 [astro-ph.CO]

66. F. Yuan, R. Narayan, Hot accretion flows around black holes. Ann. Rev. Astron. Astrophys. **52**, 529–588 (2014). https://doi.org/10.1146/annurev-astro-082812-141003. arXiv: 1401.0586 [astro-ph.HE]
67. M. Zeilig-Hess, A. Levinson, E. Nakar, Numerical simulations of AGN wind feedback on black hole accretion: probing down to scales within the sphere of influence. Mon. Not. Roy. Astron. Soc. **482**(4), 642–4653 (2019). https://doi.org/10.1093/mnras/sty3034. arXiv: 1811.02587 [astro-ph.HE]
68. S. Zhang, B. Liu, V. Bromm, Distinguishing the impact and signature of black holes from different origins in early cosmic history. Mon. Not. Roy. Astron. Soc. **528**(1), 180–197 (2024). https://doi.org/10.1093/mnras/stad3986. arXiv: 2310.01763 [astro-ph.CO]
69. F. Ziparo et al., Cosmic radiation backgrounds from primordial black holes. Mon. Not. Roy. Astron. Soc. **517**(1), 1086–1097 (2022). https://doi.org/10.1093/mnras/stac2705. arXiv: 2209.09907 [astro-ph.CO]

Chapter 14
Haloes, Other Dark-Matter Candidates, and Astrophysical Implications

Philippa S. Cole

Abstract It is possible that a multi-component dark matter model is required if primordial black holes only contribute to a fraction of the energy density in dark matter. This is increasingly more likely with respect to the case of $f_{\rm PBH} = 1$, since there is only one remaining window, on asteroid-mass scales, where PBHs can make up all of the dark matter. A mixed dark matter model can lead to interesting observables that come about due to the interactions between PBHs and the second dark matter component. This can provide unique signatures of the presence of primordial black holes and increase the prospects of detection or improvement of constraints in the mass ranges where $f_{\rm PBH} < 1$, whilst simultaneously exploring the remaining open parameter space for other dark matter candidates.

14.1 Introduction

Constraints on the abundance of primordial black holes (PBHs) have strengthened across the mass-range of PBHs that could be dark matter candidates, namely $10^{-18}\,{\rm M}_\odot \lesssim {\rm M}_{\rm PBH} \lesssim 10^4\,{\rm M}_\odot$, where $M_{\rm PBH}$ is PBH mass [22, 46]. There is only one window remaining around the mass of an asteroid, $10^{-16}\,{\rm M}_\odot \lesssim {\rm M}_{\rm PBH} \lesssim 10^{-10}\,{\rm M}_\odot$, where all of the energy density budget for dark matter could still be explained by PBHs of a single mass (or narrow distribution of masses) [74]. See Part V of this volume for details of the current status of constraints.

However, an order 1–10% fraction of the dark matter can still be explained by PBHs in much of the parameter space, which would still provide a far greater understanding of the make up of dark matter than we currently have. Furthermore, it is important to stress that even in mass ranges where only a small fraction of the dark matter can be explained by PBHs, their existence would be enlightening for how the

P. S. Cole (✉)
Dipartimento di Fisica "G. Occhialini", Universitá degli Studi di Milano-Bicocca, Piazza della Scienza 3, 20126 Milano, Italy
e-mail: philippa.cole@unimib.it

INFN, Sezione di Milano-Bicocca, Piazza della Scienza 3, 20126 Milano, Italy

C. Byrnes et al. (eds.), *Primordial Black Holes*, Springer Series in Astrophysics and Cosmology, https://doi.org/10.1007/978-981-97-8887-3_14

very first stages of the early universe evolved. This is because very specific ingredients are required from either inflationary mechanisms (see e.g. [9, 17, 26, 42, 49, 54, 73, 79] for a selection of examples) or cosmological processes such as phase transitions [58, 62] in order to produce even just one PBH [27]. Producing a specific abundance of PBHs is an extremely delicate procedure [29], and needing to match an observed abundance today would be very prescriptive for identifying the correct model of the early universe.

If it is the case that PBHs exist but only make up a fraction of the dark matter energy density budget, such that $f_{\rm PBH} = \frac{\rho_{\rm PBH}}{\rho_{\rm DM}} < 1$, then there must be more than one species of dark matter. In this chapter, the other component (or components) of dark matter will be assumed to be of a particle nature. Whilst less simple, a 'mixed' dark matter scenario offers opportunities for additional signatures owing to the interactions between PBHs and the other dark matter particles [2, 19, 31, 53, 86, 89]. This could provide additional observational probes, as well as new ways of constraining the abundance of both PBHs and particulate dark matter, which has also thus far eluded detection. From a cautionary perspective, the combined scenario could also lead to obscuring the existence of either species, not only because there would be a lower abundance of each, but also because current search strategies may not account for the presence of two species, that could result in signatures of dark matter being swept up in degeneracies [28].

14.2 Particle Companions

The two main classes of particle dark matter candidates that have been considered as companions to PBHs are weakly interacting massive particles (WIMPs) and axions, or axion-like particles (ALPs). We briefly introduce them in this section, before focusing on signatures of their possible co-existence with PBHs in the following sections.

14.2.1 WIMPs

Weakly interacting massive particles interact only via the weak force and gravity. They were postulated as a candidate that produced the correct abundance of dark matter to match cosmic microwave background observations with masses in the GeV to TeV range, i.e. at the electroweak scale, which is a natural energy scale at which to expect new physics [38, 44, 82, 85].

Briefly, dark matter particles could be produced as a thermal relic of the Big Bang such that during the early universe they would be in thermal equilibrium with all of the other particles. As the universe cools and expands, the dark matter particles would become sufficiently diffuse so that they can no longer annihilate. Their number density approaches a constant at this epoch, or 'freezes out'. Theories that produce thermal relics predict a cross-section which depends on the mass of the dark matter particle and the weak-scale gauge coupling g_{weak} which is of order 1. For particle masses in the GeV–TeV range, the cross-section which is set by m_X and g_{weak} automatically produces the correct abundance for today's dark matter energy density. This coincidence of correct abundance and energy scale is a phenomenon often dubbed the 'WIMP miracle' [39, 83].

Unfortunately, a plethora of searches for various WIMP models that employ different mediators for the interactions between the dark matter particles and the standard model have not found any evidence of their existence yet. For reviews, see [4, 84]. Furthermore, there have been various critiques of the WIMP miracle [7], decreasing their perceived novelty and broadening the space of viable dark matter candidates. Whilst this doesn't rule them out as dark matter candidates, they are certainly proving more difficult to find than first thought. Next generation experiments focusing on direct and indirect detection, as well as collider searches will continue to constrain the still-viable parameter space, see Fig. 14.1.

Nevertheless, the fact that WIMPs self-annihilate provides a unique observational signature for the combined scenario of WIMPs and PBHs, which ultimately leads to very tight constraints on their co-existence, see Sect. 14.3.1.

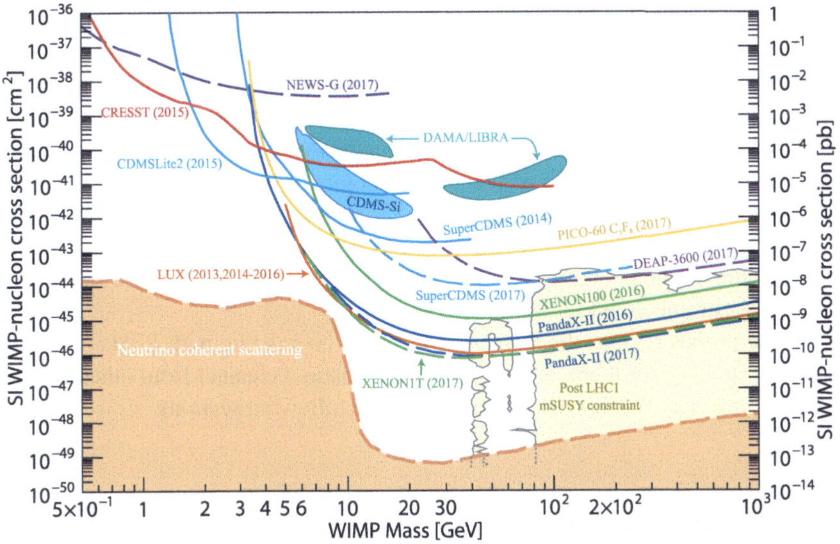

Fig. 14.1 Constraints on the WIMP cross section for spin-independent coupling as a function of WIMP mass. Reproduced under the Creative Commons by 4.0 license from [23]

14.2.2 Axions and ALPs

The quantum chromodynamics (QCD) axion was initially proposed as a particle that could explain the strong charge-parity (CP) problem [80]. In a nutshell, the electric dipole moment of the neutron is constrained to be extremely small: $|d_n| < 2.9 \times 10^{-26} e\,\mathrm{cm}$, where e is the charge of the electron. In QCD, the theory of the strong interaction between quarks which is mediated by gluons, the value of $|d_n|$ depends on the mis-alignment angle θ_{QCD} as

$$d_n \approx 3.6 \times 10^{-16} \theta_{\mathrm{QCD}} e\,\mathrm{cm}. \tag{14.1}$$

Given the strong constraint on $|d_n|$, there is seemingly a fine-tuning problem on $\theta_{\mathrm{QCD}} \lesssim 10^{-10}$, see e.g. [67, 80] for details. One solution to this fine-tuning problem is the QCD axion, which sets $\theta_{\mathrm{QCD}} = 0$ via QCD non-perturbative effects. The QCD axion can be stable and have masses in the range $m_a \sim 10^{-12} - 10^{-3}\,\mathrm{eV}$, which means that it simultaneously qualifies as a viable dark matter candidate [67].

However, many high energy physics models predict the existence of a more general class of axion-like particles (ALPs) and thus there is a much wider mass range of ALPs, $m_a \sim 10^{-33} - 10^{-3}\,\mathrm{eV}$ that can be considered as dark matter candidates [66]. As they are too light to be observed at particle accelerators, prospects for their detection tend to rely on looking for signatures of their conversion to and from photons, and their effect on astrophysical and cosmological observables. Targeted searches have found no evidence of them yet, including helioscopes such as the CERN Axion Solar Telescope (CAST) [90] which look for signatures of solar axions that have converted into x-rays, or haloscopes such as the Axion Dark Matter Experiment (ADMX) [60] which aims to convert dark matter axions into microwave photons, as well as limits from astrophysical and cosmological data-sets such as CMB observations [8], globular cluster observations [32] and white dwarf observations [33]. However the parameter space, which is usually presented in terms of their mass and their coupling to the photon $g_{a\gamma}$, is vast, and there are various regions wide open in terms of constraints, see Fig. 14.2.

Unique probes such as the signatures of ALPs co-existing with PBHs offer new ways of searching for them, and in this case, a distinct channel from other popular candidates such as WIMPs, as we will see in the following sections.

14.2.3 Other Candidates

There are many other proposals for dark matter candidates, including but not limited to self-interacting dark matter [88], warm dark matter [18], or certain scenarios of modified gravity [69]. To date they have not been considered explicitly in combination with the co-existence of PBHs, however in Sect. 14.5, we will consider the production of dark matter particles including some of these candidates from the evaporation of very light PBHs [14].

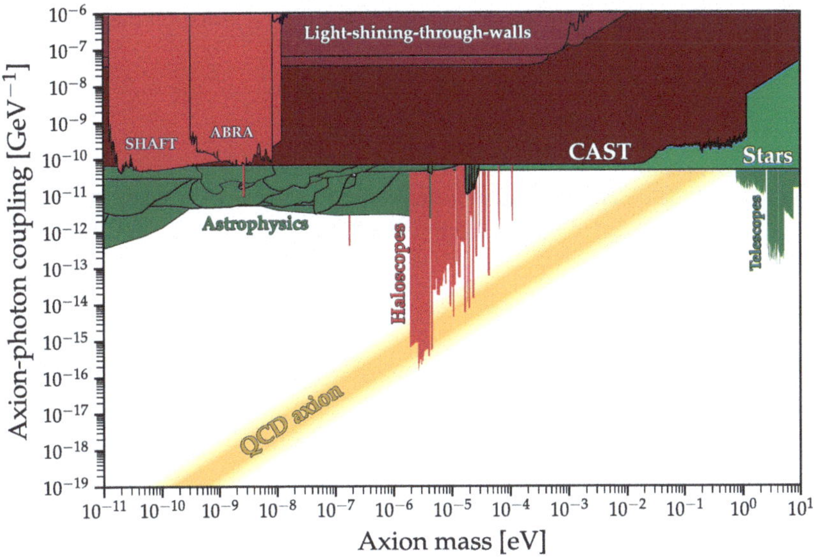

Fig. 14.2 Constraints on the axion-photon coupling as a function of axion mass. The yellow diagonal shaded region shows the region of the parameter space that corresponds to the QCD axion. Figure from [77] and further details in [78]

14.3 Formation of Dark Matter Haloes Around PBHs

If PBHs co-exist with particles of cold dark matter in the early universe, the gravitational attraction of the PBHs will overcome the decelerating expansion of the background during radiation domination. This causes PBHs to accrete the dark matter particles in the form of dense haloes. The characteristic time and radius of this process is at 'turn-around', when and where the dark matter particles stop moving away from the PBH due to the background expansion and instead start falling inwards. Following the argument in [2], an analytical approximation identifies turn-around as when the Newtonian gravitational attraction of the PBH is equal to the deceleration of the background in radiation domination. The acceleration, denoted by double dots, of a point particle at a separation r from a PBH is

$$\ddot{r} = -\frac{GM_{\mathrm{PBH}}}{r^2} + \frac{\ddot{a}}{a}r \tag{14.2}$$

where G is Newton's gravitational constant, and a is the scale factor. Then, equating the first and second term on the right hand side leads to the definition of the turn-around radius r_{TA}:

$$GM_{\mathrm{PBH}} = \frac{8\pi G}{3}\rho_{\mathrm{rad}}r_{\mathrm{TA}}^3 \tag{14.3}$$

where $\rho_{\rm rad}$ is the total background energy density during radiation domination and we have used the acceleration equation

$$\frac{\ddot{a}}{a} = -\frac{(1+3\omega)}{2}H^2 \tag{14.4}$$

with $\omega = 1/3$ the equation of state during radiation domination and H the Hubble factor, as well as the Friedmann equation in a flat universe

$$H^2 = \frac{8\pi G}{3}\rho. \tag{14.5}$$

Substituting the energy density at matter radiation equality $\rho_{\rm eq} = 2\rho_{\rm rad}$ into Eq. 14.3, demonstrates that the mass of the halo within the turn-around radius is approximately equal to the mass of the PBH at matter-radiation equality:

$$M_{\rm PBH} = \frac{4\pi}{3}\rho_{\rm eq} r_{\rm TA}^3. \tag{14.6}$$

The dark matter density profile is then found by assuming that particles on radii smaller than the turn-around radius are frozen into the dark matter halo, and hence the dark matter energy density of the halo is equal to the background radiation energy density at turn-around time. Since matter scales like a^{-3} in radiation domination, the dark matter density profile in terms of $\rho_{\rm eq}$ is

$$\begin{aligned}\rho_{\rm DM}(r) &= \rho(r(t_{\rm TA}))\\ &= \frac{\rho_{\rm eq}}{2}\left(\frac{a_{\rm TA}}{a_{\rm eq}}\right)^{-3}\\ &= \frac{\rho_{\rm eq}}{2}\left(\frac{t_{\rm TA}}{t_{\rm eq}}\right)^{-\frac{3}{2}}\\ &= \frac{\rho_{\rm eq}}{2} t_{\rm eq}^{\frac{3}{2}} (8GM_{\rm PBH})^{\frac{3}{4}} r(t_{\rm TA})^{-\frac{9}{4}}\end{aligned} \tag{14.7}$$

using $t = \frac{1}{2H}$ during radiation domination and the Friedmann equation again together with $\rho(t_{\rm TA}) \approx 3M_{\rm PBH}/4\pi r(t_{\rm TA})^3$. A numerical solution to the differential equation in Eq. 14.2 shows that in fact the turn-around radius is more accurately identified at [2]

$$GM_{\rm PBH} = 2H^2 r(t_{\rm TA})^3 \tag{14.8}$$

i.e. with an additional factor of 2 with respect to Eq. 14.3. Working this through results in a firm prediction for the dark matter density profile around PBHs of

$$\rho_{\rm DM}(r) = \frac{\rho_{\rm eq}}{2} t_{\rm eq}^{\frac{3}{2}} (2GM_{\rm PBH})^{\frac{3}{4}} r^{-\frac{9}{4}} \tag{14.9}$$

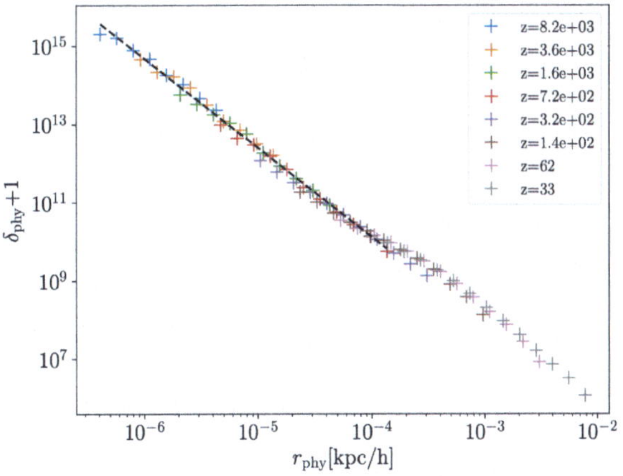

Fig. 14.3 The density profile as a function of radius of the dark matter halo surrounding a PBH. Coloured cross-markers are the results of simulations at the labelled redshift values, the dashed black line shows the analytical $\gamma_{sp} = 9/4$ prediction for the slope. Figure re-produced with permission from [2]

where we have dropped the subscript TA and it is implicit that the halo density is fixed at turn-around time.

Numerical simulations corroborate this analytical prediction [2, 36], see Fig. 14.3, for the dark matter density profile around $\mathcal{O}(10)$ $M_{\odot}$ black holes forming in radiation domination and evolved from redshift $z \sim 8000$ down to redshifts between $z \sim 30-60$. The results are expected to hold for any cold dark matter candidate, and should also be independent of the mass of the black hole, in the sense that a dense spike should form with a profile characterised by a slope of 9/4 and a normalisation as given by Eq. 14.9.

It is important to stress that if there exists a small population of PBHs accompanied by any cold dark matter particles, then dense haloes (also referred to as spikes) necessarily form around the black holes. Provided they are not significantly disrupted by, for example, close encounters or collisions with nearby stars or black holes [56], then searches for one component must take into account the presence of the other. See, for example, a recent study [51] on possible signatures of dark matter spikes around low-mass X-ray binaries that could have a primordial origin. The next sections consider signatures to be expected from the co-existence of PBHs and a sub-set of particle dark matter candidates.

14.3.1 WIMP Haloes

If the dense dark matter spikes around PBHs are made up of weakly interacting massive particles which interact amongst themselves with a given cross-section, then close to the black hole, the WIMPs will be packed close enough in that they will annihilate each other. This causes a plateau in the dark matter density profile at small radii, and also results in an annihilation signal that should be observable in gamma-rays.

The maximum density of WIMPs in the halo $\rho_{\max}$ depends on the mass m_χ and the cross-section $\langle \sigma v \rangle$ of the particles

$$\rho_{\max} = \frac{m_\chi}{\langle \sigma v \rangle t_0} \tag{14.10}$$

where t_0 is the age of the universe today. This means that there is a threshold radius of the dark matter profile, found by equating Eqs. 14.9 and 14.10, such that particles on smaller radii than the threshold will contribute to an annihilation signal, since the density of the spike exceeds the maximum value on these radii.

Assuming that the annihilation signal contributes to the diffuse gamma-ray background, constraints on the decay rate of dark matter from [3] can be used to place constraints on $f_{\rm PBH}$ for $\mathcal{O}(10)\,{\rm M}_\odot$ PBHs. For the standard WIMP scenario with $m_\chi = 100\,{\rm GeV}$ and $\langle \sigma v \rangle = 3 \times 10^{-26}\,{\rm cm}^3{\rm s}^{-1}$, $\mathcal{O}(10)\,{\rm M}_\odot$ PBHs are restricted to only making up a very small fraction of dark matter $f_{\rm PBH} \lesssim 2 \times 10^{-9}$ [2, 19, 36]. Conversely, the detection of just one PBH would fix extremely strong constraints on almost all dark matter models motivated by weak-scale extensions of the standard model [16].

This holds across the range of WIMP masses (approximately $m_\chi = 10 - 10^4\,{\rm GeV}$) and cross-sections in which there are constraints on the diffuse gamma-ray background (which depends on the decay channel assumed), and the constraint on $f_{\rm PBH}$ only varies by $\mathcal{O}(1)$ [3]. However, note that for much larger PBH masses, the number density of PBHs will be much smaller than for the case of $\mathcal{O}(10){\rm M}_\odot$ PBHs and therefore the diffuse gamma-ray background constraints do not apply. Instead, the analysis should be done with point source observations from e.g. the Fermi satellite [64]. Also for much lighter PBHs, the thermal kinetic energy of the WIMPs could alter the density profile of the dark matter depending on its mass. Therefore, the strong conclusion that WIMPs and PBHs effectively can't co-exist is only true for stellar-mass black holes and $m_\chi = 10 - 10^4\,{\rm GeV}$. Mass ranges of either the WIMPs or the PBHs that differ greatly should be investigated independently.

14.3.2 ALP Haloes and Clouds

Since axion-like particles do not self-annihilate, they are not subject to the same bounds from astrophysical observations that WIMPS are, as described in Sect. 14.3.1. This means that across the full range of masses, PBHs could be accompanied by dense spikes of axion-like particles accumulated in the manner described in Sect. 14.3, if they do not make up the entirety of the dark matter themselves. Possibilities for detecting such systems will be explored in Sect. 14.4.

However, due to the very light masses of ALPs, there is another possibility for the formation of particle clouds around PBHs. If the PBHs are spinning, then an extremely dense boson cloud can form due to superradiance [21, 61]. This occurs when the angular momentum of the black hole is extracted and transferred to the

ALPs if the Compton wavelength of the particles is comparable to the Schwarzschild radius of the black hole. The mass of the boson m_b which has a comparable Compton wavelength to the Schwarzschild radius of a PBH with mass $m_{\rm PBH}$ is given by [40]

$$m_b \sim 10^{-4}\,\mathrm{eV}\left(\frac{m_{\rm PBH}}{10^{-6}\,\mathrm{M}_\odot}\right)^{-1}. \tag{14.11}$$

Boson clouds tend to be much more dense than the spikes referred to in Sect. 14.3, and therefore could provide very distinctive signatures that will be discussed further in Sect. 14.4.

However, in many PBH scenarios it is not expected that the PBHs should possess any significant spin [72], unless they are formed during matter domination instead of radiation domination [47], or if they are formed not from the collapse of large scalar perturbations but instead from e.g. the collapse of domain walls. Even if superradiance took place during the early universe, the angular momentum of the PBH is expected to be depleted on a short time-scale and therefore its signatures may no longer be observable in the present epoch. Other possibilities for generating spin in the late universe include spinning up due to accretion [63], or that the merger of two PBHs could form a spinning remnant [10].

14.4 Searches for PBHs and Dark Matter with Gravitational Waves

One of the most promising ways of probing the nature of dark matter, be it a particle or compact object, is via gravitational wave observations [15]. Furthermore, there are distinct gravitational wave signatures for systems of PBHs surrounded by spikes or haloes of particle dark matter [56]. Improved sensitivities and wider frequency reaches of third generation detectors such as the proposed Einstein Telescope [81] and Cosmic Explorer [37] pose opportunities to simultaneously learn about two components of the dark matter. Either, through detecting PBH binaries merging in an environment of dark matter particles, or optimistically, through the signature of superradiance around a lone PBH.

14.4.1 Dephasing Signatures in the Gravitational Waveform

As described in Sect. 14.3, PBHs that do not make up all of the dark matter will necessarily have very dense dark matter particle haloes around them. If two PBHs form a binary then they will emit gravitational waves as they inspiral and eventually merge [75]. If they are of similar mass, then it is likely that within the first few orbits the dark matter haloes will be disrupted and possibly completely destroyed [56].

This means that the binary system will behave in the same way as one in vacuum, and hence the gravitational wave signal will not display evidence of the dark matter particles. This is true for heavier particle-like dark matter, however it is unclear whether this assumptions holds for light, wave-like dark matter. Simulations have shown some evidence for the persistence of the cloud of dark matter in the presence of equal-mass binaries [6], however it remains to be seen whether this is a physical effect or a consequence of the boundary conditions within the simulations that do not allow dispersion of the dark matter beyond the extent of the halo.

However, if there is a sizeable difference between the masses of the two PBHs, such that the mass ratio $q = m_2/m_1 \lesssim 10^{-2.5}$ where m_2 is the mass of the lighter PBH, then the dark matter halo is likely to survive for the duration of the inspiral, and the gravitational waveform will be altered by the presence of the dark matter [57]. This provides an opportunity for signatures of dark matter to be detected by the merging of intermediate or extreme mass ratio PBH binaries.

14.4.1.1 Dynamical Friction

As the companion (lighter) PBH drives through the dark matter halo of the primary PBH, the dark matter particles will form a wake behind the companion which slows down its velocity in the orbital direction, but increases its velocity in the radial direction. This means that the inspiral will complete in fewer cycles from a given entry frequency than an equivalent system in vacuum with the same masses. This effect is known as dynamical friction, and the difference in number of cycles is known as an accumulated dephasing of the gravitational waveform.

At Newtonian order where the orbits of the companion object are approximated as quasi-circular, this effect enters into the waveform via the change in separation of the two black holes as [35, 57]

$$\dot{r}(t) = -\frac{64G^3 M m_1 m_2}{5c^5 r(t)^3} - \frac{8\pi G^{\frac{1}{2}} m_2 \rho_{\rm DM}(r,t)\xi(r,t) r(t)^{\frac{5}{2}} \log\Lambda}{\sqrt{M} m_1} \tag{14.12}$$

where the first term is the emission of gravitational waves, and the second is the Chandrasekhar formula for dynamical friction. $M = m_1 + m_2$, ξ describes what fraction of the dark matter particles are moving more slowly than the companion object and can therefore be deflected and contribute to the drag force, $\Lambda = \sqrt{m_1/m_2}$ is the impact factor. Both the density profile of the dark matter and the fraction of particles moving more slowly than the black hole are functions of radius and time, because there is feedback from the presence of the black hole onto the spike. The initial density profile can be taken as that of Eq. 14.9, and then a prescription such as that of e.g. `HaloFeedback` [57] can be used to describe the evolution of the dark matter profile in the presence of the companion object in time. For intermediate and extreme mass ratio systems, orbits are in fact expected to be eccentric. This could have a non-negligible effect on the ability to measure the dark matter effect on the waveforms [13, 55], but a full study of more accurate waveforms including post-Newtonian effects for the regime of PBHs remains to be done.

The change in evolution of the separation of the binary is equivalent to a change in frequency evolution, which enters into the evolution of the phase of the gravitational waves:

$$\Phi(f) = \int_f^{f_f} \frac{dt}{df'} f' df'. \tag{14.13}$$

The dephasing $\delta\Phi$ is then the difference in phase between the dark matter affected system and some reference vacuum system

$$\delta\Phi = \Phi_V(f_i) - \Phi_{DM}(f_i) = 2\pi N \tag{14.14}$$

where f_i is some chosen initial frequency where the observation begins and N is the number of cycles between initial and final frequencies. Naively the vacuum system would be the equivalent system with the same mass parameters, but one could also define a 'best-fit' vacuum system that aims to minimise the dephasing and hence provide a conservative way of asserting the measurability of the dark matter environment [30].

There is also an effect on the gravitational waveform due to accretion of the dark matter onto the companion black hole, but this has been shown to be less important than the effect of dynamical friction for the mass ratios of interest [55].

14.4.1.2 Prospects for Detection with Einstein Telescope and Cosmic Explorer

With the improved sensitivities and wider frequency ranges of both proposed 3rd generation ground-based gravitational wave detectors Einstein Telescope and Cosmic Explorer, it could be possible to detect intermediate or extreme mass ratio PBH binaries and measure the properties of the dark matter clouds around them, if they exist. It has been shown that detecting PBH binaries in vacuum with small mass ratios will be possible, potentially using techniques borrowed from continuous wave search strategies, or hierarchical searches [65, 70, 71, 76]. It has also been shown that if there is an undisrupted dark matter spike around the larger of the two black holes, that with one week duration signals, it will be possible to confidently measure the density of the dark matter spike to be non-zero, as well as measure the distinctive slope of the dark matter profile. For example, for a system with $m_1 = 1\,M_\odot$ and $m_2 = 10^{-3}\,M_\odot$, the density normalisation can be measured to precision $\rho_6 = 1.7^{+0.8}_{-0.6} \times 10^{13}\,M_\odot/\mathrm{pc}^3$ and the slope to be $\gamma_{sp} = 2.22^{+0.07}_{0.05}$ where the quoted errors indicate the 68% credible intervals [28]. However, with a realistic mass function for the PBHs and respecting all other direct detection constraints on PBHs in the sub-solar mass range, the merger rate is expected to be low (~ 0.3 per year) [28], although recent studies have shown that the dark matter spike itself can have a large effect on the merger rate [52]. Future detectors will need to be lucky in order to detect these systems over their lifetimes of order years. Furthermore, various data analysis challenges such as being able to deal with these long duration signals both in terms of the search strategy and parameter estimation still need to be overcome.

14.4.1.3 Parameter Estimation Bias Between Dark Matter and Vacuum

It is possible, to some extent, to mimic the dark matter dephasing effect with a vacuum system of larger total mass (degeneracies with other parameters of the binary may also be possible). This degeneracy can be quantified by the amount of signal-to-noise ratio (SNR) lost in the mis-match between a dark matter gravitational waveform and its equivalent best-fit vacuum system, i.e. the vacuum template that does the best job of mimicking the dark matter-influenced system [28, 30]. For a range of masses, the percentage SNR loss is shown in Fig. 14.4 for both Einstein Telescope and Cosmic Explorer expected sensitivities and 1-year duration signals.

For most of the parameter space, the SNR lost if a vacuum template is used is over 50%. This is both promising in the sense that a full data analysis approach that accounts for the possibility of the presence of dark matter will be able to confidently distinguish it from vacuum, but also that these systems may go undetected if the

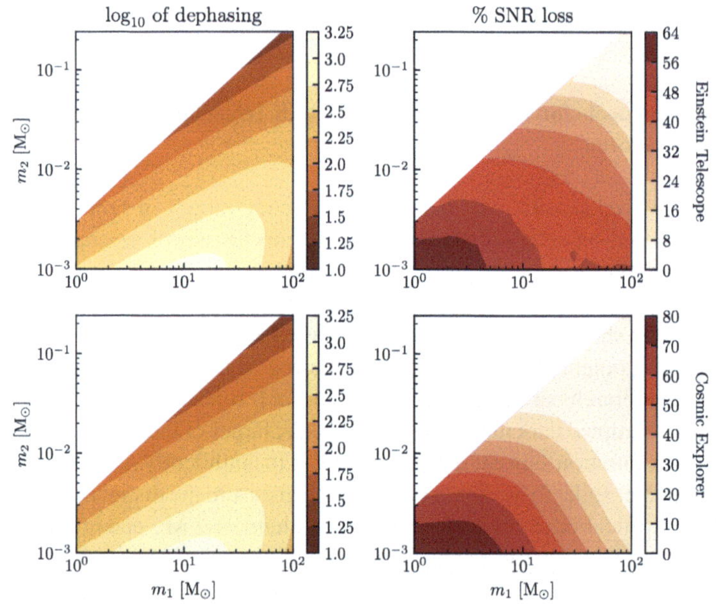

Fig. 14.4 The dephasing with respect to the best-fit vacuum system (left column) and the percentage of SNR lost with respect to the best-fit vacuum system (right column) as a function of primary m_1 and secondary m_2 mass. The top column is with respect to Einstein Telescope sensitivity and the bottom is with respect to Cosmic Explorer sensitivity. Figure reproduced with permission from [28]

analysis does not include the DM possibility in its approach. Since PBHs are expected to form dark matter haloes by necessity in this region of the parameter space where $f_{\rm PBH} = 1$ is ruled out, searching for these systems in vacuum will be likely futile. However, this only applies to binaries with a very small mass ratio where the spike is expected to have survived.

14.4.1.4 Astrophysical Versus Primordial Origin

In the case that the primary black hole is not clearly sub-solar within the measurement errors and therefore it is ambiguous as to whether it is of astrophysical or primordial origin, the observation of the presence of a dark matter spike around the primary black hole could provide distinctive distinguishing evidence. This is due to the fact that astrophysical black holes, of masses detectable with ground-based gravitational wave detectors, are not expected to form dense spikes of dark matter around them. Even in the case of intermediate mass black holes, where it is possible to invoke the idea that they formed in isolation in the centre of deep dark matter potential wells at high redshift, the expected slope of the dark matter profile is $\gamma_{\rm sp} = 7/3$ [35, 45], different to that expected for PBHs ($\gamma_{\rm sp} = 9/4$). For supermassive black holes at the centres of galaxies, it is expected that many of them will have undergone major mergers in their history, and therefore maintaining an undisrupted steep density profile is unlikely. The measurement of a dark matter profile slope of $\gamma_{\rm sp} = 9/4$ from a binary black hole inspiral would therefore be smoking gun evidence for both a primordial origin for the heavier black hole (and necessarily the lighter one since it would be sub-solar) as well as for the existence of a particle dark matter spike around it.

14.4.2 Superradiance GW Signatures

As discussed in Sect. 14.3.2, superradiance could cause a boson cloud to form around spinning PBHs. The most likely scenario for this combination to occur in the late universe such that the boson cloud is still present, as opposed to having extracted all of the angular momentum from the black hole and already decayed, is if the PBH is the remnant of a previous PBH merger [10] or if a PBH is spun up significantly by accretion [63]. There are possibilities for observing the signatures of such clouds via gravitational waves, both in the case of a single PBH, in which case the cloud itself will emit gravitational waves [43], or in the case of a PBH binary, in which case there could be evidence of ionisation of the cloud in the dephasing of the gravitational waveform [11, 12, 87], analogous to the dynamical friction discussed in Sect. 14.4.1.1. In the latter case, however, the combination of requirements: that two PBHs merge to form a spinning remnant, that a boson cloud grows via superradiance, and that then a secondary binary forms with low mass ratio (such that the ionisation effect is large enough to be observed) are challenging, so we will focus on the single PBH case.

The dominant production of gravitational waves is caused by annihilation of bosons, resulting in gravitational waves with frequencies that are twice the Compton frequency [40]:

$$f \sim 2 \times 10^6\,\mathrm{Hz}\left(\frac{m_b}{10^{-9}\,\mathrm{eV}}\right), \tag{14.15}$$

and with an amplitude h_0 (in time domain hence dimensionless) of

$$h_0 \sim 5 \times 10^{-25}\frac{1}{l}\left(\frac{\alpha}{0.1}\right)\left(\frac{\epsilon}{10^{-3}}\right)\left(\frac{d_L}{\mathrm{kpc}}\right)^{-1}\left(\frac{m_b}{10^{-9}\,\mathrm{eV}}\right)^{-1} \tag{14.16}$$

where $\alpha = Gm_{\mathrm{PBH}}m_b$ is the gravitational fine structure constant, l is the orbital angular momentum number of the bosons and ϵ is the fraction of the PBH mass which is transferred to the cloud. For superradiance to take place, $\alpha/l < 0.5$. Therefore, for example, a spinning PBH with mass $m_{\mathrm{PBH}} = 1\,\mathrm{M}_\odot$ could form a cloud of bosons with masses $m_b = 10^{-10}\,\mathrm{eV}$ via superradiance, which would produce monochromatic gravitational waves due to their self-annihilation at a frequency of $f = 2 \times 10^5\,\mathrm{Hz}$ with maximum amplitude $h_0 \sim 2.5 \times 10^{-23}$, assuming that $\alpha/l = 0.5$, $\epsilon = 10^{-3}$ and $d_L = 1\,\mathrm{kpc}$.

Current ground-based gravitational wave detectors have bandwidth up to a maximum frequency of approximately 10^4 Hz, and therefore would only be sensitive to bosons of masses $m_b < 10^{-11}\,\mathrm{eV}$, which would form clouds around spinning black holes of masses larger than $m_{\mathrm{PBH}} > 10\,\mathrm{M}_\odot$. For these masses, it would be difficult to be certain that the black hole around which the boson cloud has formed is primordial. Therefore, it will be necessary to go to the higher frequency regime, where data from various experiments such as ADMX [60] can be re-purposed for high-frequency gravitational wave searches [34]. According to recent estimates, the current sensitivities are not good enough to detect high frequency gravitational waves from the annihilation of bosons and therefore technological improvements or dedicated search strategies in this frequency regime will be required if there is hope for this channel of detection [40].

14.5 Dark Matter Particles from PBH Evaporation

PBHs evaporate due to Hawking radiation [48], and those lighter than $\mathcal{O}(10^{-18})\,\mathrm{M}_\odot$ will have completely evaporated before today. There are strong bounds on the abundance of PBHs in the mass range $10^{-24} - 10^{-16}\,\mathrm{M}_\odot$ due to the non-observation of the expected effects of their evaporation from big-bang nucleosynthesis (BBN), the extra-galactic gamma-ray background, and the cosmic microwave background (CMB) [1, 5, 20, 59]. However, black holes with masses less than $\sim 10^{-24}\,\mathrm{M}_\odot$ could have produced dark matter particles during their evaporation before BBN [25].

PBHs produce all particles with masses below the Hawking temperature T_{BH} of the black hole:

$$k_B T_{\rm BH} = \frac{1}{8\pi} \frac{(M_P c^2)^2}{M_{\rm BH} c^2} \tag{14.17}$$

where k_B is the Boltzmann constant and M_P is the Planck mass. There are two possibilities for producing stable dark matter particles from a single PBH. Either dark matter particles are emitted immediately after the formation of the PBH with number

$$N_X = \frac{g_{X,H}}{g_{*,H}} \frac{4\pi}{3} \left(\frac{M_{\rm BH}}{M_P} \right)^2 \tag{14.18}$$

that depends only on the mass of the PBH (at formation), and $g_{X,H}$, $g_{*,H}$ are the number of stable dark matter particles with masses below $k_{\rm B} T_{\rm BH}$ and total number of particles with masses below $k_{\rm B} T_{\rm BH}$ respectively. Or, dark matter particles are emitted towards the end of the lifetime of the PBH, once the temperature of the PBH has increased enough to be larger than the mass of the heavy particles with number

$$N_X = \frac{g_{X,H}}{g_{*,H}} \frac{1}{48\pi} \left(\frac{M_P}{M_{\rm X}} \right)^2 \tag{14.19}$$

that only depends on the mass of the dark matter particle, $M_{\rm X}$ [68]. There are strong constraints on such particles from requiring them to be sufficiently cold so as not to conflict with structure formation observations. However there may be ways to relax these constraints by considering a broad, and more realistic, PBH mass distribution [24] or to allow for entropy non-conservation [41]. Furthermore, there are interesting phenomenological signatures to be explored even if the entire dark matter abundance is not radiated by PBHs. For example, if PBHs are abundant enough to trigger an early-matter dominated era [50], and they produce axions as they evaporate, then the temperature at which those axions will start to oscillate will be altered [14].

14.6 Conclusion

PBHs may co-exist with a population of dark matter particles if they do not make up all of the dark matter themselves. There are interesting prospects for detection or constraints on both species due to interactions between the two. For example, weakly interacting massive particles would form dense haloes around PBHs, which on certain mass ranges, would have induced a signature in the diffuse gamma-ray background which has not been detected, thus placing constraints on large portions of the parameter space for both species if they are to co-exist. Axions or axion-like particles could also form dense haloes around PBHs which may be detectable via dephasing signatures in the gravitational waveform of PBH binaries. Or, if there exists a population of spinning PBHs along with a population of axion-like particles, then a boson cloud could form around the PBH due to superradiance. This system requires a combination of initial conditions to align, but would have very distinctive signatures in gravitational wave observations.

Alternatively, instead of co-existing, PBHs may have been the predecessors of our current dark matter population. Dark matter particles could have been produced in the early universe, before BBN, due to the evaporation of very light PBHs. Constraints on how warm dark matter can be due to structure formation observations are perhaps the most tightly constraining on this scenario, but ways of relaxing or alleviating these constraints have been considered.

Given that all but one window of the PBH mass range has been constrained such that the totality of the dark matter energy density budget cannot be explained by PBHs, i.e. $f_{\mathrm{PBH}} < 1$, it is important to consider a combined scenario for dark matter if PBHs are to play a role. Observations from the CMB, BBN and the gamma-ray background have already provided additional constraints on a sub-population of PBHs combined with WIMPs, and there are prospects for gravitational wave observations in particular to provide new constraints on distinctive properties of a multi-component model of dark matter that involves PBHs.

References

1. S.K. Acharya, R. Khatri, CMB and BBN constraints on evaporating primordial black holes revisited. J. Cosmol. Astroparticle Phys. **2020**(06), 018–018 (2020). ISBN: 1475-7516. https://doi.org/10.1088/1475-7516/2020/06/018
2. J. Adamek et al., WIMPs and stellar-mass primordial black holes are incompatible. Phys. Rev. D **100**(2), 023506 (2019). https://doi.org/10.1103/PhysRevD.100.023506. arXiv: 1901.08528 [astro-ph.CO]
3. S. Ando, K. Ishiwata, Constraints on decaying dark matter from the extragalactic gamma-ray background. JCAP **05**(024) (2015). https://doi.org/10.1088/1475-7516/2015/05/024. arXiv: 1502.02007 [astro-ph.CO]
4. G. Arcadi et al., The waning of the WIMP? A review of models, searches, and constraints. Eur. Phys. J. C **78**(3) (2018). ISBN: 1434-6052. https://doi.org/10.1140/epjc/s10052-018-5662-y
5. J. Auffinger, Primordial black hole constraints with Hawking radiation—A review. Prog. Particle Nucl. Phys. **131**, 104040 (2023). ISBN: 0146-6410. https://doi.org/10.1016/j.ppnp.2023.104040. https://www.sciencedirect.com/science/article/pii/S0146641023000212
6. J.C. Aurrekoetxea et al., The effect of wave dark matter on equal mass black hole mergers. arXiv: 2311.18156 [gr-qc]
7. H. Baer et al., Dark matter production in the early Universe: Beyond the thermal WIMP paradigm. Phys. Rep. **555**, 1–60. ISBN: 0370-1573. https://doi.org/10.1016/j.physrep.2014.10.002
8. C. Balázs et al., Cosmological constraints on decaying axion-like particles: a global analysis. J. Cosmol. Astroparticle Phys. **2022**(12), 027. ISBN: 1475-7516. https://doi.org/10.1088/1475-7516/2022/12/027
9. G. Ballesteros et al., Primordial black holes and gravitational waves from dissipation during inflation. J. Cosmol. Astroparticle Phys. **2022**(12), 006. ISBN: 1475-7516. https://doi.org/10.1088/1475-7516/2022/12/006
10. E. Barausse, L. Rezzolla, Predicting the direction of the final spin from the coalescence of two black holes. Astrophys. J. **704**(1), L40–L44. ISBN: 1538-4357. https://doi.org/10.1088/0004-637x/704/1/l40

11. D. Baumann et al., Ionization of gravitational atoms. Phys. Rev. D **105**(11), 115036 (2022). https://doi.org/10.1103/PhysRevD.105.115036. arXiv: 2112.14777 [gr-qc]
12. D. Baumann et al., Sharp signals of boson clouds in black hole binary inspirals. Phys. Rev. Lett. **128**(22), 221102 (2022). https://doi.org/10.1103/PhysRevLett.128.221102. arXiv: 2206.01212 [gr-qc]
13. N. Becker et al., Circularization versus eccentrification in intermediate mass ratio inspirals inside dark matter spikes. Phys. Rev. D **105**, 063029 (2022). https://doi.org/10.1103/PhysRevD.105.063029
14. N. Bernal, Ó. Zapata, Self-interacting dark matter from primordial black holes. J. Cosmol. Astroparticle Phys. **2021**(03), 007 (2021). ISBN: 1475-7516. https://doi.org/10.1088/1475-7516/2021/03/007
15. G. Bertone et al., Gravitational wave probes of dark matter: challenges and opportunities. In: SciPost Phys. Core **3**, 007 (2020). https://doi.org/10.21468/SciPostPhysCore.3.2.007. arXiv: 1907.10610 [astro-ph.CO]
16. G. Bertone et al., Primordial black holes as silver bullets for new physics at the weak scale. Phys. Rev. D **100**(12), 123013 (2019). https://doi.org/10.1103/PhysRevD.100.123013. arXiv: 1905.01238 [hep-ph]
17. N. Bhaumik, R.K. Jain, Primordial black holes dark matter from inflection point models of inflation and the effects of reheating. J. Cosmol. Astroparticle Phys. **2020**(01), 037–037 (2020). ISBN: 1475–7516. https://doi.org/10.1088/1475-7516/2020/01/037
18. P. Bode, J.P. Ostriker, N. Turok, Halo formation in warm dark matter models. Astrophys. J. **556**(1), 93–107 (2001). ISBN: 1538-4357. https://doi.org/10.1086/321541
19. S.M. Boucenna et al., Novel constraints on mixed dark-matter scenarios of primordial black holes and WIMPs. J. Cosmol. Astroparticle Phys. **2018**(07), 003–003 (2018). ISBN: 1475-7516. https://doi.org/10.1088/1475-7516/2018/07/003
20. M. Boudaud, M. Cirelli, Voyager 1 $e^{\pm}$ further constrain primordial black holes as dark matter. Phys. Rev. Lett. **122**(4) (2019). ISBN: 1079-7114. https://doi.org/10.1103/physrevlett.122.041104
21. R. Brito, V. Cardoso, P. Pani, *Superradiance: New Frontiers in Black Hole Physics* (Springer International Publishing, 2020). ISBN: 9783030466220. https://doi.org/10.1007/978-3-030-46622-0
22. B. Carr, F. Kuhnel, Primordial black holes as dark matter candidates. SciPost Phys. Lect. Notes **48**, 1 (2022). https://doi.org/10.21468/SciPostPhysLectNotes.48. arXiv: 2110.02821 [astro-ph.CO]
23. C. Chatterjee, *Dark Matter Self Interactions and its Impact on Large Scale Structures* (2019). arXiv: 1901.05658 [astro-ph.CO]
24. P. Chattopadhyay, A. Chaudhuri, M.Y. Khlopov, *Dark Matter from Evaporating PBH dominated in the Early Universe* (2023). arXiv: 2209.11288 [hep-ph]
25. A. Cheek et al., Evaporation of primordial black holes in the early Universe: Mass and spin distribution. Phys. Rev. D **108**, 015005 (2023). https://doi.org/10.1103/PhysRevD.108.015005
26. M. Cicoli, V.A. Diaz, F.G. Pedro, Primordial black holes from string inflation. J. Cosmol. Astroparticle Phys. **2018**(06), 034–034 (2018). ISBN: 1475-7516. https://doi.org/10.1088/1475-7516/2018/06/034
27. P.S. Cole, C.T. Byrnes, Extreme scenarios: the tightest possible constraints on the power spectrum due to primordial black holes. JCAP **02**, 019 (2018). https://doi.org/10.1088/1475-7516/2018/02/019. arXiv: 1706.10288 [astro-ph.CO]
28. P.S. Cole et al., Measuring dark matter spikes around primordial black holes with Einstein Telescope and Cosmic Explorer. Phys. Rev. D **107**(8), 0803006 (2023). https://doi.org/10.1103/PhysRevD.107.083006. arXiv: 2207.07576 [astro-ph.CO]
29. P.S. Cole et al., Primordial black holes from single-field inflation: a fine-tuning audit. JCAP **08**, 031 (2023). https://doi.org/10.1088/1475-7516/2023/08/031. arXiv: 2304.01997 [astro-ph.CO]
30. A. Coogan et al., Measuring the dark matter environments of black hole binaries with gravitational waves. Phys. Rev. D **105**(4), 043009 (2022). https://doi.org/10.1103/PhysRevD.105.043009. arXiv: 2108.04154 [gr-qc]

31. M.S. Delos, J. Silk, Ultradense dark matter haloes accompany primordial black holes. Mon. Notices Roy. Astron. Soc. **520**(3), 4370–4375 (2023). ISBN: 0035-8711. https://doi.org/10.1093/mnras/stad356. eprint: https://academic.oup.com/mnras/article-pdf/520/3/4370/49288240/stad356.pdf. URL: https://doi.org/10.1093/mnras/stad356
32. M.J. Dolan, F.J. Hiskens, R.R. Volkas, Advancing globular cluster constraints on the axion-photon coupling. J. Cosmol. Astroparticle Phys. **2022**(10), 096 (2022). ISBN: 1475-7516. https://doi.org/10.1088/1475-7516/2022/10/096. http://dx.doi.org/10.1088/1475-7516/2022/10/096
33. M.J. Dolan, F.J. Hiskens, R.R. Volkas, Constraining axion-like particles using the white dwarf initial-final mass relations. J. Cosmol. Astroparticle Phys. **2021**(09), 010 (2021). ISBN: 1475-7516. https://doi.org/10.1088/1475-7516/2021/09/010. http://dx.doi.org/10.1088/1475-7516/2021/09/010
34. V. Domcke, C. Garcia-Cely, N.L. Rodd, Novel search for high-frequency gravitational waves with low-mass axion haloscopes. Phys. Rev. Lett. **129**(4) (2022). ISBN: 1079-7114. https://doi.org/10.1103/physrevlett.129.041101. http://dx.doi.org/10.1103/PhysRevLett.129.041101
35. K. Eda et al., Gravitational waves as a probe of dark matter minispikes. Phys. Rev. D **91**(4) (2015). ISBN: 1550-2368. https://doi.org/10.1103/physrevd.91.044045. http://dx.doi.org/10.1103/PhysRevD.91.044045
36. Y.N. Eroshenko, Dark matter density spikes around primordial black holes. Astron. Lett. **42**(6), 347–356 (2016). https://doi.org/10.1134/S1063773716060013. arXiv: 1607.00612 [astro-ph.HE]
37. M. Evans et al., A Horizon Study for Cosmic Explorer: Science, Observatories, and Community.(2021). arXiv: 2109.09882 [astro-ph.IM]
38. J.L. Feng, Naturalness and the status of supersymmetry. Ann. Rev. Nucl. Particle Sci. **63**(1), 351–382 (2013). ISBN: 1545-4134. https://doi.org/10.1146/annurev-nucl-102010-130447. http://dx.doi.org/10.1146/annurev-nucl-102010-130447
39. J.L. Feng, The WIMP paradigm: theme and variations. SciPost Physics Lecture Notes (2023). ISBN: 2590-1990. https://doi.org/10.21468/scipostphyslectnotes.71. http://dx.doi.org/10.21468/SciPostPhysLectNotes.71
40. G. Franciolini, A. Maharana, F. Muia, Hunt for light primordial black hole dark matter with ultrahigh-frequency gravitational waves. Phys. Rev. D **106**(10) (2022). ISBN: 2470-0029. https://doi.org/10.1103/physrevd.106.103520. http://dx.doi.org/10.1103/PhysRevD.106.103520
41. T. Fujita et al., Baryon asymmetry, dark matter, and density perturbation from primordial black holes. Phys. Rev. D **89**(10) (2014). ISBN: 1550-2368. https://doi.org/10.1103/physrevd.89.103501. http://dx.doi.org/10.1103/PhysRevD.89.103501
42. C. Germani, T. Prokopec, *On primordial black holes from an inflection point*. (2017). arXiv: 1706.04226 [astro-ph.CO]
43. S. Ghosh, Gravitational waves from superradiant instabilities of rotating black holes. Mod. Phys. Lett. A **36**(33) (2021). ISBN: 1793-6632. https://doi.org/10.1142/s021773232130024x. http://dx.doi.org/10.1142/S021773232130024X
44. G. Francesco Giudice, Naturally speaking: the naturalness criterion and physics at the LHC. Perspect. LHC Phys. (WORLD SCIENTIFIC 2008), pp. 155–178. ISBN: 9789812779762. https://doi.org/10.1142/9789812779762_0010. http://dx.doi.org/10.1142/9789812779762_0010
45. P. Gondolo, J. Silk, Dark matter annihilation at the galactic center. Phys. Rev. Lett. **83**(9), 1719–1722 (1999). ISBN: 1079-7114. https://doi.org/10.1103/physrevlett.83.1719. http://dx.doi.org/10.1103/PhysRevLett.83.1719
46. A.M. Green, B.J. Kavanagh, Primordial Black Holes as a dark matter candidate. J. Phys. G **48**(4), 043001 (2021). https://doi.org/10.1088/1361-6471/abc534. arXiv: 2007.10722 [astro-ph.CO]
47. T. Harada et al., Spins of primordial black holes formed in the matter-dominated phase of the Universe. Phys. Rev. D **96**(8), 083517 (2017). [Erratum: Phys.Rev.D 99, 069904 (2019)]. https://doi.org/10.1103/PhysRevD.96.083517. arXiv: 1707.03595 [gr-qc]

48. S.W. Hawking, Particle creation by black holes. Commun. Math. Phys. **43**, 199–220 (1975). Ed. by G. W. Gibbons and S. W. Hawking. [Erratum: Commun.Math.Phys. 46, 206 (1976)]. https://doi.org/10.1007/BF02345020
49. M.P. Hertzberg, M. Yamada, Primordial black holes from polynomial potentials in single field inflation. Phys. Rev. D **97**(8) (2018). ISBN: 2470-0029. https://doi.org/10.1103/physrevd.97.083509. http://dx.doi.org/10.1103/PhysRevD.97.083509
50. D. Hooper, G. Krnjaic, S.D. McDermott, Dark radiation and superheavy dark matter from black hole domination. J. High Energy Phys. **2019**(8) (2019). ISBN: 1029-8479. https://doi.org/10.1007/jhep08(2019)001. http://dx.doi.org/10.1007/JHEP08(2019)001
51. A. Ireland, *Dark spiky primordial black holes*. arXiv:2406.07624 (2024)
52. P. Jangra, B.J. Kavanagh, J. M. Diego, Impact of dark matter spikes on the merger rates of primordial black holes. JCAP **11**, 069 (2023). https://doi.org/10.1088/1475-7516/2023/11/069. arXiv: 2304.05892 [astro-ph.CO]
53. K. Kadota, H. Tashiro, Radio bounds on the mixed dark matter scenarios of primordial black holes and WIMPs. J. Cosmol. Astroparticle Phys. **2022**(08), 004 (2022). https://doi.org/10.1088/1475-7516/2022/08/004. https://dx.doi.org/10.1088/1475-7516/2022/08/004
54. A. Karam et al., Anatomy of single-field inflationary models for primordial black holes. J. Cosmol. Astroparticle Phys. **2023**(03), 013 (2023). ISBN: 1475-7516. https://doi.org/10.1088/1475-7516/2023/03/013. http://dx.doi.org/10.1088/1475-7516/2023/03/013
55. T.K. Karydas, B.J. Kavanagh, G. Bertone, Sharpening the dark matter signature in gravitational waveforms I: Accretion and eccentricity evolution (2024). arXiv: 2402.13053 [gr-qc]
56. B.J. Kavanagh, D. Gaggero, G. Bertone, Merger rate of a subdominant population of primordial black holes. Phys. Rev. D **98**(2), 023536 (2018). https://doi.org/10.1103/PhysRevD.98.023536. arXiv: 1805.09034 [astro-ph.CO]
57. B.J. Kavanagh et al., Detecting dark matter around black holes with gravitational waves: Effects of dark-matter dynamics on the gravitational waveform. Phys. Rev. D **102**(8), 083006 (2020). https://doi.org/10.1103/PhysRevD.102.083006. arXiv: 2002.12811 [gr-qc]
58. K. Kawana, K-P. Xie, Primordial black holes from a cosmic phase transition: The collapse of Fermi-balls. Phys. Lett. B **824**, 136791 (2022). ISBN: 0370-2693. https://doi.org/10.1016/j.physletb.2021.136791
59. C. Keith et al., Constraints on primordial black holes from big bang nucleosynthesis revisited. Phys. Rev. D **102**(10) (2020). ISBN: 2470-0029. https://doi.org/10.1103/physrevd.102.103512. http://dx.doi.org/10.1103/PhysRevD.102.103512
60. R. Khatiwada et al., Axion dark matter experiment: detailed design and operations. Rev. Sci. Instrum. **92**(12) (2021). ISBN:1089-7623. https://doi.org/10.1063/5.0037857. http://dx.doi.org/10.1063/5.0037857
61. R. Lasenby, Black hole superradiance as a probe of ultra-light new particles. Proc. Int. Astron. Union **12**(S324), 273–278 (2016). https://doi.org/10.1017/S1743921317002319
62. M. Lewicki, P. Toczek, V. Vaskonen, Primordial black holes from strong first-order phase transitions. J. High Energy Phys. **2023**(9) (2023). ISBN: 1029-8479. https://doi.org/10.1007/jhep09(2023)092. http://dx.doi.org/10.1007/JHEP09(2023)092
63. V. De Luca et al., The evolution of primordial black holes and their final observable spins. J. Cosmol. Astroparticle Phys. **2020**(04), 052–052 (2020). ISBN: 1475-7516. https://doi.org/10.1088/1475-7516/2020/04/052. http://dx.doi.org/10.1088/1475-7516/2020/04/052
64. J.H. MacGibbon et al., Primordial black holes, in *5th International Fermi Symposium* (2015). arXiv: 1503.01166 [astro-ph.HE]
65. R. Magee et al., Methods for the detection of gravitational waves from subsolar mass ultra-compact binaries. Phys. Rev. D **98**(10). ISBN: 2470-0029. https://doi.org/10.1103/physrevd.98.103024. http://dx.doi.org/10.1103/PhysRevD.98.103024
66. D.J.E. Marsh, Axions and ALPs: a very short introduction, in *13th Patras Workshop on Axions, WIMPs and WISPs* (2018), pp. 59–74. https://doi.org/10.3204/DESY-PROC-2017-02/marsh_david. arXiv: 1712.03018 [hep-ph]
67. D.J.E. Marsh, Axion cosmology. Phys. Rep. **643**, 1–79 (2016). ISBN: 0370-1573. https://doi.org/10.1016/j.physrep.2016.06.005. http://dx.doi.org/10.1016/j.physrep.2016.06.005

68. I. Masina, Dark matter and dark radiation from evaporating primordial black holes. Eur. Phys. J. Plus **135**(7), 552 (2020). https://doi.org/10.1140/epjp/s13360-020-00564-9. arXiv: 2004.04740 [hep-ph]
69. M. Milgrom, MOND theory. Can. J. Phys. **93**(2), 107–118 (2015). ISBN: 1208-6045. https://doi.org/10.1139/cjp-2014-0211. http://dx.doi.org/10.1139/cjp-2014-0211
70. A.L. Miller et al., Gravitational wave constraints on planetary-mass primordial black holes using LIGO O3a data (2024). arXiv: 2402.19468 [gr-qc]
71. A.L. Miller et al., Probing planetary-mass primordial black holes with continuous gravitational waves. Phys. Dark Univ. **32**, 100836 (2021). ISBN: 2212-6864. https://doi.org/10.1016/j.dark.2021.100836. https://www.sciencedirect.com/science/article/pii/S2212686421000674
72. M. Mirbabayi, A. Gruzinov, J. Noreña, Spin of primordial black holes. JCAP **03**, 017 (2020). https://doi.org/10.1088/1475-7516/2020/03/017. arXiv: 1901.05963 [astro-ph.CO]
73. S.S. Mishra, V. Sahni, Primordial black holes from a tiny bump/dip in the inflaton potential. J. Cosmol. Astroparticle Phys. **2020**(04), 007–007 (2020). ISBN: 1475-7516. https://doi.org/10.1088/1475-7516/2020/04/007. http://dx.doi.org/10.1088/1475-7516/2020/04/007
74. P. Montero-Camacho et al., Revisiting constraints on asteroid-mass primordial black holes as dark matter candidates. J. Cosmol. Astroparticle Phys. **2019**(08), 031–031 (2019). ISBN: 1475-7516. https://doi.org/10.1088/1475-7516/2019/08/031. http://dx.doi.org/10.1088/1475-7516/2019/08/031
75. H. Nishikawa et al., Primordial-black-hole mergers in dark-matter spikes. Phys. Rev. D **99**(4), (2019). ISBN: 2470-0029. https://doi.org/10.1103/physrevd.99.043533. http://dx.doi.org/10.1103/PhysRevD.99.043533
76. A.H. Nitz, Yi-F. Wang, Search for gravitational waves from the coalescence of subsolar-mass binaries in the first half of advanced LIGO and Virgo's third observing run. Phys. Rev. Lett. **127**(15) (2021). ISBN: 1079-7114. https://doi.org/10.1103/physrevlett.127.151101. http://dx.doi.org/10.1103/PhysRevLett.127.151101
77. C. O'Hare. *cajohare/AxionLimits: AxionLimits*. https://cajohare.github.io/AxionLimits/. Version v1.0. July 2020. https://doi.org/10.5281/zenodo.3932430
78. C.A. J. O'Hare, Cosmology of axion dark matter. PoS COSMICWISPers 040 (2024). https://doi.org/10.22323/1.454.0040. arXiv: 2403.17697 [hep-ph]
79. O. Özsoy, G. Tasinato, Inflation and primordial black holes. Universe **9**(5), (2023). ISBN: 2218-1997. https://doi.org/10.3390/universe9050203. http://dx.doi.org/10.3390/universe9050203
80. R.D. Peccei, The strong CP problem and axions. Axions (Springer, Berlin, Heidelberg, 2008), pp. 3–17. ISBN: 9783540735182. https://doi.org/10.1007/978-3-540-73518-2_1. http://dx.doi.org/10.1007/978-3-540-73518-2_1
81. M. Punturo et al., The Einstein Telescope: a third-generation gravitational wave observatory. Class. Quant. Grav. **27**, 194002 (2010). Ed. by Fulvio Ricci. https://doi.org/10.1088/0264-9381/27/19/194002
82. L. Roszkowski, E. Maria Sessolo, S. Trojanowski, WIMP dark matter candidates and searches—current status and future prospects. Rep. Prog. Phys. **81**(6), 066201 (2018). https://doi.org/10.1088/1361-6633/aab913. https://dx.doi.org/10.1088/1361-6633/aab913
83. R.J. Scherrer, M.S. Turner, On the relic, cosmic abundance of stable, weakly interacting massive particles. Phys. Rev. D **33**, 1585–1589 (1986). https://doi.org/10.1103/PhysRevD.33.1585
84. M. Schumann, Direct detection of WIMP dark matter: concepts and status. J. Phys. G: Nucl. Particle Phys. **46**(10), 103003 (2019). ISBN: 1361-6471. https://doi.org/10.1088/1361-6471/ab2ea5
85. G. Steigman, M.S. Turner, Cosmological constraints on the properties of weakly interacting massive particles. Nucl. Phys. B **253**, 375–386 (1985). ISBN: 0550-3213. https://doi.org/10.1016/0550-3213(85)90537-1
86. H. Tashiro, K. Kadota, Constraining mixed dark-matter scenarios of WIMPs and primordial black holes from CMB and 21-cm observations. Phys. Rev. D **103**, 123532 (2021). https://doi.org/10.1103/PhysRevD.103.123532
87. G. Maria Tomaselli, T.F.M. Spieksma, G. Bertone, Dynamical friction in gravitational atoms. JCAP **07**, 070 (2023). https://doi.org/10.1088/1475-7516/2023/07/070. arXiv: 2305.15460 [gr-qc]

88. S. Tulin, H-B. Yu, Dark matter self-interactions and small scale structure. Phys. Rep. **730**, 1–57 (2018). ISBN: 0370-1573. https://doi.org/10.1016/j.physrep.2017.11.004
89. Y. Yang, X. Li, G. Li, Constraints on primordial black holes in the mixed dark matter scenarios using the ratio $(^3\mathrm{He}+\mathrm{D})/\mathrm{H}$ (2023). arXiv: 2308.07010 [astro-ph.CO]
90. K. Zioutas et al., First results from the CERN axion solar telescope. Phys. Rev. Lett. **94**, 121301 (2005). https://doi.org/10.1103/PhysRevLett.94.121301

Part IV
The Primordial Black Hole Scenario in the Gravitational-Wave Era

Chapter 15
Gravitational-Wave Observations and Primordial Black Holes

Francesco Iacovelli and Michele Maggiore

Abstract Gravitational-wave observations have the potential of allowing the identification of a population of merging primordial black-hole binaries. We provide an overview of the capabilities of present and future gravitational-wave detectors, with a special emphasis on the perspective for observing quantities that are signatures of a primordial origin.

15.1 Introduction

In this chapter we discuss the perspectives for observing PBHs and studying their properties, both individually and as a population, with current and future GW detectors.

In GW physics the last decade has been marked by the extraordinary results of the LIGO-Virgo-KAGRA Collaboration. The first detection of a binary black hole (BBH) coalescence took place on Sept. 14th, 2015 [4], and was a milestone in science. Another milestone was the observation, on Aug. 17, 2017, of the first binary neutron star (BNS) coalescence [2], and the detection and follow-up of the source in all bands of the electromagnetic spectrum [3]. Nowadays, after three observing runs, the catalog of observed compact binary coalescences contains about 90 BBH coalescences, two NSBH binaries and two BNS [6, 7]. In the currently ongoing fourth observing run, one candidate event every few days has been announced [20]. Many different fields have already been significantly impacted by these discoveries: we now have a much better understanding of the population properties of compact

F. Iacovelli (✉) · M. Maggiore
Département de Physique Théorique, Université de Genève, 24 quai Ernest Ansermet, 1211 Genève 4, Switzerland
e-mail: Francesco.Iacovelli@unige.ch

M. Maggiore
e-mail: Michele.Maggiore@unige.ch

Gravitational Wave Science Center (GWSC), Université de Genève, 1211 Geneva, Switzerland

C. Byrnes et al. (eds.), *Primordial Black Holes*, Springer Series in Astrophysics and Cosmology, https://doi.org/10.1007/978-981-97-8887-3_15

binaries (see, e.g. Ref. [8]), we start to have first constraints on deviations from GR, as well as on the expansion history of the Universe (see, e.g. Ref. [5, 9]), while the multi-messenger observation of the binary neutron-star system GW170817 and of its electromagnetic counterpart [1, 3] had important implications in relativistic astrophysics, nuclear physics, nucleosynthesis in the Universe, and cosmology.

15.2 Current and Future Ground-Based Gravitational Wave Detectors

While the results of second-generation (2G) detectors such as LIGO, Virgo and KAGRA have marked the beginning of the GW astronomy era, a new generation (3G) of ground-based detectors is currently under development. In particular, the European flagship project is Einstein Telescope (ET) [29, 30, 41], while the US community is developing the Cosmic Explorer project [22, 44]. A discussion of the ET Science Case can be found in [35], while a recent comprehensive study of the prospects for observations of compact binary coalescence and characterization at ET can be found in [32]; the perspective for multi-messenger observations with ET are discussed in detail in [46]; for CE and for a more general discussion of 3G detectors see also [17, 26, 33]. More recently, the whole science case of ET has been refined and evaluated with respect to different possible choices of design, including different options for the geometry (a single-site detector with nested interferometer in a triangular configuration, versus two L-shaped interferometers) and different sensitivity curves [18].

Figure 15.1 shows the strain sensitivities that were reached by LIGO and Virgo during the O3 run. As an example of the potential of the 2G LIGO and Virgo instruments, when pushed to their limit, we show a representative sensitivity for the so-called A# stage of the LIGO detectors. We compare these sensitivities with those expected for ET (in the two configurations currently under consideration, i.e. a triangle with 10 km arms and a 2L with 15 km arms) and CE (considering both a detector with 20 km arms and one with 40 km arms). The corresponding detection horizons, for equal-mass spinless binaries, are shown in Fig. 15.2.

From the point of view of PBHs the following aspects, in the jump from 2G to 3G detectors, are noteworthy.

1. The increased reach in distance. This is particularly significant for identifying a PBH population because, even if the redshift at which the first stars formed is not precisely known, theoretical calculations and cosmological simulations suggest that the redshift at which Pop III remnant BBHs merge is below $z \sim 30$ (see [39, 40] and refs. therein). In contrast, stellar-mass PBHs have a merger rate density that increases monotonically with redshift, up to $z \sim 10^3$ and beyond [42] (see Chap. 16). While current detectors have a reach where astrophysical BH dominates, we see from the figure that 3G detectors reach redshift as high as $z \sim 50$ (and even $z \sim 100$ for the optimal combinations of masses and orientation), where the contamination from astrophysical BH binary mergers is expected to be negligible.

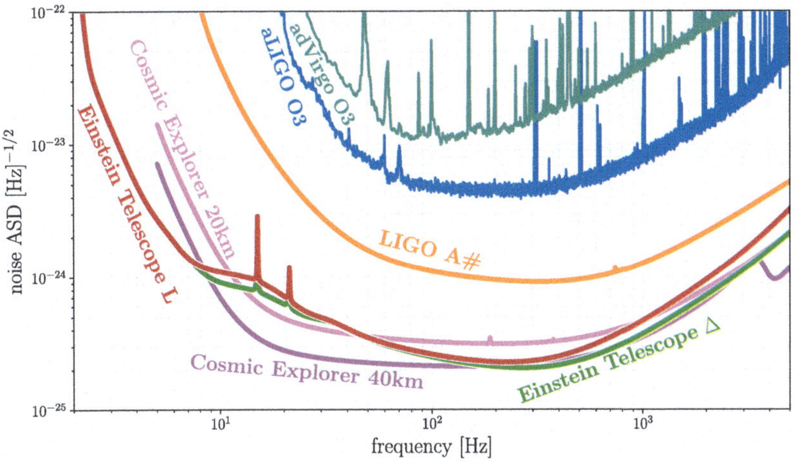

Fig. 15.1 Noise amplitude spectral densities for: the current second generation detectors LIGO and Virgo achieved during the O3 observing run; the prospects for LIGO in the so-called A# configuration; ET in the triangular configuration with 10 km arms (rescaling the spectral density by a factor 2/3 to take into account the three nested interferometers [32]); a single L-shaped interferometer (part of a 2L configuration of ET) with 15 km arms; CE with 20 km and with 40 km long arms

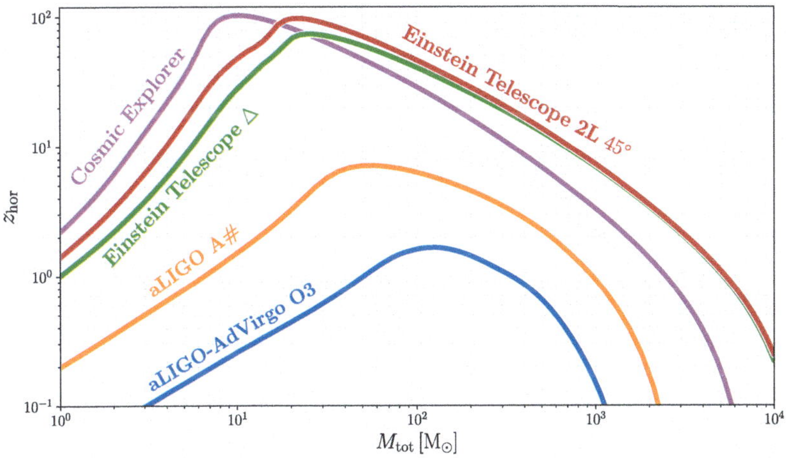

Fig. 15.2 Redshift horizons for equal-mass spinless binaries at: LIGO and Virgo during the O3 observing run; LIGO in the A# configuration; ET both in the triangular configuration with 10 km arms and as two L-shaped detector with 15 km arms at 45°; CE in a configuration consisting of two L-shaped instruments one with 40 km long arms and the other of 20 km long arms

Reaching a great distance, however, is not yet sufficient to have a smoking-gun signature of a primordial origin. It is also necessary to be able to reconstruct the redshift with an error sufficiently small, so that the support of the posterior at z smaller than a critical value, such as $z_c = 30$, is negligible. This issue was studied in [24, 36, 38, 39], where it is found that individual identification of PBHs based on the condition $z > 30$ is possible but challenging, and in any case requires a network of 3G detectors including ET and one or two CE.

2. The significant improvement in sensitivity of 3G detectors implies a corresponding improvement in the parameter estimation of coalescing binaries. From the point of view of PBHs, particularly relevant are the orbit eccentricity, and the spins of the binary components [24]. The left panel of Fig. 15.3 shows the improvement in the reconstruction of $\chi_{1,z}$ and $\chi_{2,z}$, the components of the spin parallel to the orbital angular momentum, for the two binary components (we use the label 1 to refer to the heavier object and 2 to the lighter object), assuming a population of PBHs as in [25] based on the results in [23], while the right panel shows the accuracy on the eccentricity for a source at a fixed distance $d_L = 100$ Mpc and total source-frame mass $20\text{M}_\odot$.

3. A possible smoking gun signature for PBH is a source-frame mass in the subsolar range (see Chap. 18). Figure 15.2 shows that, in this regime, even LIGO-A# cannot go significantly beyond our local Universe. In contrast, ET or CE can reach cosmological distances, up to $z \sim 1 - 2$ for a total mass of $1\text{M}_\odot$ and $z \sim 2 - 5$ for a total mass of $2\text{M}_\odot$.

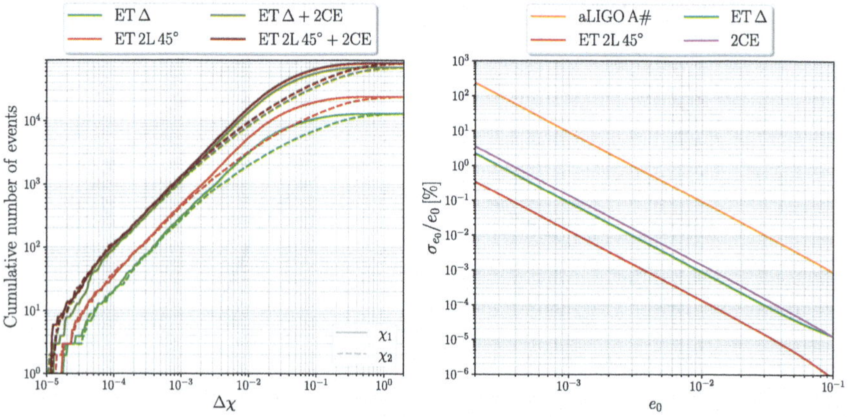

Fig. 15.3 *Left panel:* cumulative distribution of the errors on the PBH binary spin parameters obtained with the GWFAST Fisher code [31, 32] for the population used in [25] and based on [23]. We report the results for two ET configurations (10 km triangle and 15 km 2L at 45°) both alone and in a network with 2 CE. *Right panel:* Relative error on the eccentricity parameter e_0 at $f_{e_0} = 10$ Hz for a $M_{\text{tot}} = 20\ \text{M}_\odot$ spinless binary at 100 Mpc as a function of e_0. We compare the performance of aLIGO in the A# configuration, ET in the triangular and 2L configurations and 2 CE

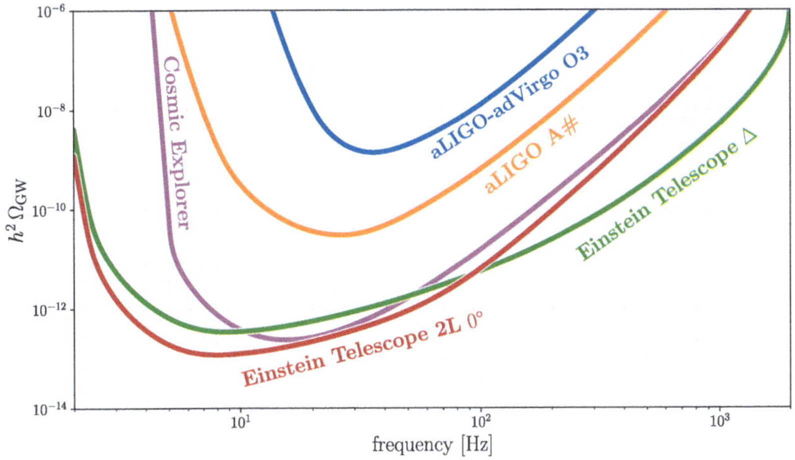

Fig. 15.4 Power-law integrated sensitivity curves to a stochastic background of GWs Ω_{GW} [48] obtained for: LIGO and Virgo during the O3 observing run; LIGO in the A# configuration; ET both in the triangular configuration with 10 km arms and as as two parallel L-shaped detector with 15 km arms; CE in a configuration consisting of two L-shaped instruments one with 40 km long arms and the other of 20 km long arms. The curves are scaled by $h = H_0/100$ km s^{-1} Mpc^{-1} and obtained assuming an observation time of 1 yr and an SNR of 1

Finally, Fig. 15.4 shows the sensitivity to stochastic background energy fraction Ω_{GW}, as expressed by the power-law integrated sensitivity curve [48]. We see that 3G detectors gain two orders of magnitudes even with respect to LIGO-A#.

15.3 Gravitational Wave Detectors in the mHz Range

The future will also be marked by pioneering experiments in completely different frequency bands as compared to the ones mentioned above, providing a new perspective on the universe.

The milliHz band will be explored by the space-based European Space Agency experiment LISA [13] that with its three spacecrafts arranged in a triangular configuration at million-km distance could collect signals from a large variety of sources: galactic binaries observations will allow to map the Milky Way through GWs, giving valuable information on the formation of such systems and their evolution through gravitational and tidal dissipation; observing the inspiral and merger of massive and supermassive BHs through cosmic time will help shedding light on their origin and evolution, and also allow us to put constraints on the expansion history of the Universe; detections of extreme mass ratio inspirals of stellar-origin BHs into massive BHs at the center of galaxies are an extremely promising way to map with exquisite

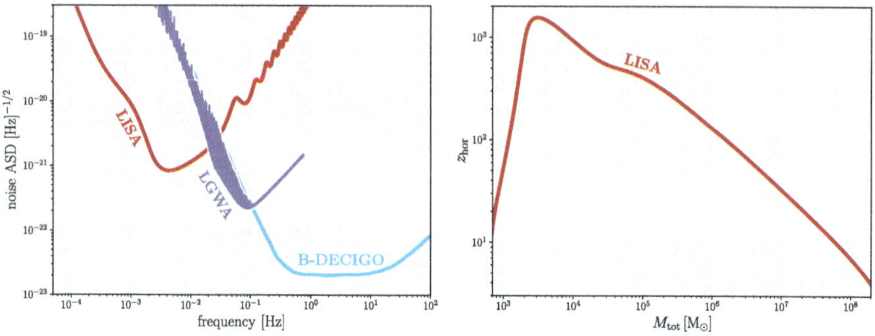

Fig. 15.5 *Left panel:* Noise amplitude spectral densities for: LISA assuming an arm length of 2.5×10^9 km and two data channels [45]; LGWA [21, 27]; B-DECIGO [34, 37]. *Right panel:* Redshift horizon for equal-mass spinless binaries at LISA assuming a mission lifetime of 4 years

precision the gravitational field around these gargantuan objects and study their environments; multiband of stellar-origin BHs in LISA and ground-based detectors would complement each other, giving e.g. the possibility to disentangle formation channels thanks to eccentricity measurements; the detection of stochastic backgrounds of cosmological origin that can be generated by several processes in the early universe can give extremely valuable information on the first instants after the Big Bang [15]. In Fig. 15.5 we report the sensitivity curve (left panel) and detection horizon for equal-mass spinless binaries (right panel) for the LISA detector, again noticing the considerable reach in terms of distance in a different mass range, giving e.g. the possibility to observe PBH binaries in a broad mass range out to extremely high redshift.

The frequency gap between LISA and ground-based detectors such as ET and CE could be covered by proposed experiments such as the Lunar Gravitational Wave Antenna (LGWA) [27], that aims at using the Moon as a GW-detector, monitoring with seismometers the vibrations of the Moon; by experiments based on atom interferometry such as MIGA [19], VLBAI [47] and AION [16] (see also [10] for review); and by the space interferometer DECIGO currently under study [34]. As an example, the sensitivity of LGWA and the so-called B-DECIGO are reported in Fig. 15.5.

15.4 Gravitational Wave Observations with Pulsar Timing Arrays

In a different frequency band, corresponding to the nanoHz, another exciting result came out in mid-2023 from pulsar timing arrays (PTAs): several experiments monitoring the arrival times of pulses emitted by galactic millisecond pulsars and searching for the correlation induced by GWs, such as the European Pulsar Timing Array (EPTA) and NANOGRAV in North America, reported evidence for the presence of

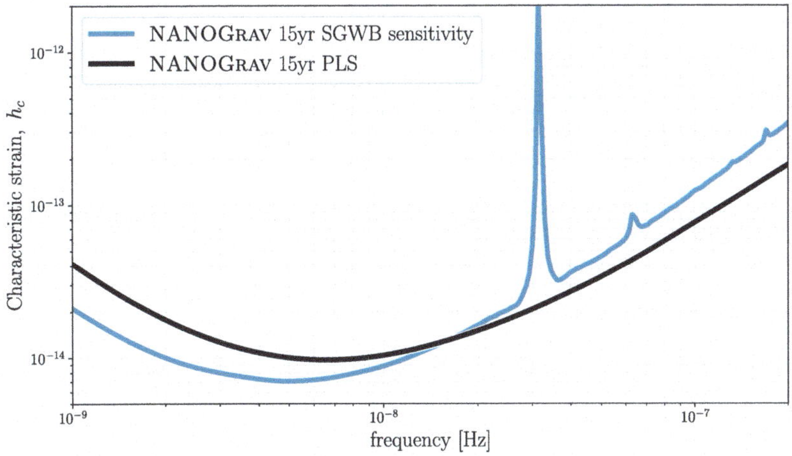

Fig. 15.6 Effective characteristic strain noise to a stochastic GW background source for the NANOGRAV 15-yr PTA data and corresponding power-law integrated sensitivity curve [48] obtained as in [11, 28]

a stochastic background of GWs [12, 14, 43, 49], in data streams spanning longer than a decade. While the origin of this signal has still to be understood, a natural candidate is offered by the incoherent superposition of signals from the inspiral of supermassive BH binaries. Another possibility is a signal of cosmological origin, produced by curvature perturbations during inflation, which can also give rise to PBHs (see Chap. 17). In Fig. 15.6 we report the effective characteristic strain noise for the data-release of 15 years of observations from the NANOGRAV collaboration in the search for a stochastic signal, and the corresponding power-law integrated sensitivity curve. Thanks to more data being collected and new facilities with outstanding accuracy expected to become available, such as the Square Kilometre Array (SKA), the accuracy of PTA experiments is expected to greatly raise in the future.

References

1. B.P. Abbott et al., Gravitational waves and gamma-rays from a binary neutron star merger: GW170817 and GRB 170817A. Astrophys. J. **848**, L13 (2017). https://doi.org/10.3847/2041-8213/aa920c. arXiv: 1710.05834 [astro-ph.HE]
2. B.P. Abbott et al., GW170817: observation of gravitational waves from a binary neutron star inspiral. Phys. Rev. Lett. **119**(16), 161101 (2017). https://doi.org/10.1103/PhysRevLett.119.161101. arXiv: 1710.05832 [gr-qc]
3. B.P. Abbott et al., Multi-messenger observations of a binary neutron star merger. Astrophys. J. Lett. **848**(2), L12 (2017). https://doi.org/10.3847/2041-8213/aa91c9. arXiv: 1710.05833 [astro-ph.HE]
4. B.P. Abbott et al., Observation of gravitational waves from a binary black hole merger. Phys. Rev. Lett. **116**(6), 061102 (2016). https://doi.org/10.1103/PhysRevLett.116.061102. arXiv: 1602.03837 [gr-qc]

5. R. Abbott et al., Constraints on the cosmic expansion history from GWTC–3. Astrophys. J. **949**(2), 76 (2023). https://doi.org/10.3847/1538-4357/ac74bb. arXiv: 2111.03604 [astro-ph.CO]
6. R. Abbott et al., GWTC-2: compact binary coalescences observed by LIGO and Virgo during the first half of the third observing run. Phys. Rev. X **11**, 021053 (2021). https://doi.org/10.1103/PhysRevX.11.021053. arXiv: 2010.14527 [gr-qc]
7. R. Abbott et al. GWTC-3: compact binary coalescences observed by LIGO and Virgo during the second part of the third observing run. Phys. Rev. X **13**(4), 041039 (2023). https://doi.org/10.1103/PhysRevX.13.041039. arXiv: 2111.03606 [gr-qc]
8. R. Abbott et al. Population of merging compact binaries inferred using gravitational waves through GWTC-3. Phys. Rev. X **13**(1), 011048 (2023). https://doi.org/10.1103/PhysRevX.13.011048. arXiv: 2111.03634 [astro-ph.HE]
9. R. Abbott et al. Tests of general relativity with GWTC-3. (2021). arXiv: 2112.06861 [gr-qc]
10. *Terrestrial Very-Long-Baseline Atom Interferometry: Workshop Summary*. (2023). arXiv: 2310.08183 [hep-ex]
11. G. Agazie et al., The NANOGrav 15 yr data set: detector characterization and noise budget. Astrophys. J. Lett. **951**(1), L10 (2023). https://doi.org/10.3847/2041-8213/acda88. arXiv: 2306.16218 [astro-ph.HE]
12. G. Agazie et al., The NANOGrav 15 yr data set: evidence for a gravitational-wave background. Astrophys. J. Lett. **951**(1), L8 (2023). https://doi.org/10.3847/2041-8213/acdac6. arXiv: 2306.16213 [astro-ph.HE]
13. P. Amaro-Seoane et al., Laser Interferometer Space Antenna (2017). arXiv: 1702.00786 [astro-ph.IM]
14. J. Antoniadis et al., The second data release from the European Pulsar Timing Array III. Search for gravitational wave signals. Astron. Astrophys. **678**, A50 (2023). https://doi.org/10.1051/0004-6361/202346844. arXiv: 2306.16214 [astro-ph.HE]
15. P. Auclair et al., Cosmology with the Laser Interferometer Space Antenna (2022). arXiv: 2204.05434 [astro-ph.CO]
16. L. Badurina et al., AION: an atom interferometer observatory and network. JCAP **05**, 011 (2020). https://doi.org/10.1088/1475-7516/2020/05/011. arXiv: 1911.11755 [astro-ph.CO]
17. S. Borhanian, B.S. Sathyaprakash, Listening to the Universe with Next Generation Ground-Based Gravitational-Wave Detectors (2022). arXiv: 2202.11048 [gr-qc]
18. M. Branchesi et al., Science with the Einstein telescope: a comparison of different designs. JCAP **07**, 068 (2023). https://doi.org/10.1088/1475-7516/2023/07/068. arXiv: 2303.15923 [gr-qc]
19. B. Canuel et al., Exploring gravity with the MIGA large scale atom interferometer. Sci. Rep. **8**, 2689 (2018). https://doi.org/10.1038/s41598-018-32165-z
20. LIGO Scientific Collaboration, GraceDB. https://gracedb.ligo.org
21. U. Dupletsa et al., gwfish: a simulation software to evaluate parameter-estimation capabilities of gravitational-wave detector networks. Astron. Comput. **42**, 100671 (2023). https://doi.org/10.1016/j.ascom.2022.100671. arXiv: 2205.02499 [gr-qc]
22. M. Evans et al., A Horizon Study for Cosmic Explorer: Science, Observatories, and Community (2021). arXiv: 2109.09882 [astro-ph.IM]
23. G. Franciolini et al., From inflation to black hole mergers and back again: Gravitational-wave data-driven constraints on inflationary scenarios with a first-principle model of primordial black holes across the QCD epoch. Phys. Rev. D **106**(12), 123526 (2022). https://doi.org/10.1103/PhysRevD.106.123526. arXiv: 2209.05959 [astro-ph.CO]
24. G. Franciolini et al., How to assess the primordial origin of single gravitational-wave events with mass, spin, eccentricity, and deformability measurements. Phys. Rev. D **105**(6), 063510 (2022). https://doi.org/10.1103/PhysRevD.105.063510. arXiv: 2112.10660 [astro-ph.CO]
25. G. Franciolini et al., Searching for primordial black holes with the Einstein Telescope: Impact of design and systematics. Phys. Rev. D **108**(4), 043506 (2023). https://doi.org/10.1103/PhysRevD.108.043506. arXiv: 2304.03160 [gr-qc]

26. I. Gupta et al., Characterizing gravitational wave detector networks: from A$^\sharp$ to cosmic explorer. Class. Quant. Grav. **41**(24), 245001 (2024). https://doi.org/10.1088/1361-6382/ad7b99. Document No. P2300019, CE Document No. P2300019-v2
27. J. Harms et al., Lunar gravitational-wave antenna. Astrophys. J. **910**(1), 1 (2021). https://doi.org/10.3847/1538-4357/abe5a7. arXiv: 2010.13726 [gr-qc]
28. J.S. Hazboun, J.D. Romano, T.L. Smith, Realistic sensitivity curves for pulsar timing arrays. Phys. Rev. D **100**(10), 104028 (2019). https://doi.org/10.1103/PhysRevD.100.104028. arXiv: 1907.04341 [gr-qc]
29. S. Hild et al., Sensitivity studies for third-generation gravitational wave observatories. Class. Quant. Grav. **28**, 094013 (2011). https://doi.org/10.1088/0264-9381/28/9/094013. arXiv: 1012.0908 [gr-qc]
30. S. Hild, S. Chelkowski, A. Freise. Pushing towards the ET sensitivity using 'conventional' technology (2008). arXiv: 0810.0604 [gr-qc]
31. F. Iacovelli et al., `GWFAST`: a Fisher information matrix Python code for third-generation gravitational-wave detectors. Astrophys. J. Suppl. **263**, 2 (2022). arXiv: 2207.06910 [astro-ph.IM]
32. F. Iacovelli et al., Forecasting the detection capabilities of third-generation gravitational-wave detectors using GWFAST. Astrophys. J. **941**(2), 208 (2022). https://doi.org/10.3847/1538-4357/ac9cd4. arXiv: 2207.02771 [gr-qc]
33. V. Kalogera et al., The Next Generation Global Gravitational Wave Observatory: The Science Book (2021). arXiv: 2111.06990 [gr-qc]
34. S. Kawamura et al., Current status of space gravitational wave antenna DECIGO and B-DECIGO. PTEP **2021**(5), 05A105 (2021). https://doi.org/10.1093/ptep/ptab019. arXiv: 2006.13545 [gr-qc]
35. M. Maggiore et al., Science case for the Einstein telescope. JCAP **03**, 050 D(2020). https://doi.org/10.1088/1475-7516/2020/03/050. arXiv: 1912.02622 [astro-ph.CO]
36. M. Mancarella, F. Iacovelli, D. Gerosa, Inferring, not just detecting: Metrics for high-redshift sources observed with third-generation gravitational-wave detectors. Phys. Rev. D **107**(10), L101302 (2023). https://doi.org/10.1103/PhysRevD.107.L101302. arXiv: 2303.16323 [gr-qc]
37. T. Nakamura et al., Pre-DECIGO can get the smoking gun to decide the astrophysical or cosmological origin of GW150914-like binary black holes. PTEP **2016**(9), 093E01 (2016). https://doi.org/10.1093/ptep/ptw127. arXiv: 1607.00897 [astro-ph.HE]
38. K.K.Y. Ng et al., Measuring properties of primordial black hole mergers at cosmological distances: Effect of higher order modes in gravitational waves. Phys. Rev. D 107.2 (2023), p. 024041. https://doi.org/10.1103/PhysRevD.107.024041. arXiv: 2210.03132 [astro-ph.CO]
39. K.K.Y. Ng et al., On the single-event-based identification of primordial black hole mergers at cosmological distances. Astrophys. J. Lett. **931**(1), L12 (2022). https://doi.org/10.3847/2041-8213/ac6bea. arXiv: 2108.07276 [astro-ph.CO]
40. K.K.Y. Ng et al., Probing multiple populations of compact binaries with third-generation gravitational-wave detectors. Astrophys. J. Lett. **913**(1), L5 (2021). https://doi.org/10.3847/2041-8213/abf8be. arXiv: 2012.09876 [astro-ph.CO]
41. M. Punturo et al.. The Einstein telescope: a third-generation gravitational wave observatory. Class. Quant. Grav. **27**, 194002 (2010). Ed. by Fulvio Ricci. https://doi.org/10.1088/0264-9381/27/19/194002
42. M. Raidal et al., Formation and evolution of primordial black hole binaries in the early universe. JCAP **02**, 018 (2019). https://doi.org/10.1088/1475-7516/2019/02/018. arXiv: 1812.01930 [astro-ph.CO]
43. D.J. Reardon et al., Search for an isotropic gravitational-wave background with the parkes pulsar timing array. Astrophys. J. Lett. **951**(1), L6 (2023). https://doi.org/10.3847/2041-8213/acdd02. arXiv: 2306.16215 [astro-ph.HE]
44. D. Reitze et al., Cosmic explorer: The U.S. contribution to gravitational-wave astronomy beyond LIGO. Bull. Am. Astron. Soc. **51**(7), 035 (2019). arXiv: 1907.04833 [astro-ph.IM]
45. T. Robson, N.J. Cornish, C. Liu, The construction and use of LISA sensitivity curves. Class. Quant. Grav. **36**(10), 105011 (2019). https://doi.org/10.1088/1361-6382/ab1101. arXiv: 1803.01944 [astro-ph.HE]

46. Samuele Ronchini et al., Perspectives for multimessenger astronomy with the next generation of gravitational-wave detectors and high-energy satellites. Astron. Astrophys. **665**, A97 (2022). https://doi.org/10.1051/0004-6361/202243705. (arXiv: 2204.01746 [astro-ph.HE])
47. D. Schlippert et al., Matter-wave interferometry for inertial sensing and tests of fundamental physics, in *Proceedings of the Eighth Meeting on CPT and Lorentz Symmetry* (World Scientific, 2020), pp. 37–40
48. E. Thrane, J.D. Romano, Sensitivity curves for searches for gravitational-wave backgrounds. Phys. Rev. D **88**(12), 124032 (2013). https://doi.org/10.1103/PhysRevD.88.124032. arXiv: 1310.5300 [astro-ph.IM]
49. H. Xu et al., Searching for the Nano-Hertz stochastic gravitational wave background with the Chinese pulsar timing array data release I. Res. Astron. Astrophys. **23**(7), 075024 (2023). https://doi.org/10.1088/1674-4527/acdfa5. arXiv: 2306.16216 [astro-ph.HE]

Chapter 16
Formation of Primordial Black Hole Binaries and Their Merger Rates

Martti Raidal, Ville Vaskonen, and Hardi Veermäe

Abstract We review the theory behind the formation of primordial black hole binaries and their merger rates. We consider the binary formation in the early and late Universe, emphasising the former as it gives the dominant contribution of the present primordial black hole merger rate. The binaries formed in the early Universe are highly eccentric so their interactions with other primordial black holes can significantly increase their coalescence times and thereby suppress the merger rate. We discuss in detail how the suppression of the merger rate arising from such interactions can be estimated and how such interactions lead to the formation of another, much harder, binary population that contributes to the present merger rate if more than 10% of dark matter consists of primordial black holes with a relatively narrow mass distribution. When the primordial abundance is below 1%, encounters between primordial black holes are unlikely and their effect on the merger rate becomes negligible.

16.1 Introduction

Gravitational wave (GW) signals created by primordial black hole (PBH) binaries provide a direct probe of hypothetical PBH populations. The PBH populations are expected to differ from astrophysical black hole (ABH) populations in aspects such

M. Raidal · V. Vaskonen · H. Veermäe (✉)
Keemilise ja Bioloogilise Füüsika Instituut, Rävala pst. 10, 10143 Tallinn, Estonia
e-mail: hardi.veermae@cern.ch

M. Raidal
e-mail: martti.raidal@cern.ch

V. Vaskonen
e-mail: ville.vaskonen@pd.infn.it

V. Vaskonen
Dipartimento di Fisica e Astronomia, Università degli Studi di Padova, Via Marzolo 8, 35131 Padova, Italy

Istituto Nazionale di Fisica Nucleare, Sezione di Padova, Via Marzolo 8, 35131 Padova, Italy

C. Byrnes et al. (eds.), *Primordial Black Holes*, Springer Series in Astrophysics and Cosmology, https://doi.org/10.1007/978-981-97-8887-3_16

as mass distribution, redshift dependence of the merger rate, spins or even spatial distribution. A good theoretical understanding of both populations is crucial for employing the full potential of GW observations. In the following, we will focus on describing the PBH binary populations and the resulting PBH merger rates.

The characteristics of the present-day PBH populations are moderately model-dependent. They can depend on the numerous ways how the PBHs could have been produced in the early Universe and potentially also on deviations from the standard cosmology, which can affect the evolution of the PBH populations after their formation. To our luck, it is not necessary to specify the details of PBH formation to estimate the properties of the present-day PBH binary populations. Only the PBH abundance, the distribution of their masses and spins as well as their spatial distribution before matter-radiation equality should be specified. This sets the initial stage of PBH structure formation as nearby PBH pairs decouple from the expansion and form the first bound structures in the Universe.

In certain scenarios, a population of merging PBH binaries can be probed by GW experiments. With enough data, these mergers can help us reconstruct the initial characteristics of PBHs in the early Universe and begin probing the early Universe by reconstructing scenarios that would produce such initial conditions. The goal of this chapter is to outline the first part of this exercise and show how an initial population of PBHs coupled to the Hubble flow in the very early Universe will give rise to a population of PBH binaries merging in the present day.

16.1.1 Binary Formation Channels

There are various channels for forming PBH binaries. These can be classified by (1) whether the binaries were formed in the early or late Universe, and (2) whether the formation included dynamics of two or more bodies. The PBH binary characteristics of each combination can be outlined as follows:

1. In the **early two-body channel**, binaries are formed from nearby PBH pairs that decouple the Hubble flow. Such binaries generally dominate the PBH binary merger rate when $f_{\rm PBH}$—the fraction of DM in PBHs—is small. However, they are very eccentric since most of their angular momentum comes from tiny tidal forces induced by fluctuations in the density of surrounding matter. This makes the merger rate susceptible to perturbations by encounters with other PBHs, so their subsequent evolution should be accounted for when estimating their present merger rate. In particular, when $f_{\rm PBH} \gtrsim 0.1$, small-scale structure formation is enhanced. This leads to greater disruption of these binaries and suppresses the corresponding merger rate.

2. In the **early three-body channel**, bound three-body systems are formed after initial compact 3-PBH configurations decouple from the Hubble flow. In such a system, a binary remains if one of the PBHs is ejected or if two of the PBHs merge. The resulting binaries are much less eccentric and, therefore, much harder[1] than the binaries from the early two-body channel. As a result, their merger rate is not significantly affected by encounters with other PBHs. However, since compact 3-PBH systems are less likely to occur in the early Universe, this channel is generally subdominant to the early two-body channel for small $f_{\rm PBH}$. However, it can dominate if $f_{\rm PBH} \approx 1$ or if the PBHs are formed in clusters.
3. In the **late two-body channel**, PBH binaries are formed from hyperbolic encounters of PBHs in DM haloes if they lose a sufficient amount of energy via GW emission. The resulting binaries are typically eccentric and merge relatively fast. Although the resulting merger rate depends on the PBH structure formation and may be boosted by spiky DM profiles or PBH clustering, such binaries tend to have a smaller merger rate than PBH binaries formed in the early Universe. Although this channel could also produce ABH-PBH binaries, their contribution is expected to be marginal [22].
4. The **late three-body channel** can be active in the environment with large PBH densities. The binaries are formed in 2 + 1 PBH encounters, where the 3rd PBH carries away enough energy for the remaining 2 to become bound. Analogously to the early three-body channel, these binaries are hard and not too eccentric. Although their formation implies an environment with relatively frequent PBH-PBH binary encounters, such encounters will not significantly affect their merger rate. This channel is generally weaker than the ones listed above. It can be enhanced if $f_{\rm PBH} \approx 1$ or if the formation of PBH structures is enhanced, i.e., by initial clustering.

In the following, we will consider each channel in detail.

16.1.2 Preliminaries

We will use geometric units $c = G = 1$ throughout this chapter. We consider a generic PBH mass function, that we define as

$$\psi(m) \equiv \frac{m}{\rho_{\rm PBH}} \frac{{\rm d}n_{\rm PBH}}{{\rm d}\ln m}\,, \tag{16.1}$$

where $\rho_{\rm PBH}$ and $n_{\rm PBH}$ denote the total energy and comoving number densities of PBHs. With this definition we get $\int \psi(m){\rm d}\ln m = 1$ and the average of a quantity X is given by

[1] A binary is considered hard when its binding energy is larger than the average kinetic energy of surrounding PBHs.

$$\langle X \rangle \equiv \int \frac{\mathrm{d}n_{\mathrm{PBH}}}{n_{\mathrm{PBH}}} X = \int \mathrm{d}\ln m \, \frac{\langle m \rangle}{m} \, \psi(m) X \,, \tag{16.2}$$

where $\langle m \rangle = \rho_{\mathrm{PBH}}/n_{\mathrm{PBH}} = 1/\int \mathrm{d}\ln m \, m^{-1} \, \psi(m)$.

We denote the masses of individual PBHs by a lowercase m and the PBHs making up the binary by m_1 and m_2. The total mass and the reduced mass of the binary are $M \equiv m_1 + m_2$ and $\mu \equiv m_1 m_2 / M$ and we often use the symmetric mass ratio $\eta \equiv \mu/M$ to quantify unequal mass binaries.

We work in the Newtonian limit. Thus, the binding energy of the binary is

$$E_{\mathrm{bin}} = \frac{\mu \dot{\mathbf{r}}^2}{2} - \frac{M\mu}{r} = \frac{M\mu}{2r_a}, \tag{16.3}$$

where r_a denotes the semimajor axis of the binary, $\mathbf{r} = \mathbf{r}_1 - \mathbf{r}_2$ is the separation of the PBHs positioned at $\mathbf{r}_1$ and $\mathbf{r}_2$, and the dot denotes time derivation. The binary's angular momentum $\mathbf{L} = \mu \mathbf{r} \times \dot{\mathbf{r}}$ will be quantified via the dimensionless quantity

$$\mathbf{j} \equiv \frac{\mathbf{L}/\mu}{\sqrt{r_a M}}, \tag{16.4}$$

which satisfies $j \equiv |\mathbf{j}| \leq 1$ and is related to the eccentricity as

$$e = \sqrt{1 - j^2}. \tag{16.5}$$

Thus, for our purposes, it is sufficient to characterize PBH binaries by four quantities, e.g., m_1, m_2, r_a and j.

To estimate the merger rate, we need to consider the formation of PBH binaries and their disruption between formation and merger. Given the PBH binary formation rate density $\mathrm{d}R_{\mathrm{b}}(t')/(\mathrm{d}m_1 \mathrm{d}m_2 \mathrm{d}j \mathrm{d}r_a)$ at time t', the merger rate density at time t is

$$\frac{\mathrm{d}R(t)}{\mathrm{d}m_1 \mathrm{d}m_2} = \int \mathrm{d}t' \mathrm{d}j \mathrm{d}r_a \, \frac{\mathrm{d}R_{\mathrm{b}}(t')}{\mathrm{d}m_1 \mathrm{d}m_2 \mathrm{d}j \mathrm{d}r_a} \delta\left(t - t' - \tau(m_1, m_2, r_a, j)\right), \tag{16.6}$$

where τ denotes the coalescence time. The coalescence time can be estimated assuming the binary will evolve only via GW emission after its formation. In the Newtonian regime, the coalescence time of eccentric binaries is [27]

$$\tau = \frac{3}{85} \frac{r_a^4 j^7}{\eta M^3}. \tag{16.7}$$

This expression would overestimate the coalescence time mostly by a factor of 1.81 when applied to almost circular binaries.[2] If the binary is disrupted[3] or evolves between formation and merger due to effects other than GW emission, e.g., accretion, then (16.6) must be modified accordingly.

16.2 PBH Binary Formation in the Early Universe

In the early Universe, PBH binaries can form when nearby PBHs decouple from the Hubble flow [26]. The formation process and the resulting merger rate depend on whether a third PBH falls into the pair. In this section, we will first lay out the derivation for the two-body formation channel and look at phenomena that may perturb these binaries and suppress this rate. Finally, we will consider the merger rate stemming from the three-body channel.

16.2.1 Dynamics of Early Binary Formation

To set up the initial configuration for the early Universe two-body channel, let us consider a pair of PBHs with masses m_1 and m_2 at a comoving separation x_0 which initially follow the Hubble flow. We assume that there are no PBHs within a comoving radius y of the pair, to guarantee that the pair evolves as a two-body system after decoupling. This initial setup is shown in Fig. 16.1. The size of the empty region will be determined later by requiring that the PBHs closest to the pair will not become gravitationally bound to the pair after decoupling from the Hubble flow and, therefore, form a bound system consisting of 3 or more bodies. Such systems will be addressed later in Sect. 16.2.3. We further assume that the positions of PBHs are uncorrelated, that is, they follow a Poisson distribution. In this case, the comoving number density of PBH configurations described above is

$$\mathrm{d}n_{\mathrm{pairs}} = \frac{1}{2} e^{-\bar{N}(y)} \mathrm{d}n(m_1) \mathrm{d}n(m_2) \mathrm{d}V(x_0), \tag{16.9}$$

---

[2] For any j, the time evolution of j and r_a of purely GW driven binaries [27] can be integrated to give,

$$\tau = \frac{3r_a^4 j^7}{85\eta M^3} \frac{F(j) - F(1)j}{\left(1 - \frac{121}{425} j^2\right)^{3480/2299} \left(1 - j^2\right)^{24/19}} \approx \frac{3r_a^4 j^7}{85\eta M^3} \frac{1 - 0.70j + 0.62j^2}{1 + 0.67j}, \tag{16.8}$$

where $F(j) \equiv F_1\left(-1/2; -1181/2299, -5/19; 1/2; j^2 121/425, j^2\right)$ with F_1 the first Appell hypergeometric function. The second approximation deviated from the exact one by less than 0.5%. The limiting case of circular binaries, $j \to 1$, gives $\tau = (5/256) r_a^4/\eta M^3$.

[3] Disruption in this context means any event that leads to a change in the orbital parameters that significantly affects the coalescence time, ionization or swapping out a constituent of the binary.

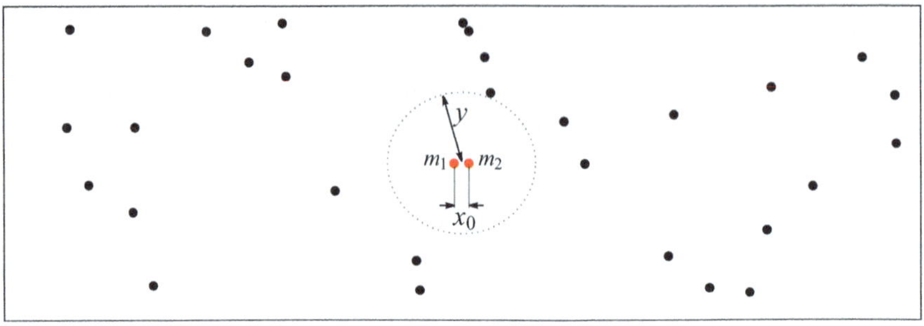

Fig. 16.1 A initial PBH configuration and its parametrization

where $\mathrm{d}n(m)$ is the comoving number density of PBHs in the mass range $(m, m + \mathrm{d}m)$, and

$$\bar{N}(y) \equiv nV(y) \tag{16.10}$$

is the expected number of PBHs in a spherical volume $V(y) \equiv (4\pi/3)y^3$ with a comoving radius y. The factor $1/2$ avoids overcounting.

The population of PBH binaries created from such initial conditions is described by the formation rate

$$\frac{\mathrm{d}R_{\mathrm{b}}(t)}{\mathrm{d}m_1 \mathrm{d}m_2 \mathrm{d}j \mathrm{d}r_a} = \int \mathrm{d}x_0 \frac{\mathrm{d}n_{\mathrm{p}}}{\mathrm{d}m_1 \mathrm{d}m_2 \mathrm{d}x_0} \frac{\mathrm{d}P(j, r_a|x_0, y)}{\mathrm{d}j \mathrm{d}r_a} \delta(t - t_{\mathrm{dc}}(x_0)), \tag{16.11}$$

where $t_{\mathrm{dc}}(x_0)$ denotes the time of decoupling of the pair from the Hubble flow and $\mathrm{d}P(j, r_a|x_0, y)/\mathrm{d}j\mathrm{d}r_a$ is the distribution orbital parameters given x_0 and y.[4] A few simplifications can be made. Since the early binaries are formed before the matter-radiation equality, we can omit the dependence on formation time and assume that $t_{\mathrm{dc}}(x_0) \approx t_0$, where t_0 is the age of the Universe.[5] Additionally, $P(j, r_a|x_0, y) \propto \delta(r_a - r_a(x_0))\, \mathrm{d}P(j|x_0, y)/\mathrm{d}j$ since r_a can be predicted from x_0 alone. Given the coalescence time (16.7), we find that the merger rate density (16.6) can be recast as

$$\begin{aligned} \mathrm{d}R_{\mathrm{E2}}^{(y)}(t) &= \int \mathrm{d}n_{\mathrm{pairs}} \mathrm{d}j \frac{\mathrm{d}P(j|x_0, y)}{\mathrm{d}j} \delta\left(t - \tau(r_a(x_0), j)\right) \\ &= \frac{1}{14\tau} \mathrm{d}n(m_1)\mathrm{d}n(m_2) \int \mathrm{d}V(x_0) e^{-\bar{N}(y)} j \frac{\mathrm{d}P(j|x_0, y)}{\mathrm{d}j}\bigg|_{j=j(\tau, r_a(x_0))}, \end{aligned} \tag{16.12}$$

where, in the second line, $j(\tau, r_a(x_0))$ is obtained by inverting Eq. (16.7).

This estimate must, however, be modified to account for the disruption of the early binaries via collisions of the binary with other PBHs in DM clusters between the binary formation and merger. We will describe this effect by the survival probability

[4] We don't show the dependence on the progenitor masses.

[5] We use $t_0 = 13.8$Gyr in numerical estimates [5].

S_L, which we take to be independent of the initial conditions. Furthermore, the exclusion of initial conditions in which the binary gets disrupted by nearby PBHs shortly after formation can be characterised by another survival probability S_E. So, the merger rate density Eq. (16.12) takes the final form

$$\boxed{dR_{E2}(t) = S_L S_E \times dR_{E2}^{(0)}(t)}, \tag{16.13}$$

where $S_E \equiv dR_{E2}^{(y)}/dR_{E2}^{(0)} \leq 1$.[6] Note that y is generally not an independent variable and can depend on various model parameters.

All in all, to proceed, we need to compute (*i*) how the semimajor axis $r_a(x_0)$ is formed during the decoupling from the Hubble flow (see Sect. 16.2.1.1), (*ii*) how spatial inhomogeneities set the distribution of initial angular momenta of the binary (see Sect. 16.2.1.2), (*iii*) what the minimal distance y is between the pair and other PBHs that would avoid disrupting the binary due to the infall of other close PBHs into the binary and how it affects S_E (see Sect. 16.2.1.3), (*iv*) estimate disruption in late DM haloes by computing the probability S_L that the binaries do not collide with other PBHs during before the merger (see Sect. 16.2.2).

16.2.1.1 Decoupling

The line element for spacetimes with small inhomogeneities in the comoving Newtonian gauge and the absence of anisotropic stress is

$$ds^2 = -(1 + 2\phi(\mathbf{x}))dt^2 + (1 - 2\phi(\mathbf{x}))a(t)^2 dx^2, \tag{16.14}$$

where the potential ϕ is determined by the total matter energy density $\rho_M(\mathbf{x})$ as $\Delta\phi = 4\pi a^2 \rho_M(\mathbf{x})$. Treating the PBHs as point masses, we obtain

$$\phi(\mathbf{x}) = -\sum_i \frac{m_i}{a|\mathbf{x}_i - \mathbf{x}|} - 4\pi\bar{\rho}_M \int \frac{d^3k}{(2\pi)^3} \frac{a^2}{k^2} e^{-i\mathbf{k}\cdot\mathbf{x}} \delta_M(\mathbf{k}), \tag{16.15}$$

where m_i denote the masses and $\mathbf{x}_i$ the comoving positions of the PBHs, $\bar{\rho}_M$ the average comoving matter energy density and $\delta_M(\mathbf{k})$ the matter density perturbation. A system of N non-relativistic PBHs, $a dx/dt \ll 1$, is described by the action

$$S^{(N)} = \int dt \left[\sum_i m_i \left(\frac{1}{2}\dot{\mathbf{r}}_i^2 + \frac{1}{2}\frac{\ddot{a}}{a}\mathbf{r}_i^2 - \phi_{ex}(\mathbf{x}) \right) + \sum_{i>j} \frac{m_i m_j}{|\mathbf{r}_i - \mathbf{r}_j|} \right], \tag{16.16}$$

[6] Later, the suppression factor will also account for particle DM fluctuations, which are neglected in $R_{E2}^{(0)}$.

where $\mathbf{r}_i \equiv a\mathbf{x}_i$ denotes the proper distance, the last term accounts for the pairwise interaction of the PBHs and the potential $\phi_{\rm ex}(\mathbf{x})$ describes the effect of the surrounding matter on the N-body system. For a PBH pair, the action can be approximated as

$$S^{(2)} \approx \int \mathrm{d}t \left[\frac{M}{2}\left(\dot{\mathbf{r}}_c^2 + \frac{\ddot{a}}{a}\mathbf{r}_c^2\right) - M\phi_{\rm ex}(\mathbf{r}_c) + \frac{\mu}{2}\left(\dot{\mathbf{r}}^2 + \frac{\ddot{a}}{a}\mathbf{r}^2 + \frac{2M}{r} - \mathbf{r}\cdot\mathbf{T}\cdot\mathbf{r}\right)\right], \tag{16.17}$$

where $\mathbf{r}$ denotes the proper separation of the PBH, $\mathbf{r}_c$ the centre of mass of the PBH pair, and $\mathbf{T}_{ij} \equiv \partial_i\partial_j\phi_{\rm ex}(\mathbf{r}_c)$ is the tidal tensor due to the external forces. In summary, the PBH pair is subject to the following forces

$$\mathbf{F}/\mu = \underbrace{\mathbf{r}\ddot{a}/a}_{\text{Hubble flow}} - \underbrace{M\hat{\mathbf{r}}/r^2}_{\text{self-gravity}} + \underbrace{(\hat{\mathbf{r}}\cdot\mathbf{T}\cdot\mathbf{r})\hat{\mathbf{r}}}_{\text{radial tidal forces}} + \underbrace{(\mathbf{r}\times(\mathbf{T}\cdot\mathbf{r}))}_{\text{tidal torque}}\times(\hat{\mathbf{r}}/r), \tag{16.18}$$

where $\hat{\mathbf{r}} \equiv \mathbf{r}/r$. The first three forces are radial, whereas the last term provides the torque that prevents the head-on collision of the PBH pair.

The angular momentum $\mathbf{L}$ of the pair is generated by the tidal torque,

$$\mathbf{L} = \int \mathrm{d}t\, \mathbf{r}\times\mathbf{F} = \mu\int \mathrm{d}t\, \mathbf{r}\times(\mathbf{T}\cdot\mathbf{r}). \tag{16.19}$$

During radiation domination, the PBHs surrounding the pair are assumed to follow the Hubble flow and the matter density perturbations are constant. This implies that the potential (16.15) scales roughly as $\phi \propto a^{-1}$. Consequently, $\mathbf{T} \propto a^{-3}$ and the tidal forces get damped fast after the pair decouples from the Hubble flow. The pair has then formed a binary with angular momentum $\mathbf{L}$ and semi-major axis r_a.

The binary decouples when the Hubble flow becomes subdominant and the second term in (16.18) takes over. To estimate when this happens, consider the overdensity in matter due to the PBH pair of comoving size x_0,

$$\delta_{\rm pair} \equiv \frac{M/2}{\rho_{\rm M} V(x_0)}, \tag{16.20}$$

which should exceed the average matter density $\delta_{\rm pair} \gg 1$ if the PBH pair is to form a binary. Such a region begins to collapse when its density becomes comparable to the density of radiation, $\rho_{\rm R}a^{-4} \approx \delta_{\rm pair}\rho_{\rm M}a^{-3}$, that is when

$$a \approx a_{\rm dc} \equiv a_{\rm eq}/\delta_{\rm pair}, \tag{16.21}$$

where $a_{\rm eq}$ denotes the scale factor during matter-radiation equality. Since $\delta_{\rm pair} \gg 1$, we expect the binary to be formed during radiation domination. To make this more precise, consider first the idealized case of a head-on collision for which, the radial equation of motion of the pair, $\ddot{r} - r\ddot{a}/a + Mr^{-2} = 0$, can be recast in terms of the comoving separation $x \equiv r/a$ as

$$x'' + \frac{a}{a_{\rm dc}} \frac{x_0^3}{x^2} = 0\,, \tag{16.22}$$

where the prime denotes derivation with respect to $\ln(a)$ and the initial conditions are given by $x(a_0) = x_0$, $x'(a_0) = 0$ with $a_0 \ll a_{\rm dc}$. The solution of the equation of motion can be expressed as $x(a) \approx x_0 \chi(a/a_{\rm dc})$, where $\chi(y)$ solves $(y\partial_y)^2\chi + y\chi^{-2} = 0$ with $\chi(y \to 0) = 1$, $\chi'(y \to 0) = 0$. The function $\chi(y)$ oscillates with an amplitude that decreases asymptotically as $0.2/y$, so the semi-major axis of the binary is

$$r_a = a x_a/2 \approx 0.1 a_{\rm dc} x_0 \approx 0.84 \rho_{\rm R} x_0^4/M\,. \tag{16.23}$$

The first root of $\chi(y)$ lies at $y \approx 0.54$ corresponding to the first close encounter of the pair at $a \approx 0.54 a_{\rm dc}$ and indicating that the binary decouples from the Hubble flow before that. In Fig. 16.2 we indicate by the dashed vertical line the moment when the self-gravity of the pair starts to dominate over the Hubble flow.

The idealised case discussed above is in excellent agreement with the radial evolution in the presence of a weak tidal field illustrated in Fig. 16.2 as well as with N-body simulations in which the tidal effects are created by dynamical PBHs surrounding the pair [30] suggesting that the angular momentum can be studied perturbatively as long as the orbit remains eccentric ($j \ll 1$) [6]. Plugging the solution of (16.22) into (16.19), we find that the dimensionless angular momentum (16.4) is

$$\mathbf{j} = \frac{1}{\sqrt{r_a M}} \int \frac{\mathrm{d}a}{a^2 H} \mathbf{x} \times (\mathbf{T} a^3 \cdot \mathbf{x}) = \frac{0.95\, x_0^3}{M} \hat{\mathbf{r}} \times (\mathbf{T} a^3 \cdot \hat{\mathbf{r}}), \tag{16.24}$$

where we used $\int_0^\infty \mathrm{d}y\, \chi(y)^2 \approx 0.3$. Note that under the above assumptions $\mathbf{T} a^3$ is constant.

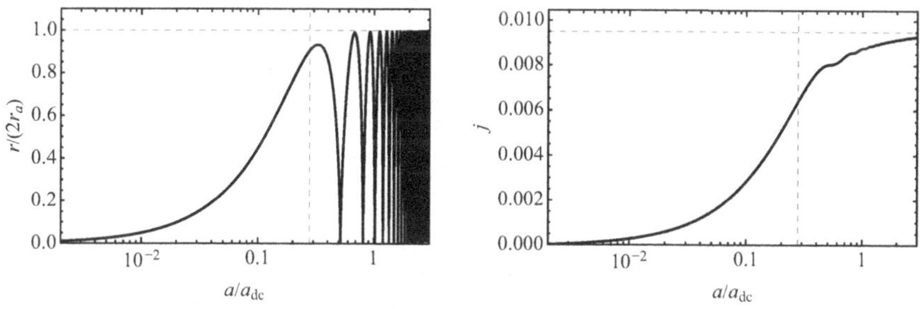

Fig. 16.2 Decoupling of the PBH pair. The left and right panels show, respectively, the evolution of the pair's separation and angular momentum in a weak tidal field. The vertical dashed lines show the moment when the self-gravity of the pair starts to dominate over the Hubble flow (see Eq. (16.18)). The horizontal dashed line in the right panel shows the approximation in Eq. (16.24)

16.2.1.2 Distribution of Angular Momenta

The tidal torque is determined by the distribution of surrounding PBHs as well as the inhomogeneities of the rest of matter. The resulting dimensionless angular momentum can thus be expressed as

$$\mathbf{j} = \mathbf{j}_{\rm PBH} + \mathbf{j}_{\rm M}, \qquad \mathbf{j}_{\rm PBH} \equiv \sum_i \mathbf{j}_{\rm l}(\mathbf{x}_i, m_i), \tag{16.25}$$

where $\mathbf{j}_{\rm l}$ is the angular momentum generated by a single PBH with mass m_i at a comoving distance x_i from the pair, and $\mathbf{j}_{\rm M}$ is the angular momentum generated by matter fluctuations. Using Eq. (16.24) with $\mathbf{T}_i = m_i(\mathbf{1} - 3\hat{\mathbf{x}}_i \otimes \hat{\mathbf{x}}_i)/x_i^3 a^{-3}$ and defining

$$j_0 \equiv 0.95\, \bar{N}(x_0)\frac{\langle m\rangle}{M} \approx 0.4\frac{f_{\rm PBH}}{\delta_{\rm pair}}, \tag{16.26}$$

where $\langle m\rangle = \rho_{\rm PBH}/n$ is the average PBH mass and we have used $\Omega_{\rm M}/\Omega_{\rm DM} = 1.2$, we get

$$\mathbf{j}_{\rm l}(\mathbf{x}_i, m_i) = j_0 \frac{m_i}{\langle m\rangle}\frac{3}{\bar{N}(x_i)}\hat{\mathbf{x}}_i \times \hat{\mathbf{r}}\,(\hat{\mathbf{x}}_i \cdot \hat{\mathbf{r}}). \tag{16.27}$$

The quantity j_0 is a reference scale for the dimensionless angular momentum. It captures its order of magnitude because torque is dominantly generated by the closest PBHs, for which $\bar{N}(x_i) \approx 1$, and the remaining terms in Eq. (16.27) will contribute an $\mathcal{O}(1)$ on average. Binaries form from PBH pairs which satisfy $\delta_{\rm pair} \gg 1$, thus we naturally expect that $j_0 \ll 1$.

To compute the distribution of angular momenta, we will consider the most common scenario in which the masses m_i and positions $\mathbf{x}_i$ as well as the matter density perturbations are statistically independent. In this case, it is convenient to use the additive property of the cumulant-generating function of $\mathbf{j}$

$$K(\mathbf{k}) \equiv \ln\left\langle e^{i\mathbf{k}\cdot\mathbf{j}}\right\rangle = K_{\rm M}(\mathbf{k}) + K_{\rm PBH}(\mathbf{k}), \tag{16.28}$$

where $K_{\rm M}(\mathbf{k}) \equiv \ln\left\langle e^{i\mathbf{k}\cdot\mathbf{j}_{\rm M}}\right\rangle$ and $K_{\rm PBH}(\mathbf{k}) \equiv \ln\left\langle e^{i\mathbf{k}\cdot\mathbf{j}_{\rm PBH}}\right\rangle$ are cumulant generating functions of $\mathbf{j}_{\rm M}$, $\mathbf{j}_{\rm PBH}$, respectively. The average is taken over all random quantities contained in the expression. The probability distribution of $\mathbf{j}$ can then be computed as

$$\frac{{\rm d}P}{{\rm d}^3 j} \equiv \langle\delta(\mathbf{j} - \mathbf{j}_{\rm M} - \mathbf{j}_{\rm PBH})\rangle = \int \frac{{\rm d}^3 k}{(2\pi)^3} e^{-i\mathbf{k}\cdot\mathbf{j} + K(\mathbf{k})}. \tag{16.29}$$

Note that the above distribution is two-dimensional since the angular momentum is perpendicular to $\hat{\mathbf{r}}$: the cumulant generating function is a function of $\mathbf{k}_\perp \equiv \mathbf{k} \times \hat{\mathbf{r}}$. Statistical isotropy of the surrounding matter further constrains it to be a function of $k_\perp$ only.

Let us first consider the matter density fluctuations. The generating function $K_{\rm M}(\mathbf{k})$ is determined by the first two cumulants if the matter density fluctuations are Gaussian, that is

$$K_{\rm M}(\mathbf{k}) = -\frac{1}{2}\left\langle (\mathbf{k}\cdot\mathbf{j})^2\right\rangle = -\frac{1}{4}\sigma_{j,\rm M}^2 k_\perp^2, \tag{16.30}$$

since the mean $\langle \mathbf{j}\rangle = 0$ must vanish due to isotropy arguments. The variance $\sigma_{j,\rm M}$ can be estimated by noting that matter density fluctuations induce a tidal field which in Fourier space reads $\mathbf{T}_{\rm M} = a^{-3}\hat{\mathbf{q}} \otimes \hat{\mathbf{q}}\, 4\pi \bar{\rho}_{\rm M}\delta_{\rm M}(q)$ by Eq. (16.15). Using Eq. (16.24) and averaging over orientations then gives [6, 30]

$$\sigma_{j,\rm M}^2 = \left\langle \mathbf{j}_{\rm M}^2\right\rangle = \left(\frac{0.95 x_0^3}{M}\right)^2 \frac{a^6}{5}\left\langle {\rm tr}(\mathbf{T}_{\rm M}\cdot\mathbf{T}_{\rm M}) - \frac{1}{3}{\rm tr}(\mathbf{T}_{\rm M})^2\right\rangle = \frac{6}{5} j_0^2 \frac{\sigma_{\rm M}^2}{f_{\rm PBH}^2}, \tag{16.31}$$

where $\sigma_{\rm M}^2 \equiv (\Omega_{\rm M}/\Omega_{\rm DM})^2 \left\langle \delta_{\rm M}^2\right\rangle$ is the rescaled variance of matter density perturbations at the time the binary is formed.[7]

The cumulant generating function $K_{\rm PBH}$ accounts for identical and statistically independent contributions from all surrounding PBHs. To send the number of surrounding PBHs to infinity, we will first consider a finite volume and then take the limit $V \to \infty$ while keeping $N/V = n$ fixed:[8]

[7] We will use $\sqrt{\left\langle \delta_{\rm M}^2\right\rangle} = 0.005$ in numerical estimates.

[8] After plugging in Eq. (16.27), the integral for $K_{\rm PBH}$ takes the form

$$K_{\rm PBH}(\mathbf{k}) = \int {\rm d}n(m) \int_{|\mathbf{x}|>y} \frac{{\rm d}\Omega}{4\pi}{\rm d}V(x)\left[\exp\left(\frac{i3mj_0}{\rho_{\rm PBH}V(x)}(\mathbf{k}_\perp\cdot\hat{\mathbf{x}})(\hat{\mathbf{x}}\cdot\hat{\mathbf{r}})\right) - 1\right]. \tag{16.32}$$

The spatial integral can be evaluated by making the change of variables $u \equiv zV(y)/V(x)$, with $z \equiv k_\perp j_0 m/(\rho_{\rm PBH}V(y))$, and taking the following steps:

$$\begin{aligned}
&-z\int \frac{{\rm d}^2\Omega}{4\pi}\int_0^z \frac{{\rm d}u}{u^2}\left[\exp\left(i3u(\hat{\mathbf{k}}_\perp\cdot\hat{\mathbf{x}})(\hat{\mathbf{x}}\cdot\hat{\mathbf{r}})\right) - 1\right] \\
&= -z\int_0^z \frac{{\rm d}u}{u^2}\int_0^\pi\int_0^{2\pi} \frac{{\rm d}\cos(\theta)\,{\rm d}\phi}{4\pi}\left[\exp\left(i\frac{3u}{2}\sin^2(\theta)\sin(2\phi)\right) - 1\right] \\
&= -z\int_0^z \frac{{\rm d}u}{u^2}\left[\frac{\pi}{2\sqrt{2}}J_{-\frac{1}{4}}\left(\frac{3u}{4}\right)J_{\frac{1}{4}}\left(\frac{3u}{4}\right) - 1\right] = {}_1F_2\left(-\frac{1}{2};\frac{3}{4},\frac{5}{4};-\frac{9z^2}{16}\right) - 1.
\end{aligned}$$

$$
\begin{aligned}
K_{\text{PBH}}(\mathbf{k}) &= \lim_{\substack{V\to\infty \\ N/V=n}} N \ln \left(\int\limits_{|\mathbf{x}|>y} \frac{\mathrm{d}^3 x}{V} \frac{\mathrm{d}n(m)}{n} e^{i\mathbf{k}\cdot\mathbf{j}_1(\mathbf{x},m)} \right) \\
&= \int \mathrm{d}^3 x\, \mathrm{d}n(m) \left(e^{i\mathbf{k}\cdot\mathbf{j}_1(\mathbf{x},m)} - 1 \right) \\
&= -\bar{N}(y) \int \frac{\mathrm{d}n(m)}{n} F\left(\frac{m}{\langle m \rangle} \frac{1}{\bar{N}(y)} j_0 k_\perp \right),
\end{aligned}
\tag{16.33}
$$

where $F(z) = {}_1F_2\left(-1/2; 3/4, 5/4; -9z^2/16\right) - 1$ and ${}_1F_2$ is the generalised hypergeometric function. It asymptotes to $F(z) \sim z - 1$ when $z \to \infty$ and $F(z) \sim 3z^2/10$ when $z \to 0$. These limits correspond to $\bar{N}(y) \ll 1$ and $\bar{N}(y) \gg 1$, respectively. In particular, the $\bar{N}(y) \gg 1$ case recovers the Gaussian limit $K_{\text{PBH}}(\mathbf{k}) \sim -\frac{1}{4}\sigma^2_{j,\text{PBH}} k_\perp^2$, with the variance given by

$$
\sigma^2_{j,\text{PBH}} = \frac{6}{5} j_0^2 \frac{\langle m^2 \rangle}{\langle m \rangle^2} \frac{1}{\bar{N}(y)}.
\tag{16.34}
$$

The variance diverges when $\bar{N}(y) \to 0$. For this reason, it is not possible to use the central limit theorem to compute the angular momentum distribution even though it arises from an infinite sum of independent identical random variables. We also remark that a simple parallel can be drawn between the variances (16.31) and (16.34) by noting that the variance of PBH density fluctuations in a volume $V(y)$ is $\langle m^2 \rangle / (\langle m \rangle^2 \bar{N}(y))$ and has thus an analogous interpretation to the factor $\sigma_{\text{M}}^2 / f_{\text{PBH}}$ in Eq. (16.31).

Due to isotropy, it is sufficient to consider the distribution of j instead of $\mathbf{j}$. The cumulant generating functions imply the angular momentum distribution (16.29) is given by

$$
\begin{aligned}
j\frac{\mathrm{d}P}{\mathrm{d}j} = \int\limits_0^\infty \mathrm{d}u\, u J_0(u) \exp\Bigg[&- \bar{N}(y) \int \frac{\mathrm{d}m}{m} \psi(m)\, F\left(u \frac{m}{\langle m \rangle} \frac{1}{\bar{N}(y)} \frac{j_0}{j} \right) \\
&- u^2 \frac{3}{10} \frac{\sigma_{\text{M}}^2}{f_{\text{PBH}}^2} \frac{j_0^2}{j^2} \Bigg].
\end{aligned}
\tag{16.35}
$$

Depending on the size of the empty region surrounding the PBH pair, this distribution interpolates between two asymptotic cases

$$
j\frac{\mathrm{d}P}{\mathrm{d}j} = \begin{cases} \dfrac{j^2/j_0^2}{(1 + j^2/j_0^2)^{3/2}}, & \bar{N}(y) \ll 1 \quad \text{(other PBHs arbitrarily close)}, \\[2ex] \dfrac{2j^2}{\sigma_j^2} e^{-j^2/\sigma_j^2}, & \bar{N}(y) \gg 1 \quad \text{(Gaussian limit)}, \end{cases}
\tag{16.36}
$$

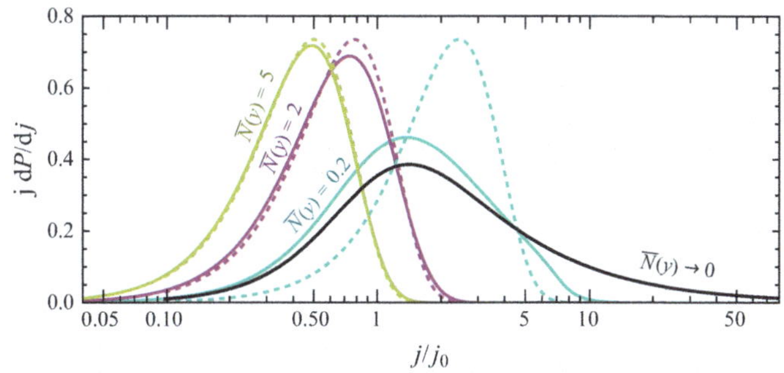

Fig. 16.3 Angular momentum distributions of PBH binaries after formation (16.35) for different $\bar{N}(y)$ are shown by solid lines. The dashed lines show the corresponding Gaussian approximation (16.36). A monochromatic mass function is assumed in the figure

where

$$\sigma_j^2 \equiv \sigma_{j,\mathrm{M}}^2 + \sigma_{j,\mathrm{PBH}}^2 = \frac{6}{5} j_0^2 \left(\frac{1}{\bar{N}(y)} \frac{\langle m^2 \rangle}{\langle m \rangle^2} + \frac{\sigma_\mathrm{M}^2}{f_\mathrm{PBH}^2} \right) \tag{16.37}$$

is the total variance of the distribution. The angular momentum distribution (16.35) together with the Gaussian approximation is shown in Fig. 16.3 for a monochromatic mass function. It can be seen that the Gaussian approximation works well already for $\bar{N}(y) = 2$, while the $\bar{N}(y) \to 0$ limit becomes a good approximation when $\bar{N}(y) \lesssim 0.2$.

The appearance of $\psi(m)$ in (16.35) implies a non-trivial dependence on the mass function. However, the dependence on $\psi(m)$ vanishes completely (16.36) in the $\bar{N}(y) \ll 1$ limit and can be reduced to the relative width of the mass distribution in the Gaussian $\bar{N}(y) \gg 1$ case.

Equation (16.35) is in excellent agreement with numerical simulations as illustrated by Fig. 16.4. It shows the angular momentum distributions obtained from N-body simulations of PBH binaries expected to merge within the age of the Universe t_0 (green region) and compares them to the distribution predicted by Eq. (16.35) (black dashed curve). Note that the distribution (16.35) gives a conditional probability assuming a fixed initial separation x_0. The corresponding distribution conditioned on a fixed coalescence time t is

$$\frac{\mathrm{d}P(j|\tau = t)}{\mathrm{d}j} \propto j_0(x_0) \frac{\mathrm{d}P(j|x_0)}{\mathrm{d}j} \bigg|_{\tau(r_a(x_0), j) = t}, \tag{16.38}$$

where we dropped prefactors independent of j. Analogously to the distribution $j\mathrm{d}P(j|x_0)/\mathrm{d}j$ that depends on j only through the combination j/j_0, it is possible to define j_τ so that $\mathrm{d}P(j|\tau)/\mathrm{d}j$ would depend on j through the combination j/j_τ only, where j_τ gives an order of magnitude estimate of the dimensionless angular momentum of initial binaries merging today. Since $j/j_0|_{\tau=t} = (j/j_\tau)^{37/16}$, we have that

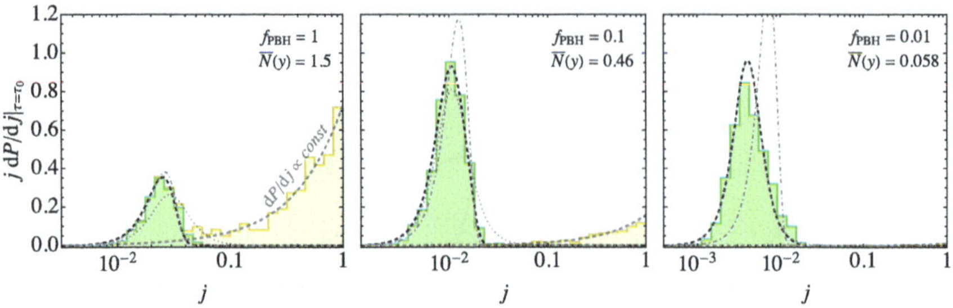

Fig. 16.4 Distribution of angular momenta at $a = 3\,a_{\rm eq}$ of PBH binaries for different $f_{\rm PBH}$ and a monochromatic mass function with $m_c = 30 M_\odot$. The initial configurations were chosen in a way that the PBH pair would form a binary that merges today if unperturbed. The perturbed and unperturbed binaries are coloured in yellow and green respectively. The thick dashed line shows the distribution (16.38) with $\bar{N}(y)$ estimated from Eq. (16.40). The thin dashed curves show limiting cases. The thick grey dashed curve compares the distribution of disrupted binaries to $\mathrm{d}P/\mathrm{d}j \propto 1$. Based on simulation data from Ref. [30]

$$j_\tau = 1.7 \times 10^{-2} \left[\frac{t}{t_0}\right]^{\frac{3}{37}} \left[\frac{M}{M_\odot}\right]^{\frac{5}{37}} (4\eta)^{\frac{3}{37}} f_{\rm PBH}^{\frac{16}{37}} . \tag{16.39}$$

The dependence on both the mass and the coalescence time is very mild, so it is safe to assume that PBHs around the stellar mass range had initial eccentricities around $\mathcal{O}(10^{-2})$ or lower when $f_{\rm PBH} \ll 1$.

16.2.1.3 Merger Rate and Binary Disruption by Nearest PBHs

The binary may be disrupted shortly after formation by becoming bound to a nearby PBH or later by interacting with PBHs in compact DM haloes, which are more likely to form in PBH cosmologies. The first possibility is accounted by an appropriate choice of $\bar{N}(y)$ and it is typically the dominant effect when $f_{\rm PBH} \ll 1$. When $f_{\rm PBH} \approx 1$, however, the binary is likely to be perturbed by later encounters due to the enhanced small-scale structure of PBH DM. This effect will be addressed separately in Sect. 16.2.2.

Since $j_0 = \mathcal{O}(10^{-2})$ for binaries merging today, the early PBH binaries tend to be extremely eccentric at formation. Since the coalescence time scales as $\tau \propto j^7$, even a $\mathcal{O}(1)$ increase in j will increase the coalescence time by several orders of magnitude. Therefore, even mild interactions can effectively remove the binary from the population of binaries merging today. It is certainly possible that some of the disrupted binaries may contribute to the present merger rate, but they must evolve from initially more compact PBH systems to merge within the age of the Universe after being perturbed. This scenario constitutes an entirely different formation channel – the early three-body channel – discussed in Sect. 16.2.3.

To determine $\bar{N}(y)$, consider collisions between the binary and the PBH initially closest to it. Let $x_{\rm NN}$ denote the initial comoving distance of the nearest neighbour. Since y sets a lower bound on viable values of $x_{\rm NN}$ for binary formation, we must find the minimal $x_{\rm NN}$ which does not lead to the disruption of the binary after the nearest neighbour decouples from the Hubble flow. This estimate can be made following the reasoning of Sect. 16.2.1.1. The region surrounding the binary corresponds to an effective matter density fluctuation $\delta_{\rm NN} \approx (M - \rho_{\rm PBH} V(x_{\rm NN}))/(\rho_{\rm M} V(x_{\rm NN}))$. Given that such configurations are expected to collapse at $a \approx a_{\rm eq}/\delta_{\rm NN}$ we estimate that, at a given redshift, the binary has collided with its neighbour if $x_{\rm NN} < y$, when

$$\bar{N}(y) \approx \frac{M}{\langle m \rangle} \frac{f_{\rm PBH}}{f_{\rm PBH} + a_{\rm eq}/a}. \tag{16.40}$$

The condition (16.40) agrees well with numerical N-body simulations of PBH binary formation in the early Universe [30]. This is illustrated in Fig. 16.5, which shows that (16.40) predicts the interface between perturbed (yellow region) and unperturbed (green region) binaries very well when $f_{\rm PBH} \ll 1$. When $f_{\rm PBH} = 1$, about half of the binaries with $x_{\rm NN} > y$ are seen disrupted. As we will show in Sect. 16.2.2, the early formation of PBH clusters in such PBH cosmologies [17] can account for this effect. The estimate Eq. (16.40) was also used when comparing numerically obtained angular momentum distributions to our analytic predictions in Fig. 16.4. In that case, an excellent agreement can be observed also when $f_{\rm PBH} = 1$.

When $a \gg a_{\rm eq}$ and for narrow mass functions, we have that $\bar{N}(y) \approx 2$ implying that the PBH pair should be positioned in an underdense region. This condition is, however, too strong when the PBH abundance is small $f_{\rm PBH} \ll 1$. In this case, it is sufficient to demand that the binary survives until the non-linear structure formation of CDM haloes sets in. Since this happens when $a/a_{\rm eq} = 1/\sigma_M$, we have that

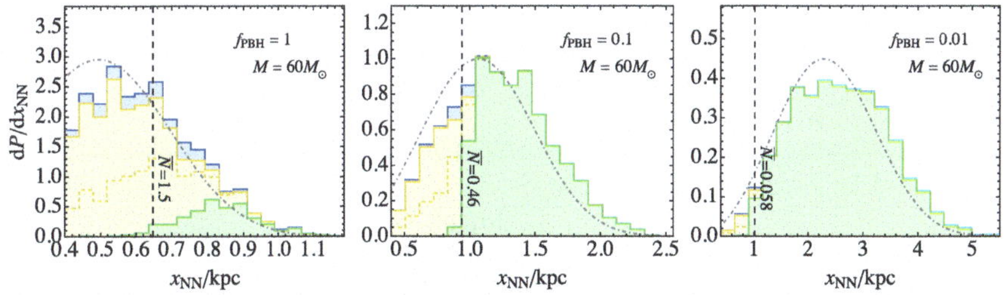

Fig. 16.5 Binary disruption at $a = 3\,a_{\rm eq}$ depending on the initial comoving distance $x_{\rm NN}$ of the PBH nearest to the binary for $30 M_\odot$ PBHs and different $f_{\rm PBH}$. The blue region indicates simulations in which the initial pair did not form a binary. The yellow region shows pairs that evolved into a binary that was later perturbed but remained bound. Binaries above the yellow dashed line swapped at least one of their components. The green region shows undisrupted binaries. The vertical dashed line indicates the estimate Eq. (16.40). The numerical simulations evolve a PBH pair estimated to evolve into a binary that would merge at the present time if not disrupted. Based on simulation data from Ref. [30]

$$\bar{N}(y) \approx \frac{M}{\langle m \rangle} \frac{f_{\rm PBH}}{f_{\rm PBH} + \sigma_{\rm M}}. \tag{16.41}$$

The binary is expected to evolve without much interference after that point since the disruption of PBH binaries in the standard late CDM haloes is very unlikely [6]. This point is illustrated further in Sect. 16.2.2, where we show that, even when considering the enhanced small-scale structures in PBH cosmologies, PBH encounters are expected to take place in relatively small haloes containing less than $\mathcal{O}(10^4)$ PBHs which are formed at high redshifts.

A potential shortcoming of the condition (16.41) is that it accounts for the masses of the surrounding PBHs on average. Thus it can fail when the mass distribution of PBHs spans several orders of magnitude. For example, it can underestimate the suppression when the nearest PBH is much heavier than the total mass of the binary since heavier PBHs are more likely to disrupt light binaries even if their distance is large. On the other hand, it can overestimate the disruption of heavy binaries by light PBHs surrounding it. In extreme cases, the surrounding binaries can be too light to significantly alter the orbit of the binary. In that case, excluding initial configurations where a light PBH is close to binary is unjustified.

Having determined the characteristics of the first binaries and constraints to their environment, we can proceed to evaluate the merger rate density (16.13) $\mathrm{d}R_{E2} = S_{\rm E} S_{\rm L} \mathrm{d}R^{(0)}_{E2}$. The unsuppressed contribution $\mathrm{d}R^{(0)}_{E2}$ is evaluated by integrating Eq. (16.12) in the $\sigma_M \to 0$ limit and using $y \to 0$ asymptotic (16.35) of the angular momentum distribution,

$$\begin{aligned} \frac{\mathrm{d}R^{(0)}_{E2}(t)}{\mathrm{d}m_1 \mathrm{d}m_2} &= \frac{1.4}{t} \left(\frac{t \eta M^{14}}{f^7_{\rm PBH} \rho^{11}_M} \right)^{\frac{3}{37}} \frac{\mathrm{d}n(m_1)\mathrm{d}n(m_2)}{\mathrm{d}m_1 \mathrm{d}m_2} \\ &\approx \frac{1.6 \times 10^6}{\mathrm{Gpc}^3 \mathrm{yr}} f_{\rm PBH}^{\frac{53}{37}} \eta^{-\frac{34}{37}} \left(\frac{M}{M_\odot} \right)^{-\frac{32}{37}} \left(\frac{t}{t_0} \right)^{-\frac{34}{37}} \frac{\psi(m_1)\psi(m_2)}{m_1 m_2}. \end{aligned} \tag{16.42}$$

The suppression factor $S_{\rm E}$ gives the remaining contribution[9]

$$\begin{aligned} S_{\rm E} = \frac{e^{-\bar{N}(y)}}{\Gamma(21/37)} \int \mathrm{d}v\, v^{-\frac{16}{37}} \exp\Bigg[&- \bar{N}(y)\langle m \rangle \int \frac{\mathrm{d}m}{m} \psi(m) F\left(\frac{m}{\langle m \rangle} \frac{v}{\bar{N}(y)} \right) \\ &- \frac{3\sigma_{\rm M}^2 v^2}{10 f^2_{\rm PBH}} \Bigg], \end{aligned} \tag{16.43}$$

with $\bar{N}(y)$ estimated from (16.41).

The suppression factor includes the effect of matter perturbations. In particular, when $f_{\rm PBH} \ll \sigma_M$, the angular momentum of the early binaries is generated mainly by matter perturbations (compare Eqs. (16.31) and (16.34)) instead of the surrounding

[9] Integrating (16.12) over x_0 and u in Eq. (16.35) can be simplified by changing x_0 to $v = u j_0 / j(\tau, x_0)$, which helps to factor out parts of the integral.

PBHs. As a result, the merger rate scales $f_{\rm PBH}^2$ instead of $f_{\rm PBH}^{53/27}$. This slightly stronger dependence causes the merger rate to drop faster when $f_{\rm PBH}$ is decreased. Since in this regime $\mathrm{d}R^{(0)}_{E2}(t) \propto \sigma_{\rm M}^{-21/74}$, a larger $\sigma_{\rm M}$, that is, more pronounced matter perturbations, will lead to a smaller merger rate.

In practice (16.43) must be evaluated numerically, which can be quite computationally expensive. A good analytic approximation can be constructed by considering the $\bar{N}(y) \to 0$ and $\bar{N}(y) \to \infty$ asymptotics of the angular momentum and noting that the following inequalities hold [30]

$$S_{\rm E,min} \leq S_{\rm E} \leq S_{\rm E,max} < 1\,, \tag{16.44}$$

where the $\bar{N}(y) \to 0$ and $\bar{N}(y) \to \infty$ limits are

$$\begin{aligned} S_{\rm E,max} &= \left(\frac{5 f_{\rm PBH}^2}{6\sigma_{\rm M}^2}\right)^{\frac{21}{74}} U\left(\frac{21}{74}, \frac{1}{2}, \frac{5 f_{\rm PBH}^2}{6\sigma_{\rm M}^2}\right), \\ S_{\rm E,min} &= \frac{\sqrt{\pi}(5/6)^{21/74}}{\Gamma(29/37)} \left[\frac{\langle m^2\rangle/\langle m\rangle^2}{\bar{N}(y)} + \frac{\sigma_{\rm M}^2}{f_{\rm PBH}^2}\right]^{-\frac{21}{74}} e^{-\bar{N}(y)}\,, \end{aligned} \tag{16.45}$$

where U denotes the confluent hypergeometric function. The factor $S_{\rm E}$ can be approximated as

$$S_{\rm E} \approx \frac{\sqrt{\pi}(5/6)^{21/74}}{\Gamma(29/37)} \left[\frac{\langle m^2\rangle/\langle m\rangle^2}{\bar{N}(y) + C} + \frac{\sigma_{\rm M}^2}{f_{\rm PBH}^2}\right]^{-\frac{21}{74}} e^{-\bar{N}(y)}\,, \tag{16.46}$$

where the factor C is fixed by demanding that $S_{\rm E} \to S_{\rm E,max}$ in the limit $\bar{N}(y) \to 0$,

$$C = f_{\rm PBH}^2 \frac{\langle m^2\rangle/\langle m\rangle^2}{\sigma_{\rm M}^2} \left\{\left[\frac{\Gamma(29/37)}{\sqrt{\pi}} U\left(\frac{21}{74}, \frac{1}{2}, \frac{5 f_{\rm PBH}^2}{6\sigma_{\rm M}^2}\right)\right]^{-\frac{74}{21}} - 1\right\}^{-1}. \tag{16.47}$$

This approximation is accurate within 7% for log-normal mass function with widths $\sigma \leq 2$. Another approximation, that is accurate within 2% but retains one numerical integral, has been constructed in Ref. [13].

16.2.2 Disruption of Early Binaries in DM Haloes

After its formation, the binary may get perturbed by encounters with a third PBH. Such encounters are likely in very dense haloes. We can estimate the probability of this by computing the rate of encounters that significantly alter the angular momentum of the binary. However, the small haloes are also unstable and experience gravothermal instabilities within the age of the Universe. This can lead to core collapse and the eventual evaporation of the PBH cluster. In the former case, the density can increase significantly and all binaries in such haloes are likely to get perturbed. In the following, we first estimate the rate of encounters that perturb the binaries and then the probability that the binary is within a halo that goes through gravothermal collapse before the binary merges.

Consider an early binary whose coalescence time initially equals the age of the Universe, $\tau = t_0$, and a third PBH of mass m_3 approaching the binary. In a hyperbolic encounter,[10] the velocity of the third body of mass m_3 at the distance of closest approach r_p is $v_p = b v_{\rm rel}/r_p$, where $v_{\rm rel}$ denotes the relative velocity of the third body and b the impact parameter that is related to r_p by $b^2 = r_c^2 + 2(M + m_3) r_c / v_{\rm rel}^2$. We can estimate the timescale of the interaction between the binary and the third body as $t_p \sim r_p/v_p$. The tidal torque caused by the third body is given by $\mathbf{T}_p = m_3(\mathbf{1} - 3\hat{\mathbf{r}}_p \otimes \hat{\mathbf{r}}_p)/r_p^3$ and the resulting change in the binary angular momentum can be estimated as $\Delta\mathbf{L} \simeq t_p \mathbf{r}_a \times (\mathbf{T}_p \cdot \mathbf{r}_a)$. This gives

$$\Delta j \simeq \frac{m_3 r_a^{3/2}}{\sqrt{M} b v_{\rm rel} r_p} . \tag{16.48}$$

The binary gets perturbed if Δj is comparable to the initial angular momentum. For the early binaries, we can estimate the initial angular momentum by j_τ given in Eq. (16.39). So, encounters with the impact parameter

$$b \lesssim \frac{[m_3(M + m_3)]^{1/3} r_a^{1/2}}{j_\tau^{1/3} M^{1/6} v_{\rm rel}} \equiv b_{\Delta j > j_\tau} \tag{16.49}$$

would perturb the binary. The corresponding interaction cross section is $\sigma_{\Delta j > j_\tau} = \pi b_{\Delta j > j_\tau}^2$.

To estimate how large the rate of such processes is, let us consider early PBH haloes of N PBHs. Such haloes form when $a_c \equiv (1 + z_c)^{-1} \approx a_{\rm eq}\sqrt{N}/f_{\rm PBH}$, where $a_{\rm eq}$ denotes the matter-radiation equality. If such haloes are virialized, the velocity dispersion is $\sigma_v^2 \approx M_{\rm H}/R$ where R denotes the virial radius and $M_{\rm H} = \langle m \rangle N/f_{\rm PBH}$ the mass of the halo.[11] The average matter density in these haloes is, following the Press-Schechter theory, given by $\rho = 3M_{\rm H}/(4\pi R^3) \approx 18\pi^2 \rho_c a_c^{-3}$, where ρ_c is the critical comoving density. So, the timescale of processes that perturb the binary in such haloes can be estimated as

$$t_p \simeq \left[\frac{\rho}{\langle m \rangle} \langle v_{\rm rel} \sigma_{\Delta j > j_\tau} \rangle\right]^{-1} \simeq 1\,{\rm Gyr} \left[\frac{N}{1000}\right]^{\frac{19}{12}} f_{\rm PBH}^{-\frac{397}{222}} \left[\frac{\langle m \rangle}{M_\odot}\right]^{-\frac{10}{111}} , \tag{16.50}$$

[10] See Sect. 16.3 for details of the hyperbolic encounters.

[11] Here we assume that the fraction of DM in PBHs in haloes matches $f_{\rm PBH}$. Ref. [17] showed that, especially in the early Universe, this fraction could be larger.

where we used Eq. (16.7) to translate r_a into τ, took $\tau = t_0$, replaced all masses by $\langle m \rangle$ and the relative velocity by the velocity dispersion σ_v. With this rough estimate, we find that for $\langle m \rangle = \mathcal{O}(10 M_\odot)$ and $f_{\rm PBH} = 1$ binaries in haloes with $N \lesssim 5800$ get perturbed before today. Moreover, by definition, the haloes need at least $N = 3$, implying that binaries are likely not perturbed when $f_{\rm PBH} \lesssim 10^{-3}$.

Next, let us estimate the effect of gravothermal instabilities of small haloes. We estimate this using the characteristic time scale of core collapse, $t_{\rm cc} \geq 18 t_{\rm r}$, where $t_{\rm r}$ denotes the relaxation time [28]:

$$t_{\rm r} = 0.065 \frac{\sigma_v^3}{m\rho \ln \Lambda} \approx 2{\rm kyr} \frac{N^{7/4}}{f_{\rm PBH}^{5/2} \ln \Lambda} . \tag{16.51}$$

By approximating the Coulomb logarithm as $\ln \Lambda \approx \ln(N/f_{\rm PBH})$ and requiring $18 t_{\rm r} < t_0$, we get that haloes with

$$N \leq 1500 f_{\rm PBH}^{10/7} \ln \Lambda^{4/7} \equiv N_c \tag{16.52}$$

experience gravothermal collapse before today (see also [4]). This indicates that, if $f_{\rm PBH} = 1$, then binaries are perturbed in haloes with $N < 5300$. All binaries survive if $f_{\rm PBH} \lesssim 0.005$. These numbers are very similar to those found above by estimating the rate of the encounters that would perturb the binaries assuming that the haloes survive.

Let us briefly consider the hardness of the early binaries. In the largest haloes in which binaries are likely disrupted according to Eq. (16.52), the velocity dispersion is

$$\sigma_v \approx 0.8 \frac{\rm km}{\rm s} f_{\rm PBH}^{\frac{1}{6}} \left[\frac{\langle m \rangle}{M_\odot}\right]^{\frac{1}{3}} N^{\frac{1}{12}} \approx 1.8 \frac{\rm km}{\rm s} f_{\rm PBH}^{\frac{2}{7}} \left[\frac{\langle m \rangle}{M_\odot}\right]^{\frac{1}{3}} . \tag{16.53}$$

On the other hand, the characteristic binding energy of binaries merging within the age of the Universe is

$$\frac{E_{\rm bin}}{\mu} \approx \left(3.4 \frac{\rm km}{\rm s} \eta^{-\frac{1}{8}} \left[\frac{j}{0.01}\right]^{\frac{7}{8}} \left[\frac{M}{M_\odot}\right]^{\frac{1}{8}}\right)^2 . \tag{16.54}$$

Therefore, the binaries can be considered hard as encounters with other PBHs are unlikely to ionise them within environments where encounters are frequent enough.

Assuming that all of the binaries in haloes of size $N \leq N_c(z)$ get perturbed, we can estimate the suppression of the merger rate by estimating the probability that the binary is in such a halo. Including both isolated haloes and subhaloes, the suppression factor, shown in Fig. 16.6, is given by [34]

$$S_{\rm L}(z) = 1 - \sum_{N=3}^{N_c(z)} \left[p_N(z_c) + \sum_{N' > N_c(z)} p_{N,N'}(z_c) \right] , \tag{16.55}$$

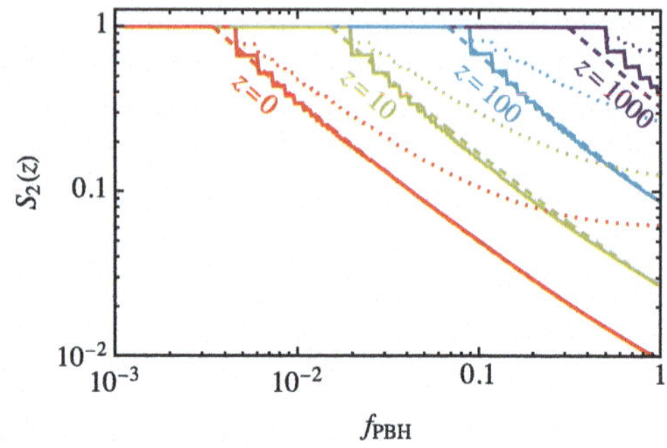

Fig. 16.6 The suppression factor of merger rate for the early two-body channel as a function of PBH abundance at different redshifts. The dashed curves show the approximation (16.60) and the dotted curves show the suppression factor neglecting the subhalos

where $z_c = f_{\rm PBH} z_{\rm eq}/\sqrt{N_c}$ denotes the redshift at which the haloes of N_c PBHs form and $z_{\rm eq}$ denotes the redshift of matter-radiation equality. The probability of the binary being part of a halo containing N PBHs at redshift z is given by

$$p_N(z) = \frac{N^{-1/2} e^{-N/N^*(z)}}{\sum_{N>2} N^{-1/2} e^{-N/N^*(z)}}, \tag{16.56}$$

and the probability of the binary being part of a subhalo containing N PBHs inside a halo of N' by

$$p_{N,N'}(z) = p_{N'}(z) \frac{N^{-1/2} e^{-N/N^*(z)}}{\sum_{N=2}^{N'} N^{-1/2} e^{-N/N^*(z)}}. \tag{16.57}$$

Here $N^*(z)$ is the characteristic number of PBHs in a halo at redshift z, and can be estimated by [17]

$$N^*(z) = \left[\ln(1+\delta^*) - \delta^*/(1+\delta^*)\right]^{-1} \tag{16.58}$$

with

$$\delta^* \approx 1.69 \left[f_{\rm PBH}\, {}_1F_2\left((1-\sqrt{21})/4,\, (1+\sqrt{21})/4;\, 1;\, -(z_{\rm eq}+1)/(z+1) \right) \right]^{-1}. \tag{16.59}$$

The numerically evaluated suppression factor $S_{\rm L}(z)$ can be approximated by [34]

$$S_{\rm L}(t) \approx \min\left\{ 1,\, 0.01 \left[\left(\frac{t}{t_0}\right)^{0.44} f_{\rm PBH} \right]^{-0.65} e^{0.03 \ln^2\left[\left(\frac{t}{t_0}\right)^{0.44} f_{\rm PBH}\right]} \right\}, \tag{16.60}$$

which, as seen in Fig. 16.6, fits the numerical estimate well. The plot also shows that for large $f_{\rm PBH}$ (e.g. $f_{\rm PBH} > 0.03$ at $z = 0$) the dominant contribution to the suppression arises from binary disruption in subhalos. We remark that this suppression factor might be affected by phenomena, such as tidal stripping, which can modify subhalo evolution. So far, DM subhaloes in PBH cosmologies have not been studied in detail and thus this term might be subject to unknown systematic uncertainties.

16.2.3 The Early Three-Body Channel: Merger Rate of Perturbed Binaries

As discussed above, a sizeable fraction of PBHs will form binaries after they decouple from the Hubble flow. However, many of them can be perturbed by interactions with other PBHs. The population of such binaries will not share the characteristics of the unperturbed binaries evolving in isolation. This was sharply illustrated by the simulated angular momentum distribution shown in Fig. 16.4, which revealed a peaked distribution for unperturbed binaries, but a nearly flat one ($\mathrm{d}P/\mathrm{d}j \approx 1$) for perturbed ones. The perturbed binaries tend to be hard so the interactions with other PBHs are likely not to break them apart. Since they are expected to be still around, many of them could contribute to the present merger rate.

Most perturbed binaries in the early Universe that would merge today would have formed from PBH pairs that initially evolved into an eccentric binary with a very short coalescence time. Let us first make a few order-of-magnitude estimates to understand the formation mechanism. Consider a compact PBH pair that decouples from the Hubble flow. Its expected initial angular momentum due to tidal forces is $j = \mathcal{O}(10^{-2})$. After being perturbed we expect that $j = \mathcal{O}(1)$, that is, the angular momentum grows by about two orders of magnitude. Assuming that this process formed a binary that merges today and given that $\tau \propto j^7$, the initial coalescence time would have been 14 orders of magnitude shorter than the age of the Universe, that is, about an hour. Such binaries would not have had time to interact with other PBHs before the merger.

It follows that perturbed binaries contributing to the present PBH merger rate are formed from PBH pairs residing in a tidal field that induces angular momenta $j \gg \mathcal{O}(10^{-2})$, which tend to be larger than for unperturbed binaries discussed in the last section. Such a tidal field implies the presence of a third PBH at a distance closer than the average separation. The presence of a nearby 3rd PBH means that we are looking at a 3-body system that eventually evolves into a binary after the third body is ejected. The binding energy of the binary is comparable to the binding energy of the 3-body system and its angular momentum can be estimated to follow the distribution

$$\frac{\mathrm{d}P(j)}{\mathrm{d}j} = \gamma j^{\gamma-1}, \tag{16.61}$$

with $\gamma \in [1, 2]$. The limiting case $\gamma = 1$ is seen in numerical simulations of early Universe binary formation [30] and is shown in Fig. 16.4 by the yellow histogram. The other limit, $\gamma = 2$, corresponds to the thermal distribution [20]. However, we note that gravitating many-body systems are unstable and will not thermalize completely, so it is not clear if a thermal $\gamma = 2$ distribution can be realized. Also, assuming that the angular momentum distribution $P(j)$ is independent of the initial angular momentum agrees with numerical studies of binary-single body collisions [12, 32].

With these considerations in mind, we can estimate the merger rate of perturbed binaries. Consider three PBHs with masses m_i, $i \in \{1, 2, 3\}$ that are of the similar. We assume that the Hubble decoupling of the 3-body system is hierarchical[12] so that the PBHs "1" and "2" at a comoving separation x will decouple first at

$$a(t_{12}) \approx a_{\rm eq} \frac{N(x)}{f_{\rm PBH}} \frac{\langle m \rangle}{m_1 + m_2}, \tag{16.62}$$

and the third PBH at a comoving distance y from the centre of mass of the pair will decouple at

$$a(t_p) \approx a_{\rm eq} \frac{N(y)}{f_{\rm PBH}} \frac{\langle m \rangle}{m_1 + m_2 + m_3}. \tag{16.63}$$

During radiation domination

$$t_{\rm p} = 65.3\,{\rm kyr} \times \left(\frac{\langle m \rangle}{m_1 + m_2 + m_3} \right)^2 \frac{N(y)^2}{f_{\rm PBH}^2}. \tag{16.64}$$

As discussed above, the coalescence time $\tau_{\rm i}$ of the binary "1 + 2" can be short. In the following, we will consider systems with

$$\tau_{\rm i} > 20 t_{\rm p}. \tag{16.65}$$

The binding energy of eccentric binaries scales as $E(t) = E_0(1 - t/\tau)^{-2}$ [27], so this condition is equivalent to demanding that the binding energy is affected by less than 11% by GW emission. We also note that, during matter-domination, Eq. (16.64) overestimates $t_{\rm p}$, so (16.65) would lead to a lower merger rate estimate.

With hierarchical collapse, the coalescence time of the initial "1 + 2" binary depends on its angular momentum j_{12} which is set mostly by the tidal torque from the third PBH. By Eqs. (16.27) and (16.26), it is given by

$$j_{12} = 1.4 \frac{m_3}{m_1 + m_2} \frac{\bar{N}(x)}{\bar{N}(y)} |\sin(2\theta)|, \tag{16.66}$$

where θ denotes the angle between the line joining the "1 + 2" pair and the line joining the center of mass of the pair and the third PBH. The order of magnitude of the angular momentum generated by the remaining PBHs is of the order of the width of the distribution (16.37), that is, $\sigma_j \approx 0.5\bar{N}(x)/\sqrt{\bar{N}(y)}$. If $\bar{N}(y) \ll 1$, then $j_{12} \gg \sigma_j$ and we can neglect the contribution from the remaining PBHs.

---

12 This assumption is likely to hold when all comoving separations are below the average, in which case hierarchical 3-body configurations would be more likely than configurations in which all distances are comparable.

In a hierarchical setup, the binding energy of the 3-body system is determined mostly by the binding energy of the initial binary. The comoving density of initial "1 + 2" binaries with per binding energy interval $(E, E + \mathrm{d}E)$ is

$$\frac{\mathrm{d}n_i(E, m_i)}{\mathrm{d}E} = \int \mathrm{d}n_3\, P(E, \tau_i \geq 20t_p | x, y, \theta, m_i), \tag{16.67}$$

where

$$\mathrm{d}n_3 = \frac{n}{2}\frac{\mathrm{d}n(m_1)}{n}\frac{\mathrm{d}n(m_2)}{n}\frac{\mathrm{d}n(m_3)}{n}\mathrm{d}\bar{N}(x)\mathrm{d}\bar{N}(y)e^{-\bar{N}(y)}\frac{\mathrm{d}\cos\theta}{2} \tag{16.68}$$

gives the density of 3-body configurations yielding the initial binary. It generalizes Eq. (16.9). The probability that the "1 + 2" pair produces a binary with binding energy E that satisfies the bound (16.65) is

$$\begin{aligned} &P(E, \tau_i \geq 20t_p | x, y, \theta, m_i) \\ &\quad = \delta(E - E(x))\Theta(\tau_\mathrm{i}(E, j_{12}) - 20t_\mathrm{p}), \end{aligned} \tag{16.69}$$

where Θ denotes the step function. The binding energy $E = m_1 m_2/2r_a$ (16.3) is determined from the initial separation x by Eq. (16.23). It gives $x^4 \approx 0.6(m_1 + m_2)m_1 m_2/(\rho_R E)$.

Consider now the integrals over x, y and θ in (16.67). The x integral can be eliminated using the δ-function. The y integral can be replaced with the integral over $\bar{N}(y)$ and the condition (16.65) can now be recast as an upper bound on $\bar{N}(y)$,

$$\begin{aligned} &\bar{N}(y) \leq \bar{N}(y)_{\max} \\ &\equiv 3800 f_{\mathrm{PBH}} E^{-\frac{37}{36}} \rho_R^{\frac{1}{4}} \frac{m_1^{\frac{11}{12}} m_2^{\frac{11}{12}} m_3^{\frac{7}{9}} (m_1 + m_2 + m_3)^{\frac{2}{9}}}{(m_1 + m_2)^{\frac{11}{36}} \langle m \rangle} |\sin(2\theta)|^{\frac{7}{9}}. \end{aligned} \tag{16.70}$$

For consistency, we will further require that $\bar{N}(y)_{\max} < 1$ at every orientation, which yields a lower bound on the binding energy

$$E > E_{\min} \equiv \langle m \rangle \left[11\frac{\mathrm{km}}{\mathrm{s}}\right]^2 \left[\frac{\langle m \rangle}{M_\odot}\right]^{\frac{18}{37}} f_{\mathrm{PBH}}^{\frac{36}{37}} \mathcal{G}(m_1, m_2, m_3) \tag{16.71}$$

where we defined the dimensionless factor

$$\mathcal{G}(m_1, m_2, m_3) = \left(\frac{2m_1^3 m_2^3}{m_1 + m_2}\right)^{\frac{11}{37}} m_3^{\frac{28}{37}} \left(\frac{m_1 + m_2 + m_3}{3}\right)^{\frac{8}{37}} \langle m \rangle^{-\frac{91}{37}} \tag{16.72}$$

that becomes unity for monochromatic mass functions. With this cut-off, we can simplify the computation and omit the exponential suppression in (16.68), so the integral over $\bar{N}(y)$ can be approximated by $\bar{N}(y)_{\max}$.

Averaging over θ then gives the initial binding energy distribution

$$\frac{\mathrm{d}n_i(E, m_i)}{\mathrm{d}E\mathrm{d}m_1\mathrm{d}m_2\mathrm{d}m_3} = \frac{\mathrm{d}n_{i,\mathrm{mono}}(E)}{\mathrm{d}E}\mathcal{F}(m_1, m_2, m_3)\prod_{i=1}^{3}\frac{\mathrm{d}n}{n\,\mathrm{d}m_i}, \tag{16.73}$$

where

$$\frac{\mathrm{d}n_{i,\mathrm{mono}}(E)}{\mathrm{d}E} = 4.0 \times 10^{10} E^{-\frac{25}{9}} f_{\mathrm{PBH}}^3 \langle m\rangle^{\frac{16}{9}} \rho_R^{\frac{3}{2}} \tag{16.74}$$

and we defined the dimensionless factor

$$\mathcal{F}(m_1, m_2, m_3) = m_1^{\frac{5}{3}} m_2^{\frac{5}{3}} m_3^{\frac{7}{9}} \left(\frac{m_1 + m_2}{2}\right)^{\frac{4}{9}} \left(\frac{m_1 + m_2 + m_3}{3}\right)^{\frac{2}{9}} \langle m\rangle^{-\frac{43}{9}} \tag{16.75}$$

that reduces to unity for monochromatic mass functions.

To convert Eq. (16.73) into the energy distribution of binary parameters after the 3rd body is ejected, we assume that a 3-body system with binding energy E' yields a binary with binding energy E

$$\frac{\partial K(E|E')}{\partial E} = \frac{\alpha}{E'} e^{-\alpha(E/E'-1)} \Theta(E - E'). \tag{16.76}$$

This approximates the behaviour observed in numerical simulations of PBH formation in the early universe for which $\alpha = 1$ gives a decent fit [30, 34]. The limiting case in which the initial binding energy does not change corresponds to the limit $\alpha \to \infty$. The Heggie-Hills law states that hard binaries tend to get harder, while soft binaries get softer [14, 15]. Thus, as the early binaries tend to be hard, decreasing α will support more sizeable energy changes and can thus model the effect of subsequent collisions with other PBHs in PBH clusters. Since we work with the assumption of narrow mass functions, we do not consider the dependence on the masses of the three interacting PBHs in Eq. (16.76).

Typically the lightest PBH will be ejected from the 3-body system, as it is more likely to be accelerated beyond the escape velocity. So, the binary will be typically composed of the two heavier PBHs. To model this, we will assume that the probability of ejecting a PBH from the system is exponentially suppressed by the relative mass $p_i(m_i) \propto \exp(-\lambda m_i/(m_1 + m_2 + m_3))$, where $\lambda \geq 0$ is a free parameter. $\lambda = 0$ corresponds to equal probability of ejection, while for $\lambda \to \infty$ the lightest PBH is always expelled. The exact shape of this function must be determined numerically. Given m_1, m_2, m_3, the binary constituent mass distribution is

$$\begin{aligned} P_m(m_1', m_2'|m_1, m_2, m_3) = {} & p_1\delta(m_1' - m_2)\delta(m_2' - m_3) \\ & + p_2\delta(m_1' - m_1)\delta(m_2' - m_3) \\ & + p_3\delta(m_1' - m_1)\delta(m_2' - m_2). \end{aligned} \tag{16.77}$$

The distribution of orbital parameters of the early binaries from the 3-body channel is

$$\frac{\mathrm{d}n_{\mathrm{E3}}(j,E,m_1',m_2')}{\mathrm{d}j\mathrm{d}E\mathrm{d}m_1'\mathrm{d}m_2'} = \frac{\mathrm{d}P(j)}{\mathrm{d}j}\int \mathrm{d}E'\mathrm{d}m_1\mathrm{d}m_2\mathrm{d}m_3\frac{\partial K(E|E')}{\partial E} \times \frac{\mathrm{d}n_i(E')}{\mathrm{d}E'\mathrm{d}m_1\mathrm{d}m_2\mathrm{d}m_3}P_m(m_1',m_2'|m_1,m_2,m_3), \tag{16.78}$$

with the angular momentum distribution given by Eq. (16.61). Since $\partial n_i(E)/\partial E \propto E^{-25/9}$, it is possible to evaluate the E' integral

$$\frac{\mathrm{d}n_{\mathrm{E3}}(j,E,m_1,m_2)}{\mathrm{d}j\mathrm{d}E} = \frac{\mathrm{d}P(j)}{\mathrm{d}j}\frac{\mathrm{d}n_{i,\mathrm{mono}}(E)}{\mathrm{d}E}\mathcal{K}(\alpha)\bar{\mathcal{F}}(m_i')\frac{\mathrm{d}n(m_1)}{n}\frac{\mathrm{d}n(m_2)}{n}, \tag{16.79}$$

where

$$\bar{\mathcal{F}}(m_1,m_2) \equiv \int \frac{\mathrm{d}n(m)}{n}\Big[(p_1(m,m_1,m_2)+p_1(m_1,m,m_2))\mathcal{F}(m,m_1,m_2) + p_3(m_1,m_2,m)\mathcal{F}(m_1,m_2,m)\Big] \tag{16.80}$$

accounts for the variation in masses and the overall factor

$$\mathcal{K}(\alpha) \equiv \alpha^{-25/9}e^{\alpha}\Gamma\,(25/9,\alpha) \tag{16.81}$$

contains the effect of hardening in binary-single PBH collisions. $\Gamma\,(a,b)$ denotes the incomplete gamma function. We will use $\alpha = 1$, which corresponds to $\mathcal{K}(1) = 4.0$. As was discussed above, accounting for hardening via later encounter would reduce α and increase the merger rate. However, as the effect of later encounters was estimated to be small [11], we will not consider this contribution. The enhancing effect of hardening might, however, become more relevant in scenarios in which PBHs are initially clustered because the probability of binary-PBH collisions is expected to increase in such cases.

When integrating over masses in Eq. (16.80) we neglected the lower bound E_{min} (16.71) on the binding energy E. To estimate E_{min} for the binary after the 3rd PBH is ejected, we make the substitution in Eq. (16.71),

$$\mathcal{G}(m_1,m_2,m_3) \to \bar{\mathcal{G}}(M) = \left(\frac{M}{2\langle m\rangle}\right)^{91/37}, \tag{16.82}$$

where M is the total mass of the binary. As the binary is more likely to consist of the heaviest PBHs, $\bar{\mathcal{G}}$ will give a stricter lower bound than what would follow from $\mathcal{G}$, i.e., $\bar{\mathcal{G}} \gtrsim \mathcal{G}$. This will in turn soften the resulting merger rate. However, we expect the effect to be small. As with $\mathcal{G}$, we have that $\bar{\mathcal{G}} = 1$ for monochromatic mass functions.

The merger rate implied by the distribution of binaries (16.79) is given by Eq. (16.6). As with R_{E2} in Eq. (16.12), we can assume that the binaries formed at the Big Bang, so that

$$\frac{\mathrm{d}R_{E3}}{\mathrm{d}\ln m_1 \mathrm{d}\ln m_2} = \int \mathrm{d}j\,\mathrm{d}E\,\frac{\mathrm{d}n_{E3}(j,E)}{\mathrm{d}j\mathrm{d}E\mathrm{d}\ln m_1\mathrm{d}\ln m_2}\delta\left(t-\tau(j,E)\right)$$
$$= R_{E3,\mathrm{mono}}\left(\frac{M}{2\langle m\rangle}\right)^{\frac{179\gamma}{259}-\frac{2122}{333}}(4\eta)^{-\frac{3\gamma}{7}-1}\bar{\mathcal{F}}(m_1,m_2)\psi(m_1)\psi(m_2), \tag{16.83}$$

where the prefactor corresponds to the merger rate for a monochromatic mass function

$$R_{E3,\mathrm{mono}} = \frac{7.9\times10^4}{\mathrm{Gpc}^3\mathrm{yr}} f_{\mathrm{PBH}}^{\frac{144\gamma}{259}+\frac{47}{37}}\left[\frac{t}{t_0}\right]^{\frac{\gamma}{7}-1}\left[\frac{\langle m\rangle}{M_\odot}\right]^{\frac{5\gamma-32}{37}}\frac{e^{3.2(1-\gamma)}\gamma}{28/9-\gamma}\mathcal{K}. \tag{16.84}$$

This merger rate depends weakly on the exact form of the condition (16.65). For instance, imposing $\tau_i > qt_p(y)$, we get $R_{E3} \propto q^{\frac{16(28-9\gamma)}{2331}}$.

We stress that the three-body merger rate estimate (16.83) holds for relatively *narrow mass functions*. A hierarchical decoupling, i.e., $a(t_{12}) \ll a(t_\mathrm{p})$ implies a relatively narrow mass function which guarantees that $m_1 + m_2$ is of the same order as m_3. If $m_3 \gg m_1 + m_2$, then the gravitational potential of the third body could dominate the "1 + 2" system before its decoupling from the Hubble flow. On the other hand, if $m_3 \ll m_1 + m_2$, the third body may not be heavy enough to disrupt the initial binary. So, extending these estimates to wider mass functions requires considering the cases involving large mass ratios separately.

16.3 PBH Binary Formation in the Late Universe

In the late Universe, PBH binaries can form dynamically through two- or three-body interactions in dense structures that have decoupled from the Hubble flow. Binary formation in two-body encounters happens through the emission of GWs, while in a three-body system, one of the bodies carries away kinetic energy from the system leaving the other two in a bound state.

We characterize the PBH number overdensities with an effective parameter $\delta_{\mathrm{eff}} > 1$. Consider first a structure that includes N PBHs. In such a structure, the average PBH number overdensity is $\delta_N = N/(V_N n_{\mathrm{PBH}}) > 1$ where $n_{\mathrm{PBH}} = \sum_N n_N N$ is the comoving average PBH number density and V_N is the volume containing the N PBHs. The PBH merger rate is $R_N = \delta_N^q(\sigma_{v,N}/\sigma_v)^b R_1$ where $q = 2$ for the two-body and $q = 3$ for the three-body channel, $\sigma_{v,N}$ is the velocity dispersion of PBHs in a halo with N PBHs, σ_v is a reference value for the velocity dispersion, $b = -11/7$ for the two-body and $b \in (-55, -47)/7$ for the three-body channel (as shown in the following) and R_1 is the merger rate assuming the background value for the PBH

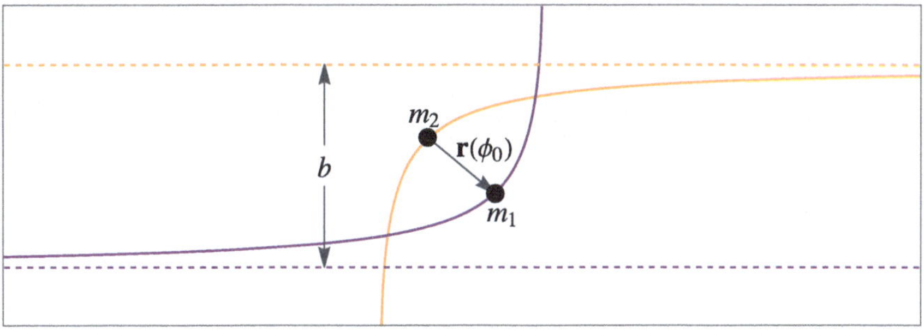

Fig. 16.7 A hyperbolic encounter of two PBHs

number density and the velocity dispersion σ_v. The total PBH merger rate is then obtained by summing over different size halos:

$$R = \sum_N n_N V_N R_N = \frac{\sum_N n_N N \delta_N^{q-1} (\sigma_{v,N}/\sigma_v)^b}{n_{\rm PBH}} R_1 \equiv \delta_{\rm eff}^{q-1} R_1, \tag{16.85}$$

where in the last step we defined the effective PBH overdensity $\delta_{\rm eff}$ that includes also the effect of the differences in the velocity dispersions in different halos. We assume that the mass distribution of PBHs in DM haloes matches their overall mass distribution.

16.3.1 Two-Body Encounters

Consider a hyperbolic encounter of two bodies of masses m_1 and m_2 with initial relative velocity $v_{\rm rel}$ and impact parameter b, as shown in Fig. 16.7. In the centre of mass frame, the energy and angular momentum of the system can be written in terms of the polar coordinates of the separation $\mathbf{r}_1 - \mathbf{r}_2 \equiv (r\cos\phi, r\sin\phi, 0)$ as

$$E = \frac{1}{2}\mu\left(\dot{r}^2 + r^2\dot{\phi}^2\right) - \frac{\mu M}{r}, \qquad \mathbf{L} = \mu r^2 \dot{\phi}\hat{z}, \tag{16.86}$$

where $M = m_1 + m_2$ and $\mu = m_1 m_2/M$. Due to conservation of energy and angular momentum, the distance r between the bodies evolves as

$$r = \frac{b\sin\phi_0}{\cos(\phi - \phi_0) - \cos\phi_0}, \qquad \dot{\phi} = -\frac{b v_{\rm rel}}{r^2}, \tag{16.87}$$

where $\phi_0 = \pi - \tan^{-1}(b v_{\rm rel}/GM)$ corresponds to the distance of closest approach.

The system loses energy by emitting GWs. The energy emitted by unit time is given by the third time derivative of the quadrupole moment tensor $Q_{ij} = m_1 r_{1,i} r_{1,j} +$

$m_2 r_{2,i} r_{2,j}$ as $\dot{E}_{\rm GW} = \dddot{Q}_{ij} \dddot{Q}^{ij}/45$. By integrating over time and taking the limit of small $v_{\rm rel}$ we get

$$E_{\rm GW} = \frac{19\pi \eta^2 M^8}{6 b^7 v_{\rm rel}^7}, \tag{16.88}$$

where $\eta = \mu/M$.

The bodies form a binary if the energy emitted in GWs is larger than the initial kinetic energy of the pair, $E_0 = \mu v_{\rm rel}^2/2$. This corresponds to the maximal impact parameter

$$b_{\rm max} = \left(\frac{19\pi \eta}{3}\right)^{1/7} M v_{\rm rel}^{-9/7}. \tag{16.89}$$

The semimajor axis and eccentricity of the resulting binary are

$$r_{a,0} = -\frac{\eta M^2}{2E_f}, \qquad e_0 = \sqrt{1 + \frac{2E_f v_{\rm rel}^2 b^2}{\eta M^3}}, \tag{16.90}$$

where $E_f \equiv E_0 - E_{\rm GW}$, and by Eq. (16.8) the binary coalescence time is

$$\begin{aligned} \tau &= \frac{19\pi M}{85 v_{\rm rel}^3} \sqrt{\frac{(b/b_{\rm max})^{21}}{1-(b/b_{\rm max})^7}} \\ &\approx 1\,{\rm day} \frac{M}{M_\odot} \left(\frac{v_{\rm rel}}{100\,{\rm km/s}}\right)^{-3} \sqrt{\frac{(b/b_{\rm max})^{21}}{1-(b/b_{\rm max})^7}}. \end{aligned} \tag{16.91}$$

This expression diverges for $b = b_{\rm max}$ since in that case the orbit is a parabola, but already for $b < 0.99 b_{\rm max}$ the square-root factor is smaller than 4 and the binaries merge very quickly.

The capture cross section is then given by $\sigma = \pi b_{\rm max}^2$. Assuming a Maxwell-Boltzmann velocity distribution $P(v) \propto v^2 \exp\left(-v^2/\sigma_v^2\right)$, that matches reasonably well the results of N-body simulations [23], we get [8, 25, 29]

$$\langle v_{\rm rel} \sigma_2 \rangle = \int {\rm d}v_{\rm rel}\, v_{\rm rel} P(v_{\rm rel}) \pi b_{\rm max}^2 \approx 2\sqrt{\pi} \left(\frac{19\pi\,\eta}{3}\right)^{2/7} \frac{M^2}{\sigma_v^{11/7}}. \tag{16.92}$$

The velocity dispersion σ_v typically takes a value in the $1 - 100\,{\rm km/s}$ range. Since the coalescence times of the binaries formed through the two-body capture process are very short, we can directly translate the capture rate into the PBH merger rate:

$$\begin{aligned}\frac{\mathrm{d}R_{\mathrm{L2}}}{\mathrm{d}\ln m_1 \mathrm{d}\ln m_2} &= \langle v_{\mathrm{rel}}\sigma_{\mathrm{mer}}\rangle \delta_{\mathrm{eff}} \frac{\mathrm{d}n_{\mathrm{PBH}}}{\mathrm{d}\ln m_1}\frac{\mathrm{d}n_{\mathrm{PBH}}}{\mathrm{d}\ln m_2} \\ &= 2\sqrt{\pi}\left(\frac{19\pi\,\eta}{3}\right)^{2/7} \frac{M^2}{\sigma_v^{11/7}} \frac{\delta f_{\mathrm{PBH}}^2 \rho_{\mathrm{DM}}^2}{m_1 m_2}\psi(m_1)\psi(m_2) \\ &\approx \frac{3.4\times 10^{-6}}{\mathrm{Gpc}^3\mathrm{yr}} f_{\mathrm{PBH}}^2\,\delta_{\mathrm{eff}}\left(\frac{\sigma_v}{\mathrm{km/s}}\right)^{-\frac{11}{7}} \eta^{-\frac{5}{7}}\psi(m_1)\psi(m_2).\end{aligned} \tag{16.93}$$

Clustered PBHs tend to have also a smaller velocity dispersion which further enhances the rate (16.93). Moreover, clustering tends to expel light PBHs from the regions of large δ_{eff} which effectively causes spatial dependence of the PBH mass function $\psi(m)$.

16.3.2 Three-Body Encounters

The merger rate of binaries formed from in three-body encounters is estimated by multiplying the rate of two-body encounters by the probability that a third body happens to be close to the pair and estimating the coalescence times of these binaries [1, 11, 18, 19, 21, 22, 24, 31].

Using (16.87) at $\phi = \phi_0$, we find that in a hyperbolic encounter, the distance between the two bodies gets smaller than r_a if the impact parameter is $b^2 < r_a^2(1 + 2M/(r_a v_{\mathrm{rel}}^2)) \equiv b_a^2$. Assuming a Maxwell-Boltzmann velocity distribution, the velocity averaged cross section for the PBHs to reach within distance r_a is

$$\langle v_{\mathrm{rel}}\sigma_2\rangle = \int \mathrm{d}v_{\mathrm{rel}}\, v_{\mathrm{rel}} P(v_{\mathrm{rel}})\pi b_a^2 \approx 2\sqrt{\pi} r_a \sigma_v \left(1 + \frac{2M}{r_a\sigma_v^2}\right). \tag{16.94}$$

The time the PBHs spend at a distance smaller than r_a can be integrated from the solution (16.87). In the limit of small v_{rel} this gives $\tau_a \approx \sqrt{8r_a^3/(9M)}$. Then, the probability that a third PBH passes the PBH pair within distance r_a from its center of mass is

$$\begin{aligned}P_3 &= \int \mathrm{d}v_3\, P(v_3)\pi r_a^2(v_3\tau_a + r_a)\delta n_{\mathrm{PBH}} \\ &= \left[\frac{4}{3}\sqrt{\frac{2\pi r_a^7}{M}}\sigma_v + \pi r_a^3\right]\delta_{\mathrm{eff}}\, n_{\mathrm{PBH}}.\end{aligned} \tag{16.95}$$

The rate of three-body encounters can be approximated as $\propto P_3\langle v_{\mathrm{rel}}\sigma_2\rangle$. The encounter leads to the formation of a binary if the semimajor axis r_a is sufficiently small, $r_a < r_{a,\mathrm{max}}$. The upper limit on r_a is estimated by setting a lower limit on the binary hardness factor κ defined by comparing the binary binding energy to the average kinetic energy of the surrounding PBHs:

$$\kappa \equiv \frac{E_{\rm bin}}{\langle m \rangle \sigma_v^2/2} = \frac{\mu M}{r_a \langle m \rangle \sigma_v^2}, \tag{16.96}$$

where $E_{\rm bin}$ is the binding energy given by Eq. (16.3). The threshold value of $\kappa > \kappa_{\rm min} = 5$ is often used in the literature (see e.g. [11, 24, 31]).

Unlike the two-body encounters, the three-body encounters produce wide binaries whose coalescence time can be even longer than the age of the Universe. To estimate the merger rate, we need to first estimate the angular momenta of the binaries. Analogously to the early universe 3-body scenarios (16.61), we take

$$\frac{{\rm d}P}{{\rm d}j} = \gamma j^{\gamma-1}, \tag{16.97}$$

where $\gamma \in [1, 2]$. The value $\gamma = 2$ corresponds to the thermal distribution [20] but Refs. [30, 33] find that the distribution could be superthermal with $\gamma < 2$.

With the above ingredients, the binary formation rate in the three-body encounters is

$$\frac{{\rm d}R_b}{{\rm d}m_1 {\rm d}m_2 {\rm d}j {\rm d}r_a} = \frac{{\rm d}P}{{\rm d}j} \frac{{\rm d}P_3 \langle v_{\rm rel}\sigma_2 \rangle}{{\rm d}r_a} \Theta(r_a - r_{a,\max}) \delta_{\rm eff} \frac{{\rm d}n_{\rm PBH}}{{\rm d}m_1} \frac{{\rm d}n_{\rm PBH}}{{\rm d}m_2}. \tag{16.98}$$

Then, using Eq. (16.6), we get the merger rate of the binaries formed in the three-body encounters:

$$\begin{aligned} \frac{{\rm d}R_{\rm L3}}{{\rm d}\ln m_1 {\rm d}\ln m_2} &= \frac{7\pi}{24} f_{\rm PBH}^3 \delta_{\rm eff}^2 \left(\frac{1360\eta\sigma_v^8 t}{3M}\right)^{\frac{\gamma}{7}} \frac{M^3 \rho_{\rm DM}^3}{\eta \langle m \rangle \sigma_v^9} \mathcal{F}\left(\frac{\langle m \rangle}{2\eta M}\kappa_{\rm min}\right) \psi(m_1)\psi(m_2) \\ &\approx \frac{1.3 \times 10^{-16} e^{-6.0(\gamma-1)}}{{\rm Gpc}^3 {\rm yr}} f_{\rm PBH}^3 \delta_{\rm eff}^2 \left(\frac{\sigma_v}{\rm km/s}\right)^{-9+\frac{8\gamma}{7}} \eta^{-1+\frac{\gamma}{7}} \left(\frac{M}{M_\odot}\right)^{3-\frac{\gamma}{7}} \left(\frac{t}{t_0}\right)^{\frac{\gamma}{7}} \\ &\times \mathcal{F}\left(\frac{\langle m \rangle}{2\eta M}\kappa_{\rm min}\right) \psi(m_1)\psi(m_2), \end{aligned} \tag{16.99}$$

where

$$\mathcal{F}(\kappa) \equiv \kappa^{-4+\frac{4\gamma}{7}} \left[\frac{6\sqrt{\pi}}{7-\gamma} + \frac{72}{63-8\gamma}\kappa^{-\frac{1}{2}} + \frac{15\sqrt{\pi}}{70-8\gamma}\kappa^{-1} + \frac{22}{77-8\gamma}\kappa^{-\frac{3}{2}}\right]. \tag{16.100}$$

Similarly, as in the case of two-body encounters, the rate is larger in the higher density environments with smaller velocity dispersion. However, the three-body channel is even more sensitive to the clustering. Moreover, the coalescence time of the binaries formed through the three-body encounters is typically very large, so the binaries that would merge in the present Universe should have formed at a high redshift in compact PBH clusters induced by their Poisson distribution.

16.4 Summary

In conclusion, the merger rate of PBH binaries consists of four components which we report here for standard PBH cosmologies:

1. **Early two-body channel**

$$\frac{\mathrm{d}R_{E2}}{\mathrm{d}\ln m_1 \mathrm{d}\ln m_2} \approx \frac{1.6\times 10^6}{\mathrm{Gpc}^3\mathrm{yr}} f_{\mathrm{PBH}}^{\frac{53}{37}} \eta^{-\frac{34}{37}} \left(\frac{M}{M_\odot}\right)^{-\frac{32}{37}} \left(\frac{t}{t_0}\right)^{-\frac{34}{37}} S_{\mathrm{L}} S_{\mathrm{E}}\, \psi(m_1)\psi(m_2)\,, \tag{16.101}$$

where S_{L} and S_{E} are suppression factors given in Eqs. (16.46) and (16.60).

2. **Early three-body channel**

$$\frac{\mathrm{d}R_{\mathrm{E3}}}{\mathrm{d}\ln m_1 \mathrm{d}\ln m_2} = \frac{7.9\times 10^4}{\mathrm{Gpc}^3\mathrm{yr}} f_{\mathrm{PBH}}^{\frac{144\gamma}{259}+\frac{47}{37}} \left[\frac{t}{t_0}\right]^{\frac{\gamma}{7}-1} \left[\frac{\langle m\rangle}{M_\odot}\right]^{\frac{5\gamma-32}{37}} \frac{e^{3.2(1-\gamma)}\gamma}{28/9-\gamma}\mathcal{K}$$
$$\times \left(\frac{M}{2\langle m\rangle}\right)^{\frac{179\gamma}{259}-\frac{2122}{333}} (4\eta)^{-\frac{3\gamma}{7}-1} \bar{\mathcal{F}}(m_1,m_2)\psi(m_1)\psi(m_2)\,, \tag{16.102}$$

where $\mathcal{K}$ and $\bar{\mathcal{F}}$ are given by Eqs. (16.81,16.80) and account for 3-body dynamics. Numerical simulations suggest that $\mathcal{K} \approx 4.0$. The second line contains the effect of extended mass functions.

3. **Late two-body channel**

$$\frac{\mathrm{d}R_{\mathrm{L2}}}{\mathrm{d}\ln m_1 \mathrm{d}\ln m_2} \approx \frac{3.4\times 10^{-6}}{\mathrm{Gpc}^3\mathrm{yr}} f_{\mathrm{PBH}}^2\, \delta_{\mathrm{eff}} \left(\frac{\sigma_v}{\mathrm{km/s}}\right)^{-\frac{11}{7}} \eta^{-\frac{5}{7}}\psi(m_1)\psi(m_2), \tag{16.103}$$

where δ_{eff} and σ_v characterize the PBH density contrast and velocity dispersion in the structures.

4. **Late three-body channel**

$$\frac{\mathrm{d}R_{\mathrm{L3}}}{\mathrm{d}\ln m_1 \mathrm{d}\ln m_2} \approx \frac{1.3\times 10^{-16} e^{-6.0(\gamma-1)}}{\mathrm{Gpc}^3\mathrm{yr}} f_{\mathrm{PBH}}^3 \delta_{\mathrm{eff}}^2 \left(\frac{\sigma_v}{\mathrm{km/s}}\right)^{-9+\frac{8\gamma}{7}}$$
$$\times\, \eta^{-1+\frac{\gamma}{7}} \left(\frac{M}{M_\odot}\right)^{3-\frac{\gamma}{7}} \left(\frac{t}{t_0}\right)^{\frac{\gamma}{7}} \mathcal{F}\left(\frac{\langle m\rangle}{2\eta M}\kappa_{\min}\right) \psi(m_1)\psi(m_2), \tag{16.104}$$

where δ_{eff} and σ_v characterize the PBH density contrast and velocity dispersion in the structures, $\mathcal{F}(\kappa)$ is given in Eq. (16.100) and γ determines the initial angular momentum distribution (16.97) of the binaries.

To demonstrate the relative importance of different PBH merger rate channels for different values of f_{PBH}, we present numerical examples of the merger rates (16.101), (16.102), (16.103), (16.104) in Fig. 16.8. For concreteness, we have chosen a monochromatic mass function with $m = 30\,M_\odot$. It is evident that for $f_{\mathrm{PBH}} \lesssim 0.1$ the early two-body channel (16.101) dominates. This channel is also

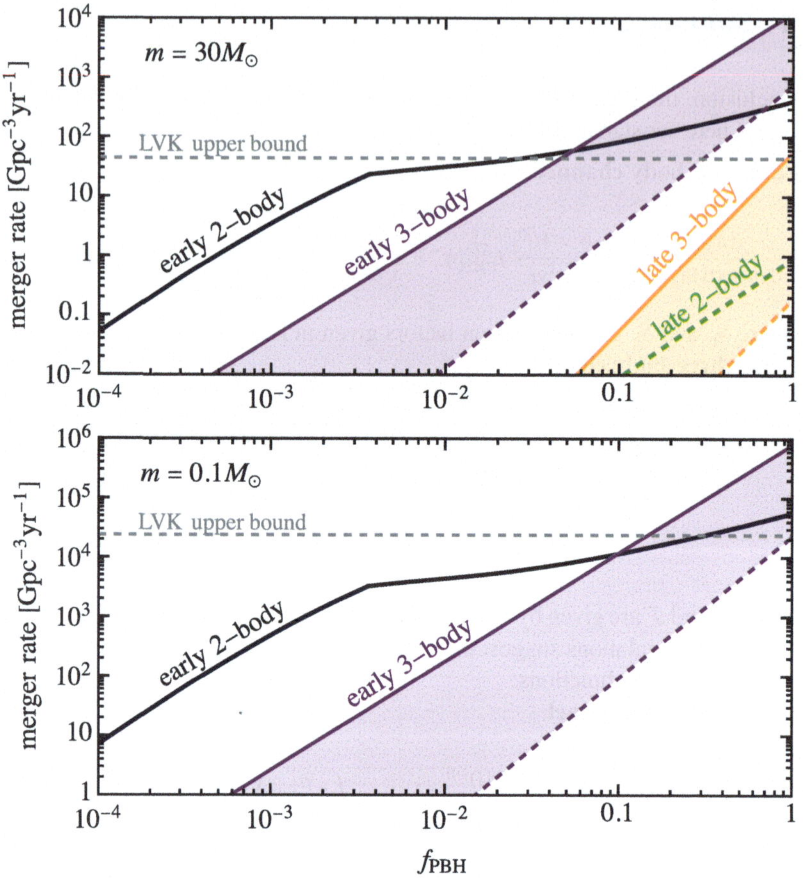

Fig. 16.8 Numerical estimates of the four channels contributing to the PBH merger rate as functions of the fraction of DM in PBHs. We assume a monochromatic mass function with $m = 30\, M_\odot$ (upper panel) and $m = 0.1\, M_\odot$ (lower panel). The bands show variations due to uncertainties in the angular momentum distribution. We show optimistic late universe merger rates with $\delta_{\rm eff} = 10^5$ for the late two-body and $\delta_{\rm eff} = 10^7$ for the late three-body channels. The gray dashed lines show the LVK upper bounds from [2] and [3]

best understood and has the smallest uncertainties. The situation may change when $f_{\rm PBH} \gtrsim 0.1$, that is, when a large fraction of DM of the Universe should be in the form of PBHs. In this case, the PBH density is large and the early three-body channel (16.102) may be the dominant one. However, the uncertainties associated with this solution are larger than in the first case. The upper (solid) and lower (dashed) boundaries of the three-body channels correspond to the merger rate with $\gamma = 1$ and $\mathcal{K} = 4$ and $\gamma = 2$ and $\mathcal{K} = 1$, respectively. The late Universe channels (16.103) and (16.104) are always expected to be subdominant. Optimistically, in Fig. 16.8 we have fixed $\sigma_v = 1\,\rm km/s$ and used $\delta_{\rm eff} = 10^5$ for the late two-body and $\delta_{\rm eff} = 10^7$

for the late three-body channel where we also show $\gamma = 1$ (upper solid) and $\gamma = 2$ (lower dashed) cases. Yet the late-time channels cannot be enhanced enough to be dominant.

The current upper bound from GW observations by LIGO-Virgo-KAGRA (LVK) observational run O3 is indicated by the gray dashed lines in Fig. 16.8. Comparing it with the predicted merger rate will implies a constraint on $f_{\rm PBH}$. However, the observational bound on the merger rate is model dependent—it is affected by the characteristics of the binary population, e.g., the mass and redshift dependencies in the merger rates. Thus, a dedicated analysis is needed for accurate GW constraints on $f_{\rm PBH}$ and Fig. 16.8 can at best indicate the order of magnitude. Given the present sensitivity to $f_{\rm PBH}$, both early channels can be relevant to observations. The early two-body channel constraints $f_{\rm PBH} < \mathcal{O}(10^{-3})$ for heavier $m \in 1 - 100 M_\odot$ PBHs, while the early three-body channel implies that the abundance sub-solar mass PBHs should not exceed $\mathcal{O}(10\%)$ [7, 10, 16].

Initial clustering of PBHs affects,[13] in particular, the early Universe binary formation channels. A rough estimate of the effect of initial clustering can be obtained by considering a constant 2-point function at the scales relevant for binary formation, $1 + \xi(\mathbf{x}) = \delta$. As this is equivalent to a local change in the PBH number density, its effect on the merger rate is a simple rescaling: $R(f_{\rm PBH}) \to \delta^{-1} R(\delta f_{\rm PBH})$. However, this scaling does not hold for later processes, such as disruption in haloes, that do not depend on the local PBH density at their formation. The merger rate of the early three-body channel is proportional to a higher power of $f_{\rm PBH}$ than that of the early two-body channel so it is also more sensitive to the initial clustering. Moreover, the initial clustering would enhance the disruption probability of the early binaries, suppressing R_{E2}. On the other hand, R_{E3} gets enhanced even more as the three-body encounters are more likely.

Acknowledgements This work was supported by the European Regional Development Fund through the CoE program grant TK202 and by the Estonian Research Council grants PRG803, PSG869, RVTT3 and RVTT7. The work of V.V. was partially supported by the European Union's Horizon Europe research and innovation program under the Marie Sklodowska-Curie grant agreement No. 101065736.

References

1. S.J. Aarseth, D.C. Heggie. The probability of binary formation by three-body encounters **53**, 259–265 (1976)
2. R. Abbott et al., Population of merging compact binaries inferred using gravitational waves through GWTC-3. Phys. Rev. X **13**(1), 011048 (2023). https://doi.org/10.1103/PhysRevX.13.011048. arXiv: 2111.03634 [astro-ph.HE]

[13] The effect of clustering will affect constraints on the PBHs. In the stellar mass range, which may be probed by terrestrial GW interferometers, it will strengthen constraints on the PBH abundance arising from the Lyman-α observations [9]. As a result, only relatively mild initial clustering is permitted in the mass range that is currently observational available.

3. R. Abbott et al., Search for subsolar-mass binaries in the first half of advanced LIGO and Virgo's third observing run (2021). arXiv: 2109.12197 [astro-ph.CO]
4. N. Afshordi, P. McDonald, D.N. Spergel, Primordial black holes as dark matter: the power spectrum and evaporation of early structures. Astrophys. J. Lett. **594**, L71–L74 (2003). https://doi.org/10.1086/378763. arXiv: astro-ph/0302035
5. N. Aghanim et al., Planck 2018 results. VI. Cosmological parameters. Astron. Astrophys. **641** (2020). [Erratum: Astron.Astrophys. 652, C4 (2021)], A6. https://doi.org/10.1051/0004-6361/201833910. arXiv: 1807.06209 [astro-ph.CO]
6. Y. Ali-Haïmoud, E.D. Kovetz, M. Kamionkowski, Merger rate of primordial black-hole binaries. Phys. Rev. D **96**(12), 123523 (2017). https://doi.org/10.1103/PhysRevD.96.123523. arXiv:1709.06576 [astro-ph.CO]
7. M. Andrés-Carcasona et al., Constraints on primordial black holes from LIGO-Virgo-KAGRA O3 events (2024). arXiv: 2405.05732 [[astro-ph.CO]
8. S. Bird et al., Did LIGO detect dark matter?. Phys. Rev. Lett. **116**(20), 201301 (2016). https://doi.org/10.1103/PhysRevLett.116.201301. arXiv: 1603.00464 [astro-ph.CO]
9. V. De Luca et al., Ruling out initially clustered primordial black holes as dark matter. Phys. Rev. Lett. **129**(19), 191302 (2022). https://doi.org/10.1103/PhysRevLett.129.191302. arXiv: 2208.01683 [astro-ph.CO]
10. G. Franciolini et al., From inflation to black hole mergers and back again: gravitational-wave data-driven constraints on inflationary scenarios with a first-principle model of primordial black holes across the QCD epoch. Phys. Rev. D **106**(12), 123526 (2022). https://doi.org/10.1103/PhysRevD.106.123526. arXiv: 2209.05959 [astro-ph.CO]
11. G. Franciolini et al., Primordial black hole mergers from three-body interactions. Phys. Rev. D **106**(8), 083529 (2022). https://doi.org/10.1103/PhysRevD.106.083529. arXiv: 2205.15340 [astro-ph.CO]
12. J.M. Fregeau et al., Stellar collisions during binary—binary and binary—single star interactions. Mon. Not. Roy. Astron. Soc. **352**, 1 (2004). https://doi.org/10.1111/j.1365-2966.2004.07914.x. arXiv: astro-ph/0401004
13. A. Hall, A.D. Gow, C.T. Byrnes, Bayesian analysis of LIGO-Virgo mergers: primordial versus astrophysical black hole populations. Phys. Rev. D **102**, 123524 (2020). https://doi.org/10.1103/PhysRevD.102.123524. arXiv: 2008.13704 [astro-ph.CO]
14. D.C. Heggie, Binary evolution in stellar dynamics. Mon. Not. Roy. Astron. Soc. **173**, 729–787 (1975). https://doi.org/10.1093/mnras/173.3.729
15. J.G. Hills, Encounters between binary and single stars and their effect on the dynamical evolution of stellar systems **80**, 809–825 (1975). https://doi.org/10.1086/111815
16. G. Hütsi et al., Two populations of LIGO-Virgo black holes. JCAP **03**, 068 (2021). https://doi.org/10.1088/1475-7516/2021/03/068. arXiv: 2012.02786 [astro-ph.CO]
17. D. Inman, Y. Ali-Haïmoud, Early structure formation in primordial black hole cosmologies. Phys. Rev. D **100**(8), 083528 (2019). https://doi.org/10.1103/PhysRevD.100.083528. arXiv: 1907.08129 [astro-ph.CO]
18. N. Ivanova et al., Formation of black-hole X-ray binaries in globular clusters. Astrophys. J. **717**, 948–957 (2010). https://doi.org/10.1088/0004-637X/717/2/948. arXiv: 1001.1767 [astro-ph.HE]
19. N. Ivanova et al., The evolution of binary fractions in globular clusters. Mon. Not. Roy. Astron. Soc. **358**, 572–584 (2005). https://doi.org/10.1111/j.1365-2966.2005.08804.x. (arXiv: astro-ph/0501131)
20. J.H. Jeans, The origin of binary systems **79**, 408 (1919). https://doi.org/10.1093/mnras/79.6.408
21. V. Korol et al., Merger rates in primordial black hole clusters without initial binaries. Mon. Not. Roy. Astron. Soc. **496**(1), 994–1000 (2020). https://doi.org/10.1093/mnras/staa1644. arXiv: 1911.03483 [astro-ph.HE]
22. K. Kritos et al., The astro-primordial black hole merger rates: a reappraisal. JCAP **05**, 039 (2021). https://doi.org/10.1088/1475-7516/2021/05/039. arXiv: 2012.03585 [gr-qc]

23. Y.-Y. Mao et al., Halo-to-halo similarity and scatter in the velocity distribution of dark matter. Astrophys. J. **764**, 35 (2013). https://doi.org/10.1088/0004-637X/764/1/35. arXiv: 1210.2721 [astro-ph.CO]
24. M. Morscher et al., The dynamical evolution of stellar black holes in globular clusters **800**(1), 9 (2015). https://doi.org/10.1088/0004-637X/800/1/9. arXiv: 1409.0866 [astro-ph.GA]
25. H. Mouri, Y. Taniguchi, Runaway merging of black holes: analytical constraint on the timescale. Astrophys. J. Lett. **566**, L17–L20 (2002). https://doi.org/10.1086/339472. (arXiv: astro-ph/0201102)
26. T. Nakamura et al., Gravitational waves from coalescing black hole MACHO binaries. Astrophys. J. Lett. **487**, L139–L142 (1997). https://doi.org/10.1086/310886. (arXiv: astro-ph/9708060)
27. P.C. Peters, Gravitational radiation and the motion of two point masses. Phys. Rev. **136**, B1224–B1232 (1964). https://doi.org/10.1103/PhysRev.136.B1224
28. G.D. Quinlan, The time scale for core collapse in spherical star clusters. New Astron. **1**, 255 (1996). https://doi.org/10.1016/S1384-1076(96)00018-8. arXiv: astro-ph/9606182
29. G.D. Quinlan, S.L. Shapiro, Dynamical evolution of dense clusters of compact stars **343**, 725 (1989). https://doi.org/10.1086/167745
30. M. Raidal et al., Formation and evolution of primordial black hole binaries in the early universe. JCAP **02**, 018 (2019). https://doi.org/10.1088/1475-7516/2019/02/018. arXiv: 1812.01930 [astro-ph.CO]
31. C.L. Rodriguez et al., Modeling dense star clusters in the milky way and beyond with the cluster Monte Carlo code. Astrophys. J. Supp. **258**(2), 22 (2022). https://doi.org/10.3847/1538-4365/ac2edf. arXiv: 2106.02643 [astro-ph.GA]
32. S. Sigurdsson, E.S. Phinney, Binary–single star interactions in globular clusters **415**, 631 (1993). https://doi.org/10.1086/173190
33. N.C. Stone, N.W.C. Leigh, A statistical solution to the chaotic, non-hierarchical three-body problem. Nature **576**(7787), 406–410 (2019). https://doi.org/10.1038/s41586-019-1833-8. arXiv: 1909.05272 [astro-ph.GA]
34. V. Vaskonen, H. Veermäe, Lower bound on the primordial black hole merger rate. Phys. Rev. D **101**(4), 043015 (2020). https://doi.org/10.1103/PhysRevD.101.043015. arXiv: 1908.09752 [astro-ph.CO]

Chapter 17
Gravitational-Wave Backgrounds Associated with Primordial Black Holes

Guillem Domènech

Abstract Primordial black hole formation requires high-density regions in the (random) density field filling the primordial universe. While only the largest (and so rarest) overdensities collapse to form primordial black holes, the rest cause large anisotropic stresses, which are the source of gravitational waves. We provide an overview of the theoretical aspects of the gravitational wave backgrounds associated with primordial black holes from large primordial fluctuations. We consider gravitational wave backgrounds associated with primordial black hole formation, primordial black hole reheating, and unresolved primordial black hole binaries. We present several graphical summaries and illustrations for the busy reader.

17.1 Introduction

Whenever there are fluctuations on top of the isotropic and homogeneous primordial universe, there is a secondary generation of gravitational waves (GWs). Kenji Tomita was the first to write about this effect in the '70s in Ref. [142], saying that "gravitational waves are induced by deformed density perturbations". Our interpretation of Ref. [142] is that by "deformed density perturbations", Tomita meant the anisotropic stress resulting from the presence of density perturbations. This will be clear later from Fig. 17.1. This effect, which we will call induced GWs, was later studied in the '90s by Matarrese, Pantano and Saez [114, 115] and by Matarrese, Mollerach and Bruni [113]. Ten years later, in 2006, we find a glimpse of the potential of induced GWs by Ananda, Clarkson, and Wands [15], where they considered an "excess power in a single mode" from the power spectrum measured by CMB observations (and also using a notation very similar to the current one). A very nice work by Baumann, Steinhardt, Takahashi, and Ichiki [32] hinted that induced GWs could be enhanced during a matter-dominated phase (see also Refs. [12, 21]). But, it was not until 2008

G. Domènech (✉)
Institute for Theoretical Physics, Leibniz University Hannover,
Appelstraße 2, 30167 Hannover, Germany
e-mail: guillem.domenech@itp.uni-hannover.de

C. Byrnes et al. (eds.), *Primordial Black Holes*, Springer Series in Astrophysics and Cosmology, https://doi.org/10.1007/978-981-97-8887-3_17

that Saito and Yokoyama [135, 136] made the connection between induced GWs and primordial black holes (PBHs). Their idea was quickly followed by Bugaev and Klimai [39–41]. For more details on the early history of induced GWs and recent developments, we refer the reader to Ref. [58] for a recent review on the topic by the author of this chapter. Other helpful reviews can be found in Refs. [24, 56, 66, 153].

Induced GWs are a crucial observable for any PBH scenario. As it will be evident by the end of this chapter, an induced GW background must be present if there is (or was) a significant fraction of PBHs in the universe. Though the opposite is not always true, the induced GW background might suggest or strongly exclude PBHs as a significant fraction of DM. This chapter will aim to introduce the main formulas and predictions for the GW backgrounds associated with PBHs, focusing on a qualitative and intuitive understanding of the physics behind such GWs and the interesting parameter space to be probed. The busy reader will find a graphical summary of the parameter space in Fig. 17.2 (for GWs associated with PBH formation) and in Fig. 17.6 (for GWs associated with PBH reheating). Their corresponding physical picture of the generation of induced GWs is respectively illustrated in Figs. 17.1 and 17.5.

The chapter is organized as follows. We first explore in Sect. 17.2 general sources of GWs in cosmology. We then revisit in Sect. 17.3 the definition of the energy density of GWs in cosmology, emphasizing potential ambiguities for induced GWs. In Sect. 17.4, we present the general formulation of induced GWs and focus on the most studied. In Sect. 17.5, we review the GWs from the PBH-dominated early universe. The GW background from PBH binaries is explored in Sect. 17.6. The chapter ends with final remarks in Sect. 17.7.

17.2 Bird's Eye View on Secondary GWs

We start with a general discussion on GWs in cosmology, with the aim of understanding secondary GWs and how they differ from other sources. We use the notation "secondary GWs" to refer to any GW sourced at second order in cosmological perturbation theory using the linear solutions to cosmic fluctuations. We will shortly see what we mean by second order. Our basic framework, which we introduce in detail below, is a perfect fluid in a perturbed FLRW universe (for the basics of cosmology, we recommend Mukhanov's [117] and Baumann's [31] books). A perfect fluid could be an adiabatic perfect fluid with a given equation of state $w = P/\rho$, where P and ρ, respectively, are the pressure and energy density of the fluid. But, it could also be a scalar field, say φ. These two cases are described with the energy-momentum tensor given by

$$T_{\mu\nu} = (\rho + P)\, u_\mu u_\nu + P g_{\mu\nu}\,, \tag{17.1}$$

where $g_{\mu\nu}$ is our spacetime metric and u_μ the normalized 4-velocity of the fluid.[1] The adiabatic perfect fluid describes, e.g., a gas of ultra-relativistic particles when $w = 1/3$. This is the situation we encounter in standard Big Bang cosmology. A gas of non-relativistic particles has $w \ll 1$, which is the case of standard cold DM. On the other hand, the scalar field can be used to describe inflation, reheating scenarios, bubble nucleation, and topological defects.

To get to the equations for GWs, we begin with Einstein Equations, namely $G_{\mu\nu} = T_{\mu\nu}$ with $G_{\mu\nu}$ the Einstein tensor, and perturb both the metric and the matter fields. For simplicity, we work in conformal time and in the shear-free slicing (also known as the Newton gauge) in which the line element reads

$$ds^2 = a^2 \left(-e^{2\Phi} d\tau^2 + e^{2\Psi} \left(e^h\right)_{ij} dx^i dx^j\right), \tag{17.3}$$

where a is the scale factor, τ is conformal time, Φ and Ψ are the gravitational potentials, h_{ij} are the transverse-traceless degrees of freedom, which respresent the GWs and we neglected the transverse degrees of freedom, i.e., vector fluctuations. Note that we used the exponential ansatz for later convenience when discussing the subtleties of second-order perturbation theory. For the matter fields, we have that

$$u_0 = -ae^{\Phi}, u_i = av_i, \rho = \rho(t) + \delta\rho, P = P(t) + \delta P, \tag{17.4}$$

where now $\rho(t)$ and $P(t)$ are the background components, i.e., homogeneous in space, and $\delta\rho$ and δP are the perturbations. Similarly, we take $\varphi \to \varphi(t) + \delta\varphi$ for the scalar field case.

With the perturbative expansion given by Eqs. (17.3) and (17.4), we may write the GW equation at second order in cosmological perturbation theory as [58, 65]

$$h''_{ij} + 2\mathcal{H}h'_{ij} - \Delta h_{ij} = \left[4\partial_i\Phi\partial_j\Phi + 2a^2(\rho + P)v_i v_j\right]^{\rm TT}, \tag{17.5}$$

where $' = d/d\tau, \mathcal{H} = a'/a, \Delta = \delta^{ij}\partial_i\partial_j$ and we expanded Einstein equations up to second order. Also, "TT" refers to the transverse-traceless component of a given tensor (see, e.g., Ref. [58] for the explicit form of the projector). There is no linear source to the GWs in Eq. (17.5) as a perfect fluid has no linear anisotropies. To derive the right-hand side of (17.5), we used that at linear order $\Psi = -\Phi$ as well as the properties of the transverse-traceless projector to write terms like $\Phi\partial_i\partial_j\Phi$ as $-\partial_i\Phi\partial_j\Phi$. It is interesting to note that in Eq. (17.5), the term with Φ comes from $G^{(2)}_{ij}$ and the term with v_i from $T^{(2)}_{ij}$, where the superscript (2) means the expansion at second order. Thus, the term with Φ can be understood as the "backreaction" of

[1] The canonical scalar field case is recovered by identifying

$$u_\mu = \frac{\partial_\mu\varphi}{\sqrt{-\partial_\alpha\varphi\partial^\alpha\varphi}}\ , \quad \rho = -\frac{1}{2}\partial_\alpha\varphi\partial^\alpha\varphi + V(\varphi), \quad P = -\frac{1}{2}\partial_\alpha\varphi\partial^\alpha\varphi - V(\varphi), \tag{17.2}$$

where $V(\varphi)$ is the potential of the scalar field.

the metric, while the term with v_i is the standard matter source to GWs. With the above equation, we may qualitatively discuss cosmic sources of GWs.

Let us start with the case in which GWs are generated on scales much smaller than the Hubble radius (also called Hubble horizon) by a "spectator" field, i.e., the field responsible for the source of GWs is a subdominant component of the universe's energy density. This includes, e.g., nucleation of bubbles due to a first-order phase transition or the formation and evolution of a network of solitons, such as cosmic strings, domain walls, oscillons, etc. In this regime, also called the weak gravity limit, one may safely neglect the contribution from Φ in Eq. (17.5). This is because, on subhorizon scales, the term containing Φ in Eq. (17.5) is suppressed with respect to the v_i contribution by the ratio of energy densities between the spectator and the dominant field. Thus, GWs are sourced by the anisotropic stress resulting from the inhomogeneous velocity flows or scalar field profiles. For instance, in the case of bubble nucleation, one of the primary sources of GWs are the acoustic density waves following bubble collisions (though the collisions themselves and later turbulences also produce GWs; see Refs. [23, 44, 134] for recent reviews). For cosmic strings, it may be better to think of Eq. (17.5) in terms of the quadrupole formula (as for a local source, the volume integral of T_{ij} is related to the second time derivative of the quadrupole). Then, a global cosmic string may be thought of as a "vibrating cylinder" with a length proportional to the Hubble radius. Its time-dependent quadrupole is then the source of GWs. There are, of course, other processes that lead to GW production, such as cosmic string loops and kinks. As the present discussion is very naive, and the above-mentioned process entirely non-linear, we refer the interested reader again to Refs. [44, 134] for recent reviews with more detailed discussions and references therein.

A somewhat similar situation applies to secondary GWs associated with PBHs. But, in contrast with previous cases, secondary GWs are generated by the dominant fluid in the universe. And, the dominant fluid in the universe determines the *curvature fluctuations*. Thus, we cannot generally neglect the term $\partial_i\Phi\partial_j\Phi$ in Eq. (17.5). This is why one says that the GWs are induced by curvature fluctuations and are called induced GWs for short. We illustrate the physical picture of the generation of induced GWs in Fig. 17.1. There, we considered a universe filled with random Gaussian superhorizon curvature fluctuations with a sharp log-normal power spectrum, peaked at $k_{\rm peak}$ and with a large enough amplitude. The left plot of Fig. 17.1 shows the 2D real space distribution of the density field, say δ, or the curvature fluctuation. We see that only a few lumps (in this case, roughly one) have a large enough amplitude to eventually collapse and form a PBH once the Hubble horizon is approximately the size of the overdensity. The right plot of Fig. 17.1 illustrates with blue arrows the velocity flow (as being proportional to $\partial_i\delta$ or $\partial_i\Phi$) and with black crosses the eigenvectors of the traceless part of T_{ij} (proportional to v_iv_j). Larger velocities have larger arrows and a lighter blue color. The size of the crosses is proportional to the eigenvalues of the traceless part of T_{ij}.[2] See how larger "sources" to the induced GWs

[2] We only consider the traceless part of T_{ij} as in 2D there are no transverse-traceless degrees of freedom. Nevertheless, we expect that a similar situation applies to a more realistic 3D picture.

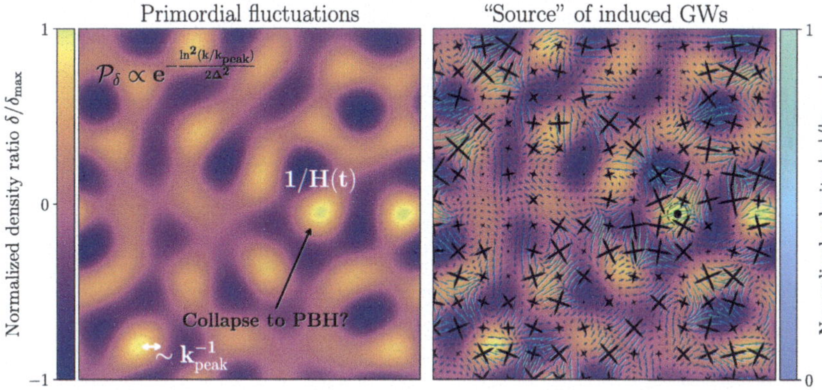

Fig. 17.1 Real space density field in 2D from Gaussian fluctuations with log-normal primordial spectrum with $\Delta = 0.01$ and $k_{\rm peak} = 0.005 L_{\rm box}^{-1}$ (i.e., the typical size of fluctuations is 20 times smaller than the box), all in arbitrary units. On the left panel, we show the density ratio and illustrate that only the rarest and largest fluctuations collapse to form PBHs. On the right panel, we show the velocity gradients and the eigenvectors times the eigenvalues of the traceless component of the energy-momentum tensor. This exemplifies that induced GWs are generated by the mean fluctuations where gradients are largest

are associated with larger velocity flows. We also see that while only one overdensity collapses to form a PBH, the rest source the induced GWs.

To put some units, we show in Fig. 17.2 the frequency of the induced GWs $f_{\rm iGWs}$ (see Eq. (17.25)), the typical mass of the PBHs $M_{\rm PBH,f}$ and the frequency related to the Innermost Stable Circular Orbit (ISCO), $f_{\rm ISCO}$, of the merger of an equal PBH binary (in the binary frame, see Eq. (17.51)). The redshift factor is the main difference between $f_{\rm iGW}$ and $f_{\rm ISCO}$. While induced GWs are generated at the same time PBHs form, the detected PBH binaries merge in the nearby universe. See how with GW detectors such as ET, LISA, and PTAs, we can indirectly probe the existence of PBH with masses ranging from $O(10^9\,{\rm g})$ to $O(M_\odot)$ with induced GWs. In fact, there may already be hints of an induced GW background at nHz frequency from PTAs [4–7, 17–19, 129, 130, 148, 156]. Interestingly, the same GW detectors can probe the mass range from $O(M_\odot)$ to $O(10^9 M_\odot)$ from the merger of PBHs binaries. In the same figure we illustrate the most interesting PBH mass windows to explain the DM [45], the planet mass candidates events in OGLE [116, 119], the BH mergers seen by the LVK collaboration [35, 70, 137] and the seeds of supermassive black holes (SMBHs) [46].

17.3 Effective Energy Density of GWs in Cosmology

Before we dig into the details of the calculations of induced GWs, we briefly revisit the definition of the energy density of GWs in cosmology, as we will recursively use it to derive predictions. We will discuss important subtleties of the definition of

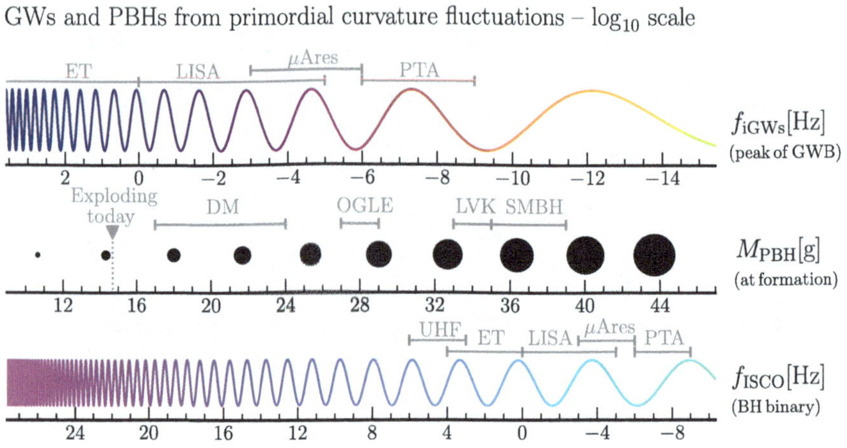

Fig. 17.2 Graphical summary of the relevant parameter space regarding GW frequency and PBH mass for the induced GWs (associated with PBH formation). In the top we show the frequency of induced GWs $\log_{10} f_{\rm iGWs}$ (17.25), in the middle we show PBH mass $\log_{10} M_{\rm PBH}$ and in the bottom we show the frequency of PBH mergers $\log_{10} f_{\rm ISCO}$ (17.51). They are all related to $M_{\rm PBH}$ by Eq. (17.24) through $k_{\rm peak} = a_{\rm f} H_{\rm f}$. We show the frequency range of current and future GW detectors such as Ultra-High-Frequency (UHF) GW detectors, ET, LISA, μ-Ares, and PTAs. We also draw the interesting PBH mass windows: PBH as all DM [45, 81], planet mass PBH to explain OGLE candidate event [116, 119], the merger of solar mass PBHs at LVK [35, 70, 137] and PBH as the seeds of SMBH [46]. Note that the PTA window for induced GW overlaps with OGLE and LVK

energy density of the induced GWs at the end of this section. The derivation of the energy-momentum tensor can be found, e.g., in Maggiore's book [108]. A detailed discussion on the subtleties is given in Ref. [63], and references therein.

17.3.1 *The Pseudo Energy Momentum Tensor of GWs*

To estimate the energy density carried by the GWs, we need to find the "backreaction" of *linear* tensor fluctuations onto the background metric. To do that, we expand Einstein equations to second order (as we did in Eq. (17.5)) and compute some kind of average. This average could be a volume and/or a time average, inside which several GW wavelengths or oscillations fit. In the case of stochastic fields, the volume/time average is equivalent to the ensemble average (by ergodicity). Expanding and taking the average, we formally find that

$$G^{(0)}_{\mu\nu}[a] \approx -\langle G^{(2)}_{\mu\nu}[a(\eta + h)]\rangle + T^{(0)}_{\mu\nu} \approx t^{\rm GW}_{\mu\nu} + T^{(0)}_{\mu\nu}\,, \tag{17.6}$$

where η refers to the Minkowski metric $\eta_{\mu\nu}$, h to h_{ij} of Eq. (17.3) and the brackets denote the chosen average. We moved the second order expansion $G^{(2)}_{\mu\nu}$ to the right-hand side so that we can interpret it as the energy-momentum tensor, which we call $t^{\rm GW}_{\mu\nu}$ in the last step. The contribution from linear fluctuations vanishes because of the linear equations of motion. After some algebra, one obtains that

$$t^{\rm GW}_{\mu\nu} = \frac{1}{4}\left\langle \partial_\mu h^{ij}\partial_\nu h_{ij} - \frac{1}{2}g_{\mu\nu}\partial_\sigma h^{ij}\partial^\sigma h_{ij}\right\rangle . \tag{17.7}$$

A fundamental assumption in deriving Eq. (17.7) is that we only consider subhorizon tensor fluctuations h_{ij}. This gives us an unambiguous notion of a slowly evolving, large-scale background and high-frequency, small-scale GWs. It is not clear how to apply a similar procedure when both scales are similar, e.g., for superhorizon tensor modes. Also, note that $t^{\rm GW}_{\mu\nu}$ is called a pseudotensor as it only transforms as a tensor at the linear level in perturbation theory. It is a priori not well-defined at higher orders.

The energy density of the GWs is given by the 00 component of Eq. (17.7), namely $a^2\rho_{\rm GW} = t^{\rm GW}_{00}$ in conformal time. Since h_{ij} is a random field, we can compute the energy density in terms of its power spectrum. Working with Fourier modes,[3] the GW energy density fraction per $\ln k$, also referred to as the spectral density, is given by

$$\Omega_{\rm GW}(k,\tau) = \frac{1}{\rho_{\rm tot}}\frac{d\rho_{\rm GW}}{d\ln k} = \frac{k^2}{12\mathcal{H}^2}\sum_\lambda \overline{\mathcal{P}_{h,\lambda}(k,\tau)}\,, \tag{17.9}$$

where $\mathcal{P}_{h,\lambda}(k,\tau)$ is the dimensionless strain power spectrum, defined by

$$\langle h_{\mathbf{k},\lambda}(\tau)h_{\mathbf{k}',\lambda'}(\tau)\rangle = (2\pi)^3\delta(\mathbf{k}+\mathbf{k}')\delta_{\lambda\lambda'}\tfrac{2\pi^2}{k^3}\mathcal{P}_{h,\lambda}(k,\tau)\,. \tag{17.10}$$

We also took an additional oscillation (time) average denoted with an overline, as the same average is used when looking for GW backgrounds. It should be noted that in deriving Eq. (17.9) we used that for subhorizon GWs $k \gg \mathcal{H}$ and that $h'_{\mathbf{k},\lambda} \sim k h_{\mathbf{k},\lambda}$. Equation (17.9) is the quantity we will use to derive and compare our predictions with observations. The energy density of GWs is related to the strain noise variance.

---

[3] Our convention is that

$$h_{ij} = \frac{1}{(2\pi)^3}\sum_\lambda \int d^3k\, h_{\mathbf{k},\lambda}(\tau)\epsilon^\lambda_{ij}(\mathbf{k})\, e^{i\mathbf{k}\cdot\mathbf{x}}\,, \tag{17.8}$$

where polarization tensors are normalized, i.e. $(\epsilon^\lambda_{ij}(\mathbf{k}))^*\epsilon^{ij,\lambda'}(\mathbf{k}) = \delta^{\lambda\lambda'}$.

17.3.2 Subtleties of Induced GWs

Before ending this section, let us point out the fundamental ambiguity in the definition (17.7) of the energy density of GWs. We can convince ourselves that the definition is ambiguous by checking whether $\rho_{\rm GW}$ is invariant under a shift of the conformal time coordinate, e.g., $\bar{\tau} = \tau + T(\tau, x^i)$, where T is an infinitesimal gauge parameter. With some effort, one finds that the tensor perturbations defined as in Eq. (17.3) transform according to [65]

$$\bar{h}_{ij} = h_{ij} + \widehat{TT}_{ij}^{kl}\left[T\, h'_{kl} + \partial_k T \partial_l T\right] + ..., \tag{17.11}$$

where "..." means higher order combinations in T and h_{ij} and we neglected vector perturbations. It turns out that what we called tensor perturbations $\bar{h}_{ij}$, in the barred time coordinate, are a mixture of tensor and scalar perturbations in the unbarred time coordinate. In other words, the definition of tensor perturbations beyond the linear order depends on the time slicing. The energy density of GWs as defined by Eq. (17.7) is certainly not invariant at second order under the time shift (17.11).

It is not so surprising that Eq. (17.7) becomes problematic at second order. First, we considered only linear perturbations, and second, there is no local notion of energy in gravity, as we can always go to a local inertial (Minkowski) frame. Although this is a general problem, it is not so problematic for GW sources in the weak gravity regime, such as cosmic strings and phase transitions (where secondary GWs from Φ are already neglected), unless one deliberately chooses a strange time coordinate. But, it may be problematic for induced GWs since they are a second-order effect from Φ. For instance, considering that the unbarred time corresponds to the time in the Newton gauge, we have that $T = T(\Phi, \tau)$, with the exact functional form depending on the chosen barred time slicing. Since $h_{ij}^{\rm iGW} \propto \partial_i \Phi \partial_j \Phi$ all the terms in the right-hand side of Eq. (17.11) can be in principle comparable. This issue was first pointed out by Ref. [87] for a matter-dominated universe, in Ref. [79] in radiation domination and in general in Ref. [141]. See also Ref. [140]. It was then argued in Refs. [54, 89] that in radiation domination, the synchronous (which should describe the frame of a GW detector) and Newton gauges give the same results at late times. Then, in Ref. [63], it was shown that such approximate gauge invariance applies to other gauges and for any cosmological background (except matter domination; see below).

The crucial point to realize, which provides a partial resolution, is that we measure GWs far from the source [63]. The same is required, e.g., to define the GWs from binaries. In cosmology, "far" from the source means that there is effectively no more production of GWs at some point in time. This is precisely the situation for induced GWs. The curvature perturbation Φ quickly decays once it enters the Hubble horizon, rendering the energy density of induced GWs practically independent of the time slicing.[4] For instance, the calculations in the Newton gauge, constant Hubble gauge, flat gauge, and synchronous gauge (after proper gauge fixing) all yield the same results. The exceptions to this statement are (i) comoving slices and (ii) a dust-dominated universe. In exception (i), one can argue that such time slices are suitable for superhorizon but not for subhorizon physics. For example, on small scales, the velocity flow can become quite non-trivial and choosing a time slice that is comoving with the fluid everywhere is undoubtedly a convoluted choice. Thus, it is not surprising that the comoving slicing might contain gauge artifacts that look like GWs. In exception (ii), the curvature perturbation is constant on all scales, so there is a constant source to the induced GWs. In this case, one must follow the GW production until the end of the dust-dominated era to have an unambiguous definition of $\Omega_{\rm GW}$. The resolution to the ambiguity lies in a systematic study of the strain measured by a GW detector at second order in cosmological perturbation theory. More details and references on the gauge issue can be found in Ref. [58].

17.4 GWs from Primordial Curvature Fluctuations

After the general discussions in Sects. 17.2 and 17.3, we focus on the mathematical formulation of induced GWs. We first take a general approach but quickly turn to the most studied case in Sect. 17.4.2. Extensions of the formalism will be discussed in Sect. 17.4.4. Details of the main calculations can be found in Refs. [57, 58, 99].

17.4.1 General Formulation

Consider that, after inflation, there are some initial curvature fluctuations Φ in the universe. Since induced GWs are sourced by the dominant fluid in the universe, the scalar component of velocity fluctuations in Eq. (17.5) is also related to the curvature perturbation via Einstein equations (the $0i$ component to be precise). For this reason, we can take a general approach to the source term in Eq. (17.5) by saying that, in Fourier modes,[5] the source term can be split into a primordial value of the curvature perturbation $\Phi^{\rm prim}_{\mathbf{k}}$ times a time-dependent transfer function. A general equation for the induced tensor modes then reads

$$h''_{\mathbf{k},\lambda} + 2\mathcal{H}h'_{\mathbf{k},\lambda} + k^2 h_{\mathbf{k},\lambda} = 4\int \frac{d^3q}{(2\pi)^3} e^{ij}_{\lambda}(k) q_i q_j f(\tau, q, |\mathbf{k}-\mathbf{q}|)\Phi^{\rm prim}_{\mathbf{q}}\Phi^{\rm prim}_{|\mathbf{k}-\mathbf{q}|}\,, \tag{17.12}$$

[4] By almost independent, we mean that the prediction for the energy density of induced GWs is independent of the time slicing up to corrections suppressed by $(k/\mathcal{H})^2$ (a ratio which evaluated today is tiny) for a restricted class of time slicings.

[5] The Fourier expansion for Φ follows the same convention as Eq. (3) for h_{ij}.

where $f(\tau, q, |\mathbf{k}-\mathbf{q}|)$ is the transfer function of the source. We will discuss particular cases later. We can also write down a formal integral solution to Eq. (17.12) by means of the Green's function method, which yields

$$h_{\mathbf{k},\lambda}(\tau) = \frac{4}{k^2}\int \frac{d^3q}{(2\pi)^3} e_\lambda^{ij}(\mathbf{k}) q_i q_j I(\tau, q, |\mathbf{k}-\mathbf{q}|)\Phi^{\rm prim}_{\mathbf{q}}\Phi^{\rm prim}_{|\mathbf{k}-\mathbf{q}|}\,, \tag{17.13}$$

where we defined for compactness

$$I(\tau, q, |\mathbf{k}-\mathbf{q}|) = k^2 \int_0^\tau d\tau'\, G_h(\tau,\tau') f(\tau', q, |\mathbf{k}-\mathbf{q}|)\,. \tag{17.14}$$

$I(\tau, q, |\mathbf{k}-\mathbf{q}|)$ is often called the kernel of induced GWs, and G_h in Eq. (17.14) is the Green's function for tensor modes which in radiation domination reads $kG_h(\tau,\tau') = \frac{a(\tau)}{a(\tau')}\sin(k(\tau-\tau'))$. The factors k^2 in Eqs. (17.13) and (17.14) are introduced for later convenience, as in this way, I is dimensionless.

With our solution, Eq. (17.13), we can compute the oscillation average of the dimensionless power spectrum that enters in the spectral density of GWs, Eq. (17.9). After some calculations, we obtain that

$$\begin{aligned}\overline{\mathcal{P}_{h,\lambda}(k,\tau)} = \frac{4k}{\pi}\int \frac{d^3q}{(2\pi)^3}\int \frac{d^3q'}{(2\pi)^3}\overline{I(\tau, q, |\mathbf{k}-\mathbf{q}|) I(\tau, q', |\mathbf{k}-\mathbf{q}'|)} \\ \times e_\lambda^{ij}(\mathbf{k}) q_i q_j e_\lambda^{ij}(-\mathbf{k}) q'_i q'_j \langle \Phi^{\rm prim}_{q}\Phi^{\rm prim}_{|\mathbf{k}-\mathbf{q}|}\Phi^{\rm prim}_{q'}\Phi^{\rm prim}_{|\mathbf{k}-\mathbf{q}'|}\rangle'\,,\end{aligned} \tag{17.15}$$

where we assumed that $\langle h_\lambda(k,\tau) h_{\lambda'}(k',\tau)\rangle$ follows the form of Eq. (17.10) so that we may subtract the factor $(2\pi)^3\delta_{\lambda\lambda'}\delta(\mathbf{k}+\mathbf{k}')$. We denote the remaining contribution to the 4-point function of Φ with a prime. Note that the oscillation average in Eq. (17.15) must be done once the tensor modes are deep inside the Hubble horizon, i.e., for $k\tau \gg 1$. We now restrict ourselves to the case of Gaussian primordial fluctuations (which is often a good approximation), where the 4-point function is determined solely by the 2-point function. In this case, we find that

$$\begin{aligned}\overline{\mathcal{P}_h(k,\tau)} &= \sum_\lambda \overline{\mathcal{P}_{h,\lambda}(k,\tau)} \\ &= 8\int_0^\infty dv \int_{|1-v|}^{1+v} du \left(\frac{4v^2-(1-u^2+v^2)^2}{4uv}\right)^2 \overline{I^2(\tau,k,v,u)}\mathcal{P}_\Phi(ku)\mathcal{P}_\Phi(kv)\,,\end{aligned} \tag{17.16}$$

where we introduced $v = q/k$ and $u = |\mathbf{k}-\mathbf{q}|/k$ for compactness.

It is worth noting that, in general, induced GWs are sourced by the typical curvature fluctuations, that is, $\langle\Phi^2\rangle^{1/2}$. In other words, the naive expectation is that $\langle\Phi^4\rangle \sim O(\langle\Phi^2\rangle^2)$, unless the probability distribution function (PDF) of Φ is highly skewed (i.e., something very far from the Gaussian quantum fluctuations expected from inflation). In contrast, PBHs form from rare curvature fluctuations, which are determined by the tail of the PDF. Thus, while PBHs are exponentially sensitive to the root mean square of curvature fluctuations, induced GWs are proportional to the fourth power. From this, we conclude that the detection of induced GWs does not generally imply a significant fraction of PBHs in the universe. However, a substantial fraction of PBHs must have a detectable induced GW signal.

17.4.2 Semi-Analytical Formula in Radiation Domination

Let us now revisit the most considered case, primordial adiabatic (curvature) fluctuations in a radiation-dominated universe. The first semi-analytical formula was derived in Refs. [69, 99] and was generalized in Refs. [57, 62]. We will discuss more possibilities in Sect. 17.4.4. In the case that one fluid dominates the universe, the $0i$ component of Einstein's equations yields

$$\Phi' + \mathcal{H}\Phi = -\frac{1}{2}a^2(\rho + P)v\,, \tag{17.17}$$

where v is defined through $v_i = \partial_i v$. Additionally, the curvature fluctuations evolve according to

$$\Phi(q\tau) = \Phi^{\rm prim}_{\mathbf{q}}\,\frac{3j_1(c_s q\tau)}{c_s q\tau}\,, \tag{17.18}$$

where j_1 is the spherical Bessel function of order 1.

Inserting the relation (17.17) between v and Φ in Eq. (17.5), and using the linear solution Eq. (17.18), we find the explicit form of $f(\tau, q, |\mathbf{k}-\mathbf{q}|)$ in Eq. (17.12). The transfer function $f(\tau, q, |\mathbf{k}-\mathbf{q}|)$ is a product of sines and cosines, and the integral $I(\tau, q, |\mathbf{k}-\mathbf{q}|)$ (17.14) can be carried out analytically. Then, combining Eqs. (17.9) and (17.16), we find a compact form for the GWs induced during standard radiation domination (that is $c_s^2 = w = 1/3$), namely

$$\Omega_{\rm GW,*} = \int_0^\infty dv \int_{|1-v|}^{1+v} du\, \mathcal{T}_{RD}(u, v)\mathcal{P}_{\mathcal{R}}(ku)\mathcal{P}_{\mathcal{R}}(kv)\,, \tag{17.19}$$

where we used Eq. (17.22) and defined

$$\mathcal{T}_{RD}(u,v) = \left(\frac{4v^2 - (1-u^2+v^2)^2}{4u^2v^2}\right)^2 \times y^2 \left\{ \frac{\pi^2}{4} y^2 \Theta[u+v-c_s^{-1}] + \left(1 - \frac{1}{2} y \ln\left|\frac{1+y}{1-y}\right|\right)^2 \right\}, \quad (17.20)$$

and

$$y = \frac{u^2+v^2-c_s^{-2}}{2uv}. \quad (17.21)$$

The meaning of the subscript "*" will be explained in the next paragraph. Note that to obtain Eq. (17.20), we took the limit $k\tau \to \infty$ of the integral (17.14). This is a very good approximation since we are interested in the GWs generated in the very early universe. So, any correction scales at least as $\mathcal{O}(1/(k\tau_{\rm eq}))$ where $\tau_{\rm eq}$ is the time of matter-radiation equality. Recall that the comoving scale that entered the Hubble horizon at that time is $k_{\rm eq} \sim 0.01\,{\rm Mpc}^{-1}$, and we are interested in scales below a parsec. For the general form of the kernel for finite $k\tau$, see Ref. [99]. We also used that

$$\Phi = \frac{3(1+w)}{5+3w}\mathcal{R} = \frac{2}{3}\mathcal{R}, \quad (17.22)$$

since predictions from inflation are given in terms of the curvature perturbation on comoving slices $\mathcal{R}$. We note that the variable y, defined in Eq. (17.21), is far from arbitrary. It is actually related to the area of the triangle composed by 1, $c_s v$ and $c_s u$ (or k, $c_s q$ and $c_s|\mathbf{k}-\mathbf{q}|$ in the original variables) and it naturally appears when integrating the product of three Bessel functions. In the end, it is a very useful variable for the extension to general constant w and c_s. In the integration range of Eq. (17.16), we have that y is bounded by $-\infty < y < 1$.

Strictly speaking, instead of taking the limit $k\tau \to \infty$, one should integrate the kernel (17.14) until today, that is at $\tau = \tau_0$. However, for the GW frequencies of interest, it is unnecessary to do the integrals across all the radiation, matter, and dark energy domination phases. We may use the fact that GWs, once inside enough the Hubble horizon, behave as radiation. From there on, we can track the energy density ratio of GWs by comparing it to that of standard radiation. Thus, if we call τ_* the time at which a given GW frequency starts behaving as radiation, we can write [58, 93]

$$.\Omega_{\rm GW,0}h^2 = 1.62\times 10^{-5}\left(\frac{\Omega_{\rm rad,0}h^2}{4.18\times 10^{-5}}\right)\left(\frac{g_\rho(T_*)}{106.75}\right)\left(\frac{g_s(T_*)}{106.75}\right)^{-4/3}\Omega_{\rm GW,*}, \quad (17.23)$$

where the subscript "$*$" means evaluation at $\tau = \tau_*$, $\Omega_{\rm rad,0}$ is energy density ratio of radiation today [8] and $h = H_0/(100\,{\rm km/s/Mpc})$. $g_\rho(T_*)$ and $g_s(T_*)$ are the effective degrees of freedom in the energy density and entropy of the standard radiation fluid. Thus, our goal is to estimate $\Omega_{\rm GW,*}$ and use Eq. (17.23) to find the amplitude of the induced GW background today. In passing, using that PBHs form with a mass given by

$$M_{\rm PBH,f} = 4\pi\gamma M_{\rm pl}^2/H_{\rm f}\,, \tag{17.24}$$

where $\gamma \sim 0.2$ is an effective parameter (see, e.g., Ref. [81, 97, 138] for reviews),[6] and that $k_{\rm peak} = a_{\rm f}H_{\rm f}$ where "f" refers to PBH formation, we can write $f_{\rm iGWs}$ of Fig. 17.2 in terms of $M_{\rm PBH,f}$ as

$$f_{\rm iGWs} \approx 1\,{\rm Hz}\left(\frac{M_{\rm PBH,f}}{10^{16}{\rm g}}\right)^{-1/2}\left(\frac{\gamma}{0.2}\right)^{1/2}\left(\frac{g_\rho(T_*)}{106.75}\right)^{1/4}\left(\frac{g_s(T_*)}{106.75}\right)^{-1/3}. \tag{17.25}$$

17.4.3 Typical Features and General Properties

Let us derive some useful and general properties of the induced GW spectrum from Eq. (17.19). This will provide us with intuition about the expected shape of the induced GW spectrum from a given primordial spectrum of curvature fluctuations, assuming that the primordial spectrum is peaked. We show a few examples of typical induced GW spectra in Fig. 17.7.

The first thing one may notice from the averaged square kernel Eq. (17.20) is that it contains a logarithm that diverges at $y = \pm 1$. The value $y = -1$ occurs when $u + v = c_s^{-1}$. The origin of such divergence is a resonant production of induced tensor modes. In fact, Eq. (17.5) is nothing but a damped harmonic oscillator with a periodic force proportional to Φ^2. So there is a resonance for $k = c_s(q + |\mathbf{k} - \mathbf{q}|)$, where k is the tensor and q the scalar wavenumbers.[7] Thus, for a peaked enough spectra, the resonance lies at

$$k_{\rm res} = 2c_s k_{\rm peak}\,. \tag{17.26}$$

[6] The value $\gamma \sim 0.2$ is based on early estimates [138], but there are recent and more accurate estimations of such value [68, 81].

[7] This intuition is only valid if the source in Fourier space (17.12) has a well-defined frequency. This only occurs when fluctuations of Φ have a typical scale, or in other words, the power spectrum of Φ is sufficiently peaked. From the point of view of the integral (17.16), the exact resonance is a zero-measure point. So if the primordial spectrum is broad, the resonance is smeared out.

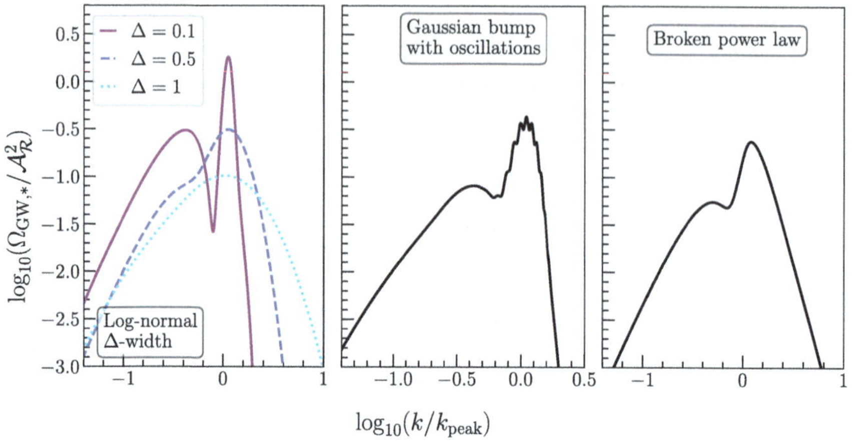

Fig. 17.3 Induced GW spectra in radiation domination computed numerically using SIGW-fast [146]. On the left panel, we show the induced GWs from a log-normal peak for various values of the logarithmic width Δ. See how, as the width increases, the resonant peak is smoothed out. On the middle panel, we show the induced GW from a Gaussian bump with $\mathcal{O}(1)$ oscillations in the primordial spectrum from Ref. [72]. Note how the oscillations in the induced GW spectrum are only $\mathcal{O}(10\%)$ modulations at the peak. On the right panel, we show the induced GW spectrum from a broken power-law primordial spectrum with k^4 in the IR and k^{-5} in the UV. See how the induced GW spectrum has a similar shape for all relatively sharp peaks. The IR tail of the induced GW spectrum in all cases goes as $k^3 \ln^2 k$ (except for the very sharp Δ case, in magenta, which initially is $k^2 \ln^2 k$)

The resonant peak is clearly visible in Fig. 17.3, where we show the induced GW spectra of a log-normal primordial spectrum with logarithmic with Δ, a Gaussian bump with oscillations and a broken power-law spectrum, respectively from left to right. On the left plot, we also plot the induced GW spectrum for various values of the width. See how as one increases Δ, the resonant peak disappears. For analytical approximations in the case of the log-normal primordial spectrum, see Ref. [126]. For the Gaussian bump with oscillations, see Refs. [36, 72, 73, 147]. And, for the broken power-law, see Refs. [22, 104, 149].

The value $y = 1$ corresponds to the limit $u \sim v \gg 1$, which is related to long wavelength tensor modes $k \ll q$. Assuming that $\mathcal{P}_\mathcal{R}(k) \propto \mathcal{A}_\mathcal{R}(k/k_{\rm peak})^n$ (with appropriate cut-off) the very low frequency, or far Infra-Red (IR), tail reads [22]

$$\Omega^{\rm IR}_{\rm GW,*}(k \ll k_{\rm peak}) \propto \begin{cases} \mathcal{P}^2_\mathcal{R}(k) & \text{for} \quad 0 < n < 3/2 \\ \left(\frac{k}{k_{\rm peak}}\right)^3 \ln^2\left(\frac{k}{k_{\rm peak}}\right) & \text{for} \quad n > 3/2 \end{cases} . \tag{17.27}$$

The first case of Eq. (17.27) is also valid for broad enough peaks.[8] This is, for example, the case of a broad log-normal for scales not too far from the peak. For scales very far from the peak, one has a clear exponential decay, and one recovers the universal k^3 scaling. The second case corresponds to sharp enough peaks. Note that for the case of a Dirac delta primordial spectrum (corresponding to the $\Delta \to 0$ limit of the log-normal) or a very sharp peak, the IR tail goes as k^2 instead of k^3. For more details on the universal IR scaling, see Ref. [42]. The logarithmic running in the IR tail seems to be a characteristic of induced GWs in a radiation dominated universe [43, 152]. However, it is not present for general values of w [62].

Lastly, we consider the value $y \gg 1$, which corresponds to $k \gg k_{\rm peak}$. Assuming the same power-law ansatz but with $n < 0$, the high frequency, or Ultra-Violet (UV), the tail is given by [22]

$$\Omega^{\rm UV}_{{\rm GW},*}(k \gg k_{\rm peak}) \propto \begin{cases} \mathcal{P}^2_{\mathcal{R}}(k) & \text{for} \quad 0 > n > -4 \\ \left(\frac{k}{k_{\rm peak}}\right)^{-4} \mathcal{P}_{\mathcal{R}}(k) & \text{for} \quad n < -4 \end{cases} . \tag{17.29}$$

From Eq. (17.29), we see that for sharply peaked primordial spectra, the UV tail is quickly suppressed. In fact, for infinitely sharp primordial spectra (i.e., a Dirac delta), there is a sharp cut-off at $k = 2k_{\rm peak}$ due to momentum conservation. Namely, one cannot induce tensor modes with a frequency higher than the source term. Nevertheless, if the primordial spectrum is not too steep, we may obtain information from the UV tail. Examples of this are the inference of the magnitude of local non-Gaussianities [22] or the presence of a second phase of slow-roll inflation after the enhancement of fluctuations [25]. We also note that the k^4 scaling of the second case in Eq. (17.29) is sensitive to the value of c_s. When $c_s \to 1$, it gradually transitions to a k^2 scaling [25]. One can see the power-law scalings (or the sharp cut-off) in the induced GW spectra of Fig. 17.3. In the IR tail, all cases decay as $k^3 \ln^2 k$. Instead, the UV tail is more sensitive to the shape of the primordial spectrum.

We end with a helpful rough estimate of the peak amplitude of the GW spectrum today from Eqs. (17.23), (17.27) and (17.29), disregarding the presence of the resonant peak. This is given by

$$\Omega^{\rm peak}_{{\rm GW},0} h^2 \approx 10^{-6}\, \mathcal{P}^2_{\mathcal{R}}(k_{\rm peak})\,, \tag{17.30}$$

which is good enough so that it also applies to a flat primordial spectrum (see Eq. (8)). From Eq. (17.30), we see that the amplitude of the primordial spectrum of fluctuations

---

[8] Looking at both Eqs. (17.27) and (17.29), we see that for broad enough spectra, a good estimate to the induced GW spectrum is given by $\Omega_{{\rm GW},*}(k) \sim \mathcal{P}^2_{\mathcal{R}}(k)$. In the limit of a flat spectrum (that is $\mathcal{P}_{\mathcal{R}} \approx$ constant), one numerically finds that [99]

$$\Omega^{\rm flat}_{{\rm GW},*} \approx 0.82\, \mathcal{P}^2_{\mathcal{R}}\,. \tag{17.28}$$

measured by CMB observations, around 10^{-9} [11], yields unobservable induced GWs. We need an enhancement of more than 4 orders of magnitude to have an observable GW signal.

17.4.4 Other Extensions

It is important to consider physics beyond the simplest model of Gaussian primordial fluctuations and a radiation-dominated universe. First, we have not probed the early universe at such small scales, so one should be as agnostic as possible. Second, there are several attractive models of inflation and reheating (and much more; see, e.g., Ref. [14] for a review) which predict departures such as non-Gaussianities of primordial fluctuation and a matter or kinetic dominated phase of reheating, respectively with $w \ll 1$ (for an oscillating moduli field) and $w \approx 1$ (in quintessential inflation models). As we shall see, the induced GW spectrum depends on the initial conditions of primordial fluctuations (adiabatic/isocurvature and Gaussian/non-Gaussian) and the model of the primordial universe, e.g., the equation of state of the universe at the time of wave generation. Below, we briefly mention four different possibilities. We illustrate the effects of these possibilities on the induced GW spectrum in Fig. 17.4.

17.4.4.1 General Equation of State of the Very Early Universe

We know the universe had to be dominated by relativistic particles of the standard model around the time of BBN. More precisely, the radiation in the universe should have thermalized at $T \sim 4\,\text{MeV}$ [83, 84, 95, 96] for a successful BBN. We also have strong evidence for a period of cosmic inflation from CMB observations [11]. However, we do not know how or when the universe was re-heated. So, it is entirely plausible that there was a re-heating phase for $T > 4\,\text{MeV}$ with a different equation of state w than radiation, e.g., a stiff fluid or a scalar field. If induced GWs (and PBHs) are generated during this w-dominated period, the shape of the induced GW spectral density will be modified. This is studied in detail in Refs. [57, 62]. The main changes are: (i) an enhancement/suppression due to the different background expansion, (ii) a different position of the resonant peak at $k_{\rm res} = 2c_s k_{\rm peak} = 2\sqrt{w}k_{\rm peak}$ and (iii) a different power-law in the low-frequency tail going as $f^{3-2|b|}$ where $b = (1 - 3w)/(1 + 3w)$. We illustrate these effects on the left of Fig. 17.4. In this way, induced GWs also probe the content of the primordial universe.

17.4.4.2 Primordial Non-Gaussianity and Trispectrum

As we discussed around Eq. (17.15), the induced GW spectrum is determined by the 4-point function of curvature fluctuations. So, strictly speaking, induced GWs are also sensitive to any departure from Gaussian fluctuations. In general, this means

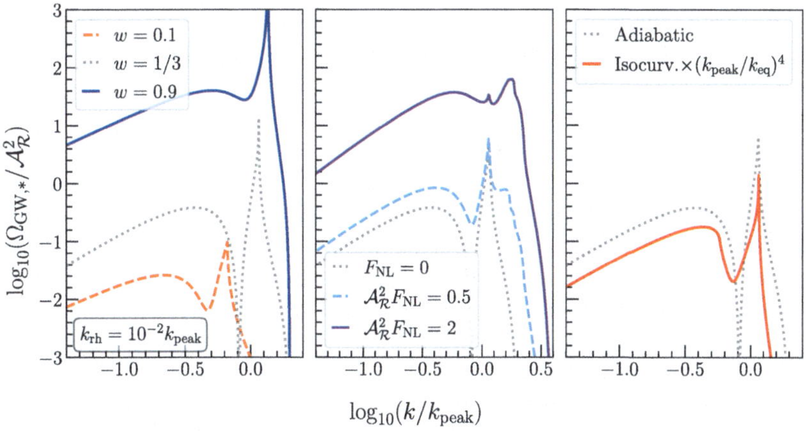

Fig. 17.4 Induced GW spectrum for different extensions assuming a Dirac delta peak in the primordial spectrum of fluctuations. The radiation domination result is shown in grey lines. On the left panel, we show the effects of softer (orange) and stiffer (blue) equations of state w from Ref. [57] (one may use SIGWfast [146] for the finite width case). See how the stiffer the w, the higher the GW spectrum and the sharper the resonant peak (which moves to higher frequencies). On the middle panel, we show the effects of local non-Gaussianity from Ref. [3] (data available at [2]). See how the effect of $F_{\rm NL}$ is to generate induced GW beyond the cut-off at $k = 2k_{\rm peak}$ from Gaussian fluctuations. On the right panel, we compare the induced GW spectrum from adiabatic and isocurvature initial conditions from Ref. [61]. Although the GW spectrum is qualitatively similar, they have a different shape of the resonant peak, and the isocurvature-induced GWs have no destructive interference at $k = \sqrt{2}c_s k_{\rm peak}$

that induced GWs probe the trispectrum of primordial fluctuations. However, if such fluctuations are generated during inflation, one expects perturbative departures from Gaussian fluctuations. And therefore, the induced spectrum would only be slightly modified compared to the Gaussian case (see, e.g., Ref. [77] for a general treatment later applied to scale-invariant fluctuations). But, one could also assume a general ansatz for primordial non-Gaussianity and study its effects, regardless of the dynamics during inflation. The most studied case is local non-Gaussianity, which is parametrized as [101]

$$\Phi(x) = \Phi_{\rm Gauss}(x) + F_{\rm NL}(\Phi^2_{\rm Gauss}(x) - \langle\Phi^2_{\rm Gauss}(x)\rangle)\,, \tag{17.31}$$

where $\Phi_{\rm Gauss}$ is the Gaussian variable. See Refs. [22, 43, 143] for early works and Ref. [3] for a precise systematic approach. As shown in the middle panel of Fig. 17.4, the local non-Gaussian contribution adds a resonant peak at $k = 3c_s k_{\rm peak}$ and dominates over the Gaussian contribution for $\mathcal{A}_{\mathcal{R}}^2 F_{\rm NL} > 1$. Non-Gaussianity does not alter the predictions for the IR tail of the induced GW spectrum, though. Note that there is no reason to stop the ansatz (17.31) at Φ^2, especially if non-Gaussianities are large, as one could include higher powers of Φ. See Ref. [1] for a nice general treatment.

As a curious possibility, we also mention that a parity-violating trispectrum could lead to chiral-induced GWs, though only as a small correction [76].

17.4.4.3 Isocurvature Initial Conditions

Observations of the CMB tell that primordial fluctuations on the largest scales are adiabatic [11]. This means that there is a time slice of uniform total energy density with only curvature perturbations but no relative number density fluctuations. But, at much smaller scales, this might not be the case. These initial conditions are called isocurvature [98, 110]. Isocurvature fluctuations do not directly source induced GWs (see Ref. [61] for the original work and Ref. [56] for a recent review). Instead, as the universe evolves, isocurvature fluctuations induce curvature fluctuations that later induce GWs. The price to pay is that the amplitude of induced GWs is suppressed by the energy density ratio of the subdominant field (responsible for isocurvature) with respect to the dominant field (responsible for curvature). For DM isocurvature fluctuations with a peaked isocurvature spectrum at $k_{\rm peak}$, the induced GW spectrum is suppressed by a factor $(k_{\rm eq}/k_{\rm peak})^4$, where $k_{\rm eq} \approx 0.01\,{\rm Mpc}^{-1}$ is the comoving scale at the standard matter-radiation equality. Interestingly, the amplitude of the isocurvature spectrum can be very large, say $\mathcal{A}_S \gg 1$,[9] and still be compatible with cosmological perturbation theory (up to the time fluctuations become non-linear and might form PBHs [124]). In this case, one has a detectable isocurvature-induced GW signal. Another interesting possibility is that the GWs are induced by the isocurvature from Poisson statistical fluctuations. This is a general signal when discrete objects or solitons exist, such as PBH, cosmic strings, oscillons, etcetera. This generic prediction has been dubbed "Universal Gravitational Waves of Solitons" in Refs. [105, 106]. In Sect. 17.5, we study in detail the GWs from PBH number density fluctuations.

17.4.4.4 Induced GWs During Inflation

We mainly discussed the GWs induced by primordial fluctuations and the different extensions after inflation. However, induced GW can also be generated during inflation as a secondary effect from vacuum fluctuations of spectator fields, such as scalar, vector, or tensor fields. The GW spectrum, in this case, is more model-dependent; therefore, we only cite the relevant literature. For scalar-induced GWs, e.g., in multifield models of inflation, we refer the reader to Refs. [34, 49, 50, 74, 88, 154]. One typical case for vector field is axion inflation with Abelian or non-Abelian gauge

[9] Large isocurvature ($S \gg 1$) implies that the matter field responsible for isocurvature fluctuations is very inhomogeneous, i.e., $\delta\rho_m \gg \rho_m$. However, as long as fluid velocities are negligible, which is the case on superhorizon scales, we may still treat the system within linear perturbation theory. In addition, since the isocurvature field is the sub-dominant field, one can check that perturbation theory holds as long as $\rho_m S \ll \rho_{\rm rad}$ (assuming radiation dominates the universe). Such a condition can also be understood as $\delta\rho_m \ll \rho_{\rm rad}$, meaning that there is no backreaction to the background geometry from isocurvature. For an extended discussion, see App. B of Ref. [61].

fields, see Refs. [27, 49, 50, 75, 109]. For more references, we refer the reader to Komatsu's review on the topic [100]. Two form fields can also source GWs at second order [71], though not at the linear order. Additional tensor fields can linearly source GWs [80]. For a review on linear GWs from metric vacuum fluctuations during inflation, see Ref. [82].

17.5 GWs from the PBH Dominated Early Universe

There is one scenario beyond the standard adiabatic primordial fluctuations, which is interesting regarding the associated GW signals. This is the case when ultra-light PBHs dominate the very early universe and evaporate before BBN, reheating the universe via Hawking radiation [16, 47, 103]. Most relevant to secondary GWs is the fact that, as shown in Refs. [90–92] (see also Ref. [125]), an epoch of dust domination (such as a gas of PBHs) may enhance the production of induced GWs. Adding that, due to the discrete nature of PBHs, there are number density fluctuations from the Poisson noise in the PBH gas [121], one finds a unique induced GW signature of the PBH-dominated universe [60, 67, 121]. Below, we focus on the induced GW signatures following Ref. [60]. The physics of BH evaporation is discussed in detail in Chap. 21.

17.5.1 The PBH Dominated Universe

For simplicity, let us assume a monochromatic PBH mass function. In that case, there are two free parameters in the PBH reheating scenario: the PBH mass at formation $M_{\rm PBH,f}$ and the initial fraction β of PBHs. The mass is proportional to the Hubble parameter at formation via Eq. (17.24). The initial fraction is related to the *mean* number density of PBHs per Hubble volume, that is

$$\beta = \frac{\rho_{\rm PBH,f}}{\rho_{\rm rad,f}} = \frac{4\pi\gamma}{3H_{\rm f}^3} n_{\rm PBH,f}\,, \tag{17.32}$$

where we used that $\rho_{\rm PBH,f} = M_{\rm PBH,f} \times n_{\rm PBH,f}$ and $\rho_{\rm rad,f} \approx 3H_{\rm f}^2 M_{\rm pl}^2$. We can also compute the mean distance between PBHs from the mean number density. The comoving wavenumber associated with this distance is given by [121]

$$k_{\rm UV} = \left(\frac{4\pi n_{{\rm PBH},f} a_{\rm f}}{3}\right)^{1/3} = \beta^{1/3}\gamma^{-1/3} k_{\rm f}\,, \tag{17.33}$$

where $k_\mathrm{f} = \mathcal{H}_\mathrm{f}$ is the comoving wavenumber that entered the horizon at formation.[10] This means that PBHs are initially displaced over super-Hubble distances.

The most interesting point is to note that, in fact, we have a gas of PBHs spread all over the universe. And, while it has on average a homogeneous distribution, there are statistical fluctuations. Namely, by chance, PBHs will be somewhere closer and somewhere sparser, leading to number density fluctuations. Since PBHs form very rarely and, therefore, are mostly distributed uniformly in space, the distribution of PBHs follows Poisson statistics [121]. The typical scale of the fluctuations is set by k_UV (17.33). The dimensionless power spectrum of PBH number density fluctuations at formation is given by [121]

$$\mathcal{P}_{\delta_{\mathrm{PBH,f}}}(k) = \frac{2}{3\pi}\left(\frac{k}{k_\mathrm{UV}}\right)^3 \Theta(k_\mathrm{UV} - k)\,, \tag{17.34}$$

where $\delta_\mathrm{PBH,f} = \delta n_\mathrm{PBH,f}/n_\mathrm{PBH,f}$ and $\Theta(x)$ is the Heaviside theta function. Note that the picture of a PBH gas is only valid if we have at least one PBH in a given volume. Thus, in the PBH gas approximation, we have a cut-off at the mean inter-PBH distance, namely at $k = k_\mathrm{UV}$. Hence, the notation "UV" to refer to the high wavenumber cut-off of the spectrum. We illustrate the PBH number density fluctuations in Fig. 17.5.

After formation, PBHs evaporate via Hawking radiation. This leads to a cosmic time-dependent PBH mass given by [91]

$$M_\mathrm{PBH}(t) \approx M_\mathrm{PBH,f}\left(1 - \frac{t}{t_\mathrm{eva}}\right)^{1/3} \quad \text{where} \quad t_\mathrm{eva} \approx \frac{M^3_\mathrm{PBH,f}}{M^4_\mathrm{pl}}\frac{160}{3.8\pi g_H(T_\mathrm{PBH})}\,. \tag{17.35}$$

The subscript "eva" refers to evaluation at evaporation. Because of PBH number density conservation, we have that $n_\mathrm{PBH} \propto a^{-3}$. This result, together with Eq. (17.35), completely determines the time evolution of ρ_PBH. Note that, during the time that Hawking evaporation is not effective, we have that $\rho_\mathrm{PBH}/\rho_\mathrm{r} \propto a$ and, eventually, PBHs might dominate the very early universe. The requirement that PBHs dominate the very early universe, which is equivalent to the requirement that PBHs evaporate after the Early PBH-radiation Equality (let us call this time "eeq"), yields

$$\beta > 6 \times 10^{-10}\left(\frac{g_H(T_\mathrm{PBH,f})}{108}\right)^{1/2}\left(\frac{\gamma}{0.2}\right)^{-1/2}\left(\frac{M_\mathrm{PBH,f}}{10^4\mathrm{g}}\right)^{-1}\,, \tag{17.36}$$

where $g_H(T_\mathrm{PBH})$ are the spin weighted degrees of freedom (which respresent the GWs) and $T_\mathrm{PBH,f}$ is the Hawking temperature of the PBH. For the standard model of particle physics, one has $g_H(T_\mathrm{PBH}) \approx 108$. The temperature of the radiation filling

[10] Note that if such PBHs formed from the collapse of primordial fluctuations, then k_f is equivalent to what we called k_peak in previous sections.

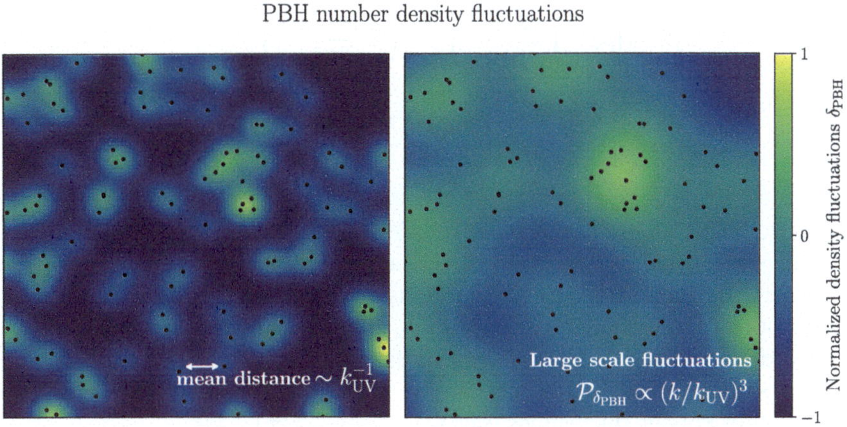

Fig. 17.5 PBH gas in the early universe. After formation, PBH are randomly distributed in space following a uniform distribution. This randomness leads to number density fluctuations. On the left panel, we show the kernel density estimate for the PBH number density in a fluid picture, smoothing on scales similar to the mean distance between PBHs. On the right panel, we show the kernel density estimate using a larger smoothing scale. See how number density fluctuations lead to large-scale density fluctuations of the PBH gas

the universe after PBH evaporation (which follows from saying that $H_{\rm eva} = 2/(3t_{\rm eva})$, where the numerical factor is that of a matter-dominated universe) is given by[11]

$$T_{\rm eva} \approx 2.8 \times 10^4\,{\rm GeV} \left(\frac{M_{\rm PBH,f}}{10^4{\rm g}}\right)^{-3/2} \left(\frac{g_H(T_{\rm PBH,f})}{108}\right)^{1/2} \left(\frac{g_\rho(T_{\rm eva})}{106.75}\right)^{-1/4} . \tag{17.37}$$

Imposing BBN bounds, namely that $T_{\rm eva} > 4\,{\rm MeV}$ [83, 84, 95, 96], yields an upper bound on the PBH mass as

$$M_{{\rm PBH},f} < 5 \times 10^8{\rm g}\,. \tag{17.38}$$

The dynamics of the PBH reheating scenario are illustrated in Fig. 17.7.

It is also useful to compute the wavenumber corresponding to the Horizon size at evaporation (e.g., relative to the Hubble radius today). This is given by[12]

[11] To derive $T_{\rm eva}$ we used that for a radiation dominated universe $3H^2M_{\rm pl}^2 = \rho_{\rm rad} = \frac{\pi^2}{30} g_\rho(T)T^4$.

[12] To derive $k_{\rm eva}$ we used that $k_{\rm eva} = a_{\rm eva}H_{\rm eva}$ and that conservation of entropy gives that $\frac{a^3}{a_*^3} = \frac{T_*^3}{T^3}\frac{g_s(T_*)}{g_s(T)}$.

$$k_{\rm eva} \approx 4.7 \times 10^{11}{\rm Mpc}^{-1}\left(\frac{M_{\rm PBH,f}}{10^4{\rm g}}\right)^{-3/2}\left(\frac{g_H(T_{\rm PBH})}{108}\right)^{1/2} \times \left(\frac{g_\rho(T_{\rm eva})}{106.75}\right)^{1/4}\left(\frac{g_s(T_{\rm eva})}{106.75}\right)^{-1/3}, \tag{17.39}$$

Since $T_{\rm eva}$ (17.37) only depends on $M_{\rm PBH,f}$ and not β, the same applies to $k_{\rm eva}$. The frequency associated with the numerical factor in Eq. (17.39), namely $4.7 \times 10^{11}{\rm Mpc}^{-1}$, is about 7.3×10^{-4} Hz. It turns out that, because of number density conservation (which implies $a \propto n_{\rm PBH}^{-1/3}$), $k_{\rm eva}$ computed in terms of $k_{\rm f}$ has the same β dependence as $k_{\rm UV}$ (17.33). This implies that the ratio $k_{\rm UV}/k_{\rm eva}$ is independent of β. And, since $k_{\rm eva}$ from (17.39) depends on $M_{\rm PBH,f}$, the frequency associated with the cut-off of Poisson fluctuations (say $f_{\rm poisson} = k_{\rm UV}/2\pi$) reads

$$f_{\rm possion} \approx 1.7 \times 10^3{\rm Hz}\left(\frac{M_{{\rm PBH},f}}{10^4{\rm g}}\right)^{-5/6}\left(\frac{g_H(T_{\rm PBH})}{108}\right)^{1/6}\left(\frac{g_\rho(T_{\rm eva})}{106.75}\right)^{1/4}\left(\frac{g_s(T_{\rm eva})}{106.75}\right)^{-1/3}. \tag{17.40}$$

This frequency is where the induced GW spectrum from PBH Poisson fluctuations peaks. We show the parameter space for $f_{\rm poisson}$, $M_{\rm PBH,f}$ and $T_{\rm eva}$ in Fig. 17.6. See how GW detectors such as the LVK collaboration, Einstein Telescope, and Cosmic Explorer can probe a wide range of masses of ultra-light PBHs if such PBHs dominate the universe.

17.5.2 GWs from PBH Number Density Fluctuations

As mentioned at the beginning of the section, PBHs are initially inhomogeneously distributed in space, according to Poisson statistics. The initial conditions of such number density fluctuations are isocurvature in nature [60, 121]. This is because as each PBH forms, it leaves a "hole" in the original radiation fluid. Thus, by energy conservation, one must have in the fluid picture that $\delta\rho_{\rm PBH} + \delta\rho_{\rm rad} = 0$ initially. This is nothing but isocurvature initial fluctuations. For instance, in the Newton gauge and on superhorizon scales, one has that $2\Phi \sim \delta\rho_{\rm tot} = \delta\rho_{\rm PBH} + \delta\rho_{\rm rad} = 0$, i.e. there are no initial curvature fluctuations. But, as the universe evolves, and the mean PBH energy density starts to overtake that of the radiation fluid (see Fig. 17.6), there is a conversion from isocurvature to curvature fluctuations. The conversion is completed once PBHs dominate the universe. Then, if we call S_i the initial isocurvature, one finds that the curvature fluctuations at PBH domination are given by $\Phi(k \ll k_{\rm eeq}) = S_i/5$ and $\Phi(k \gg k_{\rm eeq}) \approx \frac{3}{4}S_i(k/k_{\rm eeq})^{-2}$ (see e.g., Ref. [98]), where $k_{\rm eeq} = \sqrt{2}\,\beta^{2/3}\gamma^{1/3}k_{\rm UV}$ is the scale that enters the horizon at PBH-radiation equality. Scales with $k \gg k_{\rm eeq}$ are suppressed because they entered the horizon during the radiation-dominated epoch when the subhorizon curvature fluctuations decay.

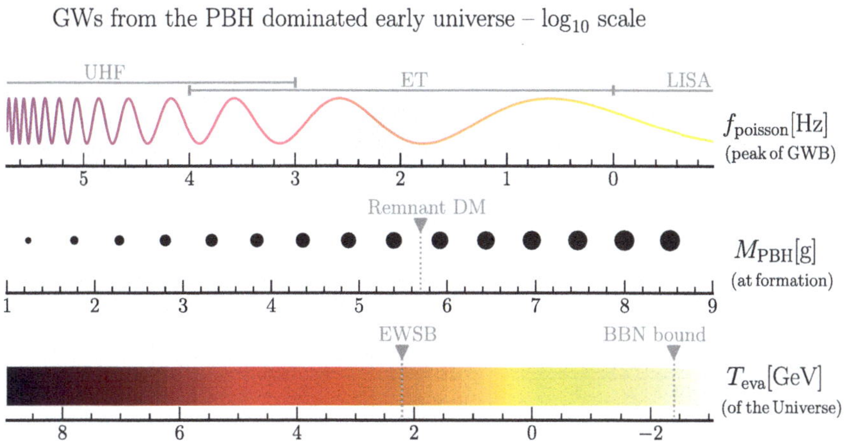

Fig. 17.6 Graphical summary of the GW signatures of the PBH-dominated universe. In the top panel, we show the frequency of the induced GWs associated with PBH reheating, $\log_{10} f_{\rm possion}$ (17.40). GW detectors in the relevant frequency range are UHF GW detectors, ET and LISA. In the middle panel, we show the corresponding PBH mass at formation, $\log_{10} M_{\rm PBH,f}$. We also show the PBH remnant scenario as DM [107], which requires $M_{\rm PBH,f} \approx 5 \times 10^5$g [64]. In the bottom panel, we give the temperature of the radiation filling the universe right after PBH evaporation $\log_{10} T_{\rm eva}$ (17.37). We show the BBN constraint $T \sim 4\,$MeV [83, 84, 95, 96] as well as the temperature corresponding to ElectroWeak Symmetry Breaking (EWSB), which is relevant for leptogenesis/baryogenesis, see e.g. Ref. [85]. Interestingly, leptogenesis via Right-handed neutrinos requires $M_{\rm PBH,f} \lesssim 10\,$g [33, 53]

During the PBH-dominated phase, curvature fluctuations remain constant on all scales due to the zero pressure of the PBH gas. This also implies that PBH density fluctuations grow. During that period, there is a constant source of induced GWs. Because of this, the induced GW spectrum for wavenumbers such that $k_{\rm eeq} < k < k_{\rm UV}$ turns out to be flat with an amplitude approximately given by [121]

$$\Omega^{\rm flat}_{\rm GW,PBH-dom} \approx 2 \times 10^{13} \left(\frac{M_{\rm PBH,f}}{10^4{\rm g}}\right)^{4/3} \beta^{16/3}\,. \tag{17.41}$$

Note that the above estimate assumes a monochromatic mass function. For the analysis with an extended mass function see Ref. [120]. However, the biggest production of induced GWs happens right after PBHs evaporate completely. The source of the enhancement is that by the time PBHs evaporate, the number of density fluctuations has grown very much. Then, the almost sudden transition from PBH to radiation domination (see in Fig. 17.7 that the transition takes less than a quarter of efold) leads to large radiation density fluctuations (this is called the "poltergeist" mechanism in Ref. [91]). Velocity fluctuations of the radiation fluid after evaporation are very large. From Eq. (17.17), we find that after evaporation $v \sim \Phi'/\mathcal{H} \propto (k/k_{\rm eva})^2$. Taking into account that the transition is not instantaneous leads to a suppression factor $\sim (k/k_{\rm eva})^{-1/3}$ [91] (that comes from the fact that PBHs lose energy according to

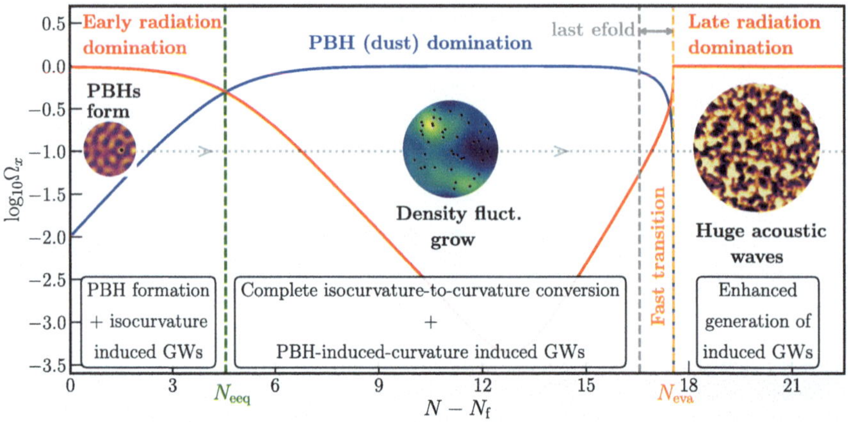

Fig. 17.7 Illustration of the universe's evolution in the PBH reheating scenario. We show $\Omega_{\rm PBH} = \rho_{\rm PBH}/\rho_{\rm tot}$ (with a blue line) and $\Omega_{\rm rad} = \rho_{\rm rad}/\rho_{\rm tot}$ (with a red line) in terms of efolds N from PBH formation $N_{\rm f}$. PBHs start to dominate the universe at $N_{\rm eeq}$ and evaporate completely at $N_{\rm eva}$. During the early radiation domination, PBHs form, but the induced GW signal associated with the formation and PBH isocurvature are respectively of too high frequency or too low amplitude. After $N_{\rm eeq}$, PBH isocurvature has been converted into curvature fluctuations and induced GWs are more efficiently produced (derived in Ref. [121]). Because the complete evaporation of PBHs is very fast, there is an enhanced production of induced GWs right after PBH evaporation [91] (estimated in Ref. [60])

$M_{\rm PBH} \propto (1 - t/t_{\rm eva})^{1/3}$). An approximation to the induced GW spectrum after PBH evaporation is given by [60, 67]

$$\Omega_{\rm GW,eva}(k) \approx \Omega^{\rm peak}_{\rm GW,eva}\left(\frac{k}{k_{\rm UV}}\right)\Theta(k_{\rm UV} - k)\,, \tag{17.42}$$

where

$$\Omega_{\rm GW,eva} \approx 10^{30}\beta^{16/3}\left(\frac{\gamma}{0.2}\right)^{8/3}\left(\frac{g_H(T_{\rm PBH})}{108}\right)^{-17/9}\left(\frac{M_{\rm PBH,f}}{10^4{\rm g}}\right)^{34/9}. \tag{17.43}$$

Note that Eq. (17.43) is larger than Eq. (17.41) by a factor $M^{22/9}_{\rm PBH,f}$. We show on the left of Fig. 17.8 the shape of the induced GW spectrum after PBH evaporation for different values of the PBH mass. On the right of Fig. 17.8, we show the bounds on β as a function of PBH mass and the corresponding amplitude of the induced GW spectrum peak (17.43).

It is important to note, though, that while the amplitude of Eq. (17.43) is very sensitive to the monochromatic mass function assumption (as a broad mass function leads to a more gradual transition to radiation domination), the amplitude of Eq. (17.41) is not as it only depends on the length PBH dominated phase (see also Ref. [120]). We can now use Eq. (17.43) to derive the current upper bounds on the initial fraction

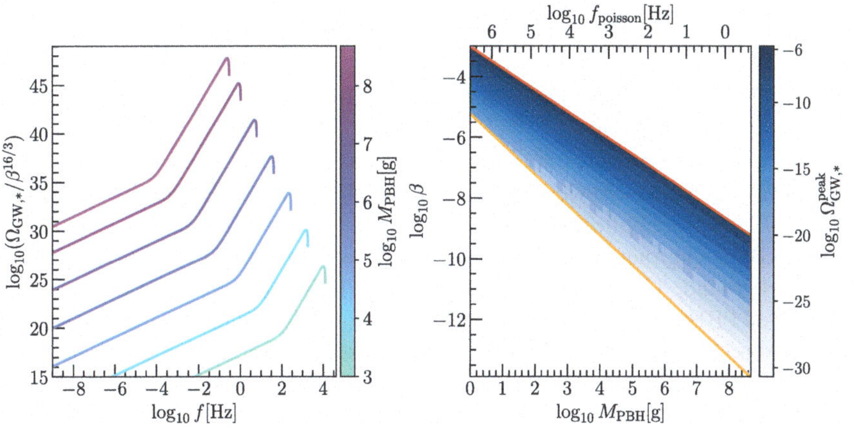

Fig. 17.8 On the left panel, we show the GW spectrum of GWs induced by PBH density fluctuations after PBH evaporation in terms of frequency for various values of $M_{PBH,f}$ (shown in color code) and normalized to $\beta^{16/3}$. The larger the mass, the higher the amplitude and the lower the peak frequency. On the right panel, we show the allowed parameter space of initial PBH fraction β requiring PBH domination (lower bound (17.36), orange line) and using BBN constraints on the effective number of relativistic species (upper bound (17.44), red line). In color code, we show the corresponding amplitude of the peak of the GW spectrum (17.43)

of PBHs using BBN bounds [20, 52] on the effective number of relativistic species (yielding $\Omega_{GW,0}h^2 \lesssim 10^{-6}$). This leads us to

$$\beta < 1.1 \times 10^{-6} \left(\frac{\gamma}{0.2}\right)^{-1/2} \left(\frac{g_H(T_{PBH})}{108}\right)^{17/48} \left(\frac{g_*(T_{eva})}{106.75}\right)^{1/16} \left(\frac{M_{PBH,f}}{10^4 g}\right)^{-17/24} . \tag{17.44}$$

We end this section by mentioning shortcomings in the GW calculations of the PBH-dominated universe. First, although curvature fluctuations remain well inside the regime of cosmological perturbation theory, density fluctuations eventually become non-linear during PBH domination. For adiabatic primordial fluctuations, one may impose a cut-off to the primordial spectrum as in Ref. [91] to ensure that all fluctuations are linear at all times. However, PBH density fluctuations are intrinsic to the model, so an arbitrary cut-off may not be as justified. More accurate predictions require more sophisticated numerical simulations. Second, the GW enhancement is highly sensitive to the monochromatic mass assumption. A broad mass function will lead to a slower transition to the late radiation-dominated regime. According to Ref. [91], the resulting induced GW spectrum is exponentially sensitive to the width of the mass function. It would be interesting to do a more detailed study. For effects of modified gravity on the evolution of PBH number density fluctuations and induced GWs, see Refs. [122, 123].

17.6 GW Background from PBH Binaries

For completeness, we dedicate the last section of this chapter to the GW background from unresolved PBH binaries. We want to mention that this short section does not do justice to such an interesting topic. The interested reader is encouraged to read Ref. [132] for a recent review (see also Refs. [131, 133] for earlier reviews) and Refs. [37, 144] for applications to PBHs.

The GW background from PBH binaries has a different nature than the induced GW background. The former results from the PBH themselves as a collection of unresolved PBH mergers, while the latter is generated by the large-scale structure of fluctuations (very few of which collapse to PBHs). The first is the local, continuous source of GWs (active until PBHs evaporate), while the second is only generated in the very early universe. The GW spectral density from unresolved PBH binaries is computed as the integrated energy flux in GWs per logarithmic frequency bin normalized to the critical energy density of the universe ρ_c [37, 132, 155]. Schematically one has that [132]

$$\Omega_{\rm GW}(f) = \frac{1}{\rho_c} f F_{\rm GW}(f)\,, \tag{17.45}$$

where $F_{\rm GW}$ is the integrated flux per frequency bin.[13] The integrated flux is nothing but an integral over time of the flux emitted per binary times the rate at which PBHs merge. The flux per binary is computed as the emitted energy in GWs per binary and per frequency bin, say $dE_{\rm GW}/df$, times the inverse area that the GWs expanded, namely $\frac{1}{4\pi r^2}\frac{dE_{\rm GW}}{df}$ where r is the distance to the source. A nice analytical fit to $dE_{\rm GW}/df$, including the inspiral and merger phases, for non-precessing binaries is provided by Refs. [9, 10] (they provide the template for the strain in frequency domain but the energy is proportional to f^2 times the strain squared). The rate at which the PBH binaries merge as a function of time (i.e., the number of mergers per unit time and unit volume) for PBHs is computed in Chapter 16. Here, we call it $\mathcal{R}_{\rm merge}(t)$. Using the cosmic redshift as a time variable, i.e., $1+z=a(t)^{-1}$, the spectral density of the GW background from PBH binaries is given by [132]

$$\Omega^{\rm binaries}_{\rm GW}(f) = \frac{f}{3H_0^2 M_{\rm pl}^2}\int_0^{z_{\rm max}} dz \frac{\mathcal{R}_{\rm merge}(z)}{(1+z)H(z)}\frac{dE_{\rm GW}}{df_s}\,, \tag{17.46}$$

where we used the notation f_s to denote the frequency in the source frame, which is related to the observed frequency by $f_s = f(1+z)$. If the GW background is composed by the merger of long-lived PBHs (i.e., $M_{\rm PBH,f} > 10^{15}\,{\rm g}$), then only the nearby binaries (in the cosmological sense) contributes to the integral. In this case, we can take the Hubble parameter of the ΛCDM universe, i.e.,

[13] To be precise, $F_{\rm GW}$ also contains an integral over the direction of observation [132]. However, we assume an isotropic background of binaries and consider the average GW energy density.

$H(z) = H_0\sqrt{\Omega_{m,0}(1+z)^3 + \Omega_{\Lambda,0}}$. The explicit expression for $dE_{\rm GW}/df_s$ for non-spinning PBHs (which is the expectation for PBH binaries at late times) reads [37, 132]

$$\frac{dE_{\rm GW}}{df_s} = \frac{(G\pi)^{2/3}\mathcal{M}_c^{5/3}}{3} f_s^{-1/3} \times \begin{cases} (1-(1.44-2.68\eta)\nu^{2/3})^2 & \text{for } f_s < f_1\,, \\ \omega_m f_s (1-1.89\nu+1.66\nu^2) & \text{for } f_1 \le f_s < f_2\,, \\ \omega_r f_s^{7/3} \frac{\nu_4^4}{4(\nu-\nu_2)^2+\nu_4^2} & \text{for } f_2 \le f_s < f_3\,, \end{cases} \tag{17.47}$$

where we defined $\nu_n \equiv (\pi G \mathcal{M} f_n)^{1/3}$, $\mathcal{M} = m_1 + m_2$ is the total mass, $\mathcal{M}_c = (m_1 m_2)^{3/5}/\mathcal{M}^{1/5}$ is the chirp mass and $\eta = m_1 m_2/\mathcal{M}^2$. The numerical parameters in Eq. (17.47) are given by

$$\begin{aligned} \nu_1^3 &= 0.07 + 0.64\eta - 0.06\eta^2 - 7.09\eta^3\,, \quad \nu_2^3 = 0.19 + 0.15\eta - 0.02\eta^2 + 2.33\eta^3\,, \\ \nu_3^3 &= 0.32 - 0.13\eta - 0.27\eta^2 + 4.92\eta^3\,, \quad \nu_4^3 = 0.09 - 0.41\eta + 1.83\eta^2 - 2.87\eta^3\,. \end{aligned} \tag{17.48}$$

Lastly, upper integration limit in Eq. (17.46) is given by $z_{\rm max} = f_3/f - 1$.

We end this section with some rough estimates for equal mass PBH binaries from Ref. [59] to develop a bit of intuition. The merger rate of equal PBH binaries evaluated today in its simplest form is given by [13, 35, 118, 137]

$$\mathcal{R}_{\rm merge} \approx 4\times 10^6\,{\rm Gpc}^{-3}{\rm yr}^{-1} \frac{{\rm f}_{\rm PBH}^2}{\left({\rm f}_{\rm PBH}^2 + \sigma_{\rm eq}^2\right)^{21/74}} \left(\frac{M_{\rm PBH}}{M_\odot}\right)^{-32/37}\,, \tag{17.49}$$

where ${\rm f}_{\rm PBH}$ is the fraction of PBHs as dark matter and $\sigma_{\rm eq}^2 \approx 2.5\times 10^{-5}$. Putting some numbers and using that only the closest binaries, which are the loudest, contribute to the peak of the GW spectrum, we find that

$$\Omega^{\rm peak}_{\rm GW,binary} h^2 \approx 1.6\times 10^{-8} \left(\frac{M_{\rm PBH,f}}{M_\odot}\right)^{5/37} \left(\frac{{\rm f}_{\rm PBH}}{0.01}\right)^{53/37} \left(1 + 0.25\left(\frac{{\rm f}_{\rm PBH}}{0.01}\right)^{-2}\right)^{-21/74}\,. \tag{17.50}$$

The peak of the GW spectrum is then roughly at the ISCO frequency, namely

$$f_{\rm ISCO} \approx 2.2\,{\rm kHz} \left(\frac{M_{\rm PBH,f}}{M_\odot}\right)^{-1}\,. \tag{17.51}$$

This concludes our estimates for the GW background from unresolved PBH binaries.

17.7 Final Remarks

The induced GW background is a decisive observable for the PBH scenario from large primordial fluctuations. In fact, it is generally expected that any mechanism that forms PBHs is always accompanied by a loud GW signal (as discussed in Sect. 17.2). The absence of an induced GW background would leave any PBH scenario in doubt. But, most important to us, any hint (or even discovery) of PBHs in a given mass range provides us with strong motivation to look for the induced GW background (see Sect. 17.4). From the GW background, we may also learn about the type of primordial density fluctuations (e.g., adiabatic/isocurvature and Gaussian/non-Gaussian) as well as the content of the very early universe (e.g., the equation of state w); see Sect. 17.4.4. We may also be able to test whether PBHs once dominated the very early universe, reheating it via Hawking radiation. And, it might give us hints of PBH remnants as DM, as we saw in Sect. 17.5. We may also be able to observe the GW background from the unresolved PBH binaries (see Sect. 17.6), providing further evidence to the PBH scenario. Besides the interesting observational consequences, induced GWs also motivate us to improve our theoretical understanding of GW energy density when dealing with second-order cosmological perturbation theory (as discussed in Sect. 17.3).

Interesting research directions that we have not discussed in this chapter are the study of statistical properties of the GW background (see, e.g. Ref. [38]) and GW background anisotropies (see, e.g., Ref. [28, 30, 51] for the general formulation and Ref. [29, 55, 102, 128, 139, 145, 150] for applications to PBHs) as additional probes/evidence of the PBH scenario. We have also not discussed high-frequency GWs from Hawking evaporation of PBHs, which might be testable by its effects on the effective number of relativistic species at CMB and BBN [20, 78, 86, 94, 111, 112]. There is also the possibility to induce GWs from scalar-tensor and tensor-tensor mixings, see e.g. Refs. [26, 48, 127, 151].

We end this chapter by pointing out some directions worth exploring. First, one must examine the extensions beyond the simplest assumptions (radiation domination and Gaussian fluctuations) in more detail. This includes exploring motivated models that modify the expansion history of the early universe and derive more accurate kernels, studying how departures from General Relativity affect the induced GW spectrum, and, perhaps, exploring different types of non-Gaussianity of primordial fluctuations. Second, it would be very interesting to derive accurate predictions using numerical simulations for the GWs from the PBH-dominated early universe, as this is a unique way to probe the existence of PBHs that evaporated long ago. Lastly, it would be remarkable from the theoretical point of view to find a rigorous solution to the gauge ambiguity of induced GWs.

Acknowledgements I would like to thank all of my collaborators from whom I learned a lot and without whom I would have probably never explored induced GWs to the extent I did. These are V. Atal, S. Balaji, P. Bari, N. Bartolo, E. Dimastrogiovanni, M. Fasiello, G. Franciolini, A. Malhotra, S. Matarrese, A. Naruko, C. Lin, S. Passaglia, S. Patil, S. Pi, S. Renaux-Petel, M. Sasaki, V. Takhistov, G. Tasinato, T. Vargas, A. Wang, J. Wang and L. Witkowski. I also thank J. Fedrow, K. Inomata,

E. Komatsu, T. Papanikolaou, S. Profumo, R. Samanta, J. Turner, C .Unal, Z. Weiner and S. Young for many stimulating discussions. This work is supported by the DFG under the Emmy-Noether program, project number 496592360.

References

1. K.T. Abe et al., Primordial black holes and gravitational waves induced by exponential-tailed perturbations (2022). arxiv:2209.13891 [astro-ph.CO]
2. P. Adshead, K. Lozanov, Z. Weiner, *ngsigw-results* (2023). https://doi.org/10.5281/ZENODO.8176107. https://zenodo.org/record/8176107
3. P. Adshead, K.D. Lozanov, Z.J. Weiner, Non-Gaussianity and the induced gravitational wave background. JCAP **10**, 080 (2021). https://doi.org/10.1088/1475-7516/2021/10/080. arxiv:2105.01659 [astro-ph.CO]
4. A. Afzal et al., The NANOGrav 15 yr data set: search for signals from new physics. Astrophys. J. Lett. **951**(1), L11 (2023). https://doi.org/10.3847/2041-8213/acdc91. arxiv:2306.16219 [astro-ph.HE]
5. G. Agazie et al., Comparing recent PTA results on the nanohertz stochastic gravitational wave background (2023). arxiv:2309.00693 [astro-ph.HE]
6. G. Agazie et al., The NANOGrav 15 yr data set: evidence for a gravitational-wave background. Astrophys. J. Lett. **951**(1), L8 (2023). https://doi.org/10.3847/2041-8213/acdac6. arxiv:2306.16213 [astro-ph.HE]
7. G. Agazie et al., The NANOGrav 15 yr data set: observations and timing of 68 millisecond pulsars. Astrophys. J. Lett. **951**(1), L9 (2023). https://doi.org/10.3847/2041-8213/acda9a. arxiv:2306.16217 [astro-ph.HE]
8. N. Aghanim et al., Planck 2018 results. VI. Cosmological parameters. Astron. Astrophys. **641** (2020). [Erratum: Astron.Astrophys. 652, C4 (2021)], A6. https://doi.org/10.1051/0004-6361/201833910. arxiv:1807.06209 [astro-ph.CO]
9. P. Ajith et al., A template bank for gravitational waveforms from coalescing binary black holes. I. Non-spinning binaries. Phys. Rev. D **77** (2008). [Erratum: Phys.Rev.D 79, 129901 (2009)], 104017. https://doi.org/10.1103/PhysRevD.77.104017. arxiv:0710.2335 [gr-qc]
10. P. Ajith et al., Inspiral-merger-ringdown waveforms for black-hole binaries with non-precessing spins. Phys. Rev. Lett. **106**, 241101 (2011). https://doi.org/10.1103/PhysRevLett.106.241101. arxiv:0909.2867 [gr-qc]
11. Y. Akrami et al., Planck 2018 results. X. Constraints on inflation. Astron. Astrophys. **641**, A10 (2020). https://doi.org/10.1051/0004-6361/201833887. arxiv:1807.06211 [astro-ph.CO]
12. L. Alabidi et al., Observable induced gravitational waves from an early matter phase. JCAP **05**, 033 (2013). https://doi.org/10.1088/1475-7516/2013/05/033. arxiv:1303.4519 [astro-ph.CO]
13. Y. Ali-Haïmoud, E.D. Kovetz, M. Kamionkowski, Merger rate of primordial black-hole binaries. Phys. Rev. D **96**(12), 123523 (2017). https://doi.org/10.1103/PhysRevD.96.123523. arxiv:1709.06576 [astro-ph.CO]
14. R. Allahverdi et al., The first three seconds: a review of possible expansion histories of the early universe (2020). arxiv:2006.16182 [astro-ph.CO]
15. K.N. Ananda, C. Clarkson, D. Wands, The cosmological gravitational wave background from primordial density perturbations. Phys. Rev. D **75**, 123518 (2007). https://doi.org/10.1103/PhysRevD.75.123518. arxiv:gr-qc/0612013
16. R. Anantua, R. Easther, J.T. Giblin, GUT-scale primordial black holes: consequences and constraints. Phys. Rev. Lett. **103**, 111303 (2009). https://doi.org/10.1103/PhysRevLett.103.111303. arxiv:0812.0825 [astro-ph]

17. J. Antoniadis et al., The second data release from the European pulsar timing array I. The dataset and timing analysis (2023). https://doi.org/10.1051/0004-6361/202346841. arxiv:2306.16224 [astro-ph.HE]
18. J. Antoniadis et al., The second data release from the European pulsar timing array III. Search for gravitational wave signals. Astron. Astrophys. **678**, A50 (2023). https://doi.org/10.1051/0004-6361/202346844. arxiv:2306.16214 [astro-ph.HE]
19. J. Antoniadis et al., The second data release from the European pulsar timing array: V. Implications for massive black holes, dark matter and the early Universe (2023). arxiv:2306.16227 [astro-ph.CO]
20. A. Arbey et al., Precision calculation of dark radiation from spinning primordial black holes and early matter-dominated eras. Phys. Rev. D **103**(12), 123549 (2021). https://doi.org/10.1103/PhysRevD.103.123549. arxiv:2104.04051 [astro-ph.CO]
21. H. Assadullahi, D. Wands, Gravitational waves from an early matter era. Phys. Rev. D **79**, 083511 (2009). https://doi.org/10.1103/PhysRevD.79.083511. arxiv:0901.0989 [astro-ph.CO]
22. V. Atal, G. Domènech, Probing non-Gaussianities with the high frequency tail of induced gravitational waves. JCAP **06**, 001 (2021). https://doi.org/10.1088/1475-7516/2021/06/001. arxiv:2103.01056 [astro-ph.CO]
23. P. Athron et al., Cosmological phase transitions: from perturbative particle physics to gravitational waves. Prog. Part. Nucl. Phys. **135**, 104094 (2024). https://doi.org/10.1016/j.ppnp.2023.104094. arxiv:2305.02357 [hep-ph]
24. E. Bagui et al., Primordial black holes and their gravitational-wave signatures (2023). arxiv:2310.19857 [astro-ph.CO]
25. S. Balaji, G. Domenech, J. Silk, Induced gravitational waves from slow-roll inflation after an enhancing phase (2022). arxiv:2205.01696 [astro-ph.CO]
26. P. Bari et al., Gravitational waves induced by scalar-tensor mixing. Phys. Rev. D **109**(2), 023509 (2024). https://doi.org/10.1103/PhysRevD.109.023509. arxiv:2307.05404 [astro-ph.CO]
27. N. Barnaby, E. Pajer, M. Peloso, Gauge field production in axion inflation: consequences for monodromy, non-Gaussianity in the CMB, and gravitational waves at interferometers. Phys. Rev. D **85**, 023525 (2012). https://doi.org/10.1103/PhysRevD.85.023525. arxiv:1110.3327 [astro-ph.CO]
28. N. Bartolo et al., Anisotropies and non-Gaussianity of the cosmological gravitational wave background. Phys. Rev. D **100**(12), 121501 (2019). https://doi.org/10.1103/PhysRevD.100.121501. arxiv:1908.00527 [astro-ph.CO]
29. N. Bartolo et al., Gravitational wave anisotropies from primordial black holes. JCAP **02**, 028 (2020). https://doi.org/10.1088/1475-7516/2020/02/028. arxiv:1909.12619 [astro-ph.CO]
30. N. Bartolo et al., Probing anisotropies of the stochastic gravitational wave background with LISA. JCAP **11**, 009 (2022). https://doi.org/10.1088/1475-7516/2022/11/009. arxiv:2201.08782 [astro-ph.CO]
31. D. Baumann. *Cosmology*. Cambridge University Press (2022). ISSN: 978-1-108-93709-2, 978-1-108-83807-8. https://doi.org/10.1017/9781108937092
32. D. Baumann et al., Gravitational wave spectrum induced by primordial scalar perturbations. Phys. Rev. D **76**, 084019 (2007). https://doi.org/10.1103/PhysRevD.76.084019. arxiv:hep-th/0703290
33. N. Bernal et al., Rescuing high-scale leptogenesis using primordial black holes. Phys. Rev. D **106**(3), 035019 (2022). https://doi.org/10.1103/PhysRevD.106.035019. arxiv:2203.08823 [hep-ph]
34. M. Biagetti, M. Fasiello, A. Riotto, Enhancing inflationary tensor modes through spectator fields. Phys. Rev. D **88**, 103518 (2013). https://doi.org/10.1103/PhysRevD.88.103518. arxiv:1305.7241 [astro-ph.CO]
35. S. Bird et al., Did LIGO detect dark matter?. Phys. Rev. Lett. **116**(20), 201301 (2016). https://doi.org/10.1103/PhysRevLett.116.201301. arxiv:1603.00464 [astro-ph.CO]

36. M. Braglia, X. Chen, D.K. Hazra, Probing primordial features with the stochastic gravitational wave background. JCAP **03**, 005 (2021). https://doi.org/10.1088/1475-7516/2021/03/005. arxiv:2012.05821 [astro-ph.CO]
37. M. Braglia, J. Garcia-Bellido, S. Kuroyanagi, Testing primordial black holes with multi-band observations of the stochastic gravitational wave background. JCAP **12**(12), 012 (2021). https://doi.org/10.1088/1475-7516/2021/12/012. arxiv:2110.07488 [astro-ph.CO]
38. M. Braglia, J. Garcia-Bellido, S. Kuroyanagi, Tracking the origin of black holes with the stochastic gravitational wave background popcorn signal (2022). arxiv:2201.13414 [astro-ph.CO]
39. E.V. Bugaev, P.A. Klimai, Bound on induced gravitational wave background from primordial black holes. JETP Lett. **91**, 1–5 (2010). https://doi.org/10.1134/S0021364010010017. arxiv:0911.0611 [astro-ph.CO]
40. E. Bugaev, P. Klimai, Constraints on the induced gravitational wave background from primordial black holes. Phys. Rev. D **83**, 083521 (2011). https://doi.org/10.1103/PhysRevD.83.083521. arxiv:1012.4697 [astro-ph.CO]
41. E. Bugaev, P. Klimai, Induced gravitational wave background and primordial black holes. Phys. Rev. D **81**, 023517 (2010). https://doi.org/10.1103/PhysRevD.81.023517. arxiv:0908.0664 [astro-ph.CO]
42. R.-G. Cai, S. Pi, M. Sasaki, Universal infrared scaling of gravitational wave background spectra. Phys. Rev. D **102**(8), 083528 (2020). https://doi.org/10.1103/PhysRevD.102.083528. arxiv:1909.13728 [astro-ph.CO]
43. R.-G. Cai, S. Pi, M. Sasaki, Gravitational waves induced by non-Gaussian scalar perturbations. Phys. Rev. Lett. **122**(20), 201101 (2019). https://doi.org/10.1103/PhysRevLett.122.201101. arxiv:1810.11000 [astro-ph.CO]
44. C. Caprini, D.G. Figueroa, Cosmological backgrounds of gravitational waves. Class. Quant. Grav. **35**(16), 163001 (2018). https://doi.org/10.1088/1361-6382/aac608. arxiv:1801.04268 [astro-ph.CO]
45. B. Carr, F. Kuhnel, Primordial black holes as dark matter: recent developments. Ann. Rev. Nucl. Part. Sci. **70**, 355–394 (2020). https://doi.org/10.1146/annurev-nucl-050520-125911. arxiv:2006.02838 [astro-ph.CO]
46. B. Carr, J. Silk, Primordial black holes as generators of cosmic structures. Mon. Not. Roy. Astron. Soc. **478**(3), 3756–3775 (2018). https://doi.org/10.1093/mnras/sty1204
47. Bernard J. Carr, Some cosmological consequences of primordial black-hole evaporations. Astrophys. J. **206**, 8–25 (1976). https://doi.org/10.1086/154351
48. Z. Chang, X. Zhang, J.-Z. Zhou, Gravitational waves from primordial scalar and tensor perturbations. Phys. Rev. D **107**(6), 063510 (2023). https://doi.org/10.1103/PhysRevD.107.063510. arxiv:2209.07693 [astro-ph.CO]
49. S.-L. Cheng, W. Lee, K.-W. Ng, Primordial black holes and associated gravitational waves in axion monodromy inflation. JCAP **07**, 001 (2018). https://doi.org/10.1088/1475-7516/2018/07/001. arxiv:1801.09050 [astro-ph.CO]
50. S.-L. Cheng, W. Lee, K.-W. Ng, Production of high stellar-mass primordial black holes in trapped inflation. JHEP **02**, 008 (2017). https://doi.org/10.1007/JHEP02(2017)008. arxiv:1606.00206 [astro-ph.CO]
51. C.R. Contaldi, Anisotropies of gravitational wave backgrounds: a line of sight approach. Phys. Lett. B **771**, 9–12 (2017). https://doi.org/10.1016/j.physletb.2017.05.020. arxiv:1609.08168 [astro-ph.CO]
52. Richard H. Cyburt et al., New BBN limits on physics beyond the standard model from ^4He. Astropart. Phys. **23**, 313–323 (2005). https://doi.org/10.1016/j.astropartphys.2005.01.005. (arxiv:astro-ph/0408033)
53. S. Datta, A. Ghosal, R. Samanta, Baryogenesis from ultralight primordial black holes and strong gravitational waves from cosmic strings. JCAP **08**, 021 (2021). https://doi.org/10.1088/1475-7516/2021/08/021. arxiv:2012.14981 [hep-ph]
54. V. De Luca et al., On the gauge invariance of cosmological gravitational waves. JCAP **03**, 014 (2020). https://doi.org/10.1088/1475-7516/2020/03/014. arxiv:1911.09689 [gr-qc]

55. E. Dimastrogiovanni et al., Enhancing gravitational wave anisotropies with peaked scalar sources. JCAP **01**, 018 (2023). https://doi.org/10.1088/1475-7516/2023/01/018. arxiv:2205.05644 [astro-ph.CO]
56. G. Domènech, Cosmological gravitational waves from isocurvature fluctuations. AAPPS Bull. **34**(1), 4 (2024). https://doi.org/10.1007/s43673-023-00109-z. arxiv:2311.02065 [gr-qc]
57. G. Domènech, Induced gravitational waves in a general cosmological background. Int. J. Mod. Phys. D **29**(03), 2050028 (2020). https://doi.org/10.1142/S0218271820500285. arxiv:1912.05583 [gr-qc]
58. G. Domènech, Scalar induced gravitational waves review. Universe **7**(11), 398 (2021). https://doi.org/10.3390/universe7110398. arxiv:2109.01398 [gr-qc]
59. G. Domènech, Were recently reported MHz events planet mass primordial black hole mergers?. Eur. Phys. J. C **81**(11), 1042 (2021). https://doi.org/10.1140/epjc/s10052-021-09853-8. arxiv:2110.00550 [astro-ph.CO]
60. G. Domènech, C. Lin, M. Sasaki, Gravitational wave constraints on the primordial black hole dominated early universe. JCAP **04**, 062 (2021). https://doi.org/10.1088/1475-7516/2021/04/062. arxiv:2012.08151 [gr-qc]
61. G. Domènech, S. Passaglia, S. Renaux-Petel, Gravitational waves from dark matter isocurvature. JCAP **03**(03), 023 (2022). https://doi.org/10.1088/1475-7516/2022/03/023. arxiv:2112.10163 [astro-ph.CO]
62. G. Domènech, S. Pi, M. Sasaki, Induced gravitational waves as a probe of thermal history of the universe. JCAP **08**, 017 (2020). https://doi.org/10.1088/1475-7516/2020/08/017. arxiv:2005.12314 [gr-qc]
63. G. Domènech, M. Sasaki, Approximate gauge independence of the induced gravitational wave spectrum (2020). arxiv:2012.14016 [gr-qc]
64. G. Domènech, M. Sasaki, Gravitational wave hints black hole remnants as dark matter. Class. Quant. Grav. **40**(17), 177001 (2023). https://doi.org/10.1088/1361-6382/ace493. arxiv:2303.07661 [gr-qc]
65. G. Domènech, M. Sasaki, Hamiltonian approach to second order gauge invariant cosmological perturbations. Phys. Rev. D **97**(2), 023521 (2018). https://doi.org/10.1103/PhysRevD.97.023521. arxiv:1709.09804 [gr-qc]
66. G. Domènech, M. Sasaki, Probing primordial black hole scenarios with terrestrial gravitational wave detectors (2024). arxiv:2401.07615 [gr-qc]
67. G. Domènech, V. Takhistov, M. Sasaki, Exploring evaporating primordial black holes with gravitational waves. Phys. Lett. B **823**, 136722 (2021). https://doi.org/10.1016/j.physletb.2021.136722. arxiv:2105.06816 [astro-ph.CO]
68. A. Escrivà, F. Kuhnel, Y. Tada, Primordial black holes (2022). arxiv:2211.05767 [astro-ph.CO]
69. J.R. Espinosa, D. Racco, A. Riotto, A cosmological signature of the SM higgs instability: gravitational waves. JCAP **09**, 012 (2018). https://doi.org/10.1088/1475-7516/2018/09/012. arxiv:1804.07732 [hep-ph]
70. G. Franciolini et al., Searching for a subpopulation of primordial black holes in LIGO-Virgo gravitational-wave data. Phys. Rev. D **105**(8), 083526 (2022). https://doi.org/10.1103/PhysRevD.105.083526. arxiv:2105.03349 [gr-qc]
71. T. Fujita et al., Inflation with two-form field: the production of primordial black holes and gravitational waves. JCAP **09**, 017 (2022). https://doi.org/10.1088/1475-7516/2022/09/017. arxiv:2202.02401 [astro-ph.CO]
72. J. Fumagalli, S. Renaux-Petel, L.T. Witkowski, Oscillations in the stochastic gravitational wave background from sharp features and particle production during inflation. JCAP **08**, 030 (2021). https://doi.org/10.1088/1475-7516/2021/08/030. arxiv:2012.02761 [astro-ph.CO]
73. J. Fumagalli, S. Renaux-Petel, L.T. Witkowski, Resonant features in the stochastic gravitational wave background. JCAP **08**, 059 (2021). https://doi.org/10.1088/1475-7516/2021/08/059. arxiv:2105.06481 [astro-ph.CO]

74. J. Fumagalli et al., Primordial gravitational waves from excited states. JHEP **03**, 196 (2022). https://doi.org/10.1007/JHEP03(2022)196. arxiv:2111.14664 [astro-ph.CO]
75. J. García-Bellido, M. Peloso, C. Unal, Gravitational waves at interferometer scales and primordial black holes in axion inflation. JCAP **12**, 031 (2016). https://doi.org/10.1088/1475-7516/2016/12/031. arxiv:1610.03763 [astro-ph.CO]
76. S. Garcia-Saenz, Y. Lu, Z. Shuai, Scalar-induced gravitational waves from ghost inflation and parity violation. Phys. Rev. D **108**(12), 123507 (2023). https://doi.org/10.1103/PhysRevD.108.123507. arxiv:2306.09052 [gr-qc]
77. S. Garcia-Saenz et al., No-go theorem for scalar-trispectrum-induced gravitational waves (2022). arxiv:2207.14267 [astro-ph.CO]
78. T.C. Gehrman et al., The primordial black holes that disappeared: connections to dark matter and MHz-GHz gravitational waves. JCAP **10**, 001 (2023). https://doi.org/10.1088/1475-7516/2023/10/001. arxiv:2304.09194 [hep-ph]
79. J.-O. Gong. Analytic integral solutions for induced gravitational waves (2019). arxiv:1909.12708 [gr-qc]
80. M.A. Gorji, M. Sasaki, Primordial-tensor-induced stochastic gravitational waves. Phys. Lett. B **846**, 138236 (2023). https://doi.org/10.1016/j.physletb.2023.138236. arxiv:2302.14080 [gr-qc]
81. A.M. Green, B.J. Kavanagh, Primordial black holes as a dark matter candidate. J. Phys. G **48**(4), 043001 (2021). https://doi.org/10.1088/1361-6471/abc534. arxiv:2007.10722 [astro-ph.CO]
82. M.C. Guzzetti et al., Gravitational waves from inflation. Riv. Nuovo Cim. **39**(9), 399–495 (2016). https://doi.org/10.1393/ncr/i2016-10127-1. arxiv:1605.01615 [astro-ph.CO]
83. S. Hannestad, What is the lowest possible reheating temperature?. Phys. Rev. D **70**, 043506 (2004). https://doi.org/10.1103/PhysRevD.70.043506. arxiv:astro-ph/0403291
84. T. Hasegawa et al., MeV-scale reheating temperature and thermalization of oscillating neutrinos by radiative and hadronic decays of massive particles. JCAP **12**, 012 (2019). https://doi.org/10.1088/1475-7516/2019/12/012. arxiv:1908.10189 [hep-ph]
85. D. Hooper, G. Krnjaic, GUT baryogenesis with primordial black holes. Phys. Rev. D **103**(4), 043504 (2021). https://doi.org/10.1103/PhysRevD.103.043504. arxiv:2010.01134 [hep-ph]
86. D. Hooper, G. Krnjaic, S.D. McDermott, Dark radiation and superheavy dark matter from black hole domination. JHEP **08**, 001 (2019). https://doi.org/10.1007/JHEP08(2019)001. arxiv:1905.01301 [hep-ph]
87. J.-C. Hwang, D. Jeong, H. Noh, Gauge dependence of gravitational waves generated from scalar perturbations. Astrophys. J. **842**(1), 46 (2017). https://doi.org/10.3847/1538-4357/aa74be. arxiv:1704.03500 [astro-ph.CO]
88. K. Inomata, Bound on induced gravitational waves during inflation era. Phys. Rev. D **104**(12), 123525 (2021). https://doi.org/10.1103/PhysRevD.104.123525. arxiv:2109.06192 [astro-ph.CO]
89. K. Inomata, T. Terada. Gauge independence of induced gravitational waves. Phys. Rev. D **101**(2), 023523 (2020). https://doi.org/10.1103/PhysRevD.101.023523. arxiv:1912.00785 [gr-qc]
90. K. Inomata et al., Enhancement of gravitational waves induced by scalar perturbations due to a sudden transition from an early matter era to the radiation era. Phys. Rev. D **100**(4), 043532 (2019). https://doi.org/10.1103/PhysRevD.100.043532. arxiv:1904.12879 [astro-ph.CO]
91. K. Inomata et al., Gravitational wave production right after a primordial black hole evaporation. Phys. Rev. D **101**(12), 123533 (2020). https://doi.org/10.1103/PhysRevD.101.123533. arxiv:2003.10455 [astro-ph.CO]
92. K. Inomata et al., Gravitational waves induced by scalar perturbations during a gradual transition from an early matter era to the radiation era. JCAP **10**, 071 (2019). https://doi.org/10.1088/1475-7516/2019/10/071. arxiv:1904.12878 [astro-ph.CO]

93. K. Inomata et al., Inflationary primordial black holes for the LIGO gravitational wave events and pulsar timing array experiments. Phys. Rev. D **95**(12), 123510 (2017). https://doi.org/10.1103/PhysRevD.95.123510. arxiv:1611.06130 [astro-ph.CO]
94. A. Ireland, S. Profumo, J. Scharnhorst, Primordial gravitational waves from black hole evaporation in standard and nonstandard cosmologies. Phys. Rev. D **107**(10), 104021 (2023). https://doi.org/10.1103/PhysRevD.107.104021. arxiv:2302.10188 [gr-qc]
95. M. Kawasaki, K. Kohri, N. Sugiyama, Cosmological constraints on late time entropy production. Phys. Rev. Lett. **82**, 4168 (1999). https://doi.org/10.1103/PhysRevLett.82.4168. arxiv:astro-ph/9811437
96. M. Kawasaki, K. Kohri, N. Sugiyama, MeV scale reheating temperature and thermalization of neutrino background. Phys. Rev. D **62**, 023506 (2000). https://doi.org/10.1103/PhysRevD.62.023506. arxiv:astro-ph/0002127
97. M.Y. Khlopov, Primordial black holes. Res. Astron. Astrophys. **10**, 495–528 (2010). https://doi.org/10.1088/1674-4527/10/6/001. arxiv:0801.0116 [astro-ph]
98. H. Kodama, M. Sasaki, Evolution of isocurvature perturbations. 2. Radiation dust universe. Int. J. Mod. Phys. A **2**, 491 (1987). https://doi.org/10.1142/S0217751X8700020X
99. K. Kohri, T. Terada, Semianalytic calculation of gravitational wave spectrum nonlinearly induced from primordial curvature perturbations. Phys. Rev. D **97**(12), 123532 (2018). https://doi.org/10.1103/PhysRevD.97.123532. arxiv:1804.08577 [gr-qc]
100. E. Komatsu, New physics from the polarized light of the cosmic microwave background. Nature Rev. Phys. **4**(7), 452–469 (2022). https://doi.org/10.1038/s42254-022-00452-4. arxiv:2202.13919 [astro-ph.CO]
101. Eiichiro Komatsu, David N. Spergel, Benjamin D. Wandelt, Measuring primordial non-Gaussianity in the cosmic microwave background. Astrophys. J. **634**, 14–19 (2005). https://doi.org/10.1086/491724. (arxiv:astro-ph/0305189)
102. J.-P. Li et al., Primordial non-Gaussianity f $_{NL}$ and anisotropies in scalar-induced gravitational waves. JCAP **10**, 056 (2023). https://doi.org/10.1088/1475-7516/2023/10/056. arxiv:2305.19950 [astro-ph.CO]
103. J.E. Lidsey, T. Matos, L. Arturo Urena-Lopez, The inflaton field as selfinteracting dark matter in the brane world scenario. Phys. Rev. D **66**, 023514 (2002). https://doi.org/10.1103/PhysRevD.66.023514. arxiv:astro-ph/0111292
104. J. Liu, Z.-K. Guo, R.-G. Cai, Analytical approximation of the scalar spectrum in the ultraslow-roll inflationary models. Phys. Rev. D **101**(8), 083535 (2020). https://doi.org/10.1103/PhysRevD.101.083535. arxiv:2003.02075 [astro-ph.CO]
105. K.D. Lozanov, M. Sasaki, V. Takhistov, Universal gravitational wave signatures of cosmological solitons (2023). arxiv:2304.06709 [astro-ph.CO]
106. K.D. Lozanov, M. Sasaki, V. Takhistov, Universal gravitational waves from interacting and clustered solitons. Phys. Lett. B **848**, 138392 (2024). https://doi.org/10.1016/j.physletb.2023.138392. arxiv:2309.14193 [astro-ph.CO]
107. Jane H. MacGibbon, Can Planck-mass relics of evaporating black holes close the universe? Nature **329**, 308–309 (1987). https://doi.org/10.1038/329308a0
108. M. Maggiore. *Gravitational Waves. Vol. 1: Theory and Experiments* (Oxford Master Series in Physics. Oxford University Press, 2007). ISBN: 978-0-19-857074-5, 978-0-19-852074-0
109. A. Maleknejad, Axion inflation with an SU(2) gauge field: detectable chiral gravity waves. JHEP **07**, 104 (2016). https://doi.org/10.1007/JHEP07(2016)104. arxiv:1604.03327 [hep-ph]
110. K.A. Malik, D. Wands, Adiabatic and entropy perturbations with interacting fluids and fields. JCAP **02**, 007 (2005). https://doi.org/10.1088/1475-7516/2005/02/007. arxiv:astro-ph/0411703
111. I. Masina, Dark matter and dark radiation from evaporating Kerr primordial black holes. Grav. Cosmol. **27**(4), 315–330 (2021). https://doi.org/10.1134/S0202289321040101. arxiv:2103.13825 [gr-qc]
112. I. Masina, Dark matter and dark radiation from evaporating primordial black holes. Eur. Phys. J. Plus **135**(7), 552 (2020). https://doi.org/10.1140/epjp/s13360-020-00564-9. arxiv:2004.04740 [hep-ph]

113. S. Matarrese, S. Mollerach, M. Bruni, Second order perturbations of the Einstein-de Sitter universe. Phys. Rev. D **58**, 043504 (1998). https://doi.org/10.1103/PhysRevD.58.043504. arxiv:astro-ph/9707278
114. Sabino Matarrese, Ornella Pantano, Diego Saez, A general relativistic approach to the non-linear evolution of collisionless matter. Phys. Rev. D **47**, 1311–1323 (1993). https://doi.org/10.1103/PhysRevD.47.1311
115. Sabino Matarrese, Ornella Pantano, Diego Saez, General relativistic dynamics of irrotational dust: cosmological implications. Phys. Rev. Lett. **72**, 320–323 (1994). https://doi.org/10.1103/PhysRevLett.72.320. (arxiv:astro-ph/9310036)
116. P. Mroz et al., No large population of unbound or wide-orbit Jupiter-mass planets. Nature **548**(7666), 183–186 (2017). https://doi.org/10.1038/nature23276. arxiv:1707.07634 [astro-ph.EP]
117. V. Mukhanov. *Physical Foundations of Cosmology* (Cambridge University Press, Cambridge, UK New York, 2005). ISBN: 978-0-511-13679-5
118. Takashi Nakamura et al., Gravitational waves from coalescing black hole MACHO binaries. Astrophys. J. Lett. **487**, L139–L142 (1997). https://doi.org/10.1086/310886. (arxiv:astro-ph/9708060)
119. H. Niikura et al., Constraints on earth-mass primordial black holes from OGLE 5-year microlensing events. Phys. Rev. D **99**(8), 083503 (2019). https://doi.org/10.1103/PhysRevD.99.083503. arxiv:1901.07120 [astro-ph.CO]
120. T. Papanikolaou, Gravitational waves induced from primordial black hole fluctuations: the effect of an extended mass function. JCAP **10**, 089 (2022). https://doi.org/10.1088/1475-7516/2022/10/089. arxiv:2207.11041 [astro-ph.CO]
121. T. Papanikolaou, V. Vennin, D. Langlois, Gravitational waves from a universe filled with primordial black holes. JCAP **03**, 053 (2021). https://doi.org/10.1088/1475-7516/2021/03/053. arxiv:2010.11573 [astro-ph.CO]
122. T. Papanikolaou et al., No constraints for f(T) gravity from gravitational waves induced from primordial black hole fluctuations. Eur. Phys. J. C **83**(1), 31 (2023). https://doi.org/10.1140/epjc/s10052-022-11157-4. arxiv:2205.06094 [gr-qc]
123. T. Papanikolaou et al., Scalar induced gravitational waves from primordial black hole Poisson fluctuations in f(R) gravity. JCAP **10**, 013 (2022). https://doi.org/10.1088/1475-7516/2022/10/013. arxiv:2112.15059 [astro-ph.CO]
124. S. Passaglia, M. Sasaki, Primordial black holes from CDM isocurvature perturbations. Phys. Rev. D **105**(10), 103530 (2022). https://doi.org/10.1103/PhysRevD.105.103530. arxiv:2109.12824 [astro-ph.CO]
125. M. Pearce et al., Gravitational wave signals from early matter domination: interpolating between fast and slow transitions (2023). arxiv:2311.12340 [astro-ph.CO]
126. S. Pi, M. Sasaki, Gravitational waves induced by scalar perturbations with a lognormal peak. JCAP **09**, 037 (2020). https://doi.org/10.1088/1475-7516/2020/09/037. arxiv:2005.12306 [gr-qc]
127. R. Picard, K.A. Malik, Induced gravitational waves: the effect of first order tensor perturbations (2023). arxiv:2311.14513 [astro-ph.CO]
128. S. Profumo, F. Yang, Anisotropy of the stochastic gravitational wave background from sub-horizon-collapsed primordial black hole mergers. Phys. Rev. D **108**(10), 103528 (2023). https://doi.org/10.1103/PhysRevD.108.103528. arxiv:2306.07454 [astro-ph.CO]
129. D.J. Reardon et al., Search for an isotropic gravitational-wave background with the Parkes pulsar timing array. Astrophys. J. Lett. **951**(1), L6 (2023). https://doi.org/10.3847/2041-8213/acdd02. arxiv:2306.16215 [astro-ph.HE]
130. D.J. Reardon et al., The gravitational-wave background null hypothesis: characterizing noise in millisecond pulsar arrival times with the Parkes pulsar timing array. Astrophys. J. Lett. **951**(1), L7 (2023). https://doi.org/10.3847/2041-8213/acdd03. arxiv:2306.16229 [astro-ph.HE]
131. T. Regimbau, The astrophysical gravitational wave stochastic background. Res. Astron. Astrophys. **11**, 369–390 (2011). https://doi.org/10.1088/1674-4527/11/4/001. arxiv:1101.2762 [astro-ph.CO]

132. T. Regimbau, The quest for the astrophysical gravitational-wave background with terrestrial detectors. Symmetry **14**(2), 270 (2022). https://doi.org/10.3390/sym14020270
133. P.A. Rosado, Gravitational wave background from binary systems. Phys. Rev. D **84**, 084004 (2011). https://doi.org/10.1103/PhysRevD.84.084004. arxiv:1106.5795 [gr-qc]
134. R. Roshan, G. White, Using gravitational waves to see the first second of the universe (2024). arxiv:2401.04388 [hep-ph]
135. R. Saito, J. Yokoyama, Gravitational wave background as a probe of the primordial black hole abundance. Phys. Rev. Lett. **102** (2009). [Erratum: Phys.Rev.Lett. 107, 069901 (2011)], 161101. https://doi.org/10.1103/PhysRevLett.102.161101. arxiv:0812.4339 [astro-ph]
136. R. Saito, J. Yokoyama, Gravitational-wave constraints on the abundance of primordial black holes. Prog. Theor. Phys. **123** (2010). [Erratum: Prog.Theor.Phys. 126, 351–352 (2011)], 867–886. https://doi.org/10.1143/PTP.126.351. arxiv:0912.5317 [astro-ph.CO]
137. M. Sasaki et al., Primordial black hole scenario for the gravitational-wave event GW150914. Phys. Rev. Lett. **117**(6) (2016). [Erratum: Phys. Rev. Lett. 121, 059901 (2018)], 061101. https://doi.org/10.1103/PhysRevLett.117.061101. arxiv:1603.08338 [astro-ph.CO]
138. M. Sasaki et al., Primordial black holes—perspectives in gravitational wave astronomy. Class. Quant. Grav. **35**(6), 063001 (2018). https://doi.org/10.1088/1361-6382/aaa7b4. arxiv:1801.05235 [astro-ph.CO]
139. G. Scelfo et al., GW$\times$LSS: chasing the progenitors of merging binary black holes. JCAP **09**, 039 (2018). https://doi.org/10.1088/1475-7516/2018/09/039. arxiv:1809.03528 [astro-ph.CO]
140. M. Sipp, B. Malte Schaefer, Scalar-induced gravitational waves in a ΛCDM cosmology. Phys. Rev. D **107**(6), 063538 (2023). https://doi.org/10.1103/PhysRevD.107.063538. arxiv:2212.01190 [astro-ph.CO]
141. K. Tomikawa, T. Kobayashi, Gauge dependence of gravitational waves generated at second order from scalar perturbations. Phys. Rev. D **101**(8), 083529 (2020). https://doi.org/10.1103/PhysRevD.101.083529. arxiv:1910.01880 [gr-qc]
142. K. Tomita, Non-linear theory of gravitational instability in the expanding universe. Progr. Theoret. Phys. **37**(5), 831–846 (1967). ISSN: 0033-068X. https://doi.org/10.1143/PTP.37.831. eprint: https://academic.oup.com/ptp/article-pdf/37/5/831/5234391/37-5-831.pdf. https://doi.org/10.1143/PTP.37.831
143. C. Unal, Imprints of primordial non-Gaussianity on gravitational wave spectrum. Phys. Rev. D **99**(4), 041301 (2019). https://doi.org/10.1103/PhysRevD.99.041301. arxiv:1811.09151 [astro-ph.CO]
144. S. Wang, T. Terada, K. Kohri, Prospective constraints on the primordial black hole abundance from the stochastic gravitational-wave backgrounds produced by coalescing events and curvature perturbations. Phys. Rev. D **99**(10) (2019). [Erratum: Phys. Rev. D 101, 069901 (2020)], 103531. https://doi.org/10.1103/PhysRevD.99.103531. arxiv:1903.05924 [astro-ph.CO]
145. S. Wang, V. Vardanyan, K. Kohri, Probing primordial black holes with anisotropies in stochastic gravitational-wave background. Phys. Rev. D **106**(12), 123511 (2022). https://doi.org/10.1103/PhysRevD.106.123511. arxiv:2107.01935 [gr-qc]
146. L.T. Witkowski, SIGWfast: a python package for the computation of scalar-induced gravitational wave spectra (2022). arxiv:2209.05296 [astro-ph.CO]
147. L.T. Witkowski et al., Expansion history-dependent oscillations in the scalar-induced gravitational wave background. JCAP **05**(05), 028 (2022). https://doi.org/10.1088/1475-7516/2022/05/028. arxiv:2110.09480 [astro-ph.CO]
148. H. Xu et al., Searching for the nano-hertz stochastic gravitational wave background with the Chinese pulsar timing array data release I. Res. Astron. Astrophys. **23**(7), 075024 (2023). https://doi.org/10.1088/1674-4527/acdfa5. arxiv:2306.16216 [astro-ph.HE]
149. W.-T. Xu et al., Gravitational waves from double-inflection-point inflation. Phys. Rev. D **101**(2), 023505 (2020). https://doi.org/10.1103/PhysRevD.101.023505. arxiv:1907.05213 [astro-ph.CO]

150. Y.-H. Yu, S. Wang, Anisotropies in scalar-induced gravitational-wave background from inflaton-curvaton mixed scenario with sound speed resonance (2023). arxiv:2310.14606 [astro-ph.CO]
151. Y.-H. Yu, S, Wang, Primordial gravitational waves assisted by cosmological scalar perturbations (2023). arxiv:2303.03897 [astro-ph.CO]
152. C. Yuan, Z.-C. Chen, Q.-G. Huang, Log-dependent slope of scalar induced gravitational waves in the infrared regions. Phys. Rev. D **101**(4), 043019 (2020). https://doi.org/10.1103/PhysRevD.101.043019. arxiv:1910.09099 [astro-ph.CO]
153. C. Yuan, Q.-G. Huang, A topic review on probing primordial black hole dark matter with scalar induced gravitational waves (2021). arxiv:2103.04739 [astro-ph.GA]
154. Z. Zhou et al., Primordial black holes and gravitational waves from resonant amplification during inflation. Phys. Rev. D **102**(10), 103527 (2020). https://doi.org/10.1103/PhysRevD.102.103527. arxiv:2010.03537 [astro-ph.CO]
155. X.-J. Zhu et al., Stochastic gravitational wave background from coalescing binary black holes. Astrophys. J. **739**, 86 (2011). https://doi.org/10.1088/0004-637X/739/2/86. arxiv:1104.3565 [gr-qc]
156. A. Zic et al., The Parkes pulsar timing array third data release (2023). arxiv:2306.16230 [astro-ph.HE]

Chapter 18
Gravitational Waves from Sub-Solar Mass Primordial Black Holes

Andrew L. Miller

Abstract Gravitational waves from inspiraling sub-solar mass compact objects would provide almost definitive evidence for the existence of primordial black holes. In this chapter, we explain why these exotic objects are interesting candidates for current and future gravitational-wave observatories, and provide detailed explanations of how they are searched for. We describe one method, matched filtering, to search for binaries with masses in $[0.01, 1]M_\odot$. Furthermore, since signals from inspiraling planetary- and asteroid-mass compact binaries ($[10^{-9}, 10^{-2}]M_\odot$) would spend hours to years in the detector frequency band, we explain the novel pattern recognition techniques that have been developed to search for them. Finally, we describe extreme mass ratio inspiral systems, and how these will be searched for in future space-based detectors. For all mass regimes, we comment on the prospects for detection.

18.1 Introduction

Sub-solar mass compact objects are broadly defined as objects more compact than white dwarfs with masses less than a solar mass. Predictions as far back as the 1930s by Chandrasekhar dictate an upper limit of the mass of a white dwarf to be $M_{\max} \sim 1.4M_\odot$ [36, 37, 50, 79, 107], thus any system with a mass larger than $M_{\max}$ would inevitably collapse to a neutron star or black hole. Since there are not any astrophysical channels to form sub-solar mass objects more compact than white dwarfs, a detection of a sub-solar mass object via gravitational waves would imply a new, unknown formation channel, one of which could be through accumulated dark matter resulting in collapse [39, 47, 100].

Data from ground-based gravitational-wave observatories has extensively been used to search looking for gravitational waves from neutron stars and black holes with

A. L. Miller (✉)
Nikhef—National Institute for Subatomic Physics, Science Park 105, 1098, XG, Amsterdam, The Netherlands

Institute for Gravitational and Subatomic Physics (GRASP), Utrecht University, Princetonplein 1, 3584, CC Utrecht, The Netherlands
e-mail: andrew.miller@nikhef.nl

C. Byrnes et al. (eds.), *Primordial Black Holes*, Springer Series in Astrophysics and Cosmology, https://doi.org/10.1007/978-981-97-8887-3_18

at least one solar mass, which has resulted in $\mathcal{O}(100)$ of them [8, 84, 86]. Already, there have been detections of unexpected systems, e.g. in the so-called "lower-mass mass gap" $[2, 5]M_\odot$ (GW190814 and GW190425) [2, 7] that are heavier than the most massive pulsar in our Galaxy ($\sim 2.14 M_\odot$) [46], though of comparable mass to that of pulsar PSR J1748-2021B and to that of the remnant of the neutron-star merger GW170817 [1]. In this mass gap, there had simply been no observations of compact objects with these masses, though some stellar evolution models could predict this gap or a continuous distribution of masses; hence, possible explanations, e.g. hierarchical mergers [17, 51, 101] or objects of primordial origin [33], were invoked. Therefore, there is not a clear explanation of this mass gap, whether there is in fact one, and thus the primordial black hole hypothesis cannot be ruled out. Furthermore, even the first detection of gravitational waves, GW150914, indicated the existence of black holes with masses of at least $\sim 25 M_\odot$ [3], which confirmed that black holes could both form in binaries and merge within a Hubble time.

Based on the unexpected physics we have already learned from "canonical" gravitational-wave sources, the currently *unknown* sources of gravitational waves could open new windows into the history of the universe and the formation of (primordial) black holes.

The range of sub-solar mass primordial black holes is vast, spanning from $\sim [10^{-18}, 1] M_\odot$. If such objects form binary systems, they will emit gravitational waves as they inspiral. The amount of time that the signals spend in-band also vary significantly as a function of the mass of the primordial black holes– lighter systems inspiral for much longer than heavier ones, since the duration scales inversely with the chirp mass of the binary—thus, different methods have been designed to cover different mass regimes. The ideal signal processing technique, the matched filter, is tractable around $\mathcal{O}(0.1) M_\odot$ and above, but becomes too computationally heavy to be used below that. Therefore, other techniques that can handle longer-duration systems, i.e. those of masses $[10^{-5}, 10^{-2}] M_\odot$, lasting hours-days, have been developed to search for such systems. We call the signals emitted by these binaries "transient continuous waves" (tCWs). Systems with primordial black holes with masses in $[10^{-10}, 10^{-5}] M_\odot$ last for durations much longer than the observation time of the detector, and thus methods developed to handle quasi-monochromatic, quasi-infinite signals can be employed here. These signals are called "continuous waves" (CWs).

In Fig. 18.1, we show the range of masses probed via gravitational waves, the kinds of methods that can be used, the durations of the signals, the expected luminosity distance assuming strain sensitivities of 10^{-26}, 5×10^{-24}, 10^{-22} at 100 Hz in each of the three regimes (CW methods, tCW methods and matched filtering, respectively) and the maximal possible sky localization we could obtain using just by using the Doppler shift induced by the relative motion of the earth and the source at one detector. The values on each color bar are meant to conceptually illustrate the different regimes in which we can search for primordial black holes and should not be interpreted as strictly corresponding to the chirp mass listed, since, for example, luminosity distance reach depends on the actual strain sensitivity of the detector, and the time to merger is affected by the choice of frequency.

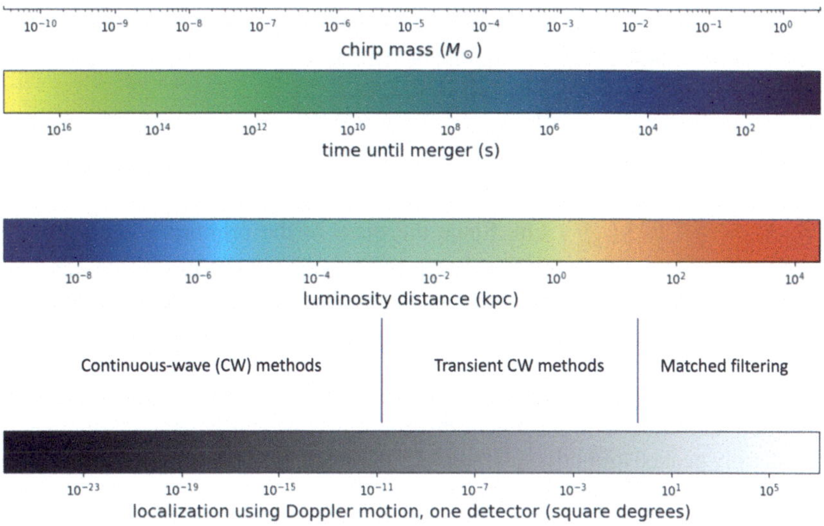

Fig. 18.1 Summary plot that shows the full sub-solar mass range that can be probed by ground-based gravitational-wave detectors, along with different methods that can be used, the expected source durations, expected distance reaches, and maximal possible sky localization

In this chapter, we will discuss the motivations for the existence of sub-solar mass objects, the gravitational-wave signatures of sub-solar mass objects, and the techniques that we can use to detect them.

18.2 Searches for GWs from [0.1, 1]$M_\odot$ Binary PBHs

18.2.1 Motivation

Below a solar mass, black holes could be primordial in origin [33]. In essence, overdensities (relative to a background density) generated in the early universe, e.g. by large primordial fluctuations during inflation [35, 54, 62, 66], through bubble nucleation or domain walls [56], or due to cosmic string loops and scalar field instabilities [45, 65], could collapse in on themselves, leading to the formation of primordial black holes [33, 38, 60, 113]. Because 84% of the matter in the Universe is currently unexplained [12], primordial black holes could comprise a fraction or all of this unknown matter. Depending on when they formed in the early universe, primordial black holes could take on a wide range of masses – as small as the Planck mass, 10^{-5} g if they formed at the Planck time (10^{-43} s), $\sim 1M_\odot$ if they formed during the QCD phase transition (10^{-5} s), or $10^5 M_\odot$ if they formed 1 s after the Big Bang [33].

The time and mass relations are obtained by equating the mean densities of a black hole of mass M and radius R and of the universe in the radiation era [111]:

$$\rho_{\rm PBH} = \frac{M}{\frac{4}{3}\pi R_s^3} = \rho_c \sim 10^6 \left(\frac{t}{1\ \rm s}\right)^{-2} \ \rm g/cm^3, \tag{18.1}$$

where $R_s = \frac{2GM}{c^2} \sim 3\left(\frac{M}{M_\odot}\right)$ km. Since the mass of the primordial black hole is roughly equivalent to the mass within the horizon, any fluctuations that enter the horizon cause over-densities and subsequent collapses into primordial black holes.

Furthermore, the thermal of the universe could enhance primordial black hole formation at different scales [34], e.g. the QCD transition significantly reduces the radiation pressure of the plasma, so that a uniform primordial enhancement stretching across the QCD scale will generate a distribution of primordial black hole masses that is sharply peaked around a solar mass [30] as well as a broader mass distribution at both larger and sub-solar masses [10].

Other physical mechanisms could result in the formation of sub-solar mass primordial black holes. For example, if dark matter is composed of particles that allow for dissipation or other chemical reactions, sub-solar mass objects could form through collapsing dark-matter halos [39, 47, 100]. Essentially, if a new "dark Hydrogen atom" exists, along with a heavy dark fermion (akin to a proton) with mass m_x, a light dark fermion (akin to an electron) with mass m_c, and a massless dark photon that mediates particle interactions in this theory with a strength determined by the dark fine structure constant α_D, the dark Chandrasekhar mass upper limit becomes [37, 100]:

$$M_{\rm DC} = 1.4 M_\odot \left(\frac{m_p}{m_x}\right)^2 \tag{18.2}$$

and goes up to $\sim 1.9 M_\odot$ for arbitrarily more complex dark-matter particle interaction models [102]. With the addition of this dark sector of particles, dark-matter structures can undergo dissipative processes such as recombination, Bremsstrahlung radiation, and collisional excitation of atoms in sufficiently dense regions [29, 96], can clump like ordinary Hydrogen and therefore collapse into compact objects [100]. Without any dark nuclear physics, e.g. dark electron or neutron degeneracy pressures, all clumps of dark matter will collapse into black holes, with a minimum mass given by Eq. 18.2. If $m_x > m_p$, sub-solar mass primordial black holes could be formed.

While extensive searches for compact objects above a solar mass have been performed by the LIGO, Virgo and KAGRA collaborations and others [8, 86], only a small effort [9, 10, 77, 81–83, 88] in terms of method design has been devoted to explore the full range of sub-solar mass objects. There are many reasons for this, that will be explored in detail in the next sub-sections: (1) the gravitational-wave amplitude falls off steeply with the masses of the two compact objects [68], (2) searches for sub-solar mass objects in $[0.1, 1] M_\odot$ are computationally expensive due to need to correlate many long-duration waveforms with the data [81], (3) searches for sub-solar mass objects below $0.1 M_\odot$ require techniques beyond the standard matched filtering ones to handle long-duration signals in the detector [13, 16, 78], and (4) other existing experiments put stringent constraints on $f_{\rm PBH}$ [58].

18.2.2 Matched Filtering Searches

Matched filtering searches correlate strain data with a bank of deterministic templates, i.e. "waveforms", that model gravitational-wave emission from two compact objects of masses m_1 and m_2 in a quasi-circular orbit in many cases, though there are exceptions [81, 82]. For each template, maximized over arrival time, phase and distance to the source, a signal-to-noise ratio ρ is computed, given by [20]:

$$\rho^2 = \frac{\langle d|h\rangle^2 + \langle d|h_{\pi/2}\rangle^2}{\langle h|h\rangle} \tag{18.3}$$

where

$$\langle a|b\rangle \equiv 4\,\mathrm{Re}\left[\int_{f_{\min}}^{f_{\max}} df \frac{\tilde{a}^*(f)\tilde{b}(f)}{S_n(f)}\right] \tag{18.4}$$

is the noise-weighted inner product, the tilde denotes the Fourier Transform of either the strain time-series d or the template time-series h, "$*$" denotes the complex conjugate, $S_n(f)$ is the frequency-dependent power spectral density of the noise, estimated empirically using, typically, a median-average [32], and $f_{\min}$ and $f_{\max}$ are the minimum and maximum frequencies of the analysis. In the stationary-phase approximation for waveforms in the frequency-domain, $\tilde{h}_{\pi/2} = i\tilde{h}$ [49]. If the data contain exactly the signal that we are searching for, i.e., $d = h$, we would obtain the optimal signal-to-noise ratio:

$$\rho^2_{\mathrm{opt}} = 4\int_{f_{\min}}^{f_{\max}} df \frac{|\tilde{h}(f)|^2}{S_n(f)} \tag{18.5}$$

Because we do not know which gravitational-wave signals are embedded in the data, we must convolve all possible waveforms with the data. In practice, since we do not have infinite computing power, we cannot actually do this, so we must construct a bank of templates across a high-dimensional parameter space in such a way as to ensure that we do not lose more than some percent of signals due to discretizing the parameter space.

Let us assume we have a template u and we correlate it with noise n plus signal s, $n + \mathcal{A}s$, where $\mathcal{A}$ quantifies the amplitude of the signal. The expectation value of matched-filter signal-to-noise ratio is [87]:

$$\begin{aligned} E[\rho^2] &= E[\langle n + \mathcal{A}s|u\rangle] = E[\langle n|u\rangle] + \mathcal{A}E[\langle s|u\rangle] \\ &= \begin{cases} \mathcal{A}, & \text{if } s = u \\ \mathcal{A}\langle s|u\rangle, & \text{otherwise,} \end{cases} \end{aligned} \tag{18.6}$$

where $E[\langle n|u\rangle] = 0$ because the noise and signal are uncorrelated. Therefore, we can read off a quantity that defines the effectiveness of our template: the "match" $\langle s|u\rangle \in [0, 1]$ between the signal s in the data and the template u.

Not all parameters of u carry equal importance in gravitational-wave searches. There are two types of parameters to consider: the *intrinsic*, i.e. phase, parameters $\boldsymbol{\lambda}$, of u, and the *extrinsic* parameters $\boldsymbol{\mu}$, e.g. the observed time of coalescence or the amplitude. The key distinction is that the full range of extrinsic parameters can be searched over simultaneously for a fixed intrinsic parameter $\boldsymbol{\lambda}$; thus, what matters for the construction of template banks are the intrinsic parameters, i.e. the *phase evolution* of a gravitational-wave signal.

We are very sensitive to changes in phase evolution, so we can maximize over the *extrinsic* parameters and place templates in the *intrinsic* parameter space $\boldsymbol{\lambda}$. Assuming that the signal matches ones of the templates in our template bank, we can write the match between templates $u(\boldsymbol{\mu}, \boldsymbol{\lambda})$ and $u(\boldsymbol{\mu} + \Delta\boldsymbol{\mu}, \boldsymbol{\lambda} + \Delta\boldsymbol{\lambda})$ as:

$$M(\boldsymbol{\lambda}, \Delta\boldsymbol{\lambda}) \equiv \max_{\mu, \Delta\mu} \langle u(\boldsymbol{\mu}, \boldsymbol{\lambda})|u(\boldsymbol{\mu} + \Delta\boldsymbol{\mu}, \boldsymbol{\lambda} + \Delta\boldsymbol{\lambda})\rangle. \tag{18.7}$$

Noting that the best-case scenario is when $\Delta\boldsymbol{\lambda} = 0$, we can Taylor expand about $\Delta\boldsymbol{\lambda}$ and write:

$$M(\boldsymbol{\lambda}, \Delta\boldsymbol{\lambda}) \approx 1 + \frac{1}{2}\left(\frac{\partial^2 M}{\partial\Delta\lambda^i \partial\Delta\lambda^j}\right)_{\Delta\lambda^k=0} \Delta\lambda^i \Delta\lambda^j. \tag{18.8}$$

We can see that a concept of "mismatch" $1 - M$ arises naturally that depends on a *metric* g_{ij}, which quantifies how close two templates should be when allowing a certain mismatch

$$g_{ij}(\boldsymbol{\lambda}) = -\frac{1}{2}\left(\frac{\partial^2 M}{\partial\Delta\lambda^i \partial\Delta\lambda^j}\right)_{\Delta\lambda^k=0} \tag{18.9}$$

So, we can see that the template mismatch equals the proper distance between them:

$$1 - M = g_{ij}\Delta\lambda^i \Delta\lambda^j. \tag{18.10}$$

Therefore, by taking derivatives of M with respect to the physical parameters of the waveform, e.g. the chirp mass, frequency, or spins, we can efficiently place templates in the template bank for a given mismatch.

Typically, this template bank is constructed across a high-dimensional parameter space with a chosen maximum mismatch between adjacent templates of a few percent; in other words, no more than a few percent of signals will be lost due to a discrete sampling of the parameter space [87].

If we return to Eq. 18.3, while $f_{\max}$ is typically taken to be $\sim$1024 Hz, a choice based primarily on the fact that signal power accumulated above 1024 Hz is negligible. $f_{\min}$ is chosen to have a reasonable computational cost. This is because the number of templates scales inversely with primordial black hole parameters (e.g. minimum total mass $M_{\min}$, f, M) in the simple case of quasi-circular orbits with a given $S_n(f)$ [87]:

$$\mathcal{N} \simeq 8.4 \times 10^6 \left(\frac{1 - MM}{0.03}\right)^{-1} \left(\frac{M_{\min}}{0.2\, M_\odot}\right)^{-2.7} \left(\frac{f_0}{70\ \mathrm{Hz}}\right)^{-2.5} \tag{18.11}$$

where MM is the "minimal match". Thus, we see that the number of templates increases as the frequency and mass decrease. Physically, this is because of the time that the binary spends in the frequency band of the detector. The time τ for a binary to reach coalescence is related to the frequency and chirp mass $\mathcal{M} \equiv \frac{(m_1 m_2)^{3/5}}{(m_1+m_2)^{1/5}}$ [68]:

$$\tau \simeq 2.18 \left(\frac{1.21 M_\odot}{\mathcal{M}}\right)^{5/3} \left(\frac{100\ \mathrm{Hz}}{f}\right)^{8/3} \mathrm{s}, \tag{18.12}$$

and hence lower-frequency signals last longer and phase mismatches therefore have more time to accumulate. In other words, ρ^2 is sensitive to smaller changes in the signal parameters, and the spacing between templates in the same parameter space must shrink to follow each potential waveform's phase evolution [87]. Thus, more templates are required to populate the parameter space at lower frequencies than at higher ones, and so $f_{\min}$ has been taken to be 45 Hz recently [9, 10], but could be taken to be lower.

While no search has yielded any viable primordial black hole candidates, upper limits on the fraction of dark matter that sub-solar mass primordial black holes could compose have been computed. To do this, analyses inject many simulated signals into the data at different distances away, randomized over spins and all other parameters, at particular chirp masses and determine the detection efficiency at a particular confidence level.

18.3 Even Lighter PBHs

18.3.1 Motivation

Primordial black holes need not lie within the region $[0.1, 1]M_\odot$, but could take on even lighter masses as well. Recently detected star and quasar microlensing events [27, 61, 80] suggest primordial black holes could take on masses $[10^{-6}, 10^{-5}]M_\odot$ and constitute around 1% of dark matter, which agrees with the unified scenarios for primordial black holes discussed in [41] but exceeds that predicted from floating planets [18]. Even Planet 9 could be a $10^{-6} M_\odot$ primordial black hole that the solar system captured long ago [98]. Though these observations provide a basis for the existence of primordial black holes, astrophysical uncertainties, e.g. those due to clustering of primordial black holes, are an unfortunate part of them [26, 31, 34, 48, 55, 109], which motivates the need for complementary probes of this mass range.

18.3.2 Signal Model

For compact objects in binary systems with chirp masses $\mathcal{M} < 0.1 M_{\odot}$, we typically assume that the spin-up of the evolution over time $\dot{f}_{\rm gw}$ follows a simple power-law evolution, which is derived from the quadrupole formula by equating the orbital energy loss due to the gravitational-wave power:

$$\dot{f}_{\rm gw} = \frac{96}{5}\pi^{8/3}\left(\frac{G\mathcal{M}}{c^3}\right)^{5/3} f_{\rm gw}^{11/3}, \tag{18.13}$$

Thus, we only consider the inspiral portion of the system's life (neglecting merger and ringdown, due to the fact that the merger frequency lies well outside of the sensitivity frequency band of our detectors) and neglect higher-order contributions to the inspiral, i.e. we only care about the Post-Newtonian 0 contribution. Higher-order Post-Newtonian terms add negligibly to the phase for such light systems [110], and primordial black holes are expected to have low spins [42]; thus, the approximations made here should be valid.

Integrating Eq. 18.13, we obtain the frequency evolution over time:

$$f_{\rm gw}(t) = f_0\left[1 - \frac{8}{3}k f_0^{8/3}(t - t_0)\right]^{-\frac{3}{8}}, \tag{18.14}$$

where $k \equiv \frac{96}{5}\pi^{8/3}\left(\frac{G\mathcal{M}}{c^3}\right)^{5/3}$, and the signal will have an amplitude of [68]:

$$h_0(t) = \frac{4}{d}\left(\frac{G\mathcal{M}}{c^2}\right)^{5/3}\left(\frac{\pi f_{\rm gw}(t)}{c}\right)^{2/3}, \tag{18.15}$$

where d is the distance to the source.

The time until coalescence, given by Eq. 18.12, varies significantly across the mass and gravitational-wave frequency parameter space, as can be seen in Fig. 18.2, based on [78]. The long durations of these signals imply that matched filtering is impractical; thus, other methods had to be developed to probe such systems.

We note that if the second term in Eq. 18.14 is small compared to the first one, Eq. 18.14 can be binomially expanded:

$$f_{\rm gw}(t) \simeq f_0\left[1 + k f_0^{8/3}(t - t_0)\right] = f_0 + \dot{f}(t - t_0), \tag{18.16}$$

which is nothing but a monochromatic gravitational-wave signal whose frequency varies slowly in time by $\dot{f}$. Physically, such signals will have $\mathcal{M} < 10^{-5} M_{\odot}$ and last for years, at least. This derivation suggests that we can apply methods to search for light primordial black holes systems used in searches for "continuous gravitational waves", quasi-monochromatic, persistent signals canonically arising from

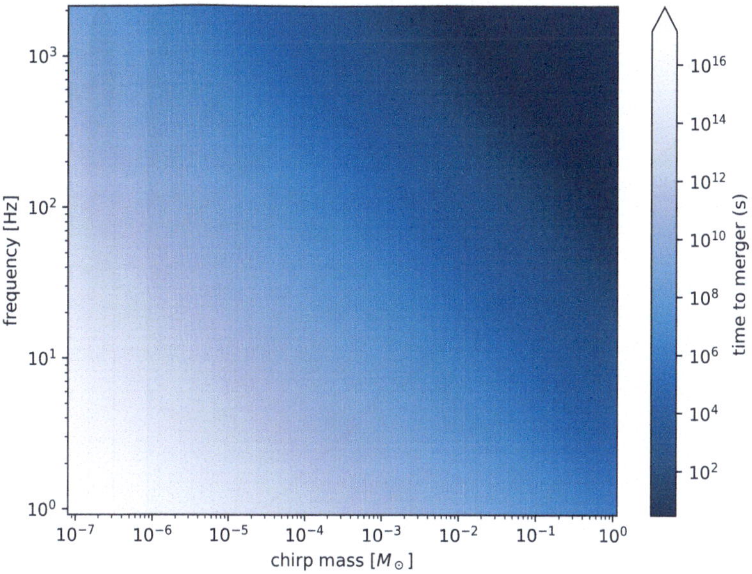

Fig. 18.2 Time before merger as a function of frequency and chirp mass. The widely distributed signal durations in the parameter space imply that different techniques are needed to probe the existence of PBHs at different masses

lumpy, rotating neutron stars [74, 90, 95, 108]. When we must use the full power-law expression, we can, instead, apply methods used in searches for "transient continuous waves", canonically arising from newborn neutron stars with large deformations [94]. While the physics of these systems are all different, the frequency evolution over time follows the same equations. The divide between the transient continuous-wave and the continuous-wave regimes is practically a function of the maximum $\dot{f}$ that continuous-wave searches probe, as well as the difference in frequency evolutions of Eqs. 18.14 and 18.16, and can be seen in Fig. 18.3. We see that continuous-wave searches are sensitive only to small chirp masses and slowly inspiraling systems, while transient continuous-wave searches could be sensitive to a much larger parameter space that matched filtering cannot probe.

18.3.3 Search Techniques

Matched filtering techniques cannot be easily applied to search for these systems, since the computational cost of correlating numerous templates with the data scales strongly with the signal duration [75]. Thus, techniques similar to those used to search for long-lived sources of continuous gravitational waves had to be developed to look for these systems. These methods are largely *semi-coherent*, meaning that they break the dataset into chunks of length T_{FFT}, analyze the data within T_{FFT} coherently (with

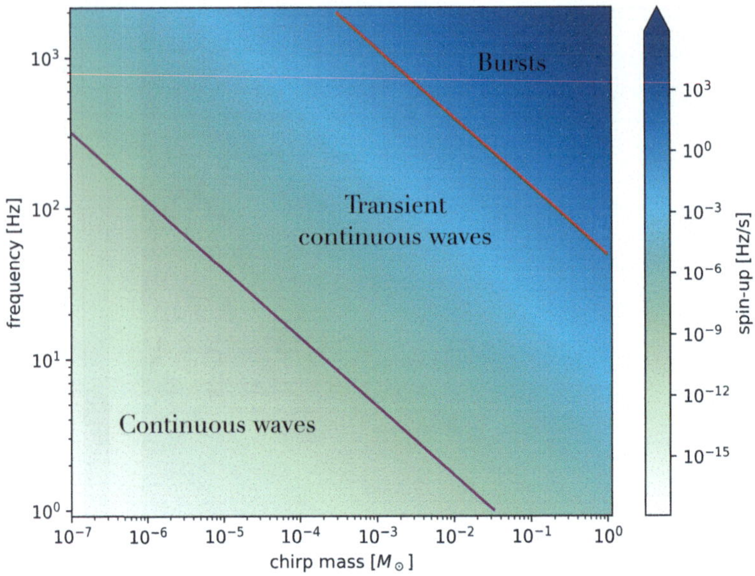

Fig. 18.3 Frequency as a function of chirp mass with spin-up colored. A magenta line representing the maximum spin-up to which continuous-wave searches have considered is also plotted [4], along with a red line indicating roughly the maximum spin-up that can be considered in a transient continuous-wave analysis (1 Hz/s). Colored points below the magenta line, meaning smaller spin-ups, represent possible masses of inspiraling PBHs that can be probed with continuous-wave methods. Transient continuous gravitational-wave methods are necessary to exhaustively constrain larger PBH chirp masses

phase information), and combine the data across different $T_{\rm FFT}$ incoherently (only the power, no phase). These techniques are much more computationally efficient than matched filtering, and can be tuned to obtain similar sensitivities by setting lower thresholds on detection statistics to make up for the degradation in sensitivity induced by performing a semi-coherent analysis [91]. They also tend to be more robust against noise disturbances and are able to deal effortlessly with non-Gaussian, non-stationary noise [19, 95].

These techniques rest on the principle that, within each $T_{\rm FFT}$, the frequency modulation of a gravitational-wave signal is confined to one frequency bin $\Delta f = 1/T_{\rm FFT}$. This means that within each $T_{\rm FFT}$, the signal is monochromatic, though over $T_{\rm obs}$, the signal will be confined to different frequency bins. The question, then, is how to choose $T_{\rm FFT}$. If $\dot{f}$ in Eq. 18.13 encapsulates the rate of change of the frequency, we would like to ensure that the following condition is met:

$$\dot{f} T_{\rm FFT} \leq \Delta f = \frac{1}{T_{\rm FFT}}. \tag{18.17}$$

Thus, in our case

$$T_{\text{FFT}} \leq \sqrt{\frac{5}{96}} \pi^{-4/3} \left(\frac{G\mathcal{M}}{c^3}\right)^{-5/6} f_{\text{gw}}^{-11/6} \tag{18.18}$$

Now, we know that f_{gw} may increase rapidly, which means that the maximum T_{FFT} decreases. But, the sensitivity of any given semi-coherent method depends on T_{FFT}. It is therefore worthwhile to consider ways to maximize the sensitivity as a function of T_{obs}, T_{FFT} and the noise power spectral density, which can be done analytically [13] or empirically [73].

Because inspiraling planetary-mass primordial black holes last for durations in between continuous-wave sources and merging compact binaries, different techniques have to be used to be optimally sensitive to these systems. While methods have been developed to search for long-lived remnants of neutron star mergers or supernovae [71, 72, 85, 104], only one has been extensively adapted to search for these kinds of primordial black holes [73] that transforms points in the time/frequency plane of the detector to lines in the frequency/chirp mass plane of the source using the Hough transform [19, 71], though others based on tracking time/frequency evolutions stochastically using the Viterbi algorithm [13, 112] and demodulating the phase evolution [16, 89], are being developed.

18.3.3.1 Hough Transform

We show one example of a technique used to search for planetary-mass primordial black holes in Fig. 18.4. Here, from the strain data $h(t)$, we take Fourier transforms of length T_{FFT} over a duration of length T_{obs}, estimate the power spectral density in each T_{FFT}, divide the square of the FFT by the estimate of the power spectral density,

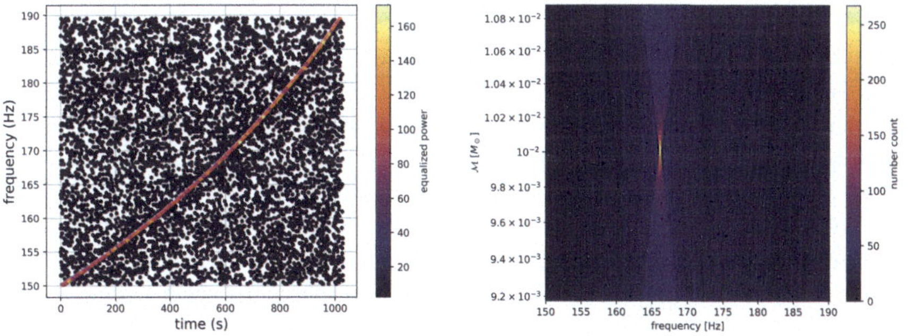

(a) (b)

Fig. 18.4 The left-hand plot shows the peakmap (time/frequency map), created with $T_{\text{FFT}} = 4$ s, of a strong, rapidly evolving signal, which is the input to the Generalized Frequency-Hough transform. The right-hand plot shows the output of the Generalized Frequency-Hough transform, which is a histogram in the $f_0/\mathcal{M}$ space of the source. The injection parameters are $h_0 = 10^{-22}$, $f_0 \simeq 165$ Hz, $\mathcal{M} = 10^{-2} M_{\odot}$. The recovered candidate is in the same bin as the injection threshold the result, and create a time/frequency map, shown in the left panel of Fig. 18.4. This time/frequency map is the input to the Generalized Frequency-Hough transform, which essentially loops over different chirp masses and sums over the times in the power law given in Eq. 18.14, and creates a two-dimensional histogram in a parameter space that relates to f_0 and $\mathcal{M}$, shown in the right panel of Fig. 18.4. This method does not actually sum the equalized power, but only a "1" when a pixel in the time/frequency plane is above a threshold, and zero otherwise. This is one example of how these semi-coherent methods deal with noise disturbances: in this case, powerful noise lines, regardless of their strength, are given weights of "1" at whatever times they appear, reducing their impact on the background.

18.3.3.2 Viterbi

Another method, used in continuous-wave all-sky searches for gravitational waves from isolated neutron stars, can be applied to search for inspiraling primordial black hole binaries. This technique relies on the Viterbi algorithm, which is a dynamical programming algorithm, to identify the most likely sequence of hidden Markov states given a set of observations [112]. Instead of computing the posterior probability that each track contains a signal and selecting that one the maximum, the method, at each time step in a set of observations, maximizes the probability to detect a gravitational-wave signal. It essentially attempts to find the "optimal path" in a two-dimensional time/frequency representation of the data without looping through all the possible tracks [70, 105, 106]. In other words, tracks at one time step that will never become the one with maximum probability are rejected immediately. Additionally, Viterbi treats the track as the result of a Markovian process, i.e. the probability calculated at a given time/frequency point (a "state") depends only on the probability computed at the previous state.

The Viterbi method is computationally cheap compared to the Hough Transform, since it operates in a model-agnostic way[1] and does not require that each possible track is computed beforehand [23–25]. The trade-off, though, is a loss of sensitivity compared to the Hough. However, Viterbi would be sensitive to primordial black holes for which the gravitational-wave waveform deviates significantly from Eq. 18.13, i.e. at higher Post-Newtonian orders required for highly asymmetric mass ratio systems [13], or additional physics, e.g. dark-matter clouds present around the primordial black holes [28, 43, 44, 63]. In Fig. 18.5, we show how the Viterbi algorithm identifies and recovers the signal track present in the spectrogram.

[1] In the case of inspiraling primordial black holes, it is imposed in the prior probability that the signal frequency must increase over time, which is the only trace of "modelling" present in the method.

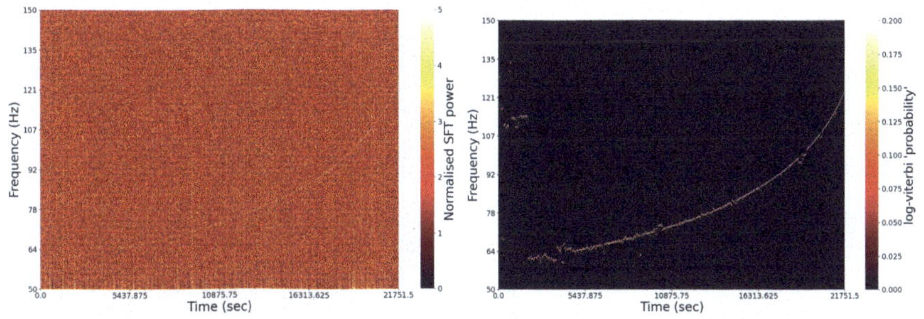

Fig. 18.5 Signal detection using the Viterbi algorithm in the case of $[\mathcal{M}_c, d_L] = [10^{-2} M_\odot, 147\text{kpc}]$. Left panel: The spectrogram of the injected signal hidden within the Gaussian noise. Right panel: The detected signal by the Viterbi algorithm. Taken from [13]

18.3.3.3 BSD-COBI

The Band-Sampled Data COmpact Binary Inspiral (BSD-COBI) has also been designed to search for binary primordial black holes [16]. This method relies on a concept known as "heterodyning", in which the gravitational-wave frequency evolution over time is demodulated out of the data. The procedure essentially applies a phase correction to the strain time series[2] to remove a signal.

In other words, consider that the strain $h(t)$ has the form

$$h(t) = \text{Re}[A(t)e^{i\phi} + n(t)] \tag{18.19}$$

where $n(t)$ is the noise, $A(t)$ is the time-varying signal amplitude, and ϕ is the phase of the signal (the integral of Eq. 18.14). At this point, we can multiply by a phase factor $e^{-i\phi}$ to heterodyne or demodulate the data

$$\begin{aligned} h^{\text{het}}(t) &= \text{Re}[h(t)e^{-i\phi}] \\ &= \text{Re}[A(t) + n(t)]. \end{aligned} \tag{18.20}$$

If the correction is done perfectly, the signal does not experience any frequency modulation over t; thus, it becomes monochromatic over its total duration. In principle, the T_{FFT} length can be increased to, at most, T_{obs}; the sensitivity after this correction, therefore, rivals that of the matched filter. However, the signal parameters have to be exactly known to achieve matched filtering sensitivity. In practice, T_{FFT} can be increased by a factor of $\mathcal{O}(1-10)$ relative to that in Eq. 18.18, which enhances the sensitivity of the search. An efficient way of performing this correction, however, is under evaluation.

In Fig. 18.6, we show the heterodyning procedure as applied to a gravitational-wave signal from an inspiraling primordial black hole binary. We see that, before the

[2] It's not the real-valued strain time series, but the so-called "reduced analytic signal" on which this correction is applied.

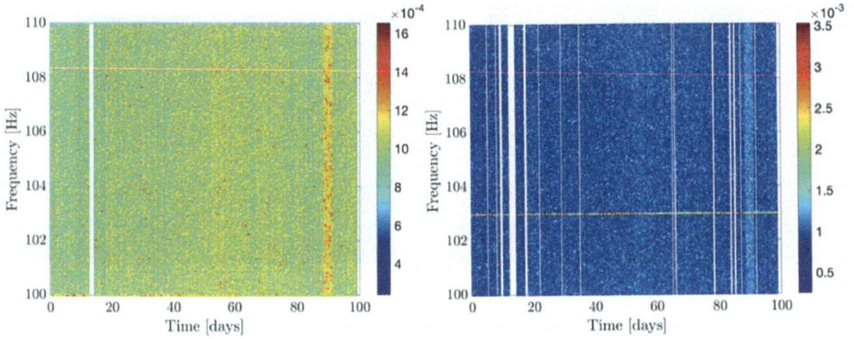

Fig. 18.6 Left: original time/frequency peakmap, before heterodyning. The signal cannot be seen. Right: time/frequency map after heterodyning. All phase modulations have been removed, so the signal appears as monochromatic. Taken from [16]

correction, the signal cannot be seen in the time/frequency map, but after heterodyning, it becomes visible. At this point, the time/frequency map will be projected onto the frequency axis, which amounts to a binary sum over time of whether the power in each time/frequency pixel is above a given threshold.

18.3.4 Constraints on Asteroid Mass Primordial Black Holes from All-Sky Searches for Continuous Waves

While planetary-mass primordial black holes have to be searched for with dedicated techniques, we can leverage existing all-sky searches for isolated, asymmetrically rotating neutron stars to place constraints on asteroid-mass primordial black holes. This portion of the parameter space corresponds to the points below the light green line in Fig. 18.3. Essentially, these searches [6] look for spinning down (but also spinning up) isolated neutron stars, whose spin-down or spin-up can be explained by having mountains or accretion, respectively. Since primordial black holes will inspiral, they will always be spinning up, hence it is the search over positive frequency drifts, up to $\dot{f}_{\rm max} \sim 2 \times 10^{-9}$ Hz/s in [6], that allow us to make some statement on very slowly inspiraling primordial black holes. The frequency range that was searched, from $\sim$[10,2000] Hz, the maximum spin-up, and the criteria that the frequency evolution be linear, $f(t) = f_0 + \dot{f}(t - t_0)$, fix the chirp masses to which we would be sensitive based on Eq. 18.13 to be $\sim [10^{-7}, 10^{-5}] M_\odot$.

The output of continuous-wave searches is often upper limits on the strain amplitude h_0 as a function of frequency, averaged over extrinsic parameters. From $h_0(f)$, we can compute the distance reach via Eq. 18.15, employing the upper limits obtained in [5, 6, 103].

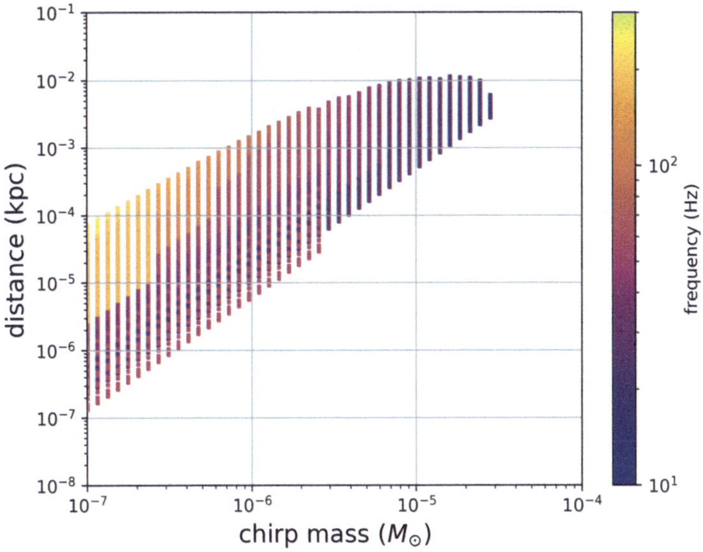

Fig. 18.7 Distance reach as a function of chirp mass with frequency on the color axis. The maximum reach is of $\mathcal{O}$(pc) for most of the planetary-chirp-mass black hole inspirals. Each point contributes to Eqs. 18.27 and 18.28 used to obtain constraints on the rates at a given chirp mass

We will now approximate theoretically the results we obtained in [76] for the distance reach as a function of chirp mass (Fig. 18.7), and the constraints on the fraction of dark-matter that primordial black holes could compose as a function of mass for equal-mass binaries, and as a function of m_2 for $m_1 = 2.5M_\odot$.

In continuous-wave semi-coherent searches, the upper limits on the strain amplitude as a function of frequency can be approximated as

$$h_0(f) = 2\frac{\sqrt{S_n(f)}}{(T_{\rm FFT}T_{\rm obs})^{1/4}}. \tag{18.21}$$

Here, $T_{\rm FFT} = 1800$ s in [5] irrespective of the frequency, though this choice is primarily a function of computational cost, and $T_{\rm obs}$ is the observation time (in O3a, 6 months).

Plugging in $h_0(f)$ into Eq. 18.15, we can obtain the distance reach to gravitational-wave systems slowly inspiraling towards one another

$$d(f) = 2\frac{(T_{\rm FFT}T_{\rm obs})^{1/4}}{\sqrt{S_n(f)}}\left(\frac{G\mathcal{M}}{c^2}\right)^{5/3}\left(\frac{\pi f_{\rm gw}(t)}{c}\right)^{2/3}, \tag{18.22}$$

From the distance reach, we can compute the space-time volume $\langle VT\rangle$, which, for nearby sources, is simply the volume of a sphere, whose radius is the distance reach, multiplied by the source duration in a particular frequency range $[f, f+\delta f]$:

$$\langle VT \rangle = \frac{4}{3}\pi d(f)^3 T \tag{18.23}$$

where $T = \max(T_{\text{obs}}, \Delta T)$, and ΔT is the time spent by the binary system in a given frequency range δf:

$$\Delta T = \frac{5}{256}\pi^{-8/3}\left(\frac{c^3}{G\mathcal{M}}\right)^{5/3}\left[f^{-8/3} - (f + \delta f)^{-8/3}\right] \tag{18.24}$$

$$\approx \frac{5}{96}\pi^{-8/3}\left(\frac{c^3}{G\mathcal{M}}\right)^{5/3}\delta f f^{-11/3} \tag{18.25}$$

where $\delta f = \dot{f}_{\max} T_{\text{obs}}$. This means that

$$\langle VT \rangle = \frac{5\pi^{1/3}\dot{f}G^4\mathcal{M}^4 T_{\text{obs}}(T_{\text{FFT}}T_{\text{obs}})^{3/4}\left(\frac{c^3}{G\mathcal{M}}\right)^{2/3}}{9c^9 f^{5/3} S_n^{3/2}(f)} \tag{18.26}$$

Now, the number of binaries detectable at a given frequency is:

$$N_{\text{bin}} = \langle VT \rangle \mathcal{R}, \tag{18.27}$$

where $\mathcal{R}$ is the formation rate density of binary primordial black holes. Summing over all possible binaries detected at each frequency

$$N_{\text{bin}}^{\text{tot}} = \sum_i N_{\text{bin}}(f_i)\,. \tag{18.28}$$

and solving for $\mathcal{R}$, we arrive at

$$\mathcal{R} = \frac{3}{4\pi}\left(\sum_i \langle VT \rangle (f_i)\right)^{-1}. \tag{18.29}$$

These rate densities are a function of each chirp mass to which we could be sensitive.

From the rate densities in Eq. 18.29, we can equate this quantity to analytical models for formation rate densities of primordial black holes for equal-mass binaries:

$$\mathcal{R} = 1.04 \times 10^{-6}\,\text{kpc}^{-3}\text{yr}^{-1} f_{\text{sup}} f(m_{\text{PBH}})^2 \left(\frac{m_{\text{PBH}}}{M_\odot}\right)^{-32/37} (f_{\text{PBH}})^{53/37}, \tag{18.30}$$

and also asymmetric-mass ratio binaries, with $m_2 \ll m_1$:

$$\mathcal{R} = 5.28 \times 10^{-7}\,\text{kpc}^{-3}\text{yr}^{-1} f_{\text{sup}} f(m_1) f(m_2) \left(\frac{m_1}{M_\odot}\right)^{-32/37} \left(\frac{m_2}{m_1}\right)^{-34/37} (f_{\text{PBH}})^{53/37}\,. \tag{18.31}$$

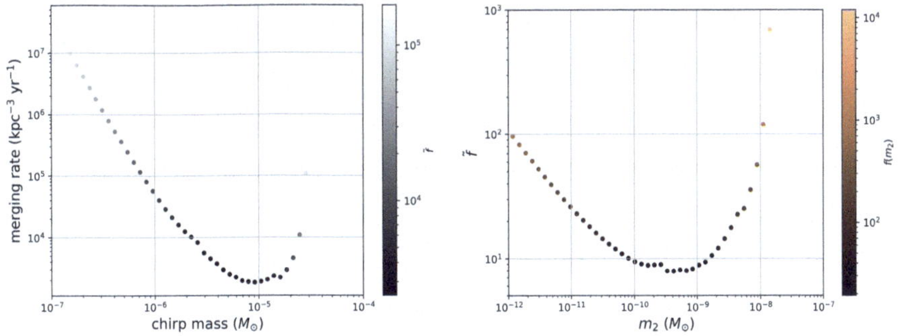

Fig. 18.8 Left: Upper limit on merging rates of planetary and asteroid chirp mass binaries. These rates do not depend on any particular PBH formation model; they only depend on the sensitivity of the continuous-wave search, i.e. the distance reach. Also shown in color are upper limits on the fraction of dark matter that primordial black holes could compose assuming rate models for equal-mass primordial black hole binaries. Right: Upper limit on the PBH abundance for asymmetric mass ratio binaries with one component at $m_1 = 2.5M_\odot$. Also shown in color is upper limit on the mass function of the smaller companion PBH assuming $f(m_1) = 1$ and $f_{\mathrm{PBH}} = 1$

We can constrain asymmetric mass-ratio binaries because the continuous-wave search is primarily sensitive to the chirp mass of the binary, so m_1 and m_2 can be chosen freely as long as $\mathcal{M}$ remains the same (assuming negligible eccentricity). Additionally, asymmetric mass-ratio binaries are more likely to occur.

Since primordial black holes are well-motivated by observations of merging black holes in the stellar-mass range, we consider systems with $m_1 = 2.5\ M_\odot$, as motivated by the QCD transition [30, 34, 40] and some mergers such as GW190425 and GW190814 [40], in a binary with a much lighter primordial black hole of mass m_2. Combining Eq. 18.29 with Eqs. 18.30 and 18.31, and denoting a model-independent parameter $\tilde{f}^{53/37} \equiv f_{\mathrm{sup}} f(m_1) f(m_2) f_{\mathrm{PBH}}^{53/37}$, we can arrive at constraints present in Fig. 18.8. These constraints, while not yet probing a physical regime, indicate that, soon, continuous-wave search could detect nearby primordial black hole systems.

18.3.5 Projected Constraints for Planetary-Mass primordial black holes

Analogously to what was described in Sect. 18.3.4, we can also constrain higher-mass with the semi-coherent methods described in Sect. 18.3.3. For systems with chirp masses $\mathcal{M} \in [10^{-5}, 10^{-2}]$, the linear approximation for the frequency/time evolution used in Sect. 18.3.4 breaks down; thus, we need to use the Generalized frqeuency-Hough transform or another semi-coherent method that can track the frequency evolution given in Eq. 18.14.

The procedure to obtain (projected) constraints for systems in this mass regime is almost identical to that in Sect. 18.3.4, except that Eq. 18.21 is replaced by one that

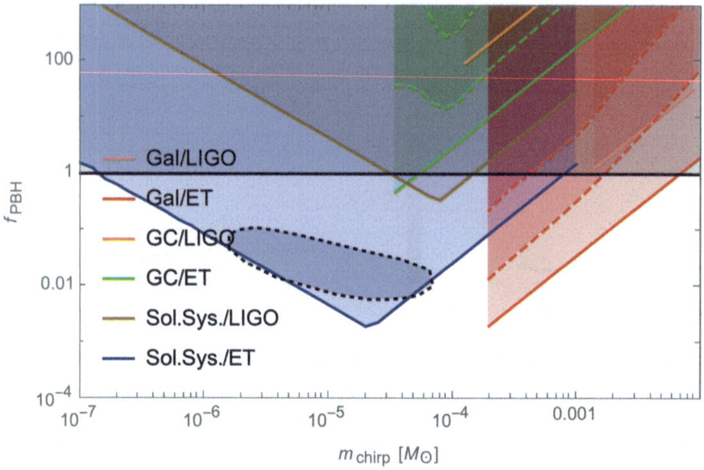

Fig. 18.9 Expected limits on the dark matter fraction made of PBHs as a function of the chirp mass, for primordial binaries in Case 1—agnostic mass function—(solid lines) and Case 2—thermal mass function—(dashed lines), and for binaries formed through tidal capture in Case 2 (dotted-dashed lines). The different colors represent the limits from galactic binaries (gal), from the galactic center (GC) and in the solar system vicinity (sol. sys.), for the expected sensitivities of advanced LIGO/Virgo and ET. The dotted elliptic region represents constraints from the Optical Gravitational Lensing Experiment (OGLE) and the Subaru Hyper Suprime-Cam (HSC) [80] for comparison. Taken from [78]

accounts for the power-law evolution of the signal that also depends on the method used, and in Eq. 18.25, the approximation that $\delta f/f \ll 1$ cannot be made. This equation can be found in [73], Eq. 29–31. Then, noting that T_{FFT} is much smaller than in the previous case (because $\dot{f}$ is much bigger), the same procedure can be applied. We show the projected constraints on f_{PBH} for equal-mass systems in Fig. 18.9.

18.3.6 Mini-EMRIs

A class of asymmetric-mass ratio binaries could be so-called mini extreme-mass ratio inspirals (mini EMRIs), which could be detected by current gravitational-wave interferometers [59]. These systems could be composed of a $10M_\odot$ or $100M_\odot$ primary object whose companion would be lighter by a factor of $\sim 10^5$, i.e. $m_2/m_1 = q = 10^{-5}$. Such systems would require the lighter object to be a primordial black hole or another exotic compact object.

Though similar to the previous discussions of asymmetric mass-ratio primordial black holes, the signal model is slightly different:

$$\frac{df}{dt} = kf^{11/3}C_f(a, f), \tag{18.32}$$

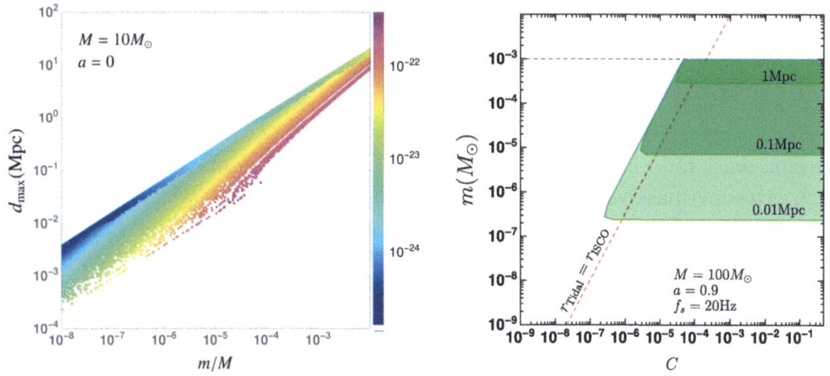

Fig. 18.10 Left: estimated distance reach at 95% confidence as a function of mass ratio, with the minimum detectable gravitational-wave amplitude colored. Right: sensitivity to a mini-EMRI consisting of a $100M_\odot$ central massive compact object and an ECO specified by its mass (vertical axis) and compactness (horizontal axis), where the color-shaded regions denote where the corresponding mini-EMRI system can emit gravitational waves and be detected by LIGO/Virgo, assuming $CR_{\rm thr} = 5$

$$h_0 = \frac{4}{d}\left(\frac{GM_c}{c^2}\right)^{5/3}\left(\frac{\pi f}{c}\right)^{2/3} C_h(a, f). \tag{18.33}$$

where these two factors, $C_f(a, f)$ and $C_h(a, f)$ characterize the change in the orbital (or gravitational-wave) frequency accounting for relativistic effects. Plots of these two quantities are shown in [59]. The luminosity distance reach for these kinds of systems varies depending on the mass ratio, and is shown in Fig. 18.10, along with projected constraints on the exotic companion as a function of its compactness.

Though the relativistic factors imply a variation in the frequency evolution over time in Eq. 18.14, if we wish to apply the Hough Transform to look for these systems, we can treat Eq. 18.32 as an "effective" power-law, so that its form resembles in Eq. 18.32 that, though not exactly $f^{11/3}$. In a real search, therefore, this would require searching over a range of power laws, in addition to a range of chirp masses and frequencies. Still, however, the system could not be too close to the plunge in order to employ this approach. A better idea may be the Viterbi method that is not bound to a particular model and could potentially pick out mini EMRI signals where $C_f(a, f) \gg 1$.

18.4 EMRIs

18.4.1 Motivation

Extreme mass ratio inspirals (EMRIs) consist of a primary mass $m_1 \sim \mathcal{O}(10^9)M_\odot$ (such as a supermassive black hole at the center of our galaxy) and a secondary mass $m_2 \sim \mathcal{O}(1)M_\odot$ (a stellar-mass compact object such as a neutron star, black hole or white dwarf). Ordinary stellar objects would be tidally disrupted by such a supermassive black hole; thus, EMRIs require that the secondary object is compact [69]. In contrast to the binary black holes that merge in less than a second in LIGO, Virgo and KAGRA, EMRIs would be visible in space-based gravitational-wave detectors at mHz frequencies, and spend years in band before the secondary object plunges into the supermassive black hole. As the secondary object inspirals into the primary one, a significant amount of gravitational-wave power will be emitted at low frequencies.

Such systems are interesting from both astrophysics and fundamental physics perspectives [21]. The detection of such sources would allow us to test whether the supermassive black holes indeed behave as Kerr black holes [22, 53], measure the mass distribution of supermassive black holes and their stellar environments [52], measure the background spacetimes of supermassive black holes [57], and measure the presence of gas around the primary object [15].

To obtain a sense of how long these signals will last, we can integrate Eq. 18.14 to obtain the number of cycles

$$N_{\rm cyc} = \int_{t_0}^{t} f(t)dt \tag{18.34}$$

$$= \int_{f_{\rm min}}^{f_{\rm max}} \frac{f}{\dot{f}} df \tag{18.35}$$

and also add a redshift-dependence $1+z$ for binaries at cosmological distances [69]

$$N_{\rm cyc} = \frac{1}{32\pi^{8/3}} \left(\frac{G\mathcal{M}}{c^3}\right)^{-5/3} \left(f_{\rm min}^{-5/3} - f_{\rm max}^{-5/3}\right) \tag{18.36}$$

$$= 10^5 \left(\frac{2\ \rm mHz}{f_{\rm min}}\right)^{5/3} \left(\frac{10^3 M_\odot}{\mathcal{M}}\right)^{5/3} \left(\frac{2}{1+z}\right)^{5/3},$$

where $\mathcal{M} = 10^3 M_\odot$ could correspond to $m_1 = 10^6 M_\odot$ and $m_2 = 10 M_\odot$.

To compare $N_{\rm cyc}$ to that of systems visible in ground-based detectors, we see that we can set $m_1 = m_2 = 10M_\odot$ ($\mathcal{M} \simeq 8.7M_\odot$) at a frequency of 10 Hz, we obtain $N_{\rm cyc} \simeq 600$ at a redshift of 1. An EMRI, therefore, spends orders of magnitude longer in the frequency band of space-based detectors compared to its duration in ground-based ones.

18.4.2 Search Techniques

As discussed in previous sections, matched filtering provides the best sensitivity of any data analysis method. However, creating EMRI waveforms may pose problems in the future. First, the number of cycles over which the template has to be reasonably accurate is much higher than in ground-based detectors. Second, because the secondary object will reach speeds close to the speed of light, Post-Newtonian expansions cannot be employed to derive the waveform of these systems, nor is numerical relativity feasible, since it only works for mass ratios up to 10^5 and $\sim 10^2$ cycles. It is possible, therefore, that semi-coherent methods discussed in Sect. 18.3.3, which operate by breaking data into smaller coherent segments that are analyzed incoherently in the time/frequency plane, may be used in the future to search for EMRIs. Furthermore, if we do wish to employ matched filtering, some approaches have already been designed, based on self-force [69], that are able to expand the waveform in powers of the (small) mass ratio m_2/m_1 of EMRIs. While all these techniques seem promising, they are in their infancy and need to be extensively studied to handle the EMRI case.

18.4.3 Detection Prospects

The number of EMRIs as a function of redshift depends on a variety of astrophysical unknowns: (1) the number of supermassive black holes formed as a function of their mass, spin and redshift, (2) the properties of the stars around the supermassive black hole: whether the supermassive black hole is surrounded by stars spatially distributed along a cusp around it (necessary for EMRI formation), how galaxy mergers over cosmic time have affected the cusp (mergers would erode the cusps on some timescale, and on a different timescale, the cusp could regrow, (3) how many plunges versus EMRIs there are, (4) the duty cycle of EMRIs, and (5) the chosen m_2 [21]. When models for each of these are chosen, the rate of EMRI formation can be derived, which can range from $\mathcal{O}(10)$ to $\mathcal{O}(10^5)$ per year up to a redshift of $z = 4.5$, of which only 10% to 50% of them would be detectable by LISA, depending on the waveform—see Table 1 of [21].

In Fig. 18.11, we show the fraction of detectable EMRI events using two waveforms in a matched filtering analysis at LISA sensitivity as a function of primary component mass and redshift, fixing $m_2 = 10M_\odot$ (M1) and $m_2 = 30M_\odot$ (M4). We can see that locally, i.e. $z < 1$, we would be able to detect close to all the EMRIs around, and we observe a fall-off with redshift.

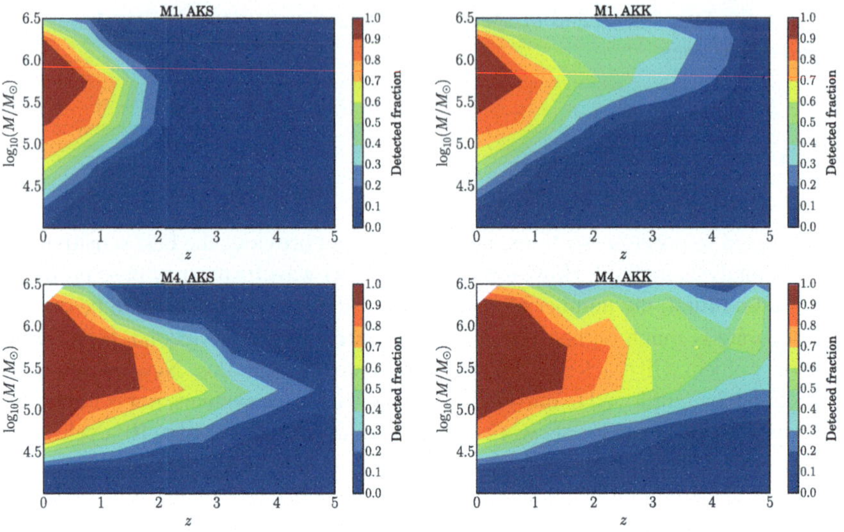

Fig. 18.11 Fraction of the intrinsic EMRI population detectable by LISA as a function of source-frame total mass and redshift, for two models of supermassive black hole mass function, spin, cusp regrowth time, etc. (M1 and M4) and with two waveforms constructed to perform a matched filtering analysis (AKS and AKK). Taken from [21]

18.5 Conclusions

We have described in great detail how searches for sub-solar mass primordial black holes from predominately current-generation ground-based work. Such a mass range also presents exciting opportunities for future observatories on the ground as well, such as Einstein Telescope and Cosmic Explorer [92, 93], and space-based ones, such as DECIGO, LISA, Taiji and TianQin [14, 64, 67, 97]. Indeed, in future ground-based detectors, the lowest detectable frequency of such inspiraling systems will be much around 2–5 Hz instead of the 10–15 Hz it currently is, which will allow a significantly longer observation time, and potentially the localization of such systems using only the relative motion of the earth and source. Furthermore, with roughly a factor of ~ 10 improvement in the luminosity distance reached, we will be able to see primordial black holes inspiraling at least in a 1000 times larger volume than we currently do. The expected exquisite sensitivity of these detectors, as well as long-duration sources, will require surmounting new challenges related to computational cost, estimation of the signal-free background, signals occurring simultaneously, and non-stationary noise. The methods presented here, especially those from tCW and CW searches, are apt at handling these problems [75], and along with matched filtering, could constitute a major probe of sub-solar mass primordial black holes in the future.

Regarding space-based detectors, such systems could easily spend years visible mHz or deci-Hz frequencies, enabling the possibility of performing multi-band gravitational-wave astronomy [99]. In essence, the source parameters could be estimated using space-based detector data, and further measurements of the inspiral could be made in future ground-based detectors. Potentially, if ultra-high frequency gravitational-wave detectors come online, the mergers of such sub-solar mass primordial black holes could be detected as well [11].

The future is therefore bright for searching for and hopefully finding gravitational waves from binary sub-solar mass primordial black holes.

References

1. B.P. Abbott et al., GW170817: observation of gravitational waves from a binary neutron star inspiral. Phys. Rev. Lett. **119**(16), 161101 (2017). https://doi.org/10.1103/PhysRevLett.119.161101. arXiv: 1710.05832 [gr-qc]
2. B.P. Abbott et al., GW190425: observation of a compact binary coalescence with total mass $\sim 3.4 M_{\odot}$. Astrophys. J. Lett. **892**(1), L3 (2020). https://doi.org/10.3847/2041-8213/ab75f5. arXiv: 2001.01761 [astro-ph.HE]
3. B.P. Abbott et al., Observation of gravitational waves from a binary black hole merger. Phys. Rev. Lett. **116**(6), 061102 (2016). https://doi.org/10.1103/PhysRevLett.116.061102. arXiv: 1602.03837 [gr-qc]
4. B.P. Abbott et al., All-sky search for continuous gravitational waves from isolated neutron stars using Advanced LIGO O2 data. Phys. Rev. D **100**(2), 024004 (2019)
5. R. Abbott et al., All-sky search for continuous gravitational waves from isolated neutron stars in the early O3 LIGO data. Phys. Rev. D **104**(8), 082004 (2021). https://doi.org/10.1103/PhysRevD.104.082004. arXiv: 2107.00600 [gr-qc]
6. R. Abbott et al., All-sky search for continuous gravitational waves from isolated neutron stars using advanced LIGO and advanced Virgo O3 data. Phys. Rev. D **106**(10), 102008 (2022). https://doi.org/10.1103/PhysRevD.106.102008. arXiv: 2201.00697 [gr-qc]
7. R. Abbott et al., GW190814: gravitational waves from the coalescence of a 23 solar mass black hole with a 2.6 solar mass compact object. Astrophys. J. Lett. **896**(2), L44 (2020). https://doi.org/10.3847/2041-8213/ab960f. arXiv: 2006.12611 [astro-ph.HE]
8. R. Abbott et al., GWTC-3: compact binary coalescences observed by LIGO and Virgo during the second part of the third observing run. Phys. Rev. X **13**(4), 041039 (2023). https://doi.org/10.1103/PhysRevX.13.041039. arXiv: 2111.03606 [gr-qc]
9. R. Abbott et al., Search for subsolar-mass binaries in the first half of advanced LIGO and Virgo's third observing run (2021). arXiv: 2109.12197 [astro-ph.CO]
10. R. Abbott et al., Search for subsolar-mass black hole binaries in the second part of advanced LIGO's and advanced Virgo's third observing run (2022). arXiv: 2212.01477 [astro-ph.HE]
11. N. Aggarwal et al., Challenges and opportunities of gravitational-wave searches at MHz to GHz frequencies. Living Rev. Rel. **24**(1), 4 (2021). https://doi.org/10.1007/s41114-021-00032-5. arXiv: 2011.12414 [gr-qc]
12. N. Aghanim et al., Planck 2018 results. VI. Cosmological parameters. Astron. Astrophys. **641** (2020). [Erratum: Astron.Astrophys. 652, C4 (2021)], A6. https://doi.org/10.1051/0004-6361/201833910. arXiv: 1807.06209 [astro-ph.CO]
13. G. Alestas et al., Applying the Viterbi algorithm to planetary-mass black hole searches (2024). arXiv: 2401.02314 [astro-ph.CO]
14. P. Amaro-Seoane et al., Laser interferometer space antenna (2017). arXiv: 1702.00786 [astro-ph.IM]

15. P. Amaro-Seoane et al., Astrophysics, detection and science applications of intermediate- and extreme mass-ratio inspirals. Class. Quant. Grav. **24**, R113–R169 (2007). https://doi.org/10.1088/0264-9381/24/17/R01. arXiv: astro-ph/0703495
16. M. Andrés-Carcasona et al., BSD-COBI: new search pipeline to target inspiraling light dark compact objects. PoS **EPS-HEP2023**, 067 (2023). https://doi.org/10.22323/1.449.0067
17. F. Antonini, S. Toonen, A.S. Hamers, Binary black hole mergers from field triples: properties, rates and the impact of stellar evolution. Astrophys. J. **841**(2), 77 (2017). https://doi.org/10.3847/1538-4357/aa6f5e. arXiv: 1703.06614 [astro-ph.GA]
18. K. Arun, C. Sivaram, O.V. Kiren, Primordial planets predominantly of dark matter. Earth Moon Planets **122**(3–4), 115–119 (2019). https://doi.org/10.1007/s11038-019-09525-4
19. P. Astone et al., Method for all-sky searches of continuous gravitational wave signals using the frequency-Hough transform. Phys. Rev. D **90**(4), 042002 (2014). https://doi.org/10.1103/PhysRevD.90.042002. arXiv: 1407.8333 [astro-ph.IM]
20. S. Babak et al., Searching for gravitational waves from binary coalescence. Phys. Rev. D **87**(2), 024033 (2013). https://doi.org/10.1103/PhysRevD.87.024033. arXiv: 1208.3491 [gr-qc]
21. S. Babak et al., Science with the space-based interferometer LISA. V: n-ratio inspirals. Phys. Rev. D **95**(10), 103012 (2017). https://doi.org/10.1103/PhysRevD.95.103012. arXiv: 1703.09722 [gr-qc]
22. L. Barack, C. Cutler, Using LISA EMRI sources to test off-Kerr deviations in the geometry of massive black holes. Phys. Rev. D **75**, 042003 (2007). https://doi.org/10.1103/PhysRevD.75.042003. arXiv: gr-qc/0612029
23. J. Bayley, G. Woan, C. Messenger, Generalized application of the Viterbi algorithm to searches for continuous gravitational-wave signals. Phys. Rev. D **100**(2), 023006 (2019). https://doi.org/10.1103/PhysRevD.100.023006. arXiv: 1903.12614 [astro-ph.IM]
24. J. Bayley, C. Messenger, G. Woan, Rapid parameter estimation for an all-sky continuous gravitational wave search using conditional varitational auto-encoders. Phys. Rev. D **106**(8), 083022 (2022). https://doi.org/10.1103/PhysRevD.106.083022. arXiv: 2209.02031 [astro-ph.IM]
25. J. Bayley, C. Messenger, G. Woan, Robust machine learning algorithm to search for continuous gravitational waves. Phys. Rev. D **102**(8), 083024 (2020). https://doi.org/10.1103/PhysRevD.102.083024. arXiv: 2007.08207 [astro-ph.IM]
26. K.M. Belotsky et al., Clusters of primordial black holes. Eur. Phys. J. C **79**(3), 246 (2019). https://doi.org/10.1140/epjc/s10052-019-6741-4. arXiv: 1807.06590 [astro-ph.CO]
27. S. Bhatiani, X. Dai, E. Guerras, Confirmation of planet-mass objects in extragalactic systems. The Astrophys. J. **885**(1), 77 (2019)
28. R. Brito, V. Cardoso, P. Pani, Superradiance: new frontiers in black hole physics. Lect. Notes Phys. **906**, 1–237 (2015). https://doi.org/10.1007/978-3-319-19000-6. arXiv: 1501.06570 [gr-qc]
29. M.R. Buckley, A. DiFranzo, Collapsed dark matter structures. Phys. Rev. Lett. **120**(5), 051102 (2018). https://doi.org/10.1103/PhysRevLett.120.051102. arXiv: 1707.03829 [hep-ph]
30. C.T. Byrnes et al., Primordial black holes with an accurate QCD equation of state. JCAP **08**, 041 (2018). https://doi.org/10.1088/1475-7516/2018/08/041. arXiv: 1801.06138 [astro-ph.CO]
31. J. Calcino, J. Garcia-Bellido, T.M. Davis, Updating the MACHO fraction of the milky way dark halowith improved mass models. Mon. Not. Roy. Astron. Soc. **479**(3), 2889–2905 (2018). https://doi.org/10.1093/mnras/sty1368. arXiv: 1803.09205 [astro-ph.CO]
32. C. Capano et al., Systematic errors in estimation of gravitational-wave candidate significance. Phys. Rev. D **96**(8), 082002 (2017). https://doi.org/10.1103/PhysRevD.96.082002. arXiv: 1708.06710 [astro-ph.IM]
33. B. Carr, The primordial black hole mass spectrum. Astrophys. J. **201**, 1–19 (1975). https://doi.org/10.1086/153853
34. B. Carr et al., Cosmic conundra explained by thermal history and primordial black holes. Phys. Dark Univ. **31**, 100755 (2021). https://doi.org/10.1016/j.dark.2020.100755. arXiv: 1906.08217 [astro-ph.CO]

35. B.J. Carr, J.E. Lidsey, Primordial black holes and generalized constraints on chaotic inflation. Phys. Rev. D **48**, 543–553 (1993). https://doi.org/10.1103/PhysRevD.48.543
36. S. Chandrasekhar, The highly collapsed configurations of a stellar mass (Second paper). Mon. Not. Roy. Astron. Soc. **95**, 207–225 (1935). https://doi.org/10.1093/mnras/95.3.207
37. S. Chandrasekhar, The maximum mass of ideal white dwarfs. Astrophys. J. **74**, 81–82 (1931). https://doi.org/10.1086/143324
38. G.F. Chapline, Cosmological effects of primordial black holes. Nature **253**(5489), 251–252 (1975). https://doi.org/10.1038/253251a0
39. J. Choquette, J.M. Cline, J.M. Cornell, Early formation of supermassive black holes via dark matter self-interactions. JCAP **07**, 036 (2019). https://doi.org/10.1088/1475-7516/2019/07/036. arXiv: 1812.05088 [astro-ph.CO]
40. S. Clesse, J. Garcia-Bellido, GW190425, GW190521 and GW190814: three candidate mergers of primordial black holes from the QCD epoch. Phys. Dark Univ. **38**, 101111 (2022). https://doi.org/10.1016/j.dark.2022.101111. arXiv: 2007.06481 [astro-ph.CO]
41. S. Clesse, J. Garcia-Bellido, Seven hints for primordial black hole dark matter. Phys. Dark Univ. **22**, 137–146 (2018). https://doi.org/10.1016/j.dark.2018.08.004. arXiv: 1711.10458 [astro-ph.CO]
42. S. Clesse, J. Garcia-Bellido, The clustering of massive primordial black holes as dark matter: measuring their mass distribution with advanced LIGO. Phys. Dark Univ. **15**, 142–147 (2017). https://doi.org/10.1016/j.dark.2016.10.002. arXiv: 1603.05234 [astro-ph.CO]
43. P.S. Cole et al., Measuring dark matter spikes around primordial black holes with Einstein telescope and cosmic explorer. Phys. Rev. D **107**(8), 083006 (2023). https://doi.org/10.1103/PhysRevD.107.083006. arXiv:2207.07576 [astro-ph.CO]
44. A. Coogan et al., Measuring the dark matter environments of black hole binaries with gravitational waves. Phys. Rev. D **105**(4), 043009 (2022). https://doi.org/10.1103/PhysRevD.105.043009. arXiv:2108.04154 [gr-qc]
45. E. Cotner, A. Kusenko, Primordial black holes from scalar field evolution in the early universe. Phys. Rev. D **96**(10), 103002 (2017). https://doi.org/10.1103/PhysRevD.96.103002. arXiv:1706.09003 [astro-ph.CO]
46. H.T. Cromartie et al., Relativistic shapiro delay measurements of an extremely massive millisecond pulsar. Nature Astron. **4**(1), 72–76 (2019). https://doi.org/10.1038/s41550-019-0880-2. arXiv: 1904.06759 [astro-ph.HE]
47. G. D'Amico et al., Massive black holes from dissipative dark matter. Mon. Not. Roy. Astron. Soc. **473**(1), 328–335 (2018). https://doi.org/10.1093/mnras/stx2419. arXiv: 1707.03419 [astro-ph.CO]
48. V. De Luca et al., The clustering evolution of primordial black holes. JCAP **11**, 028 (2020). https://doi.org/10.1088/1475-7516/2020/11/028. arXiv: 2009.04731 [astro-ph.CO]
49. S. Droz et al., Gravitational waves from inspiraling compact binaries: validity of the stationary phase approximation to the Fourier transform. Phys. Rev. D **59**, 124016 (1999). https://doi.org/10.1103/PhysRevD.59.124016. arXiv: gr-qc/9901076
50. T. Ertl et al., The explosion of helium stars evolved with mass loss (2019). https://doi.org/10.3847/1538-4357/ab6458. arXiv: 1910.01641 [astro-ph.HE]
51. G. Fragione, A. Loeb, Black hole–neutron star mergers from triples. Mon. Not. Roy. Astron. Soc. **486**(3), 4443–4450 (2019). https://doi.org/10.1093/mnras/stz1131. arXiv: 1903.10511 [astro-ph.GA]
52. J.R. Gair, C. Tang, M. Volonteri, LISA extreme-mass-ratio inspiral events as probes of the black hole mass function. Phys. Rev. D **81**, 104014 (2010). https://doi.org/10.1103/PhysRevD.81.104014. arXiv: 1004.1921 [astro-ph.GA]
53. J.R. Gair et al., Testing general relativity with low-frequency, space-based gravitational-wave detectors. Living Rev. Rel. **16**, 7 (2013). https://doi.org/10.12942/lrr-2013-7. arXiv: 1212.5575 [gr-qc]
54. J. Garcia-Bellido, A.D. Linde, D. Wands, Density perturbations and black hole formation in hybrid inflation. Phys. Rev. D **54**, 6040–6058 (1996). https://doi.org/10.1103/PhysRevD.54.6040. arXiv: astro-ph/9605094

55. J. García-Bellido, S. Clesse, Constraints from microlensing experiments on clustered primordial black holes. Phys. Dark Univ. **19**, 144–148 (2018). https://doi.org/10.1016/j.dark.2018.01.001. arXiv: 1710.04694 [astro-ph.CO]
56. J. Garriga, A. Vilenkin, J. Zhang, Black holes and the multiverse. JCAP **02**, 064 (2016). https://doi.org/10.1088/1475-7516/2016/02/064. arXiv: 1512.01819 [hep-th]
57. K. Glampedakis, S. Babak, Mapping spacetimes with LISA: inspiral of a test-body in a 'quasi-Kerr' field. Class. Quant. Grav. **23**, 4167–4188 (2006). https://doi.org/10.1088/0264-9381/23/12/013. arXiv: gr-qc/0510057
58. A.M. Green, B.J. Kavanagh, Primordial black holes as a dark matter candidate. J. Phys. G **48**(4), 043001 (2021). https://doi.org/10.1088/1361-6471/abc534. arXiv: 2007.10722 [astro-ph.CO]
59. H.-K. Guo, A. Miller, Searching for mini extreme mass ratio inspirals with gravitational-wave detectors (2022). arXiv: 2205.10359 [astro-ph.IM]
60. S. Hawking, Gravitationally collapsed objects of very low mass. Mon. Not. Roy. Astron. Soc. **152**, 75 (1971)
61. M.R.S. Hawkins, The signature of primordial black holes in the dark matter halos of galaxies. Astron. Astrophys. **633**, A107 (2020). https://doi.org/10.1051/0004-6361/201936462. arXiv: 2001.07633 [astro-ph.GA]
62. P. Ivanov, P. Naselsky, I. Novikov, Inflation and primordial black holes as dark matter. Phys. Rev. D **50**, 7173–7178 (1994). https://doi.org/10.1103/PhysRevD.50.7173
63. B.J. Kavanagh et al., Detecting dark matter around black holes with gravitational waves: effects of dark-matter dynamics on the gravitational waveform. Phys. Rev. D **102**(8), 083006 (2020). https://doi.org/10.1103/PhysRevD.102.083006. arXiv: 2002.12811 [gr-qc]
64. S. Kawamura et al., The Japanese space gravitational wave antenna DECIGO, ed. by N. Mio. Class. Quant. Grav. **23**, S125–S132 (2006). https://doi.org/10.1088/0264-9381/23/8/S17
65. M. Khlopov, B.A. Malomed, I.B. Zeldovich, Gravitational instability of scalar fields and formation of primordial black holes. Mon. Not. Roy. Astron. Soc. **215**, 575–589 (1985)
66. H. Il Kim, C.H. Lee, Constraints on the spectral index from primordial black holes. Phys. Rev. D **54**, 6001–6007 (1996). https://doi.org/10.1103/PhysRevD.54.6001
67. Z. Luo et al., A brief analysis to Taiji: science and technology. Results Phys. **16**, 102918 (2020). https://doi.org/10.1016/j.rinp.2019.102918
68. M. Maggiore, *Gravitational Waves: Volume 1: Theory and Experiments* (Gravitational Waves, OUP Oxford, 2008). ISBN: 9780198570745
69. M. Maggiore, *Gravitational Waves. Vol. 2: Astrophysics and Cosmology* (Oxford University Press, 2018). ISBN: 978-0-19-857089-9
70. A. Melatos et al., Hidden Markov model tracking of continuous gravitational waves from a binary neutron star with wandering spin. III. Rotational phase tracking. Phys. Rev. D **104**(4), 042003 (2021). https://doi.org/10.1103/PhysRevD.104.042003. arXiv: 2107.12822 [gr-qc]
71. A. Miller et al., Method to search for long duration gravitational wave transients from isolated neutron stars using the generalized frequency-Hough transform. Phys. Rev. D **98**(10), 102004 (2018). https://doi.org/10.1103/PhysRevD.98.102004. arXiv: 1810.09784 [astro-ph.IM]
72. A.L. Miller et al., How effective is machine learning to detect long transient gravitational waves from neutron stars in a real search?. Phys. Rev. D **100**(6), 062005 (2019). https://doi.org/10.1103/PhysRevD.100.062005. arXiv:1909.02262 [astro-ph.IM]
73. A.L. Miller et al., Probing new light gauge bosons with gravitational-wave interferometers using an adapted semicoherent method. Phys. Rev. D **103**(10), 103002 (2021). https://doi.org/10.1103/PhysRevD.103.103002. arXiv:2010.01925 [astro-ph.IM]
74. A.L. Miller, Recent results from continuous gravitational wave searches using data from LIGO/Virgo/KAGRA's third observing run, in *57th Rencontres de Moriond on Gravitation* (2023). arXiv: 2305.15185 [gr-qc]
75. A.L. Miller, N. Singh, C. Palomba, Enabling multimessenger astronomy with continuous gravitational waves: early warning and sky localization of binary neutron stars in the Einstein telescope. Phys. Rev. D **109**(4), 043021 (2024). https://doi.org/10.1103/PhysRevD.109.043021. [astro-ph.IMarXiv: 2309.15808[astro-ph.IM]

76. A.L. Miller et al., Constraints on planetary and asteroid-mass primordial black holes from continuous gravitational-wave searches (2021). https://doi.org/10.1103/PhysRevD.105.062008
77. A.L. Miller et al., Gravitational wave constraints on planetary-mass primordial black holes using LIGO O3a data (2024). https://doi.org/10.1103/PhysRevLett.133.111401
78. A.L. Miller et al., Probing planetary-mass primordial black holes with continuous gravitational waves. Phys. Dark Univ. **32**, 100836 (2021). https://doi.org/10.1016/j.dark.2021.100836. arXiv: 2012.12983 [astro-ph.HE]
79. B. Müller et al., Three-dimensional simulations of neutrino-driven core-collapse supernovae from low-mass single and binary star progenitors. Mon. Not. Roy. Astron. Soc. **484**(3), 3307–3324 (2019). https://doi.org/10.1093/mnras/stz216. arXiv: 1811.05483 [astro-ph.SR]
80. H. Niikura et al., Constraints on earth-mass primordial black holes from OGLE 5-year microlensing events. Phys. Rev. D **99**(8), 083503 (2019). https://doi.org/10.1103/PhysRevD.99.083503. arXiv: 1901.07120 [astro-ph.CO]
81. A.H. Nitz, Y.-F. Wang, Broad search for gravitational waves from subsolar-mass binaries through LIGO and Virgo's third observing run. Phys. Rev. D **106**(2), 023024 (2022). https://doi.org/10.1103/PhysRevD.106.023024. arXiv:2202.11024 [astro-ph.HE]
82. A.H. Nitz, Y.-F. Wang, Search for gravitational waves from the coalescence of sub-solar mass and eccentric compact binaries (2021). https://doi.org/10.3847/1538-4357/ac01d9. arXiv: 2102.00868 [astro-ph.HE]
83. A.H. Nitz, Y.-F. Wang, Search for gravitational waves from the coalescence of subsolar-mass binaries in the first half of advanced LIGO and Virgo's third observing run. Phys. Rev. Lett. **127**(15), 151101 (2021). https://doi.org/10.1103/PhysRevLett.127.151101. arXiv: 2106.08979 [astro-ph.HE]
84. A.H. Nitz et al., 4-OGC: catalog of gravitational waves from compact binary mergers. Astrophys. J. **946**(2), 59 (2023). https://doi.org/10.3847/1538-4357/aca591. arXiv: 2112.06878 [astro-ph.HE]
85. M. Oliver, D. Keitel, A.M Sintes, Adaptive transient Hough method for long-duration gravitational wave transients. Phys. Rev. D **99**(10), 104067 (2019)
86. S. Olsen et al., New binary black hole mergers in the LIGO-Virgo O3a data. Phys. Rev. D **106**(4), 043009 (2022). https://doi.org/10.1103/PhysRevD.106.043009. arXiv:2201.02252 [astro-ph.HE]
87. B.J. Owen, Search templates for gravitational waves from inspiraling binaries: choice of template spacing. Phys. Rev. D **53**, 6749–6761 (1996). https://doi.org/10.1103/PhysRevD.53.6749. arXiv : gr-qc/9511032
88. K. Sang Phukon et al., The hunt for sub-solar primordial black holes in low mass ratio binaries is open (2021). arXiv: 2105.11449 [astro-ph.CO]
89. O.J. Piccinni et al., A new data analysis framework for the search of continuous gravitational wave signals. Class. Quant. Grav. **36**(1), 015008 (2019). https://doi.org/10.1088/1361-6382/aaefb5. arXiv: 1811.04730 [gr-qc]
90. O. Juliana Piccinni, Status and perspectives of continuous gravitational wave searches. Galaxies **10**(3), 72 (2022). https://doi.org/10.3390/galaxies10030072. arXiv: 2202.01088 [gr-qc]
91. R. Prix, Gravitational waves from spinning neutron stars, in *Neutron Stars and Pulsars*, ed. by W. Becker, vol. 357 (ASSL. Springer Berlin Heidelberg, 2009). Chap. 24, pp. 651–685. ISBN: 978-3-540-76964-4. https://doi.org/10.1007/978-3-540-76965-1_24. https://dcc.ligo.org/LIGO-P060039/public
92. M. Punturo et al., The Einstein telescope: s third-generation gravitational wave observatory, ed. by F. Ricci. Class. Quant. Grav. **27**, 194002 (2010). https://doi.org/10.1088/0264-9381/27/19/194002
93. D. Reitze et al., Cosmic explorer: the U.S. contribution to gravitational-wave astronomy beyond LIGO. Bull. Am. Astron. Soc. **51**(7), 035 (2019). arXiv: 1907.04833 [astro-ph.IM]
94. K. Riles, Recent searches for continuous gravitational waves. Mod. Phys. Lett. **A32**(39), 1730035 (2017). https://doi.org/10.1142/S021773231730035X. arXiv: 1712.05897 [gr-qc]

95. K. Riles, Searches for continuous-wave gravitational radiation (2022). arXiv: 2206.06447 [astro-ph.HE]
96. E. Rosenberg, J.J. Fan, Cooling in a dissipative dark sector. Phys. Rev. D **96**(12), 123001 (2017). https://doi.org/10.1103/PhysRevD.96.123001. arXiv: 1705.10341 [astro-ph.GA]
97. W.-H. Ruan et al., Taiji program: gravitational-wave sources. Int. J. Mod. Phys. A **35**(17), 2050075 (2020). https://doi.org/10.1142/S0217751X2050075X. arXiv: 1807.09495 [gr-qc]
98. J. Scholtz, J. Unwin, What if planet 9 is a primordial black hole?. Phys. Rev. Lett. **125**(5), 051103 (2020). https://doi.org/10.1103/PhysRevLett.125.051103. arXiv: 1909.11090 [hep-ph]
99. A. Sesana, Prospects for multiband gravitational-wave astronomy after GW150914. Phys. Rev. Lett. **116**(23), 231102 (2016). https://doi.org/10.1103/PhysRevLett.116.231102. arXiv: 1602.06951 [gr-qc]
100. S. Shandera, D. Jeong, H.S. Grasshorn Gebhardt, Gravitational waves from binary mergers of subsolar mass dark black holes. Phys. Rev. Lett. **120**(24), 241102 (2018). https://doi.org/10.1103/PhysRevLett.120.241102. arXiv: 1802.08206 [astro-ph.CO]
101. K. Silsbee, S. Tremaine, Lidov-Kozai cycles with gravitational radiation: merging black holes in isolated triple systems. Astrophys. J. **836**(1), 39 (2017). https://doi.org/10.3847/1538-4357/aa5729. arXiv: 1608.07642 [astro-ph.HE]
102. D. Singh et al., Gravitational-wave limit on the Chandrasekhar mass of dark matter. Phys. Rev. D **104**(4), 044015 (2021). https://doi.org/10.1103/PhysRevD.104.044015. arXiv: 2009.05209 [astro-ph.CO]
103. B. Steltner et al., Deep Einstein@Home all-sky search for continuous gravitational waves in LIGO O3 public data. Astrophys. J. **952**(1), 55 (2023). https://doi.org/10.3847/1538-4357/acdad4. arXiv: 2303.04109 [gr-qc]
104. L. Sun, A. Melatos, Application of hidden Markov model tracking to the search for long-duration transient gravitational waves from the remnant of the binary neutron star merger GW170817. Phys. Rev. D **99**(12), 123003 (2019)
105. S. Suvorova et al., Hidden Markov model tracking of continuous gravitational waves from a binary neutron star with wandering spin. II. Binary orbital phase tracking. Phys. Rev. D **96**(10), 102006 (2017). https://doi.org/10.1103/PhysRevD.96.102006. arXiv: 1710.07092 [astro-ph.IM]
106. S. Suvorova et al., Hidden Markov model tracking of continuous gravitational waves from a neutron star with wandering spin. Phys. Rev. D **93**(12), 123009 (2016). https://doi.org/10.1103/PhysRevD.93.123009. arXiv: 1606.02412 [astro-ph.IM]
107. Y. Suwa et al., On the minimum mass of neutron stars. Mon. Not. Roy. Astron. Soc. **481**(3), 3305–3312 (2018). https://doi.org/10.1093/mnras/sty2460. arXiv: 1808.02328 [astro-ph.HE]
108. R. Tenorio, D. Keitel, A.M. Sintes, Search methods for continuous gravitational-wave signals from unknown sources in the advanced-detector era. Universe **7**(12), 474 (2021). https://doi.org/10.3390/universe7120474.arXiv: 2111.12575 [gr-qc]
109. M. Trashorras, J. García-Bellido, S. Nesseris, The clustering dynamics of primordial black boles in N-body simulations. Universe **7**(1), 18 (2021). https://doi.org/10.3390/universe7010018. arXiv: 2006.15018 [astro-ph.CO]
110. E. Velcani, Study of data analysis methods for the search of gravitational waves from primordial black hole binaries. https://opac.uniroma1.it/SebinaOpacRMS/resource/study-of-data-analysis-methods-for-the-search-of-gravitational-waves-from-primordial-black-hole-bina/RMS4652009?locale=eng. Master Thesis, Sapienza University of Rome (2022)
111. P. Villanueva-Domingo, O. Mena, S. Palomares-Ruiz, A brief review on primordial black holes as dark matter. Front. Astron. Space Sci. **8**, 87 (2021). https://doi.org/10.3389/fspas.2021.681084. arXiv: 2103.12087 [astro-ph.CO]
112. A. Viterbi, Error bounds for convolutional codes and an asymptotically optimum decoding algorithm. IEEE Trans. Inf. Theory **13**(2), 260–269 (1967)
113. Y.B. Zel'dovich, I.D. Novikov, The hypothesis of cores retarded during expansion and the hot cosmological model. Soviet Astron. AJ (Engl. Transl.) **10**, 602 (1967)

Part V
Constraints on Primordial Black Holes

Chapter 19
Overview on Current Constraints on the Primordial Black Hole Abundance

Kazunori Kohri

Abstract In this chapter, we summarise the observational constraints on PBHs with short descriptions. The master figures are given in Figs. 19.1 and 19.2. We have endeavored and put in a reminder to refer to the number of the section in which the topic is being discussed in this book.

19.1 Introduction

The abundance of PBHs is severely limited by observations. This provides useful guidance both in terms of modeling the early Universe and the possibility of PBHs becoming dark matter. In this chapter, we present the robust limits from a number of studies in terms of evaporating or stable PBHs. The criterion for selecting the lines of constraints is that the assumptions made in this section must be physically natural and the most conservative one. It should be notable that this does not mean we conclude that the lines which were not selected here are meaningless. Our goal is to make the observational bounds on the PBH tighter and tighter, and at last to discover it. Ideally, under the most conservative assumption, the adoption of restrictions by new observations should result in more stringent bounds on parameters. Thus, simply our purpose is not to intentionally impose severe restrictions on the abundance of the PBHs. That spirit must never be forgotten. The structure of this chapter is as follows:

- 19.2 Constraints on evaporating PBHs;
- 19.3 Constraints on stable PBHs;
- 19.4 Originated from constraints on curvature perturbation;
- 19.5 PBH domination and its consequences.

K. Kohri (✉)
Division of Science, National Astronomical Observatory of Japan (NAOJ), and SOKENDAI, 2-21-1 Osawa, Mitaka, Tokyo 181-8588, Japan

Theory Center, IPNS, KEK, 1-1 Oho, Tsukuba, Ibaraki 305-0801, Japan

QUP (WPI), KEK, 1-1 Oho, Tsukuba, Ibaraki 305-0801, Japan

Kavli IPMU (WPI), University of Tokyo, Kashiwa, Chiba 277-8568, Japan
e-mail: kazunori.kohri@gmail.com

C. Byrnes et al. (eds.), *Primordial Black Holes*, Springer Series in Astrophysics and Cosmology, https://doi.org/10.1007/978-981-97-8887-3_19

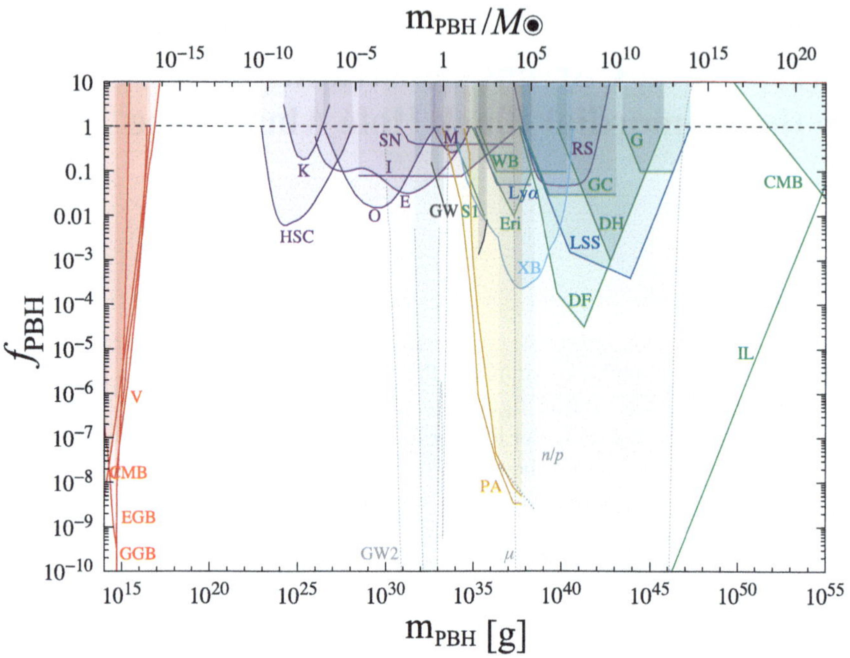

Fig. 19.1 Summary of observational constraints on the abundance of PBHs as a function of the PBH mass in g (the lower x-axis) and in the unit of the solar mass $M_{\odot}$ (the upper x-axis). The y-axis means the fraction to the mass density of CDM. The red region excludes the evaporating PBHs through the observations of gamma-ray background [23], positron by Voyager 1 [19], light elements with BBN [23], and CMB anisotropy [23]. The magenta region excludes the abundance through gravitational lensing by the projects, HSC [73, 85], OGLE [74], EROS [88], MACHO [11] and Icarus [76]. From the observations of the pulsar timing, we can constrain PBHs by the non-detection of the gravitational wave nonlinearly-produced by curvature perturbation if we assume PBHs were produced by curvature perturbation [26, 27, 81]. Through the μ-distortion of the global spectrum of CMB observations, we obtain an upper bound on curvature perturbation and the abundance of PBHs which are expected to be produced by curvature perturbation [26, 28, 51, 60] (the right grey region). The black line excludes PBHs by the non-detection of GWs by LIGO/Virgo/KAGRA [2, 10]. Because accretions of gas onto a PBH can emit radio and X-ray, observational data exclude PBHs with their masses of O(10) $M_{\odot}$ [39]. The cosmological accretions onto PBHs are severely constrained by the observations of both CMB polarization [78, 84] (the yellow ocher region denoted by PA) and the X-rays [39] (the cyan region denoted by XB). The dissipation of curvature perturbation induces dilution of baryon to photon ratio which affects BBN [48, 52, 72]. The limits from the dynamical friction (DF) [20, 97] and the large-scale structure (LSS) [4, 25, 53, 69, 71, 75] are plotted in the green and blue regions, respectively. This figure was produced by updating the source data of Fig. 10 and Fig. 11 in B. Carr, K. Kohri, Y. Sendouda, and J. Yokoyama, 2021 [26]. Thus, we encourage readers to cite Ref.[26] whenever they refer to these lines

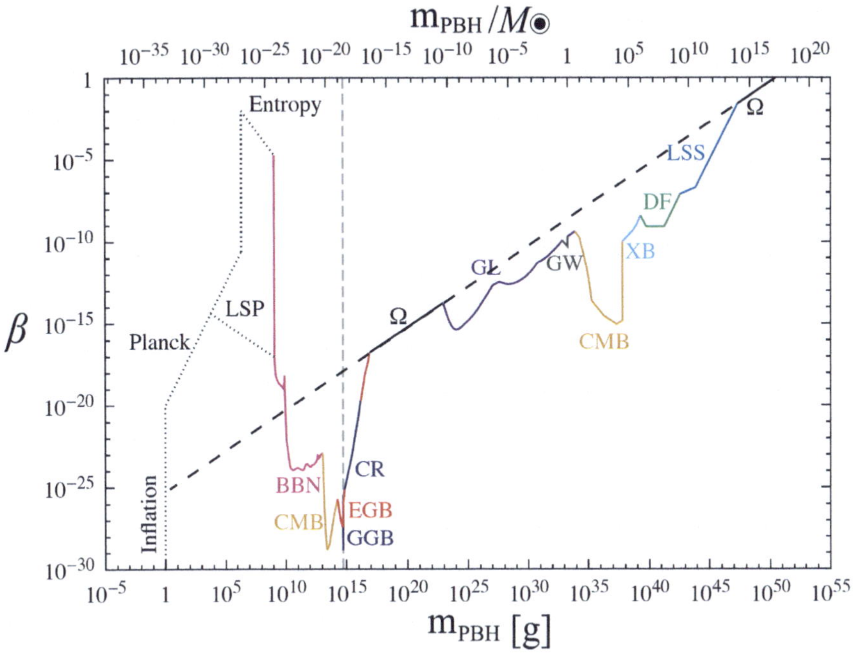

Fig. 19.2 Observational bounds on β. Their abbreviated names are as follows: Extra-galactic Gamma-ray Background (EGB), Galactic Gamma-ray Background (GGB), Cosmic-Rays (CR), Cosmic Microwave Background (CMB), Big-Bang Nucleosynthesis (BBN), the Ω parameter of PBHs to be 100% of cold dark matter (Ω), Gravitational Lensing (GL), Gravitational Wave (GW), X-ray Background (XB), Dynamical Friction (DF), Large-Scale Structure (LSS). The diagonal dashed line means the PBH can be 100% of CDM at present if we assume they are stable. This figure was produced by updating the source data of Fig. 18 in B. Carr, K. Kohri, Y. Sendouda, and J. Yokoyama, 2021 [26]

19.2 Constraints on Evaporating PBHs

19.2.1 Evaporation Through the Hawking Process

A PBH can evaporate through the Hawking process [44] by emitting radiation. The lifetime is given by

$$\tau_{\rm PBH} \sim \frac{m_{\rm PBH}^3}{m_P^4} \sim t_0 \left(\frac{m_{\rm PBH}}{5.1 \times 10^{14} {\rm g}} \right)^3 \sim 10^{64} {\rm yrs} \left(\frac{m_{\rm PBH}}{1 M_\odot} \right)^3 . \tag{19.1}$$

where $t_0 = 13.8$Gyrs is the current age of the Universe [7] with $M_\odot$ being the solar mass ($\simeq 2 \times 10^{33}$ g) and m_P being the reduced Planck mass ($\simeq 2.4 \times 10^{18}$ GeV). This means that PBHs with masses $m_{\rm PBH}$ smaller than 5.1×10^{14}g should have been evaporated by the current age of the Universe. Then, such a PBH cannot become dark matter.

According to the Hawking process, we know a black hole has a finite temperature,

$$T_{\rm PBH} \sim \frac{m_P^2}{m_{\rm PBH}} \sim 10{\rm MeV}\left(\frac{m_{\rm PBH}}{10^{15}{\rm g}}\right)^{-1} \sim 10^{-7}{\rm K}\left(\frac{m_{\rm PBH}}{1{\rm M}_\odot}\right)^{-1}, \tag{19.2}$$

which means that we may observe MeV – GeV γ-ray or cosmic rays (electron, positron, neutrino, etc.) produced by an evaporating PBH at present in principle. Thus, we can constrain the abundance of the PBHs with their masses of around 10^{15} – 10^{17} g severely by the current observations of γ-ray and/or cosmic-ray positron due to non-detections of those signals. We have relations among $\beta = \Omega_{\rm PBH}(t = t_{\rm form})$ at the formation time of the PBH ($t = t_{\rm form}$), the energy fraction to the cold dark matter (CDM) $f_{\rm PBH} = \Omega_{\rm PBH}/\Omega_{\rm CDM}$, and curvature perturbation $\mathcal{P}_\zeta$ as follows

$$\begin{aligned} f_{\rm PBH} &\sim \left(\frac{\beta}{10^{-18}}\right)\left(\frac{m_{\rm PBH}}{10^{15}{\rm g}}\right)^{-1/2} \sim \left(\frac{\beta}{10^{-8}}\right)\left(\frac{m_{\rm PBH}}{30M_\odot}\right)^{-1/2} \\ &\sim \left(\frac{m_{\rm PBH}}{10^{18}{\rm g}}\right)^{-1/2}\sqrt{\frac{\mathcal{P}_\zeta}{10^{-1}}}\exp\left[-\frac{1}{18\mathcal{P}_\zeta}\right]. \end{aligned} \tag{19.3}$$

The bounds on β is shown in Fig. 19.2. Here we assumed radiation had fully dominated after the end of the primordial inflation. By this simple relation (for more precise relations, see Refs. [12, 26, 33, 41, 58, 90, 93–96] for supplementary reading), we can transfer the bounds on β to the bounds on $f_{\rm PBH}$ or $\mathcal{P}_\zeta$ which are summarized in Fig. 19.1. Next, we introduce each constraint on the abundance of the PBHs in detail.

In Fig. 19.1, the red region excludes the evaporating PBHs by the existing observations as follows. We can constrain the abundance of PBHs through observations of gamma-ray background of isotropic components [23] (see also Refs. [13, 15] for supplementary reading), galactic gamma-ray components [16, 22, 23, 30, 62, 64], line emissions from annihilating electron-positron pairs [32, 34, 63], cosmic-ray positrons by Voyager 1 [19], etc. By Ref. [23], the authors of [23] reported the diagonal line as a function of $m_{\rm PBH}$ for the first time to constrain the longer lifetime of PBHs than the cosmic age due to the gradual (not sudden) evaporation into gamma-rays. These observations have excluded the possibility for PBHs to be dark matter for the mass ranges of $5.1 \times 10^{14}{\rm g} \lesssim m_{\rm PBH} \lesssim 10^{17}{\rm g}$.

The evaporating PBHs emit high-energy particles whose electromagnetic energy can modify the history of recombination and temperature of background baryon. This potentially affects the observations of CMB anisotropy and polarization [3, 23, 67, 77, 79, 86] and cosmological 21cm line signals [23, 29, 68]. By the non-detection of those features, we can constrain the abundance of PBHs for the masses of $10^{13.5}{\rm g} \lesssim m_{\rm PBH} \lesssim 10^{14.5}{\rm g}$.

Emissions of high-energy hadrons, gamma-rays, and charged leptons are dangerous for light elements which were produced in the BBN epoch. That is because they induce hadronic and electromagnetic showers through subsequent scatterings off the background particles, and then the produced daughter particles can destroy light elements through inelastic hadron scattering and photodissociation [23, 26, 55]. On the other hand, even if PBHs dominated before their evaporation, the reheating of the Universe was realized by the evaporation of the PBHs. The condition for the PBH domination just before the BBN is $\beta > 10^{-14} \left(m_{\rm PBH}/10^9 {\rm g}\right)^{-1}$. Then, if the evaporation occurs just before/during the beginning of the BBN, the thermalization of the background neutrinos should be modified [54] due to emissions of electromagnetic particles such as photon, electron, etc. In this case, even the effective number of neutrino species can be smaller than three [54],

$$N_{\nu,{\rm eff}} < 3, \tag{19.4}$$

due to insufficient thermalization of neutrinos because weak interaction is slow at the beginning of the BBN. Then, the lower bound on the reheating temperature is obtained to be $T_R > 1.8$ MeV by comparing theoretical predictions of BBN with observational abundances of light elements [43]. In addition, the value of the neutron to proton ratio can be also changed severely by scatterings of high-energy hadrons through strong (not weak) interaction, such as charged pions, kaons, proton, antiproton, neutron, antineutron, etc. Then, we obtain the severer bound (but still a conservative one) on the reheating temperature than that only for the electromagnetic emission [43],

$$T_R > 4\,{\rm MeV}. \tag{19.5}$$

This lower bound on reheating temperature corresponds to the straight vertical line showing $m_{\rm PBH} \sim 10^9$g in Fig. 19.2. By also considering the case for the subdomination of PBHs, in total we can exclude the mass ranges of $10^9{\rm g} \lesssim m_{\rm PBH} \lesssim 10^{13.5}$g by the destruction of the light elements by high-energy hadrons and electromagnetic particles [55]. (See also the works reconsidered by the other groups, e.g., [56] for both of hadronic and radiative emissions, and Refs. [3, 77] for only on the electromagnetic/photon emission for supplementary reading). It is notable that the crossing point with the diagonal dashed line (Ω) shows $m_{\rm PBH} \lesssim 10^{10}$g can be allowed. This limit gives a good criterion whether a daughter stable particle may contribute to a fraction of dark matter or not, which is predicted in a new theory beyond the standard model. See also Chap. 1 of this book for more details.

19.2.2 Memory Burden Effect

Some groups considered a possible mechanism which makes evaporation stop intermediately during the evaporation in the very high entropy circumstances around evaporating PBHs [36]. Then, the evaporation rate is suppressed by a power of entropy, e.g., the inverse of the entropy square. This is called the "Memory Burden" effect. By following this claim, the bounds on evaporating PBHs can be modified, and a new window $10^5\text{g} \lesssim m_{\text{PBH}} \lesssim 10^{10}\text{g}$ opens for PBHs to be dark matter [87]. See also similar works for supplementary reading [35, 42].

19.3 Constraints on Stable PBHs

19.3.1 Gravitational Lensing

Gravitational lensing can give a constraint on a PBH passing through the line of sight towards a light source. The magenta region in Fig. 19.1 excludes the abundance of PBHs through gravitational lensing by the following projects, HSC [73, 85], OGLE [70, 74], EROS [88], MACHO [11] and Icarus [76]. From Fig. 19.1, we see that the masses of $10^{-10} M_\odot \lesssim m_{\text{PBH}} \lesssim 2 \times 10^4 M_\odot$ and $6 \times 10^4 M_\odot \lesssim m_{\text{PBH}} \lesssim 10^{9.5} M_\odot$ are excluded by non-detections of lensing events caused by PBHs. See also Chaps. 1 and 23 of this book for more details.

19.3.2 Gravitational Waves from Binary PBHs

Quite recently we have had a strong motivation to study PBHs because the rate of gravitational wave (GW) detected by aLIGO [1] can be fitted by the event rate of a coalescence of binary PBHs with their masses of $\sim 30\ M_\odot$ [82, 83] for the fraction $f_{\text{PBH}} \sim O(10^{-3})$ (See also Ref. [17]). In other words, simultaneously we can constrain the number density of PBHs by observing the GWs produced by binary PBHs with arbitrary masses. The black line in Fig. 19.1 has excluded PBHs due to the non-detection of such GWs by the current sensitivities of aLIGO/AdV for $0.2 M_\odot \lesssim m_{\text{PBH}} \lesssim 1 M_\odot$, and $10 M_\odot \lesssim m_{\text{PBH}} \lesssim 3 \times 10^2 M_\odot$ [2, 10, 21, 89]. See also a future detectability of GWs from lighter binary PBHs and its consequence of a bound on the abundance [91]. See also Chap. 16 of this book for more details.

19.3.3 Constraints on Accretions Onto PBHs

From observations of radio and X-rays which are emitted through accretions of gas onto a PBH, we can obtain an upper bound on the abundance of PBHs with their masses for $O(10) - O(10^2)\ M_\odot$, which is denoted by the cyan region [39].

Considering a finite relative velocity between baryon and CDM/PBHs, a disk accretion of baryon onto a PBH must have occurred inevitably at around the high redshift from $z \sim O(10^3)$ to $z \sim O(10^2)$. The photon emitted through the accretion can change the recombination history of baryon, which is severely constrained by observations of the CMB anisotropy and polarization. The yellow ocher region in

Fig. 19.1 is excluded by the current observations about accretions onto a PBH plus its CDM halo system [78, 84] (see also Refs. [8, 9, 80] for supplementary reading as the original ideas). See also Chap. 24 of this book for more details.

19.3.4 Constraints on PBHs from Dynamical Motions and Large-Scale Structure

There are some methods to constrain PBHs dynamically by using massive objects in the Universe. Not in order to destroy a dwarf galaxy, such as Eridanus II due to dynamical friction (DF), we can constrain the mass ranges for $10^4 M_\odot \lesssim m_{\rm PBH} \lesssim 10^{12} M_\odot$ [20, 97].

Large-scale structure (LSS) constraints come from the Lyman-α forest (Lyα) and various other cosmic structures due to the fluctuation Poissonian because of the discreteness for the distribution of the massive PBHs [4, 25, 53, 69, 71, 75]. See also Chap. 1 of this book for more details.

19.4 Originated from Constraints on Curvature Perturbation

If PBHs are produced by the collapse of large density perturbation at small scales, which is without a doubt the most promising scenario, the upper bounds on density perturbation (or curvature perturbation $\mathcal{P}_\zeta$) can be transformed into the ones on the abundance of PBHs $f_{\rm PBH}$ or β. On the other words, if the PBHs were produced by the other mechanism, the following limits may be meaningless.

The dissipation of density perturbation at small scales in the early Universe induces a dilution of baryon which affects the freezeout value of neutron to proton ratio n/p in BBN. Then, the dark grey region denoted by n/p in Fig. 19.1 is excluded by comparing the theoretical predictions of light element abundances with the observations for $10^{3.5} M_\odot \lesssim m_{\rm PBH} \lesssim 10^{5.5} M_\odot$ [48, 52, 72].

Through observations of the CMB μ-distortion caused by the dissipation of density perturbation, we obtain an upper bound on curvature perturbation and, simultaneously the abundance of PBHs for $10^{4.5} M_\odot \lesssim m_{\rm PBH} \lesssim 10^{13.5} M_\odot$ [26, 28, 51, 60].

From the observations of the pulsar timing, we can constrain PBHs by the non-detection of the gravitational wave induced by curvature perturbation [26, 27, 61, 81]. Recently, a detection of the induced gravitational wave through the pulsar timing was claimed by the NANOGrav collaboration [5, 6]. See also Refs. [37, 49, 92] for the theoretical models to fit the data. However, we adopt a conservative upper limit on the induced gravitational wave or the abundance of PBHs discussed in [27]. See also Chap. 1 of this book for more details.

19.5 PBH Domination and Its Consequences

Here we discuss some scenarios in which PBHs could have dominated the Universe at some time in the early epoch. An unstable PBH whose mass is lighter than $\sim 10^9$ g with lifetime being $O(1)$ sec cannot be constrained by observations so strongly even by adopting the constraint from BBN which has a good sensitivity at around $m_{\rm PBH} \gtrsim 10^9$ g [18, 23, 26] (See also the discussions in Sect. 19.2).

For $m_{\rm PBH} \lesssim 10^9$ g In principle, an evaporating PBH may be constrained by the production of a hypothetical massive relic to be dark matter such as the lightest supersymmetric particle (LSP) or the Planck mass relic. About a non-thermal production of the LSP due to the evaporation, we have an upper bound on β in order not to exceed the observed density of cold dark matter [40, 57, 65],

$$\beta \lesssim 10^{-18} \left(\frac{m_{\rm PBH}}{10^{11}{\rm g}}\right)^{-1/2} \left(\frac{m_{\rm LSP}}{100{\rm GeV}}\right)^{-1} \tag{19.6}$$

for $m_{\rm PBH} < 10^{11}{\rm g}\,(m_{\rm LSP}/100{\rm GeV})^{-1}$ with $m_{\rm LSP}$ being the mass of LSP of the order of the weak scale. Quite recently, it was reported that a bound can be strengthened for lighter dark matter by considering a free streaming and an erase of small-scale fluctuations due to its initial high momentum [14, 66].

Additionally, about a possible production of the Planck-mass relic at the last stage of the evaporation, we have an upper bound on β in order not to exceed the density of cold dark matter,

$$\beta \lesssim 10^{-20} \left(\frac{m_{\rm PBH}}{\rm g}\right)^{3/2}, \tag{19.7}$$

for $1\,{\rm g} \lesssim m_{\rm PBH} \lesssim 10^7$ g [24].

The above bounds are based on the detailed natures of dark matter, which should be highly model-dependent. Therefore, when we omit the bounds due to such uncertainties, PBHs can be allowed to dominate the Universe even shortly at some time before they completely vanish away. Then, there will appear some exotic phenomena as follows. GWs were produced by the nonlinear effect of large curvature perturbation at small scales which produced the PBHs [50] or by a coalescence of binary PBHs before the evaporation [47]. In some models, baryogenesis through non-thermal leptogenesis may be realized successfully by right-handed neutrino produced from the huge amount of the evaporating PBHs [38]. In terms of particle physics, the electroweak vacuum may be affected by vacuum fluctuations of the Higgs field around evaporating primordial black holes [59].

Quite recently it was pointed out that very hot spots could form around evaporating primordial black holes that became dominant in energy density. This is of great interest because it indicates the possibility of producing exotic particles such as monopole and/or dark matter inside such hot spots, whose mass can be much larger than the background temperature [31, 45, 46].

Acknowledgements This work was supported by the JSPS KAKENHI, JP23KF0289 and JP24K07027, and the MEXT Grant-in-Aid for Scientific Research on Innovative Areas JP22H05270 and JP24H01825.

References

1. B.P. Abbott et al., Observation of gravitational waves from a binary black hole merger. Phys. Rev. Lett. **116**(6), 061102 (2016). https://doi.org/10.1103/PhysRevLett.116.061102. arXiv: 1602.03837 [gr-qc]
2. B.P. Abbott et al., Search for subsolar mass ultracompact binaries in advanced LIGO's second observing run. Phys. Rev. Lett. **123**(16), 161102 (2019). https://doi.org/10.1103/PhysRevLett.123.161102. arXiv: 1904.08976 [astro-ph.CO]
3. S.K. Acharya, R. Khatri, CMB and BBN constraints on evaporating primordial black holes revisited. JCAP **06**, 018 (2020). https://doi.org/10.1088/1475-7516/2020/06/018. arXiv: 2002.00898 [astro-ph.CO]
4. N. Afshordi, P. McDonald, D.N. Spergel, Primordial black holes as dark matter: the power spectrum and evaporation of early structures. Astrophys. J. Lett. **594**, L71–L74 (2003). https://doi.org/10.1086/378763. arXiv: astro-ph/0302035
5. A. Afzal et al., The NANOGrav 15 yr data set: search for signals from new physics. Astrophys. J. Lett. **951**(1), L11 (2023). https://doi.org/10.3847/2041-8213/acdc91. arXiv: 2306.16219 [astro-ph.HE]
6. G. Agazie et al., The NANOGrav 15 yr data set: evidence for a gravitational-wave background. Astrophys. J. Lett. **951**(1), L8 (2023). https://doi.org/10.3847/2041-8213/acdac6. arXiv: 2306.16213 [astro-ph.HE]
7. N. Aghanim et al., Planck 2018 results. VI. Cosmological parameters'. Astron. Astrophys. **641**, A6 (2020). https://doi.org/10.1051/0004-6361/201833910. arXiv: 1807.06209 [astro-ph.CO]
8. D. Agius et al., Feedback in the dark: a critical examination of CMB bounds on primordial black holes (2024). arXiv: 2403.18895 [hep-ph]
9. Y. Ali-Haïmoud, M. Kamionkowski, Cosmic microwave background limits on accreting primordial black holes. Phys. Rev. D **95**(4), 043534 (2017). https://doi.org/10.1103/PhysRevD.95.043534. arXiv: 1612.05644 [astro-ph.CO]
10. Y. Ali-Haïmoud, E.D. Kovetz, M. Kamionkowski, Merger rate of primordial black-hole binaries. Phys. Rev. D **96**(12), 123523 (2017). https://doi.org/10.1103/PhysRevD.96.123523. arXiv: 1709.06576 [astro-ph.CO]
11. R.A. Allsman et al., MACHO project limits on black hole dark matter in the 1-30 solar mass range. Astrophys. J. Lett. **550**, L169 (2001). https://doi.org/10.1086/319636. arXiv: astro-ph/0011506
12. K. Ando, K. Inomata, M. Kawasaki, Primordial black holes and uncertainties in the choice of the window function. Phys. Rev. D **97**(10), 103528 (2018). https://doi.org/10.1103/PhysRevD.97.103528. arXiv: 1802.06393 [astro-ph.CO]
13. A. Arbey, J. Auffinger, J. Silk, Constraining primordial black hole masses with the isotropic gamma ray background. Phys. Rev. D **101**(2), 023010 (2020). https://doi.org/10.1103/PhysRevD.101.023010. arXiv: 1906.04750 [astro-ph.CO]
14. I. Baldes et al., Non-cold dark matter from primordial black hole evaporation (2020). arXiv: 2004.14773 [astro-ph.CO]
15. G. Ballesteros, J.C.-Blázquez, D. Gaggero, X-ray and gamma-ray limits on the primordial black hole abundance from Hawking radiation. Phys. Lett. B **808**, 135624 (2020). https://doi.org/10.1016/j.physletb.2020.135624. arXiv: 1906.10113 [astro-ph.CO]
16. J. Berteaud et al., Strong constraints on primordial black hole dark matter from 16 years of INTEGRAL/SPI observations. Phys. Rev. D **106**(2), 023030 (2022). https://doi.org/10.1103/PhysRevD.106.023030. arXiv: 2202.07483 [astro-ph.HE]

17. S. Bird et al., Did LIGO detect dark matter?. Phys. Rev. Lett. **116**(20), 201301 (2016). https://doi.org/10.1103/PhysRevLett.116.201301. arXiv: 1603.00464 [astro-ph.CO]
18. A. Boccia, F. Iocco, L. Visinelli, Constraining the primordial black hole abundance through Big-Bang nucleosynthesis (2024). arXiv: 2405.18493 [astro-ph.CO]
19. M. Boudaud, M. Cirelli, Voyager 1 $e^{\pm}$ further constrain primordial black holes as dark matter. Phys. Rev. Lett. **122**(4), 041104 (2019). https://doi.org/10.1103/PhysRevLett.122.041104. arXiv: 1807.03075 [astro-ph.HE]
20. T.D. Brandt, Constraints on MACHO dark matter from compact stellar systems in ultra-faint dwarf galaxies. Astrophys. J. Lett. **824**(2), L31 (2016). https://doi.org/10.3847/2041-8205/824/2/L31. arXiv: 1605.03665 [astro-ph.GA]
21. R.-G. Cai et al., Pulsar timing array constraints on the induced gravitational waves. JCAP **10**, 059 (2019). https://doi.org/10.1088/1475-7516/2019/10/059. arXiv: 1907.06372 [astro-ph.CO]
22. B.J. Carr et al., Constraints on primordial black holes from the Galactic gamma-ray background. Phys. Rev. D **94**(4), 044029 (2016). https://doi.org/10.1103/PhysRevD.94.044029. arXiv: 1604.05349 [astro-ph.CO]
23. B.J. Carr et al., New cosmological constraints on primordial black holes. Phys. Rev. D **81**, 104019 (2010). https://doi.org/10.1103/PhysRevD.81.104019. arXiv: 0912.5297 [astro-ph.CO]
24. B. Carr, J. Gilbert, J. Lidsey, Black hole relics and inflation: limits on blue perturbation spectra. Phys. Rev. D **50**, 4853–4867 (1994). https://doi.org/10.1103/PhysRevD.50.4853. arXiv: astro-ph/9405027
25. B. Carr, J. Silk, Can graininess in the early universe make galaxies?. Astrophys. J. **268**, 1–16 (1983). https://doi.org/10.1086/160924
26. B. Carr et al., Constraints on primordial black holes. Rept. Prog. Phys. **84**(11), 116902 (2021). https://doi.org/10.1088/1361-6633/ac1e31. arXiv: 2002.12778 [astro-ph.CO]
27. Z.-C. Chen, C. Yuan, Q.-G. Huang, Pulsar timing array constraints on primordial black holes with NANOGrav 11-year dataset. Phys. Rev. Lett. **124**(25), 251101 (2020). https://doi.org/10.1103/PhysRevLett.124.251101. arXiv: 1910.12239 [astro-ph.CO]
28. J. Chluba, A. Ravenni, S.K. Acharya, Thermalization of large energy release in the early Universe. Mon. Not. Roy. Astron. Soc. **498**(1), 959–980 (2020). https://doi.org/10.1093/mnras/staa2131. arXiv: 2005.11325 [astro-ph.CO]
29. S. Clark et al., 21 cm limits on decaying dark matter and primordial black holes. Phys. Rev. D **98**(4), 043006 (2018). https://doi.org/10.1103/PhysRevD.98.043006. arXiv: 1803.09390 [astro-ph.HE]
30. A. Coogan, L. Morrison, S. Profumo, Direct detection of hawking radiation from asteroid-mass primordial black holes. Phys. Rev. Lett. **126**(17), 171101 (2021). https://doi.org/10.1103/PhysRevLett.126.171101. arXiv: 2010.04797 [astro-ph.CO]
31. S. Das, A. Hook, Black hole production of monopoles in the early universe. JHEP **12**, 145 (2021). https://doi.org/10.1007/JHEP12(2021)145. arXiv: 2109.00039 [hep-ph]
32. B. Dasgupta, R. Laha, A. Ray, Neutrino and positron constraints on spinning primordial black hole dark matter. Phys. Rev. Lett. **125**(10), 101101 (2020). https://doi.org/10.1103/PhysRevLett.125.101101. arXiv: 1912.01014 [hep-ph]
33. V. De Luca, A. Kehagias, A. Riotto, How well do we know the primordial black hole abundance: the crucial role of nonlinearities when approaching the horizon. Phys. Rev. D **108**(6), 063531 (2023). https://doi.org/10.1103/PhysRevD.108.063531. arXiv: 2307.13633 [astro-ph.CO]
34. W. DeRocco, P.W. Graham, Constraining primordial black hole abundance with the Galactic 511 keV line. Phys. Rev. Lett. **123**(25), 51102 (2019). https://doi.org/10.1103/PhysRevLett.123.251102. arXiv: 1906.07740 [astro-ph.CO]
35. G. Domènech, M. Sasaki, Gravitational wave hints black hole remnants as dark matter. Class. Quant. Grav. **40**(17), 77001 (2023). https://doi.org/10.1088/1361-6382/ace493. arXiv: 2303.07661 [gr-qc]

36. G. Dvali et al., Black hole metamorphosis and stabilization by memory burden. Phys. Rev. D **102**(10), 103523 (2020). https://doi.org/10.1103/PhysRevD.102.103523. arXiv: 2006.00011 [hep-th]
37. G. Franciolini et al., The recent gravitational wave observation by pulsar timing arrays and primordial black holes: the importance of non-gaussianities (2023). arXiv: 2306.17149 [astro-ph.CO]
38. T. Fujita et al., Baryon asymmetry, dark matter, and density perturbation from primordial black holes. Phys. Rev. D **89**(10), 103501 (2014). https://doi.org/10.1103/PhysRevD.89.103501. arXiv: 1401.1909 [astro-ph.CO]
39. D. Gaggero et al., Searching for primordial black holes in the radio and X-ray sky. Phys. Rev. Lett. **118**(24), 241101 (2017). https://doi.org/10.1103/PhysRevLett.118.241101. arXiv: 1612.00457 [astro-ph.HE]
40. A.M. Green, Supersymmetry and primordial black hole abundance constraints. Phys. Rev. D **60**, 063516 (1999). https://doi.org/10.1103/PhysRevD.60.063516. arXiv: astro-ph/9903484
41. A.M. Green et al., A new calculation of the mass fraction of primordial black holes. Phys. Rev. D **70**, 041502 (2004). https://doi.org/10.1103/PhysRevD.70.041502. arXiv: astro-ph/0403181
42. M.R. Haque et al., Quantum effects on the evaporation of PBHs: contributions to dark matter (2024). arXiv: 2404.16815 [hep-ph]
43. T. Hasegawa et al., MeV-scale reheating temperature and thermalization of oscillating neutrinos by radiative and hadronic decays of massive particles. JCAP **12**, 012 (2019). https://doi.org/10.1088/1475-7516/2019/12/012. arXiv: 1908.10189 [hep-ph]
44. S.W. Hawking, Black hole explosions. Nature **248**, 30–31 (1974). https://doi.org/10.1038/248030a0
45. M. He et al., Formation of hot spots around small primordial black holes. JCAP **01**, 027 (2023). https://doi.org/10.1088/1475-7516/2023/01/027. arXiv: 2210.06238 [hep-ph]
46. M. He et al., Thermalization and hotspot formation around small primordial black holes (2024). arXiv: 2407.15926 [hep-ph]
47. D. Hooper et al., Hot gravitons and gravitational waves from Kerr black holes in the early universe (2020). arXiv: 2004.00618 [astro-ph.CO]
48. K. Inomata, M. Kawasaki, Y. Tada, Revisiting constraints on small scale perturbations from big-bang nucleosynthesis. Phys. Rev. D **94**(4), 043527 (2016). https://doi.org/10.1103/PhysRevD.94.043527. arXiv: 1605.04646 [astro-ph.CO]
49. K. Inomata, K. Kohri, T. Terada, The detected stochastic gravitational waves and subsolar-mass primordial black holes (2023). arXiv: 2306.17834 [astro-ph.CO]
50. K. Inomata et al., Gravitational wave production right after a primordial black hole evaporation. Phys. Rev. D **101**(12), 123533 (2020). https://doi.org/10.1103/PhysRevD.101.123533. arXiv: 2003.10455 [astro-ph.CO]
51. K. Inomata et al., Inflationary primordial black holes for the LIGO gravitational wave events and pulsar timing array experiments. Phys. Rev. D **95**(12), 123510 (2017). https://doi.org/10.1103/PhysRevD.95.123510. arXiv: 1611.06130 [astro-ph.CO]
52. D. Jeong et al., Silk damping at a redshift of a billion: a new limit on small-scale adiabatic perturbations. Phys. Rev. Lett. **113**, 061301 (2014). https://doi.org/10.1103/PhysRevLett.113.061301. arXiv: 1403.3697 [astro-ph.CO]
53. A. Kashlinsky et al., Tracing the first stars with fluctuations of the cosmic infrared background. Nature **438**, 45–50 (2005). https://doi.org/10.1038/nature04143. arXiv: astro-ph/0511105
54. M. Kawasaki, K. Kohri, N. Sugiyama, Cosmological constraints on late time entropy production. Phys. Rev. Lett. **82**, 4168 (1999). https://doi.org/10.1103/PhysRevLett.82.4168. arXiv: astro-ph/9811437
55. M. Kawasaki, K. Kohri, T. Moroi, Big-bang nucleosynthesis and hadronic decay of long-lived massive particles. Phys. Rev. D **71**, 083502 (2005). https://doi.org/10.1103/PhysRevD.71.083502. arXiv: astro-ph/0408426
56. C. Keith et al., Constraints on primordial black holes from big bang nucleosynthesis revisited. Phys. Rev. D **102**(10), 103512 (2020). https://doi.org/10.1103/PhysRevD.102.103512. arXiv: 2006.03608 [astro-ph.CO]

57. M.Y. Khlopov, A. Barrau, J. Grain, Gravitino production by primordial black hole evaporation and constraints on the inhomogeneity of the early universe. Class. Quant. Grav. **23**, 1875–1882 (2006). https://doi.org/10.1088/0264-9381/23/6/004. arXiv: astro-ph/0406621
58. N. Kitajima et al., Primordial black holes in peak theory with a non-Gaussian tail. JCAP **10**, 053 (2021). https://doi.org/10.1088/1475-7516/2021/10/053. arXiv: 2109.00791 [astro-ph.CO]
59. K. Kohri, H. Matsui, Electroweak vacuum collapse induced by vacuum fluctuations of the Higgs field around evaporating black holes. Phys. Rev. D **98**(12), 123509 (2018). https://doi.org/10.1103/PhysRevD.98.123509. arXiv: 1708.02138 [hep-ph]
60. K. Kohri, T. Nakama, T. Suyama, Testing scenarios of primordial black holes being the seeds of supermassive black holes by ultracompact minihalos and CMB μ-distortions. Phys. Rev. D **90**(8), 083514 (2014). https://doi.org/10.1103/PhysRevD.90.083514. arXiv: 1405.5999 [astro-ph.CO]
61. K. Kohri, T. Terada, Semianalytic calculation of gravitational wave spectrum nonlinearly induced from primordial curvature perturbations. Phys. Rev. D **97**(12), 123532 (2018). https://doi.org/10.1103/PhysRevD.97.123532. arXiv: 1804.08577 [gr-qc]
62. M. Korwar, S. Profumo, Updated constraints on primordial black hole evaporation. JCAP **05**, 054 (2023). https://doi.org/10.1088/1475-7516/2023/05/054. arXiv: 2302.04408 [hep-ph]
63. R. Laha, Primordial black holes as a dark matter candidate are severely constrained by the galactic center 511 keV γ -ray line. Phys. Rev. Lett. **123**(25), 251101 (2019). https://doi.org/10.1103/PhysRevLett.123.251101. arXiv: 1906.09994 [astro-ph.HE]
64. R. Laha, J.B. Muñoz, T.R. Slatyer, INTEGRAL constraints on primordial black holes and particle dark matter. Phys. Rev. D **101**(12), 123514 (2020). https://doi.org/10.1103/PhysRevD.101.123514. arXiv: 2004.00627 [astro-ph.CO]
65. M. Lemoine, Moduli constraints on primordial black holes. Phys. Lett. B **481**, 333–338 (2000). https://doi.org/10.1016/S0370-2693(00)00469-X. arXiv: hep-ph/0001238
66. O. Lennon et al., Black hole genesis of dark matter. JCAP **04**, 009 (2018). https://doi.org/10.1088/1475-7516/2018/04/009. arXiv: 1712.07664 [hep-ph]
67. M. Lucca et al., The synergy between CMB spectral distortions and anisotropies. JCAP **02**, 026 (2020). https://doi.org/10.1088/1475-7516/2020/02/026. arXiv: 1910.04619 [astro-ph.CO]
68. K.J. Mack, D.H. Wesley, Primordial black holes in the dark ages: observational prospects for future 21cm surveys (2008). arXiv: 0805.1531 [astro-ph]
69. P. Mészáros, Primeval black holes and galaxy formation. Astron. Astrophys. **38**, 5–13 (1975)
70. P. Mroz et al., No massive black holes in the milky way halo (2024). arXiv: 2403.02386 [astro-ph.GA]
71. R. Murgia et al., Lyman-α forest constraints on primordial black holes as dark matter. Phys. Rev. Lett. **123**(7), 071102 (2019). https://doi.org/10.1103/PhysRevLett.123.071102. arXiv: 1903.10509 [astro-ph.CO]
72. T. Nakama, T. Suyama, J. Yokoyama, Reheating the universe once more: the dissipation of acoustic waves as a novel probe of primordial inhomogeneities on even smaller scales. Phys. Rev. Lett. **113**, 061302 (2014). https://doi.org/10.1103/PhysRevLett.113.061302. arXiv: 1403.5407 [astro-ph.CO]
73. H. Niikura et al., Microlensing constraints on primordial black holes with Subaru/HSC andromeda observations. Nature Astron. **3**(6), 524–534 (2019). https://doi.org/10.1038/s41550-019-0723-1. arXiv: 1701.02151 [astro-ph.CO]
74. H. Niikura et al., Constraints on earth-mass primordial black holes from OGLE 5-year microlensing events. Phys. Rev. D **99**(8), 083503 (2019). https://doi.org/10.1103/PhysRevD.99.083503. arXiv: 1901.07120 [astro-ph.CO]
75. M. Oguri, R. Takahashi, Probing dark low-mass halos and primordial black holes with frequency-dependent gravitational lensing dispersions of gravitational waves. Astrophys. J. **901**(1), 58 (2020). https://doi.org/10.3847/1538-4357/abafab. arXiv: 2007.01936 [astro-ph.CO]

76. M. Oguri et al., Understanding caustic crossings in giant arcs: characteristic scales, event rates, and constraints on compact dark matter. Phys. Rev. D **97**(2), 023518 (2018). https://doi.org/10.1103/PhysRevD.97.023518. arXiv: 1710.00148 [astro-ph.CO]
77. V. Poulin, J. Lesgourgues, P.D. Serpico, Cosmological constraints on exotic injection of electromagnetic energy. JCAP **1703**(03), 043 (2017). https://doi.org/10.1088/1475-7516/2017/03/043. arXiv: 1610.10051 [astro-ph.CO]
78. V. Poulin et al., CMB bounds on disk-accreting massive primordial black holes. Phys. Rev. D **96**(8), 083524 (2017). https://doi.org/10.1103/PhysRevD.96.083524. arXiv: 1707.04206 [astro-ph.CO]
79. H. Poulter et al., CMB constraints on ultra-light primordial black holes with extended mass distributions (2019). arXiv: 1907.06485 [astro-ph.CO]
80. M. Ricotti, J.P. Ostriker, K.J. Mack, Effect of primordial black holes on the cosmic microwave background and cosmological parameter estimates. Astrophys. J. **680**, 829 (2008). https://doi.org/10.1086/587831. arXiv: 0709.0524 [astro-ph]
81. R. Saito, J. Yokoyama, Gravitational wave background as a probe of the primordial black hole abundance. Phys. Rev. Lett. **102** (2009). [Erratum: Phys.Rev.Lett. 107, 069901 (2011)], 161101. https://doi.org/10.1103/PhysRevLett.102.161101. arXiv: 0812.4339 [astro-ph]
82. M. Sasaki et al., Primordial black hole scenario for the gravitational-wave event GW150914. Phys. Rev. Lett. **117**(6) (2016). [Erratum: Phys.Rev.Lett. 121, 059901 (2018)], 061101. https://doi.org/10.1103/PhysRevLett.117.061101. arXiv: 1603.08338 [astro-ph.CO]
83. M. Sasaki et al., Primordial black holes—perspectives in gravitational wave astronomy. Class. Quant. Grav. **35**(6), 063001 (2018). https://doi.org/10.1088/1361-6382/aaa7b4. arXiv: 1801.05235 [astro-ph.CO]
84. P.D. Serpico et al., Cosmic microwave background bounds on primordial black holes including dark matter halo accretion. Phys. Rev. Res. **2**(2), 023204 (2020). https://doi.org/10.1103/PhysRevResearch.2.023204. arXiv: 2002.10771 [astro-ph.CO]
85. N. Smyth et al., Updated constraints on asteroid-mass primordial black holes as dark matter. Phys. Rev. D **101**(6), 063005 (2020). https://doi.org/10.1103/PhysRevD.101.063005. arXiv: 1910.01285 [astro-ph.CO]
86. P. Stöcker et al., Exotic energy injection with ExoCLASS: application to the Higgs portal model and evaporating black holes. JCAP **03**, 018 (2018). https://doi.org/10.1088/1475-7516/2018/03/018. arXiv: 1801.01871 [astro-ph.CO]
87. V. Thoss, A. Burkert, K. Kohri, Breakdown of hawking evaporation opens new mass window for primordial black holes as dark matter candidate (2024). arXiv: 2402.17823 [astro-ph.CO]
88. P. Tisserand et al., Limits on the macho content of the galactic halo from the EROS-2 survey of the magellanic clouds. Astron. Astrophys. **469**, 387–404 (2007). https://doi.org/10.1051/0004-6361:20066017. arXiv: astro-ph/0607207
89. V. Vaskonen, H. Veermäe, Lower bound on the primordial black hole merger rate. Phys. Rev. D **101**(4), 043015 (2020). https://doi.org/10.1103/PhysRevD.101.043015. arXiv: 1908.09752 [astro-ph.CO]
90. Q. Wang et al., Primordial black holes from the perturbations in the inflaton potential in peak theory. Phys. Rev. D **104**(8), 083546 (2021). https://doi.org/10.1103/PhysRevD.104.083546. arXiv: 2111.10028 [astro-ph.CO]
91. S. Wang, T. Terada, K. Kohri, Prospective constraints on the primordial black hole abundance from the stochastic gravitational-wave backgrounds produced by coalescing events and curvature perturbations. Phys. Rev. D **99**(10) (2019). [Erratum: Phys. Rev. D 101, 069901 (2020)], 103531. https://doi.org/10.1103/PhysRevD.99.103531. arXiv: 1903.05924 [astro-ph.CO]
92. S. Wang et al., Exploring the implications of 2023 pulsar timing array datasets for scalar-induced gravitational waves and primordial black holes (2023). arXiv: 2307.00572 [astro-ph.CO]
93. C.-M. Yoo, J.-O. Gong, S. Yokoyama, Abundance of primordial black holes with local non-Gaussianity in peak theory. JCAP **09**, 033 (2019). https://doi.org/10.1088/1475-7516/2019/09/033. arXiv: 1906.06790 [astro-ph.CO]

94. C.-M. Yoo et al., Abundance of primordial black holes in peak theory for an arbitrary power spectrum. PTEP **2021**(1), 013E02 (2021). https://doi.org/10.1093/ptep/ptaa155. arXiv: 2008.02425 [astro-ph.CO]
95. C.-M. Yoo et al., Primordial black hole abundance from random Gaussian curvature perturbations and a local density threshold. PTEP **2018**(12), 123E01 (2018). https://doi.org/10.1093/ptep/pty120. arXiv: 1805.03946 [astro-ph.CO]
96. S. Young, C.T. Byrnes, M. Sasaki, Calculating the mass fraction of primordial black holes. JCAP **07**, 045 (2014). https://doi.org/10.1088/1475-7516/2014/07/045. arXiv: 1405.7023 [gr-qc]
97. S.L. Zoutendijk et al., The MUSE-faint survey: I. Spectroscopic evidence for a star cluster in Eridanus 2 and constraints on MACHOs as a constituent of dark matter. Astron. Astrophys. **635**, (2020) A107. ISSN: 1432-0746. https://doi.org/10.1051/0004-6361/201936155

Chapter 20
Positive Indications for Primordial Black Holes

F. Kühnel

Abstract Until recently, the focus of PBH research has mostly been either on formation or constraints. While being important, too little emphasis has been on the actual *detections* which could plausibly be allocated to PBHs. In fact, as we will see below, it can be said with increasing levels of confidence that those macroscopic dark matter bodies may have already been observed years ago. The present Chapter is about precisely this *Positivist Perspective*. It summarises a longer recent review article [41] by Bernard Carr, Sébastien Clesse, Juan Garía-Bellido, Michael Hawkins as well as the author, and focuses on the most important pieces of claimed evidence.

20.1 Introduction

The debate whether primordial black holes (PBHs) exist, and to which extent, is as long as their history. In fact, the very first article [160] on these compact objects claimed to exclude their existence, and has even stopped related research for years. It was only seven years later when Carr & Hawking [42] disproved the mentioned claim, which then reinvigorated PBH research. This historical turn is symptomatic for many instances in which the PBH abundance has been claimed to be strongly constrained, but later these limits got either substantially weakened or even entirely removed. For instance, it had once been claimed that femtolensing measurements exclude any notable PBH dark matter fraction in the asteroidal mass range [111]. However, some time later these constraints were reanalysed [103] and shown to disappeared entirely. Likewise, constraints originating from white-dwarf explosions had been claimed to seriously restrict the allowed abundance of similarly light PBHs [139], but a later reanalysis has completely disproved this [6]. As the third example of many, it had been claimed that microlensing with the Subaru hyper-suprime camera excluded notable

F. Kühnel (✉)
Max-Planck-Institut für Physik, Boltzmannstr. 8, 85748 Garching, Germany

Arnold Sommerfeld Center, Ludwig-Maximilians-Universität, Theresienstraße 37, 80333 München, Germany
e-mail: mail@floriankuehnel.net

C. Byrnes et al. (eds.), *Primordial Black Holes*, Springer Series in Astrophysics and Cosmology, https://doi.org/10.1007/978-981-97-8887-3_20

amounts PBH dark matter in a mass range including the window 10^{-13}–$10^{-10}\,M_{\odot}$ (see first arXiv version of Ref. [121]). This very range, however, has been completely reopened since the wavelength of the light used for gravitational lensing was (much) larger than the extent of the black holes in that interval. While not all constraint reanalyses found such trivial mistakes or unrealistic/incorrect assumptions, there is (almost) no constraint on the PBH abundance which is truly universal. All rely on assumptions, and these might not be (entirely) fulfilled. In some cases this only concerns exotic scenarios, but in most cases, it is relatively easy to evade constraints altogether; the most assumption-general constraint diagram would essentially be empty. On this note, it is now postulated that the semiclassial treatment of black hole evaporation becomes invalid at latest at the hole's half mass, possibly much earlier [60]. If confirmed, this results in a vastly prolonged lifetime and a decrease of the emission rate. While this so-called 'memory burden' effect is currently quantitatively being specified, it is clear that some of the strong bounds on light PBHs are seriously weakened and possibly disappear entirely [13, 147]—opening up an extended (and previously disregarded) PBH dark matter window below and around $10^{15}\,g$, which has been the most tightly constrained region. See chapter 19 for a summary of potential constraints.

This chapter, which is a condensed version of Ref. [41], is taking a different (*positivist*) approach by focussing entirely on reported observations for which a non-compact-body explanation is either unknown or would require substantial tuning or non-standard modifications. As spatial clustering of PBHs has a strong effect on numerous interpretations of astrophysical observations, we elaborate on its importance in the subsequent section. Then, in Sect. 20.3 we discuss numerous claimed positive evidences for PBHs originating from gravitational microlsensing observations, this being followed by those from various dynamical effects (Sect. 20.4) and gravitational waves (Sect. 20.5). Finally, in Sect. 20.6, we present a simple and natural scenario which could explain numerous of the reported pieces of claimed evidence for primordial black holes.

20.2 Clustering of PBHs

In order to interpret observations which might have been caused by PBHs, an in-depth understanding of their spatial distribution is essential. The present section is in turn devoted to the associated distinct effects of those macroscopic bodies. A first discussion has been put forward by Mészáros [113] and later many other authors studied this effect (see e.g. Refs. [35, 38, 48, 49, 68]). Additionally, the Coulomb effect of an individual PBH is also important (see Ref. [39] for a heuristic treatment). Below, we elaborate on the formation and dynamical evolution of PBH clusters.

There are two ways in which PBHs provide a source of fluctuations for objects of larger mass:

(1) via the so-called seed effect, associated with the Coulomb attraction of a *single* black hole (see e.g. Ref. [39]);
(2) via the so-called Poisson effect (see e.g. Ref. [113]), associated with the $\sqrt{N}$-fluctuation in the number of black holes.

Given a scale $\tilde{M}$ of the initial fluctuation in the matter density, and for a monochromatic distribution of PBHs of mass M, it can be shown [39] that the Poisson effect dominates for all $\tilde{M}$ if $f_{\text{PBH}} \sim 1$, while if $f_{\text{PBH}} \ll 1$, the Poisson effect dominates for $\tilde{M} > M/f_{\text{PBH}}$, and the seed effect for $\tilde{M} < M/f_{\text{PBH}}$. Indeed, the mass bound by a single seed can never exceed M/f_{PBH} because of the competition from other seeds. *We stress that these effects are inevitable.* The Poisson effect, for instance, just comes from the macroscopic nature of PBH dark matter, i.e. its discreteness.

The formation of Poisson-induced PBH clusters was studied numerically, using N-body simulations [92, 137], and analytically, using the Press–Schechter formalism and the theory of spherical collapse [9]. As an illustrative example, Fig. 20.1 shows the density distribution at redshift $z = 99$ obtained in Ref. [92] for $M = 30\,M_\odot$ and $f_{\text{PBH}} = 10^{-5}$ and 10^{-1}. As can be seen, the difference with the standard cosmological scenario, in which the Universe is still homogeneous at such a high redshift, is striking.

As pointed out by Kashlinsky [99], the contribution from Poisson fluctuations to the matter power spectrum becomes dominant on small scales. Stellar-mass PBHs lead to the high-redshift gravitational collapse of almost all small-scale perturbations into clusters with masses up to 10^6–$10^7\,M_\odot$. Figure 20.2 shows the (dimensionless) density perturbation $\Delta(k)$, being related to the power spectrum $P(k)$ through $\Delta(k)^2 = P(k)\,k^3/(2\pi^2)$, for the standard cold dark matter model with $M = 2.6\,M_\odot$ and $f_{\text{PBH}} = 1$. As can be observed, the Poisson term dominates for comoving wavemodes $k \gtrsim 100\,\text{Mpc}^{-1}$ at $z \approx 20$. For comparison, the expected spectrum for the standard cold dark matter scenario and for a primordial spectrum with a sharp

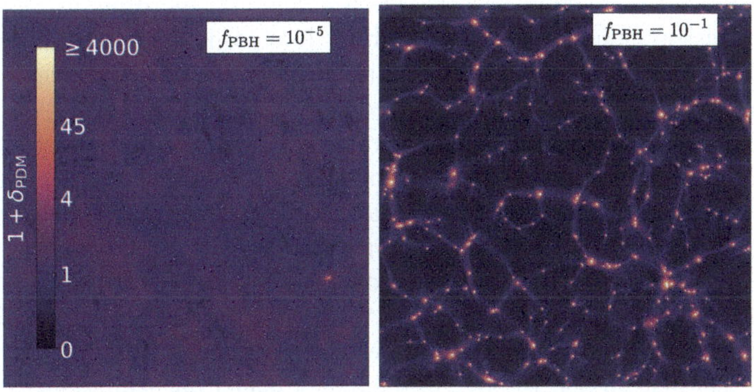

Fig. 20.1 Expected dark matter density distribution over a scale of $2\,h^{-1}$ kpc at redshift $z = 99$ as obtained from N-body simulations of Ref. [92] for $M = 30\,M_\odot$ and $f_{\text{PBH}} = 10^{-5}$ (*left*) or $f_{\text{PBH}} = 0.1$ (*right*)

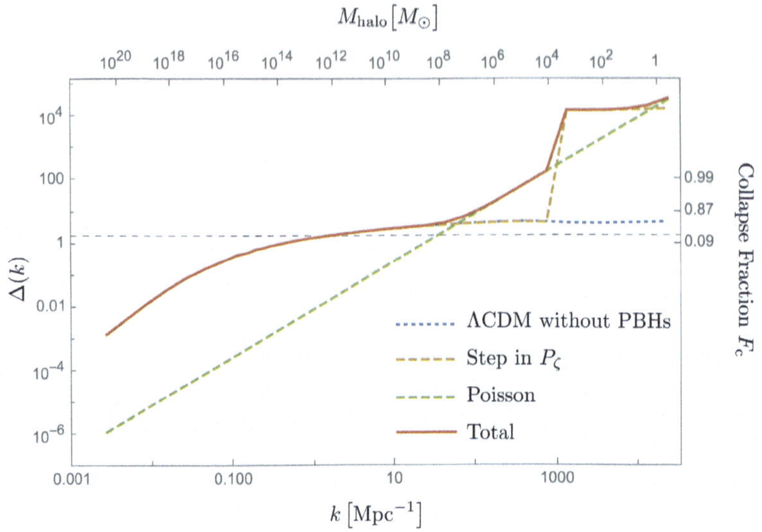

Fig. 20.2 Present dimensionless (linear) density contrast $\Delta(k)$ in the standard cold dark matter model without (dotted blue line) and with PBHs for $f_{\rm PBH} = 1$ at $M = 2.6\,M_\odot$ (solid red line), including the inevitable Poisson term in the matter power spectrum (dashed green line). For comparison, the possible power spectrum enhancement leading to PBH formation for a transition scale at $k_{\rm trans} = 10^3\,{\rm Mpc}^{-1}$ (dashed orange line) is also shown. The horizontal dashed line indicates the critical threshold for halo formation $\delta_{\rm clust} \simeq 1.686$. The upper x-axis shows the implied halo mass. The right y-axis indicates the fraction of collapsed halos, $F_c(M_c)$, being almost 1 for $k \gtrsim 100\,{\rm Mpc}^{-1}$, meaning that all PBHs are regrouped in clusters of masses 10^6–$10^7\,M_\odot$. From Ref. [41] and adapted

enhancement at a scale $k_{\rm trans} = 10^3\,{\rm Mpc}^{-1}$ to generate PBHs, is also shown. The mentioned effect modifies the spectrum below the scale at which the Poisson effect becomes dominant, but even for lower values of M or $f_{\rm PBH}$, one may have to take into account the effect of the enhanced primordial power spectrum.

Once Poisson-induced PBH clusters have formed, it takes some time for the PBH velocities and the cluster size to virialise. Afterwards, the clusters continue to expand due to dynamical heating and eventually, if they are initially sufficiently compact and dense, will completely evaporate through dynamical heating. Of course, a cluster cannot be considered in isolation. So as it expands, it is subsumed into progressively larger structures and eventually loses its integrity. Clusters larger than this minimum mass may still lose their integrity due to collisions within a larger cluster. This, however, can lead to their disruption only if the velocity dispersion within clusters increases with their mass. As long as the Poisson effect dominates, it is marginal whether the hierarchy of clustering can be destroyed. However, on mass scales greater than around $10^8\,M_\odot$, the cold dark matter fluctuations take over (see Fig. 20.2). This means that collisional disruption could occur, at least in the central region of the larger bound region.

In order to illustrate these different processes and to infer the PBH cluster mass-size relation today, Fig. 20.3 shows as broken black lines the initial cluster size, and

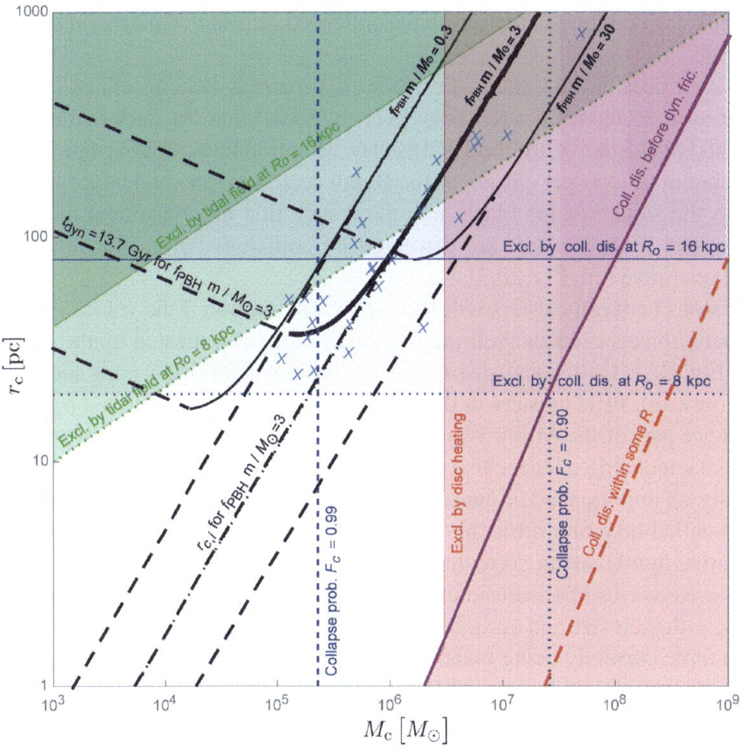

Fig. 20.3 Regions in the (M_c, r_c) plane of PBH clusters being induced by Poisson fluctuations being either excluded or favoured by (1) Galactic disc heating (red region); (2) disruption by the Galactic tidal field (green region) for a Galactocentric distance of 8 kpc (dotted line) and 16 kpc (solid line); (3) cluster collisions (blue lines); (4) dynamical friction (purple line); (5) dynamical heating (upper dashed black lines). The expected size of clusters in a scenario with $f_{\rm PBH}\, M = 3\, M_\odot$ is shown by the thick solid black line which takes into account the dilution of sub-clusters inside more massive host clusters. For comparison, the cases $f_{\rm PBH}\, M = 0.3\, M_\odot$ and $30\, M_\odot$ are also depicted (thin black lines). The lower dashed black lines indicates the initial cluster size. The vertical blue lines represent the cluster mass for which there is a probability of collapse into halos $F_c = 0.99$ and $F_c = 0.90$ (dashed and dotted lines, respectively) for Poisson fluctuations in the Press–Schechter formalism. The crosses indicate actual observations of ultra-faint dwarf galaxies, with $r_c \approx r_{1/2}$ and M_c computed from the virial theorem, using the best-fit values of the half-light-radius $r_{1/2}$ and velocity dispersion. From Ref. [41]

the region where dynamical expansion occurs within the age of the Universe, exemplifying the cases $f_{\rm PBH}\, M/M_\odot = 0.3, 3$ (relevant for the *thermal-history* model [40]) and 30 (see Sect. 20.6). The increase in size with mass approximately follows the distribution of observed ultra-faint dwarf galaxies, whose most likely half-light radius and mass are represented by crosses, indicating that they form from Poisson rather than cold dark matter fluctuations. The figure also depicts the cluster masses for which the collapse fraction $F_c(M_c) = 0.9$ and 0.99 (vertical blue lines), showing that almost all Poisson fluctuations have collapsed into such clusters on the relevant mass scale.

It is well possible that halos are made of supermassive clusters of black holes of more modest mass. Carr & Lacey [37] suggested that the dark matter could be constituted of clusters of around $10^6\,M_\odot$. The accretion luminosity is then reduced by the number of objects per cluster, as the Bondi accretion rate (and hence luminosity) scales as the square of the black hole mass such that the dynamical limits can be circumvented if the clusters are disrupted by collisions before dynamical effects become operative (c.f. Fig. 20.3).

Collisions between clusters will lead to their disruption if the velocity dispersion of the halo objects exceeds their internal dispersion, as indicated by the broken red line in Fig. 20.3. Collisional disruption is constant within the core and increases further outward. PBH clusters in this case would have been disrupted by collisions in the inner part of the Milky Way halo, but they nevertheless remain stable at the larger Galactocentric distance that is typical of ultra-faint dwarf galaxies.

Finally, an important additional component to the initial PBH clustering originates from so-called quantum diffusion [14, 20, 65, 66, 130, 153], which induces non-perturbative non-Gaußian corrections to the matter-density distribution. This leads to successive waves of enhanced curvature fluctuations inevitably generating dense clusters, with their size and mass strongly depending on the form of the non-Gaußian tails—a topic currently being under investigation. In particular, plateau-type potentials like those implied by critical Higgs inflation [64, 72] naturally generate such tails. These exponential tails also have important consequences for the early formation of massive structures, like galaxies and clusters at high redshift [66]. Although primordial non-Gaußianity at CMB scales is usually characterised by local perturbative parameters (such as $f_{\rm NL}$ or $g_{\rm NL}$), the non-Gaußian exponential tails arising from quantum diffusion during inflation cannot be described perturbatively. They are always positively squeezed with a large kurtosis and higher-order cumulants, enhancing the probability of collapse to PBHs [66].

20.3 Claimed Lensing Evidence

Gravitational lensing by PBHs has a long history (see e.g. Refs.[11, 12, 45, 71, 76, 83–89, 128, 134, 140]). Here, one has to distinguish four different effects (see Ref. [41]):

(1) multiple images of quasars produced by massive objects such as galaxies (macrolensing);
(2) brightness variations in isolated quasars induced by much less massive objects, typically in the stellar mass range (microlensing), including microlensing by an intergalactic population of compact bodies, or individual stars or halo objects within galaxies;

(3) microlensing by such objects of the images of a macrolensed quasar;
(4) microlensing of individual nearby stars by other compact bodies.

A great piece of observational support for the PBH dark matter hypothesis has already been provided by the MACHO collaboration [11, 12]. This has detected around 15 microlensing events (depending on the details of the selection criteria), being far larger than can be attributed to known stellar populations in the Milky Way or Magellanic Clouds [11]. At present, the only plausible candidates are PBHs [86], these making up around 40% of the dark matter in the Milky Way halo [86]. It is important to note that an extended PBH mass function with a peak at around a few solar masses, as being the case in the *Thermal History* model [40], beside implying an inhomogeneous population of PBHs (e.g. in the Galactic Center, dwarf spheroidals), could well yield the totality of the dark matter.

Another approach for the detection of PBHs in the Milky Way concerns the monitoring of stars in the Galactic bulge by the OGLE collaboration [158]. These observations focus on the detection of lensing events involving white dwarfs, neutron stars and black holes, using *Gaia* parallax measurements and brightness changes to estimate the mass of the lenses. This approach has led to the detection of a small sample of lenses within the mass range [2, 5] $M_\odot$, where this gap may be best explained by a population of PBHs. Figure 20.4 indicates the mass distribution of these OGLE/*Gaia* events.

Niikura et al. [123] have used data from the five-year OGLE survey of 2622 microlensing events in the Galactic bulge [119]. They have identified six ultra-short ones attributable to planetary-mass objects between 10^{-6} and 10^{-4} $M_\odot$. These would contribute about 1% of the cold dark matter, being more than expected for free-floating planets [150]. So, the most plausible explanation of these events involves PBHs. Also, there is no continuity of detections between these outliers and events associated with low-mass stars or brown dwarfs. Figure 20.5 shows the distribution of these events as a function of timescale (left panel) and also the values of $f_{\text{PBH}}(M)$ required to explain the data for a monochromatic mass function (blue band of right panel).

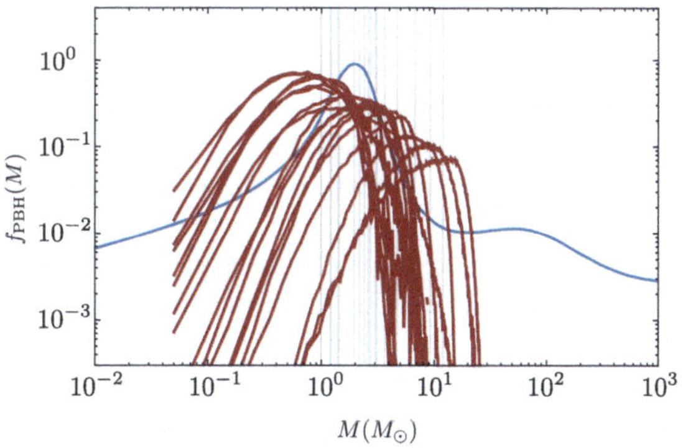

Fig. 20.4 Mass distribution of OGLE/*Gaia* events (red curves) overlaid onto the model of Ref. [40]. The mode of the distribution for each microlensing event is shown by grey vertical lines. From Ref. [73]

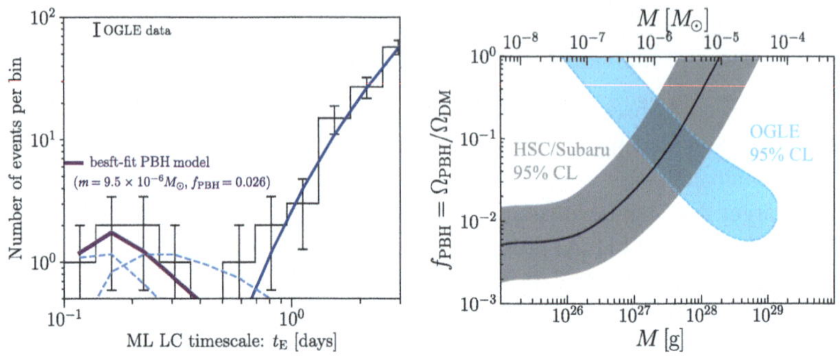

Fig. 20.5 *Left panel*: Number of observed microlensing events as a function of lightcurve (LC) timescale in days from Ref. [123]. The bold-purple solid line indicates the best-fit PBH model (with $M = 9.5 \times 10^{-6}\, M_\odot$ and $f_{\rm PBH} = 0.026$) for explaining the distribution of six ultrashort-timescale OGLE events. *Right panel*: 95% confidence level (CL) region for PBH abundance (blue-shaded region), assuming that the six observed ultrashort-timescale events in the OGLE data are due to PBHs with a monochromatic mass function. Also shown (grey band) is the 95% CL prediction for the abundance inferred from the HSC/Subaru observations. From Refs. [122, 123]

The search for microlensing of stars in the Andromeda galaxy is far more challenging because the microlensed star is unresolved and just one of many stars contributing to each CCD pixel. One must therefore use the so-called pixel-lensing technique, which involves subtracting a reference image from a target image taken at a different epoch. Several groups have attempted this. Notably, the POINT-AGAPE collaboration has detected six microlensing events [31], which has been argued to be more than could be expected from self-lensing alone. The conclusion is that is that 20% of the halo mass in the direction of Andromeda could be PBHs in the range 0.5–1 $M_\odot$.

Furthermore, Niikura et al. [121] have carried out a dense-cadence, seven-hour observation of Andromeda with the Subaru Hyper Suprime-Camera (HSC), and then used the pixel-lensing technique to search for microlensing of stars by PBHs lying in the halo of the Milky Way or Andromeda. The pointing was towards the central region of Andromeda and the observation of a single microlensing event by a compact body with mass in the range 10^{-11}–$10^{-5}\, M_\odot$ was reported. The associated light-curve is shown in Figs. 20.6, and 20.7 depicts the expected event rate for various PBH masses as a function of the full-width-at-half-maximum timescale $t_{\rm FWHM}$. Translating this into estimates for $f_{\rm PBH}$ yields the grey band in the right panel of Fig. 20.5.

Perhaps the most convincing detections of stellar-mass PBHs are made using an analysis of microlensing events in the light curves of multiply-lensed quasar systems. The first detection of such a gravitational lens [155] has revealed the quasar Q0957+561 split into two separate images by a massive galaxy along the line of sight. A photometric monitoring programme then showed that small brightness changes in one image were repeated in the second image around a year later. This has confirmed the identification of the system as a gravitational lens. However, it was only some time later when evidence as claimed that these microlenses were part of a cosmological

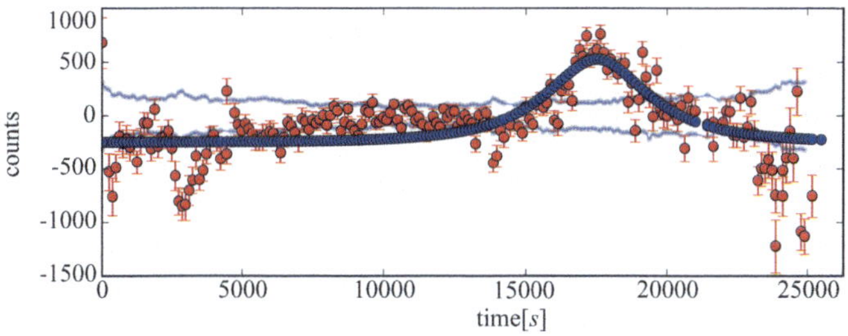

Fig. 20.6 Best-fit microlensing model (blue curve) for the light curve of the single pixel-lensing event reported by Niikura et al. [121]. Error bars denote photometric errors in the brightness measurement in the image difference at each epoch. From Ref. [121]

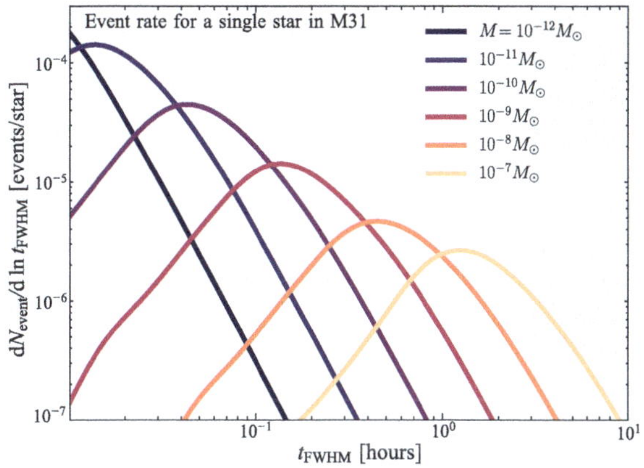

Fig. 20.7 The expected differential number of microlensing events per logarithmic interval of the full-width-at-half-maximum timescale t_{FWHM} for a single star in Andromeda. Solid lines correspond to a monochromatic PBH mass function with $f_{PBH} = 1$. The unshaded band is where the utilised data has the highest sensitivity to the lightcurves ($t_{FWHM} \simeq 0.07$–3 h). From Ref. [121]

distribution of stellar-mass compact bodies making up the dark matter [85, 86, 89]. The overall conclusion is that the probability that the observed microlensing of the quasar images can be attributed to stars is less than 10^{-4}.

Hawkins [86] has provided solid arguments why these bodies can only be PBHs making up a significant fraction of the dark matter. First of all, from combined measurements of light element abundances and the photon density, yielding an upper limit on the baryon density parameter $\Omega_b < 0.03$ [143], the bodies must be *non-baryonic*. Secondly, in order to serve as a gravitational microlense, they must be *compact enough* and in the *right mass range*. At least one of these characteristics is not shared by the suggested non-PBH compact dark matter candidates, including quark nuggets [10], their generalisations [94], as well as so-called *N*-MACHOs [59], since these are too light. Boson stars [69] constitute another class of macroscopic dark matter candidates but they are insufficiently compact (see Ref. [86] for a more detailed discussion).

20.4 Claimed Dynamical and Accretion Evidence

In this section we consider a number signatures related to dynamical and accretion effects of PBHs (see Ref. [41] for a more extensive discussion). Here, we will focus on four conundra to which PBHs could conceivably provide an explantation: *(i)* the cusp/core problem, *(ii)* the origin of the SMBH seeds, *(iii)* the cause for exploding white dwarfs, *(iv)* the source of X-ray/infrared background correlations.

Numerous dynamical effects were discussed in Ref. [43] (c.f. also Ref. [141]), these being associated with the heating of galactic disks [107], effects on dwarf galaxies [105], triggering of white dwarf explosions [78], disruption of wide binaries [19, 117, 156], globular [118] and star clusters [28], capture of PBHs by white dwarfs and neutron stars [33, 129], dynamical friction effect of spheroid stars [43], tidal distortions in galaxies [149] and the CMB dipole anisotropy [43]. Many of them involve heating or destruction of astronomical systems by the passage of nearby PBHs.

Cusp/Core Problem. Boldrini et al. [25] have used N-body simulations to explore a galactic dark matter halo comprising both PBHs and particles. As a specific example, they consider $10^7\,M_\odot$ dwarf galaxies with PBHs of mass between 25 and $100\,M_\odot$. They find that cores can form as a result of dynamical heating of the cold dark matter through PBH infall and two-body processes, provided the PBHs constitute at least 1% of the non-baryonic halo. Here, the halo shape depends upon the PBH mass distribution, which is subject to mass segregation, PBH mergers, possible slingshotting of light PBHs, and PBH interactions with the central black hole. This may resolve the cusp/core problem.

The observed rotation curve of the Milky Way indicates the presence of a supermassive central black hole of $4 \times 10^6\,M_\odot$ and a ring of dark matter at a distance of ~ 100 pc with the usual r^{-2} fall-off at larger distances. These features are predicted by detailed numerical simulations in the central region of an all-PBH halo [148]. Because the black holes scatter off each other and the central SMBH, they accumulate in a ring around the centre [74] with the density profile acquiring a hill at a distance of ~ 100 pc for a galaxy similar to our own, and of ~ 10 pc for dwarf galaxies. This configuration may be transitory but it currently seems universal for any sizes of galaxies and their halos. The profile looks like a core at large distances, so one needs precise observations of rotation curves in the inner parts of galaxies to test the model [32]. Note that several IMBHs have been observed in the central molecular zone around the Galactic centre [146]. These could be a subpopulation of massive PBHs that have migrated towards the centre due to dynamical friction.

PBHs as SMBH Seeds. The possible rôle of PBHs as seeds for the SMBHs in galactic nuclei and early structures has been highlighted by recent JWST results, as reviewed by Dolgov [57]. This shows that such structures form much earlier than expected in the standard cold dark matter scenario. For example, there is a population of galaxies with mass above $10^{10}\,M_\odot$ at redshift in the range [7.4, 9.1] with central SMBHs [106]. The *Atacama Large Millimeter/submillimeter Array* has observed an active galactic nucleus at redshift $z \sim 7$ whose large luminosity suggests the presence of a $10^9\,M_\odot$ black hole [61]. The detection of X-rays from a $4 \times 10^7\,M_\odot$ SMBH in a gravitationally-lensed galaxy at $z = 10.3$ has recently been reported [24], and it has been emphasised [57] that such observations are entirely consistent with population of PBH. Liu & Bromm [109] have also argued that unusually massive galaxies at $z > 10$ could be evidence that structure formation is accelerated by $10^9\,M_\odot$ PBHs making up a dark matter fraction of 10^{-6}–10^{-3}. Furthermore, Goulding et al. [77] suggest the presence of a heavy seeding channel for the formation of SMBHs within the first billion years of cosmic evolution. An interesting feature of these observations is that the central-black-hole-to-stellar-mass ratio is several orders of magnitude larger than observed locally, as expected if the black hole is acting as a seed. We caution the reader that the field of high-redshift observations is rapidly evolving, with novel results and their analyses/interpretations are continuously being published at the time of writing. See also Chaps. 9 and 11.

Exploding White Dwarfs. Primordial black holes can trigger explosions of white dwarfs, leading to potentially observable signatures [7, 8, 27, 44, 44, 67, 79, 142]). It is therefore exciting that some recently observed supernovæ, so-called *calcium-rich transients* [80, 135], do not trace the stellar density but are located much off-centre from their host galaxies. Furthermore, they appear to originate from white dwarfs with masses of around $0.6\,M_\odot$ [112], well below the Chandrasekhar limit, predominantly occurring in old systems. It has been argued [142] that these transient events could have been triggered by collisions with asteroidal-mass PBHs; the associated event rate being within reach of current or near-future microlensing surveys [122, 144]. The left panel of Fig. 20.8 depicts the cumulative calcium-rich-transient supernova event rate, normalised to the half-light radii of the host galaxies, as determined in Ref. [135]. As can clearly be seen, this is distinct from the corresponding results for standard type-Ia supernovæ, whose locations closely follow the stellar distribution. In contrast, the distribution of calcium-rich-transient events follows remarkably well that expected from compact dark matter/white dwarf interactions. This, together with the unique chemical properties and atypical progenitors of the calcium-rich transients, strongly supports the hypothesis that these events originate from such interactions.

The modelling of Ref. [142] suggests that PBHs with $10^{21}\,\text{g} < M < 10^{23}\,\text{g}$ and $10^{-3} < f_{\text{PBH}} < 0.1$ are plausible triggers for these events (see right panel of Fig. 20.8). Future observations will further explore this exciting possibility. Interestingly, the *Zwicky Transient Facility*, which is the largest ongoing systematic related survey, classified eight new events within its first 16 months of operation [54]. Soon the *Large Synoptic Survey Telescope* (LSST)—as part of the *Vera Rubin Observatory*—will increase the search volume by at least two orders of magnitude. Furthermore, observations with JWST should be able to resolve dwarf spheroidal galaxies for the current sample of calcium-rich transients [54], these two populations being very closely connected.

PBH Accretion: X-Ray/Infrared Backgrounds. The most compelling claimed evidence for PBHs is associated with the spatial coherence of the source-subtracted cosmic infrared and X-ray backgrounds, as shown by Kashlinsky and his collaborators [99, 100, 102, 120]. This undoubtedly requires black holes. Although these need not be primordial, the level of the infrared background suggests an overabundance

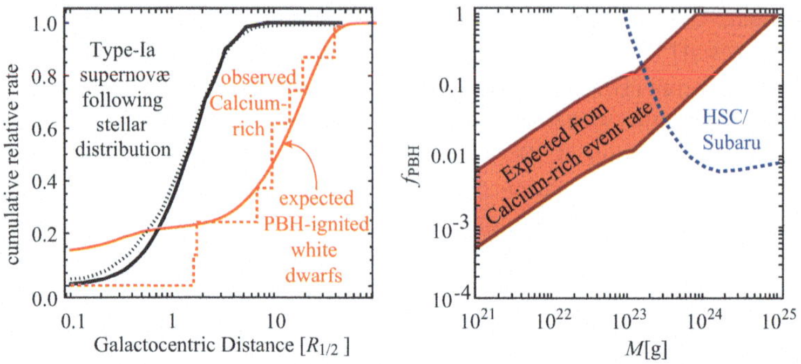

Fig. 20.8 *Left panel*: Radial distribution of galactic stars (black solid) and the expected PBH/white dwarf interactions (red solid) compared to the distributions of standard type-Ia supernovæ (black dashed) and calcium-rich transients (red dashed) within their host galaxies. *Right panel*: Primordial black hole dark matter fraction (red band) as a function of mass required to produce the observed relevant supernova rate. is the 95% CL exclusion region from stellar microlensing by the Subaru HSC [122, 144]. Figures adapted from Ref. [142]

of high-redshift halos and this could be explained by the Poisson effect discussed above if a significant fraction of the cold dark matter comprises solar-mass PBHs. In these halos, a few stars form and emit infrared radiation, while the PBHs emit X-rays due to accretion. It is challenging to find other scenarios that naturally produce such features. PBHs naturally explain both the amplitude and angular spectrum of the source-subtracted infrared anisotropies (as found by *Spitzer*). The infrared/X-ray coherence is a consistency argument but the infrared connection would apply even if there was no coherence (i.e. even if PBHs did not efficiently accrete). Kashlinksy also invokes cosmological advection at $z > 100$ in order to speed up the formation of early compact objects [101].

Hasinger [82] has analysed their cross-correlation using deep *Chandra* and *Spitzer* survey data [120], and estimated the contribution from PBH baryon accretion and stars to the cosmic X-ray and infrared backgrounds. A population of 10^{-8}–$10^{10}\,M_{\odot}$ PBHs has been argued (assuming Bondi capture and advection-dominated disc accretion) to be consistent with the residual X-ray fluctuation signal, peaking at redshifts $z \approx 17$–30. Furthermore, it has been argued that PBHs could have an important bearing on a number of other phenomena: (1) amplifying primordial magnetic fields; (2) modifying the reionisation history of the Universe (while being consistent with recent *Planck* measurements [132]); (3) impacting on X-ray heating, thereby providing a contribution to the entropy floor observed in groups of galaxies [133]; (4) certain 21-cm absorption-line features [26] which could be connected to radio emission from PBHs.

In a recent related study, Cappelutti et al. [34] have explored the high-redshift properties of PBH dark matter having a mass spectrum induced by the thermal history of the Universe (see Ref. [40]). Their main findings are: (1) a secondary peak of star formation at $z \sim 15$–20 (beyond the well-established observed peak at $z \sim 3$),

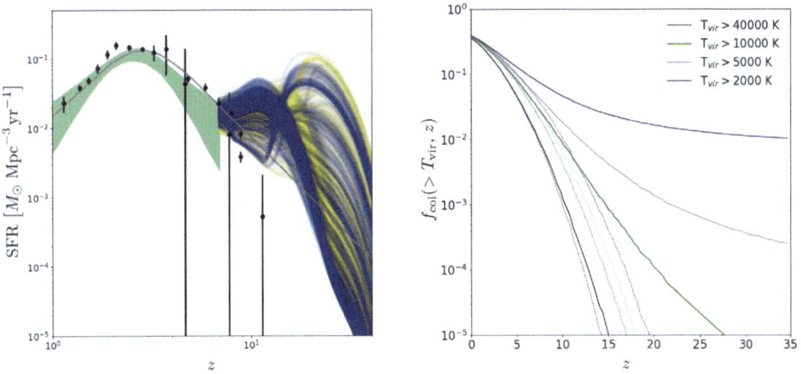

Fig. 20.9 *Left panel*: Star formation rate (SFR) density versus redshift for the *thermal-history* model [40] (see Sect. 20.6). The green band indicates compatibility with measurements from extra-galactic background light and high-redshift surveys. The grey continuous line indicates the best fit to the data. *Right panel*: Fraction of collapsed halos with mass exceeding M as a function of redshift for various virilisation temperatures $T_{\rm vir}$. Results are for both particle dark matter (dotted lines) and PBH dark matter (solid lines). From Ref. [34]

being driven by mini halos, which are likely to host the first episode of Population III star formation (c.f. Fig. 20.9); (2) a strong enhancement of the X-ray background fluctuations and the unresolved cosmic X-ray and infrared-background cross-power spectrum, with only minor effect on the cosmic infrared-background fluctuations; (3) an explanation of the integrated $1\,{\rm nW\,m^{-2}\,sr^{-1}}$ measured signal [90, 102] (which cannot be accounted for by non-PBH cosmologies) with $M > 1\,M_\odot$ PBHs; (4) the X-ray spectral-energy distribution of the cosmic X-ray- and infrared-background cross-correlation signal contains information about their production mechanism (c.f. Ref. [108]).

20.5 Claimed Gravitational-Wave Evidence

The first direct detections of gravitational waves (GWs) were in 2015 [1] and originated from the merging of black holes which were unexpectedly large (around $30\,M_\odot$) compared to those in X-ray binaries. This triggered an avalanche of interest in PBHs, not only as a source of the mergers but also as a possible solution to the dark matter problem. Eight years later, around 90 detections could have been collected in the third *Gravitational-Wave Transient Catalog* (GWTC-3) [138], and the properties of the gravitational-wave sources are still intriguing. Remarkably, *thermal-history* scenario [40] (see Sect. 20.6) is capable of explaining the merger rates, the black hole mass distribution, the existence of events in the high- and low-mass gaps for stellar remnants, the mergers with highly asymmetric masses, the reported low spins,

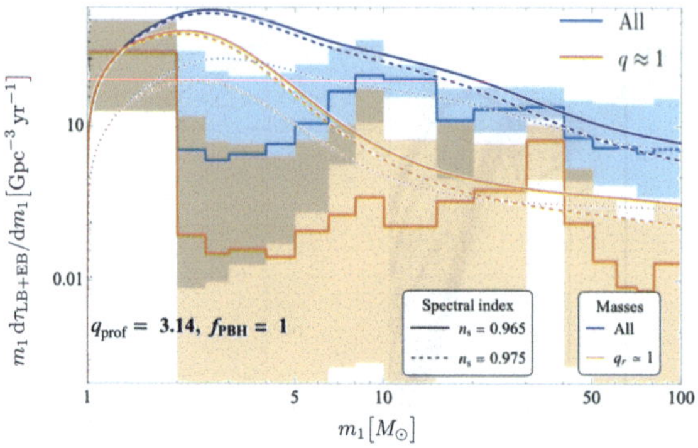

Fig. 20.10 Reconstructed rate distribution with 90% confidence intervals as a function of mass m_1 for equal-mass (orange band) and non-equal mass (blue band) mergers, using GWTC-3 data, for the so-called binned Gaußian process model with uniform ($\log m_1$, $\log m_2$) bins. Superimposed are the typical mass distributions predicted by the simulations of Ref. [62]. These include QCD effects and assume Gaußian fluctuations on PBH scales with a nearly scale-invariant power spectrum having a spectral index $n_s = 0.965$ (solid) and $n_s = 0.975$ (dotted), and $f_{\mathrm{PBH}} = 1$ for the combined merger rates of early and late binaries. The parameter q_{prof} describes the density profile. From Ref. [4]

and some subsolar black hole candidates. Figure 20.10 shows the reconstructed rate distribution confronted with predictions from the mentioned unifying scenario.

Origin of Mergers. One month after the first gravitational-wave detection, Bird et al. [21] and Clesse & García-Bellido [51] claimed that the expected PBH merging rates of binaries formed at late times in compact halos is compatible with the LVK analysis. This was later supported by other studies (see e.g. Ref. [22]).

Raidal et al. [136, 152] investigated the evolution and merging of early PBH binaries in more detail using N-body simulations, and found that the rates are highly suppressed if PBHs significantly contribute to the dark matter. We have to stress that the LVK constraints on f_{PBH} are always obtained under assumptions on their merger rates, some of which become invalid for broad mass distributions. Furthermore, for $f_{\mathrm{PBH}} > 0.1$, the rate of early PBH binaries can be heavily suppressed due to perturbations by light PBHs as well as due to matter inhomogeneities and Poisson clustering (see Sect. 20.2). In fact, late binaries formed in clusters may even dominate the merger rate (although see Chap. 25 for a less positive but more mainstream view). It may even remain possible that all LVK mergers are primordial in origin. Note, however, that this matter is currently undergoing dramatic progress and different conclusions had previously been reached.

Theoretical Model. Fig. 20.11 shows a theoretical prediction of the distribution for detections in the *thermal-history* model [40], improved by recent simulations of PBH formation at the QCD epoch [62]. As can be seen, gravitational-wave events are especially predicted between 30 $M_\odot$ and 60 $M_\odot$, which is where most observations lie. The second most likely region is around a solar mass, as expected for PBH formation at the QCD peak, since the horizon mass at that time is around this scale. For early binaries, the event distribution would have a dominant peak around solar

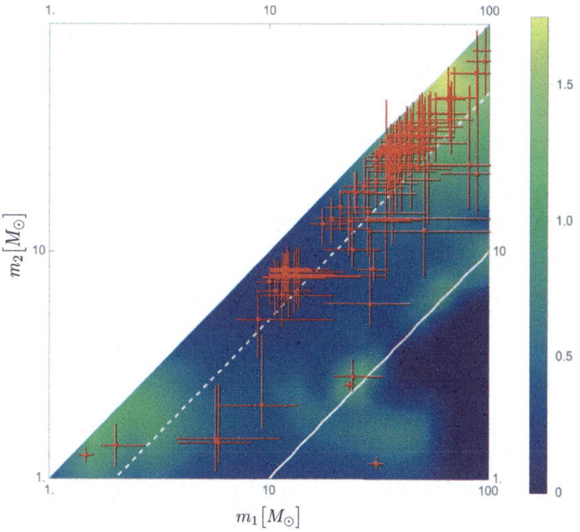

Fig. 20.11 Expected probability distribution of PBH mergers with masses m_1 and m_2 for the *thermal-history* mass function (see subsequent section) with $n_s = 0.97$, power-spectrum amplitude adjusted to yield $f_{\rm PBH} = 1$, and using the LIGO sensitivity of the third observational run. Solid and dashed white lines correspond to mass ratios $q \equiv m_2/m_1$ of 0.1 and 0.5, respectively. From Ref. [62]

mass, making it is less likely that all black holes between $20\,M_\odot$ and $100\,M_\odot$ were in early binaries. Some regions are not occupied by stellar black hole mergers but are a characteristic prediction of PBH models. These regions correspond to the pair-instability mass gap (above $60\,M_\odot$), the subsolar region, the low-mass gap (between ~ 2.5 and $\sim 5\,M_\odot$) as well as binaries with low mass ratios.

Black Holes in the Upper Mass Gap. There are numerous observations of compact objects in these regions, and GWTC-3 indicates that these exhibit solely smooth transitions in merger rates [4]. This is rather hard to explain with a stellar model but entirely natural for PBHs. The same holds true for black holes with masses larger than approximately $60\,M_\odot$ for the following reason: Above this mass, the temperature in the core of stars becomes so high during oxygen-burning that electron-positron pair production leads to a reduction of the pressure and core collapse. Below a mass of around $150\,M_\odot$, such a heavy star explodes as a remnantless supernova. Hence, stars are not expected to directly form black holes with mass between about $60\,M_\odot$ and $150\,M_\odot$. This pair-instability mass gap is well established theoretically, even if the exact range depends on various factor such as the carbon abundance at the end of helium burning, the initial angular momentum and the mass loss, rotational mass shedding, magnetic torques, disk formation during the collapse, and can be debated to a some degree (see Ref. [157]). With component masses of $66\,M_\odot$ and $85\,M_\odot$, GW190521 is the first clear observation of black holes in this mass gap. However, GWTC-3 contains several additional candidates (GW190403_051519, GW190426_190642, GW190519_153544, GW190602_175927, GW190706_222661, GW190929_012149, GW191109_010717, GW200220_061928) with components in the range from $61\,M_\odot$ to $107\,M_\odot$. A more detailed analysis of the merger rates provides statistical evidence for a smooth transition across the mass gap, showing no break [4].

Table 20.1 Reported properties (date, component masses and spins) of the subsolar triggers in Refs. [5, 131], with a (network) signal-to-noise ratio (SNR) larger than 8 and false-alarm rate (FAR) smaller then 2 yr^{-1} in the second and third observing runs of LIGO/Virgo

Date	FAR [yr^{-1}]	$m_1[M_\odot]$	$m_2[M_\odot]$	spin-1-z	spin-2-z	H SNR	L SNR	V SNR	Network SNR
2017-04-01	0.41	4.90	0.78	−0.05	−0.05	6.32	5.94	–	8.67
2017-03-08	1.21	2.26	0.70	−0.04	−0.04	6.32	5.74	–	8.54
2020-03-08	0.20	0.78	0.23	0.57	0.02	6.31	6.28	–	8.90
2019-11-30	1.37	0.40	0.24	0.10	−0.05	6.57	5.31	5.81	10.25
2020-02-03	1.56	1.52	0.37	0.49	0.10	6.74	6.10	–	9.10

Asymmetric Mergers. GWTC-3 contains information about a few additional coalescences with low mass ratios ($q < 0.25$): GW200210_092254, GW200115_042309 as well as GW191219_163121. For the latter, the secondary mass is around 1.2 $M_\odot$ and $q \simeq 0.04$, which would make it the lowest-mass neutron star ever observed with gravitational waves. The mass properties of these gravitational-wave events are shown in Table 20.1. As depicted in Fig. 20.11, one can obtain a broad range of low-mass-ratio mergers for a wide PBH mass function. Future observations can be expected to soon provide additional information on the abundance of such asymmetric binaries and their merger rates. For PBHs, one cannot exclude LVK observing even lower mass ratios (even down to $q \simeq 0.001$) using continuous-wave methods [115, 116] but specific waveforms must be assumed for such searches. For broad PBH mass distributions, asymmetric binaries also boost the stochastic gravitational-wave background to a future-observable level [18].

High Redshifts. Given the broad variety of PBH and stellar black hole scenarios, such models could accommodate very different distributions of mass, rate and spin. Therefore, it is crucial to identify a clear signature of PBHs which cannot be explained by black holes of stellar origin. There are two observations that undoubtedly point towards a PBH origin: The first is detecting a black hole merger from a redshift before the epoch of first star formation (viz. $z > 20$) [46]. For chirp masses above or around 30 $M_\odot$, such redshifts will become accessible with the third generation of ground-based gravitational-wave detectors, such as *Einstein Telescope* [110] and *Cosmic Explorer* [63]. Their expected redshift range for different values of the signal-to-noise ratio is depicted in Fig. 20.12. As indicated therein, intermediate-mass or supermassive black holes will be probed by space interferometers like LISA [17] and pulsar timing arrays.

Black Holes in the Lower Mass Gap. The second PBH-unique signature would be the detection of black holes smaller than 1 $M_\odot$. Detecting such a hole can already be achieved with the current generation of gravitational-wave detectors. The first subsolar LIGO/Virgo searches had not been successful [2, 3] but only focused on binaries with a primary component smaller than 2 $M_\odot$. Similar searches were performed independently [124, 126] (see also Refs. [125, 127] and Chap. 18).

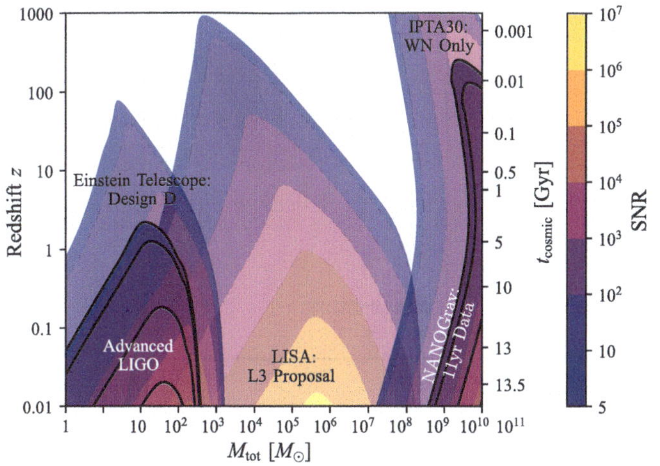

Fig. 20.12 Signal-to-Noise Ratio (SNR) as a function of source redshift and total source mass for various gravitational-wave detectors: *Einstein Telescope* (design model D), advanced LIGO, LISA (L3 proposal), IPTA30 (WN only) and NANOGrav (11yr). From Ref. [98]

Subsolar Analyses. A re-analysis of data from the second LIGO run, including binaries with a $2\,M_\odot < m_1 < 10\,M_\odot$ primary and a subsolar secondary component, has revealed four subsolar black hole candidates with high signal-to-noise ratio (> 8) and low false-alarm-rate (< 2 per year [131])—i.e. at the usual LVK thresholds for detections. Essential properties of these candidates are given in Table 20.1. Recently, three additional subsolar triggers have been reported [5] based on the same thresholds. Their properties are also shown in Table 20.1. Assuming additional tidal deformability tests exclude other known compact bodies, a primordial origin of subsolar-mass compact objects would undoubtedly be the most plausible interpretation, as all other proposed exotic compact objects are clearly less natural than PBHs.

Black Hole Spin. Another unexpected property of the first detected coalescences is their low effective spins, which is often compatible with zero. Low spins for black hole binaries are not expected in most stellar models, since these should produce a wide effective spin distribution with a peak around $\chi_{\rm eff} \approx 0.8$. On the contrary, they can be compatible with PBHs in the standard formation scenario where they form from inflationary overdensities and low spin is a generic outcome (see e.g. Ref. [56]). Furthermore, coherent accretion, previous mergers and hyperbolic encounters in dense PBH environments are expected to broaden the spin distribution [95]. In particular, for coalescences in the pair-instability mass gap, this would result in a distribution of $\chi_{\rm eff}$ centred on zero. There is some evidence for non-zero $\chi_{\rm eff}$ in the most recent gravitational-wave catalogue.

Pulsar Timing Arrays. The NANOGrav collaboration has recently reported the detection of a gravitational-wave background at nano-Hertz frequency [16] using pulsar timing arrays (PTAs). This has been confirmed by the *European Pulsar Timing Array* (EPTA) [47], Parkes [75] and *International Pulsar Timing Array* (IPTA) [15]

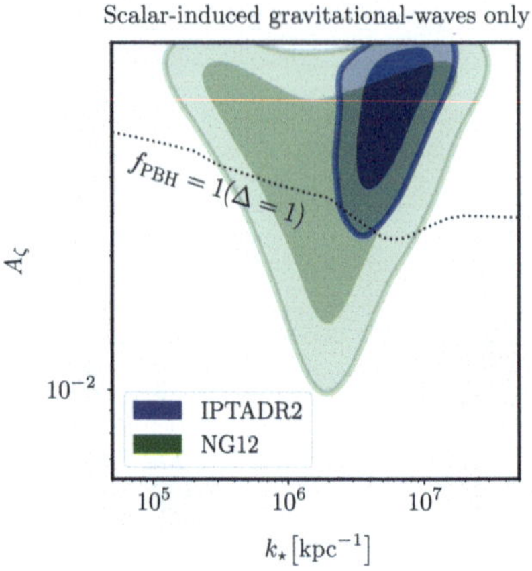

Fig. 20.13 Two-dimensional posterior distributions (1σ and 2σ contours) of the primordial power-spectrum amplitude A_ζ at peak scale k_*, assuming a log-normal shape with a width of $\Delta = 1$, from the NANOGrav 12.5-year (green) and second IPTA data release (blue) observations of PTAs. This assumes that the observed signal comes from scalar-induced gravitational waves. The dotted line represents the expected amplitude leading to $f_{\mathrm{PBH}} = 1$ on which one can identify the bumpy feature due to the QCD-epoch leading to solar-mass PBHs. From Ref. [53]

collaborations, although the gravitational-wave origin still had to be demonstrated. This claim was confirmed in June 2023, with strong evidence for the Hellings–Downs correlations [91] as predicted by the quadrupolar nature of gravitational waves. Such a background could have been sourced by the large density fluctuations needed to form PBHs. Several studies have been performed suggesting that the signal frequency would be associated with PBHs in the stellar [53, 55, 93, 104, 151, 159] or even planetary [58] mass range. The results obtained using the most recent IPTA data and a log-normal primordial power spectrum are shown in Fig. 20.13, where the required amplitude for $f_{\mathrm{PBH}} = 1$ is also estimated [53]. We should stress that, at the time of writing, there is some uncertainty in the origin of the signal.

20.6 Unified Scenario

All the claimed observational evidence for PBHs discussed in the present chapter, together with numerous additional pieces discussed in Ref. [41], might be explained with a single unified theoretical scenario, as first described in Ref. [40]. We will argue that this could naturally emerge as a consequence of the well established thermal history of the Universe. For all formation scenarios, inevitable PBH clustering (see Sect. 20.2) is another key ingredient in explaining the claimed observational evidence. One also has to consider the relationship between the primordial fluctuations on large and small cosmological scales.

At the end of inflation, reheating fills the Universe with radiation. Which, in the standard cosmological scenario, remains dominated by relativistic particles until matter-radiation equality with an energy density decreasing as the fourth power of the temperature. As time progresses, the number of relativistic degrees of freedom remains constant until around 200 GeV, when the temperature of the Universe falls to the mass thresholds of the Standard Model particles. The first particle to become non-relativistic is the top quark at 172 GeV, followed by the Higgs boson at 125 GeV, the Z boson at 92 GeV and then the W boson at 81 GeV. At a temperature of around 200 MeV, when the QCD transition occurs, protons, neutrons and pions condense out of the free light quarks and gluons. A little later, the pions become non-relativistic, followed by muons, with e^+e^- annihilation and neutrino decoupling occurring at around 1 MeV. Note that chapter 5 also discusses the impact of standard model phase transitions.

Changes in the number of relativistic degrees of freedom induce changes in the effective equation-of-state parameter $w \equiv p/\rho c^2$. As shown in Fig. 20.14, there are in turn four periods in the thermal history of the Universe when w decreases. After each of these, w increases again, but because the threshold overdensity required for PBH production δ_c is exponentially sensitive to the equation-of-state parameter, the mentioned drops modify the probability of gravitational collapse of any large curvature fluctuations. This results in pronounced features in the PBH mass function even if the primordial fluctuations have a completely uniform power spectrum.

There is a plethora of inflationary models, predicting a large variety of shapes for $\delta_{\rm rms}(M)$. Some models produce an extended plateau or dome-like feature of the power spectrum. For example, this applies for two-field models, such as hybrid inflation [52], and even for some single-field models, like critical Higgs inflation [64]. This results in a quasi-scale-invariant spectrum, $\delta_{\rm rms}(M) = A\,(M/M_\odot)^x$, with $x = (1-n_s)/4 + \alpha_s/8\ln(M/M_\odot)$. Here, the spectral index n_s, its running $\alpha_s \equiv {\rm d}\ln n_s(k)/{\rm d}\ln k$ and the amplitude A are treated as free phenomenological parameters. We choose $A = 0.0218$ for $n_s = 0.986$ and $\alpha_s = -0.0018$, in order to get an integrated abundance $f_{\rm PBH} = 1$. Following Refs. [36], the ratio of the PBH

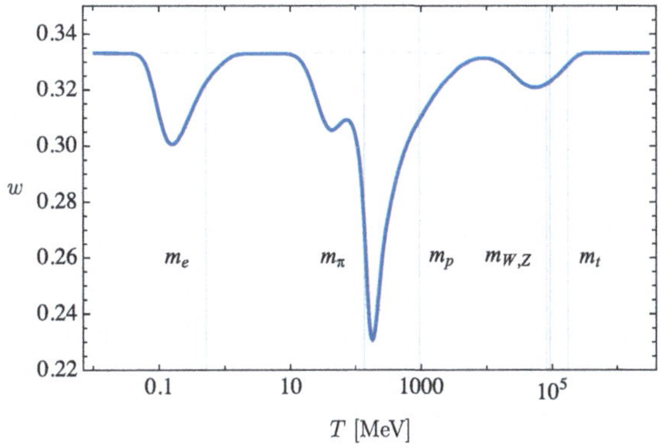

Fig. 20.14 Equation-of-state parameter w as a function of temperature T. The grey vertical lines indicate the masses of the electron, pion, proton/neutron, W/Z bosons and top quark. The grey dashed horizontal line correspond to pure radiation, i.e. to $w = 1/3$. From Ref. [40]

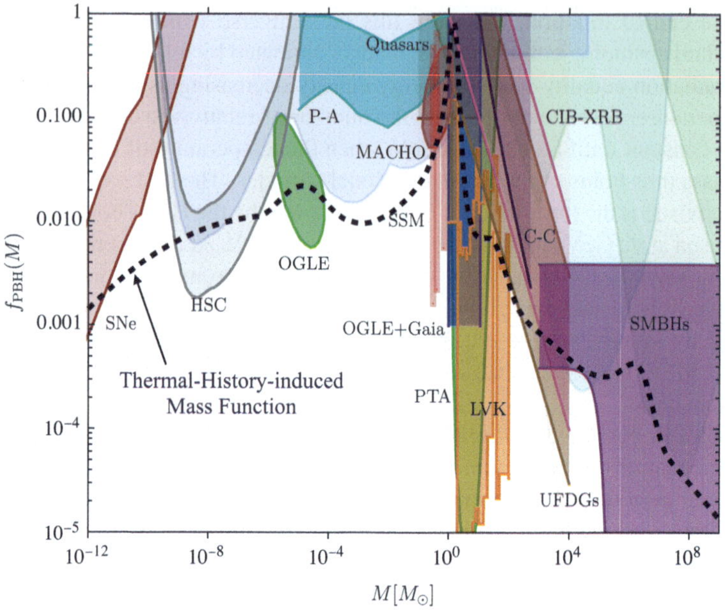

Fig. 20.15 Summary of claimed positive evidences for PBHs in terms of $f_{\rm PBH}$ values required by or consistent with the claimed detections. These come from PBH-attributed signals from supernovæ (SNe), various microlensing surveys (*Gaia*, HSC, OGLE, MACHO), POINT-AGAPE pixel-lensing (P-A), gravitational waves (LVK), ultra-faint dwarf galaxies (UFDGs), supermassive black holes (SMBHs), core/cusp (C-C) profiles for inner galactic halos, as well as correlations of the source-subtracted cosmic infrared and X-ray backgrounds (CIB-XRB). Also shown is the *thermal-history* PBH mass function (thick, dashed curve; c.f. Ref. [40]). From Ref. [41]

and the horizon mass at re-entry is assumed to be 0.8. The resulting mass function is represented by the dashed black line in Fig. 20.15, showing a dominant peak at $M \simeq 2\,M_\odot$ and three additional bumps at $10^{-5}\,M_\odot$, $30\,M_\odot$ and $10^6\,M_\odot$. Earlier works [30, 50, 96, 97] have also discussed the effect of the QCD transition on PBH formation but did not treat the other transitions. All these effects were first considered in a unified way in Ref. [40].

Depending on the running and tilt, the mentioned bumps can be more or less pronounced; small variations of the tilt and its running can dramatically change the normalisation and thus the height of the main peak. Furthermore, as demonstrated in Ref. [23], the presence of lepton-flavour asymmetries could also induce pronounced changes to the functional form of the bumps and even their location [23]. As pointed out in Refs. [114, 145, 154], scenarios with unequal lepton flavour asymmetries (before the onset of neutrino oscillations) are observationally almost unconstrained and therefore open up a whole new parameter space for the evolution of the Universe around the QCD transition. There are indications [70] that the cosmic QCD transition might even become first order if the asymmetries are sufficiently large, with a dramatic impact on the PBH mass function. Hence, PBHs are an excellent—and possibly the only conceivable—probe for the associated physics.

Figure 20.15 shows the PBH mass and dark matter fraction required to explain the various type of claimed observational evidence discussed in this chapter. The fit with the data is consistent with all the constraints if one uses a running spectral index for the power spectrum at PBH scales, as suggested in Ref. [81]. As mentioned above, we utilise a spectral tilt $n_s = 0.986$ and running $\alpha_s = -0.0018$ at PBH scales. These values are remarkably close to those measured on CMB scales. Of course, a priori, the primordial power spectrum $P(k)$ is at present essentially entirely unknown on PBH scales, but there are inflationary mechanisms (for instance, critical Higgs inflation [64, 72]) which transfer the functional form of $P(k)$ over from CMB scales, while leading to the required large increase in amplitude. This also typically comes with (non-perturbative) non-Gaußianity which plays an important rôle in evading μ-distortion constraints (cf. Ref. [29] for a recent discussion). There currently still is a lot of room for theoretical model building; the solid black curve in Fig. 20.15 should therefore be seen as an indication, but it is truly remarkable that already this simple model—with all peak scales not put in by hand by rather naturally given by the thermal history of the Universe—might explain all of the shown conundra, stressing the importance to consider the variations in the equation of state of the Universe for primordial black hole formation.

References

1. B.P. Abbott et al., Observation of gravitational waves from a binary black hole merger. Phys. Rev. Lett. **116**(6), 061102 (2016). https://doi.org/10.1103/PhysRevLett.116.061102. arXiv: 1602.03837 [gr-qc]
2. B.P. Abbott et al., Search for subsolar mass ultracompact binaries in advanced LIGO's second observing run. Phys. Rev. Lett. **123**(16), 161102 (2019). https://doi.org/10.1103/PhysRevLett.123.161102. arXiv: 1904.08976 [astro-ph.CO]
3. B. Abbott et al., Search for subsolar-mass ultracompact binaries in advanced LIGO's first observing run. Phys. Rev. Lett. **121**(23), 231103 (2018). https://doi.org/10.1103/PhysRevLett.121.231103. arXiv: 1808.04771 [astro-ph.CO]
4. R. Abbott et al., Population of merging compact binaries inferred using gravitational waves through GWTC-3. Phys. Rev. X **13**(1), 011048 (2023). https://doi.org/10.1103/PhysRevX.13.011048. arXiv: 2111.03634 [astro-ph.HE]
5. R. Abbott et al., Search for subsolar-mass black hole binaries in the second part of advanced LIGO's and advanced Virgo's third observing run (2022). https://doi.org/10.48550/ARXIV.2212.01477. arXiv:2212.01477
6. M.A. Abramowicz et al., No observational constraints from hypothetical collisions of hypothetical dark halo primordial black holes with galactic objects. Astrophys. J. **705**, 659–669 (2009). https://doi.org/10.1088/0004-637X/705/1/659. arXiv: 0810.3140 [astro-ph]
7. J. Acevedo, J. Bramante, Supernovae sparked by dark matter in white dwarfs **100**(4), 043020 (2019). https://doi.org/10.1103/PhysRevD.100.043020. arXiv: 1904.11993 [hep-ph]
8. J. Acevedo, J. Bramante, A. Goodman, Accelerating composite dark matter discovery with nuclear recoils and the Migdal effect **105**(2), 023012 (2022). https://doi.org/10.1103/PhysRevD.105.023012. arXiv: 2108.10889 [hep-ph]

9. N. Afshordi, P. McDonald, D.N. Spergel, Primordial black holes as dark matter: the power spectrum and evaporation of early structures. Astrophys. J. Lett. **594**, L71–L74 (2003). https://doi.org/10.1086/378763. arXiv: astro-ph/0302035
10. J.-E. Alam, S. Raha, B. Sinha, Quark nuggets as baryonic dark matter Astrophys. J. **513**, 572 (1999). https://doi.org/10.1086/306905. arXiv: astro-ph/9704226
11. C. Alcock et al., Binary microlensing events from the MACHO project **541**(1), 270–297 (2000). https://doi.org/10.1086/309393. arXiv: astro-ph/9907369 [astro-ph]
12. C. Alcock et al., The MACHO project first-year large Magellanic cloud results: the microlensing rate and the nature of the galactic dark halo **461**, 84 (1996). https://doi.org/10.1086/177039. arXiv: astro-ph/9506113 [astro-ph]
13. A. Alexandre, G. Dvali, E. Koutsangelas, New mass window for primordial black holes as dark matter from memory burden effect (2024). arXiv: 2402.14069 [hep-ph]
14. C. Animali, V. Vennin, Clustering of primordial black holes from quantum diffusion during inflation (2024). arXiv: 2402.08642 [astro-ph.CO]
15. J. Antoniadis et al., The international pulsar timing array second data release: search for an isotropic gravitational wave background. Mon. Not. Roy. Astron. Soc. **510**(4), 4873–4887 (2022). https://doi.org/10.1093/mnras/stab3418. arXiv: 2201.03980 [astro-ph.HE]
16. Z. Arzoumanian et al., The NANOGrav 12.5 yr data set: search for an isotropic stochastic gravitational-wave background. Astrophys. J. Lett. **905**(2), L34 (2020). https://doi.org/10.3847/2041-8213/abd401. arXiv: 2009.04496 [astro-ph.HE]
17. P. Auclair et al., Cosmology with the laser interferometer space antenna (2022). arXiv: 2204.05434 [astro-ph.CO]
18. E. Bagui, S. Clesse, A boosted gravitational wave background for primordial black holes with broad mass distributions and thermal features. Phys. Dark Univ. **38**, 101115 (2022). https://doi.org/10.1016/j.dark.2022.101115. arXiv: 2110.07487 [astro-ph.CO]
19. J.N. Bahcall, P. Hut, S. Tremaine, Maximum mass of objects that constitute unseen disk material. Astrophys. J **290**, 15–20 (1985). https://doi.org/10.1086/162953
20. M. Biagetti et al., Primordial black holes from inflation and quantum diffusion. JCAP **07**, 032 (2018). https://doi.org/10.1088/1475-7516/2018/07/032. arXiv: 1804.07124 [astro-ph.CO]
21. S. Bird et al., Did LIGO detect dark matter? Phys. Rev. Lett. **116**(20), 201301 (2016). https://doi.org/10.1103/PhysRevLett.116.201301. arXiv: 1603.00464 [astro-ph.CO]
22. S. Blinnikov et al., Solving puzzles of GW150914 by primordial black holes. JCAP **11**, 036 (2016). https://doi.org/10.1088/1475-7516/2016/11/036. arXiv: 1611.00541 [astro-ph.HE]
23. D. Bödeker et al., Lepton flavor asymmetries and the mass spectrum of primordial black holes. Phys. Rev. D **103**(6), 063506 (2021). https://doi.org/10.1103/PhysRevD.103.063506. arXiv: 2011.07283 [astro-ph.CO]
24. A. Bogdan et al., Evidence for heavy seed origin of early supermassive black holes from a $z \sim 10$ X-ray quasar (2023). arXiv: 2305.15458 [astro-ph.GA]
25. P. Boldrini et al., Cusp-to-core transition in low-mass dwarf galaxies induced by dynamical heating of cold dark matter by primordial black holes. Mon. Not. Roy. Astron. Soc. **492**(4), 5218–5225 (2020). https://doi.org/10.1093/mnras/staa150. arXiv: 1909.07395 [astro-ph.CO]
26. J. Bowman et al., An absorption profile centred at 78 megahertz in the sky-averaged spectrum **555**(7694), 67–70 (2018). https://doi.org/10.1038/nature25792. arXiv: 1810.05912 [astro-ph.CO]
27. J. Bramante, Dark matter ignition of type IA supernovae **115**(14), 141301 (2015). https://doi.org/10.1103/PhysRevLett.115.141301. arXiv: 1505.07464 [hep-ph]
28. T.D. Brandt, Constraints on MACHO dark matter from compact stellar systems in ultra-faint dwarf galaxies **824**(2), L31 (2016). https://doi.org/10.3847/2041-8205/824/2/L31. arXiv: 1605.03665 [astro-ph.GA]
29. C.T. Byrnes, J. Lesgourgues, D. Sharma, Robust μ-distortion constraints on primordial supermassive black holes from non-Gaussian perturbations (2024). arXiv: 2404.18475 [astro-ph.CO]

30. C.T. Byrnes et al., Primordial black holes with an accurate QCD equation of state JCAP **08**, 041 (2018). https://doi.org/10.1088/1475-7516/2018/08/041. arXiv: 1801.06138 [astro-ph.CO]
31. S. Calchi-Novati et al., POINT-agape pixel lensing survey of M31: evidence for a MACHO contribution to galactic halos. Astron. Astrophys. **443**, 911 (2005). https://doi.org/10.1051/0004-6361:20053135. arXiv: astro-ph/0504188
32. J. Calcino, J. Garcia-Bellido, T.M. Davis, Updating the MACHO fraction of the milky way dark halowith improved mass models. Mon. Not. Roy. Astron. Soc. **479**(3), 2889–2905 (2018). https://doi.org/10.1093/mnras/sty1368. arXiv: 1803.09205 [astro-ph.CO]
33. F. Capela, M. Pshirkov, P. Tinyakov, A comment on "Exclusion of the remaining mass window for primordial black holes..." (2014). arXiv:1401.3025. arXiv: 1402.4671 [astro-ph.CO]
34. N. Cappelluti, G. Hasinger, P. Natarajan, Exploring the High-redshift PBH- Λ CDM universe: early black hole seeding, the first stars and cosmic radiation backgrounds. Astrophys. J. **926**(2), 205 (2022). https://doi.org/10.3847/1538-4357/ac332d. arXiv: 2109.08701 [astro-ph.CO]
35. B. Carr, The statistical clustering of primordial black holes **56**(3), 377–383 (1977)
36. B. Carr, S. Clesse, J. García-Bellido, Primordial black holes from the QCD epoch: Linking dark matter, baryogenesis and anthropic selection. Mon. Not. Roy. Astron. Soc. **501**(1), 1426–1439 (2021). https://doi.org/10.1093/mnras/staa3726. arXiv: 1904.02129 [astro-ph.CO]
37. B. Carr, C. Lacey, Dark clusters in galactic halos? Astrophys. J. **316**, 23–35 (1987). https://doi.org/10.1086/165176
38. B. Carr, J. Silk, Can graininess in the early universe make galaxies? Astrophys. J. **268**, 1–16 (1983). https://doi.org/10.1086/160924
39. B. Carr, J. Silk, Primordial black holes as generators of cosmic structures. Mon. Not. Roy. Astron. Soc. **478**(3), 3756–3775 (2018). https://doi.org/10.1093/mnras/sty1204
40. B. Carr et al., Cosmic conundra explained by thermal history and primordial black holes. Phys. Dark Univ. **31**, 100755 (2021). https://doi.org/10.1016/j.dark.2020.100755. arXiv: 1906.08217 [astro-ph.CO]
41. B. Carr et al., Observational evidence for primordial black holes: a positivist perspective. Phys. Rept. **1054**, 1–68 (2024). https://doi.org/10.1016/j.physrep.2023.11.005. arXiv: 2306.03903 [astro-ph.CO]
42. B.J. Carr, S.W. Hawking, Black holes in the early Universe. Mon. Not. Roy. Astron. Soc. **168**, 399–415 (1974). https://doi.org/10.1093/mnras/168.2.399
43. B.J. Carr, M. Sakellariadou, Dynamical constraints on dark compact objects. Astrophys. J. **516**, 195–220 (1999). https://doi.org/10.1086/307071
44. H.-S. Chan et al., Delayed detonation thermonuclear supernovae with an extended dark matter component **914**(2), 138 (2021). https://doi.org/10.3847/1538-4357/abfd32. arXiv: 2012.06857 [astro-ph.HE]
45. K. Chang, S. Refsdal, Flux variations of QSO 0957 + 561 A, B and image splitting by stars near the light path **282**(5739), 561–564 (1979). https://doi.org/10.1038/282561a0
46. Z.-C. Chen, Q.-G. Huang, Distinguishing primordial black holes from astrophysical black holes by Einstein telescope and cosmic explorer **2020**(8), 039 (2020). https://doi.org/10.1088/1475-7516/2020/08/039. arXiv: 1904.02396 [astro-ph.CO]
47. S. Chen et al., Common-red-signal analysis with 24-yr high-precision timing of the European pulsar timing array: inferences in the stochastic gravitational-wave background search. Mon. Not. Roy. Astron. Soc. **508**(4), 4970–4993 (2021). https://doi.org/10.1093/mnras/stab2833. arXiv: 2110.13184 [astro-ph.HE]
48. J.R. Chisholm, Clustering of primordial black holes: basic results. Phys. Rev. D **73**, 083504 (2006). https://doi.org/10.1103/PhysRevD.73.083504. arXiv: astro-ph/0509141
49. J.R. Chisholm, Clustering of primordial black holes II: evolution of bound systems. Phys. Rev. D **84**, 124031 (2011). https://doi.org/10.1103/PhysRevD.84.124031. arXiv: 1110.4402 [astro-ph.CO]
50. S. Clesse, J. Garcia-Bellido, GW190425, GW190521 and GW190814: three candidate mergers of primordial black holes from the QCD epoch. Phys. Dark Univ. **38**, 101111 (2022). https://doi.org/10.1016/j.dark.2022.101111. arXiv: 2007.06481 [astro-ph.CO]

51. S. Clesse, J. García-Bellido, The clustering of massive primordial black holes as dark matter: measuring their mass distribution with advanced LIGO. Phys. Dark Univ. **15**, 142–147 (2017). https://doi.org/10.1016/j.dark.2016.10.002. arXiv: 1603.05234 [astro-ph.CO]
52. S. Clesse, J. García-Bellido, Massive primordial black holes from hybrid inflation as dark matter and the seeds of galaxies. Phys. Rev. D **92**(2), 023524 (2015). https://doi.org/10.1103/PhysRevD.92.023524. arXiv: 1501.07565 [astro-ph.CO]
53. V. Dandoy, V. Domcke, F. Rompineve, Search for scalar induced gravitational waves in the international pulsar timing array data release 2 and NANOgrav 12.5 years dataset (2023). arXiv: 2302.07901 [astro-ph.CO]
54. K. De et al., The zwicky transient facility census of the local universe I: systematic search for calcium-rich gap transients reveals three related spectroscopic subclasses **905**(1), 58 (2020). https://doi.org/10.3847/1538-4357/abb45c. arXiv: 2004.09029 [astro-ph.HE]
55. V. De Luca, G. Franciolini, A. Riotto, NANOGrav data hints at primordial black holes as dark matter. Phys. Rev. Lett. **126**(4), 041303 (2021). https://doi.org/10.1103/PhysRevLett.126.041303. arXiv: 2009.08268 [astro-ph.CO]
56. V. De Luca et al., The initial spin probability distribution of primordial black holes. JCAP **05**, 018 (2019). https://doi.org/10.1088/1475-7516/2019/05/018. arXiv: 1903.01179 [astro-ph.CO]
57. A. Dolgov, Tension between HST/JWST and Λ CDM Cosmology, PBH, and antimatter in the galaxy. 14th Frascati workshop on multifrequency behaviour of high energy cosmic sources (2023). arXiv:2310.00671 [astro-ph.CO]
58. G. Domènech, S. Pi, NANOGrav hints on planet-mass primordial black holes. Sci. China Phys. Mech. Astron. **65**(3), 230411 (2022). https://doi.org/10.1007/s11433-021-1839-6. arXiv: 2010.03976 [astro-ph.CO]
59. G. Dvali, E. Koutsangelas, F. Kühnel, Compact dark matter objects via N dark sectors. Phys. Rev. D **101**, 083533 (2020). https://doi.org/10.1103/PhysRevD.101.083533. arXiv: 1911.13281 [astro-ph.CO]
60. G. Dvali et al., Black hole metamorphosis and stabilization by memory burden. Phys. Rev. D **102**(10), 103523 (2020). https://doi.org/10.1103/PhysRevD.102.103523. arXiv: 2006.00011 [hep-th]
61. R. Endsley et al., ALMA confirmation of an obscured hyperluminous radio-loud AGN. Mon. Not. R. Astron. Soc. **520**(3), 4609–4620 (2023). https://doi.org/10.1093/mnras/stad266. https://doi.org/10.1093%20%2Fmnras%2Fstad266
62. A. Escrivà, E. Bagui, S. Clesse, Simulations of PBH formation at the QCD epoch and comparison with the GWTC-3 catalog. JCAP **05**, 004 (2023). https://doi.org/10.1088/1475-7516/2023/05/004. arXiv: 2209.06196 [astro-ph.CO]
63. M. Evans et al., A horizon study for cosmic explorer: science, observatories, and community (2021). arXiv: 2109.09882 [astro-ph.IM]
64. J.M. Ezquiaga, J. García-Bellido, E.R. Morales, Primordial black hole production in critical higgs inflation. Phys. Lett. B **776**, 345–349 (2018). https://doi.org/10.1016/j.physletb.2017.11.039. arXiv: 1705.04861 [astro-ph.CO]
65. J. María Ezquiaga, J. García-Bellido, Quantum diffusion beyond slow-roll: implications for primordial black-hole production. JCAP **08**, 018 (2018). https://doi.org/10.1088/1475-7516/2018/08/018. arXiv: 1805.06731 [astro-ph.CO]
66. J. María Ezquiaga, J. García-Bellido, V. Vennin, Massive galaxy clusters like El Gordo hint at primordial quantum diffusion. Phys. Rev. Lett. (2022) arXiv: 2207.06317 [astro-ph.CO]
67. P. Fayet, D. Hooper, G. Sigl, Constraints on light dark matter from core-collapse supernovae **96**(21), 211302 (2006). https://doi.org/10.1103/PhysRevLett.96.211302. arXiv: hep--ph/0602169 [hep-ph]
68. K. Freese, R. Price, D. Schramm, Formation of population III stars and galaxies with primordial planetary-mass black holes. Astrophys. J. **275**, 405–412 (1983). https://doi.org/10.1086/161542
69. K. Fujikura et al., Microlensing constraints on axion stars including finite lens and source size effects. Phys. Rev. D **104**(12), 123012 (2021). https://doi.org/10.1103/PhysRevD.104.123012. arXiv: 2109.04283 [hep-ph]

70. F. Gao, I. Oldengott, Cosmology meets functional QCD: first-order cosmic QCD transition induced by large lepton asymmetries. Phys. Rev. Lett. **128**(13), 131301 (2022). https://doi.org/10.1103/PhysRevLett.128.131301. arXiv: 2106.11991 [hep-ph]
71. J. Garcia-Bellido, M. Hawkins, Reanalysis of the MACHO constraints on PBH in the light of Gaia DR3 data (2024). arXiv: 2402.00212 [astro-ph.GA]
72. J. Garcia-Bellido, E.R. Morales, Primordial black holes from single field models of inflation. Phys. Dark Univ. **18**, 47–54 (2017). https://doi.org/10.1016/j.dark.2017.09.007. arXiv: 1702.03901 [astro-ph.CO]
73. J. García-Bellido, Primordial black holes and the origin of the matter–antimatter asymmetry. Phil. Trans. Roy. Soc. Lond. A **377**(2161) (2019). ed. by J. Dainton, p. 20190091. https://doi.org/10.1098/rsta.2019.0091
74. J. García-Bellido, Primordial black holes as dark matter
75. B. Goncharov et al., On the evidence for a common-spectrum process in the search for the nanohertz gravitational-wave background with the parkes pulsar timing array. Astrophys. J. Lett. **917**(2), L19 (2021). https://doi.org/10.3847/2041-8213/ac17f4. arXiv: 2107.12112 [astro-ph.HE]
76. R. Gott, Are heavy halos made of low mass stars—a gravitational lens test. Astrophys. J **243**, 140–146 (1981). https://doi.org/10.1086/158576
77. A.D. Goulding et al., UNCOVER: the growth of the first massive black holes from JWST/NIRSpec—spectroscopic redshift confirmation of an X-Ray luminous AGN at z = 10.1. Astrophys. J. Lett. **955**(1), L24 (2023). https://doi.org/10.3847/2041-8213/acf7c5. arXiv: 2308.02750 [astro-ph.GA]
78. P. Graham, S. Rajendran, J. Varela, Dark matter triggers of supernovae **92**(6), 063007 (2015). https://doi.org/10.1103/PhysRevD.92.063007. arXiv: 1505.04444 [hep-ph]
79. P. Graham et al., White dwarfs as dark matter detectors **98**(11), 115027 (2018). https://doi.org/10.1103/PhysRevD.98.115027. arXiv: 1805.07381 [hep-ph]
80. H. Perets et al., A faint type of supernova from a white dwarf with a helium-rich companion **465**(7296), 322–325 (2010). https://doi.org/10.1038/nature09056. arXiv: 0906.2003 [astro-ph.HE]
81. G. Hasinger, Illuminating the dark ages: cosmic backgrounds from accretion onto primordial black hole dark matter. JCAP **07**, 022 (2020). https://doi.org/10.1088/1475-7516/2020/07/022. arXiv: 2003.05150 [astro-ph.CO]
82. G. Hasinger, Illuminating the dark ages: cosmic backgrounds from accretion onto primordial black hole dark matter **2020**(7), 022 (2020). https://doi.org/10.1088/1475-7516/2020/07/022. arXiv: 2003.05150 [astro-ph.CO]
83. M.R.S. Hawkins, Evidence for microlensing by primordial black holes in quasar broad emission lines. Mon. Not. Roy. Astron. Soc. **527**(2), 2393–2402 (2023). https://doi.org/10.1093/mnras/stad3346. arXiv: 2311.08915 [astro-ph.CO]
84. M.R.S. Hawkins, Gravitational microlensing, quasar variability and missing matter. Nature **366**, 242–245 (1993). https://doi.org/10.1038/366242a0
85. M.R.S. Hawkins, SDSS J1004+4112: the case for a galaxy cluster dominated by primordial black holes. Astron. Astrophys. **643**, A10 (2020). https://doi.org/10.1051/0004-6361/202038670. arXiv: 2010.15007 [astro-ph.CO]
86. M.R.S. Hawkins, The signature of primordial black holes in the dark matter halos of galaxies. Astron. Astrophys. **633**, A107 (2020). https://doi.org/10.1051/0004-6361/201936462. arXiv: 2001.07633 [astro-ph.GA]
87. M. Hawkins, Caustic crossings in quasar light curves? Astron. Astrophys. **340**, L23 (1998). arXiv: astro-ph/9810337
88. M. Hawkins, New evidence for a cosmological distribution of stellar mass primordial black holes. Mon. Not. Roy. Astron. Soc. (2022). https://doi.org/10.1093/mnras/stac863
89. . M. Hawkins, The case for primordial black holes as dark matter. Mon. Not. Roy. Astron. Soc. **415**, 2744 (2011). https://doi.org/10.1111/j.1365-2966.2011.18890.x. arXiv: 1106.3875 [astro-ph.CO]

90. K. Helgason et al., On the physical requirements for a pre-reionization origin of the unresolved near-infrared background **455**(1), 282–294 (2016). https://doi.org/10.1093/mnras/stv2209. arXiv: 1505.07226 [astro-ph.CO]
91. R.W. Hellings, G.S. Downs, Upper limits on the isotropic gravitational radiation background from pulsar timing analysis. Astrophys. J. Lett. **265**, L39–L42 (1983). https://doi.org/10.1086/183954
92. D. Inman, Y. Ali-Haïmoud, Early structure formation in primordial black hole cosmologies. Phys. Rev. D **100**(8), 083528 (2019). https://doi.org/10.1103/PhysRevD.100.083528. arXiv: 1907.08129 [astro-ph.CO]
93. K. Inomata et al., NANOGrav results and LIGO-virgo primordial black holes in axionlike curvaton models. Phys. Rev. Lett. **126**(13), 131301 (2021). https://doi.org/10.1103/PhysRevLett.126.131301. arXiv: 2011.01270 [astro-ph.CO]
94. D. Jacobs, G. Starkman, B. Lynn, Macro dark matter. Mon. Not. Roy. Astron. Soc. **450**(4), 3418–3430 (2015). https://doi.org/10.1093/mnras/stv774. arXiv: 1410.2236 [astro-ph.CO]
95. S. Jaraba, J. García-Bellido, Black hole induced spins from hyperbolic encounters in dense clusters. Phys. Dark Universe **34**, 100882 (2021). https://doi.org/10.1016/j.dark.2021.100882. arXiv: 2106.01436 [gr-qc]
96. K. Jedamzik, Consistency of primordial black hole dark matter with LIGO/virgo merger rates. Phys. Rev. Lett. **126**(5), 051302 (2021). https://doi.org/10.1103/PhysRevLett.126.051302. arXiv: 2007.03565 [astro-ph.CO]
97. K. Jedamzik, Primordial black hole formation during the QCD epoch. Phys. Rev. D **55**, 5871–5875 (1997). https://doi.org/10.1103/PhysRevD.55.R5871. (arXiv: astro-ph/9605152)
98. A. Kaiser, S. McWilliams, Sensitivity of present and future detectors across the black-hole binary gravitational wave spectrum. Class. Quant. Grav. **38**(5), 055009 (2021). https://doi.org/10.1088/1361-6382/abd4f6. arXiv: 2010.02135 [gr-qc]
99. A. Kashlinsky, LIGO gravitational wave detection, primordial black holes and the near-IR cosmic infrared background anisotropies. Astrophys. J. Lett. **823**(2), L25 (2016). https://doi.org/10.3847/2041-8205/823/2/L25. arXiv: 1605.04023 [astro-ph.CO]
100. A. Kashlinsky et al., Tracing the first stars with fluctuations of the cosmic infrared background **438**(7064), 45–50 (2005). https://doi.org/10.1038/nature04143. arXiv: astro-ph/0511105 [astro-ph]
101. A. Kashlinsky, Cosmological advection flows in the presence of primordial black holes as dark matter and formation of first sources **126**(1), 011101 (2021). https://doi.org/10.1103/PhysRevLett.126.011101. arXiv: 2012.08047 [astro-ph.CO]
102. A. Kashlinsky et al., Looking at cosmic near-infrared background radiation anisotropies. Rev. Mod. Phys. **90**(2), 025006 (2018). https://doi.org/10.1103/RevModPhys.90.025006. arXiv: 1802.07774 [astro-ph.CO]
103. A. Katz et al., Femtolensing by dark matter revisited **2018**(12), 005 (2018). https://doi.org/10.1088/1475-7516/2018/12/005. arXiv: 1807.11495 [astro-ph.CO]
104. K. Kohri, T. Terada, Solar-mass primordial black holes explain NANOGrav hint of gravitational waves. Phys. Lett. B **813**, 136040 (2021). https://doi.org/10.1016/j.physletb.2020.136040. arXiv: 2009.11853 [astro-ph.CO]
105. S. Koushiappas, A. Loeb, Dynamics of dwarf galaxies disfavor stellar-mass black holes as dark matter **119**(4), 041102 (2017). https://doi.org/10.1103/PhysRevLett.119.041102. arXiv: 1704.01668 [astro-ph.GA]
106. I. Labbé et al., A population of red candidate massive galaxies Ë & oelig; 600 MYR after the Big Bang. Nature **616**(7956), 266–269 (2023). https://doi.org/10.1038/s41586-023-05786-2. https://doi.org/10.1038%20%2Fs41586-023-05786-2
107. C. Lacey, J. Ostriker, Massive black holes in galactic halos? Astrophys. J. **299**, 633–652 (1985). https://doi.org/10.1086/163729
108. Y. Li et al., The SPLASH and Chandra COSMOS legacy survey: the cross-power between near-infrared and X-Ray background fluctuations **864**(2), 141 (2018). https://doi.org/10.3847/1538-4357/aad55a. arXiv: 1807.10304 [astro-ph.GA]

109. B. Liu, V. Bromm, Accelerating early massive galaxy formation with primordial black holes. Astrophys. J. Lett. **937**(2), L30 (2022). https://doi.org/10.3847/2041-8213/ac927f. arXiv: 2208.13178 [astro-ph.CO]
110. M. Maggiore et al., Science case for the Einstein telescope. **2020**(3), 050 (2020). https://doi.org/10.1088/1475-7516/2020/03/050. arXiv: 1912.02622 [astro-ph.CO]
111. G.F. Marani et al., Gravitationally lensed gamma-ray bursts as probes of dark compact objects. Astrophys. J. Lett. **512**, L13 (1999). https://doi.org/10.1086/311868. arXiv: astro-ph/9810391
112. X. Meng, Z. Han, A pair of CO+ He white dwarfs as the progenitor of 2005E-like supernovae? **573**(A57) (2015). https://doi.org/10.1051/0004-6361/201424562. arXiv: 1410.8630 [astro-ph.SR]
113. P. Mészáros, Primeval black holes and galaxy formation. Astron. astrophys. **38**, 5–13 (1975)
114. M. Middeldorf-Wygas et al., Cosmic QCD transition for large lepton flavor asymmetries. Phys. Rev. D **105**(12), 123533 (2022). https://doi.org/10.1103/PhysRevD.105.123533. arXiv: 2009.00036 [hep-ph]
115. A.L. Miller et al., Constraints on planetary and asteroid-mass primordial black holes from continuous gravitational-wave searches (2021). arXiv: 2110.06188 [gr-qc]
116. A.L. Miller et al., Probing planetary-mass primordial black holes with continuous gravitational waves. Phys. Dark Univ. **32**, 100836 (2021). https://doi.org/10.1016/j.dark.2021.100836. arXiv: 2012.12983 [astro-ph.HE]
117. M. Monroy-Rodríguez, C. Allen, The end of the MACHO era- revisited: new limits on MACHO masses from halo wide binaries. Astrophys. J. **790**(2), 159 (2014). https://doi.org/10.1088/0004-637X/790/2/159. arXiv: 1406.5169 [astro-ph.GA]
118. B. Moore, An upper limit to the mass of black holes in the halo of the galaxy. Astrophys. J. Lett. **413**, L93–L96 (1993). https://doi.org/10.1086/186967 (arXiv:astro-ph/9306004)
119. P. Mróz et al., No large population of unbound or wide-orbit Jupiter-mass planets. Nature **D85**(548), 183–186. https://doi.org/10.1038/nature23276. arXiv: 1707.07634 [astro-ph.CO]
120. N. Cappelluti et al., Cross-correlating cosmic infrared and X-Ray background fluctuations: evidence of significant black hole populations among the CIB sources. Astrophys. J. **769**(1), 68 (2013). https://doi.org/10.1088/0004-637X/769/1/68. arXiv: 1210.5302 [astro-ph.CO]
121. H. Niikura et al., Microlensing constraints on primordial black holes with Subaru/HSC Andromeda observations. Nat. Astron. **3**(6), 524–534 (2019). https://doi.org/10.1038/s41550-019-0723-1. arXiv: 1701.02151 [astro-ph.CO]
122. H. Niikura et al., Microlensing constraints on primordial black holes with Subaru/HSC Andromeda observations. Nature Astronomy **3**, 524–534 (2019). https://doi.org/10.1038/s41550-019-0723-1. arXiv: 1701.02151 [astro-ph.CO]
123. H. Niikura et al., Constraints on Earth-mass primordial black holes from OGLE 5-year microlensing events. Phys. Rev. D **99**(8), 083503 (2019). https://doi.org/10.1103/PhysRevD.99.083503. arXiv: 1901.07120 [astro-ph.CO]
124. A.H. Nitz, Y.-F. Wang, Broad search for gravitational waves from subsolar-mass binaries through LIGO and Virgo's third observing run. Phys. Rev. D **106**(2), 023024 (2022). https://doi.org/10.1103/PhysRevD.106.023024. arXiv: 2202.11024 [astro-ph.HE]
125. A.H. Nitz, Y.-F. Wang, Search for gravitational waves from the coalescence of sub-solar mass and eccentric compact binaries (2021). https://doi.org/10.3847/1538-4357/ac01d9. arXiv: 2102.00868 [astro-ph.HE]
126. A.H. Nitz, Y.-F. Wang, Search for gravitational waves from the coalescence of subsolar-mass binaries in the first half of advanced LIGO and Virgo's third observing run. Phys. Rev. Lett. **127**(15), 151101 (2021). https://doi.org/10.1103/PhysRevLett.127.151101. arXiv: 2106.08979 [astro-ph.HE]
127. A.H. Nitz, Y.-F. Wang, Search for gravitational waves from high-mass-ratio compact-binary mergers of stellar mass and subsolar mass black holes. Phys. Rev. Lett. **126**(2), 021103 (2021). https://doi.org/10.1103/PhysRevLett.126.021103. arXiv: 2007.03583 [astro-ph.HE]
128. B. Paczynski, Gravitational microlensing by the galactic halo **304**, 1 (1986). https://doi.org/10.1086/164140

129. P. Pani, A. Loeb, Tidal capture of a primordial black hole by a neutron star: implications for constraints on dark matter **2014**(6), 026 (2014). https://doi.org/10.1088/1475-7516/2014/06/026. arXiv: 1401.3025 [astro-ph.CO]
130. C. Pattison et al., Quantum diffusion during inflation and primordial black holes. JCAP **10**, 046 (2017). https://doi.org/10.1088/1475-7516/2017/10/046. arXiv: 1707.00537 [hep-th]
131. K. Phukon et al., The hunt for sub-solar primordial black holes in low mass ratio binaries is open (2021). arXiv: 2105.11449 [astro-ph.CO]
132. P. Collaboration, N. Aghanim et al., Planck 2018 results VI: cosmological parameters **641**, A6 (2020). https://doi.org/10.1051/0004-6361/201833910. arXiv: 1807.06209 [astro-ph.CO]
133. T. Ponman, D. Cannon, J. Navarro, The thermal imprint of galaxy formation on X-ray clusters **397**(6715), 135–137 (1999). https://doi.org/10.1038/16410. arXiv: astro-ph/9810359 [astro-ph]
134. W. Press, J. Gunn, Method for detecting a cosmological density of condensed objects **185**, 397–412 (1973). https://doi.org/10.1086/152430
135. R. Lunnan et al., Two new calcium-rich gap transients in group and cluster environments **836**(1), 60 (2017). https://doi.org/10.3847/1538-4357/836/1/60. arXiv: 1612.00454 [astro-ph.HE]
136. M. Raidal, V. Vaskonen, H. Veermäe, Gravitational waves from primordial black hole mergers. JCAP 09, 037 (2017). https://doi.org/10.1088/1475-7516/2017/09/037. arXiv: 1707.01480 [astro-ph.CO]
137. M. Raidal et al., Formation and evolution of primordial black hole binaries in the early universe. JCAP 02, 018 (2019). https://doi.org/10.1088/1475-7516/2019/02/018. arXiv: 1812.01930 [astro-ph.CO]
138. R. Abbott et al. GWTC-3: compact binary coalescences observed by LIGO and virgo during the second part of the third observing run (2021). arXiv: 2111.03606 [gr-qc]
139. M. Roncadelli, A. Treves, R. Turolla, Primordial black holes are again on the limelight (2009). https://doi.org/10.48550/arXiv.0901.1093. arXiv: 0901.1093 [astro-ph.CO]
140. P. Schneider, Upper bounds on the cosmological density of compact objects with sub-solar masses from the variability of QSOs. Astron. Astrophys. **279**, 1–20 (1993)
141. J. Silk, Feedback by massive black holes in gas-rich dwarf galaxies. Astrophys. J. Lett. **839**(1), L13 (2017). https://doi.org/10.3847/2041-8213/aa67da. arXiv: 1703.08553 [astro-ph.GA]
142. J. Smirnov et al., White dwarfs in dwarf spheroidal galaxies: a new class of compact-dark-matter detectors (2022). arXiv: 2211.00013 [astro-ph.CO]
143. M. Smith, L. Kawano, R. Malaney, Experimental, computational, and observational analysis of primordial nucleosynthesis **85**, 219 (1993). https://doi.org/10.1086/191763
144. N. Smyth et al., Updated constraints on asteroid-mass primordial black holes as dark matter **101**(6), 063005 (2020). https://doi.org/10.1103/PhysRevD.101.063005. arXiv: 1910.01285 [astro-ph.CO]
145. M. Stuke, D. Schwarz, G. Starkman, WIMP abundance and lepton (flavour) asymmetry. JCAP **03**, 040 (2011). https://doi.org/10.1088/1475-7516/2012/03/040. arXiv:1111.3954 [astro-ph.CO]
146. S. Takekawa et al., The fifth candidate for an intermediate-mass black hole in the galactic center **890**(2), 167 (2020). https://doi.org/10.3847/1538-4357/ab6f6f. arXiv: 2002.05173 [astro-ph.GA]
147. V. Thoss, A. Burkert, K. Kohri, Breakdown of hawking evaporation opens new mass window for primordial black holes as dark matter candidate (2024). arXiv: 2402.17823 [astro-ph.CO]
148. M. Trashorras, J. García-Bellido, S. Nesseris, The clustering dynamics of primordial black boles in N-body simulations. Universe **7**(1), 18 (2021). https://doi.org/10.3390/universe7010018. arXiv: 2006.15018 [astro-ph.CO]
149. S. Van den Bergh, Collapsed objects in clusters of galaxies **224**, 891 (1969). https://doi.org/10.1038/224891a0

150. A. van Elteren et al., Survivability of planetary systems in young and dense star clusters **624**, A120 (2019). https://doi.org/10.1051/0004-6361/201834641. arXiv: 1902.04652 [astro-ph.SR]
151. V. Vaskonen, H. Veermäe, Did NANOGrav see a signal from primordial black hole formation? Phys. Rev. Lett. **126**(5), 051303 (2021). https://doi.org/10.1103/PhysRevLett.126.051303. arXiv: 2009.07832 [astro-ph.CO]
152. V. Vaskonen, H. Veermäe, Lower bound on the primordial black hole merger rate. Phys. Rev. D **101**(4), 043015 (2020). https://doi.org/10.1103/PhysRevD.101.043015. arXiv: 1908.09752 [astro-ph.CO]
153. V. Vennin, D. Wands, Quantum diffusion and large primordial perturbations from inflation (2024). arXiv: 2402.12672 [astro-ph.CO]
154. V. Vovchenko et al., Pion condensation in the early universe at nonvanishing lepton flavor asymmetry and its gravitational wave signatures. Phys. Rev. Lett. **126**(1), 012701 (2021). https://doi.org/10.1103/PhysRevLett.126.012701. arXiv: 2009.02309 [hep-ph]
155. D. Walsh, R. Carswell, R. Weymann, 0957+561 A, B: twin quasistellar objects or gravitational lens? **279**, 381–384 (1979). https://doi.org/10.1038/279381a0
156. M. Weinberg, S. Shapiro, I. Wasserman, The dynamical fate of wide binaries in the solar neighborhood **312**, 367 (1987). https://doi.org/10.1086/164883
157. S.E. Woosley, A. Heger, The pair-instability mass gap for black holes. Astrophys. J. Lett. **912**(2), L31 (2021). https://doi.org/10.3847/2041-8213/abf2c4. arXiv: 2103.07933 [astro-ph.SR]
158. L. Wyrzykowski et al., Black hole, neutron star and white dwarf candidates from microlensing with OGLE-III. Mon. Not. Roy. Astron. Soc. **458**(3), 3012–3026 (2016). https://doi.org/10.1093/mnras/stw426. arXiv: 1509.04899 [astro-ph.SR]
159. Z. Yi, Z.-H. Zhu, NANOGrav signal and LIGO-Virgo primordial black holes from Higgs inflation (2021). arXiv: 2105.01943 [gr-qc]
160. Y. Zel'dovich, I. Novikov, The hypothesis of cores retarded during expansion and the hot cosmological model. Sov. Astron. **10**, 602 (1967)

Chapter 21
Ultralight Primordial Black Holes

Stefano Profumo

Abstract The fate of ultralight black holes depends on whether or not evaporation stops at or around the Planck scale. If evaporation stops, the general expectation is that a population of Planck-scale relics will be left over, possibly including a significant fraction of electrically charged relics. If evaporation does not stop, a runaway "explosion" would occur, with significant and potentially detectable high-energy emission. Here, I review both possibilities, with an emphasis on current status and future detection prospects.

21.1 Introduction

Primordial black holes (PBHs) produced in the very early universe with mass below, approximately, $M_U \simeq 5.1 \times 10^{14}$ g (absent dark degrees of freedom, that could significantly shorten the timescale for evaporation [7]) have completed evaporation. The end products of evaporation are highly uncertain, as when $M_{\rm PBH} \to M_P$ the BH temperature $T_{\rm BH} \to M_P$ and, thus, quantum gravity effects are expected to play an important role [13]. Broadly, evaporation could in principle continue at sub-Planckian scales (corresponding to trans-Planckian temperatures), eventually consuming the BH, and $M_{\rm BH} \to 0$, or stop around, or below the Planck scale, leaving a stable relic. In addition, BHs can approach extremality as $M_{\rm PBH} \to M_P$, which could also result in the end of evaporation [30].

In this chapter I discuss the end-point of PBH evaporation. In the following Sect. 21.2 I review the possibility of stable Planck-scale relics, possibly, but not necessarily constituting a fraction, or all, of the cosmological dark matter. The following Sect. 21.3 then details how such Planck-scale relics can be directly detected with an array of experimental facilities. I then turn, in Sect. 21.4 to the opposite

S. Profumo (✉)
University of California, Santa Cruz, and Santa Cruz Institute for Particle Physics, 1156 High St, Santa Cruz, CA 95060, USA
e-mail: profumo@ucsc.edu

C. Byrnes et al. (eds.), *Primordial Black Holes*, Springer Series in Astrophysics and Cosmology, https://doi.org/10.1007/978-981-97-8887-3_21

possibility of complete evaporation and of the direct detection of the runaway high-energy emission from PBHs exploding today. The final Sect. 21.5 summarizes and concludes.

21.2 Planck-Scale Relics

The notion that Planck-scale relics of PBH evaporation could be some, or all, the cosmological dark matter (DM) has been contemplated for a long time (see e.g. [31]). As a BH mass approaches, its mass depleted by increasingly fast Hawking evaporation, the Planck scale, one out of three scenarios is conjectured to occur:

(i) a stable, Planck-scale relic is left over [31];

(ii) a naked spacetime singularity is left over [16];

(iii) evaporation continues at trans-Planckian energies and temperatures, and the singularity vanishes [23].

Any PBH formed in the early universe with a lifetime shorter than the age of the universe (as mentioned above corresponding to a mass at production of M_U) is slated to fall into one of these three categories.

In this section, I will assume that (i) or (ii) is the case. General arguments then indicate that Planck-scale relics can be either electromagnetically neutral, or carry a residual electric or magnetic charge.[1] The experimental absence of magnetic monopoles does not necessarily exclude the possibility of magnetically charged Planck relics, although, admittedly, this possibility is rather exotic [17]. I will thus entertain the possibility that a fraction $f_{\rm CPR}$ of relics be electrically charged, and denoted charged Planck-scale relics (CPR). Absent a charge, Planck scale relic are virtually entirely undetectable. Also note that there are general arguments [34] for why even if PBHs were produced with significant initial spin parameter $a = J/M$, with J the PBH's angular momentum and M its mass, evaporation rapidly and efficiently sheds angular momentum, leaving $a \to 0$ at the end of evaporation for $M < M_U$.

Note that Planck relics may have a significant initial velocity, as their momentum results from the ejection of highly energetic quanta which, around the Planck scale, have energies of the order of the Planck scale itself. The resulting relic's momentum is effectively a random walk in momentum space, dominated by the emission at the end of evaporation, when the Hawking temperature is highest. For N_q terminal emitted quanta, the momentum of the BH goes as $1/\sqrt{N_q}$. Thus, at most, calling M_R the relic mass, the relic's momentum $p \simeq M_R$ (corresponding to $N_q = 1$) and the energy $E = \sqrt{2}M_R$, thus the Lorentz factor of the relic at "production" is, at

[1] In principle dark sector charges are also possible.

most, $\gamma \simeq \sqrt{2}$ and the average initial velocity $\bar{v}_i \simeq 1/\sqrt{2}$. This corresponds to a momentum, today, $p_0 = a_i p_i$, where a_i is the scale factor at production. Structure formation limits the cosmological dark matter velocity today to be $\lesssim 5 \times 10^{-7}$ [19, 32, 40] (see also [29]), which, in turn, means that any relic produced before BBN (when the scale factor $a_{\rm BBN} \simeq 2.5 \times 10^{-10}$), as it should given the strong constraints on evaporation after BBN, is highly non-relativisitc [29].

Whether or not Planck relics are stable against discharge is a matter of some controversy. There are several reasons why a first-principles computation of the relic charge on Planck relics is problematic. First, it is unclear whether one can neglect the back-reaction of the evaporation products on the BH, and, as a consequence, whether or not the particle emission can be considered as an "individual", isolated event [34]; Second, the charge-to-mass ratio may get large enough that it could significantly affect, or even stop, the evaporation rate: recall that a Planck-mass BH reaches extremality for $Q \simeq 12\ e$. Third, it is unclear how the running of the electromagnetic coupling $\alpha_{\rm EM}$ behaves at or near the Planck scale, where several layers of new physics possibly come into play through radiative effects.

These conceptual difficulties non-withstanding, one can proceed to an analytical estimate of the relic charge on a Planck relic whose evaporation stops at some mass scale $M_R \simeq M_P$ (in what follows I always assume for definiteness $M_R = M_P$). Page [35] estimated that the probability distribution for BH charges Q is approximately Gaussian,

$$P(Q) \sim \exp\left(-4\pi\alpha(Q/e)^2\right), \tag{21.1}$$

with an rms value $Q/e = 1/\sqrt{8\pi\alpha} \simeq 2.34$. However, if the product of the BH mass and the emitted charged particle mass is small enough in Planck units, then the rms value Q/e increases up to approximately 6.

The more modern approach of Ref. [30] relaxes some assumption of the original Page calculation, such as taking into account the finite value of the BH mass times the light lepton mass, and without neglecting the BH charge in the computation of the grey body factors. The results are qualitatively similar to the original estimate, and the fraction of charged to neutral relics was found to be of order 1.

The fate of the charged relics depends on their cosmic history. It is easy to realize that direct accretion of opposite-sign particles would not discharge the BH, since the accretion cross section is approximately geometric, thus $\sigma \sim M_R^2/M_P^4 \sim 1/M_P^2$ and the accretion rate $\Gamma \lesssim n_e/M_P^2$ is always much smaller than the Hubble rate $H \sim T^2/M_P$, with T the universe's temperature, since

$$\frac{\Gamma(T)}{H(T)} \sim \frac{T^3}{M_P^2}\frac{M_P}{T^2} \sim \frac{T}{M_P} \ll 1 \text{ for } T \ll M_P. \tag{21.2}$$

Much more likely is the possibility that charged, Planck-scale relics (CPR) form bound states with opposite-sign cosmic rays. In this case, however, reionization proceeds significantly quicker [10] than for visible-sector particles [30], and thus, today, CPRs are not expected to exist in neutral bound states.

21.3 Direct Detection of Charged Planck-Scale Relics

CPRs are extremely heavy, resulting in a small direct detection event rate N, which for nominal values of the local DM density and velocity, and for a detector with effective area $A_{\rm eff}$ and efficiency ε reads

$$N \simeq 0.23\ {\rm yr}^{-1} f_{\rm CPR} \left(\frac{M_R}{M_P}\right)^{-1} \left(\frac{\rho_{\rm DM}}{0.3\ {\rm GeV/cm^3}}\right) \left(\frac{v_{\rm DM}}{300\ {\rm km/s}}\right) \left(\frac{A_{\rm eff}}{1\ {\rm m}^2}\right) \varepsilon. \quad (21.3)$$

CPRs would behave in particle detectors similarly to very heavy ions, with kinetic energies on the order of

$$K \sim \frac{1}{2} M_R v_{\rm DM}^2 \simeq 6 \times 10^{16}\ {\rm eV}\ \ (M_R \simeq M_P). \quad (21.4)$$

This should be compared with typical atomic energy scales in the 1–10^3 eV: CPRs travel through matter without significantly decreasing their kinetic energy, and essentially with no deflection, as the relative momentum transfer is also negligible.

Most of the energy deposited by the passage of a CPR through matter is delivered to nuclei and not to electrons, since $m_e \ll m_N$ and $\Delta E \sim m v_{\rm DM}^2$. Nuclear scattering of course will also produce ionization and scintillation, albeit with target-dependent quenching factors [33, 38]. Interestingly, Ref. [28] finds that the signals expected for instance in liguid Argon, ionization and scintillation, by CPRs would enable the discrimination between those tracks and those expected from other dark matter candidates. In particular, the CPRs trajectories are uniquely slated to appear as crossing the whole active medium, in every direction, producing uniform ionization and scintillation on the whole path [28].

Several classes of detectors are suitable to search for CPRs. Bubble chambers of superheated fluids such as PICO [5] are sensitive to the passage of a highly ionizing massive CPR, according to the simulations in [30], leaving a signal significantly different from that of both WIMPs and neutron backgrounds. Unfortunately, the event rate for the proposed 500 L version of PICO is too small to place significant constraints on CPRs (see Fig. 21.1).

Reference [30] estimates that the photon yield at atmsopheric fluorescence detectors such as HiRes [1] would produce a signal to noise too small to be detectable, even with very favorable assumptions on the detector's fluorescence efficiency. Similar issues apply to circumventing trigger issues in Cherenkov detectors, given that the CPRs are relatively slow.

Direct dark matter detectors are, on the other hand, ideally placed to detect CPRs: the overburdain is completely irrelevant in stopping the CPRs, and some of the largest noble gas detectors have a significant effective area and very high efficiency. Finally, so-called paleo-detectors have such long effective exposure times that they provide the best opportunity for the direct detection of CPRs.

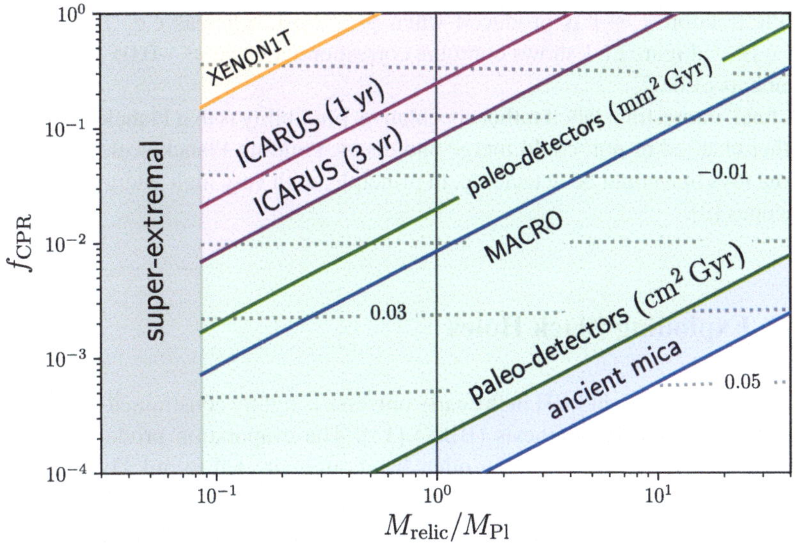

Fig. 21.1 Projected 99% CL upper limit on the mass and density of CPRs with experiments of several classes. See text for details. **Orange:** a 10 yr exposure of XENON1T [6]. **Magenta:** solid: a 3 yr exposure of ICARUS. The dashed line shows a 1 yr exposure. **Green:** solid: estimated limits from a paleo-detector with $\mathcal{E} = 1$ and a 1 cm^2 Gyr exposure. The dashed line shows a 1 mm^2 Gyr exposure. **Blue:** strongest possible limits from monopole searches, including a direct search by MACRO and a search for tracks in ancient mica [20]. **Dotted gray:** relic fractions produced assuming an initial a power-law mass function with index γ. Contours step from $\gamma = -0.05$ to $\gamma = 0.05$ from top to bottom in increments of 0.02. **Shaded gray:** region prohibited by super-extremality for a charge of $1e$. Reproduced, with permission, from Ref. [30]

Figure 21.1 shows projected, future constraints from a variety of experimental facilities. In particular, the left, grey region corresponds to masses for which the PBHs would be super-extremal for a charge $Q = 1\ e$, and the vertical line corresponds to $M_R = M_P$. Shaded regions are testable, or ruled out, by direct dark matter search experiments such as XENON1T (orange), and by neutrino experiments such as ICARUS (magenta). The green region corresponds to constraints from paleodetectors with different effective exposures. Finally, the blue region is constrained by monopole searches with MACRO and for tracks in ancient mica [20].

The dashed lines reflect the possibility that the PBHs be initially distributed according to a power-law initial mass function $M\,\mathrm{d}N/\mathrm{d}M \propto M^{\gamma-1}$. Assuming that the entirety of the cosmological dark matter is in the form of PBHs, and that black holes with mass $M > M_U$ do not lose any significant amount of mass, the mass fraction in CPRs is for $\gamma < 0$,

$$\frac{\Omega_{\text{CPR}}}{\Omega_{\text{DM}}} \approx \frac{M_{\text{Pl}} M_U^{\gamma-1} - M_{\text{Pl}}^{\gamma}}{M_U^{\gamma-1}\left[M_{\text{Pl}} - (1 - 1/\gamma) M_U\right] - M_{\text{Pl}}^{\gamma}}. \tag{21.5}$$

A CPR fraction $f \sim 1$ is produced when $\gamma \lesssim -0.1$, whereas e.g. $f \sim 10^{-2}$ for $\gamma \sim -10^{-2}$. Figure 21.1 shows contours corresponding to $\gamma = -0.05$ up to 0.05 in increments of 0.02.

A final, interesting, labeit rather speculative, possibility is that Planck-scale relics, whether charged or not, could merge and decay, emitting Planck-scale cosmic and gamma rays that could be detectable, in principle, with very high-energy cosmic ray telescopes [9].

21.4 Exploding Black Holes

The evaporation of light PBH in the early universe is tightly constrained if happening after Big Bang Nucleosynthesis (BBN) [13]. The evaporation products of PBHs evaporating *prior* to BBN, on the other hand, are expected to quickly thermalize, leaving no measurable imprint, with at least three important exceptions:

1. If the PBH energy density is comparable, or larger than the energy density of the rest of the universe, then the expansion rate is affected by evaporation, that acts as an episode of "reheating"; constraints can thus arise if such process is close enough to BBN;
2. PBH evaporation prior to BBN can, in principle, be responsible for the production of dark sector species [32], including the cosmological dark matter, and even of the baryon asymmetry (for instance via the production of right-handed heavy neutrinos whose decay is CP violating, as in the standard leptogenesis scenario [36], or via effective operators that are CP violating and couple to the $B - L$ current, as envisioned in [37]);
3. An evaporation product that is slated not to thermalize is gravitons: graviton production would manifest itself as a stochastic background of very high-frequency gravitational waves [18]. The frequency of the signal could be lowered in principle by either a non-standard cosmological history in the early universe [26], or by the existence of extra-dimensions, with an associated suppressed Planck scale [25].

PBHs produced in the early universe may, in principle, be at the end of their evaporation process today. Reference [11] calculates that for an initial mass function $\psi(M) \equiv M dN/dM$, the rate of PBH evaporating today is

$$\dot{n}_{\rm PBH} \simeq \rho_{\rm DM} \frac{\psi_i\,(M_U)}{3 t_U}, \tag{21.6}$$

with t_U the age of the universe. Depending on the mass function, constraints on the PBH abundance bound this rate to be generically rather low, at most around one event per cubic parsec per year, in the most favorable possible case of a very narrow lognormal mass function. This should be compared with the HAWC limits on the explosion rate, which at present is $\dot{n}_{\rm PBH} < 3,400\ {\rm pc}^{-3}{\rm yr}^{-1}$.

On the other hand, PBHs could arise in the *late* universe (see the discussion in Chap. 9 of this book). If that is the case, the rate of explosions is virtually unconstrained, and could be very large. As a result, PBHs could be associated with astroparticle anomalies such as the Galactic center gamma-ray excess [4], or that, as well as the AMS tentative ^{3}He events [12, 14], and the antiproton excess [15].

PBH "explosions" (as Hawking referred to them [22]) look like "backwards" gamma-ray bursts: the luminosity and photon energy grows in a runaway process, which suddenly stops as evaporation completes. Reference [11] searched for such events both in the Fermi GRB catalogue [27] and in the Fermi Large Area Telescope transient sources [8], in the case where evaporation may not have completed yet. Critically, any events potentially associated with exploding PBHs would most likely have a non-trivial proper motion in the sky, which was not observed for any of the Fermi candidate events. As a result, there is no current evidence for any PBH explosion, consistent with theoretical expectations.

Reference [11]'s results, summarized in Fig. 21.2, indicate that even the most competitive future telescope, the Cherenkov Telescope Array (CTA) [3] would only be able to observe an exploding PBH if it were closer than, approximately, 1 pc, under the best possible circumstances. As such, it may be possible to actually measure time of arrival differences with satellites positioned at macroscopic distances from one another: Observations with the Interplanetary Gamma-Ray Burst Timing Network[2]

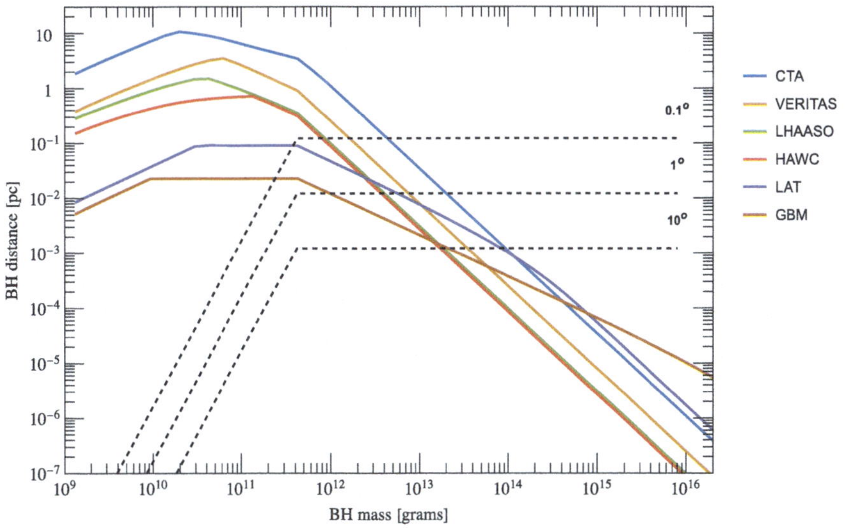

Fig. 21.2 Maximum visible distance curves with respect to black hole mass for a number of modern gamma-ray detectors. The black dashed lines indicate the distance, at a given mass, at which the proper motion during an observation up to one year long may exceed 0.1, 1 and 10 °. Reproduced with permission from [11]

[2] https://heasarc.gsfc.nasa.gov/docs/heasarc/missions/ipn.html.

(IPN) of GRB detectors could likely be the most promising search technique to identify a relatively nearby GRB-like event from a genuine, extra-galactic GRB [39]. Searches for local events are currently under way. Figure 21.2 also shows that a large proper motion (shown with black dashed lines) is expected to occur if the evaporation is not yet in its runaway, final phase.

The observation of PBH explosions could very strongly constrain the existence of otherwise "dark" (meaning not sharing any interaction with visible sector particles) degrees of freedom. BH evaporation is independent of particle interactions, and as such, as soon as a particle is kinematically accessible to BH evaporation, BHs need evaporate into it, shortening the runaway evaporation/explosion process. High-energy gamma-ray telescopes such as HAWC [2] and CTA [3] are ideally suited to search for such dark degrees of freedom in the putative light curves of evaporating PBHs [7].

A staggering possible fate of light PBH evaporation is to form "hot spots" in the early universe that would lead to a certain, finite decay rate of metastable vacua, including the electroweak vacuum; this could lead, in turn, and assuming the validity of the posited temperature profile of such hot spots, to significant constraints on the initial abundance of light PBH [21].

An additional interesting possibility is that PBH be responsible for reheating at the end of inflation, as discussed e.g. in Ref. [24]. In this scheme, ultralight PBHs are formed (e.g., from the inflaton field) after inflation, go on to dominate the energy density of the Universe, and evaporate before Big Bang Nuncleosynthesis begins. In this scenario non-gravitational couplings between the inflaton and the SM are not required, making PBH reheating and the corresponding inflation scenarios rather economic from a model building perspective.

21.5 Conclusion and Outlook

The most likely fate of light PBHs is either complete evaporation in a runaway process, or the formation of Planck-scale relics. There exist several arguments for the latter possibility, but the first is not excluded, as physics at the Planck scale is largely unknown. Here, I reviewed how the existence of such light relics, or exploding PBHs, can be tested with experiments and observations.

Unless Planck-scale relics are electrically or magnetically charged, or unless they merge and evaporate in the late universe at appreciable rates, such particles are virtually undetectable. The most competitive detectors, for relic charged Planck-scale BHs, are paleo-detectors - ancient crystals whose effective exposure time is on the order of the age of the Earth, and that can be searched for for defects created by the passage of a highly-ionizing, massive particle. Neutrino experiments and large, noble-gas direct dark matter detectors are also rather competitive.

Exploding PBHs—either coming from an initial mass function serendipitously peaked at the mass corresponding to a BH lifetime approximately equal to the age of the universe today, or formed by some non-primordial mechanism at late times—can be detected searching for what is, effectively, a "reverse gamma-ray burst": different telescopes, with effective areas peaking at vastly different energies, are sensitive to different evaporation epochs; the terminal explosive, runaway phase is most constrained by the highest-energy, largest-effective-area observatories, such as the future CTA observatory; ongoing evaporation can be effectively searched for with GRB telescopes, including the Fermi Large Area Telescope.

While no evidence of exploding PBHs has emerged so far, perhaps the most promising path ahead is to use the Interplanetary Gamma-Ray Burst Timing Network, as exploding PBHs leaving a visible imprint are slated to be much closer than, approximately, 1 pc, as opposed to cosmologically-distant GRBs.

Acknowledgements I am deeply grateful to Benjamin Lehmann, who has driven most of the the work discussed here. It is because of students like Ben that I love my job. I am also very grateful to the reviewer, Kaloian D. Lozanov, for providing numerous suggestions and corrections upon reviewing an earlier version of this paper. This material is based upon work supported in part by the U.S. Department of Energy, grant number de-sc0010107.

References

1. R.U. Abbasi et al., A Study of the composition of ultrahigh energy cosmic rays using the high resolution fly's eye. Astrophys. J. **622**, 910–926 (2005). https://doi.org/10.1086/427931, arXiv:astro-ph/0407622
2. A.U. Abeysekara et al., On the sensitivity of the HAWC observatory to gamma-ray bursts. Astropart. Phys. **35**, 641–650 (2012). https://doi.org/10.1016/j.astropartphys.2012.02.001, arXiv:1108.6034 [astro-ph.HE]
3. B.S. Acharya et al. Introducing the CTA concept. Astropart. Phys. **43**, 3–18 (2013). https://doi.org/10.1016/j.astropartphys.2013.01.007
4. M. Ackermann et al., The Fermi Galactic center GeV excess and implications for dark matter. Astrophys. J. **840**(1), 43 (2017). https://doi.org/10.3847/1538-4357/aa6cab, arXiv:1704.03910 [astro-ph.HE]
5. C. Amole et al., Improved dark matter search results from PICO-2L Run 2. Phys. Rev. D **93**(6), 061101 (2016). https://doi.org/10.1103/PhysRevD.93.061101, arXiv:1601.03729 [astro-ph.CO]
6. E. Aprile et al., Dark Matter Search Results from a One Ton-Year Exposure of XENON1T. Phys. Rev. Lett. **121**(11), 111302 (2018). https://doi.org/10.1103/PhysRevLett.121.111302, arXiv:1805.12562 [astro-ph.CO]
7. M.J. Baker, A. Thamm, Black hole evaporation beyond the standard model of particle physics. JHEP **01**, 063 (2023). https://doi.org/10.1007/JHEP01(2023)063, arXiv:2210.02805 [hep-ph]
8. L. Baldini et al., Catalog of long-term transient sources in the first 10 yr of Fermi-LAT data. Astrophys. J. Supp. **256**(1), 13 (2021). https://doi.org/10.3847/1538-4365/ac072a, arXiv:2106.00100 [astro-ph.HE]
9. A. Barrau et al., Dark matter as Planck relics without too exotic hypotheses. Phys. Rev. D **100**(12), 123505 (2019). https://doi.org/10.1103/PhysRevD.100.123505, arXiv:1906.09930 [hep-ph]
10. W.J. Boardman, *The Radiative Recombination Coefficients of the Hydrogen Atom*, vol. 9 (1964), p. 185. https://doi.org/10.1086/190101
11. X. Boluna et al., Searching for Exploding black holes. JCAP **04**, 024 (2024). https://doi.org/10.1088/1475-7516/2024/04/024, arXiv:2307.06467 [astro-ph.HE]

12. E. Carlson et al., Antihelium from dark matter. Phys. Rev. D **89**(7), 076005 (2014). https://doi.org/10.1103/PhysRevD.89.076005, arXiv:1401.2461 [hep-ph]
13. B. Carr et al., Constraints on primordial black holes. Rept. Prog. Phys. **84**(11), 116902 (2021). doi:https://doi.org/10.1088/1361-6633/ac1e31, arXiv:2002.12778 [astro-ph.CO]
14. A. Coogan, S. Profumo. Origin of the tentative AMS antihelium events. Phys. Rev. D **96**(8), 083020 (2017). https://doi.org/10.1103/PhysRevD.96.083020, arXiv:1705.09664 [astro-ph.HE]
15. M.-Y. Cui et al., Possible dark matter annihilation signal in the AMS-02 antiproton data. Phys. Rev. Lett. **118**(19), 191101 (2017). https://doi.org/10.1103/PhysRevLett.118.191101, arXiv:1610.03840 [astro-ph.HE]
16. B.S. DeWitt, Quantum field theory in curved space-time. Phys. Rept. **19**, 295–357 (1975). https://doi.org/10.1016/0370-1573(75)90051-4
17. M.D. Diamond, D.E. Kaplan, Constraints on relic magnetic black holes. JHEP **03**, 157 (2022). https://doi.org/10.1007/JHEP03(2022)157, arXiv:2103.01850 [hep-ph]
18. A.D. Dolgov, D. Ejlli, Relic gravitational waves from light primordial black holes. Phys. Rev. D **84**, 024028 (2011). https://doi.org/10.1103/PhysRevD.84.024028, arXiv:1105.2303 [astro-ph.CO]
19. T. Fujita et al., Baryon asymmetry, dark matter, and density perturbation from primordial black holes. Phys. Rev. D **89**(10), 103501 (2014). https://doi.org/10.1103/PhysRevD.89.103501, arXiv:1401.1909 [astro-ph.CO]
20. D. Ghosh, S. Chatterjea, Supermassive magnetic monopoles flux from the oldest mica samples. EPL **12**, 25–28 (1990). https://doi.org/10.1209/0295-5075/12/1/005
21. L. Hamaide et al., *Primordial Black Holes are True Vacuum Nurseries* (2023). arXiv: 2311.01869 [hep-ph]
22. S. W. Hawking. "Black hole explosions". In: *Nature* 248 (1974), pp.& Acirc; 30–31. DOI: 10.1038/248030a0
23. S.W. Hawking, Zeta function regularization of path integrals in curved space-time. Commun. Math. Phys. **55**, 133 (1977). https://doi.org/10.1007/BF01626516
24. J.C. Hidalgo, L. Arturo Urena-Lopez, A.R. Liddle, Unification models with reheating via primordial black holes. Phys. Rev. **D85**, 044055 (2012). https://doi.org/10.1103/PhysRevD.85.044055, arXiv:1107.5669 [astro-ph.CO]
25. A. Ireland, S. Profumo, J. Scharnhorst, *Gravitational Waves from Primordial Black Hole Evaporation with Large Extra Dimensions* (2023). arXiv:2312.08508 [gr-qc]
26. A. Ireland, S. Profumo, J. Scharnhorst, Primordial gravitational waves from black hole evaporation in standard and nonstandard cosmologies. Phys. Rev. D **107**(10), 104021 (2023). https://doi.org/10.1103/PhysRevD.107.104021, arXiv:2302.10188 [gr-qc]
27. A. von Kienlin et al., the fourth fermi-GBM gamma-ray burst catalog: a decade of data. Astrophys. J. **893**, 46 (2020). https://doi.org/10.3847/1538-4357/ab7a18, arXiv:2002.11460 [astro-ph.HE]
28. I. Lazanu, S. Lazanu, Mihaela Parvu, About detecting very low mass black holes in LAr detectors. JCAP **10**, 046 (2020). https://doi.org/10.1088/1475-7516/2020/10/046, arXiv:2006.09974 [hep-ph]
29. B.V. Lehmann, S. Profumo, Black hole remnants are not too fast to be dark matter. Phys. Dark Univ. **39**, 101145 (2023). https://doi.org/10.1016/j.dark.2022.101145, arXiv:2105.01627 [gr-qc]
30. B.V. Lehmann et al., Direct detection of primordial black hole relics as dark matter. J. Cosmol. Astroparticle Phys. **2019**(10), 046 (2019). ISSN:1475-7516. https://doi.org/10.1088/1475-7516/2019/10/046
31. J.H. MacGibbon, Can planck-mass relics of evaporating black holes close the universe? Nature **329**, 308–309 (1987). https://doi.org/10.1038/329308a0
32. L. Morrison, S. Profumo, Y. Yu, Melanopogenesis: dark matter of (almost) any mass and baryonic matter from the evaporation of primordial black holes weighing a ton (or less). JCAP **05**, 005 (2019). https://doi.org/10.1088/1475-7516/2019/05/005, arXiv:1812.10606 [astro-ph.CO]

33. W. Mu, X. Ji, Ionization yield from nuclear recoils in liquid-xenon dark matter detection. Astropart. Phys. **62**, 108–114 (2015). https://doi.org/10.1016/j.astropartphys.2014.07.013, arXiv:1310.2094 [physics.ins-det]
34. D.N. Page, Particle emission rates from a black hole. 2. Massless particles from a rotating hole. Phys. Rev. D **14**, 3260–3273 (1976). https://doi.org/10.1103/PhysRevD.14.3260
35. D.N. Page, Particle emission rates from a black hole. 3. Charged leptons from a nonrotating hole. Phys. Rev. D **16**, 2402–2411 (1977). https://doi.org/10.1103/PhysRevD.16.2402
36. A. Pilaftsis, The little review on leptogenesis. J. Phys. Conf. Ser. **171**, 012017 (2009). J. Bernabeu et al. (ed.). https://doi.org/10.1088/1742-6596/171/1/012017, arXiv:0904.1182 [hep-ph]
37. N. Smyth, L. Santos-Olmsted, S. Profumo, Gravitational baryogenesis and dark matter from light black holes. JCAP **03**(03), 013 (2022). https://doi.org/10.1088/1475-7516/2022/03/013, arXiv:2110.14660 [hep-ph]
38. V.I. Tretyak, Semi-empirical calculation of quenching factors for ions in scintillators. Astropart. Phys. **33**, 40–53 (2010). https://doi.org/10.1016/j.astropartphys.2009.11.002, arXiv:0911.3041 [nucl-ex]
39. T.N. Ukwatta et al., *Investigation of Primordial Black Hole Bursts using Interplanetary Network Gamma-ray Bursts* (2015). arXiv:1512.01264 [astro-ph.HE]
40. M. Viel et al., Constraining warm dark matter candidates including sterile neutrinos and light gravitinos with WMAP and the Lyman-alpha forest. Phys. Rev. D **71**, 063534 (2005). https://doi.org/10.1103/PhysRevD.71.063534, arXiv:astro-ph/0501562

Chapter 22
The Asteroid Mass Window

Peter Tinyakov

Abstract Primordial black holes are an attractive dark matter candidate, particularly if they can explain the totality of it. At primordial black hole masses below $\sim 10^{17}$ g and above $\sim 10^{23}$ g this possibility is excluded from the variety of arguments and with different confidence. The range in between, often referred to as the "asteroid mass window", currently remains unconstrained. The most promising, in our view, way to probe this mass range is to use stars as the primordial black hole detectors. If a star captures even a single primordial black hole it starts being accreted onto it and eventually gets destroyed—converted into a sub-solar mass black hole. This process may have a variety of signatures form a mere star disappearance to supernova-type explosions of a new kind. The viability of this approach depends crucially on the probability of primordial black hole capture by stars. In this chapter we summarize the existing capture mechanisms and discuss their implications for constraining the abundance of (or perhaps discovering) primordial black holes in the asteroid mass window.

22.1 Introduction

Of the whole mass range where the primordial black holes (PBHs) may potentially constitute the dark matter (DM), the window of roughly 10^{17}–10^{23} g is particularly difficult to assess. No actual constraints, only proposals exist in this window at present [11]. Below $\sim 10^{17}$ g the PBHs evaporate fast enough to either completely disappear by now or overproduce the diffuse gamma ray background which is constrained by the existing observations [10, 14]. At higher masses starting at around $\sim 10^{23}$ g (for comparison, the Moon mass is 7×10^{26} g) the PBHs start being detectable through the gravitational lensing of background stars [20, 21]. The range between these two values, spanning some 6 orders of magnitude and often referred to as the *"asteroid mass window"*, is currently unconstrained. In this mass range the PBHs have sizes of order or smaller than the atomic size and are very difficult to detect.

P. Tinyakov (✉)
Université Libre de Bruxelles (ULB), CP225 bvd. du Triomphe, 1050 Brussels, Belgium
e-mail: petr.tiniakov@ulb.be

C. Byrnes et al. (eds.), *Primordial Black Holes*, Springer Series in Astrophysics and Cosmology, https://doi.org/10.1007/978-981-97-8887-3_22

Below $\sim 10^{23}$ g the PBHs are too light to be detectable by the standard star lensing. A proposal has been made to use instead gamma-ray bursts as lensed sources [1] and look, instead of time variability, for features in the energy spectrum where PBHs, if present, would induce oscillations. However, it has been argued later that finite size of sources together with the wave optics effects will cause the decoherence resulting in no detectable spectral modulations for most of the gamma-ray bursts except, perhaps, a small fraction of rapidly varying ones [22]. Most of other proposals in the asteroid mass range turn around using stars as PBH detectors. They are based either on the capture of PBHs by stars with the subsequent star destruction [7–9, 15, 16], or on ignition of nuclear reactions in white dwarfs by a traversing PBH [19]. An overview of these proposals can be found in Ref. [26]. Here we focus on the PBH capture by stars.

As we will see below, even in the most favorable conditions a star can only capture a few PBHs over its lifetime. This is too small to have any dynamical or gravitational effects on the star. However, once captured inside the star, a PBH starts to accrete the ambient matter. If the Bondi accretion rate is assumed,[1] the accretion is sufficiently rapid even in the case of ordinary stars to be able to consume the whole star in a cosmologically short time. In the Bondi regime [5], the accretion rate by a PBH of mass m in the center of a star with the matter density ρ_* is estimated as

$$\dot{m} \simeq 4\pi \frac{G^2 m^2 \rho_*}{c_s^3}$$

where G is the gravitational constant and c_s the sound speed, and we have suppressed a numerical factor of order 1 depending on the matter equation of state. With this rate the time it takes a PBH to consume the whole star, call it accretion time $t_{\rm acc}$, is determined by the longest initial stage when $m \ll M_\odot$ and the star itself is not yet modified, while the uncertain final stages are in any case much shorter. The accretion time is therefore determined by the *initial* PBH mass and reads

$$t_{\rm acc} \simeq \frac{c_s^3}{4\pi G^2 \rho_* m} \simeq 5 \times 10^6 {\rm yr} \left(\frac{c_s}{500{\rm km/s}}\right)^3 \left(\frac{150{\rm g/cm}^3}{\rho_*}\right) \left(\frac{10^{20}{\rm g}}{m}\right).$$

We have substituted here the parameters typical of solar mass stars which have lifetimes exceeding ~ 10 Gyr. We see that for PBH masses $m \gtrsim 10^{18}$ g the accretion times are much smaller than the age of the Universe, and much smaller than the star lifetime. For compact stars such as white dwarfs and neutron stars the accretion times are much shorter.

Thus, capturing an asteroid-mass BH is fatal for stars—they get destroyed in a relatively short time. This may lead to several observable consequences like disappearance of stars, creation of (sub)solar mass black holes and supernova-type

---

[1] While not completely obvious, it is plausible that Bondi regime applies in the case of a small PBH inside the star. Nevertheless, this question requires a quantitative analysis which will be given elsewhere.

explosions of a new type. All three signatures may potentially be used to constrain (or discover) the abundance of primordial black holes. However, before discussing the signatures we need to consider how at all PBHs can be captured inside a star.

22.2 Capture Generalities

Being a part (or whole) of the DM, PBHs and stars are both concentrated in galactic halos. However, because the interactions of a PBH with matter are very weak relative to its enormous mass, it is very difficult to settle a PBH inside a star. There are basically two ways: either a PBH can be captured by a star from the local DM density, or it can get inside the star from the star birth. In either case, the process typically requires two stages:

(i) *trapping*, or getting on a star-crossing gravitationally bound orbit, and
(ii) *cooling*, i.e., gradually loosing energy until settled completely inside the star.

The second stage, cooling, is common for both mechanisms. Once the PBH is completely inside the star, the accretion phase starts.

During the cooling stage, and also during the first stage in case of the direct capture, the key quantity is the energy loss at a single star crossing. So we start this section by having a closer, more quantitative look at this quantity.

Once a PBH is on a bound star-crossing orbit it starts to lose energy at each subsequent passage through the star. This process has to be fast enough, at least it should not take longer than the star lifetime to get completely inside the star. Given the PBH mass, the only free parameter here is the initial size of the PBH orbit, depending on which there may or may not be enough time to cool. Thus, not all trapped PBHs will eventually get captured and destroy the star. The cooling process depends only on the energy loss; it will be the next topic in this section.

In the ideal case of an isolated star, once a PBH gets on a bound star-crossing orbit it will cross the star again and again until it gets absorbed inside. This is not necessarily the case for a star in a realistic environment like a galaxy, because gravitational perturbations due to neighboring stars may deviate the PBH and make it miss the star it is orbiting, so that the energy loss will halt. So, as the last general point in this section we will quantify the conditions under which the perturbers become important.

22.2.1 Energy Loss

A PBH passing through a star slows down for two reasons. First, it accretes the star matter. In the star frame the total PBH momentum does not change, while the mass grows, so the velocity decreases. This is equivalent to the drag force

$$F_{\rm acc} = -\dot{m} v$$

where $\dot{m}$ is the mass accretion rate and we have assumed a non-relativistic PBH velocity v. To estimate the mass accretion rate we note that even those PBHs that have small (inbound) asymptotic velocities get accelerated when they fall into a star, so that they cross it at a (marginally) supersonic speed. Indeed, for a sun-like star, the escape velocity from the star center is $v \sim \sqrt{3GM_*/R_*} \sim 750$ km/s, while the sound speed in the Sun center is ~ 500 km/s. As one moves away form the center towards the surface the sound speed decreases by an order of magnitude, while the escape velocity changes by $\sqrt{2/3}$ at most. For the purpose of the estimate we may therefore consider the stellar matter as a collection of free particles. Then the accretion rate is $\dot{m} = \rho_* \sigma_{\rm BH} v$ where $\sigma_{\rm BH}$ is the cross section of particle capture by a black hole,

$$\sigma_{\rm BH} = 16\pi \frac{G^2 m^2}{v^2}.$$

We have therefore

$$F_{\rm acc} = -16\pi \rho_* G^2 m^2$$

where the velocity v canceled out. Here and below we use natural units $\hbar = c = 1$.

The second force that slows down the PBH is the dynamical friction [4, 13]. This force arises because matter particles get dragged along with the passing PBH by the gravitational force; correspondingly, the PBH is slowed down. The dynamical friction force is written as follows,

$$F_{\rm dyn} = -4\pi \rho_* G^2 m^2 \frac{\ln \Lambda}{v^2} \tag{22.1}$$

where $\ln \Lambda \sim 30$ is a Coulomb logarithm. Thus, the dynamical friction is dominant over the accretion by the factor $\ln \Lambda / v^2$ which is much larger than 1, except perhaps in the case of neutron stars where $v \sim 1$ and the dynamical friction is somewhat reduced due to matter degeneracy.

As in the case of accretion, Eq. (22.1) assumes that the matter particles do not interact among themselves, which is a good approximation at supersonic PBH velocities. In our case the velocity of the PBH when passing through a star is only marginally supersonic. In this case the dynamical friction (22.1) gives a smaller force than a more accurate treatment that takes into account properties of the medium [28]. The accurate answer is some factor $\sim$2 larger than Eq. (22.1) in the case of Sun-like stars (see Fig. 4 of Ref. [27]) due to a resonance-type enhancement at velocities close to the sound speed [28]. We do not include this numerical factor in what follows.

With the force (22.1), the energy loss per single star crossing is $\Delta E \sim R_* F_{\rm dyn}$. More accurately, one has to average the energy loss over the PBH impact parameter. If one assumes for simplicity the uniform distribution of the PBH trajectories in the plane perpendicular to the PBH velocity we get, in agreement with Eq. (2) of Ref. [8]

$$E_{\rm loss} = \frac{1}{\pi R_*^2} \int ds\, dl F_{\rm dyn} = \frac{4 M_* G^2 m^2}{R_*^2 v^2} \ln \Lambda . \tag{22.2}$$

Here we have ignored the change of v^2 along the PBH trajectory within the star. We will see later, however, that the distribution of trajectories in some cases is not uniform, and Eq. (22.3) in these cases is only an order-of-magnitude estimate.

Substituting in (22.2) the escape velocity from the star surface $v_{\rm esc}^2 = 2GM_*/R_*$ we can rewrite it as follows,

$$\frac{E_{\rm loss}}{m} = \frac{2Gm}{R_*} \ln \Lambda = \frac{R_g}{R_*} \frac{m}{M_*} \ln \Lambda , \tag{22.3}$$

where $R_g = 2GM_*$ is the gravitational radius of the star. Note that this expression is only valid in the case when $v \simeq v_{\rm esc}$, which explains why the independence of $E_{\rm loss}$ of M_* is not a problem. We see from Eq. (22.3) that the more compact is the star, the larger are the energy losses, but even in the best case of a neutron star with $R_g \sim R_*/3$ the energy loss is very small for an asteroid mass PBH for which $m/M_* \lesssim 10^{-10}$.

It follows immediately from this estimate that only very slow PBHs can be trapped after one star crossing. To be trapped, the asymptotic kinetic energy of a PBH must be smaller than $E_{\rm loss}$, which translates into the condition on the asymptotic velocity v_∞ of PBH

$$v_\infty < 2 \left(\frac{Gm}{R_*} \ln \Lambda \right)^{1/2} .$$

Numerically, a PBH of mass $m = 10^{23}$ g can only get trapped if its asymptotic velocity relative to the star does not exceed some maximum value which is 0.03 km/s for the Sun and 9 km/s for a neutron star. Even in the case of a neutron star this velocity is smaller than the velocity dispersion in most astrophysical environments, with notable exception of some dwarf galaxies.

Another immediate consequence of this estimate is large initial size of the PBH bound orbit after trapping. When a PBH from the ambient DM distribution passes through a star it loses the amount of energy (22.3). If its initial kinetic energy was smaller it gets trapped. A typical energy of a freshly trapped PBH is therefore of the order $-E_{\rm loss}$. Such PBHs have radial orbits of the size

$$r_{\rm max} \sim \frac{GM_* m}{E_{\rm loss}} = \frac{M_*}{2\, m \ln \Lambda} R_* . \tag{22.4}$$

The trapped PBH continues to lose energy at each subsequent star crossing and finally gets completely absorbed in the star, provided that (i) there is enough time and (ii) the trajectory is not perturbed so that it stops passing through the star and lose energy. We come now to these two conditions.

22.2.2 Cooling Time

Since the energy loss at a single star crossing is very small, the size of the first bound orbit is very large, cf. Eq. (22.4). Outside of the star such an orbit is a Keplerian orbit with the apastron $a = r_{\max}$ and periastron $b < R_* \ll a$. One may therefore consider this orbit as radial. For such an orbit the time between the first and the next star crossing is

$$\Delta t = \pi \frac{a^{3/2}}{\sqrt{R_g}}.$$

If we assume that after the first passage the PBH total energy is $E_1 = -E_{\text{loss}}$, at the n-th step its energy is

$$E_n = -nE_{\text{loss}}.$$

Correspondingly, the size of the nth orbit is

$$a_n = \frac{r_{\max}}{n}$$

and the return time at the nth step is

$$\Delta t_n = \frac{\Delta t}{n^{3/2}}.$$

The total cooling time converges and is saturated by a few first orbits. Summing up Δt_n one finds

$$t_{\text{cool}} = \sum_{n=1} \Delta t_n = \pi \zeta(3/2) \frac{r_{\max}^{3/2}}{\sqrt{R_g}} \sim 8 \frac{r_{\max}^{3/2}}{\sqrt{R_g}}. \tag{22.5}$$

This equation is valid when the initial size of the orbit is $r_{\max}$ given by (22.4), as determined by E_{loss} of Eq. (22.3).

In the case when the initial size of the orbit a_0 is much smaller than $r_{\max}$, one may either calculate a partial sum in Eq. (22.5) and then use the asymptotics of the resulting Hurwitz zeta function, or use the continuous energy loss approximation. After the star crossing, the energy changes by $-E_{\text{loss}}$, and correspondingly the apastron of the orbit decreases by

$$\Delta a = -\frac{a^2}{GM_* m} E_{\text{loss}}.$$

Approximating da/dt by $\Delta a/\Delta t$ one obtains the differential equation for the change of the apastron

$$\frac{da}{dt} = -\frac{2E_{\text{loss}}}{\pi m \sqrt{R_g}} \sqrt{a}.$$

Solving this equation one finds the cooling time as a function of the initial apastron a_0,

$$t_{\text{cool}}(a_0) = \frac{\pi\sqrt{R_g}m}{E_{\text{loss}}}\sqrt{a_0} = \frac{\pi M_* R_*}{m\sqrt{R_g}\ln\Lambda}\sqrt{a_0}, \tag{22.6}$$

where we have substituted E_{loss} from Eq. (22.3). This equation is valid for $a \ll r_{\text{max}}$, with r_{max} given by Eq. (22.4). Note that if one formally takes $a_0 = r_{\text{max}}$ in Eq. (22.6), the result is parametrically the same as Eq. (22.5) and only differs by a numerical coefficient.

22.2.3 *Perturbers*

For a successful cooling the bound PBH trajectory must cross the star at each approach, i.e. the periastron must be smaller than the star radius. For an isolated star and PBH orbits that become bound as a result of the energy loss *inside the star* this is automatically guaranteed. However, in a realistic environment (in a galaxy) the gravitational perturbations from nearby stars may make the trajectory that originally passed through the star to miss it at the next approach. If this happens, the PBH starts orbiting the star without energy loss and the cooling will never complete. Following Ref. [16] and assuming the perturber is a nearby star of the same mass, let us derive the condition on the distance to this star which guarantees that the perturbation is not big enough to halt the cooling.

It should be noted that gravitational perturbations may also cause orbits originally not passing through the star to do so when perturbed. Therefore, the condition that we derive below may be over-conservative. This, however, requires a more detailed calculation.

For a very extended nearly radial orbit the periastron r_{min} is completely determined by the PBH angular momentum J through the relation

$$r_{\text{min}} = J^2/(m^2 R_g).$$

Requiring $r_{\text{min}} < R_*$ sets an upper bound on the angular momentum

$$J_{\text{max}}/m = \sqrt{R_* R_g}. \tag{22.7}$$

The condition that the orbit crosses the star $r_{\text{min}} < R_*$ is thus equivalent to requiring that $J < J_{\text{max}}$.

Consider the change of the PBH angular momentum under the influence of a small gravitational potential $U(\mathbf{x})$. For definiteness we suppose that the unperturbed PBH trajectory is $x(t)$, and the perturbation is in the plane (x, y). We are interested in the change of the angular momentum ΔJ accumulated over one fall from $x = r_{\text{max}}$ to $x \sim R_*$. We have

$$\Delta J/m = \int_0^{T/4} x^2(t) \frac{\partial^2 U}{\partial x \partial y} dt,$$

where T is the period of the radial Keplerian orbit with the apastron r_{max}. To estimate the integral we use the unperturbed motion $x(t)$ and assume that the perturber is far away, so that the derivative $\partial_x \partial_y U$ is approximately constant along the trajectory. This gives

$$\Delta J/m = \frac{5\pi}{16} \frac{r_{\text{max}}^{7/2}}{R_g^{1/2}} \frac{\partial^2 U}{\partial x \partial y}. \tag{22.8}$$

We have to require that $\Delta J \lesssim J_{\text{max}}$.

If the perturber is a star with the same mass at a distance $d \gg r_{\text{max}}$ which we assume to be static, the derivative of the potential is estimated as $\partial^2 U/\partial x \partial y \sim R_g/d^3$ and this condition translates into

$$r_{\text{max}} < d \left(\frac{R_*}{\alpha d} \right)^{1/7}. \tag{22.9}$$

Here α is a numerical coefficient of order 1 that depends on the relative orientation of the PBH orbit and the direction to the perturber. In view of the small power in Eq. (22.9) the numerical value of this coefficient is not so important. In a given environment the distance d can be estimated from the density of stars.

22.3 Direct Capture

With these tools at hand, consider now the direct capture in more detail. It is instructive to start by deriving the rate of trapping of PBHs by an isolated star. So, assume an isolated star at rest is placed in a DM halo of density ρ_{DM} consisting of PBHs which have Maxwell distribution in velocities with the dispersion σ,

$$dn = \frac{\rho_{\text{DM}}}{m} \left(\frac{3}{2\pi\sigma^2} \right)^{3/2} \exp\left(-\frac{3v^2}{2\sigma^2} \right) d^3 v. \tag{22.10}$$

Some of the PBHs from the halo will cross the star and lose enough energy to become gravitationally bound on orbits that cross the star, i.e., will get trapped. Averaging over the Maxwell distribution, the rate at which this happens is [29]

$$\frac{dN}{dt} = 2\sqrt{6\pi} \frac{\rho_{\text{DM}}}{\sigma m} G M_* R_* \left\{ 1 - \exp\left(-\frac{3E_{\text{loss}}}{m\sigma^2} \right) \right\} \simeq 6\sqrt{6\pi} \frac{\rho_{\text{DM}}}{\sigma^3} \frac{G M_* R_*}{m^2} E_{\text{loss}},$$

where in the last equality we have expanded the exponential assuming $3E_{\text{loss}}/(m\sigma^2) \ll 1$. Note that only the low-velocity end of the distribution contributes at small E_{loss}, so the details of the velocity distribution in fact do not matter.

If we substitute here the expression (22.3) for E_{loss} we find

$$\frac{dN}{dt} = 24\sqrt{6\pi}\frac{\rho_{\text{DM}}}{\sigma^3}G^2 M_* \ln \Lambda.$$

Remarkably, when applicable, the result does not depend neither on the PBH mass nor on the star radius. Numerically, the trapping rate reads

$$\frac{dN}{dt} = 0.5\ \text{Gyr}^{-1}\left(\frac{7\text{km/s}}{\sigma}\right)^3\left(\frac{\rho_{\text{DM}}}{100\ \text{GeV/cm}^3}\right) \tag{22.11}$$

for $M_* = M_\odot$. We stress again that this is the rate at which the PBHs get trapped (settle on bound star-crossing orbits) after the first collision with the star. While only these PBHs may repeatedly collide with the star and eventually lose enough energy to completely sink inside, not all of them will necessarily do it. Therefore, this is the upper bound; the actual capture rate can only be smaller. This expression has been derived from different arguments in Ref. [26].[2]

This bound implies that by the direct mechanism it is possible to capture a few PBHs at most during the whole lifetime of the Universe. It applies to all types of stars except perhaps neutron stars for which the approximations which we have used become marginal, so a numerical factor may arise. We will consider the case of neutron stars in more detail below.

22.3.1 Ordinary Stars

Let us show that ordinary stars are very inefficient in capturing PBHs directly. In this case the actual capture rate is much smaller than given by Eq. (22.11) because of further suppression resulting from cooling time and perturbation constraints.

Consider first the cooling time (22.5). As explained above, it is determined by the initial orbit size r_{max} which is in turn determined by the energy loss. Since the initial total energy of a PBH is positive, the smallest (most negative) value the energy may have after the star crossing is $-E_{\text{loss}}$. The larger E_{loss}, the smaller is r_{max} and the shorter the cooling time.

The energy loss varies depending on the impact parameter and is maximum for the trajectory passing through the star center. But even for such a trajectory the energy loss is only factor $3/2$ larger than the average value (22.3). With the average energy

---

[2] There is a factor 2 difference between Eq. (22.11) and Ref. [26] which we attribute to a different estimate of E_{loss}.

loss, the initial orbit size is

$$r_{\max} = 3.7 \text{ kpc} \left(\frac{10^{20} \text{ g}}{\text{m}} \right),$$

and the corresponding cooling time is

$$t_{\text{cool}} = 2 \cdot 10^{13} \text{ yr} \left(\frac{10^{20} \text{ g}}{\text{m}} \right)^{3/2}.$$

We see that the initial size of the orbit is unrealistically large except maybe at the upper end of the asteroid window $m \sim 10^{23}$g. The cooling time exceeds 10 Gyr in the whole mass window, beign comparable to it only at the upper end.

The constraint from perturbers cannot be satisfied either. As we will discuss below in Sect. 22.5, the most promising known systems maximizing the capture rate (22.11) are ultra-faint dwarf galaxies where the star-to-star distance d is of the order $d \sim 6$pc. The maximum initial orbit size has to be at least an order of magnitude smaller: making use of Eq. (22.4) with $\alpha = 1$ we find

$$r_{\max} < 0.4 \text{ pc}.$$

Again, this condition cannot be satisfied unless $m > 10^{24}$ g (cf. Eq. (22.4)), which is outside of the mass window we consider.

We conclude that direct capture does not work for ordinary Sun-like stars. Even if a fraction of PBHs escapes gravitational perturbations by chance, there is not enough time to lose the energy.

22.3.2 Neutron Stars

In the case of neutron stars there are several differences which have to be taken into account. First, a neutron star has radius only about factor ~3 larger than its Schwarzschild radius. This means the general relativity effects start becoming non-negligible. They lead, in particular, to a marginal enhancement by a factor $1/(1 - R_g/R_{\text{NS}}) \approx 1.6$ in the capture rate [24].

A PBH falling onto a neutron star accelerates to a semi-relativistic velocity $v \sim \sqrt{R_g/R_{\text{NS}}} \sim 0.6$. This velocity is similar or larger than the sound speed in the neutron star core which is most often cited within the $0.3 - 0.6$ range, but is not well known and depends on the neutron star model. For semi-relativistic velocities the dynamical friction (22.1) loses its $1/v^2$ enhancement.

Another complication is that nuclear matter is degenerate, and therefore in the approximation of non-interacting particles (neutrons) not all scattering states can be excited, but only those for which the momentum transfer exceeds the Fermi momentum. Since the dynamical friction force can be viewed as a result of the momentum transfer to scattered matter particles, less scattering means weaker friction force.

The calculation of the dynamical friction that takes into account both effects can be found in the Appendix of Ref. [8] where the contribution of the force due to the direct accretion was also included. The resulting energy loss, averaged over different impact parameters of PBH trajectories passing through the star, was found to be

$$E_{\rm loss} = \frac{4G^2 m^2 M_*}{R_*^2} \left\langle \frac{\ln \Lambda}{v^2} \right\rangle \tag{22.12}$$

where the $\langle ... \rangle$ denotes the density-weighted average over the star volume. Numerically one finds, using a realistic neutron star density profile,

$$\left\langle \frac{\ln \Lambda}{v^2} \right\rangle \simeq 15.$$

Remarkably, accounting for the degeneracy only reduces the energy loss by a factor ~ 6 as compared to the naive estimate $\ln \Lambda / v^2 \sim 83$.

Apart from the dynamical friction, one should include in the energy loss the tidal effects, excitation of the surface waves and gravity wave emission. Numerically these effects turn out not very important for PBH masses in the asteroid window. With the account of all the known energy losses, the direct capture rate by *an isolated* neutron star was calculated in Ref. [18]. The result can be written as follows,

$$\frac{dN}{dt} = 0.26\,{\rm Gyr}^{-1} \left(\frac{7{\rm km/s}}{\sigma}\right)^3 \left(\frac{\rho_{\rm DM}}{100\ {\rm GeV/cm}^3}\right) C(X), \tag{22.13}$$

where we have included the general relativistic enhancement by a factor ≈ 1.6 [24]. Here the variable X is

$$X = 0.018 \left(\frac{m}{10^{20}{\rm g}}\right) \left(\frac{7{\rm km/s}}{\sigma}\right)^2.$$

and the function $C(X) \approx 1$ in the range $X \in [0, 20]$ and drops as power law for larger values of X. Note that $X < 20$ for PBH masses up to 10^{23} g at $\sigma = 7$ km/s. The full shape of $C(X)$ can be found in Ref. [18], Fig. 3.

Let us now check the conditions coming from the cooling time and the gravitational perturbations by nearby stars. For both conditions the key quantity is the size of the initial bound orbit $r_{\rm max}$. Given the energy loss by a PBH passing through the neutron star (22.12), the initial size of the orbit can be found from Eq. (22.4) which gives

$$r_{\rm max} = \frac{R_*^2}{4Gm} \left\langle \frac{\ln \Lambda}{v^2} \right\rangle^{-1} = 0.7\,{\rm pc} \left(\frac{10^{20}{\rm g}}{\rm m}\right). \tag{22.14}$$

In terms of $r_{\rm max}$, the cooling time is given by Eq. (22.5) which contains no other parameters. This gives

$$t_{\rm cool} = 5 \cdot 10^7\,{\rm yr} \left(\frac{10^{20}{\rm g}}{\rm m}\right)^{3/2}.$$

We see that the cooling is sufficiently fast as long as $m \gtrsim 3 \cdot 10^{18}$ g, i.e. almost in the whole asteroid mass range except the lightest masses.

The condition that nearby stars do not spoil the cooling is given by Eq. (22.9). We may use the same estimate of the star-to-star distance as above, $d \sim 6$ pc, as found in some of the prominent dwarf galaxies. Making use of this value and $R_* \simeq 10$ km for the neutron star radius we find that the condition (22.9) requires

$$r_{\rm max} < 0.08\,{\rm pc}.$$

Combining this with the expression for $r_{\rm max}$ as a function of m Eq. (22.14), this condition translates to

$$m \gtrsim 10^{21}\,{\rm g}. \tag{22.15}$$

Thus, the high mass end of the asteroid mass window satisfies this condition, while the low mass end does not.

In summary, both conditions together are satisfied at the high mass end of the asteroid window, Eq. (22.15). For these masses the capture rate is given by Eq. (22.13). The rate is not vanishingly small only in environments with high DM density and low velocity dispersion, similar to dwarf galaxies where, numerically, the capture rate may reach values of 1 per a few Gyr.

22.4 Capture at Birth

Let us now turn to the second mechanism—capture of PBHs by stars at their birth. Stars form from dense gas clouds which at some point lose their stability with respect to gravitational collapse. At the initial stages these clouds do not have strong overdensities and for our purpose may be considered as uniform. As the collapse progresses a dense protostellar object develops in the center [2, 6] which starts accreting the surrounding gas, forming at the end a pre-main-sequence star. This process happens at the Kelvin-Helmholtz timescale which is much longer than the free fall time.

Star formation takes place in galaxies in the presence of DM density $\rho_{\rm DM}$ which we will assume, as above, to consist of PBHs. We will also assume that the PBHs have Maxwell distribution in velocities with the velocity dispersion σ. A typical baryonic density in a cloud of 10^6–10^7 GeV/cm^3 is much larger than the DM density in the galactic halos, so whatever happens to PBHs has no influence on the cloud dynamics.

The capture mechanism is the following. A gas cloud from which a star will form is gravitationally bound and also has some amount of PBHs gravitationally bound to it. When the cloud collapses into a pre-main-sequence star, the time-dependent gravitational potential of the collapsing baryons drags these PBHs along, so that their distribution becomes peaked at the star position. When the star is finally formed, there is a probability to find a PBH on a bound orbit either completely contained inside or passing through the star. The fist ones are already captured; the second start cooling and will be eventually captured if the cooling is successful.

Obviously, only the trajectories of PBHs that spend large time inside the contracting cloud are significantly affected by the contraction, so the PBHs that are not bound to the cloud can be neglected. The number of captured PBHs is proportional to their total number in the corresponding part of the phase space. This number in turn is proportional to the total DM density $\rho_{\rm DM}$, and to the volume of the velocity space limited from above by the escape velocity from the cloud. The escape velocity can be estimated from the baryonic density ρ_0 and radius R_0 of the cloud. These parameters are known from observations [23] and are summarized in Ref. [9]. For example, for a star of 1 $M_\odot$ the values are $R_0 = 4300$ AU and $\rho_0 = 10^6$ GeV/cm^3. For these parameters the escape velocity is $v_{\rm esc} = 0.6$ km/s. This is much smaller than the velocity dispersion in known dwarf galaxies. Therefore, the volume in the DM velocity space corresponding to trajectories bound to the cloud is proportional to $v_{\rm esc}^3/\sigma^3$. We conclude that the number of bound trajectories, and therefore the number of captured PBHs is proportional to the combination $\rho_{\rm DM}/\sigma^3$, similarly to the direct capture case.

22.4.1 Mean Captured Number: Ordinary Stars

Consider first ordinary stars because in any case compact stars—white dwarfs and neutron stars—are formed from main sequence stars and therefore pass first through this stage. So, the PBHs captured by compact stars at birth are those that they inherit from their main sequence progenitors.

In order to calculate the total number of captured PBHs one has to know the distribution of the PBHs in position and velocity after the star is formed. This distribution builds up from bound PBH trajectories during the star formation. The fact that the formation happens slowly on a time scale much longer than the free fall time implies that the motion of the bound PBHs can be described in the adiabatic approximation. This means that the details of the contraction process do not matter, only the initial and final state.

The *maximum* number of PBHs that can be captured by a star at birth can be estimated by assuming that after the star is formed the phase space of PBHs is filled with the maximum density all the way up to positive energies. From the Liouville theorem, the phase space density after the star formation can nowhere exceed the maximum phase space density before—that is, the one at low velocities, assuming the Maxwellian DM distribution. From Eq. (22.10), this phase space density is

$$\mathcal{Q} = \frac{\rho_{\rm DM}}{m} \left(\frac{3}{2\pi\sigma^2} \right)^{3/2} . \tag{22.16}$$

We are not, however, aware of the proof that this phase space *has* to be filled. It is straightforward to argue that the Liouville theorem and the angular momentum conservation together do not guarantee this. A counter-example is a very fast (instantaneous) contraction which satisfies both properties but obviously does not populate circular orbits close to the star.

It is instructive nevertheless to make an analytic estimate (the detailed calculation can be found in Ref. [27]). Starting from the distribution that has constant phase space density (22.16), we want to calculate the number of orbits that (i) are bound and (ii) have periastrons smaller than the star radius R_*. In other words, we have to integrate

$$dN = \mathcal{Q} d^3r d^3v = 8\pi^2 \mathcal{Q} r^2 v_\phi dr dv_r dv_\phi \tag{22.17}$$

over the part of the phase space that satisfies (i) and (ii). Here we have used symmetries to reduce the number of variables to three: radius r, radial velocity v_r and the modulus of the tangential velocity v_ϕ.

The shape of the Keplerian orbits is completely characterized by conserved energy E and angular momentum L,

$$\epsilon \equiv \frac{E}{m} = \frac{1}{2}(v_r^2 + v_\phi^2) - \frac{GM_*}{r}, \tag{22.18}$$

$$l \equiv \frac{L}{m} = r v_\phi. \tag{22.19}$$

At fixed r one may trade the velocities v_r and v_ϕ for ϵ and l. From Eqs. (22.18)–(22.19) the Jackobian of this transformation is

$$dv_r dv_\phi = \frac{1}{r v_r} d\epsilon dl.$$

One has therefore from Eq. (22.17)

$$dN = 8\pi^2 \mathcal{Q} \frac{l}{v_r} dr d\epsilon dl.$$

We may express the radial velocity through ϵ and l as follows,

$$v_r = \pm \sqrt{\frac{2GM_*}{r} + 2\epsilon - \frac{l^2}{r^2}}$$

where both signs are physical. We find therefore

$$dN = 8\pi^2 Q \frac{1}{\sqrt{\frac{2GM_*}{r} + 2\epsilon - \frac{l^2}{r^2}}} dr d\epsilon d(l^2). \tag{22.20}$$

The extra factor 2 in this equation is introduced to account for two signs of v_r at given E and L.

One more change of variables is needed to make the integration trivial. Since the condition (ii) is imposed on the periastron, it is useful to change the variables from ϵ and l^2 to the periastron b and apastron a. For any finite a the condition (i)—negative total energy – is then automatically satisfied. We will see at the end that most of the contributions come from very elongated orbits for which $a \gg b$. In this case one can approximate a and b as follows,

$$a = -\frac{GM_*}{\epsilon} + \mathcal{O}(b/a), \tag{22.21}$$

$$b = \frac{l^2}{2GM_*} + \mathcal{O}(b/a). \tag{22.22}$$

Recall that the energy is negative, $\epsilon < 0$. In the same approximation, one can neglect the last term under the square root in Eq. (22.20). Equations (22.20), (22.21) and (22.22) then give

$$dN = 4\pi^2 Q \frac{R_g^{3/2} r}{a^{3/2}\sqrt{ar - r^2}} dr\, da\, db,$$

where as before $R_g = 2GM_*$ is the Schwarzschild radius of the star.

We will need at the end the number of PBHs on star-crossing orbits with the apastron smaller than a given $a_{\max} \gg R_*$. This makes the integral convergent. We therefore have to integrate the above expression over r from b to a, then over a from b to $a_{\max}$ and finally over b from 0 to R_*,

$$N = 4\pi^2 Q R_g^{3/2} \int_0^{R_*} db \int_b^{a_{\max}} \frac{da}{a^{3/2}} \int_b^a \frac{r\, dr}{\sqrt{ar - r^2}}.$$

One may notice that the dominant contribution to this integral comes from the region $r, a \gg R_*$. We may therefore change the lower limit of the integration over r to 0 instead of b. The integral over r then gives

$$\int_0^a \frac{r}{\sqrt{ar - r^2}} dr = \frac{a\pi}{2},$$

so that

$$N = 2\pi^3 Q R_g^{3/2} \int_0^{R_*} db \int_b^{a_{\max}} \frac{da}{\sqrt{a}}. \tag{22.23}$$

The integral over a now becomes explicitly convergent at the lower limit and we can again replace b by 0. Then the integration gives

$$N = 4\pi^3 Q R_g^{3/2} R_* \sqrt{a_{\max}} = \frac{\rho_{\rm DM}}{m\sigma^3}(3\pi R_g)^{3/2} R_* \sqrt{2a_{\max}}. \tag{22.24}$$

This expression agrees with the one obtained in Ref. [27] for the same quantity. It also agrees with the number of PBHs that settle on the star crossing orbits calculated numerically in Ref. [16]. This implicitly confirms our starting assumption that when the star is formed, the whole phase space with $E < 0$ is filled with the maximum phase-space density.

The captured number grows infinitely as $a_{\max} \to \infty$, i.e. when larger and larger orbits are included. At some point, however, $a_{\max}$ becomes so big that either the cooling time becomes longer than the age of the star, or the perturbations from nearby stars can no more be neglected. We will see below that, in realistic conditions of dwarf galaxies, the cooling requirement is more constraining at masses below $\sim 10^{21}$ g. In order to estimate the number of captured PBHs from Eq. (22.24) in this mass range it remains, therefore, to substitute in Eq. (22.24) the maximum size of the orbit that still has time to cool and settle inside the star. The orbits that PBHs acquire after the contraction are much smaller than the critical value (22.4), so we should use Eq. (22.6) for the cooling time. Comparing Eqs. (22.24) and (22.6) we see that their dependence on a and m is the same, so that we can rewrite the captured number N as follows,

$$N = 3\sqrt{6\pi} R_g^2 \frac{\ln \Lambda}{M_*} \frac{\rho_{\rm DM}}{\sigma^3} t_{\rm cool}. \tag{22.25}$$

From this expression we conclude that when the capture is limited by the cooling time, the number of captured PBHs does not depend on the PBH mass m. Note, however, that Eq. (22.6) is approximate, so the estimate (22.25) is only valid when the former is correct.

If we substitute in Eq. (22.25) the cooling time of order of the star lifetime, $t_{\rm cool} \sim 10$ Gyr, we find that numerically

$$N = 2.3 \left(\frac{\rho_{\rm DM}}{100\,{\rm GeV/cm^3}}\right) \left(\frac{7\,{\rm km/s}}{\sigma}\right)^3 \left(\frac{t_{\rm cool}}{10\,{\rm Gyr}}\right). \tag{22.26}$$

This is very close to the result of the numerical simulations of Ref. [16] as will be discussed below. It should be stressed, however, that several approximations have been made to obtain Eq. (22.25), so the numerical agreement of Eq. (22.25) with simulations is a mere coincidence. Among these approximations we have used the *average* energy loss and assumed, when calculating the average in Eq. (22.2), that the trajectories are distributed uniformly over the impact parameter plane inside the star. We see now from Eq. (22.24) that this is not the case: if it were, the measure in Eq. (22.23) would have been $b\,db$, not db. We have also assumed that $b \ll a$. This approximation may break for small PBH masses leading to small energy loss, cf. Eq. (22.2), which would imply that only relatively compact trajectories have time to cool down.

If one does not want to rely on assumptions, numerical methods must be used to determine N. Such simulations have been performed in Refs. [7, 9, 16]. The calculation is greatly simplified by the fact that the total DM mass on bound orbits is by many orders of magnitude smaller than the star mass, so one can simulate PBH trajectories *one by one* in the same gravitational potential of the contracting cloud, each time with different initial conditions. Further simplification is due to the adiabatic character of the contraction. Because of that one may chose a convenient parameterization of the time-dependent cloud shape—all choices should give the same result as long as the contraction is slow.

The general algorithm is as follows (see Ref. [16] for details). One generates random initial conditions for PBH trajectories covering uniformly the relevant region of the phase space. For numerical efficiency one may sample only the region of negative starting energy and small angular momentum containing (almost) all the trajectories that, by the end of the contraction, may become gravitationally bound and passing through the star. Each trajectory is then evolved in the gravitational field of the contracting cloud. The trajectories that do pass through the star at the end of the contraction are selected.

Among these trajectories, the ones with the apastron too large to survive the deflections by nearby stars (see Sect. 22.2.3) are then discarded. For each of the remaining trajectories the cooling time has to be checked. It is not sufficient at this stage to rely on the average energy loss (22.3). In Ref. [16] the energy losses were calculated with the account of the actual star density profile (the Solar profile was used) and the parameters of individual trajectories. The trajectories having the cooling time longer than the star lifetime are again discarded. The remaining ones are considered successful.

The phase space region sampled by the initial conditions contains some total number and total mass of PBH. Knowing the fraction of the successful (that is, leading to capture) trajectories from simulations, one may immediately determine the mean captured number $\bar{N}$ and the mean captured mass $\bar{M} = m\bar{N}$ of PBHs. For a 1 $M_\odot$ star the result is shown in Fig. 22.1, which is the same as Fig. 1 of Ref. [16], except $\bar{N}$ is shown instead of $\bar{M}$.

On the horizontal part of the plot in the range roughly $10^{19} - 10^{21}$ g, the mean captured number is limited by the cooling time. There it is in agreement with Eq. (22.25) which predicts no dependence on the PBH mass. The drop at around 10^{21} g occurs where the requirement of not being perturbed by the nearby stars becomes more important than that of the cooling time. This part of the curve depends on the assumed ambient star density n_*; the value $n_* = 0.009$ pc^{-3} as inferred from measured parameters of the Triangulum II dwarf galaxy was used on this plot.

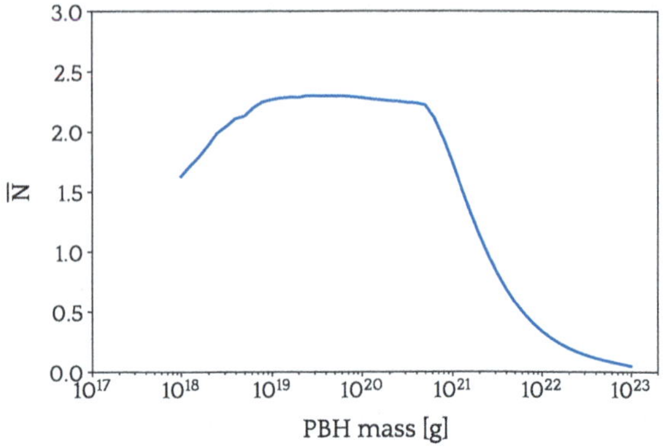

Fig. 22.1 The mean number $\bar{N}$ of PBHs captured by a star of 1 $M_\odot$ at birth as a function of the PBH mass. The calculation assumes an isolated star at rest in an environment with the PBH mean mass density of 100 GeV/cm^3 and velocity dispersion 7 km/s. The plot is based on the same simulation as Fig. 1 of Ref. [16]

At PBH masses smaller than $\sim 10^{19}$ g the mean captured number starts to decrease. This can be traced back to breaking of the approximations used to derive Eq. (22.25), in particular of the assumption that the trajectories with $a \gg b$ give the dominant contribution.

22.4.2 *Mean Captured Number: Compact Stars*

Compact stars form as a result of evolution of main sequence stars. Massive stars with masses in the range of $8M_\odot \lesssim M_* \lesssim 15M_\odot$ during their main sequence phase typically end their evolution as neutron stars, the lighter ones become white dwarfs, while the heavier ones collapse directly into black holes. The lifetimes of massive stars decrease as the mass grows, reaching about $\sim$ 10 Myr at the heavy mass end. If a main sequence star at its birth has trapped one or more PBHs on bound star crossing orbits, and if they get captured inside the progenitor star and sink to the center before it evolves into a compact remnant, they may be inherited by the latter.

This two-step process has been considered in Refs. [7, 9] where the amount of captured PBHs was estimated for neutron stars and white dwarfs in the assumption that at the first stage the PBHs do not destroy the progenitor. This assumption is not crucial for deriving constraints based on the neutron star and white dwarf disappearance because the destruction of a progenitor star has the same effect. Progenitors of various masses have been considered. The results were presented in the form of expected constraints on the fraction of PBHs in the total amount of DM that one could impose provided a population of neutron stars and/or white dwarfs is found in dwarf galaxies. Unfortunately, this is not presently the case. Note, however, that perturbations by nearby stars were not taken into account in the analysis of Refs. [7, 9].

22.5 Implications for Constraints on PBH Abundance

As already mentioned, the destruction of stars—main sequence or compact—by PBHs may potentially have several different signatures that may be used to constrain (or detect) the PBH abundance. Of these, a mere star disappearance due to their conversion into sub-solar mass black holes appears the most robust one. We will now consider this signature in more detail; other signatures will be briefly discussed in the concluding section. We will focus on main sequence stars because they are much easier to observe, while the deficit of compact stars due to their transmutation into black holes is much harder to establish experimentally. In fact, as of today, no neutron stars or white dwarfs have been detected in dwarf galaxies.

In the previous sections we have seen that PBHs of asteroid masses are very difficult to capture inside stars. All the mechanisms that we have considered require cosmologically long time and may lead to sizable probability of capture only in exceptional environments. Both the probability of the direct capture, Eq. (22.13), and the capture at birth, Eq. (22.26), are proportional to $\rho_{\rm DM}/\sigma^3$. Following Ref. [16] one may classify the existing environments according to their "merit factor" η defined as

$$\eta = \frac{\rho_{\rm DM}}{100\ {\rm GeV/cm^3}} \left(\frac{7\ {\rm km/s}}{\sqrt{2}\sigma} \right)^3 ,$$

where $\sqrt{2}$ accounts for the fact that when both PBHs and stars have the same velocity dispersion σ, the relative velocity has the dispersion $\sqrt{2}\sigma$. In the environments with $\eta = 1$ Eqs. (22.13) and (22.26) apply with $\rho_{\rm DM}$- and σ-dependent factors equal to 1.

Of all the astrophysical environments with experimentally confirmed DM density only some ultra-faint dwarf galaxies (UFDs) reach the values of η close to 1. Among the most promising UFDs are the Triangulum II ($\eta \geq 0.95$), Tucana III ($\eta \geq 0.51$), Draco II ($\eta \geq 0.39$), Segue 1 ($\eta = 0.39$) and Grus I ($\eta = 0.37$). These numbers are calculated from actually measured parameters of these galaxies [16]; the inequalities are used where only the upper bounds on the velocity dispersion exist. For comparison, the Milky Way in the solar neighborhood has $\eta \sim 2 \times 10^{-8}$.

As a word of warning, it is worth noting at this point that the identification of a cluster of stars as a dwarf galaxy having its own DM halo is often a matter of debate. Fortunately, there is a rapid progress in this direction; in particular, 4 out of 5 UFDs mentioned above have been recently confirmed as dwarf galaxies [3].

The PBH capture is a random uncorrelated process and therefore the probability to capture a given number of PBHs follows the Poisson distribution with the expectation equal to the mean captured number $\bar{N}$. For a solar mass star this quantity was calculated numerically in Ref. [16]; it is shown in Fig. 22.1 for the Trinagulum II galaxy which has $\eta \simeq 1$. In the mass range $\sim 10^{19}$–10^{21} g the mean captured number is $\bar{N} \simeq 2.2$ assuming PBHs constitute all of the DM. This corresponds to the survival probability (the probability to capture zero PBHs) $p_0 = \exp(-\bar{N}) \simeq 0.11$, meaning that the stars actually observed in this galaxy would constitute only 11% of their original number as formed about 10 Gyr ago, while 89% would have been destroyed by PBHs.

If PBHs constitute only a fraction $f_{\rm PBH} = \Omega_{\rm PBH}/\Omega_{\rm DM}$ of the DM, the mean captured number would be $f_{\rm PBH}\bar{N}$, and the survival probability would be higher for smaller $f_{\rm PBH}$. Requiring that no more than a certain fraction of stars has been destroyed translates into constraints on $f_{\rm PBH}$ [16]. The result is shown in Fig. 22.2 for two assumed values of the maximum destroyed fraction of stars equal to 50 and 20%.

It is not straightforward to infer the maximum allowed fraction of destroyed stars from present-day observations as this would involve modeling of a dwarf galaxy history. While such models exist, they have large uncertainties. A way around has been proposed in Ref. [15]; it makes use of the fact that the destruction probability grows with the mass of the star. This is illustrated in Fig. 22.3 which shows the dependence of the mean captured number of PBHs on the star mass for the case $\eta = f = 1$. This dependence is close to linear, and therefore the survival probability decreases exponentially with the star mass. Thus, the presence of PBHs, apart form reducing the number of stars, also strongly impacts the *stellar mass function* depleting its high-mass end.

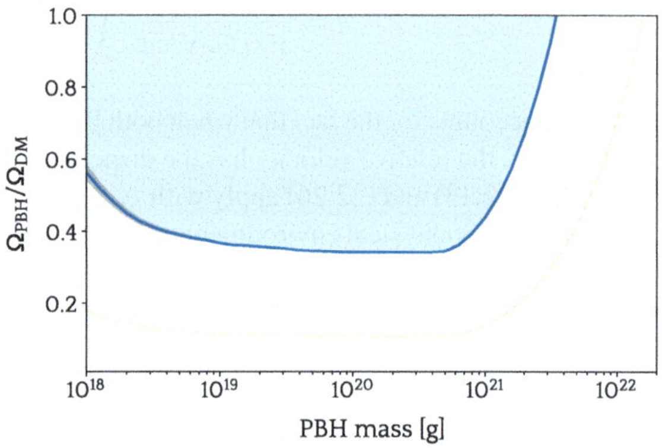

Fig. 22.2 Upper limits on the PBH abundance $\Omega_{\rm PBH}/\Omega_{\rm DM}$ as a function of the PBH mass for the Triangulum II dwarf galaxy *assuming* that no more than 50% (blue) or 20% (orange) of stars were destroyed by PBHs. Figure adopted from Ref. [16]

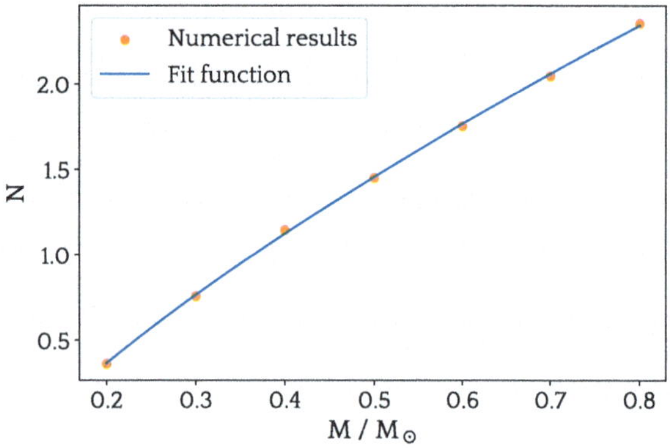

Fig. 22.3 The mean captured number $\bar{N}$ as a function of the star mass M assuming $\eta = 1$ and $f = 1$. The dots show the numerical result, the blue line represents the fit by the function $a(M/M_\odot)^b + c$ with $a = 3.76$, $b = 0.685$ and $c = -0.883$. Figure adapted from Ref. [15]

Known UFD galaxies (see Ref. [30] for a review) have from hundreds to thousands of individual stars resolved with masses typically going down to $0.5M_{\odot}$, and down to $0.2M_{\odot}$ in some cases. Moreover, it has been shown [17] that the James Webb Space Telescope can resolve stars down to $0.09M_{\odot}$ in the dwarf galaxy Draco II at a distance of ~ 20 kpc. From these data one can, in principle, reconstruct the stellar mass function with a good accuracy and in a sufficiently wide mass range. Importantly, with the already existing technology, these data may be extended to low masses $\lesssim 0.2M_{\odot}$ where the effect of PBHs should be negligible.

Modeling of the stellar mass distribution at birth, or initial mass function, is a standard tool in the galactic astrophysics [12, 25]. The UFDs are of particular interest in this context because, according to current understanding, their stars have all been formed in a single burst some 12.8 Gyr ago. The long-lived ones—those with masses below $\sim 0.8M_{\odot}$—should therefore have preserved their initial mass distribution until present.

The destruction of stars by PBHs modifies this picture. Importantly, the effect of PBHs switches on and off very sharply (exponentially) as the merit factor goes through the value $\eta \approx 1$. Thus, comparing several UFDs with η scattered around 1 should give a good handle on the parameters of the unmodified stellar mass function, as well as on its possible modifications due to the star destruction by PBHs.

Preliminary estimates of the sensitivity of this approach have been made in Ref. [15] using simulated data. A star population was generated with the Kroupa mass function with the standard parameters and the number of stars corresponding to a typical UFD, thus mimicking closely the observational data. These mock data then have been fitted to a model based on a generic mass function model (broken power law or log-normal) with a few free parameters varying in the ranges inferred from observations. The model also included the destruction of stars by PBHs controlled by one more free parameter, the fraction of PBHs f_{PBH} varying between 0 and 1. The Bayesian method was used to infer the likely values of all the parameters, including f_{PBH}. The results show that the method is sensitive enough to exclude large values of f_{PBH} down to 0.8 at more than 3σ level. It appears therefore quite feasible to exclude 100% of the DM being composed of PBHs with already existing observations.

22.6 Summary and Concluding Remarks

Primordial black holes are an appealing candidate for the dark matter. However, large parts of the potentially available mass range are constrained by various arguments, and only the so-called asteroid mass window from $\sim 10^{17}$g to $\sim 10^{23}$g remains fully allowed until now. We have argued that capture of PBHs by main sequence stars in dwarf galaxies may be a promising way to probe this mass range. Moreover, it is conceivable that this can be done with the existing technologies, if not with already existing data.

The approach that we have outlined consists, essentially, in comparing the stellar mass functions of dwarf galaxies with low and high values of the merit factor η. For that good quality data on stellar populations are required. Such data already exist for some of the galaxies with low η where stars have been resolved down to $0.2M_{\odot}$. For high η galaxies, notably for the ones listed in Sect. 22.5, the current observations go down to $0.5M_{\odot}$. This may or may not be sufficient to firmly exclude the possibility of all of the DM consisting of PBHs, $\Omega_{\mathrm{PBH}}/\Omega_{\mathrm{DM}} = 1$. Better observations of these galaxies extending to smaller star masses will be very important for this purpose.

We have focused on the most straightforward signature of the star destruction by PBHs, the mere star disappearance. But there are likely other signatures. If a sizable fraction of stars in a dwarf galaxy is destroyed by PBHs, this would create a population of sub-solar mass black holes. They are not directly visible, but may perhaps be detected in gravitational waves. It is likely that the star destruction itself is accompanied by a supernova-type energy release. If this is a short event that might have happened any time in the last 10 Gyr, we have to be very lucky to observe and identify one today, unless it has features that make these explosions unambiguously distinguishable from other supernova events. Alternatively, last stages of the star accretion onto a black hole may be relatively long and luminous, in which case we have a good chance to detect and maybe identify such an event as being due to a PHB capture.

Acknowledgements The author is indebted to N. Esser for the help with making Fig. 22.1 and for comments on the manuscript. This work is supported in part by the Institut Interuniversitaire des Sciences Nucléaires (IISN) Grant No. 4.4503.15.

References

1. A. Barnacka, J.F. Glicenstein, R. Moderski, New constraints on primordial black holes abundance from femtolensing of gamma-ray bursts. Phys. Rev. D **86**, 043001 (2012). https://doi.org/10.1103/PhysRevD.86.043001, arXiv:1204.2056 [astro-ph.CO]
2. M.R. Bate, Collapse of a molecular cloud core to stellar densities: the first three-dimensional calculations. Astrophys. J. Lett. **508**, L95 (1998). https://doi.org/10.1086/311719, arXiv:astro-ph/9810397
3. H. Baumgardt et al., Stellar mass segregation as separating classifier between globular clusters and ultrafaint dwarf galaxies. **510**(3), 3531–3545 (2022). https://doi.org/10.1093/mnras/stab3629, arXiv: 2112.04689 [astro-ph.GA]
4. J. Binney, S. Tremaine, *Galactic Dynamics* (1987)
5. H. Bondi, On spherically symmetrical accretion. Mon. Not. R. Astron. Soc. **112**, 195 (1952). https://doi.org/10.1093/mnras/112.2.195
6. Al.P. Boss, H.W. Yorke, Protoplanetary disks, mid-infrared dips, and disk gaps. Astrophys. J. **469**, 366 (1996). https://doi.org/10.1086/177786
7. F. Capela, M. Pshirkov, P. Tinyakov, Adiabatic contraction revisited: implications for primordial black holes. Phys. Rev. **D90**(8), 083507 (2014). https://doi.org/10.1103/PhysRevD.90.083507, arXiv:1403.7098 [astro-ph.CO]
8. F. Capela, M. Pshirkov, P. Tinyakov, Constraints on primordial black holes as dark matter candidates from capture by neutron stars. Phys. Rev. D **87**(12), 123524 (2013). https://doi.org/10.1103/PhysRevD.87.123524, arXiv:1301.4984 [astro-ph.CO]

9. F. Capela, M. Pshirkov, P. Tinyakov, Constraints on primordial black holes as dark matter candidates from star formation. Phys. Rev. D **87**(2), 023507 (2013). https://doi.org/10.1103/PhysRevD.87.023507, arXiv:1209.6021 [astro-ph.CO]
10. B.J. Carr et al., New cosmological constraints on primordial black holes. Phys. Rev. D **81**, 104019 (2010). https://doi.org/10.1103/PhysRevD.81.104019, arXiv:0912.5297 [astro-ph.CO]
11. B. Carr et al., Constraints on primordial black holes. Rept. Prog. Phys. **84**(11), 116902 (2021). https://doi.org/10.1088/1361-6633/ac1e31, arXiv:2002.12778 [astro-ph.CO]
12. G. Chabrier, Galactic stellar and substellar initial mass function. PASP **115**(809), 763–795 (2003). https://doi.org/10.1086/376392, arXiv:astro-ph/0304382 [astro-ph]
13. S. Chandrasekhar, Brownian motion, dynamical friction, and stellar dynamics. Rev. Mod. Phys. **21**, 383–388 (1949). https://doi.org/10.1103/RevModPhys.21.383
14. A. Coogan, L. Morrison, S. Profumo, Direct detection of hawking radiation from asteroid-mass primordial black holes. Phys. Rev. Lett. **126**(17), 171101 (2021). https://doi.org/10.1103/PhysRevLett.126.171101, arXiv:2010.04797 [astro-ph.CO]
15. N. Esser, S. De Rijcke, P. Tinyakov, The impact of primordial black holes on the stellar mass function of ultra-faint dwarf galaxies (2023). arXiv:2311.12658 [astro-ph.GA]
16. N. Esser, P. Tinyakov, Constraints on primordial black holes from observation of stars in dwarf galaxies (2022). arXiv:2207.07412 [astro-ph.HE]
17. D.R. Weisz et.al., The JWST resolved stellar populations early release science program. II. survey overview. Astrophys. J. Suppl. Ser. **268**(1), 15 (2023). https://doi.org/10.3847/1538-4365/acdcfd
18. Y. Génolini, P. Serpico, P. Tinyakov, Revisiting primordial black hole capture into neutron stars. Phys. Rev. D **102**(8), 083004 (2020). https://doi.org/10.1103/PhysRevD.102.083004, arXiv:2006.16975 [astro-ph.HE]
19. P.W. Graham, S. Rajendran, J. Varela, Dark matter triggers of supernovae. Phys. Rev. D **92**(6), 063007 (2015). https://doi.org/10.1103/PhysRevD.92.063007, arXiv:1505.04444 [hep-ph]
20. K. Griest, A.M. Cieplak, M.J. Lehner, Experimental limits on primordial black hole dark matter from the first 2 yr of kepler data. Astrophys. J. **786**(2), 158 (2014). https://doi.org/10.1088/0004-637X/786/2/158, arXiv:1307.5798 [astro-ph.CO]
21. K. Griest, A.M. Cieplak, M.J. Lehner, New limits on primordial black hole dark matter from an analysis of kepler source microlensing data. Phys. Rev. Lett. **111**, 181302 (2013). https://doi.org/10.1103/PhysRevLett.111.181302
22. A. Katz et al., Femtolensing by dark matter revisited. JCAP **12**, 005 (2018). https://doi.org/10.1088/1475-7516/2018/12/005, arXiv:1807.11495 [astro-ph.CO]
23. J.M. Kirk, D. Ward-Thompson, P. Andre, The initial conditions of isolated star formation. 6. SCUBA mapping of prestellar cores. Mon. Not. R. Astron. Soc. **360**, 1506–1526 (2005). https://doi.org/10.1111/j.1365-2966.2005.09145.x, arXiv:astro-ph/0505190
24. C. Kouvaris, WIMP annihilation and cooling of neutron stars. Phys. Rev. D **77**, 023006 (2008). https://doi.org/10.1103/PhysRevD.77.023006, arXiv:0708.2362 [astro-ph]
25. P. Kroupa, On the variation of the initial mass function. Mon. Not. R. Astron. Soc. **322**(2), 231–246 (2001). https://doi.org/10.1046/j.1365-8711.2001.04022.x, arXiv:astro-ph/0009005 [astro-ph]
26. P. Montero-Camacho et al., Revisiting constraints on asteroid-mass primordial black holes as dark matter candidates. JCAP **08**, 034 (2019). https://doi.org/10.1088/1475-7516/2019/08/031, arXiv:1906.05950 [astro-ph.CO]
27. M. Oncins et al., Primordial black holes capture by stars and induced collapse to low-mass stellar black holes. Mon. Not. R. Astron. Soc. **517**(1), 28–37 (2022). https://doi.org/10.1093/mnras/stac2647, arXiv2205.13003 [astro-ph.GA]
28. E.C. Ostriker, Dynamical friction in a gaseous medium. Astrophys. J. **513**(10, 252–258 (1999). https://doi.org/10.1086/306858, arXiv:astro-ph/9810324 [astro-ph]

29. W.H. Press, D.N. Spergel, Capture by the sun of a galactic population of weakly interacting massive particles. Astrophys. J. **296**, 679–684 (1985), by M.A. Srednicki (ed.). https://doi.org/10.1086/163485
30. J.D. Simon, The faintest dwarf galaxies. Ann. Rev. Astron. Astrophys. **57**, 375–415 (2019). https://doi.org/10.1146/annurev-astro-091918-104453, arXiv:1901.05465 [astro-ph.GA]

Chapter 23
Lensing Constraints: Substellar to Intermediate Masses

Evencio Mediavilla and Jorge Jiménez-Vicente

Abstract Gravitational microlensing is a robust tool to detect and directly measure the abundance and mass of any kind of compact object, either in our galaxy or in the extragalactic domain. Based on generic, broadly applicable arguments, it is concluded that the observed microlensing magnifications are too small and the microlensing events less frequent than the expectations for a significant population of compact objects (other than normal stars). The detection of chromatic effects of microlensing also does not support the presence of black holes. Detailed statistical studies of the observed microlensing magnifications and events frequency impose strict upper limits to the fraction of total mass of black holes ($\lesssim 1\%$) from $10^{-7} M_{\odot}$ to indefinitely large masses. These results hold even when the black holes are distributed according to a mass spectrum or are forming clusters.

23.1 Lensing Effects of Compact Objects

Since the experimental confirmation of Einstein's principle of equivalence and general theory of relativity, we know that massive objects, no matter their origin or nature, deflect light rays. This property is used to measure mass in different astrophysical and cosmological scenarios using the Gravitational Lensing theory.

Regarding the study of PBHs we are particularly interested in the detection of the possible gravitational lensing effects induced by the presence of these objects

E. Mediavilla (✉)
Instituto de Astrofísica de Canarias, Vía Láctea S/N, 38200 La Laguna, Tenerife, Spain
e-mail: emg@iac.es

Departamento de Astrofísica, Universidad de la Laguna, 38200 La Laguna, Tenerife, Spain

J. Jiménez-Vicente
Departamento de Física Teórica y del Cosmos, Universidad de Granada, Campus de Fuentenueva, 18071 Granada, Spain
e-mail: jjimenez@ugr.es

Instituto Carlos I de Física Teórica y Computacional, Universidad de Granada, 18071 Granada, Spain

C. Byrnes et al. (eds.), *Primordial Black Holes*, Springer Series in Astrophysics and Cosmology, https://doi.org/10.1007/978-981-97-8887-3_23

in galaxies and clusters of galaxies. Specifically, this chapter focuses on the consequences of lensing studies in our galaxy and, especially, in lensed quasars. Lensing allows to detect and study the presence of compact massive objects in at least two ways: globally (through macro-modeling anomalies) and locally (from microlensing effects). Macro-modeling fits the positions of the multiple images of a lensed quasar, using a suitable smooth mass distribution of the lens galaxy and predicts, among other observables, the flux-ratios between images. Consequently, positional (astrometric) and flux-ratio anomalies (greater than model allowances arising from experimental uncertainties) may indicate the presence of massive objects not considered in the modeling process. Microlensing effects[1] are caused by the granulation of the lens galaxy mass distribution in compact objects and mainly consist in the change of flux of any of the source images caused by the presence of these objects along the observer's line of sight. Thus, while microlensing is a local effect directly related to the presence of compact objects, macro-modeling anomalies can be affected by degeneracies in the description of the global mass distribution of the lens galaxy and are, consequently, more model dependent. For this reason, micro and millilensing are especially interesting in the context of PBHs although we also shall consider some inputs from macro-modeling.

From the experimental point of view, there are two approaches commonly used to study microlensing: monitoring of flux variability of lensed quasar images (light-curve method) and comparison of the continuum flux-ratio between two images with respect to their intrinsic flux-ratio, either determined from the flux of a very large region insensitive to microlensing or from a model (single-epoch method). In practical time intervals, millilensing can be studied only using the second method.

Two general properties of gravitational lensing are worthy to be highlighted as they can help to discuss the existence of massive compact objects. The first one is mass-length degeneracy (a result of the scale invariance of gravitation, [15, 42]), which has the consequence that for a source of fixed size and a population of compact objects of a given mass density, the larger the mass of the compact objects the greater the impact of microlensing. The second one is the dependence of microlensing strength with the source size (microlensing chromaticity). In principle, microlensing is sensitive to the source size relative to the Einstein radius[2] of the compact object: the smaller the source, the larger the strength of microlensing. However, in terms of the Einstein radius of the microlens, there is a low limit of the source size (point source limit) below which microlensing progressively approaches a constant value and no significant chromaticity is further present. As we comment along the chapter, none of these two properties support the existence of PBHs.

This chapter is organized as follows. For historical reasons and to introduce the main concepts, we start with galactic MACHOS studies in the Milky Way,

[1] Also millilensing effects (micro and millilensing are commonly distinguished by the strength of the effect).

[2] The Einstein radius, used as the natural scale of microlensing, increases with the square root of the microlens mass and corresponds to the radius of the region in which the magnification induced by an isolated microlens is noticeable.

Andromeda and other nearby galaxies in Sect. 23.2 and with the early work on the extension of the MACHOS experimental idea to the extragalactic domain in Sect. 23.3 (mainly focused on typical stellar masses). Above the stellar mass range, we study the very interesting (after LIGO-Virgo-KAGRA discoveries) case of the intermediate mass PBHs in Sect. 23.3.2. Below the stellar mass-range, in Sect. 23.3.3 we consider the sub-stellar domain from X-ray observations. In Sect. 23.4 we abandon the simplification of an spatially uniform distribution of single mass microlenses, to explore both, the effect of a spectrum of masses and the impact of clustering (this last study can be extrapolated to indefinitely large masses). The chapter ends with a section devoted to summarize the main results.

23.2 Galactic MACHOS ($10^{-7} M_{\odot} \lesssim M \lesssim 1 M_{\odot}$)

Massive Compact Halo Objects (MACHOs) were proposed long ago as suitable dark matter candidates constituting the bulk mass of the Milky Way halo. In a most influential paper, Paczyński [36] proposed a theoretically simple (yet extremely challenging from the observational point of view) method of detecting such compact objects using its gravitational lensing effect. The method consists on observing the temporary increase in brightness produced by the gravitational (micro)lensing effect originated by the occasional alignment of a background star with such compact objects. For the experiment to succeed, it requires a colossal observational effort, consisting in the photometric monitoring of a vast number (e.g. millions) of stars, with high cadence and for very long periods of time. Paczyński showed that the Magellanic Clouds offer the best suitable targets for this purpose. Several independent experiments followed this path, and soon succeeded in detecting the predicted microlensing events, most important being the MACHO [2], EROS [5] and OGLE [43] collaborations.

The duration of a microlensing event is dependent on the lens mass M_L. For the LMC, the rule of thumb (somewhat dependent on the model) is that the duration of a microlensing event (Einstein crossing time) is: $\langle t_E \rangle \sim 70\sqrt{M_L/M_{\odot}}$ days. These experiments have been most efficient detecting events with duration between a few days to ~100 days and are, therefore, most sensitive to lens masses in the "stellar" range $0.1 - 1 M_{\odot}$. Indeed, it is in this mass range where most events have been detected and best limits can be set, with limits for larger and lower masses being established based on the absence of excess events from what is expected. The resulting limits are dependent on the halo and galaxy model, importance of self-lensing, etc.

The main results from those microlensing experiments towards the Magellanic Clouds [cf. 3, 4, 37, 48] set abundances below 6–7% for compact objects in the halo in the mass range between roughly 0.01 to 0.5 $M_{\odot}$. The limits relax as we move to higher masses, allowing up to 20% for objects of a few solar masses. These results are somewhat model dependent, particularly on the mass of the Milky Way halo, but also on the mass and spatial distribution of the lenses [cf. 7]. In general, low mass halo models leave some more room for MACHOs in the halo than the high mass models.

Searching for larger mass objects, [6] have recently combined EROS-2 and MACHO data to obtain long time spans (up to 10.6 years) in 14.1 million objects, and searched for long duration events (100–1000 days) founding none.[3] This allowed them to establish at 95% confidence level, an upper limit fraction of 15% of the halo mass in form of compact objects in the mass range from 10 to 100 $M_\odot$ (again, higher masses are much less restricted by these procedures, allowing up to 50% for objects of 1000 $M_\odot$).

Similarly, on the low mass range, [31] used OGLE-IV microlensing observations towards the galactic bulge of about 50 million stars with very high cadence ($\sim$ 20 min). From a thorough analysis of the frequency of short events (0.1 to 2 days), they could set stringent limits of a few percent of the mass in compact objects in the planetary mass range from 10^{-5} to $10^{-2} M_\odot$. Similar results have been obtained by [33]. Even lower masses have been constrained using the Subaru-HSC survey of 87 million stars in M31 with very high cadence [32], even considering finite size effects by [10], who constrain abundances below a few percent in the range between 10^{-5} and $10^{-10} M_\odot$.

Overall, galactic microlensing experiments have established quite stringent constraints on the possibility that compact objects contribute significantly to the mass of the Milky Way Halo [cf. 30]. The results show that it is very unlikely that these compact objects are an important constituent of the halo if they are in the mass range between 10^{-7} to $1 M_\odot$. Higher masses are, nevertheless, less restricted, and although compact objects of up to 100 $M_\odot$ are not fully discarded, they cannot be the main halo constituent.

23.3 Extragalactic MACHOS

23.3.1 Stellar Mass Range ($0.1 M_\odot \lesssim M \lesssim 3 M_\odot$)

On top of the limits set on compact objects in the halo of the Milky Way, gravitational microlensing can also be used to constrain the abundance of such objects in the extragalactic realm. The idea of observing microlensing in the images of multiply lensed quasars caused by stars in the lens galaxy was first suggested by [9]. Several authors recognized the possibilities of using quasar microlensing monitoring or single-epoch snapshots of the flux-ratio anomalies induced by microlensing to estimate the optical depth of the microlenses population [26, 40, 45, 47]. A first practical attempt to implement this idea was carried out by [39], using flux-ratio anomalies of a sample of 11 lensed quasars.

A central challenge met by the early extragalactic micro-lensing studies was the determination of a robust base-line of no microlensing magnification. [28] solved the problem using the quasar emission lines, arising from a region large enough as to

[3] Two potential candidates remained uncertain and cannot be fully excluded.

be insensitive to microlensing, to measure the intrinsic flux-ratios between images in a large sample (20) of lensed quasars. The flux-ratio anomalies between images i and j attributable to microlensing are then calculated from,

$$\Delta m_{ij} = (m_i - m_j)^{cont} - (m_i - m_j)^{line.}, \tag{23.1}$$

where the continuum flux-ratio (expressed as a difference in magnitudes) emitted from a small region of a few light-days in size (according to reverberation mapping studies [11, 16, 21], and see also the discussion in [29]) and hence, strongly affected by microlensing is compared with the intrinsic flux-ratio inferred from the emission lines arising from a large region of several hundred light-days in size (almost insensitive to microlensing). Notice that if the continuum is taken at a wavelength relatively close to the emission line the anomalies estimated in this way are virtually free from extinction effects.

In Fig. 23.1 we show an updated histogram of (the absolute value of) microlensing magnifications (see [27]), which basically confirms the results by [28]: most of the measurements are concentrated at moderate magnifications, $|\Delta m| \leq 0.5$ mag, and no large magnification tail is present. Next step to estimate the abundance of the compact objects that act like microlenses is to simulate the expected microlensing for different values of the fraction of total mass in microlenses (the total mass is estimated from macro-models). This is usually performed using microlensing magnification maps from which the theoretical microlensing magnification PDFs are obtained. Considering $1 M_\odot$ compact objects, [28] compare the simulated PDFs with the observed histogram and obtain a maximum likelihood estimate of the mass fraction in microlenses between $\sim$0.05 and $\sim$0.20 when the source size ranges between $\sim$1 and $\sim$10 light-days. Alternatively, by virtue of the mass-length degeneracy, if we fix the size in 5 light-days (according to reverberation mapping estimates, see above) the results hold in the range between $\sim 0.1 M_\odot$ and $\sim 10 M_\odot$. These estimates have been confirmed by posterior studies, jointly analyzing the source size

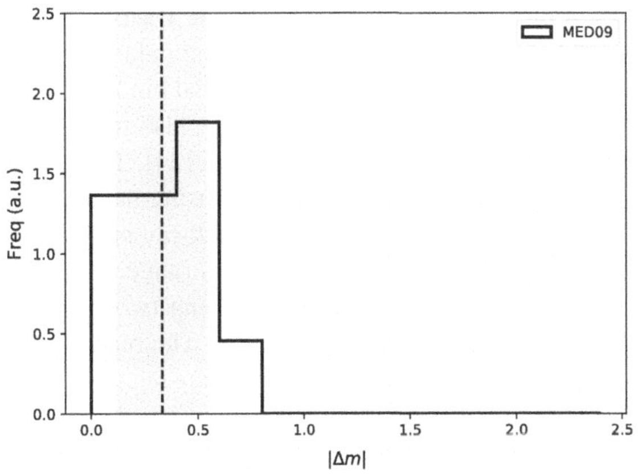

Fig. 23.1 Histogram of (absolute value of) microlensing magnifications for the UV continuum from a sample of 34 image pair measurements from 23 lensed systems. The vertical dashed line indicates the mean value and the shaded band indicates one sigma around the average

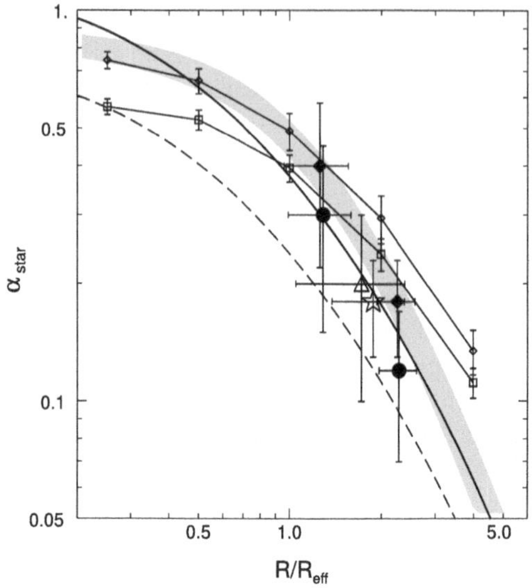

Fig. 23.2 Radial profile for the stellar mass fraction. The star (triangle) use only the X-ray (optical) data and a single radial bin for all objects. Circles (diamonds) are the estimates for the two radial bins using a logarithmic (linear) prior on size and both the X-ray and optical data. The dashed line corresponds to a simple model with a de Vaucouleurs stellar component and a total mass corresponding to a SIS with a flat rotation curve equal to the maximum rotational velocity of the stellar component. The thick line is the fiducial galaxy model from [42]. The gray band is the best fit profile for the sample of lenses analyzed by [34]. Open diamonds and squares correspond to a model using a Hernquist component for the stars, embedded in an NFW halo with (open squares) and without (open diamonds) adiabatic contraction of the dark matter, also from [34]. Figure adapted from Ref. [24]

and the mass fraction of the microlenses by [23, 24], using X-ray data [24], or based on enlarged data samples [12]. A summary of the main results about abundance of compact objects in the stellar mass range derived from gravitational lensing, is presented in Fig. 23.2, where we plot the fraction of mass in compact objects versus the radial distance to the center of the galaxy. The triangle corresponds to optical microlensing considering a single radial bin for all the objects from the study of [23] . The star corresponds to the average, also in a single radial bin, of the results of a sample of objects observed in X-rays [24]. The estimates of the fraction of mass in microlenses based on X-ray microlensing have the advantage that they are insensitive to the size of the source (that in X-ray is significantly smaller than the Einstein radius) for objects in the stellar mass range, cancelling a possible degeneracy. The diamonds are the estimates considering two radial bins for the objects combining both the X-ray and optical data [24]. The recently reported value of the fraction of mass in stars of 16% based on microlensing measurements from an enlarged sample of lenses by [12] would also be in reasonable agreement with these results. All the estimates nicely match the fiducial galaxy radial profile (continuous thick line) based on the mass fundamental plane scaling from [42] as well obtained from microlensing observations. Moreover, all these microlensing based results are in good agreement with the radial profile inferred by [34] from strong lensing models of a galaxy sample (gray band) and with a simple theoretical model (de Vaucouleurs + SIS) matching the rotational velocity of the stellar component [24].

Independent limits (albeit less stringent) have been set by [49] from gravitational lensing of type Ia supernovae (SNe Ia), who established that compact objects with masses above $0.01 M_{\odot}$ contribute less than 37% to the total mass of lens galaxies.

Thus, the main conclusion of the determination of the abundance of stellar mass MACHOS in the extragalactic domain is that quasar microlensing probably arises from the normal star population of lens galaxies and there is no statistical evidence for MACHOS in the dark halos.

23.3.2 *LIGO-Virgo-KAGRA Mass Range ($3M_{\odot} \lesssim M \lesssim 60M_{\odot}$)*

The detection of gravitational waves by the LIGO experiment produced by binary black holes mergers [1] involving masses significantly larger than originally expected for black holes of stellar origin (a typical value of $30M_{\odot}$, but with estimates as large as $60M_{\odot}$) renewed the interest in the possibility that dark matter could consist of primordial black holes in the intermediate-mass range, $1M_{\odot} \lesssim M \lesssim 1000M_{\odot}$ [8].

To analyze this possibility, [29] extend up to $100M_{\odot}$ the microlensing studies by [23, 24] focused on the stellar mass range. To do that, they fix the quasar source size to a fiducial value of 5 light-days in the optical and of 0.6 light-days in X-ray, according to the results from reverberation mapping. Then, they use the mass-length invariance of gravitational lensing to transform the joint probability distributions of the mass fraction of microlenses determined by [23, 24] in the optical and in X-rays from the (α, r) plane to the (α, M) plane. α is the surface mass density fraction in compact objects. It can be defined as $\alpha = \kappa_*/\kappa$ (where κ_* is the convergence in compact objects, and κ is the total convergence at the location of the image). The results (see Fig. 23.3) strongly constrain the compact object masses to the range $0.05M_{\odot} \lesssim M \lesssim 0.45M_{\odot}$ at the 90% confidence level. They also estimate a ~20% fraction of the total mass in compact objects. Both results are in agreement with the expected masses and abundances of the stellar component and are incompatible with the existence of any significant population of massive black holes with masses above the typical stellar ones. However, the upper value of the mass effectively probed by [29] is limited by the assumption that the broad emission lines are not affected by microlensing and can be used to define the zero microlensing baseline. If the broad line region is a few hundreds of light days large, it may be significantly microlensed by objects of $M_{BH} \gtrsim 100M_{\odot}$. Thus masses above this upper limit should be explored using other reference to set the no microlensing base-line.

The study by [29] is based on the use of a distribution of single-mass microlenses, while the real situation can be a mix of stars and BHs with very different masses. To

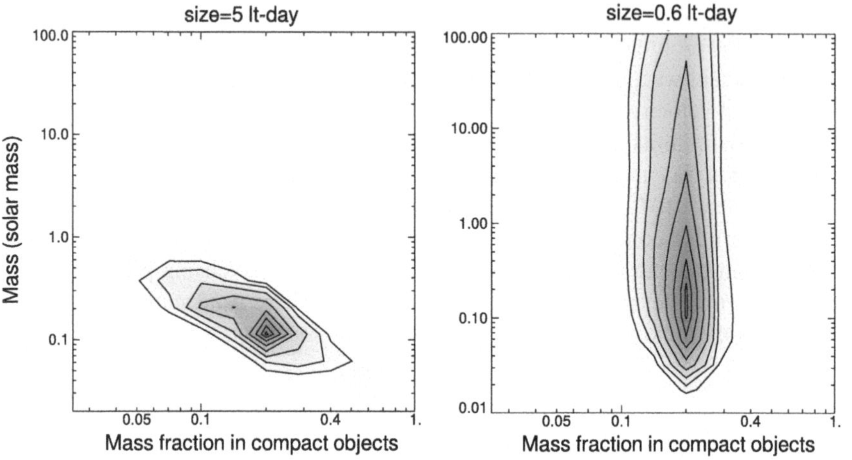

Fig. 23.3 Probability distributions for the mass and abundance of compact objects in lens galaxies from optical (left) and X-ray (right) data. Figure adapted from [29]

take into account this, [12, 14] consider a bimodal distribution of stars of $0.2M_\odot$ and BHs of $30M_\odot$. [14] present a broadly applicable argument based on the sensitivity to the size of the quasar (respect to the Einstein radius defined by the microlens mass) of the probability of observing a given microlensing magnification. They show that there is a strong difference in the predicted microlensing magnifications from a single population of stars and from a mix of stars with LIGO-Virgo-KAGRA BHs when a source of finite size (i.e. of size different from zero) is considered. The Einstein radius of a $\sim 30M_\odot$ BH is ~ 100 light-days, very much larger than the typical size of an accretion disk at the UV ($\sim$ 5 light-days, according to reverberation map results), while the Einstein ring of a $0.2M_\odot$ star is less than 10 light-days. Then, quasars approximate the point source limit for the BHs but have a size comparable to the Einstein radius of the stars. In the case of the stars, this result in a smoothing of the magnification changes at the source plane and to a substantial narrowing of the PDF of the predicted magnifications. Comparison with the observed histogram of microlensing magnifications obtained by [38], reveals that the predictions for a population of only stars match reasonably well the observed magnifications while a mix of stars and BHs predict too large magnifications (see Fig. 23.4). Consequently, if a significant part of the dark matter of the lens galaxies were in the form of BHs of the masses detected by LIGO-Virgo-KAGRA, much larger microlensing magnifications should have been regularly observed. As this is not the case, a PBHs population can safely be discarded. On the other hand, as the experimental histogram of microlensing magnifications by [38] is obtained taking as reference for zero microlensing the flux-ratios between images inferred from macro-models, it is not sensitive to the mass of the microlenses. Thus, taking into account that if a source is point like for a BH it also would be so for any BH of a larger mass, we can discard the existence of

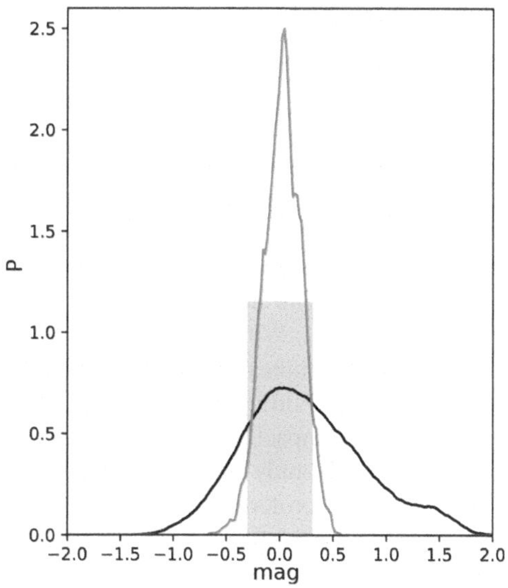

Fig. 23.4 Microlensing magnification histograms for a population of 20% stars with 0.2 $M_\odot$ plus 80% smooth dark matter (dark grey curve) and a population of 20% stars plus 80% LIGO-Virgo-KAGRA type BHs of 30 $M_\odot$ (black curve) for a source of 10 lt-days. The shaded area marks the region containing 70% of the observed microlensing magnifications by [38]. Figure adapted from [14]

a significant population not only for the intermediate mass range but also for any population of BHs of indefinitely large mass.

This result based on a general argument is quantitatively confirmed from a thorough Bayesian statistic study by [12]. They estimate at once the most likely abundances of stars and BHs, considering an additional smooth dark matter component and a quasar source size in agreement with current reverberation mapping estimates. The Bayesian study is based on the microlensing magnifications of both, the [28] sample extended to include 17 new measurements, and the [38] sample. [12] explore the $10M_\odot$ to $60M_\odot$ range for the BH mass, finding an upper limit to the abundance of BH $\lesssim$0.5% at the 68% confidence level. They also estimate a 16% contribution from the stars, in agreement with previous studies. These results are confirmed when uncertainties in source size and macrolens models are taken into account.

There is another broadly applicable rationale against the existence of a significant population of BHs with masses above $\sim 10M_\odot$. [38] find that X-ray flux-ratio anomalies are much larger than the optical ones. This result, confirmed by Jiménez-Vicente et al. [23], indicates that the effect causing the anomalies is sensitive to the differences in size between the X-ray and optical quasar sources, while they should perform as point like under the lensing action of large mass microlenses [38]. This is illustrated by Fig. 1 of [12] where we can appreciate the large sensitivity of the microlensing probability distribution corresponding to a $0.2M_\odot$ stellar population to the source size change while the probability associated to a mix of stars and $30M_\odot$ BHs is almost invariable.

23.3.3 Substellar Mass-Range ($2 \times 10^{-3} M_\odot \lesssim M \lesssim 0.1 M_\odot$)

The effect of microlensing is strongly diluted when the size of the source is much larger than the size of the Einstein radius of the microlenses, as the source *feels* a very smoothed version of the magnification structure at the source plane. Therefore, observations in the visible range (UV rest frame) corresponding to sources with a size of a few light days (which is the size of the Einstein radius of masses in the stellar range), are insensitive to lenses with substellar masses. Very compact sources are therefore needed if we are going to probe the substellar mass range of compact objects. With this idea in mind, [13] have used observations in X-rays, emitted by very compact sources of $\sim$1 lt-day, comparing them with a set of simulated models containing different proportions of stars, smoothly distributed dark matter, and substellar compact objects in the mass range from $\sim$0.1 to $2 \times 10^{-3} M_\odot$. The mass limits in this study are bounded on the upper side by the confusion with stars (as there is no way microlensing can distinguish the nature of compact objects of similar masses), and on the lower side by the Einstein radius of the lenses compared to the source size (which cannot produce significant magnifications for larger sources). The results of this study showed that the fraction of mass in form of stars is pretty well established to $\sim$12% in good agreement with expectations. For the mass in form of compact objects of substellar mass, the estimated values are around 4%, with upper limits around 10% for the masses down to 0.01 $M_\odot$. For the lowest studied masses of 0.002 $M_\odot$, whose Einstein radius is similar to the X-ray source sizes, the method is much less restrictive, with an estimated abundance of 10% (and an upper limit of 34%) of the total mass in form of this objects. Indeed, the data are fully compatible with 0% of the mass in the lens in form of objects in this mass range. In conclusion, the majority of the dark matter in these lens galaxies must be in the either in a smooth component or, at least, in the form of objects of smaller mass which cannot produce measurable microlensing efects for the X-ray sources.

Extragalactic single stars would be the optimal sources to probe low mass compact objects in the halos of galaxies via gravitational microlensing given their extremely compact source (compared to the emission from quasars in any band), but even the brightest supergiant stars are, of course, too faint to be observed at large distances. Yet, nature eventually provides unexpected opportunities, and if such star happens to be at the right spot in a galaxy behind a large cluster of galaxies (which provides very large magnifications of a few thousands), it might be magnified enough to be observed at cosmological distances. Since the first claimed observation of a single star at redshift $z = 1.5$ in this type of scenario by [25] as a transient event, a few other examples have been reported, even up to a record distance of $z = 6.2$ [46]. The transient nature of these observations in objects which are not intrinsically variable seems to indicate that an extrinsic induced variability is taking place, with gravitational microlensing by compact objects in the cluster halo being the favoured candidate. Several studies [35, 44] have carefully analyzed the shape of the light curve of such stars for a suitable set of parameters (including the amount of intra-cluster stars, the size of the star and the transverse effective velocity of the source) and have shown the huge potential of observations of these objects to restrict the mass and abundance of compact objects in the halo. In particular, given the extreme compactness of the source, these observations can probe down to very low masses. Unfortunately, present observations are scarce, and no evidence of a microcaustic crossing has been found. Moreover, there is a large uncertainty with respect to the transverse velocity of the source. Overall, limits on the abundance of objects in the mass range $10^{-5} - 10M_{\odot}$ to below 10% have been established [e.g. 35] with this procedure, but there is a large uncertainty in these limits due to the lack of knowledge on the transverse velocity. There are ongoing observational programs to discover and follow up these lensed stars (some taking benefit of the exquisite resolution and sensitivity of the James Webb Space Telescope). These observations are a promising method to be able to set more stringent limits to compact objects in the low mass range.

23.4 Mass Spectrum and Clustering Effects

In the previous sections, we have considered simple mass distributions (single or bimodal) and a reasonably (random) uniform spatial distribution of the microlenses. Nevertheless, formation theories for PBHs do not predict so simple mass and spatial distributions. PBHs should have a continuous mass distribution (as well as stars do) and, under certain circumstances, they could be spatially clustered [cf. 18]. It is therefore reasonable to wonder how the previous constraints on the abundance of PBHs from microlensing are affected when these two effects are considered. Again, there is a significant difference in the approach for galactic (based on the number of events of a certain duration) and extragalactic microlensing (based on flux ratio anomalies or light curves).

A study of galactic microlensing including an extended mass function and spatial clustering has been performed by [7]. The main conclusion of their work is that including an extended mass function and spatial clustering of the PBHs has the effect of shifting the constraints towards smaller masses, therefore leaving some more room for PBH of large mass ($\sim 10M_{\odot}$) which become much harder to detect.

In the case of extragalactic studies (mostly from quasar microlensing), we will consider the effect of the extended mass function and the spatial clustering of the lenses separately.

With respect to the mass function of microlenses, the situation is completely different for single-epoch and light curve studies. While in the latter case, the mass distribution of lenses may, in principle, leave detectable traces in a microlensing light curve, this inclusion would greatly complicate modeling (even more) to an extreme very difficult to handle. Indeed, microlensing is a quite degenerate problem (where mass and abundance of the lenses, size of the source and transverse velocity may combine in several different ways to produce similar observational results). Being able to reproduce the details of observed microlensing light curves while considering a large number of physical parameters is a very difficult problem. This is probably the reason why works dealing with these techniques have considered only a population of stars (for which we have quite a good knowledge from other sources), with none (to our knowledge) considering the contribution from a potential population of PBHs (or other population of compact objects). For the former case of single-epoch microlensing, where observations provide much more limited information, and the overall statistics of magnification is the most significant quantity, the situation is somewhat different, and several works have addressed the issue.

Several studies have shown that for continuous distributions of mass, the specific shape of the mass function is not determinant in the overall microlensing magnification statistics, with the mean mass being the most relevant parameter, except for extremely bimodal distributions with very different characteristic masses (see [41] and also discussion in [22]) which may be, indeed, the relevant case for PBHs (mixed with the stellar population of the lens). [22] already hinted that in this case, the relevant parameter determining the statistics of microlensing magnification of a mix of lenses of different masses was the geometric mean mass instead of the usual arithmetic mean mass.[4] Subsequently, [15] proved that this is also the case for pure bimodal distributions, with special application to the case of PBHs.[5] Therefore, in general, the microlensing effect of a continuous mass distribution, even extremely bimodal, could be modeled by a single mass distribution.[6] With the aim of studying the initial mass function of lens galaxies using microlensing, [22] estimated the average mass of microlenses in lens galaxies to be around 0.16 $M_{\odot}$, and were able to reproduce this average value with a plausible initial mass function with a high mass slope and a low mass cutoff fully compatible with the ones obtained by other means. That implies that the observed microlensing (both in UV rest frame continuum and X-rays) can be fully explained with the stellar population with no need for a population of other compact objects. If a population of such objects would have contributed significantly to the mass in the lens, they would have been most likely detected, by producing an estimated (geometric) mean mass significantly different of the value expected for a standard stellar population. The only situation that could affect this general statement is if the source size lies in between the Einstein radii characteristic for each of the masses in a bimodal distribution. This scenario of a bimodal mass distribution, taking into account the source size, has been explicitly modelled in the works of [12–14] whose limits are, therefore, rather general.

Finally, we need to address the issue of spatial clustering of the PBHs, and how this would affect the previous constraints on their abundance. In most studies of (single-epoch) quasar microlensing, the flux ratio anomalies are measured with respect to some baseline which is little or no affected by microlensing (cf. Equation 23.1). This baseline is usually determined by the flux ratio coming from very large sources, like

[4] For most smooth mass distributions, unless extremely wide, the arithmetic mean and the geometric mean mass are indeed very similar.

[5] The work by [20] may also open an interesting alternative path to constrain extended mass functions.

[6] A bimodal continuous distribution can be very well modeled by a two mass distribution where each mass is the characteristic (geometric) mean mass of each population.

the emission from broad lines (with typical sizes of hundreds of light days), mid infrared from the dust torus (with typical size of 1 pc) or radio emission (with a more uncertain size, but of the order of 5 pc or even larger). While these sources are much larger than the Einstein radius for lenses up to $\sim 100 M_\odot$, and are therefore little affected by microlensing by such objects, the situation may be very different if the PBHs are clustered, as in that case, the Einstein radius of the cluster as a whole, can be comparable (or even larger) than those sources which, consequently, can no longer be used as an unmicrolensed baseline. Heydenreich et al. [19] have addressed this issue by using as baseline the flux ratio prediction from the (macro) model which (although subject to quite a large uncertainty) is virtually free from microlensing effects by the clusters. They compared flux ratio anomalies in broad lines, mid infrared and radio (therefore unaffected by microlensing by stars) with numerical simulations of microlensing by clusters of PBHs in the LIGO-Virgo-KAGRA mass range (i.e. $\sim 30 M_\odot$). The modeled clusters have sizes in the range 5–15 pc and are composed of 300–3000 PBHs according to theoretical predictions [cf. 17]. They set an overall upper limit for clustered PBHs in this mass range of $\sim 3\%$ (at 68% confidence level),[7] with the highest abundances permitted for the most diluted and poorly populated clusters. Taking into account that diluted and poorly populated clusters approach the limit of uniform distribution, which has been addressed by [12, 14], we can accept the generality of the overall conclusion: PBHs in the intermediate mass range cannot be a main component of the dark matter in lens galaxies.

23.5 Summary of Constraints

As we have seen in Sect. 23.2, the galactic microlensing studies can establish rather strong limits (< few percent) for the abundance of PBHs in the halo in the low mass range from 10^{-7} to $\sim 1 M_\odot$. These constraints relax for higher masses ($> 1 M_\odot$), particularly when a mass spectrum and non uniform spatial distributions are taken into account.

In Fig. 23.5, we gather the constraints imposed by extragalactic microlensing on the fraction of mass in BHs from the study of a mixed population of BHs, stars and a smooth dark matter component. According to the mass-range, we have considered 4 regions. (i) $2 \times 10^{-3} M_\odot$ to $10^{-1} M_\odot$. As the lower end of this mass-range is approached, the microlensing magnification signatures attributable to the BHs become progressively washed out (smoothed at the scale of the X-ray source size), and the ability of microlensing to constrain the abundance of BHs diminishes, reaching a 10% for the smallest studied mass. Above $\sim 0.01 M_\odot$ a 4% limit can be established. In any case, note that the abundance inferred for the stars ($\sim 12\%$) is consistent with the expectations. (ii) In the second mass-range considered, $10^{-1} M_\odot$ to $3 M_\odot$, BHs and stars become indistinguishable, but also in this case, the consistence of the

[7] This figure can increase up to 7% if uncertainties in the macro lens models are neglected.

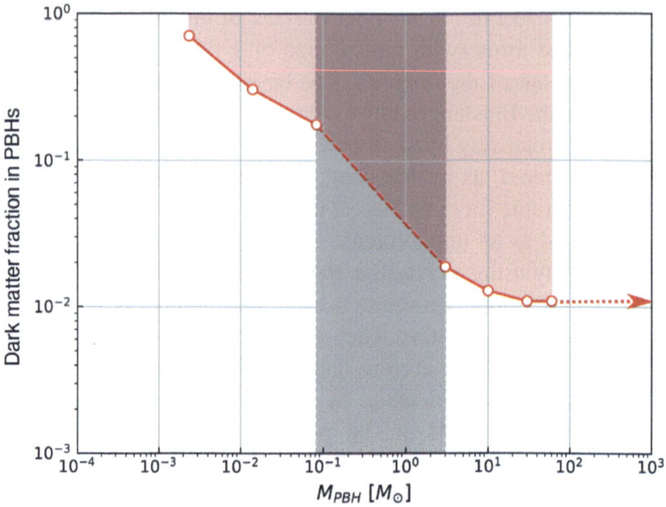

Fig. 23.5 Limits on the abundance of PBHs from quasar microlensing observations. The grey band indicates the region of confusion with stars. Limits extend to higher masses. Figure adapted from [13]

obtained fraction of mass in compact objects with the expected abundance of stars ($\sim$16%) support the exclusion of a significant BH population. (iii) In the $3M_\odot$ to $100M_\odot$ range, the upper limit on the abundance of BHs diminishes with the mass from 2 to 1%. (iv) As the BH mass increases, the point source limit is reached for the quasar continuum source and we can safely extrapolate the 1% upper limit to any larger mass, $M > 100M_\odot$. This last limit may be important to complement the constraints imposed from other grounds than microlensing in this (unlimited) mass range.

Thus, we can conclude that either from generic arguments or from detailed Bayesian calculations, observations of microlensing magnifications exclude the presence of a significant population of BHs with masses above $10^{-7}M_\odot$. Furthermore, this conclusion holds even when the BHs are distributed following a mass spectrum or appear in clusters.

Acknowledgements We thanks the anonymous referee for valuable comments, useful to improve the paper. This work is financed by grants PID2020-118687GB-C31 and PID2020-118687GB-C31, financed by MCIN/AEI/10.13039/501100011033. J. J. V. is also financed by projects FQM-108 financed by Junta de Andalucía.

References

1. B.P. Abbott et al., Observation of gravitational waves from a binary black hole merger. **116**(6), 061102 (2016). https://doi.org/10.1103/PhysRevLett.116.061102, arXiv:1602.03837 [gr-qc]
2. C. Alcock et al., Possible gravitational microlensing of a star in the large magellanic cloud. **365**(6447), 621–623 (1993). https://doi.org/10.1038/365621a0, arXiv:astro-ph/9309052 [astro-ph]
3. C. Alcock et al., The MACHO project: microlensing results from 5.7 years of large magellanic cloud observations. **542**(1), 281–307 (2000). https://doi.org/10.1086/309512, arXiv:astro-ph/0001272 [astro-ph]
4. C. Alcock et al., MACHO project limits on black hole dark matter in the 1–30 $M_{\odot}$ range. **550**(2), L169–L172 (2001). https://doi.org/10.1086/319636, astro-ph/0011506 [astro-ph]
5. E. Aubourg et al., Evidence for gravitational microlensing by dark objects in the Galactic halo. **365**(6447), 623–625 (1993). https://doi.org/10.1038/365623a0
6. T. Blaineau et al., New limits from microlensing on Galactic black holes in the mass range 10 $M_{\odot}<M<1000M_{\odot}$. **664**, A106 (2022). https://doi.org/10.1051/0004-6361/202243430, arXiv:2202.13819 [astro-ph.GA]
7. J. Calcino, J. García-Bellido, T.M. Davis, Updating the MACHO fraction of the Milky Way dark halowith improved mass models. **479**(3), 2889–2905 (2018). https://doi.org/10.1093/mnras/sty1368, arXiv:1803.09205 [astro-ph.CO]
8. B. Carr, F. Kühnel, M. Sandstad, Primordial black holes as dark matter. **94**(8), 083504 (2016). https://doi.org/10.1103/PhysRevD.94.083504, arXiv:1607.06077 [astro-ph.CO]
9. K. Chang, S. Refsdal, Flux variations of QSO 0957 + 561 A, B and image splitting by stars near the light path. **282**(5739), 561–564 (1979). https://doi.org/10.1038/282561a0
10. D. Croon et al., Subaru-HSC through a different lens: Microlensing by extended dark matter structures. Phys. Rev. D **102**, 083021 (2020). https://doi.org/10.1103/PhysRevD.102.083021
11. R. Edelson et al., Space telescope and optical reverberation mapping project. II. Swift and HST reverberation mapping of the accretion disk of NGC 5548. **806**(1), 129 (2015). https://doi.org/10.1088/0004-637X/806/1/129, arXiv:1501.05951 [astro-ph.GA]
12. A. Esteban-Gutiérrez et al., Abundance of LIGO/virgo black holes from microlensing observations of quasars with reverberation mapping size estimates. **929**(2), 123 (2022). https://doi.org/10.3847/1538-4357/ac57c5, arXiv:2203.04777 [astro-ph.CO]
13. A. Esteban-Gutiérrez et al., Constraints on the abundance of primordial black holes from X-ray quasar microlensing observations: substellar to planetary mass range. **954**(2), 172 (2023). https://doi.org/10.3847/1538-4357/ace62f, arXiv:2307.07473 [astro-ph.CO]
14. A. Esteban-Gutiérrez et al., Limiting the abundance of LIGO/virgo black holes with microlensing observations of quasars of finite size. **929**(1), L17 (2022). https://doi.org/10.3847/2041-8213/ac57c6, arXiv:2203.04758 [astro-ph.CO]
15. A. Esteban-Gutiérrez et al., The impact of the mass spectrum of lenses in quasar microlensing studies. Constraints on a mixed population of primordial black holes and stars. **904**(2), 176 (2020). https://doi.org/10.3847/1538-4357/abbdf7, arXiv:2011.05751 [astro-ph.CO]
16. M.M. Fausnaugh et al., Space telescope and optical reverberation mapping project. III. Optical continuum emission and broadband time delays in NGC 5548. **821**(1), 56 (2016). https://doi.org/10.3847/0004-637X/821/1/56, arXiv:1510.05648 [astro-ph.GA]
17. M. Gorton, A.M. Green, Effect of clustering on primordial black hole microlensing constraints. **2022**(8), 035 (2022). https://doi.org/10.1088/1475-7516/2022/08/035, arXiv:2203.04209 [astro-ph.CO]
18. A.M Green, B.J. Kavanagh, Primordial black holes as a dark matter candidate. J. Phys. G: Nuclear Particle Phys. **48**(4), 043001 (2021). https://doi.org/10.1088/1361-6471/abc534
19. S. Heydenreich, E. Mediavilla, J. Jiménez-Vicente, H. Vives-Arias, J.A. Muñoz, The abundance of clustered primordial black holes from quasar microlensing. Astron. Astrophys. **690**, A307 (2024). https://doi.org/10.1051/0004-6361/202449216
20. K. Inomata et al., Inflationary primordial black holes as all dark matter. Phys. Rev. D **96**, 043504 (2017). https://doi.org/10.1103/PhysRevD.96.043504

21. Y.-F. Jiang et al., Detection of time lags between quasar continuum emission bands based on pan-STARRS light curves. **836**(2), 186 (2017). https://doi.org/10.3847/1538-4357/aa5b91, arXiv:1612.08747 [astro-ph.HE]
22. J. Jiménez-Vicente, E. Mediavilla, The initial mass function of lens galaxies from quasar microlensing. **885**(1), 75 (2019). https://doi.org/10.3847/1538-4357/ab46b8, arXiv:1910.10509 [astro-ph.GA]
23. J. Jiménez-Vicente et al., Dark matter mass fraction in lens galaxies: new estimates from microlensing. **799**(2), 149 (2015). https://doi.org/10.1088/0004-637X/799/2/149, arXiv:1407.3955 [astro-ph.GA]
24. J. Jiménez-Vicente et al., Probing the dark matter radial profile in lens galaxies and the size of X-ray emitting region in quasars with microlensing. **806**(2), 251 (2015). https://doi.org/10.1088/0004-637X/806/2/251, arXiv:1502.00394 [astro-ph.GA]
25. P.L. Kelly et al., Extreme magnification of an individual star at redshift 1.5 by a galaxy-cluster lens. Nat. Astron. **2**, 334–342 (2018). https://doi.org/10.1038/s41550-018-0430-3, arXiv:1706.10279 [astro-ph.GA]
26. G.F. Lewis, M.J. Irwin, The statistics of microlensing light curves. II. Temporal analysis. **283**(1), 225–240 (1996). https://doi.org/10.1093/mnras/283.1.225
27. E. Mediavilla, J. Jiménez-Vicente, V. Motta, Quasar microlensing statistics and flux-ratio anomalies in lens models. **167**(4), 171 (2024). https://doi.org/10.3847/1538-3881/ad2ae0, arXiv:2404.09865 [astro-ph.CO]
28. E. Mediavilla et al., Microlensing-based estimate of the mass fraction in compact objects in lens galaxies. **706**(2), 1451–1462 (2009). https://doi.org/10.1088/0004-637X/706/2/1451, arXiv:0910.3645 [astro-ph.CO]
29. E. Mediavilla et al., Limits on the mass and abundance of primordial black holes from quasar gravitational microlensing. **836**(2), L18 (2017). https://doi.org/10.3847/2041-8213/aa5dab, arXiv:1702.00947 [astro-ph.GA]
30. M. Moniez, Microlensing as a probe of the Galactic structure: 20 years of microlensing optical depth studies. General Relat Gravitation **42**(9), 2047–2074 (2010). https://doi.org/10.1007/s10714-009-0925-4, arXiv:1001.2707 [astro-ph.GA]
31. P. Mróz et al., No large population of unbound or wide-orbit Jupiter-mass planets. Nature **D85**(548), 183–186. https://doi.org/10.1038/nature23276, arXiv:1707.07634 [astro-ph.CO]
32. H. Niikura et al., Microlensing constraints on primordial black holes with Subaru/HSC Andromeda observations. Nat. Astron. **3**, 524–534 (2019). https://doi.org/10.1038/s41550-019-0723-1, arXiv:1701.02151 [astro-ph.CO]
33. H. Niikura et al., Constraints on Earth-mass primordial black holes from OGLE 5-year microlensing events. **99**(8), 083503 (2019). https://doi.org/10.1103/PhysRevD.99.083503, arXiv:1901.07120 [astro-ph.CO]
34. M. Oguri, C.E. Rusu, E.E. Falco, The stellar and dark matter distributions in elliptical galaxies from the ensemble of strong gravitational lenses. **439**(3), 2494–2504 (2014). https://doi.org/10.1093/mnras/stu106, arXiv:1309.5408 [astro-ph.CO]
35. M. Oguri et al. Understanding caustic crossings in giant arcs: Characteristic scales, event rates, and constraints on compact dark matter. Phys. Rev. D **97**, 023518 (2018). https://doi.org/10.1103/PhysRevD.97.023518
36. B. Paczynski, Gravitational microlensing by the galactic halo. **304**, 1 (1986). https://doi.org/10.1086/164140
37. P. Tisserand et al., and EROS-2 Collaboration, Limits on the Macho content of the Galactic Halo from the EROS-2 Survey of the Magellanic Clouds. **469**(2), 387–404 (2007). https://doi.org/10.1051/0004-6361:20066017, arXiv:astro-ph/0607207 [astro-ph]
38. D. Pooley et al., X-ray and optical flux ratio anomalies in quadruply lensed quasars. I. Zooming in on quasar emission regions. **661**(1), 19–29 (2007). https://doi.org/10.1086/512115, arXiv:astro-ph/0607655 [astro-ph]
39. P.L. Schechter, J. Wambsganss, The dark matter content of lensing galaxies at 1.5 R_e, in *Dark Matter in Galaxies*, vol. 220, ed. by S. Ryder et al. (2004), p. 103. https://doi.org/10.48550/arXiv.astro-ph/0309163, arXiv:astro-ph/0309163 [astro-ph]

40. P.L. Schechter, J. Wambsganss, Quasar microlensing at high magnification and the role of dark matter: enhanced fluctuations and suppressed saddle points. **580**(2), 685–695 (2002). https://doi.org/10.1086/343856, arXiv:astro-ph/0204425 [astro-ph]
41. P.L. Schechter, J. Wambsganss, G.F. Lewis, Qualitative aspects of quasar microlensing with two mass components: magnification patterns and probability distributions. **613**(1), 77–85 (2004). https://doi.org/10.1086/422907, arXiv:astro-ph/0403558 [astro-ph]
42. P.L. Schechter et al., A calibration of the stellar mass fundamental plane at z Ë& oelig;0.5 using the micro-lensing-induced flux ratio anomalies of macro-lensed quasars. **793**(2), 96 (2014). https://doi.org/10.1088/0004-637X/793/2/96, arXiv:1405.0038 [astro-ph.GA]
43. A. Udalski et al., The optical gravitational lensing experiment. Discovery of the first candidate microlensing event in the direction of the galactic bulge. **43**, 289–294 (1993)
44. T. Venumadhav, L. Dai, J. Miralda-Escudé, Microlensing of extremely magnified stars near caustics of galaxy clusters. **850**(1), 49 (2017). https://doi.org/10.3847/1538-4357/aa9575, arXiv:1707.00003 [astro-ph.CO]
45. R.L. Webster et al., Interpreting the Light Curve of Q2237+0305. **102**, 1939 (1991). https://doi.org/10.1086/116015
46. B. Welch et al., A highly magnified star at redshift 6.2. **603**(7903), 815–818 (2022). https://doi.org/10.1038/s41586-022-04449-y, arXiv:2209.14866 [astro-ph.GA]
47. H.J. Witt, S. Mao, P.L. Schechter, On the universality of microlensing in quadruple gravitational lenses. **443**, 18 (1995). https://doi.org/10.1086/175499
48. L. Wyrzykowski et al., The OGLE view of microlensing towards the Magellanic Clouds - IV. OGLE-III SMC data and final conclusions on MACHOs. **416**(4), 2949–2961 (2011). https://doi.org/10.1111/j.1365-2966.2011.19243.x, arXiv:1106.2925 [astro-ph.GA]
49. M. Zumalacárregui, U. š Seljak, Limits on stellar-mass compact objects as dark matter from gravitational lensing of type Ia supernovae. Phys. Rev. Lett. **121**, 141101 (2018). https://doi.org/10.1103/PhysRevLett.121.141101

Chapter 24
Cosmic Microwave Background and Accretion

Pasquale Dario Serpico

Abstract Primordial black holes of stellar mass or heavier would accrete baryonic gas, which becomes denser and hotter, injecting energetic photons in the cosmological medium soon after cosmic recombination, in the so-called *dark ages*. The ionisation history of the universe would be altered, an effect cosmic microwave background temperature and polarisation anisotropies is sensitive to. The magnitude of the effect depends on the abundance and mass distribution of the primordial black holes, as well as on still uncertain aspects of accretion luminosity. We review the current understanding of this field, with an emphasis on the peculiarities of the phenomenon in the cosmological context, and present the existing constraints on the primordial black hole abundances.

24.1 Introduction

Massive bodies gravitationally accrete matter, which acquires kinetic energy at the expense of potential one; as a result of microscopic dissipative mechanisms, the accreted material heats up and emits radiation. At the same time, following energy and angular momentum losses, this material falls onto the massive body or, for a black hole (BH), it crosses the event horizon. A rich astrophysical phenomenology is associated to accretion power (see e.g. the monograph [20]). The luminosity due to the accretion of the interstellar medium by isolated stellar or planetary objects, while universal, is typically too modest to be of observational relevance. This is why accretion studies in astrophysics often focus on close binaries or episodes of large mass accretion, for instance related to a stellar disruption by a supermassive black hole. Most electromagnetic signals associated to BH detection are related to these situations.

P. D. Serpico (✉)
LAPTh, CNRS and University of Savoie Mont Blanc, 9 chemin de Bellevue, BP 110, 74941 Annecy, France
e-mail: serpico@lapth.cnrs.fr

C. Byrnes et al. (eds.), *Primordial Black Holes*, Springer Series in Astrophysics and Cosmology, https://doi.org/10.1007/978-981-97-8887-3_24

Needless to say, accretion phenomena on compact objects require the presence of compact objects in the first place. In the standard cosmological scenario, these are only present starting from the end of the dark ages, let us say at redshifts $z \lesssim 20 - 30$, after first stellar structures could form by gravitational collapse in the pristine dark matter (DM) halos. While differing in poorly known quantitative details, the accretion phenomena in these environments are qualitatively similar to the more familiar ones going on in astrophysical settings at low z.

In scenarios where sufficiently massive primordial black holes (PBHs) are present, however, it becomes of interest to study the relevance of accretion phenomena onto them in the early universe, at $z \gg 20$. In this case, the direct detection of the accretion luminosity may be impossible or impractical. Currently, the most effective tool to constrain such an exotic process is by studying the alterations it provokes on the cosmic microwave background (CMB) temperature and polarisation anisotropies. This chapter is devoted to outline the key ingredients, equations and assumptions entering these calculations, and to present current constraints to an audience presumably already familiar with the basic of the standard ΛCDM cosmology. The chapter is structured as follows: In Sect. 24.2 we recall the key arguments on why and how CMB anisotropies can be affected by such energy injection. Section 24.3 reviews the basic notions on accretion rate and luminosity, while Sect. 24.4 specialises these notions to a cosmological setting. Section 24.5 discusses *disk* accretion and luminosity in general and in the cosmological context, and Sect. 24.6 shows how to account for the presence of a DM halo around the accreting PBH. The latter treatment does not depend on the nature of DM, under the sole hypothesis that the bulk of DM is collisionless. Section 24.7 presents current constraints and a discussion on some perspectives. Natural units are used in equations, while numerical values are quoted in the most widely used astrophysical units.

24.2 Impact of Energy Injection on CMB Anisotropies

In the last couple of decades, CMB anisotropies have been repeatedly used to constrain the energy injection associated to a number of exotic processes; we just mention here seminal articles related to decaying relics [16], annihilating relics like weakly interacting massive particle dark matter [18, 47], evaporating (hence "light") PBHs [14] or accreting (hence "heavy") PBHs [39]. These putative processes have all qualitatively similar effects on the CMB, reviewed in this section; for details see e.g. [35].

Note that the energy of the injected particles, even if negligible with respect to the large CMB photon energy density, is not negligible with respect to the kinetic energy of the baryonic gas, which is cold and mostly neutral during the dark ages ($20 \lesssim z \lesssim 1000$). The main consequence of this injection is to modify the fraction of free electrons x_e, either directly, via ionisation, or in a multi-step process, via collisional excitation followed by photoionisation by a CMB photon. An indirect effect is the heating of the medium, whose temperature $T_{\rm M}$ has a feedback on the evolution of x_e. CMB anisotropies are sensitive to these effects via alterations to optical depth τ (integrated value of $x_e(z)$ over redshift) and its time dependence (or "visibility function").

Modern treatments of the relevant physics are quite detailed and involve numerous processes and transitions, see for instance refs. [6, 17, 42] at the basis of widely used codes. The key physics can be however understood within the so-called "Peebles recombination" model [33, 53], where these quantities are ruled by a system of two coupled differential equations (for details on these terms and a compact overview, we refer the reader to the appendix A of Ref. [36]):

$$\frac{\mathrm{d}x_e(z)}{\mathrm{d}z} = \frac{1}{(1+z)H(z)}(R(z) - I(z) - I_X(z)) \,,$$
$$\frac{\mathrm{d}T_\mathrm{M}}{\mathrm{d}z} = \frac{1}{1+z}\left[2T_\mathrm{M} + \beta_\mathrm{cool}(T_\mathrm{M} - T_\mathrm{CMB})\right] + K_h. \tag{24.1}$$

Here, $H(z)$ is the Hubble expansion rate (dominated by matter in the relevant redshift interval) the R and I terms are the standard recombination and ionisation rates, respectively, while $\beta_\mathrm{cool} \equiv 8\sigma_T u_\mathrm{CMB} x_e/[3(1+z)(1+x_e)H\, m_e]$ is the so-called opacity of the gas, with $u_\mathrm{CMB} = \pi^2 T_\mathrm{CMB}^4/15$ the energy density of the CMB, σ_T the Thomson cross section, m_e the electron mass. The exotic terms are represented by the effective ionisation rate $I_X(z) = I_{Xi}(z) + I_{X\alpha}(z)$ (contributed to by the direct ionisation rate I_{Xi} and excitation+ionisation rate $I_{X\alpha}$) and K_h, which describes the heating rate due to the new process. All these quantities, standard and extra terms, are in general known (non-linear) functions of x_e, and T_M. The extra terms are proportional to the deposited energy rate in each channel,

$$\left.\frac{\mathrm{d}E}{\mathrm{d}V\mathrm{d}t}\right|_{\mathrm{dep},c} , \tag{24.2}$$

which are related to the direct input of the extra injected energy rates, $\mathrm{d}E/(\mathrm{d}V\mathrm{d}t)\big|_\mathrm{inj}$, via transfer functions f_c specific to the channel c,

$$f_c(z, x_\mathrm{e}) \equiv \frac{\mathrm{d}E/(\mathrm{d}V\mathrm{d}t)\big|_{\mathrm{dep},c}}{\mathrm{d}E/(\mathrm{d}V\mathrm{d}t)\big|_\mathrm{inj}}. \tag{24.3}$$

We refer to appendix A of Ref. [36] for further definitions and more details on each coefficient.

The injection energy rates are the direct input of the model being tested. For instance, for PBHs constituting a fraction f_PBH of the DM, one has

$$\left.\frac{\mathrm{d}E}{\mathrm{d}V\mathrm{d}t}\right|_\mathrm{inj} = L n_\mathrm{pbh} = L f_\mathrm{PBH}\frac{\rho_\mathrm{DM}}{M} \tag{24.4}$$

where we are implicitly assuming a monochromatic mass spectrum at M (see Sect. 24.7 for a generalisation), and L is the associated accretion luminosity, detailed in the next sections. On the contrary, the transfer functions depend only on standard physics, but their calculation may be rather cumbersome, requiring to follow the degradation of energy and cascades of particle multiplication which are generally non-local (a particle injected at z_1 may deposit some of its energy at $z_2 < z_1$). Several calculations have been performed in the literature, see e.g. [47, 51] for early ones. Nowadays, dedicated software exists like the `EXOCLASS` package [48], based on the more modern calculations of [46], which can be directly interfaced with the Boltzmann code `CLASS` [11] to evaluate the impact of any kind of exotic electromagnetic energy injection onto CMB anisotropies. While the efforts towards more realistic treatment of energy deposition is still ongoing, see [27, 28], it is fair to say that remaining uncertainties are not the dominant ones for the problem of interest here (redshift range $100 \lesssim z \lesssim 1000$, quasi-homogeneous medium). Also note that, while conceptually the problem may be complicated by a feedback effect of the deposited energy on the deposition function itself, within current constraints this effect is at the percent level, see e.g. [28].

Once the function $x_e(z)$ is known, the impact on the CMB anisotropy angular power spectra is due to the two previously mentioned effects:

- The visibility function is altered and recombination is delayed, which (slightly) shifts the acoustic peaks and causes small wiggles at high multipoles ℓ in the differences with respect to a standard ΛCDM scenario.
- Due to overall higher integrated ionisation fraction, the optical depth is increased (all other parameters being the same). This manifests itself in a damping of temperature anisotropies and an enhanced power in the polarisation spectrum at low/intermediate ℓ's.

These effects, quantitatively depending on the accretion luminosity evolution, are illustrated in Fig. 24.1, taken from [37], for two benchmark cases motivated in Sects. 24.4, 24.5. In particular, the former effect is visible in the top panel, the latter effect in both panels.

Note that in principle, different exotic injection histories due to different processes or different accretion luminosity evolutions correspond to different $x_e(z)$ (see green vs. red lines in Fig. 24.1) and could be distinguished via a fine analysis of CMB anisotropies. Hence, if a signal were found, some constraints could be put on the PBH accretion mechanism. With current precision, however, only a rough sensitivity to extreme differences in $x_e(z)$ can be expected, and we will not indulge in these aspects any further.

24.3 Generalities on Accretion Rate and Luminosity

A particle of mass m falling without losses from rest at infinity to a distance R from a point mass M acquires kinetic energy equal to the potential one, $GMm/R = 0.5mR_S/R$, where we introduced the Schwarzschild radius $R_S \equiv 2GM$. Since some

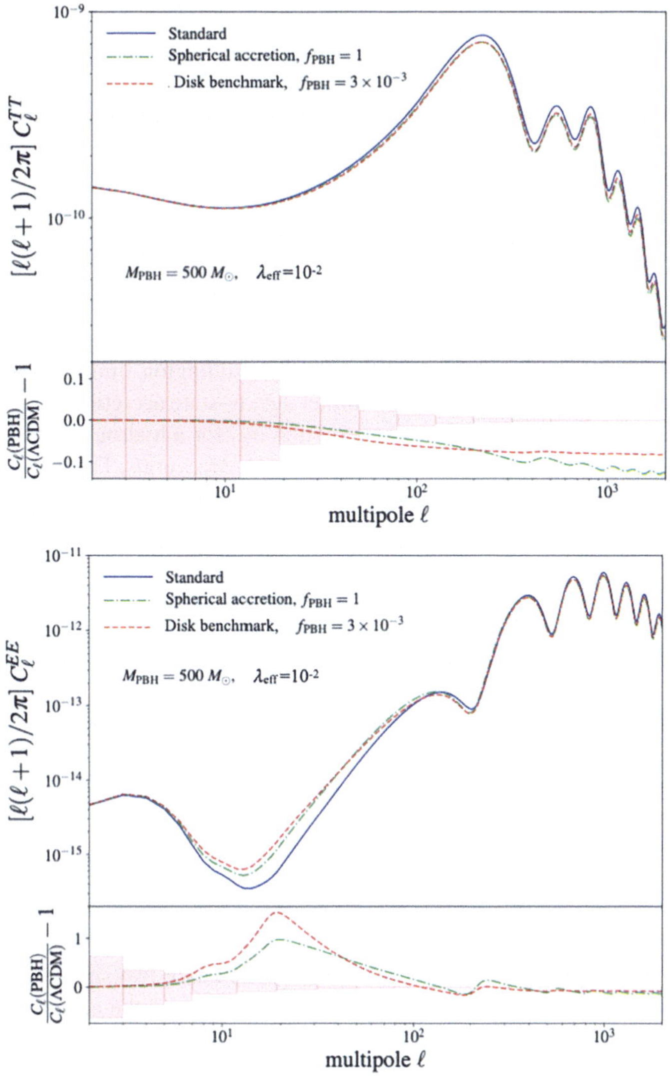

Fig. 24.1 CMB temperature (TT, top panel) and polarization (EE mode, bottom panel) angular power spectrum obtained for a monochromatic population of PBH with masses 500 $M_\odot$ depending on the accretion recipe used. See Sects. 24.4, 24.5 for a description of the two reported benchmarks. Adapted from Ref. [37]

of this energy ends up in radiation, typically via collisional/dissipative effects, an electromagnetic luminosity L is associated to accretion. The total electromagnetic luminosity L associated to mass accretion rate $\dot{M}$ is customarily written as $L = \epsilon \dot{M}$, keeping in mind that the conversion efficiency parameter ϵ is a function of $\dot{M}$ and of the BH and medium properties. If all the kinetic energy were converted into radiation, from the previous argument one would estimate $\epsilon \leq R_S/(2R) \leq 0.5$. The part of the mass accretion rate contributing to increase the BH mass hence writes $(1-\epsilon)\dot{M}$. From a macroscopic point of view, the problem is to determine ϵ and $\dot{M}$.

As a rule of thumb, only particles orbiting at or beyond the innermost stable circular orbit (ISCO) have time for radiative processes to be relevant: Once at the ISCO, the particle crosses the event horizon within a free-fall time [20]. The closest the ISCO, the higher is ϵ: A typical benchmark value used in the literature is $\epsilon \simeq 0.1$ for a Schwarzschild BH, for which $R_{\rm ISCO} = 3\,R_S$ (a slightly lower ϵ is actually expected when taking GR effects into account). For comparison, $\epsilon \simeq 0.4$ for a maximally rotating Kerr BH.

It is also worth stressing that there are physical limitations to the electromagnetic luminosity. The most famous such bound is the Eddington limit, when radiation pressure in a spherical setting prevents further steady-state accretion. In terms of the proton mass m_p and the Thomson cross section σ_T, for a hydrogen gas this condition writes $L_{\rm Edd} = 4\pi G M m_p/\sigma_T \simeq 1.3 \times 10^{38}(M/M_\odot)$ erg/s. The Eddington limit suggests that: (a) With growing BH mass, a proportionally larger limiting luminosity can be attained. (b) Steady-state accretion at larger and larger $\dot{M}$ is only possible at the expense of a decreasing efficiency ϵ, otherwise the bound would be violated.

Concerning $\dot{M}$, in most astrophysical settings of interest it is set either by the close binary system specification, or is an external input entering e.g. a fit to the data. For PBHs, we are however interested in computing it in a quasi-homogeneous cosmological background. Even the deceivingly simple problem of homogeneous accretion onto a point mass M, despite being rather old, does not have a general solution. Under steady-state hypothesis, the problem was solved by Hoyle & Littleton in 1939–40 [21–23] for an infinite and pressureless gas cloud with density ρ_∞ far away from the BH, moving at $v_{\rm rel}$, yielding

$$\dot{M}_{HL} = 4\pi\rho_\infty \frac{(GM)^2}{v_{\rm rel}^3}. \tag{24.5}$$

As shown by Bondi and Hoyle in 1944 [13], up to a factor two smaller accretion is estimated once accounting for density inhomogeneities and the accretion wake. In the opposite limit of accretion at rest in gas with pressure, Bondi found in 1952 [12] (see also [20] for a modern exposition) that

$$\dot{M}_B = 4\pi\lambda\rho_\infty \frac{(GM)^2}{c_{s,\infty}^3}, \tag{24.6}$$

where $c_{s,\infty}$ is the speed of sound in the homogeneous gas at infinity and $\lambda \sim \mathcal{O}(0.1-1)$ is the accretion eigenvalue: It is the key output of the solution of the steady-state problem involving mass conservation, Euler equation, energy conservation and the equation of state (EOS). We will see a generalisation of this problem in the cosmological framework in the next section. Note that the Bondi problem (or its relativistic generalisation [30]) yields a transonic solution for BHs, i.e. the velocity changes from subsonic at large distances to supersonic below a finite distance above the event horizon.

We can summarise both regimes of Eqs. (24.5), (24.6) and provide an interpolation between the two via the following parameterisation:

$$\dot{M} = 4\pi \lambda_{\rm eff} \rho_\infty v_{\rm rel} r_{\rm eff}^2, \tag{24.7}$$

where

$$r_{\rm eff} = \frac{GM}{v_{\rm eff}^2}, \tag{24.8}$$

where $v_{\rm eff} \simeq \sqrt{c_{s,\infty}^2 + v_{\rm PBH}^2}$, as suggested by Bondi himself [12], and $\lambda_{\rm eff} \sim 0.1 - 1$ is to be adjusted to the problem at hand. The variable $r_{\rm eff}$ thus defined is inspired by the so-called Bondi radius $r_B \equiv GM/c_{s,\infty}^2$; we see that the limit $v_{\rm eff} \to c_{s,\infty}$ in Eq. (24.7) yields indeed Eq. (24.6). Note that none of the solutions corresponding to Eqs. (24.5), (24.6) accounts for the radiation feedback onto the flow. Also, Eqs. (24.5), (24.6), (24.7) grow quadratically with the mass. In general, one does find that $\dot{M}/M$ is a growing function of the mass: As a corollary, accretion power only matters for sufficiently heavy objects, typically of stellar size or heavier.

The density and temperature profiles corresponding to a given solution can be used to compute the radiation emitted by the system, accounting for the relevant processes, such a Bremsstrahlung radiation, synchrotron radiation, etc. Once the energy-integrated emissivity (energy per unit time and volume) j is known, the luminosity can be obtained by integration, e.g. in case of spherical symmetry one has $L = \int {\rm d}r 4\pi r^2 j$. Energy-differential quantities can also be derived, of course, if needed.

24.4 Accretion in a Cosmological Setting

The first seminal studies of the PBH accretion in a cosmological context, with application to the CMB, are the articles [38, 39]. While some of the results of these works remain correct, other results have been proven incorrect and amended in [7], on which the following discussion is mostly based.

First, the Bondi problem previously introduced can be generalised to include energy cooling and momentum drag via Compton scattering (quantified via the functions $\beta_{\rm cool}$ and $\beta_{\rm drag} \equiv (1 + x_e) m_e \beta_{\rm cool}/(2m_p) \ll \beta_{\rm cool}$, respectively, named γ and β in [7]). In spherical symmetry, the relevant system of equations, i.e. mass conservation, momentum equation, heat equation and perfect gas EOS write

$$4\pi r^2 \rho |v| = \dot{M} = {\rm const} = 4\lambda\pi\rho_\infty \frac{(GM)^2}{c_{s,\infty}^3}, \tag{24.9}$$

$$v\frac{\mathrm{d}v}{\mathrm{d}r} = -\frac{GM}{r^2} - \frac{1}{\rho}\frac{\mathrm{d}P}{\mathrm{d}r} - \beta_{\mathrm{drag}}v, \tag{24.10}$$

$$v\rho^{2/3}\frac{\mathrm{d}}{\mathrm{d}r}\left(\frac{T}{\rho^{2/3}}\right) = \beta_{\mathrm{cool}}(T_{\mathrm{CMB}} - T), \tag{24.11}$$

$$P = \frac{\rho}{m_p}(1+\overline{x}_e)T, \tag{24.12}$$

for a PBH accreting hydrogen gas at rest in the cosmic frame, where the first mass-conservation equation introduces the accretion eigenvalue λ and links the accretion to asymptotic (cosmological homogeneous) density ρ_∞ and sound speed $c_{s,\infty} = \sqrt{\gamma P_\infty/\rho_\infty}$. The value of the EOS parameter are $\gamma = 5/3$ for adiabatic conditions and $\gamma = 1$ for isothermal ones. Remarkably, the authors of [7] obtain a semi-analytical solution for this system of equations, yielding the density, velocity and temperature profile versus radius r for a broad range of parameters β_{drag} and β_{cool}. These reduce to the Bondi solution (where at small r one finds $\rho \propto r^{-3/2}$ and $T \propto r^{-1}$) in the limit $\beta_{\mathrm{drag}}, \beta_{\mathrm{cool}} \to 0$.

Concerning radiative emission near the Schwarzschild radius, in the unmagnetised (or poorly magnetised) cosmological environment, it is dominated by Bremsstrahlung radiation, i.e. free-free emission in a fully-ionised thermal electron-proton plasma, where densities obey $n_e = n_p$ and the common temperature is T. The expression for the energy-integrated emissivity j is consistent with a simple estimate, writing

$$j = n_e^2\alpha\sigma_T T\mathcal{J}(X), \tag{24.13}$$

where $X \equiv T/m_e$ and $J(X)$ is a dimensionless function (typically of order ~ 10) which can be expressed within a few percent accuracy by the fit [7]

$$\mathcal{J}(X) \approx \begin{cases} \frac{4}{\pi}\sqrt{2/\pi}X^{-1/2}\left(1+5.5X^{1.25}\right), & X < 1, \\ \frac{27}{2\pi}\left[\ln(2X\mathrm{e}^{-\gamma_{\mathrm{E}}} + 0.08) + \frac{4}{3}\right], & X > 1, \end{cases} \tag{24.14}$$

including both $e-e$ and $e-p$ processes. Also, despite the problem being in general a complicated radiative transfer one, the authors of [7] bracket the efficiency parameter ϵ in the two limiting cases where ionisation of the matter either proceeds via collisional effects (if the emerging radiation field is too weak to photoionise the gas) or, in the opposite limit, via photoionisation. In both cases, one has

$$\epsilon \approx \alpha\frac{\dot{M}}{L_{\mathrm{Edd}}}\frac{T_{\mathrm{S}}}{m_p}\mathcal{J}\left(\frac{T_{\mathrm{S}}}{m_e}\right). \tag{24.15}$$

where α is the fine-structure constant and T_S the temperature at the Schwarzschild radius, predicted to be between 10^9 and 10^{11} K for the collisional and photo-ionisation case, respectively. Apart for factors of order unity, due for instance to neglecting the effect of the Helium fraction or general-relativistic corrections, the solution to this simplest formalisation of the cosmological problem is bracketed between two extremes.

24.4.1 Consistency Checks and Domains of Validity

The previous solution is a steady-state one, but the universe evolves over the dynamical timescale H^{-1}. As long as the characteristic timescale of the accretion system is shorter than the cosmological timescale, the steady-state solution is meaningful. For that, one must require [38]

$$\frac{r_B}{c_{s,\infty}} H(z) < 1 \Longrightarrow M \lesssim 10^{4.5}\, M_\odot. \tag{24.16}$$

Also, the calculation assumes that gas accretes on an isolated BH, which is valid as long as the Bondi radius is much smaller than the characteristic proper separation between PBHs, yielding [7]

$$M \lesssim 3 \times 10^4 f_{\rm PBH}^{-1/2}\, M_\odot. \tag{24.17}$$

This condition is only violated if the PBHs are subject to strong clustering. Qualitatively, neglecting clustering (as henceforth assumed) leads however to more conservative constraints, since accretion grows quadratically with the mass.

In obtaining cosmological bounds, one implicitly considers homogeneous effects of the PBH on the medium, i.e. $x_e(z)$ is not space-dependent. This makes sense as long as each PBH can ionise all of the region separating it from the nearest PBH, leading to (Appendix A in [43])

$$f_{\rm PBH} > 10^{-15} x_e^3 \frac{M_\odot}{M}\,. \tag{24.18}$$

Notice that we can use mean results ignoring the discreteness effect due to the finite number of PBHs in each patch of the CMB sky being analysed, as long as there is significantly more than a PBH in each patch of the CMB. The most stringent constraint comes from the maximum multipole used ($\ell \sim$2000), yielding [43]

$$N_{\rm PBH} \simeq \frac{5 \times 10^7}{\ell} \left(\frac{f_{\rm PBH} M_\odot}{M}\right)^{1/3} > 1\,. \tag{24.19}$$

Also note that it was explicitly checked in [7, 39] that the plasma is optically thin to both Compton scattering and free-free absorption, as long as the accretion-rate is sub-Eddington. This hypothesis is used in obtaining the estimate of Eq. (24.15). The validity of the spherical accretion regime for such moderate accretion rates was also defended in [39].

24.4.2 Challenges

If the above considerations provided a realistic and complete description of the accretion phenomenon in the cosmological framework, the robustness of the derived constraints would not be so debated in the literature. In the following, we discuss some of the subtle points omitted in the previous treatment, and still raising questions.

First of all, the solution derived assumes PBHs at rest in the cosmological frame. Reference [7] generalised it to a finite PBH velocity $v_{\rm PBH}$, by replacing $c_{s,\infty}$ with $v_{\rm eff} = \sqrt{c_{s,\infty}^2 + v_{\rm PBH}^2}$ and $T_\infty \to T_\infty + m_p v_{\rm PBH}^2/(1+x_e)$. This prescription was also followed in [39], and it is inspired by what discussed already by Bondi [12], as mentioned after Eq. (24.7). There is no strong argument supporting the consistency or the reasonable nature of this recipe, as already noted in [7]. Of course, this would be only of academic relevance if relative velocities between PBHs and the gas satisfied $v_{\rm PBH} \ll c_{s,\infty}$, but this is not what assumed in [7]. The rationale for supersonic velocities has to do with an effect discussed in [49]: At the recombination time, the sound speed in the baryonic fluid drops from relativistic to the thermal velocities of the hydrogen atoms, becoming less than the relative velocity of baryons with respect to DM computed via linear perturbation theory. Supersonic coherent flows of the baryons are expected to exist (at cosmological scales) with respect to the underlying DM potential wells and thus PBHs, as long as PBHs trace the cold DM fluid.

However, it has been questioned that such a simple picture catches the whole story. In [25], it has been argued that applying the reasoning within ΛCDM of Ref. [49] to a PBH cosmology is incorrect at small scales, where the PBH Poisson clustering is responsible for quickly driving the relative velocity of PBH and baryons to zero at the small scales relevant for accretion.

Another caveat was raised in [37] and concerns the validity of the spherical approximation, at least for the innermost accretion regions responsible for the bulk of the luminosity. For accretion to proceed spherically all the way down to the ISCO, the angular momentum of the fluid must be very small, smaller than expected by typical fluctuations in density at the accretion distance due to the small-scale power spectrum due to PBH, which is enhanced compared to ΛCDM.

Both effects just mentioned are more and more relevant, the higher $f_{\rm PBH}$ is. In the opposite limit of small $f_{\rm PBH}$, one may be reduced to a situation where the solution of [49] applies: Yet, one should then consider that when the relative motion between a BH and the gas is supersonic, an accretion disk is typically formed, as confirmed by simulations (see e.g. [31]).

Overall, we expect that when characteristic velocities between baryons and PBHs are of the order of $c_{s,\infty}$ and the universe is as homogeneous as expected in ΛCDM at small scales, the spherical solution approximation of [7] should be applicable, modulo the replacement of $v_{\rm eff}$ with $c_{s,\infty}$. This would result in bounds roughly one order of magnitude stronger than what discussed in [7]. In the opposite case, whenever baryonic angular momentum (e.g. due to gradients in density and velocities) and/or supersonic motions are relevant, we expect a disk to form in the inner region around the PBH, and the luminosity estimate would change. We quickly review the physics relevant to this case in Sect. 24.5.

The study in [7] is also conservative in another respect. When $f_{\rm PBH} \ll 1$, another component is present in the universe besides PBHs and baryons, some cold fluid responsible for the bulk of the DM. The PBH further attracts this DM, which in turn boosts the PBH capability to attract baryons. A first treatment of this effect was attempted in [39]; a more modern study, both analytical and numerical, was presented in [43]. The inclusion of this effect is dealt with in Sect. 24.6.

24.5 Accretion Disks

Unfortunately, no complete theory of disk accretion exists from first principles. A rather general classifications of different regimes matching several states seen in Nature has however been achieved, as summarised in Fig. 24.2, adapted from Ref. [52]. The two key parameters controlling most of the physics are the accretion rate (y axis in Fig. 24.2, normalised to $\dot{M}_{\rm Edd} \equiv 10 L_{\rm Edd}$) and the optical depth (for which a proxy, the vertically integrated surface density Σ in units of g/cm^2 times the dimensionless viscosity parameter α, is reported in the x-axis of Fig. 24.2).

If the disk is *optically thick*, the associated solutions are known as *cold flows*, since associated with relatively low temperatures: Baryons and electrons thermalise, attaining a radial-dependent temperature T_c which can be deduced by equating the dissipation rate per unit face area of the disk to the blackbody flux. One infers the figure of merit (e.g. see Eq. 1.10 in [20])

$$T_c \simeq 4 \times 10^7 \text{K} \left(\frac{L}{L_{\rm Edd}}\right)^{1/4} \left(\frac{M}{M_\odot}\right)^{1/4} \sqrt{\frac{3\,\text{km}}{R}}. \tag{24.20}$$

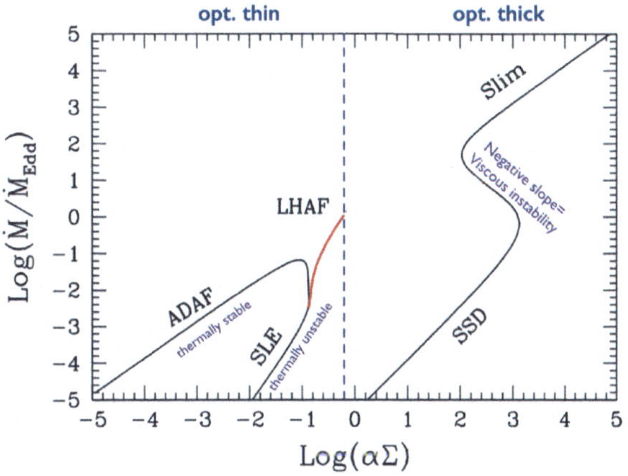

Fig. 24.2 A broad classifications of different regimes matching known accretion disk solutions, adapted from Ref. [52] by the author

Typically, the hottest regions of stellar mass BH disks in this regime do not exceed $\sim 10^7$ K. If the accretion rate is low or moderate, in the optically thick regime the standard Shakura-Sunyaev solution [44] is applicable (labelled SSD in Fig. 24.2). Its key assumption is that the dissipation happens through viscous effects parametrised by the dimensionless coefficient $\alpha \lesssim 1$, independent of other properties of the disk, so that viscous torque is proportional to the pressure. In this solution, the disk mass is negligible compared to that of the accreting BH and self-gravity of the disk can be neglected. The efficiency parameter $\epsilon \simeq 0.1$ is associated to this solution and adopted in most related literature.

This solution breaks down if accretion grows to values comparable to the Eddington scale. Other stable solutions exist at high accretion rate and high optical thickness. When the accreting gas becomes too optically thick to radiate all the locally dissipated energy, radiation is trapped and advected inward with the accretion flow. The radiative efficiency becomes lower than 10% and, as a consequence, L never significantly exceed $L_{\rm Edd}$. This alternative regime of the cold flow class is sometimes called *slim disk* (since one can still deal with it via vertically integrated equations, as in the case of the solution [44]) but it is better characterised as *optically thick advection-dominated accretion flow*.

In the opposite regime of *optically thin disk*, thermal equilibrium is not achieved and the corresponding solutions are classified as *hot flows*, since the relevant temperature scale is the virial one,

$$T_{\rm vir} \simeq \frac{GMm_p}{3R} \simeq \frac{m_p}{6}\frac{R_S}{R} \simeq 1.7 \times 10^{12}{\rm K}\frac{M}{M_\odot}\frac{R_S}{R}, \tag{24.21}$$

clearly satisfying the relation $T_{\rm vir} \gg T_c$. The state of the plasma is typically parameterised via two temperatures, one for leptons T_e and one for nuclei $T_{\rm ion}$, satisfying $T_{\rm ion} \simeq T_{\rm vir} \gg T_e$ (The first such model, dubbed SLE in Fig. 24.2, was proposed in 1976 in [45], but was realised to be thermally unstable.) In general, under the described conditions the accreting plasma forms a thick torus, and a sizeable fraction of accretion energy goes into heating the flow rather than being radiated away. This regime is known as *optically thin advection-dominated accretion flow*, or ADAF in short (bottom-left in Fig. 24.2). This regime is characterised by an efficiency $\epsilon \simeq 0.1 f(\dot{M})$, with $f \lesssim 1$ and $f' > 0$; ϵ saturates to 0.1 at large accretion rates, i.e. when approaching Eddington condition, a regime sometimes dubbed luminous hot accretion flows (LHAF in Fig. 24.2). The radiative efficiency of ADAF largely depends on the fraction of energy shared by the electrons, δ. According to [52], theoretical lower limits indicate $\delta \gg 0.01$, and fits to observations suggest $0.1 < \delta < 0.5$. For conditions where multiple stable accretion regimes can coexist, observations seem to support the conclusion that Nature selects a ADAF-type solution ("Strong ADAF principle").

Some effects have been neglected in the above schematic classification. Notably, dynamical effects due to magnetic fields, the BH spin, radiative feedback and kinetic (as opposed to fluid) effects have been neglected. While this is still an open field of research, the most important phenomenological correction to the above picture relates to the presence of outflows and jets: Some of the material falling on the BH can be actually carried away from the inner disk in the form of a non-relativistic wind or a relativistic jet, either Poynting-flux or matter dominated. In extreme cases, so-called magnetically arrested disk solutions apply. For observed active galactic nuclei (most often in the ADAF regime) the role of these outflows is relevant, suppressing the BH-associated luminosity by one order of magnitude or more compared to what expected from the Bondi solution [34]. These effects seem to be particularly important in quiet BHs, such as Sgr A* at the center of the Milky Way. In the light of this situation, a rather conservative approach was taken in works like [37, 43] when dealing with a disk model in a cosmological context: Only the low-efficiency ADAF models were considered, with a choice of $\delta = 0.1$ for the energy shared by electrons (roughly the minimum suggested value consistent with observations according to [52]) and an order-of-magnitude suppressed accretion with respect to the (cosmological) Bondi-like solution, to mimic the effects of the outflows in a way roughly consistent with observations [34].

That said, the need to go beyond the simple ADAF models in the cosmological setting and for deriving CMB anisotropy bounds is questionable, and it may well be that the prescriptions adopted in [37, 43] are overly conservative. First of all, PBHs in a cosmological setting are very different from astrophysical ones at low z: Models typically predict them to form preferentially with vanishing spin and they live in an unmagnetised medium. To mention but one important element, jets are typically Poynting flux dominated and are associated to magnetic effects as well as the BH spin. We expect these effects to be very small if not absent in the cosmological context. Also, we cannot emphasise enough that *energy is not lost*: even outflow and jet luminosities, if present, should eventually cause heating and excitation/ionisation of the cosmological medium, hence contributing to the CMB bounds even when not contributing to the BH luminosity proper. This remark is also relevant to the recent trend of applying simulation results such as [31], showing important *radiative feedback*, to the problem of cosmological bounds on PBHs, see e.g. [4, 19]. This feedback effect consists in the modification to the actual accretion rate onto the BH due to the (sizeable) modification of the medium surrounding the BH brought upon by the accretion luminosity itself, which in turn also induces a time-dependence. However, even leaving apart the reliability of the extrapolation of the medium density/temperature parameters of [31] to cosmological values, Refs. [4, 19] only consider the effects of the decreased UV/X-ray luminosity associated to the BH. Whenever significant feedback is suspected, a fraction (that should be quantified) of the kinetic energy acquired by the baryons "falling from infinity" does not disappear into the BH, but stays in the surrounding environment : either directly (as heating of the gas that never penetrates close to the BH) or indirectly, e.g. by contributing to the ionisation of a large Strömgren bubble surrounding the BH or the light it releases in the medium. While it is understandable that the energy budget associated to those is ignored in the simulations of [31], which are primarily meant to be applied to the UV/X-ray emission of astrophysical BH, they could be relevant (and should at least be discussed) for CMB constraint applications. While currently not included in Ref. [4, 19], they may be worth assessing in future studies.

Till now, we have glossed over the spectral features of the emission. In general, modern tools like EXOCLASS allow one to reliably compute the energy deposited in the cosmological medium, once the injected spectrum is known. As confirmed also by analytical approximations (see notably [7]), at least in the most relevant redshift range $300 \lesssim z \lesssim 1000$, most of the injected energy is absorbed by the medium; its repartition in ionisation, excitation, heating is only weakly dependent from the details of the spectrum. The latter can be computed rather precisely within the simplifying assumptions of [7], so that the uncertainty on the bounds related to the spectrum is definitely not the dominant one.

Even for disk models, the situation is qualitatively similar. As reported in [52], the dominant electron emission in the context of ADAF can be parameterised as (see [37])

$$\frac{\mathrm{d}L}{\mathrm{d}E} \propto \Theta(E - E_{\min})E^{-a}\exp(-E/T_s), \tag{24.22}$$

with $a \simeq 0 - 0.5$, $T_s \simeq m_e$ and $E_{\min} \simeq 10\,\mathrm{eV}$, accounting for a cut including only the fraction of the spectrum "useful" for excitation/ionisation. While details do depend on the accretion rate and BH mass, due to the above-mentioned weak dependence of the deposited energy on the spectral shape, these variations only lead to a moderate spread and uncertainty in the derived bounds, quantified in a factor $\simeq 2$ in Ref. [37].

24.6 Including Dark Matter Accretion

If PBHs *do not* constitute the totality of the DM, besides baryons they also gravitationally attract DM [8, 9, 29]. In absence of energy and angular momentum loss, as expected for most DM candidates, most of this DM does not accrete the PBH, rather forms a halo, which can however boost the PBH capability to accrete baryons.[1] Although the original Bondi problem was considering accretion onto a point particle, a natural generalization of the notion of Bondi radius for an extended distribution of mass can be written as [32]:

$$\frac{G\,M_{\rm PBH}}{r_{\rm B,eff}} - \Phi_h(M_{\rm PBH}, r_{B,\rm eff}, t) = v_{\rm eff}^2(t), \tag{24.23}$$

where $M_{\rm PBH}$ is the initial PBH mass and Φ_h the (time-dependent) gravitational potential associated to the DM halo. In [43], the problem was thus tackled by adopting Eq. (24.5), with $r_{\rm B}$ replaced by $r_{\rm B,eff}$ solution of Eq. (24.23), with the gravitational potential of the halo estimated analytically or numerically.

Analytically, one can consider the spherically symmetric case of growth of a halo of (exactly cold and dispersionless) DM, around a PBH which is the only center of

[1] One implication is that the $\dot{M}$ associated to both baryons and DM is insufficient, once integrated down to $z \sim 100$, to significantly alter M. For the purposes of this chapter, we can consider $f_{\rm PBH} \simeq$ const.

attraction in the whole universe. The time evolution of a mass-shell at position r can be obtained by solving the following differential equation,

$$\frac{\mathrm{d}^2 r}{\mathrm{d}t^2} = -\frac{4G}{\pi} 3r \left[\rho_{\mathrm{PBH}} + \sum_i (\rho_i + 3p_i) \right], \tag{24.24}$$

where ρ_i and p_i are the energy density and pressure of the component "i" (radiation, matter, and dark energy, if effective at the redshift considered), respectively, and we defined the energy density of the PBH as $\rho_{\mathrm{PBH}} = 3\, M_{\mathrm{PBH}}/(4\pi r^3)$. The physical radius r is represented by $r = a(t)x$ where $a(t)$ is the scale factor normalised as $a_0 = 1$ at present, and x is the co-moving coordinate. At each time t, the DM halo mass is the DM density integrated up to the radius r_s defined by $\mathrm{d}r_s/\mathrm{d}t = 0$. This approach yields the following key results [29, 43]:

- A time evolution given by

$$M_{\mathrm{halo}} \simeq \left(\frac{3000}{1+z} \right) M_{\mathrm{PBH}}\,. \tag{24.25}$$

- A density profile proportional to $\propto r^{-3}$ down to small distances, where a free-fall profile $r^{-3/2}$ takes over.

Equation (24.25) only yields an upper limit to the DM halo mass. At least at late times, the growth breaks down, e.g. due to tidal effects of nearby PBHs and other halos. It is also worth noting that the radial profile crucially depends on the free-fall boundary condition at the center; not accounting for the DM angular momentum is expected to be a too crude approximation. Resorting to numerical simulations, in [43] it was found that as long as f_{PBH} is not too large and the redshift is not too low (evolution was followed down to $z = 99$), halos accrete onto isolated PBHs, and the profile is independent of f_{PBH}. Over about two decades in radius, at early times the profile matches the $r^{-2.25}$ power-law predicted by [9], confirming the conjecture advanced in Ref. [38], Sec. 4; at late times, the profile evolves towards the slightly steeper $r^{-2.5}$. This power-law profile is consistent with those found in independent numerical simulations by [1], which also found a smooth transition to standard NFW-like profile of DM halos at large radii. The Poisson's equation for the gravitational potential $\phi_i = (4\pi G)^{-1} \nabla^{-2} \rho_i$ is solved in Fourier space for PBHs and DM separately. The unknown $r_{\mathrm{B,eff}}$ follows from plugging the potential thus found in Eq. (24.23).

At high redshifts, the PBH is much more relevant than the DM halo and $r_{\mathrm{B,eff}}$ is small, close to the naive estimate r_{B} for the naked PBH. When the halo eventually exceed the PBH mass, its further evolution is largely independent of the PBH contribution to the potential. The numerical results lead to an estimated halo mass which is about 60% of the simple result of Eq. (24.25), with a similar scaling with redshift, although with a different mass profile. Note that it is safe to neglect the "ordinary" DM halos feedback onto the halos growing around PBH, since the former ones only grow at much later times (typically $z \lesssim 30$ in a ΛCDM cosmology) than those of concern for us. *A fortiori*, the feedback of the baryons can also be neglected. Also, note that most baryons are still unbound to halos, and their ratio to the DM in the growing halos around PBH is much smaller than the baryon to DM cosmological density ratio of $\sim 15\%$. Hence, objections on the realism of power law DM density profiles around BH surviving in the current universe [50] do not apply to the pristine configurations considered here.

A simple semi-analytical model that matches these results and interpolates between limiting behaviours was proposed in [43]. In the specific case of a point-like potential due to the PBH plus the power-law matter distribution around it, with density $\rho(r) \propto r^{-\nu}$ up to a distance r_h and total mass M_h, Eq. (24.23) rewrites

$$\frac{v_{\rm eff}^2(z) r_{B,\rm eff}}{G M_{\rm PBH}} = 1 + \frac{M_h}{M_{\rm PBH}} \left\{ \Theta(r_{B,\rm eff} - r_h) + \frac{\Theta(r_h - r_{B,\rm eff})}{1-p} \left[\left(\frac{r_b}{r_h}\right)^p - p \left(\frac{r_{B,\rm eff}}{r_h}\right) \right] \right\}, \tag{24.26}$$

where $p \equiv 3 - \nu$, and M_h and r_h depend from $M_{\rm PBH}$ and z. Reference [43] adopted Eq. (24.25) for the halo mass M_h, assuming it is extending up to the turnaround radius, defined as (see e.g. Sect. 4 in Ref. [38])

$$r_h \simeq 58\,{\rm pc}\,(1+z)^{-1} \left(\frac{M_h(M_{\rm PBH}, z)}{M_\odot}\right)^{1/3}. \tag{24.27}$$

Equation (24.26) admits either the solution ("Bondi radius" associated to the halo mass)

$$r_{\rm B,eff} = \frac{G(M_{\rm PBH} + M_h)}{v_{\rm eff}^2} \simeq \frac{G M_h}{v_{\rm eff}^2} \equiv r_{{\rm B},h}\,, \tag{24.28}$$

which holds if $r_h < r_{{\rm B},h}$; otherwise, if $r_h > r_{{\rm B},h}$, neglecting the PBH mass one has

$$r_{\rm B,eff} \simeq r_h \left[(1-p)\frac{r_h}{r_{{\rm B},h}} + p \right]^{\frac{1}{p-1}} \leq r_h. \tag{24.29}$$

Note that Eq. (24.29) tends to $r_{{\rm B},h}$ when $p \to 0$, as expected: When the DM halo profile is very steep and/or the halo is very compact, as far as accreting baryons are concerned they simply see a BH whose effective mass is the sum of the PBH and the DM halo mass. If the halo is fluffy or large, only a fraction of the mass of the halo contributes to the accretion. In any case, the condition $r_{\rm B,eff} \geq r_{\rm B,PBH}$ must hold. In Ref. [43], this model with the choice $p = 0.75$ suggested by simulations was thus used to asses the impact of PBH accretion onto the CMB.

24.7 Constraints and Perspectives

In the limit of a monochromatic mass distribution for PBHs, Fig. 24.3 (from [43]) reports the bound for the "conservative" ADAF disk model described in Sect. 24.5 (top panel), and the spherical accretion model of [7] (with $v_{\rm eff} = c_{s,\infty}$) described in Sect. 24.4 (bottom panel). In each case, light shaded regions represent the excluded parameter space when including the DM halo effect. Here are the datasets from which these bounds are derived: *Planck* 2018 high-ℓ and low-ℓ TT, EE and lensing likelihood [2, 3]; the isotropic BAO measurements from 6dFGS at $z = 0.106$ [10] and from the MGS galaxy sample of SDSS at $z = 0.15$ [40]; the anisotropic BAO and the growth function $f\sigma_8(z)$ measurements from the CMASS and LOWZ galaxy samples of BOSS DR12 at $z = 0.38$, 0.51, and 0.61 [5]; the Pantheon supernovae dataset [41], which includes measurements of the luminosity distances of 1048 SNe Ia in the redshift range $0.01 < z < 2.3$. The non-CMB datasets are essentially instrumental in breaking some degeneracies. For technical details on the derivation of the bounds and the analysis chain, see [43]. A reasonably conservative statement is that the totality of DM in the form of PBH (i.e. $f_{\rm PBH} = 1$) is excluded for $M \gtrsim 4\, M_\odot$, with a factor ~ 5 systematic uncertainty dominated by the poorly known accretion luminosity. The

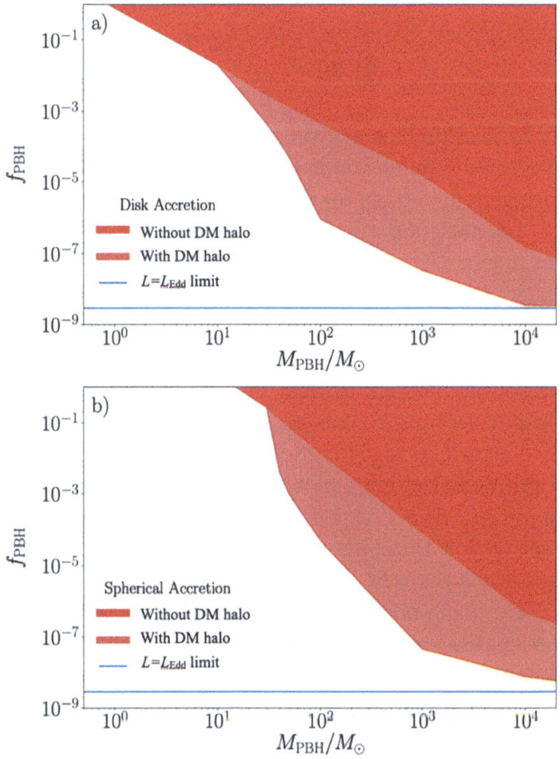

Fig. 24.3 Bounds on the abundance of PBH assuming disk accretion (top panel) or spherical accretion (bottom panel). We show the results with (light-shaded) and without (dark-shaded) the formation of a DM halo. The horizontal line shows the limiting bound assuming constant emission at Eddington luminosity. Adapted from Ref. [43]; see text for details

constraints on $f_{\rm PBH}$ improve more than linearly with mass up to hundreds of solar masses: The scaling derived in [37] with the mass as well as the accretion eigenvalue was $f_{\rm PBH}^{\rm lim} \propto (M\,\lambda_{\rm eff})^{-1.6}$. The dependence is stronger if the DM halo effect is taken into account, as it should. At least from $\sim 50\,M_\odot$ to $\sim 10^3\,M_\odot$, the CMB anisotropy bounds are nominally the tightest known to date. If extrapolated to high masses, they saturate eventually at $f_{\rm PBH} \simeq 3\times 10^{-9}$ around $10^4\,M_\odot$, when the luminosity is expected to approach the Eddington one. The reliability of the bounds in this range is however doubtful, since the accretion physics extrapolation (see Sects. 24.4, 24.5) as well as some of the cosmological assumptions (see Sect. 24.4.1) break down.

Further, there is no indication that assuming a monochromatic mass distribution leads to over-conservative bounds. Matching the trend typically observed in other cases where a broader mass function is used [15, 26], also CMB bounds are stable or improve for broad mass functions. In [37], by recasting the monochromatic bounds according to the technique suggested in [15], it was checked that the constraints derived on the monochromatic mass function also apply to the average mass value of a broad, log-normal mass distribution; the broad distribution is more tightly constrained when its width covers a decade or more in mass space. More recently, in [24], the CMB anisotropy bound was also computed for a very broad and multi-peaked mass function $\psi(M)$. This was done by "brute force", integrating numerically the bounds obtained for each mass value, i.e. by using deposition energies

$$\left.\frac{\mathrm{d}E}{\mathrm{d}V\mathrm{d}t}(z)\right|_{\rm dep,c} = \int \mathrm{d}M f_c(z,M)\psi(M)\left.\frac{\mathrm{d}E}{\mathrm{d}V\mathrm{d}t}\right|_{\rm inj}, \tag{24.30}$$

or by recasting the monochromatic bounds, as done in [37]. For the case considered, the "exact" numerical bounds for a very broad mass function are about 50% more stringent than if approximately computed via the prescription of [15].

The most important phenomenological consequence of these bounds is to provide an independent argument against the possibility that the bulk of the surprisingly heavy BH population (a few tens $M_\odot$ to $\sim 100 M_\odot$) inferred by LIGO and Virgo constitutes a significant fraction of the DM. However, the possibility that some of these objects is of primordial origin is still marginally consistent with the CMB anisotropy bounds, at least for some assumptions on accretion. As a side remark, despite the tight bounds at large masses and the fact that some assumptions become shakier in this range (See Sect. 24.4.1), the CMB angular power spectra do not exclude a primordial origin hypothesis for the supermassive black hole, if combined with a significant accretion phase, although this hypothesis is challenged by other arguments (see Sect. V.B in [43] as well as Ref. [24]) and no convincing self-consistent model has been proposed, yet.

The major uncertainties still plaguing these constraints are not the statistical errors on the data, rather the theory systematics related to accretion physics, as discussed at length. Based on current knowledge, the bounds discussed here should rather err on the conservative side. Also, we have only considered PBH accretion in the dark ages, stopping integration at $z \sim 100$, where cosmological linear perturbation theory is reliable. Accretion certainly continues once halos become self-gravitating and capable of retaining significant amount of baryons, but extrapolation of the cosmological solutions discussed here cannot be considered educated guesses, since density, velocities, and accretion properties will be dominated by the astrophysical conditions in the (proto)halos. It is worth keeping in mind, however, that if a significant growth of the PBH mass in the dark ages is excluded, this is not necessarily the case at lower redshift (although such a situation may raise more phenomenological problems than it solves). The bounds from CMB anisotropies must be implicitly thought as bounds on the primordial value of $f_{\rm PBH}$.

The most important improvement in this field is expected from dedicated numerical simulations. Note that some recent analysis [19] has attempted to directly use the results of hydrodynamical simulations, notably those of Ref. [31], to describe a supersonically moving accreting BH. Reassuringly, for masses between 1 and 100 $M_\odot$, the fiducial model of Ref. [31] does lead to constraints bracketed by the spherical ones of [7] and the disk one of Ref. [37], obtaining even stronger constraints at $M \gtrsim 10^2\,M_\odot$.[2] As discussed in Sect. 24.5, however, the main limitation of these exercises is in accounting for all other channels of energy into which the, "kinetic" accretion luminosity is leaking into, which is potentially relevant, given the claim of significant feedback. Qualitatively, this should obviously improve the CMB bounds, but a quantitative analysis is currently missing.

Some interesting avenue for the future would be to refine the model of [7] for steady-state, spherical accretion in an unmagnetised medium of cosmological density at rest with respect to the BH, but including radiative transfer and possibly GR effects. This would collapse the results currently bracketed between the collisional and photoionisation regime into a well-defined benchmark family of solutions. Eventually, it would be important to extend those simulation to the case of relative motion between the PBH and the gas (notably in the supersonic regime). The energy carried by outflows and jets, if present, should also be estimated. Since the cosmological medium does not share a number of complications with the astrophysical medium (think of the lack of large inhomogeneities and magnetisation, vanishingly low metallicities, absence of dust grains, etc.), a determination of universal laws of the accretion rate and luminosity could be within reach. From the cosmological theory side, clarifying the expectation for the power spectrum and the relative motion between PBHs and baryons at very small scales not only in ΛCDM, but also in a cosmology dominated by PBHs, is another intriguing direction for improvement.

Acknowledgements I would like to thank all the colleagues who have helped me deepening this fascinating topic, at the interplay of cosmology, astrophysics and particle physics. Special thanks to my co-authors on this subject, F. Calore, S. Clesse, G. Franco Abellán, J. Iguaz, D. Inman, K. Kohri, and V. Poulin.

---

[2] The authors of Ref. [19] also present results combining the accretion extracted from Ref. [31] with values for the radiative efficiency taken from the analysis of Ref. [7], yielding more conservative bounds. We note however that this choice does not match any result or prescription in the literature, and combines ingredients which appear mutually incompatible (for instance, the solution of Ref. [7] assumes spherical symmetry, while the configuration in Ref. [31] leads to disk formation).

References

1. J. Adamek et al., WIMPs and stellar-mass primordial black holes are incompatible. Phys. Rev. D **100**(2), 023506 (2019). https://doi.org/10.1103/PhysRevD.100.023506. arXiv: 1901.08528 [astro-ph.CO]
2. N. Aghanim et al., Planck 2018 results. V. CMB power spectra and likelihoods. Astron. Astrophys. **641**, A5 (2020). https://doi.org/10.1051/0004-6361/201936386. arXiv: 1907.12875 [astro-ph.CO]
3. N. Aghanim et al., Planck 2018 results. VI. Cosmological parameters. Astron. Astrophys. **641**, A6 (2020). [Erratum: Astron. Astrophys. **652**, C4 (2021)]. https://doi.org/10.1051/0004-6361/201833910. arXiv: 1807.06209 [astro-ph.CO]
4. D. Agius et al., Feedback in the dark: a critical examination of CMB bounds on primordial black holes (2024). arXiv: 2403.18895 [hep-ph]
5. S. Alam et al., The clustering of galaxies in the completed SDSS-III Baryon oscillation spectroscopic survey: cosmological analysis of the DR12 galaxy sample. Mon. Not. Roy. Astron. Soc. **470**(3), 2617–2652 (2017). https://doi.org/10.1093/mnras/stx721. arXiv: 1607.03155 [astro-ph.CO]
6. Y. Ali-Haimoud, C.M. Hirata, HyRec: a fast and highly accurate primordial hydrogen and helium recombination code. Phys. Rev. D **83**, 043513 (2011). https://doi.org/10.1103/PhysRevD.83.043513. arXiv: 1011.3758 [astro-ph.CO]
7. Y. Ali-Haimoud, M. Kamionkowski, Cosmic microwave background limits on accreting primordial black holes. Phys. Rev. D **95**(4), 043534 (2017). https://doi.org/10.1103/PhysRevD.95.043534. arXiv: 1612.05644 [astro-ph.CO]
8. V.S. Berezinsky, V.I. Dokuchaev, Y.N. Eroshenko, Formation and internal structure of superdense dark matter clumps and ultracompact minihaloes. JCAP **11**, 059 (2013). https://doi.org/10.1088/1475-7516/2013/11/059. arXiv: 1308.6742 [astro-ph.CO]
9. E. Bertschinger, Self-similar secondary infall and accretion in an Einstein-de Sitter universe. Astrophys. Acirc. J. **58**, 39 (1985). https://doi.org/10.1086/191028
10. F. Beutler et al., The 6dF galaxy survey: baryon acoustic oscillations and the local hubble constant. Mon. Not. Roy. Astron. Soc. **416**, 3017–3032 (2011). https://doi.org/10.1111/j.1365-2966.2011.19250.x. arXiv: 1106.3366 [astro-ph.CO]
11. D. Blas, J. Lesgourgues, T. Tram, The cosmic linear anisotropy solving system (CLASS) II: approximation schemes. JCAP **07**, 034 (2011). https://doi.org/10.1088/1475-7516/2011/07/034. arXiv: 1104.2933 [astro-ph.CO]
12. H. Bondi, On spherically symmetrical accretion. Mon. Not. Roy. Astron. Soc. **112**, 195 (1952). https://doi.org/10.1093/mnras/112.2.195
13. H. Bondi, F. Hoyle, On the mechanism of accretion by stars. Mon. Not. Roy. Astron. Soc. **104**, 273 (1944)
14. B.J. Carr et al., New cosmological constraints on primordial black holes. Phys. Rev. D **81**, 104019 (2010). https://doi.org/10.1103/PhysRevD.81.104019. arXiv: 0912.5297 [astro-ph.CO]
15. B. Carr et al., Primordial black hole constraints for extended mass functions. Phys. Rev. D **96**(2), 023514 (2017). https://doi.org/10.1103/PhysRevD.96.023514. arXiv: 1705.05567 [astro-ph.CO]
16. X.-L. Chen, M. Kamionkowski, Particle decays during the cosmic dark ages. Phys. Rev. D **70**, 043502 (2004). https://doi.org/10.1103/PhysRevD.70.043502. arXiv: astro-ph/0310473
17. J. Chluba, R.M. Thomas, Towards a complete treatment of the cosmological recombination problem. Mon. Not. Roy. Astron. Soc. **412**, 748 (2011). https://doi.org/10.1111/j.1365-2966.2010.17940.x. arXiv: 1010.3631 [astro-ph.CO]
18. M. Cirelli, F. Iocco, P. Panci, Constraints on dark matter annihilations from reionization and heating of the intergalactic gas. JCAP **10**, 009 (2009). https://doi.org/10.1088/1475-7516/2009/10/009. arXiv: 0907.0719 [astro-ph.CO]

19. G. Facchinetti, M. Lucca, S. Clesse, Relaxing CMB bounds on primordial black holes: the role of ionization fronts. Phys. Rev. D **107**(4), 043537 (2023). https://doi.org/10.1103/PhysRevD.107.043537. arXiv: 2212.07969 [astro-ph.CO]
20. J. Frank, A. King, D.J. Raine, *Accretion Power in Astrophysics*, 3rd edn. (2002)
21. F. Hoyle, R.A. Lyttleton, On the accretion of interstellar matter by stars. Proc. Camb. Philos. Soc. **36**(3), 325 (1940). https://doi.org/10.1017/S0305004100017369
22. F. Hoyle, R.A. Lyttleton, The effect of interstellar matter on climatic variation. Proc. Camb. Philos. Soc. **35**(3), 405 (1939). https://doi.org/10.1017/S0305004100021150
23. F. Hoyle, R.A. Lyttleton, The evolution of the stars. Proc. Camb. Philos. Soc. **35**(4), 592 (1939). https://doi.org/10.1017/S0305004100021368
24. J.I. Juan, P.D. Serpico, G.F. Abellán, The QCD phase transition behind a PBH origin of LIGO/Virgo events? JCAP **07**(07), 009 (2022). https://doi.org/10.1088/1475-7516/2022/07/009. arXiv: 2204.07027 [astro-ph.CO]
25. A. Kashlinsky, Cosmological advection flows in the presence of primordial black holes as dark matter and formation of first sources. Phys. Rev. Lett. **126**(1), 011101 (2021). https://doi.org/10.1103/PhysRevLett.126.011101. arXiv: 2012.08047 [astro-ph.CO]
26. F. Kühnel, K. Freese, Constraints on primordial black holes with extended mass functions. Phys. Rev. D **95**(8), 083508 (2017). https://doi.org/10.1103/PhysRevD.95.083508. arXiv: 1701.07223 [astro-ph.CO]
27. H. Liu et al., Exotic energy injection in the early universe i: a novel treatment for low-energy electrons and photons. Phys. Rev. D **108**(4), 043530 (2023). https://doi.org/10.1103/PhysRevD.108.043530. arXiv: 2303.07366 [astro-ph.CO]
28. H. Liu et al., Exotic energy injection in the early Universe. II. CMB spectral distortions and constraints on light dark matter. Phys. Rev. D **108**(4), 043531 (2023). https://doi.org/10.1103/PhysRevD.108.043531. arXiv: 2303.07370 [astro-ph.CO]
29. K.J. Mack, J.P. Ostriker, M. Ricotti, Growth of structure seeded by primordial black holes. Astrophys. J. **665**, 1277–1287 (2007). https://doi.org/10.1086/518998. arXiv: astro-ph/0608642
30. F.C. Michel, Accretion of matter by condensed objects. Astrophys. Space Sci. **15**(1), 153–160 (1972). https://doi.org/10.1007/BF00649949
31. K.H. Park, M. Ricotti, Accretion onto black holes from large scales regulated by radiative feedback III: enhanced luminosity of intermediate mass black holes moving at supersonic speeds. Astrophys. J. **767**, 163 (2013). https://doi.org/10.1088/0004-637X/767/2/163. arXiv: 1211.0542 [astro-ph.CO]
32. K.H. Park et al., Bulge-driven fueling of seed black holes. Astrophys. J. **818**(2), 184 (2016). https://doi.org/10.3847/0004-637X/818/2/184. arXiv: 1512.03434 [astro-ph.GA]
33. P.J.E. Peebles, Recombination of the primeval plasma. Astrophys. J. **153**, 1 (1968). https://doi.org/10.1086/149628
34. S. Pellegrini, Nuclear accretion in galaxies of the local universe: clues from Chandra observations. Astrophys. J. **624**, 155–161 (2005). [Erratum: Astrophys. J. **636**(564) (2005)]. https://doi.org/10.1086/429267. arXiv: astro-ph/0502035
35. V. Poulin, J. Lesgourgues, P.D. Serpico, Cosmological constraints on exotic injection of electromagnetic energy. JCAP **1703**(03), 043 (2017). https://doi.org/10.1088/1475-7516/2017/03/043. arXiv: 1610.10051 [astro-ph.CO]
36. V. Poulin, P.D. Serpico, J. Lesgourgues, Dark matter annihilations in halos and high-redshift sources of reionization of the universe. JCAP **12**, 041 (2015). https://doi.org/10.1088/1475-7516/2015/12/041. arXiv: 1508.01370 [astro-ph.CO]
37. V. Poulin et al., CMB bounds on disk-accreting massive primordial black holes. Phys. Rev. D **96**(8), 83524 (2017). https://doi.org/10.1103/PhysRevD.96.083524. arXiv: 1707.04206 [astro-ph.CO]
38. M. Ricotti, Bondi accretion in the early universe. Astrophys. J. **662**, 53–61 (2007). https://doi.org/10.1086/516562. arXiv: 0706.0864 [astro-ph]
39. M. Ricotti, J.P. Ostriker, K.J. Mack, Effect of primordial black holes on the cosmic microwave background and cosmological parameter estimates. Astrophys. J. **680**, 829 (2008). https://doi.org/10.1086/587831. arXiv: 0709.0524 [astro-ph]

40. A.J. Ross et al., The clustering of the SDSS DR7 main Galaxy sample – I. A 4% distance measure at $z = 0.15$. Mon. Not. Roy. Astron. Soc. **449**(1), 835–847 (2015). https://doi.org/10.1093/mnras/stv154. arXiv: 1409.3242 [astro-ph.CO]
41. D.M. Scolnic et al., The complete light-curve sample of spectroscopically confirmed SNe Ia from Pan-STARRS1 and cosmological constraints from the combined pantheon sample. Astrophys. J. **859**(2), 101 (2018). https://doi.org/10.3847/1538-4357/aab9bb. arXiv: 1710.00845 [astro-ph.CO]
42. S. Seager, D.D. Sasselov, D. Scott, A new calculation of the recombination epoch. Astrophys. J. Lett. **523**, L1–L5 (1999). https://doi.org/10.1086/312250. arXiv: astro-ph/9909275
43. P.D. Serpico et al., Cosmic microwave background bounds on primordial black holes including dark matter halo accretion. Phys. Rev. Res. **2**(2), 023204 (2020). https://doi.org/10.1103/PhysRevResearch.2.023204. arXiv: 2002.10771 [astro-ph.CO]
44. N.I. Shakura, R.A. Sunyaev, Black holes in binary systems: observational appearance. Astron. Astrophys. **24**, 337–355 (1973)
45. S.L. Shapiro, A.P. Lightman, D.M. Eardley, A two-temperature accretion disk model for Cygnus X-1: structure and spectrum. Astrophys. J. **204**, 187–199 (1976). https://doi.org/10.1086/154162
46. T.R. Slatyer, Indirect dark matter signatures in the cosmic dark ages II: ionization, heating and photon production from arbitrary energy injections. Phys. Rev. D **93**(2), 023521 (2016). https://doi.org/10.1103/PhysRevD.93.023521. arXiv: 1506.03812 [astro-ph.CO]
47. T.R. Slatyer, N. Padmanabhan, D.P. Finkbeiner, CMB constraints on WIMP annihilation: energy absorption during the recombination epoch. Phys. Rev. D **80**, 043526 (2009). https://doi.org/10.1103/PhysRevD.80.043526. arXiv: 0906.1197 [astro-ph.CO]
48. P. Stöcker et al., Exotic energy injection with ExoCLASS: application to the higgs portal model and evaporating black holes. JCAP **03**, 018. (2018) https://doi.org/10.1088/1475-7516/2018/03/018. arXiv: 1801.01871 [astro-ph.CO]
49. D. Tseliakhovich, C. Hirata, Relative velocity of dark matter and baryonic fluids and the formation of the first structures. Phys. Rev. D **82**, 083520 (2010). https://doi.org/10.1103/PhysRevD.82.083520. arXiv: 1005.2416 [astro-ph.CO]
50. P. Ullio, H.S. Zhao, M. Kamionkowski, A dark matter spike at the galactic center? Phys. Rev. D **64**, 043504 (2001). https://doi.org/10.1103/PhysRevD.64.043504. arXiv: astro-ph/0101481
51. M. Valdes, C. Evoli, A. Ferrara, Particle energy cascade in the intergalactic medium. Mon. Not. Roy. Astron. Soc. **404**, 1569–1582 (2010). https://doi.org/10.1111/j.1365-2966.2010.16387.x. arXiv: 0911.1125 [astro-ph.CO]
52. F. Yuan, R. Narayan, Hot accretion flows around black holes. Ann. Rev. Astron. Astrophys. **52**, 529–588 (2014). https://doi.org/10.1146/annurev-astro-082812-141003. arXiv: 1401.0586 [astro-ph.HE]
53. Y.B. Zeldovich, V.G. Kurt, R.A. Sunyaev, Recombination of hydrogen in the hot model of the universe. Zh. Eksp. Teor. Fiz. **55**, 278–286 (1968)

Chapter 25
Confronting Primordial Black Holes with LIGO-Virgo-KAGRA and the Einstein Telescope

Zu-Cheng Chen and Alex Hall

Abstract The detection of gravitational waves from binary black hole coalescences by the LIGO-Virgo-KAGRA Collaboration has raised fundamental questions about the genesis of these events. In this chapter, we explore the possibility that primordial black holes, proposed candidates for dark matter, may serve as the progenitors of the binary black holes observed by LIGO-Virgo-KAGRA. Employing a Bayesian analysis, we constrain the primordial black hole model using the LIGO-Virgo-KAGRA third gravitational wave transient catalog, revealing that stellar-mass primordial black holes cannot dominate cold dark matter. Considering a mixed population of astrophysical black holes and primordial black holes, we determine that approximately 1/4 of the detectable events in the third gravitational wave transient catalog can be attributed to primordial black hole binaries. We also forecast detectable event rate distributions for primordial black hole and astrophysical black hole binaries by the third-generation ground-based gravitational wave detectors, such as the Einstein Telescope, offering a potential avenue to distinguish primordial black holes from astrophysical black holes based on their distinct redshift evolutions.

25.1 Introduction

In recent years, the detection of gravitational waves (GWs) from compact binary coalescences by the LIGO-Virgo-KAGRA (LVK) Collaboration has ushered in a new era of GW astronomy [2, 8, 9]. The latest release of the third GW Transient Catalog (GWTC-3) by LVK reports approximately 90 GW events detected during their first three observing runs, with a majority identified as binary black hole (BBH) mergers exhibiting a broad mass distribution. The heaviest event, GW190521 [6],

Z.-C. Chen (✉)
Department of Physics, Hunan Normal University, Yuelu District, China
e-mail: zuchengchen@hunnu.edu.cn

A. Hall
Institute for Astronomy, University of Edinburgh, Edinburgh, UK
e-mail: ahall@roe.ac.uk

C. Byrnes et al. (eds.), *Primordial Black Holes*, Springer Series in Astrophysics and Cosmology, https://doi.org/10.1007/978-981-97-8887-3_25

involves component masses $m_1 = 85^{+21}_{-14} M_{\odot}$ and $m_2 = 66^{+17}_{-18} M_{\odot}$, residing within the upper black hole mass gap believed to originate from pulsation pair-instability supernovae [100]. The lower cutoff of this mass gap is currently estimated to be around $50 \pm 4 M_{\odot}$ [34, 63, 64, 99, 100]. Even after accounting for measurement uncertainties, this observation strongly suggests that the primary mass of GW190521 lies within the mass gap, challenging the idea of a direct stellar progenitor origin [23] within the conventional stellar evolution scenario for astrophysical black holes (ABHs).

As an alternative explanation for the observed BBHs, primordial black holes (PBHs) have gained attention alongside conventional ABHs. Formed in the early Universe through the gravitational collapse of primordial density fluctuations [47, 74] (see also Chap. 2), PBHs serve as a plausible explanation for LVK BBHs [37, 124] and are considered candidates for cold dark matter (CDM) [45] and seeds for galaxy formation [32, 83]. Moreover, the recent pulsar timing observation [28] of the eccentric binary millisecond pulsar, PSR J0514−4002E, suggests that its companion could potentially be a PBH [51]. The formation of PBHs is inevitably associated with the generation of GWs induced by primordial scalar perturbations [29, 30, 41, 53, 59, 123, 141–143] (see also Chap. 17). These induced GWs may be detectable with pulsar timing array experiments [17, 24, 53, 55, 79, 86, 87, 90, 114, 136, 137, 139].

In this chapter, we explore the plausibility of PBHs as an explanation for the BBH events observed by LVK (see Ref. [27] for a recent review). The observed rate of GW events puts a strong constraint on models in which a significant proportion of CDM is in PBHs. Despite this, a subdominant population of PBHs is permitted by the data, and such a population may be able to account for features in the source population that are not easily explained by ABHs. Our analysis indicates that the abundance of stellar-mass PBHs in CDM, f_{PBH}, cannot exceed $\mathcal{O}(10^{-3})$. Furthermore, the GW data suggest that not all BBH events detected so far can be solely explained by a single formation channel, whether astrophysical [145] or primordial [73, 133]. We, therefore, also present a comprehensive Bayesian inference study of the GWTC-3 catalog, incorporating a mixed ABH + PBH model.

We also explore the potential for distinguishing between ABHs and PBHs using upcoming third-generation ground-based GW detectors, such as the Einstein Telescope (ET) [112]. The population properties of PBH binaries differ significantly from those of ABH binaries. Notably, the merger rate of PBH binaries increases with redshift (z), while the merger rate of ABH binaries, which follows the star formation rate, peaks around $z \sim 2$ and then rapidly decreases. This characteristic feature can serve as a distinguishing factor between PBHs and ABHs [49, 103].

25.2 PBH Models and Bayesian Methodology

In this section, we begin with an overview of the calculation of the merger rate density of PBH binaries by considering the four most commonly used PBH mass functions in the literature. Then, we present the Bayesian inference framework employed in data analyses to constrain the various PBH models.

25.2.1 *Merger Rate Density Distribution of PBH Binaries*

In this subsection, we outline the calculation of the PBH merger rate density. Throughout this chapter, our focus is solely on the scenario that PBH binaries are formed in the early Universe. For the late Universe scenario and a comprehensive discussion of the merger rate density of PBH binaries, readers are directed to Chap. 15.

The sample of BBHs observed by LVK implies a broad distribution of BH masses, prompting the consideration of an extended mass function for PBHs. We impose the normalization condition on the probability distribution function of PBH mass, $P(m)$, ensuring that

$$\int_0^{\infty} P(m)\,\mathrm{d}m = 1. \tag{25.1}$$

Assuming that the fraction of CDM in PBHs is $f_{\mathrm{PBH}} \equiv \Omega_{\mathrm{PBH}}/\Omega_{\mathrm{CDM}}$, then the fraction of PBHs in the non-relativistic matter within the mass interval $(m, m+\mathrm{d}m)$ is [48]

$$f_{\mathrm{CDM}} f_{\mathrm{PBH}}\, P(m)\,\mathrm{d}m, \tag{25.2}$$

where the coefficient $f_{\mathrm{CDM}} \approx 0.85$ accounts for the fraction of CDM in the non-relativistic matter, encompassing both CDM and baryons.

We can now proceed to estimate the merger rate density of PBH binaries. We assume that PBHs are initially randomly distributed, following a spatial Poisson distribution in the early Universe, a condition that holds when they decouple from the cosmic background evolution [21, 104, 124]. Pairs of PBHs acquire angular momentum due to the gravitational torque exerted by other PBHs, ultimately culminating in the formation of a PBH binary upon decoupling from cosmic expansion. Subsequently, the binary undergoes the emission of gravitational radiation, ultimately resulting in a merger GW event potentially detectable by GW detectors. The merger rate density of a PBH binary with masses m_1 and m_2 at cosmic time t, $\mathcal{R}(t, m_1, m_2)$, is expressed as [50, 89, 113]

$$\mathcal{R}(t, m_1, m_2) = \frac{1.6\times 10^6}{\mathrm{Gpc}^3\,\mathrm{yr}} \left(\frac{t}{t_0}\right)^{-\frac{34}{37}} f_{\mathrm{PBH}}^{\frac{53}{37}} \eta^{-\frac{34}{37}} \left(\frac{M}{M_\odot}\right)^{-\frac{32}{37}} P(m_1)P(m_2), \tag{25.3}$$

where t_0 is the present cosmic time, $M = m_1 + m_2$ is the total mass, and $\eta \equiv m_1 m_2/(m_1 + m_2)^2$. The contribution of hierarchical mergers [88] has been neglected, given its subdominance as constrained by GWTC [91, 135]. Additionally, the total merger rate can be obtained by integrating the component masses as

$$R(t) = \int \mathcal{R}(t, m_1, m_2)\, \mathrm{d}m_1\, \mathrm{d}m_2. \tag{25.4}$$

We will frequently switch between cosmic time t and redshift z. The cosmic time t and redshift z are related by

$$t(z) = \int_z^\infty \frac{dz'}{H(z')(1+z')}, \tag{25.5}$$

where $H(z)$ is the Hubble parameter. We neglect contributions from radiation and neutrinos, given the limited sensitivity of current ground-based GW detectors to a small redshift range. When converting between luminosity distances, times, and redshifts, we adopt the best-fit cosmological model of Planck 2018 [18].

To constrain the PBH scenario, one must place some a priori constraints on the form of the PBH mass function, $P(m)$. The form of the mass function is sensitive to the details of PBH formation. Throughout this chapter we consider four PBH mass functions frequently employed in the literature: the lognormal, power-law (with a lower-mass cutoff), broken power-law (BPL), and critical collapse (CC) distributions. In Fig. 25.1, we provide a visual illustration of these four PBH mass distributions. In the rest of the subsection, we will introduce these mass functions and briefly describe their key properties.

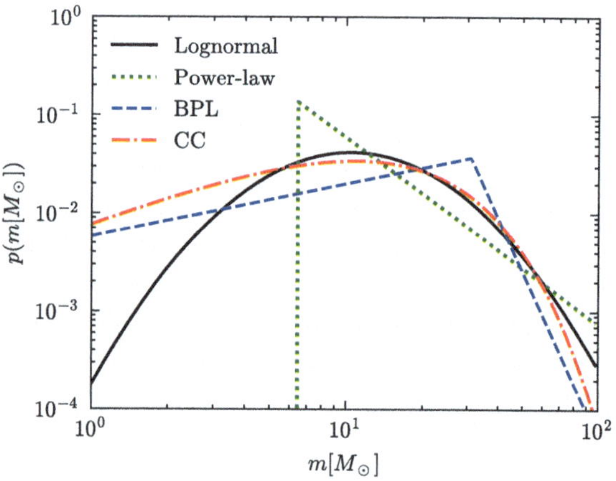

Fig. 25.1 An illustrative depiction presenting the four PBH mass distributions explored in this chapter: lognormal, power-law, BPL, and CC functional forms. All four PBH mass functions have been normalized to unity, satisfying Eq. (25.1)

25.2.1.1 Lognormal Mass Function

A lognormal PBH mass function is defined by [62]

$$P(m) = \frac{1}{\sqrt{2\pi}\sigma m} \exp\left(-\frac{\ln^2(m/M_c)}{2\sigma^2}\right), \tag{25.6}$$

where M_c is the median mass and σ characterizes the width of the mass distribution. The lognormal mass function proves to be a versatile approximation for a wide range of extended mass distributions, especially when PBHs originate from a smooth, symmetric peak in the inflationary power spectrum under the slow-roll approximation [46, 72, 80]. A PBH population with a lognormal mass function therefore requires at least three parameters to model the binary merger rate, $\Lambda = \{M_c, \sigma, f_{\rm PBH}\}$.

25.2.1.2 Power-Law Mass Function

The next model we consider is a PBH mass function characterized by the power-law form [44]

$$P(m) = \frac{\alpha - 1}{M_{\rm min}} \left(\frac{m}{M_{\rm min}}\right)^{-\alpha}, \tag{25.7}$$

where $M_{\rm min}$ denotes a lower-mass cutoff and α is the power-law index. We require $\alpha > 1$ such that $\int_{M_{\rm min}}^{\infty} P(m) dm$ will not be divergent. The power-law mass function typically emerges from a broad or flat power spectrum of curvature perturbations [57] during the radiation-dominated era [45, 46]. In this case, three hyperparameters, $\Lambda = \{M_{\rm min}, \alpha, f_{\rm PBH}\}$, need to be determined through the analysis of GW data.

25.2.1.3 Broken Power-Law Mass Function

A refinement of the power-law model is to consider a BPL form [60] given by

$$P(m) = \left(\frac{m_*}{\alpha_1 + 1} + \frac{m_*}{\alpha_2 - 1}\right)^{-1} \begin{cases} (\frac{m}{m_*})^{\alpha_1}, & m < m_* \\ (\frac{m}{m_*})^{-\alpha_2}, & m > m_* \end{cases}, \tag{25.8}$$

where m_* is the peak mass of $mP(m)$. Here, $\alpha_1 > 0$ and $\alpha_2 > 1$ are two power-law indices. The BPL mass function is a generalization of the power-law form and can be realized if PBHs are formed by vacuum bubbles nucleating during inflation via quantum tunneling [60]. In this case, four hyperparameters, $\Lambda = \{m_*, \alpha_1, \alpha_2, f_{\rm PBH}\}$, need to be determined through the analysis of GW data.

25.2.1.4 Critical Collapse Mass Function

Lastly, we consider a PBH mass function with the CC form [43, 71, 106, 138]

$$P(m) = \frac{\alpha^2 m^\alpha}{M_{\rm f}^{1+\alpha}\,\Gamma(1/\alpha)} \exp\left(-(m/M_{\rm f})^\alpha\right), \tag{25.9}$$

where α is a universal exponent related to the CC of radiation, and $M_{\rm f}$ is a mass scale approximately of the order of the horizon mass at the collapse epoch [43]. This mass spectrum does not exhibit a lower mass cutoff; however, it experiences exponential suppression beyond the mass scale of $M_{\rm f}$. The CC mass function is closely related to a power spectrum of density fluctuations resembling a δ-function [43, 71, 106, 138]. In this case, three hyperparameters, $\Lambda = \{M_{\rm f}, \alpha, f_{\rm PBH}\}$, need to be determined through the analysis of GW data.

25.2.2 Bayesian Formalism

In the PBH context, population studies using the Bayesian formalism have become standard practice [48, 52, 54, 58, 66, 73, 78, 91, 146]. This subsection gives an overview of the Bayesian formalism employed in the analysis of GW data to infer model parameters. We refer to [127] for a pedagogical introduction to Bayesian inference in GW astronomy. With the recent accumulation of almost 100 GW observations from the LVK Collaboration [10], the study of GW sources is swiftly transitioning to the domain of "population inference" [1, 12, 13, 48, 49, 52, 54, 91, 140, 146]. This involves addressing selection biases introduced by the detectors' sensitivity, particularly concerning GW frequency and binary parameters like masses and redshift. Beyond correcting for selection bias, GW population inference faces the challenge of dealing with uncertain source parameter measurements due to noise in the heterogeneous data. This requires a specialized statistical framework to reconstruct population properties, with current techniques often leveraging hierarchical Bayesian inference [96, 131]. To exemplify how the machinery of BBH population analysis works in detail, in the remainder of this chapter we will employ a hierarchical Bayesian approach to deduce BBH population parameters from GW data, accounting for the uncertainty in estimating individual event parameters.

The merger rate density, as expressed in Eq. (25.3), corresponds to measurements made in the BBH source frame of reference. The source frame is the coordinate system attached to the astrophysical source of GW, while the detector frame is the coordinate system associated with the GW detector. In other words, the source frame is linked to the astrophysical properties of the GW source, while the detector frame is associated with the measurements recorded by the GW detector. To facilitate analysis, a conversion into the detector frame is required using the transformation given by

$$\mathcal{R}_{\rm pop}(\theta|\Lambda) = \frac{1}{1+z}\frac{dV_{\rm c}}{dz}\mathcal{R}(\theta|\Lambda), \tag{25.10}$$

where z is the cosmological redshift, and $\theta \equiv \{z, m_1, m_2\}$ constitutes the parameter set defining individual BBH events. In this chapter, we focus on the redshift and mass distributions and ignore the spin distribution. The collection of parameters Λ encompasses $f_{\rm PBH}$ and parameters derived from the mass function $P(m)$. Additionally, $dV_{\rm c}/dz$ represents the differential comoving volume. The incorporation of the factor $1/(1+z)$ serves to convert time increments from the source frame to the detector frame.

Given the observed data $\mathbf{d} = \{d_1, d_2, \cdots, d_{N_{\rm obs}}\}$, comprising $N_{\rm obs}$ BBH merger events, we formulate the total number of events as an inhomogeneous Poisson process, leading to the likelihood [92, 96, 127]

$$\mathcal{L}(\mathbf{d}|\Lambda) \propto N_{\rm exp}^{N_{\rm obs}} e^{-N_{\rm exp}} \prod_{i=1}^{N_{\rm obs}} \frac{\int \mathcal{L}(d_i|\theta)\,\mathcal{R}_{\rm pop}(\theta|\Lambda)d\theta}{\xi(\Lambda)}, \tag{25.11}$$

where $N_{\rm exp} \equiv N_{\rm exp}(\Lambda)$ represents the expected number of detections over the observation timespan. Here, $\mathcal{L}(d_i|\theta)$ denotes the likelihood for the individual ith GW event, derived from the posterior of the individual event by reweighing with the prior on θ. Furthermore, $\xi(\Lambda)$ quantifies selection biases for a population with parameters Λ and is defined as

$$\xi(\Lambda) = \int P_{\rm det}(\theta)\,\mathcal{R}_{\rm pop}(\theta|\Lambda)\,{\rm d}\theta, \tag{25.12}$$

where $0 < P_{\rm det}(\theta) < 1$ denotes the detection probability [107], a function of the source parameters θ.

In practice, the estimation of $\xi(\Lambda)$ is achieved using simulated injections [11]. Then, $\xi(\Lambda)$ in Eq. (25.12) is approximated through a Monte Carlo integral over found injections [12] as

$$\xi(\Lambda) \approx \frac{1}{N_{\rm inj}} \sum_{j=1}^{N_{\rm found}} \frac{\mathcal{R}_{\rm pop}(\theta_j|\Lambda)}{p_{\rm draw}(\theta_j)}, \tag{25.13}$$

where $N_{\rm inj}$ stands for the total number of injections, $N_{\rm found}$ is the count of successfully detected injections, and $p_{\rm draw}$ represents the probability density function from which the injections are drawn. Utilizing the posterior samples from each event, we compute the hyper-likelihood (25.11) as

$$\mathcal{L}(\mathbf{d}|\Lambda) \propto N_{\rm exp}^{N_{\rm obs}} e^{-N_{\rm exp}} \prod_{i=1}^{N_{\rm obs}} \frac{1}{\xi(\Lambda)} \left\langle \frac{\mathcal{R}_{\rm pop}(\theta|\Lambda)}{\pi(\theta)} \right\rangle, \tag{25.14}$$

where $\langle \cdots \rangle$ denotes the weighted average over posterior samples of θ. The denominator $\pi(\theta)$ refers to the priors on source parameters used to construct the posterior of individual events.

According to Bayes's theorem, the posterior distribution for the hyperparameter Λ is expressed as

$$p(\Lambda|\mathbf{d}) = \frac{\mathcal{L}(\mathbf{d}|\Lambda)\pi(\Lambda)}{\mathcal{Z}}, \tag{25.15}$$

$\pi(\Lambda)$ denotes the prior distribution for Λ, and $\mathcal{Z}$ is a normalization factor referred to as the "evidence"

$$\mathcal{Z} \equiv \int \mathrm{d}\Lambda \, \mathcal{L}(\mathbf{d}|\Lambda)\pi(\Lambda). \tag{25.16}$$

In this chapter, we integrate the PBH population distribution (25.3) into the `ICAROGW` [101] package to calculate the likelihood function (25.14). We utilize the `dynesty` [125] sampler, accessed through `Bilby` [26, 119], to perform sampling across the parameter space.

The evidence, $\mathcal{Z}$, defined above plays a crucial role in model selection, addressing the question of which model is statistically favored by the data and to what extent. The Bayes factor, quantifying the ratio of evidence between two exclusive models, $\mathcal{M}_2$ and $\mathcal{M}_1$, serves as a quantifiable measure for model selection scores. It is expressed as

$$\mathrm{BF}_1^2 \equiv \frac{\mathcal{Z}_2}{\mathcal{Z}_1}, \tag{25.17}$$

with $\mathcal{Z}_2$ and $\mathcal{Z}_1$ denoting the evidence for $\mathcal{M}_2$ and $\mathcal{M}_1$, respectively. The evidence, being the average of the likelihood over the prior volume, naturally incorporates Occam's razor principle, which states that, all else being equal, a model with fewer parameters is preferred. A substantial value of BF_1^2 indicates a strong preference for model $\mathcal{M}_2$ over $\mathcal{M}_1$. Table 25.1 presents an interpretation of the Bayes factor in the model comparison [81].

Table 25.1 An interpretation of the Bayes factor in quantifying the model selection scores as given by Ref. [81]

BF	ln BF	Strength of evidence
< 1	< 0	Negative
$1 - 3$	$0 - 1$	Not worth more than a bare mention
$3 - 20$	$1 - 3$	Positive
$20 - 150$	$3 - 5$	Strong
> 150	> 5	Very strong

25.3 Constraints from the GWTC-3 Catalog

In this section, we first present the constraints on the population parameters using GWTC-3 for the individual scenarios of purely ABH population, purely PBH population, and a combined model containing both ABH and PBH populations.

As described above, the GWTC provides samples from the posterior distribution of parameters describing each GW event. These include both *intrinsic* parameters, such as source-frame component masses and spins, and *extrinsic* parameters such as matched filter signal-to-noise (SNR) and merger times. Bayesian inference of population models can then be performed by approximating the integral over source parameters θ in the likelihood function Eq. (25.11) with the Monte Carlo estimator Eq. (25.14).

In the subsection, we will describe constraints from GWTC-3 [10] (incorporating its previous iterations, see Refs. [2, 8]) assuming the observed BBH merger events are drawn from a population of solely ABHs, solely PBHs, and finally a mixed ABH+PBH population.

25.3.1 Constraints on the ABH Population

Population inference has been performed extensively with the GWTC-3 data and its predecessors in the ABH context [2, 12, 13, 120, 121, 129]. Detailed modelling of the ABH merging binary population is challenging due to the plurality of possible formation channels and astrophysical uncertainties (for recent reviews, see Refs. [94, 95]). Partly for this reason, ABH population studies have primarily taken an agnostic approach to model inference, for example, by considering simple empirical parametric merger rate models.

The best fitting empirical parametric models are those that can accommodate the primary features visible in the distribution of detector-frame chirp mass, as this is typical a well constrained parameter for a BBH merger event. The empirical distribution is qualitatively similar to that of the source-frame total mass, shown in the top panel of Fig. 25.3; a peak in the merger rate for total masses around 20 $M_{\odot}$, relatively few sources in the range 30 – 50 $M_{\odot}$, a broad peak between 50 $M_{\odot}$ and 80 $M_{\odot}$, and a gradual suppression in the rate of sources with higher mass. While models with more parameters may be able to provide superior fits to the data, overfitting incurs penalties for such models when compared against simpler models in Bayesian model selection.

While detailed predictions of the merger rate density as a function of source masses and redshift are challenging in the ABH context, stellar physics makes several predictions for the black hole population of varying robustness to physical assumptions. The most secure of these is the lower limit on the mass of stellar black holes, thought to be equivalent to the maximum neutron star mass, around 2 $M_{\odot}$ [108, 116, 130].

For black holes just heavier than this, a mass gap spanning the range 2–5 $M_{\odot}$ can develop [35], although this is subject to uncertainty based on the timescale for instability growth and supernova launch [67]. The discovery of the event GW190814, with a secondary component mass of 2.6 $M_{\odot}$ [7] directly challenges the existence of this mass gap [144]. Continuing up the mass spectrum, a second mass gap is expected between roughly 65–120 $M_{\odot}$ from pulsational pair-instability supernova theory [134]. The location of the upper mass gap is subject to substantial uncertainty [63, 98], and the gap may be populated through dynamical formation channels [85]. Even if the mass gap exists in the spectrum of newly formed stellar black holes, multi-generational mergers could populate the gap if the local stellar density is high enough [68, 118]. The detection of GW190521, with component masses of roughly 85 $M_{\odot}$ and 66 $M_{\odot}$ [6], challenges the existence of the upper mass gap [14], and the population analysis of the GWTC-3 performed by the LVK Collaboration found no evidence for a suppression of the merger rate above 60 $M_{\odot}$ [12]. Finally, beyond the upper limit of the upper mass gap, searches for merging binaries with component black holes in the Intermediate Mass Black Hole (IMBH) regime, have so far yielded negative results [16]. Well-fitting parametric ABH mass distributions are usually endowed with power law tails with upper mass cutoffs to represent this high-mass suppression, but the uncertainty in the formation mechanism of IMBHs makes a physical explanation challenging.

Whilst the distribution of chirp masses is the most constraining feature of the data for population studies, additional information comes from the distribution of less well-measured source parameters, in particular the spins, redshifts, and mass ratios of the binary systems. ABH models do not make strong predictions for the distributions of these quantities; instead, constraints from the GWTC provide key observational hints at the evolutionary pathways for BBH events. We summarise the key findings from these distributions in the ABH context below.

- Spin constraints are most succinctly summarised by the parameter $\chi_{\rm eff}$, the projection of a component mass-weighted total spin angular momentum vector onto the total angular momentum vector, which is typically measured with better precision than either spin component individually [19]. The LVK Collaboration has found evidence for a broadening of the distribution of $\chi_{\rm eff}$ above 30 $M_{\odot}$, and a negative correlation between spins and the mass ratio $q \equiv m_2/m_1$, where m_1 is the primary component (i.e. the heavier of the two component masses), at 97.5% credibility [12].
- The redshift dependence of the BBH merger rate is well fit by a power law model having merger rate per comoving volume $\mathcal{R}(z) \propto (1+z)^{\kappa}$, with $\kappa = 2.9^{+1.7}_{-1.8}$ at 90% credibility [12]. Constraints on $\mathcal{R}(z)$ are summarized in Fig. 25.4. The redshift dependence is consistent with the Madau-Dickinson evolution of the global star formation rate [93].
- Reference [12] finds that the distribution of mass ratio is well fitted by a power law in q with exponent $\beta_q = 1.1^{+1.7}_{-1.3}$ at 90% credibility. The constraints on q are relatively broad compared with the prior for most events, making the mass ratio a relatively less powerful discriminator between models than proxies for the total

mass. As is visible from Fig. 25.2, most of the systems in GWTC-3 have mass ratios compatible with unity, although it should be noted that extreme mass ratios ($q \ll 0.01$) are relatively less detectable for the LVK detector network at fixed total mass (see, e.g. Fig. 3 of Ref. [73]).

In the context of constraining the possibility of a PBH population from GWTC-3, one should keep in mind that posterior constraints on CBC source parameters can be sensitive to the adopted priors. When constraining populations models, one effectively replaces these priors with a distribution for the parameters conditioned on the source population model. This model will typically have its own free parameters ("hyperparameters") that themselves require priors to be specified. As stressed in Ref. [36], the influence of the prior can strongly impact the source parameter posteriors for some GWTC events. Results from population-informed priors for empirical parametric ABH populations are presented alongside the default "uninformative" priors in Ref. [12]. The use of these alternative priors can impact, for example, conclusions drawn about the existence of the lower mass gap. The results discribed above and presented in Figs. 25.2, 25.3, and 25.4 assume the uninformative priors adopted in Ref. [12].

25.3.2 PBH Constraints Assuming a Single Population

We employ GW events from GWTC-3, excluding those with false alarm rates exceeding 1 yr^{-1} and secondary component masses below $3M_{\odot}$ to mitigate potential contamination from events involving neutron stars following Ref. [58]. This selection yields a total of 69 GW events from GWTC-3 that satisfy these criteria, and the corresponding posterior samples for these BBH events are publicly accessible through Ref. [15]. In the analyses, we use combined posterior samples obtained from the IMRPhenom [111, 126] and SEOBNR [102, 109] waveform families.

Employing the Bayesian inference framework outlined in Sect. 25.2.2, we perform parameter estimations for four distinct PBH mass functions, as described in Sect. 25.2.1. A comprehensive summary of the parameters and their corresponding prior distributions is provided in Table 25.2. Utilizing data from 69 BBH events from GWTC-3 and employing hierarchical Bayesian inference, we obtain posterior distributions for the hyperparameters, as illustrated in Fig. 25.5. Additionally, the posterior predictive distributions (PPD) for the four PBH mass functions are depicted in Fig. 25.6.

The PPD updates the prior for parameters θ based on data $\mathbf{d}$. Recall the hyperposterior $p(\Lambda|\mathbf{d})$ reflects our post-measurement understanding of hyperparameters shaping the mass distribution $P(\theta)$. The PPD addresses: given this hyper-posterior, what is the distribution of $P(\theta)$? It signifies the probability that the next event possesses true parameter values θ based on our knowledge of population hyperparameters Λ:

$$p_{\Lambda}(\theta|\mathbf{d}) = \int \mathrm{d}\Lambda\; p(\Lambda|\mathbf{d})\, P(\theta|\Lambda). \tag{25.18}$$

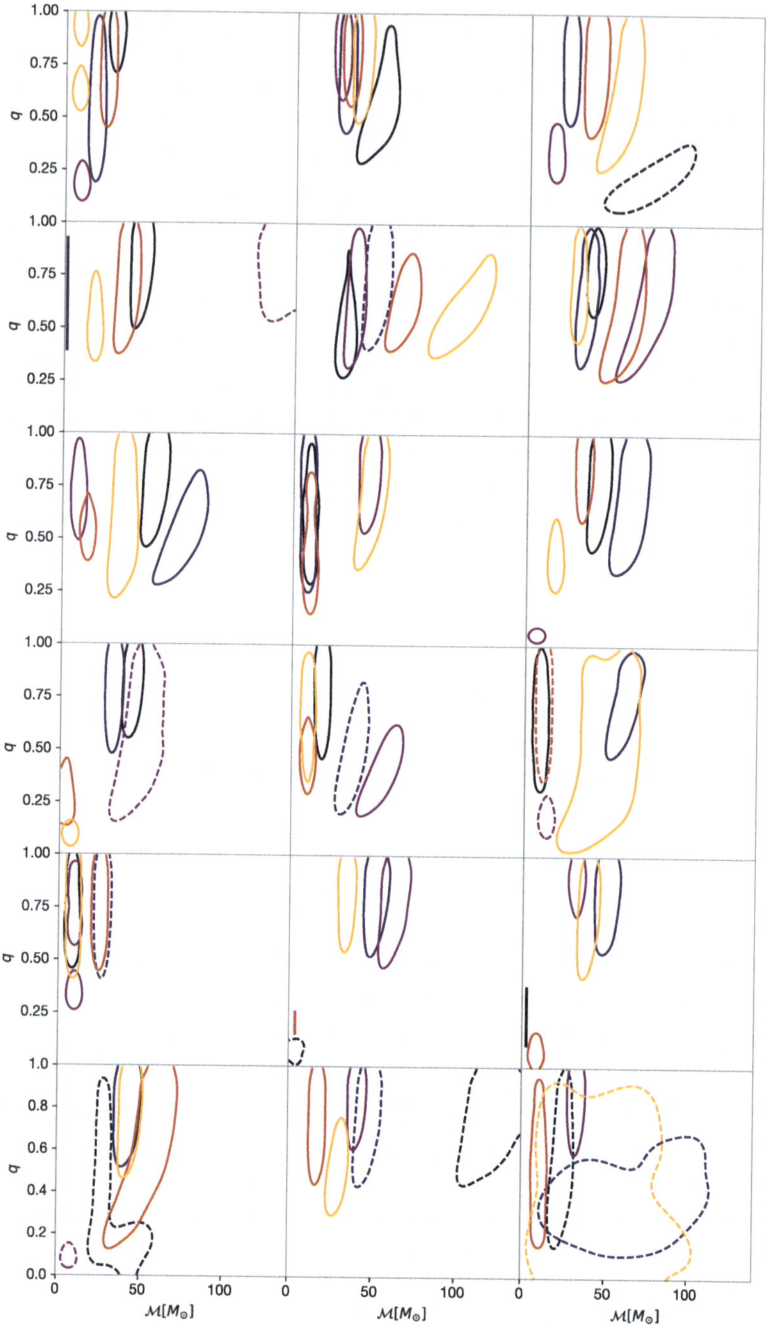
1.00
0.75
0.50
0.25
0.0
0.2
0.4
0.6
0.8
1.0
q
0
50
100
$\mathcal{M}[M_\odot]$

◀**Fig. 25.2** Source parameter constraints for the 90 compact binary coalescences (CBCs) that constitute GWTC-2.1 and GWTC-3. Each contour encloses 68% of the posterior probability, and we show constraints projected into the plane of mass ratio q and detector-frame chirp mass $\mathcal{M}$. Sources have been placed into groups of five according to their detection time, with the earliest detections (i.e. those detected in O1) in the top-left panel and latest in the bottom-right (i.e. those detected in O3b). Hotter, more luminant colours denote sources with a later detection date within its group of five. Sources with dashed contours have a false alarm rate larger than 1 per year, namely FAR $> 1\,\text{yr}^{-1}$, and are hence excluded from the population analysis of [12]. We have also excluded the binary neutron star event GW170817, but included GW200105_162426, which was excluded from GWTC-3 due to its low p_{astro} but included in the populations analysis of Ref. [12]

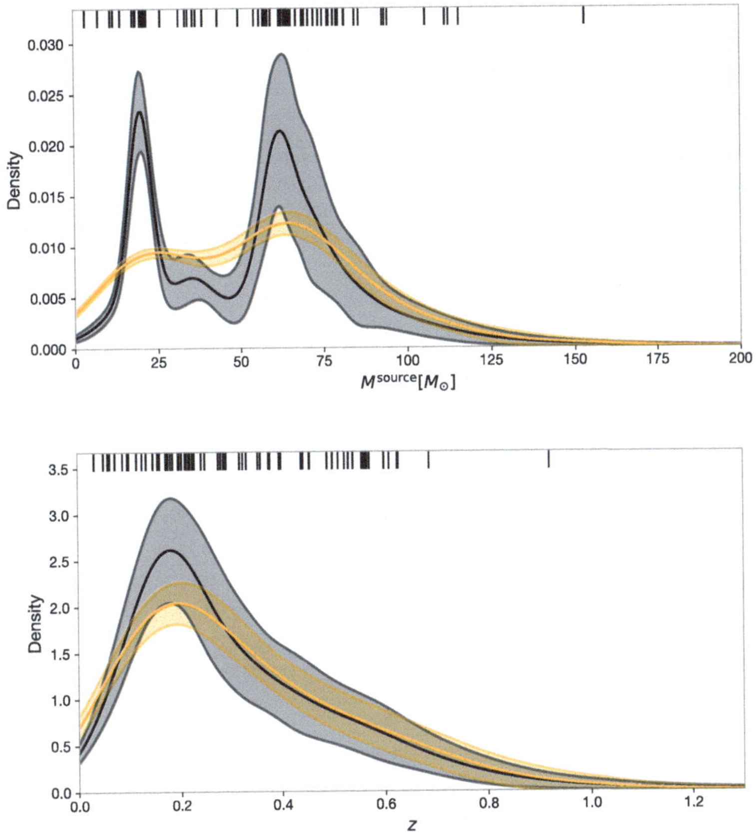

Fig. 25.3 Distribution of the source-frame total mass (top panel) and redshift (bottom panel) of CBCs in the GWTC catalog, excluding GW170817 and events with FAR $> 1\,\text{yr}^{-1}$. Posterior medians are shown as tick marks at the top of the figure. The black curve shows an adaptive Gaussian KDE constructed with `awkde` (https://github.com/mennthor/awkde) and using the algorithm of Ref. [122], the orange curve shows a standard Gaussian KDE with a global bandwidth set using Scott's rule. Both curves are mean KDEs over bootstrap resamples from the source posteriors, shaded regions are 2σ confidence intervals constructed from these resamples

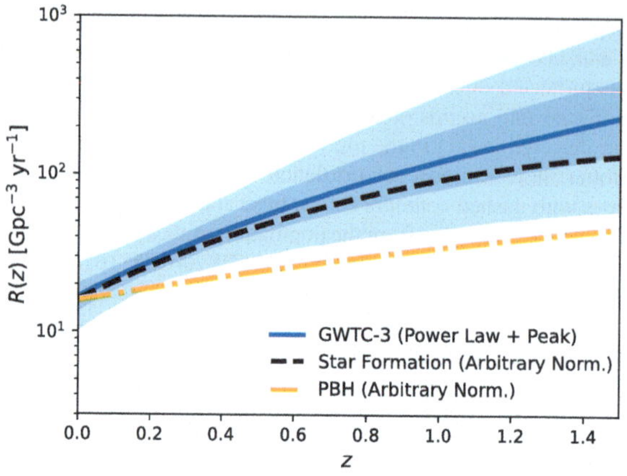

Fig. 25.4 Comoving merger rate density of BBH mergers in GWTC-3 [12]. We show the median (blue solid curve) and 50% and 90% credibility intervals (shaded regions). Overplotted is a fit to the global star formation rate density from Ref. [93] (black dashed curve) and the PBH merger rate model from [113] at fixed source mass and neglecting the suppression factor S (orange dot-dashed curve), both with an arbitrary normalisation

Table 25.2 Model parameters and their corresponding prior distributions employed in our Bayesian parameter estimation for the four PBH mass function models. Here, $\mathcal{U}$ and $\log\mathcal{U}$ refer to uniform and log-uniform distributions, respectively

Parameter	Description	Prior
$f_{\rm PBH}$	Abundance of PBH in CDM	$\log\mathcal{U}(-4, 0)$
Lognormal PBH mass function		
M_c	Central mass in $M_\odot$	$\mathcal{U}(5, 50)$
σ	Mass width	$\mathcal{U}(0.1, 2)$
Power-law PBH mass function		
$M_{\rm min}$	Lower mass cutoff in $M_\odot$	$\mathcal{U}(3, 10)$
α	Power-law index	$\mathcal{U}(1.05, 4)$
BPL PBH mass function		
m_*	Peak mass in $M_\odot$	$\mathcal{U}(5, 50)$
α_1	First power-law index	$\mathcal{U}(0, 3)$
α_2	Second power-law index	$\mathcal{U}(1, 10)$
CC PBH mass function		
M_f	Horizon mass scale in $M_\odot$	$\mathcal{U}(1, 50)$
α	Universal exponent	$\mathcal{U}(0, 5)$

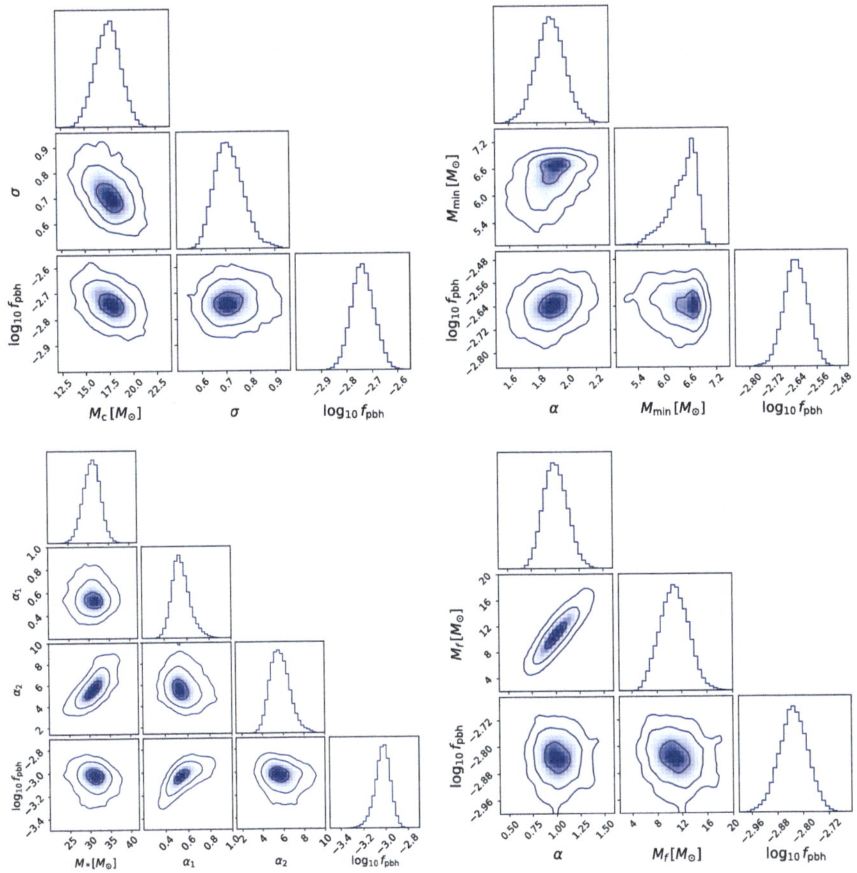

Fig. 25.5 Marginalized posterior distributions for hyperparameters inferred from GWTC-3, with each panel corresponding to a specific mass function: lognormal (top left), power-law (top right), BPL (bottom left), and CC (bottom right). Contours denote the 1σ, 2σ, and 3σ credible regions, respectively

The subscript Λ distinguishes the PPD from the posterior $p(\theta|\mathbf{d})$. In the hyper-posterior sample version, the PPD is expressed as

$$p_\Lambda(\theta|\mathbf{d}) = \frac{1}{n_s}\sum_{k}^{n_s} P(\theta|\Lambda_k), \tag{25.19}$$

where k iterates over n_s hyper-posterior samples. While the PPD best estimates the appearance of $P(\theta)$, it doesn't convey potential variability in $P(\theta)$ due to uncertainty in Λ. Effectively conveying this variability may involve overplotting multiple realizations of $P(\theta|\Lambda_k)$, where Λ_k is a randomly chosen hyper-posterior sample.

Fig. 25.6 Posterior predictive distributions for the four PBH mass functions using GWTC-3 catalog

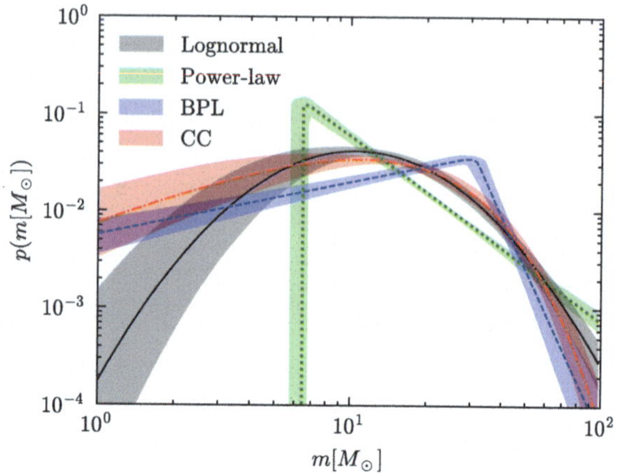

For the lognormal PBH mass function model, our analysis yields hyperparameter values with median estimates and 90% equal-tailed credible intervals: $M_c = 17.3^{+2.2}_{-2.0} M_\odot$, $\sigma = 0.71^{+0.10}_{-0.08}$, and $f_{\rm PBH} = 1.8^{+0.3}_{-0.3} \times 10^{-3}$. It's noteworthy that the derived M_c value is larger than that obtained from GWTC-1 in Ref. [135] due to the inclusion of heavier black holes in GWTC-3. Moreover, utilizing Eq. (25.4), we deduce the local merger rate as $R(t_0) = 41^{+16}_{-12}{\rm Gpc}^{-3}\,{\rm yr}^{-1}$.

For the power-law PBH mass function model, our analysis yields $M_{\rm min} = 6.5^{+0.3}_{-0.8} M_\odot$, $\alpha = 1.9^{+0.2}_{-0.2}$, and $f_{\rm PBH} = 2.3^{+0.3}_{-0.3} \times 10^{-3}$. Notably, the inferred α value is smaller than that from GWTC-1 in Ref. [135] due to the inclusion of more massive BHs in GWTC-3. Using Eq. (25.4), we estimate the local merger rate as $R(t_0) = 48^{+15}_{-12}{\rm Gpc}^{-3}$, ${\rm yr}^{-1}$.

For the BPL PBH mass function model, our analysis yields $m_* = 31.1^{+1.8}_{-2.1} M_\odot$, $\alpha_1 = 0.54^{+0.08}_{-0.06}$, $\alpha_2 = 5.6^{+0.9}_{-0.8}$, and $f_{\rm PBH} = 0.9^{+0.1}_{-0.1} \times 10^{-3}$. Applying Eq. (25.4), we estimate the local merger rate as $R(t_0) = 46^{+15}_{-11}{\rm Gpc}^{-3}\,{\rm yr}^{-1}$.

For the CC PBH mass function model, our analysis yields $M_{\rm f} = 10.8^{+3.7}_{-3.6} M_\odot$, $\alpha = 1.0^{+0.2}_{-0.2}$, and $f_{\rm PBH} = 1.5^{+0.2}_{-0.2} \times 10^{-3}$. Using Eq. (25.4), we estimate the local merger rate as $R(t_0) = 49^{+26}_{-16}{\rm Gpc}^{-3}\,{\rm yr}^{-1}$.

Across all four mass functions, our inferences suggest that the abundance of PBHs in CDM, $f_{\rm PBH}$, is on the order of $\mathcal{O}(10^{-3})$. These results align with previous estimations [21, 48–50, 52, 54, 124, 135, 146], supporting the conclusion that stellar-mass PBHs do not dominate the composition of CDM. While the late-time clustering of PBHs may impact the observed merger rate, our inferred values for $f_{\rm PBH}$ consistently remain below 3×10^{-3} for all four PBH mass functions. Therefore, this clustering effect can be safely neglected according to Ref. [78].

We also calculate the Bayes factors ($\rm BF_{PL}$) between models with different PBH mass functions, using the model with the power-law mass function as the fiducial model. From Table 25.3, we see that $\rm BF^{LG}_{PL}$ attains the highest value, suggesting that the lognormal mass function provides the most optimal fit to GWTC-3 among the

Table 25.3 Bayes factors, BF_{PL}, comparing models with various PBH mass functions to the model with a power-law PBH mass function

	Lognormal	Power-law	BPL	CC
BF_{PL}	166	1	49	139

four mass functions. Note that the Bayes factors presented in Table 25.3 should be interpreted with caution. The hierarchical nature of our analysis and the different parameterizations in each mass model mean that the Bayes factors can be affected by the choice of priors on the hyperparameters. As a result, the nominal values of the Bayes factors may be subject to specific prior choices, and they should be considered as a qualitative comparison between the models rather than a definitive quantitative assessment.

While the results presented in this subsection suggest a lognormal or CC mass function is preferred over power law or BPL forms, it is important to note that none of the considered mass functions provides a particularly good fit to the observed BBH mass distribution. As discussed in Sect. 25.3.1 and emphasized in Refs. [65, 73], the observed distribution has features, such as multiple peaks, that are not captured by any of our assumed PBH mass functions. This fact presents a stark challenge to the PBH scenario in the BBH context, and motivates the consideration of mixed ABH+PBH populations, to which we now turn.

25.3.3 PBH Constraints Assuming Mixed Populations

Previous investigations suggest the possibility that the BBH mergers observed by the LVK Collaboration may originate from both ABHs and PBHs [58, 65, 66, 73, 78, 133]. In this section, we explore a hybrid model, referred to as the mixed ABH+PBH model, wherein BBHs can arise from both ABH and PBH channels.

We adopt the lognormal mass function for the PBH component, as it is a reasonable fit to the mass function resulting from a peak in the primordial curvature perturbation. For the ABH component, we employ a phenomenological model following Ref. [5]. The combined merger rate, $\mathcal{R}_{\rm total}(m_1, m_2, z)$, is the sum of the ABH merger rate ($\mathcal{R}_{\rm ABH}$) and the PBH merger rate ($\mathcal{R}_{\rm PBH}$), given by

$$\mathcal{R}_{\rm total}(z, m_1, m_2) = \mathcal{R}_{\rm ABH}(z, m_1, m_2) + \mathcal{R}_{\rm PBH}(z, m_1, m_2), \tag{25.20}$$

where $\mathcal{R}_{\rm PBH}(z, m_1, m_2)$ is given by Eq. (25.3). Specifically, the ABH merger rate is expressed as

$$\mathcal{R}_{\rm ABH}(z, m_1, m_2) = R_{0,\rm ABH}\, \pi(z)\, \pi(m_1)\, \pi(m_2|m_1), \tag{25.21}$$

where $R_{0,\rm ABH}$ represents the local merger rate of ABH binaries.

The redshift distribution $\pi(z)$ is parameterized as

$$\pi(z|\gamma, \kappa, z_{\rm p}) = \left[1 + \left(1 + z_{\rm p}\right)^{-\gamma-\kappa}\right] \frac{(1+z)^{\gamma}}{1 + \left[(1+z)/(1+z_{\rm p})\right]^{\gamma+\kappa}}, \tag{25.22}$$

where γ and k denote the slopes of two power-law regimes before and after a turning point $z_{\rm p}$. This choice of parameterization is motivated by the potential correlation between the binary formation rate and the star formation rate [42, 93].

The primary mass distribution $\pi(m_1) \equiv \pi\left(m_1|m_{\rm min}, m_{\rm max}, \alpha, \lambda_{\rm g}, \mu_{\rm g}, \sigma_{\rm g}\right)$ is chosen to follow the "Power Law + Peak" model [12], which is a combination of a power-law and Gaussian component, given by

$$\pi(m_1) = \left[(1-\lambda_{\rm g})\mathcal{P}(m_1|m_{\rm min}, m_{\rm max}, -\alpha) + \lambda_{\rm g}\mathcal{G}(m_1|\mu_{\rm g}, \sigma_{\rm g})\right] S(m_1, m_{\rm min}, \delta_m), \tag{25.23}$$

where $\mathcal{P}(x|x_{\rm min}, x_{\rm max}, \alpha)$ represents a truncated power-law with slope α between the bounds $x_{\rm min}$ and $x_{\rm max}$, given by

$$\mathcal{P}(x|x_{\rm min}, x_{\rm max}, \alpha) \propto \begin{cases} x^{\alpha} & (x_{\rm min} \leqslant x \leqslant x_{\rm max}) \\ 0 & \text{otherwise.} \end{cases} \tag{25.24}$$

The Gaussian distribution, $\mathcal{G}(x|\mu, \sigma)$, is characterized by mean μ and standard deviation σ:

$$\mathcal{G}(x|\mu, \sigma) = \frac{1}{\sigma\sqrt{2\pi}} \exp\left[-\frac{(x-\mu)^2}{2\sigma^2}\right]. \tag{25.25}$$

The parameter $\lambda_{\rm g}$ serves as the ratio parameter between the power-law component $\mathcal{P}$ and the Gaussian component $\mathcal{G}$. Additionally, the sigmoid-like window function, $S(m_1, m_{\rm min}, \delta_m)$, introduces a smoothing rise within the interval $(m_{\rm min}, m_{\rm min} + \delta_m)$:

$$S(m, m_{\rm min}, \delta_m) = \begin{cases} 0 & (m < m_{\rm min}) \\ \left[f\left(m - m_{\rm min}, \delta_m\right) + 1\right]^{-1} & (m_{\rm min} \le m < m_{\rm min} + \delta_m) \\ 1 & (m \ge m_{\rm min} + \delta_m)\,, \end{cases} \tag{25.26}$$

where $f(m', \delta_m)$ is defined by

$$f(m', \delta_m) = \exp\left(\frac{\delta_m}{m'} + \frac{\delta_m}{m' - \delta_m}\right). \tag{25.27}$$

The secondary mass distribution $\pi(m_2)$ is modeled with a truncated power-law with slope β between a minimum mass $m_{\min}$ and the primary mass m_1 as

$$\pi(m_2|m_1, m_{\min}, \alpha) = \mathcal{P}(m_2|m_{\min}, m_1, \beta)\, S(m_2, m_{\min}, \delta_m). \tag{25.28}$$

The priors for the parameters associated with the lognormal PBH mass function align with those presented in Table 25.2. Concurrently, the priors for parameters in the ABH model are outlined in Table 25.4. Using the data from 69 BBHs obtained from GWTC-3 and employing hierarchical Bayesian inference, we derive the posteriors for the hyperparameters within the mixed ABH+PBH model, as illustrated in Fig. 25.7.

Our analysis yields hyperparameter values with median estimates and 90% equal-tailed credible intervals: $M_c = 18.1^{+26.7}_{-11.6} M\odot$, and $\sigma = 0.98^{+0.87}_{-0.75}$. Meanwhile, we obtain an upper limit of $f_{\rm PBH} \lesssim 9.7 \times 10^{-4}$ at the 95% credible interval. The inclusion of the ABH channel introduces notably larger uncertainties in the inferred values for M_c and σ, with a substantial reduction of the upper limit on $f_{\rm PBH}$, as expected. Additionally, we determine the local merger rate for ABHs to be $R_{0,\rm ABH} = 8.8^{+6.2}_{-3.7} {\rm Gpc}^{-3}{\rm yr}^{-1}$.

Table 25.4 Model parameters and their corresponding prior distributions employed in our Bayesian parameter estimation for the ABH model

Parameter	Description	Prior
γ	Slope of the power-law regime for the rate evolution before the point z_p	$\mathcal{U}(0, 12)$
κ	Slope of the power-law regime for the rate evolution after the point z_p	$\mathcal{U}(0, 6)$
z_p	Redshift turning point between the power-law regimes with γ and κ	$\mathcal{U}(0, 4)$
α	Spectral index for the power-law of the primary mass distribution	$\mathcal{U}(1.5, 12)$
β	Spectral index for the power-law of the mass ratio distribution	$\mathcal{U}(-4, 12)$
$m_{\min}$ $[M_\odot]$	Minimum mass of the power-law component of the primary mass distribution	$\mathcal{U}(2, 10)$
$m_{\max}$ $[M_\odot]$	Maximum mass of the power-law component of the primary mass distribution	$\mathcal{U}(50, 200)$
λ_g	Fraction of the model in the Gaussian component	$\mathcal{U}(0, 1)$
μ_g $[M_\odot]$	Mean of the Gaussian component in the primary mass distribution	$\mathcal{U}(20, 50)$
σ_g $[M_\odot]$	Width of the Gaussian component in the primary mass distribution	$\mathcal{U}(0.4, 10)$
δ_m $[M_\odot]$	Range of mass tapering at the lower end of the mass distribution	$\mathcal{U}(0, 10)$

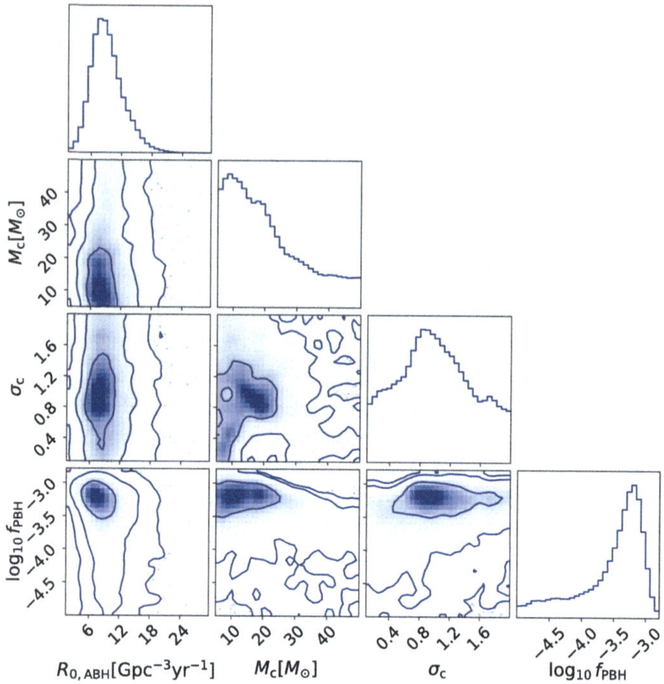

Fig. 25.7 One and two-dimensional marginalized posteriors for some of the hyperparameters in the mixed ABH+PBH model

Within the mixed ABH+PBH model, our analysis indicates that the fraction of detectable events originating from PBH binaries in GWTC-3 is $f_P \equiv N^{det}_{PBH}/(N^{det}_{PBH} + N^{det}_{ABH}) = 24.5^{+30.6}_{-17.3}\%$. This finding is consistent with the results obtained in previous studies [58, 146]. Despite the substantial uncertainty associated with f_P, it implies that at least a subset of BBHs in GWTC-3 can be attributed to the PBH channel. While our analysis suggests that the data allows for a non-zero contribution from PBHs, this does not imply conclusive evidence for PBH mergers.

So far, we have parameterised the PBH population in terms of its mass function. An alternative approach, demonstrated in Refs. [65, 146], is to drill down further and parameterise the curvature power spectrum that gives rise to PBHs. This approach is a potentially powerful way to constrain inflationary models directly from GW source catalogs, but will require further progress in developing accurate models of the ABH binary population to realize its full potential.

To conclude this subsection, we have seen that GW data are not consistent with models in which all of the sources are PBH binaries. Such models are unable to reproduce the detailed features of the black hole mass distribution that GWTC-3 has revealed. Mixed populations on the other hand do allow for a non-zero PBH contribution to the observed population, but this result should be treated with extreme caution due to the large uncertainties associated with ABH population models. While a PBH component could explain the presence of objects in the ABH mass gaps, such as GW190814, it remains to be seen whether the existence of these gaps is a strong enough prediction of stellar physics that PBHs can be considered a leading candidate for the observed GW sources.

25.3.4 Systematic Errors from Astrophysical Modelling Uncertainties

We have presented a simplified analysis of the GWTC-3 catalog to constrain the contribution of PBHs to the observed BBH mergers. However, it is essential to acknowledge the limitations of our approach and discuss the potential impact of astrophysical modelling uncertainties on our results.

Our analysis assumes a single formation channel for ABH mergers, described by a phenomenological model. This simplification may not fully capture the complexity of the various astrophysical processes that could lead to the formation of BBHs. In reality, there are several possible formation channels for ABHs, each with its own set of assumptions and uncertainties [94, 97]. These include isolated binary evolution [31, 33], dynamical formation in dense stellar environments [25, 117], and hierarchical mergers [69, 84]. Neglecting the diversity of these formation channels in our analysis could potentially lead to an overestimation of the PBH contribution.

Moreover, even within each formation channel, there are numerous uncertainties in astrophysical modelling. For example, in the isolated binary evolution scenario, the outcome of the binary system depends on various factors such as the initial mass function, stellar metallicity, binary fraction, and the efficiency of common envelope ejection [94, 97]. These uncertainties can significantly impact the predicted mass and redshift distributions of ABH mergers [94, 97].

The issue of astrophysical modelling uncertainties is particularly relevant for events like GW190521 [6], which lies in the upper mass gap, where the formation mechanisms are not yet fully understood. The seemingly positive measurement of $f_{\rm PBH}$ in our analysis may be driven by the lack of knowledge of the formation mechanisms for such high-mass events [66]. Alternative astrophysical scenarios, such as hierarchical mergers [69, 84] or stellar mergers in young star clusters [61, 85], could potentially explain the observed inconsistencies without invoking PBHs.

To mitigate these systematic errors, a comprehensive analysis should incorporate more comprehensive astrophysical modelling, taking into account results from population synthesis codes and considering multiple formation channels. Population synthesis codes, such as StarTrack [33], MOBSE [70], and COSMIC [40], provide a detailed description of the ABH population by modelling the evolution of massive binary systems. By including these simulated populations in the analysis, we can better distinguish between the contributions of ABHs and PBHs to the observed merger events.

In conclusion, the systematic errors arising from astrophysical modelling uncertainties pose a significant challenge in constraining the PBH contribution to BBH mergers. To draw robust conclusions about the role of PBHs in the observed GW events, future analyses should incorporate more comprehensive astrophysical modelling, considering multiple formation channels and the latest results from population synthesis simulations.

25.4 Prospects for Constraints with the Einstein Telescope

The second-generation GW detectors, Advanced LIGO and Advanced Virgo, have significantly advanced our understanding of the Universe. Despite their ongoing improvements in sensitivity, these detectors are anticipated to encounter limitations in the future, primarily due to constraints imposed by their hosting infrastructures. Consequently, the exploration of new GW detectors is underway to overcome these limitations. One such initiative is the Einstein Telescope (ET) [75, 76, 112], envisioned as a third-generation GW observatory in Europe. ET represents a novel research infrastructure designed to accommodate a detector with the capability to observe the entire Universe through GWs. It is conceptualized as a multi-interferometer observatory with the ambitious goal of enhancing sensitivity by a factor of ten compared to previous-generation detectors. See Chap. 15 for an overview of various GW detectors.

This section delves into the exploration and projection of constraints on PBHs and the potential to distinguish them from ABHs using third-generation ground-based GW detectors such as ET. These advanced detectors are expected to surpass the capabilities of the current LVK detector network, potentially detecting on the order of $\mathcal{O}(10^5)$ BBH events annually [115, 132].

25.4.1 Constraints on the PBH Abundance from the ET

The detection of PBHs serves to constrain the PBH abundance parameter f_{PBH}. Conversely, the absence of PBH detection will place upper limits on the merger rate of PBH binaries, consequently constraining the PBH abundance. The detectable event rate of PBH binaries hinges on both the merger rate and the sensitivity of GW detectors. In this subsection, under the assumption of PBHs having a monochromatic mass distribution, we proceed to estimate the detectable limits on the abundance of PBHs through a targeted search conducted by ET.

The expected number of detections, N_{exp}, can be estimated by [50, 82]

$$N_{\text{exp}} = \int R(z) \frac{\mathrm{d}VT}{\mathrm{d}z} \mathrm{d}z, \tag{25.29}$$

where $\mathrm{d}VT/\mathrm{d}z$ characterizes the spacetime sensitivity of a GW detector as a function of redshift and incorporates the selection effects inherent to that particular detector. See also Eq. (25.11). Generally, $\mathrm{d}VT/\mathrm{d}z$ depends on the properties θ (e.g. masses and spin) of a binary and is defined as [3, 4]

$$\frac{\mathrm{d}VT}{\mathrm{d}z} = \frac{\mathrm{d}V_c}{\mathrm{d}z}\frac{T_{\mathrm{obs}}}{1+z}P_{\mathrm{det}}(\theta), \tag{25.30}$$

where T_{obs} is the observing time, and the denominator $1 + z$ accounts for the converting of cosmic time from source frame to detector frame due to cosmic redshift.

Considering a scenario in which ET does not detect any PBH binaries, the observatory will establish an upper limit on the merger rate of PBH binaries. The 90% credible level upper limit on the binary merger rate is determined using the loudest event statistic formalism [38]:

$$R_{90} = \frac{2.303}{VT}, \tag{25.31}$$

where

$$VT = \int \frac{\mathrm{d}VT}{\mathrm{d}z}\mathrm{d}z. \tag{25.32}$$

To compute VT, we employ the semi-analytical approximation from [3, 4], neglecting the effect of spins for BHs and utilizing the `IMRPhenomPv2` waveform to simulate the BBH templates. Additionally, we set a single-detector SNR threshold $\rho_{\mathrm{th}} = 8$ as a detection criterion, corresponding approximately to a network threshold of 12.

The detectable limits for f_{PBH} through the targeted search using data from LVK's first three runs (O1-O3) and ET are illustrated in Fig. 25.8. Our assumptions include the absence of viable PBH binary candidates during ET observations, and ET operating at full duty for one year. Given the ongoing debate regarding whether the observed super-solar mass BBHs are of PBH origin or not, we represent the detectable limits for super-solar mass PBHs with dashed lines. Notably, if no PBH binaries are detected, ET is anticipated to constrain f_{PBH} at the order of $\mathcal{O}(10^{-4})$ for $1\,M_{\odot}$, which is two orders of magnitude more stringent than that from LVK O1-O3.

25.4.2 Distinguishing PBHs from ABHs

Currently, the second-generation GW detectors, such as LIGO, Virgo, and KAGRA, are limited to observing BBHs at low redshifts ($z < 1$), but future third-generation GW detectors like ET will extend the reach to much higher redshifts ($z \geq 10$). In this subsection, we will explore how ET, can effectively distinguish between PBHs and ABHs based on their distinct event rate distributions at high redshifts.

We have seen in Eq. (25.3) that the merger rate of PBH binaries increases with redshift z, specifically $\mathrm{R}(z) \propto t(z)^{-34/37}$ [50]. This relationship is independent of

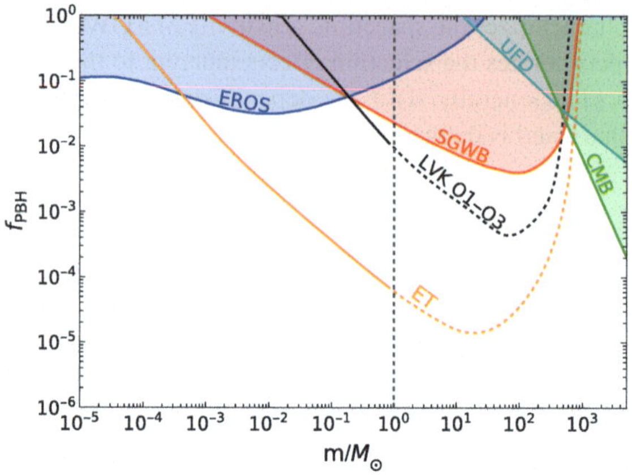

Fig. 25.8 Constraints on the PBH abundance, f_{PBH}, characterized by a monochromatic mass distribution. These constraints are from both the non-detection of SGWBs during LVK O1-O3 runs and the null result of targeted searches for PBH systems. The vertical gray line at $1M_\odot$ signifies that the constraints from the targeted search are applicable solely to sub-solar mass PBHs, as it remains uncertain whether any of the BBHs detected by LVC are composed of PBH binaries. The black and orange curves are the results of the targeted search from the LVK O1-O3 runs and the ET, respectively, assuming an observation time of one year for ET. The red curve represents the updated upper bound of f_{PBH} constrained by the non-detection of SGWBs in LIGO O1-O3 searches. Additionally, results from other experiments are included: EROS/MACHO microlensing (EROS) [128], dynamical heating of ultra-faint dwarf galaxies (UFD) [39], and accretion constraints derived from CMB observations [20, 22, 56, 77, 110]

both the abundance and mass function of PBHs. In contrast, the merger rate of ABH binaries is thought to mimic the global star formation rate history; an initially rise with z, peaking at a low redshift, and then rapidly declining as described by Eq. (25.22). A comparison of the merger rates of PBH binaries and ABH binaries as a function of redshift is illustrated in Fig. 25.9. The distinct features of the redshift distribution of the merger rates may be used to quantify the constraints on f_{PBH} [105].

The redshift-dependent observable events' number density of a GW detector can be calculated using

$$\frac{\mathrm{d}N_{\mathrm{exp}}}{\mathrm{d}z} = \int \mathrm{d}m_1 \mathrm{d}m_2\, \mathcal{R}(z, m_1, m_2)\, \frac{\mathrm{d}VT}{\mathrm{d}z}, \tag{25.33}$$

where $\frac{\mathrm{d}VT}{\mathrm{d}z}$ is defined in Eq. (25.30). Integrating over redshift z yields the total number of observable events, N_{exp},

$$N_{\mathrm{exp}} = \int \mathrm{d}z \frac{\mathrm{d}N_{\mathrm{exp}}}{\mathrm{d}z}. \tag{25.34}$$

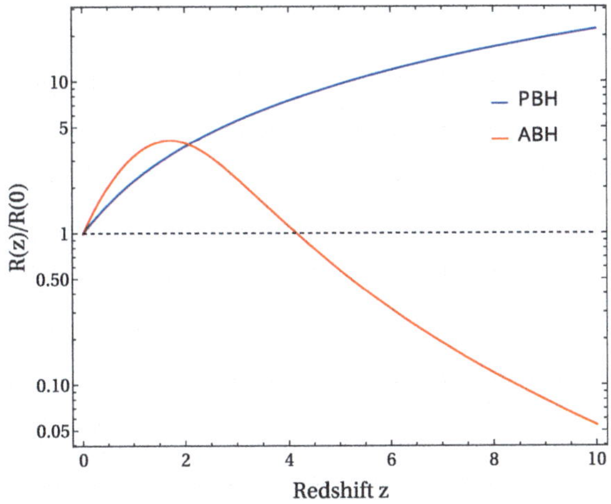

Fig. 25.9 Merger rates of PBH binaries and ABH binaries as a function of redshift. The redshift dependence of the ABH binary merger rate follows Eq. (25.22), while the PBH binary merger rate is proportional to $t(z)^{-34/37}$. For a comparison with observational data, Fig. 25.4 shows the best-fit merger rate densities out to $z = 1.4$, assuming an ABH model based on the Madau-Dickinson global star formation history [93]

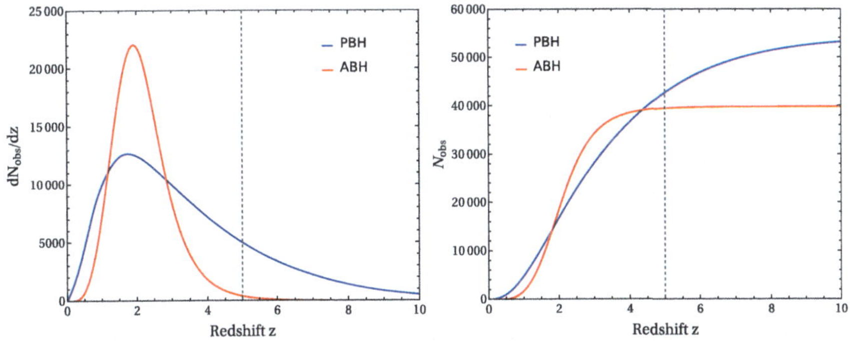

Fig. 25.10 **Left panel**: Redshift distribution of the expected number of observable BBHs, $\mathrm{d}N_{\mathrm{exp}}/\mathrm{d}z$, for ET. We have fixed the model parameters to their best-fit values from Sect. 25.3. **Right panel**: Redshift distribution of the total number of observable BBHs, N_{exp}, for ET

It's worth noting that Eq. (25.29) is a special case of Eq. (25.34) when $P(m_1) = P(m_2) = \delta(m)$. Figure 25.10 illustrates the expected number of observable BBHs $\mathrm{d}N_{\mathrm{exp}}/\mathrm{d}z$ and the total number of observable BBHs N_{exp}, as a function of redshift for ET.

Third-generation GW detectors like ET are anticipated to detect $\mathcal{O}(10^5)$ BBH mergers annually and extend their reach to much higher redshifts. The distinct redshift distributions of $\mathrm{d}N_{\mathrm{exp}}/\mathrm{d}z$ for PBH binaries and ABH binaries provide a complementary tool to distinguish between these two BBH formation models. Particularly, for ABH models, the contribution to the expected number of observable BBHs from high redshifts ($z > 5$) may be negligible, leading to the total number of observable events, $N_{\rm exp}$, approaching a constant at $z > 5$. However, for PBH binaries, the contribution from higher redshifts cannot be ignored, causing $N_{\rm exp}$ to continue increasing when $z > 5$. Therefore, an "excess" of the total number of observable events after redshift $z = 5$ could potentially indicate a population of PBHs as shown in Fig. 25.10.

It is crucial to highlight a fundamental difference between PBH and ABH merger scenarios when considering the next-generation forecast. PBH mergers are thought to have formed in the early Universe, where the environment and physical principles are arguably cleaner and better understood compared to the complex astrophysical environments in which ABH mergers occur. As a result, theoretical predictions of the merger rate history for PBH scenarios are generally more robust and subject to fewer uncertainties. In contrast, ABH mergers are strongly influenced by the formation history of their stellar progenitors and the associated astrophysical environments, which are still poorly constrained. This leads to significant uncertainties in the predicted merger rate history for ABH scenarios. Therefore, a comprehensive analysis that takes into account both the redshift distribution and the associated uncertainties in the merger rate predictions will be essential to effectively distinguish between PBH and ABH models using third-generation GW detectors like ET.

25.5 Summary

In summary, our exploration of the origins of BBH events detected by the LVK collaboration has led us to consider PBHs as potential progenitors, given their candidacy as dark matter constituents. Through a Bayesian analysis utilizing the LVK GWTC-3 catalog, we find that the stellar-mass PBHs are unlikely to constitute the entirety of CDM.

Our investigation extends to a mixed population of ABHs and PBHs, revealing that approximately one-fourth of the detectable events in GWTC-3 may be attributed to PBH binaries. This nuanced understanding of the BBH population sheds light on the coexistence of different astrophysical sources.

Looking ahead, we anticipate the third-generation GW detector, such as the ET, to provide valuable insights. By forecasting detectable event rate distributions for PBH and ABH binaries, this advanced detector offers a potential avenue to distinguish between PBHs and ABHs based on their distinct redshift evolutions. This promises to be a crucial step in refining our understanding of the various sources contributing to the rich tapestry of GW signals observed by GW detectors.

Acknowledgements ZCC is supported by the National Natural Science Foundation of China (Grant No. 12247176 and No. 12247112) and the innovative research group of Hunan Province under Grant No. 2024JJ1006. AH is supported by a Royal Society University Research Fellowship. For the purpose of open access, the author has applied a Creative Commons Attribution (CC BY) license to any Author Accepted Manuscript version arising from this submission.

References

1. B.P. Abbott et al., Binary black hole population properties inferred from the first and second observing runs of advanced Ligo and advanced Virgo. Astrophys. J. Lett. **882**(2), L24 (2019). https://doi.org/10.3847/2041-8213/ab3800. arXiv: 1811.12940 [astro-ph.HE]
2. B.P. Abbott et al., GWTC-1: a gravitational-wave transient catalog of compact binary mergers observed by LIGO and Virgo during the first and second observing runs. Phys. Rev. X **9**(3), 031040 (2019.https://doi.org/10.1103/PhysRevX.9.031040. arXiv: 1811.12907 [astro-ph.HE]
3. B.P. Abbott et al., Supplement: the rate of binary black hole mergers inferred from advanced LIGO observations surrounding GW150914. Astrophys. J. Suppl. **227**(2), 14 (2016). https://doi.org/10.3847/0067-0049/227/2/14. arXiv: 1606.03939 [astro-ph.HE]
4. B. Abbott et al., The rate of binary black hole mergers inferred from advanced LIGO observations surrounding GW150914. Astrophys. J. Lett. **833**(1), L1 (2016). https://doi.org/10.3847/2041-8205/833/1/L1. arXiv: 1602.03842 [astro-ph.HE]
5. R. Abbott et al., Constraints on the cosmic expansion history from GWTC–3. Astrophys. J. **949**(2), 76 (2023). https://doi.org/10.3847/1538-4357/ac74bb. arXiv: 2111.03604 [astro-ph.CO]
6. R. Abbott et al., GW190521: a binary black hole merger with a total mass of $150 M_{\odot}$. Phys. Rev. Lett. **125**(10), 101102 (2020). https://doi.org/10.1103/PhysRevLett.125.101102. arXiv: 2009.01075 [gr-qc]
7. R. Abbott et al., GW190814: gravitational waves from the coalescence of a 23 solar mass black hole with a 2.6 solar mass compact object. Astrophys. J. Lett. **896**(2), (2020), p. L44. https://doi.org/10.3847/2041-8213/ab960f. arXiv: 2006.12611 [astro-ph.HE]
8. R. Abbott et al., GWTC-2: compact binary coalescences observed by LIGO and Virgo during the first half of the third observing run. Phys. Rev. X **11**, 021053 (2021). https://doi.org/10.1103/PhysRevX.11.021053. arXiv: 2010.14527 [gr-qc]
9. R. Abbott et al., GWTC-3: compact binary coalescences observed by LIGO and Virgo during the second part of the third observing run. arXiv: 2111.03606 [gr-qc]
10. R. Abbott et al., GWTC-3: compact binary coalescences observed by LIGO and Virgo during the second part of the third observing run. Phys. Rev. X **13**(4), 041039 (2023). https://doi.org/10.1103/PhysRevX.13.041039. arXiv: 2111.03606 [gr-qc]
11. R. Abbott et al., GWTC-3: compact binary coalescences observed by LIGO and Virgo during the second part of the third observing run — O3 search sensitivity estimates. https://doi.org/10.5281/zenodo.5546676
12. R. Abbott et al., Population of merging compact binaries inferred using gravitational waves through GWTC-3. Phys. Rev. X **13**(1), 011048 (2023). https://doi.org/10.1103/PhysRevX.13.011048. arXiv: 2111.03634 [astro-ph.HE]
13. R. Abbott et al., Population properties of compact objects from the second LIGO-Virgo gravitational-wave transient catalog. Astrophys. J. Lett. **913**(1), L7 (2021). https://doi.org/10.3847/2041-8213/abe949. arXiv: 2010.14533 [astro-ph.HE]
14. R. Abbott et al., Properties and astrophysical implications of the 150 $M_{\odot}$ binary black hole merger GW190521. Astrophys. J. Lett. **900**(1), L13 (2020). https://doi.org/10.3847/2041-8213/aba493. arXiv: 2009.01190 [astro-ph.HE]
15. R. Abbott et al., The population of merging compact binaries inferred using gravitational waves through GWTC-3 - Data release. Version v1. https://doi.org/10.5281/zenodo.5655785
16. R. Abbott et al., Search for intermediate-mass black hole binaries in the third observing run of Advanced LIGO and Advanced Virgo. Astron. Astrophys. **659**, A84 (2022). https://doi.org/10.1051/0004-6361/202141452. arXiv: 2105.15120 [astro-ph.HE]
17. G. Agazie et al., The NANOGrav 15 yr data set: evidence for a gravitational-wave background. Astrophys. J. Lett. **951**(1), L8 (2023). https://doi.org/10.3847/2041-8213/acdac6. arXiv: 2306.16213 [astro-ph.HE]

18. N. Aghanim et al., Planck 2018 results. VI. Cosmological parameters. Astron. Astrophys. **641**, A6 (2020). [Erratum: Astron.Astrophys. 652, C4 (2021)], A6. https://doi.org/10.1051/0004-6361/201833910. arXiv: 1807.06209 [astro-ph.CO]
19. P. Ajith et al., Inspiral-merger-ringdown waveforms for black-hole binaries with non-precessing spins. Phys. Rev. Lett. **106**, 241101 (2011). https://doi.org/10.1103/PhysRevLett.106.241101. arXiv: 0909.2867 [gr-qc]
20. Y. Ali-Haïmoud, M. Kamionkowski, Cosmic microwave background limits on accreting primordial black holes. Phys. Rev. D **95**(4), 043534 (2017). https://doi.org/10.1103/PhysRevD.95.043534. arXiv: 1612.05644 [astro-ph.CO]
21. Y. Ali-Haïmoud, Ely D. Kovetz, M. Kamionkowski . Merger rate of primordial black-hole binaries. Phys. Rev. D **96**(12), 123523 (2017). https://doi.org/10.1103/PhysRevD.96.123523. arXiv: 1709.06576 [astro-ph.CO]
22. D. Aloni , K. Blum, R. Flauger, Cosmic microwave background constraints on primordial black hole dark matter. JCAP 05 (2017), p. 017. https://doi.org/10.1088/1475-7516/2017/05/017. arXiv: 1612.06811 [astro-ph.CO]
23. O. Anagnostou , M. Trenti, A. Melatos, Repeated mergers of black hole binaries: implications for GW190521. Astrophys. J. **941**(1), 4 (2022). https://doi.org/10.3847/1538-4357/ac9d95. arXiv: 2010.06161 [astro-ph.HE]
24. J. Antoniadis et al., The second data release from the European Pulsar Timing Array III. Search for gravitational wave signals. Astron. Astrophys. 678 (2023), A50. https://doi.org/10.1051/0004-6361/202346844. arXiv: 2306.16214 [astro-ph.HE]
25. F. Antonini , M. Gieles, A. Gualandris, Black hole growth through hierarchical black hole mergers in dense star clusters: implications for gravitational wave detections. Mon. Not. Roy. Astron. Soc. 486.4 (2019), pp. 5008–5021. https://doi.org/10.1093/mnras/stz1149. arXiv: 1811.03640 [astro-ph.HE]
26. G. Ashton et al., BILBY: a user-friendly Bayesian inference library for gravitational-wave astronomy. Astrophys. J. Suppl. **241**(2), 27 (2019). https://doi.org/10.3847/1538-4365/ab06fc. arXiv: 1811.02042 [astro-ph.IM]
27. E. Bagui et al., Primordial black holes and their gravitational-wave signatures. arXiv: 2310.19857 [astro-ph.CO]
28. E.D. Barr et al., A pulsar in a binary with a compact object in the mass gap between neutron stars and black holes. Science **383**(6680), 275–279 (2024). https://doi.org/10.1126/science.adg3005. arXiv: 2401.09872 [astro-ph.HE]
29. N. Bartolo et al., Primordial black hole dark matter: LISA serendipity. Phys. Rev. Lett. **122**(21), 211301 (2019). https://doi.org/10.1103/PhysRevLett.122.211301. arXiv: 1810.12218 [astro-ph.CO]
30. N. Bartolo et al., Testing primordial black holes as dark matter with LISA. Phys. Rev. D **99**(10), 103521 (2019). https://doi.org/10.1103/PhysRevD.99.103521. arXiv: 1810.12224 [astro-ph.CO]
31. Simone S. Bavera et al., The impact of mass-transfer physics on the observable properties of field binary black hole populations. Astron. Astrophys. **647**, A153 (2021). https://doi.org/10.1051/0004-6361/202039804. arXiv: 2010.16333 [astro-ph.HE]
32. R. Bean, J. Magueijo, Could supermassive black holes be quintessential primordial black holes? Phys. Rev. D **66**, 063505 (2002). https://doi.org/10.1103/PhysRevD.66.063505. (arXiv: astro-ph/0204486)
33. K. Belczynski et al., Evolutionary roads leading to low effective spins, high black hole masses, and O1/O2 rates for LIGO/Virgo binary black holes. Astron. Astrophys. **636**, A104 (2020). https://doi.org/10.1051/0004-6361/201936528. arXiv: 1706.07053 [astro-ph.HE]
34. K. Belczynski et al., The effect of pair-instability mass loss on black hole mergers. Astron. Astrophys. **594**, A97 (2016). https://doi.org/10.1051/0004-6361/201628980. arXiv: 1607.03116 [astro-ph.HE]
35. K. Belczynski et al., Missing black holes unveil the supernova explosion mechanism. Astrophys. J. 757 (2012), p. 91. https://doi.org/10.1088/0004-637X/757/1/91. arXiv: 1110.1635 [astro-ph.GA]

36. S. Bhagwat et al., The importance of priors on LIGO-Virgo parameter estimation: the case of primordial black holes. JCAP 01 (2021), p. 037. https://doi.org/10.1088/1475-7516/2021/01/037. arXiv: 2008.12320 [astro-ph.CO]
37. S. Bird et al., Did LIGO detect dark matter? Phys. Rev. Lett. **116**(20), 201301 (2016). https://doi.org/10.1103/PhysRevLett.116.201301. arXiv: 1603.00464 [astro-ph.CO]
38. R. Biswas et al., The loudest event statistic: general formulation, properties and applications. Class. Quant. Grav. 26 (2009). [Erratum: Class.Quant.Grav. 30, 079502 (2013)], p. 175009. https://doi.org/10.1088/0264-9381/26/17/175009. arXiv: 0710.0465 [gr-qc]
39. T.D. Brandt, Constraints on MACHO dark matter from compact stellar systems in ultra-faint dwarf galaxies. Astrophys. J. Lett. 824.2 (2016), p. L31. https://doi.org/10.3847/2041-8205/824/2/L31. arXiv: 1605.03665 [astro-ph.GA]
40. K. Breivik et al., COSMIC variance in binary population synthesis. Astrophys. J. **898**(1), 71 (2020). https://doi.org/10.3847/1538-4357/ab9d85. arXiv: 1911.00903 [astro-ph.HE]
41. R-g. Cai, S. Pi, M. Sasaki, Gravitational waves induced by non-gaussian scalar perturbations. Phys. Rev. Lett. **122**(20), 201101 (2019). https://doi.org/10.1103/PhysRevLett.122.201101. arXiv: 1810.11000 [astro-ph.CO]
42. T. Callister et al., Shouts and murmurs: combining individual gravitational-wave sources with the stochastic background to measure the history of binary black hole mergers. Astrophys. J. Lett. **896**(2), L32 (2020). https://doi.org/10.3847/2041-8213/ab9743. arXiv: 2003.12152 [astro-ph.HE]
43. B.J. Carr et al., Constraints on primordial black holes from the Galactic gamma-ray background. Phys. Rev. D **94**(4), 044029 (2016). https://doi.org/10.1103/PhysRevD.94.044029. arXiv: 1604.05349 [astro-ph.CO]
44. B. Carr, The Primordial black hole mass spectrum. Astrophys. J. **201**, 1–19 (1975). https://doi.org/10.1086/153853
45. B. Carr, F. Kühnel, M. Sandstad, Primordial black holes as dark matter. Phys. Rev. D **94**(8), 083504 (2016). https://doi.org/10.1103/PhysRevD.94.083504
46. B. Carr et al., Primordial black hole constraints for extended mass functions. Phys. Rev. D **96**(2), 023514 (2017). https://doi.org/10.1103/PhysRevD.96.023514. arXiv: 1705.05567 [astro-ph.CO]
47. B.J. Carr, S.W. Hawking, Black holes in the early Universe. Mon. Not. Roy. Astron. Soc. **168**, 399–415 (1974). https://doi.org/10.1093/mnras/168.2.399
48. Z-C. Chen , F. Huang, Q-G. Huang, Stochastic gravitational-wave background from binary black holes and binary neutron stars and implications for LISA. Astrophys. J. **871**(1), 97 (2019). https://doi.org/10.3847/1538-4357/aaf581. arXiv: 1809.10360 [gr-qc]
49. Z-C. Chen , F. Huang, Q-G. Huang, Distinguishing primordial black holes from astrophysical black holes by einstein telescope and cosmic explorer. JCAP **08**, 039 (2020). https://doi.org/10.1088/1475-7516/2020/08/039. arXiv: 1904.02396 [astro-ph.CO]
50. Z-C. Chen , F. Huang, Q-G. Huang, Merger rate distribution of primordial-black-hole binaries. Astrophys. J. **864**(1), 61 (2018). https://doi.org/10.3847/1538-4357/aad6e2. arXiv: 1801.10327 [astro-ph.CO]
51. Z-C. Chen , L. Liu, Is PSR J0514E in a PBH-NS binary? (2024). arXiv: 2401.12889 [astro-ph.HE]
52. Z-C. Chen , C. Yuan, Q-G. Huang, Confronting the primordial black hole scenario with the gravitational-wave events detected by LIGO-Virgo. Phys. Lett. B **829**, 137040 (2022). https://doi.org/10.1016/j.physletb.2022.137040. arXiv: 2108.11740 [astro-ph.CO]
53. Z-C. Chen , C. Yuan, Q-G. Huang, Pulsar Timing Array Constraints on Primordial Black Holes with NANOGrav 11-Year Dataset Phys. Rev. Lett. **124**(25), 251101 (2020). https://doi.org/10.1103/PhysRevLett.124.251101. arXiv: 1910.12239 [astro-ph.CO]
54. Z-C. Chen et al., Constraints on primordial-black-hole population and cosmic expansion history from GWTC-3. JCAP **03**, 024 (2023). https://doi.org/10.1088/1475-7516/2023/03/024. arXiv: 2205.11278 [astro-ph.CO]
55. Z-C. Chen et al., Probing the speed of scalar-induced gravitational waves with pulsar timing arrays (2024). arXiv: 2401.09818 [gr-qc]

56. L. Chen, Q-G. Huang, K. Wang, Constraint on the abundance of primordial black holes in dark matter from Planck data. JCAP **12**, 044 (2016). https://doi.org/10.1088/1475-7516/2016/12/044. arXiv: 1608.02174 [astro-ph.CO]
57. V. De Luca, G. Franciolini, A. Riotto, On the primordial black hole mass function for broad spectra. Phys. Lett. B **807**, 135550 (2020). https://doi.org/10.1016/j.physletb.2020.135550. arXiv: 2001.04371 [astro-ph.CO]
58. V. De Luca et al., Bayesian evidence for both astrophysical and primordial black holes: mapping the GWTC-2 catalog to third-generation detectors. JCAP **05**, 003 (2021). https://doi.org/10.1088/1475-7516/2021/05/003. arXiv: 2102.03809 [astro-ph.CO]
59. V. De Luca et al., On the gauge invariance of cosmological gravitational waves. JCAP **03**, 014 (2020). https://doi.org/10.1088/1475-7516/2020/03/014. arXiv: 1911.09689 [gr-qc]
60. H. Deng, A possible mass distribution of primordial black holes implied by LIGO-Virgo. JCAP **04**, 058 (2021). https://doi.org/10.1088/1475-7516/2021/04/058. arXiv: 2101.11098 [astro-ph.CO]
61. U.N. Di Carlo et al., Binary black holes in young star clusters: the impact of metallicity. Mon. Not. Roy. Astron. Soc. **498**(1), 495–506 (2020). https://doi.org/10.1093/mnras/staa2286. arXiv: 2004.09525 [astro-ph.HE]
62. A. Dolgov, J. Silk, Baryon isocurvature fluctuations at small scales and baryonic dark matter. Phys. Rev. D **47**, 4244–4255 (1993). https://doi.org/10.1103/PhysRevD.47.4244
63. R. Farmer et al., Mind the gap: The location of the lower edge of the pair instability supernovae black hole mass gap (2019). https://doi.org/10.3847/1538-4357/ab518b. arXiv: 1910.12874 [astro-ph.SR]
64. R. Farmer et al., Constraints from gravitational wave detections of binary black hole mergers on the $^{12}C\,(\alpha,\gamma)^{16}O$ rate. Astrophys. J. Lett. **902**(2), L36 (2020). https://doi.org/10.3847/2041-8213/abbadd. arXiv: 2006.06678 [astro-ph.HE]
65. G. Franciolini et al., From inflation to black hole mergers and back again: Gravitational-wave data-driven constraints on inflationary scenarios with a first-principle model of primordial black holes across the QCD epoch. Phys. Rev. D **106**(12), 123526 (2022). https://doi.org/10.1103/PhysRevD.106.123526. arXiv: 2209.05959 [astro-ph.CO]
66. G. Franciolini et al., Searching for a subpopulation of primordial black holes in LIGO-Virgo gravitational-wave data. Phys. Rev. D **105**(8), 083526 (2022). https://doi.org/10.1103/PhysRevD.105.083526. arXiv: 2105.03349 [gr-qc]
67. C.L. Fryer et al., Compact Remnant Mass Function: Dependence on the Explosion Mechanism and Metallicity. Astrophys. J. **749**, 91 (2012). https://doi.org/10.1088/0004-637X/749/1/91. arXiv: 1110.1726 [astro-ph.SR]
68. D. Gerosa , M. Fishbach, Hierarchical mergers of stellar-mass black holes and their gravitational-wave signatures. Nature Astron. **5**(8), 749–760 (2021). https://doi.org/10.1038/s41550-021-01398-w. arXiv: 2105.03439 [astro-ph.HE]
69. D. Gerosa, S. Vitale, E. Berti, Astrophysical implications of GW190412 as a remnant of a previous black-hole merger. Phys. Rev. Lett. **125**(10), 101103 (2020). https://doi.org/10.1103/PhysRevLett.125.101103. arXiv: 2005.04243 [astro-ph.HE]
70. N. Giacobbo , M. Mapelli, M. Spera, Merging black hole binaries: the effects of progenitor's metallicity, mass-loss rate and Eddington factor. Mon. Not. Roy. Astron. Soc. **474**(3), 2959–2974 (2018). https://doi.org/10.1093/mnras/stx2933. arXiv: 1711.03556 [astro-ph.SR]
71. A.D. Gow, C.T. Byrnes, A. Hall, Accurate model for the primordial black hole mass distribution from a peak in the power spectrum. Phys. Rev. D **105**(2), 023503 (2022). https://doi.org/10.1103/PhysRevD.105.023503. arXiv: 2009.03204 [astro-ph.CO]
72. A.M. Green, Microlensing and dynamical constraints on primordial black hole dark matter with an extended mass function. Phys. Rev. D **94**(6), 063530 (2016). https://doi.org/10.1103/PhysRevD.94.063530. arXiv: 1609.01143 [astro-ph.CO]
73. A. Hall, A.D. Gow, C.T. Byrnes, Bayesian analysis of LIGO-Virgo mergers: Primordial vs. astrophysical black hole populations. Phys. Rev. D **102**, 123524 (2020). https://doi.org/10.1103/PhysRevD.102.123524. arXiv: 2008.13704 [astro-ph.CO]

74. S. Hawking, Gravitationally collapsed objects of very low mass. Mon. Not. Roy. Astron. Soc. **152**, 75 (1971)
75. S. Hild et al., Sensitivity Studies for Third-Generation Gravitational Wave Observatories. Class. Quant. Grav. **28**, 094013 (2011). https://doi.org/10.1088/0264-9381/28/9/094013. arXiv: 1012.0908 [gr-qc]
76. S. Hild, S. Chelkowski, A. Freise, Pushing towards the ET sensitivity using 'conventional' technology (2008). arXiv: 0810.0604 [gr-qc]
77. B. Horowitz, Revisiting primordial black holes constraints from ionization history (2016). arXiv: 1612.07264 [astro-ph.CO]
78. G. Hütsi et al., Two populations of LIGO-Virgo black holes. JCAP **03**, 068 (2021). https://doi.org/10.1088/1475-7516/2021/03/068. arXiv: 2012.02786 [astro-ph.CO]
79. J-H. Jin et al., Confronting sound speed resonance with pulsar timing arrays (2023). arXiv: 2307.08687 [astro-ph.CO]
80. K. Kannike et al., Single Field Double Inflation and Primordial Black Holes. JCAP **09**, 020 (2017). https://doi.org/10.1088/1475-7516/2017/09/020. arXiv: 1705.06225 [astro-ph.CO]
81. R.E. Kass, A.E. Raftery, Bayes factors. J. Am. Stat. Assoc. **90**(430), 773–795 (1995). https://doi.org/10.1080/01621459.1995.10476572
82. B.J. Kavanagh, D. Gaggero, G. Bertone, Merger rate of a subdominant population of primordial black holes. Phys. Rev. D **98**(2), 023536 (2018). https://doi.org/10.1103/PhysRevD.98.023536. arXiv: 1805.09034 [astro-ph.CO]
83. M. Kawasaki , A. Kusenko , T.T. Yanagida, Primordial seeds of supermassive black holes. Phys. Lett. B **711**, 1–5 (2012). https://doi.org/10.1016/j.physletb.2012.03.056. arXiv: 1202.3848 [astro-ph.CO]
84. C. Kimball et al., Evidence for Hierarchical Black Hole Mergers in the Second LIGO–Virgo Gravitational Wave Catalog. Astrophys. J. Lett. **915**(2), L35 (2021). https://doi.org/10.3847/2041-8213/ac0aef. arXiv: 2011.05332 [astro-ph.HE]
85. K. Kremer et al., Populating the upper black hole mass gap through stellar collisions in young star clusters. Astrophys. J. **903**(1), 45 (2020). https://doi.org/10.3847/1538-4357/abb945. arXiv: 2006.10771 [astro-ph.HE]
86. L. Liu, Z-C. Chen, Q-G. Huang, Implications for the non-Gaussianity of curvature perturbation from pulsar timing arrays (2023). arXiv: 2307.01102 [astro-ph.CO]
87. L. Liu, Z-C. Chen, Q-G. Huang, Probing the equation of state of the early Universe with pulsar timing arrays (2023). arXiv: 2307.14911 [astro-ph.CO]
88. L. Liu, Z-K. Guo, R-G. Cai, Effects of the merger history on the merger rate density of primordial black hole binaries. Eur. Phys. J. C **79**(8), 717 (2019). https://doi.org/10.1140/epjc/s10052-019-7227-0. arXiv: 1901.07672 [astro-ph.CO]
89. L. Liu, Z-K. Guo, R-G. Cai, Effects of the surrounding primordial black holes on the merger rate of primordial black hole binaries. Phys. Rev. D **99**(6), 063523 (2019). https://doi.org/10.1103/PhysRevD.99.063523. arXiv: 1812.05376 [astro-ph.CO]
90. L. Liu, Y. Wu, Z-C. Chen, Simultaneously probing the sound speed and equation of state of the early Universe with pulsar timing arrays (2023). arXiv: 2310.16500 [astro-ph.CO]
91. L. Liu et al., Constraining the merger history of primordial-black-hole binaries from GWTC-3. Phys. Rev. D **107**(6), 063035 (2023). https://doi.org/10.1103/PhysRevD.107.063035. arXiv: 2210.16094 [astro-ph.CO]
92. T.J. Loredo, Accounting for source uncertainties in analyses of astronomical survey data. AIP Conf. Proc. **735**(1), 195–206 (2004). Ed. by Rainer Fischer, Roland Preuss, and Udo von Toussaint. https://doi.org/10.1063/1.1835214. arXiv: astro-ph/0409387
93. P. Madau, M. Dickinson, Cosmic Star Formation History. Ann. Rev. Astron. Astrophys. **52**, 415–486 (2014). https://doi.org/10.1146/annurev-astro-081811-125615. arXiv: 1403.0007 [astro-ph.CO]
94. I. Mandel, F.S. Broekgaarden, Rates of compact object coalescences. Living Rev. Rel. **25**(1), 1 (2022). https://doi.org/10.1007/s41114-021-00034-3. arXiv: 2107.14239 [astro-ph.HE]

95. I. Mandel, A. Farmer, Merging stellar-mass binary black holes. Phys. Rept. **955**, 1–24 (2022). https://doi.org/10.1016/j.physrep.2022.01.003. arXiv: 1806.05820 [astro-ph.HE]
96. I. Mandel, W.M. Farr, J.R. Gair, Extracting distribution parameters from multiple uncertain observations with selection biases. Mon. Not. Roy. Astron. Soc. **486**(1), 1086–1093 (2019). https://doi.org/10.1093/mnras/stz896. arXiv: 1809.02063 [physics.data-an]
97. M. Mapelli, *Formation Channels of Single and Binary Stellar-mass Black Holes* (Springer, 2020), pp. 1–65
98. M. Mapelli et al., Impact of the Rotation and Compactness of Progenitors on the Mass of Black Holes. Astrophys. J. 888 (2020), p. 76. https://doi.org/10.3847/1538-4357/ab584d. arXiv: 1909.01371 [astro-ph.HE]
99. P. Marchant, T. Moriya, The impact of stellar rotation on the black hole mass-gap from pair-instability supernovae. Astron. Astrophys. **640**, L18 (2020). https://doi.org/10.1051/0004-6361/202038902. arXiv: 2007.06220 [astro-ph.HE]
100. P. Marchant et al., Pulsational pair-instability supernovae in very close binaries (2018)https://doi.org/10.3847/1538-4357/ab3426. arXiv: 1810.13412 [astro-ph.HE]
101. S. Mastrogiovanni et al., On the importance of source population models for gravitational-wave cosmology. Phys. Rev. D **104**(6), 062009 (2021). https://doi.org/10.1103/PhysRevD.104.062009. arXiv: 2103.14663 [gr-qc]
102. A. Matas et al., ligned-spin neutron-star–black-hole waveform model based on the effective-one-body approach and numerical-relativity simulations. Phys. Rev. D **102**(4), 043023 (2020). https://doi.org/10.1103/PhysRevD.102.043023. arXiv: 2004.10001 [gr-qc]
103. S. Mukherjee, J. Silk, Can we distinguish astrophysical from primordial black holes via the stochastic gravitational wave background? Mon. Not. Roy. Astron. Soc. **506**(3), 3977–3985 (2021). https://doi.org/10.1093/mnras/stab1932. arXiv: 2105.11139 [gr-qc]
104. T. Nakamura et al., Gravitational waves from coalescing black hole MACHO binaries. Astrophys. J. Lett. **487**, L139–L142 (1997). https://doi.org/10.1086/310886. arXiv: astro-ph9708060
105. K.K.Y. Ng et al., Constraining high-redshift stellar-mass primordial black holes with next-generation ground-based gravitational-wave detectors. Astrophys. J. Lett. **933**(2), L41 (2022). https://doi.org/10.3847/2041-8213/ac7aae. arXiv: 2204.11864 [astro-ph.CO]
106. J.C. Niemeyer, K. Jedamzik, Near-critical gravitational collapse and the initial mass function of primordial black holes. Phys. Rev. Lett. **80**, 5481–5484 (1998). https://doi.org/10.1103/PhysRevLett.80.5481. (arXiv: astro-ph/9709072)
107. R. O'Shaughnessy, V. Kalogera, K. Belczynski, Binary compact object coalescence rates: the role of elliptical galaxies. Astrophys. J. **716**, 615–633 (2010). https://doi.org/10.1088/0004-637X/716/1/615. arXiv: 0908.3635 [astro-ph.CO]
108. J.R. Oppenheimer, G.M. Volkoff, On massive neutron cores. Phys. Rev. **55**, 374–381 (1939). https://doi.org/10.1103/PhysRev.55.374
109. S. Ossokine et al., Multipolar effective-one-body waveforms for precessing binary black holes: construction and validation. Phys. Rev. D **102**(4), 044055 (2020). https://doi.org/10.1103/PhysRevD.102.044055. arXiv: 2004.09442 [gr-qc]
110. V. Poulin et al., CMB bounds on disk-accreting massive primordial black holes. Phys. Rev. D **96**(8), 083524 (2017). https://doi.org/10.1103/PhysRevD.96.083524. arXiv: 1707.04206 [astro-ph.CO]
111. G. Pratten et al., Computationally efficient models for the dominant and subdominant harmonic modes of precessing binary black holes. Phys. Rev. D **103**, 104056 (2021). https://doi.org/10.1103/PhysRevD.103.104056. arXiv: 2004.06503 [gr-qc]
112. M. Punturo et al., The Einstein Telescope: A third-generation gravitational wave observatory. Class. Quant. Grav. **27**, 194002 (2010). Ed. by Fulvio Ricci. https://doi.org/10.1088/0264-9381/27/19/194002
113. M. Raidal et al., Formation and evolution of primordial black hole binaries in the early Universe. JCAP **02**, 018 (2019). https://doi.org/10.1088/1475-7516/2019/02/018. arXiv: 1812.01930 [astro-ph.CO]

114. D.J. Reardon et al., Search for an isotropic gravitational-wave background with the parkes pulsar timing array. Astrophys. J. Lett. **951**(1), L6 (2023). https://doi.org/10.3847/2041-8213/acdd02. arXiv: 2306.16215 [astro-ph.HE]
115. T. Regimbau et al., Digging deeper: observing primordial gravitational waves below the binary black hole produced stochastic background. Phys. Rev. Lett. **118**(15), 151105 (2017). https://doi.org/10.1103/PhysRevLett.118.151105. arXiv: 1611.08943 [astro-ph.CO]
116. L. Rezzolla , E.R. Most, L.R. Weih, Using gravitational-wave observations and quasi-universal relations to constrain the maximum mass of neutron stars. Astrophys. J. Lett. **852**(2), L25 (2018). https://doi.org/10.3847/2041-8213/aaa401. arXiv: 1711.00314 [astro-ph.HE]
117. C.L. Rodriguez, A. Loeb, Redshift evolution of the black hole merger rate from globular clusters. Astrophys. J. Lett. **866**(1), L5 (2018). https://doi.org/10.3847/2041-8213/aae377. arXiv: 1809.01152 [astro-ph.HE]
118. C.L. Rodriguez et al., Black holes: The next generation—repeated mergers in dense star clusters and their gravitational-wave properties. Phys. Rev. D **100**(4), 043027 (2019). https://doi.org/10.1103/PhysRevD.100.043027. arXiv: 1906.10260 [astro-ph.HE]
119. I.M. Romero-Shaw et al., Bayesian inference for compact binary coalescences with bilby: validation and application to the first LIGO–Virgo gravitational-wave transient catalogue. Mon. Not. Roy. Astron. Soc. **499**(3), 3295–3319 (2020). https://doi.org/10.1093/mnras/staa2850. arXiv: 2006.00714 [astro-ph.IM]
120. J. Roulet, M. Zaldarriaga, Constraints on binary black hole populations from LIGO–Virgo detections. Mon. Not. Roy. Astron. Soc. **484**(3), 4216–4229 (2019). https://doi.org/10.1093/mnras/stz226. arXiv: 1806.10610 [astro-ph.HE]
121. J. Roulet et al., Binary black hole mergers from LIGO/Virgo O1 and O2: population inference combining confident and marginal events. Phys. Rev. D **102**(12), 123022 (2020). https://doi.org/10.1103/PhysRevD.102.123022. arXiv: 2008.07014 [astro-ph.HE]
122. J. Sadiq, T. Dent, D. Wysocki, Flexible and fast estimation of binary merger population distributions with an adaptive kernel density estimator. Phys. Rev. D **105**(12), 123014 (2022). https://doi.org/10.1103/PhysRevD.105.123014. arXiv: 2112.12659 [gr-qc]
123. R. Saito, J. Yokoyama, Gravitational wave background as a probe of the primordial black hole abundance. Phys. Rev. Lett. **102**, 161101 (2009). [Erratum: Phys.Rev.Lett. 107, 069901 (2011)]. https://doi.org/10.1103/PhysRevLett.102.161101. arXiv: 0812.4339 [astro-ph]
124. M. Sasaki et al., Primordial Black Hole Scenario for the Gravitational-Wave Event GW150914. Phys. Rev. Lett. **117**(6), 061101 (2016). [Erratum: Phys.Rev.Lett. 121, 059901 (2018)]. https://doi.org/10.1103/PhysRevLett.117.061101. arXiv: 1603.08338 [astro-ph.CO]
125. J.S. Speagle, dynesty: a dynamic nested sampling package for estimating Bayesian posteriors and evidences. Mon. Not. Roy. Astron. Soc. **493**(3), 3132–3158 (2020). https://doi.org/10.1093/mnras/staa278. arXiv: 1904.02180 [astro-ph.IM]
126. J.E. Thompson et al., Modeling the gravitational wave signature of neutron star black hole coalescences. Phys. Rev. D **101**(12), 124059 (2020). https://doi.org/10.1103/PhysRevD.101.124059. arXiv: 2002.08383 [gr-qc]
127. E. Thrane , C. Talbot, An introduction to Bayesian inference in gravitational-wave astronomy: parameter estimation, model selection, and hierarchical models. Publ. Astron. Soc. Austral. **36**, e10 (2019). [Erratum: Publ.Astron.Soc.Austral. 37, e036 (2020)]. https://doi.org/10.1017/pasa.2019.2. arXiv: 1809.02293 [astro-ph.IM]
128. P. Tisserand et al., Limits on the macho content of the galactic halo from the EROS-2 survey of the magellanic clouds. Astron. Astrophys. **469**, 387–404 (2007). https://doi.org/10.1051/0004-6361:20066017. arXiv: astro-ph/0607207
129. V. Tiwari, Exploring Features in the Binary Black Hole Population. Astrophys. J. **928**(2), 155 (2022). https://doi.org/10.3847/1538-4357/ac589a. arXiv: 2111.13991 [astro-ph.HE]
130. R.C. Tolman, Static solutions of Einstein's field equations for spheres of fluid. Phys. Rev. **55**, 364–373 (1939). https://doi.org/10.1103/PhysRev.55.364
131. S. Vitale et al., Inferring the properties of a population of compact binaries in presence of selection effects (2020). https://doi.org/10.1007/978-981-15-4702-7-1. arXiv: 2007.05579 [astro-ph.IM]

132. S. Vitale et al., Measuring the star formation rate with gravitational waves from binary black holes. Astrophys. J. Lett. **886**(1), L1 (2019). https://doi.org/10.3847/2041-8213/ab50c0. arXiv: 1808.00901 [astro-ph.HE]
133. K.W.K. Wong et al., Constraining the primordial black hole scenario with Bayesian inference and machine learning: the GWTC-2 gravitational wave catalog. Phys. Rev. D **103**(2), 023023 (2021). https://doi.org/10.1103/PhysRevD.103.023026. arXiv: 2011.01865 [gr-qc]
134. S.E. Woosley, Pulsational pair-instability supernovae. Astrophys. J. **836**(2), 244 (2017). https://doi.org/10.3847/1538-4357/836/2/244. arXiv: 1608.08939 [astro-ph.HE]
135. Y. Wu, Merger history of primordial black-hole binaries. Phys. Rev. D **101**(8), 083008 (2020). https://doi.org/10.1103/PhysRevD.101.083008. arXiv: 2001.03833 [astro-ph.CO]
136. H. Xu et al., Searching for the Nano-Hertz Stochastic Gravitational Wave Background with the Chinese Pulsar Timing Array Data Release I Res. Astron. Astrophys. **23**(7), 075024 (2023). https://doi.org/10.1088/1674-4527/acdfa5. arXiv: 2306.16216 [astro-ph.HE]
137. Z. Yi, Z-Q. You, Y. Wu, Model-independent reconstruction of the primordial curvature power spectrum from PTA data (2023). arXiv: 2308.05632 [astro-ph.CO]
138. J. Yokoyama, Cosmological constraints on primordial black holes produced in the near critical gravitational collapse. Phys. Rev. D **58**, 107502 (1998). https://doi.org/10.1103/PhysRevD.58.107502. arXiv: gr-qc/9804041
139. Z-Q. You, Z. Yi, Y. Wu, Constraints on primordial curvature power spectrum with pulsar timing arrays (2023). arXiv: 2307.04419 [gr-qc]
140. Z-Q. You et al. Constraints on peculiar velocity distribution of binary black holes using gravitational waves with GWTC-3 (2023). arXiv: 2306.12950 [astro-ph.CO]
141. C. Yuan, Z-C. Chen, Q-G. Huang, Log-dependent slope of scalar induced gravitational waves in the infrared regions. Phys. Rev. D **101**(4), 043019 (2020). https://doi.org/10.1103/PhysRevD.101.043019. arXiv: 1910.09099 [astro-ph.CO]
142. C. Yuan, Z-C. Chen, Q-G. Huang, Probing primordial–black-hole dark matter with scalar induced gravitational waves. Phys. Rev. D **100**(8), 081301 (2019). https://doi.org/10.1103/PhysRevD.100.081301. arXiv: 1906.11549 [astro-ph.CO]
143. C. Yuan, Z-C. Chen, Q-G. Huang, Scalar induced gravitational waves in different gauges. Phys. Rev. D **101**(6), 063018 (2020). https://doi.org/10.1103/PhysRevD.101.063018. arXiv: 1912.00885 [astro-ph.CO]
144. M. Zevin et al., Exploring the lower mass gap and unequal mass regime in compact binary evolution. Astrophys. J. Lett. **899**, L1 (2020). https://doi.org/10.3847/2041-8213/aba74e. arXiv: 2006.14573 [astro-ph.HE]
145. M. Zevin et al., One Channel to Rule Them All? Constraining the Origins of Binary Black Holes Using Multiple Formation Pathways. Astrophys. J. **910**(2), 152 (2021). https://doi.org/10.3847/1538-4357/abe40e. arXiv: 2011.10057 [astro-ph.HE]
146. L-M. Zheng et al., Towards a reliable reconstruction of the power spectrum of primordial curvature perturbation on small scales from GWTC-3. Phys. Lett. B **838**, 137720 (2023). https://doi.org/10.1016/j.physletb.2023.137720. arXiv: 2212.05516 [astro-ph.CO]

Chapter 26
LIGO-Virgo-KAGRA Constraints on Primordial Black Holes from Stochastic Gravitational-Wave Background Searches

Alba Romero-Rodríguez and Sachiko Kuroyanagi

Abstract Primordial black holes may have left an imprint in the form of a stochastic gravitational wave background throughout their evolution in the history of the Universe. This chapter highlights two types of stochastic gravitational wave background: those generated by scalar curvature perturbations associated with primordial black hole formation in the early Universe and those composed of ensembles of gravitational waves emitted by primordial black hole binaries. After describing detection methods and a brief introduction on Bayesian inference, we discuss current constraints imposed by LIGO-Virgo-KAGRA observations through the non-detection of the stochastic gravitational wave backgrounds and discuss their physical implications.

26.1 Introduction

A stochastic gravitational wave background (SGWB) offers an exciting opportunity to explore primordial black holes (PBHs). The SGWB is a random and persistent background of gravitational waves (GWs) coming from all directions. It is commonly referred to as 'stochastic' because it can only be characterized statistically. The sources can be either cosmological or astrophysical. The former serves as a unique probe of the primordial Universe since it can propagate directly from the early epoch of inflation. The latter is formed by an ensemble of unresolved astrophysical events,

A. Romero-Rodríguez (✉)
Chargèe de recherche du Centre National de La Recherche Scientifique (CNRS), France, LAPP, IN2P3 (Institut de Physique Nuclèaire et de Physique des Particules), Univ. Savoie Mont Blanc, CNRS, Laboratoire d' Annecy de Physique des Particules - IN2P3, 74940 Annecy, France
e-mail: romero@lapp.in2p3.fr

S. Kuroyanagi
Instituto de Física Teórica UAM-CSIC, Universidad Autónoma de Madrid, 28049 Madrid, Spain
e-mail: sachiko.kuroyanagi@csic.es

Department of Physics, Nagoya University, Furo-cho Chikusa-ku, Nagoya 464-8602, Japan

C. Byrnes et al. (eds.), *Primordial Black Holes*, Springer Series in Astrophysics and Cosmology, https://doi.org/10.1007/978-981-97-8887-3_26

and is also important for acquiring information on astrophysical source populations at high redshift, as well as for verifying the existence of PBH binaries.

This chapter aims to discuss the implications for PBHs that can be obtained from SGWB observations. In particular, we review the constraints on PBH models derived from the nondetection in the most recent observation, namely Observation run 3 (O3) of the LIGO-Virgo-KAGRA (LVK) detector network. We focus on two different types of SGWBs: one arising from scalar curvature perturbations associated with PBH formation and the other generated from the collective GWs emitted by PBH binaries.

The first source is often termed the scalar-induced GW background (SIGWB). If there is a mechanism to amplify small-scale density perturbations in the early Universe, PBHs can form due to the collapse of extremely dense regions. In the scenario where primordial curvature fluctuations are amplified during inflation, PBHs form soon after the corresponding modes enter the Hubble radius. Associated with this process, SGWBs are sourced by the second-order terms of scalar perturbations, which is predicted in the cosmological perturbation theory [15, 23, 47, 76, 77, 94, 105]. In this case, the peak scale is determined by the detailed mechanism of enhancing scalar perturbations during inflation, and it corresponds to the frequency of GWs, denoted by f, as

$$f = \frac{ck}{2\pi} = 25 \left(\frac{k}{1.6 \times 10^{16}\,\mathrm{Mpc}^{-1}} \right) \mathrm{Hz} \simeq 5.8 \times 10^{-9} \alpha^{1/2} \left(\frac{M_{\mathrm{PBH}}}{M_{\odot}} \right)^{-1/2} \mathrm{Hz}, \tag{26.1}$$

where k is the scale of the curvature perturbations, $M_{\odot} \simeq 2 \times 10^{33}$g denotes the solar mass, $\alpha \equiv M_{\mathrm{PBH}}/M_H$, typically of order unity, is used to account for the deviation between the PBH mass M_{PBH} and the horizon mass M_H.[1] GW experiments can probe scales much smaller than those accessible with current cosmological observations. Thus, they offer a unique opportunity to explore different epochs of inflation.

Second, the SGWB originates from the collective GWs emitted by PBH binaries. In this case, the SGWB consists of an ensemble of a large number of weak, independent, and unresolved GW sources [40, 86, 106]. Differences between primordial and astrophysical black holes typically emerge at high redshifts as PBHs form much earlier than astrophysical black holes. The SGWB provides a unique opportunity to probe such high redshifts and may offer hints to discriminate between astrophysical and primordial scenarios [16, 24, 81, 82, 112]. Additionally, the PBH mass function can be very different from astrophysical ones, and this difference could be prominently reflected in the spectral shape [20, 26]. This SGWB exhibits a peak at the merger frequency (redshifted), which is determined by the mass of PBHs. Assuming equal-mass binaries, the energy spectrum has a peak roughly at

[1] Here, we have assumed the simplest scenario in which PBHs form with a fixed fraction of the horizon mass. However, to be precise, simulations of PBH formation show that the PBH mass depends on both the scale and amplitude of the perturbation from which it forms.

$$f \simeq 8.3 \times 10^3 \left(\frac{M_{\rm PBH}}{M_\odot}\right)^{-1} \text{Hz}. \tag{26.2}$$

The spectrum at low frequencies is attributed to the inspiral phase, which can be tested by the LVK detectors for solar mass PBHs.

As observed in Eqs. (26.1) and (26.2), the two SGWB are generated by different mechanisms at different times and emerge in completely different frequency bands with different mass dependencies. Therefore, they can be investigated in multi-wavelength GW experiments [110], including ground-based/space-borne interferometers, pulsar timing, etc. In this chapter, we focus on constraints obtained by the current LVK detectors, which explore two different PBH masses: assuming $f = 25$ Hz, we are probing PBHs with a mass of $5.0 \times 10^{-20} M_\odot$ with SIGWB, while for PBH binaries, the corresponding redshifted PBH mass is $300 M_\odot$ (due to the low-frequency inspiral spectrum mentioned earlier, it can also access lower mass PBHs).

We aim to provide an overview of LVK SGWB constraints, hoping to serve as a useful guide for individuals interested in deriving constraints on their theoretical models from the data. The chapter is organized as follows: first, in Sect. 26.2, we briefly summarize the standard detection methods of SGWB, which involve cross-correlation between data from two or more interferometers, as well as the basics of Bayesian inference, which is often used to provide constraints on model parameters. Subsequently, in Sect. 26.3, we describe the current constraints on the SIGWB and provide their implications for PBH scenarios. In Sect. 26.4, we discuss the SGWB from PBH binaries and its constraints by the current LVK observations. We conclude in Sect. 26.5 describing future prospects.

26.2 Analysis Techniques for the SGWB Search in the LIGO-Virgo-Kagra Collaboration

26.2.1 *Basics and Assumptions*

GWs can be described as a small tensor perturbation around the Minkowski background metric, i.e., $g_{ab} = \eta_{ab} + h_{ab}$ where $\eta_{ab} = \text{diag}(-1, 1, 1, 1)$ and the tensor perturbations h_{ab} satisfy the transverse traceless conditions. They are often expressed as a superposition of plane waves by decomposing into their Fourier modes using the two independent polarization states

$$h_{ab}(t, \mathbf{x}) = \sum_{A=+,\times} \int_{-\infty}^{\infty} \mathrm{d}f \int_{S^2} \mathrm{d}\hat{\Omega}\, h_A(f, \hat{\Omega}) e^{2\pi i f (t - \hat{\Omega}\cdot\mathbf{x}/c)} e^A_{ab}(\hat{\Omega}), \tag{26.3}$$

where f is the GW frequency. The unit vector $\hat{\Omega}$ represents the direction of propagation of the wave and is related to the wave vector as $\mathbf{k} = 2\pi f \hat{\Omega}/c$. Each plane wave is characterized by the Fourier amplitude $h_A(f, \hat{\Omega})$ and the phase $2\pi i f(t - \hat{\Omega}\cdot\mathbf{x}/c)$. The metric perturbations $h_{ab}(t, \mathbf{x})$ are real numbers, so the plane wave components satisfy $h_A(-f, \hat{\Omega}) = h_A^*(f, \hat{\Omega})$. The polarization tensors $e_{ab}^A(\hat{\Omega})$ denote the two independent polarizations of GWs. Although the choice of polarization basis is not unique and may be chosen based on the specific problem, here we use the most common basis to define the plus and cross polarizations ($A = +, \times$)

$$e_{ab}^{+}(\hat{\Omega}) = \hat{m}_a\hat{m}_b - \hat{n}_a\hat{n}_b, \tag{26.4}$$

$$e_{ab}^{\times}(\hat{\Omega}) = \hat{m}_a\hat{n}_b + \hat{n}_a\hat{m}_b, \tag{26.5}$$

where $\hat{\Omega}$, $\hat{m}$ and $\hat{n}$ are the unit vectors (see Fig. 26.1), defined as

$$\hat{\Omega} = \cos\phi\sin\theta\hat{x} + \sin\phi\sin\theta\hat{y} + \cos\theta\hat{z}, \tag{26.6}$$

$$\hat{m} = \sin\phi\hat{x} - \cos\phi\hat{y}, \tag{26.7}$$

$$\hat{n} = \cos\phi\cos\theta\hat{x} + \sin\phi\cos\theta\hat{y} - \sin\theta\hat{z}. \tag{26.8}$$

They satisfy $e_{ab}^A(\hat{\Omega})e^{A',ab}(\hat{\Omega}) = 2\delta^{AA'}$.

The metric perturbations h_{ab} and the Fourier amplitudes $h_A(f, \hat{\Omega})$ are random variables. Hence, the stochastic background is characterized statistically. Due to the central limit theorem, the SGWB is often assumed to follow a Gaussian distribution. This assumption arises from the consideration that, in most cases of cosmological backgrounds, a very large number of Hubble patches in which GWs are produced are observed across the resolution area of the detector. For astrophysical backgrounds, it holds if it comprises a sufficiently large number of overlapped independent sources. The assumption of Gaussianity simplifies the analysis because the entire probability density function can be characterized by the quadratic expectation value, i.e., $\langle h_A^*(f, \hat{\Omega})h_{A'}(f', \hat{\Omega}')\rangle$. Note that a SGWB has a zero mean, indicating $\langle h_A(f, \hat{\Omega})\rangle = 0$.

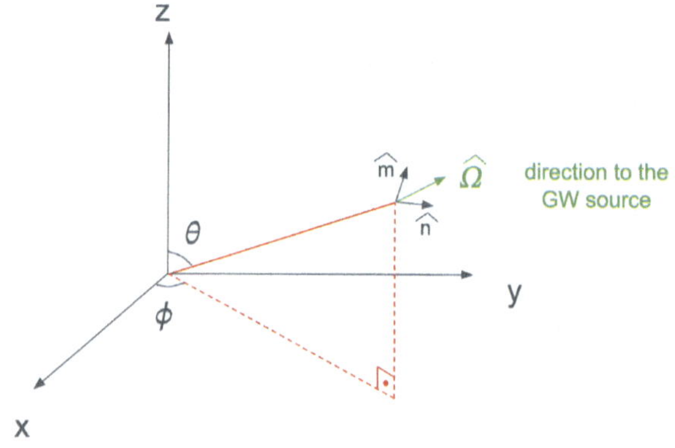

Fig. 26.1 The coordinate system consists of the unit vectors $\hat{\Omega}, \hat{m}, \hat{n}$, which define the polarization tensors $e_{ab}^A(\hat{\Omega})$. The unit vector $\hat{\Omega}$ points in the direction of the GW propagation. The other two unit vectors lie perpendicular to $\hat{\Omega}$

Background Information - Random variables
The statistical properties of a random variable X are determined by its probability distribution $p_X(x)$. The n-th order moments of the distribution are defined by $\langle X^n \rangle \equiv \int_{-\infty}^{\infty} \mathrm{d}x x^n p_X(x)$. The first moment, $\langle X \rangle$, is the mean value of X and is usually referred to as μ. The second moment, $\langle X^2 \rangle$, is related to the variance of the distribution as $\sigma^2 = \langle X^2 \rangle - \langle X \rangle^2$.

If X is a Gaussian random variable, then its probability distribution can be expressed as

$$p_X(x) = \frac{1}{\sqrt{2\pi}\sigma} e^{-\frac{(x-\mu)^2}{2\sigma^2}}, \tag{26.9}$$

which implies that X is entirely described by its first two momenta. In this case, all the higher-order moments are either zero or can be expressed as a sum of products of the first two momenta (Isserlis' theorem [59]).

In this chapter, we only focus on a Gaussian, stationary, unpolarized, and isotropic SGWB. This assumption is valid in the case of the SIGWB. The SIGWB signal is intrinsically non-Gaussian, unlike other cosmological sources, because its source is second-order in the curvature perturbation. However, the propagation effects of the GWs across the perturbed Universe suppress the level of non-Gaussianity to an unobservable level [21, 22]. In the case of the SGWB from PBH binaries (and in general astrophysical backgrounds), it can deviate from this assumption and the SGWB can exhibit non-Gaussianity [27], anisotropy [111], and polarizations [107]. In fact, these additional statistics contain rich information on the black hole population and help differentiate between different PBH models.

Another major assumption that we make in this section concerns Gaussian, stationary, and uncorrelated detector noise. Real-world detectors experience numerous non-Gaussian transient noises, often termed glitches, along with non-stationary noise characterized by slow variations in detector sensitivity. The stochastic analysis in the LVK [1, 5–7] is conducted to ensure that the Gaussianity and stationarity of noise are consistently maintained through preprocessing the strain data with the procedure known as *gating* and implementing the so-called *delta-sigma cut*. During the gating process, loud glitches are eliminated by applying an inverse Tukey window to instances where the data exhibits a significant excess [79]. Residual non-Gaussian noises and non-stationarities are also addressed through the delta-sigma cut by monitoring the standard deviation of each time-series segment and removing all times where segments vary too much compared to adjacent segments. See Refs. [89, 90] for details of data conditioning.

Furthermore, specific sources of correlated noise exist, deviating from our idealized framework. Noteworthy are narrow-frequency noise artifacts such as correlations between the electronics mains frequency (e.g., 60 Hz in LIGO and 50 Hz in Virgo) and data sampling referenced to GPS clocks [41]. These known noise lines are typically *notched out*, meaning that the values of the spectra at the affected frequency bins are removed from the analysis. Of particular concern are Schumann resonances, electromagnetic phenomena originating from lightning strikes that persist within the Earth's ionosphere, resulting in coherent oscillations in magnetometer readings at ground-based GW detectors. These resonances, primarily below 50 Hz, present a significant challenge due to their potential coupling with mirror suspension systems, electric cables, and electronics. A noise budget for magnetic correlations was constructed in the observing runs of LIGO-Virgo data [7, 103, 104]. Despite the current findings indicating no significant evidence for correlated magnetic noise, future sensitivities demand vigilant consideration of these resonances [61]. One of the solutions to this challenge could be GW geodesy [30, 60], a new robust method that enables us to discern between potential correlated noises and genuine SGWB signals by requiring true observations to be consistent with the known geometry of the detector network.

26.2.2 Gaussian, Stationary, Unpolarized, and Isotropic SGWB

The SGWB, being Gaussian with zero mean, isotropic, unpolarized, and stationary allows us to write the 2-point correlation function as

$$\langle h_A^*(f, \hat{\Omega}) h_{A'}(f', \hat{\Omega}') \rangle = \frac{1}{16\pi} S_h(f) \delta_{AA'} \delta(f - f') \delta^2(\hat{\Omega}, \hat{\Omega}'), \tag{26.10}$$

where $S_h(f)$ is called the strain power spectral density (PSD) and satisfies $S_h(f) = S_h(-f)$. The SGWB is often characterized by the dimensionless energy density parameter

$$\Omega_{\mathrm{GW}}(f) = \frac{1}{\rho_{c,0}} \frac{\mathrm{d}\rho_{\mathrm{GW}}}{\mathrm{d}(\ln f)}, \tag{26.11}$$

where $\rho_{c,0} = 3H_0^2 c^2/(8\pi G)$ is the critical energy density of the Universe today with H_0 being the Hubble parameter at present time. The energy density of GWs, ρ_{GW}, is expressed as

$$\rho_{\mathrm{GW}} = \frac{c^2}{32\pi G} \langle \dot{h}_{ab}(t, \mathbf{x}) \dot{h}^{ab}(t, \mathbf{x}) \rangle. \tag{26.12}$$

By substituting (26.3), we obtain

$$\begin{aligned}\rho_{\rm GW} =& \frac{c^2}{32\pi G} \sum_{A,A'=+,\times} \int_{-\infty}^{\infty} \mathrm{d}f \int_{-\infty}^{\infty} \mathrm{d}f' \int_{S^2} \mathrm{d}\hat{\Omega} \int_{S^2} \mathrm{d}\hat{\Omega}' \langle h_A^*(f,\hat{\Omega}) h_{A'}(f,\hat{\Omega})\rangle \\ &\times 4\pi^2 f f' e^{-2\pi i f(t-\hat{\Omega}\cdot\mathbf{x}/c)} e^{2\pi i f'(t-\hat{\Omega}'\cdot\mathbf{x}/c)} e_{ab}^{A}(\hat{\Omega}) e^{A',ab}(\hat{\Omega}') \\ =& \frac{c^2}{128G} \sum_{A,A'=+,\times} \int_{-\infty}^{\infty} \mathrm{d}f \int_{-\infty}^{\infty} \mathrm{d}f' \int_{S^2} \mathrm{d}\hat{\Omega} \int_{S^2} \mathrm{d}\hat{\Omega}' S_h(f) \delta_{AA'} \delta(f-f') \delta^2(\hat{\Omega},\hat{\Omega}') \\ &\times f f' e^{-2\pi i f(t-\hat{\Omega}\cdot\mathbf{x}/c)} e^{2\pi i f'(t-\hat{\Omega}'\cdot\mathbf{x}/c)} e_{ab}^{A}(\hat{\Omega}) e^{A',ab}(\hat{\Omega}') \\ =& \frac{\pi c^2}{8G} \int_{-\infty}^{\infty} \mathrm{d}f\, f^2 S_h(f) = \frac{\pi c^2}{4G} \int_{0}^{\infty} \mathrm{d}f\, f^2 S_h(f). \end{aligned} \tag{26.13}$$

In the third step, under the assumption of an unpolarized background, we have used $\sum_{+,\times} e_{ab}^{A}(\hat{\Omega}) e^{A,ab}(\hat{\Omega}) = 4$. Furthermore, assuming an isotropic background, the solid angle integral yields a factor of 4π. In the last step, we have replaced the frequency integration range as $\int_{-\infty}^{\infty} \mathrm{d}f = 2\int_0^{\infty} \mathrm{d}f$. Substituting this result into Eq. (26.11), we obtain[2]

$$\Omega_{\rm GW}(f) = \frac{2\pi^2}{3H_0^2} f^3 S_h(f). \tag{26.14}$$

This equation implies that, because of the f^3 dependence, given two detectors with the same strain sensitivity, the one at lower frequencies is sensitive to lower values of $\Omega_{\rm GW}(f)$. Consequently, for instance, in the case of LVK O3 analysis, the most sensitive frequency band for stochastic search lies in the relatively low-frequency range of 30 – 40Hz, while the noise floor is observed above 100Hz when sensitivity is plotted in terms of the usual strain noise amplitude.

Now that the energy density spectrum has been introduced, Fig. 26.2 displays examples of the spectrum of the SGWB for different cosmological models, alongside the current best upper limits and future expected GW detector sensitivities. Many generation mechanisms predict the existence of a SGWB across a wide range of frequencies, emphasizing the importance of multi-band GW observations in the future. Current ground-based detectors have limited sensitivity to probe possible early Universe signals. However, future detectors such as the Laser Interferometer Space Antenna (LISA) and third-generation detectors like the Einstein Telescope (ET) could investigate specific sources such as inflation, cosmic strings, and phase transitions.

[2] Note that some literature (e.g. Ref. [71]) has a factor of 2 difference in the definition of Eq. (26.10), resulting in a factor of 2 difference in $\Omega_{\rm GW}$. We follow the definition in Refs. [14, 91], which is commonly used in LVK stochastic papers.

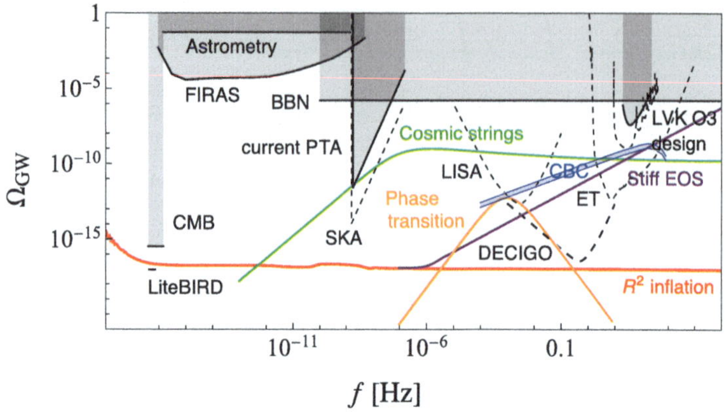

Fig. 26.2 SGWB spectra for several cosmological models (colored solid curves) are compared to current upper limits (shaded in gray) and expected sensitivities of future experiments (black dashed lines). As examples of cosmological sources, we show GW signals from inflation (assuming R^2 inflation [101]) and its modification by a stiff Equation of State [53]. Note that the amplitude and spectral behavior of these signals heavily depend on the model parameters, and the spectra shown here represent just one example of this variability. We also depict the GW spectrum of cosmic string loops [44, 45] (assuming tension $G\mu = 10^{-12}$) and the electroweak phase transition [67]. Current constraints include the Advanced-LIGO O3 upper bound [7], constraints based on big bang nucleosynthesis (BBN) and cosmic microwave background (CMB) observations [114], pulsar timing array [12], astrometric measurement by Gaia satellite [62], CMB μ-distortion constraints by COBE/FIRAS [65], and CMB temperature and polarization observation [10]. Future expected sensitivities include the final sensitivity of Advanced-LIGO [3], ET [2], DECIGO [64], LISA [39], SKA [46], and LiteBIRD [54]. We assume 3 years of observation for interferometer experiments and 20 years for the SKA. The blue-shaded band indicates the expected amplitude of the SGWB due to the cosmic population of compact binary coalescences (CBCs), based on the observed individual events in the O3 catalog [8]. Note that the expected spectral amplitude is extrapolated to the LISA frequency band assuming the $f^{2/3}$ dependence of the inspiral phase [18, 70]. However, the lower frequencies could be modified by the effects of eccentricity and precession at the time of binary formation [48, 77]

26.2.3 SGWB Detection Method: Cross-Correlation Technique

In this subsection, we outline the framework for the standard detection method. Given the stochastic nature of a SGWB, distinguishing between a signal and local detector noise poses a significant challenge. Thus, SGWB searches often involve cross-correlating data from pairs of interferometers[3] [37, 50, 80]. For a detailed review, we refer the reader to Ref. [14, 91].

[3] Note that LISA (and possibly ET, whose design is still under consultation) uses the so-called Time Delay Interferometry (TDI) variables for data processing. In this method, six laser links are transformed into a null channel containing mainly noise and two orthogonal signal channels, where the noise is suppressed. In this case, two orthogonal signal channels are not correlated (as they effectively correspond to two L-shape detectors; one of which is rotated by $\pi/4$ radians with

Let us define the output of a detector $s_i(t)$ as the sum of the detector's noise $n_i(t)$ and a GW signal $h_i(t)$, where the label i denotes each detector,

$$\begin{aligned} s_1(t) =& n_1(t) + h_1(t), \\ s_2(t) =& n_2(t) + h_2(t). \end{aligned} \tag{26.15}$$

The basic concept is that, when we correlate the outputs of the two detectors and integrate them over time, the noise terms cancel out because they are zero-mean random variables $\langle n_i(t) \rangle = 0$ and ideally have no correlation with each other $\langle n_i(t) n_j(t) \rangle \propto \delta_{ij}$. As a simple example, if we have two co-located and co-aligned detectors, so that $h_1(t) = h_2(t) = h(t)$, we can extract the signal in the following way:

$$\begin{aligned} \int_{-T/2}^{T/2} \mathrm{d}t \; s_1(t) s_2(t) &= \int_{-T/2}^{T/2} \mathrm{d}t \left(n_1(t) n_2(t) + h_1(t) n_2(t) + n_1(t) h_2(t) + h(t)^2 \right) \\ &\simeq \int_{-T/2}^{T/2} \mathrm{d}t \; h(t)^2 = T \langle h(t)^2 \rangle. \end{aligned} \tag{26.16}$$

In reality, the situation is more complex. The LVK analysis employs the optimal filtering method, specifically designed to detect a particular spectral shape of a SGWB while accounting for the locations of the detectors. It is equivalent to constructing the cross-correlation estimator for an observation period T using the output of the two detectors s_1 and s_2 as

$$Y = \int_{-T/2}^{T/2} \mathrm{d}t \int_{-T/2}^{T/2} \mathrm{d}t' s_1(t) s_2(t) Q(t, t'). \tag{26.17}$$

The function $Q(t, t')$ is known as a filter function, chosen to maximize the signal-to-noise ratio (SNR) for given detector locations and the spectral shape of the GW signal and detector noise, as we will see shortly. Since we have to determine the spectral shape of the SGWB, the Fourier transform of this function is sometimes referred to as a "template". As described in the following section, to constrain a particular model, one has to prepare a set of spectral shapes predicted by theory (a "template bank" in the terminology of matched filtering search) and find the most probable template by iteratively calculating the estimator of Eq. (26.17) for different templates.

Noise and SGWB are assumed to be stationary during the observation time T, so the optimal choice of the filter function depends only on the time difference $\Delta t = t - t'$ and $Q(t, t')$ can be written as $Q(t - t')$. Note that the cross-correlation is maximum when two detectors are co-located and co-aligned, and the filter function has a peak at $t = t'$. It falls off rapidly to zero when Δt is much larger than the light

respect to the other [42]), so the auto-correlation of the two signal channels is used to search for a SGWB [99].

travel time between the two sites. Using this fact, one can extend the integration range of t' to $[-\infty, \infty]$ and perform the Fourier transformation as

$$Y = \int_{-\infty}^{\infty} \mathrm{d}f \int_{-\infty}^{\infty} \mathrm{d}f' \delta_{\mathrm{T}}(f - f') s_1^*(f) s_2(f') Q(f'), \tag{26.18}$$

where $\delta_{\mathrm{T}}(f - f')$ is a finite time approximation to the Dirac delta function,

$$\delta_{\mathrm{T}}(f - f') = \int_{-T/2}^{T/2} \mathrm{d}t \, e^{-2\pi i (f - f')t} = \frac{\sin(\pi T(f - f'))}{\pi(f - f')}. \tag{26.19}$$

For $f = f'$, we have $\delta_{\mathrm{T}}(0) \to T$, and we get the usual Dirac delta function for $T \to \infty$.

The expectation value of Y, denoted by $\langle Y \rangle$, has an associated variance given by $\sigma_{\mathrm{Y}}^2 = \langle Y^2 \rangle - \langle Y \rangle^2$. Thus, we define the signal-to-noise ratio as

$$\mathrm{SNR}^2 := \frac{\langle Y \rangle^2}{\sigma_Y^2}. \tag{26.20}$$

To calculate $\langle Y \rangle$, we need to evaluate $\langle s_1^*(f) s_2(f') \rangle$. Using Eq. (26.15) and assuming uncorrelated random noises, this expectation value can be approximated as

$$\begin{aligned} \langle s_1^*(f) s_2(f') \rangle &= \langle n_1^*(f) n_2(f') + h_1^*(f) n_2(f') + n_1^*(f) h_2(f') + h_1^*(f) h_2(f') \rangle \\ &\simeq \langle h_1^*(f) h_2(f') \rangle. \end{aligned} \tag{26.21}$$

Now let us remove the assumption of co-located and co-aligned detectors. The GW signal at detector i can be expressed as the contraction between the metric and the detector $h_{\mathrm{i}}(t) = h_{ab}(t, \mathbf{x}_{\mathrm{i}}) d^{\mathrm{ab}}(t, \mathbf{x}_{\mathrm{i}})$, where $\mathbf{x}_{\mathrm{i}}$ is the location of the detector. The vector $d^{\mathrm{ab}}(t, \mathbf{x}_{\mathrm{i}})$ denotes the response of the detector i located at $\mathbf{x}_{\mathrm{i}}$ and time t. When the round-trip light travel time in the detector arms is short compared to the period of the GWs (which is the case for the LVK detectors), it can be expressed in terms of the unit vectors along the two laser arms of the detector, $\hat{X}(t, \mathbf{x}_{\mathrm{i}})$ and $\hat{Y}(t, \mathbf{x}_{\mathrm{i}})$,

$$d^{\mathrm{ab}}(t, \mathbf{x}_{\mathrm{i}}) \simeq \frac{1}{2}\Big(\hat{X}^{\mathrm{a}}(t, \mathbf{x}_{\mathrm{i}}) \hat{X}^{\mathrm{b}}(t, \mathbf{x}_{\mathrm{i}}) - \hat{Y}^{\mathrm{a}}(t, \mathbf{x}_{\mathrm{i}}) \hat{Y}^{\mathrm{b}}(t, \mathbf{x}_{\mathrm{i}})\Big). \tag{26.22}$$

Then the Fourier amplitude is expressed as

$$h_{\mathrm{i}}(f; t) = \sum_{A=+,\times} \int_{S^2} \mathrm{d}\hat{\Omega} \, h_A(f, \hat{\Omega}) e^{-2\pi i f \hat{\Omega} \cdot \mathbf{x}_{\mathrm{i}}/c} e_{ab}^{\mathrm{A}}(\hat{\Omega}) d^{\mathrm{ab}}(t, \mathbf{x}_{\mathrm{i}}). \tag{26.23}$$

In the equation, we see the contraction of the polarization tensor with the detector response tensor at location $\mathbf{x}_{\mathrm{i}}$ and time t, which is often defined as $F_{\mathrm{i}}^{\mathrm{A}}(\hat{\Omega}, t) := e_{ab}^{\mathrm{A}}(\hat{\Omega}) d^{\mathrm{ab}}(t, \mathbf{x}_{\mathrm{i}})$ and called the *detector pattern functions*. Here the t index is kept to emphasize the fact that the analysis is done per segment of time. In reality, the direction of the detectors changes due to the Earth's motion, and this must be carefully considered when the SGWB is anisotropic. In the following, we omit this dependence by focusing on the isotropic case, where detectors always receive the same contributions from the SGWB regardless of changes in the direction of the detector over time.

Then, Eq. (26.21) is now expressed as

$$\begin{aligned}
&\langle s_1^*(f)s_2(f')\rangle \\
&\simeq \sum_{A,A'=+,\times} \int_{S^2} \mathrm{d}\hat{\Omega} \int_{S^2} \mathrm{d}\hat{\Omega}' \langle h_A^*(f,\hat{\Omega})h_{A'}(f',\hat{\Omega}')\rangle\, e^{2\pi i f\hat{\Omega}\cdot\mathbf{x}_1/c} e^{-2\pi i f\hat{\Omega}'\cdot\mathbf{x}_2/c} F_1^A(\hat{\Omega})F_2^{A'}(\hat{\Omega}') \\
&= \frac{3H_0^2}{32\pi^3 f^3}\Omega_{\mathrm{GW}}(f)\delta(f-f') \sum_{A,A'=+,\times} \int_{S^2} \mathrm{d}\hat{\Omega}\, e^{2\pi i f\hat{\Omega}\cdot\Delta\mathbf{x}/c} F_1^A(\hat{\Omega})F_2^{A'}(\hat{\Omega}),
\end{aligned} \tag{26.24}$$

where we have defined $\Delta\mathbf{x} := \mathbf{x}_1 - \mathbf{x}_2$, and have used Eqs. (26.10) and (26.14) in the second step. Then the expectation value of the estimator Y can be expressed as

$$\langle Y\rangle = \frac{3H_0^2}{20\pi^2}T\int_{-\infty}^{\infty} \mathrm{d}f \frac{\Omega_{\mathrm{GW}}(|f|)}{|f|^3}\gamma_{12}(|f|)Q(f), \tag{26.25}$$

where $\gamma_{12}(f)$ is a purely geometrical quantity known as the *overlap reduction function* (ORF), given by

$$\gamma_{12}(f) := \frac{5}{8\pi}\sum_{A,A'=+,\times}\int_{S^2} \mathrm{d}\hat{\Omega} e^{2\pi i f\hat{\Omega}\cdot\Delta\mathbf{x}/c} F_1^A(\hat{\Omega})F_2^{A'}(\hat{\Omega}). \tag{26.26}$$

The pre-factor $5/(8\pi)$ is chosen so that $\gamma_{12}(f) = 1$ for co-aligned and co-located detectors.

The next step consists of calculating the variance σ_Y^2. Assuming the weak-signal limit, i.e. the detector noise is much larger than the SGWB signal $h_i(t) \ll n_i(t)$, we can approximate $\sigma_Y^2 = \langle Y^2\rangle - \langle Y\rangle^2 \simeq \langle Y^2\rangle$.[4] For a Gaussian random variable, all cumulants vanish aside from the 2-point cumulants. Thus, using the assumption of uncorrelated random noises, the 4-point correlation function of the noise is reduced to

$$\begin{aligned}
&\langle n_1^*(f)n_2(f')n_1^*(f'')n_2(f''')\rangle \\
&= \langle n_1^*(f)n_2(f')\rangle\langle n_1^*(f'')n_2(f''')\rangle + \langle n_1^*(f)n_1^*(f'')\rangle\langle n_2(f')n_2(f''')\rangle \\
&\quad + \langle n_1^*(f)n_2(f''')\rangle\langle n_2(f')n_1^*(f'')\rangle \\
&\simeq \langle n_1^*(f)n_1^*(f'')\rangle\langle n_2(f')n_2(f''')\rangle.
\end{aligned} \tag{26.27}$$

[4] See Ref. [69], for the case of strong-signal regime, where we can no longer neglect the $\langle Y\rangle^2$ term.

The time-series noise $n_i(t)$ has real numbers and its Fourier transform satisfies $n_i^*(f) = n_i(-f)$. The same holds for the filter function. Using this, we obtain

$$\begin{aligned}
\sigma_Y^2 &\simeq \int_{-\infty}^{\infty} df \int_{-\infty}^{\infty} df' \int_{-\infty}^{\infty} df'' \int_{-\infty}^{\infty} df''' \langle n_1^*(f) n_1(-f'')\rangle \langle n_2^*(-f') n_2(f''')\rangle \\
&\qquad \times \delta_T(f - f')\delta_T(f'' - f''')Q(f')Q(f''') \\
&= \frac{1}{4}\int_{-\infty}^{\infty} df \int_{-\infty}^{\infty} df' P_1(|f|)P_2(|f'|)\delta_T^2(f - f')Q(f)Q^*(f') \\
&= \frac{T}{4}\int_{-\infty}^{\infty} df\, P_1(|f|)P_2(|f|)|Q(f)|^2,
\end{aligned} \tag{26.28}$$

where, in the second step, we have used the definition of the one-sided noise power spectrum[5]

$$\langle n_i^*(f) n_i(f')\rangle = \frac{1}{2}\delta(f - f')P_i(|f|). \tag{26.29}$$

In the last equality, $\delta_T(0) = T$ is used.

Now we define an inner product between functions $A(f)$ and $B(f)$ as

$$(A, B) = \int_{-\infty}^{\infty} df\, A^*(f)B(f)P_1(|f|)P_2(|f|). \tag{26.30}$$

Then the expectation value of the cross-correlation estimator in Eq. (26.25) can be written as

$$\langle Y\rangle = \frac{3H_0^2 T}{20\pi^2}\left(Q(|f|), \frac{\Omega_{GW}(|f|)\gamma_{12}(|f|)}{f^3 P_1(|f|)P_2(|f|)}\right). \tag{26.31}$$

Similarly, the variance of Y, Eq. (26.28), can be written as

$$\sigma_Y^2 = \frac{T}{4}(Q(|f|), Q(|f|)). \tag{26.32}$$

Therefore, the SNR is described as

$$\mathrm{SNR}^2 = \frac{\langle Y\rangle^2}{\sigma_Y^2} = \left(\frac{3H_0^2}{10\pi^2}\right)^2 T \frac{\left(Q(|f|), \frac{\Omega_{GW}(|f|)\gamma_{12}(|f|)}{f^3 P_1(|f|)P_2(|f|)}\right)^2}{(Q(|f|), Q(|f|))}. \tag{26.33}$$

The filter function $Q(|f|)$ is chosen such that the SNR is maximized. This happens when $Q(|f|)$ is

$$Q(|f|) = \lambda \frac{\Omega_{GW}(|f|)\gamma_{12}(|f|)}{f^3 P_1(|f|)P_2(|f|)}, \tag{26.34}$$

[5] Frequencies are positive, leading to a factor of $1/2$ in the definition of the noise power spectrum. The absolute value over f is used to emphasize $f \in \mathcal{R}^+$.

where λ is a overall normalization constant. Substituting Eq. (26.34) into Eq. (26.33) results in

$$\mathrm{SNR} = \frac{3H_0^2\sqrt{T}}{10\pi^2}\left(\int_{-\infty}^{\infty} \mathrm{d}f \frac{\Omega_{\mathrm{GW}}^2(|f|)\gamma_{12}^2(|f|)}{|f|^6 P_1(|f|)P_2(|f|)}\right)^{1/2}. \tag{26.35}$$

Major implications deriving from this equation are explained in what follows.

- The SNR increases proportionally to $\sqrt{T}$.
- The GW signal amplitude $\Omega_{\mathrm{GW}}(|f|)$ in the numerator is multiplied by the ORF, introduced in Eq. (26.26). It quantifies the reduction in sensitivity of the cross-correlation due to the response of the detectors as well as their separation and orientation. In Fig. 26.3, we plot the ORF between the LIGO and Virgo interferometers.
- The noise power spectra $P_i(|f|)$ in the denominator represent the detector noise. A smaller $P_i(|f|)$ leads to better detector sensitivity, resulting in an increased SNR. This naturally suppresses noisy frequencies from the analysis.
- The filter function to maximize the SNR includes the spectral shape $\Omega_{\mathrm{GW}}(|f|)$, which is not known a priori. Therefore, we need to prepare a set of filters (templates) and apply them iteratively to find the maximum SNR.

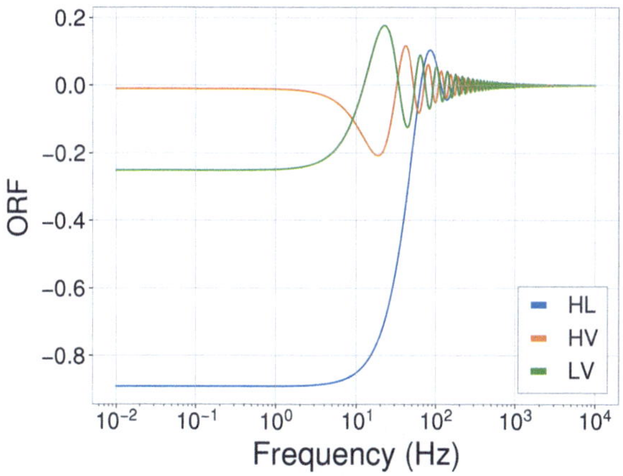

Fig. 26.3 Overlap reduction function (ORF) between the three LIGO-Virgo interferometers, where 'H' indicates Hanford, 'L' denotes Livingston, and 'V' represents Virgo. The curves are calculated by substituting actual relative positions and orientations of detector pairs (see e.g., Ref. [96] for parameter values). The ORF starts to oscillate and decay rapidly above the characteristic frequency $f_c = c/(2\Delta\mathbf{x})$. This implies that nearby detectors in a pair lose less sensitivity at low frequencies and are more sensitive to a SGWB. The cases in which the ORF is negative as $f \to 0$ are due to the fact that the two interferometers considered for the calculation are rotated by 90° with respect to each other

Background information
In the LVK stochastic analysis, the spectrum is usually modeled using a power law form

$$\Omega_{\rm GW}(f) = \Omega_\alpha \left(\frac{f}{f_{\rm ref}}\right)^\alpha , \qquad (26.36)$$

where α denotes the spectral index, $f_{\rm ref}$ is the reference frequency, and Ω_α represents the amplitude of the SGWB at $f_{\rm ref}$. Typically, $f_{\rm ref}$ is set to 25 Hz, as this frequency roughly corresponds to the peak sensitivity of the LIGO-Virgo network.

The typical values of the spectral index α are 0, 2/3, or 3. For example, $\alpha = 0$ characterizes cosmic strings at the high-frequency end [97] and the slow-roll inflation (although its amplitude is much smaller than the sensitivity of LVK). The value $\alpha = 2/3$ corresponds to the inspiral phase of compact binary coalescences (CBC) [31]. Finally, when a SGWB arises from the overlap of uncorrelated sources with a typical time scale of $\Delta\tau$, it results in white noise (frequency-independent) strain power at the low-frequency tail $f < 1/\Delta\tau$, corresponding to $\alpha = 3$. This phenomenon can be observed in phase transition origins [92] or an ensemble of burst-like astrophysical sources such as supernovae [75].

The CBC background is the most likely source in the LVK frequency band [4]. Therefore, whenever searching for other sources, CBCs must be simultaneously fitted using this spectrum, as we also do in the following analysis sections. The power law spectrum with $\alpha = 2/3$, which is typical of the inspiral phase, provides a good approximation as long as we consider the LVK sensitivities, while it may no longer be applicable for future detectors, and thus the contributions from the merger and ringdown phases must be included.

26.2.4 Narrowband Analysis

As seen in the formulas derived in the previous subsection, it is more convenient to work in the frequency domain rather than in the time domain. Working in the time domain requires phase information, which cannot be theoretically predicted in the case of the SGWB, where random fluctuations come from all directions. Thus, especially for constructing the filter function that requires information of the SGWB signal, the frequency domain is simpler and more practical. Therefore, in real data analysis, we first construct the cross-spectral density using the Fourier-transformed data $s_i(f)$ as

$$C_{12}(f) \equiv \frac{2}{T} s_1^*(f) s_2(f'). \qquad (26.37)$$

It is important to note that in actual searches, the data is divided into short segments, typically 192 seconds each. This approach allows us to handle gaps in data due to interruptions in data collection and address potential non-stationarities in the detector noise over both short and long timescales. Each segment is analyzed individually first, and then the information from each bin is combined [90]. Additionally, the cross-spectral density is estimated through coarse graining by averaging over neighboring frequencies to reduce fluctuations in the PSD estimates [102].

In practical terms, measured data are discretely sampled and the analysis is first performed in discrete bins of frequency. This approach is called *narrowband analysis*. The narrowband definition of the estimator can be obtained from Eq. (26.18) by reducing the bandwidth of the analysis to δf

$$\hat{Y}_f = \delta f \cdot 2\mathrm{Re}[s^*_{1,f} s_{2,f}]Q_f = \delta f \cdot T\ \mathrm{Re}[C_{12,f}]Q_f. \tag{26.38}$$

Here "Re" stands for the real part of a complex number, which arises due to our focus on positive frequencies, and comes from utilizing the relation $\int_{-\infty}^{\infty} \mathrm{d}f\ s_1(f)s_2^*(f) = 2\int_0^{\infty} \mathrm{d}f\ \mathrm{Re}[s_1(f)s_2^*(f)]$. The subscript f denotes the value at a frequency bin f and is used when we want to emphasize that the value is obtained from data and is discrete, while the variables with the notation (f) can be calculated theoretically without data. Additionally, we use a hat to indicate that the value is constructed from the data rather than being the true value. With this notation, as derived in Eq. (26.34), the optimal filter Q_f is rewritten as

$$Q_f = \lambda \frac{\Omega_{\mathrm{GW},f}\gamma_{12}(f)}{f^3 P_{1,f} P_{2,f}}. \tag{26.39}$$

By rewriting Eq. (26.25), the expectation value of $\hat{Y}_f$ is given by

$$\langle \hat{Y}_f \rangle = \frac{3H_0^2}{10\pi^2} T\ \delta f \frac{\Omega_{\mathrm{GW},f}}{f^3} \gamma_{12}(f) Q_f. \tag{26.40}$$

The factor 2 difference in the coefficient is again due to the fact that we work with positive frequencies. Let us take the overall normalization constant λ to satisfy $\langle \hat{Y}_f \rangle = \Omega_{\mathrm{GW},f}$. With this choice of normalization, using Eqs. (26.39) and (26.40), we obtain

$$\lambda^{-1} = \delta f \frac{3H_0^2}{10\pi^2} T \frac{\Omega_{\mathrm{GW},f}\gamma_{12}^2(f)}{f^6 P_{1,f} P_{2,f}}. \tag{26.41}$$

By substituting Eqs. (26.39) and (26.41) to (26.38), we obtain the final form of the estimator,

$$\hat{Y}_f = \frac{\mathrm{Re}[C_{12,f}]}{\gamma_{12}(f) S_0(f)}, \tag{26.42}$$

where we have defined

$$S_0(f) \equiv \frac{3H_0^2}{10\pi^2}\frac{1}{f^3}. \tag{26.43}$$

It is important to note that, by setting the normalization to ensure $\langle \hat{Y}_f \rangle = \Omega_{\mathrm{GW},f}$, the mean value of the estimator $\hat{Y}_f$ directly corresponds to the value of the spectral amplitude of the SGWB at a frequency f. From Eq. (26.28), we can see that the associated variance is given by

$$\sigma_{\hat{Y}_f}^2 = \frac{1}{2T\Delta f}\frac{P_{1,f}P_{2,f}}{\gamma_{12}^2(f)S_0^2(f)}. \tag{26.44}$$

Finally, we aim to utilize the information from all frequency bins, as described in Sect. 26.2.3, which is called *broadband analysis*. The full broadband statistics can be derived by combining the narrowband statistics of all frequency bins,

$$\hat{Y} \equiv \frac{\sum_f H^2(f)\sigma_{\hat{Y}_f}^{-2}\hat{Y}_f}{\sum_f H^2(f)\sigma_{\hat{Y}_f}^{-2}}, \tag{26.45}$$

$$\sigma_{\hat{Y}}^{-2} \equiv \sum_f H^2(f)\sigma_{\hat{Y}_f}^{-2}. \tag{26.46}$$

By weighting with the variance $\sigma_{\hat{Y}_f}^2$, this approach assigns less weight to measurements with higher noise levels at frequencies where variances are larger, and it is optimal in the sense of being an unbiased, minimal variance estimator (for further details, see Sect. 6.1 of Ref. [91]). Furthermore, the rescaling function $H(f)$ is applied to reweight the estimate of the spectrum, optimizing the statistic for a specific shape. This means that the analysis is tailored to seek a specific spectral shape that is theoretically prepared. For example, in the case of the power law form, Eq. (26.36), the function is given by

$$H_{\mathrm{ref},\alpha}(f) = \left(\frac{f}{f_{\mathrm{ref}}}\right)^{\alpha}. \tag{26.47}$$

This function may be a set of templates to be examined using Bayesian inference to determine the most probable model, as elaborated in the subsequent subsection.

In the LVK collaboration, we use a user-friendly Python-based package for the search of an isotropic SGWB in ground-based interferometer data called **pygwb** [55, 90]. Its primary purpose is to construct the optimal estimator and variance of the SGWB, as described by Eqs. (26.45) and (26.46). Additionally, it facilitates Bayesian parameter estimation to constrain SGWB models. It is worth mentioning that in practical searches, the estimators $\hat{Y}_f$ calculated for each time series segment are combined for the entire observation period before averaging across frequency bins. While we do not describe the details here, interested readers can find a detailed explanation in Sect. 3.3 of Ref. [90].

26.2.5 Bayesian Inference

The current LVK stochastic analysis employs a hybrid approach that combines frequentist and Bayesian analysis techniques [73]. Specifically, certain frequentist statistics, namely $\hat{Y}_f$ and $\sigma^2_{\hat{Y}_f}$ combined for the entire observation period, are calculated as described in the previous subsection. Subsequently, their frequency series are utilized as the fundamental input data for a Bayesian analysis to calculate posterior probability distributions of model parameters, typically to provide upper limits on the spectral amplitude of the SGWB. The hybrid frequentist-Bayesian analysis method has been utilized over the years to establish upper limits on GWBs with various amplitudes and spectral shapes [5–7]. Reference [78] has demonstrated that this hybrid approach does not lose information compared to a fully Bayesian search, which produces posterior distributions from the full time-series data rather than from the frequency series of frequentist statistics.

The Bayesian approach offers advantages over frequentist methods, as with a proper choice of the prior, the posterior distribution provides more information about the true parameter within the identified region. This frequentist-Bayesian approach allows us to conveniently explore arbitrary spectra of the SGWB and is often used for parameter estimation to provide constraints on a certain cosmological model.

Background Information - Quick summary of Bayesian inference

In Bayesian inference, probabilities are related to knowledge about an event (or signal). For example, when we want to infer the amplitude of a SGWB (a), the data (d) are known and it is the value of a that is uncertain. The relevant probability is that the amplitude has a certain value, given the data. This probability distribution is the posterior $p(a|d)$. This is unlike in frequentist inference, where the uncertainty is intrinsic to the data, and the relevant probability is that of observing the data given a signal of amplitude a. This probability is known as the likelihood, represented by $p(d|a)$. If we assume that N time-series data samples $d = \{d_j\}$ is the sum of the signal a and white, zero-mean noise n_j, i.e. $d_j = a + n_j$, with variance σ, the likelihood function can be defined as

$$p(d|a) = \frac{1}{(2\pi)^{N/2}\sigma^N} \exp\left[-\frac{1}{2\sigma^2}\sum_j (d_j - a)^2\right]. \qquad (26.48)$$

The likelihood and the posterior distribution are related by the Bayes' theorem, given by

$$p(a|d) = \frac{p(d|a)p(a)}{p(d)}, \qquad (26.49)$$

where $p(a)$ is the prior probability distribution for the amplitude. The prior represents the knowledge about the range and distribution of the parameter in the model (the SGWB amplitude in this example). The normalization factor $p(d)$ is the marginalised likelihood or evidence, obtained by integrating over the model parameters, a in this case, $p(d) = \int p(d|a)p(a)\mathrm{d}a$.

Utilizing the estimator and the variance in each frequency bin, $\hat{Y}_f$ and $\sigma^2_{\hat{Y}_f}$, constructed from the data, the likelihood function that is used in the LVK collaboration is [73]

$$p(\{\hat{Y}_f\}|\Theta) \propto \exp\left[-\sum_f \frac{(\hat{Y}_f - Y(f|\Theta))^2}{2\sigma^2_{\hat{Y}_f}}\right], \tag{26.50}$$

where $Y(f|\Theta)$ represents the true value of Y, determined by an assumed model and dependent on a set of model parameters Θ. We remind the reader that the mean value of the estimator $\hat{Y}_f$ directly corresponds to the value of the spectral amplitude of the SGWB at a frequency f, as mentioned in the previous section. Thus, $Y(f|\Theta)$ is sometimes written as $\Omega_{\rm GW}(f|\Theta)$.

In our searches, we always compute the Bayes factor, a measure of the relative evidence provided by the data for two competing hypotheses. This factor facilitates the comparison of how effectively each hypothesis aligns with the observed data, thereby quantifying the strength of support for one hypothesis over the other. Given two hypotheses $\mathcal{H}_0$ and $\mathcal{H}_1$, the hypothesis of having a signal and the hypothesis of only having noise in the data, the Bayes factor is defined as

$$\mathcal{B} = \frac{p(d|\mathcal{H}_1)}{p(d|\mathcal{H}_0)}, \tag{26.51}$$

where d represents the data and $p(d|\mathcal{H}_i)$ is the evidence of model $\mathcal{H}_i$. It is customary to present the logarithm of the Bayes factor. For positive values, hypothesis $\mathcal{H}_1$ is preferred over $\mathcal{H}_0$. In the case of the search for a SGWB, we typically compute the Bayes factor between the hypothesis of a signal being present and the hypothesis of only noise being present in the data. Another example is the comparison between the hypothesis of the existence of only general relativity (GR) polarization (tensor polarization) and non-GR polarizations (additional scalar/vector polarizations) [32].

Background information - Bayesian credible interval
While all the information from a Bayesian inference run is encoded in the posterior distribution, the posterior mean and probability intervals are concise ways of expressing the results of an analysis. A Bayesian confidence interval (CI), also referred to as a credible interval (CI), represents the degree of belief about an event. It is defined by the area under the posterior distribution between two parameter values with the range containing a particular percentage of probable values. If the data are insufficient to determine the model parameter,

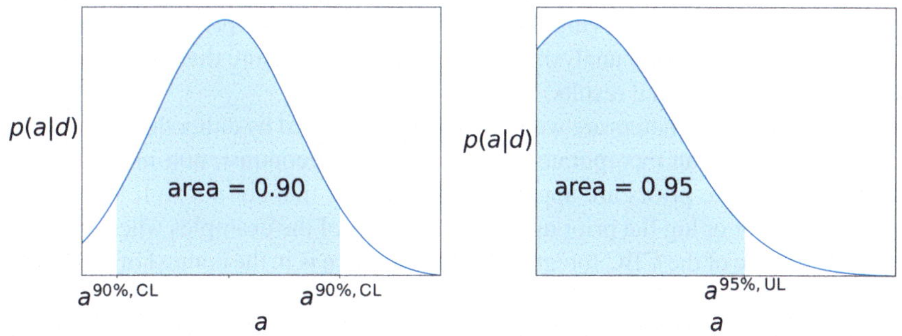

Fig. 26.4 Graphical representations of a Bayesian 90% credible interval (CI) and 95% upper limit (UL), assuming that the posterior is dependent on a parameter a. **Left:** The area below the posterior $p(a|d)$ colored in light blue represents the CI, which coincides with 90% of the total area. The two boundaries of the interval are commonly determined by the so-called Highest Density Interval, which is defined to contain the required area such that all points within the interval have a higher probability density than points outside the interval. In other words, the probability density at the two boundaries is the same. **Right:** The area below the posterior $p(a|d)$ colored in light blue coincides with a 95% of the total area. The lower bound coincides with the lowest possible value a can take, as determined by the prior distribution. The value of the parameter a at the upper bound corresponds to the 95% UL

we can obtain a Bayesian upper limit (UL), where one end of the interval is determined by the smallest value that the parameter can take.

Practically speaking, the CI or UL is obtained by integrating the posterior distribution on a parameter until xx% of the area under the posterior is achieved. The value of the parameter for which this occurs will be the UL on the parameter at the xx% confidence level. In literature, we often find 90% and 95% CI. The 90% is preferred for its stability, as insufficient sampling points may lead to instability in the tail distribution. The 95% is chosen primarily motivated by its intuitive relationship with the standard deviation (2σ). It is recommended to have a sample size of at least 10^4 for precise computation of the 95% CI [68, 72]. The definition for the CI and UL is represented graphically in Fig. 26.4.

Another noteworthy aspect is that the 68 and 95% CL do not correspond to the notion of 1 and 2 σ regions in two dimensions. In this case, within the 1σ region, the Gaussian distribution contains 39% of the volume, not 68%. Therefore, it is important to carefully choose the wording of "σ" or "%" when displaying the results of parameter estimation using 2-dimensional contour plots. The default corner plot output of **pygwb** utilizes the σ notation.

This section concludes with a brief discussion on priors. Prior distributions represent the beliefs or knowledge about the parameters of a model before observing the data. These distributions encode information available before data collection and, as mentioned earlier, are updated through Bayes' theorem, Eq. (26.49), to form posterior distributions after observing the data. The choice of the prior distribution is a critical aspect of Bayesian analysis, and it is essential to ensure that this choice does not unduly affect the final results.

When certain parameters are well known or constrained by either theory or experiment, their priors can incorporate this information. In contrast, non-informative or weakly informative priors are applied for parameters that are not well known. For instance, a linear or log flat prior might be used. One of the examples where the prior on the amplitude of the CBC foreground is informative is in the context of estimating astrophysical foreground amplitude in SGWB parameter estimation. This is because direct detections of individual CBC events help update the parameters of the CBC population, leading to better estimates of the spectral amplitude [6]. As a result, more constrained priors on this parameter can be established.

26.2.6 Calibration Uncertainties in the Search for the GWB

The data obtained from ground-based GW detectors requires calibration to translate the digital output of the detector into a relative displacement of the test masses within the detectors. This calibration process introduces both statistical uncertainties and systematic errors, which are commonly referred to as calibration uncertainties [109]. For an in-depth examination of calibration uncertainties in the Advanced-LIGO detectors, the reader is referred to Ref. [100].

The Bayesian analysis for a SGWB has to account for these calibration uncertainties [113]. We do this by introducing an unknown calibration factor Λ, which has a mean value of 1 and a variance of ε^2. Thus the probability distribution for Λ is given by

$$p(\Lambda) \propto \exp\left[-\frac{(\Lambda-1)^2}{2\varepsilon^2}\right]. \tag{26.52}$$

Note that we impose the constraint that Λ must be positive, and $\Lambda = 1$ corresponds to perfect amplitude calibration.

The likelihood function for the Bayesian analysis, introduced in Eq. (26.50), is constructed using the narrowband estimator $\hat{Y}_f$ along with its associated error $\sigma_{\hat{Y}_f}$. With the inclusion of calibration uncertainties, if the calibration is imperfect ($\Lambda \neq 1$), the set of estimators $\{\hat{Y}_f\}$ represents measurements of $\Lambda Y(f|\Theta)$, where $Y(f|\Theta)$ is a model described by parameters Θ. Consequently, the likelihood is modified as follows,

$$p(\{\hat{Y}_f\}|\Theta, \Lambda) \propto \exp\left[-\sum_f \frac{(\hat{Y}_f - \Lambda Y(f|\Theta))^2}{2\sigma^2_{\hat{Y}_f}}\right], \tag{26.53}$$

and the posteriors are obtained by marginalizing over Λ,

$$p(\{\hat{Y}_f\}|\Theta) = \int \mathrm{d}\Lambda\, p(\{\hat{Y}_f\}|\Theta, \Lambda) p(\Lambda). \tag{26.54}$$

Given our assumption of the Gaussian distribution for Λ as stated in Eq. (26.52), we can analytically perform the integration, which is implemented in the **pygwb** package [90].

26.3 Constraints on a Scalar-Induced Gravitational Wave Background

In this section, we present an overview of the latest constraints derived from the initial three observing runs conducted by the LVK collaboration regarding a SIGWB [58, 63, 93], which is closely associated with the formation of PBH [94]. This section is based on the results presented in Ref. [93].

Among the processes for deriving constraints from LVK data, a crucial initial decision involves modeling the spectral shape of the curvature power spectrum to be explored in the analysis. Typically, an agnostic approach is adopted, and no specific inflation model is selected. It is customary to opt for a log-normal shape for the peak in the curvature power spectrum

$$\mathcal{P}_\zeta(k) = \frac{A}{\sqrt{2\pi}\,\Delta} \exp\left[-\frac{\ln^2(k/k_*)}{2\Delta^2}\right], \tag{26.55}$$

where the peak in the curvature power spectrum is defined by its position k_* with its width controlled by the parameter Δ and its amplitude characterized by the integrated power A. In the $\Delta \to 0$ limit, the spectrum reduces to a Dirac delta function $\mathcal{P}_\zeta(k) = A\delta(\ln(k/k_*))$.

Given the curvature perturbation spectrum, the energy density spectrum for the SIGWB is calculated using the approximated analytical expression [49, 66]

$$\Omega_{\mathrm{GW}}(k)h^2 = 1.62 \times 10^{-5} \left(\frac{\Omega_{\mathrm{R},0}h^2}{4.18 \times 10^{-5}}\right) \left(\frac{g_*}{106.75}\right) \left(\frac{g_{*,s}}{106.75}\right)^{-4/3} \\ \times \frac{1}{12} \int_{-1}^{1} \mathrm{d}x \int_{1}^{\infty} \mathrm{d}y\, \mathcal{P}\Big(k\frac{y-x}{2}\Big) \mathcal{P}\Big(k\frac{x+y}{2}\Big) F(x, y), \tag{26.56}$$

where $\Omega_{\mathrm{R},0}$ is the present value of the energy density fraction of radiation, h is the dimensionless Hubble constant, and g_* and $g_{*,s}$ are the effective number of degrees of freedom for energy density and entropy density, respectively. The function $F(x, y)$ is given by

$$F(x, y) = \frac{288(x^2 + y^2 - 6)^2(x^2 - 1)^2(y^2 - 1)^2}{(x - y)^8(x + y)^8} \times \left[\left(x^2 - y^2 + \frac{x^2 + y^2 - 6}{2}\ln\left|\frac{y^2 - 3}{x^2 - 3}\right|\right)^2 + \frac{\pi^2}{4}(x^2 + y^2 - 6)^2\theta(y - \sqrt{3})\right]. \tag{26.57}$$

The LIGO-Virgo network exhibits its sensitivity in the frequency range between $10 - 500$Hz, corresponding to wavenumbers between approximately 10^{16} and 10^{17} Mpc^{-1}. These scales re-entered the Hubble radius when temperatures were above 10^8 GeV, so we can adopt $g_* = g_{*,s} = 106.75$ within the Standard Model paradigm.

Figure 26.5 illustrates the GWB spectrum for different values of the width parameter Δ, with fixed values of A and f_*. It can be observed that, for larger Δ, the peak is smoothed out, and the spectral amplitude decreases. The integrated amplitude A serves as the normalization of the SGWB amplitude, and the entire spectrum scales as $\Omega_{\rm GW}(f) \propto A^2$. The peak scale $k_* = 2\pi f_*/c$ determines the frequency at which the SGWB spectrum peaks. The spectrum exhibits a peak at approximately the same wavenumber as the curvature power spectrum. The entire spectrum is shifted to lower or higher frequencies without changing the spectrum shape when a different value of k_* is used.

To derive constraints on the SIGWB from LVK data, we conduct a Bayesian search using the methodology outlined in Sect. 26.2.5. Specifically, we utilize data products obtained from the stochastic isotropic analysis of the initial three observational runs of the LVK network, employing the former stochastic pipeline (similar to **pygwb**). In addition to the SIGWB signal, we include the CBC foreground, $\Omega_{\rm GW,CBC}(f) = \Omega_{\rm ref}(f/f_{\rm ref})^{2/3}$ with $f_{\rm ref} = 25$Hz, in the analysis, and estimate the background amplitude parameter $\Omega_{\rm ref}$ simultaneously. Therefore, the Bayesian search involves four parameters $\Theta = (\Omega_{\rm ref}, A, k_*, \Delta)$, with their respective priors described in Table 26.1. The priors on the integrated power A and the peak wavenumber k_* are selected so that the resulting signals could be detectable by the LVK net-

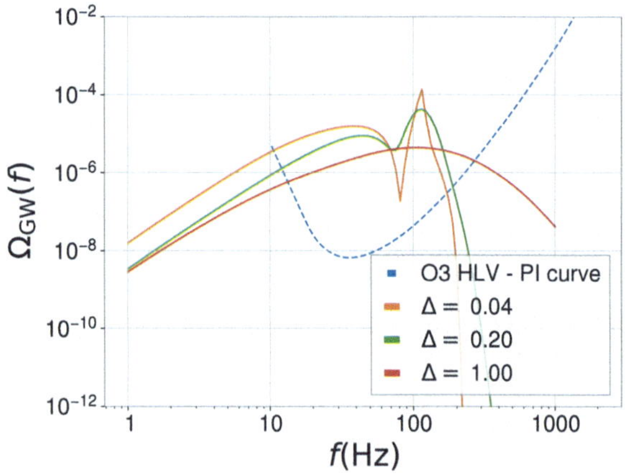

Fig. 26.5 Example of the SGWB spectrum. Different curves correspond to different values of the width parameter $\Delta = 0.05, 0.2, 1$. The other parameters are fixed as $A = 1$ and $f_* = ck_*/(2\pi) = 100$ Hz. Additionally, we plot the O3 HLV power-law integrated sensitivity curve as a dashed blue line

Table 26.1 Summary of prior distributions used in the Bayesian analysis. We have four free parameters: $\Omega_{\rm ref}$ is the amplitude of the CBC background at 25 Hz, A is the integrated power of the log-normal distribution of the curvature power spectrum, k_* is the peak position, and Δ is the width

Parameter	Prior
$\Omega_{\rm ref}$	LogUniform $(10^{-10}, 10^{-7})$
A	LogUniform $(10^{-3}, 10^{0.5})$
k_*/Mpc^{-1}	LogUniform $(10^{13}, 10^{21})$
Δ	LogUniform (0.05, 5)

work. Furthermore, the width parameter Δ is strategically chosen to encompass both narrow and wide peaks. Additionally, the prior on $\Omega_{\rm ref}$ is informed by prior estimates of the CBC background [4].

The posterior distributions for the model parameters obtained from the Bayesian analysis are shown in Fig. 26.6. The posterior on $\Omega_{\rm ref}$ allows to obtain an UL at 95% CL of 6.0×10^{-9}, which is consistent with the UL obtained in the O3 agnostic isotropic GWB search [7]. Data excludes a part of the parameter space in k_* and A. Exclusions at 95% CL are obtained in the region where the LIGO-Virgo interferometers have the highest sensitivity, $k_* \in [10^{16}, 10^{17}]\text{Mpc}^{-1}$. The posterior of Δ shows no preference for any range of values. Finally, we have obtained a Bayes factor of $\log \mathcal{B}^{\rm CBC+PBH}_{\rm noise} = -0.8$, which indicates that there is no evidence for a GWB signal.

The 2-dimensional contour plot as well as the 1-dimensional posterior distributions in Fig. 26.6 are produced by *marginalizing over* the other parameters. Marginalization involves integrating the posterior probability over other parameters. For example, consider that we have the joint posterior distribution $p(\theta_1, \theta_2|d)$ and we are interested in the constraint on parameter θ_1. Then the marginalized distribution of θ_1 is obtained by $p(\theta_1|d) = \int p(\theta_1, \theta_2|d)\, \mathrm{d}\theta_2$, and the integration range is determined by the prior range of θ_2.

It is important to note that such marginalized constraints are *prior dependent*, especially when the parameters lack precise determination and the posterior distribution fails to converge to zero at the edge of the priors. For instance, the integrated power A remains unconstrained for excessively large and small k_* values due to the limited frequency coverage of the LVK sensitivity. Consequently, the marginalized constraint on A weakens with a larger prior range of k_* because we assign a greater weight to such unconstrained regions by expanding the integration range of k_*. Therefore, instead of discussing the marginalized constraints, we rerun the Bayesian search by fixing the values of Δ and k_* to obtain ULs on A. The results for different combinations of Δ and k_* are summarized in Table 26.2. The most stringent ULs are obtained for $k_* = 10^{17}\text{Mpc}^{-1}$, corresponding to $\sim$ 100 Hz, where the interferometers have their highest sensitivity. For large widths, $\Delta \gg 1$, the spectrum becomes flat and the UL on A becomes independent of the scale. The most stringent UL is obtained for a narrow peak $\Delta = 0.05$ and $k_* = 10^{17}$ Mpc $^{-1}$. In all these

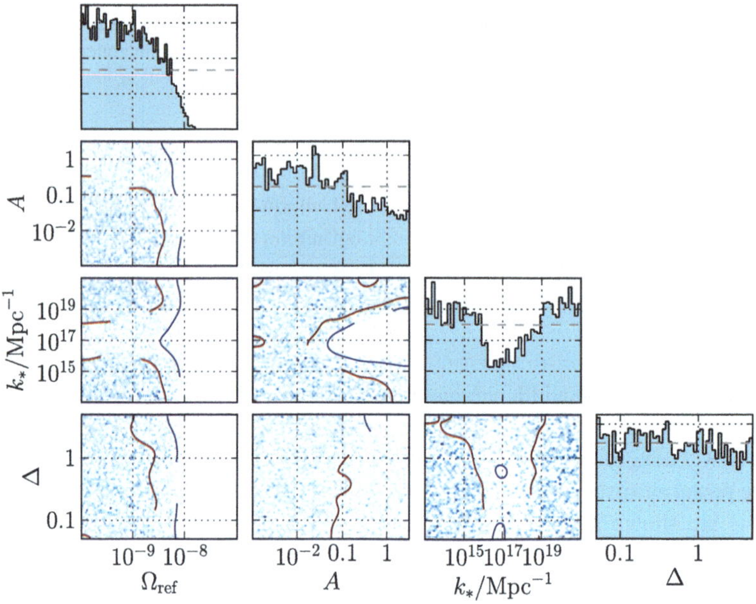

Fig. 26.6 Figure retrieved from Ref. [92]. Posterior distributions on the parameters of the curvature power spectrum describing the peak of the curvature perturbations. The 2D subplots show red and blue lines indicating exclusions at 68% and 95% confidence level

Table 26.2 Upper limits on the integrated power A of the curvature power spectrum at 95% CL for fixed values of the peak position k_* and width Δ

	$k_* = 10^{15}\,\mathrm{Mpc}^{-1}$	$k_* = 10^{17}\,\mathrm{Mpc}^{-1}$	$k_* = 10^{19}\,\mathrm{Mpc}^{-1}$
$\Delta = 0.05$	2.1	0.02	1.4
$\Delta = 0.2$	2.2	0.03	1.6
$\Delta = 1$	1.6	0.05	1.8
$\Delta = 5$	0.2	0.2	0.3

runs, the ULs at 95% CL on $\Omega_{\rm ref}$ are between 5.5×10^{-9} and 6.6×10^{-9}, which is consistent with the isotropic results from the first three observing runs.

In Fig. 26.7, the LVK bounds of the curvature power spectrum (shaded red area) are compared with other cosmological constraints for a very narrow peak ($\Delta \to 0$) in the left panel and for a relatively wide spectrum ($\Delta = 1$) in the right panel. LVK bounds exhibit heightened stringency compared to indirect constraints from BBN/CMB (shaded blue area) within the $10^{16} - 10^{19}\mathrm{Mpc}^{-1}$ region. Despite this, the PBH formation constraints (green solid lines) emerge as the most stringent across all scales. It is noteworthy, however, that with the anticipated sensitivities of Advanced-LIGO (depicted by the dashed red line) and ET (depicted by the dashed orange line), these bounds are expected to surpass even those derived from PBH formation.

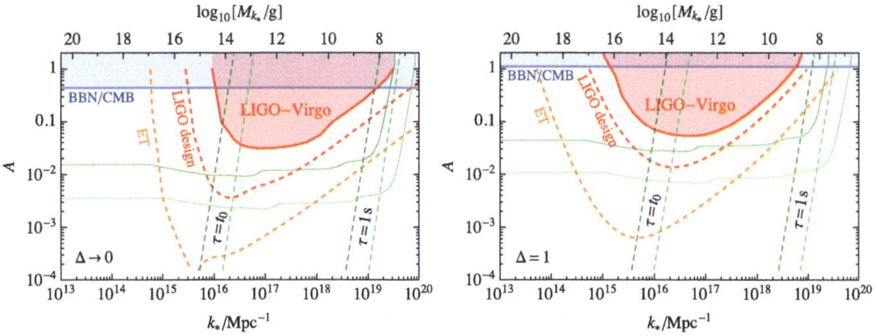

Fig. 26.7 Figures retrieved from Ref. [93]. The bounds set on A from the LVK O3 data (shaded red region) are compared with other observational constraints (**Left:** for $\Delta \to 0$. **Right:** for $\Delta = 1$.). The bottom and top horizontal axis represent the peak scale of the curvature perturbation k_* and the PBH masses associated with the scale, respectively. The shaded blue region are indirect bounds by BBN/CMB on the SGWB abundance. The bounds from PBH formation are calculated for two benchmark cases: $\{\kappa = 3.0, \delta_c = 0.65\}$ (solid dark green) and $\{\kappa = 11.0, \delta_c = 0.45\}$ (solid light green). Further details can be found in the main text. The evaporation timescales of PBHs ($\tau = t_0$ and 1 s) are also indicated with dashed green lines for reference, where t_0 is the present time. Lastly, the dashed red line represents the bound expected from LIGO-design sensitivity, while the dashed orange line denotes that of ET

Background information - Relation between PBH mass and scale of curvature perturbations

PBHs are produced by critical collapse when large curvature perturbations of scale k re-entered the Hubble radius. The mass of the PBH is determined by the horizon mass, M_k, which is given by [34]

$$M_k \approx 1.4 \times 10^{13} M_\odot \left(\frac{k}{\text{Mpc}^{-1}}\right)^{-2} \left(\frac{g_*}{106.75}\right)^{1/2} \left(\frac{g_{*,s}}{106.75}\right)^{-2/3} . \qquad (26.58)$$

The PBH mass M_{PBH} then follows the scaling law [108]

$$M_{\text{PBH}} = \kappa M_k \left(\delta_m - \delta_c\right)^\gamma , \qquad (26.59)$$

where δ_m is the density contrast, which is related to the curvature perturbations δ_ζ by $\delta_m = \delta_\zeta - 3\delta_\zeta^2/8$. The value of the parameter κ depends on the procedure used to smooth the primordial perturbations [115]. The value $\gamma = 0.36$ is the universal critical exponent during radiation domination [36]. Finally, to explain the critical collapse, it is customary to introduce the density contrast induced by curvature fluctuations at some scale, denoted by δ. If the density contrast exceeds the critical value δ_c, the region collapses gravitationally and forms a PBH after the fluctuation re-enters Hubble radius. An estimate of the value of

δ_c was first obtained by Carr in Ref. [34], where he studied the Jeans mass of the fluctuation. This study led to $\delta_c = 1/3$. More recent numerical studies have shown that the value of δ_c is a bit higher.

For the two benchmark cases to produce the green curves in Fig. 26.7, the critical value for the density contrast is chosen to be 0.45 and 0.65, which are paired with $\kappa = 11.0$ and $\kappa = 3.0$, respectively. The difference in the estimate of PBH abundance between the two benchmark cases reflects the uncertainties in the calculation of the PBH formation. To calculate the bounds of A obtained from constraints in PBH formation, the PBH abundance from the peak in the curvature power spectrum was calculated following the procedure in Ref. [116].

Background information - Other observational constraints

Here we provide brief descriptions of other observational constraints shown in Fig. 26.7:

- *BBN/CMB indirect SGWB constraints*: BBN and CMB can indirectly constrain the amplitude of the SGWB. In the case of BBN, these indirect bounds emerge from the fact that introducing an additional radiation energy component, including the SGWB, influences the expansion rate of the Universe. This impact on the cosmic expansion can be examined through the analysis of light element abundances produced during the BBN epoch [71]. Similarly, the presence of a SGWB at the time of CMB decoupling introduces alterations to both the observed CMB and matter power spectra in a manner identical to massless neutrinos [98]. Recent joint CMB+BBN analysis implies that $\Delta N_{\rm eff} < 0.136$ at 2σ [114], which leads to $\Omega_{\rm GW}h^2 < 1.53 \times 10^{-6}$. This provides mild constraints on the curvature perturbation spectral amplitude applicable in a wide range of frequencies $f > 2 \times 10^{-11}$ Hz, which corresponds to $M_{\rm PBH} < 10^{38}$ g.
- *Constraints on evaporating PBHs:* When the PBH mass is below 5×10^{14} g, PBHs evaporate through Hawking radiation by today and cannot explain the abundance of dark matter in the present cosmological epoch. Particularly, in the PBH mass range of $10^{14} - 10^{15}$ g, the abundance of PBHs strongly radiating in the present Universe is strictly constrained by extragalactic γ-ray background observations. Furthermore, the PBH mass range of $10^{13} - 10^{14}$ g is constrained by the CMB, as PBH evaporations affect CMB observations in various ways. Additionally, lower PBH mass ranges, $10^9 - 10^{13}$ g, are constrained by BBN, as extra radiation from evaporating PBHs alter light element abundances. In Fig. 26.7 we illustrate the collective envelope of these constraints [33, 35] with green curves. For reference, we also show green dashed lines labeled as $\tau = t_0$ (indicating PBHs that evaporate today)

and $\tau = 1$ s (representing those that completely evaporated before BBN), between which constraints of the CMB and BBN should lie.

In summary, the Bayesian analysis does not show evidence for a GWB signal, though it does allow the exclusion of certain regions within the model's parameter space. Constraints have been established based on the peak's width, its position, and the integrated power, surpassing the constraints from BBN and CMB observations in the range of approximately $10^{15} < k/\text{Mpc}^{-1} < 10^{18}$. These constraints, achieving $A \simeq 0.02$ for a narrow peak centered at $k \simeq 10^{17}\text{Mpc}^{-1}$, remain formidable, rivaling the restrictions arising from the PBH abundance associated with such peaks in the curvature power spectrum. However, we find that current ground-based experiments at their design performance, and the future Einstein Telescope, will reach the required sensitivity to set even more stringent upper limits, providing a very powerful probe of the standard formation mechanism of very light PBHs.

Finally, we note that, although we have been discussing the constraints assuming a Gaussian distribution of the primordial curvature perturbations, many inflation models predicting PBHs exhibit non-Gaussian distributions [17, 28, 51, 52], which changes the SGWB constraints. As studied in the literature [9, 11, 29, 43, 85, 117], non-Gaussianity enhances or suppresses the SGWB spectral amplitude. In particular, when we consider the quadratic local-type non-Gaussianity often parameterized by the F_{NL} parameter, the overall spectral amplitude is enhanced, with a more pronounced elevation at high frequencies. This leads to stronger constraints on the curvature perturbation amplitude A for the large non-Gaussian case. Reference [58] discussed constraints using LVK O3 data assuming local-type non-Gaussianity with F_{NL} parameterization and found that there is a strong degeneracy in the parameters of A and F_{NL}. In the limit of large non-Gaussianity ($F_{\text{NL}} \gtrsim 4$), a 95% CL upper limit is found to be $F_{\text{NL}}A \leq 0.13, 0.09, 0.10$ at fixed scales of 10^{16}, $10^{16.5}$, 10^{17} Mpc^{-1}, respectively, for $\Delta \to 0$.

26.4 Constraints on SGWB from PBH Binaries

In this section, we review the constraints on a SGWB from PBH binaries. PBHs have sufficient time to form binary systems after their creation and eventually emit strong GWs when they merge. Those that occur near us can be detected as individual events, while those at high redshift cannot be resolved individually; instead, the ensemble of them can be detected as a SGWB.

In contrast to the SGWB described in the previous section, predicting the amplitude of the SGWB in the case of the PBH binary background is not so robust due to several uncertainties. These uncertainties stem from factors such as the binary formation mechanism, the probability of disruption by a third body, and additional waveform-altering factors like eccentricity, spin, and precession, etc. (we refer the reader to Ref. [19] for an overview). Consequently, even if constraints on the SGWB amplitude are obtained from data, translating them to PBH parameters involves a lot of theoretical uncertainties. Therefore, it is essential to approach the results with caution and carefully consider the assumptions that are made. In this section, we describe the most basic calculation method and briefly review the constraints obtained with the LVK) data in the literature.

The energy density of the SGWB generated by PBH binaris can be derived by integrating single contributions over redshift z and the two binary component masses, m_1 and m_2, as follows [86]

$$\Omega_{\rm GW}(f) = \frac{f}{\rho_{c,0}} \int \mathrm{d}z' \int \mathrm{d}\ln m_1 \int \mathrm{d}\ln m_2 \, \frac{1}{(1+z')H(z')} \times \frac{\mathrm{d}^2\tau_{\rm merg}(z',\, m_1,\, m_2)}{\mathrm{d}\ln m_1 \mathrm{d}\ln m_2} \frac{\mathrm{d}E_{\rm GW}(f_s)}{\mathrm{d}f_s}, \tag{26.60}$$

where $f_s = f(1+z)$ is the redshifted source frame frequency, with z being redshift, $H(z)$ is the Hubble expansion rate, and $\tau_{\rm merg}$ is the differential merger rate per unit time, comoving volume, and mass interval.

In the non-spinning limit, the single source energy spectrum is given by the following phenomenological expression [13]

$$\frac{\mathrm{d}E_{\rm GW}(f)}{\mathrm{d}f} = \frac{(G\pi)^{2/3}}{3}\mathcal{M}_c^{5/3} f^{-1/3} \times \begin{cases} \left(1+\alpha_2 u^2\right)^2 & \text{for} \quad f < f_1, \\ w_1 f \left(1+\epsilon_1 u + \epsilon_2 u^2\right)^2 & \text{for} \quad f_1 \le f < f_2, \\ w_2 f^{7/3} \frac{f_4^4}{(4(f-f_2)^2+f_4^2)^2} & \text{for} \quad f_2 \le f < f_3, \end{cases} \tag{26.61}$$

where $\mathcal{M}_c = (m_1 m_2)^{3/5}/(m_1+m_2)^{1/5}$ is the chirp mass and $u = c^{-1}[\pi (m_1+m_2)Gf]^{1/3}$. The frequencies with index $i = 1, 2, 3, 4$ are given by $f_i = c^3 u_i^3 [\pi(m_1+m_2)G]^{-1}$ with

$$\begin{aligned} u_1^3 &= 0.066 + 0.6437\eta - 0.05822\eta^2 - 7.092\eta^3, \\ u_2^3 &= 0.185 + 0.1469\eta - 0.0249\eta^2 + 2.325\eta^3, \\ u_3^3 &= 0.3236 - 0.1331\eta - 0.2714\eta^2 + 4.922\eta^3, \\ u_4^3 &= 0.0925 - 0.4098\eta + 1.829\eta^2 - 2.87\eta^3, \end{aligned} \tag{26.62}$$

where $\eta = m_1 m_2/(m_1+m_2)^2$ is the symmetric mass ratio. The other fitting parameters are given by $\alpha_2 = -323/224 + 451\eta/168$, $\epsilon_1 = -1.8897$, $\epsilon_2 = 1.6557$, and

$$\begin{aligned} w_1 &= f_1^{-1} \frac{[1+\alpha_2 u_1^2]^2}{[1+\epsilon_1 u_1 + \epsilon_2 u_1^2]^2}, \\ w_2 &= w_1 f_2^{-4/3} [1+\epsilon_1 u_2 + \epsilon_2 u_2^2]^2. \end{aligned} \tag{26.63}$$

The merger rate is a key factor to see the model dependence of PBHs. It depends on the population of black holes and the binary formation scenario, and their details affect the spectral shape. There are two major binary formation channels [20, 87], see Chap. 16 for more details. One is the so-called early binary formation, where a binary is formed by a pair of closely spaced PBHs with the surrounding third object acting on the pair via a tidal force that generates the angular momentum of the binary [84, 95]. In this case, the merger rate is given by

$$\frac{\mathrm{d}^2 \tau_{\mathrm{merg}}(z, m_1, m_2)}{\mathrm{d}\ln m_1 \, \mathrm{d}\ln m_2} = 3.4 \times 10^6 f_{\mathrm{sup}}(m_1, m_2, f_{\mathrm{PBH}}) f_{\mathrm{PBH}}^{53/37} \psi(m_1)\psi(m_2) \times \left(\frac{t}{t_0}\right)^{-34/37} \left(\frac{m_1 + m_2}{M_\odot}\right)^{-32/37} \left(\frac{m_1 m_2}{(m_1+m_2)^2}\right)^{-34/37} \mathrm{yr}^{-1}\mathrm{Gpc}^{-3}, \tag{26.64}$$

where $f_{\mathrm{sup}}(m_1, m_2, f_{\mathrm{PBH}})$ is a supression factor that takes into account several physical effects (see Ref. [57, 88] for details), $f_{\mathrm{PBH}} \equiv \Omega_{\mathrm{PBH}}/\Omega_{\mathrm{DM}}$ is the fraction of PBHs in dark matter, and $\psi(m)$ is the PBH mass function defined by

$$\psi(m) \equiv \frac{1}{\rho_{\mathrm{PBH}}} \frac{\mathrm{d}\rho_{\mathrm{PBH}}}{\mathrm{d}\ln m}. \tag{26.65}$$

The value of the suppression factor is still not clearly understood, but a plausible range is $0.001 < f_{\mathrm{sup}} < 0.1$ when considering a monochromatic PBH mass function [19].

Another channel is the so-called late binary formation, where PBH binaries are formed by tidal capture in very dense halos during the matter-dominated era [25, 38, 74]. In this case, the merger rate is given by

$$\frac{\mathrm{d}^2 \tau_{\mathrm{merg}}(z, m_1, m_2)}{\mathrm{d}\ln m_1 \, \mathrm{d}\ln m_2} = R_{\mathrm{clust}} f_{\mathrm{PBH}}^2 \frac{(m_1+m_2)^{10/7}}{(m_1 m_2)^{5/7}} \psi(m_1)\psi(m_2) \quad \mathrm{yr}^{-1}\mathrm{Gpc}^{-3}, \tag{26.66}$$

where R_{clust} is a scaling factor that depends on the PBH clustering properties and velocity distribution. A rough estimation, using a simplified model where the PBH density is uniformly enhanced within the halo, yields $R_{\mathrm{clust}} \sim 10^2 - 10^3$ [19]. However, this estimate has not been thoroughly refined by considering factors such as dynamical heating, the merger and disruption history of PBH clusters, and the radial distribution of PBHs within a cluster. Additionally, the effects of an extended mass function are likely to be complex.

Now, with predictions for the merger rate given by Eq. (26.64) or (26.66), and the energy spectrum for single sources provided by Eq.(26.61), we can calculate the SGWB spectrum by substituting them into Eq. (26.60). In the left panel of Fig. 26.8, we show the SGWB spectrum of the PBH binaries, assuming a monochromatic mass function for different values of PBH masses and considering both early and late binary formation. We can see that SGWB from early formation channel generically dominates the one from late formation channel when the PBH mass is $M_{\mathrm{PBH}} < O(10) M_\odot$. Roughly speaking, the PBH mass determines the peak frequency of the SGWB. The single source emits the strongest GWs at the merger,

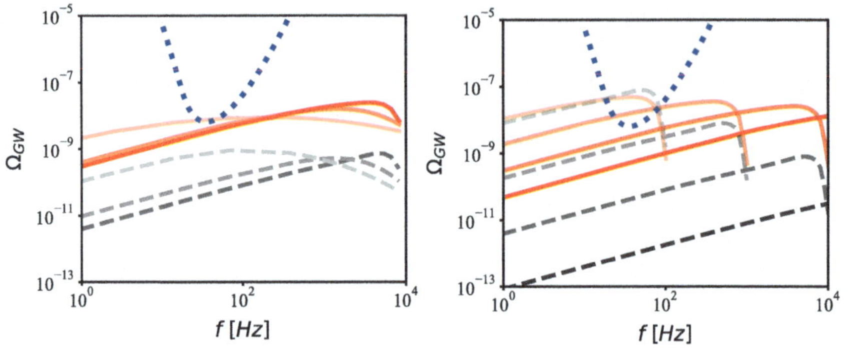

Fig. 26.8 **Left:** Spectra of the SGWB from PBH binaries plotted assuming monochromatic mass function for different values of PBH masses (lines correspond to $M = 100, 10, 1, 0.1 M_\odot$ from left to right). Solid red curves are calculated assuming early binary formation channel with $f_{\rm sup} = 0.002$. Dashed black curves are calculated assuming late binary formation channel with $R_{\rm clust} = 10^2$. $f_{\rm PBH} = 1$ is assumed for both cases. **Right** SGWB spectra for different width of the lognormal mass function (lines correspond to $\sigma = 2, 1, 0.2$ from left to right; we assume $\mu = 1 M_\odot$). Solid red and dashed black curves corresponds early and late formation channels, respectively. O3 sensitivity is also plotted with the dotted blue curve for comparison

corresponding to the peak frequency f_2 in Eq. (26.61). For binary PBHs of equal mass, it is given by $f_2 = 8286(M_\odot/M_{\rm PBH})$ Hz. In the right panel, we depict the spectrum assuming a lognormal Gaussian distribution of the PBH mass function $\psi(m) = 1/(\sqrt{2\pi}\sigma)\exp[-(\ln(m/\mu)^2)/(2\sigma^2)]$ for different width parameters σ. We observe that a wider mass function smooths out the peak of the spectrum.

As far as we consider PBH masses smaller than $\mathcal{O}(10)M_\odot$, it is most likely that we will observe the $\Omega_{\rm GW} \propto f^{2/3}$ behavior of the inspiral phase with ground-based detectors. Therefore, the astrophysical CBC background, jointly considered with the SGWB signal in the previous section, effectively represents the PBH binary background. The most conservative 95% CL upper limit on $\Omega_{\rm ref}$ obtained in the previous section is $\Omega_{\rm ref} < 6.6 \times 10^{-9}$ at 25Hz. For a monochromatic mass function, it is straightforward to show that the spectral amplitude during the inspiral phase depends on $\Omega_{\rm GW} \propto M_{\rm PBH}^{89/111}$ for early binary formation and $\propto M_{\rm PBH}^{5/3}$ for late binary formation. Therefore, the bound on $\Omega_{\rm ref}$ can provide upper limits on the following combinations of parameters

$$f_{\rm sup} f_{\rm PBH}^{53/37} (M_{\rm PBH}/M_\odot)^{89/111} \lesssim 2.6 \quad \text{for early binary formation} \tag{26.67}$$

$$R_{\rm clust} f_{\rm PBH}^{2} (M_{\rm PBH}/M_\odot)^{5/3} \lesssim 200 \quad \text{for late binary formation} \tag{26.68}$$

Note that, for $M_{\rm PBH} > \mathcal{O}(10)M_\odot$, this argument is not applicable as the spectrum no longer follows the $\Omega_{\rm GW} \propto f^{2/3}$ behavior.

When we have observations with better sensitivity across a wider frequency range, we must be cautious about deviations from the spectral shape of $\Omega_{\rm GW} \propto f^{2/3}$, and more precise considerations are needed. Reference [83] performed parameter estimation for a SGWB from PBH binaries using LVK data. As seen in their work, with the current sensitivity of the experiments, we are unable to obtain useful constraints on the model parameters, but third-generation detectors will be a powerful tool to extract information on PBH populations [81].

26.5 Conclusions

In this chapter, we have explored potential signals of PBHs in SGWB observations. We have delved into two distinct types of signals: those associated with PBH formation in the early stages and signals arising from the binaries of PBHs. We have discussed their current constraints, focusing on observations from the recent LVK O3 run. These two observables probe completely different epochs of the Universe; one provides access to the very early radiation-dominated phase, while the other probes PBHs in the late Universe. The scales of PBHs probed by the LVK detectors are also very different: the former imposes limits on PBHs of $10^{-24} M_\odot$ to $10^{-18} M_\odot$, while the latter exhibits sensitivity to $\mathcal{O}(10^{1\div2}) M_\odot$.

Present constraints already provide meaningful limits, and even stronger constraints will become available with enhanced sensitivity in ongoing and upcoming observation runs, O4 and O5. Moreover, third-generation detectors, such as the Einstein Telescope (ET) and the Cosmic Explorer (CE), are expected to significantly increase sensitivity, reaching $\Omega_{\rm GW} \sim 10^{-13}$ and expanding the frequency range. Additionally, while not explored in detail here, space-based gravitational wave detectors like LISA, TianQin, Taiji, and DECIGO can offer intriguing constraints in a similar manner. Furthermore, pulsar timing arrays offer another avenue to probe SGWB at nano-Hz frequencies. In such cases, the corresponding PBH mass tends to shift toward higher masses because of the lower targeting frequencies of these experiments. All the advancements in gravitational wave probes will enable us to explore a much broader parameter space of PBH models. We conclude with Fig. 26.9, where we present the anticipated future explorations for PBHs across a wide range of mass scales, reflecting the immense potential of future gravitational wave experiments to explore new physics.

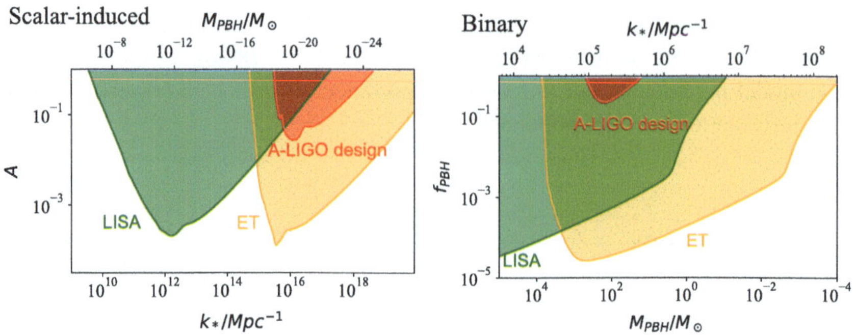

Fig. 26.9 Accessible parameter space with future experiment sensitivities **Left**: for the spectral amplitude of the curvature power spectrum A, achievable through the SIGWB in the future is shown as functions of both k_* and $M_{\rm PBH}$, and **Right**: for the fraction of PBH $f_{\rm PBH}$ through the PBH binary background (assuming the early binary formation channel with $f_{\rm sup} = 0.002$). For both cases, monochromatic PBH mass function is assumed. The curves are obtained assuming 1 year of observation by Advanced-LIGO design sensitivity, ET, and LISA, with a conservative detection threshold of SNR = 8

References

1. J. Aasi et al., Searching for stochastic gravitational waves using data from the two colocated LIGO Hanford detectors. Phys. Rev. D **91**, 022003 (2 Jan 2015). https://doi.org/10.1103/PhysRevD.91.022003. https://link.aps.org/doi/10.1103/PhysRevD.91.022003
2. B.P. Abbott et al., Exploring the sensitivity of next generation gravitational wave detectors. Classical Quantum Gravity **34**(4), 044001 (2017). https://doi.org/10.1088/1361-6382/aa51f4. https://doi.org/10.1088/1361-6382/aa51f4
3. B.P. Abbott et al., GW150914: implications for the stochastic gravitational-wave background from binary black holes. Phys. Rev. Lett. **116**, 131102 (13 Mar 2016). https://doi.org/10.1103/PhysRevLett.116.131102. https://link.aps.org/doi/10.1103/PhysRevLett.116.131102
4. B.P. Abbott et al., GW170817: implications for the stochastic gravitational-wave background from compact binary coalescences. Phys. Rev. Lett. **120**, 091101 (9 Feb 2018). https://doi.org/10.1103/PhysRevLett.120.091101. https://link.aps.org/doi/10.1103/PhysRevLett.120.091101
5. B.P. Abbott et al., Search for the isotropic stochastic background using data from Advanced LIGO's second observing run. Phys. Rev. D **100**, 061101 (6 Sept 2019). https://doi.org/10.1103/PhysRevD.100.061101. https://link.aps.org/doi/10.1103/PhysRevD.100.061101
6. B.P. Abbott et al., Upper limits on the stochastic gravitational-wave background from advanced LIGO's first observing run. Phys. Rev. Lett. **118**, 121101 (12 Mar 2017). https://doi.org/10.1103/PhysRevLett.118.121101. https://link.aps.org/doi/10.1103/PhysRevLett.118.121101
7. R. Abbott et al., Upper limits on the isotropic gravitational-wave background from advanced LIGO and advanced virgo's third observing run. Phys. Rev. D **104**, 022004 (2 July 2021). https://doi.org/10.1103/PhysRevD.104.022004. https://link.aps.org/doi/10.1103/PhysRevD.104.022004
8. R. Abbott et al., Population of merging compact binaries inferred using gravitational waves through GWTC-3. Phys. Rev. X **13**(1), 011048 (2023). https://doi.org/10.1103/PhysRevX.13.011048. arXiv: 2111.03634 [astro-ph.HE]
9. K.T. Abe et al., Primordial black holes and gravitational waves induced by exponential-tailed perturbations (2022). arXiv: 2209.13891 [astro-ph.CO]

10. P.A.R. Ade et al., Improved constraints on primordial gravitational waves using planck, WMAP, and BICEP/Keck observations through the 2018 observing season. Phys. Rev. Lett. **127**(15), 151301 (2021). https://doi.org/10.1103/PhysRevLett.127.151301. arXiv: 2110.00483 [astro-ph.CO]
11. P. Adshead, K.D. Lozanov, Z.J. Weiner, Non-gaussianity and the induced gravitational wave background. JCAP **10**, 080 (2021). https://doi.org/10.1088/1475-7516/2021/10/080. arXiv: 2105.01659 [astro-ph.CO]
12. G. Agazie et al., Comparing recent PTA results on the nanohertz stochastic gravitational wave background (2023). arXiv: 2309.00693 [astro-ph.HE]
13. P. Ajith et al., Inspiral-merger-ringdown waveforms for black-hole binaries with non-precessing spins. Phys. Rev. Lett. **106**, 241101 (2011). https://doi.org/10.1103/PhysRevLett.106.241101. arXiv: 0909.2867 [gr-qc]
14. B. Allen, J.D. Romano, Detecting a stochastic background of gravitational radiation: signal processing strategies and sensitivities. Phys. Rev. D **59**, 102001 (10 Mar 1999). https://doi.org/10.1103/PhysRevD.59.102001. https://link.aps.org/doi/10.1103/PhysRevD.59.102001
15. K.N. Ananda, C. Clarkson, D. Wands, The cosmological gravitational wave background from primordial density perturbations. Phys. Rev. D **75**, 123518 (2007). https://doi.org/10.1103/PhysRevD.75.123518. (arXiv: gr-qc/0612013)
16. V. Atal et al., Constraining changes in the merger history of BH and PBH binaries with the stochastic gravitational wave background. Phys. Rev. D **105**(12), 123522 (2022). https://doi.org/10.1103/PhysRevD.105.123522. arXiv: 2201.12218 [astro-ph.CO]
17. V. Atal et al., PBH in single field inflation: the effect of shape dispersion and non-Gaussianities. JCAP **05**, 022 (2020). https://doi.org/10.1088/1475-7516/2020/05/022. arXiv: 1908.11357 [astro-ph.CO]
18. S. Babak et al., Stochastic gravitational wave background from stellar origin binary black holes in LISA. JCAP **08**, 034 (2023). https://doi.org/10.1088/1475-7516/2023/08/034. arXiv: 2304.06368 [astro-ph.CO]
19. E. Bagui et al., Primordial black holes and their gravitational-wave signatures (2023). arXiv: 2310.19857 [astro-ph.CO]
20. E. Bagui, S. Clesse, A boosted gravitational wave background for primordial black holes with broad mass distributions and thermal features. Phys. Dark Univ. **38**, 101115 (2022). https://doi.org/10.1016/j.dark.2022.101115. arXiv: 2110.07487 [astro-ph.CO]
21. N. Bartolo et al., Primordial black hole dark matter: LISA serendipity. Phys. Rev. Lett. **122**(21), 211301 (2019). https://doi.org/10.1103/PhysRevLett.122.211301. arXiv: 1810.12218 [astro-ph.CO]
22. N. Bartolo et al., Testing primordial black holes as dark matter with LISA. Phys. Rev. D **99**(10), 103521 (2019). https://doi.org/10.1103/PhysRevD.99.103521. arXiv: 1810.12224 [astro-ph.CO]
23. D. Baumann et al., Gravitational wave spectrum induced by primordial scalar perturbations. Phys. Rev. D **76**, 084019 (2007). https://doi.org/10.1103/PhysRevD.76.084019. (arXiv: hep--th/0703290)
24. S.S. Bavera et al., Stochastic gravitational-wave background as a tool for investigating multi-channel astrophysical and primordial black-hole mergers. Astron. Astrophys. **660**, A26 (2022). https://doi.org/10.1051/0004-6361/202142208. arXiv: 2109.05836 [astro-ph.CO]
25. S. Bird et al., Did LIGO detect dark matter? Phys. Rev. Lett. **116**(20), 201301 (2016). https://doi.org/10.1103/PhysRevLett.116.201301. arXiv: 1603.00464 [astro-ph.CO]
26. M. Braglia, J. Garcia-Bellido, S. Kuroyanagi, Testing Primordial Black Holes with multi-band observations of the stochastic gravitational wave background. JCAP **12**(12), 012 (2021). https://doi.org/10.1088/1475-7516/2021/12/012. arXiv: 2110.07488 [astro-ph.CO]
27. M. Braglia, J. Garcia-Bellido, S. Kuroyanagi, Tracking the origin of black holes with the stochastic gravitational wave background popcorn signal (2022). arXiv: 2201.13414 [astro-ph.CO]

28. Y.-F. Cai et al., Revisiting non-gaussianity from non-attractor inflation models. JCAP **05**, 012 (2018). https://doi.org/10.1088/1475-7516/2018/05/012. arXiv: 1712.09998 [astro-ph.CO]
29. R.-g. Cai, S. Pi, M. Sasaki, Gravitational waves induced by non-gaussian scalar perturbations. Phys. Rev. Lett. **122**(20), 201101 (2019). https://doi.org/10.1103/PhysRevLett.122.201101. arXiv: 1810.11000 [astro-ph.CO]
30. T.A. Callister, M.W. Coughlin, J.B. Kanner, Gravitational-wave geodesy: a new tool for validating detection of the stochastic gravitational-wave background. Astrophys. J. **869**(2), L28 (2018). https://doi.org/10.3847/2041-8213/aaf3a5. https://doi.org/10.3847/2041-8213/aaf3a5
31. T. Callister et al., Limits of astrophysics with gravitational-wave backgrounds. Phys. Rev. X **6**, 031018 (3 Aug 2016). https://doi.org/10.1103/PhysRevX.6.031018. https://link.aps.org/doi/10.1103/PhysRevX.6.031018
32. T. Callister et al., Polarization-based tests of gravity with the stochastic gravitational-wave background. Phys. Rev. X **7**(4), 041058 (2017). https://doi.org/10.1103/PhysRevX.7.041058. arXiv: 1704.08373 [gr-qc]
33. B.J. Carr et al., New cosmological constraints on primordial black holes. Phys. Rev. D **81**, 104019 (2010). https://doi.org/10.1103/PhysRevD.81.104019. arXiv: 0912.5297 [astro-ph.CO]
34. B. Carr, The Primordial black hole mass spectrum. Astrophys. J. **201**, 1–19 (1975). https://doi.org/10.1086/153853
35. B. Carr et al., Constraints on primordial black holes. Rept. Prog. Phys. **84**(11), 116902 (2021). https://doi.org/10.1088/1361-6633/ac1e31. arXiv: 2002.12778 [astro-ph.CO]
36. M.W. Choptuik, Universality and scaling in gravitational collapse of a massless scalar field. Phys. Rev. Lett. **70**, 9–12 (1 Jan 1993). https://doi.org/10.1103/PhysRevLett.70.9. https://link.aps.org/doi/10.1103/PhysRevLett.70.9
37. N. Christensen, Measuring the stochastic gravitational radiation background with laser interferometric antennas. Phys. Rev. D **46**, 5250–5266 (1992). https://doi.org/10.1103/PhysRevD.46.5250
38. S. Clesse, J. García-Bellido, Detecting the gravitational wave background from primordial black hole dark matter. Phys. Dark Univ. **18**, 105–114 (2017). https://doi.org/10.1016/j.dark.2017.10.001. arXiv: 1610.08479 [astro-ph.CO]
39. M. Colpi et al., LISA definition study report (2024). arXiv: 2402.07571 [astro-ph.CO]
40. N.J. Cornish, J.D. Romano, When is a gravitational-wave signal stochastic? Phys. Rev. D **92**, 042001 (2015)
41. P.B. Covas et al., Identification and mitigation of narrow spectral artifacts that degrade searches for persistent gravitational waves in the first two observing runs of advanced LIGO. Phys. Rev. D **97**(8), 082002 (2018). https://doi.org/10.1103/PhysRevD.97.082002. arXiv: 1801.07204 [astro-ph.IM]
42. C. Cutler, Angular resolution of the LISA gravitational wave detector. Phys. Rev. D **57**, 7089–7102 (1998). https://doi.org/10.1103/PhysRevD.57.7089. (arXiv: gr-qc/9703068)
43. L.V. Dall'Armi et al., Circular polarization of the astrophysical gravitational wave background (2023). arXiv: 2301.08205 [astro-ph.CO]
44. T. Damour, A. Vilenkin, Gravitational wave bursts from cosmic strings. Phys. Rev. Lett. **85**, 3761–3764 (2000). https://doi.org/10.1103/PhysRevLett.85.3761. (arXiv: gr-qc/0004075)
45. T. Damour, A. Vilenkin, Gravitational wave bursts from cusps and kinks on cosmic strings. Phys. Rev. D **64**, 064008 (2001). https://doi.org/10.1103/PhysRevD.64.064008. (arXiv: gr-qc/0104026)
46. P.E. Dewdney et al., The square kilometre array (2009). https://doi.org/10.1109/JPROC.2009.2021005
47. G. Domènech, Scalar induced gravitational waves review. Universe **7**(11), 398 (2021). https://doi.org/10.3390/universe7110398. arXiv: 2109.01398 [gr-qc]
48. D.J. D'Orazio, J. Samsing, Black hole mergers from globular clusters observable by LISA II: resolved eccentric sources and the gravitational wave background. Mon. Not. Roy. Astron.

Soc. **481**(4), 4775–4785 (2018). https://doi.org/10.1093/mnras/sty2568. arXiv: 1805.06194 [astro-ph.HE]
49. J.R. Espinosa, D. Racco, A. Riotto, A cosmological signature of the SM higgs instability: gravitational waves. JCAP **09**, 012 (2018). https://doi.org/10.1088/1475-7516/2018/09/012. arXiv: 1804.07732 [hep-ph]
50. E.E. Flanagan, The sensitivity of the laser interferometer gravitational wave observatory (LIGO) to a stochastic background, and its dependence on the detector orientations. Phys. Rev. D **48**, 2389–2407 (1993). https://doi.org/10.1103/PhysRevD.48.2389. (arXiv: astro-ph/9305029)
51. J. García-Bellido, M. Peloso, C. Unal, Gravitational wave signatures of inflationary models from primordial black hole dark matter. JCAP **09**, 013 (2017). https://doi.org/10.1088/1475-7516/2017/09/013. arXiv: 1707.02441 [astro-ph.CO]
52. J. García-Bellido, M. Peloso, C. Unal, Gravitational waves at interferometer scales and primordial black holes in axion inflation. JCAP **12**, 031 (2016). https://doi.org/10.1088/1475-7516/2016/12/031. arXiv: 1610.03763 [astro-ph.CO]
53. M. Giovannini, Gravitational waves constraints on postinflationary phases stiffer than radiation. Phys. Rev. D **58**, 083504 (1998). https://doi.org/10.1103/PhysRevD.58.083504. (arXiv: hep-ph/9806329)
54. M. Hazumi et al., LiteBIRD: JAXA's new strategic L-class mission for all-sky surveys of cosmic microwave background polarization. Proc. SPIE Int. Soc. Opt. Eng. **11443**, 114432F (2020). https://doi.org/10.1117/12.2563050. arXiv: 2101.12449 [astro-ph.IM]
55. https://pypi.org/project/pygwb
56. https://dcc.ligo.org/P2000546/public
57. G. Hütsi et al., Two populations of LIGO-Virgo black holes. JCAP **03**, 068 (2021). https://doi.org/10.1088/1475-7516/2021/03/068. arXiv: 2012.02786 [astro-ph.CO]
58. R. Inui et al., Constraints on non-gaussian primordial curvature perturbation from the LIGO-Virgo-KAGRA third observing run (2023). arXiv: 2311.05423 [astro-ph.CO]
59. L. Isserlis, On a formula for the product-moment coefficient of any order of a normal frequency distribution in any number of variables. Biometrika **12**(1–2), 134–139 (1918). https://doi.org/10.1093/biomet/12.1-2.134
60. K. Janssens et al., Gravitational-wave geodesy: defining false alarm probabilities with respect to correlated noise. Phys. Rev. D **105**, 082001 (8 Apr 2022). https://doi.org/10.1103/PhysRevD.105.082001. https://link.aps.org/doi/10.1103/PhysRevD.105.082001
61. K. Janssens et al., Impact of schumann resonances on the einstein telescope and projections for the magnetic coupling function. Phys. Rev. D **104**, 122006 (12 Dec 2021). https://doi.org/10.1103/PhysRevD.104.122006. https://link.aps.org/doi/10.1103/PhysRevD.104.122006
62. S. Jaraba et al., Stochastic gravitational wave background constraints from Gaia DR3 astrometry. Mon. Not. Roy. Astron. Soc. **524**(3), 3609–3622 (2023). https://doi.org/10.1093/mnras/stad2141. arXiv: 2304.06350 [astro-ph.CO]
63. S.J. Kapadia et al., Search for the stochastic gravitational-wave background induced by primordial curvature perturbations in LIGO's second observing run. Astrophys. J. Lett. **910**(1), L4 (2021). https://doi.org/10.3847/2041-8213/abe86e. arXiv: 2009.05514 [gr-qc]
64. S. Kawamura et al., Current status of space gravitational wave antenna DECIGO and B-DECIGO. PTEP **2021**(5), 05A105 (2021). https://doi.org/10.1093/ptep/ptab019. arXiv: 2006.13545 [gr-qc]
65. T. Kite, A. Ravenni, S.P. Patil, J. Chluba, Bridging the gap: spectral distortions meet gravitational waves. Mon. Not. Roy. Astron. Soc. **505**(3), 4396–4405 (2021). arXiv:2010.00040. https://doi.org/10.1093/mnras/stab1558
66. K. Kohri, T. Terada, Semianalytic calculation of gravitational wave spectrum nonlinearly induced from primordial curvature perturbations. Phys. Rev. D **97**, 123532 (12 June 2018). https://doi.org/10.1103/PhysRevD.97.123532. https://link.aps.org/doi/10.1103/PhysRevD.97.123532
67. A. Kosowsky, M.S. Turner, Gravitational radiation from colliding vacuum bubbles: envelope approximation to many bubble collisions. Phys. Rev. D **47**, 4372–4391 (1993). https://doi.org/10.1103/PhysRevD.47.4372. (arXiv: astro-ph/9211004)

68. J.K. Kruschke, *Doing Bayesian Data Analysis: A Tutorial with R, JAGS, and Stan*, 2nd edn., pp. 1–759 (2014). https://doi.org/10.1016/B978-0-12-405888-0.09999-2
69. H. Kudoh et al., Detecting a gravitational-wave background with next-generation space interferometers. Phys. Rev. D **73**, 064006 (2006). https://doi.org/10.1103/PhysRevD.73.064006. (arXiv: gr-qc/0511145)
70. L. Lehoucq et al., Astrophysical uncertainties in the gravitational-wave background from stellar-mass compact binary mergers. Mon. Not. Roy. Astron. Soc. **526**(3), 4378–4387 (2023). https://doi.org/10.1093/mnras/stad2917. arXiv: 2306.09861 [astro-ph.HE]
71. M. Maggiore, Gravitational wave experiments and early universe cosmology. Phys. Rept. **331**, 283–367 (2000). https://doi.org/10.1016/S0370-1573(99)00102-7. (arXiv: gr-qc/9909001)
72. D. Makowski, M. Ben-Shachar, D. Lüdecke, BayestestR: describing effects and their uncertainty, existence and significance within the bayesian framework. J. Open Source Softw. **4**(40), 1541 (2019). https://doi.org/10.21105/joss.01541
73. V. Mandic et al., Parameter estimation in searches for the stochastic gravitational-wave background. Phys. Rev. Lett. **109**, 171102 (17 Oct 2012). https://doi.org/10.1103/PhysRevLett.109.171102. https://link.aps.org/doi/10.1103/PhysRevLett.109.171102
74. V. Mandic, S. Bird, I. Cholis, Stochastic gravitational-wave background due to primordial binary black hole mergers. Phys. Rev. Lett. **117**(20), 201102 (2016). https://doi.org/10.1103/PhysRevLett.117.201102. arXiv: 1608.06699 [astro-ph.CO]
75. S. Marassi, R. Schneider, V. Ferrari, Gravitational wave backgrounds and the cosmic transition from Population III to Population II stars. Mon. Not. R. Astron. Soc. **398**(1), 293–302 (2009). ISSN: 0035-8711. https://doi.org/10.1111/j.1365-2966.2009.15120.x. https://academic.oup.com/mnras/article-pdf/398/1/293/3829554/mnras0398-0293.pdf
76. S. Matarrese, O. Pantano, D. Saez, General relativistic dynamics of irrotational dust: cosmological implications. Phys. Rev. Lett. **72**, 320–323 (1994). https://doi.org/10.1103/PhysRevLett.72.320. (arXiv: astro-ph/9310036)
77. S. Matarrese, S. Mollerach, M. Bruni, Second order perturbations of the Einstein-de Sitter universe. Phys. Rev. D **58**, 043504 (1998). https://doi.org/10.1103/PhysRevD.58.043504. arXiv: astro-ph/9707278
78. A. Matas, J. Romano, Frequentist versus Bayesian analyses: cross-correlation as an approximate sufficient statistic for LIGO-Virgo stochastic background searches. Phys. Rev. D **103** (2021). https://doi.org/10.1103/PhysRevD.103.062003
79. P.F. Michelson, On detecting stochastic background gravitational radiation with terrestrial detectors. Mon. Not. R. Astron. Soc. **227**(4), 933–941 (1987). ISSN: 0035-8711. https://doi.org/10.1093/mnras/227.4.933. https://academic.oup.com/mnras/article-pdf/227/4/933/3926536/mnras227-0933.pdf
80. S. Mukherjee, M.S.P. Meinema, J. Silk, Prospects of discovering subsolar primordial black holes using the stochastic gravitational wave background from third-generation detectors. Mon. Not. Roy. Astron. Soc. **510**(4), 6218–6224 (2022). https://doi.org/10.1093/mnras/stab3756. arXiv: 2107.02181 [astro-ph.CO]
81. S. Mukherjee, J. Silk, Can we distinguish astrophysical from primordial black holes via the stochastic gravitational wave background? Mon. Not. Roy. Astron. Soc. **506**(3), 3977–3985 (2021). https://doi.org/10.1093/mnras/stab1932. arXiv: 2105.11139 [gr-qc]
82. S. Mukherjee, J. Silk, Time-dependence of the astrophysical stochastic gravitational wave background. Mon. Not. Roy. Astron. Soc. **491**(4), 4690–4701 (2020). https://doi.org/10.1093/mnras/stz3226. arXiv: 1912.07657 [gr-qc]
83. T. Nakamura et al., Gravitational waves from coalescing black hole MACHO binaries. Astrophys. J. Lett. **487**, L139–L142 (1997). https://doi.org/10.1086/310886. (arXiv: astro-ph/9708060)
84. G. Perna et al., Fully non-gaussian scalar-induced gravitational waves (2024). arXiv: 2403.06962 [astro-ph.CO]
85. E.S. Phinney, A practical theorem on gravitational wave backgrounds (2001). arXiv: astro-ph/0108028

86. M. Raidal, V. Vaskonen, H. Veermäe, Gravitational waves from primordial black hole mergers. JCAP **09**, 037 (2017). https://doi.org/10.1088/1475-7516/2017/09/037. arXiv: 1707.01480 [astro-ph.CO]
87. M. Raidal et al., Formation and evolution of primordial black hole binaries in the early universe. JCAP **02**, 018 (2019). https://doi.org/10.1088/1475-7516/2019/02/018. arXiv: 1812.01930 [astro-ph.CO]
88. T. Regimbau, The astrophysical gravitational wave stochastic background. Res. Astron. Astrophys. **11**, 369–390 (2011). https://doi.org/10.1088/1674-4527/11/4/001. arXiv: 1101.2762 [astro-ph.CO]
89. A.I. Renzini et al., pygwb: a python-based library for gravitational-wave background searches. Astrophys. J. **952**(1), 25 (2023). https://doi.org/10.3847/1538-4357/acd775. arXiv: 2303.15696 [gr-qc]
90. J. Romano, N. Cornish, Detection methods for stochastic gravitational-wave backgrounds: a unified treatment. Living Rev. Relativ. **20** (2017). https://doi.org/10.1007/s41114-017-0004-1
91. A. Romero et al., Implications for first-order cosmological phase transitions from the Third LIGO-Virgo observing run. Phys. Rev. Lett. **126**(15), 151301 (2021). https://doi.org/10.1103/PhysRevLett.126.151301. arXiv: 2102.01714 [hep-ph]
92. A. Romero-RodrÍguez et al., Search for a scalar induced stochastic gravitational wave background in the third LIGO-Virgo observing run. Phys. Rev. Lett. **128**, 051301 (5 Feb 2022). https://doi.org/10.1103/PhysRevLett.128.051301. https://link.aps.org/doi/10.1103/PhysRevLett.128.051301
93. R. Saito, J. Yokoyama, Gravitational wave background as a probe of the primordial black hole abundance. Phys. Rev. Lett. **102**, 161101 (2009). [Erratum: Phys. Rev. Lett. **107**, 069901 (2011)]. https://doi.org/10.1103/PhysRevLett.102.161101. arXiv: 0812.4339 [astro-ph]
94. M. Sasaki et al., Primordial black hole scenario for the gravitational-wave event GW150914. Phys. Rev. Lett. **117**(6), 061101 (2016). [Erratum: Phys. Rev. Lett. **121**, 059901 (2018)]. https://doi.org/10.1103/PhysRevLett.117.061101. arXiv: 1603.08338 [astro-ph.CO]
95. N. Seto, A. Taruya, Polarization analysis of gravitational-wave backgrounds from the correlation signals of ground-based interferometers: measuring a circular-polarization mode. Phys. Rev. D **77**, 103001 (2008). https://doi.org/10.1103/PhysRevD.77.103001. arXiv: 0801.4185 [astro-ph]
96. X. Siemens, V. Mandic, J. Creighton, Gravitational-wave stochastic background from cosmic strings. Phys. Rev. Lett. **98**, 111101 (11 Mar 2007). https://doi.org/10.1103/PhysRevLett.98.111101. https://link.aps.org/doi/10.1103/PhysRevLett.98.111101
97. T.L. Smith, E. Pierpaoli, M. Kamionkowski, A new cosmic microwave background constraint to primordial gravitational waves. Phys. Rev. Lett. **97**, 021301 (2006). https://doi.org/10.1103/PhysRevLett.97.021301. (arXiv: astro-ph/0603144)
98. T.L. Smith et al., LISA for cosmologists: calculating the signal-to-noise ratio for stochastic and deterministic sources. Phys. Rev. D **100**(10), 104055 (2019). [Erratum: Phys. Rev. D **105**, 029902 (2022)]. https://doi.org/10.1103/PhysRevD.100.104055. arXiv: 1908.00546 [astro-ph.CO]
99. A.A. Starobinsky, A new type of isotropic cosmological models without singularity. Phys. Lett. B **91** (1980). Ed. by I.M. Khalatnikov, V.P. Mineev, pp. 99–102. https://doi.org/10.1016/0370-2693(80)90670-X
100. L. Sun et al., Characterization of systematic error in advanced LIGO calibration. Classical Quantum Gravity **37**(22), 225008 (2020). https://doi.org/10.1088/1361-6382/abb14e. https://doi.org/10.1088/1361-6382/abb14e
101. C. Talbot et al., Inference with finite time series: observing the gravitational universe through windows. Phys. Rev. Res. **3**, 043049 (4 Oct 2021). https://doi.org/10.1103/PhysRevResearch.3.043049. https://link.aps.org/doi/10.1103/PhysRevResearch.3.043049
102. E. Thrane, N. Christensen, R.M.S. Schofield, Correlated magnetic noise in global networks of gravitational-wave detectors: observations and implications. Phys. Rev. D **87**, 123009 (12 June 2013). https://doi.org/10.1103/PhysRevD.87.123009. https://link.aps.org/doi/10.1103/PhysRevD.87.123009

103. E. Thrane et al., Correlated noise in networks of gravitational-wave detectors: subtraction and mitigation. Phys. Rev. D **90**, 023013 (2 July 2014). https://doi.org/10.1103/PhysRevD.90.023013. https://link.aps.org/doi/10.1103/PhysRevD.90.023013
104. K. Tomita, Non-linear theory of gravitational instability in the expanding universe. Prog. Theor. Phys. **37**(5), 831 (1967). https://doi.org/10.1143/PTP.37.831
105. C. Unal, Imprints of primordial non-gaussianity on gravitational wave spectrum. Phys. Rev. D **99**(4), 041301 (2019). https://doi.org/10.1103/PhysRevD.99.041301. arXiv: 1811.09151 [astro-ph.CO]
106. N. van Remortel, K. Janssens, K. Turbang, Stochastic gravitational wave background: Methods and implications. Prog. Part. Nucl. Phys. **128**, 104003 (2023). https://doi.org/10.1016/j.ppnp.2022.104003. arXiv: 2210.00761 [gr-qc]
107. V. Vaskonen, H. Veermäe, Did NANOGrav see a signal from primordial black hole formation? Phys. Rev. Lett. **126**, 051303 (5 Feb 2021). https://doi.org/10.1103/PhysRevLett.126.051303. https://link.aps.org/doi/10.1103/PhysRevLett.126.051303
108. S. Vitale et al., Physical approach to the marginalization of LIGO calibration uncertainties. Phys. Rev. D **103**, 063016 (6 Mar 2021). https://doi.org/10.1103/PhysRevD.103.063016. https://link.aps.org/doi/10.1103/PhysRevD.103.063016
109. S. Wang, T. Terada, K. Kohri, Prospective constraints on the primordial black hole abundance from the stochastic gravitational-wave backgrounds produced by coalescing events and curvature perturbations. Phys. Rev. D **99**(10), 103531 (2019). [Erratum: Phys. Rev. D **101**, 069901 (2020)]. https://doi.org/10.1103/PhysRevD.99.103531. arXiv: 1903.05924 [astro-ph.CO]
110. S. Wang, V. Vardanyan, K. Kohri, Probing primordial black holes with anisotropies in stochastic gravitational-wave background. Phys. Rev. D **106**(12), 123511 (2022). https://doi.org/10.1103/PhysRevD.106.123511. arXiv: 2107.01935 [gr-qc]
111. S. Wang et al., Constraints on the primordial black hole abundance from the first advanced LIGO observation run using the stochastic gravitational-wave background. Phys. Rev. Lett. **120**(19), 191102 (2018). https://doi.org/10.1103/PhysRevLett.120.191102. arXiv: 1610.08725 [astro-ph.CO]
112. J. Whelan et al., Treatment of calibration uncertainty in multi-baseline cross-correlation searches for gravitational waves. J. Phys.: Conf. Ser. **484** (2012). https://doi.org/10.1088/1742-6596/484/1/012027
113. T.-H. Yeh et al., Probing physics beyond the standard model: limits from BBN and the CMB independently and combined. JCAP **10**, 046 (2022). https://doi.org/10.1088/1475-7516/2022/10/046. arXiv: 2207.13133 [astro-ph.CO]
114. S. Young, The primordial black hole formation criterion re-examined: Parametrisation, timing and the choice of window function. Int. J. Mod. Phys. D **29** (2019). https://doi.org/10.1142/S0218271820300025
115. S. Young, I. Musco, C.T. Byrnes., Primordial black hole formation and abundance: contribution from the non-linear relation between the density and curvature perturbation. JCAP **11**, 012 (2019). https://doi.org/10.1088/1475-7516/2019/11/012. arXiv: 1904.00984 [astro-ph.CO]
116. C. Yuan, Q.-G. Huang, Gravitational waves induced by the local-type non-Gaussian curvature perturbations. Phys. Lett. B **821**, 136606 (2021). https://doi.org/10.1016/j.physletb.2021.136606. arXiv: 2007.10686 [astro-ph.CO]
117. Y. Zhao, Y. Lu, Stochastic gravitational wave background and eccentric stellar compact binaries. Mon. Not. Roy. Astron. Soc. **500**(1), 1421–1436 (2020). https://doi.org/10.1093/mnras/staa2707. arXiv: 2009.01436 [astro-ph.HE]

Part VI
Outlook

Chapter 27
The Future of Primordial Black Holes: Open Questions and Roadmap

Antonio Riotto and Joseph Silk

Abstract We close this book by discussing some of the open questions and the roadmap in the physics of primordial black holes. Black holes are the only dark matter candidate that is known to actually exist. Their conjectured primordial role is admittedly based on hypothesis rather than fact, most straightforwardly as a simple extension to the standard models of inflation, or even, in homage to quantum physics, more controversially via a slowing-down of Hawking evaporation. Regardless of one's stance on the theoretical basis for their existence, the possibility of primordial black holes playing a novel role in dark matter physics and gravitational wave astronomy opens up a rich astrophysical phenomenology that we lay out in this brief overview.

27.1 Introduction

Primordial black holes (PBHs) provide an attractive way to resolve the dark matter problem. Unlike essentially all other dark matter particle candidates, BHs actually exist, and there are several robust prospects for determining whether there is a primordial component over a broad mass range. Theoretically PBHs may populate the

A. Riotto
Département de Physique Théorique, Université de Genève, 24 quai Ansermet, Genève 4, 1211 Geneva, Switzerland
e-mail: antonio.riotto@unige.ch

Gravitational Wave Science Center (GWSC), Université de Genève, 1211 Geneva, Switzerland

J. Silk (✉)
Institut d'Astrophysique de Paris, UMR 7095, CNRS, Sorbonne Universit'e, 98bis Boulevard Arago, 75014 Paris, France
e-mail: silk@iap.fr

Department of Physics and Astronomy, The Johns Hopkins University, Baltimore, MD 21218, USA

Beecroft Institute for Particle Astrophysics and Cosmology, Department of Physics, University of Oxford, Oxford OX1 3RH, UK

C. Byrnes et al. (eds.), *Primordial Black Holes*, Springer Series in Astrophysics and Cosmology, https://doi.org/10.1007/978-981-97-8887-3_27

universe in a humongous range of masses from $10^{-20}\,M_\odot$ to $10^{20}\,M_\odot$. They could be the whole dark matter for masses around $10^{-12}\,M_\odot$. Moreover even if PBHs are dark matter-subdominant, they could be responsible for some of the currently observed mergers of BHs in the solar mass range or they may play key roles as rare IMBH (Intermediate Mass BHs) in the range $(10^3 - 10^6)M_\odot$ that may seed both SMBH (Supermassive Black Holes) observed in the mass range $(10^6-10^{10})M_\odot$ and massive galaxies in the early universe.

PBHs formed very early and hence may be ubiquitous at the highest redshifts accessible by direct imaging, that is $z \lesssim 100$, when there was little astrophysical competition. This provides us with the possibility of finding truly unique high redshift signatures of PBHs, for example via gravitational wave interferometers such as Einstein Telescope and Cosmic Explorer that will probe the existence of binaries up to such redshift, as well as potential EMRI signals up to $z \lesssim 20$.

Well before these telescopes are in action, there will be 21cm dark ages telescopes destined for the lunar far side, such as LUSEE-Night and DSL to probe PBH accretion imprints on diffuse hydrogen absorption against the CMB to $z \lesssim 100$, both scheduled for deployment in 2026. And on a similar time-scale we anticipate LIGO-class gravitational wave interferometers to make significant inroads on the possible existence of solar mass black holes as well as black holes populating the "forbidden" region in the pair instability gap, at masses $\gtrsim 50\,M_\odot$. Such observations would provide further evidence for the existence of PBHs.

We review some of these arguments in this closing chapter, with particular attention to the open questions, challenges and future opportunities.

27.2 Some Open Questions

How well do we know the PBH abundance? And over what mass range? How well can we ascertain the PBH clustering properties and therefore the GW signal from binary mergers via LIGO and future terrestrial GW interferometers? And can we use the vicinities of SMBH and IMBH, most likely rich in dark matter and where PBH abundances are in consequence likely to be enhanced, as potential sources of EMRI signals for experiments such as LISA? We briefly discuss such open theoretical and observational questions below.

27.2.1 What Is the Abundance of PBHs?

The formation of a PBH in the early universe is a rare event and knowing the precise formation probability represents, even nowadays, a challenge. This issue is of fundamental importance as the formation probability is one of the key ingredients to calculate the current PBH abundance and mass function which not only enter in many observables, such as the merger rate of BH binaries, but which are also routinely used to express current constraints from various observations.

To illustrate the difficulty in assessing the precise abundance of PBHs, let us consider the most standard scenario in the literature where the PBHs are formed by the collapse of large inhomogeneities generated during the inflationary stage. The latter delivers a stochastic quantity, the curvature perturbation ζ on superhorizon scales, whose properties are known statistically. As we mentioned, PBHs are rare events which require a given observable, e.g. the density contrast δ or the compaction function C, to be above a typically large critical value. This implies that the calculation of the PBH abundance requires going beyond linear perturbation theory and the knowledge of the tail of the probability where the latter is typically far from being a simple Gaussian distribution. Nonlinearities enter into the game even when the initial curvature perturbation is Gaussian (because of the nonlinear relation between the overdensity/compaction function and the curvature perturbation itself. The logic chain is (in a radiation-dominated universe)

$$\zeta(r) \rightarrow \text{non-Gaussian compaction function} \rightarrow C(r) \tag{27.1}$$

where

$$C(r) = C_1(r) - \frac{3}{8}C_1^2(r), \quad C_1 = -\frac{4}{3}r\zeta'(r). \tag{27.2}$$

The presence of nonlinearities at the various stages of the calculation makes the latter cumbersome. One more complication comes from the fact that the critical threshold C_c to form a PBH depends on the shape of the initial perturbation, specifically on the curvature around the maximum r_m of the compaction function

$$C_c = C_c(q), \quad q = -\frac{r_m^2 C''(r_m)}{4C(r_m)}, \tag{27.3}$$

where primes indicate derivatives with respect to the radius and we are assuming spherical symmetry. The formation probability therefore needs the knowledge of the exact probability. Given an initial profile of the curvature perturbation, one needs therefore to compute the corresponding critical profile for the compaction function. The latter determines the position of the maximum r_m in units of the typical momentum scale $k_\star$ of the power spectrum of the curvature perturbation and the correlators derived from the compaction function, which in turn determine its non-Gaussian statistics. Even though the dependence on the profile may be accounted for integrating over all possible values of the curvature of the compaction function at its maximum in real space or realizing that in some cases the broadest profiles count [48], the position of the maximum r_m is typically calculated only for the average profile of the curvature perturbation and this automatically introduces some intrinsic and unavoidable uncertainties due to the statistical nature of the observables. A possible, albeit incomplete, way to proceed might be to find an observable whose critical value does not depend on its profile, e.g. the volume average of the compaction function, as suggested in Ref. [32] and whose origin is explained in Ref. [54]

$$\overline{C}(r_m) = \frac{3}{R_m^3} \int_0^{R_m} \mathrm{d}x\, x^2\, C(x), \quad R_m = r_m\, e^{\zeta(r_m)}, \tag{27.4}$$

but no calculation exists in the literature so far about its probability. Even without the need of mentioning other possible problems, e.g. dealing with the nonlinearities entering in the radiation transfer function when perturbations re-enter the horizon [25] or the need to deal with the phenomenon of operator mixing [38], we believe the calculation of the PBH abundance in the standard scenario is still an open issue which deserves further study. The same is true in alternative mechanisms to generate the PBHs, e.g. in supercooled first-order phase transitions [35] where the threshold criterion does not suffice.

27.2.2 What Is the Effect of PBH Clustering?

Another open issue worth mentioning is the clustering of PBHs (see Chap. 2 for a basic introduction). It may significantly affect the merger rates of coalescing binaries and, consequently, the gravitational wave signal measured by the LIGO-Virgo-KAGRA collaboration and by future experiments. Indeed, even if isolated PBH binaries are not significantly affected by encounters with a third PBH, PBH binaries residing in PBH clusters might suffer such interactions, thus modifying the expected merger rate. If PBHs cluster or not is therefore an important question to address. A significant PBH clustering might also modify constraints arising from microlensing observations and from the cosmic microwave background. Clustering might also be relevant in determining PBH spins. While the gravitational collapse of a spherical overdensity during radiation domination generates nearly spinless PBHs [24], they may acquire a large spin in the presence of accretion [22]. Indeed, if PBHs cluster they can more easily merge to form binaries at late times, so that the resulting PBH coming from such a merger will have a non-zero value even when no accretion is present.

Even though PBHs in the standard scenario formed through the collapse of large overdensities are initially not clustered [2, 26], PBH clusters form not long after matter-radiation equality [51] if $f_{\rm PBH}$ is large. Current N-body simulations are limited to large redshifts $z \sim 10^2$ and a chief challenge is how to extrapolate to lower redshifts relevant for some observational constraints. Analytical insight is certainly called for in understanding the complex evolution of the PBH population.

PBHs, even if rare, are potentially important for seeding galaxy and even quasar formation at very high redshift [17]. These would be intermediate mass PBHs that enhanced local growth by gravitational instability of cold dark matter. In such a scenario, both massive galaxies and SMBH would form at epochs much earlier than in the standard model of structure formation. Moreover SMBH formation would precede or be contemporaneous with massive galaxy formation, as actually favored by recent JWST observations [72].

27.2.3 What Fraction of the Currently Observed GW Events Can Be ascribed to PBHs?

Thanks to the many binary BH events detected so far with many more expected in the next few years, gravitational-wave astronomy is about to become a population study. Currently hierarchical Bayesian analysis combining astrophysical formation models and PBHs may only constrain the fraction of a putative subpopulation of PBHs in the data (see for instance Refs. [39, 43] and Chap. 25). The exciting possibility that PBHs are contributing to current observations will be verified only by further reducing uncertainties in astrophysical and primordial formation models. This is another fundamental open issue.

27.2.4 Are PBHs the Dark Matter?

The idea that PBHs can comprise most of the dark matter is one of the main motivations for studying PBH. Unfortunately, observational constraints eliminate this possibility for most of the range of possible PBH masses, with a notable exception around asteroid mass PBHs, spanning several decades in mass around $\sim 10^{-12}\,\mathrm{M}_\odot$ down to the limit $\sim 10^{-16}\,\mathrm{M}_\odot$, obtained from the isotropic x-ray and soft gamma ray background observational limits on fluxes produced by PBHs currently undergoing Hawking evaporation [49].

In the standard formation scenario of PBHs (see Chap. 9 for alternatives), it is unavoidable that gravitational waves are generated with a frequency which today is in the mHz range, exactly where the LISA mission has maximum sensitivity [10]. The scenario of PBHs as dark matter can be therefore tested in the future by LISA by measuring the GW two-point correlator. The fact that the asteroid mass range studied in Chap. 22 is still unconstrained is due to the fact that the microlensing constraints are ineffective around the value $10^{-11}\,M_\odot$ under which the geometric optics approximation is no longer valid and the constraints from the presence of neutron stars in globular clusters is based on extreme assumptions about the dark matter density. It is of tantamount importance to come out with possible ideas for constraining or identifying PBH in the asteroid mass range.

One promising approach is that PBH capture leads to neutron star conversion to BH. This could occur in dense star clusters that contain DM, as may be the case for nuclear star clusters, and NS conversion would occur in the case of capture of such "endoparasitic" PBHs for PBH masses larger than $\sim 10^{-11}\,M_\odot$ [71]. Such a phenomenon may produce a deficit of pulsars in our galactic center as possibly observed [27]. Another interesting avenue is to observe the number of massive main-sequence stars in ultra-faint dwarfs which should be suppressed if the dark matter is made of asteroid-mass PBHs [33], making measured the mass distribution of stars depleted in the high-mass range. Additionally, microlensing of X-ray pulsars, due to the higher photon energies and smaller source sizes, may be able to constrain the abundance of PBHs in a portion of the open window [8, 74] with future AstroSat, Athena, AXIS, eXTP, and STROBE-X experiments.

27.2.5 Ultralight PBHs as DM

Ultralight PBHs as studied in Chap. 21 below the usual Hawking radiation limit mentioned above provide an intriguing DM candidate, provided that Hawking radiation can be suppressed. Even rare early production of such stable objects deep in the radiation-dominated era guarantees the possibility of a dark matter contribution in the matter-dominated era. Acquisition of near-extremality soon after formation provides an intriguing means of stabilizing ultralight PBHs that form early. One example is that of low mass five-dimensional PBHs that initially act like four-dimensional PBHs, Hawking radiating down to the radius of the extra dimension where their effective temperature is effectively zero, to attain a stable mass [69] that is limited to $\gtrsim 10^9$gm from BBN considerations. Such ultralight PBHs are generated in higher dimensional scenarios [5] to provide a dark matter population of ultralight PBHs [4]. Other scenarios for producing near-extremal PBH include formation of maximally charged [60] or rotating PBHs [40].

There is also a window for ultralight PBHs all the way down to of order the Planck mass. Considerations of stochasticity in Hawking evaporation of Planck scale PBHs allow survival of extremal Kerr PBHs that are very rare at formation, but subsequently dominant at late epochs as a DM component [76]. Extremal charged PBHs down to Planck mass scales can avoid Schwinger discharge processes at late epochs, and generate transient high energy gamma ray or neutrino signals as nearby mergers occur of oppositely charged PBHs [9].

The quantum gravity phenomenon of so-called memory burden suppression [29] provides another route to suppressing Hawking radiation. One scenario results in ultralight PBHs undergoing Hawking evaporation after formation until about half the initial mass is lost, then leaving a stable relic that is a DM candidate, [1] on the small mass scales where evaporation is not BBN-constrained, $\lesssim 10^9$ gm. If there is continuing leakage of Hawking radiation after the onset of memory suppression [77], diffuse gamma ray background limits allow the ultralight PBH window to be somewhat more restricted.

27.3 The PBH Roadmap

Numerous future experiments, ranging from approved projects to those which are entering their detailed design phase, are set to provide significantly enhanced constraints on PBH scenarios. They include terrestrial laser interferometers (Einstein Telescope, Cosmic Explorer) [28], and space laser interferometers (LISA, DECIGO [53], TianQin and Taiji [41]) as well as atomic interferometers on the ground (with demonstrators under construction and extensions planned to km scales (AION MAGIS, MAGIA, MIGA, ELGA, ZAIGA)) and in space (AEDGE) [30]. Furthermore, there are designs for lunar interferometer/seismometer arrays (LGWA, GLOC) [45].

The science goals focus on bridging frequency gaps in the LIGO-Virgo-KAGRA frequency range that has so far led to hundreds of detections of BH mergers in the $(10 - 100) M_\odot$ range by going both to higher and lower frequencies. The new science that can be explored ranges from the masses and equation of state of neutron stars as well as multimessenger signals from core collapse supernovae to kpc distances (kHz to Hz range) to SMBH and IMBH detections and BH inspiralling (Hz to mHz, and eventually μHz). One would ultimately approach the pulsar timing array sensitivities (NANOGrav, SKA) at nanohertz frequencies to enable exploration of first order very early universe phase transitions, and presence of loop decays in a network of cosmic strings [31].

We next address various "smoking-gun" signatures that may identify PBHs, notably including high redshift mergers and sub-solar mass PBHs.

27.3.1 *High Redshift Mergers*

The PBH model predicts a binary merger rate density which grows monotonically with redshift [3, 23, 70]. See Chap. 16 for details. Focusing on binaries generated at early epochs, one has a well-defined time evolution with redshift

$$R^{\mathrm{EB}}(z) \propto \left(\frac{t(z)}{t(z=0)}\right)^{-34/37}, \tag{27.5}$$

extending up to redshifts $z \gtrsim \mathcal{O}(10^3)$. Since mergers from BHs of astrophysical origin switch off at redshifts $z \gtrsim 30$ [11, 44, 50, 55–57, 61, 62, 73, 75], detecting merger redshifts at $z \gtrsim 30$ will represent smoking guns for primordial binaries [21, 59, 67]. Of course, large measurement uncertainties on the inferred luminosity distance due to the low SNR at 3G detectors [37, 64, 65, 67] may bring relatively large errors, but recent studies show that identifying merger rate evolution at redshift larger than $\mathcal{O}(10)$ will allow to constrain PBH populations up to abundances as low as $f_{\mathrm{PBH}} \approx 10^{-5}$ [66] in the solar mass range, even accounting for the contamination of Pop III binaries, see Fig. 27.1.

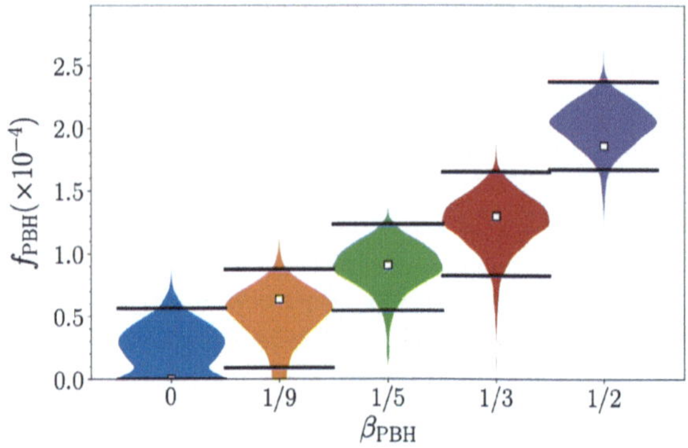

Fig. 27.1 Posteriors of $f_{\rm PBH}$ given a lognormal PBH mass function centered at $M_c = 30\,M_\odot$ and variance equal to 0.3 at $\beta_{\rm PBH} = 0, 1/9, 1/5, 1/3,$ and $1/2$, being $\beta_{\rm PBH}$ the fraction of the PBH merger rate to the total merger rate. The solid black lines indicate the 95% CIs. From Ref. [66]

27.3.2 Sub-Solar PBHs

The detection of sub-solar BHs in a compact binary merger is regarded as one of the smoking gun signatures of PBHs (Chap. 18). Indeed, BHs of astrophysical origin are not expected. The sub-solar mass range may be however populated by other compact objects like neutron stars, white dwarfs, or exotic compact objects [15] (e.g. boson stars [42]).

GW signatures of any population of PBH below the astrophysical production limit provide a powerful statistical constraint on any solar or subsolar component, as evidenced by the sharp decline observed in O3 events. There could of course be rare events generated by solar or subsolar PBHs that might contribute to DM at the $\mathcal{O}(10\%)$ level. Most interestingly, early universe physics predicts early dips in the sound speed, most prominently at the QCD transition, and thereby can potentially boost the formation of $\sim \mathrm{M}_\odot$ PBHs [16]. One key ingredient is therefore to distinguish PBHs from other compact objects, e.g. through the measurements of the tidal disruption and tidal deformability Λ. A recent analysis [20] has shown that for subsolar neutron-star binaries, the O4 and O5 projected sensitivities would allow measuring the effect of tidal disruption on the waveform in a large portion of the parameter space, also constraining the tidal deformability at $\mathcal{O}(10\%)$ level, thus excluding a primordial origin of the binary. Viceversa, for subsolar PBH binaries, model-agnostic upper bounds on the tidal deformability can rule out neutron stars or more exotic competitors. Assuming events similar to the sub-threshold candidate SSM200308 reported in LVK O3b data are PBH binaries, O4 projected sensitivity would allow ruling out the presence of neutron-star tidal effects at $\approx 3\sigma$ C.L., thus strengthening the PBH hypothesis. Future experiments would lead to even stronger ($> 5\sigma$) conclusions on potential discoveries of this kind, see Fig. 27.2.

PBHs of lunar mass, a regime where hints of unaccounted-for microlensing events have been reported by the OGLE survey of the galactic bulge [68], could be sufficiently numerous that nearby binaries might be detectable with future generation detectors [52].

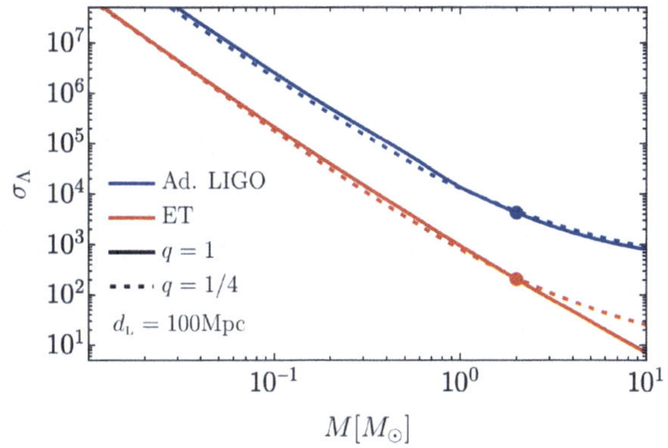

Fig. 27.2 Precision of the deformability parameter Λ at both Ad. LIGO and ET. The binary is assumed to have spinless components and negligible eccentricity and deformability, as predicted by the PBH scenario. The solid (dashed) line indicates the result for mass ratios 1 (1/4). From Ref. [37]

27.3.3 Plugging the Pair Instability Gap with PBH?

An equally powerful mass challenge for PBH arises from the pair instability gap. This is generated by the catastrophic core collapse of massive stars that results from pair-instability SNe, and predicts a deficit of astrophysical BH between $\sim 50\, M_\odot$ and $\sim 120\, M_\odot$. The full LIGO-Virgo O3 data set reveals a definitive decline in merger rates above $\sim 40\, M_\odot$ [19] but with several candidates filling this gap at high credibility [79]. This merger rate feature is plausibly explained in the astrophysical scenario by second and third generation mergers, with predicted spin distributions providing a possible discriminant [6].

Could PBHs populate the pair instability gap? A smoking gun signature would be a feature in the BH mass spectrum, most likely indicative of a new formation mechanism.

27.3.4 PBH Eccentricity

Even though PBH binaries are formed with large eccentricity at high redshift, they have enough time to circularize before the GW signal can enter the observation band of current and future detectors. This implies that an observation of a non-zero eccentricity e would rule out the interpretation as a primordial binary formed in the early Universe, while it may still be compatible with a PBH binary formed in the late-time universe [18, 80].

27.3.5 PBH Spin

The PBHs formed in the early universe through the standard scenario of the collapse of large overdensities inherit characteristic mass-spin correlations induced by accretion effects [22]. Using this criterion for determining the possible primordial nature of individual GW events would require reducing uncertainties on the accretion model and studied at the population level [36].

Figure 27.3 presents a flowchart from Ref. [37] that one might follow to assess the primordial nature of a binary merger based on measurements of the redshift z, eccentricity e, tidal deformability Λ, component masses m, and dimensionless spins χ.

27.3.6 Future Gamma-Ray Telescopes

PBHs have a range of potentially interesting direct gamma ray signals associated with Hawking evaporation. These include the 511 keV gamma-ray line, produced by electron-positron pair-annihilation, where positrons originate from black hole evaporation. The INTEGRAL detection of the Galactic Bulge provides one of the strongest bounds attainable with present observations, and should be greatly improved upon by future MeV gamma-ray telescopes such as GECCO [58], as well as AMEGO [34] and ASTROGRAM [63], among others [12].

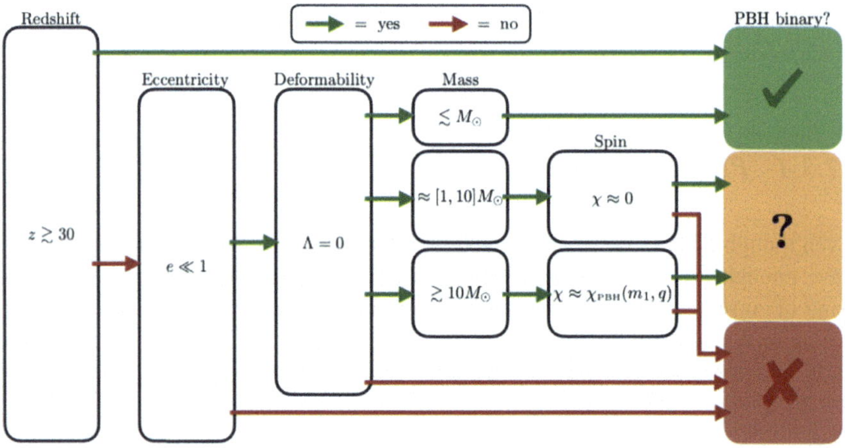

Fig. 27.3 Flowchart from Ref. [37] to assess the primordial nature of a binary merger based on measurements of the redshift z, eccentricity e, tidal deformability Λ, component masses m, and dimensionless spins χ. Each arrow indicates if the condition in the box is met (green) or violated (red)

Exploding PBHs, the late-time limit of Hawking evaporation for small PBHs provides intriguing candidates for future very high energy gamma ray telescopes that may address transient phenomena, that pertain especially to the new frontier of ultralight PBHs.

The Hawking temperature characterizes the effective temperature as inversely proportional to PBH mass and is constrained as discussed previously by soft gamma ray constraints at the $\sim 10^{17}$gm PBH mass scale, incorporating appropriate spectral corrections due to nonthermal aspects of the complex emission processes [7]. Once the allowable PBH mass range is extended to much lower masses due to slowing or suppression of Hawking radiation, as proposed in recent discussions of near-extremal and higher dimensional PBHs, intriguing prospects arise for high energy gamma ray experiments [1, 5, 40, 47], as well as for exploring direct detection [60] or merger events [9], where constraints at much higher gamma ray energies could become important.

Such gamma ray experiments include HAWC (100 GeV–50 TeV) and CTA (20 GeV–300 TeV) [13] at TeV scales, and at PeV-scale photon energies LHAASO [14] and SWGO [46]. Such searches will be closely coordinated with multimessenger projects spanning next generation underground gravitational wave telescopes and fast radio burst observatories (CHORD) [78].

27.4 Conclusions

Support for the existence of the PBH hypothesis comes from a range of arguments spanning dark matter physics, inflationary models, astrophysics, gravitational microlensing and quantum cosmology (Chap. 20). None are yet conclusive but the cumulative support for the existence of PBHs seems reasonably compelling. More to the point however is that there is a rich astrophysical agenda of open questions that contribute to a future road map for a plethora of research projects.

Two of the most fundamental PBH probes that distinguish them from astrophysical counterparts are their redshift and mass distributions. This combination provides an armoury of smoking guns that will inevitably enrich the next generation of gravitational wave and gravitational microlensing experiments. Confirmation will require improved statistics as single events may not be conclusive. It is only too easy to come up with rare events that can masquerade as PBH candidates, for example by appealing to higher order generations of merging events or to non-Gaussianity. We have closed this book by laying out a possible future road map for definitively assessing whether PBHs can present a viable candidate for dark matter, or possibly contribute to understanding other astrophysical or even quantum cosmology anomalies.

References

1. A. Alexandre, G. Dvali, E. Koutsangelas, New mass window for primordial black holes as dark matter from memory burden effect (2024). arXiv: 2402.14069 [hep-ph]
2. Y. Ali-Haïmoud, Correlation function of high-threshold regions and application to the initial small-scale clustering of primordial black holes. Phys. Rev. Lett. **121**(8) (2018). https://doi.org/10.1103/physrevlett.121.081304. https://doi.org/10.1103%2Fphysrevlett.121.081304
3. Y. Ali-Haïmoud, E.D. Kovetz, M. Kamionkowski, Merger rate of primordial black-hole binaries. Phys. Rev. D **96**(12), 123523 (2017). https://doi.org/10.1103/PhysRevD.96.123523. arXiv: 1709.06576 [astro-ph.CO]
4. L.A. Anchordoqui, I. Antoniadis, D. Lust, Dark dimension, the swampland, and the dark matter fraction composed of primordial near-extremal black holes. Phys. Rev. D **109**(9), 095008 (2024). https://doi.org/10.1103/PhysRevD.109.095008. arXiv: 2401.09087 [hep-th]
5. L.A. Anchordoqui, I. Antoniadis, D. Lüst, Dark dimension, the swampland, and the dark matter fraction composed of primordial black holes. **106**(8), 086001 (2022). https://doi.org/10.1103/PhysRevD.106.086001. arXiv: 2206.07071 [hep-th]
6. F. Antonini et al., Coalescing black hole binaries from globular clusters: mass distributions and comparison to gravitational wave data from GWTC-3. **522**(1), 466–476 (2023). https://doi.org/10.1093/mnras/stad972. arXiv: 2208.01081 [astro-ph.HE]
7. J. Auffinger, Primordial black hole constraints with Hawking radiation—a review. Prog. Part. Nucl. Phys. **131**, 104040 (2023). ISSN: 0146-6410. https://doi.org/10.1016/j.ppnp.2023.104040. http://dx.doi.org/10.1016/j.ppnp.2023.104040
8. Y. Bai, N. Orlofsky, Microlensing of X-ray pulsars: a method to detect primordial black hole dark matter. Phys. Rev. D **99**(12), 123019 (2019). https://doi.org/10.1103/PhysRevD.99.123019. arXiv: 1812.01427 [astro-ph.HE]
9. Y. Bai, N. Orlofsky, Primordial extremal black holes as dark matter. Phys. Rev. D **101**(5) (2020). ISSN: 2470-0029. https://doi.org/10.1103/physrevd.101.055006. http://dx.doi.org/10.1103/PhysRevD.101.055006
10. N. Bartolo et al., Testing primordial black holes as dark matter with LISA. Phys. Rev. D **99**(10), 103521 (2019). https://doi.org/10.1103/PhysRevD.99.103521. arXiv: 1810.12224 [astro-ph.CO]
11. K. Belczynski et al., On the likelihood of detecting gravitational waves from Population III compact object binaries. Mon. Not. Roy. Astron. Soc. **471**(4), 4702–4721 (2017). https://doi.org/10.1093/mnras/stx1759. arXiv: 1612.01524 [astro-ph.HE]
12. S. Bird et al., Snowmass2021 cosmic frontier white paper: primordial black hole dark matter. Phys. Dark Universe **41**, 101231 (2023). ISSN: 2212-6864. https://doi.org/10.1016/j.dark.2023.101231. http://dx.doi.org/10.1016/j.dark.2023.101231
13. X. Boluna et al., Searching for exploding black holes (2024). arXiv: 2307.06467 [astro-ph.HE]
14. Z. Cao et al., Ultrahigh-energy photons up to 1.4 petaelectronvolts from 12 γ-ray Galactic sources. Nature **594**(7861), 33–36 (2021). https://doi.org/10.1038/s41586-021-03498-z
15. V. Cardoso, P. Pani, Testing the nature of dark compact objects: a status report. Living Rev. Rel. **22**(1), 4 (2019). https://doi.org/10.1007/s41114-019-0020-4. arXiv: 1904.05363 [gr-qc]
16. B.J. Carr et al., Observational evidence for primordial black holes: a positivist perspective **1054**, 1–68 (2024). https://doi.org/10.1016/j.physrep.2023.11.005. arXiv: 2306.03903 [astro-ph.CO]
17. B. Carr, J. Silk, Primordial black holes as generators of cosmic structures. Mon. Not. R. Astron. Soc. **478**(3), 3756–3775 (2018). ISSN: 1365-2966. https://doi.org/10.1093/mnras/sty1204. http://dx.doi.org/10.1093/mnras/sty1204
18. I. Cholis et al., Orbital eccentricities in primordial black hole binaries. Phys. Rev. D **94**(8), 084013 (2016). https://doi.org/10.1103/PhysRevD.94.084013. arXiv: 1606.07437 [astro-ph.HE]
19. The LIGO Scientific Collaboration et al., The population of merging compact binaries inferred using gravitational waves through GWTC-3 (2022). arXiv: 2111.03634 [astro-ph.HE]

20. F. Crescimbeni et al., Primordial black holes or else? tidal tests on subsolar mass gravitational-wave observations (2024). https://doi.org/10.48550/arXiv.2402.18656. arXiv: 2402.18656 [astro-ph.HE]
21. V. De Luca et al., Bayesian evidence for both astrophysical and primordial black holes: mapping the GWTC-2 catalog to third-generation detectors. JCAP **05**, 003 (2021). https://doi.org/10.1088/1475-7516/2021/05/003. arXiv: 2102.03809 [astro-ph.CO]
22. V. De Luca et al., Constraints on primordial black holes: the importance of accretion. Phys. Rev. D **102**(4), 043505 (2020). https://doi.org/10.1103/PhysRevD.102.043505. arXiv: 2003.12589 [astro-ph.CO]
23. V. De Luca et al., Primordial black holes confront LIGO/Virgo data: current situation. JCAP **06**, 044 (2020). https://doi.org/10.1088/1475-7516/2020/06/044. arXiv: 2005.05641 [astro-ph.CO]
24. V. De Luca et al., The initial spin probability distribution of primordial black holes. JCAP **05**, 018 (2019). https://doi.org/10.1088/1475-7516/2019/05/018. arXiv: 1903.01179 [astro-ph.CO]
25. V. De Luca, A. Kehagias, A. Riotto, How well do we know the primordial black hole abundance: the crucial role of nonlinearities when approaching the horizon. Phys. Rev. D **108**(6), 063531 (2023). https://doi.org/10.1103/PhysRevD.108.063531. arXiv: 2307.13633 [astro-ph.CO]
26. V. Desjacques, A. Riotto, Spatial clustering of primordial black holes. Phys. Rev. D **98**(12), 123533 (2018). https://doi.org/10.1103/PhysRevD.98.123533. arXiv: 1806.10414 [astro-ph.CO]
27. J. Dexter, R.M. O'Leary, The peculiar pulsar population of the central parsec. Astrophys. J. **783**(1), L7 (2014). ISSN: 2041-8213. https://doi.org/10.1088/2041-8205/783/1/l7. http://dx.doi.org/10.1088/2041-8205/783/1/L7
28. G. Domènech, M. Sasaki, Probing primordial black hole scenarios with terrestrial gravitational wave detectors (2024). arXiv: 2401.07615 [gr-qc]
29. G. Dvali, A microscopic model of holography: survival by the burden of memory (2018). https://doi.org/10.48550/arXiv.1810.02336. arXiv: 1810.02336 [hep-th]
30. J. Ellis et al., Probing supermassive black hole seed scenarios with gravitational wave measurements (2023). arXiv: 2312.02983 [astro-ph.CO]
31. J. Ellis et al., What is the source of the PTA GW signal? (2023). arXiv: 2308.08546 [astro-ph.CO]
32. A. Escrivà, C. Germani, R.K. Sheth, Universal threshold for primordial black hole formation. Phys. Rev. D **101**(4), 044022 (2020). https://doi.org/10.1103/PhysRevD.101.044022. arXiv: 1907.13311 [gr-qc]
33. N. Esser, S. De Rijcke, P. Tinyakov, The impact of primordial black holes on the stellar mass function of ultra-faint dwarf galaxies (2023). arXiv: 2311.12658 [astro-ph.GA]
34. H. Fleischhack, AMEGO-X: MeV gamma-ray astronomy in the multimessenger era (2021). arXiv: 2108.02860 [astro-ph.IM]
35. M.M. Flores, A. Kusenko, M. Sasaki, Revisiting formation of primordial black holes in a supercooled first-order phase transition (2024). arXiv: 2402.13341 [hep-ph]
36. G. Franciolini, P. Pani, Searching for mass-spin correlations in the population of gravitational-wave events: the GWTC-3 case study. Phys. Rev. D **105**(12), 123024 (2022). https://doi.org/10.1103/PhysRevD.105.123024. arXiv: 2201.13098 [astro-ph.HE]
37. G. Franciolini et al., How to assess the primordial origin of single gravitational-wave events with mass, spin, eccentricity, and deformability measurements. Phys. Rev. D **105**(6), 063510 (2022). https://doi.org/10.1103/PhysRevD.105.063510. arXiv: 2112.10660 [astro-ph.CO]
38. G. Franciolini et al., Renormalized primordial black holes (2023). arXiv: 2311.03239 [astro-ph.CO]
39. G. Franciolini et al., Searching for a subpopulation of primordial black holes in LIGO-Virgo gravitational-wave data. Phys. Rev. D **105**(8), 083526 (2022). https://doi.org/10.1103/PhysRevD.105.083526. arXiv: 2105.03349 [gr-qc]

40. J.A. de Freitas Pacheco, J. Silk, Primordial rotating black holes. Phys. Rev. D **101**(8), 083022 (2020). https://doi.org/10.1103/PhysRevD.101.083022. arXiv: 2003.12072 [astro-ph.CO]
41. Y. Gong, J. Luo, B. Wang, Concepts and status of Chinese space gravitational wave detection projects. Nat. Astron. **5**(9), 881–889 (2021). ISSN: 2397-3366. https://doi.org/10.1038/s41550-021-01480-3. http://dx.doi.org/10.1038/s41550-021-01480-3
42. H.-K. Guo, K. Sinha, C. Sun, Probing boson stars with extreme mass ratio inspirals. JCAP **09**, 032 (2019). https://doi.org/10.1088/1475-7516/2019/09/032. arXiv: 1904.07871 [hep-ph]
43. A. Hall, A.D. Gow, C.T. Byrnes, Bayesian analysis of LIGO-Virgo mergers: Primordial versus astrophysical black hole populations. hys. Rev. D **102**, 123524 (2020). https://doi.org/10.1103/PhysRevD.102.123524. arXiv: 2008.13704 [astro-ph.CO]
44. T. Hartwig et al., Gravitational waves from the remnants of the first stars. Mon. Not. Roy. Astron. Soc. **460**(1), L74–L78 (2016). https://doi.org/10.1093/mnrasl/slw074. arXiv: 1603.05655 [astro-ph.GA]
45. J. van Heijningen et al., The payload of the lunar gravitational-wave antenna. J. Appl. Phys. **133**(24), 244501 (2023). https://doi.org/10.1063/5.0144687. arXiv: 2301.13685 [gr-qc]
46. J. Hinton, SWGO Collaboration, The southern wide-field gamma-ray observatory: status and prospects, in *37th International Cosmic Ray Conference*, p 23 (2022). https://doi.org/10.22323/1.395.0023. arXiv: 2111.13158 [astro-ph.IM]
47. F. Hoyle, R.A. Lyttleton, On the accretion of interstellar matter by stars. Proc. Camb. Philos. Soc. **36**(3), 325 (1940). https://doi.org/10.1017/S0305004100017369
48. A. Ianniccari et al., The primordial black hole abundance: the broader, the better (2024). arXiv: 2402.11033 [astro-ph.CO]
49. J. Iguaz, P.D. Serpico, T. Siegert, Isotropic x-ray bound on primordial black hole dark matter. Phys. Rev. D **103**(10) (2021). ISSN: 2470-0029. https://doi.org/10.1103/physrevd.103.103025. http://dx.doi.org/10.1103/PhysRevD.103.103025
50. K. Inayoshi et al., Formation pathway of Population III coalescing binary black holes through stable mass transfer. Mon. Not. Roy. Astron. Soc. **468**(4), 5020–5032 (2017). https://doi.org/10.1093/mnras/stx757. arXiv: 1701.04823 [astro-ph.HE]
51. D. Inman, Y. Ali-Haïmoud, Early structure formation in primordial black hole cosmologies. Phys. Rev. D **100**)(8), 083528 (2019). https://doi.org/10.1103/PhysRevD.100.083528. arXiv: 1907.08129 [astro-ph.CO]
52. K.T. Inoue, T. Tanaka, Gravitational waves from sub-lunar-mass primordial black-hole binaries: a new probe of extradimensions. Phys. Rev. Lett. **91**(2) (2003). ISSN: 1079-7114. https://doi.org/10.1103/physrevlett.91.021101. http://dx.doi.org/10.1103/PhysRevLett.91.021101
53. S. Kawamura et al., urrent status of space gravitational wave antenna DECIGO and B-DECIGO. PTEP **2021**(5), 05A105 (2021). https://doi.org/10.1093/ptep/ptab019. arXiv: 2006.13545 [gr-qc]
54. A. Kehagias, D. Perrone, A. Riotto, Why the universal threshold for primordial black hole formation is universal (2024). arXiv: 2405.05208 [astro-ph.CO]
55. T. Kinugawa, T. Nakamura, H. Nakano, Chirp mass and spin of binary black holes from first star remnants. Mon. Not. Roy. Astron. Soc. **498**(3), 3946–3963 (2020). https://doi.org/10.1093/mnras/staa2511. arXiv: 2005.09795 [astro-ph.HE]
56. T. Kinugawa et al., Possible indirect confirmation of the existence of Pop III massive stars by gravitational wave. Mon. Not. Roy. Astron. Soc. **442**(4), 2963–2992 (2014). https://doi.org/10.1093/mnras/stu1022. arXiv: 1402.6672 [astro-ph.HE]
57. T. Kinugawa et al., The detection rate of inspiral and quasi-normal modes of Population III binary black holes which can confirm or refute the general relativity in the strong gravity region. Mon. Not. Roy. Astron. Soc. **456**(1), 1093–1114 (2016). https://doi.org/10.1093/mnras/stv2624. arXiv: 1505.06962 [astro-ph.SR]
58. M. Korwar, S. Profumo, Updated constraints on primordial black hole evaporation. J. Cosmol. Astropart. Phys. **2023**(05), 054 (2023). ISSN: 1475-7516. https://doi.org/10.1088/1475-7516/2023/05/054. http://dx.doi.org/10.1088/1475-7516/2023/05/054

59. S.M. Koushiappas, A. Loeb, Maximum redshift of gravitational wave merger events. Phys. Rev. Lett. **119**(22), 221104 (2017). https://doi.org/10.1103/PhysRevLett.119.221104. arXiv: 1708.07380 [astro-ph.CO]
60. B.V. Lehmann et al., Direct detection of primordial black hole relics as dark matter. J. Cosmol. Astropart. Phys. **2019**(10), 046–046 (2019). ISSN: 1475-7516. https://doi.org/10.1088/1475-7516/2019/10/046. http://dx.doi.org/10.1088/1475-7516/2019/10/046
61. B. Liu, V. Bromm, Gravitational waves from Population III binary black holes formed by dynamical capture. Mon. Not. Roy. Astron. Soc. **495**(2), 2475–2495 (2020). https://doi.org/10.1093/mnras/staa1362. arXiv: 2003.00065 [astro-ph.CO]
62. B. Liu, V. Bromm, The Population III origin of GW190521. Astrophys. J. Lett. **903**(2), L40 (2020). https://doi.org/10.3847/2041-8213/abc552. arXiv: 2009.11447 [astro-ph.GA]
63. M. Mallamaci et al., All-Sky-ASTROGAM: a MeV companion for multimessenger astrophysics. PoS **ICRC2019**, 579 (2019). https://doi.org/10.22323/1.358.0579
64. M. Martinelli et al., Dancing in the dark: detecting a population of distant primordial black holes. JCAP **08**(08), 006 (2022). https://doi.org/10.1088/1475-7516/2022/08/006. arXiv: 2205.02639 [astro-ph.CO]
65. K.K.Y. Ng et al., Measuring properties of primordial black hole mergers at cosmological distances: Effect of higher order modes in gravitational waves. Phys. Rev. D **107**(2), 024041 (2023). https://doi.org/10.1103/PhysRevD.107.024041. arXiv: 2210.03132 [astro-ph.CO]
66. K.K.Y. Ng et al., Constraining high-redshift stellar-mass primordial black holes with next-generation ground-based gravitational-wave detectors. Astrophys. J. Lett. **933**(2), L41 (2022). https://doi.org/10.3847/2041-8213/ac7aae. arXiv: 2204.11864 [astro-ph.CO]
67. K.K.Y. Ng et al., On the single-event-based identification of primordial black hole mergers at cosmological distances. Astrophys. J. Lett. **931**(1), L12 (2022). https://doi.org/10.3847/2041-8213/ac6bea. arXiv: 2108.07276 [astro-ph.CO]
68. H. Niikura et al., Constraints on earth-mass primordial black holes from OGLE 5-year microlensing events. Phys. Rev. D **99**(8) (2019). ISSN: 2470-0029. https://doi.org/10.1103/physrevd.99.083503. http://dx.doi.org/10.1103/PhysRevD.99.083503
69. J.A. de Freitas Pacheco et al., Quasiextremal primordial black holes are a viable dark matter candidate. Phys. Rev. D **107**(12) (2023). ISSN: 2470-0029. https://doi.org/10.1103/physrevd.107.123525. http://dx.doi.org/10.1103/PhysRevD.107.123525
70. M. Raidal et al., Formation and evolution of primordial black hole binaries in the early universe. JCAP **02**, 018 (2019). https://doi.org/10.1088/1475-7516/2019/02/018. arXiv: 1812.01930 [astro-ph.CO]
71. C.B. Richards, T.W. Baumgarte, S.L. Shapiro, Accretion onto a small black hole at the center of a neutron star. Phys. Rev. D **103**(10) (2021). ISSN: 2470-0029. https://doi.org/10.1103/physrevd.103.104009. http://dx.doi.org/10.1103/PhysRevD.103.104009
72. J. Silk et al., Which came first: supermassive black holes or galaxies? insights from JWST (2024). arXiv: 2401.02482 [astro-ph.GA]
73. N. Singh et al., Exploring compact binary populations with the einstein telescope. Astron. Astrophys. **667**, A2 (2022). https://doi.org/10.1051/0004-6361/202142856. arXiv: 2112.04058 [astro-ph.HE]
74. M. Tamta, N. Raj, P. Sharma, Breaking into the window of primordial black hole dark matter with x-ray microlensing (2024). arXiv: 2405.20365 [astro-ph.HE]
75. A. Tanikawa et al., Merger rate density of Population III binary black holes below, above, and in the pair-instability mass gap. Astrophys. J. **910**(1), 30 (2021). https://doi.org/10.3847/1538-4357/abe40d. arXiv: 2008.01890 [astro-ph.HE]
76. Q. Taylor et al., Extremal Kerr black hole dark matter from Hawking evaporation. Phys. Rev. D **109**(10), 104066 (2024). https://doi.org/10.1103/PhysRevD.109.104066. arXiv: 2403.04054 [gr-qc]
77. V. Thoss, A. Burkert, K. Kohri, Breakdown of hawking evaporation opens new mass window for primordial black holes as dark matter candidate (2024). arXiv: 2402.17823 [astro-ph.CO]

78. K. Vanderlinde et al., The Canadian hydrogen observatory and radio-transient detector (CHORD). Can. Range Plan Astron. Astrophys. White Pap. **2020**, 28 (2019). https://doi.org/10.5281/zenodo.3765414. arXiv: 1911.01777 [astro-ph.IM]
79. D. Wadekar et al., New black hole mergers in the LIGO-Virgo O3 data from a gravitational wave search including higher-order harmonics (2023). arXiv: 2312.06631 [gr-qc]
80. Y.-F. Wang, A.H. Nitz, Prospects for detecting gravitational waves from eccentric subsolar mass compact binaries. Astrophys. J. **912**(1), 53 (2021). https://doi.org/10.3847/1538-4357/abe939. arXiv: 2101.12269 [astro-ph.HE]

The manufacturer's authorised representative in the EU is Springer Nature Customer Service Centre GmbH, Europaplatz 3, 69115 Heidelberg, Germany. If you have any concerns regarding our products, please contact ProductSafety@springernature.com

Printed and bound by CPI Group (UK) Ltd, Croydon, CR0 4YY

01/10/2026

02225929-0020